# TIMBER CONSTRUCTION MANUAL

# TIMBER CONSTRUCTION MANUAL

### FOURTH EDITION
1994

## AMERICAN INSTITUTE OF TIMBER CONSTRUCTION

Englewood, Colorado

A WILEY-INTERSCIENCE PUBLICATION

## JOHN WILEY & SONS, INC.

New York • Chichester • Brisbane • Toronto • Singapore

*Library of Congress Cataloging in Publication Data:*

Timber construction manual / American Institute of Timber
    Construction. — 4th ed.
        p.    cm.
    "A Wiley-Interscience publication."
    Includes bibliographical references and index.
    ISBN 0-471-30970-2 (alk. paper)
    1. Building, Wooden—Handbooks, manuals, etc.   I. American
Institute of Timber Construction.
TA666.T47   1994
694—dc20                                          93-28233

Printed in the United States of America

10 9 8 7 6 5 4 3 2 1

# PREFACE

The first edition of the American Institute of Timber Construction (AITC) *Timber Construction Manual* was published in 1966. Changes in the wood products industry, technological advances, and improvements in the structural timber fabricating industry have necessitated several revisions of the *Manual.*

New lumber sizes and revisions in grading requirements for lumber and glued laminated timber were reflected in the second edition published in 1974. The third edition was published in 1985 to reflect new information on timber design methods. This edition of the manual contains the latest design procedures used for timber construction.

The American Institute of Timber Construction has developed this *Timber Construction Manual* for convenient reference by architects, engineers, contractors, teachers, the laminating and fabricating industry, and all others having a need for reliable up-to-date technical data and recommendations on engineered timber construction. The information and the recommendations herein are based on the most reliable technical data available, and reflect the commercial practices found to be most practical. Their application results in structurally sound construction.

The *Manual* has been arranged primarily for convenient use by designers, detailers, and fabricators of engineered timber construction. To avoid repetition, material which pertains to more than one area will be found in only one section. Suitable cross references are made in the other pertinent sections.

Information of an engineering textbook nature, such as derivations of equations, is not included since the purpose of the *Manual* is to present data for design and construction application by those familiar with engineering procedures.

The work of the preparation of the *Timber Construction Manual* was guided by the AITC Technical Advisory Committee, and was carried out by AITC staff engineers and the engineers and technical representatives of AITC member firms.

Suggestions for the improvement of this manual will be welcomed and will receive consideration in the preparation of future editions.

The *Timber Construction Manual* has been adopted by the American Institute of Timber Construction as its official recommendation.

**While these data have been prepared in accordance with recognized engineering principles and are based on the most accurate and reliable technical data available, they should not be used or relied upon for any general or specific application without competent professional examination and verification of their accuracy, suitability, and applicability by a licensed**

professional engineer, designer, or architect. By the publication of this manual, AITC intends no representation or warranty, expressed or implied, that the information contained herein is suitable for any general or specific use or is free from infringement of any patent or copyright. Any user of this information assumes all risk and liability arising from such use.

# PREFACE TO THIRD EDITION

This Third Edition has been prepared to update the AITC Timber Construction Manual to reflect current timber design methods. Part I of the manual contains general design data and construction information. Part II contains information on loads and the design of structural elements and their fastenings. Part III contains reference information and AITC recommended standards and specifications for engineered timber construction.

The work of the preparation of the *Timber Construction Manual* was guided by the AITC Technical Advisory Committee and was carried out by AITC staff engineers and by engineers and technical representatives of AITC member firms.

Suggestions for the improvement of this manual will be welcomed and will receive consideration in the preparation of future editions.

The *Timber Construction Manual* has been adopted by the American Institute of Timber Construction as its official recommendation.

The American Institute of Timber Construction has developed this *Timber Construction Manual* for convenient reference by architects, engineers, contractors, teachers, and the laminating and fabricating industry, and all others having need for up-to-date technical data and recommendations on engineered timber construction.

# PREFACE TO SECOND EDITION

The first edition of the AITC *Timber Construction Manual* was published in 1966. Changes in the wood products industry and technological advances and improvements in the structural timber fabricating industry have necessitated this revised edition of the *Manual*.

New lumber sizes and revisions in grading requirements for lumber and glued laminated timber are reflected in this second edition. Improved and refined design procedures are also incorporated.

The *Timber Construction Manual* was prepared by the AITC engineering staff with the guidance of the Institute's Technical Advisory Committee. The valuable assistance provided from many sources in developing technical data for the *Manual* is gratefully acknowledged.

# PREFACE TO FIRST EDITION

In recent years, technical developments and the establishment of an engineered timber fabricating and laminating industry have had a profound effect on construction. Long clear spans of timber trusses, girders, arches, and decking are now commonplace. Engineered timber is widely used in such diversified construction as schools, churches, commercial buildings, industrial buildings, residences and farm buildings, highway and railway bridges, towers, theater screens, ships, and military and marine installations.

Modern practices combine engineering, quality control, and careful grading with the use of proper working stresses, dependable adhesives, and efficient mechanical fastenings to produce reliable construction. Laminating with strong, durable adhesives permits the manufacture of curved and variable shaped members and thus increases the versatility of timber construction.

The American Institute of Timber Construction is a nonprofit, technical, industrial association of manufacturers and fabricators who may design, plant-laminate, fabricate, assemble, and erect load-carrying sawn and glued timber framing and decking for roofs and other structural parts of schools, churches, commercial, industrial, and other buildings, and for other structures such as bridges, towers, and marine installations.

The American Institute of Timber Construction has developed this *Timber Construction Manual* for convenient reference by architects, engineers, contractors, teachers, the laminating and fabricating industry, and all others having a need for reliable, up-to-date technical data and recommendations on engineered timber construction. The information and the recommendations herein are based on the most reliable technical data available and reflect the commercial practices found to be most practical. Their application results in structurally sound construction.

The *Manual* has been arranged primarily for convenient use by designers, detailers, and fabricators of engineered timber construction. To avoid repetition, material which pertains to more than one area will be found in only one section. Suitable cross references are made in the other pertinent sections.

Information of an engineering textbook nature, such as derivations of formulae, is not included, since the purpose of the *Manual* is to present data for design and construction application by those familiar with engineering procedures.

Part I of the *Manual* contains design data and construction information. Part II contains AITC recommended standards and specifications which will aid the designer in preparing plans and specifications for engineered timber construction.

Material has been compiled from many sources. Where it has been possible to identify the author of the material reproduced, it is used with the author's permission.

Every precaution has been taken to assure that all the data and information included are as accurate as possible. However, the Institute cannot assume responsibility for errors or omissions resulting from the use of this *Manual* in the preparation of plans or specifications. The Institute does not prepare engineering plans.

The work of the preparation of the *Timber Construction Manual* was guided by the AITC Technical Advisory Committee and was carried out by AITC staff engineers and by engineers and technical representatives of AITC member firms.

Suggestions for the improvement of this *Manual* will be welcomed and will receive consideration in the preparation of future editions.

The *Timber Construction Manual* has been adopted by the American Institute of Timber Construction as its official recommendation.

# CONTENTS

xi

## Part III   REFERENCE

# GENERAL NOMENCLATURE

The following abbreviations and symbols are in general use throughout this manual. Deviations from these notations are indicated where they occur.

## ABBREVIATIONS

| | |
|---|---|
| AASHTO | American Association of State Highway and Transportation Officials |
| AREA | American Railway Engineers Association |
| AITC | American Institute of Timber Construction |
| ANSI | American National Standards Institute |
| APA | American Plywood Association |
| ASCE | American Society of Civil Engineers |
| ASTM | American Society for Testing and Materials |
| AWPA | American Wood-Preservers' Association |
| Btu | British thermal unit |
| COV | Coefficient of variation |
| DL | Dead load (psf) |
| EL | Earthquake load (psf) |
| EMC | Equilibrium moisture content (%) |
| FPL | Forest Products Laboratory, U.S. Forest Service |
| ft, $ft^2$, $ft^3$ | Feet, square feet, cubic feet |
| hr | Hour |
| in., $in.^2$, $in.^3$, $in.^4$ | Inches, square inches, cubic inches, inches to the fourth power |
| in.-lb | Inch-pounds |
| k | Kip (one thousand pounds) |
| KD | Kiln dried |
| lb | Pound |
| LL | Live load (psf) |
| MC | Moisture content (%) |
| min | Minimum |

| MEL | Machine evaluated lumber |
| MSR | Machine stress rated |
| NA | Neutral axis |
| NDS | *National Design Specification for Wood Construction*® |
| o.c. | On centers |
| °F | Degrees Fahrenheit |
| pcf | Pounds per cubic foot |
| plf | Pounds per lineal foot |
| psf | Pounds per square foot |
| psi | Pounds per square inch |
| SL | Snow load (psf) |
| TL | Total load (psf) |
| USDA | United States Department of Agriculture |
| WL | Wind load (psf) |

## SYMBOLS

| $A$ | Area of cross section (in.$^2$) |
| $a$ | Dimension of member (in.) |
| $a$ | Horizontal distance to load for bracket columns (in.) |
| $A_c$ | Area of concrete footing in pole design (ft$^2$) |
| $A_m$ | Gross cross-sectional area of main wood member(s) (in.$^2$) |
| $A_n$ | Cross-sectional area of notched member (in.$^2$) |
| $A_s$ | Area of steel member (in.$^2$) |
| $A_s$ | Sum of gross cross-sectional areas of side member(s) (in.$^2$) |
| $b$ | Breadth (width) of rectangular member (in.) |
| $b$ | Smaller side of beam or column before exposure to fire (in.) |
| $b$ | Width of column flange (in.) |
| $C$ | Compressive force (lb) |
| $C$ | Pole circumference at point of maximum moment (in.) |
| $C$ | Thermal conductance (Btu/hr ft.$^2$ °F) |
| $c$ | Distance from neutral axis to outer surface of beam (in.) |
| $C_b$ | Bearing area factor |
| $C_c$ | Curvature factor |
| $C_{co}$ | Seasoning conditioning factor for round timber piles |
| $C_{cs}$ | Critical section factor for round timber piles |
| $C_D$ | Duration-of-load factor |
| $C_d$ | Penetration depth factor for connections |

| | |
|---|---|
| $C_{di}$ | Diaphragm factor for nailed connections |
| $C_{dt}$ | Constant for tapered beam deflection |
| $C_{eg}$ | End grain factor for connections |
| $C_F$ | Size factor |
| $C_f$ | Form factor |
| $C_{fu}$ | Flat use factor |
| $C_g$ | Group action factor for connections |
| $C_H$ | Shear stress adjustment factor |
| $C_I$ | Interaction stress factor |
| $C_L$ | Beam stability factor |
| $C_M$ | Wet service factor |
| $C_M$ | Steel stress coefficient for bridge dowel design (psi) |
| $C_P$ | Column stability factor |
| $C_p$ | Ponding magnification factor |
| $C_R$ | Steel stress coefficient for bridge dowel design (psi) |
| $C_r$ | Reduction factor for double-tapered curved beams |
| $C_r$ | Repetitive member factor for dimension lumber |
| $C_{sp}$ | Single pile factor for timber piles |
| $C_{st}$ | Metal side plate factor for 4 in. shear plate connections |
| $C_T$ | Buckling stiffness factor for dimension lumber |
| $C_t$ | Temperature factor |
| $C_{tn}$ | Toe-nail factor for nailed connections |
| $C_u$ | Untreated factor for timber piles |
| $C_V$ | Volume factor for structural glued laminated timber |
| $C_x$ | Spaced column fixity factor |
| $C_y$ | Factor for tapered beam deflection |
| $C_\Delta$ | Geometry factor for connections |
| $C_{\Delta e}$ | Edge distance factor |
| $C_{\Delta n}$ | End distance factor |
| $C_{\Delta s}$ | Spacing factor |
| $D$ | Diameter (in.) |
| $d$ | Bridge dowel diameter (in.) |
| $d$ | Depth of bending member (in.) |
| $d$ | Least dimension of rectangular compression member (in.) |
| $d$ | Larger side of beam or column before exposure to fire (in.) |
| $d$ | Pennyweight of nail or spike |
| $d_b$ | Arch depth at base (in.) |
| $d_c$ | Depth of cross section at centerline (in.) |

| | |
|---|---|
| $d_{cb}$ | Approximate centerline depth for pitched and tapered curved beams (in.) |
| $d_{crt}$ | Minimum centerline depth due to radial tension for pitched and tapered curved beams (in.) |
| $d_e$ | Effective depth of member at a connection (in.) |
| $d_{eb}$ | Factor for calculating depth of pitched and tapered curved beams (in.) |
| $d_{eff}$ | Approximate effective centerline deflection for pitched and tapered curved beams (in.) |
| $d_n$ | Depth of remaining at a notch (in.) |
| $d_1, d_2$ | Cross-sectional dimensions of rectangular compression member in planes of lateral support (in.) |
| $D_H$ | Diameter of hole for pole design (ft) |
| $d_t$ | Depth of tangent point (in.) |
| $E, E'$ | Tabulated and allowable modulus of elasticity (psi) |
| $E_m$ | Modulus of elasticity of main member (psi) |
| $E_s$ | Modulus of elasticity of side member (psi) |
| $e$ | Eccentricity of load (in.) |
| $f_b$ | Bending stress (psi) |
| $F_b, F_b'$ | Tabular and allowable bending design value (psi) |
| $F_{bE}$ | Critical buckling design value for bending members (psi) |
| $F_{bx}, F_{bx}'$ | Tabular and allowable design value in bending about the $x-x$ axis (psi) |
| $F_{by}, F_{by}'$ | Tabular and allowable design value in bending about the $y-y$ axis (psi) |
| $F_{b1}'$ | Allowable edgewise bending design value $F_{bx}'$ (psi) |
| $F_{b2}'$ | Allowable flatwise bending design value $F_{by}'$ (psi) |
| $f_{b1}$ | Edgewise bending stress $F_{bx}$ (psi) |
| $f_{b2}$ | Flatwise bending stress $F_{by}$ (psi) |
| $f_c$ | Compression parallel to grain stress (psi) |
| $F_c, F_c'$ | Tabular and allowable compression parallel to grain design value (psi) |
| $f_{c\perp}$ | Compression perpendicular to grain stress (psi) |
| $F_{c\perp}, F_{c\perp}'$ | Tabular and allowable design value in compression perpendicular to grain (psi) |
| $F_{cE}$ | Critical buckling design value for compression members (psi) |
| $F_{cE1}, F_{cE2}$ | Critical buckling design value for compression member in planes of lateral support (psi) |

| | |
|---|---|
| $F_e$ | Dowel bearing strength (psi) |
| $F_{em}$ | Dowel bearing strength of main member (psi) |
| $F_{es}$ | Dowel bearing strength of side member (psi) |
| $F_{e\parallel}$ | Dowel bearing strength parallel to grain for bolt or lag screw connections (psi) |
| $F_{e\perp}$ | Dowel bearing strength perpendicular to grain for bolt or lag screw connections (psi) |
| $F_{e\theta}$ | Dowel bearing strength at an angle to grain for bolt or lag screw connections (psi) |
| $f_g$ | End grain in bearing perpendicular to grain (psi) |
| $F_g, F_g'$ | Tabular and allowable end grain in bearing parallel to grain (psi) |
| $f_o$ | Reference stress for pitched and tapered curved beams (psi) |
| $f_r$ | Radial stress in curved bending member (psi) |
| $f_{rt}$ | Radial tension stress (psi) |
| $F_{rt}, F_{rt}'$ | Tabular and allowable design value in radial tension (psi) |
| $f_s$ | Torsional stress (psi) |
| $f_t$ | Tension parallel to grain stress (psi) |
| $F_t, F_t'$ | Tabular and allowable tension parallel to grain stress (psi) |
| $F_v, F_v'$ | Tabular and allowable design value shear parallel to grain stress (horizontal shear) (psi) |
| $f_v$ | Shear parallel to grain stress (horizontal shear) (psi) |
| $F_{yb}$ | Bending yield strength of fastener |
| $F_\theta'$ | Allowable bearing design value at an angle to grain (psi) |
| $G$ | Specific gravity |
| $G$ | Shear modulus (modulus of rigidity) (psi) |
| $g$ | Gauge of screw |
| $h$ | Height of crown of arch (ft) |
| $h_a$ | Height of apex for pitched and tapered curved beams (in.) |
| $h_s$ | Height of soffit at midspan for pitched and tapered curved beams (in.) |
| $I$ | Initial moisture content (below 30%) (%) |
| $I$ | Moment of inertia (in.$^4$) |
| $I_K I_G$ | Ratio of moment of inertia of knots to moment of inertia of gross cross section |
| $K$ | Constant for bridge deck design |
| $K$ | Bending stress factor for pitched and tapered curved beams |
| $k$ | Change in member thickness for arch deflection (%) |
| $k$ | Thermal conductivity (Btu in./hr ft$^2$ °F) |

| | |
|---|---|
| $K_D$ | Diameter coefficient for wood screw, nail, and spike connections |
| $K_L$ | Loading condition coefficient for glued laminated beams |
| $K_M$ | Moisture content coefficient for sawn lumber truss compression chord |
| $K_r$ | Radial stress factor |
| $K_T$ | Truss compression chord coefficient for sawn lumber |
| $K_{bE}$ | Euler buckling coefficient for beams |
| $K_{cE}$ | Euler buckling coefficient for columns |
| $K_e$ | Buckling length coefficient for compression members |
| $K_f$ | Column stability coefficient for bolted and nailed built-up columns |
| $K_r$ | Radial stress coefficient |
| $K_t$ | Temperature coefficient |
| $K_v$ | Shear coefficient |
| $K_x$ | Spaced column fixity coefficient |
| $K_\theta$ | Angle to grain coefficient for bolt and lag screw connections |
| $K_1, K_2$ | Coefficients for truss deflection |
| $L$ | Span (ft) |
| $L$ | Span length of bending member (ft) |
| $L$ | Distance between points of lateral support of compression members (ft) |
| $L_c$ | Length from tip of pile to critical section (ft) |
| $L_c$ | Length between tangent points for pitched and tapered curved beams (ft) |
| $l$ | Span length of bending member (in.) |
| $l$ | Distance between points of lateral support of compression member (in.) |
| $l_b$ | Bearing length (in.) |
| $l_c$ | Clear span (in.) |
| $L_e$ | Effective length for shear (ft) |
| $l_e$ | Effective length of compression member (in.) |
| $l_e$ | Effective span length of bending member (in.) |
| $l_{e1}, l_{e2}$ | Effective length of compression member in planes of lateral support (in.) |
| $l_e/d$ | Slenderness ratio of compression member |
| $l_m$ | Length of bolt in wood main member (in.) |
| $l_n$ | Length of notch (in.) |

| | |
|---|---|
| $l_s$ | Total length of bolt in wood side member(s) (in.) |
| $l_t$ | Length of tapered leg for pitched and tapered curved beams (in.) |
| $l_u$ | Laterally unsupported span length of bending member (in.) |
| $l_1, l_2$ | Distances between points of lateral support of compression member in planes 1 and 2 (in.) |
| $l_3$ | Distance from center of spacer block to centroid of group of split ring or shear plate connectors in end block for a spaced column (in.) |
| $M$ | Moment capacity (in. lb) |
| $m$ | Final moisture content (below 30%) (%) |
| $m.c.$ | Moisture content of wood based on oven-dry weight of wood (%) |
| $M_D$ | Moment capacities for dowel bridge design (in. lb) |
| $M_s$ | Bending moment due to unit load (in. lb) |
| $M_y$ | Total secondary moment for dowel bridge design (in. lb) |
| $N$ | Fastener value for angle with direction of grain (lb) |
| $n$ | Number of dowels for bridge deck design |
| $n$ | Number of fasteners in a row |
| $N, N'$ | Nominal and allowable lateral design value at an angle to grain for a single split ring connector unit or shear plate connector unit (lb) |
| $P$ | Total concentrated load or axial load (lb) |
| $P$ | Design wheel load for bridge design (lb) |
| $P$ | Allowable passive soil pressure for poles (psf) |
| $P$ | Fastener value for load acting parallel to grain |
| $p$ | Depth of fastener penetration into wood member (in.) |
| $P, P'$ | Nominal and allowable withdrawal design value for fastener, lbs per inch of penetration |
| $Q$ | Fastener value for load acting perpendicular to grain (lb) |
| $Q$ | Statical moment of area about the neutral axis (in.$^3$) |
| $Q, Q'$ | Nominal and allowable lateral design value perpendicular to grain for a single split ring connector unit or shear plate connector unit (lb) |
| $R$ | Radius of curvature of inside face of lamination (in.) |
| $R_H$ | Horizontal reaction (lb) |
| $r$ | Radius of gyration (in.) |
| $R_B$ | Slenderness ratio of bending member |
| $R_D$ | Shear capacities for dowel bridge design (lb) |

| | |
|---|---|
| $R_m$ | Radius of curvature of centerline of curved member (in.) |
| $R_T, R_1, R_2,$ | Thermal resistance (hr ft$^2$ °F/Btu) |
| $R_v$ | Vertical reaction (lb) |
| $R_y$ | Total secondary shear for dowel bridge design (lb) |
| $S$ | Section modulus (in.$^3$) |
| $s$ | Effective bridge deck span (in.) |
| $s$ | Length of arch segment (in.) |
| $s$ | Center-to-center spacing between adjacent fasteners in a row (in.) |
| $S_B$ | Allowable soil-bearing capacity for poles (psf) |
| $S_m$ | Shrinkage from initial moisture condition to final moisture content $m$ (%) |
| $S_o$ | Total shrinkage from Table 2.3 (%) |
| $S_0, S_1, S_3, S_4$ | Allowable lateral soil-bearing pressure for poles (psf) |
| $T$ | Temperature (°F) |
| $T$ | Applied torque (in. lb) |
| $T$ | Tensile force (lb) |
| $t$ | Thickness (in.) |
| $t$ | Bridge deck thickness (in.) |
| $t$ | Fire resistance rating (min.) |
| $t$ | Thickness of column flange (in.) |
| $t$ | Thickness of lamination (in.) |
| $t_m$ | Thickness of main member (in.) |
| $t_s$ | Thickness of side member (in.) |
| $U$ | Overall heat transfer coefficient (Btu/hr ft$^2$ °F) |
| $u$ | Force in truss member caused by unit load (lb) |
| $V$ | Shear force (lb) |
| $W$ | Total uniform load (lb) |
| $w$ | Uniform load (pounds per unit length) |
| $W$ | Total load of 1 in. of water (lb/in.) |
| $W, W'$ | Nominal and allowable withdrawal design value for fastener, lbs per inch of penetration |
| $X$ | distance (ft) |
| $x$ | Distance (in.) |
| $x$ | Distance from beam support face to load (in.) |
| $x$ | Horizontal location (ft) |
| $y$ | Vertical location (ft) |
| $y$ | Wall height of arch (ft) |

| | |
|---|---|
| $Z$ | Dimensionless factor for 1 hr. fire rating |
| $Z, Z'$ | Nominal and allowable lateral design value for a single fastener connection (lb) |
| $Z_\parallel$ | Nominal lateral design value for a single bolt or lag screw connection with all wood members loaded parallel to grain (lb) |
| $Z_\perp$ | Nominal lateral design value for a single bolt or lag screw wood-to-metal connection with wood member(s) loaded perpendicular to grain (lb) |
| $Z_{m\perp}$ | Nominal lateral design value for a single bolt or lag screw wood-to-wood connection with main member loaded perpendicular to grain and side member loaded parallel to grain (lb) |
| $Z_{s\perp}$ | Nominal lateral design value for a single bolt or lag screw wood-to-wood connection with main member loaded parallel to grain and side member loaded perpendicular to grain (lb) |
| $\alpha$ | Angle measure (degrees) |
| $\alpha_r$ | Radial coefficient of thermal expansion |
| $\alpha_t$ | Tangential coefficient of thermal expansion |
| $\Delta_H$ | Horizontal movement (in.) |
| $\Delta$ | Deflection (in.) |
| $\Delta_c$ | Centerline deflection (in.) |
| $\theta$ | Angle measure (degrees) |
| $\pi$ | Pi |
| $\delta_{PL}$ | Proportional limit stress for bridge dowel design (psi) |
| $\gamma$ | Load slip modulus for a connection (lb/in.) |
| $\phi$ | Angle measure (degrees) |
| $\Omega$ | Coefficient of variation |

Other notations are explained where they occur.

# Part I

# GENERAL

# CHAPTER 1

# DESIGN CONSIDERATIONS IN THE USE OF STRUCTURAL TIMBER

## 1.1  INTRODUCTION

The American Institute of Timber Construction (AITC) has developed this *Timber Construction Manual* to provide state-of-the-art technical data and recommendations on engineered timber construction.

Chapter 1 includes basic information related to the use of structural timber framing. Topics include economy, durability, fire safety, erection, and detailing. With an understanding of these areas, the designer can more effectively utilize the advantages of wood construction. The unique characteristics of wood, design information, and standard practices are covered in subsequent chapters.

This manual applies to two types of engineered timber construction—sawn lumber and structural glued laminated timber (glulam). Sawn lumber is the product of lumber mills and is produced from many species. Glued laminated timber members are produced in laminating plants by gluing together dry lumber, normally of 2-in. or 1-in. nominal thickness, under controlled conditions of temperature and pressure. Members with a wide variety of sizes, profiles, and lengths can be produced having superior characteristics of strength, serviceability, and appearance. Glued laminated timber members are manufactured from several species, primarily Douglas Fir-Larch and Southern Pine, but also lesser amounts of Hem-Fir, Western Woods, Alaskan Cedar, and California Redwood are used. For cost-efficient design of structural glued laminated timber members, stress values as determined by the design should be specified rather than a particular species or stress combination.

## 1.2  DESIGNING FOR ECONOMY

The economic success of the construction of a project may be greatly influenced by design. Important elements of design include, but are not limited to, the layout of the framing, proper selection of materials and design of all components, ease of construction, serviceability for the intended use, and durability.

The best economy in timber construction is generally realized when standard-size members can be utilized in a repetitious arrangement. However, timber framing, especially glued laminated timber, can be custom fabricated to provide a nearly infinite variety of unique but cost-effective architectural forms and arrangements.

### 1.2.1    Standard Sizes and Grades

The use of standard sizes and grades will result in maximum economy. For glued laminated members, standard sizes as given in Table 8.2 are generally most economical. Any length, up to the maximum length limited by transportation and handling restrictions, is available. For sawn lumber, the sizes given in Table 8.1 are more economical than special sizes and special lengths. Lengths are generally available in even 2-ft increments, and there is a limit on the maximum length normally available from local suppliers.

### 1.2.2    Standard Connection Details

Specially designed connecting hardware should be avoided. A great variety of steel connecting devices have proved their effectiveness in permanent construction. Typical connection details are given in *Typical Construction Details*, AITC 104(1), included in Chapter 8. Also, see fastener manufacturers' catalogs.

### 1.2.3    Framing Systems, Sawn Timbers, and Glued Laminated Timbers

A great variety of structural timber framing systems are available. The relative economy of any one system over another will depend on the particular requirements of a specific job. Consideration of the overall structure, intended use, geographical location, required configuration, and other factors play an important part in determining the framing system to be used on a job. Table 1.1 may be used for preliminary design purposes to determine the economical span ranges for various timber framing systems. It must be emphasized that the table is to be used for preliminary planning purposes only. All systems require a more extensive analysis for final design.

The following additional considerations, when applied to timber framing system design, tend to reduce costs. Joints as simple and as few in number as practicable should be used. Splices should be so placed as to minimize design, fabrication, and erection problems. Unnecessary variations in members should be avoided; that is, the identical member design should be used repetitively where practical and the number of variations kept to a minimum. Certain roof profiles will affect the amount and type of load on a structure and may, therefore, affect economy. Better economy usually results from specifying the required design values rather than the lumber grades to be used. Judicious use of multiple continuous spans or cantilever systems with suspended spans tends to balance positive and negative moments and may lower costs.

### 1.2.4    Appearance Grades for Glued Laminated Timber

An additional consideration related to economy of design is appearance. AITC has developed specifications for three standard appearance grades of glued laminated members. These are given in *Standard Appearance Grades for Structural Glued Laminated Timber*, AITC 110 (2), included in Chapter 8. These appearance grades

## TABLE 1.1

### Economical Span Ranges for Various Timber Framing Systems

| Type of System | Economical Span Range (ft) | Considerations |
|---|---|---|
| **A. Primary Framing Systems** | | |
| *Roof Framing Systems* | | |
| *Beams* | | Beam systems are frequently used where a low-pitched roof shape is desired |
| Simple spans | | |
| Straight beams | | |
| Glued laminated | 10–100 | |
| Sawn | 6–32 | |
| I Joists | 12–40 | |
| Tapered beams | 25–100 | |
| Double tapered-pitched beams | 25–100 | |
| Curved beams | 25–100 | |
| Cantilevered systems | | |
| Glued laminated | Up to 90 | |
| Sawn | Up to 24 | Usually more economical than simple spans when span is over 40 ft |
| Continuous spans | | |
| Glued laminated | 10–32 | |
| Sawn | Up to 16 | |
| Girders | 40–80 | |
| *Arches* | | |
| Three-hinged arches | | For relatively high rise applications |
| Gothic | 40–90 | |
| Tudor | 20–120 | Provides required vertical wall frame |
| A-Frame | 20–100 | |
| Three centered | 40–250 | |
| Parabolic | 40–250 | |
| Radial | 40–250 | |
| Two-hinged arches | | For relatively low rise applications |
| Radial | 50–200 | |
| Parabolic | 50–200 | |

*(continued)*

**TABLE 1.1**   (*Continued*)

| Type of System | Economical Span Range (ft) | Considerations |
| --- | --- | --- |
| *Trusses (Heavy)* | | Provide openings for passage of wiring, piping, etc. |
| Flat or parallel chord | 50–150 | Low roof profile |
| Triangular or pitched | 50–90 | For pitched roofs requiring flat surfaces |
| Bowstring (continuous chord) | 50–200 | Provide greatest clearance with least wall height |
| Carrying | 40–60 | |
| *Trusses (light)* | | Most light trusses commonly used within these ranges are based on proprietary connections and fabrication methods |
| Flat or parallel chord | 20–50 | |
| Triangular or pitched | 20–75 | |
| *Tied arches* | 50–200 | Good where no ceiling is wanted; give clear open appearance for low-rise curve; normally more expensive than bowstring; buttress not required |
| *Dome structures* | 50–500+ | |
| *Beams* | *Floor Framing Systems* | |
| Simple span | | |
|   Glued laminated | 6–40 | |
|   Sawn | 6–20 | |
|   I Joists | 12–30 | |
| Continuous | 25–40 | |

**B. Secondary Framing Systems**

| | *Roof Framing Systems* | |
| --- | --- | --- |
| *Sheathing and decking* | | |
| 1-in. sheathing applied directly to primary system | 1–4 | Check deflection on spans greater than 32 in. |
| 2-in. roof deck applied directly to primary system | 6–10 | Check deflection on spans greater than 8 ft |
| 3-in. roof deck applied directly to primary system | 8–15 | 2-, 3-, and 4-in. decking provide good insulation, fire resistance, appearance; easy to erect |

**TABLE  1.1**  (*Continued*)

| Type of System | Economical Span Range (ft) | Considerations |
|---|:---:|---|
| *Sheathing and decking   (Continued)* | | |
| 4-in. roof deck applied directly to primary system | 12–20 | |
| Plywood or structural panel sheathing applied directly to primary system | 1–4 | |
| Stressed skin panels | 8–40 | |
| Joists with sheathing | 16–24 | |
| Purlins with sheathing | 16–36 | |
| Beams | 20–40 | |
| *Floor Framing Systems* | | |
| *Plank decking* | | Floor and ceiling in one |
| Edge to edge | 4–16 | |
| Wide face to wide face | 4–16 | |
| Joists with sheathing | 10–24 | |

are not related to strength. It is more economical to specify the finish or appearance grade best suited for each job than to require the ''premium'' appearance grade for all jobs.

## 1.3  DESIGNING FOR PERMANENCE

Permanent timber structures should be built not only to be structurally adequate, but also to be durable with a minimum of maintenance. With proper design details, construction procedures, and usage, wood is a permanent construction material. Certain conditions affect durability and maintenance costs. If proper consideration is given to these in the design phases of a project, there will be greater assurance that the structure will be durable and that maintenance costs will be minimal. Untreated wood has a proven performance of indefinitely long service if it is kept below 20% moisture content. When wood is exposed to the weather and not properly protected by a roof, eave overhang, or similar covering, or subjected to other conditions of free water or high relative humidity where decay is possible, a preservative treatment is required unless the heartwood of a naturally decay resistant species such as Redwood or Cedar is used. See *Wood Handbook* (3) for a listing of domestic woods that are resistant or very resistant to heartwood decay. The need for preservatively treated wood is a design consideration based on the conditions intended for the wood in service. See *Evaluation, Maintenance and Upgrading of Wood Structures* (4) for information on existing structures.

### 1.3.1    Wood–Moisture Relationships

Dimensional changes in wood result primarily from a gain or loss of moisture. Once the moisture content (MC) has been lowered to the fiber saturation point (approximately 25–30%), further loss of moisture results in wood shrinkage. This shrinkage continues almost linearly down to 0% moisture content. Eventually, wood assumes a condition near equilibrium with its environment. Wood shrinks and swells most significantly in a direction perpendicular to the grain. Consideration must be given for changes in dimension if moisture changes can occur.

Wood absorbs water in free liquid form and as a water vapor, and gives off water again when the humidity of the surrounding air is lowered. If wood becomes wet after being installed dry, its swelling results in increased dimensions and sometimes in distortion and twisting. If installed wet, wood may dry and shrink in service with resulting checking, movement, or distortion as a function of the final moisture content.

### 1.3.2    Significance of Checking

Checking is the result of rapid lowering of surface moisture content combined with differential moisture contents of the inner and outer portions of the piece. As wood loses moisture content to the surrounding atmosphere, the outer cells of the members lose moisture at a more rapid rate than do the inner cells. As the outer cells try to shrink, they are restrained by the inner portion of the member, which has a higher moisture content. The more rapid the rate of drying, the greater will be the differential in shrinkage between the outer and inner fibers and the higher will be the shrinkage stresses. Some species exhibit more checking than others because they may gain or lose moisture faster or have higher shrinkage rates.

In sawn lumber and timber, controlling the rate of drying will avoid or minimize checking. Many monumental buildings with great historical value have been constructed in such a way as to exploit the natural beauty of large sawn timbers. In these structures, seasoning checks are accepted as an inherent characteristic of the material.

One of the principal advantages of glued laminated timber members is their freedom from major checking; however, seasoning checks may occur in glued laminated members for the same reasons as in sawn timbers, but generally the range of moisture content permitted by industry standards approximates the moisture content in normal-use conditions, thereby minimizing checking that might occur. Moisture content of lumber at the time of gluing is, therefore, of great importance in the control of checking in service. However, serious rapid changes in moisture content after gluing will result in shrinkage or swelling of the wood and may develop stresses in both glued joints and wood that will cause checking. If glued laminated timbers are not carefully protected during shipping, storage, and erection, they may pick up moisture during this period. Subsequent drying may result in more checking than would have occurred had the members remained at the moisture content existing at the time of manufacture. Differentials in the shrinkage rate of individual laminations tend to concentrate shrinkage stresses at or near glue lines. The presence of wood fiber separation indicates adequate glue bond and not delamination. Additional information is contained in AITC Technical Note 11 (5).

### 1.3.2.1 Structural Considerations

**Glued Laminated Timber.** In general, checks have little effect on the strength of laminated members. Glued laminated members are made from laminations thin enough to season readily without developing checks on the edges. Checks or splits that occur in drying generally are on the wide faces, and do not materially affect the shear strength of bending members loaded perpendicular to the wide faces of the laminations (horizontally laminated beams), which is the most common use of laminated members. If bending members are designed with the load applied parallel to the wide faces of the laminations (vertically laminated members), the occurrence of checks on the wide faces of laminations may affect the shear strength of a beam. For this reason, the design value in shear, $F_{vy}$, for vertically laminated members has been reduced as shown in the tables in AITC 117—Design (6) included in Chapter 8 of this manual. Seasoning checks in bending members affect the shear parallel to grain value only. They are usually not of structural importance unless the checks are significant in depth, occur in the midheight of the member near the support, and shear governs the design of the members. All these factors must be considered in appraising checks from a structural viewpoint. In general, the reduction in shear strength is directly proportional to the ratio of depth of check to the width of the bending member. Minor checking may be disregarded, as there is an ample factor of safety in design values. For more information, see *Evaluation of Checking in Glued Laminated Timbers*, AITC Technical Note 18 (7).

**Sawn Lumber and Timber.** Checks affect the shear parallel to grain strength of lumber, and in establishing design values, checks are anticipated by a large reduction factor applied to test values in recognition of stress concentrations at the ends of the checks. The published design values for parallel to grain shear for the lumber grades are again adjusted for the amount of checking permissible in the various grades at the time of the grading. Because the strength properties of wood increase with dryness, checks may enlarge somewhat with increased dryness from the time of shipment without decreasing shear strength materially. Actually, a fully seasoned timber may be checked in excess of the grade limitations without affecting its adequacy in the structure because the grading rules are set up with anticipation that some checking beyond the grade limitations at the moisture content graded may occur with seasoning. This is particularly true if loads producing shear stresses are low. Furthermore, only checks that occur near the supports and in the midhalf of the depth of the member are important to shear strength.

Even though the grading rules would exclude such a piece, a column may be checked nearly through without having its strength seriously reduced because the main consideration is that the member act as a unit without splitting entirely into two parts because the $l/d$ ratio would then be reduced.

Cross-grain checks and splits that tend to run out the side of a piece or excessive checks and splits that tend to enter connection areas may be serious and may require repair or replacement. Details for minimizing and controlling the effects of checking in connection areas should be incorporated into the design details. To avoid excessive splitting between rows of bolts due to shrinkage during the seasoning period of timbers, the rows of bolts should not be spaced more than 5 in. apart or a saw kerf, terminating in a bored hole, should be provided between the lines of bolts. Alternatively, the use of two splice plates, one for each row of bolts, rather than a single splice plate will lessen the probability of splitting. Whenever possible, maximum end distances for connections should be specified to minimize the effect of checks running into the joint area. Some designs require stitch bolts or fully threaded lag screws in members with multiple connections loaded at an angle to the grain. Stitch bolts, which should be kept tight, will reinforce pieces where checking is excessive.

The final decision about whether or not shrinkage checks are detrimental to the strength requirements of any particular structure should be made by a competent engineer experienced in timber construction.

### 1.3.3    Protection

#### 1.3.3.1    Glued Laminated Timber

The application of selected sealers, paints, and similar protective measures to the surface of the members can retard, but not prevent, checking under extreme conditions. Laminated members for use under conditions that may produce checking more serious than slight surface checks should receive protective coatings or other protection should be required in accordance with *Recommended Practice for Protection of Structural Glued Laminated Timber during Transit, Storage and Erection*, AITC 111 (8), included in Chapter 8.

#### 1.3.3.2    Sawn Lumber and Timber

Because lumber loses moisture approximately 10 times faster through the end grain than through flat grain or vertical grain faces, the application of sealer to the end grain will help to minimize the shrinkage checks or splits. The extent of checking is related to the steepness of the moisture gradient between the surfaces and the interior of the piece. Unseasoned lumber should be protected from hot sun and dry winds. Because attic spaces are frequently higher in temperature and lower in humidity than the surrounding atmosphere, they should be properly vented to prevent lumber from rapid drying and resultant checking.

### 1.3.4    Wood-Destroying Organisms and Their Control

#### 1.3.4.1    Mold, Stain, and Decay

Decay of wood is caused by low forms of plant life (fungi) that develop and grow from microscopic spores that are present wherever wood is used [see *Wood Handbook* (3)]. The fungi convert wood substance into food. If deprived of any one of the four essentials of life (food, air, moisture, or favorable temperature), decay growth is prevented or stopped and the wood remains sound, retaining its existing strength with no further deterioration. Wood will not be attacked by fungi if it is submerged in water, thereby excluding air, kept continuously below 20% moisture or maintained at temperatures below freezing or much above 100°F. Growth can begin or resume whenever climatic conditions are favorable. The *Standard for Preservative Treatment of Structural Glued Laminated Timber*, AITC 109 (9), included in Chapter 8, provides information about preservative treatment of glued laminated timber. For treatment of sawn lumber, see the American Wood-Preservers' Association *Book of Standards* (10).

Molds and stains are confined largely to sapwood and are of various colors.

Molds generally do not stain the wood, but produce surface blemishes varying from white or light colors to black that can often be brushed off. Fungus stains penetrate the sapwood and normally cannot be removed by scraping or sanding. Molds and stains should not be considered as stages of decay because these fungi do not attack the wood substance appreciably. For most uses in which appearance is not a factor, wood strength is practically unimpaired by stains and molds. Ordinarily, their only effect is confined to those properties that determine shock resistance or toughness. The only danger is that the early stages of decay may also be hidden in the discolored areas of molds or stain.

Decay-producing fungi, under conditions that favor their growth, may attack both heartwood and sapwood. Heartwood is more resistant to attack than sapwood. Fresh surface growths are usually fluffy or cottony, seldom powdery like the surface growths of molds. The early stages of decay are often accompanied by a discoloration of the wood, which is more evident on freshly exposed surfaces of unseasoned wood than on dry wood. However, many fungi produce early stages of decay that are similar in color to that of normal wood or they give the wood a water-soaked appearance. Later stages of decay are easy to recognize because the wood has undergone definite changes in color and texture.

Decay from fungi reduces the strength of wood. In later stages, all decay fungi seriously reduce the strength of wood and also its fire resistance.

Brown, crumbly rot, in the dry condition, is sometimes called dry rot, but the term is incorrect because decay fungi must have some source of moisture for development, even though the wood may have subsequently become dry.

In some cases, wood submerged in water may be attacked by bacteria, resulting in a slow decay process [see *Wood Handbook* (3)].

**Prevention and Control of Decay.**   Timber to be used where conditions favorable to the growth of decay-producing organisms are unavoidable should be pressure treated with a wood preservative, unless the heartwood of naturally decay-resistant species is available and is considered adequate in view of the decay hazard encountered. See *Wood Handbook* (3) for a list of woods resistant or very resistant to decay. Where serious decay problems are present in buildings, they are almost always signs of faulty design or construction or of lack of reasonable maintenance. Design and construction principles that will assure long service and avoid decay hazards in buildings include

1.   positive site and building drainage,

2.   adequate separation of wood from known moisture sources, and

3.   ventilation and condensation control in enclosed spaces.

Building sites should always be graded to provide positive drainage away from foundation walls. All stumps, wood debris, stakes, or wood concrete forms should be removed from the immediate vicinity of a building before backfilling and before placing floor or slabs on grade.

All exposed wood surfaces and adjoining areas should be pitched to assure rapid runoff of water. Construction details that tend to trap moisture in or near wood members should be avoided. The prevention of decay in timber structures rests largely in designs that prevent the entrance and retention of rainwater.

A fairly wide roof overhang with gutters and downspouts that are properly designed is desirable.

Adequate separation of wood from known sources of moisture is necessary to prevent excessive absorption of moisture and to facilitate periodic inspection. When it is impossible or impractical to provide adequate separation, it is recommended that preservatively treated wood or the heartwood of naturally durable species be used.

Wood in contact with concrete near the ground should be protected by a moistureproof membrane such as heavy asphalt paper. In many cases, preservative treatment of the wood in actual contact with the concrete is advisable and required by building codes. Girder and joist openings in masonry walls should be big enough to assure that there will be an air space around the sides and ends of these wood members, and if the members are below the outside soil level, moistureproofing of the outer face of the walls is essential.

Unventilated and inaccessible spaces under buildings should be avoided. Wetting of the wood by condensation may result in serious decay damage. A crawl space with at least an 18-in. clearance should be left under wood joists and girders. Condensation can be minimized by providing openings on opposite foundation walls for cross ventilation. Laying heavy roll roofing or polyethylene membrane with lapped joints on the soil provides an effective moisture barrier.

Porches, breezeways, patios, and other appurtenances may present a decay hazard that cannot be fully avoided by construction practices. Therefore, it is recommended that preservatively treated wood be used for members exposed to decay hazards in these applications.

Sheathing papers on the cold surface of the walls should be of a ''breathing'' or vapor-permeable type. Vapor barriers, if installed, should be near the warm face of insulated walls and ceilings [see ASHRAE recommendations (11)]. Attic ventilation should be provided.

Where highly humid conditions are present in buildings, as in textile mills, pulp and paper mills, controlled atmosphere plants or storage buildings, and enclosed pools and shower rooms, preservatively treated wood should be used unless the heartwood of naturally decay resistant species is available and considered adequate in view of the decay hazard encountered. When the humid condition involves chemical fumes as well as water vapor, the compatibility of the environment with the treatment and the wood itself should be investigated.

To supplement good design and construction practices, periodic inspections of a structure will provide assurance that decay-preventive measures are being maintained and that additional decay hazards are not present. These inspections should reveal indications of moisture penetration or condensation, and if detected, corrective measures should be taken to avoid significant damage.

### 1.3.4.2  Insects

In terms of economic loss, the most destructive insect to attack wood buildings is the subterranean termite. In certain localities, nonsubterranean termites are also very destructive. Other insects attack timber buildings, but ordinarily, these occurrences are rather rare and their damage is slight. In many cases, these insects can be controlled by the methods used for termites.

**Subterranean Termites.**  The occurrence of subterranean termites and the damage caused by them is much greater in southern states than in northern states, where lower temperatures do not favor their development (see Fig. 1.1). However, damage to individual buildings may be just as great in northern states as in southern states.

Subterranean termites develop and maintain colonies in the ground from which they build their tunnels through the earth and around obstructions to get at the wood they need for food. Each colony shuts itself off and lives in the dark; but unless they have a constant source of moisture, the termites will die. The worker members of the colony cause the destruction of wood. At certain seasons of the year, male and female winged forms swarm from the colony, fly a short time, lose

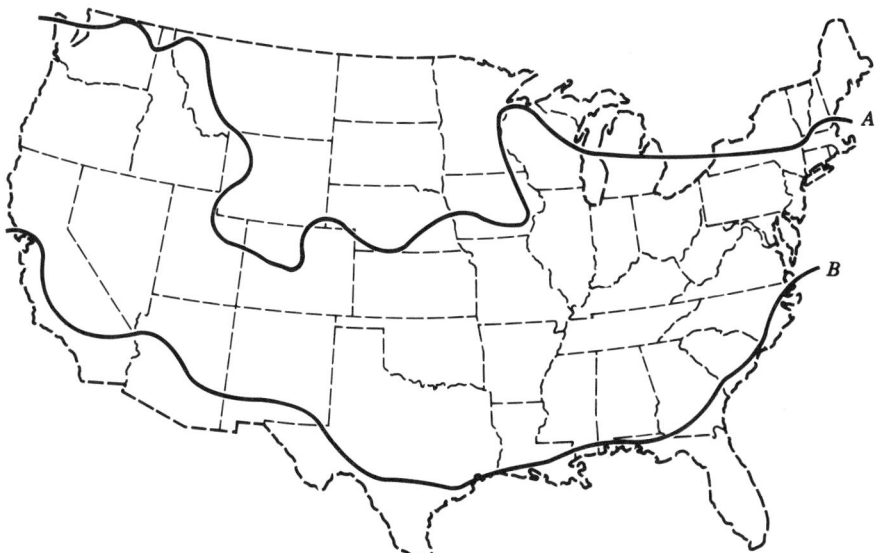

**FIGURE 1.1**   Limits of Termite Damage. *A*, the northern limit of recorded damage done by subterranean termites in the United States; *B*, the northern limit of damage done by dry wood or nonsubterranean termits. Source: *Wood Handbook*, U.S. Department of Agriculture Handbook No. 72, 1987.

their wings, mate and, if they succeed in locating suitable places, start new colonies. The appearance of "flying ants" or of their shed wings is an indication that a termite colony may be near and causing serious damage. Not all flying ants are termites; therefore, suspected insects should be identified before money is spent for their eradication.

Subterranean termites do not establish themselves in buildings by being carried there in lumber, but by entering from the ground nests after the building has been constructed. They must maintain contact with a source of moisture such as the soil. Telltale signs of the presence of termites are the earthen tubes, or runways, built by these insects over the surfaces of foundation walls to reach the wood above. Another sign is the swarming of winged adults early in the spring or fall. In wood itself, the termites make galleries that follow the grain, leaving a shell of sound wood to conceal their activities. Because the galleries seldom show on the wood surface, probing with an ice pick or knife is advisable if the presence of termites is suspected.

*Protection and Control.*    Where subterranean termites are prevalent, the best protection is to build so as to prevent their gaining access to the building. The foundations should be of preservatively treated wood, concrete, or other solid material through which the termites cannot penetrate. With brick, stone, or concrete blocks, cement mortar should be used, for termites can work through some other kinds of mortar.

Wood that is not impregnated with an effective preservative must be kept away from the ground. If there is a basement, it should preferably be floored with concrete. Posts supporting the first-floor beams must be thoroughly treated if they bear directly on the ground or on wood blocking. Untreated posts should rest on concrete piers extending at least 1 in. above the basement floor if that floor is of concrete; if the basement floor is of earth, the concrete piers should extend at least 18 in. above it. If the earth is not excavated beneath the building, and if the floor is of wood construction, the floor joists and other woodwork, unless adequately treated with preservative, should be kept at least 18 in. above the earth and good ventilation should be provided beneath the floor.

Moisture condensation on the floor joists and subflooring, which may cause conditions favorable to decay and thus make the wood more attractive to termites, can be avoided by covering the soil with a waterproof membrane. For buildings with concrete slab floors laid over the soil or directly over a sand or gravel fill, the soil under the floor should be treated before the concrete is placed. Furthermore, when insulation containing cellulose is to be used as a filler in expansion joints, it should be impregnated with a chemical toxic to termites. Sealing the top $\frac{1}{2}$ in. of expansion joints and other openings will provide additional protection from ground-nesting termites.

All concrete forms, stakes, stumps, and waste wood should be removed from the building site at the time of construction because they are possible sources of infestation. Generally, precautions that are effective against decay are also helpful against subterranean termites. Where protection is needed in addition to that obtained by structural methods, the soil adjacent to the foundation walls and piers beneath the building should be thoroughly treated with a recognized insecticide. To control termites already in a building, break contact between the termite colony in the soil and the wood. This can be done by mechanically blocking the runways from soil to wood, by treating the soil, or by both of these methods. Possible reinfestations should be guarded against by frequent inspections for the telltale signs previously given.

The Formosa termite has become established in the United States. It is more active and voracious than the subterranean termites native to the United States. However, the conventional methods of protection appear to be effective.

**Nonsubterranean Termites.** Nonsubterranean termites have been found only in a narrow strip of territory extending from central California around the southern edge of the continental United States to Virginia, and also in the West Indies and Hawaii (see Fig. 1.1). Their damage is confined to an area in southern California, to parts of southern Florida, notably Key West, and to Hawaii. The nonsubterranean dry-wood termites are fewer in number, and the depredations are not rapid, but if they are allowed to work unmolested for a few years, they can occasionally ruin timbers with their tunnelings.

*Protection and Control.* In the principal damage areas, careful examination of wood is needed to avoid the occurrence of infestations during the construction of a building. All exterior wood can be protected by placing fine-mesh screen over all holes in the walls or roof of the building. If a building is found to be infested by dry-wood termites, badly damaged wood must be replaced. Further termite activity can be arrested by approved chemical treatments applied under proper supervision which will provide for safety of people, domestic animals, and wildlife. Where practical, fumigation is another method of destroying insects.

**Other Wood-Inhabiting Insects.** Large wood-boring beetles and wood wasps infect green wood, but may complete their development in seasoned wood. They do not reinfest dry wood. They occur in all forest areas. The borers in timber or lumber can be killed by heating the wood to a center temperature of 130°F for 1 hr or by fumigation. If infested wood is used in constructing a building, the emerging adult borers will chew $\frac{1}{8}$-$\frac{1}{2}$-in. holes to the surface, penetrating insulation, vapor barriers, siding, or interior surface materials. The surface holes can be plugged and the damaged spots finished or refinished to make them inconspicuous. Because the borers do not reinfest dry wood, extermination treatments are not required in buildings.

Powder-post beetles can infest and reinfest dry wood, reducing it to floury sawdust. The *Lyctus* powder-post beetles, which are encountered most frequently, attack large-pored hardwoods. Their attacks may be recognized by tunnels packed with floury sawdust and numerous emergence holes $\frac{1}{32}$-$\frac{1}{8}$ in. in diameter. Heat or fumigation treatments will kill the beetles, but will not prevent reinfestation. It can be prevented by a surface application of an approved insecticide in a light-oil solution. Any finishing material that plugs the surface holes of wood will also immunize the wood from *Lyctus* attack. Usually, infestations in buildings result from the use of infested wood, and insecticidal treatment or fumigation may be needed to eliminate them.

Carpenter ants chew nesting galleries in wood. The principal species are large dark-colored ants, and many individuals in the colony are $\frac{1}{2}$ in. long. They exist throughout the United States. Because the ants require a nearly saturated atmosphere in their nest, an ant infestation may indicate a moisture problem in the wood that could also result in decay damage. Ant infestations may be controlled by insecticides.

**Marine Borers.**   Fixed or floating wood structures in salt or brackish water are subject to attack by marine borers. Marine borers include shipworms such as *Teredo* and *Bankia*, the pholads *Martesia* and *Zylophaga*, and *Limnoria* and *Sphaeroma*. Almost all attack wood as free-swimming organisms in the early part of their lives. Shipworms and pholads bore an entrance hole generally at the waterline, attach themselves, and grow in size as they bore tunnels into the wood. *Limnoria* and *Sphaeroma* generally burrow just below the surface of the wood.

*Protection and Control.*   For areas where shipworm and pholad attacks are known or expected and where *Limnoria* attack is not expected, the wood should be pressure treated with a creasote and/or creosote-coal tar solution. For areas where *Limnoria* and pholad attack are known or expected, a dual treatment of waterborne salts and creosote is recommended. Where *Limnoria* attack is known or expected and where pholads are absent, either a dual treatment or waterborne salts preservatives may be used.

### 1.3.5  Effects of Temperature on Wood

The effect of temperature on the dimensional stability of wood is discussed in Chapter 2. Temperature also has an important effect on the strength of wood.

Design values given in tables in Chapter 8 apply to glued laminated timber or sawn lumber used under ordinary temperature conditions. Some reduction of these values may be necessary for members subjected to elevated temperatures for repeated or prolonged periods of time, especially where the high temperature is associated with a high moisture content in the wood.

The temperature effect on strength is immediate, and its magnitude depends on the moisture content of the wood and, when the temperature is elevated, on the time of exposure. When wood is exposed to temperatures above normal for a limited period, and the temperature is not excessive, the wood can be expected to recover essentially all its original strength when the temperature is reduced to normal. However, when wood is exposed to temperatures above normal for a limited period and is expected to carry design loads during this period, reduction in design values should be considered. Experiments indicate that dry wood (12% MC) can probably be exposed to temperatures up to nearly 150°F for a year or more without an important permanent loss in most of its strength properties, but while heated, its strength will be temporarily reduced compared to its strength at normal temperature. However, if wood is heated to temperatures of up to 150°F for extended periods of time, adjustment of design values may be necessary as indicated in Table 4.11.

Tests of wood conducted at about −300°F show that the important strength properties of dry wood in bending and in compression, including stiffness and shock resistance, are much higher at the extremely low temperature than at normal temperatures.

The tabular design values of wood are based on ordinary ranges of temperatures occasionally heated to 150°F. Wood increases in strength when cooled below normal and decreases when heated. A temporary decrease in strength occurs when wood is heated up to 150°F. When the wood cools, it will regain its strength. The change in properties will be greater if the moisture content is low. In some geographical locations, fairly high temperatures are commonly experienced, but the accompanying relative humidity is ordinarily quite low. Wood exposed to such conditions generally has a low moisture content, and the immediate effect of the high temperature is not significant.

When wood is exposed to temperatures of 150°F or higher for extended periods of time, it is permanently weakened, even though the temperature is subsequently reduced and the wood is used at normal temperatures. The permanent or nonrecoverable strength loss depends on a number of factors, including the moisture content and temperature of the wood, the heating medium and time of exposure, and to some extent on the species and the size of the piece. In special cases, timbers that are exposed to elevated temperatures for extended lengths of time may need adjustments to the design values. Refer to Appendix C, ANSI/NF₀PA NDS— 1991, *National Design Specification for Wood Construction* (12).

Glued laminated members are normally cured at temperatures of less than 150°F. Therefore, no reduction in their design values due to temperature effects during manufacturing is necessary.

Adhesives used under standard specifications for structural glued laminated members (i.e., casein, resorcinol resin, phenol resin, and melamine resin adhesives) are not affected substantially by the high temperatures that char wood. The use of other adhesives that might deteriorate at lower temperatures is not permitted by standard specifications for structural glued laminated timber. Low temperatures appear to have no significant effect on the strength of glue joints.

### 1.3.6 Effects of Chemical Processes or Stored Chemicals on Wood

Wood is often superior to many other common construction materials in its resistance to chemical attack. For this reason, wood is used for storage buildings and containers for many chemicals and in processing plants in which structural members are subjected to spillage, leakage, or condensation of chemicals.

Wooden tanks are commonly employed for the storage of water or chemicals that deteriorate other materials. Experience has shown that the heartwood of Cypress, Douglas Fir-Larch, Southern Pine, and California Redwood is the most suitable for water tanks and that the heartwood of the first three of these species is most suitable for tanks when resistance to chemicals in appreciable concentrations is an important factor. The heartwood of each of the four species combines moderate to high resistances to water penetration with moderate to high natural resistance to decay and hydrolysis. Structural members, floors, stairways, catwalks, and docks are frequently exposed to spillage or leakage of chemicals. Volatile chemicals may attack roof supports, ventilating ducts, and stacks in chemical processing plants. Wood should be considered for these applications.

Chemical actions of three general types may affect the strength of wood. The first causes swelling and the resultant weakening of the wood. Liquids such as water, alcohols, and some other organic liquids swell wood. This action is almost completely reversible; hence, if the swelling liquid or solution is removed by evaporation or by extraction followed by evaporation of the solvent, the original dimensions and strength are practically restored. Liquids such as petroleum oils and creosote do not swell wood. The second type of action brings about permanent changes in the wood, such as hydrolysis of the cellulose by acids or acid salts. The third type of action, which is also permanent, involves delignification of the wood and dissolving of hemicelluloses by alkalies.

Experience and available data indicate species and conditions where wood is equal or superior to other materials in resisting the degradative action of chemicals. In general, heartwood of such species as Cypress, Douglas Fir-Larch, Southern Pine, California Redwood, Maple, and White Oak is quite resistant to attack by dilute mineral and organic acids. Oxidizing acids, such as nitric acid, have a greater degradative action than nonoxidizing acids. Alkaline solutions are more destructive than acidic solutions, and hardwoods are more susceptible to attack by both acids and alkalies than softwoods.

Highly acidic salts tend to hydrolyze wood when present in high concentrations. Even relatively low concentrations of such salts have shown signs that the salt may migrate to the surface of railroad ties, which are occasionally wet and dried in a hot, arid region. This migration, combined with the high concentrations of salt relative to the small amount of water present, causes an acidic condition sufficient to make wood brittle.

Iron salts, which develop at points of contact with plates, bolts, and the like, have a degradative action on wood, especially in the presence of moisture. In addition, iron salts probably precipitate toxic extractives and thus lower the natural decay resistance of wood. The softening and discoloration of wood around corroded iron fastenings is a commonly observed phenomenon; it is especially pronounced in acidic woods, such as Oak, and in woods such as California Redwood, which contain considerable tannin and related compounds. The oxide layer formed on iron is transformed through reaction with wood acids into soluble iron salts, which not only degrade the surrounding wood, but probably catalyze the further corrosion of the metal. The action is accelerated by moisture; oxygen may also play an important role in the process. This effect is not encountered with well-dried wood used in dry locations. Under damp-use conditions, it can be avoided or minimized by using corrosion-resistant fastenings.

Many substances have been employed as impregnants to enhance the natural resistance of wood to chemical degradation. One of the more economical treatments involves pressure impregnation with a viscous coke-oven coal tar to retard liquid penetration. Acid resistance of wood is increased by impregnation with phenolic resin solutions followed by appropriate drying and curing. Treatment with furfuryl alcohol has been used to increase resistance to alkaline solutions. Another procedure involves massive impregnation with a monomeric resin, such as methyl methacrylate, followed by polymerization.

## 1.4  DESIGNING FOR FIRE SAFETY

Neither building materials alone, nor building features alone, nor detection and fire extinguishing equipment alone can provide adequate safety from fire in buildings. A proper combination of these should provide the necessary degree of protection for the occupants and for the property. Some of the more important design considerations that should be investigated are listed.

General considerations:

1.  Use, occupancy, or activity taking place in the building.
2.  Fire stopping, draft stopping, and the elimination or proper protection of concealed spaces.
3.  Separation of areas in which hazardous processes or operations take place, such as boiler rooms and workshops.
4.  Location of structure within the property lines.
5.  Height of structure, allowable floor areas, automatic alarms, and sprinkler systems.

Safety to life considerations:

1.  The number, size, type, and accessibility of exit ways (particularly stairways) and their distance from each other.
2.  The installation of automatic alarm systems.
3.  Enclosure of stairwells and use of self-closing fire doors.
4.  Interior finishes that will assure that the surfaces will not spread flame at hazardous rates.
5.  Ventilation systems or self-contained breathing apparatus.

Safety to property considerations:

1.  The installation of automatic sprinkler systems.
2.  Proper placement of firewalls and proper protection of openings in them.
3.  Interior finishes to assure that surfaces will not spread flame at hazardous rates.
4.  Roof venting equipment or provision for draft curtains.

Protection of the occupants of a building and of the property itself can be achieved by taking advantage of the fire-endurance properties of wood and by giving careful attention to the above-mentioned details that make a building firesafe, that is, fire-safe rather than so-called fireproof.

Wood, when exposed to heat and/or flame, forms a self-insulating surface layer of char. Although the surface chars, the undamaged inner wood below the char retains its strength and will support loads equivalent to the capacity of the uncharred section. Very often, heavy timber members will retain their structural integrity through long periods of fire exposure and still remain serviceable after the surface has been cleaned and refinished. The fire endurance and excellent performance of heavy timber are attributable to the size of the wood members and to the slow rate at which the charring penetrates.

Timber, unlike most other construction materials, is not appreciably distorted by high temperatures; therefore, it is not likely that walls will be displaced or pushed over by expanding members, as often happens with some other materials.

Size of timber members is of particular importance with respect to fire endurance. Building codes specify minimum nominal dimensions for exposed heavy timber structural members. For fire-resistance classification purposes, buildings of wood construction are generally classified as heavy timber construction, ordinary construction, or wood frame construction. *Heavy timber construction* is that type in which fire resistance is attained by placing limitations on the minimum size, thickness, or composition of all load-carrying wood members; by avoiding concealed spaces under floors or roofs; by using approved fastenings, construction details, and adhesives; and by providing the required degree of fire resistance in the exterior and interior walls. *Ordinary construction* has exterior masonry walls and wood framing members of sizes smaller than heavy timber sizes. *Wood frame construction* has wood-framed walls and structural framing of sizes smaller than heavy timber sizes. Depending on the occupancy of a building or the hazard of the operations within it, a building of wood frame or ordinary construction may have its members covered with fire-resistive coverings. All of the nationally recognized building codes and most other codes recognize heavy timber construction by allowing larger building areas and uses for which ordinary construction and wood frame is not permitted. The requirements for heavy timber construction are contained in *Standard for Heavy Timber Construction*, AITC 108 (13), in Chapter 8.

### 1.4.1 Fire-Rated Timber Construction

In some instances, a specific fire rating is required by code for structural members in a building. Common stud walls, when covered with a fire-rated gypsum board, can provide a rating of 1 hr or more. Larger members may also be covered with gypsum board to increase the resistance to fire. The fire rating of wood members may be calculated by taking into account the depth, breadth, and number of sides exposed to fire and the ratio of the actual load on a member to the allowable design load. For instance, a member loaded to 50% of its design load has a greater fire rating than a member fully loaded. The following formulas may be used to calculate the fire-resistance rating of wood members, both sawn and glued laminated timber. This method is based on a procedure contained in *A Method for Assessing the Fire Resistance of Laminated Beams and Columns* (14) by T. T. Lie. The fire-resistance rating $t$, in minutes, of beams and columns can be calculated by use of the following equations.

For beams exposed to fire on four sides,

$$t + 2.54\ Zb\left(4 - \frac{2b}{d}\right) \tag{1-1}$$

For beams exposed to fire on three sides,

$$t = 2.54\ Zb\left(4 - \frac{b}{d}\right) \tag{1-2}$$

For columns exposed to fire on four sides,

$$t = 2.54\ Zd\left(3 - \frac{d}{b}\right) \tag{1-3}$$

For columns exposed to fire on three sides,

$$t = 2.54\ Zd\left(3 - \frac{d}{2b}\right) \tag{1-4}$$

where   $Z$ = dimensionless load factor from Fig. 1.2,
   $b$ = breadth (width) of a beam or larger side of a column before fire exposure (in.), and
   $d$ = larger side of a beam or smaller side of a column before exposure to fire (in.).

Eq. (1-4) is valid only if the unexposed face is the smaller side of a column. There are no experimental data to justify this equation when one of the larger faces is not exposed to the fire. If the column is recessed into a wall, Eq. (1-4) may be used, but the full dimensions of the column apply.

The value of the effective length factor $K_e$ is obtained from Table 5.4. In Fig. 1.2, $l_e$ is the unsupported length of the column or compression member, $l$, multiplied by $K_e$.

The allowable design loads on beams and columns are determined by procedures in Chapter 6. Also, an example of calculation of fire-resistance rating is included in Chapter 6. Fastenings for fire-resistance-rated members are discussed in Chapter 6. For additional information on calculation of fire resistance, see Refs. (15) and (16). For more information on this procedure and requirements for fastenings, see Ref. (17).

### 1.4.2  Fire Safety in Buildings

Regulations governing the construction of buildings have evolved over the years mainly to protect property against the ravages of fire. The life safety and fire protection provisions found in modern building codes provide what are considered minimum requirements for fire safety based on experience accumulated from fires and on research in the fire protection field. Such provisions become law when adopted by local, state, or regional jurisdictions. Typical provisions are found in all nationally recognized model building codes.

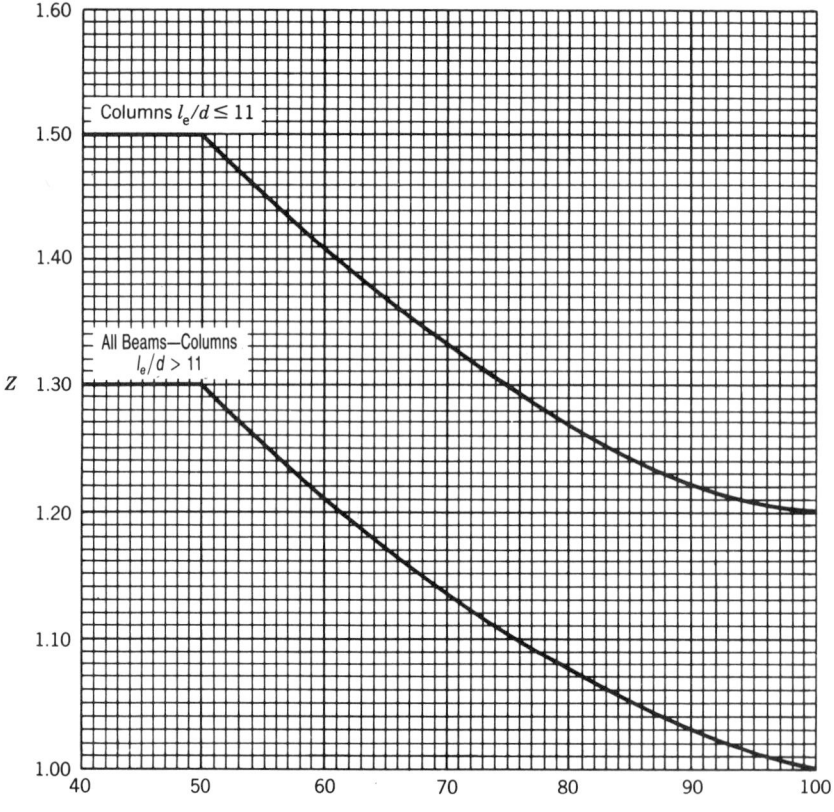

**FIGURE 1.2**    Dimensionless Load Factor, $Z$, as a function of $l_e/d$.

The extent of structural fire endurance desirable beyond building code requirements is a matter of economic balance between the risk of loss and the available fire protection facilities. Experience indicates that too much emphasis on the fire resistance of the structural frame results in a false sense of security. The emphasis is wrongly placed on so-called fireproof buildings rather than on fire-safe buildings. Good building design, adequate structural endurance, and automatic fire detection equipment and extinguishing facilities may provide fire safety at reasonable cost.

Safety from fire begins with the fire prevention measures that are taken. They include a properly designed building, including the electrical and mechanical systems. However, the fire record over the years has shown that most fires have their beginning in the contents of the building rather than in the structure itself. For this reason, fire prevention is related largely to housekeeping and maintenance, which are in turn directly associated with building administration. Thus, preventing a fire from starting has little or nothing to do with the structural materials used in the building, and the word *fireproof* as related to structural materials often deludes the building planner and the occupant by giving them an illusory feeling of safety. Because of the combustible nature of the contents of a building, no occupied building is fireproof.

Even with the best of prevention measures, fires will and do start in buildings, perhaps through unforeseeable conditions and events or uncontrollable acts of nature. The first consideration then is the safety of the occupants of the building and, after their evacuation, the safety of the fire fighters and then the protection of the building and adjacent property.

The most important protection factors are the prompt detection of a fire, immediate alarm, and the rapid extinguishment of the fire. The fact that there are people in a building is no guarantee that a fire will be discovered promptly. The promptness with which a fire is discovered and the speed with which it is extinguished have an important bearing on life safety and on the extent of damage both to the contents and to the structure. Automatic detectors and extinguishing systems can provide a high degree of protection, but they should not be considered a substitute for other protection measures such as the proper number, location, and type of exit ways. The Life Safety Code, promulgated by the National Fire Protection Association, is widely used as a guide for safety to life from fire in buildings.

## 1.5 TIMBER DETAILING

*Typical Construction Details*, AITC 104 (1), is included in Chapter 8 of this manual as a suggested guide for designers to use in their work and to give greater assurance of quality with a minimum of maintenance for engineered timber structures. Also, see Chapter 7 for recommendations on connection design and detailing.

The architect or engineer must select, modify, and design the details best suited to fit the requirements of a particular job. In doing so, the designer should keep in mind ease and economy of assembly and erection, as well as minimizing maintenance requirements and increasing durability through design and construction practices. Allowances should be made for differential movements resulting from dimensional changes, especially shrinkage perpendicular to grain. Whenever possible, connections should be designed to avoid putting wood members in tension perpendicular to grain. Moisture barriers, flashing, and other features should be used where necessary to avoid moisture or water traps. Durable structures result when such details are incorporated with other good design and construction, and protective treatment and coatings for wood and construction hardware.

When not otherwise required, the overall side clearance between nonbearing surfaces of steel assemblies and timber members should be not less than 1/8 in. (1/16 in. each side). Unless shown and required on the shop details, welds should not be located where they will interfere with the assembly of the connection. Hardware and steel accessories should be painted or coated to prevent staining of timbers during assembly and erection in work where appearance is a factor.

## 1.6 HANDLING, STORAGE, ERECTION, AND SEASONING OF STRUCTURAL TIMBER

The erection of structural timber framing requires experienced erection crews and adequate lifting equipment to protect lives and property and to assure that the framing is not improperly assembled or damaged during handling. The unloading and storage of structural timber framing before erection also demands care and good judgment. It is suggested that a shipment of structural timber framing, on receipt at the job site, be checked for tally and damage. The following general precautions apply.

### 1.6.1 Precautions During Unloading

Structural timber framing is subject to surface marring and damage when not properly handled and protected. At the erection site, the following precautions are suggested:

1.  Lift members or roll them on dollies or rollers out of railroad cars; do not drag or drop them. Unload trucks by lifting from the truck; do not dump or drop members.
2.  If unloading with lifting equipment, use wide fabric or plastic belts or other slings that will not mar wood. If chains or cables are used, provide protective blocking or padding to sharp edges or sharp corners.
3.  Guard against soiling, dirt, footprints, or abrasions. If members are wrapped, avoid tearing or damaging the protective material.

### 1.6.2 Precautions During Storage

If structural timber framing is to be stored before erection, it should be placed on blocks well off the ground, and individual members should be separated by strips so that air may circulate around all four sides. The top and all sides of storage piles should be covered with moisture-resistant material. Clear polyethylene films should not be used because wood members are subject to bleaching from sunlight. Individual wrappings should be slit or punctured on the lower side to permit drainage of water that may have accumulated. Water-resistant wrapping used for the in-transit protection of glued laminated members should be left intact until the members are enclosed within the building. If wrapping has to be removed at certain connection points during the erection, it should be replaced after the connection is made. If it is impractical to replace the wrapping, all of it should be removed to avoid the nonuniform appearance caused by sun and weather exposure.

### 1.6.3 Precautions During Erection

#### 1.6.3.1 Assembly

Trusses are usually shipped partially or completely disassembled and are assembled on the ground at the site before erection. Arches, which are generally shipped in halves, may be assembled on the ground or connections may be made after the half arches are in position. When trusses and arches are assembled on the ground at the site, they should be assembled on level blocking to permit connections to be fitted properly and tightened securely without damage. The end compression joints should be brought into full bearing and compression plates installed where specified.

Before erection, the assembly should be checked for prescribed overall dimensions, prescribed camber, and accuracy of anchorage connections. Erection should be planned and executed in such a way that the close fit and neat appearance of joints and the structure as a whole will not be impaired.

Anchor bolts should be checked prior to the start of erection. Before erection begins, all supports and anchors should be complete, accessible, and free of obstructions. The weights and balance points of the structural timber framing should be determined before lifting begins so that proper equipment and lifting methods may be employed. When long members or timber trusses of long span are raised from a flat to a vertical position preparatory to lifting, stresses entirely different from the normal design stress may be introduced. The magnitude and distribution of these stresses will vary, depending on such factors as the weight, dimensions, and type of member. A competent rigger should consider these factors in determining how much suspension and stiffening, if any, is required and where it could be located.

### 1.6.3.2  Bracing

All framing must be true and plumbed. Permanent bracing is bracing so designed and installed as to form an integral part of the final structure. Erection bracing is bracing installed to hold the framing in a safe position until sufficient permanent bracing is in place to provide full stability. Proper and adequate temporary erection bracing is introduced whenever necessary to take care of all loads to which the structure may be subjected during erection, including equipment and its operation. This bracing is left in place as long as may be required for safety. Part or all of the permanent bracing may also act as erection bracing. Erection bracing serves to plumb the framing during erection and gives it adequate stability to receive purlins, joists, and roofing materials. It may include sway bracing, guy ropes, tieing off framing nearest to end walls, steel tie rods with turnbuckle take-ups, struts, shoes, and similar items. As erection progresses, bracing is securely fastened in place to take care of all dead load, erection stresses, and normal weather conditions. Excessive concentrated construction loads, such as bundles of sheathing, piles of purlins, roofing, or other materials, should be avoided.

### 1.6.3.3  Final Alignment

Final tightening of alignment bolts should not be completed until the structure has been properly aligned.

### 1.6.3.4  Removal of Temporary Bracing

Temporary erection bracing should be removed only after diaphragms and permanent bracing are installed, the structure has been properly aligned, and connections and fastenings have been finally tightened. Retightening of connections prior to final completion or closing in of inaccessible connections is recommended.

### 1.6.3.5 Field Connections

The joining, holding, and welding of steel connections in the field are performed according to the requirements for shop work of such operations, except where such requirements apply to shop conditions only. Steel connections should comply with the specifications of the American Institute of Steel Construction (18) and the American Welding Society (19).

### 1.6.3.6 Protection of Field Cuts

All field cuts of timbers should be coated with an approved moisture seal if the member was initially coated unless otherwise specified. All field framing is done in accordance with the requirements of shop practice except where such requirements apply to shop conditions only. If timber framing has been pressure treated, field framing after treatment must be avoided or at least, insofar as possible, held to a minimum. When field cuts in pressure-treated material are unavoidable, additional treatment should be provided in accordance with AWPA Standard M4 (10).

### 1.6.3.7 Protection Against Moisture

During erection operations, all timber framing that requires moisture content control, whether sawn or glued laminated timbers, should be protected against moisture pickup. Any fabricated structural materials to be stored for an extended period of time before erection should, insofar as is practicable, be assembled into subassemblies for storage purposes.

### 1.6.3.8 Seasoning Period

Heat should not be fully turned on as soon as the structure is enclosed; otherwise, excessive checking may occur due to rapid lowering of the relative humidity in the building. A gradual seasoning period at moderate temperature should be provided.

## 1.7 REFERENCES

1. American Institute of Timber Construction, *Typical Construction Details*, AITC 104, Englewood, CO, 1984.

2. American Institute of Timber Construction, *Standard Appearance Grades for Structural Glued Laminated Timber*, AITC 110, Englewood, CO, 1984.

3. United States Department of Agriculture, Forest Service, Forest Products Laboratory, *Wood Handbook: Wood as an Engineering Material*, Agriculture Handbook No. 72, Madison, WI, 1987.

4. American Society of Civil Engineers, *Evaluation, Maintenance and Upgrading of Wood Structures*, New York, NY, 1982.

5. American Institute of Timber Construction, *Checking in Glued Laminated Timbers*, Technical Note No. 11, Englewood, CO, 1987.

6. American Institute of Timber Construction, *Standard Specifications for Structural Glued Laminated Timber of Softwood Species*, AITC 117—Design, Englewood, CO, 1993.

7. American Institute of Timber Construction, *Evaluation of Checking in Glued Laminated Timbers*, Technical Note No. 18, Englewood, CO, 1991.

8. American Institute of Timber Construction, *Recommended Practice for Protection of Structural Glued Laminated Timber During Transit, Storage and Erection*, AITC 111, Englewood, CO, 1979.

9. American Institute of Timber Construction, *Standard for Preservative Treatment of Structural Glued Laminated Timber*, AITC 109, Englewood, CO, 1990.

10. American Wood-Preservers' Associaton, *Book of Standards*, Stevensville, MD, 1992.

11. American Society of Heating, Refrigerating and Air-Conditioning Engineers, Inc., *ASHRAE Fundamentals Handbook*, Atlanta, GA, 1989.

12. American Forest and Paper Association, ANSI/NF₀PA NDS® 1991, *National Design Specification® for Wood Construction*, Washington, DC, 1991.

13. American Institute of Timber Construction, *Standard for Heavy Timber Construction*, AITC 108, Englewood, CO, 1993.

14. T. T. Lie, "A Method for Assessing the Fire Resistance of Laminated Beams and Columns." National Research Council of Canada, Division of Building Research, DBR 718, Ottawa, Ontario, 1977.

15. United States Department of Agriculture, Forest Service, Forest Products Laboratory, *Strength Validation and Fire Endurance of Glued Laminated Beams*, Madison, WI, 1985.

16. Warnock-Hersey, *Report of Standard Fire Endurance Test, Nominal $8\frac{3}{4}in.$ by $16\frac{1}{2}$ in. Glulam Wood Beam*, Report No. WH1-694-0069, Severna Park, MD, 1982.

17. American Institute of Timber Construction, *Calculation of Fire Resistance of Glued Laminated Timbers*, Technical Note No. 7, Englewood, CO, 1993.

18. American Institute of Steel Construction, *Manual of Steel Construction*, 9th ed., Chicago, IL, 1989.

19. American Welding Society, *Structural Welding Code*, D1.1, Miami, FL, 1983.

# CHAPTER 2

# PHYSICAL STRUCTURE OF WOOD

## 2.1 INTRODUCTION

Wood is a cellular organic material made up principally of cellulose, which comprises the structural units (cells), and lignin, which cements the structural units together. It also contains hemicelluloses, extractives, and ash-forming minerals. Wood cells are hollow, and they vary from about 0.04 to 0.33 in. in length and from 0.0004 to 0.0033 in. in diameter. Most cells are elongated and are oriented vertically in the growing tree, but some, called rays, are oriented horizontally and extend from the bark toward the center or pith of the tree.

### 2.1.1  Hardwoods and Softwoods

Trees are divided into two broad classes: hardwoods, which have broad leaves, and softwoods or conifers, which have needlelike or scalelike leaves. Most hardwoods shed their leaves at the end of each growing season, and most softwoods are evergreens. The terms *hardwood* and *softwood* are often misleading because they do not directly indicate the hardness or softness of wood. Some hardwoods are softer than certain softwoods and some softwoods are harder than some hardwoods.

### 2.1.2  Heartwood and Sapwood

The cross section of a tree shows several distinct zones: the bark; a light-colored zone called sapwood; an inner zone, generally of darker color, called heartwood; and, at the center, the pith.

A tree increases in diameter by adding new layers of cells from the pith outward. For a time, this new layer functions as living cells that conduct sap and store food, but eventually, as the tree increases in diameter, cells toward the center become inactive and serve only as support for the tree. The inactive inner layer is the heartwood; the outer layer containing living cells is the sapwood. There is no consistent difference between the weight and strength properties of heartwood and sapwood. In some species, heartwood is significantly more resistant to decay fungi than is sapwood, although there is a great range in the durability of heartwood among the various species.

### 2.1.3   Annual Rings

In climates where temperature limits the growing season of a tree, each annual increment of growth usually is readily distinguishable. Such an increment is known as an annual growth ring, or annual ring, and consists of an earlywood (springwood) and a latewood (summerwood) band.

### 2.1.4   Earlywood (Springwood) and Latewood (Summerwood)

In many wood species, large, thin-walled cells are formed in the spring when growth is greatest, whereas smaller, thicker-walled cells are formed later in the year. The areas of fast growth are called earlywood, and the areas of slower growth, latewood. In annual rings, the inner, lighter-colored area is the earlywood and the outer, darker layer is the latewood.

Latewood contains more solid wood substance than does earlywood and, therefore, is denser and stronger. The proportion of width of latewood to width of annual ring as well as the number of rings per inch is used as one of the visual measures of the quality and strength of wood for some species.

### 2.1.5   Grain and Texture

The terms *grain* and *texture* are used in many ways to describe the characteristics of wood and, in fact, do not have a definite meaning. Grain often refers to the width of the annual rings, as in "close-grained" or "coarse-grained." Sometimes it indicates whether the fibers are parallel to or at an angle with the sides of the pieces, as in "straight-grained" or "cross-grained." Texture usually refers to the fineness of wood structure rather than to the annual rings. When these terms are used in connection with wood, the meaning intended should be defined.

### 2.1.6   Moisture Content

A tree develops in the presence of moisture, and throughout its life the tree remains moist or "green." The amount of moisture in a living tree varies among species, in individual trees within the same species, in different parts of the same tree, and between heartwood and sapwood.

Moisture content (MC) is the weight of the water contained in wood, expressed as a percentage of the weight of the oven-dry wood. (An oven-dry condition is reached when no further loss of weight is experienced on subsequent oven drying.) As wood loses moisture, the water in the cell cavity is evaporated first. The condition at which the water in the cell cavity has been evaporated but the cell wall is still saturated is known as the fiber saturation point. The fiber saturation point is quite variable between species and among individual pieces of the same species, but averages about 30% MC. Further drying of wood below its fiber saturation point results in loss of moisture from the cell walls, which, in turn, causes shrinkage of the wood.

Wood in use gives off or takes on moisture from the surrounding atmosphere with changes in temperature and relative humidity until it attains a balance relative to the atmospheric conditions. The moisture content at this point of balance is known as the equilibrium moisture content (EMC). At constant temperature, the equilibrium moisture content depends entirely on the relative humidity of the atmosphere surrounding the wood and the hygroscopicity of the wood or contained materials. (See Table 2.1, which, for practical purposes, applies to the wood of any species.)

### 2.1.7 Growth Characteristics

Wood contains certain natural growth characteristics such as knots, slope of grain, compression wood, and shakes, which may, depending on their size, number, and location in a structural member, adversely affect the strength properties of that member. These characteristics are discussed in detail in *Wood Handbook* (1) by the U.S. Forest Products Laboratory. Structural grading rules take into account the effects of these growth characteristics on the strength of wood in establishing design values for lumber and glued laminated timber.

### 2.1.8 Directional Properties

Wood is nonisotropic because of the orientation of its cells and the manner in which it increases in diameter. It has different mechanical properties with respect to its three principal axes of symmetry: longitudinal, radial, and tangential (see Fig. 2.1). Strength and elastic properties corresponding to these three axes may be used in design.

The difference between properties in the radial and tangential directions is seldom of practical importance in most structural designs; for structural purposes, it is usually sufficient to differentiate only between properties parallel and perpendicular to the grain.

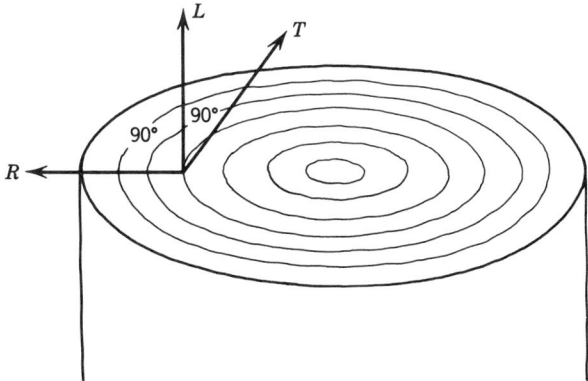

**FIGURE 2.1** The Three Principal Axes of Wood. *L*, longitudinal (parallel to grain); *R*, radial (perpendicular to grain, radial to annual rings); *T*, tangential (perpendicular to grain, tangential to annual rings).

## TABLE 2.1

## Moisture Content of Wood in Equilibrium With Stated Dry-Bulb Temperature and Relative Humidity[a]

| Temperature Dry-bulb, °F | Relative Humidity, Percent | | | | | | | | | | | | | | | | | | | |
|---|---|---|---|---|---|---|---|---|---|---|---|---|---|---|---|---|---|---|---|---|
| | 5 | 10 | 15 | 20 | 25 | 30 | 35 | 40 | 45 | 50 | 55 | 60 | 65 | 70 | 75 | 80 | 85 | 90 | 95 | 98 |
| 30 | 1.4 | 2.6 | 3.7 | 4.6 | 5.5 | 6.3 | 7.1 | 7.9 | 8.7 | 9.5 | 10.4 | 11.3 | 12.4 | 13.5 | 14.9 | 16.5 | 18.5 | 21.0 | 24.3 | 26.9 |
| 40 | 1.4 | 2.6 | 3.7 | 4.6 | 5.5 | 6.3 | 7.1 | 7.9 | 8.7 | 9.5 | 10.4 | 11.3 | 12.3 | 13.5 | 14.9 | 16.5 | 18.5 | 21.0 | 24.3 | 26.9 |
| 50 | 1.4 | 2.6 | 3.6 | 4.6 | 5.5 | 6.3 | 7.1 | 7.9 | 8.7 | 9.5 | 10.3 | 11.2 | 12.3 | 13.4 | 14.8 | 16.4 | 18.4 | 20.9 | 24.3 | 26.9 |
| 60 | 1.3 | 2.5 | 3.6 | 4.6 | 5.4 | 6.2 | 7.0 | 7.8 | 8.6 | 9.4 | 10.2 | 11.1 | 12.1 | 13.3 | 14.6 | 16.2 | 18.2 | 20.7 | 24.1 | 26.8 |
| 70 | 1.3 | 2.5 | 3.5 | 4.5 | 5.4 | 6.2 | 6.9 | 7.7 | 8.5 | 9.2 | 10.1 | 11.0 | 12.0 | 13.1 | 14.4 | 16.0 | 17.9 | 20.5 | 23.9 | 26.6 |
| 80 | 1.3 | 2.4 | 3.5 | 4.4 | 5.3 | 6.1 | 6.8 | 7.6 | 8.3 | 9.1 | 9.9 | 10.8 | 11.7 | 12.9 | 14.2 | 15.7 | 17.7 | 20.2 | 23.6 | 26.3 |
| 90 | 1.2 | 2.3 | 3.4 | 4.3 | 5.1 | 5.9 | 6.7 | 7.4 | 8.1 | 8.9 | 9.7 | 10.5 | 11.5 | 12.6 | 13.9 | 15.4 | 17.3 | 19.8 | 23.3 | 26.0 |
| 100 | 1.2 | 2.3 | 3.3 | 4.2 | 5.0 | 5.8 | 6.5 | 7.2 | 7.9 | 8.7 | 9.5 | 10.3 | 11.2 | 12.3 | 13.6 | 15.1 | 17.0 | 19.5 | 22.9 | 25.6 |
| 110 | 1.1 | 2.2 | 3.2 | 4.0 | 4.9 | 5.6 | 6.3 | 7.0 | 7.7 | 8.4 | 9.2 | 10.0 | 11.0 | 12.0 | 13.2 | 14.7 | 16.6 | 19.1 | 22.4 | 25.2 |
| 120 | 1.1 | 2.1 | 3.0 | 3.9 | 4.7 | 5.4 | 6.1 | 6.8 | 7.5 | 8.2 | 8.9 | 9.7 | 10.6 | 11.7 | 12.9 | 14.4 | 16.2 | 18.6 | 22.0 | 24.7 |
| 130 | 1.0 | 2.0 | 2.9 | 3.7 | 4.5 | 5.2 | 5.9 | 6.6 | 7.2 | 7.9 | 8.7 | 9.4 | 10.3 | 11.3 | 12.5 | 14.0 | 15.8 | 18.2 | 21.5 | 24.2 |
| 140 | 0.9 | 1.9 | 2.8 | 3.6 | 4.3 | 5.0 | 5.7 | 6.3 | 7.0 | 7.7 | 8.4 | 9.1 | 10.0 | 11.0 | 12.1 | 13.6 | 15.3 | 17.7 | 21.0 | 23.7 |
| 150 | 0.9 | 1.8 | 2.6 | 3.4 | 4.1 | 4.8 | 5.5 | 6.1 | 6.7 | 7.4 | 8.1 | 8.8 | 9.7 | 10.6 | 11.8 | 13.1 | 14.9 | 17.2 | 20.4 | 23.1 |

16%    19%

[a]Numbers to the right of the heavy line represent wet-use service conditions (16% or greater for glued laminated timber and greater than 19% for sawn timbers shaded).

## 2.2 WEIGHT AND SPECIFIC GRAVITY OF COMMERCIAL LUMBER SPECIES

Table 2.2 gives average weights, in pounds per cubic foot, of various commercial lumber species based on their weight and volume at 12% and 20% MC. It also gives specific gravities based on weight oven dry and volume at 12% MC and when green.

For design purposes, it is satisfactory to use weights at 12% MC for glued laminated timber and sawn timber that will remain dry in service. The weight for unseasoned sawn material when it is installed and before it dries in service should be taken as 20% MC for relatively small sizes of less than 3 in. in thickness. The weight based on the equilibrium moisture content with allowance for preservative treatments, if applied, should be employed for unprotected members in exterior or other wet conditions of use. Adhesives used in laminating do not have an appreciable effect on the weight.

### 2.2.1 Effect of Moisture Content

A charge in moisture content will result in a corresponding change in the weight of the wood. Values for weights at 12 and 20% MC are given in Table 2.2. For other moisture contents, the weight may be determined by the designer if this value is critical by using the formulas given in the Appendix to ASTM D2395 (2) or Fig. 1 of ASTM D 2395 (2).

### 2.2.2 Effect of Treatments

Some preservative treatments may add significantly to the weight of wood, depending on the retentions obtained and, in the case of waterborne salts, on the extent to which the wood is seasoned after treatment. Weight increases due to preservative salts are small because retentions of the dry salt are in the range 0.30–1.25 pcf. However, some salt treatments are hygroscopic, and wood treated with these salts may tend to increase in moisture content, and thus increase in weight during service conditions. In pressure treatments with oil-borne preservatives, retentions are higher, and the weight increase may be from 5 to 20 pcf or more. For specific values of recommended retentions for various preservatives in different species, see *Standard for Preservative Treatment of Structural Glued Laminated Timber*, AITC 109 (3), included in Chapter 8.

## 2.3 DIMENSIONAL STABILITY

### 2.3.1 Effect of Temperature

Wood, like most other solids, expands on heating and contracts on cooling. In most structural designs, the increase in size of wood for a rise in temperature is negligible, and as a result, the secondary stresses due to temperature changes may, in most cases, be neglected. This increase in size is important only in certain structures that are subjected to considerable temperature changes or in members with very long spans.

# TABLE 2.2

## Weights and Specific Gravities of Commercial Lumber Species

| Species[a] | Specific Gravity, $G$, Based on Oven-Dry Weight and Volume | | Weight (pcf) Based on Weight and Volume | |
| --- | --- | --- | --- | --- |
| | At 12% MC[b] | When Green[c] | At 12% MC | At 20% MC |
| *Softwoods* | | | | |
| Balsam Fir | 0.34 | 0.32 | 23.8 | 24.7 |
| California Redwood | 0.36 | 0.34 | 25.2 | 26.2 |
| Coast Sitka Spruce | 0.41 | 0.38 | 28.7 | 29.2 |
| Coast Species | 0.44 | 0.41 (0.36–0.55) | 30.8 | 32.2 |
| Douglas Fir-Larch | 0.49 | 0.45 (0.45–0.48) | 34.3 | 35.2 |
| Douglas Fir-Larch (North) | 0.50 | 0.46 (0.45–0.55) | 35.0 | 36.0 |
| Douglas Fir South | 0.46 | 0.43 | 32.2 | 33.7 |
| Eastern Hemlock | 0.42 | 0.39 | 29.4 | 30.7 |
| Eastern Hemlock–Tamarack | 0.43 | 0.40 (0.39–0.44) | 30.1 | 31.5 |
| Eastern Hemlock–Tamarack (North) | 0.45 | 0.42 (0.40–0.48) | 31.5 | 33.0 |
| Eastern Softwoods | 0.38 | 0.36 (0.32–0.49) | 26.6 | 27.7 |
| Eastern Spruce | 0.38 | 0.36 (0.33–0.38) | 26.6 | 27.7 |
| Eastern White Pine | 0.37 | 0.35 | 25.9 | 27.0 |
| Eastern White Pine (North) | 0.38 | 0.36 | 26.6 | 27.7 |
| Eastern Woods | 0.40 | 0.37 (0.32–0.49) | 28.0 | 28.5 |
| Engelmann Spruce–Alpine Fir | 0.34 | 0.32 (0.31–0.33) | 23.8 | 24.7 |
| Engelmann Spruce–Lodgepole Pine | 0.38 | 0.36 (0.33–0.39) | 26.6 | 27.7 |
| Hem-Fir | 0.42 | 0.39 (0.35–0.42) | 29.4 | 30.7 |
| Hem-Fir (North) | 0.43 | 0.40 (0.36–0.41) | 30.1 | 31.5 |
| Idaho White Pine | 0.37 | 0.35 | 25.9 | 27.0 |
| Lodgepole Pine | 0.42 | 0.39 | 29.4 | 30.7 |
| Mountain Hemlock | 0.45 | 0.42 | 31.5 | 33.0 |
| Mountain Hemlock–Hem-Fir | 0.42 | 0.39 (0.35–0.42) | 29.4 | 30.7 |

| Species | | | | |
|---|---|---|---|---|
| Northern Pine | 0.46 | 0.43 (0.40–0.47) | 32.2 | 33.7 |
| Northern Species | 0.41 | 0.38 (0.31–0.55) | 28.7 | 29.2 |
| Northern White Cedar | 0.30 | 0.29 | 21.0 | 22.5 |
| Ponderosa Pine | 0.42 | 0.39 | 29.4 | 30.7 |
| Ponderosa Pine–Sugar Pine | 0.40 | 0.37 (0.34–0.39) | 28.0 | 28.5 |
| Ponderosa Pine–Lodgepole Pine | 0.42 | 0.39 | 29.4 | 30.7 |
| Red Pine | 0.45 | 0.42 | 31.5 | 33.0 |
| Sitka Spruce | 0.41 | 0.38 | 28.7 | 29.2 |
| Southern Pine | 0.52 | 0.48 (0.47–0.54) | 36.4 | 37.5 |
| Spruce–Pine–Fir | 0.41 | 0.38 (0.33–0.42) | 28.7 | 29.2 |
| Virginia Pine–Pond Pine | 0.52 | 0.48 (0.46–0.51) | 36.4 | 37.5 |
| Western Cedars[d] | 0.35 | 0.33 (0.31–0.42) | 24.5 | 25.5 |
| Western Cedars (North) | 0.34 | 0.32 (0.31–0.42) | 23.8 | 24.7 |
| Western Hemlock | 0.45 | 0.42 | 31.5 | 33.0 |
| Western Hemlock (North) | 0.44 | 0.41 | 30.8 | 32.2 |
| Western White Pine | 0.38 | 0.36 | 26.6 | 27.7 |
| White Woods (Western Woods) | 0.41 | 0.38 (0.31–0.42) | 28.7 | 29.2 |
| *Hardwoods* | | | | |
| Aspen | 0.37 | 0.35 (0.35–0.36) | 25.9 | 27.0 |
| Black Cottonwood | 0.32 | 0.30 | 22.4 | 23.2 |
| Cottonwood | 0.40 | 0.37 | 28.0 | 28.2 |
| Northern Aspen | 0.40 | 0.37 (0.37–0.39) | 28.0 | 28.5 |
| Yellow Poplar | 0.43 | 0.40 | 30.1 | 31.5 |

[a] Species listed are those commonly used for structural purposes as listed in Supplement to the *National Design Specification* (4).

[b] Obtained by conversion from specific gravity when green by using Figure 1 ASTM D 2395, *Standard Test Methods for Specific Gravity of Wood and Wood-Base Materials* (2).

[c] Unseasoned condition, obtained from Tables 1 and 2, ASTM D 2555, *Standard Methods for Establishing Clear Wood Strength Values* (5). Species groups show weighted average specific gravities for the groups based on standing timber volumes shown in Tables 4 and 5, ASTM D 2555. The range of average specific gravities of species included in the species grouping is shown to indicate the variation to be expected.

[d] For Alaska Cedar glued laminated timber, the specific gravities are 0.45 at 12% and 0.42 when green, and weights are 31.5 and 33.0.

The coefficient of linear thermal expansion differs in wood's three structural directions. In the longitudinal direction (parallel to grain), this coefficient appears to be independent of specific gravity and species and ranges from about $1.7 \times 10^{-6}$ to $2.5 \times 10^{-6}$ per $1°F$ for oven-dry wood for both hardwoods and softwoods. This value is about one-tenth to one-third of those for other common structural materials. For that reason, consideration must be given to the differential thermal expansion of materials used in conjunction with wood.

Coefficients of linear thermal expansion across the grain (radial and tangential) are proportional to wood's specific gravity $G$. These coefficients range from about 5 to more than 10 times greater than the parallel-to-grain coefficients. The radial coefficient of thermal expansion, $\alpha_r$, can be approximated by the equation

$$\alpha_r = (32G + 9.9) \times 10^{-6} \tag{2-1}$$

per $1°F$ for oven-dry specific gravities in the range of about 0.1–0.8. The tangential coefficient of thermal expansion, $\alpha_t$, over the same oven-dry specific gravity range is approximated by the equation

$$\alpha_t = (33G + 18.4) \times 10^{-6} \tag{2-2}$$

per $1°F$. Coefficients of thermal expansion vary slightly within the temperature range of $-60$ to $+130°F$, but for most practical purposes can be considered constant.

Wood containing moisture reacts to varying temperature differently than does dry wood. As moist wood is heated, it tends to expand due to temperature, but shrinks due to loss of moisture. Unless the wood is very dry initially (3 or 4% MC or less), shrinkage due to moisture loss on heating will exceed thermal expansion, resulting in a negative net dimensional change due to heating. At the intermediate moisture levels (about 8–20%), wood will expand when first heated, but will shrink gradually to a volume smaller than the initial volume as the wood gradually loses moisture while in the heated condition. In the longitudinal direction, where dimensional change due to moisture change is very small, shrinkage still predominates over thermal expansion unless the wood is initially very dry.

### 2.3.2   Effect of Moisture Content

Between the fiber saturation point and zero moisture content, wood shrinks as it loses moisture and swells as it absorbs moisture. Above the fiber saturation point, there is no dimensional change with variation in moisture content. The amount of shrinkage and swelling differs in the tangential, radial, and longitudinal directions of the piece. Engineering design should consider shrinkage and swelling in the detailing and use of timbers.

Shrinkage occurs when the moisture content is reduced to a value below the fiber saturation point (for purpose of dimensional change, commonly assumed to be 30% MC) and is proportional to the amount of moisture lost below this point. Swelling occurs when the moisture content is increased until the fiber saturation point is reached. The total swelling is equal numerically to the total shrinkage.

Shrinking and swelling are expressed as percentages based on the green dimensions of the wood. Wood shrinks most in a direction tangent to the annual growth rings and somewhat less in the radial direction or across these rings. In general, shrinkage is greater in heavier pieces than in lighter pieces of the same species and greater in hardwoods than in softwoods.

As a piece of green or wet wood dries, the surface is reduced to a moisture content below the fiber saturation point much sooner than is the interior. Thus, the piece may show some shrinkage before the average moisture content reaches the fiber saturation point.

Table 2.3 gives the average tangential, radial, and volumetric shrinkage values for various species during drying from the green condition to 0% MC. Because the faces of the pieces of lumber are seldom so oriented that the annual growth rings are exactly tangent and radial to the faces of the piece, it is customary in determining cross-sectional dimensional changes to use an intermediate or average value between the tangential and radial values. The values in Table 2.3 can be used to estimate shrinkage by using the equation

$$S_m = \frac{S_0(I - m)}{30} \tag{2-3}$$

where $S_m$ = shrinkage from initial moisture condition to final moisture content $m(\%)$,

$S_0$ = total shrinkage from Table 2.3 (%),

$m$ = final moisture content (below 30%) (%), and

$I$ = initial moisture content (below 30%) (%).

Values for longitudinal shrinkage with a change in moisture content are not tabulated in Table 2.3 because they are ordinarily negligible. The total longitudinal shrinkage of commonly used species from fiber saturation to oven-dry condition usually ranges from 0.1 to 0.2% of the green dimension. Abnormal longitudinal shrinkage may occur in compression wood, wood with steep slope of grain, and exceptionally lightweight wood of any species.

Because there is considerable variation in shrinkage for any species, it is difficult to predict the shrinkage of an individual piece of wood. The values given in Table 2.3 may be used to estimate the average shrinkage of a quantity of pieces.

There is a normal tendency for more flat grain lumber to be used in glued laminated timber than edge grain lumber. Therefore, the shrinkage in the direction perpendicular to the glue line is more comparable to radial shrinkage than tangential shrinkage. Shrinkage can be estimated by using Eq. (2-3) and Table 2.3. A rule of thumb commonly used is to assume that 1% shrinkage occurs for every 5% change in moisture content.

The effects of dimensional changes due to a change in moisture content have been considered in the development of the details shown in *Typical Construction Details*, AITC 104 (6), in Chapter 8.

**TABLE 2.3**

**Average Shrinkage Values of Wood Based on Dimensions when Green**

| Species[a] | Percentage of Shrinkage from Green to Oven-Dry Moisture Content[b] | | |
|---|---|---|---|
| | Radial | Tangential | Volumetric |
| *Softwoods* | | | |
| Balsam Fir | 2.9 | 6.9 | 11.2 |
| California Redwood | 2.2 | 4.9 | 7.0 |
| Douglas Fir-Larch[c] | 4.6 (3.8–4.8) | 7.6 (6.9–9.1) | 12.0 (10.7–14.0) |
| Eastern Hemlock | 3.0 | 6.8 | 9.7 |
| Eastern Hemlock–Tamarack[c] | 3.1 (3.0–3.7) | 6.9 (6.8–7.4) | 10.2 (9.7–13.6) |
| Eastern Softwoods[c] | 3.2 (2.1–4.1) | 7.1 (6.1–7.8) | 10.5 (8.2–13.6) |
| Eastern White Pine | 2.1 | 6.1 | 8.2 |
| Engelmann Spruce-Alpine Fir[c] | 3.4 (2.6–3.8) | 7.2 (7.1–7.4) | 10.5 (9.4–11.0) |
| Engelmann Spruce-Lodgepole Pine[c] | 4.1 (3.8–4.3) | 6.9 (6.7–7.1) | 11.1 (11.0–11.1) |
| Hem-Fir[c] | 3.9 (3.5–4.5) | 7.9 (7.0–9.2) | 11.7 (9.8–13.0) |
| Idaho White Pine | 4.1 | 7.4 | 11.8 |
| Lodgepole Pine | 4.3 | 6.7 | 11.1 |
| Mountain Hemlock | 4.4 | 7.1 | 11.1 |
| Mountain Hemlock-Hem-Fir[c] | 3.9 (3.3–4.5) | 7.7 (7.0–9.2) | 11.8 (9.8–13.0) |
| Northern White Cedar | 2.2 | 4.9 | 7.2 |
| Ponderosa Pine-Sugar Pine[c] | 3.8 (2.9–3.9) | 6.1 (5.6–6.2) | 9.5 (7.9–9.7) |
| Ponderosa Pine-Lodgepole Pine[c] | 4.0 (3.9–4.3) | 6.3 (6.2–6.7) | 10.2 (9.7–11.1) |
| Sitka Spruce | 4.3 | 7.5 | 11.5 |
| Southern Pine[c] | 4.8 (4.6–5.4) | 7.6 (7.4–7.7) | 12.3 (12.1–12.3) |
| Virginia Pine-Pond Pine[c] | 4.5 (4.2–5.1) | 7.1 (7.1–7.2) | 11.6 (11.2–11.9) |
| Western Cedars[c] | 2.8 (2.4–4.6) | 5.2 (5.0–6.9) | 7.2 (6.8–10.1) |
| Western Hemlock | 4.2 | 7.8 | 12.4 |
| White Woods (Western Woods)[c] | 4.0 (2.9–4.5) | 7.2 (5.6–9.2) | 11.1 (7.9–13.0) |
| *Hardwoods* | | | |
| Aspen[c] | 3.5 (3.3–3.5) | 7.0 (6.7–7.4) | 11.6 (11.5–11.8) |
| Eastern Cottonwood | 3.9 | 9.2 | 13.9 |
| Yellow Poplar | 4.6 | 8.2 | 12.7 |

[a]Species listed are those commonly used for structural purposes as listed in Supplement to the *NDS* (4).

[b]Values are from Table 3.5, *Wood Handbook* (1).

[c]Species values for combinations are the weighted average based on the standing timber volumes shown in Table 4, ASTM D2555, *Standard Methods for Establishing Clear Wood Strength Values* (5). The range of values within the combination is shown in parentheses.

## 2.4  THERMAL INSULATING PROPERTIES

Thermal insulation, commonly referred to simply as insulation, is the inherent characteristic of a substance to prevent or retard the transfer of heat through its body. Heat transfer can occur in several ways: solid conduction, gas conduction, convection, and radiation. These mechanisms may occur separately or in combination. Wood is considered to be a very good insulator due to its physical structure. Each wood fiber or cell contains a void, which in living wood is used to transfer or store sap. The sawn and dried wood thus contains microscopic voids or air spaces that retard heat transfer.

Thermal conductivity $k$ is defined as the rate of heat transmission by conduction only through a unit area of an infinite slab, in a direction perpendicular to the slab surface, when the temperature gradient between the two surfaces of the slab is unity. This definition gives rise to several possible values of thermal conductivity for the same material. The first and most common, based on a temperature gradient of $1°F/in.$, has units of

$$\frac{\text{Btu}}{\text{hr ft}^2 \ °F/\text{in.}} \quad \text{or} \quad \frac{\text{Btu}}{\text{hr ft}^2 \ °F}$$

The second, based on a temperature gradient of $1°F/ft$, has units of

$$\frac{\text{Btu}}{\text{hr ft}^2 \ °F/\text{ft}} \quad \text{or} \quad \frac{\text{Btu}}{\text{hr ft} \ °F}$$

The relationship between the two values for a given material is

$$k \ (\text{Btu in.}/\text{hr ft}^2 \ °F) = 12 \ k \ (\text{Btu}/\text{hr ft} \ °F)$$

The thermal conductivity of wood varies with (1) the direction of grain, (2) specific gravity, (3) moisture content, (4) extractives present in the wood, and (5) growth characteristics such as knots, slope of grain, seasoning checks, and growth rings.

Thermal conductivity is approximately the same in radial and tangential directions, but is about $2\frac{1}{2}$ times greater along the grain. To determine thermal conductivity value across the grain for wood at various moisture contents and specific gravities, use the equation

$$k = G(1.39 + 0.028 \ \text{MC}) + 0.165 \tag{2.4}$$

where  $k$ = thermal conductivity (Btu in./hr ft$^2$ °F),
  $G$ = specific gravity based on volume at the existing moisture content and oven-dry weight, and
  MC = the existing moisture content (%).

A similar term, thermal conductance $C$, is defined as the rate of heat transmission through a unit area of a particular body or assembly having defined surfaces when the unit average temperature difference is established between the surfaces. Thermal conductance values are measured across the total thickness of the material or assembly of materials, whatever that thickness may be.

A direct measure of the insulating value of a material or assembly of materials is the reciprocal of thermal conductance and is termed thermal resistance, $R$. Thermal resistance values are most commonly used to indicate the insulating qualities of a material or a type of construction. Tables 2.4–2.6 show approximate thermal design values for some common building materials. Thermal resistance values for plane constructions, such as flat walls or roofs that have essentially one-dimensional heat flow, can be approximated as

$$R_T = R_1 + R_2 + R_3 + \cdots + R_n \qquad (2\text{-}5)$$

where   $R_T$ = total thermal resistance of the assembly (hr ft$^2$ °F/Btu) and $R_1, R_2, \cdots$ = individual resistances of the components in the assembly (hr ft$^2$ °F/Btu).

The overall heat transfer coefficient or overall transmittance of the assembly is defined by

$$U = \frac{1}{R_T} \qquad (2\text{-}6)$$

with $U$ having the same units as conductance.

Normally, the first and last resistances in Eq. (2-5) are due to the thin film of air that tends to adhere to the solid surfaces of the structure. These surface resistances are the reciprocals of the surface conductance ($k$). Surface conductance (or film coefficient) has the same units as $C$, but the symbol $k$ is used to distinguish the fluid-film conductance from solid-body conductance. The value of $k$, and therefore film resistance, depends on the orientation of the surface, air movement, direction of heat flow, and reflectivity of the surface.

The thermal resistance of the enclosed air spaces depends on orientation of the space, direction of heat flow, thickness of space, mean air temperature in the space, and temperature difference across the space. Values of thermal resistance for $3\frac{1}{2}$-in.-thick air spaces are given in Table 2.5. A typical determination of approximate total thermal resistance and overall heat transfer coefficient for a frame wall is shown in Fig. 2.2.

$$R_T = \frac{1}{h_i} + \frac{L_1}{k_1} + R_a + \frac{L_2}{k_2} + \frac{1}{h_0} \qquad (2\text{-}7)$$

$$U = \frac{1}{R_T}$$

where   $h_i$ = inside surface conductance (Btu/hr ft$^2$ °F),
$h_0$ = outside surface conductance (Btu/hr ft$^2$ °F),
$L_1, L_2$ = individual thickness of materials (in.),
$k_1, k_2$ = individual thermal conductivities of the material (Btu in./hr ft$^2$ °F),
$R_a$ = thermal resistance of air space (hr ft$^2$ °F/Btu), and,
$U$ = overall heat transfer coefficient (Btu/hr ft$^2$ °F).

## TABLE 2.4

### Thermal Design Values: Surface Conductances and Resistances[a]

| Position of Surface | Direction of Heat Flow | Nonreflective $\epsilon = 0.90$[b] | | Reflective $\epsilon = 0.20$[c] | | Reflective $\epsilon = 0.05$[d] | |
|---|---|---|---|---|---|---|---|
| | | $h_i$ | R | $h_i$ | R | $h_i$ | R |
| Still air | | | | | | | |
| Horizontal | Upward | 1.63 | 0.61 | 0.91 | 1.10 | 0.76 | 1.32 |
| Sloping—45° | Upward | 1.60 | 0.62 | 0.88 | 1.14 | 0.73 | 1.37 |
| Vertical | Horizontal | 1.46 | 0.68 | 0.74 | 1.35 | 0.59 | 1.70 |
| Sloping—45° | Downward | 1.32 | 0.76 | 0.60 | 1.67 | 0.45 | 2.22 |
| Horizontal | Downward | 1.08 | 0.92 | 0.37 | 2.70 | 0.22 | 4.55 |
| Moving Air (Any position) | | | | | | | |
| 15 mph wind (for winter) | Any | 6.00[e] | 0.17 | | | | |
| 7.5 mph wind (for summer) | Any | 4.00[e] | 0.25 | | | | |

[a]Reprinted with permission of the American Society of Heating, Refrigerating and Air-Conditioning Engineers, Inc., Atlanta, GA (7).

[b]Emittance typical of building materials such as wood, paper, masonry, nonmetallic paints, and regular glass.

[c]Emittance typical of aluminum-coated paper or bright galvanized steel.

[d]Emittance typical of bright aluminum foil.

[e]These values are $h_o$.

## TABLE 2.5

### Thermal Design Values: Thermal Resistances (R) of Air Spaces[a, b]

| Position of Air Space | Direction of Heat Flow | $\epsilon = 0.03$[c] | $\epsilon = 0.82$[d] |
|---|---|---|---|
| Horizontal | Upward | 2.84 | 0.80 |
| Sloping—45° | Upward | 3.18 | 0.82 |
| Vertical | Horizontal | 3.69 | 0.85 |
| Sloping—45° | Downward | 4.81 | 0.90 |
| Horizontal | Downward | 10.07 | 1.00 |

[a]Excerpted with permission of the American Society of Heating, Refrigerating and Air-Conditioning Engineers, Inc., Atlanta, GA (7).

[b]Values are for 3.5 in. dead air space, 90°F mean temperature, 10°F temperature difference.

[c]Effective emittance $\epsilon$ typical of air space surfaces of bright aluminum foil.

[d]Effective emittance $\epsilon$ typical of air space surfaces of wood, paper, masonry, or nonmetallic paints.

**TABLE 2.6**

**Thermal Design Values: Typical Building Materials**[a]

| Material | Conductivity $k$ per in. Thickness | Conductance $C$ for Thickness Listed | Resistance $R$ Per in. Thickness $(1/k)$ | Resistance $R$ For Thickness Listed $(1/C)$ |
|---|---|---|---|---|
| Wood | | | | |
| Hardwoods: maple, oak | 1.09–1.25 | | 0.80–0.92 | |
| Softwoods: fir, pine | 0.95–1.12 | | 1.25 | |
| Insulation | | | | |
| Blanket, mineral fiber 3 to 4 in. | | 0.091 | | 11 |
| Boards | | | | |
| Polystyrene, extruded | 0.20 | | 5.00 | |
| Polystyrene, molded beads | 0.26 | | 3.85 | |
| Glass fiber | 0.25 | | 4.00 | |
| Polyurethane | 0.16 | | 6.25 | |
| Mineral fiber | 0.29 | | 3.45 | |
| Loose fill | | | | |
| Cellulosic insulation | 0.30 | | 3.33 | |
| Perlite, expanded | 0.37 | | 2.70 | |
| Mineral fiber, $7\frac{1}{2}$ to 10 in | | | | 22 |
| Vermiculite, exfoliated | 0.47 | | 2.13 | |
| Concrete | | | | |
| Sand and gravel aggregate | | | | |
| 140 lb/ft³ | 12.0 | | 0.08 | |
| Lightweight aggregate | | | | |
| 100 lb/ft³ | 3.7 | | 0.27 | |

| Material | | | | |
|---|---|---|---|---|
| Masonry units | | | | |
| Brick, common, 120 lb/ft$^3$ | 5.0 | | 0.20 | 1.11 |
| Brick, face, 130 lb/ft$^3$ | 5.4–9.0 | | 0.19–0.11 | 2.00 |
| Concrete blocks | | | | |
| Sand and gravel aggregate, 8 in. | | 0.90 | | |
| Lightweight aggregate, 8 in. | | 0.50 | | 0.45 |
| Building boards | | | | |
| Plywood | 0.80 | | | |
| Gypsum board, $\frac{1}{2}$ in. | | 2.22 | 1.25 | |
| Particle board, medium density | 0.94 | | 1.06 | |
| Building membranes | | | | |
| Vapor-permeable felt | | 16.70 | | 0.06 |
| Vapor-seal plastic film | | | | negl. |
| Plastering materials | | | | |
| Cement plaster, sand aggregate | 5.0 | | 0.20 | |
| Gypsum plaster, sand aggregate | 5.6 | | 0.18 | |
| Roofing | | | | |
| Asbestos cement shingles | | 4.76 | | 0.21 |
| Asbestos roll roofing | | 6.50 | | 0.15 |
| Asphalt shingles | | 2.27 | | 0.44 |
| Built-up roofing, $\frac{3}{8}$ in. | | 3.00 | | 0.33 |
| Wood shingles | | 1.06 | | 0.94 |
| Siding Materials | | | | |
| Asbestos cement shingles | | 4.75 | | 0.21 |
| Wood shingles, 16 in., $7\frac{1}{2}$ in. exposure | | 1.15 | | 0.87 |
| Asphalt insulating siding | | 0.69 | | 1.46 |

(*continued*)

**TABLE 2.6** (*Continued*)

| Material | Conductivity $k$ per in. Thickness | Conductance $C$ for Thickness Listed | Resistance $R$ | |
|---|---|---|---|---|
| | | | Per in. Thickness $(1/k)$ | For Thickness Listed $(1/C)$ |
| Hardboard siding, $\frac{7}{16}$ in. | | 0.49 | | 0.67 |
| Wood, drop, $1 \times 8$ in. | | 1.27 | | 0.79 |
| Wood, bevel, $\frac{1}{2} \times 8$ in. lapped | | 1.23 | | 0.81 |
| Wood, bevel, $\frac{3}{4} \times 10$ in. lapped | | 0.95 | | 1.05 |
| Wood, plywood, $\frac{3}{8}$ in. lapped | | 1.59 | | 0.59 |
| Aluminum or steel over sheathing | | 1.61 | | 0.61 |

[a]Reprinted with permission of the American Society of Heating, Refrigerating and Air-Conditioning Engineers, Inc., Atlanta, GA (7).

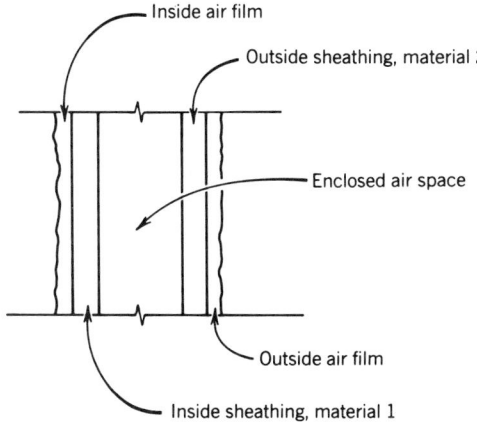

**FIGURE 2.2**  Typical wall section for determining total thermal resistance.

Tables 2.7 and 2.8 show typical examples of numerical calculations of the above type, but in tabular form. Note that in these examples, the overall heat transfer coefficient is calculated for both the area between framing ($U_i$) and for the area backed by framing members ($U_s$). A weighted average is then taken to determine an average value of $U$ for the actual construction. For more detailed information, refer to *ASHRAE Fundamentals Handbook* (7).

## 2.5  ACOUSTICAL PROPERTIES

The acoustical properties of a material or composite construction are determined by its sound insulation and sound absorption abilities. Sound insulation abilities are measured in terms of the reduction in intensity of sound when it passes through a barrier. Sound absorption refers to the amount of incident sound on a surface that is not reflected by that surface.

### 2.5.1  Sound Insulation

Sound insulating values for materials of construction are related to the sound transmission loss for the construction measured in decibels at various frequencies. The insulating values depend on the intensity of sound on the opposite face of a barrier, the mass and stiffness of the material making up a barrier, and the nature of its design and fastenings. Sound insulation may be designed into a barrier by using materials with a high mass per unit area, using materials with low rigidity, or using barriers that employ air spaces and that avoid rigid ties from one face to the other. Like most common construction materials, wood alone does not provide good sound insulation, but when combined with other materials in typical constructions, will provide a structural unit of satisfactory sound-insulating ability.

TABLE 2.7

**Examples of Calculations for Thermal Resistance and Overall Heat Transfer Coefficient**[a]

COEFFICIENTS OF TRANSMISSION $(U)$ OF FRAME WALLS[b]

| | 1 | | 2 | |
| --- | --- | --- | --- | --- |
| | Resistance $R$ | | | |
| Construction 1 | Between Framing | At Framing | Between Framing | At Framing |
| 1. Outside surface (15-mph wind) | 0.17 | 0.17 | 0.17 | 0.17 |
| 2. Siding, wood, 0.5 × 8 in. lapped (average) | 0.81 | 0.81 | 0.81 | 0.81 |
| 3. Sheathing, 0.5 in. asphalt impregnated | 1.32 | 1.32 | 1.32 | 1.32 |
| 4. Nonreflective air space, 3.5 in. (50°F mean; 10°F temperature difference) | 1.01 | — | 11.00 | — |
| 5. Nominal 2 × 4-in. wood stud | — | 4.35 | — | 4.35 |
| 6. Gypsum wallboard, 0.5 in. | 0.45 | 0.45 | 0.45 | 0.45 |
| 7. Inside surface (still air) | 0.68 | 0.68 | 0.68 | 0.68 |
| Total Thermal Resistance (R) | $R_i = 4.44$ | $R_s = 7.78$ | $R_i = 14.43$ | $R_s = 7.78$ |

Construction 1: $U_i = 1/4.44 = 0.225$; $U_s = 1/7.78 = 0.128$. With 20% framing (typical of 2 × 4-in. studs at 16-in. o.c.), $U_{av} = (0.8)(0.225) + (0.2)(0.128) = 0.206$.

Construction 2: $U_i = 1/14.43 = 0.069$; $U_s = 0.128$. With framing unchanged, $U_{av} = (0.8)(0.069) + (0.2)(0.128) = 0.081$.

COEFFICIENTS OF TRANSMISSION ($U$) OF PITCHED ROOFS[c,d]

| Construction 1 (Reflective Air Space) | 1 Heat Flow Up (Winter) | | 2 Heat Flow Down (Summer) | |
|---|---|---|---|---|
| | Between Rafters | At Rafters | Between Rafters | At Rafters |
| 1. Inside surface (still air) | 0.62 | 0.62 | 0.76 | 0.76 |
| 2. Gypsum wallboard 0.5 in., foil backed | 0.45 | 0.45 | 0.45 | 0.45 |
| 3. Nominal 2 × 4-in. ceiling rafter | — | 4.35 | — | 4.35 |
| 4. 45° slope reflective air space, 3.5 in. (50°F mean, 30°F temperature difference) | 2.17 | — | 4.33 | 4.33 |
| 5. Plywood sheathing, $\frac{5}{8}$ in. | 0.77 | 0.77 | 0.77 | 0.77 |
| 6. Felt building membrane | 0.06 | 0.06 | 0.06 | 0.06 |
| 7. Asphalt shingle roofing | 0.44 | 0.44 | 0.44 | 0.44 |
| 8. Outside surface (15 mph wind) | 0.17 | 0.17 | 0.25[e] | 0.25[e] |
| Total Thermal Resistance ($R$) | $R_i = 4.68$ | $R_s = 6.86$ | $R_i = 7.06$ | $R_s = 7.08$ |

Construction 1: $U_i = 1/4.68 = 0.214$; $U_s = 1/6.86 = 0.146$. With 10% framing (typical of 2-in. rafters at 16-in. o.c.),
$U_{av} = (0.9)(0.214) + (0.1)(0.146) = 0.207$.
Construction 2: $U_i = 1/7.06 = 0.142$; $U_s = 1/7.08 = 0.141$. With framing unchanged, $U_{av} = (0.9)(0.142) + (0.1)(0.141) = 0.142$.

## TABLE 2.7 (Continued)

| Construction 3 (Nonreflective Air Space) | 3 Heat Flow Up (Winter) | | 4 Heat Flow Down (Summer) | |
|---|---|---|---|---|
| | Between Rafters | At Rafters | Between Rafters | At Rafters |
| 1. Inside surface (still air) | 0.62 | 0.62 | 0.76 | 0.76 |
| 2. Gypsum wallboard, 0.5 in. | 0.45 | 0.45 | 0.45 | 0.45 |
| 3. Nominal 2 × 4-in. ceiling rafter | — | 4.35 | — | 4.35 |
| 4. 45° slope, nonreflective air space, 3.5 in. (50°F mean; 10°F temperature difference) | 0.96 | — | 0.90[f] | — |
| 5. Plywood sheathing, $\frac{5}{8}$ in. | 0.77 | 0.77 | 0.77 | 0.77 |
| 6. Felt building membrane | 0.06 | 0.06 | 0.06 | 0.06 |
| 7. Asphalt shingle roofing | 0.44 | 0.44 | 0.44 | 0.44 |
| 8. Outside surface (15-mph wind) | 0.17 | 0.17 | 0.25[e] | 0.25[e] |
| Total Thermal Resistance ($R$) | $R_i = 3.47$ | $R_s = 6.86$ | $R_i = 3.63$ | $R_s = 7.08$ |

Construction 3: $U_i = 1/3.47 = 0.288$; $U_s = 1/6.86 = 0.146$. With 10% framing (typical of 2-in. rafters at 16-in. o.c.),
$U_{av} = (0.9)(0.288) + (0.1)(0.146) = 0.274.$

Construction 4: $U_i = 1/3.63 = 0.275$; $U_s = 1/7.08 = 0.141$. With framing unchanged, $U_{av} = (0.9)(0.275) + (0.1)(0.141) = 0.262.$

[a] Reprinted with permission of the American Society of Heating, Refrigerating and Air-Conditioning Engineers, Inc., Atlanta, GA (7).

[b] Coefficients expressed in Btu per (hour) (square foot) (degree Fahrenheit difference in temperature between air on two sides) and are based on outside wind velocity of 15 mph. Replace air space with 3.5-in. R-11 blanket insulation for construction 2.

[c] Coefficients are expressed in Btu per (hour) (square foot) (degree Fahrenheit difference in temperature between the air on the two sides) and are based on an outside wind velocity of 15 mph for heat flow upward and 7.5 mph for heat flow downward.

[d] Pitch of roof—45°.

[e] 7.5 mph wind.

[f] Air space value at 90°F mean, 10°F temperature difference.

# TABLE 2.8

## Approximate Thermal Insulation Design Values for Typical Plywood Roof Deck Systems

### No Insulation

| Construction | Heat Flow Up (Winter) | Heat Flow Down (Summer) |
|---|---|---|
| 1. Outside surface (15-mph wind) | 0.17 | 0.25 |
| 2. $\frac{3}{8}$ in. B.U. roofing | 0.33 | 0.33 |
| 3. $\frac{1}{2}$ in. plywood | 0.62 | 0.62 |
| 4. Inside surface (still air) | 0.61 | 0.92 |
| Total Thermal Resistance ($R$) | 1.73 | 2.12 |
| $U = 1/R$ | 0.578 | 0.472 |

### Reflective Insulation[a]

| Construction | Heat Flow Up (Winter) | | Heat Flow Down (Summer) | |
|---|---|---|---|---|
| | Between Framing | At Framing | Between Framing | At Framing |
| 1. Outside surface (15-mph wind) | 0.17 | 0.17 | 0.25 | 0.25 |
| 2. $\frac{3}{8}$ in. B.U. roofing | 0.33 | 0.33 | 0.33 | 0.33 |
| 3. $\frac{1}{2}$ in. plywood | 0.62 | 0.62 | 0.62 | 0.62 |
| 4. $3\frac{1}{2}$ in. air space[b] | 2.01 | — | 8.17 | — |
| 5. $2 \times 4$ in. stiffener | — | 4.35 | — | 4.35 |
| 6. Aluminum paper | — | — | — | — |
| 7. Inside surface (still air) | 0.61 | 0.61 | 0.92 | 0.92 |
| Total Thermal Resistance ($R$) | 3.74 | 6.08 | 10.29 | 6.47 |

Winter: $U_i = 1/3.74 = 0.267$, $U_s = 1/6.08 = 0.164$; $U_{avg} = (0.92)(0.267) + (0.08)(0.164) = 0.259$.

Summer: $U_i = 1/10.29 = 0.097$, $U_s = 1/6.47 = 0.155$; $U_{avg} = (0.92)(0.097) + (0.08)(0.155) = 0.102$.

Diagram 1 labels: 3/8" B.U. roof; Outside air; 1/2" plywood; Inside air; 2 × 4 Stiffeners 24" o.c. (typical)

Diagram 2 labels: 3/8" B.U. roofing; Outside air; 3½" Air space; Reflective side; Inside air; 2 × 4 Stiffeners 24" o.c. 8% Framing[d]; 1/2" Plywood; Paper[a]

2-53

## TABLE 2.8 (Continued)

### FLEXIBLE INSULATION

| Construction | Heat Flow Up (Winter) | | Heat Flow Down (Summer) | |
|---|---|---|---|---|
| | Between Framing | At Framing | Between Framing | At Framing |
| 1. Outside surface (15-mph wind) | 0.17 | 0.17 | 0.25 | 0.25 |
| 2. $\frac{3}{8}$ in. B.U. roofing | 0.33 | 0.33 | 0.33 | 0.33 |
| 3. $\frac{1}{2}$ in. plywood | 0.62 | 0.62 | 0.62 | 0.62 |
| 4. R11 insulation | 11.0 | — | 11.0 | — |
| 5. 2 × 4 in. stiffener | — | 4.35 | — | 4.35 |
| 6. Aluminum paper | — | — | — | — |
| 7. Inside surface (still air) | 0.61 | 0.61 | 0.92 | 0.92 |
| Total Thermal Resistance ($R$) | 12.73 | 6.08 | 13.12 | 6.47 |

Winter: $U_i = 1/12.73 = 0.078$, $U_s = 1/6.08 = 0.164$; $U_{av} = (0.92)(0.078) + (0.08)(0.164) = 0.085$.

Summer: $U_i = 1/13.12 = 0.076$, $U_s = 1/6.47 = 0.155$; $U_{av} = (0.92)(0.076) + (0.08)(0.155) = 0.082$.

| Construction | Heat Flow Up (Winter) | | Heat Flow Down (Summer) | |
|---|---|---|---|---|
| | Between Framing | At Framing | Between Framing | At Framing |
| 1. Outside surface (15-mph wind) | 0.17 | 0.17 | 0.25 | 0.25 |
| 2. $\frac{3}{8}$ in. B.U. roofing | 0.33 | 0.33 | 0.33 | 0.33 |
| 3. $\frac{3}{4}$ in. plywood | 0.93 | 0.93 | 0.93 | 0.93 |
| 4. $R19$ insulation | 19.0 | — | 19.0 | — |
| 5. $2 \times 6$ in. stiffener | — | 6.88 | — | 6.88 |
| 6. Aluminum paper | — | — | — | — |
| 7. Inside surface (still air) | 1.32 | 1.32 | 4.55 | 4.55 |
| Total Thermal Resistance ($R$) | 21.75 | 9.63 | 25.06 | 12.94 |

Winter: $U_i = 1/21.75 = 0.046$, $U_s = 1/9.63 = 0.104$; $U_{av} = (0.92)(0.046) + (0.08)(0.104) = 0.051$.

Summer: $U_i = 1/25.06 = 0.040$, $U_s = 1/12.94 = 0.077$; $U_{av} = (0.92)(0.040) + (0.08)(0.077) = 0.043$.

RIGID INSULATION

| Construction | Heat Flow Up (Winter) | Heat Flow Down (Summer) |
|---|---|---|
| 1. Outside surface (15-mph wind) | 0.17 | 0.25 |
| 2. $\frac{3}{8}$ in. B.U. roofing | 0.33 | 0.33 |
| 3. Insulation, $R = 8.33$ | 8.33 | 8.33 |
| 4. $\frac{3}{4}$ in. plywood | 0.93 | 0.93 |
| 5. Inside surface (still air) | 0.61 | 0.92 |
| Total Thermal Resistance ($R$) | 10.37 | 10.76 |
| $U = 1/R$ | 0.096 | 0.093 |

3/8" B.U. roofing
Outside air
3/4" Plywood
Reflective paper
Inside air
R19 Insulation
2 × 6 Stiffeners
24" o.c. 8% Framing

8% Framing
3/8" B.U. roofing
Outside air
R = 8.33 Rigid insulation
Inside air
3/4" Plywood
2 × 4 Stiffeners 8% Framing

2-55

## TABLE 2.8 (Continued)

### HEAVY TIMBER ROOF DECK SYSTEMS

Type of Rigid Insulation

| Type of Deck | None | | Polystyrene or Fiberglass Thickness (in.) (k = 0.25) | | | | Polystyrene Thickness (in.) (k = 0.20) | | | | Polyurethane Thickness (in.) (k = 0.14) | |
| | | | 1 | | 1½ | | 1 | | 1½ | | 1 | |
| | Winter | Summer | Winter | Summer | Winter | Summer | Winter | Summer | Winter | Summer | Winter | Summer |
| 1⅛ in. thick plywood | | | | | | | | | | | | |
| $U$ | 0.397 | 0.344 | 0.154 | 0.145 | 0.117 | 0.112 | 0.133 | 0.126 | 0.100 | 0.096 | 0.104 | 0.100 |
| $R$ | 2.52 | 2.91 | 6.49 | 6.90 | 8.55 | 8.93 | 7.52 | 7.94 | 10.00 | 10.42 | 9.62 | 10.00 |
| 2 in. nominal wood deck (1½ in. actual) | | | | | | | | | | | | |
| $U$ | 0.335 | 0.296 | 0.143 | 0.136 | 0.111 | 0.107 | 0.125 | 0.119 | 0.095 | 0.092 | 0.099 | 0.095 |
| $R$ | 2.99 | 3.38 | 6.99 | 7.35 | 9.01 | 9.35 | 8.00 | 8.40 | 10.53 | 10.87 | 10.10 | 10.53 |

| | | | | | | | | | | | | |
|---|---|---|---|---|---|---|---|---|---|---|---|---|
| 3 in. nominal wood deck (2½ in. actual) | $U$ | 0.236 | 0.216 | 0.121 | 0.116 | 0.098 | 0.094 | 0.108 | 0.104 | 0.085 | 0.082 | 0.088 | 0.085 |
| | $R$ | 4.24 | 4.63 | 8.26 | 8.62 | 10.20 | 10.64 | 9.26 | 9.62 | 11.76 | 12.20 | 11.36 | 11.76 |
| 4 in. nominal wood deck (3½ in. actual) | $U$ | 0.182 | 0.170 | 0.105 | 0.101 | 0.087 | 0.084 | 0.095 | 0.092 | 0.077 | 0.075 | 0.079 | 0.077 |
| | $R$ | 5.49 | 5.88 | 9.52 | 9.90 | 11.49 | 11.90 | 10.53 | 10.87 | 12.99 | 13.33 | 12.66 | 12.99 |

[a]Aluminum-coated paper, reflective inside only.

[b]50°F mean, 30°F temperature differential, $\epsilon = 0.05$. Venting may be required for high-humidity conditions.

[c]Verify $k$ value of polystyrene with manufacturer.

[d]Framing occupies 8% of the roof surface.

### 2.5.2  Sound Absorption

The effectiveness of any material with respect to absorbing sound is given by its absorption coefficient at the frequency range specified. The sound absorption coefficient for a material is used to determine the total magnitude of the absorption property of the material expressed in units of sabins. Thus, the total sound absorption capacity of a wall, ceiling, floor, and so forth in sabins is determined by multiplying the sound absorption coefficient of the material involved by its total area.

Sound absorption values for any material may be compared to that of an open window, which is assumed to be the most complete absorber of sound. An open window is given a sound absorption coefficient of unity. A piece of hairfelt of equal size may absorb half as much sound as the open window and, therefore, has a coefficient of 0.50. Sound absorption coefficients are measured for various frequencies of sound, and the absorption ability of a material may vary widely with differences in frequency. Sound absorption values for wood vary with moisture content, direction of grain, and density.

An important acoustical property of wood is that it absorbs low-frequency sounds more rapidly than high-frequency sounds.

### 2.6  ELECTRICAL PROPERTIES

The most important electrical properties of wood are its resistance to the passage of an electric current and its dielectric properties. The electrical resistance of wood is utilized in electric moisture meters used to determine moisture content. The dielectric properties of wood are utilized in the high-frequency curing of adhesives in glued laminated members and in the seasoning of wood.

The electrical resistance of wood varies with moisture content, density, direction of travel of the current with respect to the direction of the grain, and temperature. It varies greatly with moisture content, especially below the fiber saturation point, decreasing with an increase in moisture content. Electrical resistance varies inversely with the density of wood, although this effect is slight compared to the variance due to moisture content and is greater across the grain than along it. The electrical resistance of wood approximately doubles for each drop in temperature of 22.5°F, but this relationship varies considerably with the level of moisture content. There is also a variation in electrical resistance between species, which is possibly caused by minerals or electrolytes in the wood itself or dissolved in the water present in the wood.

Metallic salts, such as used in preservative and fire-retardant treatments, may lower the electrical resistance of the wood considerably. Use of wood containing such salts should be avoided for applications where electrical resistance is critical and in processes involving dielectric heating. Electric moisture meters may give erroneous readings for such wood.

Wood at a low moisture content is normally classified as an electric insulator, or dielectric, rather than as a conductor. A dielectric can be heated by using it as the medium between electrodes carrying charges of oscillating high-frequency electricity or by placing it in an electric field of like nature. The dielectric constant of wood (the ratio of the capacitance of a wood condenser to the capacitance of a similar condenser employing a vacuum) is proportional to density at a given moisture content. The dielectric constant parallel to grain is significantly greater than the corresponding constant perpendicular to grain. The constant also decreases with an increase in frequency of the oscillating current.

The power factor of wood is the ratio of the power absorbed in the wood per cycle of oscillation of an electric current to the total apparent power stored in the wood during that cycle. The power factor generally increases with an increase in moisture content. The parallel-to-grain power factors are usually greater than the corresponding perpendicular-to-grain factors.

## 2.7 COEFFICIENT OF FRICTION

The coefficient of friction depends on the moisture content of the wood and surface roughness. It varies little with species, except for those species that contain abundant oily or waxy extractives.

Coefficients of static friction for wood on unpolished steel have been reported to be approximately 0.70 for dry wood and 0.40 for green wood. Coefficient of static friction for smooth wood on smooth wood are 0.60 for dry wood and 0.83 for green wood.

It is not usual practice to use friction to resist forces in the design of timber connections.

## 2.8 PROPERTIES OF SECTIONS

### 2.8.1 Sawn Lumber and Timber

Table 8.1 lists sizes and properties of sections for sawn lumber and timber, and includes values for net cross-sectional area, section modulus, moment of inertia, weight per lineal foot, and board measure per lineal foot.

The use of standard sizes is more economical than the use of special sizes or lengths. Lengths are generally available in even 2-ft increments. There is a limit on the maximum length normally available.

### 2.8.2 Glued Laminated Timber

Table 8.2 gives section properties for structural glued laminated timbers based on industry-recommended standard thicknesses and widths.

The most efficient and economical production of glued laminated structural members results when standard lumber sizes are used for the laminates. Industry recommended practices uses nominal 2-in.-thick lumber of standard nominal width to produce straight members and curved members where the radius of cur-

**TABLE 2.9**

**Net Depth Glued Laminated Timber**

| No. Laminations | Net Depth of Member (in.) | | |
|---|---|---|---|
| | Nominal 1-in. Laminations[a] | Nominal 2-in. Laminations | |
| | $\frac{3}{4}$ in. Actual | $1\frac{1}{2}$ in. Actual | $1\frac{3}{8}$ in. Actual |
| 4 | 3 | 6 | $5\frac{1}{2}$ |
| 5 | $3\frac{3}{4}$ | $7\frac{1}{2}$ | $6\frac{7}{8}$ |
| 6 | $4\frac{1}{2}$ | 9 | $8\frac{1}{4}$ |
| 7 | $5\frac{1}{4}$ | $10\frac{1}{2}$ | $9\frac{5}{8}$ |
| 8 | 6 | 12 | 11 |
| etc. | etc. | etc. | etc. |

[a]Based on laminations being surfaced to actual $\frac{3}{4}$ in. thickness. When laminations are resurfaced to less than $\frac{3}{4}$ in. thickness, depth is the thickness times the number of laminations.

vature is within the bending radius limits for that thickness of the species. Nominal 1-in.-thick boards are normally used when the bending radius is too sharp to permit use of nominal 2-in.-thick laminations. These standard practices are subject to modification for specific job requirements and plant procedures. The use of nominal 1-in. and 2-in.-thick laminations will generally be the most economical, and therefore, conformance with these usual thicknesses is recommended for all normal uses. Exceptions should be made only when the shape of the structure requires nonstandard laminations.

For laminating, dimension lumber of 2 in. nominal thickness is normally surfaced to a net thickness ranging from $1\frac{1}{2}$ to $1\frac{3}{8}$ in. The section properties included in Table 8.2 are based on $1\frac{1}{2}$ in. and $1\frac{3}{8}$ in. thickness. However, other thicknesses of laminations may be used to achieve the desired member depth. Finished depths of members are normally increments of these net thicknesses, as illustrated in the Table 2.9. Nominal 1-in.-thick boards usually are surfaced to $\frac{3}{4}$ in. or less net thickness.

The use of laminations of special thicknesses because of the bending radius of the member or the mixing of thicknesses for special purposes may result in net finished depths that may differ from those indicated in Table 2.9.

It is necessary to surface the sides of laminated members after gluing to remove the glue squeeze-out and to provide a uniformly smooth surface. Therefore, the net finished width of the glued laminated member is less than the net finished width of industry standard boards and dimension. Normal standard net finished widths for glued laminated structural members are as follows. Other finished widths may be used to meet the size requirements of a design or to meet other special requirements.

| Nominal width, in. | 3 | 4 | | 6 | | 8 | 10 | 12 | 14 | 16 |
|---|---|---|---|---|---|---|---|---|---|---|
| Net finished width, in. (western softwoods) | $2\frac{1}{8}$ | $3\frac{1}{8}$ | | $5\frac{1}{8}$ | | $6\frac{3}{4}$ | $8\frac{3}{4}$ | $10\frac{3}{4}$ | $12\frac{1}{4}$ | $14\frac{1}{4}$ |
| Net finished width, in. (Southern Pine) | $2\frac{1}{8}$ | 3 or $3\frac{1}{8}$ | | 5 or $5\frac{1}{8}$ | | $6\frac{3}{4}$ | $8\frac{1}{2}$ | $10\frac{3}{4}$ | $12\frac{1}{4}$ | $14\frac{1}{4}$ |

Industry standards permit the following tolerances at the time of manufacture:

1. Width—Plus or minus $\frac{1}{16}$ in. of the specified width.
2. Depth—Plus $\frac{1}{8}$ in. per ft of specified depth, minus $\frac{3}{16}$ in., or minus $\frac{1}{16}$ in. per ft of specified depth, whichever is the larger.
3. Length—Up to 20 ft, plus or minus $\frac{1}{16}$ in.; over 20 ft, plus or minus $\frac{1}{16}$ in. per 20 ft of length or fraction thereof.
4. Squareness—The cross section of all glued laminated structural members shall be square within plus or minus $\frac{1}{8}$ in. per ft of specified depth of member unless a specially shaped section is specified.
5. Camber or Straightness—The tolerances are applicable at the time of manufacture without allowances for dead load deflection. Up to 20 ft, the tolerances is plus or minus $\frac{1}{4}$ in. Over 20 ft, increase tolerance $\frac{1}{8}$ in. per each additional 20 ft or fraction thereof, but not to exceed $\frac{3}{4}$ in. The tolerances are intended for use with straight or slightly cambered members, and are not applicable to curved members such as arches.

## 2.9  REFERENCES

1. United States Department of Agriculture, Forest Service, Forest Products Laboratory, *Wood Handbook: Wood as an Engineering Material*, Agriculture Handbook No. 72, Madison, WI, 1987.
2. American Society for Testing and Materials, *Standard Test Methods for Specific Gravity of Wood and Wood-Base Materials*, ASTM D 2395, Philadelphia, PA, 1983.
3. American Institute of Timber Construction, *Standard for Preservative Treatment of Structural Glued Laminated Timber*, AITC 109, Englewood, CO, 1990.
4. American Forest & Paper Association, ANSI/NF$_0$ PA NDS®-1991, *National Design Specification® for Wood Construction*, Washington, DC, 1991.
5. American Society for Testing and Materials, *Standard Methods for Establishing Clear Wood Strength Values*, ASTM D 2555, Philadelphia, PA, 1988.
6. American Institute of Timber Construction, *Typical Construction Details*, AITC 104, Englewood, CO, 1984.
7. American Society of Heating, Refrigerating and Air-Conditioning Engineers, Inc., *ASHRAE Fundamentals Handbook*, Atlanta, GA, 1989.

# Chapter 3

# MECHANICAL PROPERTIES OF WOOD

## 3.1   DESIGN VALUES

The tabular design values for wood are given in Chapter 8. Tables 8.3–8.7 contain design values for sawn lumber. Table 8.8 contains bearing design values parallel to grain (end grain in bearing) for both sawn lumber and glued laminated timber. Design values for softwood glued laminated timber are given in AITC 117-93 *Design—Laminating Specifications* (1). The design values for hardwoods are given in AITC 119-94 *Laminating Specifications for Hardwoods* (2). Fastening design values are given in Chapter 7. These design values are based on loads of normal duration and the moisture service condition specified.

### 3.1.1   Adjustment Factors

Other conditions may exist that require adjustment to the tabular values. The tabular design values are denoted as $F_b$, $F_t$, $F_v$, $F_{c\perp}$, $F_c$, $E$, and $F_g$. After all applicable adjustment factors have been applied, the allowable design values are $F_b'$, $F_t'$, $F_v'$, $F_{c\perp}'$, $F_c'$, $E'$, and $F_g'$. See Table 4.6 for applicability of adjustment factors. Adjustment factors are explained in Chapter 4.

## 3.2   LUMBER GRADING AND DESIGN VALUES

Lumber, as it is sawn from a log, is quite variable in its mechanical properties. Individual pieces may differ in strength by as much as several hundred percent. For simplicity and economy in use, pieces of lumber of similar mechanical properties can be placed in a single class known as a grade. The properties of a particular grade depend on the sorting criteria used and on additional factors independent of the sorting criteria.

Grades and their associated design properties are found in grading rules published by rules-writing grading agencies, and these design values are also included in ANSI/NF$_o$PA NDS®—1991, *National Design Specification® for Wood Construction* (3). Not all grades listed are available in commercial quantities, and availability varies with market conditions and by geographical area. The designer should check on availability before specifying.

### 3.2.1   Visual Grading

Visual grading is the oldest lumber-grading method. It is based on the premise that mechanical properties of lumber differ from mechanical properties of clear wood because of naturally occurring characteristics that can be seen and judged by eye. These visual characteristics are used to sort the lumber into grades and include, but are not limited to, density, slope of grain, size and location of knots, checks and splits.

The design values of visually graded timbers, decking, and some species and grades of dimension lumber are based on procedures in *Standard Practices for Establishing Structural Grades and Related Properties for Visually Graded Lumber*, ASTM D245 (4). Design values for most species and grades of visually graded dimension lumber, lumber with widths of 2–4 in. are based on an in-grade testing program of many pieces of as-graded lumber of various widths on species as described in *Standard Practice for Establishing Allowable Properties for Visually-Graded Dimension Lumber from In-Grade Tests of Full Size Specimens*, ASTM D 1990 (5).

The lumber design values based on the ASTM 245 procedure are determined as follows:

The derivation of mechanical properties of visually graded lumber is based on clear-wood properties and on the presence of lumber characteristics allowed by the visual sorting criteria, as well as on the effects of size, duration of load, and moisture content. Clear-wood properties are determined in accordance with the American Society for Testing and Materials, *Standard Methods for Establishing Clear Wood Strength Values*, ASTM D 2555 (6).

Each strength property of a piece of lumber is derived from the product of the clear-wood strength for the species and the limiting strength ratio. The strength ratio is a hypothetical ratio of the strength of a piece of lumber with visible strength reducing characteristics to its strength if these characteristics are absent. Adjustments of clear-wood strength values for such factors as density, slope of grain, knots, shakes, checks and splits, size, variability, duration of load, and moisture content are given in ASTM D 245 (4).

In visually graded lumber, the modulus of elasticity $E$ assigned is an estimate of the mean modulus of the lumber grade. The average $E$ value for clear wood of the species, as given in ASTM D 2555 (6), is used as a base. The clear-wood value is multiplied by "quality factors" representing the reduction in $E$ that occurs by lumber grade. This procedure is outlined in ASTM D 245 (4).

The design values based on the in-grade testing of lumber are calculated based on testing of large numbers of different sizes and species of dimension lumber. The procedure for testing and calculating design values is given in ASTM D 1990 (5).

### 3.2.2   Machine Stress-Rated (MSR) Grading or *E*-Rated Grading

MSR grading is based on an observed relationship between modulus of elasticity $E$ and bending strength, tensile strength, or other strength properties. The $E$ value of lumber is thus a major sorting criterion for this method of grading.

The $E$ used as a sorting criterion for mechanical properties can be measured in a variety of ways. Usually, the apparent $E$ or a stiffness-related deflection is measured. Most MSR grading machines in the United States are designed to measure the flatwise bending stiffness that occurs in any span of approximately 4 ft and to use this measurement in the sorting criterion.

Each grade derived from MSR lumber relates allowable strength in bending, compression, and tension parallel to grain to the $E$ levels by which the grade is identified. As in visual grading, the $E$ value assigned to a grade for design represents the average for the grade. However, the stress-rating machines are adjusted so that the coefficient of variation of $E$ for an MSR grade is less than for a visual grade.

In machine stress rating, strengths for shear parallel to grain and compression perpendicular to grain are handled in relationship to clear-wood properties and visual grading characteristics.

Commercial machine stress-rating practices in the United States combine edge characteristic size restriction (such as knots, knot holes, burls, distorted grain, or decay) with $E$ as a predictor of grade properties. It has been shown that these concepts used together provide a better strength prediction than either alone. Design values for MSR lumber are given in Table 8.5.

### 3.2.3  Machine Evaluated Lumber (MEL)

Machine Evaluated Lumber is lumber that has been nondestructively evaluated by *American Lumber Standards (ALS) Board of Review* approved mechanical grading equipment. The MEL machine evaluates each piece, and sorts and marks the material into various strength classifications. MEL lumber is also required to meet certain visual requirements.

Design values for MEL are given in the Supplement to the ANSI/NF$_o$PA NDS—1991, *National Design Specification for Wood Construction* (3).

### 3.2.4  Design Values for Sawn Lumber

The design values for the most commonly used species and grades of visually graded lumber are tabulated in Tables 8.3–8.7. These tables are for different sizes and species. For design values of additional species or species groups, see *Design Values for Construction*, a supplement to the ANSI/NF$_o$PA NDS—1991, *National Design Specification for Wood Construction (3)*. The availability of particular grades and species should be determined prior to the use of the design values.

### 3.2.5  Design Values for Bearing on End Grain

The tabular design values for bearing stress parallel to grain $F_g$ on a rigid surface are given in Table 8.8. These design values apply to the net area in bearing.

When the design value for end grain in bearing exceeds 75% of the values given in Table 8.8, the bearing should be on a metal plate or strap or on other rigid material of adequate strength.

The values in Table 8.8 apply to end-to-end bearing of compression members, for example, as in a column, provided there is adequate lateral support and the end cuts are accurately squared and parallel. When a rigid insert between ends of members loaded in end grain bearing is required, it should be of not less than 20-gage metal plate or equivalent and should be snugly placed between abutting ends.

## 3.3   TREATMENTS—SAWN LUMBER

### 3.3.1   Preservative Treatments

The design values tabulated in Tables 8.3–8.8 apply to sawn lumber pressure impregnated with preservatives by an approved process as recommended by the *American Wood-Preservers' Association* (AWPA) (7). For sawn lumber pressure impregnated with preservative salts to the heavy retentions required for "marine" exposure as defined by the AWPA, the duration-of-load factor for impact does not apply.

### 3.3.2   Fire-Retardant Treatments

It is recommended that the effects of fire-retardant chemical treatments on strength be considered, and that information on these effects be obtained from the company providing the treating and redrying service.

## 3.4   STRUCTURAL GLUED LAMINATED TIMBER DESIGN VALUES

### 3.4.1   Establishment of Design Values

The term *structural glued laminated timber* refers to an engineered, stress-rated product of a timber laminating plant comprising assemblies of suitably selected and prepared wood laminations bonded together with adhesives. The grain of all laminations is approximately parallel longitudinally. The separate laminations do not exceed 2 in. in net thickness. They may be comprised of pieces end joined to form any length, of pieces placed or glued edge to edge to make wider ones, or of pieces bent to curved form during gluing. The design values for glued laminated timber are established by following the procedures given in *Standard Method for Establishing Stresses for Structural Glued Laminated Timber (Glulam)*, ASTM D 3737 (8). The following discussion covers the method used to establish design values in bending. (See ASTM D 3737 (8) for procedures for establishing other design values.)

The bending strength of horizontally laminated timbers (bending members that are loaded perpendicular to the wide faces of the laminations) depends on the position of various grades of laminations within the member. High-grade laminations are normally placed in the outer portions of the member where the higher strength is effectively used; lower-grade laminations are placed in the inner portion where lower strength will not greatly affect the overall strength of the member.

Also, due to the random dispersal of knots and other strength-reducing characteristics (the laminating effect), improved strength is obtained.

Evaluations of timbers vertically laminated from dimension lumber (members loaded parallel to the wide faces of the laminations) indicate that the effect of the lamination grade and number of laminations may be accounted for by procedures recommended in USDA Forest Service Research Paper FPL 333, *Bending Strength of Vertically Glued Laminated Beams with One to Five Plies* (9), and in ASTM D 3737 (8).

The principal determinants of bending strength in glued laminated structural members are the clear-wood strength and grade characteristics such as knots and slope of grain. The effects of these grade characteristics, called grade factors, are not cumulative; that is, the lowest of the grade factors controls the strength. Where other factors such as curvature or size of member are applicable, they are applied in addition to those for knots and slope of grain.

### 3.4.2 Knots

The effect of knots on the bending strength and stiffness of laminated timbers depends on their frequency within a cross section and their size and position with respect to the neutral axis of the member. The sum of the moments of inertia of areas occupied by knots within 6 in. of a critical cross section is represented by the symbol $I_K$. The moment of inertia for the full cross section of the member is represented by $I_G$. Tests of laminated timbers containing knots in various concentrations established a relationship between the $I_K/I_G$ ratio and bending strength of the members. The procedures for calculating the $I_K/I_G$ ratio are given in USDA Forest Service Research Paper FPL 292, *Improved Utilization of Lumber in Glued Laminated Beams* (10), and in ASTM D 3737 (8).

The $I_K/I_G$ ratio and its relationship to bending strength yields reliable results if specially selected tension laminations are used. AITC has established these tension lamination grades for various design value levels, and they are included in AITC 117—Manufacturing (11).

### 3.4.3 Slope of Grain

The strength of wood decreases with increasing slope of grain. For glued laminated timbers, it is possible to vary the slope-of-grain requirements for laminations at different points in the depth of a beam in accordance with the stress requirements. That is, steeper slopes of grain may be permitted in the interior laminations than in the outer laminations.

### 3.4.4 End Joints

For most glued laminated timbers manufactured, pieces of lumber must be joined end to end to provide laminations of sufficient length because of the size of the member. Both plane scarf joints and finger joints can be manufactured with adequate strength for structural glued laminated timber. The required end joint strength is monitored by physical testing procedures contained in industry standards.

Butt joints generally can transmit no tensile stress, and can transmit compressive stress only after considerable deformation. They also cause concentrations of both shear stress and longitudinal stress. For these reasons, they are not permitted for use in structural glued laminated timbers.

### 3.4.5   Design Values

Design values for structural glued laminated timber are given in Tables 1 and 2 of *Standard Specifications for Structural Glued Laminated Timber of Softwood Species,* AITC 117—Design (1), included in Chapter 8.

When curved members are subject to bending moment, radial stresses are induced in a direction parallel to the radius of curvature of the centerline of the member (perpendicular to grain). If the moment decreases the curvature (tends to straighten the member), the stress is tension; if it increases the curvature (makes the member more sharply curved), the stress is compression. For these members, when the moment induces a stress in tension across the grain, this tensile stress perpendicular to grain is limited to one-third the design value in horizontal shear for Douglas Fir-Larch, Douglas Fir South, Hem-Fir, and Western Woods and Canadian softwood species for wind or earthquake loadings and for Southern Pine for all conditions of loading. The limit is 15 psi for Douglas Fir-Larch, Douglas Fir South, Hem-Fir, and Western Woods and Canadian softwood species for other types of loads. For Douglas Fir-Larch, Douglas Fir South, Hem-Fir, or Western Woods and Canadian softwood species where the calculated stress exceeds this 15-psi value, mechanical reinforcing designed in accordance with the procedure in Chapter 5 (or an equivalent design method) should be used and should be sufficient to resist all radial tension stresses. In no case shall the calculated radial stress exceed one-third of the design value in shear parallel to grain. When mechanical reinforcing is used, the maximum moisture content of the laminations at the time of manufacture should not exceed 12%.

When the moment is in a direction causing a stress in compression across the grain, this stress is limited to the design value in compression perpendicular to the grain for the species involved.

### 3.4.6   Design Values for Bearing Parallel to Grain

Design values for bearing on the end grain, $F_g$, of sawn lumber and structural glued laminated timbers are based on the clear-wood strength of the species and are given in Table 8.8. Note that the design values are shown for both wet service conditions and dry service conditions, and use of the adjustment factor for moisture $C_M$ is not necessary.

## 3.5   TREATMENTS—GLUED LAMINATED TIMBER

### 3.5.1   Preservative Treatments

Design values tabulated in Tables 1 and 2 of AITC 117—Design (1) are applicable to glued laminated timbers pressure impregnated by processes and preservatives as recommended by the American Wood-Preservers' Association (7).

The provisions in *Standard for Preservative Treatment of Structural Glued Laminated Timber*, AITC 109 (12), included in Chapter 8 also apply.

### 3.5.2 Fire-Retardant Treatments

The design values for fire-retardant treated structural glued laminated timber treated before or after gluing are dependent on the species and treatment combinations involved. The effect on strength must be determined for each treatment; however, indications are that a 10–25% reduction in the design values in bending may be expected. The manufacturer of the treatment should be contacted for more specific information on stress adjustments for all design values.

## 3.6 REFERENCES

1. American Institute of Timber Construction, *Standard Specification for Structural Glued Laminated Timber of Softwood Species*, AITC 117—Design, Englewood, CO, 1993.

2. American Institute of Timber Construction, *Standard Specification for Hardwood Structural Glued Laminated Timber*, AITC 119, Englewood, CO, 1994.

3. American Forest and Paper Association, ANSI/NF$_o$PA NDS—1991, *National Design Specification for Wood Construction*, Washington, DC.

4. American Society for Testing and Materials, *Standard Methods for Establishing Structural Grades and Related Allowable Properties for Visually Graded Lumber*, ASTM D 245, Philadelphia, PA, 1992.

5. American Society for Testing and Materials, *Standard Practice for Establishing Allowable Properties of Visually-Graded Dimension Lumber from In-Grade Tests of Full-Size Specimens*, ASTM D1990, Philadelphia, PA, 1991.

6. American Society for Testing and Materials, *Standard Methods for Establishing Clear Wood Strength Values*, ASTM D 2555, Philadelphia, PA, 1988.

7. American Wood-Preservers' Association, *Book of Standards*, Stevensville, MD, 1984.

8. American Society for Testing and Materials, *Standard Method for Establishing Stresses for Structural Glued Laminated Timber (Glulam)*, ASTM D 3737, Philadelphia, PA, 1993.

9. United States Department of Agriculture, Forest Service, Forest Products Laboratory, *Bending Strength of Vertically Glued Laminated Beams with One to Five Plies*, Research Paper FPL 333, Madison, WI, 1979.

10. United States Department of Agriculture, Forest Service, Forest Products Laboratory, *Improved Utilization of Lumber in Glued Laminated Beams*, Research Paper FPL 292, Madison, WI, 1977.

11. American Institute of Timber Construction, *Standard Specifications for Structural Glued Laminated Timber of Softwood Species*, AITC 117—Manufacturing, Englewood, CO, 1993.

12. American Institute of Timber Construction, *Standard for Preservative Treatment of Structural Glued Laminated Timber*, AITC 109, Englewood, CO, 1990.

# Part II

# DESIGN

# CHAPTER 4

# LOADS, DEFLECTION, TABULAR VALUES, AND ADJUSTMENT FACTORS

## 4.1   LOADS

Buildings or other structures and all parts thereof should be designed to safely support all loads, including dead loads, that may reasonably be expected to affect the structure during its service life. These loads should be as stipulated by the governing building code or, in the absence of such a code, the loads, forces, and combination of loads should be in accordance with accepted engineering practice for the geographical area under consideration.

### 4.1.1   Dead Loads

*Dead load* is defined as the vertical load due to all permanent structural and nonstructural components of a structure, such as walls, floors, roofs, partitions, stairways, and fixed service equipment.

During the life of a building or other structure, additional loads may be applied that actually become dead loads, but in all probability are not treated as such in the original design. If it is anticipated that such loads will be added at a later date, modifications should be made in the original design. For example, consideration should be given to the possible reroofing of the structure, and allowance appropriate to the type of roofing involved should be made.

The actual weights of the various materials and constructions should be used in design if this information is available. In the absence of such information, values satisfactory to the building official should be used. Table 8.15 gives the approximate weight of various construction and other materials as a guide to the designer.

### 4.1.2   Live Loads

*Live load* is defined as the load superimposed by the use and occupancy of the building or other structure, not including the wind load, snow load, earthquake load, or dead load.

### 4.1.3   Floor Live Loads

In the absence of a governing building code, the minimum uniformly distributed live loads or concentrated loads in *Minimum Design Loads for Buildings and Other Structures*, ASCE 7-88 (1), are recommended.

### 4.1.4   Roof Live Loads

The minimum roof live loads should be as stipulated by the governing building code. Roof live loads used in design should represent the designer's determination of the particular service requirements for the structure, but in the absence of a governing code, in no case should they be less than the recommended minimum.

Minimum roof live loads for flat, pitched, or curved roofs are recommended in ASCE 7-88 (1). Roofs should be designed to resist either the tabulated minimum live loads, applied as balanced or full unbalanced, or the snow load, whichever produces the greater stress. Roofs to be used for special purposes should be designed for appropriate anticipated loads.

### 4.1.5   Ground Snow Loads

Snow loads vary widely throughout the United States. Factors affecting snow load accumulation on roofs include climatic variables, geographic location, roof exposure, roof slope, roof thermal condition, snow drifting, and sliding snow. Snowfall varies from year to year, and either a mean recurrence interval must be established for design purposes or design should be based on the maximum recorded snow load for which data are available. Snow loads should be as stipulated by the governing building code, but in the absence of such a code, snow loading used for design should be based on local experience or the use of accepted snow load maps.

Although the analysis of roofs for snow loading is complex because of the many variables involved, recent technical data have provided the designer with sufficient information to make a realistic analysis.

Roof design loads are obtained from ground snow loads. Several researchers and agencies have measured ground snow load distribution and plotted appropriate isogram maps depicting these loads. Data from the National Weather Service are used in preparing those maps. Snow loads for mountainous areas, especially in the Western states, should be established based on local experience. Actual snow pack of over 700 psf has been recorded in isolated regions, and some of the inhabited regions have snow loads ranging up to 300 psf.

### 4.1.6   Roof Snow Loads

Roof snow loads may be based on a determination of the ground snow load as specified by the governing building code or the roof snow loads may be specified in the code. Maximum snow load maps may also be used to determine the actual snow loads to be expected on the roof surface. As previously indicated, the roof snow load is a function of various factors.

These factors can be accounted for in design by applying appropriate snow load coefficients to the basic ground snow loads. Specific snow load coefficients have been developed to relate roof snow load to ground snow load based on comprehensive surveys of actual conditions.

For the design of both ordinary and multiple series roofs, either flat, pitched, or curved, minimum roof snow loads may be determined by multiplying the ground snow load by the appropriate coefficients based on the roof geometry and climatic conditions. Specific coefficients for these roof configurations are given in ASCE 7-88 (1). Because unbalanced loading can occur as the result of drifting, sliding, melting, and refreezing or physical removal of snow, structural roof members should be designed to resist increased loading in certain areas or other unbalanced loading conditions. For example, the roof should resist the full snow load as defined above distributed over the entire roof area, the full snow load distributed on any one portion of the area, and dead load only on the remainder of the area, depending on which load produces the greatest stress on the member considered.

Duration of load is the cumulative time during which the maximum design load is on the structure over its entire life. A 2-month duration is generally recognized as the proper design level for snow loads. Although some snow remains on roofs for periods exceeding 2 months in a single year, such snow loads seldom approach the full design load. In geographical areas where near-maximum snow load remains on the roof each winter for long periods of time or in buildings that are unheated, have heavily insulated roofs, or are used for cold storage, the use of the 2-month duration-of-load increase may not be advisable.

### 4.1.7 Wind Loads

Much research has been conducted to evaluate wind effects on various structures, and has resulted in the establishment of design coefficients that account for building shape and wind direction. In addition, extensive studies of basic wind velocities related to geographical location have resulted in the development of detailed wind speed maps for the United States. Other studies of surface resistance relative to the degree of land development and gust characteristics at a given location have provided a method for a further refinement of the basic wind velocity and its effect on structures. Additional work relates the dynamic behavior of structures to wind forces that are attributable to gusting and turbulence.

Sources of wind load analysis information include ASCE 7-88 (1), which contains a number of other references on this subject.

The duration of maximum wind load is relatively short during the life of a structure. Ten minutes is the generally accepted load duration resulting in a duration of load adjustment factor of 1.6. Note that this is a duration of load adjustment and should not be confused with probability factors.

### 4.1.8 Basic Wind Speed

Basic wind speeds for observed air flows in open, level country at a height of 33 ft above the ground have been developed. For the design of most permanent structures, a basic wind speed with a 50-yr mean recurrence interval should be applied. However, if in the judgment of the engineer or authority having jurisdiction the structure presents an unusually high degree of hazard to life and property in case of failure, the basic wind speed may be modified appropriately. Similarly, for temporary structures or structures having negligible risk of human life in case of failure, the design wind speed may be reduced.

Basic wind speeds are adjusted for height and site exposure conditions. In addition, gust response factors are applied to account for the response of the structure or building to the fluctuating nature of wind. Because of the complex nature of this subject, the designer is referred to the appropriate sections of ASCE 7-88 (1).

### 4.1.9   Wind Pressure

To determine the design wind pressure distribution acting on a building or structural element, the calculated velocity pressure is multiplied by an appropriate pressure coefficient. These pressure coefficients thus define the wind pressure acting normally on the surface of a building or element thereof, and are dependent on the external shape of the structure and its orientation with the wind. Pressure coefficients are considered to be either positive, representing an inward pressure, or negative, indicating an outward force on the structural element being analyzed. Therefore, depending on orientation to the wind and existence of openings, a building or element may be subjected to a pressure difference between opposite sides or faces, and it is thus the total resultant wind pressure that must be accounted for in design.

### 4.1.10   Minimum Roof Load Combinations

In designing, the most severe realistic distribution, concentration, and combination of roof loads and forces should be taken into consideration. Table 4.1 gives examples of roof load design combinations that should be checked and examples of the application of the duration-of-load factor for different load combinations.

**TABLE 4.1**

**Examples of Load Combinations for Design**[a]

| Windward Side | Leeward Side | Duration of Load | Duration of Load Factor $C_D$ |
|---|---|---|---|
| DL | DL | permanent | 0.90 |
| DL + LL | DL + LL | 7 days | 1.25 |
| DL[b] | DL + LL[b] | 7 days | 1.25 |
| DL + SL | DL + SL | 2 months | 1.15 |
| DL + $\frac{1}{2}$SL[c] | DL + SL[c] | 2 months | 1.15 |
| DL + WL | DL + WL | 10 minutes | 1.6 |
| DL + EL | DL + EL | 10 minutes | 1.6 |

[a] Symbols: DL = dead load, LL = live load, SL = snow load, WL = wind load, EL = earthquake load. Special configurations, locations, and occupancies of structures may require investigation of (a) full unbalanced SL or (b) combinations of WL with LL and DL, or WL with partial SL and DL. It may be assumed that wind and earthquake loads will not occur simultaneously.

[b] Full unbalanced loading.

[c] Half unbalanced loading.

Each type of load should be determined individually, and the effect of all possible combinations should be investigated. All possibilities must be investigated for unsymmetrical buildings. Special consideration should be given to structures of great span and/or height and to trusses bearing on very long columns. In addition to analyzing the structure for the combinations of loading shown in Table 4.1, the designer should check to see that all loading combinations as required by the governing building code have been met.

Structural members or systems should be designed to resist the stresses caused by partially or fully unbalanced live or snow loads and wind loads, including uplift, in combination with dead loads on the member or system. Such loading may result in reversal of stresses or stresses greater in some portion than the stresses produced when the entire roof is fully loaded. The occupancy or use of the structure, its configuration, heating and insulating considerations, local conditions, or other considerations may also cause partial or full unbalanced loading on the structure.

In order to determine the critical load combination and associated duration-of-load factor, each load combination may be divided by the applicable duration-of-load factors. The highest value determines the critical combination of loads. The following example illustrates this method.

**Example.**   A beam is to be designed for the following loads. Determine the critical combination of loads and applicable duration-of-load factor.

$$DL = 10 \text{ psf} \quad C_D = 0.9$$

$$LL = 12 \text{ psf} \quad\quad = 1.25$$

$$SL = 20 \text{ psf} \quad\quad = 1.15$$

$$WL = \phantom{0}5 \text{ psf} \quad\quad = 1.6$$

$$DL/0.9 = 11.1 \text{ psf}$$

$$(DL + LL)/C_D = (10 + 12)/1.25 = 17.6 \text{ psf}$$

$$(DL + SL)/C_D = (10 + 20)/1.15 = 26.1 \text{ psf}$$

$$(DL + WL)/C_D = (10 + 5)/1.6 = 9.4 \text{ psf}$$

$$(DL + WL + SL/2)/C_D = (10 + 5 + 20/2)/1.6 = 15.6 \text{ psf}$$

$$(DL + SL + WL/2)/C_D = (10 + 20 + 5/2)/1.6 = 20.3 \text{ psf}$$

For these loads and the loading combinations checked, the combination of dead load plus snow load with a duration-of-load factor of 1.15 is critical.

### 4.1.11   Earthquake Loads

In areas subject to earthquake shocks, every building or other structure and every portion thereof should be so designed and constructed as to resist stresses produced by lateral forces as provided by the governing building code regulations. Sources of earthquake load design information include the current *Uniform Building Code* (2) by the International Conference of Building Officials, *Recommended*

*Lateral Force Requirements* (3) by the Seismology Committee of the Structural Engineers Association of California, and *Recommended Provisions for the Development of Seismic Regulations for New Buildings*: 1991 Edition by the Building Seismic Safety Council (4).

### 4.1.12   Highway Loading

As indicated by the excellent performance of timber highway bridges over the years, wood bridges meet all the general bridge design requirements for modern highway traffic loads. In addition, wood offers the important advantage of being able to absorb impact stresses to the degree that impact loads may be neglected in the design of wood highway structures; however, they must be considered in the metal-to-metal connections. Highway design loads and their application should be in accordance with *Standard Specifications for Highway Bridges*, adopted by the American Association of State Highway and Transportation Officials (AASHTO) (5). Except for bridge railing loading, which is a short-time load, normal duration of loading is currently being used for design.

### 4.1.13   Railway Loading

Timber railway bridges and trestles have been long and extensively used in this country. With the advent of pressure preservative treatments, a service life of 50 years is commonplace. In the design of timber bridges and trestles, the recommendations in *Manual of Recommended Practice* of the American Railway Engineering Association (AREA) (6) are ordinarily applied. Recognizing the increased strength of wood under quickly applied loading, the AREA specifications permit omission of impact loads in determining stresses in the wood. Impact must be considered, however, in metal-to-metal connections. It should be pointed out that AREA specifications assume that all railway bridges and trestles are under "long-time" loading.

### 4.1.14   Crane Beam Loading

Timber is often used for crane beams and girders because of its ability to absorb impact forces. In designing crane beams, wheel loads recommended by crane manufacturers should be used if available.

ASCE 7-88 (1) recommends that for the design of all crane runways, the design loads should be increased for impact as

1.  a vertical force equal to 25% of the maximum wheel load;
2.  a lateral force equal to 20% of the weight of trolley and lifted load only, applied one-half at the top of each rail; and
3.  a longitudinal force of 10% of the maximum wheel loads of the crane applied at the top of the rail.

Some crane manufacturers specify wheel loads that include a percentage for impact. This factor should be taken into account in designing because it is unnecessary to add impact to wheel loads for timber designs unless there is a steel connection through which the load must pass or unless impact exceeds 100% of other normal loads. Impact must be considered for metal-to-metal connections. Normal duration of loading should be used for design.

Crane runway rails should be designed to prevent undue vertical and lateral deflection in accordance with crane manufacturers' recommendations.

### 4.1.15   Cyclic Loading

Tests to date indicate that wood is less sensitive to repeated loads than are crystalline structural materials such as metals. In general, the fatigue strength of wood is higher in proportion to the ultimate strength values than are the endurance limits for most structural metals. In present design practice, no factor is applied for fatigue in deriving working stress values for wood, nor is it considered necessary to do so. For these reasons, normal duration of load is used for cyclic loading design.

Where estimation of repetitions of design or near-design loads are indicated to approximate one million or more cycles, the designer should investigate the possibility of fatigue failures in shear or tension perpendicular to grain. When shear governs design and fatigue is a distinct possibility, shear stresses should be reduced 10% or changes should be made on the basis of a detailed analysis as indicated by examination of USDA Forest Products Laboratory Report No. 2236, *Fatigue Resistance of Quarter-Scale Bridge Stringers in Flexure and Shear* (7) and/or ASCE Paper No. 2470, *Design Considerations for Fatigue in Timber Structures* (8). Analyses made on the research available indicate that bending failure due to fatigue is not usual and need not be considered in ordinary design situations. Tension perpendicular to grain loading of wood should be avoided, especially when cyclic loading is expected.

### 4.1.16   Vibration

Vibration is closely related to impact and cyclic loading. The effects of vibration can usually be neglected in timber structures, except in those portions made of materials for which impact or vibration forces may be critical or where the vibration may be objectionable to the human occupancy. However, vibration may cause loosening of threaded connections used in timber structures. If there is a possibility of fatigue failure due to vibration, it should be considered, and care should be taken to avoid notches, eccentric connections, and similar design conditions.

### 4.1.17   Vibrating Loads

Experience has shown that vibrating mechanical equipment, such as air conditioning equipment, results in additional loading to the structural members and their connections. As an interim measure pending further research, the following procedure is recommended:

> In determining the design dead load, double the weight of the vibrating mechanical equipment.

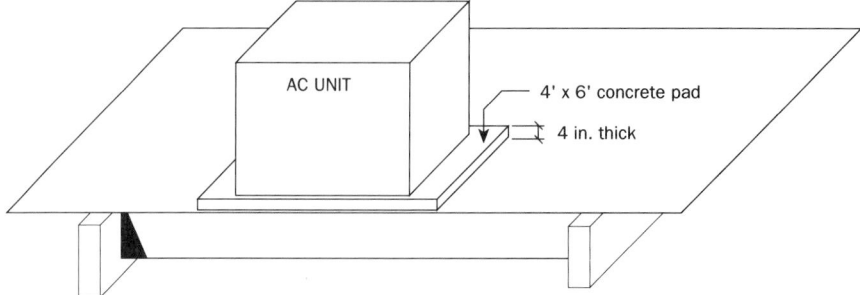

**FIGURE 4.1**    Air conditioning unit on roof causing vibrating load.

The weight of the support pad shall be included as a part of the dead load of the structure.

Check the dead load stresses by using duration-of-load factor, $C_D = 0.90$.

All vibrating equipment should be equipped with vibration isolation devices in the equipment base.

**Example:**    Determine the design dead load associated with vibrating equipment as shown in Fig. 4.1.

Given: Weight of concrete = 150 pcf, Weight of AC unit = 1,000 lb

Vibrating equipment dead load,

$$DL_{VE} = 2 \times (\text{weight of AC Unit}) + \text{weight of concrete}$$

$$= 2(1000) + (6)(4)\,(4/12)(150)$$

$$DL_{VE} = 3200 \text{ lb}$$

This modified dead load is used in design calculations.

### 4.1.18    Blast Loads

Data on the design of structures resistant to blast loading such as may be caused by nuclear weapon explosions are beyond the scope of this manual. Such data may be found in the American Society of Civil Engineers Manual No. 42, *Design of Structures to Resist Nuclear Weapons Effects* (9).

## 4.2    DEFLECTION

Deflection as related to design is primarily a function of serviceability rather than strength except for ponding as discussed in 5.4. Deflection should be limited to prevent a deflected part from interfering with another part of a structure, reduce vibration, and prevent an unsightly appearance of a sagging structure. Both vertical deflection and horizontal deflection are of concern to the designer. Deflection limitations must be in compliance with the governing building code. The deflection limitations listed in Tables 4.3 and 4.4 are recommendations of the American Institute of Timber Construction to provide greater serviceability of the product. They may be more restrictive than required by codes.

### 4.2.1.   Deflection of Beams

A member of given span and stiffness will deflect, on application of load, in direct proportion to the load applied and by an amount approximating that computed by standard engineering equations. Table 4.2 may be used to simplify the calculation of the deflection of beams under more than one type of loading by combining the equations as given in the table for the various types of loading (see 4.1 for the effects of long-time loading). The following example illustrates the use of Table 4.2.

**Example.**   Determine deflection (1) at free end of overhang and (2) at centerline of supported span.

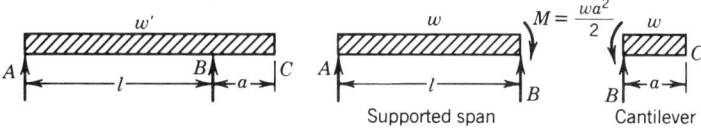

1.   The cantilever portion has an initial slope $\theta_B$ at its left end equal to the slope at the right end of the supported portion, or

$$\theta_B = \theta_M + \theta_w = \frac{Ml}{3EI} + \frac{wl^3}{24EI} = \frac{wa^2l}{6EI} - \frac{wl^3}{24EI}$$

$$\theta_B = \frac{wl}{24EI}(4a^2 - l^2)$$

Deflection at the free end: $\Delta_c = \Delta_w + a\theta_B$, or

$$\Delta_c = \frac{wa^4}{8EI} + \frac{wal}{24EI}(4a^2 - l^2) = \frac{wa}{24EI}(3a^3 + 4a^2l - l^3)$$

2.   Deflection at centerline of supported span:

$$\Delta_{\mathcal{C}} = \Delta_w + \Delta_M = \frac{5wl^4}{384EI} - \frac{Ml^2}{16EI}$$

$$= \frac{5wl^4}{384EI} - \frac{wa^2l^2}{32EI} = \frac{wl^2}{384EI}(5l^2 - 12a^2)$$

### 4.2.2   Deflection Limitations

Conditions often require that a beam not exceed certain deflection limitations. Tables 4.3 and 4.4 give deflection limits for timber beams as recommended by AITC.

For special uses, such as beams supporting vibrating machinery or carrying moving loads, more severe limitations may be required.

## TABLE 4.2
## Beam Deflection Equations for Various Types of Loading

| Point on Beam $A \quad \mathcal{C} \quad B$ | Slope of Tangent to Elastic Curve $\theta_A$ | $\theta_B$ | Deflection $\Delta_x$ | $\Delta_{max}$ and $\Delta_{\mathcal{C}}$ |
|---|---|---|---|---|
| 1 | $\dfrac{Pb(l^2 - b^2)}{6lEI}$ | $\dfrac{Pab(2l - b)}{6lEI}$ | When $x < a$: $\dfrac{Pbx}{6lEI}(l^2 - x^2 - b^2)$  When $x > a$: $\dfrac{Pb}{6lEI}\left[\dfrac{l}{b}(x-a)^3 + (l^2 - b^2)x - x^3\right]$ | $\Delta_{max}\left(\text{at } x = \sqrt{\dfrac{l^2 - b^2}{3}}\right)$: $\dfrac{Pb(l^2 - b^2)^{3/2}}{9\sqrt{3}lEI}$  $\Delta_{\mathcal{C}}$ (if $a > b$): $\dfrac{Pb}{48EI}(3l^2 - 4b^2)$ |
| 2 | $\dfrac{Pl^2}{16EI}$ | | When $x < l/2$: $\dfrac{Px}{48EI}(3l^2 - 4x^2)$ | $\dfrac{Pl^3}{48EI}$ |
| 3 | $\dfrac{Pa(l-a)}{2EI}$ | | When $x < a$: $\dfrac{Px}{6EI}(3la - 3a^2 - x^2)$  When $(l-a) > x > a$: $\dfrac{Pa}{6EI}(3lx - 3x^2 - a^2)$ | $\dfrac{Pa}{24EI}(3l^2 - 4a^2)$ |

| | | | | |
|---|---|---|---|---|
| 4 | | $\dfrac{5Pl^2}{32EI}$ | When $x < l/4$: $\dfrac{Px}{32EI}(5l^2 - 8x^2)$<br>When $l/2 > x > l/4$: $\dfrac{P}{384EI}(72l^2x - 48lx^2 - 32x^3 - l^3)$ | $\dfrac{19Pl^3}{384EI}$ |
| 5 | | $\dfrac{wl^3}{24EI}$ | $\dfrac{wx}{24EI}(l^3 - 2lx^2 + x^3)$ | $\dfrac{5wl^4}{384EI}$ |
| 6 | | $\dfrac{wa^2(l+b)^2}{24lEI}$  $\dfrac{wa^2(2l^2 - a^2)}{24lEI}$ | When $x < a$: $\dfrac{wx}{24lEI}\left[a^2(l+b)^2 - 2ax^2(l+b) + lx^3\right]$<br>When $x > a$: $\dfrac{wa^2(l-x)}{24lEI}(4xl - 2x^2 - a^2)$ | $\Delta_{\max}\left(\text{at } x = l - \sqrt{\dfrac{2l^2 + a^2}{6}}\right)$:<br>$\dfrac{wa^2(2l^2 - a^2)}{36lEI}\sqrt{\dfrac{2l^2 + a^2}{6}}$<br>$\Delta_\ell$ (if $a < b$):<br>$\dfrac{wa^2}{96lEI}(3l^2 - 2a^2)$ |
| 7 | | $\dfrac{wa^2}{12EI}(3l - 2a)$ | When $x < a$: $\dfrac{wx}{24EI}\left[2a^2(3l - 2a) + x^2(x - 4a)\right]$<br>When $x > a$: $\dfrac{wa^2}{24EI}(6lx - 6x^2 - a^2)$ | $\dfrac{wa^2}{48EI}(3l^2 - 2a^2)$ |

TABLE 4.2 (Continued)

| Point on Beam | Slope of Tangent to Elastic Curve | | Deflection | |
|---|---|---|---|---|
| A  ℓ  B | $\theta_A$ | $\theta_B$ | $\Delta_x$ | $\Delta_{max}$ and $\Delta_\ell$ |
| 8 | $\dfrac{w}{24EI}\left[l^3 - 2a^2(3l-2a)\right]$ | | When $x < a$: $\dfrac{wx}{24EI}\left[l^3 - 2x^2(l-2a) - 2a^2(3l-2a)\right]$ <br> When $x > a$: $\dfrac{w}{24EI}\left[a^4 + l^3x - x^3(2l-x) - 6a^2x(l-x)\right]$ | $\dfrac{w}{384EI}(l^2-4a^2)(5l^2-4a^2)$ |
| 9 | $\dfrac{Ml}{3EI}$ | $\dfrac{Ml}{6EI}$ | $\dfrac{Mx}{6lEI}(l-x)(2l-x)$ | $\Delta_{max}\left(\text{at } x=l-\dfrac{l}{\sqrt{3}}\right):$ <br> $\dfrac{Ml^2}{9\sqrt{3}EI}$ <br> $\Delta_\ell = \dfrac{Ml^2}{16EI}$ |
| 10 | $\dfrac{Ml}{6EI}$ | $\dfrac{Ml}{3EI}$ | $\dfrac{Mlx}{6EI}\left(1 - \dfrac{x^2}{l^2}\right)$ | $\Delta_{max}\left(\text{at } x=\dfrac{l}{\sqrt{3}}\right):$ <br> $\dfrac{Ml^2}{9\sqrt{3}EI}$ <br> $\Delta_\ell = \dfrac{Ml^2}{16EI}$ |

| A ℄ B | $\theta_B$ (at free end) | $\Delta_x$ | $\Delta_{max}$ (at free end) |
|---|---|---|---|
| 11 | $\dfrac{Pa^2}{2EI}$ | When $x < a$: $\dfrac{Px^2}{6EI}(3a - x)$ <br> When $x > a$: $\dfrac{Pa^2}{6EI}(3x - a)$ | $\dfrac{Pa^2}{6EI}(3l - a)$ |
| 12 | $\dfrac{Pl^2}{2EI}$ | $\dfrac{Px^2}{6EI}(3l - x)$ | $\dfrac{Pl^3}{3EI}$ |
| 13 | $\dfrac{wl^3}{6EI}$ | $\dfrac{wx^2}{24EI}(x^2 + 6l^2 - 4lx)$ | $\dfrac{wl^4}{8EI}$ |
| 14 | $\dfrac{Ml}{EI}$ | $\dfrac{Mx^2}{2EI}$ | $\dfrac{Ml^2}{2EI}$ |

**TABLE 4.3**

**Recommended Deflection Limitations**

| Use Classification | Applied Load Only | Applied Load + Dead Load |
|---|---|---|
| Roof beams | | |
| Industrial | $l/180$ | $l/120$ |
| Commercial and institutional | | |
| Without plaster ceiling | $l/240$ | $l/180$ |
| With plaster ceiling | $l/360$ | $l/240$ |
| Floor beams | | |
| Ordinary usage[a] | $l/360$ | $l/240$ |
| Highway bridge stringers | $l/300$ | |
| Railway bridge stringers | $l/300$ to $l/400$ | |

[a] Ordinary usage classification for floors is intended for construction in which walking comfort and minimized plaster cracking are the main considerations. These recommended deflection limits may not eliminate all objections to vibrations such as in long spans approaching the maximum limits or for some office and institutional applications where increased floor stiffness is desired. For these usages, the deflection limitations in Table 4.4 have been found to provide additional stiffness.

**TABLE 4.4**

**Deflection Limitations for Uses Where Increased Floor Stiffness is Desired**

| Use Classification | Applied Load Only | Applied Load + $K$(Dead Load)[a] |
|---|---|---|
| Floor Beams | | |
| Commercial, Office and Institutional | | |
| Floor joists, spans to 26 ft[b] | | |
| LL $\leq$ 60 psf | $l/480$ | $l/360$ |
| 60 psf < LL < 80 psf | $l/480$ | $l/360$ |
| LL $\leq$ 80 psf | $l/420$ | $l/300$ |
| Girders, spans to 36 ft[b] | | |
| LL $\leq$ 60 psf | $l/480$[c] | $l/360$ |
| 60 psf < LL < 80 psf | $l/420$[c] | $l/300$ |
| LL $\leq$ 80 psf | $l/360$[c] | $l/240$ |

[a] $K = 1$ except for seasoned members where $K = 0.5$. Seasoned members for this usage are defined as having a moisture content of less than 16% at the time of installation.

[b] For girder spans greater than 36 ft and joist spans greater than 26 ft, special design considerations may be required such as more restrictive deflection limits and vibration considerations that include the total mass of the floor.

[c] Based on reduction of live load as permitted by the code.

## TABLE 4.5

### Recommended Minimum Camber for Glued Laminated Timber Beams

| | |
|---|---|
| Roof beams[a] | $1\frac{1}{2}$ times dead load deflection |
| Floor beams[b] | $1\frac{1}{2}$ times dead load deflection |
| Bridge beams[c] | |
| Long span | 2 times dead load deflection |
| Short span | 2 times dead load $+ \frac{1}{2}$ of applied load deflection |

[a]Roof beams—the minimum camber of $1\frac{1}{2}$ times dead-load deflection will produce a nearly level member under dead load alone after plastic deformation has occurred. Additional camber is usually provided to improve appearance and/or provide necessary roof drainage. Roof beams should have a positive slope or camber equivalent to $\frac{1}{4}$ in. per foot of horizontal distance between the level of the drain and the high point of the roof, in addition to the minimum camber, to avoid the ponding of water. In addition, on long spans, level roof beams may not be desirable because of the optical illusion that the ceiling sags. This condition may also apply to floor beams in multistory buildings.

[b]Floor beams—the minimum camber of $1\frac{1}{2}$ times dead-load deflection will produce a nearly level member under dead load alone after plastic deformation has occurred. For warehouse or similar floors where live load may remain for long periods, additional camber should be provided to give a level floor under the permanently applied load.

[c]Bridge beams—bridge members are normally cambered for dead load only on multiple spans to obtain acceptable riding qualities.

## 4.3   CAMBER

Camber is built into a structural glued laminated timber member by introducing a curvature, either circular or parabolic, opposite to the anticipated deflection movement. Camber recommendations (see Table 4.5) vary with design criteria for various conditions of use. In addition, recommendations for camber depend on whether the member is of simple, continuous, or cantilever span, whether roof drainage is to be provided by the camber, and other factors. Reverse camber may be required in continuous and cantilever spans to permit adequate drainage.

Camber is usually designated in custom members by specifying the amount of camber required in inches and the location. For a simple span beam, the camber is usually specified for the midpoint. The location for cantilevered or continuous beams depends on the shape of the deflected structure.

Camber for noncustom or stock beams is usually designated by the manufacturer as radius of curvature. Typically long billets are made and sold with a camber of a constant radius. There is no industry standard for radius of curvature, but it usually varies between 1600 and 2000 ft. The longer billets are cut into shorter pieces for use. The amount of camber in inches of these shorter pieces can be obtained from Fig. 4.2.

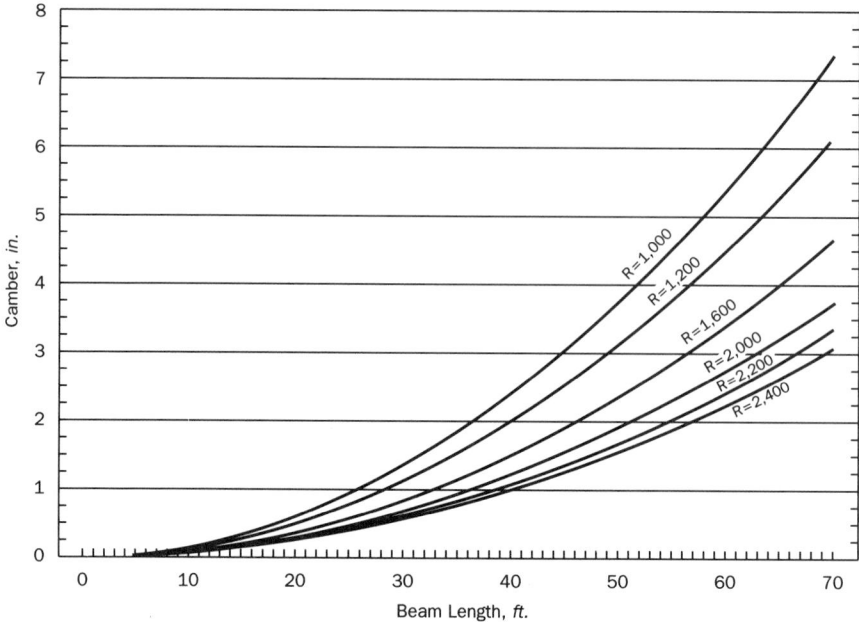

**FIGURE 4.2**    Camber for various radii.

**Example.**    A stock beam (noncustom) is manufactured with a 1600 ft radius. What is the camber at midspan of a 30 ft long beam cut from a 60 ft billet?

From Fig. 4.2, camber = 0.85 in.

The camber can also be determined by calculation using the equation included in Fig. 4.2:

$$\text{Camber} = r - 1/2 \sqrt{4r^2 - c^2}$$

$$\text{Camber} = 1600 - 1/2 \sqrt{4(1600)^2 - 30^2} \,(12) = 0.843 \text{ in.}\quad \text{Use } \tfrac{7}{8} \text{ in.}$$

When camber in inches is specified for a simple beam, the beam is made with the specified camber at the midspan on the tension side of the beam. It is either of parabolic or circular shape. For the radii of the curvatures used commonly, the two shapes are almost identical.

**Example.**    A 100 ft long beam has 4 in. camber specified. What is the camber at the $\frac{1}{4}$ point?

a.    Assuming a parabolic shape,

$$\text{Camber} = 4 - \left(\frac{25}{50}\right)^2 (4) = 3.0 \text{ in.}$$

b. Assuming a circular shape from geometry,

$$\text{radius} = \frac{4b^2 + c^2}{8b} = \frac{4\left(\dfrac{4}{12}\right)^2 + 100^2}{8\left(\dfrac{4}{12}\right)} = 3750 \text{ ft}$$

where

$b$ = the rise = camber (in.), and
$c$ = the chord length = length of the beam (ft),

$$b = \left(\frac{4}{12}\right) - 3750 + \sqrt{3750^2 + 25^2}\,(12) = 3.000 \text{ in.}$$

The normal use of camber in beams is to offset the anticipated dead load deflection and leave a slight crown in the member, which is desirable for sake of appearance. Additional camber can also be used to improve drainage. See 5.4 on ponding for more details.

## 4.4 TABULAR DESIGN VALUES

The tabular design values in Tables 8.3–8.7 for sawn lumber, AITC 117—93 Design (11) for glued laminated timber of softwood species and AITC 119—94 (12) for glued laminated timber of hardwood species, are defined as the maximum allowable stresses based on standard conditions as set forth in the tables. These tabular design values are designated as $F_b$ for bending, $F_c$ for compression parallel to grain, $F_v$ for shear, and so on. The symbol $F$ denotes a tabular design value for a given strength property, and the lower-case subscript denotes the strength property. For some conditions, these values must be adjusted before they can be used in design. The adjustment factors are denoted by $C$ and a subscript letter indicating the reason for adjustment, such as $C_F$ for size factor, $C_V$ for volume factor, and $C_D$ for duration of load. Some of these adjustment factors are applicable to all strength properties, while others are applicable to specific strength properties; for example, the curvature factor $C_C$ is applicable to bending in curved members only.

When all applicable adjustment factors have been applied to the tabular design value, the allowable value to use in design is denoted by $F'$, such as $F'_b$ for bending, $F'_c$ for compression parallel to grain, and $F'_v$ for shear parallel to grain.

Calculations involving columns and beams frequently require an interim design value which is the tabular design value adjusted by all applicable adjustment factors, except the modification for slenderness or size. These values are designated as $F^*$, such as $F^*_c$ and $F^*_b$. They are used only as interim values in calculations and are explained when they occur.

The actual unit stresses of a member under load are denoted by the same letters as the tabular design value, except that a lower case $f$ is used, such as $f_b$, $f_c$, $f_v$, etc.

### 4.4.1   Glued Laminated Timber

The tabular design values for glued laminated timber are included in AITC 117—Design (11) for softwoods which is included in Chapter 8 and AITC 119 (12) for hardwoods. These values are based on laterally braced straight members with an average moisture content of 12%. The design value for bending members is based on 12 in. depth, $5\frac{1}{8}$ in. width, and 21 ft length. When other conditions exist, appropriate adjustment factors as explained in the following pages must be applied.

Glued laminated timbers may have different tabular design values when loaded in bending about the $x$–$x$ axis and the $y$–$y$ axis, depending on the grades and arrangement of grades of lumber within the member. Therefore, Tables 1 and 2 in AITC 117—Design (11) have strength properties listed for loading in bending about both the $x$–$x$ and $y$–$y$ axes. The tabular design values when bending is about the $x$–$x$ axis are denoted by the subscript $x$ such as $F_{bx}$, $F_{vx}$, and $E_x$. For bending about the $y$–$y$ axis, the subscript $y$ is used. When no subscripts are used, the tabular design values about the $x$–$x$ axis are assumed. For clarity in specifying, it is recommended that the subscripts be used, particularly when the member is loaded in bending about the $y$–$y$ axis. Note that the subscripts $xx$ and $yy$ are sometimes used to indicate stresses about the $x$–$x$ and $y$–$y$ axes. The single-letter subscript is the preferred form.

When glued laminated timbers are loaded in flexure perpendicular to the wide faces of the laminations, the orientation of the $x$–$x$ and $y$–$y$ axis is as shown in Fig. 4.3(a). When the timbers are loaded parallel to the wide faces of the laminations, the orientation is as shown in Fig. 4.3(b). Occasionally a glued laminated timber

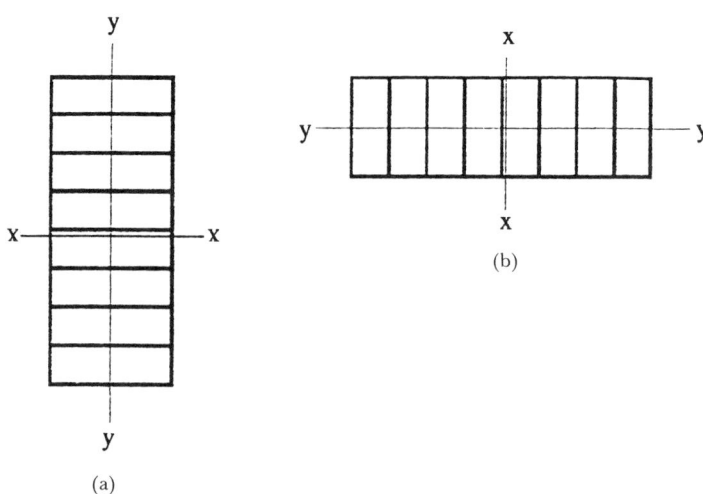

**FIGURE 4.3**   Orientation of axes.

may consist of only a few laminations, in which case, the $y$–$y$ axis could become the stronger axis. But this occurs so seldom that the common usage is to designate the axis as shown in Fig. 4.3(b) for all cases where the loading in flexure is parallel to the wide faces of the laminations.

### 4.4.2  Sawn Lumber

Sawn lumber is available as visually graded lumber in all sizes and mechanically graded lumber which includes machine stress-rated lumber (MSR) and machine evaluated lumber (MEL) with 2 in. nominal thickness. The design values for visually graded lumber are contained in Tables 8.3, 8.4, 8.6, and 8.7. These design values for the most commonly used species and grades of lumber were taken from ANSI/NF$_o$PA NDS—1991, *National Design Specification for Wood Construction* (10) and were current at the time of publication of this manual. Additional species and grades of lumber are given in the NDS. The design values for dimension size lumber are based to a large part on an in-grade testing program conducted on U.S. and Canadian species. The results of this testing have been evaluated in accordance with ASTM D1990 (21) and are included in five tables in the NDS (10). For current information on grades and design values, the designer is referred to the latest NDS, the standard grading rules, or the governing building code for current design values. See Refs. (13)–(19).

### 4.4.3  Visually Graded Lumber

Visually graded lumber is available in four categories:

1. **Visually Graded Dimension Lumber.** Visually graded dimension lumber is lumber that is from 2 to 4 in. thick and 2 in. and wider in width. A further subdivision of dimension lumber is permitted by the National Grading Rule for Dimension Lumber based on width and use categories. See Tables 8.3–8.7 for additional details or consult the NDS (10) or the current Standard Grading Rules for the species (13)–(19). Table 8.3 contains base design values for dimension lumber (lumber 2 to 4 in. thick) for all species except Southern Pine. The in-grade test program referred to previously showed that bending, tension parallel to grain, and compression parallel to grain vary with grades, width, and thickness. Table 8.3a includes the size factor $C_F$ to be applied to the base values.

Table 8.4 contains design values for Southern Pine. Similar adjustment factors for grade, width, and thickness also apply to Southern Pine, but adjustments have been included in the published values of the tables which are set up in grades, widths, and thicknesses, and the size factor is not applicable except for a few grades.

2. **Visually Graded Timbers 5 in. × 5 in. and Larger.** This category is further divided into two categories according to use: "Post & Timbers" and "Joists and Planks." Table 8.6 contains design values for species 5 in. × 5 in. and larger. The size factor $C_F$ is applied to bending stresses only for members over 12 in. deep.

3. **Visually Graded Decking.** This category is for lumber that is intended for use as decking. It is from 2 to 4 in. thick and 6 in. and wider in width. Some species have width limitations of 8 in. Table 8.7 contains design values for deck-

ing. The design values are based on a 4 in. depth and may be increased by the size factor $C_F$ for 2 and 3 in. thick decking.

Since decking is used as a bending member stressed primarily in bending only, the design values $F_b$, $F_b C_r$, $F_{c\perp}$, and $E$ are published. Shear seldom, if ever, controls the design and is not published.

   4.   **Visually Graded Boards.**   Board grades of lumber are generally graded for appearance and are not rated for stress. However, stress-rated boards in thicknesses of nominal $1-1\frac{1}{2}$ in. can be obtained when graded according to the dimension lumber grade rules for the species. See footnotes in Tables 8.3 and 8.4.

### 4.4.4   Mechanically Graded Lumber

There are two categories of mechanically graded lumber, Machine Stress Rated (MSR) and Machine Evaluated Lumber (MEL). The mechanically graded lumber is available in sizes 2 in. and less in thickness and 6 in. and wider. Table 8.5 is set up for mechanically graded lumber which is limited to lumber 2 in. and less in thickness. The effect of width and grade is included in the published design values.

### 4.4.5   Bearing Design Values Parallel to Grain

The bearing design values parallel to grain (sometimes referred to as end grain bearing) are not included in the design tables for either sawn lumber or glued laminated timber but are included in Table 8.8. These tabulated design values are subject to the applicable adjustment factors included in Table 4.6.

## 4.5   ADJUSTMENT FACTORS

The tabular design values must be multiplied by adjustment factors before use if the design conditions are other than those for which the tables were developed.

Many of the adjustment factors are common to both sawn lumber and glued laminated timber. Others are for use with only one of the types of wood products. Since the tabular design values for the different categories of sawn lumber are based on different procedures, a table of adjustment factors for that category is shown for each table, 8.3–8.7. Table 4.6 lists the applicability of all adjustment factors for both sawn lumber and glued laminated timber.

### 4.5.1   Duration of Load

The design values tabulated in Tables 8.3–8.7, AITC 117—Design (11), and AITC 119—Hardwoods (12) are for normal duration of loading. Normal load duration anticipates fully stressing a member to the full design value by the application of the full design load for a duration of approximately 10 years (applied either continuously or cumulatively). If the member is designed to be fully stressed by maximum design loads for long-term loading conditions (greater than 10 years either continuously or cumulatively), the duration of load factor is 0.90.

For other durations of load, either continuously or intermittently applied, that result in a cumulative effect, the appropriate factor $C_D$ taken from Table 4.7 or

## TABLE 4.6
### Applicability of Adjustment Factors

| | LOAD DURATION FACTOR | WET SERVICE FACTOR | TEMPERATURE FACTOR | BEAM STABILITY FACTOR [1] | SIZE FACTOR [2] | VOLUME FACTOR [1,3] | FLAT USE FACTOR [4] | REPETITIVE MEMBER FACTOR [5] | CURVATURE FACTOR [6] | FORM FACTOR | COLUMN STABILITY FACTOR | SHEAR STRESS FACTOR [7] | BUCKLING STIFFNESS FACTOR [8] | BEARING AREA FACTOR |
|---|---|---|---|---|---|---|---|---|---|---|---|---|---|---|
| $F_b' = (F_b)$ | $(C_D)$ | $(C_M)$ | $(C_t)$ | $(C_L)$ | $(C_F)$ | $(C_V)$ | $(C_{fu})$ | $(C_r)$ | $(C_c)$ | $(C_f)$ | • | • | • | • |
| $F_t' = (F_t)$ | $(C_D)$ | $(C_M)$ | $(C_t)$ | • | $(C_F)$ | • | • | • | • | • | • | • | • | • |
| $F_v' = (F_v)$ | $(C_D)$ | $(C_M)$ | $(C_t)$ | • | • | • | • | • | • | • | • | $(C_H)$ | • | • |
| $F_{c\perp}' = (F_{c\perp})$ | • | $(C_M)$ | $(C_t)$ | • | • | • | • | • | • | • | • | • | • | $(C_b)$ |
| $F_c' = (F_c)$ | $(C_D)$ | $(C_M)$ | $(C_t)$ | • | $(C_F)$ | • | • | • | • | • | $(C_P)$ | • | • | • |
| $E' = (E)$ | • | $(C_M)$ | $(C_t)$ | • | • | • | • | • | • | • | • | • | $(C_T)$ | • |
| $F_g' = (F_g)$ | $(C_D)$ | • | $(C_t)$ | • | • | • | • | • | • | • | • | • | • | • |

*Source:* ANSI/NF₀PA NDS—1991, *National Design Specification for Wood Construction*, AFPA, 1250 Connecticut Ave., NW, Washington, DC 20036.

1. The beam stability factor, $C_L$, shall not apply simultaneously with the volume factor, $C_V$, for glued laminated timber bending members (see 4.5.6). Therefore the lesser of these adjustment factors shall apply.

2. The size factor, $C_F$, shall apply only to visually graded sawn lumber members and to round timber bending members (see 4.5.6).

3. The volume factor, $C_V$, shall apply only to glued laminated timber bending members (see 4.5.5).

4. The flat use factor, $C_{fu}$, shall apply only to dimension lumber bending members 2" to 4" thick (see 4.5.5) and to glued laminated timber bending members 2" to 4" thick (see Table 4.9).

5. The repetitive member factor, $C_r$, shall apply only to dimension lumber bending members 2" to 4" thick (see 4.5.10).

6. The curvature factor, $C_c$, shall apply only to curved portions of glued laminated timber bending members (see 4.5.11).

7. Shear design values parallel to grain, $F_v$, for sawn lumber members shall be permitted to be multiplied by the shear stress factors, $C_H$, specified in Tables 8.3–8.6.

8. The buckling stiffness factor, $C_T$, shall apply only to 2" × 4" or smaller sawn lumber truss compression chords subjected to combined flexure and axial compression when $\frac{3}{8}$" or thicker plywood sheathing is nailed to the narrow face (see 4.5.15).

**TABLE 4.7**

**Frequently Used Load Duration Factors, $C_D{}^a$**

| Load Duration | $C_D$ | Typical Design Loads |
|---|---|---|
| Permanent | 0.9 | Dead Load |
| Ten years (Normal) | 1.0 | Occupancy Live Load |
| Two months | 1.15 | Snow Load |
| Seven days | 1.25 | Construction Load (Roof Included) |
| Ten minutes | 1.6 | Wind/Earthquake Load |
| Impact[b] | 2.0 | Impact Load |

[a] Load duration factors shall not apply to modulus of elasticity, $E$, nor to compression perpendicular to grain design values, $F_{c\perp}$, based on a deformation limit.

[b] The impact load duration factor shall not apply to structural members pressure-treated with water-borne preservatives to the heavy retentions required for "marine" exposure (see Ref. 20), nor to structural members pressure-treated with fire retardant chemicals. The impact load duration factor shall not apply to connections.

determined from Fig. 4.4 should be applied to adjust the tabulated design values. However, these duration-of-load adjustments, including the 0.90 reduction for permanent loading, are not applicable to modulus of elasticity or compression perpendicular to grain. These adjustment factors are applicable to the adjustment of the design values for mechanical fastenings when the wood (not the strength of the metal fastening) controls the load capacity.

If loads of different durations are applied simultaneously, the size of member required is determined for the total of all loads applied at the design value modified by the factor for the load of shortest duration in the combination. In like manner, but neglecting the load of shortest duration, the size of member required to support the remaining loads at the design value adjusted by the factor for the load of next shortest duration is determined. By repeating this procedure for all the remaining loads, the size of member required for the controlling duration-of-load condition is obtained. When the permanently applied load is less than or equals 90% of the total normal load (including the permanently applied load), the normal loading condition will control the size of member required.

### 4.5.2  Wet Service Factor

Most strength properties of wood increase as moisture content decreases. For practicality of design, the tabular design values for most wood properties have been set for dry conditions of use and are adjusted for wet service. In some cases, both wet service and dry service design values are published.

#### 4.5.2.1  Glued Laminated Timber

Glued laminated timber tabular design values are based on wood at a maximum moisture content of 16% with an average of approximately 12%. The wet service factors for all strength properties except bearing parallel to grain, $F_g$, are included in Table 4.8 and Table 1 of AITC 117—Design.

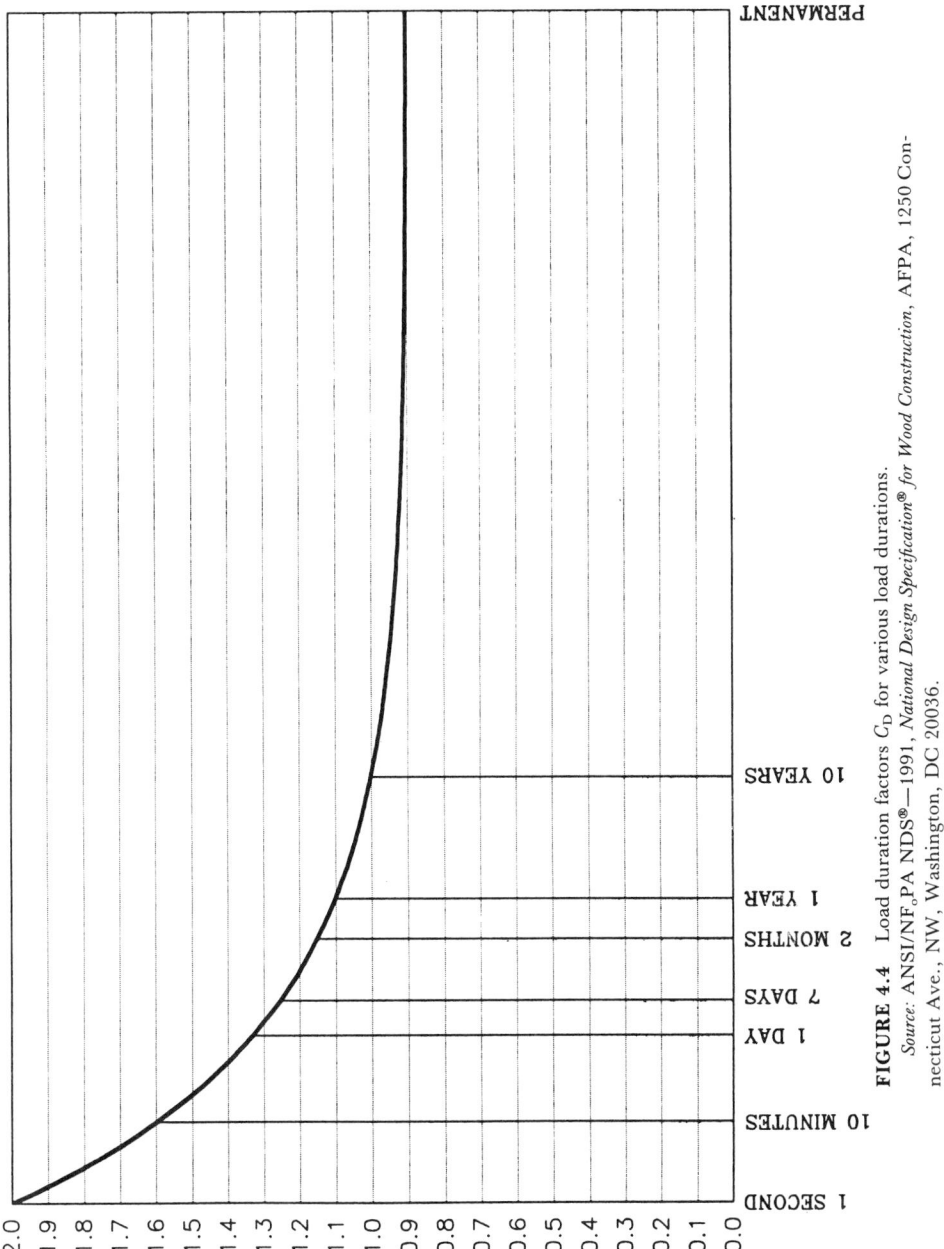

**FIGURE 4.4** Load duration factors $C_D$ for various load durations.

*Source:* ANSI/NF₀PA NDS®—1991, *National Design Specification® for Wood Construction,* AFPA, 1250 Connecticut Ave., NW, Washington, DC 20036.

**TABLE 4.8**

**Moisture Content Factors $C_M$**

| Strength Property | $F_b$ | $F_t$ | $F_c$ | $F_{c\perp}$ | $F_v$ | $E$ | $F_{rt}$ | $F_g$ |
|---|---|---|---|---|---|---|---|---|
| Glued Laminated Timber–All Species Wet conditions of use (MC greater than 16%) | 0.80 | 0.80 | 0.73 | 0.53 | 0.875 | 0.833 | 0.875 | 0.57 |
| Sawn Lumber Wet conditions of use MC greater than 19% Dimension lumber | 0.85[a] | 1.00 | 0.8[b] | 0.67 | 0.97 | 0.9 | — | —[d] |
| 5 in. × 5 in. and larger[c] | 1.00 | 1.00 | 0.91 | 0.67 | 1.00 | 1.00 | — | —[d] |
| Decking wet conditions of use all species except Southern Pine[c] | 0.85 | — | — | 0.67 | — | 0.90 | — | — |

[a] When $(F_b)(C_F)$ for dimension lumber of all species $\leq 1150$ psi, $C_M = 1.0$.

[b] When $(F_c)(C_F)$ for dimension lumber of all species except Southern Pine $\leq 750$ psi, $C_M = 1.0$; when $F_c$ for visually graded Southern Pine $\leq 750$ psi, $C_M = 1$.

[c] Southern Pine (use tabulated design values for wet service conditions).

[d] Wet service factors have already been applied to the bearing parallel to grain design values $F_g$ based on specified moisture service conditions.

[e] For Southern Pine use tabulated design values for wet service conditions.

### 4.5.2.2 Sawn Lumber

The tabular design values for dimension lumber (2 to 4 in. thick) and mechanically graded lumber are given in Tables 8.3–8.5. They are based on dry service conditions of 19% moisture content or less.

The adjustment factors for moisture content, $C_M$, for wet service conditions are given in Table 4.8 and Tables 8.3–8.5.

Both the dry service and wet service use conditions for Dense Select Structural 86, Dense Select Structural 72, and Dense Select Structural 65 Southern Pine are included in the tabular design values, and use of $C_M$ for these grades is not required.

The tabular design values for visually graded timbers, 5 in. × 5 in. and larger, for all species except Southern Pine and Mixed Southern Pine are based on use where moisture content will not exceed 19% in service. For wet service conditions where the moisture content exceeds 19%, the adjustment factors for wet service conditions, $C_M$, are given in Tables 4.8 and 8.6. The design values published for Southern Pine are for wet service conditions, and no adjustments are necessary. These values are also used for dry service conditions.

The tabular design values for decking in Table 8.7 are based on a dry service use not to exceed 19%. When used under wet service conditions exceeding 19%, the wet service factors $C_M$ included in Tables 4.8 and 8.7 must be used for all species except Southern Pine where the tabular design values are listed for both dry service conditions and wet service conditions.

### 4.5.3  Effect of Size or Volume

Research has indicated that design values are affected by the size or volume of a member as reflected by its dimensions: thickness, width and length for sawn lumber, and width, depth, and length for glued laminated timber.

In sawn lumber, which is usually limited in length, the design values have been set so that no adjustment for length is needed. However, adjustments are needed for the effect of thickness and width for bending, tension, and compression parallel to grain. These adjustments are accomplished in several ways, as shown in the tables for the various lumber categories. When a sawn lumber member is loaded flatwise (load applied perpendicular to the wide face), the depth is less than that for which the design value was based on, and a flat use factor $C_{fu}$ is applied. The tabular design values for Southern Pine dimension are listed by width and thickness, and no further adjustment is necessary. Machine graded lumber has the effect of size included in the grading process, and sawn timbers have the effect of size included in the design values, except for members over 12 in. deep which are adjusted by the size factor $C_F$. The size effect in bending is based on a member being loaded about its major $x$–$x$ axis.

The volume factor $C_V$ is used for glued laminated timber loaded in bending about the $x$–$x$ axis. It is calculated by Eq. (4-2). Note that the depth, width, and length are included in this factor. The standard beam for which the volume factor is calculated is 12 in. deep, 5.125 in. wide, and 21 ft long, loaded with a uniform load.

When a glued laminated beam is loaded in bending about the $y$–$y$ axis, a flat use factor $C_{fu}$ is used because the standard beam for which tabular design values are calculated is 12 in. deep for all species.

### 4.5.4  Size Factor, $C_F$

The size factor $C_F$ is used with sawn lumber. Since the size factor is unique to grade, thickness, and width, tables for size effect values are included with the tables of tabular design values for sawn lumber as shown below:

| | |
|---|---|
| Dimension Lumber (all species except Southern Pine) | Table 8.3 |
| Dimension Lumber Southern Pine | Table 8.4 |
| Mechanically Graded Lumber | Table 8.5 |
| Visually Graded Timbers (5 in. × 5 in. and larger) | Table 8.6 |
| Visually Graded Decking | Table 8.7 |

**Examples.**

1.  Determine the size factor, $C_F$, for bending of a 4 in. × 10 in. select structural grade of Hem-Fir.

    From Table 8.3, $C_F$ for bending = 1.2

2. Determine the size factor, $C_F$, for a 4 in. × 12 in. dense select structural grade of Southern Pine loaded in bending.

   From Table 8.4, $C_F$ for bending = 1.1

3. Determine the size factor, $C_F$, for a 6 in. × 14 in. No. 1 grade Douglas Fir-Larch timber.

   From Table 8.5, $C_F = (12/d)^{1/9} = (12/13.5)^{1/9} = 0.99$

## 4.5.5    Flat Use Factor, $C_{fu}$

The bending design values for sawn lumber loaded about the $x$–$x$ axis (loaded on edge) are obtained by adjusting the tabular design value by the appropriate size factor $C_F$. When lumber is loaded about the $y$–$y$ axis (flatwise), these tabular design values in bending are further multiplied by the flat use factor, $C_{fu}$. The flat use factor is not applied to decking as shown in Table 8.7 because the design value in bending $F_b$ for decking is based on a decking 4 in. thick (except for Redwood). The size factor $C_F$ is used to adjust bending values for nominal 2 in. and 3 in. decking. (See Table 8.7.)

**Examples.**

1. Determine the design value in bending for a 4 in. × 10 in. select structural grade of Douglas Fir-Larch loaded about the $y$–$y$ axis (flatwise bending).

   From Table 8.3, the tabular $F_b = 1450$ psi

   $C_F = 1.2$, $C_{fu} = 1.1$

   Bending design stress = $(1450)(1.2)(1.1) = 1910$ psi

2. Determine the bending design stress of a 4 in. × 12 in. dense select structural grade of Southern Pine.

   From Table 8.4, $C_F = 1.1$, $C_{fu} = 1.1$, $F_b = 2050$ psi

   Bending stress = $(2050)(1.1)(1.1) = 2480$ psi

3. Determine the bending design stress of a 6 in. × 14 in. No. 1 SR grade of Southern Pine loaded about the $y$–$y$ axis (flatwise bending).

   There is no flatwise bending factor for sawn timbers 5 in. × 5 in. and larger. Use the same value used for bending about the $x$–$x$ axis, which is $F_b C_F$.

   From Table 8.6, $F_b = 1350$ psi, $C_F = (12/d)^{1/9}$

   Design bending stress, $F_b' = F_b(C_F) = 1350(12/13.5)^{1/9} = 1330$ psi

The flat use size factor $C_{fu}$ is also used for glued laminated timber loaded in bending parallel to the wide faces of the laminations (the $y$–$y$ axis). The tabular design values in bending for members loaded in bending parallel to the wide faces of the laminations (bending about the $y$–$y$ axis) are based on members with laminations 12 in. wide. These values must be adjusted for members other than 12 in. deep by the flat use factor $C_{fu}$. Values of $C_{fu}$ for members with laminations less than 12 in. wide are shown in Table 4.9.

When the width of the laminations is greater than 12 in. as may occur in members with multiple piece laminations, $C_{fu}$ may be obtained by use of the equation in footnote a Table 4.9.

**TABLE 4.9**

**Flat Use Factor, $C_{fu}$, for Glued
Laminated Timber**

| Member Dimensions Parallel to Wide Faces of Laminations in. | Flat Use[a] Factor $C_{fu}$ |
|---|---|
| $10\frac{3}{4}$ or $10\frac{1}{2}$ | 1.01 |
| $8\frac{3}{4}$ or $8\frac{1}{2}$ | 1.04 |
| $6\frac{3}{4}$ | 1.07 |
| $5\frac{1}{8}$ or 5 | 1.10 |
| $3\frac{1}{8}$ or 3 | 1.16 |
| $2\frac{1}{2}$ | 1.19 |

[a] Values for $C_{fu}$ are rounded values from the equation $(12/d)^{1/9}$ where $d$ is the dimension of the wide faces of the laminations.

### 4.5.6  Volume Factor, $C_V$

The effect of size on the bending strength of glued laminated timber is a function of the volume as determined by the depth, width, and length of the member. Research has shown that the effect of volume can be accounted for by the Eq. (4-2) for members loaded about the $x$-$x$ axis. See Ref. (21). $F_{bx}$ should be multiplied by the volume factor $C_V$ obtained from the following equation:

$$C_V = K_L \left(\frac{5.125}{b}\right)^{1/x} \left(\frac{12}{d}\right)^{1/x} \left(\frac{21}{L}\right)^{1/x} \leq 1.0 \qquad (4\text{-}2)$$

where  $b$ = width (breadth) of bending member (in.). For multiple piece width layups, $b$ = width of widest piece in the layup. Thus, $b \leq 10.75$ in.

$d$ = depth of bending member (in.),

$L$ = length of bending member between points of zero moment (ft),

$x$ = 20 for Southern Pine,

$x$ = 10 for Western species, and

$K_L$ = loading condition coefficient shown in Table 4.10.

For convenience, Eq. (4-2) can also be written with the exponents $1/x$ written as 0.1 for Western species and 0.05 for Southern Pine.

Note that the volume factor does not apply to bending about the $y$-$y$ axis $F_{by}$. For convenience, the equation can be written as

$$C_V = K_L (1291.5/V)^{1/x} \qquad (4\text{-}3)$$

where $K_L$ and $x$ are defined in Eq. (4-2) and $V$ is a nondimensional volume determined by multiplying $b$ in inches by $d$ in inches and $L$ in feet.

## TABLE 4.10

### Loading Condition Coefficients

| Single Span Beam | $K_L$ |
|---|---|
| Concentrated load at midspan | 1.09 |
| Uniformly distributed load | 1.0 |
| Two equal concentrated loads at 1/3 points of span | 0.96 |

| Continuous Beam or Cantilever | |
|---|---|
| All loading conditions | 1.0 |

When $K_L$ is unity, it is often omitted in writing the volume factor equation.

The volume factor, $C_v$, is not accumulative with the beam stability factor, $C_L$, because the volume effect is associated with failure on the tension side of a bending member and the stability factor is associated with buckling on the compression side of a bending member.

For rectangular bending members with variable cross section along their length, such as tapered beams, $d$ for determination of the volume factor should be taken as that depth at which the stresses are being analyzed.

**Examples.**

1. Determine the volume factor $C_V$ for a 6.75 in. $\times$ 36 in. $\times$ 40 ft glued laminated timber made of Douglas Fir-Larch.

$$C_V = K_L \left(\frac{5.125}{b}\right)^{1/10} \left(\frac{12}{d}\right)^{1/10} \left(\frac{21}{L}\right)^{1/10}$$

$$= (1) \left(\frac{5.125}{6.75}\right)^{1/10} \left(\frac{12}{36}\right)^{1/10} \left(\frac{21}{40}\right)^{1/10} = 0.82$$

Using Eq. (4-3)

$$C_V = K_L (1291.5/V)^{1/x} = (1)[(1291.5)/(36)\,(6.75)\,(40)]^{1/10} = 0.82$$

Both equations give the same answer.

2. Determine the volume factor $C_V$ for an 8.75 in. $\times$ 41.25 in. $\times$ 50 ft glued laminated timber beam made of Southern Pine with two equal concentrated loads at the third points.

$$K_L = 0.96 \text{ (from Table 4.10)}$$

$$C_V = K_L \left(\frac{5.125}{b}\right)^{1/20} \left(\frac{12}{d}\right)^{1/20} \left(\frac{21}{L}\right)^{1/20}$$

$$C_V = (0.96) \left(\frac{5.125}{8.75}\right)^{1/20} \left(\frac{12}{41.25}\right)^{1/20} \left(\frac{21}{50}\right)^{1/20} = 0.88$$

### 4.5.7    Temperature

The ANSI/NF$_o$PA NDS—1991, *National Design Specification for Wood Construction* (10) indicates that the design values for sawn lumber tabulated in Tables 8.3–8.8 are applicable to members used under ordinary ranges of temperature and occasionally heated in use to temperatures up to 150°F. These same effects apply to glued laminated timber. Wood increases in strength when cooled below normal temperatures and decreases in strength when heated. Members heated in use to temperatures up to 150°F will return essentially to original strength when cooled. Prolonged temperatures above 150°F may result in permanent loss of strength. When a member is exposed to sustained elevated temperatures up to 150°F, the temperature factor $C_t$ must be applied to the design values as shown in Table 4.11. More information on the effect of temperature on design values is given in *Wood Handbook* (22) by the U.S. Forest Products Laboratory and Appendix C of *National Design Specification for Wood Construction* (10).

### 4.5.8    Temperature Factor, $C_t$

Most wood is used under conditions where temperature is not a significant factor. When an elevated temperature of a member is anticipated for an extended period of time or when elevated temperatures are expected to occur simultaneously with maximum design loads, a reduction in design values may be necessary.

**Example.**    Determine the temperature factor for $F_b$ when a sustained temperature between 100° and 125° exists in a dry location.

$$\text{From Table 4.11, } C_t = 0.8$$

### 4.5.9    Beam Stability Factor, $C_L$

The beam stability factor, $C_L$, is applied to the tabular design value in bending of beams. Calculation of the beam stability factor is covered in 5.4.11. If the depth of a beam is equal or less than the width, $C_L$ is equal to 1. If the beam is laterally supported at the ends and along the compression flange as described in 5.4.11, $C_L$ is also equal to 1. When these conditions do not exist, $C_L$ is calculated by Eq.

**TABLE 4.11**

**Temperature Factors, $C_t$**

| | In Service | $C_t$ | | |
| | Moisture | | | |
| Design Values | Conditions[1] | T≤100°F | 100°F<T≤125°F | 125°F<T≤150°F |
|---|---|---|---|---|
| $F_t$, E | Wet or Dry | 1.0 | 0.9 | 0.9 |
| $F_b$, $F_v$, $F_c$, | Dry | 1.0 | 0.8 | 0.7 |
| and $F_{c\perp}$ | Wet | 1.0 | 0.7 | 0.5 |

1. Wet and dry service conditions for sawn lumber and glued laminated timber are specified in 4.1.4 and 5.1.5.

Source: ANSI/NF$_o$PA NDS®—1991, *National Design Specification for Wood Construction*, AFPA, 1250 Connecticut Ave., NW, Washington, DC. 20036.

(5-10) using Tables 5.3 and Fig. 5.5. See examples in 5.4.11 for determining $C_L$. This factor applies to both sawn lumber and glued laminated timber.

$C_L$ is not accumulative with $C_V$ for glued laminated timber, but is accumulative with $C_F$ for sawn lumber. It is, however, accumulative with $C_I$ for tapered members when the taper is on the compression side. Tapering on the tension side of a beam is not recommended.

### 4.5.10    Repetitive Member Factor, $C_r$

This factor is 1.15 and is used to adjust the design value in bending for sawn dimension lumber when such members are used as joists, truss chords, rafters, studs, planks, and decking, or similar members in contact or not spaced more than 24 in. on centers. There must be at least three such members, and they must be joined by floor, roof, or other load distributing elements capable of supporting the design load. See Tables 8.3–8.5. The repetitive member factor is not used with decking because the factor has already been included in establishing the design values for decking. Also, it does not apply to glued laminated timber.

### 4.5.11    Curvature Factor, $C_c$

The curvature factor is used to adjust the design value in bending for curved members only. It takes into account the difference in extreme outer fiber stress between a curved member and a straight prismatic member, as well as any residual stresses that may remain in a lamination that has been bent to the stated curvature. It is not used in straight beams or cambered beams because it has very little effect at the radii of curvature used for cambering. It is not applied in the design of pitched and tapered curved beams because this effect is accounted for in the $K_\theta$ factor. The effect is less than 1% for members with $1\frac{1}{2}$-in.-thick laminations and a radius of 56 ft or more.

Stress is induced when laminations are bent to curved forms. Although much of this stress is quickly relieved, some remains and tends to reduce the strength of a curved member. Also, in a curved member, the extreme outer fiber bending stress is greater than that of a straight prismatic member subjected to the same moment. Therefore, to account for these factors, the bending design value $F_b$ must be adjusted by multiplying by the curvature factor $C_c$:

$$C_c = 1 - 2000 \left(\frac{t}{R}\right)^2 \tag{4-4}$$

where    $t$ = thickness of lamination (in.) and
$\phantom{where}$    $R$ = radius of curvature of inside face of lamination (in.).

The ratio $t/R$ may not exceed $1/100$ for hardwoods and Southern Pine nor $1/125$ for softwoods other than Southern Pine. The curvature factor is not applied to design values in the straight portion of a member regardless of curvature in other portions.

Fig. 4.5 may be used to determine curvature factors for several different lamination thicknesses.

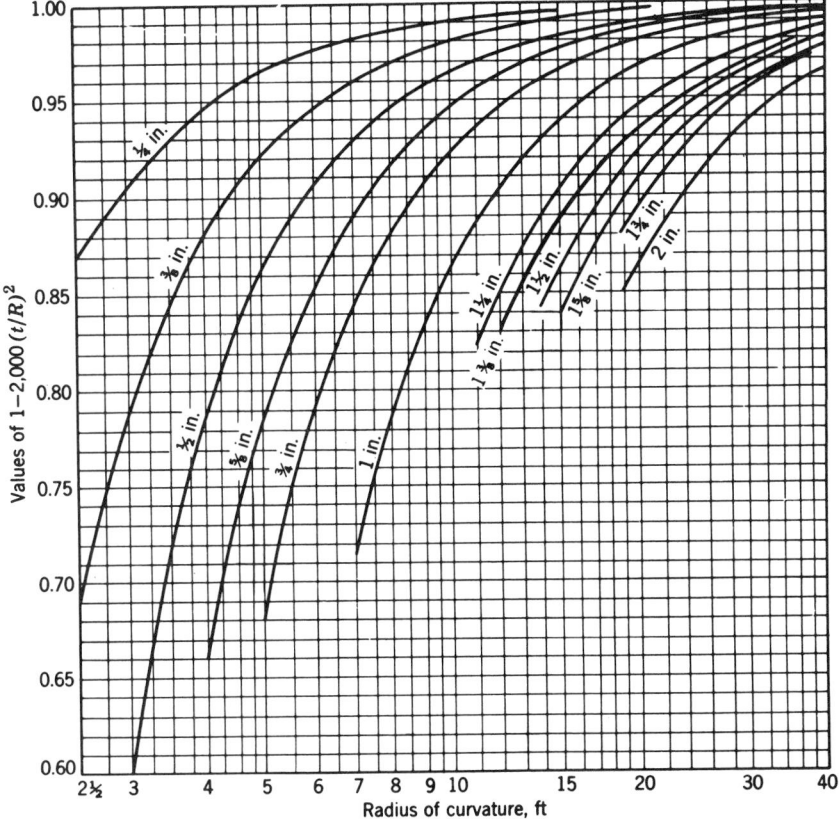

**FIGURE 4.5**   Curvature factors.

## 4.5.12   Form Factor, $C_f$

For bending members with circular or square cross sections loaded in the plane of the diagonal, the design value in bending should be adjusted by the factors shown in Table 4.12.

Adjustment of the tabular design bending values by the form factor results in a diamond-shaped member or circular-shaped member having the same moment capacity of a square member or circular-shaped member of the same cross section.

**TABLE 4.12**

**Form Factors, $C_f$**

|                  | $C_f$ |
|------------------|-------|
| Round Section    | 1.18  |
| Diamond Section  | 1.414 |

The form factor applies only to the tabular design values in bending for round wood members and rectangular or square members loaded on a diagonal. It is cumulative with all other applicable adjustment factors, including size factor and volume factor.

### 4.5.13    Column Stability Factor, $C_P$

The column stability factor, $C_P$, is a function of the dimensions, design value in compression, and modulus of elasticity. $C_P$ is determined by use of Eq. (5-14) in 5.2. When a column is supported throughout its length, $C_P = 1.0$. See 5.2 for additional information and examples of use.

### 4.5.14    Shear Stress Factor, $C_H$

The shear stress factor, $C_H$, may be used with sawn lumber when the size of the split, check, or shake is known and no further increase is anticipated. The design values in shear parallel to grain for the various grades have been set to allow for the occurrence of certain size splits, checks, or shakes. When these characteristics in a sawn timber are smaller than allowed, the design value in shear parallel to grain may be adjusted upward as shown in the adjustment factors included in Tables 8.3–8.6. Limitations on the size of split, checks, or shakes on most species have been set so that the design values are $\frac{1}{2}$ of the solid unchecked piece without splits or shakes. The shear factor $C_H$ is multiplied by the tabulated design values multiplied by other applicable adjustment factors.

$$f_v \leq F_V C_D C_M C_t C_H$$

The limitation of shear reducing factors for some grades or Redwood, mixed Southern Pine, and Southern Pine have been set so that the design values are greater than $\frac{1}{2}$ of the solid unchecked or unsplit piece without shakes. Therefore, the design values of these grades have to be reduced to obtain a 50% shear ratio before applying $C_H$, as shown in Tables 8.3, 8.4, and 8.6.

For Southern Pine

$$f_v \leq (90 \text{ psi}) \, C_D C_M C_t C_H$$

For Redwood

$$f_v \leq (80 \text{ psi}) \, C_D C_M C_t C_H$$

Values of $C_H$ are included in Tables 8.3–8.6.

The designer should use caution when applying this factor, and each piece has to be examined. If additional drying is anticipated, additional checking may occur. This factor is very useful in analyzing existing structures. It is also used for the alternate design procedures for shear included in 5.4.4.

### 4.5.15    Buckling Stiffness Factor, $C_T$

The buckling stiffness factor, $C_T$, is used with 2 × 4 sawn compression chords of wood trusses sheathed with $\frac{3}{8}$ in. or thicker plywood when they are subject to combined bending and compression loads and used under dry conditions. The plywood must be applied with good nailing practice. The factor is applied to the modulus of elasticity used in column stability calculations for the sawn 2 × 4 or smaller compression chords.

$$C_T = 1 + \frac{K_M l_e}{K_T E} \qquad (4\text{-}5)$$

in which $l_e$ = effective column length of truss compression chord (in.),

$K_M$ = 2300 for wood seasoned to 19% moisture content or less at the time of plywood attachment,

$K_M$ = 1200 for unseasoned or partially seasoned wood at the time of plywood attachment,

$K_T$ = 0.59 for visually graded lumber and machine evaluated lumber (MEL),

$K_T$ = 0.82 for products with $COV_E \leq 0.11$, and

when $l_e > 96$ in., $C_T$ shall be calculated based on $l_e = 96$ in.

### 4.5.16 Bearing Area Factor, $C_b$

For bearings less than 6 in. in length and not nearer than 3 in. to the end of a member, the maximum design load per square inch is obtained by multiplying the tabular design values in compression perpendicular to grain by the bearing area factor, $C_b$, which may be obtained from Eq. (4-6).

$$C_b = \frac{l_b + 0.375}{l_b} \qquad (4\text{-}6)$$

where $l_b$ = the length in bearing parallel to grain of the wood (in.).

The factor $C_b$ can also be obtained from Table 4.13.

### 4.5.17 Interaction Stress Factor, $C_I$

The interaction stress factor applies to the bending stress of taper cut glued laminated timber bending members. It is not applied to arches or pitched and tapered curved beams. When the tapered cut is on the compression side of a bending member, it is not cumulative with the volume factor $C_V$. However, it is cumulative when the taper cut is on the tension side. (Taper cuts on the tension side of beams are not recommended.) It is cumulative with the beam stability factor $C_L$ when the taper cut is on the compression side. See 5.6 for additional information.

### TABLE 4.13

### Bearing Area Factor $C_b$

| Length of Bearing, in. | $\frac{1}{2}$ | 1 | $1\frac{1}{2}$ | 2 | 3 | 4 | 6 or more |
|---|---|---|---|---|---|---|---|
| $C_b$ | 1.75 | 1.38 | 1.25 | 1.19 | 1.13 | 1.10 | 1.09 |

### 4.5.18    Other Adjustment Factors

The preceding discussion covers adjustment factors usually applicable to timber design. For special cases, such as poles and piles, other adjustment factors may be necessary.

## 4.6    APPLICATION OF ADJUSTMENT FACTORS

The following examples illustrate the application of adjustment factors to various design values:

Bending:
 Sawn lumber $F'_b = F_b C_D C_F C_L C_M C_f C_t C_{fu}$
 Glued laminated timber $F'_b = F_b C_D C_L C_c C_M C_I C_f C_t C_V C_{fu}$
Tension parallel to grain:
 Sawn lumber $F'_t = F_t C_D C_M C_t C_F$
 Glued laminated timber $F'_t = F_t C_D C_M C_t$
Compression parallel to grain:
 Sawn lumber $F'_c = F_c C_D C_P C_M C_f C_F$
 Glued laminated timber $F'_c = F_c C_D C_P C_M C_t$
Compression perpendicular to grain:
 Sawn lumber $F'_{c\perp} = F_{c\perp} C_M C_t$
 Glued laminated timber $F'_{c\perp} = F_{c\perp} C_M C_t$
Shear parallel to grain:
 Sawn lumber $F'_v = F_v C_D C_M C_t C_H$
 Glued laminated timber $F'_v = F_v C_D C_M C_t$
End grain in bearing:
 Sawn lumber $F'_g = F_g C_D C_t$
 Glued laminated timber $F'_g = F_g C_D C_t$
Modulus of elasticity:
 Sawn lumber $E' = E C_M C_t$
 Glued laminated timber $E' = E C_M C_t$
Radial tension:
 Glued laminated timber $F'_{rt} = F_{rt} C_D C_M C_t$

Note $C_V$ and $C_L$ are not applied cumulatively.

The symbol for the adjusted design value is shown with a prime (such as $F'_b$), and the tabular value is shown without the prime such as $F_b$. In many cases, the adjustment factors are unity and are usually not shown in design calculations. It is customary to show only those applicable adjustment factors that change the tabular design values. See Table 4.6 for applicability of adjustment factors.

**Example: Bending.**    Determine the allowable design value in bending for a straight simply supported glued laminated beam $5\frac{1}{8}$ in. × 30 in. Southern Pine beam 35 ft long that is continuously laterally supported along the top and at the ends. It is used in a dry location at normal temperature to support a snow load.

The dead load is less than $0.90/1.15$ times the total load; therefore, the duration of load for snow controls.

$$F'_b = F_b C_D C_V C_L C_c C_M C_I C_f C_t$$

$$C_L, \ C_c, \ C_M, \ C_I, \ C_f, \text{ and } C_t = 1$$

$$C_D = 1.15 \ (\text{Table } 4.7)$$

$$C_V = 0.93 \ [\text{determined by Eq. (4-2)}] \text{ and}$$

$$F'_b = F_b \ (1.15) \ (0.93) = 1.07 \ F_b$$

**Example: Bending.**    Determine the allowable design value in bending for a simply supported curved beam $6\frac{3}{4}$ in. × 27 in. × 40 ft made of Douglas Fir that is continuously laterally supported along the top and at the ends. The radius of curvature, $R$, is 35 ft and the thickness of laminations, $t$, is 1.5 in. It is used in a wet location at normal temperature. The maximum bending stress is caused by wind. The beam has been made from preservatively treated lumber to prevent decay and will be used under a wet service condition.

$$F'_b = F_b C_D C_V C_L C_c C_M C_I C_f C_t$$

$$C_L, \ C_I, \ C_f, \text{ and } C_t = 1$$

$$C_D = 1.6, \ K_L = 1$$

$$C_V = K_L \left(\frac{5.125}{b}\right)^{1/10} \left(\frac{12}{d}\right)^{1/10} \left(\frac{21}{L}\right)^{1/10}$$

$$\quad = \left(\frac{5.125}{6.75}\right)^{1/10} \left(\frac{12}{27}\right)^{1/10} \left(\frac{21}{40}\right)^{1/10} = 0.84$$

$$C_c = 1 - 2000 \ (t/R)^2 = 1 - 2000 \ [1.5/(35)\,(12)]^2 = 0.97$$

$$C_M = 0.80 \ (\text{from Table 1, AITC 117—Design or Table 4.8})$$

$$F'_b = F_b \ (1.6) \ (0.97) \ (0.80) \ (0.84) = 1.04 \ F_b$$

## 4.7  REFERENCES

1. American Society of Civil Engineers, *Minimum Design Loads for Buildings and Other Structures*, ASCE 7-88, New York, NY, 1988.

2. International Conference of Building Officials, *Uniform Building Code*, Whittier, CA, 1991.

3. Structural Engineers Association of California, *Recommended Lateral Force Requirements*.

4. Building Seismic Safety Council, *Recommended Provisions for the Development of Seismic Regulations for New Buildings*, Washington, DC.

5. American Association of State Highway and Transportation Officials, *Standard Specifications for Highway Bridges*, Washington, DC, 1987.

6.  American Railway Engineering Association, *Manual of Recommended Practice*, Washington, DC, 1969.

7.  United States Department of Agriculture, Forest Service, Forest Products Laboratory, *Fatigue Resistance of Quarter-Scale Bridge Stringers in Flexure and Shear*, Report No. 2236, Madison, WI, 1962.

8.  American Society of Civil Engineers, *Design Considerations for Fatigue in Timber Structures, Journal of Structural Division*, ASCE Paper No. 2470, Vol. 86, pp. 15–23, New York, NY, 1960.

9.  American Society of Civil Engineers, *Design of Structures to Resist Nuclear Weapons Effects*, Manual No. 42, New York, NY, 1985.

10. American Forest & Paper Association, *National Design Specification for Wood Construction*, Washington, DC, 1991.

11. American Institute of Timber Construction, *Standard Specifications for Structural Glued Laminated Timber of Softwood Species*, AITC 117—Design, Englewood, CO, 1993.

12. American Institute of Timber Construction, *Standard Specifications for Hardwood Glued Laminated Timber*, AITC 119, Englewood, CO, 1994.

13. Northeastern Lumber Manufacturers Association (NELMA), Cumberland Center, ME 04021.

14. Northern Softwood Lumber Bureau (NSLB), Cumberland Center, ME 04021.

15. Redwood Inspection Service (RIS), Novato, CA 94949.

16. Southern Pine Inspection Bureau (SPIB), Pensacola, FL 32504.

17. West Coast Lumber Inspection Bureau (WCLIB), *Standard No. 17 Grading Rules for West Coast Lumber*, Portland, OR 97223.

18. Western Wood Products Association (WWPA), Portland, OR 97204.

19. National Lumber Grades Authority (NLGA), Vancouver, B.C., Canada V6E 2E9.

20. American Wood Preservers Association, *AWPA Book of Standards*, Stevensville, MD, 1991.

21. ASTM D1990-91, *Standard Practice for Establishing Allowable Properties for Visually-Graded Dimension Lumber from In-Grade Tests of Full-Size Specimens*, Philadelphia, PA, 1991.

22. American Institute of Timber Construction, *Use of a Volume Effect Factor in the Design of Glued Laminated Timber Beams*, Technical Note No. 21, Englewood, CO, 1991.

23. United States Department of Agriculture, Forest Service, Forest Products Laboratory, *Wood Handbook: Wood as an Engineering Material*, Agriculture Handbook No. 72, Madison, WI, 1987.

# CHAPTER 5

# DESIGN OF STRUCTURAL ELEMENTS

## 5.1 INTRODUCTION

### 5.1.1 General

This chapter covers the basic types of structural members. Under the headings herein are contained general information on the design features of the member, tabular data, a typical design procedure, and, in most cases, a design example.

The procedures presented have been found to be most suited to the particular member in question; however, other procedures may be used if substantiated by tests or by sound engineering principles. The examples are illustrative only. In actual designs, all the conditions that might be expected to have a bearing on the ability of the member or structure to support all anticipated loads safely should be considered.

In general, in the example calculations, no rounding was done during intermediate steps to obtain the final answer. For purposes of illustration, intermediate steps were recorded using rounded figures, which may result in slightly different answers when the rounded intermediate figures are used.

In most cases, the actual design can be more efficiently performed by use of a computer or programmable calculator. However, hand calculations are used throughout the manual to illustrate all of the steps of calculation required for design.

### 5.1.2 Abbreviations and Symbols

The abbreviations and symbols used in this section are contained in the List of Symbols immediately following the Contents. Deviations from these notations are identified where they occur.

## 5.2 TABULAR DESIGN VALUES

The tabular design values in Tables 8.3–8.7 for sawn lumber, AITC 117—Design (1) for glued laminated timber of softwood species and AITC 119 (2) for hardwood species are defined as the maximum allowable stresses based on standard conditions as set forth in the tables. These tabular design values are designated as $F_b$ for bending, $F_c$ for compression parallel to grain, $F_v$ for shear parallel to grain, and so on. The symbol $F$ denotes a tabular design value for a given strength property and the lower case subscript denotes the strength property. For some conditions, these values must be adjusted before they can be used in design. The adjustment factors are denoted by $C$ and a subscript letter indicating the reason for modification, such as $C_V$ for volume factor and $C_D$ for duration of load. Some of these adjustment factors are applicable to all strength properties, while others are applicable to specific strength properties; for example, the curvature factor $C_C$ is applicable to bending in curved members only.

When all applicable adjustment factors have been applied to the tabular design value, the allowable value to use in design is denoted by $F'$, such as $F'_b$ for bending, $F'_c$ for compression parallel to grain, and $F'_v$ for shear parallel to grain.

The actual unit stresses of a member under load are denoted by the same letters as the tabular design value, except that a lower case $f$ is used, such as $f_b$, $f_c$, $f_v$, etc.

Some calculations, such as in calculating the beam stability factor, require an intermediate design value without all of the adjustments. In such cases, an asterisk is used for the design value such as $F_b^*$. These values are used only as interim values in the calculations and are explained where they occur.

### 5.2.1   Glued Laminated Timber

The tabular design values for glued laminated timber in AITC 117—Design (1) and AITC 119 (2) are based on dry conditions of use, a depth of 12 in., a width of $5\frac{1}{8}$ in., a length of 21 ft, and loads of normal duration. These values are based on an average moisture content of 12%. The adjustment factors for moisture content for glued laminated timber are different from those used for sawn lumber, which has several bases for moisture content modifications. For glued laminated timber, wet service adjustment factors are listed at the bottoms of the columns of design values in Tables 1 and 2 of AITC 117—Design (1), in Table 5.1 for softwoods, and AITC 119 (2) for hardwoods.

Glued laminated timbers may have different tabular design values when loaded in bending about the $x$-$x$ and the $y$-$y$ axis, depending on the grades and arrangement of grades of lumber within the member. Therefore, Tables 1 and 2 in AITC 117—Design (1) have strength properties listed for loading in bending about both the $x$-$x$ and $y$-$y$ axes. The tabular design values when bending is about the $x$-$x$ axis are denoted by the subscript $x$ such as $F_{bx}$, $F_{vx}$, and $E_x$. For bending about the $y$-$y$ axis, the subscript $y$ is used. When no subscripts are used, the tabular design values about the $x$-$x$ axis are assumed. For clarity in specifying, it is recommended that the subscripts be used, particularly when the member is loaded in bending about the $y$-$y$ axis. Note that the subscripts $xx$ and $yy$ are sometimes used to indicate stresses about the $x$-$x$ and $y$-$y$ axes. The single-letter subscript is the preferred form.

### 5.2.2   Sawn Lumber

Tabular design values for sawn lumber are available in various sizes shown in Tables 8.3–8.7. Tables 8.3 and 8.4 include dimension lumber with nominal thickness of 2–4 in. Table 8.6 includes lumber of 5 in. × 5 in. dimension and larger. Table 8.5 includes mechanically graded lumber consisting of machine stress-graded lumber (MSR) and machine evaluated lumber (MEL). These are available only in nominal 2 in. or less thicknesses. Design values for decking are included in Table 8.7 for decking 2–4 in. thick.

Adjustment factors for sawn lumber are not always the same. Therefore, adjustment factors are included with each table of sawn lumber. These design values in Tables 8.3–8.7 were taken from ANSI/NF₀PA NDS®—1991, *National Design Specification® for Wood Construction* (3) and were current at the time of publication of this manual. For current information on grades and design values, the designer is referred to the latest NDS® or the governing building code, which may include the applicable tabular design values.

Machine evaluated lumber (MEL), in addition to MSR lumber, is also being produced by some companies. The designer should check on availability of this lumber prior to specifying.

## 5.3  ADJUSTMENTS TO TABULAR DESIGN VALUES

Some of the adjustment factors, such as duration of load, are applicable to all strength properties (except modulus of elasticity and compression perpendicular to grain), but other adjustment factors may have different values for different strength properties or are applicable only to a specific property. In general, all applicable adjustment factors are cumulative, but in some cases, the more restrictive of two adjustment factors apply such as the volume factor $C_V$ and the beam stability factor $C_L$ for bending members.

A discussion of adjustment factors is contained in Chapter 4. A summary of the applicability of adjustment factors is shown in Table 4.6. Additional comments on adjustment factors are given where they occur in the design of components.

### 5.3.1  Other Adjustment Factors

The preceding discussion covers adjustment factors usually applicable to timber design. For special cases such as poles and piles, other adjustment factors are also necessary.

### 5.3.2  Application of Adjustment Factors

The following examples illustrate the application of adjustment factors to various design values for sawn lumber:

Bending $F_b' = F_b\, C_D\, C_F\, C_L\, C_M\, C_{fu}\, C_t\, C_r\, C_f$
Tension parallel to grain $F_t' = F_t\, C_D\, C_M\, C_t\, C_F$
Compression parallel to grain $F_c' = F_c\, C_D\, C_P\, C_M\, C_t\, C_F$
Shear parallel to grain $F_v' = F_v\, C_D\, C_M\, C_t\, C_H$
Bearing parallel to grain $F_g' = F_g\, C_D\, C_t$
Modulus of elasticity $E' = E C_M\, C_t\, C_T$

The symbol for the adjusted design value is shown with a prime (such as $F_b'$) and the tabular design value is shown without the prime (such as $F_b$). In many cases, the adjustment factors are unity and are usually not shown in design calculations. It is customary to show only those applicable adjustment factors that change the tabular design values (see Table 4.6). For instance, there are eight adjustment factors for bending shown in the preceding paragraph. In most cases, not more than two or three are applicable.

### 5.3.3    Adjustment Factor $C_V$ for Cantilevered and Continuous Beams

The length $L$ to use in the volume factor equation is the length in feet between points of zero moment as illustrated. For continuous and cantilevered beams, use the longest $L$ measured. The depth for a tapered member is the depth at which bending stress is being calculated and the length is the distance between points of zero moment.

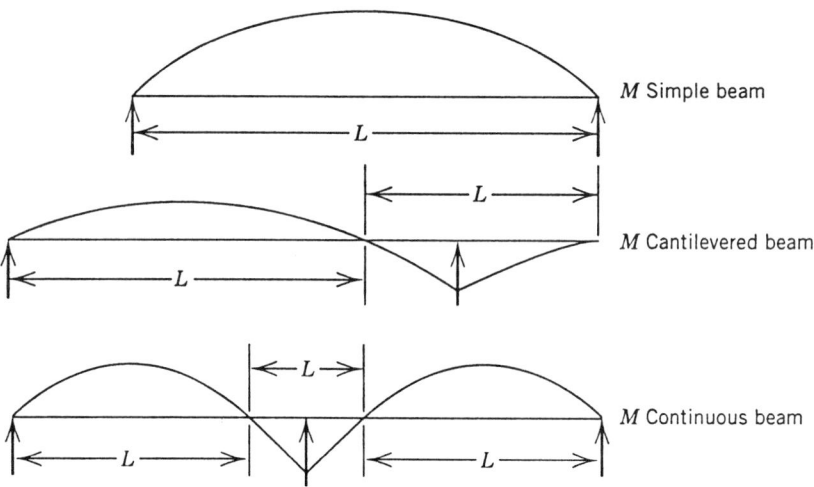

### 5.4    BEAMS AND BENDING MEMBERS

Wood members subjected to bending as the principle method of loading are usually designated as joists, purlins, beams, or girders. The term "beam" is often used to designate any of these members. Wood planks and decking are also subjected to bending. Beams are classified according to structural use: simple beams, cantilever beams, or continuous beams. Glued laminated timber beams are also classified by shape: straight, single-tapered, double-tapered, curved, pitched and tapered curved beams. Glued laminated timber beams may be cambered.

Most sawn lumber or timber beams are used in simple-span applications, but they may also be used as continuous or cantilevered beams.

Glued laminated timber beams are readily obtainable in long lengths, and for many applications, beams cantilevered over a support are economical. In some cases, continuous beams are more suitable. Simple-span beams can provide long clear spans that are limited only by manufacturing and shipping restrictions.

The design of both sawn and glued laminated timber beams should be in accordance with established engineering practice. The following structural analysis should be taken into account in design:

1. Bending including strength and lateral stability
2. Deflection
3. Shear parallel to grain (horizontal shear)
4. Compression perpendicular to grain
5. Radial stress in curved members

### 5.4.1 Bending Stress

Bending stresses are determined by standard engineering equations usually found in handbooks for the loading conditions most frequently encountered in design. Chapter 8 contains equations for various loading and support conditions for simple, continuous, or cantilevered beam design. Tables 8.1 and 8.2 contain section properties for sawn lumber and glued laminated timber, respectively. Where the moment $M$ and the section modulus $S$ of a straight or relatively straight prismatic member are known, the bending stress $f_b$ can be calculated by the flexure equation

$$f_b = \frac{M}{S} \tag{5-1}$$

The bending stress obtained from this equation must be modified for tapered, curved, and pitched and tapered curved beams as explained in the appropriate portion of this chapter. The tabular design values for bending as well as shear parallel to grain, compression perpendicular to grain, and modulus of elasticity are given in Tables 8.3–8.7 for sawn lumber. Glued laminated timber tabular design values are given in AITC 117—Design (1) for structural glued laminated timber made of softwood species and AITC 119 for glued laminated timber made of hardwoods (2).

Adjustments of these tabular design values are necessary when the conditions of use are other than stated in the tables.

### 5.4.2 Deflection

Deflection is a serviceability consideration in some structures such as warehouses and similar construction. It may not be an important consideration provided the design is such that ponding is prevented. In other structures, deflection may be a prime consideration in design.

For methods of calculations and recommendations for deflection and camber, see Chapter 4.

### 5.4.3 Shear Parallel to Grain (Horizontal Shear)

The terminology "shear parallel to grain" is replacing the term "horizontal shear" because it is more descriptive to the shear under consideration. However, the term "horizontal shear" has enjoyed a long history of use, and many existing references contain this term. The two are synonymous.

The general equation for calculating shear parallel to grain stress, $f_v$ in a beam is

$$f_v = \frac{VQ}{Ib} \tag{5-2}$$

where  $V$ = vertical shear force (lb),
$Q$ = statical moment of area above or below neutral axis about neutral axis (in.$^3$),
$I$ = moment of inertia of section (in.$^4$), and
$b$ = width of beam at neutral axis (in.).

For a rectangular beam, the equation becomes

$$f_v = \frac{3V}{2bd} \tag{5-3}$$

where  $d$ = depth of beam (in.) and
$b$ = width of beam (in.).

The shear parallel to grain stress as determined by these equations, except in the case of checked sawn beams, may not exceed the adjusted design value in shear parallel to grain, $F_v'$. The equation should not be used when the beam end is notched, when the beam is supported by fastenings at the end such as bolts, or when the beam supports hanging loads near the end (see Chapter 7 for further information). Note the following when calculating the reaction, $V$:

(a)   All loads within a distance from either support equal to the depth of the beam may be neglected when the loads are applied to one surface of the beam and the beam is fully supported by the opposite surface (see Fig. 5.1).

(b)   When there is a single moving load, or one moving load that is considerably greater than the others, place that load at a distance from the support equal to the depth of the beam; place any other loads in their normal relation. When there are two or more moving loads of approximately

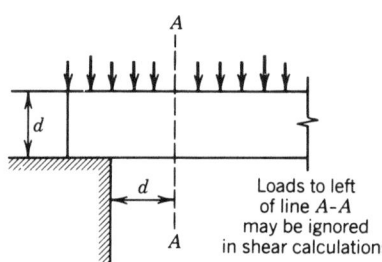

**FIGURE 5.1**   Loads within depth, $d$, from support.

equal weight and in proximity, place the loads in the position that produces the maximum value for $V$, neglecting any loads within a distance from the support equal to the depth of the beam.

### 5.4.4 Shear Parallel to Grain in Checked Sawn Beams

The design values in shear for sawn lumber contained in Tables 8.3–8.6 include allowances for checks, end splits, and shakes. These design values are recommended for use with Eq. (5-3). The ANSI/NF$_0$PA NDS—1991, *National Design Specification for Wood Construction* (3) contains an alternate procedure for shear design that considers a checked sawn beam as if the shear were resisted by two beams, that is, the upper and lower halves of the checked beam. See Appendix E of NDS (3).

### 5.4.5 End-Notched Members

Normally, beams should not be notched or tapered on the tension side. If it becomes necessary to notch a bending member at its end at a support on the tension side, the shear parallel to grain $f_v$ should be computed by the equation

$$f_v = \frac{3R_v}{2bd_e}\left(\frac{d}{d_e}\right) \tag{5-4}$$

where $R_v$ = vertical reaction (lb),
  $b$ = width of beam (in.),
  $d$ = depth of beam (in.), and
  $d_e$ = depth of beam less the depth of the notch (in.) (see Fig. 5.2).

In calculating the vertical reaction $R_v$ for use in Eq. (5-4), the loads within a depth $d_e$ from the face of the support may be neglected. The notching of a bending member on the tension side results in a decrease in strength caused by stress concentrations around the notch, as well as a reduction of the area resisting the shear forces. The notch induces tension perpendicular to grain stresses which interact with the shear parallel to grain creating a splitting tendency. For these reasons, the notching of large glued laminated timbers is not recommended, and it should be limited to smaller wood members. The equation given above is an empirical equation developed for the condition of a square-cornered end notch, and the ratio of the depth of the notch to the depth of the beam should be limited to 1:10. The

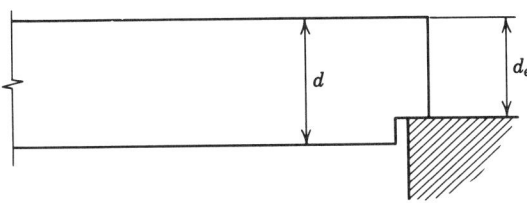

**FIGURE 5.2** End-notched beam (notched on tension side).

designer should also consider reducing the stress concentration that occurs when a member is notched by using a gradual tapered notch configuration in lieu of a square-cornered notch. Also, the designer should consider the use of reinforcement such as full threaded lag screws to resist the tendency to split at the notch. Notching on the tension side of simple beams away from the ends is not recommended.

When a beam is notched or beveled on its upper (compression) side at the ends, a less severe condition from the standpoint of stress concentrations is realized. If such a notch is square cornered, as illustrated in Fig. 5.3, the shear should be checked by the equation

$$f_v = \frac{3R_v}{2b\left[d - \left(\dfrac{d - d_e}{d_e}\right)e\right]} \tag{5-5}$$

where $e$ is the distance the notch extends inside the inner edge of the support (in.). See Fig. 5.3. If $e$ exceeds $d_e$, the preceding equation is not used; instead, the shear strength is computed by using only the depth of beam below the notch, $d_e$. In no case should a notch on the upper side of a beam exceed 40% of the total depth of a beam. If the end of a beam is beveled (as shown by the dotted line in Fig. 5.3), $d_e$ is measured from the inner edge of the support to the bevel. The same equation and considerations as for beams notched on the upper side apply.

### 5.4.6    Shear Perpendicular to Grain (Vertical Shear)

It is ordinarily not necessary to compute or check the strength of wood beams in shear perpendicular to grain.

### 5.4.7    Compression Perpendicular to Grain

The stress induced in compression perpendicular to grain, $f_{c\perp}$, at beam reactions or from loads applied to members (except as adjusted in the next paragraph) should not exceed the adjusted compression perpendicular to grain design values, $F'_{c\perp}$. These adjusted design values apply to bearings of any length at the ends of a member and to all bearings 6 in. or more in length at any other location. When calculating the bearing area at the ends of members, it is traditional to make no allowance for the fact that as the member bends, the end rotation will result in increased pressure on the inner edge of the bearing.

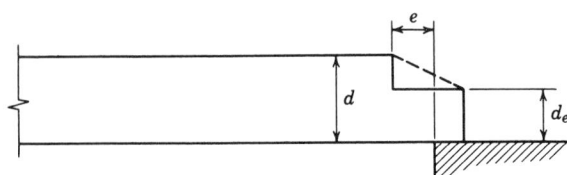

**FIGURE 5.3**    End-notched beam (notched on compression side).

For bearings less than 6 in. in length and not nearer than 3 in. to the end of a member, the maximum load per square inch is obtained by multiplying the tabular design values in compression perpendicular to grain by the bearing factor:

$$C_b = \frac{l_b + 0.375}{l_b} \tag{5-6}$$

where $l_b$ is the length of bearing parallel to the grain of the wood (in.).

This equation gives the following multiplying factors for the indicated lengths of bearing on such small areas as plates and washers (see Table 5.1). In using the preceding equation and Table 5.1 for round bearing areas, such as washers, use a length of bearing equal to the diameter.

The tabular design values for compression perpendicular to grain in Tables 8.3–8.7 for sawn lumber and in AITC 117—Design (1) for glued laminated timber are based on a deformation limit of 0.04 in. obtained when tested in accordance with the Standard ASTM D 143 (4) for compression perpendicular to grain. In special applications where deformation may be critical, the use of a reduced compression perpendicular to the grain design value may be appropriate.

When a more restrictive deformation of 0.02 in. is desired for special design considerations, the following equation may be used for glued laminated timber.

$$F_{c\perp (0.02)} = 0.73 F_{c\perp} \tag{5-7}$$

where $F_{c\perp (0.02)}$ = compression perpendicular to grain at 50% of deformation limit associated with tabulated $F_{c\perp}$ values (0.02 in.) (psi) and

$F_{c\perp}$ = compression perpendicular to grain at 0.04 in. deformation limit (psi).

For sawn lumber, NDS (3) recommends that 5.6 psi be added to the value calculated by Eq. (5-7).

Prior to 1982 for sawn lumber and 1983 for glued laminated timber, the compression perpendicular to grain design values were based on the observed proportional limit. These design values were lower than the new values, but were subject to modification for duration of load. The design values for compression perpendicular to grain in Tables 8.3–8.7 and in AITC 117—Design (1) should not be adjusted for duration of load.

The design values for compression perpendicular to grain for glued laminated timber are generally lower than those for sawn lumber for the same deformation limit. This is based, in part, on the consideration of the larger sizes of glued lam-

**TABLE 5.1**

**Bearing Area Factors, $C_b$**

| Length of bearing, in., $l_b$ | 1/2 | 1 | 1 1/2 | 2 | 3 | 4 | 6 or more |
|---|---|---|---|---|---|---|---|
| Factor, $C_b$ | 1.75 | 1.38 | 1.25 | 1.19 | 1.13 | 1.10 | 1.00 |

inated timber, the length of bearing, and the method used to derive the design values.

### 5.4.8   Angle of Load to Grain (Hankinson Formula)

The angle of load to grain is the angle between the direction of the resultant load acting on a member and the longitudinal axis of the member. In this manual, tabular design values in compression are given for bearing parallel to grain, $F_g$, and compression perpendicular to grain, $F_{c\perp}$.

Design values for $F_g$ may be obtained in Table 3.1 for sawn lumber and in Annex A to AITC 117—Design (1) for glued laminated timber. To obtain design values in compression at any angle of load to grain, the Hankinson formula may be used when the loaded surface is perpendicular to the direction of load (see Fig. 5.4). The Hankinson formula for determining the design value in compression perpendicular to the inclined surface at an angle of load to grain

$$F_\theta' = \frac{F_g' F_{c\perp}'}{F_g' \sin^2 \theta + F_{c\perp}' \cos^2 \theta} \tag{5-8}$$

where   $F_\theta'$ = design value in compression acting perpendicular to the inclined surface (psi),

$F_g'$ = design value in bearing parallel to grain, including all applicable adjustment factors (psi),

$F_{c\perp}'$ = design value in compression perpendicular to grain (psi) including all applicable adjustment factors [note: $F_{c\perp}$ is not adjusted for load duration factor $(C_D)$], and

$\theta$ = angle between the direction of load and the direction of the grain (degrees).

When the resultant load is at an angle between 0° and 90° with the surface being considered, the angle $\theta$ is the angle between the direction of grain and the direction of the load component.

The allowable design loads $F_g'$ and $F_{c\perp}'$, which include all applicable adjustment factors, must be used in Hankinson's formula rather than using tabular design values and adjusting $F_\theta$ after application of Hankinson's formula because $F_{c\perp}$ is not adjusted for duration of load.

The Hankinson formula may also be solved graphically through the use of the

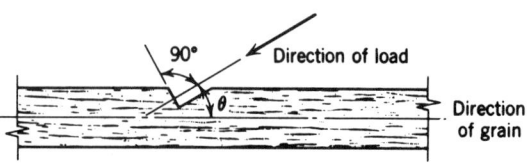

**FIGURE 5.4**   Angle of load to grain.

nomograph in Fig. 7.2. In this nomograph, set $F'_g = P'$, $F'_{c\perp} = Q'$, and $F'_\theta = N'$.

### 5.4.9 Lateral Stability

To prevent sidewise buckling in bending and compression members and to permit the members to carry maximum loads, it is usually necessary to provide lateral stability at intervals. The length of the intervals will depend on the dimensions of the member.

### 5.4.10 Sawn Beams

For sawn rectangular beams and joists, the approximate lateral stability rules based on nominal dimensions and listed in Table 5.2 may be applied by the designer. When adequately stabilized purlins or joists are set between bending members, the depth of the member below the bottom of the purlins or joist is used as the dimension $d$ in determining the depth–breadth ratio $(d/b)$. For more precise engineering analysis of the lateral stability of sawn bending members, the lateral stability calculations in 5.4.11 may be applied.

When joists have depth–breadth ratios of 8 or 9 (such joists have been in use for many years) and the compression edge is continuously supported by a deck, the end bridging prevents torsional rotation. In such cases, intermediate bridging is provided to distribute concentrated loads to adjacent joists.

**TABLE 5.2**

**Approximate Lateral Stability Rules for Sawn Beams**

| Depth–Breadth Ratio[a] | Rule |
|---|---|
| 2 : 1 or less | No lateral bracing is required |
| 3 : 1 to 4 : 1 | The ends shall be held in position, as by full-depth solid blocking, bridging, hangers, nailing or bolting to other framing members, or other acceptable means |
| 5 : 1 | One edge shall be held in line for its entire length |
| 6 : 1 | Bridging, full-depth solid blocking, or cross bracing shall be installed at intervals not exceeding 8 ft unless both edges are held in line or unless the compression edge of the member is supported throughout its length to prevent lateral displacement, as by adequate sheathing or subflooring, and the ends at points of bearing have lateral stability to prevent rotation |
| 7 : 1 | Both edges shall be held in line for their entire length |

[a]Based on nominal dimensions. If a bending member is subject to both flexure and compression parallel to grain, the depth–breadth ratio may be as much as 5 : 1 if one edge is held firmly in line. If under all combinations of load the unbraced edge of the member is in tension, the ratio for the beam may be 6 : 1.

Economy in wood beam design usually favors a deep and narrow section. Such a section increases the likelihood of lateral buckling of the compression edge. In the past, many building codes contained rules governing depth–breadth ratios to minimize such lateral buckling tendencies. The limitation of the depth-to-breadth ratio in a beam to provide lateral stability is believed to be an inefficient method of preventing lateral buckling in wood beams. The following design procedure takes into account the unsupported length of the compression side of the bending member, as well as the stability of the compression side of the member.

The allowable design value in bending $F_b'$ is determined by multiplying the tabular design value $F_b$ by all applicable adjustment factors, including the beam stability factor $C_L$.

### 5.4.11    Beam Stability Factor

The beam stability factor $C_L$ is based on a procedure developed by Hooley and Madsen in *Lateral Stability of Glued Laminated Beams* (5).

This procedure was based on the same buckling concept then in use for columns with short beams, intermediate beams, and long beams.

When the Ylinen continuous column equation was adopted for columns, the procedure for determining the column stability factor $C_P$ was modified to use in calculating the beam lateral stability factor $C_L$.

No lateral support is required when the depth of a bending member does not exceed its width; when the designer chooses to use the rules for supporting sawn lumber bending members in Table 5.2, $C_L = 1.0$; and when the compression edge of a bending member is supported throughout its length and the ends at points of bearing are held to prevent rotation, $C_L = 1.0$. Recent tests (Ref. 6) have shown that supporting the ends to prevent rotation is not necessary when strap hangers that are unsupported laterally at the bottom are used.

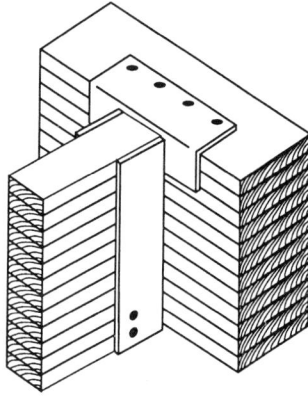

The design value in bending $F_b'$ of a member is calculated by applying all of the applicable adjustment factors to the tabular design value $F_b$. See Table 4.6 for applicability. In order to calculate the adjustment factor $C_L$, an interim design value in bending $F_b^*$ is needed where:

$F_b^*$ = $F_b$ multiplied by all applicable adjustment factors except $C_L$, $C_V$, and $C_{fu}$

Another value $F_{bE}$ is required for calculating $C_L$ [see Eq. (5-10)]. This is the Euler critical buckling design value for bending members.

The first step in design is to determine the unsupported length $l_u$ of the member, and then by use of Table 5.3 and Fig. 5.5, determine the effective length of the member $l_e$.

Next, determine the slenderness ratio $R_B$.

$$R_B = \sqrt{\frac{l_e/d}{b^2}} \qquad (5\text{-}9)$$

where  $l_e$ = the effective length (in.)
  $d$ = depth (in.), and
  $b$ = width (breadth) (in.).

The beam lateral stability factor is then calculated as follows:

$$C_L = \frac{1 + (F_{bE}/F_b^*)}{1.9} - \sqrt{\left(\frac{1 + F_{bE}/F_b^*}{1.9}\right)^2 - \left(\frac{F_{bE}/F_b^*}{0.95}\right)} \qquad (5\text{-}10)$$

where  $F_b^*$ = tabular bending design value multiplied by all applicable adjustment factors except $C_{fu}$, $C_V$, and $C_L$

  $F_{bE} = \dfrac{K_{bE}E'}{R_B^2}$

  $K_{bE}$ = 0.438 for wood products of $COV_E > 0.11$
  $K_{bE}$ = 0.609 for wood products with $COV_E \leq 0.11$

Machine stress rated (MSR) lumber and glued laminated timber with 6 or more laminations have a coefficient of variation of $E$ of 0.11 or less. Visually graded lumber has a $COV_E > 0.11$ unless a different $COV_E$ is designated on the MEL grade stamp.

The adjustment factor $C_{fu}$ is applied to members stressed in bending about the $y$-$y$ axis (whose breadth is greater than the depth), in which case $C_L = 1.0$. Therefore, calculation of $C_L$ for these members is unnecessary.

The volume factor $C_V$ and the beam stability factor $C_L$ are not accumulative. The smaller of the two calculated values apply.

## 5.4.12  Unsupported Length

When the compression edge of a beam is supported throughout its length to prevent its lateral displacement and the ends at points of bearing have lateral support to prevent lateral rotation, the effective length may be taken as zero and $C_L = 1$.

When lateral support is provided to prevent rotation at the points of bearing, but no other support to prevent lateral rotation or lateral displacement is provided

## TABLE 5.3

### Effective Length, $l_e$, for Bending Members

| CANTILEVER[a] | when $l_u/d < 7$ | when $l_u/d \geq 7$ |
|---|---|---|
| uniformly distributed load | $l_e = 1.33\ l_u$ | $l_e = 0.90\ l_u + 3d$ |
| concentrated load at unsupported end | $l_e = 1.87\ l_u$ | $l_e = 1.44\ l_u + 3d$ |

| SINGLE SPAN BEAM[a] | when $l_u/d < 7$ | when $l_u/d \geq 7$ |
|---|---|---|
| uniformly distributed load | $l_e = 2.06\ l_u$ | $l_e = 1.63\ l_u + 3d$ |
| concentrated load at center with no intermediate lateral support | $l_e = 1.80\ l_u$ | $l_e = 1.37\ l_u + 3d$ |
| concentrated load at center with lateral support at center | $l_e = 1.11\ l_u$ | |
| two equal concentrated loads at 1/3 points with lateral support at 1/3 points | $l_e = 1.68\ l_u$ | |
| three equal concentrated loads at 1/4 points with lateral support at 1/4 points | $l_e = 1.54\ l_u$ | |
| four equal concentrated loads at 1/5 points with lateral support at 1/5 points | $l_e = 1.68\ l_u$ | |
| five equal concentrated loads at 1/6 points with lateral support at 1/6 points | $l_e = 1.73\ l_u$ | |
| six equal concentrated loads at 1/7 points with lateral support at 1/7 points | $l_e = 1.78\ l_u$ | |
| seven or more equal concentrated loads, evenly spaced, with lateral support at points of load application | $l_e = 1.84\ l_u$ | |
| equal end moments | $l_e = 1.84\ l_u$ | |

Source: ANSI/NF$_o$PA NDS®—1991, *National Design Specification® for Wood Construction*, AFPA, 1250 Connecticut Ave., N.W., Washington, DC 20036.

[a] For single span or cantilever bending members with loading conditions not specified in Table 5.3:

$l_e = 2.06\ l_u$      when $l_u/d < 7$

$l_e = 1.63\ l_u + 3d$      when $7 \leq l_u/d \leq 14.3$

$l_e = 1.84\ l_u$      when $l_u/d > 14.3$

(a)  uniformly distributed load

(b)  concentrated load at center
       with no intermediate
          lateral support

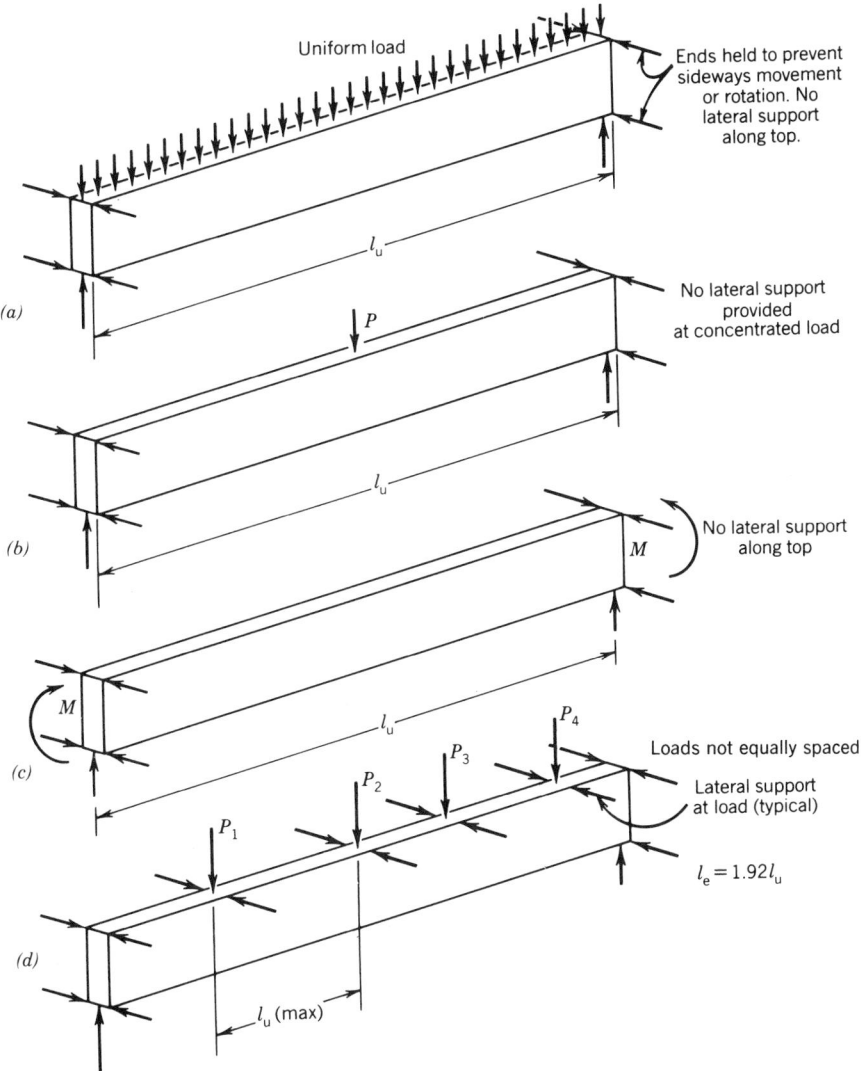

**FIGURE 5.5a–d**   Various conditions of lateral support.

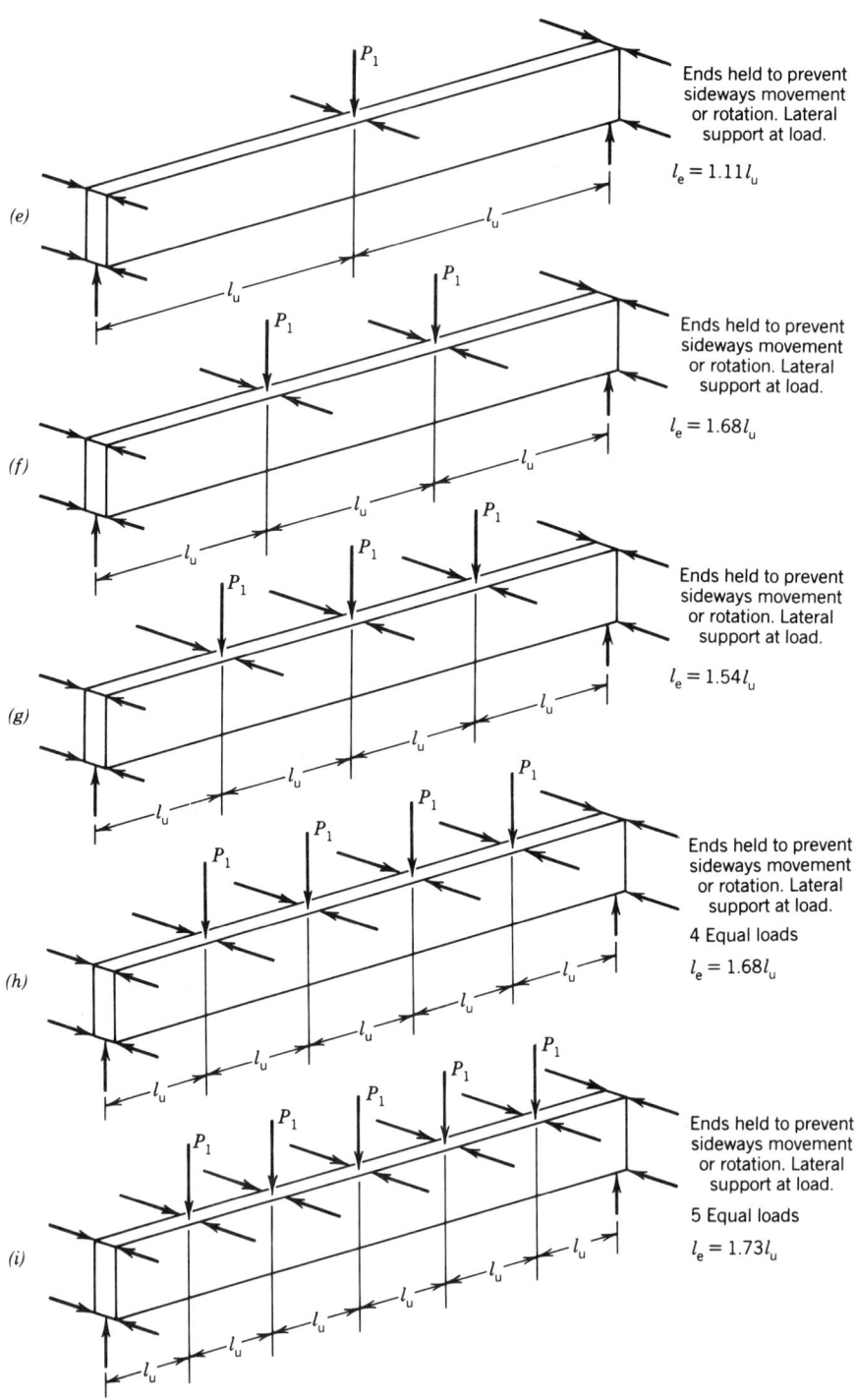

(e)

Ends held to prevent sideways movement or rotation. Lateral support at load.

$l_e = 1.11 l_u$

(f)

Ends held to prevent sideways movement or rotation. Lateral support at load.

$l_e = 1.68 l_u$

(g)

Ends held to prevent sideways movement or rotation. Lateral support at load.

$l_e = 1.54 l_u$

(h)

Ends held to prevent sideways movement or rotation. Lateral support at load.

4 Equal loads

$l_e = 1.68 l_u$

(i)

Ends held to prevent sideways movement or rotation. Lateral support at load.

5 Equal loads

$l_e = 1.73 l_u$

**FIGURE 5.5e–i**   Various conditions of lateral support.

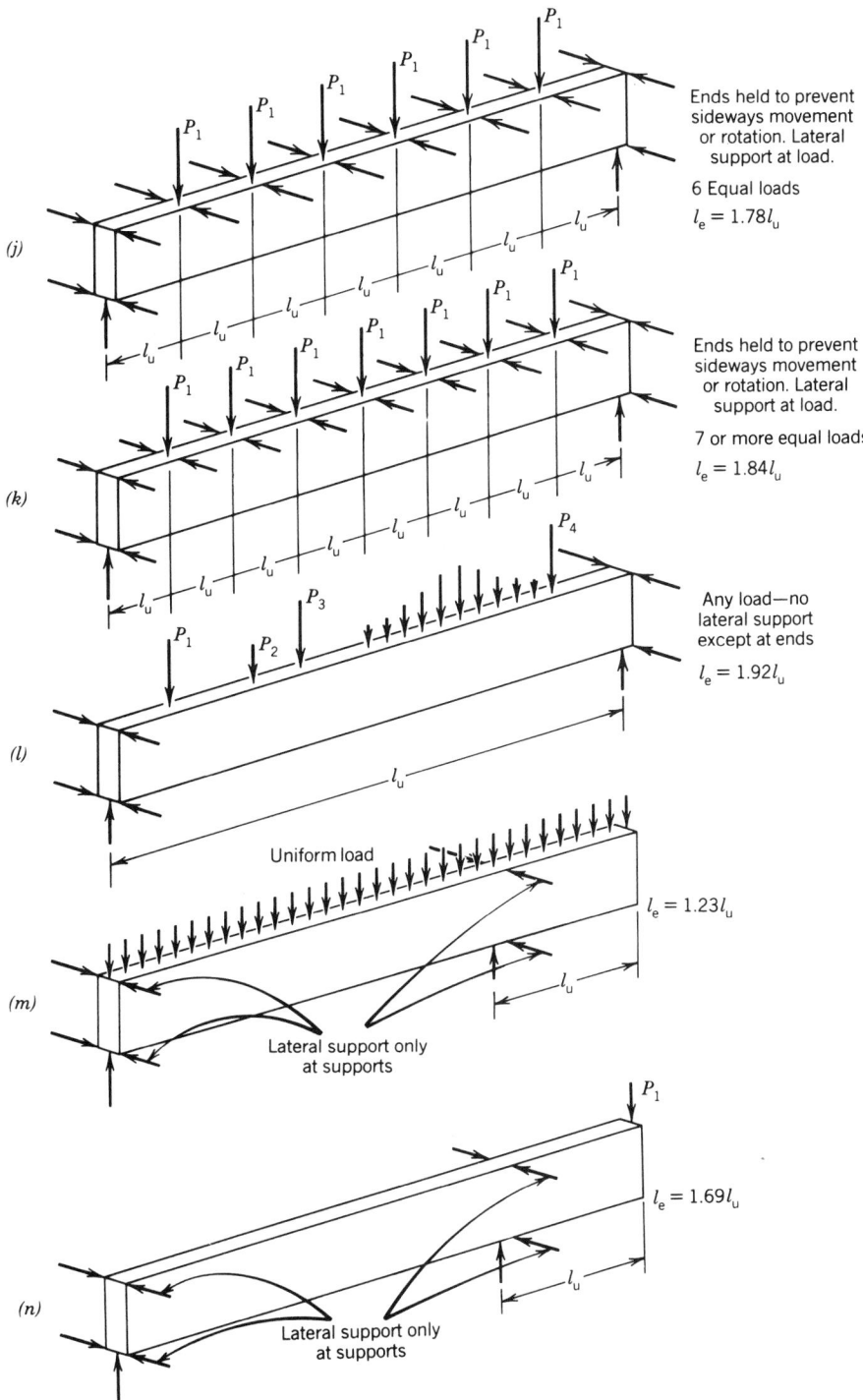

Ends held to prevent sideways movement or rotation. Lateral support at load.

6 Equal loads

$l_e = 1.78l_u$

(j)

Ends held to prevent sideways movement or rotation. Lateral support at load.

7 or more equal loads

$l_e = 1.84l_u$

(k)

Any load—no lateral support except at ends

$l_e = 1.92l_u$

(l)

Uniform load

$l_e = 1.23l_u$

(m)

Lateral support only at supports

$l_e = 1.69l_u$

(n)

Lateral support only at supports

**FIGURE 5.5j–n**   Various conditions of lateral support.

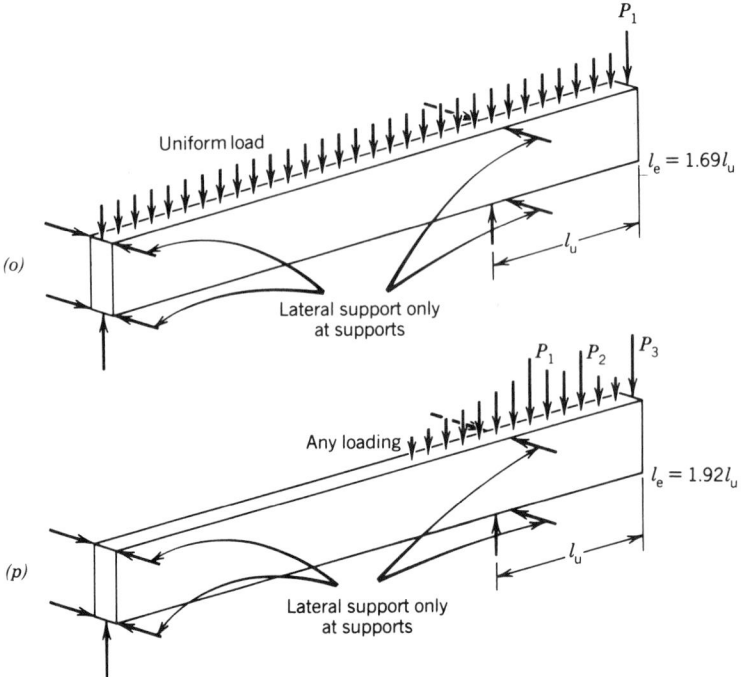

**FIGURE 5.5o–p**    Various conditions of lateral support.

throughout the length of a beam, the unsupported length is the distance between such points of bearing. See Fig. 5.5 for other conditions of support. Fig. 5.6 illustrates the unsupported length for cantilever beam conditions.

### 5.4.13    Design Values for Slender Beams

The tabular design value in bending $F_b$ must be multiplied by all applicable adjustment factors as shown in Table 4.6, including the beam stability factor $C_L$ when applicable.

The design value $F'_{bx}$ for glued laminated timber is equal to either $F^*_{bx} C_L$ or $F^*_{bx} C_V$; $C_L$ is not cumulative with $C_V$, and the smaller factor controls.

The unsupported length of a bending member $l_u$ is the distance between lateral supports of the member located at either the ends or at intermediate points as shown in Figs. 5.5 and 5.6. The effective span length $l_e$ is determined in accordance with Fig. 5.6.

When the beams are provided with lateral bracing to prevent both rotational and lateral displacement at intermediate points as well as at the ends, the unsupported length may be taken as the distance between such points of intermediate lateral bracing. If lateral displacement is not prevented at these points of intermediate support, the unsupported length must be defined as the full distance be-

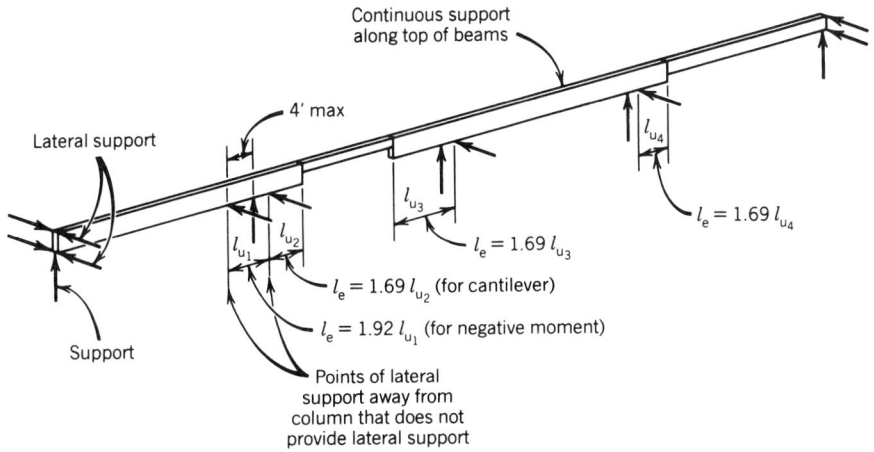

**FIGURE 5.6**    Effective lengths for various cantilevered support conditions.

tween points of bearing with adequate provision against rotation or as the full length of the cantilever.

Some of the means available for preventing lateral rotation of a beam at its points of bearing are to anchor the bottom of the beam to the wall or pilaster and the top of the beam to the parapet, to attach the roof diaphragm to the wall, and to provide a girt at the top of the wall and diagonal rod bracing for beams on wood columns with open sidewalls.

Connection of the roof sheathing to the compression side of the member is a good method of achieving lateral stability for a beam. A plywood diaphragm nailed to the beam is an example of such support. When plank decking is used, nailing patterns are most important so that nail couples will be created. A nailing pattern that contains only one nail per deck plank and no nails between planks does not provide a system with adequate lateral stability. If a wood deck is to supply such support, each piece must be securely nailed directly to the beam and to adjacent pieces to provide adequate diaphragm action. If adjacent deck planks are nailed to each other so that little or no differential movement can occur between planks, the decking will act as a diaphragm. If other kinds of decks are to provide such support, they must supply equivalent rigidity as a diaphragm.

When lateral movement of the compression edge is prevented by a continuous support, and the ends at points of bearing are held to prevent lateral rotation, there is no danger of lateral buckling and the design values require no reduction. Also, there is no need to limit the depth–breadth ratio to 5 or 6. When the depth of a beam does not exceed its breadth, no lateral stability is required, and the design value in bending is determined by applying the other applicable modifying factors to the tabular design values. When the depth of a beam exceeds the breadth, bracing must be provided at the points of bearing, and it must be so arranged

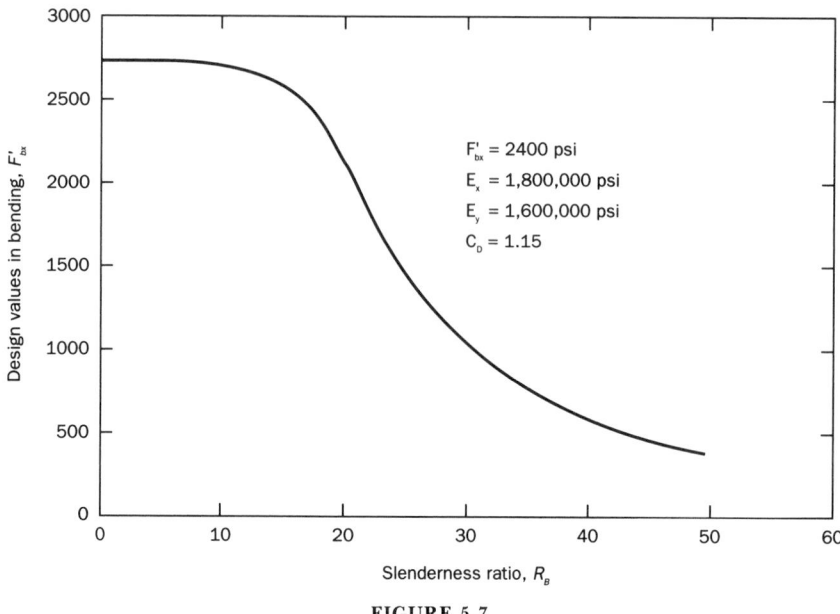

**FIGURE 5.7**

as to prevent lateral rotation of the beam at those points in a plane perpendicular to its longitudinal axis.

Recent research (6) has shown that simple-span glued laminated timber beams may be suspended by straps at the ends and carry the same amount of vertical load as a member with the bottom support fixed to prevent rotation. Such hangers have been successfully used on relatively deep and thin glued laminated timber purlins. Before using, the designer should consider uplift or other requirements of design that may require restraint of the hanger.

**Examples.**    The following three examples represent members with varying slenderness ratios. For each example, assume that the design situation requires a simple-span glued laminated timber beam having a clear span of 60 ft. A preliminary analysis indicates that the required bending moment is

$$M = 1,500,000 \text{ in.-lb}$$

The roof load is primarily snow and the duration of load factor $C_D = 1.15$. The member is used under dry service conditions and at normal temperatures. A 24F Douglas Fir-Larch glued laminated timber combination is used with an $E_x = E'_x = 1,800,000$ psi, $E_y = E'_y = 1,600,000$ psi, $F^*_b = (2400) 1.15 = 2760$ psi.

**Example 1.**    Assume the main beam is braced by joists placed 24 in. on centers with the only mode of lateral buckling taking place between the joists. For this

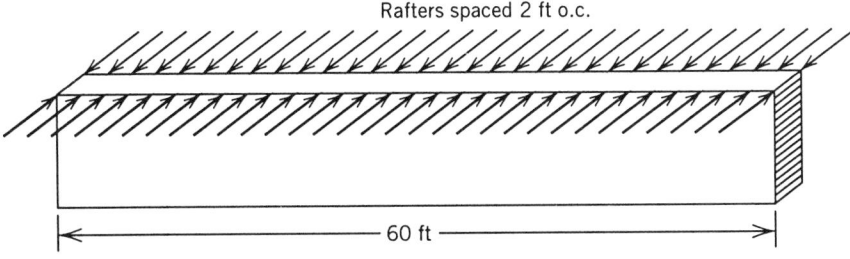

Rafters spaced 2 ft o.c.

60 ft

case, the effective length $l_e$ can be determined by the use of Table 5.3 and Fig. 5.5 for uniformly distributed load because this condition closely approximates a uniform load.

Assume depth $d = 30$ in., width $b = 5\frac{1}{8}$ in., and $S = 768.8$ in.[3]

$$l_u/d = \frac{24}{30} = 0.8$$

$$l_e = 2.06 l_u = (2.06)(24) = 49.44 \text{ in.}$$

$$R_B = \sqrt{\frac{l_e d}{b^2}} = \sqrt{\frac{(49.44)(30)}{5.125^2}} = 7.51$$

$$F_{bE} = \frac{K_{bE} E'_y}{R_B^2} = \frac{(0.609)(1,600,000)}{7.51^2} = 17,255 \text{ psi}$$

Note: Buckling is about the $y$-$y$ axis and $E'_y$ was used.

$$C_L = \frac{1 + F_{bE}/F_b^*}{1.9} - \sqrt{\left[\frac{1 + F_{bE}/F_b^*}{1.9}\right]^2 - \frac{F_{bE}/F_b^*}{0.95}}$$

$$F_{bE}/F_b^* = \frac{17,255}{2760} = 6.252$$

$$C_L = \frac{1 + 6.252}{1.9} - \sqrt{\left[\frac{1 + 6.252}{1.9}\right]^2 - \frac{6.252}{0.95}} = 0.991$$

The value of $C_L$ for $l_u = 2$ ft is very close to 1 (0.991). Unless extreme accuracy is necessary, the value of $C_L$ for lateral supports spaced 2 ft apart, $C_L$ can be considered to be equal to 1.0.

$C_V$ for a Douglas Fir beam $5\frac{1}{8}$ in. $\times$ 30 in. $\times$ 60 ft is

$$C_V = \left(\frac{5.125}{5.125}\right)^{1/10} \left(\frac{12}{30}\right)^{1/10} \left(\frac{21}{60}\right)^{1/10} = 0.822$$

$$0.822 < 0.991, \ C_V \text{ controls}$$

$$F'_b = F^*_b C_V = (2760)(0.822) = 2267 \text{ psi}$$

$$S_{\text{req'd}} = \frac{M}{F'_b} = \frac{1{,}500{,}000}{2267} = 661.7 \text{ in.}^3 < 768.8 \text{ in.}^3$$

Try $5\frac{1}{8}$ in. $\times$ $28\frac{1}{2}$ in. beam. $S = 693.8$ in.$^3$

$$C_V = \left(\frac{5.125}{5.125}\right)^{1/10} \left(\frac{12}{28.5}\right)^{1/10} \left(\frac{21}{60}\right)^{1/10} = 0.826$$

$$F'_b = (2760)(0.826) = 2279 \text{ psi}$$

$$S_{\text{req'd}} = \frac{1{,}500{,}000}{2279} = 658.2 \text{ in.}^3 < 693.8 \qquad \text{O.K.}$$

A check of a 27 in. deep beam indicated it was too small.

**Example 2.**   Assume the 60 ft Douglas Fir-Larch main beam is laterally braced by purlins spaced 60 in. o.c. with the only mode of lateral buckling taking place between the purlins.

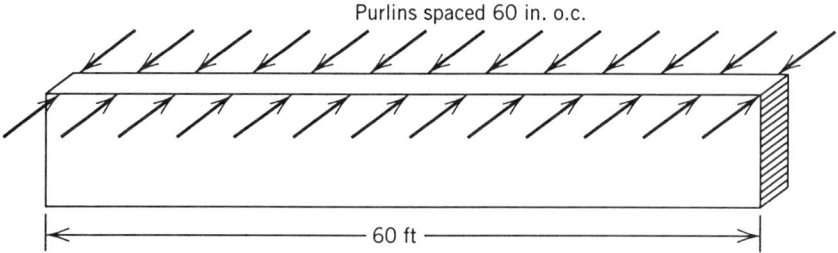

Purlins spaced 60 in. o.c.

60 ft

From Table 5.3, $l_e = 1.84l_u = 110.4$ in.

$$R_B = \sqrt{\frac{l_e d}{b^2}} = \sqrt{\frac{(110.4)(28.5)}{5.125^2}} = 10.94$$

$$F_{bE} = \frac{K_{bE} E'}{R_B^2} = \frac{(0.609)(1{,}600{,}000)}{10.94^2} = 8134 \text{ psi}$$

$$C_L = \frac{1 + 8134/2760}{1.9} - \sqrt{\left[\frac{1 + 8134/2760}{1.9}\right]^2 - \frac{8134/2760}{0.95}} = 0.976$$

$$C_V \text{ for Example 1} = 0.826 < 0.976, \, C_V \text{ controls}$$

$$F'_b = (2760)(0.826) = 2279 \text{ psi}$$

The beam size for this example is the same as for beam in Example 1. The volume factor frequently controls the design values in bending for glued laminated timber beams.

**Example 3.** Assume the 60 ft Douglas Fir-Larch beam is loaded with varying loads along its length that develop a moment of 1,500,000 in.-lb, but the beam is braced only at a point 25 ft from the left end, $l_u = 35$ ft.

Assume beam depth = 30 in.

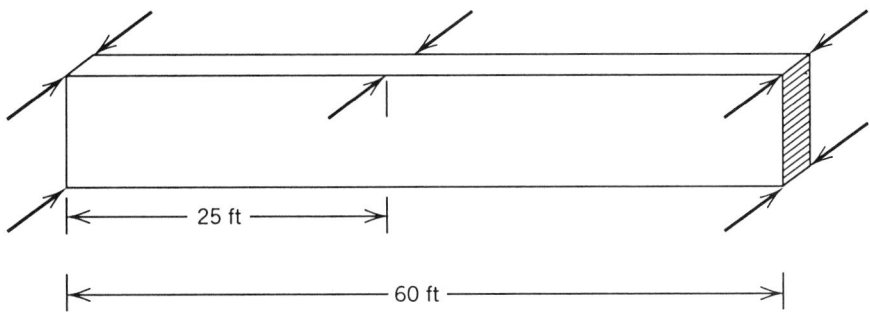

From Fig. 5.5, $l_e = (1.63)(l_u) + 3d = (1.63)(35)(12) + (3)(30) = 775$ in.

$$R_B = \sqrt{\frac{l_e d}{b^2}} = \sqrt{\frac{(775)(30)}{5.125^2}} = 29.74 < 50 \qquad \text{O.K.}$$

$$F_{bE} = \frac{K_{bE} E'}{R_B^2} = \frac{(0.609)(1,600,000)}{29.74^2} = 1101 \text{ psi}$$

$$C_L = \frac{1 + 1101/2760}{1.9} - \sqrt{\left[\frac{(1 + 1101/2760)}{1.9}\right]^2 - \frac{1101/2760}{0.95}} = 0.387$$

$$0.387 < 0.822, \ C_L \text{ controls}$$

$$F_b' = F_b^* C_L = (2760)(0.387) = 1068 \text{ psi}$$

$$S_{req'd} = \frac{M}{F_b'} = \frac{1,500,000}{1068} = 1404 \text{ in.}^3 > 768.8 \text{ in.}^3$$

The $5\frac{1}{8} \times 30$ in. beam is inadequate, and a larger beam must be used. A deeper $5\frac{1}{8}$ in. wide beam could be tried, but the $d/b$ ratio increases rapidly and a wider beam appears to be more appropriate where buckling is involved. Try a $6\frac{3}{4} \times 27$ in. beam. $S = 820.1$ in.[3]

$$R_B = \sqrt{\frac{(765.6)(27)}{6.75^2}} = 21.30$$

$$F_{bE} = \frac{(0.609)(1,600,000)}{21.30^2} = 2148 \text{ psi}$$

$$C_L = \frac{1 + 2148/2760}{1.9} - \sqrt{\left[\frac{1 + 2148/2760}{1.9}\right]^2 - \frac{2148/2760}{0.95}} = 0.698$$

$$C_V = \left(\frac{5.125}{6.75}\right)^{1/10} \left(\frac{12}{27}\right)^{1/10} \left(\frac{21}{60}\right)^{1/10} = 0.808$$

$$0.698 < 0.808, \; C_L \text{ controls}$$

$$F_b' = (2760)(0.698) = 1926 \text{ psi}$$

$$S_{\text{req'd}} = \frac{1,500,000}{1926} = 778.8 \text{ in.}^3 < 820.1 \text{ in.}^3 \qquad \text{O.K.}$$

Use a $6\frac{3}{4}$ in. $\times$ 27 in. beam.

## 5.5  TORSION

Torsional stresses are sometimes induced in timber members. Torsional loads are forces that are offset from the shear center of the section under consideration, which cause the members to twist. For square and rectangular members, the shear center is the same as the centroid of the cross section. The torsional stress in a member is the stress that resists the tendency of a torsional load to twist the member.

Some examples of designs where torsion might be significant are beams with heavy eccentric side loads, bridge stringers resisting guard rail loads, and utility towers.

The torsional stress $f_s$ in a rectangular section is

$$f_s = \frac{T(3a + 1.8b)}{8a^2b^2} \tag{5-11}$$

where  $f_s$ = maximum torsional stress at midpoint of each long side (psi),
    $T$ = applied torque (in.-lb),
    $a$ = one-half long side dimension (in.), and
    $b$ = one-half short side dimension (in.).

For a square section, the torsional stress is

$$f_s = \frac{4.8T}{s^3} \tag{5-12}$$

where  $s$ = dimension of side of square (in.).

There is limited information available on design values to be used when analyzing torsion in timber members. Until further research is done in this area, it is recommended that the design value in torsion, $F_s$, be limited to the same value recommended for radial stress.

If high torsional stresses are anticipated, the structural system should be modified to resist rotation and torsional stresses. Torsional twisting of girders may be resisted by fastening supported beams to the girders with connections that will resist the twist of the girder. These types of connections should be designed to account for shrinkage of the wood while providing the necessary rigidity.

**Example.** A $6\frac{3}{4}$ × 30-in. Douglas Fir-Larch glued laminated timber member is subjected to a torque $T$ of 5000 in.-lb. The design value for radial tension stress $F_{rt}$ is 15 psi.

$$2a = 30 \text{ in.} \qquad a = 15 \text{ in.}$$

$$2b = 6.75 \text{ in.} \qquad b = 3.375 \text{ in.}$$

$$f_s = \frac{T(3a + 1.8b)}{8a^2 b^2}$$

$$= \frac{5000\,[(3)(15) + (1.8)(3.375)]}{(8)(15)^2\,(3.375)^2}$$

$$= 12.4 \text{ psi} < F_{rt}\ (15 \text{ psi}) \qquad \text{O.K.}$$

## 5.6 DESIGN OF GLUED LAMINATED TIMBER BEAMS

### 5.6.1 Straight or Slightly Cambered Beams

Glued laminated timber beams should be designed with the ends held in place and the compression side supported so that no reduction for lack of lateral stability is required for simple-span beams. Cantilevered beams should be designed, proportioned, and braced so that the reduction for lateral stability is minimized. When it is not possible to provide this stability, the beam stability factor $C_L$ should be applied. Also, beams should be provided with adequate roof slope to drain the roof and prevent ponding.

For cost-efficient glued laminated timber beams, the stress values determined by design should be specified rather than a particular species or stress combination. It is recommended that the tabular bending stress be selected first and a trial size be determined based on bending. Other tabular design values should be checked based on this trial size. If the required tabular design value for any of these is higher than that which can be reasonably obtained, the size of the member should be increased. Sometimes this increased size will permit the use of a lower tabular design value in bending than was originally assumed. Table 1, AITC 117—Design (1), contains design values for laminating combinations for members intended to be used to resist bending loads.

These values are based on the bending strength of the combination of a given species. For instance, combination 24F–V3 SP is a combination made of visually graded Southern Pine with a tabular value in bending $F_{bx} = 2400$ psi; combination 24F–E1 WS is a combination made of E-rated Douglas Fir which is included as the category *Western Species* in the tables. The combinations listed have tabular design values in bending ranging from 1600 to 3000 psi. The designer should check on the availability of glued laminated timber with tabular design values in bending greater than 2400 psi.

### 5.6.2    Design Procedure for Glued Laminated Timber Roof Beams

The following steps are recommended in the design of a glued laminated timber beam:

1.    *Select the most practical framing system.* Simple-span beams of up to 50 ft in length are sufficiently economical to permit framing without interior columns or bearing walls. Where desired, long spans are competitive in cost with other construction materials. However, the cost is usually greater than that of systems using interior supports. For spans over 50 ft, systems using cantilever beams are often the most practical. Spacing of beams usually varies from 15 to 25 ft center to center with greater spacing possible with small glued laminated timber purlins or light trusses in place of conventional sawn purlins or joist framing. Fig. 5.8 shows a few examples of cantilever patterns that can be used in determining preliminary framing systems. However, the actual length of the cantilevered portion of the beam will depend on the ratio of the live load to dead load because most building codes require the investigation of maximum bending stresses that occur both when all spans are fully loaded with dead load plus live, snow, or ponding loads and also when only alternate spans are fully loaded and dead load alone is applied on unloaded spans.

2.    *Determine the design loads.*

(a)    Live load includes all applied loads superimposed by the occupancy and the use of the structure, except wind, snow, ponding, earthquake, or dead loads. If applied loads act continuously, they should be treated as permanent loads and considered to be the same as a dead load. Minimum live loads are usually established by the building codes.

Under the provisions of many building codes, certain reductions in live loads are allowed based on the tributary area of loading supported by the glued laminated timber. This is an important consideration that must be recognized during the preliminary design stage.

(b)    Dead load includes the weight of the glued laminated timber plus the weight of the appropriate portion of the structure that it supports as well as permanently applied loads including fixed service equipment such as heating, ventilating, and air-conditioning equipment.

(c)    Snow loads (where applicable) are those designated by the governing building code.

(d)    Wind loads (where applicable) are those designated by the governing building code.

For additional information on loads, see Chapter 4.

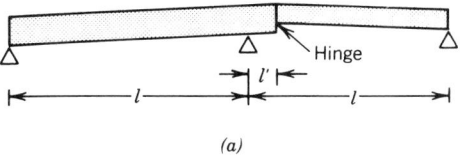

(a)

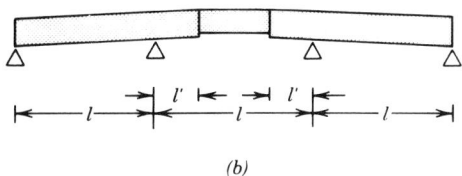

(b)

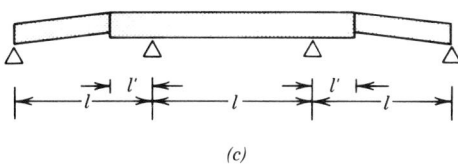

(c)

**FIGURE 5.8** Cantilever beam system. (a) Two-span cantilever system: Cantilevered beam extends over center support with the length of cantilever $l'$ equal to approximately $0.20 \times$ main support spacing $l$, (b) Three-span cantilever system: End members cantilevered over intermediate column supports and carrying the suspended beam. Length of cantilevers, $l'$ equal to approximately $0.25 \times$ main support spacing $l$, (c) Three-span cantilever system: Center member double-cantilevered over intermediate column supports and carrying the suspended wall beams. Length of cantilevers, $l'$ equal to approximately $0.17 \times$ main support spacing $l$.

3. *Select design values in bending.* Glued laminated timber beams are manufactured from several species of lumber, and consequently vary in availability regionally throughout the United States. As a general rule, availability can be increased by specifying the stresses required unless there are specific reasons for selecting a given species or combination. Table 1, AITC 117—Design (1), contains a list of combinations and corresponding design values that may be available. In general, glued laminated timber can be readily obtained with tabular design values in bending up to 2400 psi with design values of 2600 and 3000 psi available in some species. The design values in Table 1 are based on members subjected to loads of normal duration, dry conditions of use, normal temperature, members 12 in. in depth, members laterally supported, and straight prismatic members.

Combinations with a design value in bending of 2600 and 2800 psi can be obtained in Southern Pine in all sizes. Combinations with design values in bending 3000 psi can be obtained in nominal 4 in. and 6 in. widths. The designer should check on the availability of the higher strength combinations before specifying.

In most cases, the only stress adjustment factors necessary are duration of load

$C_D$ and volume factor $C_V$. Adjustment factors for other conditions are given in Chapter 4.

When glued laminated timbers are used under other conditions, adjustment to design values by applicable adjustment factors must be made.

4. *Determine size required by bending.* Use customary engineering procedures to determine the bending moment. Then, using the selected tabular design value in bending multiplied by applicable adjustment factors, determine the required section modulus. The actual size used depends on several factors. Members with the large depth–width ratios provide greater efficiency. However, practical considerations such as lateral stability, volume factor, excessive wall height, handling, and architectural considerations may limit the actual depth used. A normal relationship of depth to width for beams is between 4:1 and 5:1. Beams with an upper limit of approximately 9:1 can be used when a review of construction details permit such use. Thus, $5\frac{1}{8}$-in.-wide beams with depths up to $25\frac{1}{2}$ in. are very common, and depths up to a 36 in. may be used where adequate bracing is provided. Occasionally, beam depths are determined by architectural considerations rather than structural requirements. In those cases, where the bending stresses are very low, a larger depth–width ratio may be possible. Length limitations vary with the manufacturer's facilities, shipping conditions between the plant and job site, and site construction limitations. Beams up to 100 ft are used. When longer beams are needed, manufacturing and shipping conditions should be investigated.

5. *Determine the required tabular design value for shear parallel to grain $F_v$.* Determine the end reaction associated with shear $V$ and the required tabular design value for shear parallel to grain, $F_v$.

6. *Determine the required tabular design value for compression perpendicular to grain $F_{c\perp}$.* Determine the area in bearing, and then calculate the required tabular design value in compression perpendicular to grain. If the design value is higher than what may be reasonably available, the bearing area needs to be increased. Note that the duration-of-load adjustment factor $C_D$ does not apply to compression perpendicular to grain.

7. *Determine the required tabular design value for modulus of elasticity $E$.* Recommended deflection limitations for glued laminated timber roof beams are included in Table 4.3. The recommended deflection limitations are independent of the required camber. These are minimum recommendations, and it may be necessary under some circumstances to provide greater stiffness by increasing the member size.

Once the deflection limitation has been determined, the required tabular design value for $E$ is determined by use of the appropriate deflection equation as given in Table 4.2.

8. *Determine camber requirements.* Glued laminated beams can be manufactured with camber to compensate for the dead load deflection. For roof beams, this camber is in addition to any curvature that may used to provide a slope for drainage. Due to creep under long-term applied loads, the usual practice is to provide a minimum camber equal to 1.5 times dead-load deflection.

9. *Consider detailing requirements including roof drainage.* The plan for roof drainage should be detailed showing drainage paths. Downspouts and/or scuppers should be designed and detailed to handle the anticipated flow of water. In order to provide positive effective drainage and prevent ponding, a minimum slope of $\frac{1}{4}$ in./ft after dead-load deflection has taken place should be designed into the roof system (see Table 4.5). This slope must be considered when proposed camber and anticipated deflection are calculated. Additional slope or camber may be required where camber or deflection problems may occur. One method of improving drainage through scuppers and downspouts is to slope the ends of the beams with a tapered cut on the top face. Varying beam seat elevations is another effective means to improve drainage. See Section 5.15 for more information on ponding and the procedures necessary to prevent ponding.

10. *Specification.* The specification for glued laminated timber should include the following:

Size (width, depth, length)
Species
Camber
Appearance grade
Required design values:
    Bending $F_b$
    Shear $F_v$
Compression perpendicular to grain
    $F_{c\perp}$ (tension face)
    $F_{c\perp}$ (compression face)
Modulus of elasticity $E$

**Example.** Design a Southern Pine glued laminated timber beam to support the roof for an industrial building 40 × 60 ft as shown for dry conditions of use.

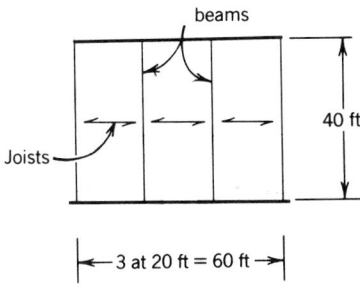

1. A simple joist framing system is selected with glued laminated timber beams spaced 20 ft apart spanning the 40-ft direction. Beams will be laterally braced at the ends and held in place along the top (compression side) by the roof decking. Therefore, no adjustment for lateral stability is required ($C_L = 1.0$).

2. Determine design loads. The controlling building code requires a mini-

mum roof live load of 20 psf. A 40% reduction in live load is permitted when the tributary area of a member exceeds 600 ft².

(a)   Assumed loads:

LL on joists = 20 psf (no snow)
Beam tributary area (40)(20) = 800 ft² > 600 ft²
LL on glued laminated timber beam, use (20)(1 − 0.40) = 12 psf (code reduction in load for tributary area)

$$DL = \quad 6.0 \text{ psf (roofing)}$$
$$+ \; 1.6 \text{ psf} \; (\tfrac{1}{2} \text{ in. plywood)}$$
$$+ \; 1.7 \text{ psf } (2 \times 10 \text{ joists at 24 in. on center)}$$
$$+ \; 4 \text{ psf (mechanical system)}$$
$$+ \; \underline{1.7} \text{ psf (glulam beam)}$$

Total DL = 15.0 psf
Total load = LL + DL
$$= 12 + 15 = 27 \text{ psf}$$
$w_{TL} = (27)(20) = 540 \text{ plf}$
$w_{DL} = (15)(20) = 300 \text{ plf}$

Note that some designers prefer to include additional dead load as a precaution to cover unanticipated loads or reroofing. However, this additional load should not be included in uplift calculations.

3.   Select design value in bending. Try $F_b$ = 2400 psi; duration of load factor $C_D$ = 1.25 for 7 day loading.

$$F_{bx}^* = (2400)(1.25) = 3000 \text{ psi}$$

4.   Determine size required by bending:

$$M = \frac{wL^2}{8} = \frac{(540)(40)^2(12)}{8} = 1{,}296{,}000 \text{ in.-lb}$$

Assume $C_V = 0.90$

$$S_{(req'd)} = \frac{M}{F_{bx}^* C_V} = \frac{1{,}296{,}000}{(3000)(0.90)} = 480 \text{ in.}^3$$

From Table 8.2 assuming a beam width of 5 in., try a 5 × 24¾ in. beam.

$$S = 510.5 \text{ in.}^3 \quad I = 6317 \text{ in.}^4 \quad A = 123.8 \text{ in.}^2$$

Determine $C_V$

$$C_V = \left(\frac{5.125}{b}\right)^{1/20} \left(\frac{12}{d}\right)^{1/20} \left(\frac{21}{L}\right)^{1/20}$$

$$= \left(\frac{5.125}{5}\right)^{1/20} \left(\frac{12}{24.75}\right)^{1/20} \left(\frac{21}{40}\right)^{1/20} = 0.935$$

$$S_{(req'd)} = \frac{1,296,000}{(3000)(0.935)} = 462.0 \text{ in.}^3 < 510.5 \text{ in.}^3 \quad \text{O.K.}$$

Use $5 \times 24\frac{3}{4}$ in. beam.

5. Determine required tabular design value for shear parallel to grain $F_v$. The effective length $L_e$ for determining shear parallel to grain is

$$L_e = L - \frac{2d}{12} = 40 - \frac{(2)(24.75)}{12} = 35.875 \text{ ft}$$

$$V = \frac{wL_e}{2} = \frac{(540)(35.875)}{2} = 9686 \text{ lb}$$

$$f_v = \frac{3V}{2A}, \quad F_v(\text{required}) = \frac{f_v}{C_D}$$

$$F_v(\text{required}) = \frac{(3)(9686)}{(2)(130.7)(1.25)} = 94 \text{ psi}$$

6. Determine the required tabular design value for compression perpendicular to grain, $F_{c\perp}$. Tension face: a typical commercial beam seat provides a length of 5 in.

$$\text{Area in bearing} = A = (5)(5) = 25 \text{ in.}^2$$

$$\text{Load on bearing plate} = P = \frac{wL}{2} = \frac{(540)(40)}{2} = 10,800 \text{ lb}$$

$$F_{c\perp}(\text{required}) = \frac{P}{A} = \frac{10,800}{25} = 432 \text{ psi (tension face)}$$

Compression face: the loads from the joists are transferred through joist hangers. A commercial joist hanger with a $1\frac{5}{16} \times 3\frac{1}{2}$-in. top plate to transfer the load will be used.

$$\text{Area in bearing} = A = (1.3125)(3.5) = 4.59 \text{ in.}^2$$

$$\text{Load on plate} = P = \frac{wL}{2} = \frac{(2)(27 - 1.7)(20)}{2} = 506 \text{ lb}$$

$$F_{c\perp}(\text{required}) = \frac{P}{A} = \frac{506}{4.59} = 110 \text{ psi (compression face)}$$

7. Determine the required tabular design value for modulus of elasticity $E$. The recommended deflection limitations for an industrial building from Table 4.3 are:

For live load plus dead load,

$$\Delta_{(LL+DL)} = \frac{l}{120} = \frac{(40)(12)}{120} = 4.0 \text{ in.}$$

For live load only,

$$\Delta_{LL} = \frac{l}{180} = \frac{(40)(12)}{180} = 2.67 \text{ in.}$$

$$\frac{LL}{TL} = \frac{12}{27} = 0.444$$

$$\frac{LL}{TL} \Delta_{(LL + DL)} = (0.444)(4.0) = 1.78 \text{ in.} < 2.67 \text{ in.}$$

Live load plus dead load controls.
For a uniform load,

$$\Delta = \frac{5wL^4}{384EI} \qquad E(\text{required}) = \frac{5wL^4}{384I\Delta}$$

$$E(\text{required}) = \frac{(5)(540)(40)^4(12)^3}{(384)(6317)(4.0)} = 1,231,000 \text{ psi}$$

The lowest tabular Modulus of Elasticity design value for a readily available 24F visual combination is 1,700,000 psi. Therefore, an $E$ of 1,700,000 psi will be specified and used to calculate dead-load deflection for camber.

8.  Determine camber requirements:

$$\Delta_{DL} = \frac{5w_{DL}L^4}{384EI} = \frac{(5)(300)(40)^4(12)^3}{(384)(1,700,000)(6317)} = 1.61 \text{ in.}$$

$$\text{Camber} = 1.5\Delta_{DL} = (1.5)(1.61) = 2.41 \text{ in. (use } 2\tfrac{1}{2} \text{ in.)}$$

9.  Roof drainage—The simplest method of roof drainage with straight beams in a building of this type is to drain the water to one side by varying the elevation of the beam seats so that a slope of at least $\tfrac{1}{4}$ in./ft is obtained. If the slope of the roof drainage is in the direction of the longitudinal axis of the glued laminated timbers, the difference in elevation is $(40)(\tfrac{1}{4}) = 10.0$ in.

10. Specification. Two 5 in. $\times$ $24\tfrac{3}{4}$ in. $\times$ 40 ft Southern Pine beams (note that length should be the exact out to out dimensions required)

Camber = $2\tfrac{1}{2}$ in.
Appearance grade—Industrial.
The members shall be made from a Southern Pine laminating combination that provides tabular design values equal to or exceeding the following:
Bending $F_{bx}$ = 2400 psi
Shear parallel to grain $F_v$ = 90 psi
Compression perpendicular to grain
    Tension face $F_{c\perp}$ = 650 psi
    Compression face $F_{c\perp}$ = 560 psi
Modulus of elasticity $E$ = 1,700,000 psi

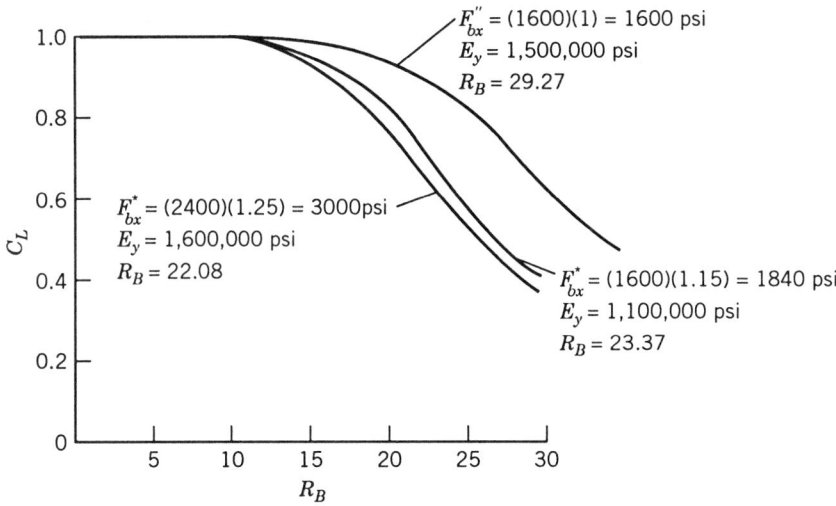

**FIGURE 5.9** Relationship of lateral stability factor $C_L$ to $R_B$.

Note: These values are lowest values for a 24$F$ Southern Pine combination meeting the required design values.

### 5.6.3 Designing for Lateral Stability

Usually, the most efficient method of roof framing is to provide continuous support to the glued laminated timbers with sheathing, decking, or joists and purlins spaced closely enough to provide lateral support. When this cannot be done, such as when purlins are installed over the top of the beam without direct connection between beam and roof sheathing, the lateral stability of the beam must be investigated.

See Figs. 5.5 and 5.6 for determining $l_e$ and Eq. (5-10) for determining the lateral stability factor $C_L$.

Fig. 5.9 shows the relationship between $R_B$ and $C_L$ for various values of $F_{bx}^*$ and $E_y$. The curves for other relationships will, in most cases, fall between the upper and lower curves. Note that the factors $C_V$ and $C_L$ are not cumulative, and the smaller of the two controls for bending about the $x$ axis. In many cases, $C_V$ is less than $C_L$ and no reduction for increased distances between points of support is required.

**Example.** Assume that all conditions and requirements of the previous example are the same, except that the roof system consists of nominal 4-in. wide purlins spaced 8 ft apart on top of the beams. Assume no change in dead loads. From Table 5.3 and Fig. 5.5, use condition $h$ as being closest to design application (four equal concentrated loads).

$$l_e = 1.68 l_u = (1.68)(8)(12) = 161.3 \text{ in.}$$

$$R_B = \sqrt{\frac{l_e d}{b^2}} = \sqrt{\frac{(161.3)(24.75)}{(5)^2}} = 12.64$$

$$F^*_{bx} = (2400)(1.25) = 3000 \text{ psi}$$

$$E'_y = 1,400,000 \text{ psi}$$

$E_y$ is used because lateral buckling is about the $y$–$y$ axis. Use Eq. (5-10) to calculate $C_L$.

$$C_L = \frac{1 + F_{bE}/F^*_b}{1.9} - \sqrt{\left[\frac{1 + (F_{bE}/F^*_b)}{1.9}\right]^2 - \frac{F_{bE}/F^*_b}{0.95}}$$

$$K_{bE} = 0.609$$

$$F_{bE} = \frac{K_{bE} E'}{R_B^2} = \frac{(0.609)(1,400,000)}{(12.64)^2} = 5340 \text{ psi}$$

$$C_L = \frac{1 + 5340/3000}{1.9} = \sqrt{\left[\frac{1 + (5340/3000)}{1.9}\right]^2 - \frac{5340/3000}{0.95}} = 0.946$$

From the previous example, $C_V = 0.935 < 0.946$, $C_V$ controls. In this case, changing the roof framing system resulted in the same size member because the volume factor still controlled design. Use a $5 \times 24\frac{3}{4}$ in. member.

### 5.6.4   Roof Beams with Tapered End Cuts

Taper cuts on the top at the end of beams are sometimes used to improve drainage, to provide extra head for downspouts and scuppers, to facilitate discharge of water, and to reduce the height of the wall. This sloping cut is commonly made

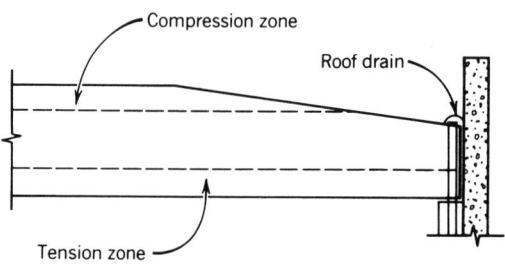

through the compression zone laminations exposing the lower strength core lamination and results in a member with a lower design value in bending than listed for the combination in Table 1, AITC 117—Design (1). The design values of $F_{bx}$, $E_x$, and $F_{c\perp}$ for taper cuts up to one-half the depth of the member can be obtained from Table 5.4. The variation in design values for combinations with the same design value in bending is caused by the use of various grades or species in the core of the combination. The value of $F_v$ remains unchanged. The values of $E_x$ and $F_{c\perp}$ are usually less because of the removal of some of the higher-grade lumber in the compression zone.

The laminated beam is usually designed as a straight prismatic member assuming full depth throughout its length, and the design values specified are based on this assumption. The effect of the sloping end cut is then checked, and required design values are modified if necessary. The length of taper cuts vary depending on the length of span, the roof-framing systems, or other requirements. Sloping cuts from 4 to 8 ft long are commonly used. The depth of the end cut usually depends on the drainage requirements. However, the remaining cross section at the end of the member must be able to resist the shear stresses. The sloping end cut is then checked by application of the interaction stress factor $C_I$ or use of the interaction equation [see Eq. (5-31)].

**TABLE 5.4**

**Design Values for Members with Taper Cuts on Compression Side**

| Combination Symbol | $F_{bx}^a$(psi) | $E_x^a$(million psi) | $F_{c\perp}^b$(psi) |
|---|---|---|---|
| VISUALLY GRADED WESTERN SPECIES | | | |
| 16F-V1 | 1350 | 1.3 | 255 |
| 16F-V2 | 1600 | 1.3 | 375 |
| 16F-V3 | 1600 | 1.5 | 560 |
| 16F-V4 | 850 | 1.3 | 255 |
| 16F-V6 | 1600 | 1.5 | 560 |
| 16F-V7 | 1600 | 1.3 | 375 |
| 20F-V1 | 1400 | 1.3 | 255 |
| 20F-V2 | 1900 | 1.5 | 375 |
| 20F-V3 | 2000 | 1.6 | 560 |
| 20F-V7 | 2000 | 1.6 | 560 |
| 20F-V8 | 2000 | 1.6 | 560 |
| 20F-V9 | 1900 | 1.5 | 375 |
| 20F-V10 | 1950 | 1.5 | 375 |
| 20F-V12 | 1850 | 1.3 | 470 |
| 22F-V1 | 1400 | 1.3 | 255 |
| 22F-V3 | 2200 | 1.7 | 560 |
| 22F-V8 | 2200 | 1.7 | 560 |

**TABLE 5.4**    (*Continued*)

| Combination Symbol | $F^a_{bx}$(psi) | $E^a_x$(million psi) | $F^b_{c\perp}$(psi) |
|---|---|---|---|
| VISUALLY GRADED WESTERN SPECIES (*Cont.*) | | | |
| 22F-V10 | 1900 | 1.5 | 500 |
| 24F-V1 | 1450 | 1.3 | 255 |
| 24F-V2 | 1900 | 1.5 | 375 |
| 24F-V4 | 2200 | 1.7 | 560 |
| 24F-V5 | 2100 | 1.6 | 375 |
| 24F-V8 | 2200 | 1.7 | 560 |
| 24F-V10 | 2000 | 1.6 | 375 |
| 24F-V11 | 2100 | 1.5 | 500 |
| E-RATED WESTERN SPECIES | | | |
| 16F-E1 | 1400 | 1.3 | 255 |
| 16F-E2 | 1500 | 1.3 | 375 |
| 16F-E3 | 850 | 1.6 | 560 |
| 16F-E6 | 1500 | 1.6 | 560 |
| 16F-E7 | 1500 | 1.3 | 375 |
| 20F-E1 | 1400 | 1.3 | 255 |
| 20F-E2 | 1900 | 1.5 | 375 |
| 20F-E3 | 2000 | 1.6 | 560 |
| 20F-E6 | 2000 | 1.6 | 560 |
| 20F-E7 | 1850 | 1.5 | 375 |
| 24F-E1 | 2300 | 1.8 | 560 |
| 24F-E2 | 2000 | 1.6 | 375 |
| 24F-E3 | 2000 | 1.6 | 375 |
| 24F-E4 | 2300 | 1.7 | 560 |
| 24F-E5 | 2300 | 1.8 | 560 |
| 24F-E6 | 1450 | 1.3 | 255 |
| 24F-E10 | 2300 | 1.8 | 560 |
| 24F-E11 | 2000 | 1.6 | 375 |
| 24F-E13 | 2300 | 1.7 | 560 |
| 24F-E14 | 2300 | 1.8 | 560 |
| 24F-E15 | 1800 | 1.5 | 375 |
| 24F-E17 | 1400 | 1.3 | 315 |
| 24F-E18 | 2100 | 1.7 | 560 |
| 24F-E20 | 1800 | 1.5 | 350 |
| VISUALLY GRADED SOUTHERN PINE | | | |
| 16F-V2 | 1500 | 1.4 | 560 |
| 16F-V3 | 1550 | 1.4 | 560 |
| 16F-V5 | 1500 | 1.4 | 560 |
| 20F-V2 | 1600 | 1.5 | 560 |
| 20F-V3 | 2000 | 1.4 | 560 |
| 20F-V4 | 1150 | 1.4 | 470 |

## TABLE 5.4 (*Continued*)

| Combination Symbol | $F_{bx}^{a}$ (psi) | $E_{x}^{a}$ (million psi) | $F_{c\perp}^{b}$ (psi) |
|---|---|---|---|
| VISUALLY GRADED SOUTHERN PINE (*Cont.*) | | | |
| 20F-V5 | 1550 | 1.5 | 560 |
| 22F-V1 | 2100 | 1.5 | 560 |
| 22F-V2 | 2100 | 1.4 | 560 |
| 22F-V3 | 1600 | 1.5 | 560 |
| 22F-V4 | 1200 | 1.5 | 470 |
| 22F-V5 | 2100 | 1.5 | 560 |
| 24F-V1 | 1650 | 1.6 | 560 |
| 24F-V3 | 2300 | 1.6 | 560 |
| 24F-V4 | 1200 | 1.5 | 470 |
| 24F-V5 | 2300 | 1.6 | 560 |
| 26F-V1 | 2300 | 1.6 | 560 |
| 26F-V2 | 2400 | 1.8 | 650 |
| 26F-V3 | 2400 | 1.8 | 560 |
| 26F-V4 | 2400 | 1.8 | 560 |
| E-RATED SOUTHERN PINE | | | |
| 16F-E1 | 1600 | 1.6 | 560 |
| 16F-E3 | 1600 | 1.6 | 560 |
| 20F-E1 | 2000 | 1.6 | 560 |
| 20F-E3 | 2000 | 1.6 | 560 |
| 22F-E1 | 2200 | 1.7 | 560 |
| 22F-E3 | 2100 | 1.6 | 560 |
| 24F-E1 | 2200 | 1.7 | 560 |
| 24F-E2 | 2400 | 1.7 | 560 |
| 24F-E4 | 2200 | 1.7 | 560 |
| 28F-E1 | 2600 | 1.8 | 560 |
| 28E-E2 | 2600 | 1.8 | 560 |
| 30F-E1 | 2600 | 1.8 | 560 |
| 30F-E2 | 2600 | 1.8 | 560 |
| CALIFORNIA REDWOOD | | | |
| B-16F-V1 | 1600 | 1.1 | 315 |

[a]Design value applicable to members that have up to one-half the depth on the compression side removed by taper cutting. Values are for normal duration of load, dry conditions of use, and 12 in. or less depth.

[b]Design value in compression perpendicular to grain for the core laminations of the combinations.

**Example.**    Redesign the beam selected in the previous example of the joist roof system for a tapered end cut 6 ft long with a depth of cut $7\frac{1}{2}$ in. on the end as shown:

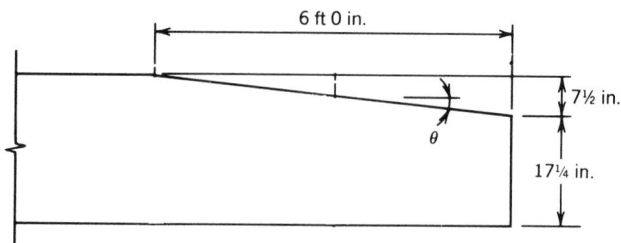

6 ft 0 in.

$7\frac{1}{2}$ in.

$17\frac{1}{4}$ in.

$\theta$

1.    Determine the required design value in shear.

$$L_e = 40 - \frac{(2)(17.25)}{12} = 37.125 \text{ ft}$$

$$V = \frac{(540)(37.125)}{2} = 10,024 \text{ lb}$$

$$F_{v(\text{required})} = \frac{3V}{2AC_D} = \frac{(3)(10,024)}{(2)(5)(17.25)(1.25)} = 139 \text{ psi}$$

2.    Determine the interaction factor $C_I$.

$$\tan \theta = \frac{7.5}{(6)(12)} = 0.1045$$

Assume trial design values in bending, shear, and compression perpendicular to grain for the tapered surface. $F_v = 200$ psi; $F_b = 1600$ psi; and $F_{c\perp} = 560$ psi for the tapered slope. This will permit most of the 24$F$ visually graded Southern Pine combinations to be used. From Table 5.5, $C_I = 0.7687$. [If Table 5.5 does not contain the combination of design values selected, $C_I$ can be calculated by use of the interaction equation (5-31)].

3.    Check bending stresses at Sections A and B.

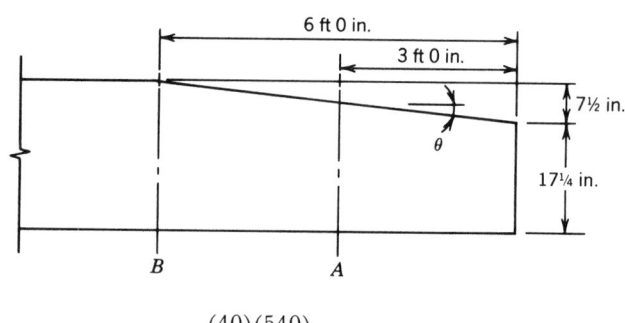

6 ft 0 in.

3 ft 0 in.

$7\frac{1}{2}$ in.

$\theta$

$17\frac{1}{4}$ in.

$B$

$A$

$$R = \frac{(40)(540)}{2} = 10,800 \text{ lb}$$

*Section A:*

$$M_A = (10,800)(3)(12) - \frac{(540)(3)^2(12)}{2} = 359,640 \text{ in.-lb}$$

Depth at $A = d = 24.75 - \frac{7.5}{2} = 21.0$ in.

$$S_A = \frac{bd^2}{6} = \frac{(5)(21.0)^2}{6} = 367.5 \text{ in.}^3$$

$$f_{bA} = \frac{M_A}{S_A} = \frac{359,640}{367.5} = 979 \text{ psi}$$

The volume factor $C_V$ for Section A-A is calculated on the basis of the depth $d$ at A-A, the width of the beam, and the entire length of the beam. Assume a Southern Pine species will be used.

$$C_V = \left(\frac{5.125}{b}\right)^{1/20} \left(\frac{12}{d}\right)^{1/20} \left(\frac{21}{L}\right)^{1/20}$$

$$= \left(\frac{5.125}{5}\right)^{1/20} \left(\frac{12}{21.0}\right)^{1/20} \left(\frac{21}{40}\right)^{1/20} = 0.943$$

Since $0.943 > 0.767$, $C_I$ controls ($C_I$ and $C_V$ are not cumulative)

$$F'_{bx} = F^*_{bx} C_I = (1600)(1.25)(0.767) = 1534 \text{ psi} > 979 \text{ psi}\quad \text{O.K.}$$

*Section B:*

$$M_B = (10,800)(6)(12) - \frac{(540)(6)^2(12)}{2} = 660,960 \text{ in.-lb}$$

$$S_B = 510.5 \text{ in.}^3 \text{ (from Table 8.2)}$$

$$f_{bB} = \frac{M_B}{S_B} = \frac{660,960}{510.5} = 1295 \text{ psi}$$

$$C_V \text{(for Section B-B)} = \left(\frac{5.125}{5.0}\right)^{1/20} \left(\frac{12}{24.75}\right)^{1/20} \left(\frac{21}{40}\right)^{1/20} = 0.935$$

$0.935 > 0.767$, $C_I$ controls.

$$F'_{bx} = F^*_{bx} C_I = 2000(0.767) = 1534 \text{ psi} > 1294 \text{ psi}\quad \text{O.K.}$$

Tapering on the end does not significantly affect the deflection, and the same deflection calculations used for an untapered beam may be used.

Camber $= 2\frac{1}{2}$ in.
Appearance grade $=$ Industrial

## TABLE 5.5

### Interaction Stress Factor $C_I$ as a Function of $\tan \theta^a$

| $\tan \theta$ | $C_I^b$ | $C_I^c$ | $C_I^d$ | $C_I^e$ | $C_I^f$ | $C_I^g$ | $C_I^h$ | $C_I^i$ | $C_I^j$ | $C_I^k$ | $C_I^l$ |
|---|---|---|---|---|---|---|---|---|---|---|---|
| 0.02 | 0.960 | 0.972 | 0.962 | 0.966 | 0.977 | 0.972 | 0.968 | 0.981 | 0.982 | 0.979 | 0.987 |
| 0.025 | 0.940 | 0.958 | 0.942 | 0.949 | 0.964 | 0.957 | 0.952 | 0.970 | 0.972 | 0.968 | 0.981 |
| 0.03 | 0.910 | 0.941 | 0.920 | 0.928 | 0.950 | 0.940 | 0.933 | 0.958 | 0.960 | 0.955 | 0.972 |
| 0.035 | 0.891 | 0.922 | 0.896 | 0.906 | 0.933 | 0.921 | 0.911 | 0.944 | 0.947 | 0.940 | 0.963 |
| 0.04 | 0.864 | 0.902 | 0.870 | 0.882 | 0.915 | 0.900 | 0.889 | 0.928 | 0.932 | 0.924 | 0.952 |
| 0.045 | 0.837 | 0.880 | 0.843 | 0.857 | 0.896 | 0.878 | 0.865 | 0.912 | 0.917 | 0.907 | 0.941 |
| 0.05 | 0.809 | 0.857 | 0.815 | 0.832 | 0.876 | 0.855 | 0.840 | 0.894 | 0.900 | 0.889 | 0.928 |
| 0.055 | 0.781 | 0.835 | 0.788 | 0.806 | 0.856 | 0.832 | 0.815 | 0.876 | 0.882 | 0.870 | 0.915 |
| 0.06 | 0.753 | 0.811 | 0.761 | 0.781 | 0.835 | 0.809 | 0.791 | 0.857 | 0.864 | 0.850 | 0.901 |
| 0.065 | 0.727 | 0.788 | 0.735 | 0.756 | 0.813 | 0.785 | 0.766 | 0.838 | 0.846 | 0.830 | 0.887 |
| 0.07 | 0.701 | 0.766 | 0.709 | 0.731 | 0.792 | 0.762 | 0.742 | 0.819 | 0.827 | 0.810 | 0.872 |
| 0.075 | 0.676 | 0.743 | 0.685 | 0.707 | 0.771 | 0.740 | 0.718 | 0.800 | 0.809 | 0.790 | 0.857 |
| 0.08 | 0.652 | 0.721 | 0.661 | 0.684 | 0.751 | 0.718 | 0.696 | 0.781 | 0.790 | 0.771 | 0.842 |
| 0.085 | 0.629 | 0.700 | 0.638 | 0.662 | 0.730 | 0.696 | 0.674 | 0.762 | 0.772 | 0.751 | 0.827 |
| 0.09 | 0.607 | 0.679 | 0.616 | 0.640 | 0.710 | 0.676 | 0.652 | 0.743 | 0.753 | 0.732 | 0.811 |
| 0.095 | 0.586 | 0.659 | 0.595 | 0.619 | 0.691 | 0.656 | 0.632 | 0.725 | 0.735 | 0.714 | 0.796 |

| 0.100 | 0.566 | 0.640 | 0.576 | 0.600 | 0.672 | 0.636 | 0.612 | 0.707 | 0.718 | 0.695 | 0.781 |
|-------|-------|-------|-------|-------|-------|-------|-------|-------|-------|-------|-------|
| 0.105 | 0.548 | 0.621 | 0.557 | 0.581 | 0.654 | 0.618 | 0.593 | 0.689 | 0.701 | 0.678 | 0.765 |
| 0.110 | 0.530 | 0.604 | 0.539 | 0.563 | 0.637 | 0.600 | 0.576 | 0.672 | 0.684 | 0.661 | 0.750 |
| 0.115 | 0.513 | 0.586 | 0.522 | 0.546 | 0.620 | 0.583 | 0.558 | 0.656 | 0.667 | 0.644 | 0.736 |
| 0.120 | 0.497 | 0.570 | 0.506 | 0.530 | 0.604 | 0.566 | 0.542 | 0.640 | 0.652 | 0.628 | 0.721 |
| 0.125 | 0.482 | 0.554 | 0.491 | 0.514 | 0.588 | 0.551 | 0.526 | 0.624 | 0.636 | 0.612 | 0.707 |
| 0.150 | 0.416 | 0.485 | 0.424 | 0.447 | 0.518 | 0.482 | 0.458 | 0.554 | 0.566 | 0.542 | 0.640 |
| 0.200 | 0.325 | 0.384 | 0.331 | 0.351 | 0.413 | 0.381 | 0.360 | 0.446 | 0.458 | 0.435 | 0.529 |

[a] Applicable when taper cut is on compression side only.

[b] $F_b = 2400$ psi; $F_v = 165$ psi; $F_{c\perp} = 560$ psi.

[c] $F_b = 2400$ psi; $F_v = 200$ psi; $F_{c\perp} = 560$ psi.

[d] $F_b = 2200$ psi; $F_v = 155$ psi; $F_{c\perp} = 375$ psi.

[e] $F_b = 2200$ psi; $F_v = 165$ psi; $F_{c\perp} = 560$ psi.

[f] $F_b = 2200$ psi; $F_v = 200$ psi; $F_{c\perp} = 560$ psi.

[g] $F_b = 2000$ psi; $F_v = 165$ psi; $F_{c\perp} = 560$ psi.

[h] $F_b = 2000$ psi; $F_v = 155$ psi; $F_{c\perp} = 375$ psi.

[i] $F_b = 2000$ psi; $F_v = 200$ psi; $F_{c\perp} = 560$ psi.

[j] $F_b = 1600$ psi; $F_v = 165$ psi; $F_{c\perp} = 560$ psi.

[k] $F_b = 1600$ psi; $F_v = 155$ psi; $F_{c\perp} = 375$ psi.

[l] $F_b = 1600$ psi; $F_v = 200$ psi; $F_{c\perp} = 560$ psi.

Bending $F_{bx}$ = 2400 psi
Shear parallel to grain $F_v$ = 200 psi
Tension face $F_{c\perp}$ = 560 psi
Compression face $F_{c\perp}$ = 560 psi
Modulus of elasticity $E$ = 1,700,000 psi

In addition, the interior core laminations shall provide the following minimum design values along the sloping cut:

Bending $F_{bx}$ = 1600 psi
Compression face $F_{c\perp}$ = 560 psi
Modulus of elasticity $E$ = 1,500,000 psi

4.   Review assumed design values. If the assumed design values along the tapered cut are too high or too low, other assumptions can be made and a new size determined. The required compression perpendicular to grain stress under the joist hangers is 110 psi, and the design value of $F_{c\perp}$ of 560 psi is adequate. The modulus of elasticity is slightly reduced in the tapered section, but because the length of taper is small, the effect is negligible and can be ignored (when a fully tapered beam is used, a reduced $E$ value may be necessary).

5.   Check specification. Tapering the beam at the ends requires a change in the specification of $F_v$ and the additional provision for design values along the sloping cut as follows:

Specification—Two 5 in. × $24\frac{3}{4}$ in. × 40 ft Southern Pine beams taper cut on one end as shown on the drawing.

Camber = $2\frac{1}{2}$ in.
Appearance grade = Industrial
Bending $F_{bx}$ = 2400 psi
Shear parallel to grain $F_v$ = 200 psi
Compression perpendicular to grain
Tension face $F_{c\perp}$ = 560 psi
Compression face $F_{c\perp}$ = 560 psi
Modulus of elasticity $E_x$ = 1,700,000 psi
                        $E_y$ = 1,500,000 psi

In addition, the interior core laminations shall provide the following minimum design values along the sloping cut:

Bending $F_{bx}$ = 1600 psi
Compression perpendicular to grain $F_{c\perp}$ = 560 psi

## 5.6.5   Design Procedure for Glued Laminated Timber Floor Beams

The following steps are recommended for the design of floor beams. They are similar to those for the design of roof beams, except that the order of calculations is different. This is because deflection is more critical in the design of floor beams and the check of modulus of elasticity is recommended after the size required by bending is determined. Table 4.3 contains the traditional deflection limitations for floor beams that meet most code requirements. When the length of spans increase and when a more rigid floor system is desired for certain uses such as commercial, office, and institutional buildings, these limits may not provide the desired stiffness and freedom from vibration. The deflection limitations for floors in Table 4.4 are suggested for such uses. Requirements for determination of floor performance vary with the intended use, and the design requirements are usually subjective. The deflection limitations included in Table 4.4 have been found by experience to provide floors that meet the more restrictive requirements. In addition to considering the deflection limitations of the glued laminated timber girders and joists, attention must also be given to the material spanning between the joists. The recommended $1\frac{1}{8}$-in. plywood on joists spaced 2 ft on centers as shown in the example is more than required by codes but is recommended in order to provide additional stiffness.

1.   Select the most practical framing system. Simple-span girders with spans of up to 35 or 40 ft are practical for floor-framing systems that provide a reasonably stiff floor for use in commercial, office, or institutional applications. Girders may be spaced up to about 26 ft on center with glued laminated timber joists spanning between them. These joists may be spaced 2 ft on centers with $1\frac{1}{8}$-in.-thick plywood and lightweight concrete topping over the joists. Added stiffness is obtained by gluing the plywood to both the joists and girders. This provides extra stiffness; however, the deflection limitations shown in Table 4.4 were based on excluding the stiffness provided by the plywood from the computations. Cantilevered girder or joist systems are generally not recommended for floor systems because of difficulties in cambering the members as well as avoiding transmission of floor vibrations between sections of the building.

2.   Determine the design loads. Minimum floor live loads are generally specified by the applicable building code. In some cases, it may be advisable to design the floor for live loads higher than the minimums recommended by the code in order to obtain a more rigid floor. Many building codes provide for reductions in live loads based on the tributary area and dead load/live load ratio.

3.   Select design value in bending. For glued laminated timber girders, design value in bending, an $F_b$ of 2400 or 2200 psi is recommended. Because of the many glued laminated timber joists used in floor systems, it is recommended that they be designed with a 16F laminating combination.

The design values in Table 1, AITC 117—Design (1), are based on members subjected to loads of normal duration, dry conditions of use, and normal temperatures and members that are laterally supported. All of these conditions usually apply to floor beams and no adjustments for other conditions will be needed. The design values are also based on members 12 in. deep. For deeper members, the volume factor $C_V$ should be applied.

4.   Determine size required by bending. An adjusted section modulus that includes the volume factor can be calculated by dividing the bending moment by the design value in bending selected in step 3.

5.    Determine the required tabular design value for modulus of elasticity $E$. Recommended limitations for deflection of floor beams are included in Tables 4.3 and 4.4. These recommendations are minimums, and it may be necessary to apply more restrictive limitations where stiffer floors are required.

6.    Determine the required tabular design value for shear parallel to grain $F_v$.

7.    Determine the required tabular design value for compression perpendicular to grain, $F_{c\perp}$.

8.    Determine camber requirements.

9.    Specification.

**Example.**    Design a glued laminated timber joist and girder system using Western Species for the second floor bay of an office building 48 ft × 35 ft as shown:

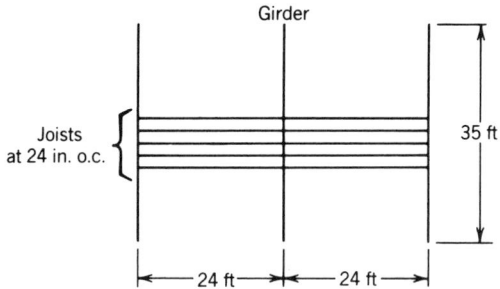

1.    The framing system selected is for girders spanning 35 ft and joists 24 in. on centers framing into each side of the girder. The structural floor consists of $1\frac{1}{8}$-in. plywood topped with lightweight concrete.

2.    Determine design loads.

Dead Loads on joists (see Table 8.15)

| | |
|---|---:|
| Carpet | 1 psf |
| Lightweight concrete topping ($\frac{3}{4}$ in.) | $6\frac{1}{2}$ psf |
| $1\frac{1}{8}$-in. plywood | $3\frac{1}{2}$ psf |
| Joist (glued laminated timber) | 6 psf |
| Mechanical allowance | 4 psf |
| Suspended ceiling | 2 psf |
| Partition allowance | 20 psf |
| For joist design, DL = | 43 psf |
| Add 2 psf for girder weight | 2 psf |
| For girder design, DL = | 45 psf |
| Live load on joists | 50 psf |

The applicable building code permits the following live load-reduction, $R$, for the girder:

$$R = r(A - 150) = (0.0008)[(35)(24) - 150] = 0.552 = 55.2\%$$

$$R = 23.1 \left( 1 + \frac{DL}{LL} \right) = (23.1)\left( 1 + \frac{45}{50} \right) = 43.9\%$$

$$R_{max} = 40\%$$

$$LL(girder) = (50)(1 - 0.40) = 30 \text{ psf}$$

where  $R$ = live load reduction in %
  $r$ = rate of reduction (0.08% for floors), and
  $A$ = area supported by member (ft$^2$).

3.  Select design value in bending for joists. Because of the close spacing and for economy in design, try 1600 psi.

$$F^*_{bx} = F_{bx} = 1600 \text{ psi}$$

4.  Determine size required by bending:

$$M = \frac{wL^2}{8} = \frac{(93)(2)(24)^2(12)}{8} = 160{,}700 \text{ in.-lb}$$

$C_L = 1.0$, floor sheathing provides continuous support of the compression side of the joists.

Assume the volume factor $C_V = 0.95$

$$S(\text{required}) = \frac{160{,}700}{(1600)(0.95)} = 95.42 \text{ in.}^3$$

From Table 8.2, assuming a $2\frac{1}{2}$ in. width, try a $2\frac{1}{2} \times 16\frac{1}{2}$-in. joist:

$$S = 113.4 \text{ in.}^3 \quad I = 935.9 \text{ in.}^4 \quad A = 41.25 \text{ in.}^2$$

Check volume factor. For Western Species:

$$C_V = \left( \frac{5.125}{2.5} \right)^{1/10} \left( \frac{12}{16.5} \right)^{1/10} \left( \frac{21}{24} \right)^{1/10} = 1.027$$

$$1.027 > 1 \text{ (use 1.0)}$$

$$S(\text{required}) = \frac{160{,}700}{1600} = 100.4 \text{ in.}^3$$

$$100.4 \text{ in.}^3 < 113.4 \text{ in.}^3$$

$$\text{Use } 2\frac{1}{2} \times 16\text{-}\frac{1}{2} \text{ in. joist.}$$

5.  Check the stiffness of the joists without considering the composite behavior between the joist and the plywood floor sheathing due to gluing of the plywood to the joists. The recommended deflection limitations from Table 4.4 are: for live load $+ \frac{1}{2}$ dead load, $\Delta = l/360 = (24)(12)/360 = 0.8$ in.; for live load only, $\Delta = l/480 = (24)(12)/480 = 0.60$ in.

$$\frac{LL}{LL + \frac{1}{2}DL} = \frac{50}{50 + (\frac{1}{2})(23)} = 0.81$$

Note that the partition load is not included in the dead load. Since $(0.81)(0.8 \text{ in.})$ = 0.65 in. > 0.60 in., live load only controls.

$$E_{(required)} = \frac{5wL^4}{384I\Delta_{LL}} = \frac{(5)(50)(2)(24)^4(12)^3}{(384)(935.9)(0.60)}$$

$$= 1{,}329{,}000 \text{ psi (use } 1{,}400{,}000 \text{ psi or higher)}$$

6. Determine the required tabular design value for shear parallel to grain $F_v$.

$$L_e = L - \frac{2d}{12} = 24 - \frac{(2)(16.5)}{12} = 21.25 \text{ ft}$$

$$V = \frac{wL_e}{2} = \frac{(93)(2)(21.25)}{2} = 1976 \text{ lb}$$

$$f_v = \frac{3V}{2A} = \frac{(3)(1976)}{(2)(41.25)} = 72 \text{ psi}$$

$$F_{v(required)} = f_v = 75 \text{ psi}$$

7. Determine the required tabular design value for compression perpendicular to grain, $F_{c\perp}$. A typical commercial joist hanger has a $2\frac{1}{2}$-in. bearing length.

$$\text{Joist reaction} = \frac{wL}{2} = \frac{(93)(2)(24)}{2} = 2232 \text{ lb}$$

$$F_{c\perp \text{ (required)}} = \frac{2232}{(2.5)(2.5)} = 357 \text{ psi}$$

There is no critical bearing stress on the compression face of the joist.

8. Determine camber requirements. Camber recommendations given in Table 4.3 for floor beams are for ordinary usage, and in some cases result in excessive camber. For the type of floor system in this example, the camber recommendation is $1\frac{1}{4}$ times the dead load deflection, which does not include the 20-psf partition load.

$$\Delta_{DL} = \frac{(5)(23)(2)(24)^4(12)^3}{(384)(1{,}500{,}000)(935.9)} = 0.24 \text{ in.}$$

$$\text{Camber} = (0.24)(1.25) = 0.31 \text{ in. (use } \tfrac{3}{8} \text{ in.)}$$

9. Specification for joist:

$$\text{Size} = 2\tfrac{1}{2} \times 16\tfrac{1}{2} \text{ in.}, \qquad \text{Camber} = \tfrac{3}{8} \text{ in.}$$

$$F_{bx} = 1600 \text{ psi}, \qquad F_v = 90 \text{ psi}$$

$$F_{c\perp \text{ (tension face)}} = 375 \text{ psi} \qquad F_{c\perp \text{ (compression face)}} = 375 \text{ psi}$$

$$E_x = 1,500,000 \text{ psi}$$

Start at step 3 of the procedure for the design of the girder.

3.   Select design value in bending for girder. Try $F_{bx} = 2400$ psi.

$$F_{bx}^* = F_{bx} = 2400 \text{ psi}$$

4.   Determine size required by bending. (Use reduced live load allowed by code.)

$$w = \text{DL} + \text{LL} = 45 + 30 = 75 \text{ psf}$$

$$M = \frac{wL^2}{8} = \frac{(75)(24)(35)^2(12)}{8} = 3,308,000 \text{ in.-lb}$$

Assume $C_V = 0.80$

$$S_{(\text{required})} = \frac{M}{F_b C_V} = \frac{3,308,000}{(2400)(0.80)} = 1723 \text{ in.}^3$$

Try a $6\frac{3}{4} \times 39$ in. girder.

$$C_V = \left(\frac{5.125}{6.75}\right)^{1/10} \left(\frac{12}{39}\right)^{1/10} \left(\frac{21}{35}\right)^{1/10} = 0.822$$

$$S_{(\text{required})} = \frac{3,308,000}{(2400)(0.822)} = 1678 \text{ in.}^3$$

From Table 8.2:

$$S = 1711 \text{ in.}^3, \ I = 33,370 \text{ in.}^4, \ A = 263.3 \text{ in.}^2$$

5.   Determine the required tabular design value for modulus of elasticity $E$. The recommended deflection limitations from Table 4.4 are: for live load $+ \frac{1}{2}$ (dead load), $\Delta = l/360 = (35)(12)/360 = 1.17$ in.; for live load only, $\Delta = l/480 = (35)(12)/480 = 0.875$ in.

$$\frac{\text{LL}}{\text{LL} + \frac{1}{2}\text{DL}} = \frac{30}{30 + (0.5)(45)} = 0.57$$

$$(0.57)(1.17) = 0.67 \text{ in.} < 0.875 \text{ in.}$$

Live load $+ \frac{1}{2}$ dead load controls.

$$E_{(\text{required})} = \frac{5[30 + (0.5)(45)](24)(35)^4(12)^3}{(384)(33,370)(1.17)} = 1,093,000 \text{ psi}$$

Use $E = 1,700,000$ psi because this is the lowest $E$ value readily available for 24F combinations.

6.  Determine the required tabular design value for shear parallel to grain, $F_v$.

$$L_e = L - \frac{2d}{12} = 35 - \frac{(2)(39)}{12} = 28.5 \text{ ft}$$

$$V = \frac{(75)(24)(28.5)}{2} = 25{,}650 \text{ lb}$$

$$f_v = \frac{3V}{2A} = \frac{(3)(25{,}650)}{(2)(263.3)} = 146 \text{ psi}$$

$$F_{v\text{(required)}} = f_v = 146 \text{ psi}$$

7.  Determine the required tabular design value for compression perpendicular to grain, $F_{c\perp}$. A typical commercial beam hanger has an 8-in. bearing length.

$$\text{Reaction} = \frac{(75)(24)(35)}{2} = 31{,}500 \text{ lb}$$

$$F_{c\perp \text{(required)}} = f_{c\perp} = \frac{(31{,}500)}{(6.75)(8)} = 583 \text{ psi}$$

8.  Determine camber requirements. (Do not camber for the 20-psf partition load.)

$$\Delta_{\text{DL}} = \frac{(5)(25)(24)(35)^4(12)^3}{(384)(1{,}700{,}000)(33{,}370)} = 0.36 \text{ in.}$$

$$\text{Camber} = (1.25)(0.36) = 0.45 \text{ in. (use } \tfrac{1}{2} \text{ in.)}$$

9.  Specification for girder:

$$\text{Size} = 6\tfrac{3}{4} \times 39 \text{ in.} \qquad \text{Camber} = \tfrac{1}{2} \text{ in.}$$

$$F_{bx} = 2400 \text{ psi} \qquad F_v = 150 \text{ psi}$$

$$F_{c\perp \text{(tension face)}} = 650 \text{ psi} \qquad F_{c\perp \text{(compression face)}} = 500 \text{ psi}$$

$$E_x = 1{,}700{,}000 \text{ psi}$$

## 5.6.6   Design of Sawn Beams

Sawn lumber and timber beams are usually used for simple spans consisting of straight members. Section properties are given in Table 8.1, and tabular design values are in Tables 8.3–8.7. The same general design considerations used for straight glued laminated timber beams are also applicable to sawn timber, except that sawn timbers cannot be cambered. Bending design values are tabulated for single member use (see NDS). As a general rule, however, sawn members may contain a small amount of curvature prior to erection, and the crown of the member should be turned up.

Size effect is taken into account in most of the grades, sizes, and species of sawn lumber. There are differences in the method which size effects are presented. Table 8.3 for all species of dimension lumber, except Southern Pine, contains base values and a table of $C_F$ values for various widths. Table 8.4, for Southern Pine, contains design values for widths up to 12 in. and a $C_F$ of 0.9 for all widths over 12 in., except for dense structural 86, dense structural 72, and dense structural 65 dimension lumber.

The size factor equation

$$C_F = (12/d)^{1/9}$$

is used for these grades for depths $d$ greater than 12 in.

This equation is also used for timbers $5 \times 5$ in. and larger included in Table 8.6 whose depth exceeds 12 in.

The size factor for machine stress rated lumber shown in Table 8.5 is included in the tabular values, and no further adjustment for size is needed.

When dimension lumber 2–4 in. thick is used flatwise as in a plank, the flat use factor $C_{fu}$ may be applied.

See the adjustment factors at the beginning of each table of design values for sawn lumber for additional information (see Tables 8.3–8.7).

Lateral support can also be taken into account by the application of approximate rules to prevent lateral rotation or lateral displacement as shown in Table 5.2.

The designer should check on the availability of the larger sizes of sawn timbers prior to design.

**Examples.**    Analyze a $4 \times 16$-in. select structural Douglas Fir-Larch sawn timber roof beam under the following conditions. The roof deck is applied directly to the beams $C_L = 1.0$, the beams are seated in hangers with $3.5 \times 4$-in. bearing area, the ends are restrained to prevent lateral rotation, and the slope of the roof is $\frac{3}{8}$ in./ft with adequate downspouts.

$$L = 20 \text{ ft}$$

$$\text{Spacing} = 10 \text{ ft}$$

$$\text{SL} = 25 \text{ psf}$$

$$\text{DL} = 10 \text{ psf}$$

$$C_D = 1.15 \text{ (snow load)}$$

$$C_F = 1.0 \text{ (from Table 8.3)}$$

$$F_b = 1450 \text{ psi (from Table 8.3)}$$

$$F'_b = (1450)(1.15)(1.0) = 1668 \text{ psi}$$

$$F'_v = (95)(1.15) = 109 \text{ psi}$$

$$F'_{c\perp} = 625 \text{ psi}$$

$$E' = 1,900,000 \text{ psi}$$

$$\Delta_{SL} = \frac{l}{240} = 1 \text{ in.}$$

$$\Delta_{TL} = \frac{l}{180} = 1.33 \text{ in.}$$

1.   Determine actual bending stress $f_b$:

$$w_{DL} = (\text{spacing})(\text{DL}) = (10)(10) = 100 \text{ plf}$$
$$w_{SL} = (\text{spacing})(\text{SL}) = (10)(25) = 250 \text{ plf}$$
$$w_{TL} = w_{DL} + w_{SL} \qquad\qquad\quad = \overline{350 \text{ plf}}$$

For a 4 × 16-in. member: $A = 53.38$ in., $S = 135.7$ in.$^3$ $I = 1034$ in.$^4$ from Table 8.1.

$$M = \frac{wL^2}{8} = \frac{(350)(20)^2(12)}{8} = 210,000 \text{ in.-lb.}$$

$$f_b = \frac{M}{S} = \frac{210,000}{135.7} = 1548 \text{ psi} < F'_b = 1668 \text{ psi} \qquad \text{O.K.}$$

2.   Determine actual shear parallel to grain stress $f_v$:

$$\text{Effective span } L_e = L - 2d = 20 - \frac{(2)(15.25)}{12} = 17.46 \text{ ft}$$

$$\text{Design shear at end } V = \frac{(17.46)(350)}{2} = 3055 \text{ lb}$$

$$f_v = \frac{3V}{2A} = \frac{(3)(3055)}{(2)(53.38)} = 86 \text{ psi} < F'_v = 109 \text{ psi} \qquad \text{O.K.}$$

3.   Determine the actual deflection of the beam:

$$\text{Deflection for snow load, } = \Delta_{SL} = \frac{5wL^4}{384EI}$$

$$= \frac{(5)(250)(20)^4(12)^3}{(384)(1,900,000)(1034)}$$

$$= 0.46 \text{ in.} < 1 \text{ in.} \qquad \text{O.K.}$$

$$\text{Deflection for total load} = (0.46)\tfrac{350}{250} = 0.64 \text{ in.} < 1.33 \text{ in.} \qquad \text{O.K.}$$

4.   Determine actual compression perpendicular to grain stress, $f_{c\perp}$, at the ends of the beam. Beam reaction:

$$R_v = \frac{(20)(350)}{2} = 3500 \text{ lb}$$

$$f_{c\perp} = \frac{3500}{(3.5)(4)} = 250 \text{ psi} < 625 \text{ psi} \qquad \text{O.K.}$$

5.  Check lateral stability. Ratio of depth to width (nominal size) $= \frac{16}{4}$ or $4:1$. This complies with criteria in Table 5.2 for the lateral stability of sawn members, and no other support is necessary. It also complies with lateral stability requirements because the unsupported length is zero and the ends are restrained from rotation.

6.  Ponding. Roof slope of $\frac{3}{8}$ in./ft with adequate roof drainage should prevent the accumulation of water, and ponding analysis is not required.

## 5.7 MEMBERS SUBJECTED TO BENDING ABOUT BOTH AXES

When members are subject to bending about both axes, the following equation must be used:

$$\frac{f_{bx}}{F'_{bx}} + \frac{f_{by}}{F'_{by}[(1 - f_{bx}/F_{bE})^2]} \le 1 \tag{5-13}$$

where:  $f_{bx}$ = actual bending stress about the $x$–$x$ axis (psi),
$f_{by}$ = actual bending stress about the $y$–$y$ axis (psi),
$F'_{bx}$ = tabular bending design value multiplied by all applicable adjustment factors (psi),
$F'_{by}$ = tabular bending design value multiplied by all applicable adjustment factors (psi), and
$F_{bE}$ = $K_{bE}E'/R_B^2$ calculated as shown in Eq. (5-10) (psi).

For lumber, bending about the $x$–$x$ axis is always edgewise bending and bending about the $y$–$y$ axis is always flatwise bending.

For glued laminated timber, the $x$–$x$ axis is defined as the axis perpendicular to the wide faces of the laminations; the $y$–$y$ axis is defined as the axis parallel to the wide faces of the laminations. When the depth $d_y$ is greater than $d_x$, the equation is written as follows:

$$\frac{f_{by}}{F'_{by}} + \frac{f_{bx}}{F'_{bx}[(1 - f_{by}/F_{bE})^2]} \le 1$$

For example, if three $2 \times 12$'s are glued together to form a $4\frac{1}{2} \times 10\frac{3}{4}$ in. glued laminated timber, $d_x = 4\frac{1}{2}$ in. and $d_y = 10\frac{3}{4}$ in.

**Example 1.** A $24F$ glued laminated timber Southern Pine beam 30 ft long is to be used to support a uniform dead load of 400 plf and a vertical live load of 800 plf. In addition, the beam is subjected to a horizontal wind load $P$ of 4000 lb located at the midpoint of the member which can act in either direction. The member is to be used under normal temperature and dry service conditions and is not laterally braced between the end supports.

$$\text{Live load}, \; C_D = 1.25$$

$$\text{Wind load}, \; C_D = 1.6$$

Determine trial size.
Moment about x-x axis

$$M_{DL} = \frac{wL^2}{8} = \frac{(400)(30)^2(12)}{8} = 540,000 \; \text{in.-lb}$$

$$M_{LL} = \left(\frac{800}{400}\right)(540,000) = 1,080,000 \; \text{in.-lb}$$

$$\text{Total } M_x = 1,620,000 \; \text{in.-lb}$$

The design is accomplished by trial and error. First, a size is estimated and calculations are made to determine the adequacy of the assumed size. To aid in estimation, a section modulus is calculated for the dead load and live load using $C_D$ of 1.6 since it is the duration of load factor for the shortest duration of load when all loads are considered simultaneously. Note, however, that the size of the member calculated must be checked for dead load and live load only using a $C_D$ of 1.25.

Assume $C_V = 0.85$

$$S_{x(required)} = \frac{M}{F_b C_D C_V} = \frac{1,620,000}{(2400)(1.6)(0.85)} = 496 \; \text{in.}^3$$

Moment about y-y axis

$$M_{WL} = \frac{PL}{4} = \frac{(4000)(30)(12)}{4} = 360,000 \; \text{in.-lb}$$

Assume width of beam = 8.5 in., $F_{by}$ = 1600 psi, from Table 4.9, $C_{fu}$ = 1.04

$$S_{y(required)} = \frac{360,000}{(1600)(1.04)(1.6)} = 135 \; \text{in.}^3$$

Try a member with an $S_x$ arbitrarily set at approximately 2 × 496 in.$^3$ and an $S_y$ of approximately 2 × 135 in.$^3$.
Try an $8\frac{1}{2}$ × $28\frac{7}{8}$ in. beam.

$S_x$ = 1181 in.$^3$, $I_x$ = 17,050 in.$^4$, $S_y$ = 347.7 in.$^3$, and $I_y$ = 1478 in.$^4$

$$f_{bx} = \frac{M_x}{S_x} = \frac{1,620,000}{1181} = 1372 \; \text{psi}$$

$$f_{\text{by}} = \frac{M_{\text{y}}}{S_{\text{y}}} = \frac{360{,}000}{347.7} = 1035 \text{ psi}$$

$$F'_{\text{bx}} = F_{\text{bx}} C_{\text{D}} C_{\text{V}} \text{ or } F_{\text{bx}} C_{\text{D}} C_{\text{L}}$$

$$C_{\text{D}} = 1.6 \text{ (wind load)}$$

$$
\begin{aligned}
C_{\text{V}} &= \left(\frac{5.125}{b}\right)^{1/20} \left(\frac{12}{d}\right)^{1/20} \left(\frac{21}{L}\right)^{1/20} \\
&= \left(\frac{5.125}{8.5}\right)^{1/20} \left(\frac{12}{28.875}\right)^{1/20} \left(\frac{21}{30}\right)^{1/20} = 0.917
\end{aligned}
$$

The beam is supported at the ends to prevent rotation.

$$l_{\text{u}} = 30 \text{ ft}$$

$$l_{\text{e}} = 1.63 l_{\text{u}} + 3d \text{ from Table 5.3}$$

$$l_{\text{e}} = 1.63(30)(12) + (3)(28.875) = 673.4 \text{ in.}$$

$$R_{\text{B}} = \sqrt{\frac{l_{\text{e}} d}{b^2}} = \sqrt{\frac{(673.4)(28.875)}{8.5^2}} = 16.40 < 50 \qquad \text{O.K.}$$

$$K_{\text{bE}} = 0.609$$

$$E' = E'_{\text{y}} = 1{,}600{,}000 \text{ psi}$$

$$F_{\text{bE}} = \frac{K_{\text{bE}} E'}{R_{\text{B}}^2} = \frac{(0.609)(1{,}600{,}000)}{16.40^2} = 3620 \text{ psi}$$

$$F^*_{\text{bx}} = 2400(1.6) = 3840 \text{ psi}$$

$$C_{\text{L}} = \frac{1 + (F_{\text{bE}}/F^*_{\text{bx}})}{1.9} - \sqrt{\left(\frac{1 + [F_{\text{bE}}/F^*_{\text{bx}}]}{1.9}\right)^2 - \frac{(F_{\text{bE}}/F^*_{\text{bx}})}{0.95}}$$

$$C_{\text{L}} = \frac{1 + (3620/3840)}{1.9} - \sqrt{\left(\frac{1 + [3620/3840]}{1.9}\right)^2 - \frac{3620/3840}{0.95}} = 0.792$$

$$0.792 < 0.917, \ C_{\text{L}} \text{ controls}$$

$$F'_{\text{bx}} = F^*_{\text{bx}} C_{\text{L}} = (3840)(0.792) = 3041 \text{ psi}$$

In considering loading about the $y$–$y$ axis, $b_{\text{y}} = 28.875$ in. and $d_{\text{y}} = 8.5$ in., $b/d = 28.875/8.5 = 3.40 > 1$; therefore, $C_{\text{L}} = 1$. $C_{\text{V}}$ is not applicable to members loaded parallel to the wide faces of the laminations, but $C_{\text{fu}}$ applies.

$$C_{\text{fu}} = 1.04 \text{ (from Table 4.9)}$$

$$F'_{by} = F_{by} C_D C_{fu} = (1600)(1.6)(1.04) = 2662 \text{ psi}$$

$$\frac{f_{bx}}{F'_{bx}} + \frac{f_{by}}{F'_{by}\left(1 - \left(\dfrac{f_{bx}}{F_{bE}}\right)\right)^2} \leq 1.0$$

$$\frac{1372}{3041} + \frac{1035}{(2662)\left(1 - \left(\dfrac{1372}{3620}\right)\right)^2} = 1.08 > 1 \qquad \text{N.G.}$$

Similar calculations show that an $8\frac{1}{2}$ in. $\times$ $30\frac{1}{4}$ in. beam is satisfactory. Use an $8\frac{1}{2}$-in. $\times$ $30\frac{1}{4}$-in. beam. A $10\frac{1}{2}$-in.-wide beam could also be used if the $30\frac{1}{4}$-in. depth is too large. Calculations show that a $10\frac{1}{2}$-in. $\times$ $23\frac{3}{8}$-in. beam is also satisfactory.

## 5.8   COLUMNS AND COMPRESSION MEMBERS

The term *column* is generally applied to all compression members including truss members, posts, or other structural components stressed in compression. Columns are divided into three general types consisting of (1) simple columns, (2) spaced columns, and (3) built-up columns. Simple wood columns consist of a single piece of sawn lumber, post, timber, pole, or glued laminated timber. Spaced columns consist of two or more individual members of sawn lumber or glued laminated timbers with their longitudinal axes parallel, separated at their ends and midpoints by blocking, and joined at the ends by connectors capable of developing the required shear resistance. Built-up columns consist of two or more pieces of lumber placed side by side and joined with mechanical fasteners.

### 5.8.1   Effective Column Length

The unbraced length of a column or compression member is the distance between two points along its length between which the member is assumed to buckle. For a laterally unsupported simple column with assumed pinned ends, the effective length $l_e$ is equal to the total length of the column. For columns with different degrees of fixity at the ends or with intermediate lateral support, as illustrated in Table 5.6 and Fig. 5.10, the effective length varies. The $K_e$ value should be determined by good engineering judgment following the guidelines in Table 5.6. The effective buckling length factors $K_e$ are shown as theoretical $K_e$ values and the recommended values of $K_e$ to use for design. The recommended values of $K_e$ for design take into account the lack of perfect fixity in use. However, when compression members depend on the rigidity of other in-plane members entering the joint to provide fixity and the combined stiffness of these members is relatively small compared to the unbraced column segments, $K_e$ can exceed the values shown in Table 5.6. Determine the effective length by multiplying $K_e$ by the length of the member between points of lateral support. Columns with intermediate bracing points may have different theoretical $K_e$ factors.

**TABLE 5.6**

**Effective Column Length for Various End Conditions**

| | | | | | | |
|---|---|---|---|---|---|---|
| Buckling Modes | | | | | | |
| Theoretical $K_e$ value | 0.5 | 0.7 | 1.0 | 1.0 | 2.0 | 2.0 |
| Recommended design $K_e$ when ideal conditions approximated | 0.65 | 0.8 | 1.2 | 1.0 | 2.1 | 2.4 |

End condition code

Rotation fixed, translation fixed

Rotation free, translation fixed

Rotation fixed, translation free

Rotation free, translation free

Source: ANSI/NF₀PA NDS—1991, *National Design Specification for Wood Construction*, AFPA, 1250 Connecticut Ave., NW, Washington, DC 20036.

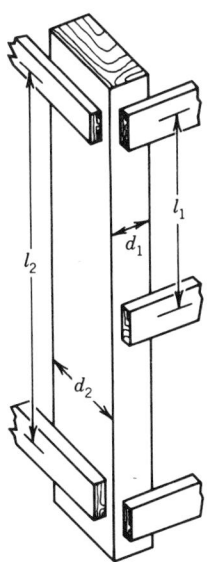

**FIGURE 5.10**   Simple solid column [from National Design Specification (3)]. $l_1$ and $l_2$ are the distances between points of lateral support of column in planes 1 and 2 (in.); $d_1$ and $d_2$ are the dimensions of column in planes of lateral support (in.).

The column equations which follow are based on pin end conditions with the ends fixed against translation. It is assumed that the column buckling curve approximates the shape of a sine wave. These equations are commonly used for columns with square cut ends which is less critical than the pin end condition. Conditions are sometimes encountered in design where the restraint is more or less than the standard conditions.

### 5.8.2   Continuous Column Design Equation

The 1991 edition of NDS (3) contains a new continuous column equation based on a procedure first proposed by Ylinen in 1956, a method of determining the buckling stress and the required cross sectional area for axially loaded straight columns in the elastic and inelastic range.

This procedure eliminates the need for classification of short, intermediate, and long columns traditionally used for design of compression members. The inclusion of the $c$ factor in the equation allows results of testing various materials to be conveniently included.

The older 4th power curve used to define the capacity of the intermediate length column compared to the short column was developed by testing larger size columns. These tests were reported by Newlin and Gahagan in *Tests of Large Timber Columns and Presentation of the Forest Products Laboratory Column Formula* (7). Subsequent tests on smaller members indicate that this equation is unconservative for smaller size members. The use of the $c$ factor of 0.80 best fit these new data for sawn lumber and 0.90 (Fig. 5.11) was used for glued laminated timber. The value of $c = 0.85$ for poles and piles is based on experience with these types of wood members.

The curve of the continuous column equation is very close to that of the older 4th power parabolic equation when $c = 0.957$ and lies below for lower $c$ values. Thus, the compression design value $F'_c$ will be slightly lower for the values of $c$ = 0.90, 0.85, and 0.80. Fig. 5.11 shows this relationship.

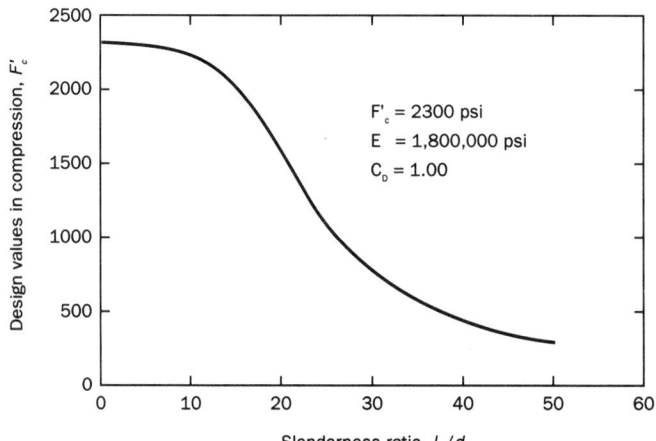

**FIGURE 5.11**   Column relationships.

There are no "short", "intermediate", or "long" columns in the continuous column equation. Theoretically, the $C_P$ factor is 1.0 at $l_e/d = 0$. For practical purposes, however, the reduction of strengths at the $l_e/d$ ratios up to about 3 can be ignored because the difference in allowable stress is small.

The design value $F_c'$ is calculated by multiplying the tabular design value $F_c$ by all applicable adjustment factors, including the column stability factor $C_P$.

The column stability factor $C_P$ is determined as follows:

When the column is laterally braced throughout its length in both directions, the column stability factor is equal to 1.0. When the column is not braced throughout its length, $C_P$ for a solid column of rectangular cross section must be calculated by the following equation:

$$C_P = \frac{1 + (F_{cE}/F_c^*)}{2c} - \sqrt{\left[\frac{1 + (F_{cE}/F_c^*)}{2c}\right]^2 - \frac{F_{cE}/F_c^*}{c}} \qquad (5\text{-}14)$$

where   $F_c^*$ = tabular compression design value multiplied by all applicable adjustment factors except $C_P$ (psi),

$F_{cE} = \dfrac{K_{cE}E'}{(l_e/d)^2}$  (5-15),

$K_{cE}$ = 0.3 for visually graded lumber,

$K_{cE}$ = 0.418 for products with $COV_E \leq 0.11$,

   $c$ = 0.8 for sawn lumber,

   $c$ = 0.85 for round timber piles,

   $c$ = 0.90 for glued laminated timber,

   $l_e = K_e l$ (in.),

$K_e$ = buckling length coefficient (Table 5.6),

$l_e d$ = the larger of the ratios of $l_{e1}/d_1$ or $l_{e2}/d_2$ (see Fig. 5.10), and

$E'$ = $E$ in the direction that buckling takes place multiplied by all applicable adjustment factors.

The solution for $C_P$ is cumbersome when done by hand calculations, but it can be obtained easily by use of a programmable calculator, a computer spreadsheet, or any one of a number of computer programs developed for wood design. The following solutions are shown in hand calculation form, although the actual solution was by use of a computer spreadsheet. Intermediate values in the calculations have been rounded, and the final value was rounded from the computer output. When the examples are checked by hand calculations, a slight discrepancy in the answer may occur if the intermediate rounded values are used in the calculations.

### 5.8.3   Design of Columns

The following examples illustrate the design of sawn lumber and glued laminated timber columns.

**Example 1(a).**   Determine the axial snow load capacity $P$ of a 10 ft long 6 in. × 6 in. No. 1 Dense Southern Pine column. The column is located in a dry location. The top and bottom of the column are held to prevent translation, and no lateral support is provided along the length.

1. Determine tabular design values from Table 8.6.

$$F_c = 975 \text{ psi} \qquad E = 1,600,000 \text{ psi}$$

For snow load, $C_D = 1.15$ from Table 4.7

2. Determine $F_c^*$ and $E'$

$$F_c^* = F_c C_D = (975)(1.15) = 1121 \text{ psi}$$

$$E' = E = 1,600,000 \text{ psi}$$

3. Dimensions

$$b = 5.5 \text{ in.,} \quad d = 5.5 \text{ in.,} \quad l = 10 \text{ ft}$$

4. Determine $l_e$

$$l_e = K_e l$$

$$K_e \text{ from Table 5.6} = 1.0,$$

$$l_e = (1.0)(12)(10) = 120 \text{ in.}$$

5. Determine $(l_e/d)^2$

$$l_e/d = 120/5.5 = 21.81 < 50 \quad \text{O.K.}$$

$$(l_e/d)^2 = (21.81)^2 = 476.03$$

6. Determine input values $c$, $K_{cE}$, and $F_{cE}$

$$c = 0.8 \text{ for sawn lumber [from Eq. (5-14)]}$$

$$K_{cE} = 0.3 \text{ for visually graded lumber [from Eq. (5-14)]}$$

$$F_{cE} = \frac{K_{cE}E'}{(l_e/d)^2} = \frac{(0.3)(1,600,000)}{476.03} = 1008 \text{ psi}$$

7. Determine $C_P$

$$C_P = \frac{1 + (F_{cE}/F_c^*)}{2c} - \sqrt{\left[\frac{1 + (F_{cE}/F_c^*)}{2c}\right]^2 - \frac{F_{cE}/F_c^*}{c}}$$

$$C_P = \frac{1 + (1008/1121)}{(2)(0.8)}$$

$$- \sqrt{\left[\frac{1 + (1008/1121)}{(2)(0.8)}\right]^2 - \frac{1008/1121}{0.8}} = 0.653$$

8. Determine $F_c'$

$$F_c' = F_c^* C_P = (1121)(0.653) = 732 \text{ psi}$$

9. Determine $P$

$$P = F_c' bd = 732(5.5)(5.5) = 22,160 \text{ lb}$$

**Example 1(b).** Determine the capacity of the same size column 20 ft long.

$$\frac{l_e}{d} = \left(\frac{(1)(20)12}{5.5}\right) = 43.64 < 50 \qquad \text{O.K.}$$

$$\left(\frac{l_e}{d}\right)^2 = 43.64^2 = 1904$$

$$F_{cE} = \frac{K_{cE}E'}{(l_e/d)^2} = \frac{(0.3)(1,600,000)}{1904} = 252 \text{ psi}$$

$$C_P = \frac{1 + (252/1121)}{(2)(0.8)}$$

$$- \sqrt{\left[\frac{1 + (252/1121)}{(2)(0.8)}\right]^2 - \frac{252/1121}{0.8}} = 0.213$$

$$F_c' = F_c^* C_p = (1121)(0.213) = 239 \text{ psi}$$

$$P = F_c' bd = (239)(5.5)^2 = 7230 \text{ lb}$$

**Example 1(c).** Determine the capacity of the same size column 5 ft long.

$$\frac{l_e}{d} = \frac{(1)(5)(12)}{5.5} = 10.91, \quad \left(\frac{l_e}{d}\right)^2 = 10.91^2 = 119$$

$$F_{cE} = \frac{(0.3)(1,600,000)}{119} = 4033 \text{ psi}$$

$$C_P = \frac{1 + 4033/1121}{(2)(0.8)} - \sqrt{\left[\frac{1 + 4033/1121}{(2)(0.8)}\right]^2 - \frac{4033/1121}{0.8}} = 0.934$$

$$F_c' = (1121)(0.934) = 1047 \text{ psi}$$

$$P = F_c' bd = (1047)(5.5)(5.5) = 31,690 \text{ lb}$$

Note that $C_P$ in 1(c) is 0.934 with $l_e/d$ equal approximately to 11. This shows that the strength of this column determined by the continuous column equation is approximately 7% less than when designed as a short column with an $l_e/d$ of 11 using the previous column design procedure.

**Example 2.** Determine the axial snow load capacity, $P$, of an 18 ft long $6\frac{3}{4}$-in. × $8\frac{1}{4}$-in. column made from combination 16F–V1 Southern Pine. (Note: Laminating combinations in Table 2, AITC 117—Design, combinations are customarily used for columns, but this combination from Table 1 was selected to illustrate the application of $E_x$ and $E_y$.) The top and bottom of the column are held to prevent translation, and lateral support is provided at midheight about the $y$-$y$ axis. The column is used in a dry location.

1.    Determine tabular design values from Table 1, AITC 117—Design included in Chapter 8.

$$F_c = 1450 \text{ psi}, \ E_x = 1,400,000 \text{ psi}, \ E_y = 1,300,000 \text{ psi}$$

$$C_D = 1.15 \text{ from Table 4.7}$$

$$E_{\text{axial}} = 1,300,000 \text{ psi (same as } E_y)$$

From Table 8.8, bearing parallel to grain across the full cross section, $F_g$, = 1690 > 1450 psi. Bearing parallel to grain does not control. It generally controls only when ends of columns are not in full bearing.

2.    Determine $F_c^*$ and $E'$

$$F_c^* = F_c C_D = (1450)(1.15) = 1668 \text{ psi}$$

$E_{\text{axial}}$ will be used

$$E' = E_{\text{axial}} = 1,300,000 \text{ psi}$$

3.    Dimensions $d_1 = 8.25$ in., $b_1 = 6.75$ in., $d_2 = 6.75$ in., $b_2 = 8.25$ in.

$$l_1 = 18.0 \text{ ft}, \ l_2 = 9 \text{ ft}$$

4.    Determine $l_e$

$$l_e = K_e l$$

$$K_e = 1.0 \text{ from Table 5.7}$$

$$l_{e1} = (1.0)(18)(12) = 216 \text{ in.}$$

$$l_{e2} = (1.0)(9)(12) = 108 \text{ in.}$$

5.    Determine $l_e/d$ and $(l_e/d)^2$

$$\frac{l_{e1}}{d_1} = \frac{216}{8.25} = 26.18, \quad \left(\frac{l_{e1}}{d_1}\right)^2 = 685.5$$

$$\frac{l_{e2}}{d_2} = \frac{108}{6.75} = 16.0, \quad \left(\frac{l_{e2}}{d_2}\right)^2 = 256.0$$

$$\left(\frac{l_{e1}}{d_1}\right)^2 \text{ controls}$$

$$\frac{l_{e1}}{d_1} = 26.18 < 50 \qquad \text{O.K.}$$

6.    Determine the input values $c$, $K_{cE}$, and $F_{cE}$.

For glued laminated timber, $c = 0.9$, $K_{cE} = 0.418$ [from Eq. (5-14)]

$$F_{cE} = \frac{K_{cE}E'}{(l_{e1}/d_1)^2} = \frac{(0.418)(1,300,000)}{685.5} = 793 \text{ psi}$$

7. Determine $C_P$

$$C_P = \frac{1 + (F_{cE}/F_c^*)}{2c} - \sqrt{\left[\frac{1 + (F_{cE}/F_c^*)}{2c}\right]^2 - \frac{F_{cE}/F_c^*}{c}}$$

$$C_P = \frac{1 + (793/1667)}{(2)(0.9)} - \sqrt{\left[\frac{1 + (793/1667)}{(2)(0.9)}\right]^2 - \frac{793/1667}{0.9}} = 0.441$$

8. Determine $F_c'$

$$F_c' = F_c^* C_P = (1667)(0.441) = 735 \text{ psi}$$

9. Determine $P$

$$P = F_c'(b_1)(d_1) = (735)(6.75)(8.25) = 40,920 \text{ lb}$$

## 5.8.4 Round Columns

A round column is a simple solid column with a circular cross section. Its load carrying capacity equals that of a square column with the same cross-sectional area. For design purposes, it is sometimes convenient to determine the required size of a square column and use a circular column of the same cross-sectional area. The diameter of the equivalent round column, $D$, is $1.128d$, where $d$ is the dimension of a side of a square column with the same cross-sectional area. It is frequently more convenient to work with the diameter of round columns rather than converting to equivalent square columns. Pole sizes are generally designated by the circumference from which the diameter is readily obtained. The radius of gyration $r$ of a round column is $D/4$ and is convenient to use in the column equations.

The radius of gyration, $r$, is also equal to $d/\sqrt{12} = 0.289d$, where $d$ is the side of a square column. The maximum $l_e/d$ ratio of 50 converts to an $l_e/r$ by multiplying $l_e/d$ by $\sqrt{12}$,

$$l_e/d\sqrt{12} = 50\sqrt{12} = 173$$

which is customarily rounded to 175.

All adjustment factors applicable to sawn columns are used. The column stability factor, $C_P$, is calculated in the usual manner, except $F_{cE}$ which is determined from the following equation to account for using $r$ rather than $d$ in the slenderness ratio.

$$F_{cE} = \frac{12K_{cE}E'}{(l_e/r)^2} \tag{5-15}$$

Round poles and piles are more frequently used than sawn lumber or glued laminated timbers machined to a round shape. Therefore, examples of design are included in 6.3 and 6.4 on poles and piles.

### 5.8.5   Tapered Columns

Tapered columns may be tapered from the larger cross section toward one end or both ends. They are designed as simple columns with the least dimension, $d$, of columns of rectangular cross section or the minimum diameter, $D$, of round columns. These dimensions are modified to determine the slenderness ratio, $l_e/d$ or $l_e/r$, to use in design for columns of rectangular cross section tapered at one end or both ends. The representative dimension, $d$, for each face of the column is derived as follows:

$$d = d_{min} + (d_{max} - d_{min})\left[a - 0.15\left(1 - \frac{d_{min}}{d_{max}}\right)\right] \qquad (5\text{-}16)$$

where  $d_{min}$ = the minimum dimension for that face of the column (in.),
$d_{max}$ = the maximum dimension for that face of the column (in.), and
$a$ = the values as shown for various support conditions

| | |
|---|---:|
| Large end fixed, small end unsupported or simply supported | $a = 0.70$ |
| Small end fixed, large end unsupported or simply supported | $a = 0.30$ |
| Both ends simply supported | |
| Tapered toward one end | $a = 0.50$ |
| Tapered toward both ends | $a = 0.70$ |

For all other support conditions

$$d = d_{min} + (d_{max} - d_{min})(1/3) \qquad (5\text{-}17)$$

Tapered round columns are designed on the basis of a square column of the same cross-sectional area and having the same degree of taper. The design capacity of a tapered column may also be used with the slenderness ratios of $l_e/r$ rather than $l_e/d$ as shown in Eq. (5-15).

### 5.8.6   Eccentrically Loaded Columns

The preceding procedures for designing columns are based on concentric loading. When columns are eccentrically loaded, the loading is considered as combined bending and axial compression and is discussed in 5.9, "Combined Bending and Axial Loading."

### 5.8.7   Spaced Columns

Spaced columns have increased load-carrying capacity in a direction perpendicular to the wide faces of the individual members (parallel to dimension $d_1$, Fig. 5.12) due to the fixity at the ends, which provides shear resistance that tends to make the column function as a unit in this direction. The load capacity in the direction parallel to the wide faces of the pieces (parallel to dimension $d_2$ as shown in Fig. 5.12) is not increased due to the fixity and the load capacity in this direction

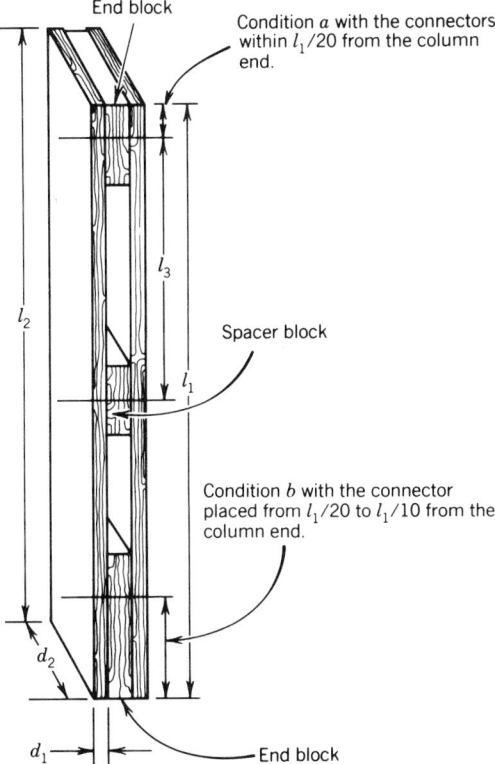

**FIGURE 5.12** Spaced column, connector joined [from the NDS (3)] $l_1$ and $l_2$ are the distances from center to center of lateral supports of continuous spaced columns, and from end to end of simple spaced columns (in.); $l_3$ is the distance from center of connectors, in end blocks, to center of spacer block (in.); $d_1$ is the dimension of the narrowest side of individual member (in.); and $d_2$ is the dimension of wide face of individual member (in.).

Source: ANSI/NF$_0$PA NDS—1991, *National Design Specification for Wood Construction*, AFPA, 1250 Connecticut Ave, NW, Washington, DC 20036.

is the sum of the load capacities of the individual members designed as simple solid columns using $l_2/d_2$.

The individual members in a spaced column are considered to act together to carry the total column load. Each member is designed separately on the basis of its $l_e/d$ ratio. Because of end fixity developed in spaced columns, a greater $l_e/d_1$ ratio than allowed for simple solid columns is permitted. These limitations for individual members of a spaced column are

1. $l_1/d_1$ shall not exceed 80, where $l_1$ is the distance between lateral supports that provide restraint perpendicular to the wide faces of the individual members,

2. $l_2/d_2$ shall not exceed 50, where $l_2$ is the distance between lateral supports that provide restraint parallel to the wide faces of the individual members, and

3.  $l_3/d_1$ is limited to 40 where $l_3$ is the distance between the centroid of connectors in an end block and the center of the spacer block.

The effective length of a spaced column, $l_e$, should be determined by good engineering practice. The lengths $l_1$ and $l_2$ should be multiplied by the buckling length coefficient $K_e$ in Table 5.6, but in no case should the length be less than the actual length of the column.

The capacity of a spaced column, $P$, is the capacity of the individual members.

$$P = 2F_c' (d_1)(d_2)$$

where    $F_c' = F_c^* C_P$.

The smallest $F_c'$ is determined by using the slenderness ratios $l_2/d_2$ or $l_1/d_1$. $F_c^*$ is the tabular design value multiplied by all applicable adjustment factors except $C_P$.

The dimensions $l_1$, $l_2$, $l_3$, $d_1$, and $d_2$ are shown in Fig. 5.12. The design values for spaced columns are a function of the fixity condition $a$ and $b$ as shown in Fig. 5.12 and the slenderness ratio of the individual members of the spaced column.

Because the following design procedure is based on tests of spaced columns using split rings for the connections at the ends, it is recommended that split rings be used for these connections.

When different grades, species, or thicknesses of members are used, the lesser value of $F_c'$ determined for either member is applied to both members. Connectors are not required for a single spacer block located in the middle tenth of the column length $l_2$. Connectors are required for multiple spacer blocks, and the distance between two adjacent blocks may not exceed one-half the distance between centers of connectors in the end blocks.

When spaced columns are used as truss compression members, panel points that are restrained laterally are considered as the ends of the spaced column. The portion of the web members between the individual pieces of which the spaced column is comprised may be considered as the end blocks in the end of the column (see Fig. 5.12). The total load capacity determined by using the spaced column equations should be checked against the sum of the load capacities of the individual members taken as simple solid columns without regard to fixity; their greater $d$ and the $l$ between lateral supports that provide restraint in a direction parallel to the greater $d$ should be used.

The design values for $F_c'$ determined by the equations should not exceed the tabular design values for compression parallel to grain, $F_c$, for the species, as multiplied by applicable adjustment factors.

Spacer and end block thickness should not be less than that of the individual members of the spaced column, nor should thickness, width, and length of spacer and end block be less than required for connectors of a size and number capable of carrying the computed load.

To obtain spaced column action, the connectors in mutually contacting surfaces of end blocks and individual members at each end of a spaced column must

### TABLE 5.7

### End Spacer Block Constant, $K_s$

| Species Group | Constant |
|---|---|
| A | $K_s = 9.555 \, (l_1/d_1 - 11) \leq 468$ |
| B | $K_s = 8.14 \, (l_1/d_1 - 11) \leq 399$ |
| C | $K_s = 6.73 \, (l_1/d_1 - 11) \leq 330$ |
| D | $K_s = 5.32 \, (l_1/d_1 - 11) \leq 261$ |

Source: ANSI/NF$_o$PA NDS-1991, *National Design Specification for Wood Construction*, AFPA, 1250 Connecticut
Ave., NW, Washington, DC 20036.

be of a size and number to provide a load capacity equal to the required cross-
sectional area of one of the individual members times the appropriate constant
$K_s$ from Table 5.7. For spaced columns that are part of a truss system, the connectors required by joint design should be checked against the values obtained
by using the constants from Table 5.7. Species groupings for use in Table 5.7
are shown in 7.4.

For calculating the spaced column stability factor $C_P$, the value of $F_{cE}$ is determined as follows:

$$F_{cE} = \frac{K_{cE}K_x E'}{(l_e/d)^2} \tag{5-18}$$

where

$K_{cE}$, $E'$ and $l_e/d$ are defined previously,
$K_x = 2.5$ for fixity condition "*a*", and
$K_x = 3.0$ for fixity condition "*b*"

**Example.**   Design a two member spaced column using No. 1 Douglas Fir-Larch
(Group B species) to meet the following requirements:

Two month (snow load) duration of load, $C_D = 1.15$
Dry conditions of use, $C_M = 1.0$
Used under normal temperatures, $C_t = 1.00$

See Table 5.7 for end spacer blocks constants.

$$P = 39,000 \text{ lb}$$

$$F_c = 1450 \text{ psi}, \ E' = E = 1,700,000 \text{ psi from Table 8.3}$$

Try 3 × 8 in. size

Note: $F_c = 1450$ psi for 8 in. width with $C_F = 1.05$
Length, $L = 12$ ft, $l_1 = l_2 = (12)(12) = 144$ in.

$F'_c = F_c C_D C_P C_F$

$F^*_c = F_c C_D C_F = (1450)(1.15)(1.05) = 1750 \text{ psi}$

$K_e = 1$

$l_{1e} = l_{2e} = 1(144) = 144 \text{ in.}$

$c = 0.8$ and $K_{cE} = 0.3$ for visually graded lumber

Buckling parallel to $d_1$ (perpendicular to wide face of an individual member) Assume side pieces used are $3 \times 8$

$$d_1 = 2.5 \text{ in.}, \ d_2 = 7.25 \text{ in.}, \ A = 18.12 \text{ in.}^2$$

$$l_{1e}/d_1 = \frac{144}{2.5} = 57.6 < 80 \qquad \text{O.K.}$$

Check end fixity distance from end of column to center of 4-in. split ring connector in the end blocks $x = 12$ in.     O.K.

$$\frac{l_1}{20} = \frac{144}{20} = 7.2 \text{ in.} \qquad \frac{l}{10} = \frac{144}{10} = 14.4 \text{ in.}$$

$$\frac{l_2}{d_1} = \frac{[(12 - 2) - 3]12}{2.5} = 33.6 < 40 \qquad \text{O.K.}$$

Column has a fixity condition $b$ since $l/20 < x < l/10$,

$$K_x = 3$$

$$F_{cE1} = \frac{K_{cE} K_x E}{(l_e/d)^2} = \frac{(0.3)(3)1,700,000}{(144/2.5)^2} = 461 \text{ psi}$$

Using Eq. (5-14)

$$C_P = \frac{1 + (461/1750)}{(2)(0.8)} - \sqrt{\left[\frac{1 + 461/1750}{(2)(0.8)}\right]^2 - \frac{461/1750}{0.8}} = 0.247$$

$$F'_c = F^*_c C_P = (1750)(0.247) = 433 \text{ psi}$$

Required $A = \dfrac{39,000}{433} = 90.1 \text{ in.}^2 > 2 (18.1) = 36.2 \text{ in.}^2 \quad \text{No good}$

Try $2\ 4 \times 8$ members

$$A = (2)(3.5)(7.25) = 50.75 \text{ in.}^2$$

$$l_{1e}/d_1 = \frac{144}{3.5} = 41.14$$

$$F_{cE1} = \frac{(0.3)(3)(1,700,000)}{(41.14)^2} = 904 \text{ psi}$$

$$F^*_c = (1450)(1.15)(1.05) = 1750 \text{ psi}$$

$$C_P = \frac{1 + (904/1750)}{(2)(0.8)} - \sqrt{\left[\frac{1 + 904/1750}{(2)(0.8)}\right]^2 - \frac{904/1750}{0.8}}$$

$$C_P = 0.445$$

$$F_c' = (1750)(0.445) = 779 \text{ psi}$$

$$P = F_c'A = (779)(50.75) = 39,514 \text{ lb} > 39,000 \text{ lb} \qquad \text{O.K.}$$

Check buckling in plane parallel to wide faces of members (parallel to $d_2$)

$$l_e/d = \frac{144}{7.25} = 19.9$$

$$F_{cE2} = \frac{(0.3)(1,700,000)}{(19.9)^2} = 1293 \text{ psi}$$

At this point, it is obvious that buckling in a plane perpendicular to the wide faces of the members (parallel to $d_1$) controls since $F_{cE2} > F_{cE1}$. However, the full calculations are shown.

$$C_P = \frac{1 + (1293/1750)}{(2)(0.8)} - \sqrt{\left[\frac{1 + (1293/1750)}{(2)(0.8)}\right]^2 - \frac{(1293/1750)}{0.8}}$$

$$C_P = 0.579$$

$$F_c' = (1750)(0.579) = 1014 \text{ psi} > 779 \text{ psi}$$

Buckling in a plane parallel to $d_1$ controls

Check the connectors for spaced column action. From Table 5.7, the end spacer block constant $K_s$ for Group B species is:

$$K_s = (8.14)(l_1/d_1 - 11) \le 399$$

$$K_s = (8.14)(41.14 - 11) = 245 < 399, \text{ use } 245$$

Total connector load $= K_sA$

$$K_sA = (245)(3.5)(7.25) = 6217 \text{ lb}$$

Allowable load for one 4 in. split ring in Group B species from Table 7.35 is

$$(5260)(1.15) = 6050 \text{ lb} < 6217 \text{ lb}$$

Use two 4 in. split rings.

## 5.8.8 Built-Up Columns

Built-up columns of two or more pieces of wood fastened together with mechanical fasteners such as nails, bolts, or other fasteners will not carry as much load as a solid column of the same size because of shear distortion that can occur in the fastening joints. The nailing of cover plates to the edge of the built-up layers

(*a*) Boxed solid core

(*b*) Edges fastened together
with cover plates

| *l/d* ratio | Strength of single piece column (%) |
|:---:|:---:|
| 6 | 82 |
| 10 | 77 |
| 14 | 71 |
| 18 | 65 |
| 22 | 74 |
| 26 | 82 |

**FIGURE 5.13**   Built-up columns; mechanically fastened; data from *Wood Handbook* (8).

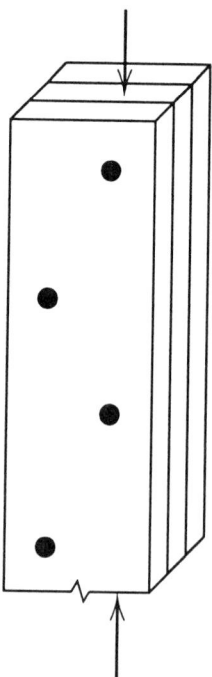

**FIGURE 5.14**   Built-up column.

improves the load-carrying capacity, according to the *Wood Handbook* (8). (See Fig. 5.13.) For more details on the design of built-up columns of the type shown in Fig. 5.14, see Part XV, *National Design Specification®* (3).

### 5.8.9   Columns with Flanges

Glued laminated and built-up columns are usually square or rectangular, but can also be made with flanges. Because of fabrication and handling difficulties, these shapes are not common, and it is recommended that the designer check with the fabricator prior to designing a column with flanges. The strength of flanged columns is somewhat less than would be determined by use of the standard column equation because the thin outstanding flanges are subject to buckling and there is incomplete shear transfer when mechanical fastenings are used. An equation that can be converted to a design equation by dividing by 2.74 (the same factor used for reducing the ultimate buckling strength determined by the Euler equation) can be found in the *Wood Handbook* (8). A design value, $F'_c$, for the flanges of *H* and *I* shapes (shown in Fig. 5.16) can be obtained from Fig. 5.15 or calculated

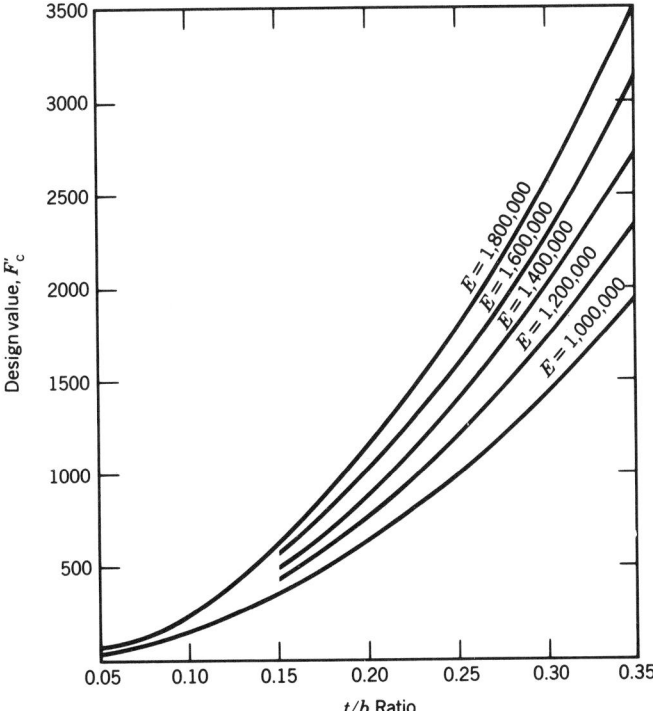

**FIGURE 5.15**   Design values for columns with outstanding flanges. $F_c$ determined by Eq. (5-19) or from figure should not exceed the design value for the grade or combination.

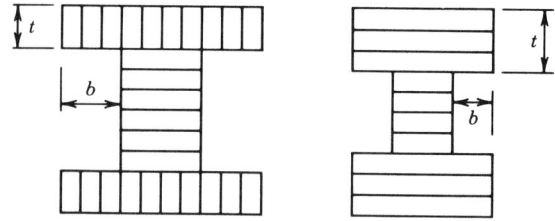

**FIGURE 5.16**    Glued laminated columns with outstanding flanges.

as follows:

$$F'_c = 0.016E' \left(\frac{t}{b}\right)^2 \tag{5-19}$$

where    $E'$ = tabular modulus of elasticity (psi) adjusted by applicable adjustment factors

$t$ = thickness of the flange (in.), and

$b$ = width of outstanding flange (in.).

The design value for the entire column cross section must not be exceeded after application of the appropriate column equation.

Fig. 5.15 may be used to determine design values, $F'_c$, for flanges of columns for various $E$ and thickness–breadth, $t/b$, ratios. The reduction of design values varies with modulus of elasticity, and in general, no reduction is necessary for $t/b$ ratios of more than 0.35.

The column equation must be modified to reflect the radius of gyration of the shape being designed. The tabular design value for compression parallel to grain $F_c$ to use in determining $F'_c$ is that shown for four or more laminations of the combination being used for the arrangements shown in Fig. 5.16. When the flanges contain fewer than four laminations and the $t/b$ is one or more, the tabular design value of $F_c$ for four or more laminations may be used. When $t/b$ is less than 1 and the flanges contain fewer than four laminations, the tabular design value to use is that for two or three laminations.

## 5.9    COMBINED BENDING AND AXIAL STRESSES

A combination of bending and axial stresses must be taken into account in design by use of the combined bending and axial stresses equations.

### 5.9.1    Combined Bending and Axial Tension

For a combination of bending and axial tension stresses, the equations for checking strength are:

$$\frac{f_t}{F'_t} + \frac{f_b}{F^*_b} \leq 1 \tag{5-20}$$

and

$$\frac{f_b - f_t}{F_b^{**}} \le 1 \tag{5-21}$$

where  $f_t$ = actual tension stress parallel to grain (psi),
$F_t'$ = tabular tension design value parallel to grain (psi) multiplied by applicable adjustment factors,
$f_b$ = actual bending stress (psi),
$F_b^*$ = tabular design value in bending (psi) multiplied by all applicable adjustment factors except $C_L$, and
$F_b^{**}$ = tabular design value in bending (psi) multiplied by all applicable adjustment factors except $C_V$.

The volume factor applies to $F_{bx}$ on the tension side of a glued laminated timber bending member; therefore, it is always applied to a member subjected to combined tension and bending about the $x$-$x$ axis. When bending is about the $y$-$y$ axis, $F_{by}$ should be used and adjusted by the flat use factor.

The beam stability factor $C_L$ affects the compression side and is offset by the axial tension stresses in the member. The bending stress, $f_b$, minus the axial tension stress, $f_t$, should not exceed $F_b^{**}$. Therefore, the stability check Eq. (5-21) should be used.

The designer should design a tension member by determining the required tabular design value for both tension and bending. However, the specified design values should be realistic in that combinations are reasonably available that can fulfill the stress requirements. If bending is the predominant stress, the combinations in Table 1, AITC 117—Design (1), can be used as guides. If tension is the predominant stress, one of the combinations in Table 2, AITC 117—Design (1), may be more appropriate than those in Table 1 because these combinations were developed primarily for axial stresses.

**Example.**   Select a 10-ft-long glued laminated timber tension member made of Western Species to resist a 30,000-lb axial tension load and a moment of 75,000 in.-lb about the $x$-$x$ axis. No lateral support is provided between the ends of the member. Design for normal duration of load and dry condition of use. Ends of member are laterally supported to prevent lateral rotation.

$$F_{by} = 2100 \text{ psi} \quad F_{bx} = 2000 \text{ psi}$$

$$F_t = 1450 \text{ psi} \quad E_x = E_y = 1,800,000 \text{ psi}$$

Use all adjustment factors as unity for the determination of a trial size.

$$F_t' = F_t = 1450 \text{ psi}$$

$$E' = E = 1,800,000 \text{ psi (either } E_x \text{ or } E_y)$$

Assume a trial size of $5\frac{1}{8} \times 10\frac{1}{2}$ in.

$$A = 53.81 \text{ in.}^2 \quad S = 94.17 \text{ in.}^3$$

$$K_L = 1.0$$

$$C_V = \left(\frac{5.125}{5.125}\right)^{1/10} \left(\frac{12}{10.5}\right)^{1/10} \left(\frac{21}{10}\right)^{1/10} = 1.09, \text{ use } C_V = 1$$

$$F_{bx}^* = F_{bx} C_V = (2000)(1) = 2000 \text{ psi}$$

$$f_t = \frac{30,000}{53.81} = 557 \text{ psi}$$

$$f_{bx} = \frac{75,000}{94.17} = 796 \text{ psi}$$

Check lateral support. From Fig. 5.5$c$ for equal end moments,

$$l_e = 1.84 l_u = (1.84)(10)(12) = 220.8 \text{ in.}$$

$$R_B = \sqrt{\frac{l_e d}{b^2}} = \sqrt{\frac{(220.8)(10.5)}{(5.125)^2}} = 9.4$$

Determine $C_L$

$$C_L = \frac{1 + F_{bE}/F_{bx}}{1.9} - \sqrt{\left[\frac{1 + F_{bE}/F_{bx}}{1.9}\right]^2 - \frac{F_{bE}/F_{bx}}{0.95}}$$

$$K_{bE} = 0.609 \text{ [from Eq. (5-10)]}$$

$$F_{bE} = \frac{K_{bE} E'}{R_B^2} = \frac{(0.609)(1,800,000)}{9.4^2} = 12,419 \text{ psi}$$

$$C_L = \frac{1 + 12,419/2000}{1.9} - \sqrt{\left[\frac{1 + 12,419/2000}{1.9}\right]^2 - \frac{12,419/2000}{0.95}} = 0.990$$

$$\frac{f_t}{F_t'} + \frac{f_b}{F_{bx}^*} = \frac{557}{1450} + \frac{796}{2000} = 0.782 < 1 \qquad \text{O.K.}$$

$$F_{bx}^{**} = F_{bx} C_L = (2000)(0.99) = 1980 \text{ psi}$$

$$\frac{f_{bx} - f_t}{F_{bx}^{**}} \le 1.0, \quad \frac{(796 - 557)}{1980} = 0.12 < 1 \qquad \text{O.K.}$$

A $5\frac{1}{8} \times 10\frac{1}{2}$-in. member is satisfactory.

If the length of the member had been 30 ft rather than 10 ft,

$$l_e = (1.84)(30)(12) = 662.4 \text{ in.}$$

$$R_B = \sqrt{\frac{(662.4)(10.5)}{5.125^2}} = 16.27$$

$$F_{bE} = \frac{K_{bE}E'}{R_B^2} = \frac{(0.609)(1,800,000)}{(16.27)^2} = 4140 \text{ psi}$$

$$C_L = \frac{1 + 4140/2000}{1.9} - \sqrt{\left[\frac{1 + 4140/2000}{1.9}\right]^2 - \frac{4140/2000}{0.95}} = 0.959$$

$$C_V = \left(\frac{5.125}{5.125}\right)^{1/10} \left(\frac{12}{10.5}\right)^{1/10} \left(\frac{21}{30}\right)^{1/10} = 0.978$$

$$\frac{557}{1450} + \frac{796}{(2000)(0.978)} = 0.791 \leq 1 \qquad \text{O.K.}$$

$$\frac{f_b - f_t}{F_{bx}^{**}} = \frac{796 - 557}{(2000)(0.959)} = 0.125 < 1 \qquad \text{O.K.}$$

### 5.9.2   Combined Bending and Axial Compression

When a member is subjected to both bending and axial compression, the combination of the compressive stresses and the tensile stresses caused by bending results in a smaller absolute value of stress on the tension side than on the compression side. This affects the manner in which the volume factor $C_V$ and the beam stability factor $C_L$ are applied. The size of member affects the bending strength of the member because the ability to resist tension on the tension side of a bending member decreases as the size increases. This is accounted for in design by the application of the volume factor $C_V$. When a member is subjected to both axial compressive and bending stresses, the compressive stresses reduce the net stress on the tension side. When $C_V < 1$ and the axial compressive stress $f_c$ is equal to or larger than $F_b^* (1 - C_V)$, the stress on the tension side has already been reduced and no further reduction by the volume factor is needed. Note that $F_b^* = F_b$ multiplied by all applicable adjustment factors except $C_L$ and $C_V$ does not apply. When the axial compressive stress $f_c$ is less than $(F_b) (1 - C_V)$, the design value in bending, $F_b'$, should not exceed that determined as follows:

$$F_b' = F_b^* C_V + f_c, \text{ but in no case greater than } F_b^*$$

where   $F_b^*$ = tabular design value in bending (psi) adjusted by all applicable adjustment factors except $C_L$ and $C_V$ and other terms are as previously noted.

The beam stability factor $C_L$ is applied to prevent buckling of the compression side of a member subjected to bending. Thus, reductions for lateral stability (compression effect) and volume (tension effect) are not cumulative. In beams with combined axial compression and bending stresses, the design value in bending, $F_b'$, should not exceed $F_b^* C_L$ or $(F_b^* C_V + f_c)$, whichever is smaller. When the interaction factor $C_I$ is involved, it is cumulative with $C_L$ when the taper cut is on the compression side. In this case, the design value in bending $F_b'$, is calculated as follows:

$$F_b' = F_b^* C_L C_I \text{ or } F_b^* C_V, \text{ whichever is smaller}$$

Members subject to an axial compression force and bending about one or more axes are designed so that the following apply:

$$\left[\frac{f_c}{F_c'}\right]^2 + \frac{f_{b1}}{F_{b1}'[1 - (f_c/F_{cE1})]}$$

$$+ \frac{f_{b2}}{F_{b2}'[1 - (f_c/F_{cE2}) - (f_{b1}/F_{bE})^2]} \leq 1 \qquad (5\text{-}22)$$

where   $f_c$ = actual applied compressive stress (psi),
$\quad f_{b1}$ = actual edgewise bending stress about the $x$-$x$ axis (psi),
$\quad f_{b2}$ = actual flatwise bending stress about the $y$-$y$ axis (psi),
$\quad d_1$ = wide face dimension (depth about the $x$-$x$ axis) (in.),
$\quad d_2$ = narrow face dimension (depth about the $y$-$y$ axis) (in.),
$\quad F_c'$ = tabulated design value in compression parallel to grain multiplied by all applicable adjustment factors (psi),
$\quad F_{b1}' = F_{bx}'$ = tabulated design value in bending about the $x$-$x$ axis multiplied by all applicable adjustment factors (psi),
$\quad F_{b2}' = F_{by}'$ = tabulated design value in bending about the $y$-$y$ axis multiplied by all applicable adjustment factors (psi),
$\quad F_{cE1} = \dfrac{K_{cE} E'}{(l_{e1}/d_1)^2}$ for either uniaxial or biaxial bending.

Where notations for columns are defined in 5.8.2:

$$F_{cE2} = \frac{K_{cE} E'}{(l_2/d_2)^2} \text{ for biaxial bending}$$

$$F_{bE} = \frac{K_{bE} E'}{R_B^2} \text{ for biaxial bending.}$$

For compression stress, $f_{c1}$ must be less than $F_{cE1}$ for either uniaxial or biaxial bending, $f_{c2}$ must be less than $F_{cE2}$ for biaxial bending, and $f_{b1}$ must be less than $F_{bE}$ for biaxial bending. The subscript 1 used for $f_{b1}$ and $F_{b1}$ indicates bending about the $x$-$x$ axis, and the subscript 2 indicates bending about the $y$-$y$ axis in biaxial bending.

When bending exists only about the $x$-$x$ axis, the third term of Eq. (5-21) drops out and the equation becomes:

$$\left[\frac{f_c}{F_c'}\right]^2 + \frac{f_{b1}}{F_{b1}'[1 - (f_c/F_{cE1})]} \leq 1 \qquad (5\text{-}23)$$

When bending exists about the $y$-$y$ axis, only the second term drops out because $f_{b1}$ becomes zero, and the equation becomes:

$$\left[\frac{f_c}{F_c'}\right]^2 + \frac{f_{b2}}{F_{b2}'[1 - (f_c/F_{cE2})]} \leq 1 \tag{5-24}$$

The effective column lengths $l_{e1}$ and $l_{e2}$ are determined by use of Table 5.6, and the effective length of bending members is determined by the procedures in Figs. 5.5 and 5.6. The load duration factor $C_D$ associated with the shortest time duration shall be determined in accordance with Table 4.7 or Fig. 4.4 as explained in 4.6.1. All applicable load combinations must be evaluated to determine the critical load combination.

**Example.** In a member subjected to both a bending and compression load, determine $F_b'$ when volume factor $C_V = 0.80$ and the beam stability factor $C_L = 0.90$. Loads are primarily snow ($C_D = 1.15$). Assume a 24F combination is being used; $F_b = 2400$ psi. Actual compression stress, $f_c = 1400$ psi:

$$F_b^* = F_b C_D = (2400)(1.15) = 2760 \text{ psi}$$

$$F_b^*(1 - C_V) = (2760)(1 - 0.80) = 552 \text{ psi} < f_c, 1400 \text{ psi}$$

Therefore

$$F_b' = F_b^* C_V + f_c = (2760)(0.8) + 1400 = 3608 \text{ psi}$$

$$3608 > 2760, \text{ use } 2760 \text{ psi}$$

or

$$F_b' = F_b^* C_L = (2760)(0.9) = 2484 \text{ psi}$$

Use $F_b' = 2480$ psi.

Glued laminated arches, discussion beginning in 5.13, are another example of members subjected to a combination of axial compression and bending stresses. In most cases, the value of $f_c$ is large enough so that no volume factor $C_V$ need be applied to arches, but a check for applicability should be made as shown in the example in 5.13.3.

**Example 1.** Select a 20-ft-long glued laminated timber simple span beam to resist a 240 plf uniform total load acting perpendicular to the wide faces of the laminations of the member and an axial compressive load of 20,000 lb. The member is braced perpendicular to the plane of the $y$-$y$ axis at the ends and throughout its length at the top. Assume normal duration of load and dry conditions of use. For a trial design, use $F_{bx} = 2000$ psi, $F_c = 1450$ psi, $F_v = 200$ psi, $E_x = 1,500,000$ psi, $E_y = 1,400,000$ psi. Try a 5 × 12⅜-in. Southern Pine member: $A = 61.88$ in.$^2$, $S = 127.6$ in.$^3$, $F_{bx}^* = F_{bx}$, $E_x' = E_x$, and $E_y' = E_y$.

The beam is braced along the top and at the ends. Determine $f_c$ and $F_c'$

$$f_c = \frac{20,000}{61.88} = 323 \text{ psi}$$

$$F_c^* = F_c = 1450 \text{ psi}$$

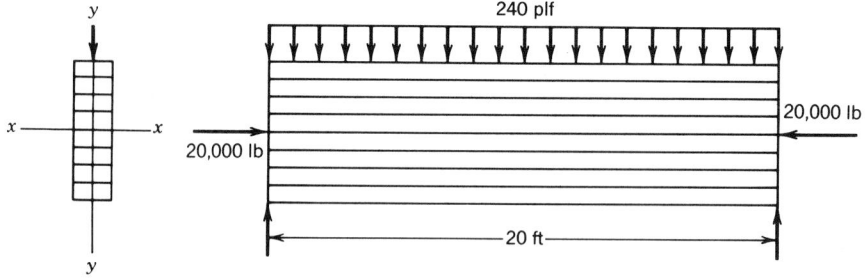

Member is braced from buckling about the $y$–$y$ axis,

$$K_e = 1, \ l_e \doteq 20(1)(12) = 240 \text{ in.}$$

$$\frac{l_e}{d} = \frac{240}{12.375} = 19.39$$

$$F_{cEx} = \frac{K_{cE}E'_x}{(l_e/d)^2} = \frac{(0.418)(1,500,000)}{(19.39)^2} = 1667 \text{ psi}$$

$$C_P = \frac{1 + F_{cE}/F^*_c}{2c} - \sqrt{\left[\frac{1 + F_{cE}/F^*_c}{2c}\right]^2 - \frac{F_{cE}/F^*_c}{c}}$$

$$C_P = \frac{1 + 1667/1450}{(2)(0.9)} - \sqrt{\left[\frac{1 + 1667/1450}{(2)(0.9)}\right]^2 - \frac{1667/1450}{0.9}} = 0.808$$

$$F'_c = F^*_c C_P = (1450)(0.808) = 1172 \text{ psi}$$

Determine $f_{bx}$ and $F'_{bx}$

$$M = \frac{wL^2}{8} = \frac{(240)(20)^2(12)}{8} = 144,000 \text{ in.-lb}$$

$$f_b = \frac{M}{S} = \frac{144,000}{127.6} = 1128 \text{ psi}$$

$$C_V = \left(\frac{5.125}{5}\right)^{1/20}\left(\frac{21}{20}\right)^{1/20}\left(\frac{12}{12.375}\right)^{1/20} = 1.00$$

$$F'_b = F^*_b(C_V) = 2000(1) = 2000 \text{ psi}$$

$$\left[\frac{f_c}{F'_c}\right]^2 + \frac{f_{bx}}{F'_{bx}(1 - f_c/F_{cE1})}$$

$$\left(\frac{323}{1172}\right)^2 + \frac{1128}{2000(1 - 323/1667)} = 0.776 < 1 \qquad \text{O.K.}$$

Use $5 \times 12\frac{3}{8}$-in. beam.

Check shear.

$$L_e = L - 2d = 20 - (2)(1.03) = 17.94 \text{ ft}$$

$$V = \frac{(17.94)(240)}{2} = 2153 \text{ lb}$$

$$f_v = \frac{3V}{2A} = \frac{(3)(2153)}{(2)(61.88)} = 52 \text{ psi}$$

A 20F Southern Pine combination with the lowest design value in shear (90 psi) could be used.

Use any combination of Southern Pine glued laminated timber that provides the following design values or greater.

$$F_{bx} = 2000 \text{ psi} \quad F_c = 1450 \text{ psi} \quad F_v = 90 \text{ psi}$$

$$E_x = 1,500,000 \text{ psi} \quad E_y = 1,400,000 \text{ psi}$$

Note that $E_y$ was not used in this example, but will be used in Example 2.

**Example 2.**  Assume that the member in the preceding example was not braced along the top, and all other conditions are the same except that the uniform load is 200 plf. The lack of top bracing creates a tendency for the member to buckle in bending and compression about the $y$-$y$ axis, and a wider beam ($6\frac{3}{4}$ in.) with the same depth ($12\frac{3}{8}$ in.) will be assumed for a trial size.

$$A = 83.53 \text{ in.}^2 \quad S = 172.3 \text{ in.}^3$$

1.  Determine $f_c$ and $f_{bx}$

$$f_c = \frac{P}{A} = \frac{20,000}{83.53} = 240 \text{ psi}$$

$$f_{bx} = \frac{M_x}{S_x} = \frac{(200)(20)^2(12)}{(8)(172.3)} = 698 \text{ psi}$$

2.  Determine $C_V$

$$C_V = \left(\frac{5.125}{b}\right)^{1/20} \left(\frac{12}{d}\right)^{1/20} \left(\frac{21}{L}\right)^{1/20}$$

$$= \left(\frac{5.125}{6.75}\right)^{1/20} \left(\frac{12}{12.375}\right)^{1/20} \left(\frac{21}{20}\right)^{1/20} = 0.987$$

3.  Determine $C_L$

   The member is unbraced, but ends are held in place.

$$l_u = (20)(12) = 240 \text{ in.}, \quad l_u/d = \frac{240}{12.375} = 19.39 \text{ in.}$$

$$l_e = 1.63 l_u + 3d \text{ (from Table 5.3)}$$

$$l_e = 1.63(240) + (3)(12.375) = 428.3 \text{ in.}$$

$$R_B = \sqrt{\frac{l_e d}{b^2}} = \sqrt{\frac{(428.3)(12.375)}{6.75^2}} = 10.78 < 50 \qquad \text{O.K.}$$

$$F_{bE} = \frac{(K_{bE})(E_y')}{R_B^2} = \frac{(0.609)(1,400,000)}{10.78^2} = 7329 \text{ psi}$$

$$F_{bx}^* = (2000)(C_D) = 2000 \text{ psi}$$

$$C_L = \frac{1 + F_{bE}/F_{bx}^*}{1.9} - \sqrt{\left[\frac{1 + F_{bE}/F_{bx}^*}{1.9}\right]^2 - \frac{F_{bE}/F_{bx}^*}{0.95}}$$

$$C_L = \frac{1 + 7329/2000}{1.9} - \sqrt{\left[\frac{1 + 7329/2000}{1.9}\right]^2 - \frac{7329/2000}{0.95}} = 0.982$$

$$0.982 < 0.987, \ C_L \text{ controls}$$

$$F_{bx}' = F_{bx}^* C_L = (2000)(0.982) = 1964 \text{ psi}$$

4.  Determine $C_P$ and $F_c'$

Member is unbraced between ends. Compression buckling is about $y$–$y$ axis.

$$K_e = 1, \ l_e = (20)(12)(1) = 240 \text{ in.}$$

$$l_e/d = \frac{240}{6.75} = 35.56 < 50 \qquad \text{O.K.}$$

$$F_c^* = F_c C_D = (1450)(1) = 1450 \text{ psi}$$

$$K_{cE} = 0.418, \ c = 0.9 \ [\text{from Eq. (5-14)}]$$

$$F_{cE} = \frac{K_{cE} E_y'}{(l_e/d)^2} = \frac{(0.418)(1,400,000)}{35.56^2} = 463 \text{ psi}$$

$$C_P = \frac{1 + F_{cE}/F_c^*}{2c} - \sqrt{\left[\frac{1 + F_{cE}/F_c^*}{2c}\right]^2 - \frac{F_{cE}/F_c^*}{c}}$$

$$C_P = \frac{1 + (463/1450)}{(2)(0.9)}$$

$$- \sqrt{\left[\frac{1 + (463/1450)}{(2)(0.9)}\right]^2 - \frac{463/1450}{0.9}} = 0.306$$

$$F_c' = F_c^* C_P = (1450)(0.306) = 443 \text{ psi}$$

5.   Check the combined bending and axial compression equation.

$$\left[\frac{f_c}{F'_c}\right]^2 + \frac{f_{bx}}{F'_{bx}(1 - f_c/F_{cE1})} \le 1$$

$$F_{cE1} = \frac{K_{cE}E'_x}{(l_1/d_1)^2} = \frac{(0.418)(1,500,000)}{(240/12.375)^2} = 1667 \text{ psi}$$

$$\left[\frac{240}{443}\right]^2 + \frac{698}{1964(1 - 240/1667)} = 0.706 < 1 \qquad \text{O.K.}$$

A check of a $6\frac{3}{4} \times 11$ in. beam using the same procedures shows that the combined stress equation is less than 1.

Use a $6\frac{3}{4} \times 11$ in. Southern Pine beam

### 5.9.3   Members Subjected to a Compression Load and Bending about Both Axes

The basic equation for combined bending and axial loads [Eq. (5-22)] is used.

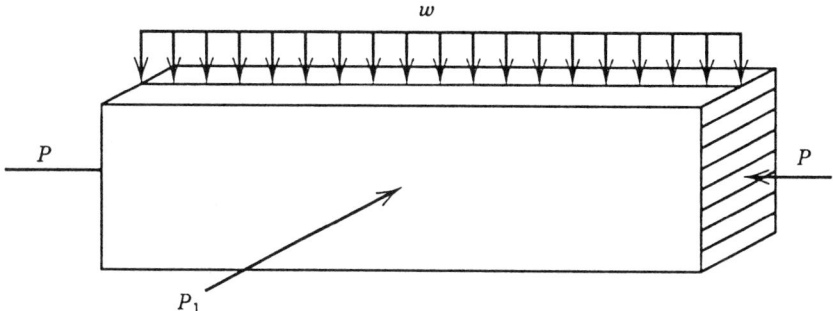

**Example.**   A 30-ft long glued laminated timber beam used in a dry location is loaded about the $x$–$x$ axis with a uniform load consisting of a dead load of 200 plf and snow load of 400 plf. A side load $P_1$ consisting of a 2500 lb wind load is applied at the midpoint. In addition, the member is loaded in compression with a 30,000 lb load $P$ consisting of a 10,000 lb dead load and a 20,000 lb snow load. The member is supported at the ends to prevent rotation, but is not braced along its length. The supports provide an adequate area so that $F_{c\perp}$ is not exceeded.

For the trial design, use a 24F combination of visually graded western species that provides the following tabular design values:

$F_{bx} = 2400$ psi, $F_{by} = 1500$ psi, $F_c = 1450$ psi, $F_{vx} = 155$ psi, $F_{vy} = 70$ psi

$E_x = 1,700,000$ psi, $E_y = 1,500,000$ psi, $E_{axial} = 1,500,000$ psi

Assume an $8\frac{3}{4} \times 24$-in. beam (Western Species).

$A = 210.0$ in.$^2$, $S_x = 840.0$ in.$^3$, $S_y = 306.3$ in.$^3$ (from Table 8.2)

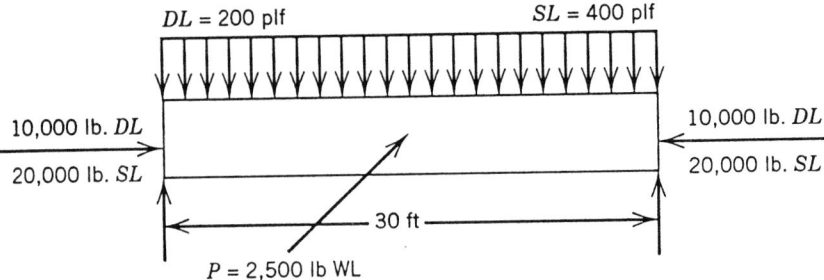

1. Determine $C_D$

$C_D$ associated with the load of the shortest duration can be used. Since $f_{by}$ is caused by wind stress and both $f_{bx}$ and $f_c$ are caused by snow load and dead load, each load must be considered. The duration of load adjustment factor for bending about the $x$-$x$ axis and the compressive load $C_D = 1.15$. Dead load alone cannot control because it is less than 90% of the total. The member must be proportioned so that it can resist the bending about the $x$-$x$ axis using a $C_D$ of 1.15 when the wind is not blowing. When the wind is blowing, $F_{bx}$, $F_c$, and $F_{by}$ can all be adjusted by $C_D = 1.6$.

For the first trial size, the assumption is made that $C_D = 1.6$ applies to all loads.

$$F^*_{bx} = (2400)(1.6) = 3840 \text{ psi}, \quad F^*_{by} = (1500)(1.6) = 2400 \text{ psi}$$

$$F^*_c = (1450)(1.6) = 2320 \text{ psi}, \quad F'_v = (155)(1.6) = 248 \text{ psi}$$

$$E'_x = E_x = 1,700,000 \text{ psi}, \quad E'_y = E_y = 1,500,000 \text{ psi}$$

2. Determine $f_c$, $f_{bx}$, and $f_{by}$

$$f_c = \frac{P}{A} = \frac{30,000}{210} = 143 \text{ psi}$$

$$f_{bx} = \frac{M_x}{S_x} = \frac{wL^2}{8S_x} = \frac{(600)(30)^2(12)}{(8)(840)} = 964 \text{ psi}$$

$$f_{by} = \frac{M_y}{S_y} = \frac{PL}{4S_y} = \frac{(2500)(30)(12)}{(4)(306.3)} = 735 \text{ psi}$$

3. Determine $C_V$

Bending about $x$-$x$ axis for Western Species

$$C_V = \left(\frac{5.125}{b}\right)^{1/10} \left(\frac{12}{d}\right)^{1/10} \left(\frac{21}{L}\right)^{1/10}$$

$$= \left(\frac{5.125}{8.75}\right)^{1/10} \left(\frac{12}{24}\right)^{1/10} \left(\frac{21}{30}\right)^{1/10} = 0.853$$

3840 $(1 - 0.853) = 564$ psi $> 143$ psi, therefore $C_V$ does apply as follows: $F'_{bx} = F^*_b C_V + f_c$.

For bending about $y$-$y$ axis, $C_V$ does not apply. When members are stressed in bending about $y$-$y$ axis, $C_{fu}$ applies. From Table 4.9, $C_{fu}$ for 8.75 in. depth = 1.04.

4.  Determine $C_L$ for bending about $x$-$x$ axis.

$F^*_{bx} = (2400)(1.6) = 3840$ psi

$K_{bE} = 0.609$ [from Eq. (5-10)]

$l_u = 30(12) = 360$ in.

$l_u/d = 360/24 = 15$

$l_e = 1.63 l_u + 3d = 1.63(360) + 3(24) = 658.8$ in.

$$R_B = \sqrt{\frac{l_e d}{b^2}} = \sqrt{\frac{(658.8)(24)}{8.75^2}} = 14.37$$

$$F_{bE} = \frac{K_{bE} E'}{R_B^2} = \frac{(0.609)(1,500,000)}{14.37^2} = 4423 \text{ psi}$$

$$C_L = \frac{1 + (F_{bE}/F^*_{bx})}{1.9} - \sqrt{\left[\frac{1 + (F_{bE}/F^*_{bx})}{1.9}\right]^2 - \frac{F_{bE}/F^*_{bx}}{0.95}}$$

$$C_L = \frac{1 + 4423/3840}{1.9} - \sqrt{\left[\frac{1 + (4423/3840)}{1.9}\right]^2 - \frac{4423/3840}{0.95}} = 0.868$$

5.  Determine $F'_{bx}$, use the minimum of:

$$F'_{bx} = F^*_{bx} C_V + f_c = (3840)(0.853) + 143 = 3419 \text{ psi}$$

or $F'_{bx} = F^*_{bx} C_L = 3840(0.868) = 3333$ psi; use $F'_{bx} = 3333$ psi

6.  Determine $F'_{by}$

$C_L$ for bending about $y$-$y$ axis is 1.

$$F'_{by} = F_{by}(C_D)(C_{fu}) = 1500(1.6)(1.04) = 2496 \text{ psi}$$

7.  Determine $C_P$

$K_e = 1 \quad l_e = (1)(30)(12) = 360$ in.

$K_{cE} = 0.418, \ c = 0.9$ [from Eq. (5-14)]

$$F_{cEy} = \frac{K_{cE} E'_y}{(l_e/d)^2} = \frac{(0.418)(1,500,000)}{(360)/8.75^2} = 370 \text{ psi}$$

$$C_P = \frac{1 + (F_{cE}/F_c^*)}{2c} - \sqrt{\left[\frac{1 + F_{cE}/F_c^*}{2c}\right]^2 - \frac{F_{cE}/F_c^*}{c}}$$

$$C_P = \frac{1 + 370/2320}{2(0.9)} - \sqrt{\left[\frac{1 + 370/2320}{(2)(0.9)}\right]^2 - \frac{370/2320}{0.9}} = 0.157$$

8.  Determine $F_c'$

$$F_c' = (F_c)(C_D)(C_P) = (1450)(1.6)(0.157) = 364 \text{ psi}$$

9.  Determine summation of ratios

$$\left(\frac{f_c}{F_c'}\right)^2 + \frac{f_{bx}}{F_{bx}'[1 - (f_c/F_{cE1})]} + \frac{f_{by}}{F_{by}'[1 - f_c/F_{cE2} - (f_{bx}/F_{bE})^2]} \le 1$$

$$F_{cE1} = F_{cEx} = \frac{(0.418)(1{,}700{,}000)}{\left(\frac{(30)(12)}{24}\right)^2} = 3158 \text{ psi}$$

$$F_{cE2} = F_{cEy} = 370 \text{ psi}$$

$$\left(\frac{143}{364}\right)^2 + \frac{964}{3333\left[1 - \frac{143}{3158}\right]} + \frac{735}{2496\left[1 - \frac{143}{370} - \left(\frac{964}{4423}\right)^2\right]} = 0.977 \quad \text{O.K.}$$

10.  Check with wind load omitted.

When wind load is not acting, the combined stress equation is:

$$\left(\frac{f_c}{F_c'}\right)^2 + \frac{f_{bx}}{F_{bx}'[1 - (f_c/F_{cE1})]}$$

$f_c = 143$ psi, $f_{bx} = 964$ psi (previously calculated)

$C_V = 0.853$ (previously calculated)

$F_{bx}^* = (F_{bx}C_D) = (2400)(1.15) = 2760$ psi

$F_{bE} = 4423$ psi (previously calculated)

$$C_L = \frac{1 + 4423/2760}{1.90} - \sqrt{\left[\frac{1 + 4423/2760}{1.90}\right]^2 - \frac{4423/2760}{0.95}} = 0.935$$

$F_{bx}' = (2760)(0.853) + 143 = 2497$ psi

or $F_{bx}' = (2760)(0.935) = 2581$ psi    Use $F_{bx}' = 2497$ psi

$F_{cE} = 370$ psi (previously calculated)

$F_c^* = (1450)(1.15) = 1667$ psi

$$C_P = \frac{1 + 370/1667}{(2)(0.9)} - \sqrt{\left[\frac{1 + 370/1667}{(2)(0.9)}\right]^2 - \frac{370/1667}{0.9}} = 0.216$$

$$F_c' = (1667)(0.216) = 360 \text{ psi}$$

$$\left(\frac{143}{360}\right)^2 + \frac{964}{2497\,[1 - (143/4423)]} = 0.549 < 1 \qquad \text{O.K.}$$

### 5.9.4 Columns with Side Loads and Eccentricity

Although members used as beam columns subject to combined axial and bending loading are usually designed in the same manner, members designed as columns required special comment because of the manner in which bending moment may be applied. Combined axial and bending loads may occur in any one of the following bending conditions:

1. Axial compression plus applied end moment
2. Concentric end load and side loads
3. Eccentric end load only
4. Combined end loads, side loads, and eccentricity
5. Loads applied at brackets on columns

The design of columns in each case is usually a trial and error process in which a tentative size is selected and analyzed. If the size is too large or too small, adjustments in size are made until the design criteria are satisfied.

The basic equation for columns subjected to eccentric loads, side loads, or a combination of some or all end loads is:

$$\left(\frac{f_c}{F_c'}\right)^2 + \frac{f_{b1} + f_c(6e_1/d_1)[1 + 0.234(f_c/F_{cE1})]}{F_{b1}'[1 - (f_c/F_{cE1})]}$$

$$+ \frac{f_{b2} + f_c(6e_2/d_2)\left(1 + 0.234(f_c/F_{cE2}) + 0.234\left[\dfrac{f_{b1} + f_c(6e_1/d_1)}{F_{bE}}\right]^2\right)}{F_{b2}'\left(1 - (f_c/F_{cE2}) - \left[\dfrac{f_{b1} + f_c(6e_1/d_1)}{F_{bE}}\right]^2\right)} \le 1.0$$

$$\tag{5-25}$$

See Section 5.95 for explanation of symbols.

### 5.9.5 Eccentric Load

When eccentric end load only exists, the general equation reduces to:

$$\left[\frac{f_c}{F_c'}\right]^2 + \frac{f_c(6e_1/d_1)[1 + 0.234(f_c/F_{cE1})]}{F_{b1}'[1 - (f_c/F_{cE1})]}$$

$$+ \frac{f_c(6e_2/d_2)\left(1 + 0.234(f_c/F_{cE2}) + 0.234\left[\dfrac{f_c(6e_1/d_1)}{F_{bE}}\right]^2\right)}{F_{b2}'\left(1 - (f_c/F_{cE2}) - \left[\dfrac{f_c(6e_1/d_1)}{F_{bE}}\right]^2\right)} \le 1.0 \qquad (5\text{-}26)$$

where   $f_c < F_{cE1} = K_{cE}E'/(l_{e1}/d_1)^2$ for either uniaxial or biaxial bending

$$\text{AND}$$

$f_c < F_{cE2} = K_{cE}E'/(l_{e2}/d_2)^2$ for biaxial bending

$$\text{AND}$$

$f_{b1} < F_{bE} = K_{bE}E'/(R_B)^2$ for biaxial bending

$f_c$ = compression stress parallel to grain due to axial load (psi),

$f_{b1}$ = edgewise bending stress due to side loads on narrow face only (psi),

$f_{b2}$ = flatwise bending stress due to side loads on wide face only (psi),

$F_c'$ = allowable compression design value parallel to grain that would be permitted if axial compressive stress only existed, determined in accordance with 4.6 and 5.8 (psi),

$F_{b1}'$ = allowable edgewise bending design value that would be permitted if edgewise bending stress only existed, determined in accordance with 4.6 and 5.4 (psi),

$F_{b2}'$ = allowable flatwise bending design value that would be permitted if flatwise bending stress only existed, determined in accordance with 4.6 and 5.4 (psi),

$K_{cE}$ = 0.3 for visually graded lumber and machine evaluated lumber (MEL),

$K_{cE}$ = 0.418 for products with $COV_E \le 0.11$,

$K_{bE}$ = 0.438 for visually graded lumber and machine evaluated lumber (MEL),

$K_{bE}$ = 0.609 for products with $COV_E \le 0.11$,

$R_B$ = slenderness ratio of bending member (see 5.4.11),

$d_1$ = wide face dimension (in.),

$d_2$ = narrow face dimension (in.),

$e_1$ = eccentricity, measured parallel to wide face from centerline of column to centerline of axial load (in.), and

$e_2$ = eccentricity, measured parallel to narrow face from centerline of column to centerline of axial load (in.).

$l_{e1}$ and $l_{e2}$ are determined in accordance with 5.4.14, $F_{cE1}$ and $F_{cE2}$ are determined in accordance with Eq. (5-18), and $F_{bE}$ is determined in accordance with Eq. (5-10).

Eqs. (5-24) and (5-25) take into account the *P*-$\Delta$ effect of the end load caused by buckling in both the *x-x* and *y-y* planes. If the loading is such that bending occurs about only one axis, either the second or third term of the equation will drop out, depending on the direction of the bending moment remaining. Eqs. (5-25) and (5-26), as printed in the NDS® (3) and as included in this manual, show the subscript "1" for bending about the *x-x* axis and the subscript "2" for bending about the *y-y* axis. The notations $f_{b1}$ and $F_{b1}$ are frequently shown in the literature as $f_{bx}$ and $F_{bx}$. In a like manner, $f_{b2}$ and $F_{b2}$ are shown as $f_{by}$ and $F_{by}$.

The value of $F_b'$ is determined by multiplying $F_b$ by all applicable adjustment factors, except that $C_V$ and $C_L$ are not applied cumulatively. When axial compressive stress, $f_c$, equals or exceeds $F_b^*(1 - C_V)$, $C_V$ does not apply because the net stress on the face of the member where bending creates a tensile stress will not exceed $F_b^* C_V$. For information on the application of $C_L$, see 5.4.11.

There is some question about using the reduced value of $F_b'$ based on lateral buckling in bending when column buckling is about a different axis. Most references use the more conservative approach of modifying $F_b$ by the beam stability factor, $C_L$, regardless of the direction of column buckling. This procedure is used in the examples in this manual. When members are designed primarily as compression members, the depth–breadth ratio is usually close to 1, and the effect of using the more conservative approach is small. When the depth–breadth ratio is larger, in the range of 3–7, as is the case with members designed primarily as beams, the effect of including $C_L$ in the equation is more significant, but the beam buckling and the column buckling are more likely to be about the same axis. $C_L$ is not usually applied in the design of arches.

**Example.**   Check the design of a 5 $\times$ 9-$\frac{5}{8}$ in. glued laminated timber column for the loading conditions illustrated. (*Note:* A 5-in. width is used for Southern Pine). The column is braced at both ends in the $x$ and $y$ directions. Pinned ends are assumed in the design. For purposes of illustration, the dry-use stresses of Southern Pine combination 16F-V3 will be used. In practice, a combination from Table 2, AITC 117—Design would probably be more efficient.

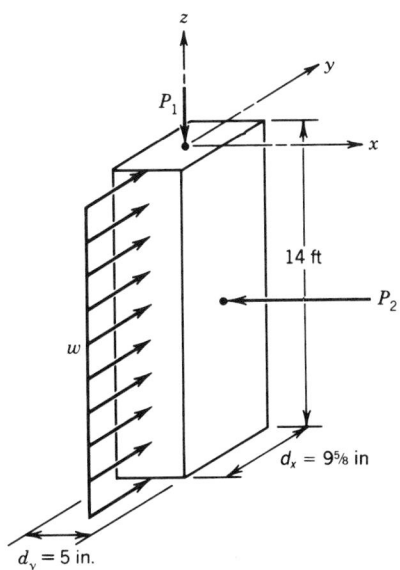

$F_c = 1450$ psi
$F_{bx} = 1600$ psi
$F_{by} = 1450$ psi
$E_x = E_x' = 1,400,000$ psi

$E_y = E'_y = 1,300,000$ psi
$P_1 = 10,000$ lb (dead load + snow load)
$P_2 = 600$ lb (wind load)
$w = 150$ plf (wind load only)

Section properties: $A = 48.12$ in.$^2$, $S_x = 77.20$ in.$^3$, $S_y = 40.10$ in.$^3$

1. Determine stresses from applied loads:

$$f_c = \frac{P_1}{A} = \frac{10,000}{48.12} = 208 \text{ psi}$$

$$f_{bx} = \frac{M_x}{S_x} = \frac{wL^2}{8S_x} = \frac{(150)(14)^2(12)}{(8)(77.2)} = 571 \text{ psi}$$

$$f_{by} = \frac{M_y}{S_y} = \frac{P_2 L}{4S_y} = \frac{(600)(14)(12)}{(4)(40.10)} = 628 \text{ psi}$$

2. Determine $F'_c$.

Assume duration of load $C_D = 1.6$ applies.

$$F^*_c = F_c C_D = (1450)(1.6) = 2320 \text{ psi}$$

$$K_e = 1$$

$$\left(\frac{l_e}{d_x}\right) = \frac{(14)(12)}{9.625} = 17.45 \quad \left(\frac{l_e}{d_y}\right) = \frac{(14)(12)}{5} = 33.6$$

$$c = 0.9 \quad K_{cE} = 0.418 \text{ [from Eq. (5-14)]}$$

Since the column is unsupported and compression buckling is about the $y$-$y$ axis, $E'_y$ and $l_e/d$ about the $y$-$y$ axis will be used.

$$F_{cEy} = \frac{K_{cE} E'_y}{(l_e/d)^2} = \frac{(0.418)(1,300,000)}{(33.6)^2} = 481 \text{ psi}$$

$$C_P = \frac{1 + F_{cE}/F^*_c}{2c} - \sqrt{\left[\frac{1 + F_{cE}/F^*_c}{2c}\right]^2 - \frac{F_{cE}/F^*_c}{c}}$$

$$C_P = \frac{1 + 481/2320}{1.8} - \sqrt{\left[\frac{1 + 481/2320}{1.8}\right]^2 - \frac{481/2320}{0.9}} = 0.202$$

$$F'_c = F^*_c C_P = (2320)(0.202) = 469 \text{ psi}$$

3. Determine $F^*_{bx}$.

$$F^*_{bx} = (1600)(1.6) = 2560 \text{ psi}$$

$$K_{bE} = 0.609 \text{ [from Eq. (5-10)]}$$

$$l_u = 14 \times 12 = 168 \text{ in.}, \quad l_u/d = \frac{168}{9.625} = 17.46 > 7$$

$$l_e = 1.63l_u + 3d = (1.63)(168) + (3)(9.625)$$

$$= 302.7 \text{ in. (Table 5.3 and Fig. 5.5)}$$

$$R_B = \sqrt{\frac{l_e d}{b^2}} = \sqrt{\frac{(302.7)(9.625)}{5^2}} = 10.80$$

$$F_{bE} = \frac{K_{bE}E_y'}{R_B^2} = \frac{(0.609)(1,300,000)}{10.80^2} = 6793 \text{ psi}$$

$$C_L = \frac{1 + F_{bE}/F_{bx}^*}{1.9} - \sqrt{\left[\frac{1 + F_{bE}/F_{bx}^*}{1.9}\right]^2 - \frac{F_{bE}/F_{bx}^*}{0.95}}$$

$$C_L = \frac{1 + (6793/2560)}{1.9} - \sqrt{\left[\frac{1 + 6793/2560}{1.9}\right]^2 - \frac{6793/2560}{0.95}} = 0.972$$

$$F_{bx}' = F_{bx}^* C_L = (2560)(0.972) = 2488 \text{ psi}$$

4.  Determine $F_{by}'$. $C_{fu} = 1.1$.

$$F_{by}^* = F_{by}C_D C_{fu} = (1450)(1.6)(1.1) = 2552 \text{ psi}$$

Since the depth-to-breadth ratio is less than 1 in bending about the $y$–$y$ axis,

$$C_L = 1,$$

$$F_{by}' = F_{by}^* = 2552 \text{ psi.}$$

5.  The general equation for combined bending and axial loads Eq. (5-22) will be used. Note that in the equation, $f_{b1}, f_{b2}, F_{b1}$, and $F_{b2}$ are $f_{bx}, f_{by}, F_{bx}$, and $F_{by}$, respectively, and $d_1$ and $d_2$ are $d_x$ and $d_y$, respectively.

$$\left(\frac{f_c}{F_c'}\right)^2 + \frac{f_{b1}}{F_{b1}'[1 - (f_c/F_{cE1})]} + \frac{f_{b2}}{F_{b2}'[1 - (f_c/F_{cE2}) - (f_{b1}/F_{bE})^2]} \leq 1.0$$

$$F_{cE1} = F_{cEx} = \frac{K_{cE}E_x'}{(l_e/d_x)^2} = \frac{(0.418)(1,400,000)}{(17.45)^2} = 1922 \text{ psi}$$

$$\left(\frac{208}{469}\right)^2 + \frac{571}{2488\left[1 - \left(\dfrac{208}{1922}\right)\right]} + \frac{628}{2552\left(1 - \dfrac{208}{481} - \dfrac{571}{6793}\right)}$$

$$= 0.892 \leq 1 \text{ O.K.}$$

6.  Check snow load and dead load only. The bending no longer applies.

$$f_c = \frac{P}{A} = 207 \text{ psi}$$

$$F_c^* = F_c C_D = (1450)(1.15) = 1668 \text{ psi}$$

$$C_P = \frac{1 + 481/1668}{1.8} - \sqrt{\left(\frac{1 + (481/1668)}{1.8}\right)^2 - \frac{481/1668}{0.9}} = 0.278$$

$$F_c' = F_c^* C_P = (1668)(0.278) = 463 \text{ psi} > 208 \text{ psi} \qquad \text{O.K.}$$

In most wood structures loaded with axial and bending loads, bending occurs about only one of the axes, and one of the bending terms in the general combined stress equation drops out. The following example illustrates the design of a column subjected to bending about one axis.

**Example.**   Check the design of a $6\frac{3}{4} \times 10\frac{1}{2}$-in. glued laminated timber column for the loading and condition illustrated. The column is braced in the $x$ and $y$ directions at both the top and bottom. Use Douglas Fir Combination 2 and wet conditions of use.

$F_c = 1900 \text{ psi}$
$F_{bx} = 1700 \text{ psi}$
$E_x = E_y = 1,700,000 \text{ psi}$
$P = 17,000 \text{ lb (including 7,000 lb dead load and 10,000 lb snow load)}$
$w = 200 \text{ plf (wind load only)}$

Section properties: $A = 70.88 \text{ in.}^2 \quad S_x = 124.0 \text{ in.}^3$

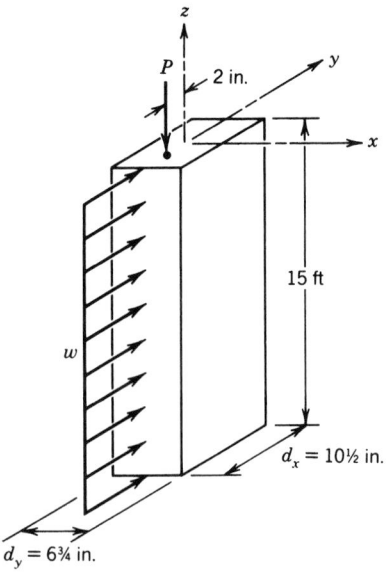

1.   Check column for total combination of loads, $C_D$ (wind) $= 1.6$, $C_D$ (snow) $= 1.15$, $C_M$ (bending) $= 0.80$, $C_M$ (compression) $= 0.73$, $C_M$ (modulus of elasticity) $= 0.833$. Assume wind load controls.

2.   Determine $f_c$ and $F'_c$.

$$f_c = \frac{P}{A} = \frac{17{,}000}{70.88} = 240 \text{ psi}$$

$$F^*_c = F_c C_D C_M = (1900)(1.6)(0.73) = 2219 \text{ psi}$$

$$E' = E C_M = (1{,}700{,}000)(0.833) = 1{,}420{,}000 \text{ psi}$$

$$c = 0.9 \quad K_{cE} = 0.418 \text{ [from Eq. (5-14)]}$$

$$K_e = 1, \frac{l_e}{d} = \frac{K_e l}{d} = \frac{1(15 \times 12)}{6.75} = 26.67$$

$$F_{cE} = \frac{K_{cE} E'}{(l_e/d)^2} = \frac{(0.418)(1{,}420{,}000)}{(26.67)^2} = 835 \text{ psi}$$

$$C_P = \frac{1 + F_{cE}/F^*_c}{2c} - \sqrt{\left[\frac{1 + F_{cE}/F^*_c}{2c}\right]^2 - \frac{F_{cE}/F^*_c}{c}}$$

$$C_P = \frac{1 + 835/2219}{1.8} - \sqrt{\left[\frac{1 + 835/2219}{1.8}\right]^2 - \frac{835/2219}{0.9}} = 0.356$$

$$F'_c = F^*_c C_P = (2219)(0.356) = 790 \text{ psi}$$

3.   Determine $f_{bx}$ and $F'_b$.

$$f_{bx} = \frac{M}{S_x} = \frac{wL^2}{8S_x} = \frac{(200)(15)^2(12)}{(8)(124.0)} = 544 \text{ psi}$$

$$C_V = \left(\frac{5.125}{b}\right)^{1/10} \left(\frac{12}{d}\right)^{1/10} \left(\frac{21}{L}\right)^{1/10}$$

$$C_V = \left(\frac{5.125}{6.75}\right)^{1/10} \left(\frac{12}{10.5}\right)^{1/10} \left(\frac{21}{15}\right)^{1/10} = 1.02, \text{ use } 1.0$$

Determine $l_e$.

$$l_u/d = \frac{(15)(12)}{10.5} = 17.14 > 7$$

$$l_e = 1.63 l_u + 3d = 1.63(15)(12) + (3)(10.5) = 325 \text{ in.}$$

$$R_B = \sqrt{\frac{l_e d}{b^2}} = \sqrt{\frac{(325)(10.5)}{(6.75)^2}} = 8.65 < 50 \qquad \text{O.K.}$$

$$F_{bx}^* = F_b C_D C_M = (1700)(1.6)(0.80) = 2176 \text{ psi}$$

$$K_{bE} = 0.609$$

$$F_{bE} = \frac{K_{bE} E'}{(R_B)^2} = \frac{(0.609)(1,420,000)}{(8.65)^2} = 11,550 \text{ psi}$$

$$C_L = \frac{1 + F_{bE}/F_{bx}^*}{1.9} - \sqrt{\left[\frac{1 + F_{bE}/F_{bx}^*}{1.9}\right]^2 - \frac{F_{bE}/F_{bx}^*}{0.95}}$$

$$C_L = \frac{1 + 11,550/2176}{1.9} - \sqrt{\left[\frac{1 + 11,550/2176}{1.9}\right]^2 - \frac{11,550/2176}{0.95}} = 0.989$$

$0.989 < 1.0$, $C_L$ controls.

$$F_{bx}' = F_{bx}^* C_L = (2176)(0.989) = 2151 \text{ psi}$$

4. Check for unity by using column and eccentric load equation.

$$F_{cE1} = F_{cEx} = \frac{(0.418)(1,420,000)}{(17.14)^2} = 2020 \text{ psi}$$

$$\left(\frac{f_c}{F_c'}\right)^2 + \frac{f_{bx} + f_c(6e_1/d_1)[1 + 0.234(f_c/F_{cE1})]}{F_{bx}'[1 - (f_c/F_{cE1})]} \leq 1$$

$$\left(\frac{240}{790}\right)^2 + \frac{544 + [(240)(6)(2)/10.5]\left[1 + 0.234\left(\frac{240}{2020}\right)\right]}{2152[1 - (240/2020)]}$$

$$= 0.527 < 1 \qquad \text{O.K.}$$

5. Try a $6\frac{3}{4}$ in. × 9 in. column.

$$A = 60.75 \text{ in.}^2 \quad S = 91.12 \text{ in.}^3$$

$$f_c = \frac{17,000}{60.75} = 280 \text{ psi}$$

$$F_c' = 790 \text{ psi}$$

Bending:

$$f_{bx} = \frac{wL^2}{8S_x} = \frac{(200)(15^2)(12)}{(8)(91.12)} = 741 \text{ psi}$$

$$F_{bx}^* = 2176 \text{ psi}$$

$l_u = (15)(12) = 180$ in.

$l_u/d = 180/9 = 20$

$l_e = 1.63 l_u + 3d = 1.63(180) + 3(9) = 320.4$ in.

$$R_B = \sqrt{\frac{(20.4)(9)}{(6.75)^2}} = 7.96$$

$$F_{bE} = \frac{(0.609)(1{,}420{,}000)}{(7.96)^2} = 13{,}664 \text{ psi}$$

$$C_L = \frac{1 + 13{,}664/2176}{1.9} - \sqrt{\left[\frac{1 + 13{,}664/2176}{1.9}\right]^2 - \frac{13{,}644/2176}{0.95}} = 0.991$$

$$F'_{bx} = F^*_b C_L = (2176)(0.991) = 2156 \text{ psi}$$

$$l_e/d = \frac{(15)(12)}{9} = 20$$

$$F_{cE1} = \frac{(0.418)(1{,}420{,}000)}{20^2} = 1484 \text{ psi}$$

$$\left(\frac{280}{790}\right)^2 + \frac{741 + [(280)(6)(2)/9]\left[1 + 0.234\left(\dfrac{280}{1484}\right)\right]}{2156[1 - (280/1484)]} = 0.771 < 1 \text{ O.K.}$$

Use $6\frac{3}{4}$ in. × 9 in. member.
A check of the column for dead load and snow load only indicates that a $6\frac{3}{4}$ in. × 9 in. column is more than adequate.

### 5.9.6   Columns with Side Bracket

Exact solutions of columns loaded on a side bracket are complicated and laborious. For these reasons, special design procedures which are safe are recommended when eccentric load is applied through a bracket some distance from the top as shown in Fig. 5.17 rather than at the top. It is assumed that the bracket load $P$ applied at distance $a$ from the center of the column is replaced by the same load $P$ centrally applied at the top of the columns plus a side load $P_s$ applied at midheight, $l/2$. This condition is shown in Fig. 5.18.

The load $P_s$ is determined by the empirical equation

$$P_s = \frac{3Pal_p}{l^2} \tag{5-27}$$

where   $P_s$ = assumed horizontal side load, placed midheight of the column (lb),
   $P$ = actual load on bracket (lb),
   $a$ = horizontal distance from the bracket load to center of column (in.),

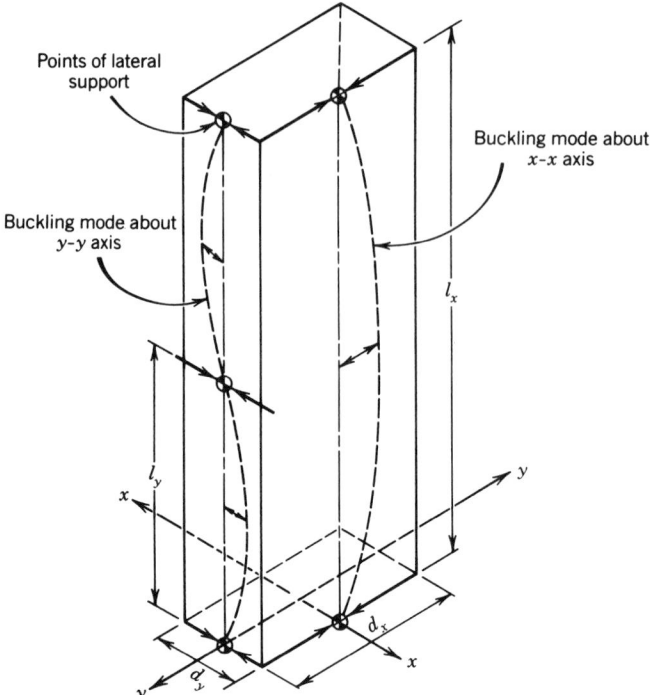

**FIGURE 5.17**    Column notations for combined bending and compression.

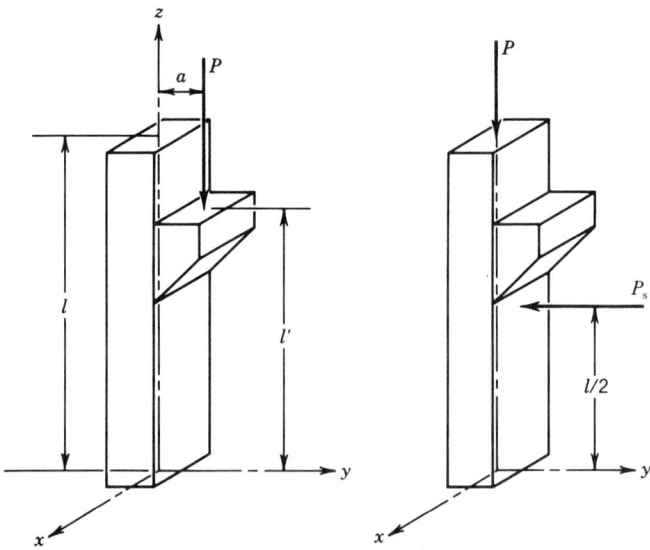

**FIGURE 5.18**    Column with load applied at side bracket.

$l_p$ = distance from point of application of load on the bracket to the farthest end of the column (in.), and

$l$ = length of column (in.).

The assumed axially applied load $P$ should be added to other concentric column
loads, and the calculated horizontal side load should be used to determine the
flexural stress

$$f_b = \frac{P_s l}{4S} = \frac{3al_p P}{4lS} \tag{5-28}$$

where $S$ is the section modulus and the other terms are as defined previously, with
the actual flexural stress $f_b$ determined by Eq. (5-28). The solution is obtained
by revising Eq. (5-22) as follows:

$$\left[ \frac{f_c}{F'_c} \right]^2 + \frac{f_b}{F'_{b1}[1 - (f_c/F_{cE1})]} \le 1 \tag{5-29}$$

This equation for combined axial compression and bending assumes that the column is of such a size that the combination of loads is critical. The designer is
cautioned that this empirical method is to be used only in checking the combined
axial and bending stress, and should not be used to determine shear or horizontal
reaction. These should be determined by the usual method of statics.

**Example.** Determine the size of a sawn wood column 12 ft long to carry a 14,000-
lb concentric axial load consisting of a 10,000 lb snow load and a 4,000 lb dead
load and a permanent equipment load of 2000 lb mounted on a side bracket as
shown. Use Douglas Fir-Larch, Posts and Timbers, No. 1 grade.

$$F_b = 1200 \text{ psi} \quad F_c = 1000 \text{ psi} \quad E = 1,600,000 \text{ psi}$$

$$a = 24 \text{ in.} \quad l = 144 \text{ in.} \quad l_p = 108 \text{ in.}$$

The top of the bracket is 36 in. below the top of the column; duration-of-load factor
$C_D = 1.15$ for snow load; dry conditions of use factor $C_M = 1.00$.

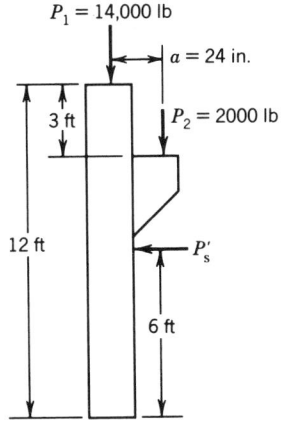

$P_1 = 14,000$ lb

$a = 24$ in.

3 ft

$P_2 = 2000$ lb

12 ft

$P'_s$

6 ft

End fixity condition is assumed to be pin end at both ends ($K_e = 1.0$). Try a 6 × 10-in. column with the 10-in. direction in the plane of the side load:

$$A = 52.25 \text{ in.}^2, \ S = 82.73 \text{ in.}^3, \ F_c^* = (1000)(1.15) = 1150 \text{ psi},$$

$$F_b^* = (1200)(1.15) = 1380 \text{ psi}, \ E' = E = 1,600,000 \text{ psi}$$

### Bending:

Replace the load on the bracket $P_2$ with centrally applied load $P$ at the top of column and place load $P_s$ at midheight.

$$P_s = \frac{3al_pP_2}{l^2} = \frac{(3)(24)(108)(2000)}{(12)^2} = 750 \text{ lb}$$

Determine $f_b$ due to assumed side load $P_s$.

$$f_b = \frac{3al_pP}{4lS} = \frac{(3)(24)(108)(2000)}{(4)(144)(51.56)} = 326 \text{ psi}$$

Determine $F_b'$.

$$l_u/d = 144/7.5 = 19.2 > 7$$

$$l_e = 1.63l_u + 3d = 1.63(144) + 3(9.5)$$

$$l_e = 263.2 \text{ in.}$$

$$R_B = \sqrt{\frac{l_ed}{b^2}} = \sqrt{\frac{(263.2)9.5}{(5.5)^2}} = 9.09$$

$$K_{bE} = 0.438$$

$$F_{bE} = \frac{K_{bE}E'}{R_B^2} = \frac{(0.438)(1,600,000)}{(9.09)^2} = 8481 \text{ psi}$$

$$C_L = \frac{1 + F_{bE}/F_b^*}{1.9} - \sqrt{\left[\frac{1 + F_{bE}/F_b^*}{1.9}\right]^2 - \frac{F_{bE}/F_b^*}{0.95}}$$

$$C_L = \frac{1 + 8481/1380}{1.9} - \sqrt{\left[\frac{1 + 8481/1380}{1.9}\right]^2 - \frac{8481/1380}{0.95}} = 0.990$$

$$F_b' = F_bC_DC_L = (1200)(1.15)(0.990) = 1367 \text{ psi}$$

### Compression:

Determine $F_c'$.
The maximum $l/d$ ratio applies in determining $F_c'$.

$$K_e = 1, \ l_e = K_el = (1)(12)(12) = 144 \text{ in.}$$

$$\frac{l_e}{d} = \frac{144}{5.5} = 26.18$$

$$c = 0.8, \ K_{cE} = 0.3 \ [\text{from Eq. (5-14)}]$$

$$F_{cE} = \frac{K_{cE}E'}{(l_e/d)^2} = \frac{(0.3)(1,600,000)}{(26.18)^2} = 700 \ \text{psi}$$

$$C_P = \frac{1 + F_{cE}/F_c^*}{2c} - \sqrt{\left[\frac{1 + F_{cE}/F_c^*}{2c}\right]^2 - \frac{F_{cE}/F_c^*}{c}}$$

$$C_P = \frac{1 + 700/1150}{(2)(0.8)} - \sqrt{\left[\frac{1 + 700/1150}{(2)(0.8)}\right]^2 - \frac{700/1150}{0.8}} = 0.505$$

$$F_c' = F_c^* C_P = (1150)(0.505) = 581 \ \text{psi}$$

$$f_c = \frac{P}{A} = \frac{14,000 + 2000}{52.25} = 306 \ \text{psi}$$

$$\left[\frac{f_c}{F_c'}\right]^2 + \frac{f_b}{F_{b1}'(1 - f_c/F_{cE1})}$$

$$\left[\frac{306}{581}\right]^2 + \frac{326}{1367(1 - 306/700)} = 0.702 \qquad \text{O.K.}$$

A check of the next smaller size, a 6 × 8, shows that the smaller column would be overstressed.
Check for dead load only.

**Bending:**

$$f_b = 326 \ \text{psi}$$

$$F_{bx}^* = F_b C_D = (1200)(0.9) = 1080 \ \text{psi}$$

$$F_{bE} = \text{Same as previously calculated} = 8481 \ \text{psi}$$

$$C_L = \frac{1 + 8481/1080}{1.9} - \sqrt{\left[\frac{1 + 8481/1080}{1.9}\right]^2 - \frac{1 + 8481/1080}{0.9}}$$

$$= 0.993$$

$$F_b' = (1200)(0.90)(0.993) = 1072 \ \text{psi}$$

**Compression:**

$$f_c = \frac{P}{A} = \frac{6000}{52.25} = 115 \ \text{psi}$$

$$F_c^* = F_c C_D = (1000)(0.90) = 900 \ \text{psi}$$

$F_{cE}$ = Same as previously calculated = 700 psi

$$C_P = \frac{1 + 700/900}{(2)(0.8)} - \sqrt{\left[\frac{1 + 700/900}{(2)(0.8)}\right]^2 - \frac{700/900}{0.8}} = 0.599$$

$F_c' = F_c^* C_P = (900)(0.599) = 539$ psi

$$\left(\frac{115}{539}\right)^2 + \frac{326}{1072(1 - 115/700)} = 0.409 \qquad \text{O.K.}$$

Use 6 × 10 No. 1 Post and Timbers grade Douglas Fir-Larch.

**Example.**  Determine the adequacy of an 8 × 8-in. pin-end column subjected to the following conditions:

Load $P$ is a load of 10,000 lb on the bracket consisting of 7000 lb snow load and 3000 lb dead load $C_D = 1.15$.

Assume dry conditions of use, $C_M = 1.0$.

Southern Pine No. 1 Dense

$$F_b = 1550 \text{ psi} \quad F_c = 975 \text{ psi} \quad E = 1,600,000 \text{ psi}$$

$$S = 70.31 \text{ in.}^3 \quad A = 56.25 \text{ in.}^2$$

$$F_c^* = 975(1.15) = 1121 \text{ psi} \quad F_b^* = (1550)(1.15) = 1782 \text{ psi}$$

Top of bracket at point of load application is located 2 ft below top of column. Analyze for combined stresses due to concentric load $P$ and side load $P_s$.

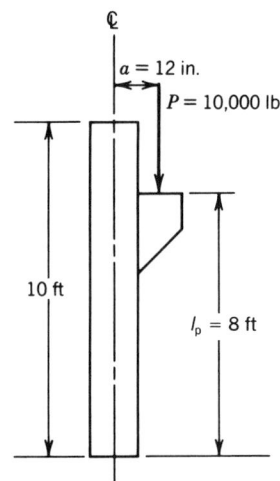

Assume 10,000 lb load is placed concentrically on column and a side load $P_s$ is placed at midheight of column.

$$a = 12 \text{ in.,} \quad l_p = (8)(12) = 96 \text{ in.,} \quad l = (10)(12) = 120 \text{ in.}$$

$$P_s = \frac{3al_pP}{l^2} = \frac{(3)(12)(96)(10,000)}{(120)^2} = 2400 \text{ lb}$$

$$f_b = \frac{P_s l}{4S} = \frac{(2400)(120)}{(4)(70.31)} = 1024 \text{ psi}$$

**Bending:**

$b/d$ ratio $= 1$, $C_L = 1$

$$F_b' = F_b^* = 1782 \text{ psi}$$

**Compression:**

$$f_c = \frac{P}{A} = \frac{10,000}{56.25} = 178 \text{ psi}$$

$$l_e/d = \frac{120}{7.5} = 16$$

$$c = 0.8 \quad K_{cE} = 0.3 \text{ [from Eq. (5-14)]}$$

$$F_{cE} = \frac{K_{cE}E'}{(l_e/d)^2} = \frac{(0.3)(1,600,000)}{16^2} = 1875 \text{ psi}$$

$$C_P = \frac{1 + F_{cE}/F_c^*}{2c} - \sqrt{\left[\frac{1 + F_{cE}/F_c^*}{2c}\right]^2 - \frac{F_{cE}/F_c^*}{c}}$$

$$C_P = \frac{1 + 1875/1121}{1.6} - \sqrt{\left[\frac{1 + 1875/1121}{1.6}\right]^2 - \frac{1875/1121}{0.8}} = 0.834$$

$$F_c' = F_c^* C_P = (1121)(0.834) = 935 \text{ psi}$$

$$\left[\frac{f_c}{F_c'}\right]^2 + \frac{f_b}{F_b'[1 - (f_c/F_{cE1})]} \leq 1$$

$$\left[\frac{178}{935}\right]^2 + \frac{1024}{1782[(1 - 178/1875)]} = 0.673 < 1 \quad \text{O.K.}$$

$$f_b = \frac{3al_pP}{4lS} = \frac{(3)(12)(96)(2400)(120)}{(4)(27.73)} = 2597 \text{ psi}$$

### 5.9.7    End Eccentricity

The column design equations customarily used are based on pin-end conditions. In practice, most columns are square cut. This increases the capacity due
to the extra amount of fixity at the ends.

In actual practice, the loads transmitted to columns from beams may be eccentric, especially when a column supports the end of a beam. Also, a slight error
in fabrication can result in the load being eccentrically applied. Usually this causes
no problem when the eccentricity is small because the partial fixity of square cut
ends tends to compensate for the eccentricity. Where the design is critical, the
designer may wish to check for some assumed eccentricity and make allowances
for it in design. The following example illustrates the effect of 1 in. of eccentricity
in a typical wood column.

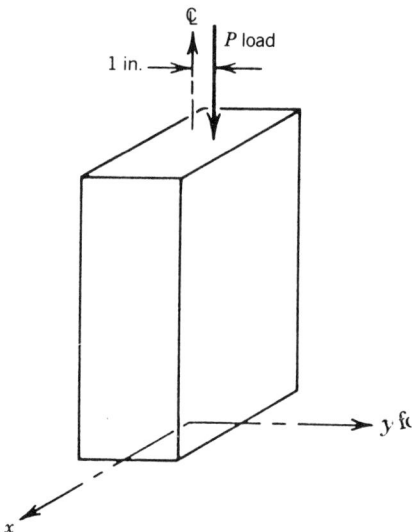

**Example.**   Column size is 6 in. × 6 in. Southern Pine No. 1 Dense grade.

$F_b$ = 1550 psi   $F_c$ = 975 psi   $E$ = 1,600,000 psi
$A$ = 20.25 in.$^2$   $S$ = 27.73 in.$^3$   Length = 13 ft
Axial load $P$ = 14,000 lb
Normal duration of load $C_D$ = 1.0
Dry conditions of use $C_M$ = 1.0
$F_b^*$ = $F_b$ = 1550 psi, $F_c^*$ = $F_c$ = 975 psi, $E'$ = $E$ = 1,600,000 psi

First, based on concentric load, determine compression:

$$f_c = P/A = \frac{14,000}{30.25} = 463 \text{ psi}$$

$$K_e = 1, \, l_e = 13(12) = 156 \text{ in.}$$

$$l_e/d = \frac{156}{5.5} = 28.36$$

$$c = 0.8, \, K_{cE} = 0.3$$

$$F_{cE} = \frac{K_{cE}E'}{(l/d)^2} = \frac{(0.3)(1,600,000)}{28.36^2} = 597 \text{ psi}$$

$$C_P = \frac{1 + (F_{cE}/F_c^*)}{2c} - \sqrt{\left[\frac{1 + (F_{cE}/F_c^*)}{2c}\right]^2 - \frac{F_{cE}/F_c^*}{c}}$$

$$= \frac{1 + (597/975)}{1.6} - \sqrt{\left[\frac{1 + (597/975)}{1.6}\right]^2 - \frac{597/975}{0.8}} = 0.507$$

$$F_c' = F_c^* C_P = (975)(0.507) = 495 \text{ psi}$$

$$463 \text{ psi} < 495 \text{ psi} \qquad \text{O.K.}$$

A 6 in. × 6 in. column based on assumption of pin-end support with concentric load is satisfactory. Square cut ends will result in additional fixity at the ends which will take care of minor amounts of eccentricity.

The effect of 1 in. eccentricity based on the pin-end assumption will be examined.

$f_c = 463$ psi (same as previously calculated)
$F_c' = 495$ psi (same as previously calculated)
$F_{cE} = 597$ psi

**Bending:**
$F_b^* = 1550$ psi
$F_{b1}' = F_b^* C_L = 1550(1) = 1550$ psi   $C_L = 1.0$ since $b/d = 1.0$

$$\left(\frac{f_c}{F_c'}\right)^2 + \frac{f_c(6e_1/d_1)[1 + 0.234(f_c/F_{cE1})]}{F_{b1}'[1 - (f_c/F_{cE1})]}$$

$$\left(\frac{463}{495}\right)^2 + \frac{(463)(6)(1)(5.5)\left[1 + 0.234\left(\dfrac{463}{597}\right)\right]}{1550\left[1 - \left(\dfrac{463}{597}\right)\right]} = 2.59 \qquad \text{N.G.}$$

The column is considerably overstressed based on assumption of pin-end supports. Try 6 in. × 8 in. column.

$$A = 41.25 \text{ in.}^2 \quad S = 51.56 \text{ in.}^3$$

$$f_c = P/A = \frac{14,000}{41.25} = 339 \text{ psi}$$

$$F'_c = 495 \text{ psi}$$

Determine $F'_b$

$K_{bE} = 0.438$

$$R_B = \sqrt{\frac{l_e d}{b^2}} = \sqrt{\frac{(156)(7.5)}{5.5^2}} = 6.22$$

$$F_{bE} = \frac{K_{bE}E'}{R_B^2} = \frac{(0.438)(1,600,000)}{(6.22)^2} = 18,114 \text{ psi}$$

$$C_L = \frac{1 + 18,114/1550}{1.9} - \sqrt{\left[\frac{1 + 18,114/1550}{1.9}\right]^2 - \frac{18,114/1550}{0.95}} = 0.995$$

$$F'_b = F^*_b(C_L) = (1550)(0.995) = 1542 \text{ psi}$$

Check column by Eq. (5-26) for 1 in. eccentricity.

$$\left(\frac{339}{495}\right)^2 + \frac{(339)(6)(1)/(7.5)[1 + 0.234(339/597)]}{1542[1 - (339/597)]} = 0.93 \qquad \text{O.K.}$$

A 6 in. × 8 in. column is satisfactory based on pin-end supports.

## 5.10  TAPERED STRAIGHT BEAMS

Glued laminated beams are often tapered to meet architectural requirements, to provide pitched roofs, to facilitate drainage, and to lower wall height requirements at the end supports. Fig. 5.19 illustrates the most common forms of these beams.

It is recommended that any sawn taper cuts be made only on the compression face of tapered beams. The laminations should be parallel to the tension face of the beam.

Camber may be provided in beams that are taper cut on the compression face by providing camber on the tension face similar to that in nontapered beams and by sawing camber into the compression face. For single-tapered straight beams, the minimum camber based on dead-load deflection may be built into the tension

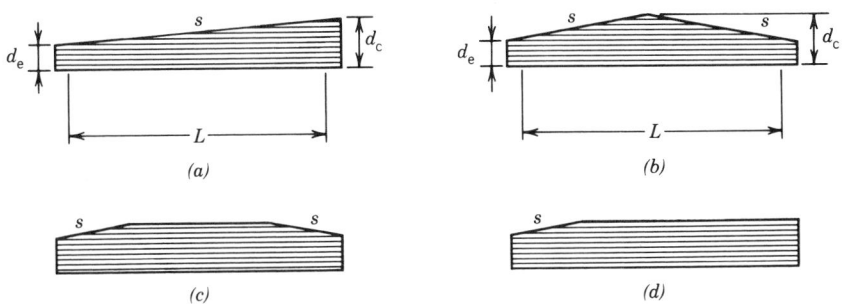

**FIGURE 5.19** Simple-span single-tapered straight beams. (*a*) Single-tapered straight; (*b*) double-tapered straight; (*c*) tapered both ends straight; (*d*) tapered one end straight.

face as well as sawn into the tapered face. For double-tapered straight beams, the minimum camber is usually built into the tension face, and if this camber is over 2-in., one-fourth the centerline camber is sawn into the tapered face. The designer should specify the required camber both for the tension face and the sawn compression face. Camber is usually not provided by the fabricator for custom glued laminated timber unless specified by the designer. For non-custom beams, camber is usually provided unless specified otherwise.

The design methods for tapered beams are based on the procedures contained in *Deflection and Stresses of Tapered Wood Beams* (9). Those procedures presented are based on the Bernoulli–Euler theory of bending and beams of isotropic material. This results in an approximate solution for wood beams, which is considered satisfactory for design purposes.

### 5.10.1  Stresses in Tapered Beams of Constant Width

When the top and bottom of a beam are not parallel to each other (a variable depth along the length of the beam is created by a sawn surface), consideration must be given to the combined effects of bending, compression, tension, and shear parallel to grain, and also to compression or tension perpendicular to grain. An illustration of the distribution of the shear stresses existing in a tapered beam is shown in Fig. 5.20.

This analysis involves an interaction equation when stresses, $f_x, f_y$, and $f_{xy}$ occur simultaneously to reduce the beam capacity below the capacity that would result from considering each stress separately.

In (Ref. 9), the basic theory was expanded to verify the applicability of an interaction equation in predicting the strength of tapered timber bending members. Such an equation, expressing the effects of combined stresses, is also used for materials other than wood and takes the form

$$\frac{f_x^2}{F_x^2} + \frac{f_y^2}{F_y^2} + \frac{f_{xy}^2}{F_{xy}^2} \leq 1 \tag{5-30}$$

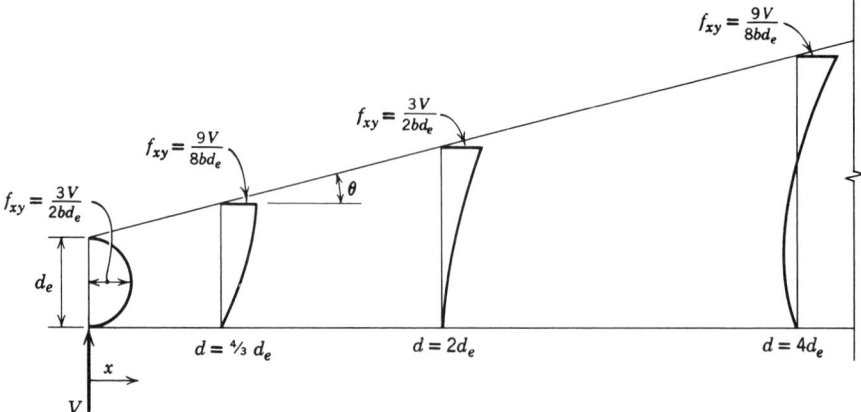

**FIGURE 5.20**   Distribution of shear stresses in a simply supported beam under concentrated loads.

where   $f_x$ = actual bending stress (psi),
   $f_y$ = actual compression or tension stress perpendicular to grain (psi)
   (compression if the taper cut is on the compression face and tension
   if the taper cut is on the tension face) (note that taper cutting on the
   tension face is not recommended),
   $f_{xy}$ = actual shear stress (psi),
   $F_x$ = tabular design value in bending (psi), $F_b$, multiplied by all applicable adjustment factors except $C_V$, $C_I$, and $C_L$,
   $F_y$ = tabular design value in compression perpendicular to grain (psi),
   $F_{c\perp}$, multiplied by all applicable adjustment factors (when the taper
   cut is on the tension side (not recommended), $F_y = F_{rt}$), and
   $F_{xy}$ = tabular design value in shear parallel to grain (psi), $F_v$, multiplied
   by all applicable adjustment factors.

For tapered beams, the actual stresses at the tapered edge can be expressed as
a function of the bending stress $f_x$ as follows:

$$f_{xy} = f_x \tan \theta, \, f_y = f_x \tan^2 \theta \text{ or } f_y = f_{xy} \tan \theta$$

where $\tan \theta$ = slope of tapered face.
   By assuming an optimum design (i.e., the interaction stress equation set equal
to 1), it is possible to write this equation as

$$\frac{f_x^2}{F_b^2} + \frac{f_x^2}{F_v^2} \tan^2 \theta + \frac{f_x^2}{F_{c\perp}^2} \tan^4 \theta = 1$$

Solving this equation for $f_x$,

$$f_x = F_b C_I$$

where $C_{\mathrm{I}}$ = interaction stress factor:

$$C_{\mathrm{I}} = \left[ \frac{1}{1 + (F_{\mathrm{b}} \tan \theta / F_{\mathrm{v}})^2 + (F_{\mathrm{b}} \tan^2 \theta / F_{\mathrm{c}\perp})^2} \right]^{1/2} \qquad (5\text{-}31)$$

To account for interaction stresses in design, the design value in bending, $F_{\mathrm{b}}$, is multiplied by $C_{\mathrm{I}}$:

$$F_{\mathrm{b}}' = F_{\mathrm{b}}^* C_{\mathrm{I}}$$

where   $F_{\mathrm{b}}^*$ = tabular design value in bending (psi) multiplied by all applicable adjustment factors except $C_{\mathrm{L}}$ and $C_{\mathrm{V}}$.

For a taper cut on the compression side of a beam, $C_{\mathrm{I}}$ is cumulative with adjustments for lateral stability $C_{\mathrm{L}}$, but is not cumulative with $C_{\mathrm{V}}$. Although taper cutting on the tension side of beams is not recommended, $C_{\mathrm{I}}$ in such cases is cumulative with $C_{\mathrm{V}}$, but not $C_{\mathrm{L}}$. When the taper cut is on the tension side, the design value obtained by the use of $C_{\mathrm{I}}$ and other applicable adjustment factors is then compared to the actual calculated bending stress $f_{\mathrm{b}}$ to determine if the section is adequate. Because $C_{\mathrm{I}}$ is dependent only on the slope of the tapered face and the material properties, tabular values of $C_{\mathrm{I}}$ as a function of $\tan \theta$ have been developed for commonly used species-bending combinations and are given in Table 5.5.

In the calculation of stresses and the interaction of stresses, it is necessary to know where the maximum stress conditions will occur. The shear stress $f_{\mathrm{xy}}$ and the perpendicular to grain stress $f_{\mathrm{y}}$ are functions of the bending stress $f_{\mathrm{x}}$. Determination of the location of the highest bending stress also gives the location of the highest values of $f_{\mathrm{xy}}$ and $f_{\mathrm{y}}$. The location of the point of maximum stress can be obtained by simple calculation or by inspection in some members. In other members, it may be necessary to divide the members into finite sections and calculate the maximum bending stress at each cross section.

Unless otherwise specified, tapered glued laminated timber beams will have the same grades on the taper cut as required for the compression side of the combination being used. The tabular design values required for calculating $C_{\mathrm{I}}$ ($F_{\mathrm{bx}}$, $F_{\mathrm{vx}}$, and $F_{\mathrm{c}\perp}$) can be obtained from Table 1, AITC 117—Design (1), included in Chapter 8. Tapered beams can also be manufactured by cutting through the compression zone as illustrated for straight beam with end cuts. If this is done, the tabular design values in Table 5.4 for $F_{\mathrm{bx}}$, $F_{\mathrm{vx}}$, $F_{\mathrm{c}\perp}$, and $E_{\mathrm{x}}$ should be used to calculate $C_{\mathrm{I}}$. Other design values in Table 1, AITC 117—Design (1), remain unchanged. There must be a clear understanding between the designer and the manufacturer when this process is used.

The depth $d$ of tapered beams where maximum bending stress occurs for uniformly loaded single-tapered straight beams ($a$ in Fig. 5.21) or symmetrical double tapered beams ($b$ in Fig. 5.21) can be determined by the equation

$$d = 2d_{\mathrm{e}} \frac{d_{\mathrm{e}} + l \tan \theta}{2d_{\mathrm{e}} + l \tan \theta} \qquad (5\text{-}32)$$

## 5.10.2  Depth at Maximum Stress, Uniform Load

The preceding equation can be converted to the following forms for easier computations.

For single-tapered straight beams (Fig. 5.21a):

$$d = 2d_e \frac{d_c}{d_e + d_c} \tag{5-33}$$

For symmetrical double-tapered straight beams (Fig. 5.21b):

$$d = \frac{d_e}{d_c}(2d_c - d_e) \tag{5-34}$$

## 5.10.3  Bending Stress, Uniform Load

The bending stress $f_x$ at the point of maximum stress for both types of beams shown in Fig. 5.21 can be determined by the following equation for a uniform load:

$$f_x = \frac{3Wl}{4bd_e(d_e + l\tan\theta)} \tag{5-35}$$

where    $W$ = total uniform load (lb) and
$\quad\quad\quad l$ = span (in.).

For single-tapered straight beams (Fig. 5.21a), the equation reduces to

$$f_x = \frac{3Wl}{4bd_cd_e} \tag{5-36}$$

For symmetrical double-tapered straight beams (Fig. 5.21b), the equation converts to

$$f_x = \frac{3Wl}{4bd_e(2d_c - d_e)} \tag{5-37}$$

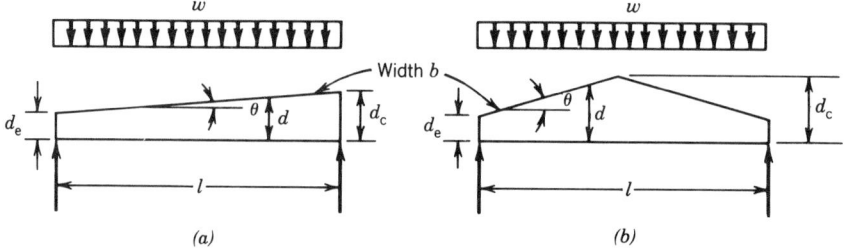

**FIGURE 5.21**    Tapered beams: (a) single-tapered straight beam; (b) double-tapered straight beam.

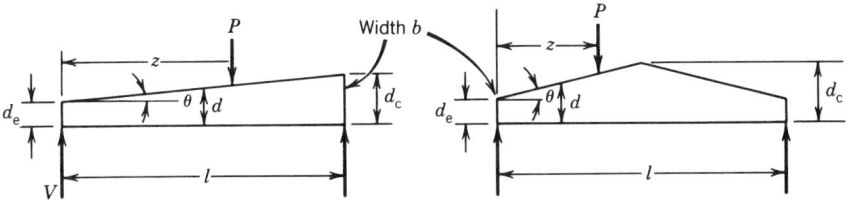

**FIGURE 5.22**   Depth at which maximum bending stress occurs.

### 5.10.4   Depth at Maximum Stress, Concentrated Load

For a beam loaded with a single concentrated load as shown in Fig. 5.22, the depth at which maximum stress occurs can be determined by the equation

$$d = 2d_e \tag{5-38}$$

when $d$ lies within the region 0 to $z$.

When $d$ as determined from this equation lies outside of the region 0 to $z$, the depth at which maximum bending stress occurs is at the load point. If the depth $d$ under the load point is such that $4/3d_e \leq d < 2d_e$, the maximum shear at the section also occurs on the tapered surface. If the depth $d$ under the load point is such that $d_e < d < 4/3d_e$, the maximum value of shear stress lies within the beam, but because the stresses along the taper cut are those considered in the interaction equation, the value of shear stress as determined by the equation $f_{xy} = f_x \tan \theta$ is still applicable for use in the interaction equation.

### 5.10.5   Stresses at Tapered Surface

The stresses at the tapered surface can be determined by the equations

$$f_x = \frac{6M}{bd^2}, f_{xy} = f_x \tan \theta, \text{ and } f_y = f_x \tan^2 \theta \tag{5-39}$$

where   $M$ = moment at point where depth $d$ occurs (in.-lb),
   $b$ = width at point where depth $d$ occurs (in.), and
   $d$ = depth where maximum bending stress occurs (in.).

When a tapered member contains loads or combinations of loads other than a uniform load or a single concentrated load or if a nonsymmetrical double-tapered straight member is used, the simplified equations for determining the depth at which the maximum stress occurs do not apply. In these cases, it is generally necessary to determine, $f_b$ ($f_x$) at various points along the member to determine the maximum bending stress. Once the maximum bending stress $f_x$ has been obtained, $f_y$ and $f_{xy}$ can be calculated for that cross section and the effect of their interaction determined.

### 5.10.6    Deflection

The shear deflection in tapered members is larger than in prismatic members, and the resulting total deflection including both bending deflection and shear deflection is slightly larger than that obtained by the customary methods of calculating deflection in prismatic members. (See Ref. 9 for further information.) Acceptable accuracy in determining the deflection of tapered members can be obtained by determining the deflection of an equivalent prismatic member. The depth of an equivalent member of constant cross section of the same width that will have the same deflection as a tapered beam can be determined by the equation

$$d = C_{dt}d_e \qquad (5\text{-}40)$$

where    $d_e$ = depth of ends of symmetrical double-tapered beam or smaller end depth of single-tapered beam (in.) and

$C_{dt}$ = constant derived from relationship of equations for deflection of tapered beams and straight prismatic beams.

For a uniformly symmetrical double-tapered beam as shown in Fig. 5.23b,

$$C_{dt} = 1 + 0.66C_y \quad \text{when } 0 < C_y \le 1 \qquad (5\text{-}41)$$

$$C_{dt} = 1 + 0.62C_y \quad \text{when } 1 < C_y \le 3 \qquad (5\text{-}42)$$

$$C_y = \frac{d_c - d_e}{d_e} \qquad (5\text{-}43)$$

For a uniformly loaded single tapered straight beam as shown in Fig. 5.23a,

$$C_{dt} = 1 + 0.46C_y \quad \text{when } 0 < C_y \le 1.1 \qquad (5\text{-}44)$$

$$C_{dt} = 1 + 0.43C_y \quad \text{when } 1.1 < C_y \le 2 \qquad (5\text{-}45)$$

where the terms are as previously defined.

The calculation of the deflection of tapered beams with other than uniform loads or single loads at the midpoint becomes somewhat complicated. In many cases, the resulting moment diagram approximates the moment diagram for a uniform load, and a close approximation can be obtained by assuming the load to be uniform. When this cannot be done, it is recommended that one of the other methods of determining deflection such as the virtual work (dummy load) method be used.

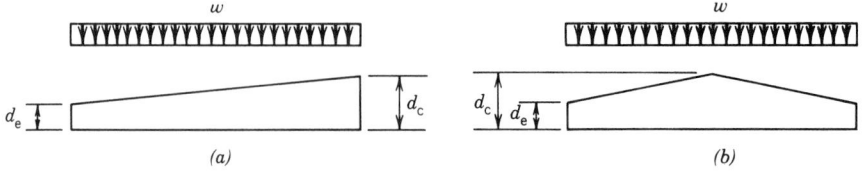

**FIGURE 5.23**    Notations for end and center depths.

Theoretically, both the shear deflection $\Delta_s$ and the bending deflection $\Delta_b$ should be determined. The total deflection $\Delta$ is the summation of the two.

$$\Delta = \Delta_s + \Delta_b$$

The tabular design values for modulus of elasticity already contain small elements of shear deflection because they are based on the deflection of a member with an $l/d$ of 21. In most cases, sufficient accuracy in deflection can be obtained by using only the tabular modulus of elasticity in deflection calculations for bending deflection $\Delta_b$ and ignoring the shear deflection $\Delta_s$ because it has been largely compensated for by the lower tabular design values for $E$.

If it is desired to calculate both $\Delta_b$ and $\Delta_s$, the modulus of elasticity $E$ in the equation for bending deflection $\Delta_b$ should be 5% larger than the tabular design value of $E$. The shear deflection can be approximated by the equation

$$\Delta_s = \frac{3Wl}{20Gbd_e} \tag{5-46}$$

where $G$ = shear modulus (psi) (approximately equal to $0.06E$) and other terms are as previously defined. See Chapter 4 for a detailed discussion on the variability of modulus of elasticity.

**Example.**   Design a double-tapered straight glued laminated timber roof beam to meet the following requirements. Roof decking is applied directly to beams to provide lateral stability and ends are held in place. Assume dry conditions of use and the following parameters:

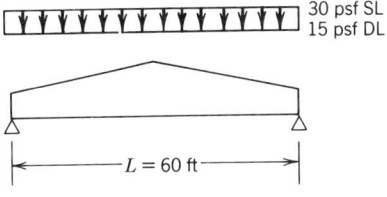

Use Western Species
$L = 60$ ft
Spacing $= 16$ ft
Roof Slope $= 1:12$
$F_b = 2400$ psi
$F_v = 165$ psi
$F_{c\perp}$ (top) $= 560$ psi
$F_{c\perp}$ (bottom) $= 650$ psi
$E = 1,700,000$ psi
$\Delta_{TL} = l/180 = (60)(12)/180 = 4$ in.

Camber $= 1\frac{1}{2}\Delta_{DL}$
SL $= 30$ psf   $C_D = 1.15$
DL $= 15$ psf
$F_b^* = (2400)(1.15) = 2760$ psi
$F_v' = (165)(1.15) = 190$ psi
$F_{c\perp}'$ (top) $= 560$ psi
$F_{c\perp}'$ (bottom) $= 650$ psi
$E' = 1,700,000$ psi

1.  Determine end depth: Assume effective span for shear $L_e = 55$ ft and width $b = 5\frac{1}{8}$ in.

$$V = \frac{wL_e}{2} = \frac{(30 + 15)(16)(55)}{2} = 19,800 \text{ lb}$$

$$\text{End depth} = d_e = \frac{3V}{2bF'_v} = \frac{(3)(19,800)}{(2)(5.125)(190)} = 30.5 \text{ in. (use 30 in.)}$$

Check $l_e$:

$$L_e = 60 - \frac{(2)(30)}{12} = 55 \text{ ft} \qquad \text{O.K.}$$

2.  Determine trial centerline depth: Centerline depth $= d_c = d_e +$ (roof slope)(span/2)

$$d_c = 30 + \left(\frac{1}{12}\right)\left(\frac{60}{2}\right)(12) = 60 \text{ in.}$$

3.  Check maximum deflection: Determine depth $d$ of equivalent prismatic beam that has the same deflection as the tapered beam:

$$d = d_e C_{dt}$$

where   $d_e = $ depth at end (in.),
$\qquad C_{dt} = $ multiplying factor obtained from the following equation for a uniformly loaded double-tapered straight beam.
$\qquad C_{dt} = 1 + 0.66C_y$
$\qquad C_y = (d_c - d_e)/d_e = (60 - 30)/30 = 1$
$\qquad C_{dt} = 1 + (0.66)(1) = 1.66$
$\qquad d = d_e C_{dt} = (30)(1.66) = 49.8 \text{ in.}$

$$I \text{ of equivalent beam} = \frac{(5.125)(49.8)^3}{12} = 52,750 \text{ in.}^4$$

$$\Delta_{TL} = \frac{5wL^4}{384EI} = \frac{(5)(45)(16)(60)^4(12)^3}{(384)(1,700,000)(52,750)} = 2.34 \text{ in.} < 4 \text{ in.} \qquad \text{O.K.}$$

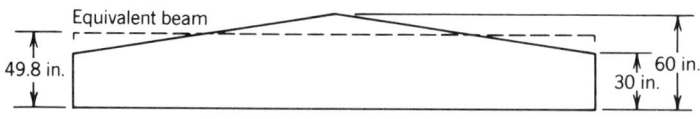

Equivalent beam

49.8 in.

60 in.
30 in.

4.  Check the actual stresses: Determine depth at which maximum stresses occur:

$$d = \frac{d_e}{d_c}(2d_c - d_e) = \frac{30}{60}(120 - 30) = 45 \text{ in.}$$

$$C_V = \left(\frac{5.125}{b}\right)^{1/10} \left(\frac{12}{d}\right)^{1/10} \left(\frac{21}{L}\right)^{1/10}$$

$$C_V = \left(\frac{5.125}{5.125}\right)^{1/10} \left(\frac{12}{45}\right)^{1/10} \left(\frac{21}{60}\right)^{1/10} = 0.789$$

Assuming volume factor controls:

$$F_b' = F_b^* C_V = (2760)(0.789) = 2177 \text{ psi}$$

$$f_x = \frac{3Wl}{4bd_e(2d_c - d_e)}$$

$$= \frac{(3)(720)(60)(60)(12)}{(4)(5.125)(30)(120 - 30)} = 1686 \text{ psi} < 2177 \text{ psi} \qquad \text{O.K.}$$

Actual shear stress at point where $d = 45$ in.:

$$f_{xy} = f_x \tan \theta = 1686 \left(\frac{1}{12}\right) = 140 \text{ psi} < 190 \text{ psi} \qquad \text{O.K.}$$

Actual compression perpendicular to grain stress:

$$f_y = f_x \tan^2 \theta = 1686 \left(\frac{1}{12}\right)^2 = 12 \text{ psi} < 560 \text{ psi} \qquad \text{O.K.}$$

5. Check the effects of combined stresses at the point of maximum stress ($d = 45$ in.) by either the interaction equation or use of the interaction stress factor $C_I$. (Both methods are shown for purposes of illustration.) Using interaction equation:

$$F_x = F_b^* \qquad F_y = F_{c\perp}' \qquad F_{xy} = F_v'$$

$$\frac{f_x^2}{F_x^2} + \frac{f_y^2}{F_y^2} + \frac{f_{xy}^2}{F_{xy}^2} \leq 1$$

$$\frac{(1686)^2}{(2760)^2} + \frac{(12)^2}{(560)^2} + \frac{(140)^2}{(190)^2} = 0.92 \qquad \text{O.K.}$$

Using interaction stress factor $C_I$:

$$\tan \theta = \frac{1}{12} = 0.083$$

$$C_I = 0.638 \text{ (from Table 5.5)}$$

$$C_V = 0.789$$

$$C_I < C_V \text{ and } C_I \text{ controls.}$$

$$F_b' = F_b^* C_I = (2760)(0.638) = 1760 \text{ psi} > 1686 \text{ psi} \qquad \text{O.K.}$$

6.  Determine camber:

$$\text{Dead load deflection} = \Delta_{\text{DL}} = \Delta_{\text{TL}}\frac{W_{\text{DL}}}{W_{\text{TL}}}$$

$$\Delta_{\text{DL}} = 2.34\left(\frac{240}{720}\right) = 0.78 \text{ in.}$$

Centerline camber $= 1.5\Delta_{\text{DL}} = (1.5)(0.78) = 1.17$ in.    Use $1\frac{1}{4}$ in.

No compression face camber is required because the camber is less than 2 in. The assumed design values are satisfactory. For greater choices of available laminating combinations, the designer should recheck the design using lower design values for compression perpendicular to grain and shear.

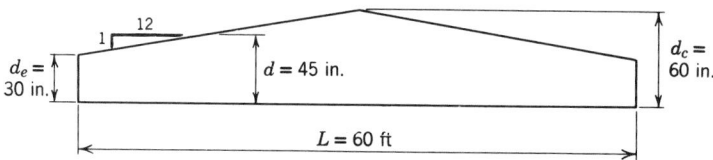

## 5.11    CURVED MEMBERS

### 5.11.1    Minimum Bending Radii $_1 = 12 ! 6 "$ no tangent
$2 = 25! 6"$ no tangent

The recommended minimum radii of curvature $R$ for curved structural glued laminated timbers are 9 ft 4 in. for Western Species and 7 ft 0 in. for Southern Pine for a lamination thickness $t$ of $\frac{3}{4}$ in.; and 27 ft 6 in. for a lamination thickness of $1\frac{1}{2}$ in. for all species. If required for architectural or other design considerations, other radii of curvature may be used with these thicknesses, and other radius-thickness combinations may be used provided that the ratio $t/R$ does not exceed $\frac{1}{100}$ for hardwoods and Southern Pine, or $\frac{1}{125}$ for other softwoods.

### 5.11.2    Curvature Factor

Because of stresses induced in bending laminations to a required curvature, the allowable flexural stress in curved laminated members is less than in straight members. Therefore, the allowable unit stress in bending for a curved member must be adjusted by multiplying by the curvature factor $C_C$ given in Eq. (4-4) as follows:

$$C_C = 1 - 2000\left(\frac{t}{R}\right)^2$$

where    $t =$ thickness of lamination (in.) and
$R =$ radius of curvature (bending radius) of the lamination (in.).

The curvature factor $C_C$ should not be applied to stresses in straight portions of a member, regardless of curvature elsewhere in the member. See Fig. 4.4 for a graphical solution of the curvature factor equation.

## 5.11.3 Radial Tension and Compression

When curved members are subjected to a bending moment $M$, radial stresses are set up in a direction parallel to the radius of curvature $R$ of the centerline of the member (perpendicular to grain). If the moment increases the radius of curvature (causes the member to become straighter), the stress is tension; if it decreases the radius (causes the member to become more sharply curved), the stress is compression.

The equation for computing the actual stress $f_r$ in members of constant cross section is:

$$f_r = \frac{3M}{2R_m bd} \tag{5-47}$$

where $M$ = bending moment (in.-lb),
$b$ = width of rectangular member (in.),
$d$ = depth of rectangular member (in.), and
$R_m$ = radius of curvature of centerline of member (in.).

For these members, when $M$ causes stress in tension across the grain (radial tension), the tensile stress shall be limited to:

### Radial Tension Design Values

| Species | Loading other than Wind or Earthquake<br>psi | Wind or Earthquake loading<br>psi |
|---|---|---|
| Alaska Cedar | 15 | 63 |
| California Redwood | 42 | 42 |
| Canadian Spruce Pine | 15 | 53 |
| Douglas Fir-Larch | 15 | 55 |
| Douglas Fir-South | 15 | 55 |
| Eastern Spruce | 15 | 48 |
| Hem-Fir | 15 | 52 |
| Softwood Species | 15 | 47 |
| Southern Pine | 67 | 67 |

These values are subject to adjustment for duration of load.

If the calculated stress exceeds the applicable design value indicated, the design should be reevaluated by changing the geometry of the section by increasing the radius of curvature or by changing the pitch to increase the centerline depth of the member. As an alternate procedure for Douglas Fir-Larch, Douglas Fir South, Hem-Fir, Eastern Spruce, Alaska Cedar, Canadian Spruce Pine, and Softwood Species, mechanical reinforcement may be designed sufficient to resist the full magnitude of the calculated radial tension stress in accordance with the recommendations given in 5.12.2. When radial reinforcement is used, it resists shrinkage that may occur between the time of installation of the reinforcement and the time the member reaches equilibrium. If the shrinkage is excessive, this restraint may cause cracking in the member. To minimize this condition where members are to be used in dry conditions, the member should be manufactured from laminations with a maximum moisture content of 12%. With radial reinforcement, the calculated radial tension stress $f_{rt}$ is limited to 55 psi for Douglas Fir-Larch and Douglas Fir South, 52 psi for Hem-Fir, 48 psi for Eastern Spruce, 63 psi for Alaska Cedar, 53 psi for Canadian Spruce Pine, and 47 psi for Softwood Species subject to adjustment for duration of load.

When the moment is in a direction causing a stress in compression across the grain, this stress shall be limited to the design value in compression perpendicular to the grain, $F_{c\perp}$ for the species involved.

When designing a curved bending member of variable cross section such as a pitched and tapered curved beam, the radial stress $f_r$ is computed by the equation

$$f_r = K_r C_r \frac{6M}{bd_c^2} \tag{5-48}$$

This equation is sometimes written as

$$f_r \text{ or } f_{rt} = K_r C_r f_o$$

where

$$f_o = \frac{6M}{bd_c^2}$$

$K_r$ = radial stress factor obtained from Fig. 5.24 or calculated using the polynomial approximations from Table 5.10,

$M$ = bending moment at midspan (in.-lb),

$b$ = width of cross section (in.),

$d_c$ = depth of cross section at centerline (in.), and

$C_r$ = reduction factor, a function of the shape of the member obtained from Figs. 5.31–5.34.

The design values in radial tension, $F_{rt}$, for pitched and tapered curved beams are the same as the design values given for members of constant cross section.

The radial stress factor $K_r$ varies with the ratio of the slope of the bottom face to the slope of the top face, $\phi_B/\phi_T$. To reduce the design complexity, $K_r$ has been selected for one $\phi_B/\phi_T$ ratio. This results in a slightly conservative design for other $\phi_B/\phi_T$ ratios.

$$K_r = A + B\left(\frac{d_c}{R_m}\right) + C\left(\frac{d_c}{R_m}\right)^2 \tag{5-49}$$

where    $A, B, C$ = are dimensionless factors from Table 5.8,

$d_c$ = depth of centerline of member (in.), and

$R_m$ = radius of curvature of centerline of member (in.).

The values of factors $A$, $B$, and $C$ are also shown graphically in Fig. 5.25.

### 5.11.4  Determination of Radial Tension Stresses

$K_r$ in Eq. (5-48) and Eq. (5-49) is defined as the radial stress factor and accounts for the effect of the key geometric variables of a pitched and tapered curved beam,

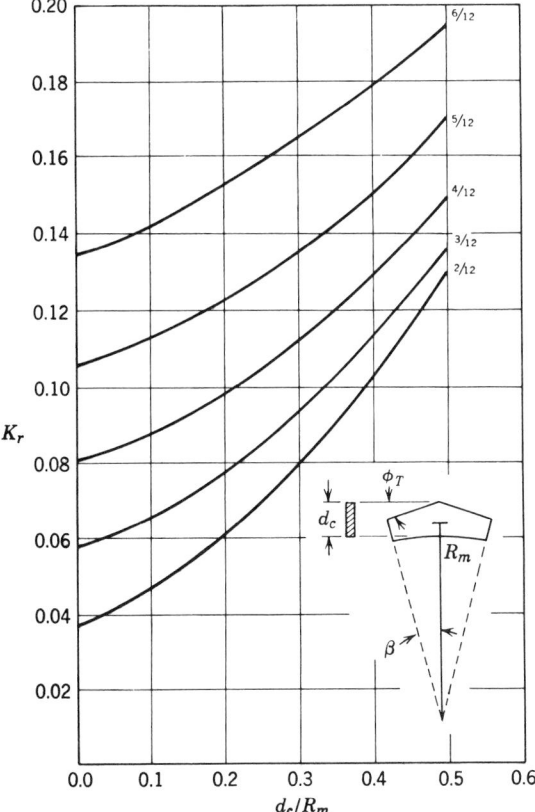

**FIGURE 5.24** Determination of $K_r$.

**TABLE 5.8**

**Polynomial Approximation to $K_r^a$**

| Angle of Upper Tapered Surface, $\phi_T$ (degrees) | Factors | | |
|:---:|:---:|:---:|:---:|
| | $A$ | $B$ | $C$ |
| 2.5 | 0.0079 | 0.1747 | 0.1284 |
| 5.0 | 0.0174 | 0.1251 | 0.1939 |
| 7.5 | 0.0279 | 0.0937 | 0.2162 |
| 10.0 | 0.0391 | 0.0754 | 0.2119 |
| 15.0 | 0.0629 | 0.0619 | 0.1722 |
| 20.0 | 0.0893 | 0.0608 | 0.1393 |
| 25.0 | 0.1214 | 0.0605 | 0.1238 |
| 30.0 | 0.1649 | 0.0603 | 0.1115 |

[a] For intermediate values of $\phi_T$, use straight-line interpolation between calculated values of $K_r$, as illustrated for $K_\theta$ in the design example for Douglas Fir-Larch beams.

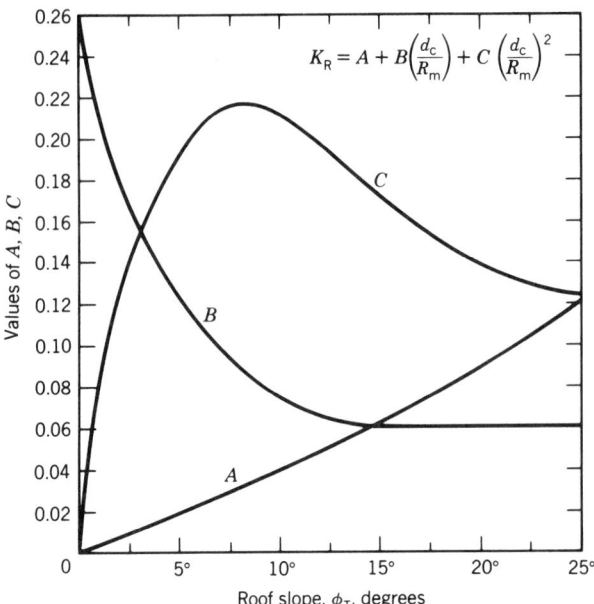

**FIGURE 5.25**    Roof slope versus *A*, *B*, and *C*. From *Timber Design Manual*, Laminated Timber Institute of Canada.

namely the top slope, $\phi_T$, the bottom slope, $\phi_B$, the ratio of $d_c/R_m$, and the span length, *L*, as they affect the radial stress distribution in these members. While a series of design curves were initially generated for different ratios of $\phi_B/\phi_T$, only one set of curves, as shown in Fig. 5.24, were used in order to simplify the design procedure. The curves on this figure are based on the assumption of $\phi_B/\phi_T = 1.0$, and are slightly conservative for all other cases of $\phi_B/\phi_T$ less than 1.0.

In addition, the generation of the $K_r$ curves was also based on the assumption that the member was subjected to pure bending. It can be shown from equilibrium equations that for a given bending moment, the case of pure bending will result in higher values of radial stresses as compared to other more common loading conditions such as a uniformly distributed load. Since most structural roof members such as pitched and tapered curved beams are seldom designed for the case of pure bending, the $C_r$ factor in Eq. (5-48) and Eq. (5-49) was introduced to allow the designer to reduce the $K_r$ value to account for uniform loading conditions.

The curves presented in Figs. 5.31–5.34 for the determination of $C_r$ are based on the member geometry for a range of $l/l_c$ from 1.0 to 4.0. For cases of $l/l_c$ greater than 4.0 or for loading conditions other than uniformly distributed, the designer can account for other loading conditions and member geometry in determining $C_r$ by using the following adjustments:

For $l/l_c \geq 8.0$, use $C_r = 1.0$

For values of $4.0 < l/l_c < 8.0$, straight line interpolation may be used between values obtained from the graph for $l/l_c = 4.0$ (Fig. 5.34) and 1.0 as determined for $l/l_c \geq 8.0$.

**TABLE 5.9**

**Adjustment Factor for Ratios**
**of *l/c***

| *l/c* | $C_r$ Adjustment Factor |
|---|---|
| 1.0 | 0.75 |
| 2.0 | 0.80 |
| 3.0 | 0.85 |
| ≥ 4.0 | 0.90 |

For equal concentrated loads positioned at $\frac{1}{3}$ points of the span, use the value of $C_r$ as determined for a uniform load case increased by a factor of 1.05 for all cases of $l/l_c$.

For point loads located at midspan, adjust the value of $C_r$ as determined for the uniformly distributed load case by the factors in Table 5.9.

The adjustment factors for loading conditions other than uniformly distributed loads are approximate values based on a ratio of the area of the moment diagrams in the curved portion of the pitched and tapered curved member for equal maximum moment conditions. For other more complex conditions, the designer may wish to determine these adjustment factors by comparing the area of the moment diagram in the curved portion of the pitched and tapered curved beam with the area for an equivalent uniformly distributed loading condition.

## 5.12   CURVED BEAMS

A number of different types of curved beams can be manufactured from glued laminated timbers. The most commonly used are shown in Fig. 5.26.

Pitched and tapered curved glued laminated timber beams are among the most popular types of structural roof members where a sloping roof and maximum interior clearance are desired. In these beams, as shown in Fig. 5.26, the top edge slopes from the apex at the centerline toward the supports at an angle to the horizontal $\theta_T$ and the lower edge is curved between the tangent points P.T. and slopes at an angle $\theta_B$ between the tangent point and the supports. The end portion of the member is usually tapered, but may be of constant cross section between the end and the tangent point. The tangent points are usually located near the quarter points, but the location can vary from a point located as close to the center as the minimum radius of curvature permits to a point at the end support. The procedure illustrated applies only to symmetrical pitched and tapered curved beams which are uniformly supported along the top edge.

The design of pitched and tapered curved beams is similar to that of straight prismatic beams, with the exception that radial tension stress $f_{rt}$ is induced and the distribution of bending stresses about the assumed neutral axis is different

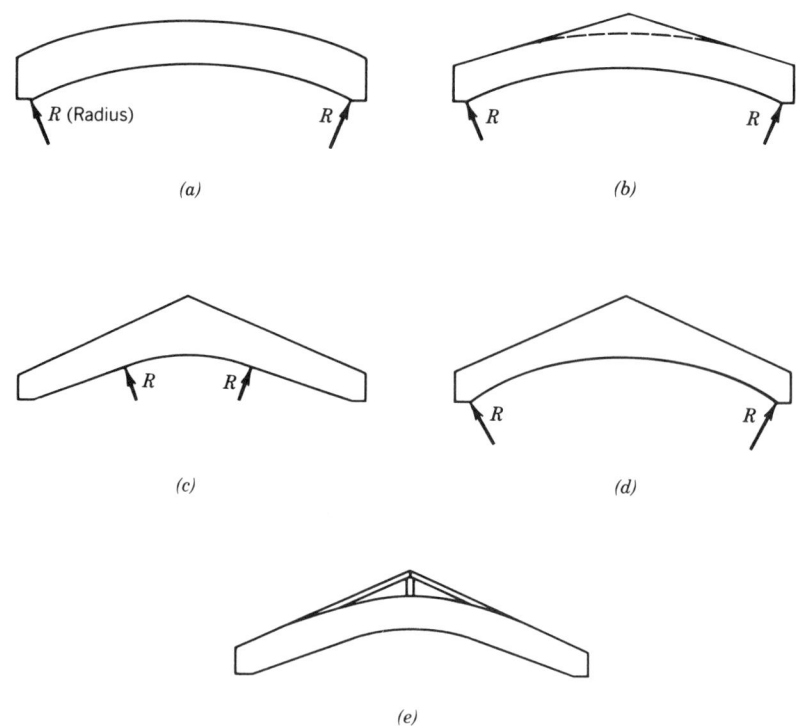

**FIGURE 5.26**   Simple span curved beam forms. (*a*) Curved beam with constant cross section; (*b*) Pitched and tapered curved beams, with constant cross section and mechanically attached haunch; (*c*) Pitched and tapered curved beams (tangent ends); (*d*) Pitched and tapered curved beams (constant curvature); (*e*) Pitched and tapered curved beams, with constant cross section and framed haunch.

from a prismatic member. Radial tension stresses were recognized by T. R. C. Wilson in Technical Bulletin No. 691 (10). The equation $f_{rt} = 3M/2R_{bd}$ was derived on the basis of an analogy to a pressure vessel. It is applicable to members of constant cross section only. The procedure included in this manual is based on subsequent research.

The bending and radial stresses in pitched and tapered curved beams are affected by the variable shape of the section, and their exact determination is complex. The procedures presented in this manual are based on a simplification of those contained in *Behavior and Design of Double-Tapered Pitched and Curved Glulam Beams* (11). The effect of interaction of stresses on the top tapered cut of these types of members is usually very small and can be ignored in beams of the usual configurations. Because the factors for determining bending and radial tension stresses are a function of the geometric configuration of the member, they cannot be accurately determined until the final size and shape are known. Therefore, the procedure presented in this manual is a trial-and-error method whereby a trial size is determined and adjusted until the correct size is obtained. Other methods may be used provided the design criteria are satisfied. The stress in bending, $f_b$, is increased by the bending stress factor $K_\theta$ due to the shape of the member.

$$f_b = K_\theta \frac{6M}{bd_c^2} \tag{5-50}$$

where   $M$ = moment at midspan (in.-lb),
     $b$ = width (in.),
     $d_c$ = depth at midspan (in.),
     $K_\theta = D + E(d_c/R_m) + F(d_c/R_m)^2$
$D, E, F$ = dimensionless factors from Table 5.10 and the other terms are as defined previously.

The deflection is also affected by the shape of the member, and the following equation represents a close approximation based on test data:

$$\Delta_c = \frac{5Wl^3}{32E'bd_{eb}^3} \tag{5-51}$$

where   $\Delta_c$ = deflection at midspan (in.),
     $W$ = total uniform load (lb),
     $l$ = span (in.),
     $E'$ = modulus of elasticity (psi) multiplied by appropriate adjustment factors,
     $b$ = width (in.),
     $d_{eb} = (d_e + d_c)(0.5 + 0.735 \tan \phi_T) - 1.41(d_c) \tan \phi_B$, (in.),
     $\phi_B$ = slope of bottom (soffit), at ends (degrees),
     $\phi_T$ = slope of top (degrees), and $d_c$ and $d_e$ are shown in Fig. 5.27.

Other methods of determining deflection such as the virtual work method may also be used.

**TABLE 5.10**

**Coefficients for Determining $K_\theta^a$**

| $\phi_T$ (degrees) | $D$ | $E$ | $F$ |
|:---:|:---:|:---:|:---:|
| 2.5 | 1.042 | 4.247 | −6.201 |
| 5 | 1.149 | 2.036 | −1.825 |
| 10 | 1.330 | 0.0 | 0.927 |
| 15 | 1.738 | 0.0 | 0.0 |
| 20 | 1.961 | 0.0 | 0.0 |
| 25 | 2.625 | −2.829 | 3.538 |
| 30 | 3.062 | −2.594 | 2.440 |

[a] For intermediate values of $K_\theta$, use straight-line interpolation between the values of $\phi_T$ (see examples).

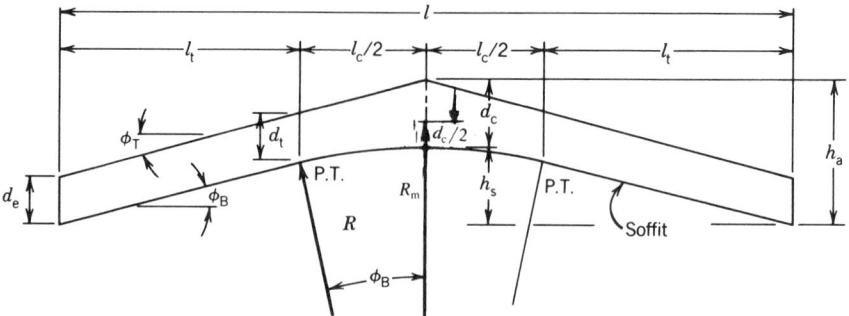

**FIGURE 5.27**    Pitched and tapered curved beam.

### 5.12.1  Design Procedures for Pitched and Tapered Curved Beams

The design procedure is based on a mathematical solution of the beam geometry (see Fig. 5.27) and eliminates the need for a graphical solution. Other methods may be used provided the requirements in steps 7 and 8 are met. When reinforcement is needed, step 10 must be followed.

1.  Determine the minimum end depth $d_e$.

The following steps are based on a uniformly loaded symmetric beam. When point loads are used, the factor $C_r$ is adjusted by coefficients in Table 5.9.

The distance, $d_e$, is the vertical dimension at the end. The depth at the end for determining shear parallel to grain $d_e'$ can be set at $d_e \cos \phi_B$ with very small error.

The shear $V$ at the end perpendicular to the grain of the bottom lamination is $V \cos \phi_B$ with very small error.

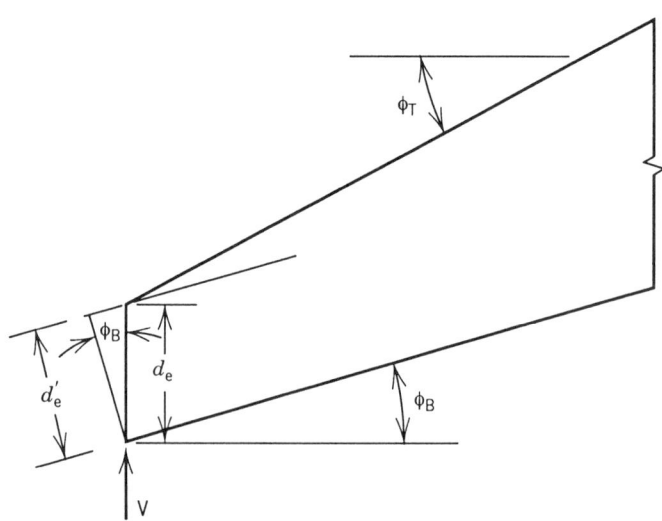

The parallel to grain shear stress at the end is determined as follows:

$$f_v = \frac{3V \cos \phi_B}{2bd_e \cos \phi_B} = \frac{3V}{2bd_e} \tag{5-52}$$

The $\cos \phi_B$'s in the equation cancel, and for convenience, $d_e$ will be used in the equation:

$$d_e = \frac{3V}{2bF_v'} \tag{5-53}$$

where  $V$ = shear based on effective span (lb),
   $b$ = width (in.), (assumed), and
   $F_v'$ = tabular design value in shear (psi) multiplied by the applicable adjustment factors.

2.   Determine approximate centerline depth $d_{cb}$ required to meet bending stress limitations:

$$d_{cb} = \sqrt{\frac{6MD}{bF_b'}} \tag{5-54}$$

where  $D$ = coefficient obtained from Table 5.10,
   $M$ = maximum bending moment (in.-lb), and
   $F_b'$ = tabular design value in bending (psi) multiplied by applicable adjustment factors (including $C_V$ or $C_L$).

3.   Determine approximate centerline depth $d_{c\Delta}$ to meet deflection limitations $(\Delta_{max})$ as shown in Table 4.3. Determine approximate effective centerline depth $d_{eff}$ for calculating $d_{c\Delta}$

$$d_{eff} = \sqrt[3]{\frac{Wl^3}{6.4E' b\Delta_{max}(\cos \phi_T)^3}} \tag{5-55}$$

where  $W$ = total uniform load (lb),
   $l$ = span (in.),
   $E'$ = tabular design value of $E$ (psi), multiplied by applicable adjustment factors,
   $b$ = width (in.), and
   $\phi_T$ = slope of top face (degrees).

The approximate centerline depth is

$$d_{\Delta c} = 2d_{eff} - d_e \tag{5-56}$$

4.   Determine the trial minimum centerline depth $d_{crt}$ due to the radial tension stress limitation.

$$d_{crt} = \sqrt{\frac{6MK_r}{bF_{rt}'}} \tag{5-57}$$

where   $K_r$ = value obtained from Fig. 5.24,
   $R_m$ = radius of curvature at neutral axis (in.),
   $M$ = maximum bending moment (in.-lb),
   $F_{rt}'$ = design value for radial tension (psi) multiplied by applicable adjustment factors, and
   $b$ = width (in.).

The determination of $K_r$ requires an estimate of a value for $d_c/R_m$. It is recommended that the first trial value of $K_r$ be determined by taking $d_c$ as the greater of the trial values of $d_{cb}$ (from step 2) or $d_{c\Delta}$ (from step 3). A trial value for $R_m$ can be taken as the span in inches.

If the design value of radial tension for Douglas Fir-Larch, Douglas Fir South, Hem-Fir, Eastern Spruce, Alaska Cedar, Canadian Spruce Pine or Softwood Species is exceeded, the member should be reproportioned or mechanical reinforcing sufficient to resist all of the radial tension load should be used. When radial reinforcement is used, the design value for radial stress, $F_{rt}'$, may not exceed one-third the tabular design value in shear multiplied by applicable adjustment factors for the species (55 psi for Douglas Fir-Larch and Douglas Fir South, 52 psi for Hem-Fir, 48 psi for Eastern Spruce, 63 psi for Alaska Cedar, 53 psi for Canadian Spruce Pine, and 47 psi for Softwood Species). As an alternative to mechanical reinforcement, the beam may be manufactured with the peaked portion (haunch) not glued to the remainder of the beam. Since this is a nonstructural part of the member, the peaked area may also be replaced by open framing. In these cases, the member may be designed as a curved beam of constant cross section. If $d_{crt}$ is greater than the height of the apex point, $h_a$ (see step 5), or is impractically large for unreinforced Douglas Fir-Larch, Douglas Fir South, Hem-Fir, Eastern Spruce, Alaska Cedar, Canadian Spruce Pine, or Softwood Species, recalculate $d_{crt}$ based on $F_{rt}'$ for radial reinforcement. In no case should $F_{rt}'$ exceed one-third the design value in shear parallel to grain, with or without mechanical reinforcement.

5.   Determine the height of the apex point, $h_a$, the height of the soffit at midspan, $h_s$, a trial bottom slope and soffit radius $R$ as follows:

$$h_a = d_e + \frac{l}{2} \tan \phi_T \tag{5-58}$$

where   $l$ = span length (in.), and
   $\phi_T$ = top slope (degrees).

$$h_s = h_a - d_c \tag{5-59}$$

where   $d_c$ = largest of $d_{cb}$, $d_{c\Delta}$, or $d_{crt}$.

From Fig. 5.28, determine the maximum bottom slope, $\phi_{B\ max}$. If $\phi_{B\ max}$ is greater than the top slope, $\phi_T$, set $\phi_{B\ max} = \phi_T$ (in which case the beam becomes a pitched curved beam as shown in Fig. 5.26). From Fig. 5.29, determine the smallest possible value of the bottom slope, $\phi_{B\ min}$. If the value for $\phi_{B\ min}$ cannot

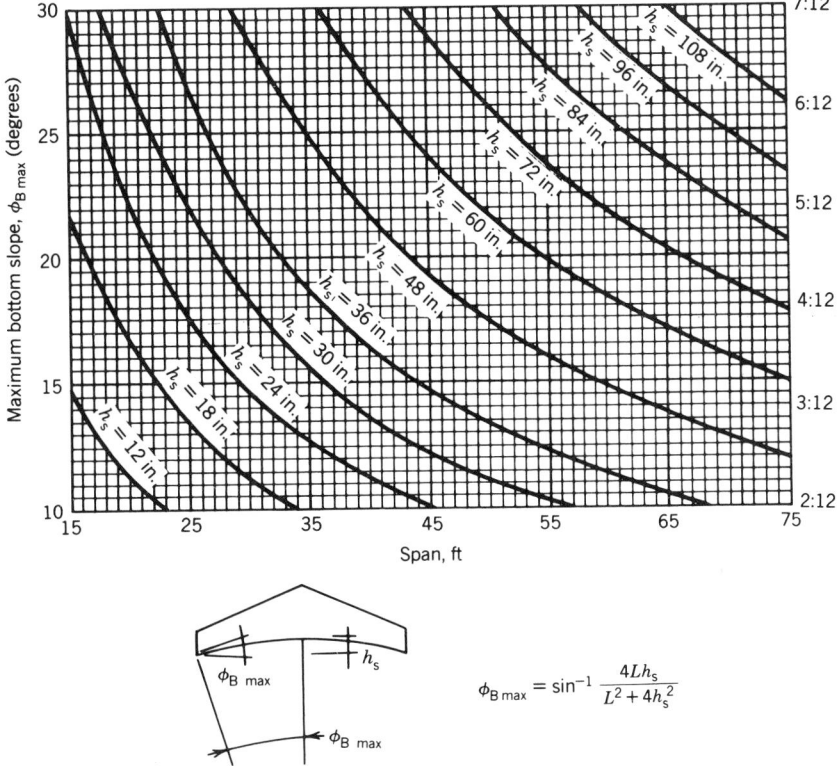

**FIGURE 5.28**   Graph for determining the largest possible bottom slope ($\phi_{B\,max}$).

be obtained from Fig. 5.29 (because of the curves not extending far enough) or if the value of $\phi_{B\,min}$ obtained from Fig. 5.29 is greater than $\phi_T$, then use $\frac{3}{4}$-in. laminations and determine $\phi_{B\,min}$ from Fig. 5.30.

Choose the trial bottom slope that is between the maximum and minimum permitted as:

$$\phi_B = 0.45\,(\phi_{B\,max} + \phi_{B\,min})\ \text{(but not less than } \phi_{B\,min}) \qquad (5\text{-}60)$$

$$R = \frac{h_s - \dfrac{l}{2}\tan\phi_B}{1 - \cos\phi_B - \sin\phi_B \tan\phi_B} \qquad (5\text{-}61)$$

6.   Determine the following values using the trial bottom slope $\phi_B$ and soffit radius $R$ determined in step 5. The following notations are used in the calculations:

(a) $l_t = l/2 - R\sin\phi_B$ = length of tapered leg (in.),

(b) $l/l_c = l/(l - 2l_t)$ = ratio of span length to distance between tangent points,

(c) $d_t' = [d_e + l_t(\tan\phi_T - \tan\phi_B)][\cos\phi_T/\cos(\phi_T - \phi_B)],$ \qquad (5\text{-}62)

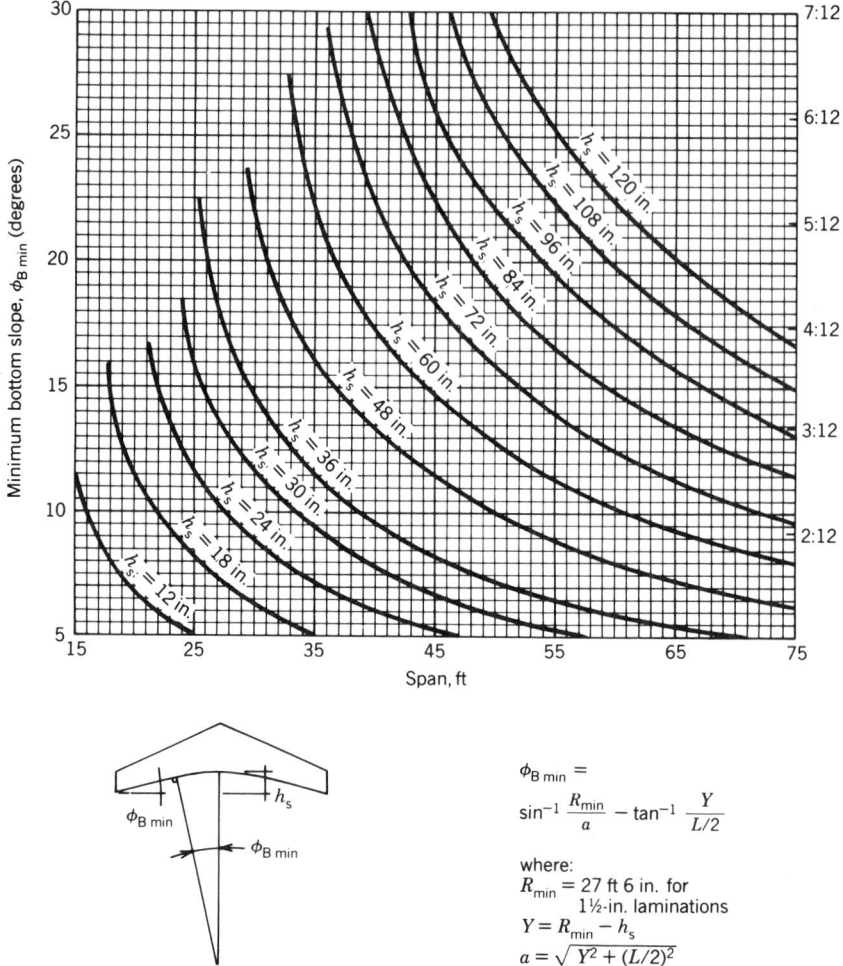

**FIGURE 5.29**  Graph for determining the smallest possible bottom slope ($\phi_{B\,min}$) when using 1½-in. laminations (Western Species).

(d) $f_0 = 6M/bd_c^2$ = reference stress at centerline used for convenience in the calculations where $M$ = maximum bending moment at centerline (in.-lb).

7.  Check deflection at the centerline $\Delta_c$.

$$d_{eb} = (d_e + d_c)(0.5) + 0.735 \tan \phi_T - 1.41 d_c \tan \phi_B \qquad (5\text{-}63)$$

$$\Delta_c = \frac{5 W l^3}{32 E' \, b d_{eb}^3} \qquad (5\text{-}64)$$

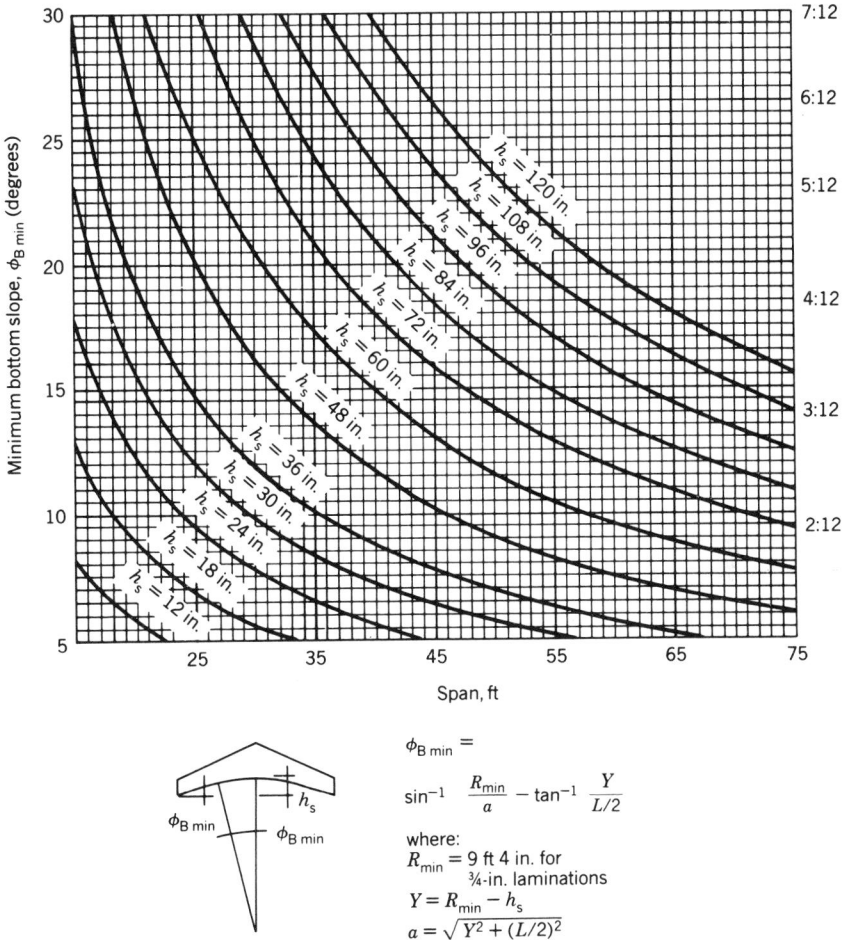

**FIGURE 5.30** Graph for determining the smallest possible bottom slope ($\phi_{B\,min}$) when using ¾-in. laminations (Western Species).

8. Check stresses.

(a) Determine extreme fiber bending stress at the apex centerline. From Table 5.10, determine bending stress factor $K_\theta$ corresponding to $\phi_T$ and $d_c/R_m$:

$$K_\theta = D + E(d_c/R_m) + F(d_c/R_m)^2 \tag{5-65}$$

$$f_b = K_\theta f_0 \tag{5-66}$$

The stress $f_b$ must be no larger than the design value in bending, $F_b'$.

(b) Determine extreme fiber bending stress at the tangent point:

$$f_{bt} = 6M_t/bd_t'^2 \tag{5-67}$$

where $M_t$ = bending moment at a distance $L_t$ from the end (in.-lb)

$f_{bt}$ must be no larger than $F'_b$. Note the depth $d_t$ is the depth at the tangent point measured perpendicular to the bottom lamination.

(c) Check the extreme fiber stress at intervals along the straight portion (leg) of pitched and tapered curved beams. Usually checking at about 4 points along the leg is sufficient, but in members with extremely long legs, more points may need to be checked. For a symmetric pitched and tapered beam with uniformly distributed load only, the distance from the end of the member to the point of maximum bending stress can be determined from the following relationship.

$$x = [d_e \cos \phi_B (L/2)]/[(L/2)\{\tan (\phi_T - \phi_B)/\cos \phi_B\} + d_e(\cos \phi_B)] \quad (5\text{-}68)$$

The depth $d'_x$ perpendicular to the bottom lamination at any point $x$ can be determined by the following equation:

$$d'_x = [d_e + x(\tan \phi_T - \tan \phi_B)][\cos \phi_T/\cos(\phi_T - \phi_B)] \quad (5\text{-}69)$$

where $x$ is distance from the end of member (in.).

The moment $M_x$ at $x$ distance from the end is determined by the following equation:

$$M_x = 0.5w/12 \, [x - 0.5d'_x(\tan \phi_B)][l - \{x - 0.5d'_x(\tan \phi_B)\}] \quad (5\text{-}70)$$

where   $M_x$ = moment at $x$ distance from end of member (in.-lb),
        $w$ = uniformly distributed load (plf),
        $d'_x$ = depth of member perpendicular to soffit lamination at the point being evaluated (in.),
        $l$ = total member span (in.), and
        $x$ = horizontal distance from the end of the beam.

$F'_x$ equals $F^*_b$ multiplied by $C_V$ or $C_I$ whichever controls.

(d) Check the maximum radial stress. Determine the radial stress factor $K_r$ corresponding to $d_c/R_m$ for the given $\phi_T$ from Fig. 5.16. Determine the reduction factor $C_r$ corresponding to $l/l_c$ for the given $\phi_T$ from Figs. 5.31, 5.32, 5.33, or 5.34.

$$f_{rt} = K_r C_r f_o \quad (5\text{-}71)$$

For values of $l/l_c$ between those shown in the figures, use straight-line interpolation.

The applied radial tension stress $f_{rt}$ must be no larger than the design value in radial tension $F'_{rt}$.

*Note:* If either the deflection or the radial stress limitations cannot be met, redesign by increasing depth $d_c$. Redesign requires going through steps 5–8 again. Steps 8a and 8b can be skipped in the redesign if the bending stress limitations were satisfied in the earlier trial geometry. If only the bending stress at the tangent point does not meet the limitations, revise the geometry by decreasing the angle $\phi_B$ (maintain $\phi_B \geq \phi_{Bmin}$). Redesign requires going through steps 5, 6, 8b, and 8c only (because the magnitude of the maximum deflection and the bending stresses

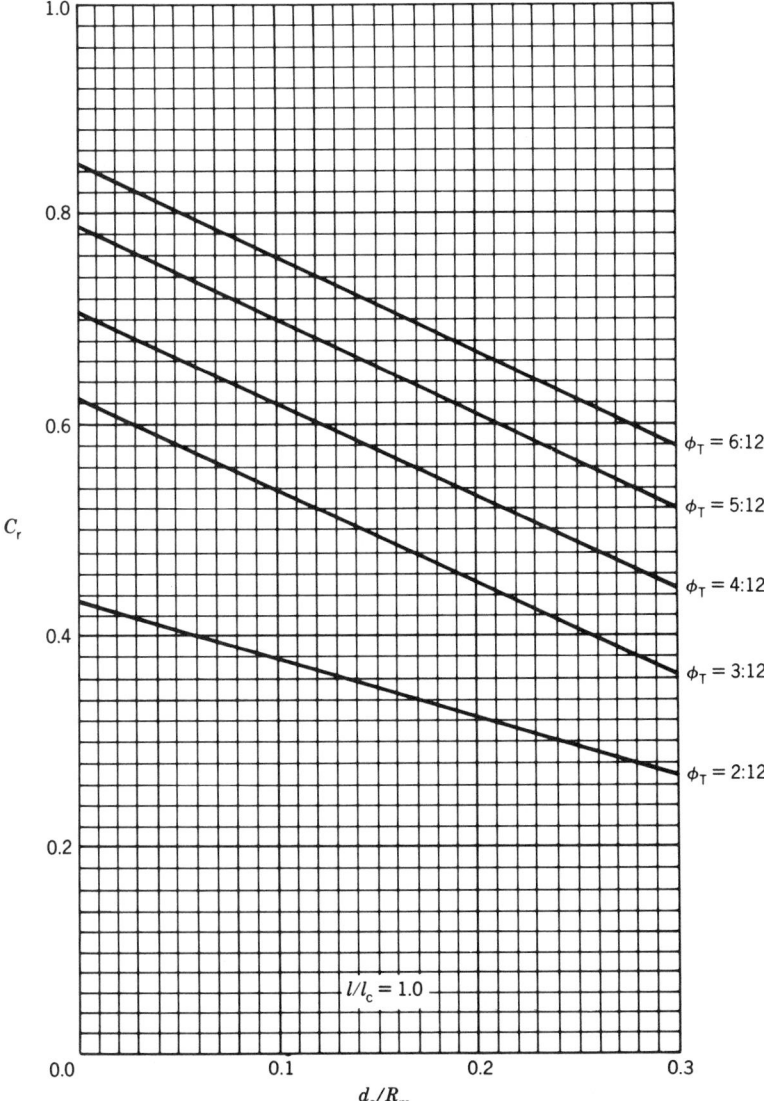

**FIGURE 5.31** Reduction factor $C_r$, to obtain radial stress for uniformly distributed loading case.

at the centerline decrease due to this revision in geometry). If the revised geometry is not adequate to satisfy the limitation, increase the depth ($d_c$) and redesign.

9. Determine the horizontal movement at the supports, $\Delta_H$, using the equation

$$\Delta_H = \frac{2h\Delta_c}{l} \qquad (5\text{-}72)$$

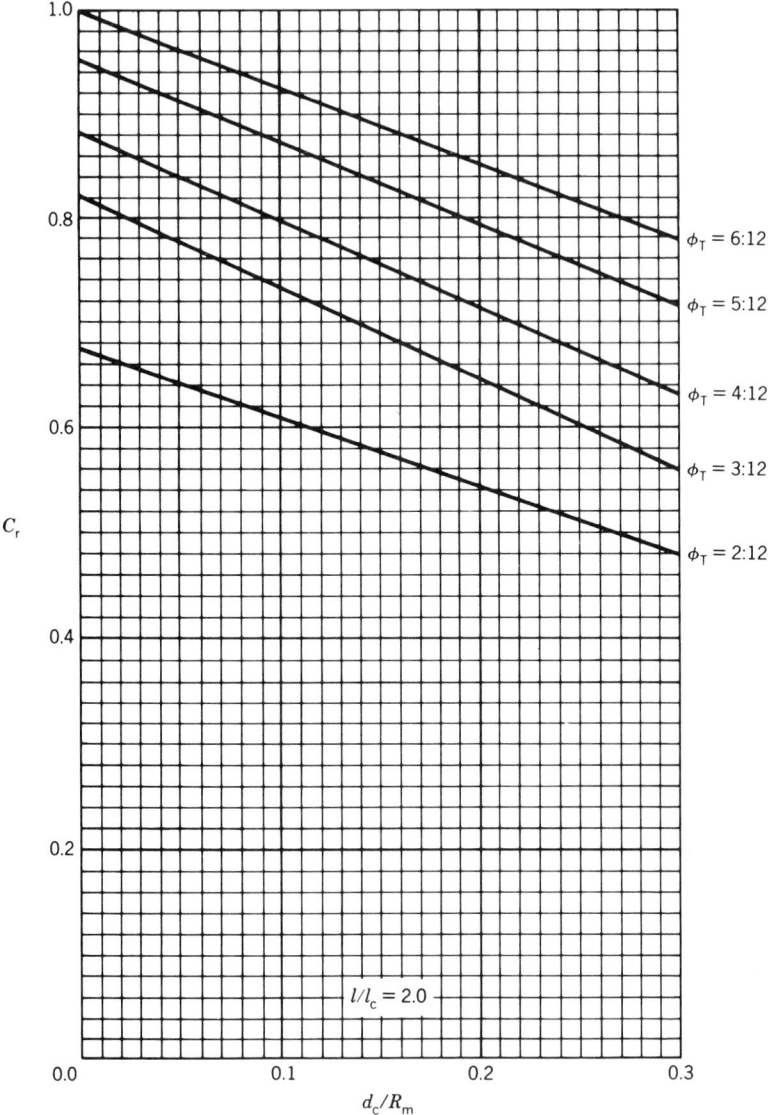

**FIGURE 5.32**   Reduction factor $C_r$, to obtain radial stress for uniformly distributed loading case.

where   $h$ = rise in centroidal axis from end to center (in.)($h = h_a - d_c/2$),
  $l$ = span length (in.), and
  $\Delta_c$ = centerline deflection as determined in step 7.

Note that horizontal movement should be provided for by the use of a slotted connection and an antifriction pad.

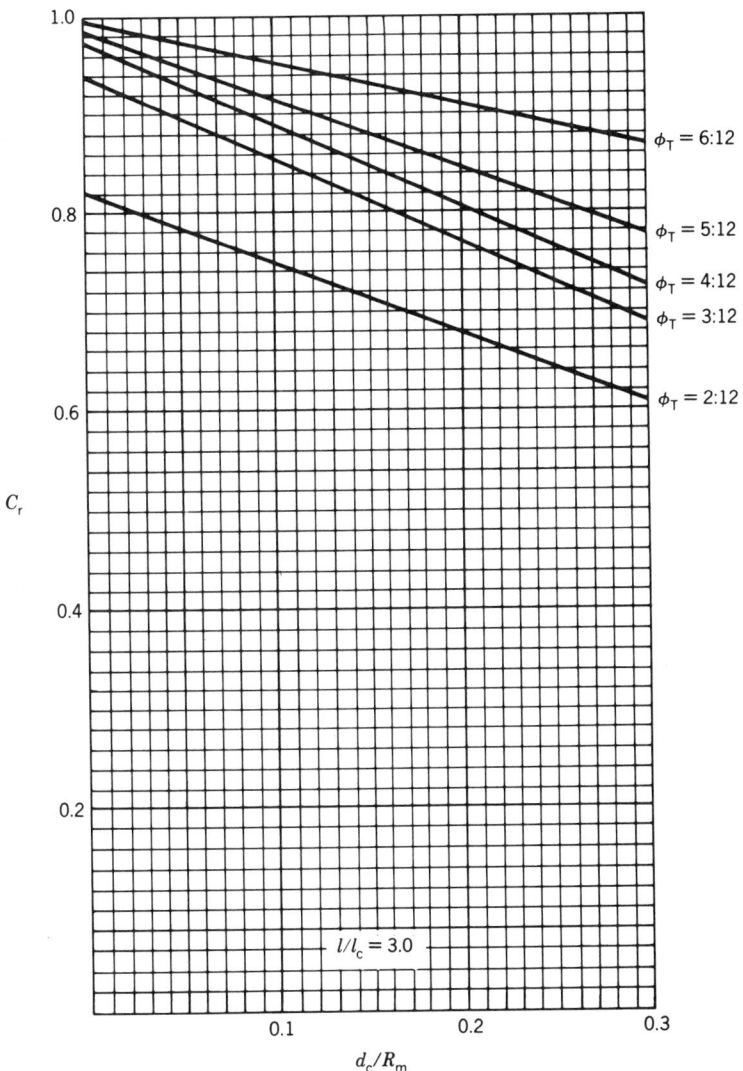

**FIGURE 5.33** Reduction factor $C_r$, to obtain radial stress for uniformly distributed loading case.

10. Design the radial reinforcement in accordance with the recommended procedures given in the next section if required by the analysis of radial stresses in step 8c.

### 5.12.2 Radial Reinforcement

If the radial tension stress in step 8c exceeds 15 psi multiplied by applicable adjustment factors for Douglas Fir-Larch, Douglas Fir South, Hem-Fir, Eastern Spruce, Alaska Cedar, Canadian Spruce Pine, or Softwood Species, the design

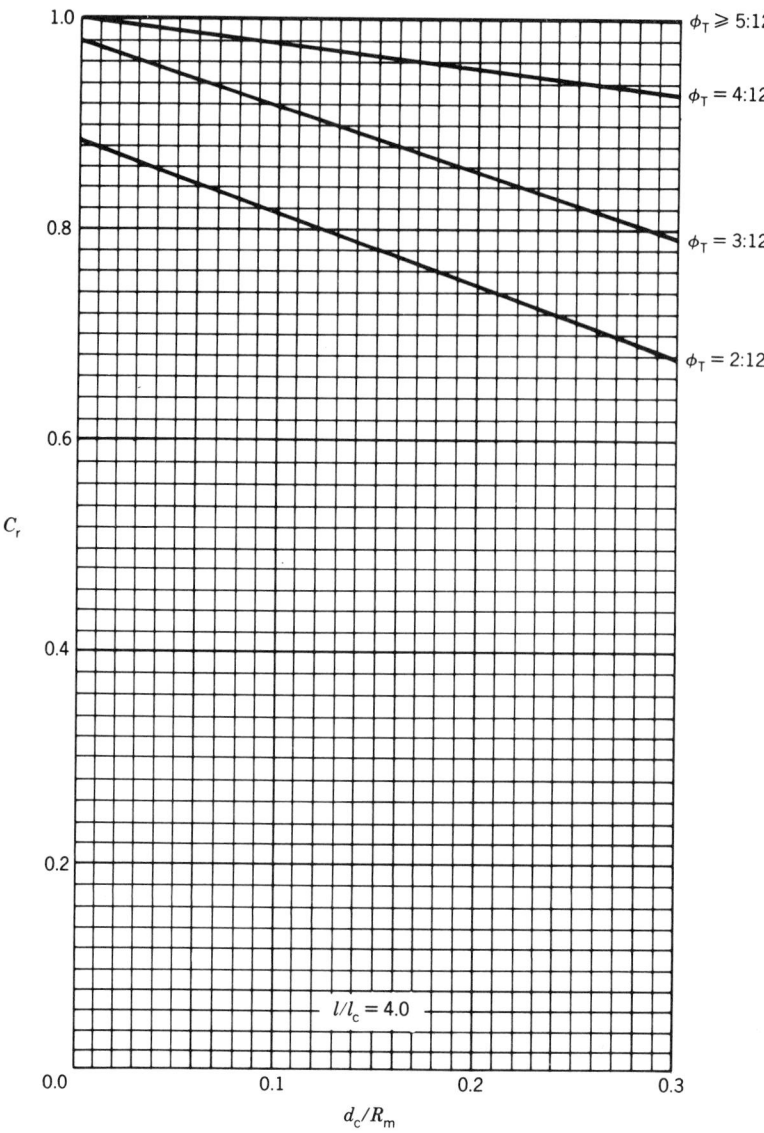

**FIGURE 5.34**    Reduction factor $C_r$, to obtain radial stress for uniformly distributed loading case.

procedure is to use radial reinforcement sufficient to resist the full magnitude of radial tension force. This reinforcement should be designed on the basis of sound engineering principles. Any type of reinforcement such as lag screws mechanically attached to wood or deformed steel bars bonded by adhesive that will effectively transfer the radial tension stresses between the wood and the reinforcement throughout the entire depth of embedment may be used. The method of bonding or attaching the reinforcement to the wood should be of a durable quality and capable of developing the required tensile strength of the reinforcement. Radial reinforcement is not used for Southern Pine.

When radial reinforcement is used, it tends to restrain shrinkage due to moisture loss. Because of this, it is recommended that the moisture content of the laminations prior to gluing not exceed 12% for dry conditions of use.

A typical design is to utilize fully threaded lag screws as radial tension reinforcement. The following comments provide general installation recommendations and design guidelines for the use of lag screw reinforcement.

(a)   The shop-installed radial tension reinforcing to be used is lag screws threaded full length. (Although the lag screws are specified to be threaded full length, manufacturing the lag screws requires a small length of unthreaded shank.) These lag screws should be shop installed in a prebored hole from the top of the member on the width centerline of the member. They should be used only in the curved or radially stressed portion of the member. The lag screws should be installed normal to the axis of the lamination (90° to the direction of the glue line) at the section where the lag screw is located. No washer should be used under the head of the lag screw. It is desirable, for structural reasons, not to countersink the head of the lag screw into the top of the beam. If it is necessary to install the lag screw in such a way that the top surface is smooth following installation, the head of the lag screw can be sawn off flush with the member or it can be countersunk, but the countersunk hole should only be large enough to install the head flush with the top of the beam.

(b)   The prebored lead hole for the threaded portion of the lag screw should have a diameter not greater than 85% of the shank diameter for species with a specific gravity of 0.50 or greater, and 80% for species with a specific gravity of less than 0.50 and a depth equal to the length of the threaded portion.

The design values for lag screws in Table 5.11 for Douglas Fir-Larch are 85% of the design values for lag screws in Table 7.24 to account for the larger hole, which is necessary for the longer lag screws. For other species, use 85% of the design value in Table 7.24 for lag screws used for radial reinforcement.

(c)   The length of the full thread of the lag screw should extend from the top (head end of the lag screw) of the beam to not less than 2 in. or not more than 3 in. from the soffit of the beam. Lag screw lengths should be in multiples of 1 in.

(d)   In general, lag screws from $\frac{5}{8}$ to 1 in. in diameter should be limited to a maximum length of 60 in. and the $\frac{5}{8}$-in.-diameter lag screws to a maximum length of 30 in. If greater lengths of reinforcement are required, larger diameter lag screws should be specified.

(e)   The fully threaded lag screws should be designed to take the entire radial tension stress developed in the member with no radial tension stress carried by the wood. The lag screws should be so spaced along the width centerline throughout the curved portion of the member as to carry the entire radial tension force.

(f)   The magnitude of the radial tension force to be carried by each lag screw should not exceed either of the following:

## TABLE 5.11

### Lag Screw Reinforcement Data for Douglas Fir-Larch

| | | Steel | Wool | | |
|---|---|---|---|---|---|
| Lag Screw Shank Diameter $D$ (in.) | Net Area at Root of Thread (in.$^2$) | Allowable TensionLoad (lb) for a Unit Stress of 20,000 psi[a] | Allowable Withdrawal Load (lb/in.) of Threaded Portion[b] | | |
| | | | Normal 1.00[c] | Snow 1.15[c] | 7-Day 1.25[c] |
| 1/4 | 0.0235 | 470 | 225 | 259 | 281 |
| 5/16 | 0.0405 | 810 | 266 | 306 | 332 |
| 3/8 | 0.0552 | 1105 | 305 | 350 | 381 |
| 7/16 | 0.0845 | 1690 | 342 | 393 | 425 |
| 1/2 | 0.108 | 2160 | 378 | 435 | 472 |
| 9/16 | 0.149 | 2970 | 410 | 472 | 512 |
| 5/8 | 0.174 | 3485 | 447 | 514 | 558 |
| 3/4 | 0.263 | 5265 | 513 | 589 | 641 |
| 7/8 | 0.366 | 7330 | 576 | 662 | 720 |
| 1 | 0.478 | 9555 | 636 | 731 | 795 |
| 1 1/8 | 0.618 | 12,360 | 695 | 799 | 869 |
| 1 1/4 | 0.804 | 16,090 | 752 | 865 | 940 |

[a]Rounded to nearest 5 lb.

[b]Based on Douglas Fir-Larch; specific gravity of 0.50 based on weight and volume when oven dry. For other species, see Table 7.24. These values are 85% of the values in Table 7.24 because of the larger lead hole used for long lag screws used for radial reinforcement.

[c]Load duration factors.

1. Maximum allowable tension in lag screws on net area at root of thread— The maximum tension load to be carried by each lag screw is limited to a unit stress of 20,000 psi on the net area at the root of the thread. Table 5.11 lists the net area at the root of threads and the allowable tension at a unit stress of 20,000 psi.

2. Maximum allowable tension in lag screws on allowable thread holding in wood (allowable withdrawal)—The maximum tension to be carried by each lag screw should not exceed the allowable withdrawal thread holding of the lag screw threads in the wood. The effective embedded thread length to be used in this determination is the embedded thread length from the neutral axis of the member to the point end of the lag screw or the thread length from the neural axis of the member to the head end of the lag screw, whichever is the lesser. The allowable unit withdrawal load for the lag screw threads in the wood should be in accordance with the values given in Table 5.11.

(g)   The section modulus is reduced somewhat by the hole bored for the radial reinforcement. In many cases, the effect is small but the reduction should be considered in design. The reduced section modulus is determined by subtracting only that portion of the lag screw hole below the neutral axis, because the lag screw completely fills the hole on the compression side. Consider the diameter of the hole as being equal to the shank diameter of the lag screw.

### 5.12.3   Design of Pitched and Tapered Curved Beams Made of Douglas Fir-Larch

The design of a pitched and tapered curved beam made of Douglas Fir-Larch is illustrated by the following example.

**Example: Douglas Fir-Larch Beam.**   Design a pitched and tapered curved glued laminated timber roof beam to meet the following requirements. Decking is applied to the top of the beam, providing adequate lateral support.

Species: Douglas Fir-Larch

$L = 60$ ft

Spacing $= 16$ ft

Roof slope $= 3:12$ $(14.04°)$

Snow load $= 30$ psf

$(C_D = 1.15)$

Dead load $= 15$ psf

$F_b^* = (2400)(1.15) = 2760$ psi

$F_v' = (165)(1.15) = 190$ psi

$F_{rt}' = (15)(1.15) = 17.2$ psi

$E' = 1,800,000$ psi

$\Delta_{TL(max)} = \dfrac{l}{180}$ for total load

1.   Determine end depth $d_e$. Use $b = 6\frac{3}{4}$ in. Calculate end reaction $R_v$:

$$R_v = \frac{wL}{2} = \frac{(45)(16)(60)}{2} = 21,600 \text{ lb}$$

Calculate trial end depth $d$:

$$d = \frac{3R_v}{2bF_v'} = \frac{(3)(21,600)}{(2)(6.75)(190)} = 25.3 \text{ in.}$$

$$L_e = L - 2d = 60 - 2\left(\frac{25.3}{12}\right) = 55.8 \text{ ft (use 56 ft)}$$

$$V = \frac{wL_e}{2} = \frac{(45)(16)(56)}{2} = 20,160 \text{ lb}$$

$$d_e = \frac{3V}{2bF_v'} = \frac{(3)(20,160)}{(2)(6.75)(190)} = 23.6 \text{ in. (use 24 in.)}$$

2.   Determine approximate trial centerline depth $d_{cb}$ from the bending stress limitation:

Assume $C_V = 0.75$:

$$F_b' = (2400)(1.15)(0.75) = 2070 \text{ psi}$$

From Table 5.10, $D$ for $10° = 1.330$ and $D$ for $15° = 1.738$.

$$D \text{ for } 14.04° = 1.330 + \frac{(1.738 - 1.330)(14.04 - 10)}{15 - 10} = 1.66$$

$$M = \frac{wl^2}{8} = \frac{(45)(16)(60)^2(12)}{8} = 3{,}890{,}000 \text{ in.-lb}$$

$$d_{cb} = \sqrt{\frac{6MD}{bF_b'}} = \sqrt{\frac{(6)(3{,}890{,}000)(1.66)}{(6.75)(2070)}} = 52.6 \text{ in.}$$

3.  Determine trial minimum centerline depth $d_{c\Delta}$ from the deflection limitation:

$$\Delta_{\max} = \frac{l}{180} = \frac{(60)(12)}{180} = 4 \text{ in.}$$

$$d_{\text{eff}} = \sqrt[3]{\frac{Wl^3}{6.4E'b\Delta_{\max}(\cos \phi_T)^3}}$$

$$= \sqrt[3]{\frac{(45)(16)(60)(60)^3(1728)}{(6.4)(1{,}800{,}000)(6.75)(4)(0.9701)^3}} = 38.5 \text{ in.}$$

4.  Determine trial minimum centerline depth $d_{crt}$ from radial stress limitation. Use $d_c = 56$ in. and $R_m = L = 60$ ft; $d_c/R_m = 56/(60)(12) = 0.08$. From Fig. 5.24, $K_r = 0.064$

$$d_{crt} = \sqrt{\frac{6MK_r}{bF_{rt}'}} = \sqrt{\frac{(6)(3{,}890{,}000)(0.064)}{(6.75)(17.2)}} = 113 \text{ in.}$$

This is a very large depth. Therefore, to reduce the depth, the beam will be designed for radial reinforcement, using a maximum radial tension stress for Douglas Fir-Larch of $F_v/3 = 165/3 = 55$ psi; $F_{rt}' = (55)(1.15) = 63.25$ psi.

$$d_{crt} = \sqrt{\frac{(6)(3{,}890{,}000)(0.064)}{(6.75)(63.25)}} = 59.1 \text{ in. (use 60 in.)}$$

5.  Determine height of apex $h_a$, trial bottom slope $\phi_B$, and soffit radius $R$:

$$h_a = d_e + \left(\frac{l}{2}\right)\tan \phi_T = 24 + \left[\frac{(60)(12)}{2}\right]\left(\frac{3}{12}\right) = 114 \text{ in.}$$

$$d_c = 60 \text{ in.}$$

$$h_s = h_a - d_c = 114 - 60 = 54 \text{ in.}$$

From Fig. 5.28, by interpolation, $\phi_{B\,\max} = 17.1° > \phi_T$. Therefore, set $\phi_{B\,\max} = 14.0°$. From Fig. 5.29, by interpolation, $\phi_{B\,\min} = 9.2°$
Choose trial bottom slope

$$\phi_B = 0.45(\phi_{B\,\max} + \phi_{B\,\min}) = (0.45)(14.0 + 9.2) = 10.4°$$

$$R = \frac{h_s - (l/2)\tan \phi_B}{1 - \cos \phi_B - \sin \phi_B \tan \phi_B} = \frac{54 - (360)(0.184)}{1 - 0.984 - (0.181)(0.184)} = 707 \text{ in.}$$

6.   Find the following values based on the values of $\phi_B$ and $R$ determined in step 5:

(a) $l_t = l/2 - R \sin \phi_B = (30)(12) - (707)(\sin 10.4°) = 232$ in.

(b) $l_c = l - 2l_t = (60)(12) - (2)(232) = 256$ in.
$l/l_c = 720/256 = 2.81$

(c) $d_t = d_e + l_t(\tan \phi_T - \tan \phi_B) = 24 + 232(\tan 14.04° - \tan 10.4°) = 39.4$ in.

(d) $R_m = d_c/2 + R = 60/2 + 707 = 737$ in.
$d_c/R_m = 60/737 = 0.81$

(e) $f_0 = 6M/bd_c^2 = (6)(3,890,000)/(6.75)(60)^2 = 960$ psi (reference stress at centerline).

7.   Check maximum deflection:

$$d_{eb} = (d_e + d_c)(0.5 + 0.735 \tan \phi_T) - 1.41d_c \tan \phi_B$$

$$= (24 + 60)[0.5 + (0.735)(0.250)] - (1.41)(60)(0.184) = 41.9 \text{ in.}$$

$$\Delta_c = \frac{5Wl^3}{32E'bd_{eb}^3}$$

$$= \frac{(5)(45)(16)(60)(720)^3}{(32)(1,800,000)(6.75)(41.9)^3} = 2.83 \text{ in.} < 4.0 \text{ in.} \quad \text{O.K.}$$

8.   Check stresses:

(a) Bending stress at centerline: From Table 5.10, determine $K_\theta$ (for $\phi_T$ of 10° and $d_c/R_m$ of 0.081) $= 1.33 + 0.0(0.081) + (0.927)(0.081) = 1.41$ and $K_\theta$ (for $\phi_T$ of 15° and $d_c/R_m$ of 0.081) $= 1.738$. Then, by straight-line interpolation, $K_\theta$ for $\phi_T$ of 14.0° and $d_c/R_m$ of 0.081:

$$K_\theta = 1.41 + \frac{(1.738 - 1.41)(4.0)}{5} = 1.67$$

$$f_b = K_\theta f_0 = (1.67)(960) = 1600 \text{ psi}$$

$$C_V = \left(\frac{5.125}{b}\right)^{1/10} \left(\frac{12}{d}\right)^{1/10} \left(\frac{21}{L}\right)^{1/10}$$

$$C_V = \left(\frac{5.125}{6.75}\right)^{1/10} \left(\frac{12}{60}\right)^{1/10} \left(\frac{21}{60}\right)^{1/10} = 0.746$$

$$F_b' = F_b^* C_V = (2760)(0.746) = 2058 \text{ psi} > 1600 \text{ psi} \quad \text{O.K.}$$

(b) Bending stress at tangent point: determine depth at tangent point

$$d_t' = [d_e + l_t(\tan \phi_T - \tan \phi_B)][\cos \phi_T/\cos (\phi_T - \phi_B)]$$

$$d_t' = [24 + 232(\tan 14.04° - \tan 10.4°)]$$

$$\cdot [\cos 14.04°/\cos (14.04° - 10.4°)] = 38.21 \text{ in.}$$

$$M_t = \frac{wll_t}{2} - \frac{wl_t^2}{2} = \frac{(45)(16)(720)(232)}{2(12)} - \frac{(45)(16)(232)^2}{2(12)}$$

$$= 3{,}396{,}000 \text{ in.-lb}$$

$$f_{bt} = \frac{6M_t}{bd_t'^2} = \frac{(6)(3{,}396{,}000)}{(6.75)(38.21)^2} = 2068 \text{ psi}$$

Determine volume factor:

$$C_V = \left(\frac{5.125}{6.75}\right)^{1/10} \left(\frac{12}{38.21}\right)^{1/10} \left(\frac{21}{60}\right)^{1/10} = 0.780$$

$$F_b' = F_b^* C_V = (2760)(0.780) = 2152 \text{ psi} > 2068 \text{ psi} \qquad \text{O.K.}$$

(c) Radial stress at centerline: From Fig. 5.23, $K_r = 0.064$. For $l/l_c = 2.81$, from Figs. 5.24 and 5.25 by interpolation,

$$C_r = 0.85$$

$$f_{rt} = K_r C_r f_0 = (0.064)(0.85)(960) = 52 \text{ psi}$$

$$F_{rt}' = (55)(1.15) = 63 \text{ psi} > 52 \text{ psi} \qquad \text{O.K.}$$

Check stresses at 3 places along tapered section between tangent point and end of beam. Use geometry to calculate depth of cross sections.

The tangent point depth is 39.3 in. measured in a vertical direction. Determine the depth perpendicular to the face lamination $d_t'$. Draw a line perpendicular to the face lamination at the tangent point. The perpendicular crosses the neutral axis at a point $d_t/2 \sin \phi_B$ to the left of the $d_t$ line. The moment will be measured

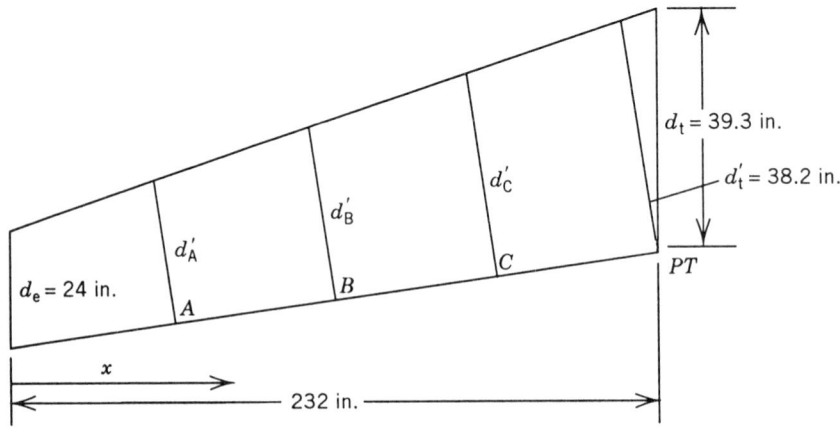

at this point. Determine $d_t'$ by Eq. (5-69). Divide the tapered portion into 4 parts with points $A$, $B$, and $C$ and draw lines perpendicular to the soffit lamination at these points. Then calculate moments and stresses as shown in the summary table.

| Point | $x$ in. | Depth[a] $d_x'$ in. | Moment[b] $M$ in.-lb | $f_b^c$ psi | $F_b'^d$ psi | $f_b/F_b'$ |
|-------|---------|---------------------|----------------------|-------------|--------------|------------|
| $A$   | 58      | 27.05               | 1,107,000            | 1345        | 2230         | 0.603      |
| $B$   | 116     | 30.77               | 2,060,000            | 1934        | 2200         | 0.879      |
| $C$   | 174     | 34.49               | 2,814,000            | 2102        | 2175         | 0.967      |
| $PT$  | 232     | 38.21               | 3,369,000            | 2051        | 2153         | 0.953      |

[a] Calculated from Eq. (5-69).
[b] Calculated from Eq. (5-70).
[c] Calculated as $f_b = 6M/b(d_x')^2$.
[d] $F_b' = F_b C_D C_V$.

The tapered portion of the beam is not overstressed.

9.   Determine horizontal movement at supports:

$$h = h_a - \frac{d_c}{2} = 114 - \frac{60}{2} = 84 \text{ in.}$$

$$\Delta_H = \frac{2h\Delta_c}{l} = \frac{(2)(84)(2.83)}{720} = 0.66 \text{ in.}$$

10.   Design radial reinforcement

Radial force per inch $= f_{rt}b = (52.2)(6.75) = 352 \text{ lb/in.}$

Assume $\frac{3}{4}$-lag screws used for radial reinforcement. From Table 5.11,

Allowable steel tension load $= 5265$ lb
Allowable withdrawal load per in. of thread $= 589$ lb
Effective thread penetration $= (d_c/2) - 2 = (60/2) - 2 = 28$ in.
Allowable withdrawal load $= (589)(28) = 16,492$ lb $> 5265$ (use 5265 lb)
Maximum spacing (controlled by steel) $= (5265/352) = 15$ in.

Use $\frac{3}{4}$-in. lag screws at 15 in. on centers, symmetric about centerline; nine lag screws on each side, starting $7\frac{1}{2}$ in. on each side of centerline; lag screws to extend to within 2 in. of the soffit.

Recheck bending stresses at reduced section caused by lag screw hole at the centerline and tangent points. Locate the neutral axis assuming that the portion of the reinforcement in the tension zone of the beam does not carry stress:

$$c = \frac{bd^2 - b'c^2 + b'a^2}{2(bd - b'c + b'a)}$$

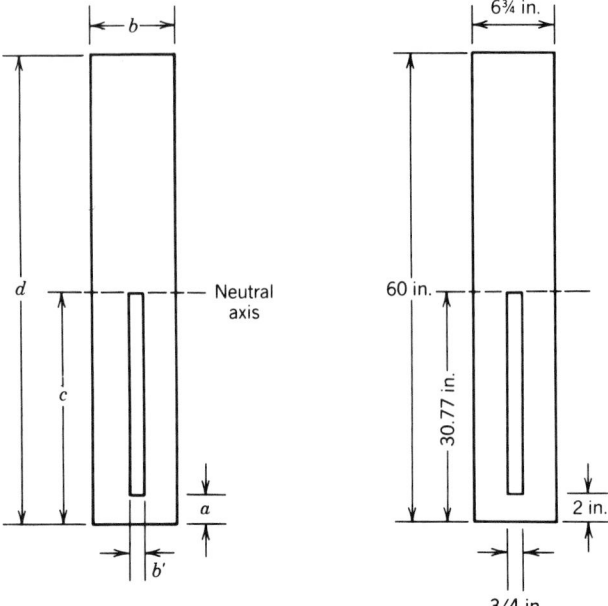

Using the actual dimensions at the centerline and tangent point depths and solving the quadratic equation,

$c = 30.77$ in. (centerline)

$c = 20.11$ in. (tangent point)

$$I_{CL} = \frac{(6.75)(60)^3}{12} + (6.75)(60)(0.77)^2 - \frac{(0.75)(28.77)^3}{3} = 115{,}800 \text{ in.}^4$$

$F_b' = (2400)(1.15)(0.746) = 2058$ psi

$$f_0 = \frac{(3{,}890{,}000)(30.77)}{115{,}800} = 1033 \text{ psi}$$

$f_b = 1.67 f_0 = (1.67)(1033) = 1725$ psi      O.K.

$$I_{PT} = \frac{(6.75)(39.3)^3}{12} + (6.75)(39.3)(0.46)^2 - \frac{(0.75)(18.11)^3}{3} = 32{,}760 \text{ in.}^4$$

$F_b' = (2400)(1.15)(0.780) = 2153$ psi

$$f_b = \frac{Mc}{I} = \frac{(3{,}396{,}000)(20.11)}{32{,}760} = 2085 \text{ psi}      \text{O.K.}$$

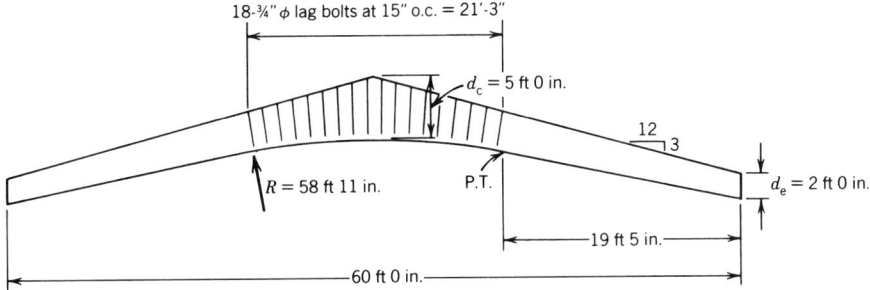

### 5.12.4 Design of Pitched and Tapered Curved Beams Made of Southern Pine

The design of a pitched and tapered curved beam made of Southern Pine is illustrated in the following example.

**Example: Southern Pine Beam.** Design a Southern Pine pitched and tapered curved glued laminated timber roof beam to meet the following requirements. Assume decking is applied to the top of the beam providing adequate lateral support:

Species: Southern Pine
$L = 45$ ft

Spacing $= 14$ ft

Roof slope $= 4:12$ $(18.4°)$

Snow load $= 30$ psf

$C_D = 1.15$

Dead load $= 15$ psf

$F_b^* = (2400)(1.15) = 2760$ psi

$F_v' = (200)(1.15) = 230$ psi

$F_{rt}' = (200/3)(1.15) = 77$ psi

$E_x = 1,800,000$ psi

$\Delta_{max} = l/180$ for total load

1.   Determine end depth $d_e$. Assume $b = 5$ in.
Calculate end reaction $R_v$:

$$R_v = \frac{wL}{2} = \frac{(45)(14)(45)}{2} = 14,175 \text{ lb}$$

Calculate trial end depth $d$

$$d = \frac{3R_v}{2bF_v'} = \frac{(3)(14,175)}{(2)(5)(230)} = 18.0 \text{ in. Use 18.0 in.}$$

$$L_e = L - 2d = 45 - \frac{(2)(18)}{12} = 42 \text{ ft Use 42 ft}$$

$$V = \frac{wL_e}{2} = \frac{(45)(14)(42)}{2} = 13,230 \text{ lb}$$

$$d_e = \frac{3V}{2bF_v'} = \frac{(3)(13,230)}{(2)(5)(230)} = 17.26 \text{ in. (use 17 in.)}$$

2.  Determine approximate centerline depth $d_{cb}$ from the bending stress limitation:

$$M = \frac{wL^2}{8} = \frac{(45)(14)(45)^2(12)}{8} = 1,914,000 \text{ in.-lb}$$

Assume $C_V = 0.85$.

$$F'_b = (2400)(1.15)(0.85) = 2346 \text{ psi}$$

From Table 5.10, $D$ for $15° = 1.738$ and $D$ for $20° = 1.961$.

$$D \text{ for } 18.4° = 1.738 + \frac{(1.961 - 1.738)(3.4)}{5} = 1.89$$

$$d_{cb} = \sqrt{\frac{6MD}{bF'_b}} = \sqrt{\frac{(6)(1,914,000)(1.89)}{(5)(2346)}} = 43.0 \text{ in.}$$

3.  Determine minimum centerline depth $d_{c\Delta}$ from the deflection limitation.

$$\Delta_{max} = \frac{l}{180} = \frac{(45)(12)}{180} = 3 \text{ in.}$$

$$d_{eff} = \sqrt[3]{\frac{Wl^3}{6.4E'b\Delta_{max}(\cos \phi_T)^3}}$$

$$= \sqrt[3]{\frac{(45)(14)(45)(540)^3}{(6.4)(1,800,000)(5)(3)(0.945)^3}} = 31.3 \text{ in.}$$

$$d_{c\Delta} = 2d_{eff} - d_e = (2)(31.3) - 17 = 45.0 \text{ in.}$$

4.  Determine minimum centerline depth $d_{crt}$ from the radial stress limitation. Assume $d_c = 45$ in. and $R_m = L = 45$ ft; $d_c/R_m = 45/(45)(12) = 0.08$. From Fig. 5.24, $K_r = 0.086$.

$$d_{crt} = \sqrt{\frac{6MK_r}{bF'_{rt}}} = \sqrt{\frac{(6)(1,914,000)(0.086)}{(5)(77)}} = 50.6 \text{ in.}$$

5.  Determine height of apex $h_a$, trial bottom slope $\phi_B$, and soffit radius $R$:

$$h_a = d_e + \frac{l}{2}\tan \phi_T = 17 + \frac{(45)(12)(0.333)}{2} = 107 \text{ in.}$$

$$d_c = 50 \text{ in.}$$

$$h_s = h_a - d_c = 107 - 50 = 57 \text{ in.}$$

From Fig. 5.28, by interpolation, $\phi_{B\,max} = 23.7° > \phi_T$; therefore, set $\phi_{B\,max} = \phi_T = 18.4°$. From Fig. 5.29, by interpolation, $\phi_{B\,min} = 13.9°$. Note that the graphs in Figs. 5.29 and 5.30 are based on $R_{min}$ values for Douglas Fir-Larch.

For Southern Pine, the $R_{min}$ value for $1\frac{1}{2}$-in. laminations is 18 ft and for $\frac{3}{4}$-in. laminations is 7 ft. These values may be used in the equations on the graphs to determine values for $\phi_{B\,min}$ for Southern Pine. For this example, the value of $\phi_{B\,min}$ is determined as follows:

$$Y = R_{min} - h_s = (18)(12) - 57 = 159 \text{ in.}$$

$$a = \sqrt{Y^2 + (l/2)^2} = \sqrt{(159)^2 + (270)^2} = 313 \text{ in.}$$

$$\phi_{B\,min} = \sin^{-1}\left(\frac{R_{min}}{a}\right) - \tan^{-1}\left(\frac{Y}{l/2}\right)$$

$$= \sin^{-1}\left(\frac{216}{313}\right) - \tan^{-1}\left(\frac{159}{270}\right) = 13.1°$$

Choose the trial bottom slope:

$$\phi_B = 0.45(\phi_{B\,max} + \phi_{B\,min}) = (0.45)(18.4 + 13.1) = 14.2°$$

$$R = \frac{h_s - l/2 \tan \phi_B}{1 - \cos \phi_B - \sin \phi_B \tan \phi_B} = \frac{57 - (270)(0.253)}{1 - 0.969 - (0.245)(0.253)} = 365 \text{ in.}$$

6.  Find the following values based on the values of $\phi_B$ and $R$ determined in step 5:

(a)  $l_t = l/2 - R \sin \phi_B = (22.5)(12) - (365)(\sin 14.2°) = 181$ in.

(b)  $l_c = l - 2l_t = (45)(12) - (2)(181) = 178$ in.; $l/l_c = 540/178 = 3.03$

(c)  $d_t' = [d_e + l_t(\tan \phi_T - \tan \phi_B)][\cos \phi_T/\cos(\phi_T - \phi_B)] = [17 + (181)(\tan 18.43° - \tan 14.2°)][\cos 18.43°/\cos (18.43° - 14.2°)] = 29.95$ in.

(d)  $R_m = d_c/2 + R = 50/2 + 365 = 390$ in.; $d_c/R_m = 50/390 = 0.13$

(e)  $f_0 = 6M/bd_c^2 = (6)(1,914,000)/(5)(50)^2 = 919$ psi (reference stress at centerline).

7.  Check maximum deflection:

$$d_{eb} = (d_e + d_c)(0.5 + 0.735 \tan \phi_T) - 1.41 d_c \tan \phi_B$$

$$= (17 + 50)[0.5 + (0.735)(0.333)] - (1.41)(50)(0.253) = 32.1 \text{ in.}$$

$$\Delta_c = \frac{5Wl^3}{32E' bd_{eb}^3}$$

$$= \frac{(5)(45)(14)(45)(540)^3}{(32)(1,800,000)(5)(32.1)^3} = 2.35 \text{ in.}$$

8.  Check stresses:

(a)  Bending stress at centerline: Determine $K_\theta$ using Eq. (5-65) and Table 5.10.

$$K_\theta \text{ (for } \phi_T \text{ of } 15° \text{ and } \frac{d_c}{R_m} \text{ of } 0.13) = 1.738$$

$$K_\theta \left( \text{for } \phi_T \text{ of } 20° \text{ and } \frac{d_c}{R_m} \text{ of } 0.13 \right) = 1.961$$

$$K_\theta \left( \text{for } \phi_T \text{ of } 18.4° \text{ and } \frac{d_c}{R_m} \text{ of } 0.13 \right) = 1.738 + \frac{(1.961 - 1.738)(3.4)}{5}$$

$$= 1.89$$

$$f_b = K_\theta f_0 = (1.89)(919) = 1736 \text{ psi}$$

$$C_V = \left( \frac{5.125}{b} \right)^{1/20} \left( \frac{12}{d} \right)^{1/20} \left( \frac{21}{L} \right)^{1/20}$$

$$= \left( \frac{5.125}{5} \right)^{1/20} \left( \frac{12}{50} \right)^{1/20} \left( \frac{21}{45} \right)^{1/20} = 0.897$$

$$F_b' = F_b C_D C_V = (2400)(1.15)(0.897) = 2477 \text{ psi} \qquad \text{O.K.}$$

(b)   Check bending stress at tangent point. Also check bending stress at three places along the tangent portion of the beam.

$$M_t = \frac{wLl_t}{2} - \frac{wl_t^2}{2}$$

$$= \frac{(45)(14)(45)(181)}{2} - \frac{(45)(14)(181)^2}{2(12)}$$

$$= 1,706,000 \text{ in.-lb}$$

$$f_{bt} = \frac{6M}{b(d_t')^2} = \frac{(6)(1,706,000)}{(5.0)(29.95)^2} = 2282 \text{ psi}$$

$$C_V = \left( \frac{5.125}{5.0} \right)^{1/20} \left( \frac{12}{29.95} \right)^{1/20} \left( \frac{21}{45} \right)^{1/20} = 0.921$$

$$F_b' = F_b C_D C_V = (2400)(1.15)(0.921) = 2540 \text{ psi}$$

$$2540 > 2282 \text{ psi} \qquad \text{O.K.}$$

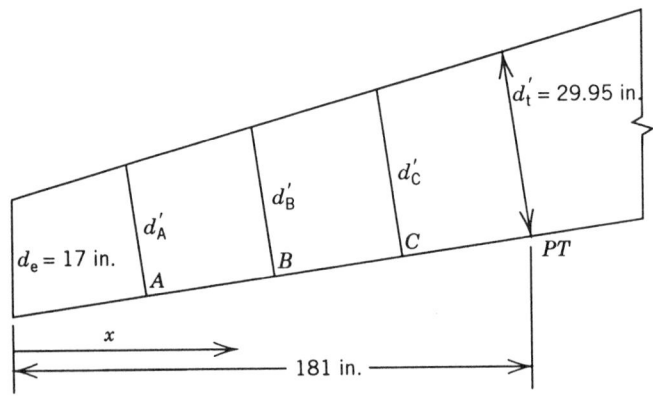

Determine depths perpendicular to soffit laminations at points $A$, $B$, and $C$. Then determine moments, $f_b$ and $F'_b$ as shown in the following table:

| Point | $x$ in. | $d'^{\,a}_x$ in. | Moment[b] in.-lb | $f_b{}^c$ psi | $F'_b{}^d$ psi | $f_b/F'_b$ |
|---|---|---|---|---|---|---|
| $A$ | 45 | 19.62 | 558,000 | 1739 | 2594 | 0.670 |
| $B$ | 91 | 23.06 | 1,040,000 | 2346 | 2575 | 0.911 |
| $C$ | 136 | 26.51 | 1,417,000 | 2420 | 2556 | 0.947 |
| $PT$ | 181 | 29.95 | 1,688,000 | 2258 | 2542 | 0.888 |

[a]Calculated from Eq. (5-69).
[b]Calculated from Eq. (5-70).
[c]Calculated as $f_b = 6M/b(d'_x)^2$.
[d]Calculated as $F'_b = F_b C_D C_V$.

The tapered portion of the beam is not overstressed.
   (c)   Check radial stress at centerline. From Fig. 5.24,

$$K_r \doteq 0.09 \qquad l/l_c = 3.03$$

From Fig. 5.33, $C_r = 0.86$.

$$f_{rt} = K_r C_r f_0 = (0.09)(0.86)(919) = 71 \text{ psi} < 77 \qquad \text{O.K.}$$

9.   Determine horizontal movement at supports:

$$h = h_a - \frac{d_c}{2} = 107 - \frac{50}{2} = 82 \text{ in.}$$

$$\Delta_H = \frac{2h}{l} \Delta_c = \frac{(2)(82)(2.35)}{540} = 0.71 \text{ in.}$$

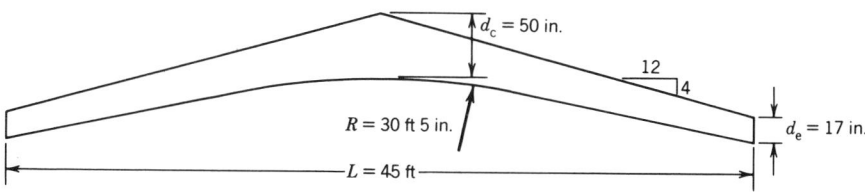

$d_c = 50$ in.   12   4   $R = 30$ ft 5 in.   $d_e = 17$ in.   $L = 45$ ft

## 5.12.5   Alternative Design Procedure for Determining Centerline Depth and $\phi_B$

An alternative procedure which reduces computation time has been developed for determining initial trial values of the centerline depth ($d_c$) and bottom slope ($\phi_B$) for pitched, tapered, and curved glued laminated timber bending members. This procedure is applicable only to members which satisfy the following general design limitations:

1.   Design loads are uniformly distributed.

2.  The depth of the member at the end, $d_e$, is equal to or greater than two times the width of the member, $b$.

3.  The radius of curvature of the member satisfies industry guidelines based on species and thickness of laminations used.

The calculation procedure is based on one of two geometry options to be selected by the designer. These options are as follows:

Option (a): determine $d_c$ and $\phi_B$ based on optimum beam geometry having the smallest curved segment

Option (b): determine $d_c$ and $\phi_B$ based on the length of the curved segment being equal to $1/4$th of the span

For Option (a), use the following equations:

$$d_c = A_1 + A_2 b + A_3 l + A_4(w)/F_b^* + A_5(w)(\phi_T)/F_b^* \qquad (5\text{-}73)$$
$$+ A_6(F_v')(l)(\phi_T)/E$$

$$\phi_B = B_1 + B_2 \phi_T + B_3(b)(\phi_T)/l + B_4(F_b^*)(\phi_T)/F_v' \qquad (5\text{-}74)$$

where

$A_1 \cdots A_6$ and $B_1 \cdots B_4$ are coefficients from Part A and Part B, Table 5.12,

$\quad b =$ width of beam (in.),
$\quad l =$ span (in.),
$\quad \phi_T =$ top slope (in degrees),
$\quad F_b^* =$ allowable bending stress adjusted for $C_D$ (psi),
$\quad F_v' =$ allowable shear stress adjusted for $C_D$ (psi),
$\quad E =$ modulus of elasticity (psi), and
$\quad w =$ total uniform load on beam (plf).

For Option (b), use the following equations:

$$d_c = C_1 + C_2 l + C_3(l)(\phi_T) - C_4(w)(\phi_T)/E + C_5(F_b^*)(b)(\phi_T)/E \quad (5\text{-}75)$$

$$\phi_B = D_1 + D_2 \phi_T + D_3(b)(\phi_T)/(l) + D_4(F_b^*)(\phi_T)/F_v' \qquad (5\text{-}76)$$

where

$C_1 \cdots C_5$ and $D_1 \cdots D_4$ are coefficients from Parts C and D, Table 5.12.

Determine values of $d_c$ and $\phi_B$ for the Southern Pine pitched and tapered curved beam design example.

(a)   $d_c = -0.7135 + (-3.2418)(5) + (0.0673)(540)$
$\qquad + (43.7746)(630/2760) + (2.7628)(630)(18.3)/2760$
$\qquad + (7.5358)(230)(540)(18.4)/1.8 \times 10^6 = 50.2$ in.
$\quad \phi_B = -3.664 + (0.6759)(18.4) + (13.3698)(5)(18.4)/540$
$\qquad + (0.0151)(2760)(18.4)/230 = 14.47°$

**TABLE 5.12**

**Coefficients for Determining Beam Geometry for Alternate Method**

| | Part A $d_c$ | | Part B $\phi_b$ |
|---|---|---|---|
| $A_1$ | $-0.7135$ | $B_1$ | $-3.664$ |
| $A_2$ | $-3.2418$ | $B_2$ | $0.6759$ |
| $A_3$ | $0.0673$ | $B_3$ | $13.3698$ |
| $A_4$ | $43.7746$ | $B_4$ | $0.0151$ |
| $A_5$ | $2.7628$ | | |
| $A_6$ | $7.5358$ | | |
| | Part C $d_c$ | | Part D $\phi_B$ |
| $C_1$ | $-0.9656$ | $D_1$ | $-3.5208$ |
| $C_2$ | $0.0520$ | $D_2$ | $0.6592$ |
| $C_3$ | $0.0019$ | $D_3$ | $12.4217$ |
| $C_4$ | $3445.7$ | $D_4$ | $0.0162$ |
| $C_5$ | $-119.23$ | | |

(b)   $d_c = -0.9656 + (0.0520)(540) + (0.0019)(540)(18.4)$
$\qquad + (3445.7)(630)(18.4)/1.8 \times 10^6$
$\qquad - (119.2342)(2760)(5)(18.4)/1.8 \times 10^6 = 51.0$ in.
$\phi_B = -3.5208 + (0.6592)(18.4) + (12.4217)(5)(18.4)/540$
$\qquad + (0.0162)(2760)(18.4)/230 = 14.38°$

Note that for either option, the values of $d_c$ and $\phi_B$ are approximately the same as the corresponding values determined using the more complicated procedure. (The remaining steps of the design procedure are the same as presented previously.)

### 5.12.6   Curved Beams of Constant Cross Section with Attached Haunch

For curved beams of constant cross section, the radial tension is calculated by Eq. (5-47). These beams may be of constant cross section and constant curvature as shown in Fig. 5.35a, of constant cross section but with tangent ends (Fig. 5.35b), or of constant cross section with tangent ends with the haunch attached using mechanical fastenings (Fig. 5.35c). The configuration shown in Fig. 5.35c is sometimes used to reduce radial tension stresses in Western species beams by eliminating the need for mechanical reinforcement. One of the more common shapes is shown in Fig. 5.36. The member is manufactured with a loose haunch that is attached mechanically. The curved portion is of constant cross section, and the tangent ends are tapered.

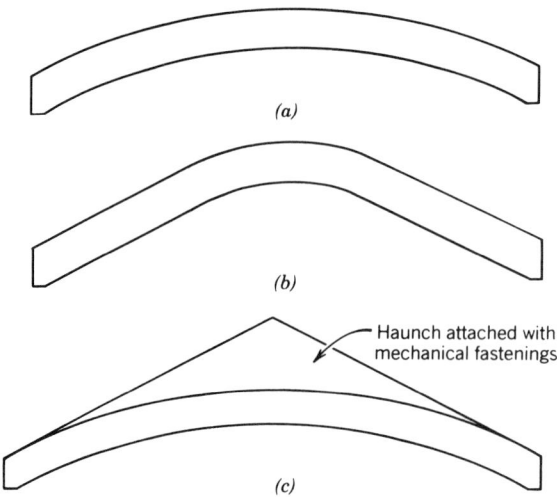

**FIGURE 5.35**    Curved beams of constant cross section.

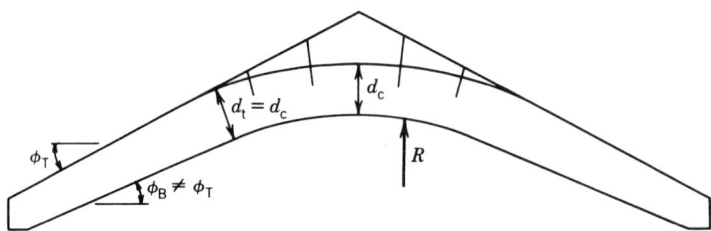

**FIGURE 5.36**    Pitched and tapered curved beam with attached haunch.

Radial tension in the curved portion can be determined by Eq. (5-47) for members of constant cross section

$$f_r = \frac{3M}{2Rbd}$$

which can be written as

$$f_{rt} = \frac{3M}{2R_m bd} \tag{5-77}$$

where terms are described in Eq. (5-47).

The fiber stress in bending in the curved portion is

$$f_b = \frac{M}{S} = \frac{6M}{bd_c^2}$$

The tabular design value in bending, $F_b$, is multiplied by the curvature factor $C_C$ in addition to other applicable adjustment factors. The tapered portion should be considered in the same manner as a straight tapered beam where $\theta = \phi_T - \phi_B$. The tabular design value for bending in this portion of the member is further multiplied by the interaction factor $C_I$ in addition to other applicable adjustment factors. Note that $C_V$ and $C_I$ are not cumulative and whichever is smaller should be used.

### 5.12.7   Design of Pitched and Tapered Curved Beam with Attached Haunch

The design of pitched and tapered curved beams with attached haunch is illustrated by the following example.

**Example.**   Design a pitched and tapered curved glued laminated timber roof beam with attached haunch to meet the following requirements:

Species: Douglas Fir-Larch     $F_b = 2400$ psi,
                               $F_b^* = (2400)(1.15) = 2760$ psi
$L = 60$ ft                    $F_v = 165$ psi, $F_v' = (165)(1.15) = 190$ psi
Spacing $= 16$ ft             $E_x = 1,800,000$ psi, $E_x' = 1,800,000$ psi
Roof slope $= 3:12$          $E_y = 1,600,000$ psi, $E_y' = 1,600,000$ psi
Camber $= 1.5\ \Delta_{DL}$   $G = 0.06E = 108,000$ psi
                               $G' = 108,000$ psi
Snow load $= 30$ psf          $\Delta_{TL} = l/180 = (60)(12)/180 = 4$ in.
Dead load $= 15$ psf          $F_{rt} = 15$ psi, $F_{rt}' = (15)(1.15) = 17.2$ psi

The haunch is to be mechanically attached, and no radial reinforcement is to be used.

1.   Determine end depth $d_e$. Assume $b = 6\frac{3}{4}$ in.
Calculate end reaction $R_v$:

$$R_v = \frac{wL}{2} = \frac{(45)(16)(60)}{2} = 21,600 \text{ lb}$$

Calculate trial end depth $d$:

$$d = \frac{3R_v}{2bF_v'} = \frac{(3)(21,600)}{(2)(6.75)(190)} = 25.30 \text{ in.}$$

$$L_e = 60 - 2\left(\frac{25.30}{12}\right) = 55.8 \text{ ft}$$

$$V = \frac{wL_e}{2} = \frac{(45)(16)(55.8)}{2} = 20,082 \text{ lb}$$

$$d_e = \frac{3V}{2bF_v'} = \frac{(3)(20,082)}{(2)(6.75)(190)} = 23.5 \text{ in. (use 24 in.)}$$

2.   Determine the approximate trial centerline depth $d_{cb}$ from the bending stress limitation:

$$M = \frac{wL^2}{8} = \frac{(45)(16)(60)^2(12)}{8} = 3{,}888{,}000 \text{ in.-lb}$$

$$F'_b = F_b C_D C_V C_C$$

$$C_D = 1.15 \text{ (snow load)}$$

$$C_V = \text{Assume } 0.85$$

$$C_C = 1 - 2000 \left[\frac{(1.5)}{(60)(12)}\right]^2 = 0.99 \text{ (use radius } R = \text{span} = 60 \text{ ft)}$$

$$F'_b = (2400)(1.15)(0.85)(0.99) = 2326 \text{ psi}$$

$$d_{cb} = \sqrt{\frac{6M}{bF'_b}} = \sqrt{\frac{(6)(3{,}888{,}000)}{(6.75)(2326)}} = 38.5 \text{ in.}$$

3.    Determine trial minimum centerline depth $d_{c\Delta}$ from deflection limitation. For preliminary design, assume deflection is 10% larger than for a straight beam with the same depth as the constant cross section.

$$\Delta = \frac{(1.1)(5)wl^4}{384E'(bd_{c\Delta}^3/12)}$$

$$\Delta_{max} = 4 \text{ in.}$$

$$d_{c\Delta} = \sqrt[3]{\frac{5.5wl^4}{384\,\Delta_{max}\,E'\,b}} = \sqrt[3]{\frac{(5.5)(45)(16)(720)^4}{(384)(4)(1{,}800{,}000)(6.75)}} = 38.5 \text{ in.}$$

4.    Determine the trial minimum centerline depth $d_{crt}$ due to radial tension stress. Assume radius = span = 60 ft.

$$d_{crt} = \frac{3M}{2R_m bF'_{rt}} = \frac{(3)(3{,}888{,}000)}{(2)(60)(12)(6.75)(17.2)} = 69.80 \text{ in.}$$

At this point, a decision needs to be made as to whether the depth is acceptable or whether to increase the radius or the width to reduce the depth. A direct solution to the problem is difficult, and a trial-and-error approach that gives due consideration to the geometry involved is suggested. The depth $d_{crt}$ is inversely proportional to both the radius and the width. Usually, changing the radius should be investigated first. Consideration should also be given to the final shape of the member for aesthetic as well as structural considerations. The minimum depth required to satisfy both bending stress and deflection requirements is 38.5 in.

From an appearance standpoint, the tangent point is usually located near the quarter point. For convenience of calculations, the point of tangency is assumed to be 15 ft from the support measured along the bottom face of the member. The depth of the member at the tangent points and curved portion of the member,

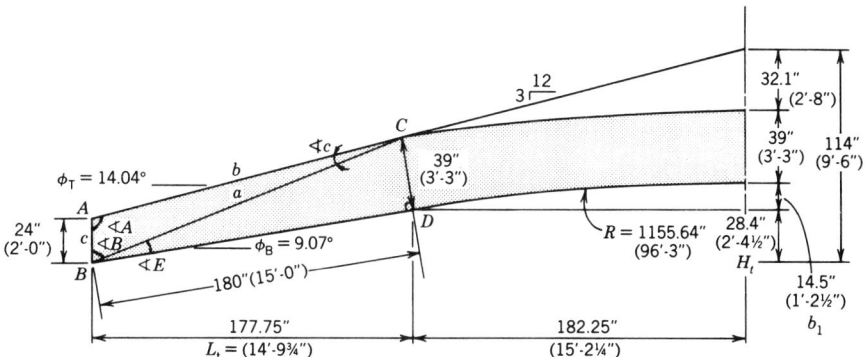

**FIGURE 5.37**   Geometry for mechanically attached haunch (trial size).

$d_t'$, is assumed to be 39 in. From these assumptions, the critical dimensions and radius for this configuration are determined. Fig. 5.37 gives critical dimensions calculated from these assumed dimensions.

(a)   Calculate angle $A$:

$$A = 90° + \phi_T = 104.04°$$

(b)   Calculate the length of side $a$:

$$a = \sqrt{(15)^2 + (3.25)^2} = 15.35 \text{ ft}$$

(c)   From triangle $ABC$:

Calculate angle $C$:

$$C = \sin^{-1}\left(\frac{d_e(\sin A)}{a}\right) = \sin^{-1}\left(\frac{24(\sin 104.4°)}{15.35(14)}\right)$$

$$C = 7.26°$$

$$B = 180° - A - C = 180° - 104.04° - 7.26° = 68.70°$$

(d)   Calculate angle $E$:

$$E = \tan^{-1}\left(\frac{3.25}{15}\right) = 12.23°$$

(e)   Calculate $\phi_B$:

$$\phi_B = 90° - B - E = 90° - 68.70° - 12.23° = 9.07°$$

(f)   Calculate the horizontal distance from the support to the tangent point, $l_t$:

$$l_t = (15)(12) \cos \phi_B = (15)(12) \cos 9.07° = 177.66 \text{ in.}$$

(g)  Calculate the vertical distance from the support to the tangent point, $H_t$:

$$H_t = (15)(12) \sin \phi_B = (15)(12) \sin 9.07° = 28.44 \text{ in.}$$

(h)  Calculate the chord distance between tangent points, $l_c$:

$$l_c = L - 2l_t = (60)(12) - (2)(177.66) = 364.68 \text{ in.}$$

(i)  Calculate rise, $b_1$:

$$b_1 = \frac{l_c}{2} \tan \frac{\phi_B}{2} = \frac{364.68}{2} \tan \frac{9.07}{2} = 14.47 \text{ in.}$$

(j)  Calculate the radius of the soffit, $R$

$$R = \frac{4b_1^2 + l_c^2}{8b_1} = \frac{(4)(14.47)^2 + (364.68)^2}{(8)(14.47)} = 1156 \text{ in.} = 96.35 \text{ ft}$$

The radius determined for this assumed location of tangent point and depth is 96.35 ft. The required $d_{crt}$ for a 96.35 ft radius is:

$$R_m = R + \frac{d_c}{2} = 96.35 + \frac{3.25}{2} = 97.97 \text{ ft}$$

$$d_{crt} = \frac{(3)(3,888,000)}{(2)(97.97)(12)(6.75)(17.25)} = 42.6 \text{ in.}$$

Changing the depth to 42.6 in. will decrease $\phi_B$ and change the location of point of tangency. After several trials, a tangent depth of 40.5 in. and a radius of 103 ft are determined which satisfies the radial stress limit. The point of tangency maintained at 15 ft from the end measured along the soffit $PT$ is located 14 ft + $10\frac{1}{8}$ in. from the end measured horizontally.

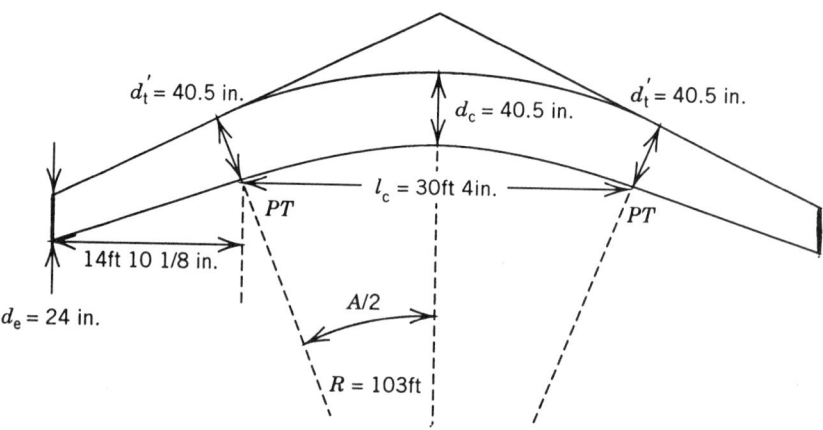

Note $d'_t$ as shown in the sketch is perpendicular to the bottom lamination and will also be designated as $d'_t$ in these calculations. From the geometry of the circle with $c = 29$ ft and $R = 103$ ft, angle $A/2$ and $\phi_B$ are calculated to be 8.61°.

5.   Check deflection: Assume deflection is 10% higher than that determined for a beam of constant cross section.

$$\Delta_{TL} = \frac{(1.1)(5)wl^4}{384E'bd^3/12} = \frac{(1.1)(5)(45)(16)(720)^4}{(384)(1,800,000)(6.75)(40.5)^3/12}$$

$$= 3.43 \text{ in.} < 4 \text{ in.} \qquad \text{O.K.}$$

6.   Check stresses:

$$f_b = \frac{6M}{bd^2} = \frac{(6)(3,888,000)}{(6.75)(40.5)^2} = 2106 \text{ psi}$$

$$C_C = 1 - 2000 \left[ \frac{1.5}{(103)(12)} \right]^2 = 0.997$$

$$C_D = 1.15 \text{ (snow load)}$$

$$C_V = \left( \frac{5.125}{6.75} \right)^{1/10} \left( \frac{12}{40.5} \right)^{1/10} \left( \frac{21}{60} \right)^{1/10} = 0.776$$

Note: The volume factor $C_V$ calculated applies to all of the curved portion of the beam between the tangent points.

The spacing between points of connection of the haunch to the beam must be considered in determining the lateral stability of the beam. The lateral stability equations used in this example are those intended for use with straight members. However, they should be applied for the lateral stability check of pitched and tapered curved beams with a detached haunch. Assume connections are made 5 ft on either side of the centerline.

$$l_u = 10 \text{ ft} \qquad l_u/d = \frac{120}{40.5} = 2.96 < 7$$

$$l_e = 2.06 l_u = (2.06)(10)(12) = 247.2 \text{ in.}$$

$$R_B = \sqrt{\frac{l_e d}{b^2}} = \sqrt{\frac{(247.2)(40.5)}{(6.75)^2}} = 14.83$$

$$F_b = (2400)(0.997)(1.15) = 2752 \text{ psi}$$

$$K_{bE} = 0.609$$

$$F_{bE} = \frac{(K_{bE})(E'_y)}{R_B^2} = \frac{(0.609)(1,600,000)}{(14.83)^2} = 4431 \text{ psi}$$

$$C_L = \frac{1 + F_{bE}/F_b^*}{1.9} - \sqrt{\left[\frac{1 + (F_{bE}/F_b^*)}{1.9}\right]^2 - \frac{F_{bE}/F_b^*}{0.95}}$$

$$C_L = \frac{1 + 4984/2752}{1.9} - \sqrt{\left[\frac{1 + (4984/2752)}{1.9}\right]^2 - \frac{4984/2752}{0.95}}$$

$C_L = 0.948 > 0.776$, $C_V$ controls

$$F_b' = F_b^* C_V = (2752)(0.776) = 2134 \text{ psi} > 2106 \text{ psi} \qquad \text{O.K.}$$

Haunches are usually attached with connections closer together and $C_V$ almost always controls.

Check interaction of stresses in the straight tapered portion of the beam. Determine $\theta$:

$$\theta = \phi_T - \phi_B = 14.04° - 8.61° = 5.43°$$

$$\tan 5.43° = 0.095$$

From Table 5.5, $C_I = 0.586$. Check stress at tangent point.

$$C_I < C_V$$

$$F_b' = F_b C_D C_I$$

$$F_b' = (2400)(1.15)(0.586) = 1618 \text{ psi}$$

$$M = (21,600)(14.835)(12) - \frac{(720)(14.835)^2(12)}{2} = 2,894,000 \text{ in.-lb}$$

$$S = \frac{(6.75)(40.5)^2}{6} = 1846 \text{ in.}^3$$

$$f_b = \frac{M}{S} = \frac{2,894,000}{1846} = 1568 < 1618 \text{ psi} \qquad \text{O.K.}$$

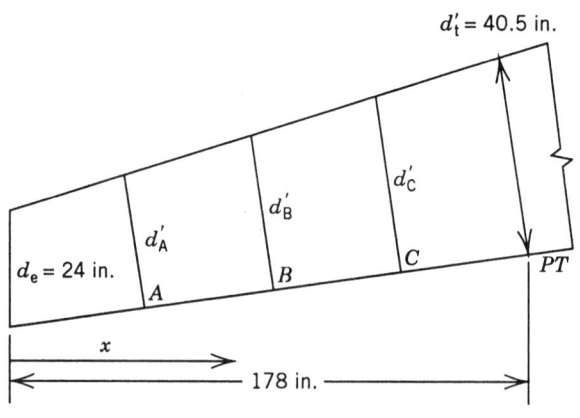

The portion of the beam from the end is tapered as shown in the preceding sketch. Check the stress at three points: $A$, $B$, and $C$. Calculate the depths perpendicular to the soffit face at each point and determine moments, $f_b$ and $F'_b$, for each point as shown in the following table:

| Point | $x$ in. | $d'_x$ in.[a] | Moment in.-lb[b] | $f_b$ psi[c] | $F'_b$ psi[d] | $f_b/F'_b$ |
|-------|---------|--------------|------------------|--------------|---------------|-----------|
| $A$ | 44.51 | 27.67 | 863,000 | 1001 | 1618 | 0.619 |
| $B$ | 89.01 | 31.96 | 1,646,000 | 1432 | 1618 | 0.886 |
| $C$ | 133.52 | 36.25 | 2,312,000 | 1564 | 1618 | 0.967 |
| $PT$ | 178.02 | 40.54 | 2,861,000 | 1547 | 1618 | 0.956 |

[a]Calculated from Eq. (5-69)
[b]Calculated from Eq. (5-70). Note that $f_b$ is slightly smaller than previously calculated because point for which moment was calculated was where the depth line for $d'_x$ crosses the neutral axis.
[c]Calculated from equation $f_b = 6M/b(d'_x)^2$.
[d]Calculated from equation $F'_b = F^*_b$ multiplied by $C_1$ or $C_V$, whichever controls.

The tapered portion of the beam is not overstressed.
Check radial tension:

$$R_m = R + \frac{d_c}{2} = 103 + \frac{40.5}{(12)(2)} = 104.7 \text{ ft}$$

$$f_{rt} = \frac{3M}{2R_m bd} = \frac{(3)(3,888,000)}{(2)(104.7)(12)(6.75)(40.5)} = 17.0 < 17.2 \text{ psi} \qquad \text{O.K.}$$

Check deflection: Since no input values have changed, the deflection of 3.43 in. calculated in Sept 5 has not changed. If a more accurate deflection is required, one of the energy equations can be used.

7.   $h = 114 - 33 - 40.5/2 = 60.75 \text{ in.}$

$$\Delta_H = \frac{2h\Delta_c}{l} = \frac{(2)(60.75)(3.43)}{(60)(12)} = 0.58 \text{ in. (use 0.6 in.)}$$

The haunch should be designed to be mechanically attached to the beam as shown in Fig. 5.36. This portion carries no bending load, and may be of all L3 grade Douglas Fir-Larch (Western Species Combination 1), which is sufficient to transfer the load from the roof decking to the beam. The top is attached by $\frac{1}{2}$-in. lag screws with depth of penetration of the lag screw loaded in tension as shown in the following calculations:

Maximum allowable load in tension = 2160 lb (see Table 5.11)
Withdrawal load for a $\frac{1}{2}$-in. lag screw = 378 lb/in. of penetration of point (see Table 7.24)
Depth of penetration = $2160/378 = 5.7$ in.; use 6 in.

Spacing of lag screws is arbitrary. A spacing of approximately 4 ft should be satisfactory. Heads should be countersunk as they may interfere with the placement of decking, purlins, or roof joists.

*Note:* This member contains more wood and is deeper at the tangent points and the centerline than the beam designed with radial reinforcement. The tangent depth, $d_t$, is 40.5 in. compared to 39.3 in., and the centerline depth, $d_c$, as shown in Fig. 5.37 is 74.1 in. compared to 60 in. This results in approximately 14 in. less head room at the centerline. However, in some cases, it may be more economical and more desirable to use a beam without reinforcement.

## 5.13    ARCHES

### 5.13.1    Three-Hinged Arches

Three-hinged arches, as the name implies, are hinged at each support and at the crown or peak. They may take the shapes shown in Fig. 5.38.

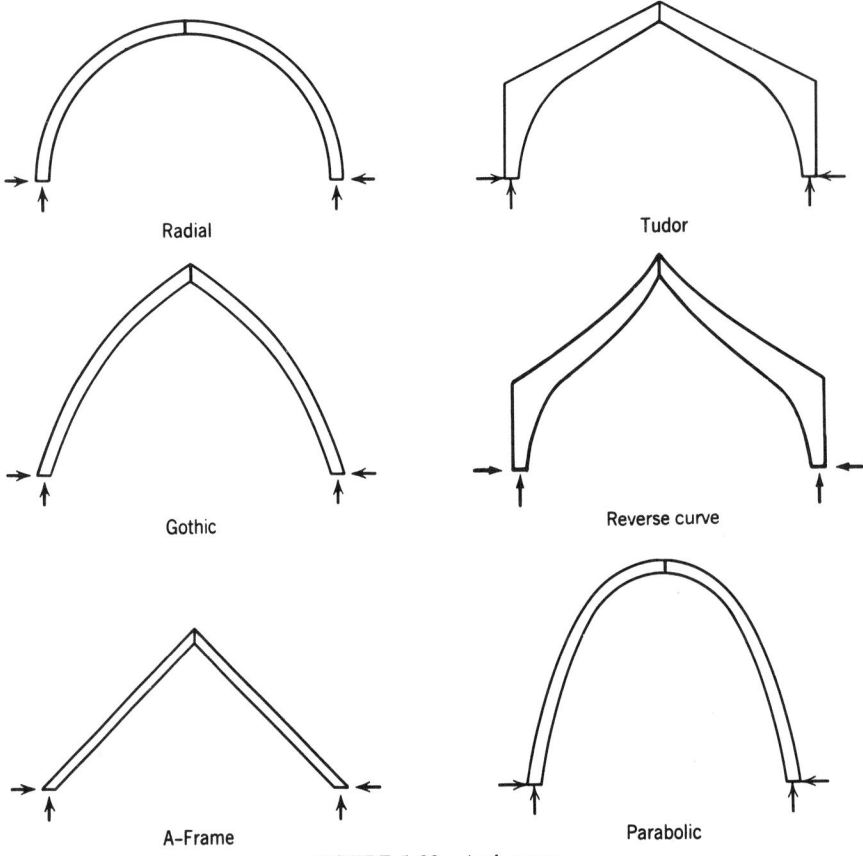

**FIGURE 5.38**    Arch types.

## 5.13.2 Design Procedure

The following procedures may be used to design simple three-hinged arches under usual loading conditions. These procedures are essentially trial-and-error methods, and revision of original values may be necessary. These procedures are given for a three-hinged tudor arch. The design of other three-hinged arches is similar.

Either a graphical or mathematical procedure can be used in the design process. The graphical solution works very well with a trial-and-error procedure, whereas the mathematical procedure is useful where advantage can be taken of modern calculating methods including computer technology.

Many building codes require a more realistic application of wind loads than is shown in the wind load diagram in Fig. 5.40. The graphical solution to wind loads applied perpendicular to the roof surface is laborious. One method for a graphical solution is to divide the outside surface of the arch into short segments, and to assume that the wind load over each segment is applied as a point load perpendicular to the arch surface. This load is then resolved into its vertical and horizontal components. An equilibrium polygon for the series of vertical loads is constructed, and another is constructed for the horizontal loads. The reactions and moments of the vertical and horizontal loads are determined separately and then combined. The solution for wind loads applied perpendicular to the outside surface of an arch usually can be determined best by mathematical rather than graphical methods.

### 5.13.3 Graphical Solution

1. Determine the dead, live, or snow loads and wind loads and the combinations of loads that should be considered. The roof loads may be assumed to be uniform if roof decking is applied directly to the arch or they may be concentrated at purlin points. See Chapter 4 for dead, live, snow, and wind load recommendations.

2. Lay out to a convenient scale the arch outline indicating external architectural dimensions (see Fig. 5.39a). Radius of curvature is also an architectural dimension, and it should not be less than the minimums recommended for various

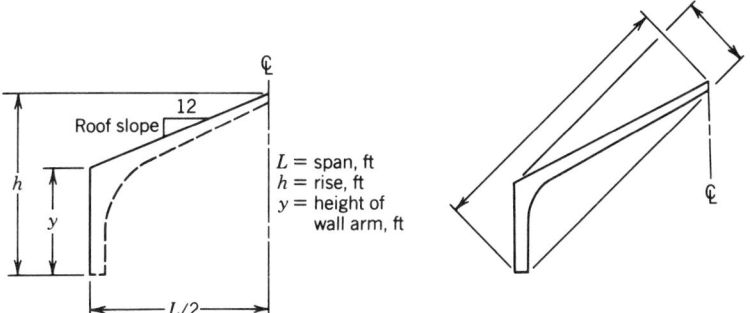

**FIGURE 5.39** (*a*) Arch dimensions; (*b*) critical dimensions for shipping.

lamination thicknesses in 5.11.1. In the case of arches with concentrated loads applied at purlin points, note the locations of the load.

3.  Calculate and tabulate the approximate reactions for various combinations of loading conditions. Fig. 5.40 indicates some of the more common loading conditions. Other loading conditions to be checked are given in Table 4.7.

4.  Determine minimum section size at base, $bd_b$, by the equation

$$bd_b = \frac{3R_H}{2F_v'} \tag{5-78}$$

where    $b$ = arch width (in.),
$d_b$ = arch depth at base (in.),
$R_H$ = horizontal reaction (lb), as determined in step 3, and
$F_v'$ = design value in shear parallel to grain (psi) multiplied by applicable adjustment factors.

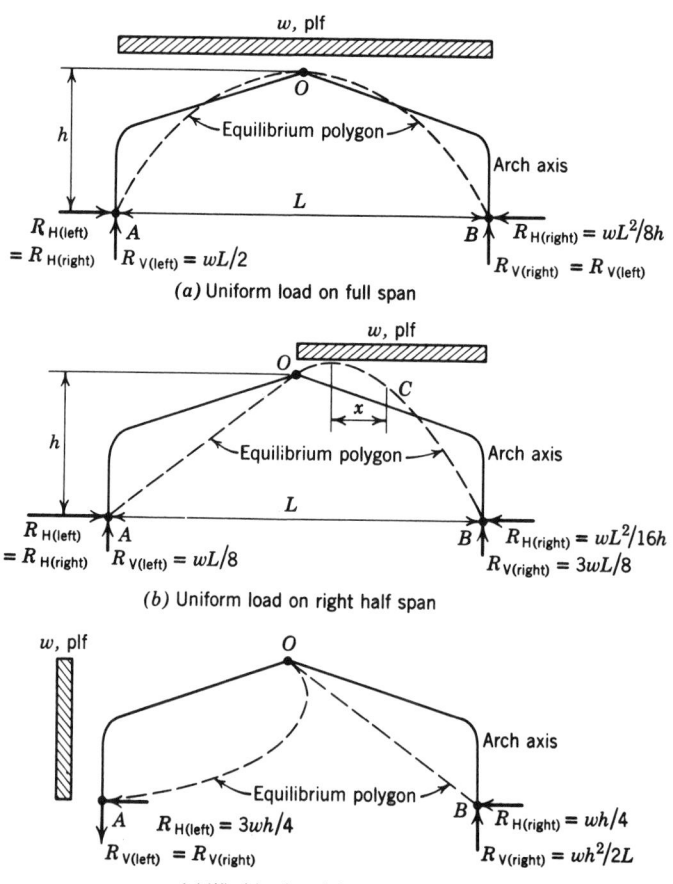

**FIGURE 5.40**    Three-hinged arch reactions and equilibrium polygons.

5.   As a first trial in the trial-and-error process, determine approximate depths of arch at crown, $d_c$, and tangent points, $d_t$, by the equations

$$d_c = 1.5b \quad \text{and} \quad d_t = 1.5d_b$$

The section at the crown is often proportioned for architectural appearance or to match the depth of purlins that frame into the arch at this point. In no case should the crown depth be less than the arch width. For design simplicity, the upper and lower tangent point depths are often assumed to be equal.

Note the shipping heights or widths for the tudor arch configuration may be a controlling factor, and the designer should check with the manufacturer to determine the applicable limitations (see Fig. 5.39*b*).

6.   If a graphical solution of the arch is used to determine the design moments and section properties along the arch axis, the following procedures are recommended.

Lay out the arch to scale, using the approximate dimensions determined in steps 4 and 5 and the assumed radius of curvature $R$. Locate the arch axis at the midpoint of the arch depth. Determine the effective span $L'$ and effective rise $h'$ (see Fig. 5.41). Finally, determine the design moments through the use of equilibrium polygons.

Equilibrium polygons are moment diagrams for a specified loading condition drawn to scale and in a position so that they pass through the hinges. Equilibrium polygons for each loading condition are illustrated in Fig. 5.40. For loading condition (*a*) in Fig. 5.40, uniform load on full span, the equilibrium polygon is a parabola with vertex at the crown hinge $O$. For condition (*b*), uniform load on right-half span, the equilibrium polygon is a straight line from base hinge $A$ to $O$ and a parabola from $O$ to base hinge $B$ with its vertex $L/8$ to the right and $h/8$

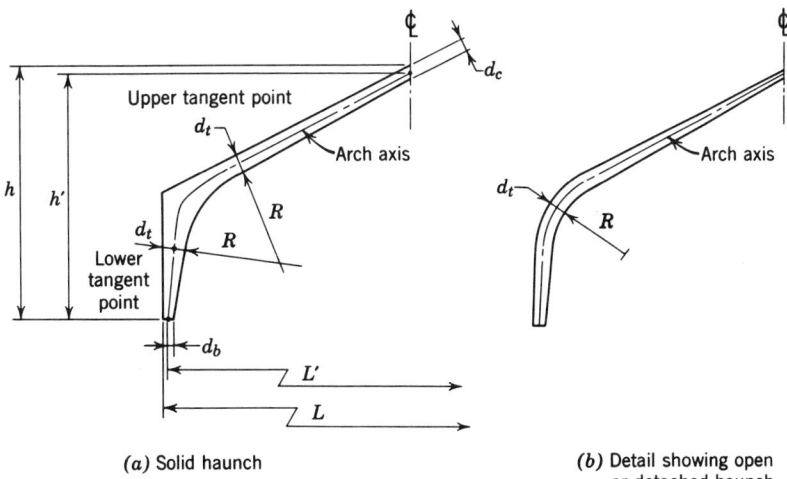

(*a*) Solid haunch

(*b*) Detail showing open or detached haunch

**FIGURE 5.41**   Arch details.

above $O$. For condition (*c*), uniform wind load on left vertical projection, the equilibrium polygon is a parabola from $A$ to $O$ with its vertex $L/16$ to the right and $h/4$ below $O$ and a straight line from $O$ to $B$.

*Note:* When the governing building code requires that wind loads be applied perpendicular to the roof surface, the graphical solution described in 5.13.2 should be used or else a mathematical solution is required.

Parabolas may easily be constructed by applying the fact that the perpendicular offset from a tangent to a parabola varies directly as the square of the tangent distances. For example, the vertical offset $y$ to point $C$ from the horizontal tangent through the vertex of the equilibrium polygon for loading conditions (*b*) is equal to $(h/8)[x/(L/8)]^2$.

The advantages of the equilibrium polygons are that they provide a means of determining the moment at any point on the arch axis and a means of determining by inspection the points of maximum moment and the signs of the moments; however, the maximum moment point may not be at the point of maximum stress in a tapered roof arm or vertical leg. When stresses approach the allowable design values, several points along the roof arm or vertical leg must be checked to determine the location of maximum stress.

For vertical loads, loading conditions (*a*) and (*b*) in Fig. 5.40, the moment at any point may be determined by multiplying the scaled length of the vertical line between the arch axis and the equilibrium polygon by the horizontal reaction $R_H$ as determined in step 3. Moments are positive if the equilibrium polygon lies above the arch axis and negative if it lies below.

For horizontal loads, loading condition (*c*) in Fig. 5.40, the moment at any point is equal to the vertical reaction ($R_v$ from step 3) times the scaled length of the horizontal line between the arch axis and the equilibrium polygon. Moments are positive at a point on the arch axis if the point is between the equilibrium polygon and the load. They are negative if the equilibrium polygon and the load are to the same side of the point. Thus, for the wind from the left as illustrated in Fig. 5.40, loading condition (*c*), moments in the left half of the arch are positive and those in the right half are negative.

7.  Recalculate horizontal reactions $R_H$ for vertical loads and vertical reactions $R_V$ for horizontal loads using the effective span and rise from step 6 in the appropriate equation from step 3.

8.  Check the arch section at several points along the arch axis for combined bending and axial compression. Points to be checked include the upper and lower tangent points and at least two points on the roof arm; and when the haunch is not glued integrally, additional points between the upper and lower tangent points should be checked. Manufacturing and shipping requirements that would determine whether or not the haunch will be integrally glued should be coordinated with potential manufacturers. More points on the roof arm should be checked if the two points investigated have stresses approaching the allowable values. One or more points on the vertical leg should also be checked.

The combined bending and axial compression Eq. (5-23) was developed for relatively straight members where the $P\Delta$ effect is factored into the bending portion of the equation. Arches are essentially curved members with built in curvature that places the arch axis a considerable distance from the force lines of the equilibrium polygon. The $P\Delta$ effect is usually considerably smaller than the moment developed by multiplying the force in the equilibrium polygon times the distance to the arch axis (see Fig. 5.40).

In view of the fact that the Eqs. (5-22) and (5-23) were not developed for curved members, it is recommended that the combined stress equation to be used for arches is:

$$\frac{f_c}{F'_c} + \frac{f_{bx}}{F'_{bx}} \leq 1 \tag{5-79}$$

where  $f_c$ = axial compressive stress at the point under consideration (psi) $(f_c = P'/A)$,

$A$ = cross-sectional area at point under consideration (in.$^2$),

$P'$ = axial compression at point under consideration (lb) (determined by laying out loads and reactions to scale, as in Fig. 5.42),

$F'_c$ = design value in compression parallel to grain if compression alone existed (psi) (this value is based on the maximum $l_e/d$ ratio of the member),

$f_{bx}$ = bending stress (psi), and

$F'_{bx}$ = design value in bending if flexural stress about the x-x axis alone existed (psi).

This equation has a long experience of successful use and has been published in previous editions of this manual.

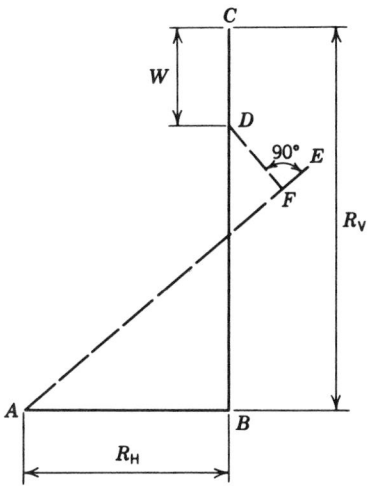

**FIGURE 5.42**  Force diagram. (1) Lay out lines to scale: $AB$ = horizontal reaction (steps 3 and 7); $BC$ = vertical reaction (steps 3 and 7); $CD$ = sum of vertical loads to left of point under consideration. (2) Draw line $AE$ parallel to the tangent to the arch axis at the point under consideration. (3) Draw line through $D$ perpendicular to line $AE$ = axial compression at point under consideration = $P'$, and $DF$ = shear perpendicular to arch axis at point under consideration.

When $f_c > F^*_{bx}(1 - C_V)$, the volume factor $C_V$ is not applicable and $F'_{bx} = F^*_{bx}$.

When $f_c < F^*_{bx}(1 - C_V)$, $F'_{bx} = F^*_{bx}C_V + f_c$.

See 5.9 for a discussion of members subjected to bending and axial compression.

If the combined stress equation value exceeds 1, it will be necessary to revise the original trial sections approximated in step 5 and to repeat the subsequent steps.

9.   Lateral stability considerations. When one edge of the arch is braced by decking fastened directly to the arch or braced at frequent intervals, as by girts or roof purlins, the ratio of tangent point depth $d_t$ to breadth $b$ of the arch based on actual dimensions should not exceed 6. When such lateral bracing is lacking, the ratio should not exceed 5, and the arch should be checked for column action in the lateral (width) direction.

The least dimension $d$ in the slenderness ratio $l_e/d$ for lateral column action is the breadth $b$ of the arch. For the roof arm of the arch, the column length $l_e$ in the slenderness ratio is the distance between points of lateral support to the roof. For the wall arm, the unsupported length is either the height of the wall arm ($y$ in Fig. 5.39$a$) or the distance between points of lateral support to the wall. The beam stability factor, $C_L$, is not usually applied to arches.

10.   Check radial stress. Radial stresses in arches seldom control the design because, for most loading conditions, the radial stress is compressive. However, for some arch configurations such as tudor arches with steep roof slopes, the wind loads may cause high radial tension stresses. Note that the design value in radial tension for wind loads for all species including Douglas Fir-Larch, Douglas Fir South, Hem-Fir, Eastern Spruce, Alaska Cedar, Canadian Spruce Pine, and Softwood Species is one-third the design value in shear adjusted by the duration-of-load factor of 1.6. When the arch is of constant cross section, the radial stress can be checked by Eq. (5-47):

$$f_r = \frac{3M}{2R_m bd}$$

where   $f_r$ = radial stress (psi) (tension $f_{rt}$, or compression $f_{rc}$),
$M$ = maximum bending moment (in.-lb),
$R_m$ = radius of curvature at arch axis (in.),
$b$ = arch width (in.), and
$d$ = arch depth at point of maximum moment (in.).

Limitations on allowable radial stress values are given in 5.11.3.

When the arch is of variable cross section such as tudor-type arches, the radial stresses can be approximated by modifying the procedure used for pitched and tapered curved beams of variable cross section as shown in Eq. (5-48). Assume that the wall arm of the arch represents one-half of a symmetrical beam with a moment at the centerline equal to the moment at the haunch of the arch as shown in the following drawing:

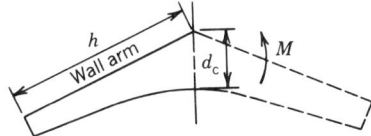

11.  Deflection of arches. Specific deflection limitations for arches have not been generally established because this is a function of the intended use. Vertical as well as horizontal deflection of arches must be considered by the designer of the building. This is especially important when adjoining construction will not tolerate the anticipated movement of the arch.

Determine deflection due to bending and change in moisture content. The deflection at any point in an arch due to bending may be determined by the equation

$$\Delta_a = \frac{s}{E} \sum \frac{Mm}{I} \tag{5-80}$$

where  $\Delta_a$ = deflection (in.),

   $s$ = length of segments along arch axis (in.) (each arch half should be divided into equal segments of length along the arch axis),

   $E$ = modulus of elasticity (psi),

   $M$ = moment at midpoint of each segment $s$ (in.-lb) under the loading condition for which the deflection is desired,

   $m$ = moment at midpoint of each segment $s$ (in.-lb) under a unit load applied at the point where the deflection is desired and in the direction in which the magnitude of the deflection is desired, and

   $I$ = moment of inertia of a section perpendicular to the arch axis at midpoint of each segment $s$ (in.$^4$).

Deflection in curved members caused by changes in moisture content can be significant and should be considered by the designer when it is critical. An explanation of this deflection is contained in *Fabrication and Design of Glued Laminated Wood Structural Members*, Technical Bulletin No. 1069, (12) and AITC Technical Note No. 2 (13). Most arches are manufactured at a slightly higher moisture content than the equilibrium moisture content in service. This change in moisture content, although small, can have a significant effect on deflection. The determination of deflection due to moisture content is complex because the exact moisture content of members at the time of fabrication is difficult to determine. The moisture content at the time of fabrication can vary as much as 5%. The equilibrium moisture content of the member in service will also vary depending on a number of factors. An approximation of the deflection can be obtained by assuming a reasonable decrease in moisture content such as 3 or 4%. The deflection is caused by a change in the thickness in the curved portion of a member. When the moisture content changes, the thickness $t$ changes and the lengths $L$ and $l$ remain relatively unchanged. Therefore, the angle $\alpha$ must change to accom-

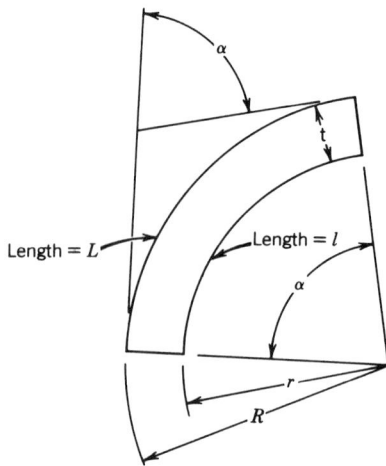

**FIGURE 5.43**    Notations for deflection.

modate the change in thickness (Fig. 5.43). The change in angle $\alpha$ can be approximated as $q\alpha$ where $q$ is equal to $-k$ and $k$ is equal to the change in thickness. Slightly greater accuracy can be obtained by using the equation

$$q = \frac{-k\alpha}{1 + k} \qquad (5\text{-}81)$$

Calculation of deflection due to a change in moisture content involves a number of approximations, and unless accurate data are available, the value of $k$ for Douglas Fir-Larch and Southern Pine can be taken as 0.2% shrinkage for each 1% change in moisture content.

**Example.**    Determine the change in the angle for a 4% decrease in moisture content when $\alpha$ is 60°.

$$k = (-4)(0.2) = -0.8\%$$

$$q = -k = -(-0.8) = +0.8\%$$

$$q\alpha = (60)(0.008) = 0.48°$$

When the change in $\alpha$ is known, the deflection at the crown of tudor arches, $\Delta_m$, can be determined by a combination of graphics and calculations or can be approximated very closely by the following equation

$$\Delta_m = \tan (q\alpha)\left[\frac{L'}{2}\right]\left(1 - \frac{h - y}{h}\right) \qquad (5\text{-}82)$$

where    $q\alpha$ = change in interior angle $\alpha$ due to shrinkage (degrees),
$\quad\quad\quad L'$ = span of arch measured from hinge to hinge (ft),
$\quad\quad\quad h$ = height of crown of arch (ft), and
$\quad\quad\quad y$ = wall height (ft).

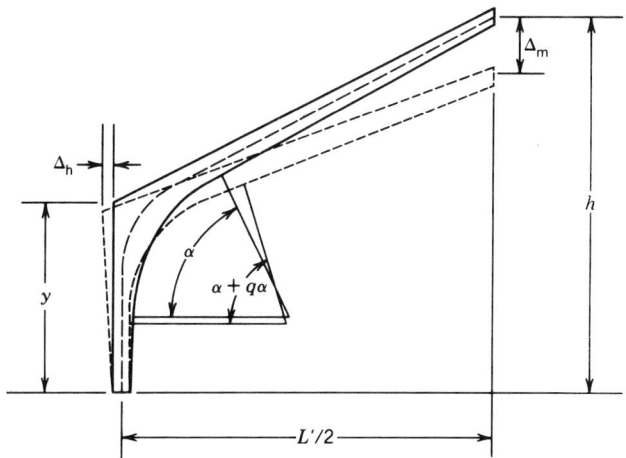

**FIGURE 5.44**   Deflection due to drying.

When the crown of the arch deflects due to a decrease in moisture content, the sides of the arch deflect outward (Fig. 5.44). The deflection of a Tudor type arch at the top of the wall, $\Delta_h$, can be closely approximated by the equation

$$\Delta_h = \tan q\alpha \, \frac{(h - y)(y)}{h} \qquad (5\text{-}83)$$

where terms are as defined previously.

**Example.**   Design a three-hinged glued laminated tudor arch made of Southern Pine to meet the following conditions using a graphical solution.

Spacing = 15 ft
Radius of curvature = $R = 10.5$ ft
Thickness of lamination = $t = \frac{3}{4}$ in.
Curvature factor = $C_C = 1 - 2000(t/R)^2$

$$C_C = 1 - 2000 \left[ \frac{0.75}{(10.5)(12)} \right]^2 = 0.929$$

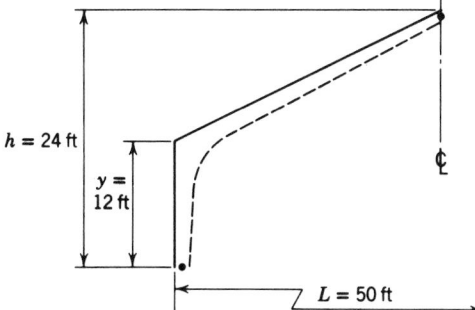

Dry conditions of use, $C_M = 1.0$

$F_b = 2400$ psi            Wall is laterally supported at midheight and at top

$F_c = 1500$ psi            Decking is applied directly to roof arm

$F_{c\perp} = 560$ psi            DL $= 13.5$ psf (estimated)

$F_v = 200$ psi            DL converted to psf of horizontal projection $= 15$ psf

$E_x = 1,800,000$ psi            SL $= 25$ psf ($C_D = 1.15$)

$E_y = 1,600,000$ psi            WL $= 20$ psf of vertical projection. (When governing code requires wind load acting perpendicular to the arch surface, see discussion in 5.13.2)

1. Application of loads:

$$w_{DL} = (\text{spacing})(DL) = (15)(15) = 225 \text{ plf}$$

$$w_{SL} = (\text{spacing})(SL) = (15)(25) = \underline{375} \text{ plf}$$

$$\text{Total vertical load} \qquad = 600 \text{ plf}$$

2. Arch layout, see sketch.

3. Calculate reactions.

Determine reactions for all loading conditions by use of equations of statics (Table 5.13). The initial calculations for reactions are based on the outside dimensions of the arch and will be recalculated when effective dimensions are determined.

4. Determine section at base.

$$\text{Maximum } R_H = 7810 \text{ lb (DL + full SL)}$$

$$bd_b = \frac{3R_H}{2F_v} = \frac{(3)(7810)}{(2)(230)} = 50.9 \text{ in.}^2$$

Try $5\frac{1}{8}$ in. $\times$ $10\frac{1}{2}$ in. section, $A = 53.8$ in.$^2$

5. Approximate crown and tangent point depths.

$$d_c = 1.5b = (1.5)(5.125) = 7.7 \text{ in. (use } 7\frac{3}{4} \text{ in.)}$$

$$d_t = 1.5d = (1.5)(10.5) = 15.75 \text{ in. (use } 15\frac{3}{4} \text{ in.)}$$

6. Lay out arch to scale and determine effective dimensions. Construct equilibrium polygons. (See Fig. 5.45.)

$$L' = 49.2 \text{ ft}$$

$$h' = 23.6 \text{ ft}$$

7. Recalculate reactions (Table 5.14).

**TABLE 5.13**

**Reactions Based on Trial Size**

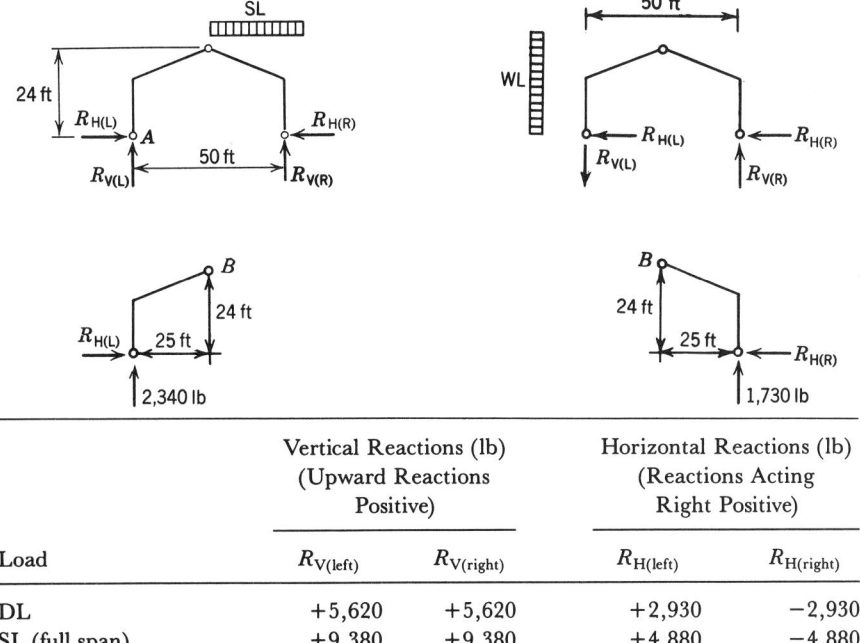

| | Vertical Reactions (lb) (Upward Reactions Positive) | | Horizontal Reactions (lb) (Reactions Acting Right Positive) | |
|---|---|---|---|---|
| Load | $R_{V(\text{left})}$ | $R_{V(\text{right})}$ | $R_{H(\text{left})}$ | $R_{H(\text{right})}$ |
| DL | +5,620 | +5,620 | +2,930 | −2,930 |
| SL (full span) | +9,380 | +9,380 | +4,880 | −4,880 |
| DL + SL (full span) | +15,000 | +15,000 | +7,810 | −7,810 |
| SL (right half) | +2,340 | +7,030 | +2,440 | −2,440 |
| DL + SL (right half) | +7,960 | +12,650 | +5,370 | −5,370 |
| WL (from left) | −1,730 | +1,730 | −5,400 | −1,800 |
| DL + WL (from left) | +3,890 | +7,350 | −2,470 | −4,730 |

8.   Check arch sections. Using the equilibrium polygons, determine moments
     at points indicated in Fig. 5.45 and also at corresponding points on right
     half of arch as shown in Table 5.15.

Tabulate maximum moments (Table 5.16) for each point along with the corresponding thrust and shear as determined from Figs. 5.46 and 5.47. Properties
of sections with depths scaled from Fig. 5.45 are given in Table 5.17.
Check each point using the combined stress equation [Eq. (5-79)]:

$$\frac{f_c}{F'_c} + \frac{f_{bx}}{F'_{bx}} \leq 1$$

Points 1 and 1′—DL + full SL: Lateral unbraced length $= l = 6$ ft-0 in. $= 72$
in.; assume $K_e = 1$, $l_e = K_e l = 72$ in.

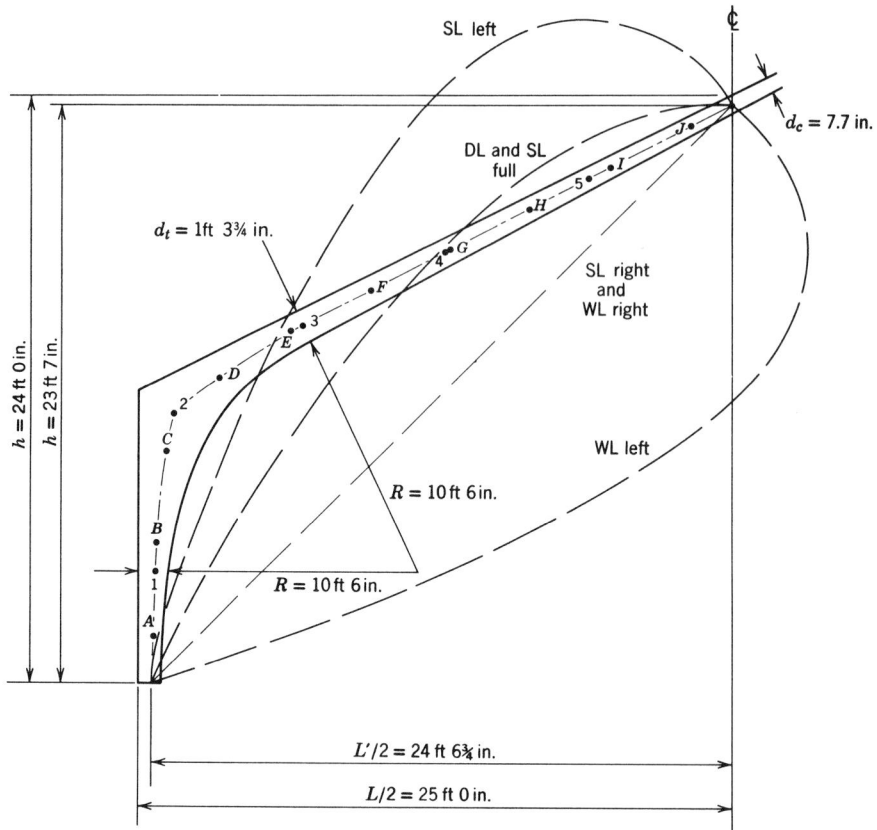

**FIGURE 5.45**   Equilibrium polygons for various loads.

**Table 5.14**

**Reactions Based on Recalculations**

| Load | Vertical Reactions (lb) (Upward Reactions Positive) | | Horizontal Reactions (lb) (Reactions Acting Right Positive) | |
|---|---|---|---|---|
| | $R_{V(\text{left})}$ | $R_{V(\text{right})}$ | $R_{H(\text{left})}$ | $R_{H(\text{right})}$ |
| DL | +5620 | +5620 | +2880 | −2880 |
| SL (full span) | +9380 | +9380 | +4800 | −4800 |
| DL + SL (full span) | +15,000 | +15,000 | +7680 | −7680 |
| SL (right half) | +2340 | +7030 | +2520 | −2520 |
| DL + SL (right half) | +7960 | +12,650 | +5400 | −5400 |
| WL (from left) | −1700 | +1700 | −5400 | −1800 |
| DL + WL (from left) | +3920 | +7320 | −2520 | −4680 |

## TABLE 5.15

### Calculation of Moments

(a) DEAD LOAD ONLY AND DEAD LOAD PLUS SNOW LOAD ON FULL SPAN

| Point | Vertical Distance between Arch Axis and Equilibrium Polygon | Horizontal Reaction(lb) | | Moment (in.-lb) | |
|---|---|---|---|---|---|
| | | DL | DL + SL | DL | DL + SL |
| 1 and 1' | 4 ft 5 in. = 53 in. | 2,880 | 7,680 | −152,600 | −407,000 |
| 2 and 2' | 9 ft 3 in. = 111 in. | 2,880 | 7,680 | −319,700 | −852,500 |
| 3 and 3' | 3 ft 8 in. = 44 in. | 2,880 | 7,680 | −126,700 | −337,900 |
| 4 and 4' | 3 in. = 3 in. | 2,880 | 7,680 | +860 | +23,000 |
| 5 and 5' | 1 ft 7 in. = 19 in. | 2,880 | 7,680 | +54,700 | +145,900 |

(b) SNOW LOAD ON RIGHT-HALF SPAN

| Point | Vertical Distance between Arch Axis and Equilibrium Polygon | Horizontal Reaction (lb) | Moment (in.-lb) | |
|---|---|---|---|---|
| | | | SL only | DL + SL |
| Left | | | | |
| 1 | 4 ft 6 in. = 54 in. | 2,520 | −136,000 | −288,600 |
| 2 | 10 ft 1 in. = 121 in. | 2,520 | −304,900 | −624,600 |
| 3 | 8 ft 5 in. = 101 in. | 2,520 | −254,500 | −381,200 |
| 4 | 5 ft 7 in. = 67 in. | 2,520 | −168,800 | −177,400 |
| 5 | 2 ft 9 in. = 33 in. | 2,520 | −83,200 | −137,900 |
| Right | | | | |
| 5' | 6 ft 5 in. = 77 in. | 2,520 | +194,000 | +248,700 |
| 4' | 6 ft 9 in. = 81 in. | 2,520 | +204,100 | +212,700 |
| 3' | 1 ft 7 in. = 19 in. | 2,520 | +47,900 | −78,800 |
| 2' | 7 ft 2 in. = 86 in. | 2,520 | −216,700 | −536,400 |
| 1' | 3 ft 9 in. = 45 in. | 2,520 | −113,400 | −266,000 |

(c) WIND LOAD FROM LEFT

| Point | Horizontal Distance between Arch Axis and Equilibrium Polygon | Vertical Reaction (lb) | Moment (in.-lb) | |
|---|---|---|---|---|
| | | | WL only | DL + WL |
| Left | | | | |
| 1 | 12 ft 5 in. = 149 in. | 1,700 | +253,300 | +100,700 |
| 2 | 22 ft 10 in. = 274 in. | 1,700 | +465,800 | +146,100 |
| 3 | 20 ft 5 in. = 245 in. | 1,700 | +416,500 | +289,800 |
| 4 | 15 ft 2 in. = 182 in. | 1,700 | +309,400 | +318,000 |
| 5 | 8 ft 4 in. = 100 in. | 1,700 | +170,800 | +224,700 |
| Right | | | | |
| 5' | 2 ft 10 in. = 34 in. | 1,700 | −57,800 | −3,100 |
| 4' | 5 ft 9 in. = 69 in. | 1,700 | −117,300 | −108,700 |
| 3' | 8 ft 8 in. = 104 in. | 1,700 | −176,800 | −303,500 |
| 2' | 10 ft 5 in. = 125 in. | 1,700 | −212,500 | −532,200 |
| 1' | 4 ft 7 in. = 55 in. | 1,700 | −93,500 | −246,100 |

## TABLE 5.16

### Maximum Moment, Thrust, and Shear

| Point | Loading Condition | Maximum Moment $M$ (in.-lb) | Thrust $P'$ (lb) | Shear Perpendicular to Arch Axis (lb) |
|---|---|---|---|---|
| Left | | | | |
| 1 | DL + full SL | −407,000 | 15,240 | 6,480 |
| 2 | DL + full SL | −852,500 | 16,000 | 1,140 |
| 3 | DL + SL right | −381,200 | 7,640 | 3,240 |
| 4 | DL + WL left | +318,000 | 2,680 | 220 |
| 5 | DL + WL left | +224,700 | 2,100 | 1,500 |
| Right | | | | |
| 5′ | DL + SL right | +248,700 | 5,360 | 1,360 |
| 4′ | DL + SL right | +212,700 | 6,960 | 1,860 |
| 3′ | DL + full SL | −337,900 | 11,640 | 6,160 |
| 2′ | DL + full SL | −852,500 | 16,000 | 1,140 |
| 1′ | DL + full SL | −407,000 | 15,240 | 6,480 |

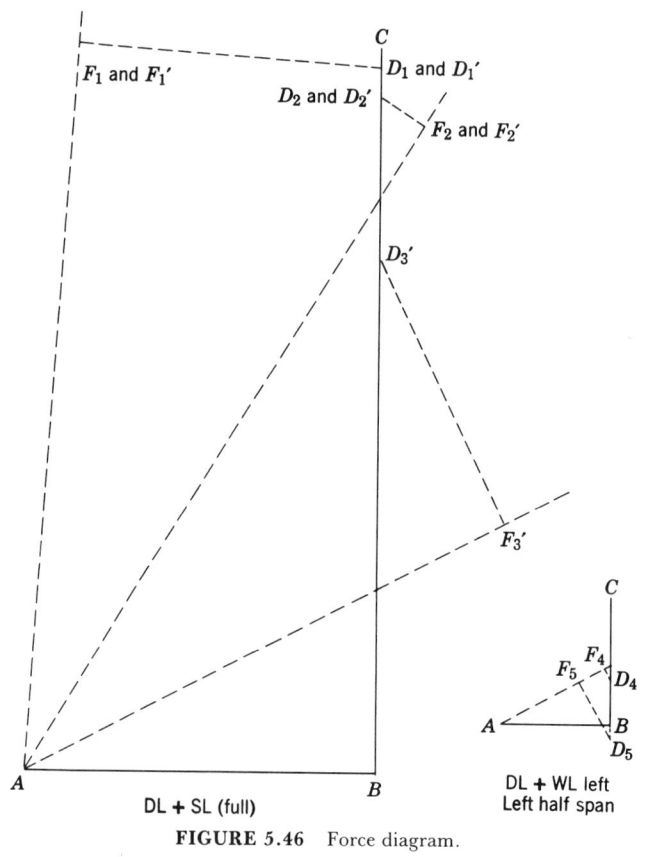

**FIGURE 5.46**   Force diagram.

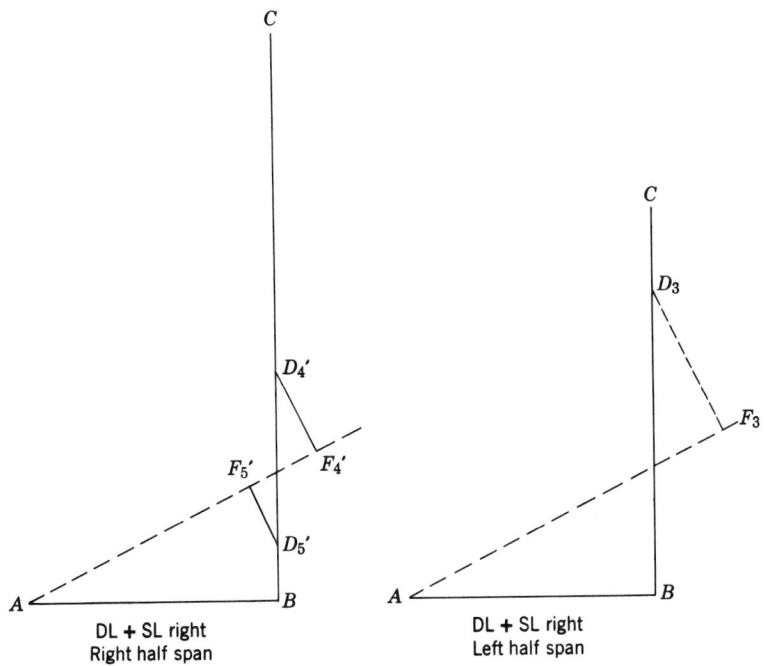

**FIGURE 5.47** Force diagram.

**TABLE 5.17**

**Properties of Sections**

| Point | Depth $d$ (in.)[a] | Area $A$ (in.$^2$) | Section Modulus $S$ (in.$^3$) | Volume Factor $C_V$ |
|---|---|---|---|---|
| 1 and 1' | 15.75 | 80.72 | 211.9 | 0.973 |
| 2 and 2' | 43.0 | 220.4 | 1579 | 0.880 |
| 3 and 3' | 15.75 | 80.72 | 211.9 | 0.973 |
| 4 and 4' | 14.0 | 71.75 | 167.4 | 0.985 |
| 5 and 5' | 11.0 | 56.38 | 103.4 | 1.00 |

[a]Scaled from Fig. 5.45.

$$\frac{l_e}{d} = \frac{72}{5.125} = 14.05$$

$$F_c^* = F_c C_D = (1500)(1.15) = 1725 \text{ psi}$$

$$E_y' = E_y = 1,600,000 \text{ psi}$$

Determine the column stability factor as follows:

From Eq. (5-14), $K_{cE} = 0.418$

$$F_{cE} = \frac{K_{cE} E_y'}{(l_e/d)^2} = \frac{(0.418)(1,600,000)}{(14.05)^2} = 3388 \text{ psi}$$

$c = 0.9$

$$C_P = \frac{1 + F_{cE}/F_c^*}{2c} - \sqrt{\left[\frac{1 + (F_{cE}/F_c^*)}{2c}\right]^2 - \frac{F_{cE}/F_c^*}{C}}$$

$$C_P = \frac{1 + 3388/1725}{(2)(0.9)} - \sqrt{\left[\frac{1 + 3388/1725}{(2)(0.9)}\right]^2 - \frac{3388/1725}{0.9}} = 0.919$$

$$F_c' = F_c^* C_P = (1725)(0.919) = 1586 \text{ psi}$$

$$f_c = \frac{P'}{A} = \frac{15,240}{80.7} = 189 \text{ psi}$$

$$F_b^* = F_b C_D C_C = (2400)(1.15)(0.929) = 2564 \text{ psi}$$

Determine volume factor $C_V$. The volume factor in Eq. (4-2):

$$C_V = K_L \left(\frac{5.125}{b}\right)^{1/x} \left(\frac{12}{d}\right)^{1/x} \left(\frac{21}{L}\right)^{1/x}$$

is modified for arches by using the $l_e/d$ ratio $= 21$. $L = l_e/12$; $l_e = 21d$; $L = 21d/12$. Substituting $21d$ in Eq. (4-2) results in the equation:

$$C_V = K_L \left(\frac{5.125}{b}\right)^{1/x} \left(\frac{12}{d}\right)^{2/x} \tag{5-84}$$

For Southern Pine

$$C_V = K_L \left(\frac{5.125}{b}\right)^{1/20} \left(\frac{12}{d}\right)^{2/20}, \quad K_L = 1 \text{ for uniform load} \tag{5-85}$$

For Point 1

$$C_V = \left(\frac{5.125}{5.125}\right)^{1/20} \left(\frac{12}{15.75}\right)^{2/20} = 0.973$$

$C_V$ for other points are shown in Table 5.17.
Check for applicability of volume factor:

$$F_b^*(1 - C_V) = (2564)(1 - 0.973) = 69 \text{ psi} < f_c \ (189 \text{ psi})$$

Volume factor is not applicable.

$$F_b' = F_b^* = 2564 \text{ psi}$$

$$f_b = \frac{M}{S} = \frac{407{,}000}{211.9} = 1920 \text{ psi}$$

$$\frac{189}{1586} + \frac{1920}{2564} = 0.87 < 1 \qquad \text{O.K.}$$

Points 2 and 2'—laterally braced by decking:

$$F_c' = F_c C_D = (1500)(1.15) = 1725 \text{ psi}$$

$$f_c = \frac{P'}{A} = \frac{16{,}000}{80.72} = 198 \text{ psi}$$

Check for applicability of volume factor:

$$F_b^* = 2564 \text{ psi (same as point 1)}$$

$$F_{bx}^*(1 - C_V) = (2564)(1 - 0.88) = 307 \text{ psi} > f_c = 198 \text{ psi}$$

$$F_{bx}' = F_{bx}^* C_V + f_c = (2564)(0.88) + 198 = 2455 \text{ psi}$$

$$f_{bx} = \frac{M}{S} = \frac{852{,}500}{1579} = 540 \text{ psi}$$

$$\frac{f_c}{F_c'} + \frac{f_{bx}}{F_{bx}'} \le 1.0, \quad \frac{198}{1725} + \frac{540}{2455} = 0.33 \le 1.0 \qquad \text{O.K.}$$

Point 3—DL + SL right:

The arch is braced laterally by the decking, and no reduction for buckling is necessary at this point, $C_P = 1.0$.

$$F_c' = F_c^*(C_P) = 1725(1) = 1725 \text{ psi}$$

$$f_c = \frac{P'}{A} = \frac{7640}{80.72} = 95 \text{ psi}$$

$$F_b^* = 2564 \text{ (same as point 1)}$$

Check for applicability of volume factor:

$$F_b^*(1 - C_V) = (2564)(1 - 0.973) = 69 \text{ psi} < f_c \text{ (95 psi)}$$

Volume factor is not applicable.

$$F_b' = F_b^* = 2564 \text{ psi}$$

$$f_b = \frac{M}{S} = \frac{381{,}200}{211.9} = 1800 \text{ psi}$$

$$\frac{95}{1725} + \frac{1800}{2564} = 0.76 < 1 \qquad \text{O.K.}$$

Point 3′—DL + full SL

$$F_c' = 1725 \text{ psi}$$

$$f_c = \frac{P'}{A} = \frac{11,640}{80.72} = 144 \text{ psi}$$

$$F_b^* = 2564 \text{ psi (same as point 1)}$$

$$C_V = 0.973 \text{ (same as point 1)}$$

Check for applicability of volume factor:

$$F_b^*(1 - C_V) = (2564)(1 - 0.973) = 69 \text{ psi} < f_c = 144 \text{ psi}$$

Volume factor is not applicable.

$$F_b' = 2564 \text{ psi (same as point 3)}$$

$$f_b = \frac{M}{S} = \frac{337,900}{211.9} = 1595 \text{ psi}$$

$$\frac{144}{1725} + \frac{1595}{2564} = 0.71 < 1.0 \qquad \text{O.K.}$$

Point 4—DL and WL from left:
  Point 4 is laterally braced by decking.

$$F_c' = F_c C_D = (1500)(1.60) = 2400 \text{ psi}$$

$$f_c = \frac{P'}{A} = \frac{2680}{71.75} = 37 \text{ psi}$$

Check for applicability of volume factor:

$$F_b^* = F_b C_D = (2400)(1.60) = 3840 \text{ psi}$$

$$F_b^*(1 - C_V) = (3840)(1 - 0.985) = 59 \text{ psi} > f_c = 37 \text{ psi}$$

Volume factor is applicable when $f_c < F_b^*(1 - C_V)$.

$$F_b' = F_b^* C_V + f_c$$

$$F_b' = (3840)(0.985) + 37 = 3820 \text{ psi}$$

$$f_b = \frac{M}{S} = \frac{318,000}{167.4} = 1900 \text{ psi}$$

$$\frac{37}{2400} + \frac{1900}{3820} = 0.51 < 1 \qquad \text{O.K.}$$

Point 4′—DL and SL right:

$$F_c' = F_c C_D = (1500)(1.15) = 1725 \text{ psi}$$

$$f_c = \frac{P'}{A} = \frac{6960}{71.75} = 97 \text{ psi}$$

Check for applicability of volume factor:

$$C_V = 0.985 \text{ (same as point 4)}$$

$$F_b^* = F_b C_D = (2400)(1.15) = 2760 \text{ psi}$$

$$F_b^*(1 - C_V) = (2760)(1 - 0.985) = 41 \text{ psi} < 97 \text{ psi}$$

Volume factor is not applicable.

$$F_b' = F_b^* = 2760 \text{ psi}$$

$$f_b = \frac{M}{S} = \frac{212,700}{167.4} = 1270 \text{ psi}$$

$$\frac{97}{1725} + \frac{1270}{2760} = 0.512 < 1 \qquad \text{O.K.}$$

Point 5—DL + WL left:
  Point 5 is laterally braced by decking $C_P = 1$

$$f_c = \frac{P'}{A} = \frac{2100}{56.38} = 37 \text{ psi}$$

$$d = 11 \text{ in.} \quad C_V = 1$$

$$F_b' = F_b^* = 3840 \text{ psi}$$

$$f_b = \frac{M}{S} = \frac{224,700}{103.4} = 2174 \text{ psi}$$

$$\frac{37}{2400} + \frac{2174}{3840} = 0.58 < 1 \qquad \text{O.K.}$$

Point 5′—DL and SL right:

$$F_c' = F_c C_D = 1500(1.15) = 1725 \text{ psi}$$

$$f_c = \frac{5360}{56.38} = 95 \text{ psi}$$

$$d = 11 \text{ in.} \quad C_V = 1$$

$$F_b' = F_b^* = 2760 \text{ psi}$$

$$f_b = \frac{248,700}{103.4} = 2406 \text{ psi}$$

$$\frac{f_c}{F'_c} + \frac{f_b}{F'_b} = \frac{95}{1725} + \frac{2406}{2760} = 0.93 < 1 \qquad \text{O.K.}$$

The depths of the arch at the crown and base are initially set based on shear requirements. Shear will seldom govern along the length of the arch. A sample calculation is shown for point 1.

$$F'_v = (200)(1.15) = 230 \text{ psi}$$

$$f_v = \frac{3V}{2A} = \frac{(3)(6480)}{(2)(80.72)} = 120 \text{ psi} < 230 \text{ psi} \qquad \text{O.K.}$$

9.   Check arch for lateral stability in regard to tangent point depth–width ratio

$$\frac{d_t}{b} = \frac{15.75}{5.125} = 3.07 < 6 \qquad \text{O.K.}$$

10.   Check radial stress.

$$d_c = d_2 = 43 \text{ in.}$$

$$R_m = R + \frac{d_c}{2} = (10.5)(12) + \frac{43}{2} = 147.5 \text{ in.}$$

$$\frac{d_c}{R_m} = \frac{43}{147.5} = 0.292$$

Consider the arch leg and haunch as $\frac{1}{2}$ of a pitched and tapered curved beam as shown in the sketch:

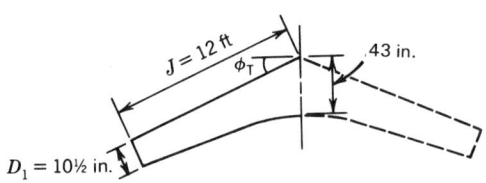

$$\phi_T = 90° - \frac{90° + SL}{2}$$

where   $SL$ = Slope of the arch roof (degrees)
$SL$ = Arc tan $\frac{12}{25}$ = 25.64°
$\phi_T$ = 90° − (90° + 25.64°/2) = 32.2°

The value of $l/l_c$ is approximately 2. The slope of the beam is $7\frac{1}{2}$ : 12. Use $\phi_T$ in Fig. 5.32 = 6 : 12.
The value of $C_r$ is between 0.8 and 1.0 in the figure. Use $C_r$ = 1.0 because the

member is not uniformly loaded. The maximum moment is 852,500 in.-lb, which results in radial compression.

$$f_{rc} = K_r C_r \frac{6M}{bd_c^2} = \frac{(0.165)(1)(6)(852,500)}{(5.125)(43)^2} = 89 \text{ psi}$$

$$F_{rc} \text{ for core laminations} = 560 \text{ psi}$$

$$F'_{rc} = F_{rc} = 560 \text{ psi} > 89 \text{ psi}$$

The largest positive moment at point 2 = +146,100 in.-lb (DL + WL).

$$f_{rt} = \frac{(0.165)(1)(6)(146,100)}{(5.125)(43)^2} = 15 \text{ psi (radial tension)}$$

$$F'_{rt} \text{ for wind load} = \frac{F_v C_D}{3} = \frac{200(1.6)}{3} = 107 \text{ psi} \quad 107 \text{ psi} > 15 \text{ psi} \quad \text{O.K.}$$

11. Check deflection. Vertical deflection at the apex and horizontal deflection at the haunch are usually considered in design. The calculations for deflection at the haunch are shown. Deflection at the apex is determined in a similar manner by placing the unit load at the apex.

Determine the outward horizontal deflection at the haunch under full snow load plus dead load. Figure 5.48 shows application of unit load for determining outward horizontal deflection at haunch and reactions due to unit load (determined by statics).

Divide each half arch axis into ten segments, $s$, 45.4 in. long. The midpoints of each segment on the left half arch axis are indicated as points $A$–$J$ on Fig. 5.45.

Moments at each midpoint may be determined by using the equilibrium polygon for DL + full SL for actual load moments, $M$, and using the equilibrium polygons for wind load right for unit load moments, $m$.

A summary of calculations to determine $\Sigma MmI$ are shown in Table 5.18.

$$\Delta_a = \frac{s}{E} \Sigma \frac{Mm}{I}$$

$$\Delta_a = \frac{(45.4)(13,910)}{1,800,000} = 0.351 \text{ in.}$$

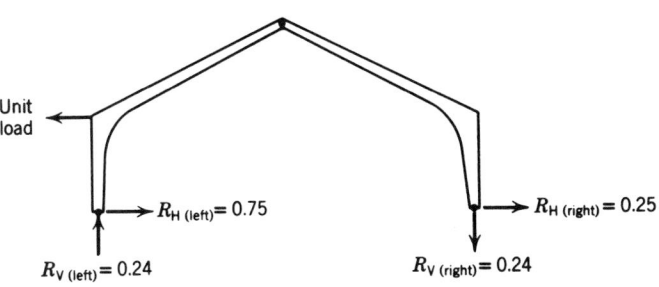

**FIGURE 5.48** Reactions for unit loading.

**TABLE 5.18**

**Summary of Deflection Calculations**

| Point | Moment, $M$, in.-lb | Unit Load Moment, $m$, in. | Moment of Inertia, $I$, in.$^4$ | $Mm/I$ |
|-------|--------------------|--------------------------|----------------------------------|--------|
| **Left** | | | | |
| $A$ | $-176{,}600$ | $-15.8$ | 940 | $+2{,}970$ |
| $B$ | $-514{,}600$ | $-43.2$ | 2,100 | $+10{,}590$ |
| $C$ | $-783{,}400$ | $-61.2$ | 12,700 | $+3{,}780$ |
| $D$ | $-668{,}200$ | $-64.8$ | 8,400 | $+5{,}150$ |
| $E$ | $-384{,}000$ | $-58.3$ | 1,700 | $+13{,}170$ |
| $F$ | $-138{,}200$ | $-52.3$ | 1,400 | $+5{,}160$ |
| $G$ | $+\phantom{0}30{,}700$ | $-43.2$ | 1,200 | $-1{,}100$ |
| $H$ | $+130{,}600$ | $-32.6$ | 740 | $-5{,}750$ |
| $I$ | $+138{,}200$ | $-20.9$ | 430 | $-6{,}720$ |
| $J$ | $+\phantom{0}76{,}800$ | $-7.2$ | 310 | $-1{,}780$ |
| **Right** | | | | |
| $J'$ | $+\phantom{0}76{,}800$ | $+2.2$ | 310 | $+540$ |
| $I'$ | $+138{,}200$ | $+7.0$ | 430 | $+2{,}250$ |
| $H'$ | $+130{,}600$ | $+11.8$ | 740 | $+2{,}080$ |
| $G'$ | $+30{,}700$ | $+16.1$ | 1,200 | $+410$ |
| $F'$ | $-138{,}200$ | $+21.1$ | 1,400 | $-2{,}080$ |
| $E'$ | $-384{,}000$ | $+25.4$ | 1,700 | $-5{,}740$ |
| $D'$ | $-668{,}200$ | $+28.8$ | 8,400 | $-2{,}290$ |
| $C'$ | $-783{,}400$ | $+26.4$ | 12,700 | $-1{,}630$ |
| $B'$ | $-514{,}600$ | $+16.6$ | 2,100 | $-4{,}070$ |
| $A'$ | $-176{,}600$ | $+5.5$ | 940 | $-1{,}030$ |

$$\Sigma Mm/I = +13{,}910$$

Summary of actual section sizes:

Base depth $= 10\frac{1}{2}$ in.
Crown depth $= 7\frac{3}{4}$ in.
Tangent depths $= 15\frac{3}{4}$ in.
Arch width $= 5\frac{1}{8}$ in.

### 5.13.4  Mathematical Solution

The mathematical or algebraic solution of arch design allows advantage to be taken of modern calculating methods including computer technology. It is a very useful procedure because a computer program can be developed for use with a number of designs. However, when only one or two designs are being considered, the graphical method is usually faster unless a program is readily available. The mathematical solution also can be readily used when wind loads are applied per-pendicular to the outer surfaces of the arch.

For additional information of the mathematical solution, see AITC Technical Note 23, *Mathematical Solution of a Three-Hinged Arch* (14).

### 5.13.5   Circular Arches

Three-hinged circular or parabolic arches are designed in a manner similar to that for three-hinged tudor arches. They will usually require slightly more wood than two-hinged arches, but are frequently more economical because of the elimination of the moment splice required for most two-hinged arches.

### 5.13.6   Two-Hinged Arches

Two-hinged arches are hinged at each base and are statically indeterminate. They may have a profile of any shape with any combination of straight or curved sections with constant or variable section depth. When shipping restrictions limit the size of member that can be transported, it is common to utilize moment splices to reduce component dimensions. The horizontal thrust at the base must be resisted by some adequate means, such as tie rods, abutments, or foundations. After the reactions, moments, shears, and axial forces have been determined, the part of the design for determining the required arch section is similar to that for the three-hinged arch.

### 5.13.7   Design Procedure

Two-hinged arches are statically indeterminate to the first degree. The usual design procedure is to determine the horizontal reactions by one of the energy methods and complete the solution by statics. Usually, the left hinge of the arch is considered the origin of a coordinate system and horizontal distances $x$ and vertical distances $y$ are measured from this point. Sometimes the configuration is such that an algebraic solution of the location of $x$ and $y$ coordinates of various points along the arch axis is difficult and a graphical solution may be preferred.

When arches are of constant cross section, constant $E$, and symmetrical about the centerline, both $E$ and $I$ drop out of the energy equations and symmetry reduces the number of calculations. The design is further simplified when the shape of the arch axis is an arc of a circle or a parabola because an algebraic solution is more readily obtainable. The effect of tie rod elongation, differential settlement, or spread of abutments is usually small and is commonly neglected in design. Tie rod elongation can change values in larger arches unless the elongation is compensated for by initial shortening of the rod with a provision for sliding bearings to permit movement of the arch base. The following procedure shows the steps required for the design of a two-hinged arch.

1.   Lay out one-half of the arch axis to a convenient scale, and divide it into any number of equal divisions (the more divisions used, the more precise the results will be). Determine and tabulate the $x$ and $y$ distances for each midpoint of a division (see Fig. 5.47). Tabulate the values for $y^2$.

2.    Determine vertical reactions $R_v$ for the various loading conditions by the summation of vertical forces equal to zero or by taking moments about one hinge. For balanced loading, $R_v$ = one-half the total live or snow and dead loads on the arch.

3.    The arch is made statically determinate by assuming that the left horizontal reaction $R_H$ (left) has been removed and the left support is free to move. For vertical loads, the entire arch can be considered as a simple-span beam. Compute and tabulate the bending movement $M_s$ at each division point for dead loads and balanced and unbalanced live or snow loads. For wind loads, $M_s$ is determined by use of the principles of statics.

4.    Multiply the $M_s$ values by the corresponding $y$ values and tabulate.

5.    Compute the horizontal reactions $R_H$ for the various loadings by the equation

$$R_H = \frac{\Sigma\,(M_s y)}{\Sigma\,(y^2)} \tag{5-86}$$

This equation neglects the effect of the rod elongation or spread in abutments because these movements are usually small.

6.    Determine and tabulate bending moment $M$ on the arch at each division point using the following equation. For dead and vertical loads:

$$M = M_s - R_H y \tag{5-87}$$

The procedure for calculating wind loads depends on the code requirements for distribution of wind load. When the projected area method is used, moments are determined by Eq. (5-87). When the wind is from the left and with the sign convention for moments used, $M_s$ is negative and $R_H y$ is positive. Eq. (5-87) is then written as $M = R_H y - M_s$.

In the absence of a governing building code, ASCE 7-88 is recommended for determination of wind loads as discussed in Chapter 4.

7.    Determine the axial thrust $P'$ at each division point.

8.    On the basis of the assumption of full lateral support and using the following equations, determine a trial size required at each division point.

$$A = \frac{P'}{F_c^*} \qquad S = \frac{M}{F_b^*}$$

where   $A$ = cross-sectional area (in.$^2$),
$\quad\quad\quad P'$ = axial thrust as determined in step 7 (lb),
$\quad\quad\quad F_c^*$ = design value in compression parallel to grain (psi) multiplied by applicable adjustment factors except $C_P$,
$\quad\quad\quad S$ = section modulus (in.$^3$),
$\quad\quad\quad M$ = bending moment as determined in step 6 (in.-lb), and
$\quad\quad\quad F_b^*$ = design value in bending (psi) multiplied by applicable adjustment factors except $C_L$ and $C_V$.

On the basis of these requirements and industry standard depths and widths [AITC 117—Design (1)], determine the required arch depths $d$ and widths $b$ at each division point. When one edge of the arch is braced by decking fastened directly to the arch or braced at frequent intervals, as by girts or purlins, the depth–breadth ratio $d/b$ of the arch based on actual dimensions should not exceed 6. When such lateral bracing is lacking, the ratio should not exceed 5.

9.   Check the section for combined bending and axial compression by Eq. (5-79) as follows:

$$\frac{f_c}{F_c'} + \frac{f_{bx}}{F_{bx}} \leq 1$$

See *Comments* in 5.13.3 for development of this equation.

where   $f_c$ = axial compression parallel to grain stress induced by the axial loads (psi),

$f_{bx}$ = bending stress about the $x$-$x$ axis (psi),

$F_c'$ = design value in compression parallel to grain (psi) multiplied by applicable adjustment factors, and

$F_{bx}'$ = design value in bending about the $x$-$x$ axis (psi) multiplied by applicable adjustment factors.

Depending on the loading condition, the length for determining the slenderness ratio, $(l_e/d)_x$, may be either of the following straight-line distances and both should be checked:

(a)   hinge to point of contraflexure on the same half of the arch or

(b)   point of contraflexure to point of contraflexure.

In either case, $d$ is the arch depth. For determining the length for calculating the slenderness ratio, $(l_e/d)_y$, the distance between purlins should be used for $l_u$. If the combined stress equation exceeds 1, it will be necessary to revise the chosen section size and repeat this step.

10.   Check the section for shear stress at the base and at the point of maximum shear, using Eq. (5-3):

$$f_v = \frac{3V}{2A}$$

where   $f_v$ = actual shear stress (psi),

$V$ = shear perpendicular to arch axis (lb), and

$A$ = cross-sectional area (in.$^2$).

This value may not exceed the design value in shear, $F_v'$, where $F_v' = F_v$ multiplied by all applicable adjustment factors.

11.   Check radial stress $f_r$ by Eq. (5-47):

$$f_r = \frac{3M}{2R_m bd}$$

where  $f_r$ = radial stress (psi) (tension $f_{rt}$, or compression $f_{rc}$),
  $M$ = maximum bending moment (in.-lb),
  $R_m$ = radius of curvature at arch axis (in.), and
  $A$ = cross-sectional area (in.$^2$).

Limitations on radial design values are given in 5.11.3.

12.   Determine the deflection due to bending. The deflection at any point in the arch may be determined by the following equation:

$$\Delta_s = \frac{s}{EI} \Sigma \, Mm \qquad (5\text{-}88)$$

where  $\Delta_a$ = deflection (in.),
  $s$ = length of segments along arch axis (in.) (any number of equal-length segments may be used; however, the greater the number of segments, the greater the accuracy),
  $E$ = modulus of elasticity (psi),
  $I$ = moment of inertia of a section perpendicular to the arch axis (in.$^4$),
  $M$ = moment at the midpoint of each segment $s$ (in.-lb) under the loading conditions for which deflection is desired, and
  $m$ = moment at midpoint of each segment $s$ (in.) under a unit load applied at point where deflection is desired and in the direction in which magnitude of deflection is desired.

To facilitate deflection calculations, tabulate the values.

### 5.13.8   Constant-Radius Arches

The following equations will assist in the design of two-hinged, constant-section, constant-radius arches.

Step 1:

$$x_n = R(\sin A - \sin \alpha_n) \qquad (5\text{-}89)$$

$$y_n = R(\cos \alpha_n - \cos A) \qquad (5\text{-}90)$$

where  $\alpha$ = the central angle to the point under consideration (degrees)
  $x_n, y_n$ = $x$ and $y$ coordinates corresponding with central angle $\alpha_n$ (see Fig. 5.48),
and other terms are as defined in Fig. 5.48.

Step 3:
Dead load:

$$M_s = R_{V(\text{left})}x = w_{DL}Ry + w_{DL}R^2 \sin \alpha_n (A - \alpha_n) \text{ (for left half span)}   (5\text{-}91)$$

where  $w_{DL}$ = uniform dead load along arch axis (plf) and
  $A - \alpha_n$ = difference between tangent angle and central angle to point under consideration (radians).

Balanced live or snow load:

$$M_s = R_{V(\text{left})}x - \tfrac{1}{2}w_{SL}x^2 \text{ (for left half span)} \qquad (5\text{-}92)$$

where  $w_{SL}$ = uniform snow load or live load along horizontal projection (plf).

Unbalanced snow load on right half span:

$$M_{s(\text{left})} = R_{V(\text{left})}x \qquad (5\text{-}93)$$

$$\text{(simple moments in left or unloaded half span)}$$

$$M_{s(\text{right})} = R_{V(\text{right})}(L - x) - \tfrac{1}{2}w_{SL}(L - x)^2 \qquad (5\text{-}94)$$

$$\text{(simple moments in right or loaded half span)}$$

Wind load from the left (projected area method):

$$M_{s(\text{left})} = -R_{V(\text{left})}x - \tfrac{1}{2}w_{WL}y^2 \qquad (5\text{-}95)$$

$$\text{(simple moments in left half span)}$$

$$M_{s(\text{right})} = R_{V(\text{right})}(L - x) - R_{H(\text{right})}y \qquad (5\text{-}96)$$
$$\text{(simple moments in right half span)}$$

where  $w_{WL}$ = uniform wind load along vertical projection (plf) .
and other terms are as shown in Fig. 5.49.

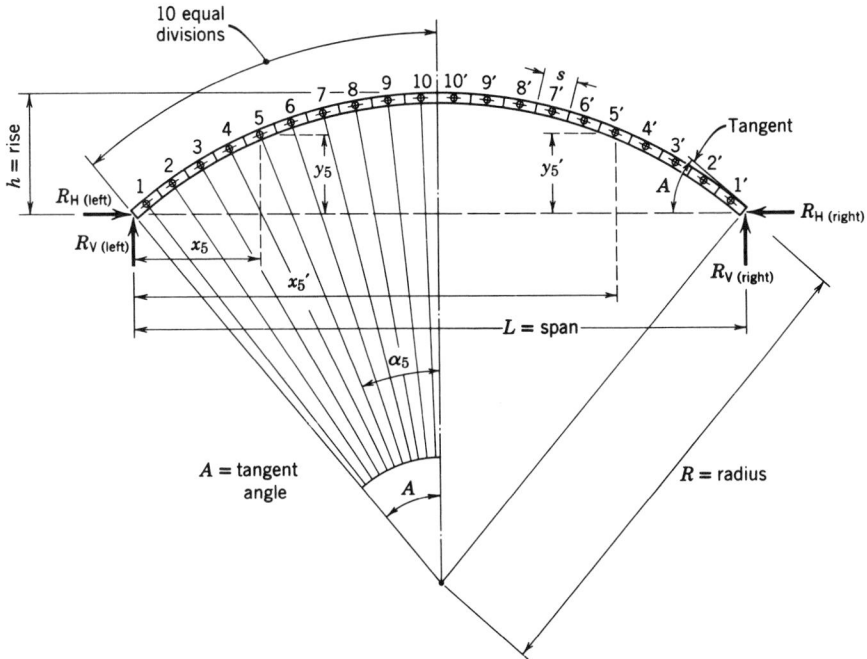

**FIGURE 5.49**  Two-hinged arch notations.

Step 7: At the point of maximum moment (which may be located from tabulation in step 6), the axial thrust $P'$ is determined by the equation

$$P' = R_H \cos \alpha_n + (R_V - \Sigma\, wx) \sin \alpha_n \qquad (5\text{-}97)$$

where    $R_H$ = horizontal reaction for loading causing maximum moment (lb) (for maximum moment occurring in left half span, use $R_{H(left)}$; in right half span, use $R_{H(right)}$

$R_V$ = vertical reaction for the loading causing maximum moment (lb),

$\alpha_n$ = central angle to point of maximum moment (degrees), and

$\Sigma wx$ = sum of loads to left (for point on left half span) or to right (for point on right half span) of point of maximum moment (lb).

Step 10: Check shear. The maximum horizontal and vertical reactions occur under dead load plus full snow load; therefore, check shear at base for this condition.

$$V = (R_{VDL} + R_{SL(left)}) \cos A - (R_{HDL} + R_{HSL(left)}) \sin A \qquad (5\text{-}98)$$

$$f_v = \frac{3V}{2A}$$

The maximum shear at the crown occurs under dead load (DL) and unbalanced snow load (UL).

$$V = R_{VDL} + R_{VUL(left)} - \Sigma\, w_{DL}x \text{ (snow load on right)} \qquad (5\text{-}99)$$

**Example.**   Design a constant-section, constant-radius glued laminated two-hinged arch to meet the following conditions:

Species is Douglas Fir-Larch
$L = 86$ ft                              Loads:
Spacing $= 17$ ft                       DL $= 10$ psf
$R = 64$ ft (radius of centerline of     SL $= 20$ psf
  arch)                                  Wind loading for 70-mph wind as
                                         shown

Assuming the arch is on buttresses 15 ft high, the wind loading is shown in Fig. 5.50(*a*). By inspection, it is obvious that when wind load stresses are added to dead load stresses, this loading will not control. Therefore, the minimum horizontal load of 10 psf acting on the vertical projection is used in design as shown in (*b*).

Lamination thickness $t = 1\frac{1}{2}$ in.
Curvature factor $C_C = 1 - 2000(t/R)^2$
   $= 1 - 2000[(1.5)/(64)(12)]^2 = 0.992$
Duration of load factor $C_D = 1.15(DL + SL), 1.6(DL + WL)$
$F_b^* = F_b C_D C_C = (2400)(1.15)(0.992) = 2740$ psi
$F_c^* = F_c C_D = (1650)(1.15) = 1900$ psi
$F_{c\perp}' = F_{c\perp} = 560$ psi

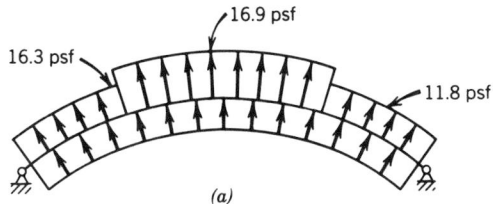

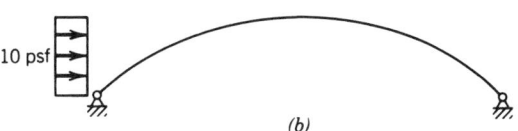

**FIGURE 5.50**   Wind load on arch.

$$F'_{rt} = F_{rt}C_D = (15)(1.15) = 17.25 \text{ psi}$$
$$F'_v = F_v C_D = (165)(1.15) = 190 \text{ psi}$$
$$E'_x = E_x = 1,800,000 \text{ psi}, \quad E'_y = E_y = 1,600,000 \text{ psi}$$

Lateral support provided by $3 \times 12$ purlins spaced at intervals of 9.57 ft measured along the arch.

   1.   Determine the tangent angle $A$ using the dimensions and geometry shown in Fig. 5.51.

$$\sin A = \frac{L/2}{R} = \frac{43}{64} \qquad A = 42.21°$$

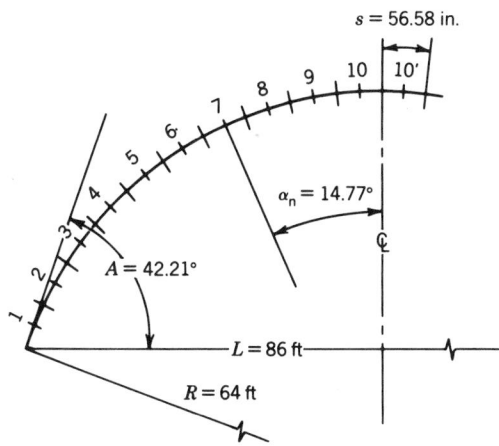

**FIGURE 5.51**   Geometry of arch.

Divide the arch half shown in Fig. 5.51 into 10 equal parts. Determine length of arc, $s$.

$$s = \frac{A2\pi R}{(360)(10)} = \frac{(42.21°)(2)(\pi)(64)(12)}{(360)(10)} = 56.58 \text{ in.}$$

Determine and tabulate $\alpha_n$, $x$, $y$, $y^2$. (See Table 5.19.)

2.  Determine vertical reactions.

Dead load:

$$R_{V(\text{left})} = R_{V(\text{right})} = \frac{w_{DL} \text{ (length of arch axis)}}{2}$$

$$= \frac{(10)(17)(94.3)}{2} = 8016 \text{ lb}$$

### TABLE 5.19[a]

### Tabulation of Input Values

| Point | $\alpha_n$ (degrees) | (rads) | $\sin \alpha_n$ | $\cos \alpha_n$ | $x$(ft) | $y$(ft) | $y^2$(ft$^2$) |
|---|---|---|---|---|---|---|---|
| Left | | | | | | | |
| 1 | 40.10 | 0.69990053 | 0.64414 | 0.76491 | 1.77 | 1.55 | 2.41 |
| 2 | 35.88 | 0.62622679 | 0.58609 | 0.81024 | 5.49 | 4.45 | 19.83 |
| 3 | 31.66 | 0.55255305 | 0.52486 | 0.85119 | 9.41 | 7.07 | 50.03 |
| 4 | 27.44 | 0.47887931 | 0.46078 | 0.88751 | 13.51 | 9.40 | 88.33 |
| 5 | 23.22 | 0.40520557 | 0.39421 | 0.91902 | 17.77 | 11.41 | 130.3 |
| 6 | 19.00 | 0.33153183 | 0.32549 | 0.94554 | 22.17 | 13.11 | 171.9 |
| 7 | 14.77 | 0.25785809 | 0.25501 | 0.96694 | 26.68 | 14.48 | 209.7 |
| 8 | 10.55 | 0.18418435 | 0.18314 | 0.98309 | 31.28 | 15.51 | 240.7 |
| 9 | 6.33 | 0.11051061 | 0.11029 | 0.99390 | 35.94 | 16.21 | 262.7 |
| 10 | 2.11 | 0.03683687 | 0.03683 | 0.99932 | 40.64 | 16.55 | 274.0 |
| Right | | | | | | | |
| 10' | | | | | 45.36 | 16.55 | 274.0 |
| 9' | | | | | 50.06 | 16.21 | 262.7 |
| 8' | | | | | 54.72 | 15.51 | 240.7 |
| 7' | | | | | 59.32 | 14.48 | 209.7 |
| 6' | | | | | 63.83 | 13.11 | 171.9 |
| 5' | | | | | 68.23 | 11.41 | 130.3 |
| 4' | | | | | 72.49 | 9.40 | 88.33 |
| 3' | | | | | 76.59 | 7.07 | 50.03 |
| 2' | | | | | 80.51 | 4.45 | 19.83 |
| 1' | | | | | 84.23 | 1.55 | 2.41 |
| | | | | | | $\Sigma y^2 =$ | 2899.8 ft$^2$ |

[a]Angle $A = 42.21° = 0.73674$ rad; $\sin A = 0.671875$; $\cos A = 0.740665$.

Balanced snow load:

$$R_{V(left)} = R_{V(right)} = w_{SL}\frac{L}{2} = \frac{(20)(17)(86)}{2} = 14,620 \text{ lb}$$

Unbalanced snow load (full snow load on right half span, no snow load on left half span): Note that some codes may require the combination of full snow load on one half with one-half snow load on the other half, which may be more critical than full unbalanced loading on one half.

By the sum of the moments about the left hinge, $\Sigma M_{Left} = 0$,

$$R_{V(right)} = \frac{3\,w_{SL}L}{8} = \frac{(3)(20)(17)(86)}{8} = 10,965 \text{ lb}$$

$$R_{V(left)} = \frac{w_{SL}L}{2} - R_{V(right)} = \frac{(20)(17)(86)}{2} - 10,965 = 3655 \text{ lb}$$

Wind load acting from the left:

$$\text{Arch rise} = \frac{L}{2}\tan\frac{A}{2} = 43 \tan\frac{42.21°}{2} = 16.60 \text{ ft}$$

Assume arch is 20 in. deep. Add 10 in. for one-half depth of arch and 6 in. for decking and roofing:

$$h = 16.60 + \frac{16}{12} = 17.93 \text{ ft}$$

By the sum of the moments about left hinge, $\Sigma M_{Left} = 0$,

$$R_{V(right)} = \frac{w_{WL}h^2}{2L} = \frac{(10)(17)(17.93)^2}{(2)(86)} = 318 \text{ lb}$$

$$R_{V(left)} = -R_{V(right)} = -318 \text{ lb (acts downward)}$$

Steps 3 and 4. Calculate and tabulate values of $M_s$ and $M_s y$ (see Table 5.20). By sum of the horizontal forces = 0,

$$R_{H(right)} = w_{WL}h = (10)(17)(17.93) = 3050 \text{ lb}$$

5.  Compute horizontal reactions

$$R_H = \frac{\Sigma M_s y}{\Sigma y^2}$$

$$R_{HDL} = \frac{28,241,200}{2899.8} = 9740 \text{ lb}$$

$$R_{HSL} = \frac{53,727,760}{2899.8} = 18,530 \text{ lb}$$

$$R_{HUL} = \frac{26,863,890}{2899.8} = 9260 \text{ lb}$$

## TABLE 5.20
### Summary of Moment Calculations

| Point | Dead Load $M_s$ | Dead Load $M_{s,y}$ | Balanced SL $M_s$ | Balanced SL $M_{s,y}$ | Unbalanced SL $M_s$ | Unbalanced SL $M_{s,y}$ | Wind Load $M_s$ | Wind Load $M_{s,y}$ |
|---|---|---|---|---|---|---|---|---|
| | | | | Simple Beam Moments (ft-lb) | | | | |
| **Left** | | | | | | | | |
| 1 | 13,850 | 21,460 | 25,340 | 39,280 | 6,480 | 10,040 | 767 | 1,190 |
| 2 | 40,690 | 181,060 | 75,140 | 334,370 | 20,090 | 89,420 | 3,430 | 15,260 |
| 3 | 65,820 | 465,320 | 122,520 | 866,220 | 34,440 | 243,500 | 7,240 | 51,190 |
| 4 | 88,750 | 834,270 | 166,490 | 1,564,990 | 49,450 | 464,800 | 11,810 | 110,980 |
| 5 | 109,300 | 1,247,100 | 206,120 | 2,351,780 | 65,040 | 742,090 | 16,720 | 190,740 |
| 6 | 126,920 | 1,663,870 | 240,570 | 3,153,860 | 81,140 | 1,063,770 | 21,660 | 283,950 |
| 7 | 141,300 | 2,046,060 | 269,050 | 3,895,870 | 97,650 | 1,413,950 | 26,310 | 380,910 |
| 8 | 152,410 | 2,363,830 | 290,980 | 4,513,080 | 114,480 | 1,775,660 | 30,400 | 471,420 |
| 9 | 159,770 | 2,589,900 | 305,860 | 4,957,940 | 131,540 | 2,132,270 | 33,760 | 547,310 |
| 10 | 163,610 | 2,707,730 | 313,380 | 5,186,490 | 148,740 | 2,461,690 | 36,200 | 599,200 |
| **Right** | | | | | | | | |
| 10' | | | | | 164,640 | 2,724,800 | 37,520 | 621,000 |
| 9' | | | | | 174,320 | 2,825,660 | 37,980 | 615,670 |
| 8' | | | | | 176,490 | 2,737,430 | 37,330 | 578,970 |
| 7' | | | | | 171,400 | 2,481,910 | 35,650 | 516,240 |
| 6' | | | | | 159,430 | 2,090,080 | 32,910 | 431,460 |
| 5' | | | | | 141,080 | 1,609,700 | 29,130 | 332,350 |
| 4' | | | | | 117,040 | 1,100,190 | 24,360 | 228,950 |
| 3' | | | | | 88,080 | 622,730 | 18,560 | 131,200 |
| 2' | | | | | 55,050 | 244,960 | 11,820 | 52,590 |
| 1' | | | | | 18,870 | 29,240 | 4,160 | 6,450 |

$\Sigma M_{s,y} = 14,120,600$
$\times 2 = 28,241,200$

$\Sigma M_{s,y} = 26,863,800$
$\times 2 = 53,727,600$

$\Sigma M_{s,y} = 26,863,890$

$\Sigma M_{s,y} = 6,167,030$

$$R_{\text{HWL(left)}} = \frac{6,164,390}{2899.8} = 2130 \text{ lb (acting to left)}$$

$$R_{\text{HSL(right)}} = w_{\text{WL}}h - R_{\text{HWL(left)}}$$

$$= (10)(17)(17.93) - 2130 = 920 \text{ lb (acting to left)}$$

6. Actual moment values are tabulated in Table 5.21.

7. Determine axial thrust. From step 6, the point of maximum moment occurs at point 5 under dead load on full span and full snow load on the right half span (unbalanced snow load). At point 5,

$$P' = [R_{\text{HDL(left)}} + R_{\text{HUL(left)}}] \cos 23.22°$$

$$+ [R_{\text{VDL(left)}} + R_{\text{VUL(left)}} - \Sigma wx] \sin 23.22°$$

$$\Sigma wx = Rw_{\text{DL}}(A - \alpha_n)$$

$$= (64)(10)(17)(0.73674 - 0.40520) = 3610 \text{ lb}$$

**TABLE  5.21**

**Actual Moment Values**

| | | | | Actual Moments (ft-lb) | | | |
|---|---|---|---|---|---|---|---|
| Point | DL | SL | $\frac{1}{2}$SL | WL | DL + SL | DL + $\frac{1}{2}$SL | DL + WL |
| Left | | | | | | | |
| 1 | −1250 | −3380 | −7880 | +2530 | −4630 | −9130 | +1280 |
| 2 | −2650 | −7320 | −21,130 | +6040 | −9970 | −23,780 | +3390 |
| 3 | −3040 | −8490 | −31,060 | +7800 | −11,530 | −34,100 | +4760 |
| 4 | −2810 | −7690 | −37,630 | +8190 | −10,500 | −40,440 | +5380 |
| 5 | −1830 | −5310 | −40,660 | +7550 | −7140 | −42,490 | +5720 |
| 6 | −770 | −2360 | −40,310 | +6230 | −3130 | −41,080 | +5460 |
| 7 | +260 | +740 | −36,490 | +4490 | +1000 | −36,230 | +4750 |
| 8 | +1340 | +3580 | −29,200 | +2590 | +4920 | −27,860 | +3930 |
| 9 | +1880 | +5490 | −18,630 | +710 | +7370 | −16,750 | +2590 |
| 10 | +2410 | +6710 | −4580 | −1000 | +9120 | −2170 | +1410 |
| Right | | | | | | | |
| 10′ | +2410 | +6710 | +11,320 | −2320 | +9120 | +13,730 | +90 |
| 9′ | +1880 | +5490 | +24,150 | −3500 | +7370 | +26,030 | −1620 |
| 8′ | +1340 | +3580 | +32,800 | −4340 | +4920 | +34,140 | −3000 |
| 7′ | +260 | +740 | +37,260 | −4850 | +1000 | +37,520 | −4590 |
| 6′ | −770 | −2360 | +37,980 | −5020 | −3130 | +37,210 | −5790 |
| 5′ | −1830 | −5310 | +35,380 | −4860 | −7140 | +33,550 | −6690 |
| 4′ | −2810 | −7690 | +29,960 | −4360 | −10,500 | +27,150 | −7170 |
| 3′ | −3040 | −8490 | +22,580 | −3520 | −11,530 | +19,540 | −6560 |
| 2′ | −2650 | −7320 | +13,830 | −2350 | −9970 | +11,170 | −5000 |
| 1′ | −1250 | −3380 | +4510 | −860 | −4630 | +3260 | −2110 |

$$P' = (9740 + 9260)(0.91902)$$
$$+ (8015 + 3660 - 3610)(0.39421) = 20,640 \text{ lb}$$

8.   Determine section size

$$A = \frac{P'}{F_c^*} = \frac{20,640}{1900} = 10.86 \text{ in.}^2$$

$$S = \frac{M}{F_b^*} = \frac{(42,490)(12)}{(2740)} = 186 \text{ in.}^3$$

Try a $5\frac{1}{8} \times 16\frac{1}{2}$-in. section, $A = 84.56 \text{ in.}^2$, $S = 232.5 \text{ in.}^3$

9.   Check for combined stresses

(a)   Compression: Consider buckling about the $x$-$x$ axis. From step 6, point of contraflexure for this loading condition occurs near the centerline of the arch. Chord distance between base of arch to center of arch is

$$\frac{L/2}{\cos(A/2)} = 46.1 \text{ ft} = 553 \text{ in.}$$

$$\left(\frac{l_e}{d}\right)_x = \frac{553}{16.5} = 33.5$$

Consider buckling between purlins equally spaced at 9.57 ft, $l_u = 114.8$ in. A value of $K_e$ slightly less than unity could be justified because of continuity of support. However, $K_e = 1$ was used in the example.

$$l_e = K_e l_u = (1)(114.8) = 114.8 \text{ in.}$$

$$\left(\frac{l_e}{d}\right)_y = \frac{114.8}{5.125} = 22.4 < 33.5 \quad (l_e/d)_x \text{ governs}$$

$K_{cE}$ for glued laminated timber = 0.418.

$$c = 0.9$$

$$F'_{cE} = \frac{K_{cE}E'}{(l/d)^2} = \frac{0.418(1,800,000)}{(33.5)^2} = 670 \text{ psi}$$

$$C_P = \frac{1 + F_{cE}/F_c^*}{2c} - \sqrt{\left[\frac{1 + F_{cE}/F_c^*}{2c}\right]^2 - \frac{F_{cE}/F_c^*}{c}}$$

$$C_P = \frac{1 + 670/1900}{(2)(0.9)} - \sqrt{\left[\frac{1 + 670/1900}{(2)(0.9)}\right]^2 - \frac{670/1900}{0.9}} = 0.336$$

$$F'_c = F_c^* C_P = (1900)(0.336) = 638 \text{ psi}$$

$$f_c = \frac{P'}{A} = \frac{20,640}{84.56} = 244 \text{ psi}$$

Determine volume factor $C_V$

$$C_V = \left(\frac{5.125}{5.125}\right)^{1/10} \left(\frac{12}{16.5}\right)^{2/10} = 0.938$$

$$F_{bx}^* (1 - C_V) = 2740(1 - 0.938) = 169 \text{ psi} < f_c \text{ (244 psi)}$$

Volume factor is not applicable.

Distance between purlins = 114.8 in.

(b)  Bending (check for buckling about $y$–$y$ axis):

For a beam with four equal loads (Table 5.3).

$$l_e = 1.68l_u = (1.68)(114.8) = 192.9 \text{ in.}$$

$$R_B = \sqrt{\frac{l_e d}{b^2}} = \sqrt{\frac{(192.9)(16.5)}{5.125^2}} = 11.01$$

$K_{bE} = 0.609$ [from Eq. (5-10)]

$$F_{bE} = \frac{K_{bE} E_y'}{R_B^2} = \frac{(0.609)(1,600,000)}{(11.01)^2} = 8040 \text{ psi}$$

$$C_L = \frac{1 + F_{bE}/F_b^*}{1.9} - \sqrt{\left[\frac{1 + F_{bE}/F_b^*}{1.9}\right]^2 - \frac{F_{bE}/F_b^*}{0.95}}$$

$$C_L = \frac{1 + 8040/2740}{1.9} - \sqrt{\left[\frac{1 + 8040/2740}{1.9}\right]^2 - \frac{8040/2740}{0.95}} = 0.976$$

$$F_{bx}' = F_{bx}^* C_L = (2740)(0.976) = 2672 \text{ psi}$$

$$f_{bx} = \frac{M}{S} = \frac{(42,490)(12)}{232.5} = 2193 \text{ psi}$$

$$\frac{f_c}{F_c'} + \frac{f_{bx}}{F_{bx}'} \leq 1$$

$$\frac{244}{638} + \frac{2193}{2672} = 0.38 + 0.82 = 1.20 \text{ (overstressed)}$$

Repeating the same procedure using a $5\frac{1}{8} \times 19\frac{1}{2}$-in. section,

$$C_V = \left(\frac{5.125}{5.125}\right)^{1/10} \left(\frac{12}{19.5}\right)^{2/10} = 0.907$$

$$f_c = \frac{20,640}{99.94} = 207 \text{ psi}$$

$$(F_{bx}^*) (1 - C_V) = 2740(1 - 0.907) = 253 \text{ psi}$$

253 psi > 207 psi Volume factor applies.

When considering volume factor

$$F_b' = F_b^* C_V + f_c = 2740(0.907) + 207 = 2692 \text{ psi}$$

Check for $F_b'$ determined by lateral stability factor, the lower value controls. From Table 5.3 for a beam with four equal loads

$$l_e = 1.68 l_u = (1.68)(114.8) = 192.9 \text{ in.}$$

$$R_B = \sqrt{\frac{l_e d}{b^2}} = \sqrt{\frac{(192.9)(19.5)}{5.125^2}} = 11.97$$

$$K_{bE} = 0.609 \text{ [from Eq. (5-10)]}$$

$$F_{bE} = \frac{(0.609)(1,600,000)}{11.97^2} = 6801 \text{ psi}$$

$$C_L = \frac{1 + 6801/2740}{1.9} - \sqrt{\left(\frac{1 + 6801/2740}{1.9}\right)^2 - \frac{6801/2740}{0.95}} = 0.969$$

$$F_{bx}' = F_b^* C_L = (2740)(0.969) = 2655 \text{ psi} < 2692 \text{ psi}$$

Lateral stability ($C_L$) controls.

$$f_b = \frac{M}{S} = \frac{(42,490)(12)}{324.8} = 1570 \text{ psi}$$

Determine $F_c'$

$$A = 99.94 \text{ in.}^2 \quad S = 324.8 \text{ in.}^3$$

$$I = 3167 \text{ in.}^4$$

$$\left(\frac{l_e}{d}\right)_x = \frac{553}{19.5} = 28.4, \quad \left(\frac{l_e}{d}\right)_y = 22.4, \quad \left(\frac{l_e}{d}\right)_x \text{ controls}$$

$$K_{cE} = 0.418 \text{ [from Eq. (5-14)]}$$

$$c = 0.9$$

$$F_{cE} = \frac{K_{cE} E_x'}{(l/d)^2} = \frac{(0.418)(1,800,000)}{(553/19.5)^2} = 936 \text{ psi}$$

$$C_P = \frac{1 + 936/1900}{(2)(0.9)} - \sqrt{\left[\frac{1 + 936/1900}{(2)(0.9)}\right]^2 - \frac{936/1900}{0.9}} = 0.454$$

$$F_c' = F_c^* C_P = (1900)(0.454) = 863 \text{ psi}$$

$$\frac{207}{863} + \frac{1570}{2655} = 0.83 < 1.0 \qquad \text{O.K.}$$

10.   Check for shear. Maximum horizontal and vertical reactions occur under dead load plus full snow load; therefore, check shear at the base for this condition.

$$V = (R_{\text{VDL}} + R_{\text{VSL}}) \cos A - (R_{\text{HDL}} + R_{\text{HSL}}) \sin A$$

$$= (8015 + 14{,}620)(0.740665) - (9740 + 18{,}530)(0.671875)$$

$$= -2230 \text{ lb (minus sign indicates opposite direction)}$$

$$f_{\text{v}} = \frac{3V}{2A} = \frac{(3)(2230)}{(2)(99.94)} = 33 \text{ psi} < 190 \text{ psi} \qquad \text{O.K.}$$

At the crown of the arch, the point of maximum shear occurs under dead load plus unbalanced snow load.

$$V = R_{\text{VDL}} + R_{\text{VUL(left)}} - \Sigma w_{\text{DL}} x \text{ (snow load on right)}$$

$$\Sigma w_{\text{DL}} x = (64)(10)(17)(0.73674) = 8015 \text{ lb}$$

$$V = 8015 + 3665 - 8015 = 3665 \text{ lb}$$

$$f_{\text{v}} = \frac{3V}{2A} = \frac{(3)(3665)}{(2)(99.94)} = 55 \text{ psi} < 190 \text{ psi} \qquad \text{O.K.}$$

Check $d/b$ ratios with new size. Deduct the depth of the purlin from the depth of the arch.

$$\frac{d}{b} = \frac{19.5 - 11.25}{5.125} = 1.61 < 5 \qquad \text{O.K.}$$

Check for column buckling between purlins. Purlins are equally spaced at 9.57 ft = 114.8 in.

$$\frac{d}{b} = \frac{19.5}{5.125} = 3.80 < 5 \qquad \text{O.K.}$$

$$d = 5.125 \text{ in. (width is smaller, so it is used)} \qquad (l_{\text{e}}/d)_{\text{y}} = \frac{114.8}{5.125} = 22.4$$

$$F_{\text{cE}} = \frac{K_{\text{cE}} E_{\text{y}}'}{(l/d)_{\text{y}}^2} = \frac{(0.418)(1{,}600{,}000)}{(22.4)^2} = 1333 \text{ psi}$$

$$C_{\text{P}} = \frac{1 + (1333/1900)}{(2)(0.9)} - \sqrt{\left[\frac{1 + 1333/1900}{(2)(0.9)}\right]^2 - \frac{1333/1900}{0.9}} = 0.607$$

$$F_{\text{c}}' = F_{\text{c}}^* C_{\text{P}} = (1900)(0.607) = 1154 \text{ psi}$$

$$1154 \text{ psi} > 863 \text{ psi} \qquad \text{O.K.}$$

11.   Check radial stress.

$$R = \text{radius of curvature} = (64)(12) = 768 \text{ in.}$$

Maximum moment is negative; therefore, radial stress is compressive.

$$F'_{rc} = F_{c\perp} = 560 \text{ psi}$$

$$f_{rc} = \frac{3M}{2RA} = \frac{(3)(42,490)(12)}{(2)(768)(99.94)} = 10 \text{ psi} < 560 \text{ psi} \quad \text{O.K.}$$

Maximum positive moment occurs at point $7' = 37,520$ ft-lb (DL $+ \frac{1}{2}$SL)

$$F'_{rt} = (15)(1.15) = 17.25 \text{ psi}$$

$$f_{rt} = \frac{3M}{2RA} = \frac{(3)(37,520)(12)}{(2)(768)(99.94)} = 8.8 \text{ psi} < 17.25 \text{ psi} \quad \text{O.K.}$$

12.    Determine vertical deflection at the centerline under unbalanced snow load. Using the layout from step 1 of this example, $s = 56.58$ in. Moments due to unbalanced snow loads were determined in step 6 and are retabulated. Moments due to the unit load are calculated in the same manner. They are also tabulated as in Table 5.18.

$$R_{H(\text{left})} = \frac{\Sigma M_s y}{\Sigma y^2} = \frac{2840.8}{2899.8} = 0.97965$$

$$m = m_s - R_{H(\text{left})} y$$

A summary of calculations for determining $\Sigma Mm$ is shown in Table 5.22. At centerline,

$$\Delta_a = \frac{s}{EI} \Sigma Mm = \frac{(56.58)(184,900)(144)}{(1,800,000)(3167)} = 0.26 \text{ in.}$$

## 5.14   MOMENT SPLICES

The use of glued laminated timber has reduced the need for moment splices in timber beams and girders, but such splices are sometimes used in arches and rigid frames where transportation requirements make the transportation of full-size structural frames impractical or uneconomical. Moment splices should be located at points of minimum moment whenever possible. They must be designed to resist bending moment, shear, stress reversal, uplift, and axial forces by various means, such as tension or compression straps located on the sides or edges and compression plates between members. A means of holding the two sections in alignment must also be provided.

Close tolerances are required in the fabrication of moment splices. Consideration should be given to inelastic as well as elastic deformation in the joint. For sawn lumber sections, and for many smaller glued laminated timbers, the simplest form of moment connection is a plywood splice plate (or pair of plates). Such plates can be glued or nailed but are also frequently used in conjunction with bolts and shear connectors.

**TABLE 5.22**

Summary of Deflection Calculations

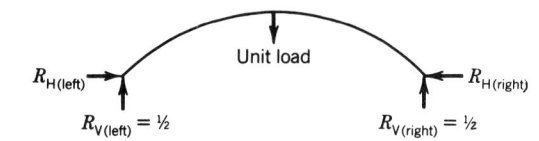

| Point | Unit Load $m_s$ | $m_s y$ | $m$ | Moments $M$<br>DL + $\frac{1}{2}$ SL (ft-lb) | $Mm$ |
|---|---|---|---|---|---|
| **Left** | | | | | |
| 1 | 0.885 | 1.3718 | −0.63344 | − 9,130 | +5,780 |
| 2 | 2.745 | 12.215 | −1.61440 | −23,780 | +38,390 |
| 3 | 4.705 | 33.264 | −2.22105 | −34,100 | +75,740 |
| 4 | 6.755 | 63.497 | −2.4536 | −40,440 | +99,220 |
| 5 | 8.885 | 101.38 | −2.29270 | −42,490 | +97,420 |
| 6 | 11.085 | 145.32 | −1.75808 | −41,080 | +72,220 |
| 7 | 13.335 | 193.16 | −0.84518 | −36,230 | +30,620 |
| 8 | 15.64 | 242.58 | +0.44579 | −27,860 | −12,420 |
| 9 | 17.97 | 291.29 | +2.0900 | −16,750 | −35,010 |
| 10 | 20.32 | 336.30 | +4.1070 | −2,170 | −8,910 |
| | | 1420.38 | | | |
| | | | | | |
| **Right** | | | | | |
| 10′ | | | +4.1070 | +13,730 | +56,390 |
| 9′ | | | +2.0900 | +26,030 | +54,400 |
| 8′ | | | +0.44579 | +34,140 | +15,220 |
| 7′ | | | +0.84518 | +37,520 | −31,710 |
| 6′ | | | −1.75808 | +37,210 | −65,420 |
| 5′ | | | −2.2927 | +33,550 | −76,920 |
| 4′ | | | −2.4536 | +27,150 | −66,620 |
| 3′ | | | −2.22105 | +19,540 | −43,400 |
| 2′ | | | −1.61440 | +11,170 | −18,030 |
| 1′ | | | −0.63344 | +3,260 | −2,060 |
| | $\Sigma m_s y = (2)(1420.38)$ | | | | $\Sigma Mm = +184,900$ |
| | $= 2840.8$ | | | | |

A typical moment splice for glued laminated timber is illustrated in the construction details in AITC 104 (15) in Chapter 8. In this splice, compression stress is taken in end bearing between the two sections. If the unit stress for end grain in bearing parallel to grain $f_g$ as determined by the loading conditions and splice location exceeds 75 % of the design value in end grain bearing $F_g'$, a snug fitting metal bearing plate not thinner than twenty gage and not deeper than one-half the depth of the member should be installed between the abutting ends. See Annex A, AITC 117—Design (1) for design values for end grain in bearing.

Tension stress is taken across the splice by means of steel straps and shear plates as illustrated or by wood splice plates and split rings. Connector capacity must be adequate to carry the tension stress. Additional side straps and shear plates are required to keep the sides and top of the members in position and to take the stress reversals from erection loads or wind uplift. Shear connections must be provided to transfer the maximum design shear through the splice connection.

Consideration should be given to the use of separate side plates for each row of connectors in order to minimize secondary stresses due to shrinkage effects in the member. See Chapter 7, "Fastenings and Connections," for spacing of rows.

The design of moment-resisting splices of this type is affected by the inelastic slip in the connections that can modify the usual assumptions on which elastic design with materials of different moduli of elasticity are based. The recommendations in this section consider this inelastic slip. If the assumption is made that the force on the compression side of the member is resisted entirely by the wood and the tension force is resisted by the steel splice plates and no inelastic slip exists between the wood and steel, a transformed section can be assumed as shown.

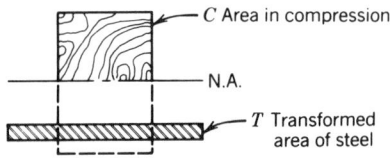

When the moment is very small compared to the compressive force (less than $Pd/6$) where $P$ is the axial thrust and $d$ is the depth of the member at the splice, the entire cross section is in compression (see Fig. 5.52). This condition may exist on circular or parabolic shaped arches which are subjected to uniform loads only. When wind loads or unbalanced loads are taken into account, the moment increases and the resultant of the compressive forces moves away from the neutral axis. The area under compression starts to decrease when the resultant has moved more than one-sixth of the depth of the member from the neutral axis. The moment and the axial thrust can be shown as an equivalent loading of $P'e$ and $P'$ where $P'$ is equal to the axial thrust and $P'e$ is equivalent to the moment (see Fig. 5.51).

When the ratio of the moment to axial thrust is high, $e$ may be greater than $d/2$ and $P'$ acts outside of the cross section as shown (see Fig. 5.51). Each loading condition should be examined so that the different ratios of bending moment to axial thrust can be checked.

In case (*a*) and some case (*b*) conditions (Fig. 5.52), the structure would be stable provided that its members are held in line, a method for transferring shear forces is provided, and end grain bearing is not exceeded. However, in both cases, a splice is needed to provide continuity so that the assumption of a continuous

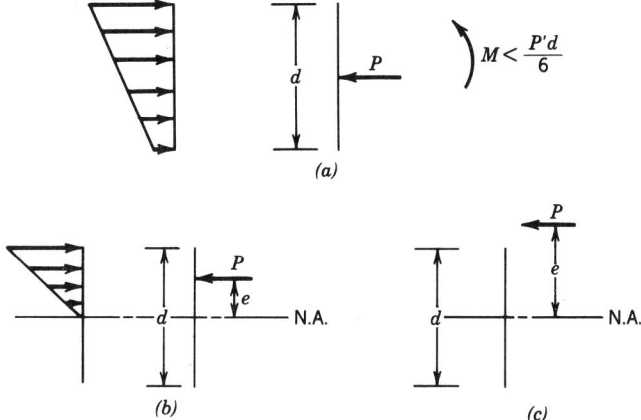

**FIGURE 5.52** Forces of splice.

member in a two-hinged arch design is validated. In case (*c*), the structure is unstable and a moment splice is required for stability as well as continuity.

The inelastic slip in mechanical fastenings, crushing of wood fibers in end grain bearing, and inaccuracies in fabrication result in errors in location of the neutral axis, and some designers use arbitrary methods of locating the area under compression such as using the entire area above the neutral axis. Others prefer to use the transformed section approach.

A conservative approach is to design the splice so that it can resist the entire applied moment. The tension force in the splice is taken by a steel strap and the compressive force by end bearing of the wood. As shown in Fig. 5.53, the maximum end grain in bearing, $f_g$, is the summation of the axial compressive stress $f_c$ and the compressive stress in bending caused by the splice $f_b$.

Another method is to consider the compression stress and bending stress assuming the splice is a cross section of the member.

The combined stresses in Fig. 5.54(*c*) show that a small triangle of tension stresses would exist in a solid cross section. In a splice, the resultant of these stresses can be replaced with a steel strap to resist the tension force. The tension force $T$ is calculated based on using the steep strap fastened $\frac{2}{3}$ of the distance $c$ between the neutral surface and the face of the member. However, it can be placed closer to the bottom face if space permits, or placed on the face.

Moment-resisting splices can also be designed so that the compression force

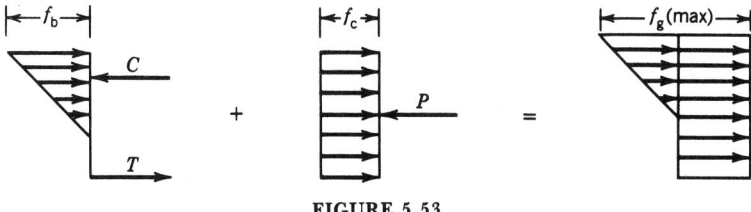

**FIGURE 5.53**

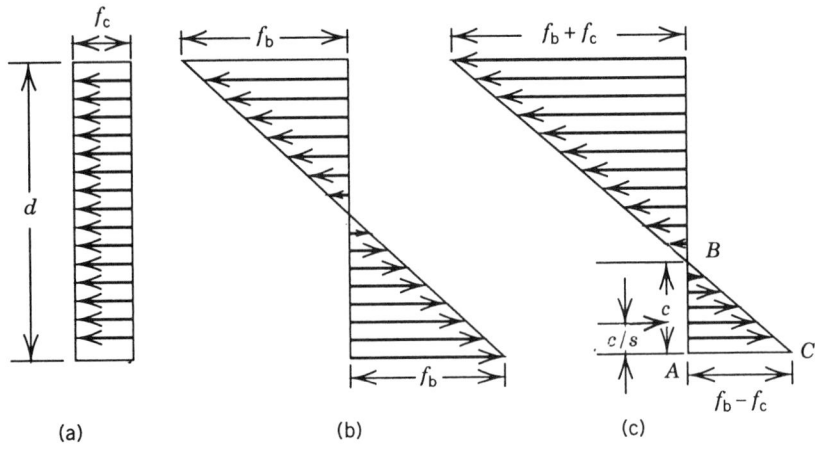

(a)          (b)          (c)

is carried by splice plates, in which case the compression splice plates must be designed as compression members. These types require more material and fabrication than those where the ends of the member abut in compression. However, the location of the resultant of the compressive force is more precise.

Inelastic slip in the tension splice plates can be offset by using a tension-type splice with bolts and nuts that can be tightened to eliminate the slip, as shown in Fig. 5.55. These also require extra materials and fabrication, but the location of forces to resist the moment can be determined more accurately with this system and the slippage that occurs in the tension connection can be compensated for by tightening the bolts.

Another type of splice can be used with the splice plates on the bottom of the

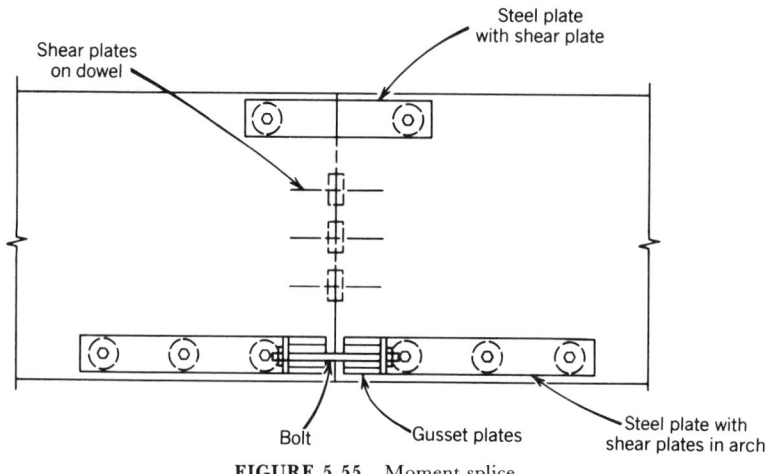

**FIGURE 5.55**    Moment splice.

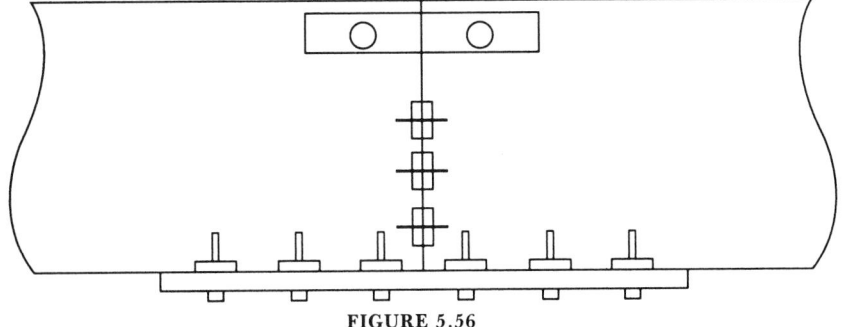

**FIGURE 5.56**

arch as shown in Fig. 5.56. This places the tension force $T$ resisting the moment farther from the compression force $P$ creating a longer moment arm $x$, producing greater efficiency.

Also, the shear plates are located in the bottom lamination that is quite often dense wood, depending on the species and combination used. The shear plate design values are higher for dense Southern Pine and Douglas Fir-Larch.

**Example.** Design a moment splice of the type illustrated in *Typical Construction Details*, AITC 104 (15), for a two-hinged arch with a radius of 60 ft. At the location of the splice, positive moment is 255,000 in.-lb caused by DL + SL. The maximum axial compression is 27,000 lb caused by DL + SL and the maximum shear is 5200 lb caused by DL + SL (on half the span). The maximum negative moment is 15,000 in.-lb caused by DL + WL. The glued laminated timber is combination 24F–V3 Southern Pine.

$$C_D = 1.15 \text{ (snow load)} \quad C_D = 1.6 \text{ (wind load)}$$

Calculate design value $F'_g$ for snow loading:

$$E_x = 1,800,000 \text{ psi}$$

$$F'_g = (2300)(1.15) = 2700 \text{ psi}$$

The section size is 5 in. $\times$ $17\frac{7}{8}$ in.

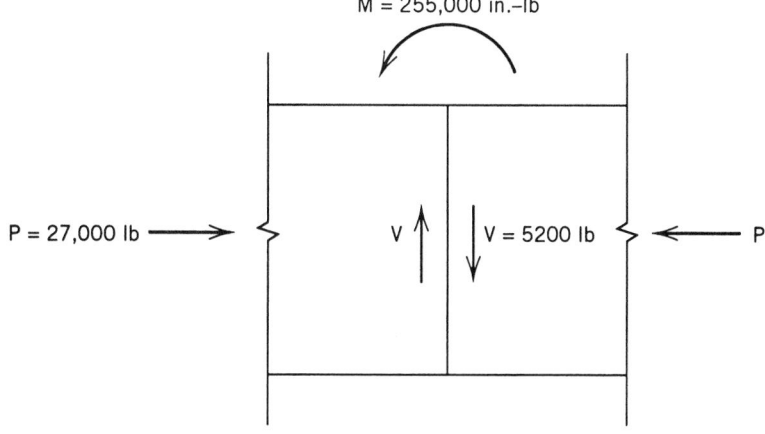

Assume the member at the point of the splice is a continuous member and determine $f_b$ and $f_c$

$$f_c = \frac{P}{A} = \frac{27,000}{89.38} = 302 \text{ psi}$$

$$f_b = \frac{M}{S} = \frac{255,000}{266.3} = 958 \text{ psi}$$

The combination of stresses is illustrated in Fig. 5.57. The neutral surface is 6.12 in. above the bottom of the member. The tension force created in triangle *ABC* is

$$T = \frac{(6.12)(656)}{2} = 2006 \text{ lb}$$

This triangle of force can be replaced by 2006 lb tension force in a steel strap fastened to the member $2\frac{1}{8}$ in. above the face with appropriate fasteners. Load required to be resisted by one tension strap $2\frac{1}{8}$ in. from the bottom is $2006/2 = 1003$ lb.

Several types of fasteners can be used—bolts, lag screws, or shear plates. From Table 7.18, one $\frac{5}{8}$ in. diameter bolt with $\frac{1}{4}$-in. steel side plates has a capacity of 2350 lb.

$$Z' = (2350)(1.15) = 2702 \text{ lb} > 1003 \text{ lb}$$

From Table 7.23, a $\frac{5}{8}$-in. diameter lag screw has a capacity of 1190 lb.

$$P = 5\text{-in.} - E - \tfrac{1}{4}\text{-in.} = 5\text{-in.} - \tfrac{3}{8}\text{-in.} - \tfrac{1}{4}\text{-in.} = 4\text{-}\tfrac{3}{8} \text{ in.}$$

$$C_d = \frac{P}{8D} = \frac{4.375}{(8)0.625} = 0.875$$

$$Z' = ZC_DC_d = (1190)(1.15)(0.875) = 1197 \text{ psi}$$

If lag screws are used in the side face, the opposing lag screws on the opposite face should be offset by the parallel-to-grain spacing of $4D = 2\frac{1}{2}$ in.

If the steel strap is placed on the face as shown in Fig. 5.56, a $\frac{5}{8}$ in. × 6 in. lag screw could be used. This also results in greater efficiency of the steel strap.

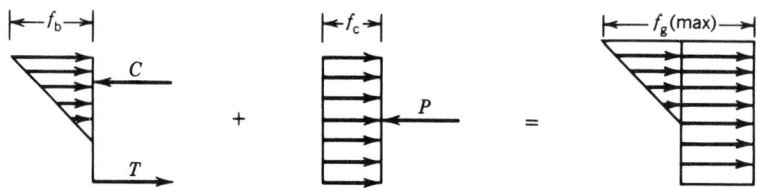

**FIGURE 5.57** Combination of stresses.

The net area of the steel strap required is

$$A_n = \frac{1197}{22,000} = 0.05 \text{ in.}^2$$

Assume one $\frac{5}{8}$ in. diameter bolt and a steel strap $\frac{1}{4}$ in. thick will be used.

Try a 2 in. $\times$ $\frac{1}{4}$ in. steel strap which provides adequate edge distance for a $\frac{5}{8}$ in. diameter bolt.

See *AISC Manual of Steel Construction* (16) for design of the steel.

$$A_s = (2)(0.25) - \left(\frac{11}{16}\right)(0.25) = 0.328 \text{ in.}^2$$

$$(0.328)(22,000) = 7216 \text{ lb} > 1197 \text{ lb} \qquad \text{O.K.}$$

The moment resisting splice requires two 2 in. $\times$ $\frac{1}{4}$ in. steel straps on each side of the splice with a $\frac{5}{8}$-in. diameter bolt located 2 in. from the face. If unusual handling stresses are anticipated, two bolts can be used on each side of the strap.

To resist stress reversal and provide resistance to handling stresses, a second strap is placed 2 in. from the opposite face. Use a 2 in. $\times$ $\frac{1}{4}$ in. steel strap and $\frac{5}{8}$ in. diameter bolt for this connection also.

Determine the shear resistance requirements. Maximum shear occurs with full dead load and $\frac{1}{2}$ snow load.

For a $2\frac{5}{8}$-in. shear plate in end grain, use 60% of the value tabulated for perpendicular to grain loading in Table 7.33 adjusted for duration of loading. Edge distance is $2\frac{1}{2}$ in., which is greater than the $1\frac{3}{4}$-in. minimum edge distance for full load.

$$C_{De} = 1.0$$

$$P' = (1990)(1.15)(0.60) = 1370 \text{ lb}$$

Try four $2\frac{5}{8}$-in. diameter shear plates in a row loaded perpendicular to grain. Minimum spacing for full allowable load perpendicular to grain is $4\frac{1}{4}$ in.

$A_s = A_m$ (both are greater than 64 in.$^2$)
$A_1 / A_2 = 1.0$
Adjustment factor $C_g = 0.99$ (from Table 7.4)
$P' = (1370)(0.99) = 1356 \text{ lb}$ (average allowable load per connector)
Number of shear plates required $= V/P' = 5200/1356 = 3.83$, use 4.
Percentage $= 3.83/4 (100) = 96\%$
Use $4\frac{1}{4}$-in. spacing.
Use loaded edge distance $= 3\frac{1}{16}$ in. $> 2\frac{3}{4}$ in. $C_{\Delta e} = 1.0$
Dowel length—Use standard penetration required for lag screw used with a shear plate as shown in Table 7.30 and add 1 in. on each end.
Standard penetration $= 5$ times lag screw diameter.
Length of dowel $= [(5)(0.625) + 1](2) = 8.25 \text{ in.}$

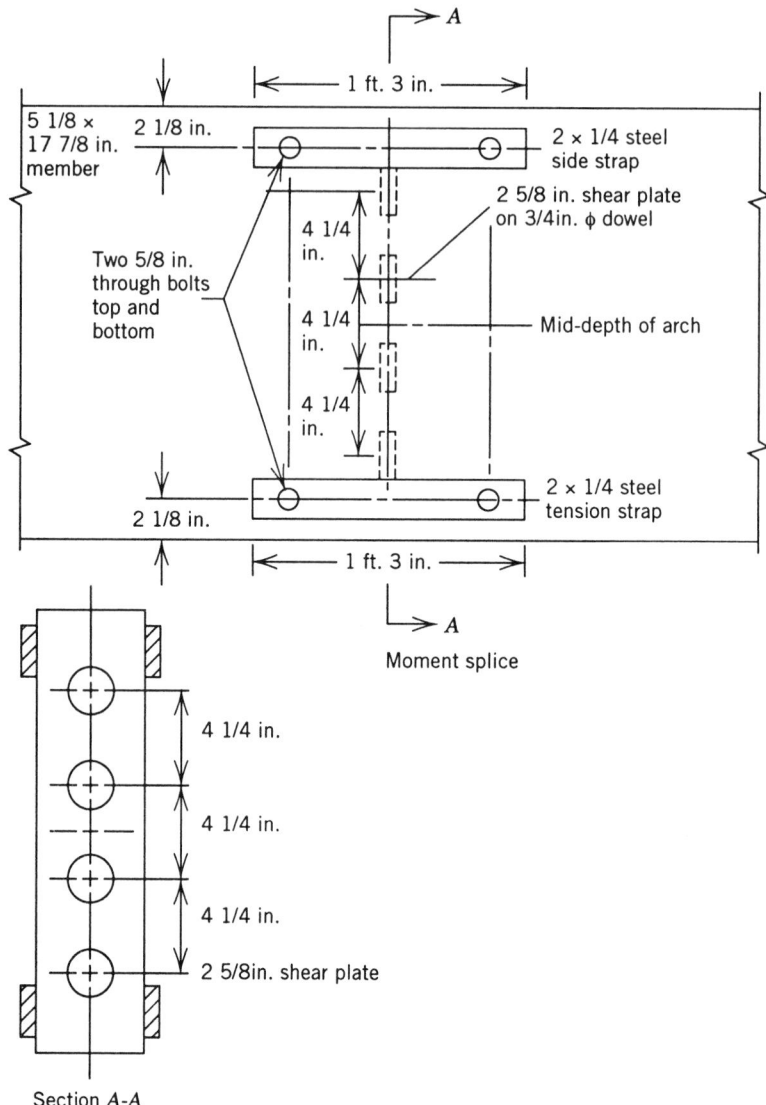

Moment splice

Section A-A

The preceding example illustrated the principle involved in designing a moment splice of the type used for the midpoint of a two-hinged arch where the ratio of the bending stress to the compressive stress is typically small and the tension force to be resisted is also small. In such cases, judgment is necessary to design the splice to resist handling stresses. In other cases, such as designing a splice for a tudor-type arch where the splice is located away from the crown, the ratio of bending stress to compressive stress is relatively high, resulting in a large tension force to be resisted. In this case, the splice designed to develop the moment is usually sufficient to resist handling stresses as well.

## 5.15 PONDING ON ROOFS

When there is the possibility of water ponding, which may cause excessive loads and additional progressive deflection, each component of the roof system, including decking, purlins, beams, girders, or other principal structural supports, should be designed accordingly. Continuous or cantilevered components should be designed for balanced or unbalanced load, whichever produces the more critical condition. Adequate drainage capacity and proper construction details should be provided.

Ponding is usually a greater problem on roofs with long-span structural members than for roofs with short-span beams because the "spring constant" of a long span system is likely to be larger. The spring constant of a roof can be expressed as the deflection in inches per arbitrary unit of load. It is convenient to express spring constant as inches of deflection per 5 lb of load per square foot because the weight of a 1 in. depth of water is approximately equal to a load of 5 psf. (The actual weight of a 1 in. depth of water is 5.2 psf.) Roofs with large tributary drainage areas are, in general, more susceptible to ponding than roofs with smaller areas.

Ponding problems may occur in all parts of the country. They appear to be greater in areas where small design live loads are used. Ponding failures have occurred in semiarid regions because rainstorms of high intensity occur, although the annual rainfall in these regions is small. Problems have also been found in colder climates. Roofs in these areas are generally designed for large snow loads, which reduces the spring constant of the system, but unusual weather conditions can occur that may result in the blockage of drain paths by packed snow or ice. In such cases, the ponding load must be added to that of the snow or ice on the roof.

The most effective method of preventing problems due to ponding is to provide adequate roof slope and drainage (see Table 4.5). It is recommended that roof surfaces have a minimum residual slope or curvature of not less than $\frac{1}{4}$ in./ft of horizontal distance between the level of the drain and any point of the roof to minimize the ponding of water. The minimum slope should remain after long-term deflection of the members has taken place. In most cases, this should provide for adequate drainage. In selecting the roof slope, the designer should take into account the unevenness of the type of roofing used, the combined effect of all members within the framing system, and the possibility of strong winds blowing the water up-slope and the effect of member shrinkage on drainage. The designer should also make sure that all elements of the roof maintain the minimum slope. This is sometimes overlooked in side or end bays where uncambered members are parallel to the slope.

Many ponding problems are caused by inadequate drainage. Consideration should be given to providing adequate drainage to handle the maximum intensity rainfall expected. (Maximum storm intensity for 50- or 100-yr storms are commonly used.) The location of drains is critical. They should be located away from points that remain at high levels when the roof deflects, such as at columns. Quite often, gravel stops will cause a dam that can prevent drainage. Improper sizing and placement of scuppers and roof drainage can allow water to accumulate to depths not anticipated in the design. Wood members may shrink somewhat in reaching moisture equilibrium in a structure, causing movement of the roof. Therefore, it is preferable to use flexible connections for roof drains so that they may move with the roof. After the roof is completed and the building is in use, an ongoing maintenance program for checking the roof drains should be established. Roof drains should be kept clean of debris or opened if they are frozen over to assure their proper function.

Camber is normally recommended for glued laminated timbers so that they do not appear to be sagging after the application of dead load. Experience has shown that, in order to offset inelastic deflection or ''permanent set'' as well as the calculated deflection, a camber of $1\frac{1}{2}$ times the dead load deflection is appropriate. This camber is intended to provide a relatively straight beam. Where it is used to obtain the required slope to prevent ponding, additional curvature is required. Also, when curvature is used to provide slope for drainage, the middle third of the member may be relatively flat. This condition should be considered in design. Extreme care should also be taken when camber is used to compensate for dead-load deflection in a cantilever beam system. Too little as well as too much camber is to be avoided.

Beam deflection should not exceed the recommended limitations given in Table 4.3. Building codes vary in their requirements for deflection limitations. Deflection limitations should not be considered as being adequate for preventing a potential ponding condition or objectionable vibration. They merely set a minimum stiffness that should be used under certain conditions. The recommended limitations given in Table 4.3 are primarily for appearance purposes. In addition, they provide a reasonable minimum stiffness where ponding or vibration are not problems.

It is strongly recommended that roofs be provided with adequate slope and drainage to prevent ponding. However, if such slope and drainage cannot be provided, the roofs must be designed to carry the added load. The complete analysis and design of a roof system subject to ponding is very time consuming because the deflection of all elements of the roof must be considered. When the deflection of purlins, stiffeners, and roof sheathing is significant, the analysis and design is best accomplished by use of a computer. The examples shown in this section illustrate the calculations involved in the analysis of beams intended to support ponding loads. When roof slope is not adequate to prevent ponding, the designer should use one of the several design methods available to make sure that the roof is stiff enough to prevent the condition. The following are some suggested considerations in the use of these methods.

a.   The moduli of elasticity of wood contained in the tables of design values are average values. All strength properties of wood, including modulus of elasticity, are variable. This variability is expressed by the coefficient of variation $COV_E$, which is a measure of the extent of the variation. ANSI/NF$_o$PA

NDS—1991, *National Design Specification for Wood Construction* (3) published by the American Forest & Paper Association lists the coefficients of variation for modulus of elasticity, $COV_E$, values of wood products as follows:

| | $COV_E$ |
|---|---|
| Visually graded sawn lumber | 0.25 |
| Machine stress-rated sawn lumber (MSR) | 0.11 |
| Glued laminated timber (six or more laminations) | 0.10 |
| Machine evaluated lumber (MEL) | 0.15 |

This variation should be considered when designing for ponding because members with lower $E$ values deflect more, creating the potential for larger ponding loads than would occur by using average $E$ values in calculations. The variations in the modulus of elasticity of wood approach a "normal distribution" from a statistical standpoint. If a designer wishes to know the estimated value of modulus of elasticity at some percentile of the total population, such as the fifth percentile (19 pieces out of 20 will have a higher $E$ than the estimated value), it can be calculated as follows:

$$E_{0.05} = E - 1.645E\ COV_E \qquad (5\text{-}100)$$

where  $E_{0.05}$ = estimated modulus of elasticity at the lower fifth percentile (psi),
  $E$ = average modulus of elasticity (psi),
  $COV_E$ = coefficient of variation of $E$, and
  1.645 = a statistical constant for determination of values at the fifth percentile based on a normal distribution of data (constants for other percentiles may be obtained from standard statistical tables).

**Example.**  The average $E$ of a grade of visually graded sawn lumber is 1,800,000 psi; the $E$ value at the fifth percentile is

$$E_{0.05} = 1,800,000 - (1.645)(1,800,000)(0.25) = 1,060,000 \text{ psi}$$

b.  The inelastic deflection, creep, or permanent set of glued laminated timber averages approximately one-half of the calculated dead-load deflection. This deflection can be offset by camber, as previously stated. The permanent set of unseasoned sawn lumber often used for secondary framing is approximately equal to the calculated dead-load deflection. Therefore, the total long-time deflection is approximately 2 times the calculated initial elastic deformation. The additional ponding load resulting from the permanent set of all members within a particular framing system should be considered in the design analysis.

c.  The deflection of all elements in a roof system should be considered. In a typical panelized wood roof system, this includes the glued laminated timber beams, the purlins, the subpurlins (or stiffeners), and the plywood roof sheathing.

When low-pitched roofs have insufficient slope for a drainage (less than $\frac{1}{4}$ in./ft), the stiffness of supporting members should be such that a 5-psf load will cause no more than $\frac{1}{2}$ in. deflection (17). A simplified approach is based on the assumption that the roof structure between the supporting members is relatively stiff and only a small amount of deflection will occur between supporting members. This method permits some additional load due to deflection of purlins, stiffeners, or sheathing as well as some variation in the modulus of elasticity, and where drainage is adequate and no unusual conditions occur, this rule of thumb works reasonably well.

Where the deflection of the secondary roof framing system between the supporting members is large enough to produce significant ponding loads, a complete analysis should be made. It must be recognized that some water has to accumulate to the start deflection. This accumulation can be from a number of causes—roughness of the surface, blowing winds, high gravel stops, drains placed slightly above the level of the roof, and so on.

**Example.**    A low-pitched roof is supported by simple-span glued laminated timber beams that have spring constants of $\frac{1}{2}$ in. deflection for a uniform load of 5 psf. A $\frac{1}{2}$-in.-high gravel stop is around the perimeter of the roof. No drains are provided. The deflection of a secondary framing system between the beams is small. Assume that $\frac{1}{2}$ in. of water, held by the gravel stops, is spread uniformly over the roof. The load from this water causes $\frac{1}{4}$ in. of deflection, which in turn causes an additional load of water to accumulate, resulting in $\frac{1}{8}$ in. of deflection and so forth until equilibrium is reached at about $\frac{1}{2}$ in. of deflection. The total depth of water at midspan is $\frac{1}{2} + \frac{1}{2} = 1$ in., or approximately a 5-psf load. Thus, a roof system limited to $\frac{1}{2}$ in. of deflection for a 5-psf load should reach equilibrium before the ponding load becomes excessive.

The ponding load caused by a deflection of 1 in. in a member with equal height supports is somewhat less than 5 psf because the load caused by the deflection is not uniform but in the approximate shape of a curve, defined by the equation

$$\Delta = \Delta_{\text{max}} \sin \frac{\pi x}{L} \tag{5-101}$$

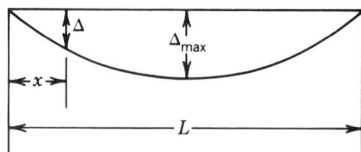

Actual measurements of pools of ponded water on roofs indicate that large variations from the theoretical deflection curve can be expected and that a close approximation can be obtained by assuming the shape of the deflection curve to be parabolic. A parabolic load is roughly equivalent to a uniform load with a depth equal to 0.83 times the depth of the parabola. This method can be used to estimate deflections of individual portions of a roof system.

**Example.**    For the roof system used in the previous example, the deflection of the supporting member is calculated as follows: The $\frac{1}{2}$-in. depth uniform load of water causes a deflection of $\left(\frac{1}{2}\right)(0.5) = 0.25$ in. where the beam deflects $\frac{1}{2}$ in. per inch depth of water. This 0.25-in. deflection in turn results in a parabolic load equivalent to a uniform load with a depth of $(0.83)(0.25) = 0.208$ in., which causes an additional deflection of $(0.208)(0.5) = 0.104$ in. The complete calculations are shown as follows:

Deflection for 0.5 in. uniform depth of water $\qquad\qquad\qquad = 0.25$ in.
Deflection for 0.25 in. parabolic load $\quad = (0.83)(0.25)(0.5) = 0.104$ in.
Deflection for 0.104 in. parabolic load $\quad = (0.83)(0.104)(0.5) = 0.043$ in.
Deflection for 0.043 in. parabolic load $\quad = (0.83)(0.043)(0.5) = 0.018$ in.
Deflection for 0.018 in. parabolic load $\quad = (0.83)(0.018)(0.5) = \underline{0.007}$ in.
$\qquad\qquad\qquad\qquad\qquad\qquad\qquad\qquad\qquad\qquad\qquad\qquad 0.42$ in.

At equilibrium, this results in a total depth of water of 0.92 in. weighing approximately 4.8 psf.

This method can be used to calculate the deflection of all portions of a roof system by using a series of successive calculations.

More accurate calculations can be performed by using computer programs that take into account the deflection of all elements of the roof structure.

### 5.15.1  Magnification Factor

The effect of ponding of water on a low-pitched roof system supported by sawn beams or glued laminated beams without camber is to increase deflections and stresses. This increase can be expressed by the equation

$$M_F = \frac{1}{1 - W'l^3/\pi^4EI} \qquad\qquad (5\text{-}102)$$

where  $M_F$ = factor for multiplying stresses and deflections under existing loads to determine stresses and deflections under existing loads plus ponding and is known as the magnification factor,
$W'$ = total load of 1-in. depth of water on the roof area supported by the beam or deck (lb/in.),
$l$ = span length of beam or deck (in.),
$E$ = modulus of elasticity of beam or deck (psi), and
$I$ = moment of inertia of beam or deck (in.$^4$).

The value of $E$ frequently used in design calculations is the tabular design value, which is an average $E$ of a grade of lumber or laminating combination. If the $E$ at the lower fifth percentile is desired for use in Eq. (5-102), Eq. (5-100) may be used to determine $E_{0.05}$.

It is noted that as the term $W'l^3/\pi^4EI$ approaches unity, the magnification factor approaches infinity and the deflection and bending stresses are increased indefinitely. The derivation of this magnification factor was based on an approximate analysis verified closely by experiment (18). The analysis assumed elastic behavior and did not account for stresses or deflections caused by creep. The factor applies to each element of a roof system, including decking, purlins, beams, and girders. Design values as modified for duration of loading, size or volume factor, and other applicable factors, and deflection limits may not be exceeded after application of the magnification factor to existing stresses and deflections.

As a further illustration of the potential effect of ponding as it relates to increasing bending stress and deflection, Fig. 5.58 is a graphic presentation of the magnification factor as a function of span length, deflection criteria, and ratio of ponding load to design load. By examining this plot, it can readily be seen that for a given span and deflection criteria, roofs designed for a relatively low value of dead load plus applied live load (i.e., a high $W'/W$ ratio, where $W'$ is as previously defined and $W$ is the total design dead load plus live load) will be most susceptible to possible ponding problems.

The use of the magnification factor is a simplified approach that should be used with judgment. It is valid only for roofs and single members that are flat and do not contain camber (disregarding camber gives a conservative answer). It disregards the deflection of secondary members. The magnification factor will increase stresses caused by dead and live loads assumed to be acting at the time ponding can occur. There are no uniformly recognized criteria for determination of loads to be magnified, and the designer should make a reasonable judgment

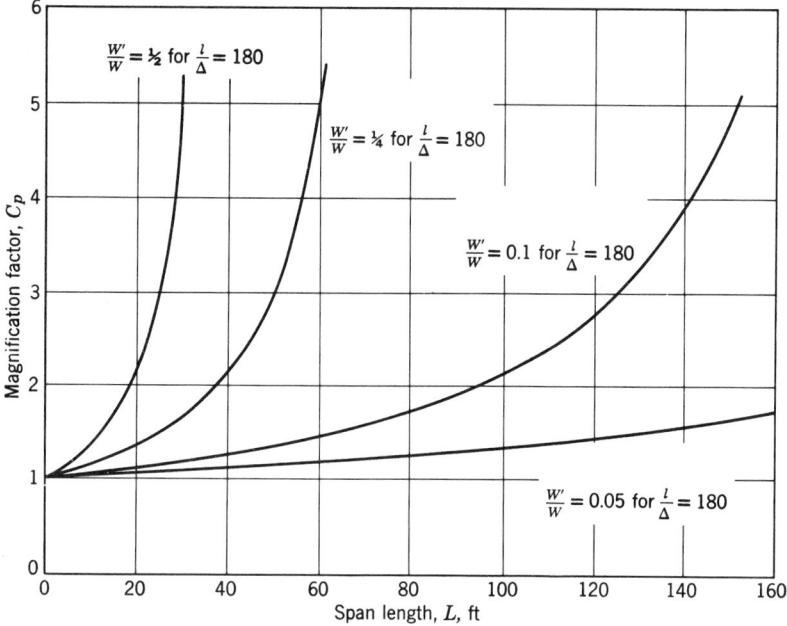

**FIGURE 5.58**  Magnification factor, $M_F$, relating deflection or bending values under ponding to values without ponding, for simply supported beams under initial uniformly distributed load.

as to what snow, water, or other load is to be included. If any conservatism is obtained by using full design load, it could be counterbalanced to some extent by this omission. Theoretically, all subsystems in the roof should be investigated. This is a very laborious calculation, and methods making use of computers are suggested where greater accuracy is desired.

**Examples.** The following examples illustrate the application of ponding analysis. They are based on a simplified approach for ponding calculations that can be used when the effect of the deflection of the secondary roof-framing members on ponding is small.

To illustrate the effect of the magnification factor on flat roof design, two examples have been selected: (1) a beam designed for dead load plus snow load, and (2) a beam with a 2-in.-high gravel stop designed for dead load plus live load.

**Example 1.** Check an $8\frac{1}{2} \times 35\frac{3}{4}$-in. glued laminated timber simple-span beam under the following conditions:

Species: Southern Pine

$$L = 62 \text{ ft} \qquad F_b^* = (2400)(1.15) = 2760 \text{ psi}$$

$$\text{Spacing} = s = 16 \text{ ft} \qquad F_v' = (200)(1.15) = 230 \text{ psi}$$

$$A = 303.9 \text{ in.}^2 \qquad E' = 1,800,000 \text{ psi}$$

$$S = 1811 \text{ in.}^3 \qquad I = 32,360 \text{ in.}^4$$

Allowable total-load deflection $= l/120 = (62)(12)/120 = 6.20$ in.

The beam is continuously braced along its top surface.

Determine the volume factor using Eq. (4-2)

$$C_V = \left(\frac{5.125}{b}\right)^{1/20} \left(\frac{12}{d}\right)^{1/20} \left(\frac{21}{L}\right)^{1/20}$$

$$C_V = \left(\frac{5.125}{8.50}\right)^{1/20} \left(\frac{12}{35.75}\right)^{1/20} \left(\frac{21}{62}\right)^{1/20} = 0.875$$

(See 4.6 for determination of volume factor $C_V$.)

$$F_b' = F_b^* C_V = (2760)(0.875) = 2413 \text{ psi}$$

Duration of load factor $C_D = 1.15$ (from Table 4.7).

| | | |
|---|---|---|
| SL = 25 psf | $w_{SL}$ = 400 plf | $f_b$ = 2445 psi (dead load + snow load) |
| DL = 23 psf | $w_{DL}$ = 368 plf | $f_v$ = 106 psi (dead load + snow load) |
| TL = 48 psf | $w_{TL}$ = 768 plf | $\Delta_{TL}$ = 4.38 in. |

Check for the effects of ponding:

$$\Delta \text{ (for 5-psf load)} = (4.38)\frac{5}{48} = 0.456 \text{ in.} < 0.5 \text{ in. (spring constant) O.K.}$$

This simplified check does not take into account that snow load may be acting at the same time ponding occurs. If maximum snow load is assumed to be acting at the same time, a further check should be made by use of the magnification factor $M_F$:

$$M_F = \frac{1}{1 - W'L^3/\pi^4 EI}$$

$$W' = 5.2Ls = (5.2)(62)(16) = 5160 \text{ lb}$$

$$MF = \left[1 - \frac{(5160)(62)^3(1728)}{\pi^4(1,800,000)(32,360)}\right]^{-1} = 1.60$$

Multiply the calculated stresses for the member under consideration by $M_F$ to determine the effects of ponding in combination with dead and snow load:

$$f_b M_F = (2445)(1.60) = 3912 \text{ psi} > 2400 \text{ psi} \qquad \text{N.G.}$$

$$f_v M_F = (106)(1.60) = 170 \text{ psi} < 230 \text{ psi} \qquad \text{O.K.}$$

$$\Delta_{TL} M_F = (4.38)(1.60) = 7.01 \text{ in.} > 6.20 \text{ in.} \qquad \text{N.G.}$$

In this example, it can be seen that the beam is overstressed in bending, and that the deflection limit is exceeded as the result of ponding in combination with the design loads; therefore, if it is not possible to eliminate the cause of ponding, a larger section is required.

Try an $8\frac{1}{2} \times 41\frac{1}{4}$-in. section:

$$A = 350.6 \text{ in.}^2 \quad I = 49.720 \text{ in.}^4$$

$$C_V = \left(\frac{5.125}{8.50}\right)^{1/20} \left(\frac{12}{41.25}\right)^{1/20} \left(\frac{21}{62}\right)^{1/20} = 0.868$$

$$f_b = 1837 \text{ psi} \quad F_b' = (2400)(1.15)(0.868) = 2397 \text{ psi}$$

$$f_v = 91 \text{ psi}$$

$$\Delta_{TL} = 2.85 \text{ in.}$$

$$M_F = \left[1 - \frac{(5160)(62)^3(1728)}{\pi^4(1,800,000)(49,720)}\right]^{-1} = 1.32$$

Effects of ponding:

$$f_b M_F = (1837)(1.32) = 2429 \text{ psi} > 2400 \text{ (consider O.K.)}$$

$$f_v M_F = (91)(1.32) = 120 \text{ psi} < 230 \qquad \text{O.K.}$$

$$\Delta_{TL} M_F = (2.85)(1.32) = 3.77 \text{ in.} < 6.20 \text{ in.} \qquad \text{O.K.}$$

The larger section is adequate.

**Example 2.**   A $6\frac{3}{4} \times 31\frac{1}{2}$-in. glued laminated timber simple-span beam with a 2-in. high gravel stop is used under the following conditions:

The beam is continuously braced along its top surface.
Species = Douglas Fir-Larch.

$$L = 58 \text{ ft} \qquad F_b^* = (2400)(1.25) = 3000 \text{ psi}$$

$$\text{Spacing} = 16 \text{ ft} \qquad F_v' = (165)(1.25) = 206 \text{ psi}$$

$$A = 212.6 \text{ in.}^2 \qquad E' = E = 1{,}800{,}000 \text{ psi}$$

$$S = 1116 \text{ in.}^3 \qquad I = 17{,}580 \text{ in.}^4$$

Allowable live load deflection $\quad = l/240 = (58)(12)/240 = 2.90 \text{ in.}$

$$C_V = \left(\frac{5.125}{6.75}\right)^{1/10} \left(\frac{12}{31.5}\right)^{1/10} \left(\frac{21}{58}\right)^{1/10} = 0.798$$

Duration-of-load factor $C_D = 1.25$

$$LL = 12 \text{ psf} \qquad w_{LL} = 192 \text{ plf}$$

$$DL = \underline{15} \text{ psf} \qquad w_{DL} = \underline{240} \text{ plf}$$

$$TL = 27 \text{ psf} \qquad w_{TL} = 432 \text{ plf}$$

For the loading conditions shown, the following stresses and deflections were calculated:

$$f_b = 1953 \text{ psi} \quad f_v = 80 \text{ psi} \quad \Delta_{LL} = 1.54 \text{ in.}$$

$$\Delta \text{ (for 5-psf load)} = (1.54)\frac{5}{12} = 0.64 \text{ in.} > 0.5 \text{ in.}$$

Since the $6\frac{3}{4} \times 31\frac{1}{2}$-in. section has insufficient stiffness to meet the criterion that a 5-psf load will cause not more than $\frac{1}{2}$-in. deflection, a stiffer section must be selected.

Try a $6\frac{3}{4} \times 34\frac{1}{2}$-in. section:

$$A = 232.9 \text{ in.}^2 \quad S = 1339 \text{ in.}^3$$

$$C_V = \left(\frac{5.125}{6.75}\right)^{1/10} \left(\frac{12}{34.5}\right)^{1/10} \left(\frac{21}{58}\right)^{1/10} = 0.791$$

For the loading conditions shown, the following stresses and deflections were calculated:

$$f_b = 1628 \text{ psi} \qquad I = 23{,}100 \text{ in.}^4$$

$$\Delta_{LL} = 1.18 \text{ in.} \qquad F_b' = (2400)(1.25)(0.791) = 2372 \text{ psi}$$

$$f_v = 73 \text{ psi}$$

$$\Delta \text{ (for 5-psf load)} = (1.18)\frac{5}{12} = 0.49 \text{ in.} < 0.5 \text{ in.} \qquad \text{O.K.}$$

This simplified check works satisfactorily when only a small amount of water accumulates to start the deflection process. When a larger amount of water may be present due to inadequate drainage, high gravel stops, and so on; a further check by use of the magnification factor $M_F$ should be made.

A 2-in. depth of water caused by the gravel stop will be assumed. It will also be assumed that the design live load (12 psf) is not acting simultaneously with the ponding load. The roof is flat with no slope.

A 2-in. depth of water over the roof will cause a load of $(2/12)(62.4)(16) = 166$ plf. The total load on the roof beam is equal to the dead load plus the water load, or

$$w_{TL} = 240 + 166 = 406 \text{ plf}$$

$$f_b = (1628)\frac{406}{432} = 1530 \text{ psi}$$

$$f_v = (73)\frac{406}{432} = 68 \text{ psi}$$

$$\Delta_{LL} = (1.18)\frac{406}{432} = 1.10 \text{ in.}$$

$$W' = 5.2Ls = (5.2)(58)(16) = 4825 \text{ lb}$$

$$M_F = \left[1 - \frac{(4825)(58)^3(1728)}{\pi^4(1,800,000)(23,100)}\right]^{-1} = 1.67$$

Effects of ponding:

$$f_b M_F = (1530)(1.67) = 2555 \text{ psi} > 2372 \qquad \text{N.G.}$$

$$f_v M_F = (68)(1.67) = 114 \text{ psi} < 206 \qquad \text{O.K.}$$

$$\Delta_{LL} M_F = (1.10)(1.67) = 1.84 \text{ in.} < 2.90 \qquad \text{O.K.}$$

The member is slightly overstressed in bending due to ponding. Either reduce the height of the gravel guard to $1\frac{3}{4}$ in. or use the next larger size, $6\frac{3}{4} \times 36$ in. (In some cases, the live load may occur simultaneously with the ponding. If so, the live load should also be included in determining the total design load.)

## 5.16   REFERENCES

1. American Institute of Timber Construction, *Standard Specifications for Structural Glued Laminated Timber of Softwood Species*, AITC 117—Design, Englewood, CO, 1993.

2. American Institute of Timber Construction, *Standard Specification for Hardwood Glued Laminated Timber*, AITC 119, Englewood, CO, 1994.

3. American Forest & Paper Association, *National Design Specification for Wood Construction*, Washington, DC, 1991.

4. American Society for Testing and Materials, *Standard Methods of Testing Small Clear Specimens of Timber*, ASTM D 143, Philadelphia, PA, 1983.

5.  Hooley, R. F. and B. Madsen, *Lateral Stability of Glued Laminated Beams*, Journal of Structural Division, Proceedings of the Structural Division, American Society of Civil Engineers, Vol. 90 ST8:1964.

6.  American Institute of Timber Construction, *Wood Design—A Commentary on the Performance of Deep and Narrow Resawn Glued Laminated Purlins*, Englewood, CO, 1993.

7.  United States Department of Agriculture, Forest Service, *Test of Large Timber Columns and Presentation of the Forest Products Laboratory Column Formula*, Technical Bulletin 167, U.S. Forest Products Laboratory, Madison, WI, 1930.

8.  United States Department of Agriculture, Forest Service, Forest Products Laboratory, *Wood Handbook: Wood as an Engineering Material*, Agriculture Handbook No. 72, Madison, WI, 1955.

9.  United States Department of Agriculture, Forest Service, Forest Products Laboratory, *Deflection and Stresses of Tapered Wood Beams*, Research Paper FPL 34, Madison, WI, 1965.

10. United States Department of Agriculture, Forest Service, Forest Products Laboratory, *The Glued Laminated Wooden Arch*, Technical Bulletin No. 691, Madison, WI, 1939.

11. Colorado State University, Civil Engineering Department, *Behavior and Design of Double-Tapered Pitched and Curved Glulam Beams*, Structural Research Report No. 16, Ft. Collins, CO, 1976.

12. United States Department of Agriculture, Forest Service, Forest Products Laboratory, *Fabrication and Design of Glued Laminated Wood Structural Members*, Technical Bulletin No. 1069, Madison, WI, 1954.

13. American Institute of Timber Construction, *Deflection of Arches*, Technical Note No. 2, Englewood, CO, 1992.

14. American Institute of Timber Construction, Mathematical Solution of 3-Hinged Arch Design, Technical Note 22, Englewood, CO, 1993.

15. American Institute of Timber Construction, *Typical Construction Details*, AITC 104-84, Englewood, CO, 1984.

16. American Institute of Steel Construction, *Manual of Steel Construction*, Chicago, IL, 1991.

17. Haussler, R. W., *Roof Deflection Caused by Rainwater Pools*, Civil Engineering, ASCE Vol. 32, No. 10, Oct. 1962, 89.

18. E. W. Kuenzi and B. Bohannan, *Increases in Deflection and Stresses Caused by Ponding of Water on Roofs*, Forest Products Journal, 14(9) 421–424, September 1964.

# CHAPTER 6

# DESIGN OF STRUCTURAL SYSTEMS

## 6.1  INTRODUCTION

This chapter contains recommended procedures for the design of structural systems using information developed in the preceding chapter on the design of components. The methods of design illustrated are those in common use. However, other suitable methods based on sound engineering practice may be used.

## 6.2  TRUSSES

The subject of timber truss design is quite broad. This discussion is limited to basic design procedures and highlighting of features unique to timber truss construction.

### 6.2.1  Truss Types

The types of timber trusses most commonly built are (see Fig. 6.1): (a) parallel chord; (b) pitched or triangular; (c) bowstring; (d) camelback; and (e) special, such as crescent, scissors, sawtooth, lenticular, and king or queen post trusses.

Some roof shapes are dictated by architectural considerations, such as a sawtooth roof. Generally, however, the designer selects the truss type for best overall economy, along with the necessary clear span, clearances, and aesthetic appearance desired for the structure. The maximum economical span for any given type of timber truss will vary with the materials available, loading conditions, ratio of labor to material cost, and fabrication methods. Table 1.1 gives span ranges for various primary truss-framing systems. Also, span ranges for various secondary framing systems are given, which can be used to help determine truss spacings.

Truss members are designated in three categories: top chord, bottom chord, and webs (all interior vertical or diagonal members between the top and bottom chords are called webs). Joints at which members intersect and connect are called panel points. If flat or parallel chord trusses are used as roof trusses, adequate pitch must be provided for drainage. (See Chapter 5.)

Pratt trusses have the advantage that for normal loading, the longer (diagonal) webs are in tension, whereas the shorter (vertical) webs are in compression. Flat

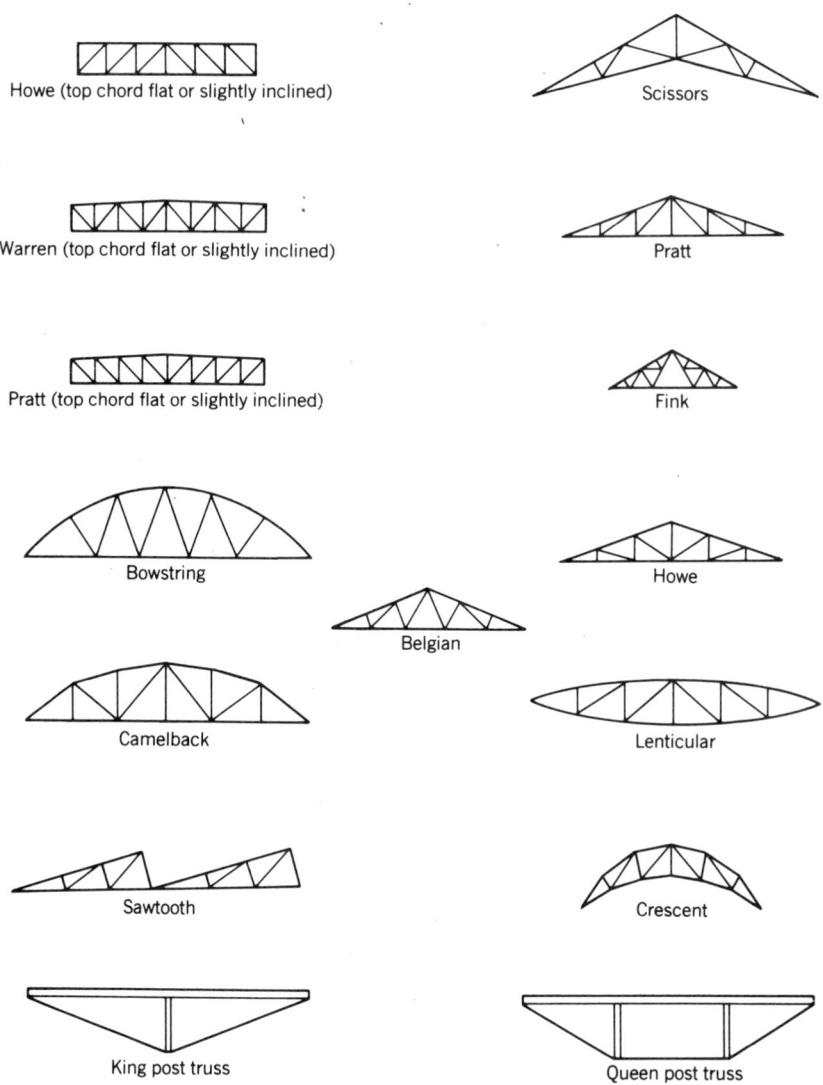

**FIGURE 6.1**    Truss types.

or low-pitched Pratt roof trusses can be used for clear spans up to 120 ft, with the most economy in spans 80 ft and under.

Pitched Pratt and Belgian trusses can be used for clear spans up to 100 ft. They become more economical than Fink trusses when the pitch is less than 25° and the span is over 50 ft. The roof pitch should be a minimum of 3 : 12.

Fink trusses can be used for clear spans up to 80 ft, with roof pitches 25° or more from the horizontal. Optimum span range is 40–60 ft.

Bowstring trusses are most economical for spans from 80 to 150 ft, but may range from 50 to 200 ft. Usually, the radius of the top chord is equal to the span. Modified bowstring trusses, where the center portion of the top chord is straight instead of curved, can be used for spans up to 300 ft. The chords and heel connections take the major stresses; the web stresses under uniform loading conditions are negligible, and for unbalanced loading, web stresses are comparatively light. Bowstring trusses, because of low web stresses, have light webs and web connections. Chord stresses are nearly equal throughout their length and, with constant cross section, chords have full-length economy.

Bowstring trusses usually have the upper chord shaped to the form of a circular arc or parabola. With the normal depth–span ratios used, the variation between the parabola and the circular arc is very slight and is not sufficient to change stresses materially. Circular curves are customarily used. In addition to normal axial and bending stresses, upper chords of bowstring trusses are subject to moments due to the eccentricity of their curved shape.

Scissor trusses can be efficiently designed for clear spans up to 80 ft, and are used to provide more height clearance toward the midspan. They are primarily used in churches, gymnasiums, and various types of assembly halls.

Bridge trusses can be designed for economy with clear spans up to 200 ft.

Table 6.1 lists aids for determining economical truss dimensions without a great deal of preliminary calculations. When these ratios of depth to span are lower, larger forces and members usually result. Also, these trusses may deflect excessively under applied loads.

Chords and webs in all truss types may be constructed as single-leaf, double-leaf, or multileaf members. Single-leaf chords are also known as monochords. The most common arrangements of trusses are monochord trusses with single-leaf chords and webs and those with double-leaf chords having single-leaf webs located between the chord leaves. Webs may be attached to sides of chords or may be in the same plane and attached thereto with steel straps or gussets. Truss members may be sawn lumber or glued laminated timber. The use of steel rods or other steel shapes for members in timber trusses is acceptable if they fulfill all conditions of design and service.

Timber truss web systems should be selected for convenience of connection and economy. Web locations and panel point spacings may be dictated by selection of secondary purlin framing so as to minimize chord bending stresses. In parallel chord trusses, diagonal webs should be sloped between 45° and 60° from

### TABLE 6.1

#### Recommended Timber Truss Depth–Span Ratios

| | |
|---|---|
| Flat or parallel chord | 1/8 to 1/10 |
| Triangular or pitched | 1/6 or deeper |
| Bowstring | 1/6 to 1/8 |

the horizontal for greatest economy. Bowstring trusses use panel lengths from 8 to 14 ft, depending on the truss span.

Monochord trusses may require steel gusset plate connections with through bolts alone or in combination with shear plates. In multileaf trusses, connections can often be made more easily by using through bolts and shear plates or split rings. When split-ring or shear plate connectors are required, it is usually best to have the wood trusses fabricated in a shop.

Glued laminated timber provides many features desirable for truss designs. Glued laminated timber can be made in almost any shape, size, or length, and provides higher design values than do sawn timbers. Sawn timbers are limited in maximum length and cross-sectional size and are prone to checking. However, sawn members may provide cost savings where they will serve. Thus, depending on truss span and loading requirements, trusses can be made of all sawn timber, all glued laminated timber, or a combination of both and may include some steel tension members. If sawn members are used in conjunction with glued laminated timber, care must be taken to match widths at connections and provide for differential shrinkage.

### 6.2.2    Deflection

The deflection $\Delta_a$ at any point in a truss may be determined by the method of virtual work. In this method, a unit load that acts in the direction of the desired component of movement is applied at the point being investigated. The force caused in each truss member by this unit load is determined. Deflection is then determined by the summation of the effects in each member of the unit load and actual design loads, using the equation

$$\Delta_a = \sum \frac{Pul}{AE} \tag{6-1}$$

where    $P$ = axial force load in a truss member caused by design loads (lb),
$\quad\quad\quad u$ = force in a truss member caused by a unit load (dimensionless),
$\quad\quad\quad l$ = length of truss member (in.),
$\quad\quad\quad A$ = cross-sectional area of truss member (in.$^2$), and
$\quad\quad\quad E$ = modulus of elasticity (psi).

The effects of truss deflection should be considered in the design of columns and walls and in the working of truss-supported doors or other building fixtures.

### 6.2.3    Camber

Truss camber (Fig. 6.2) should be such that total load deflection does not produce a sag below a straight line between points of support. Camber may be determined by using Eq. (6-1) to determine the dead-load elastic deformation to which the inelastic deformation (or slip) is added, test data, the recommendations given in CSA Standard 086 (1), or the following empirical formula recommended by TECO/Lumberlock (2) for multileaf, timber-connected trusses:

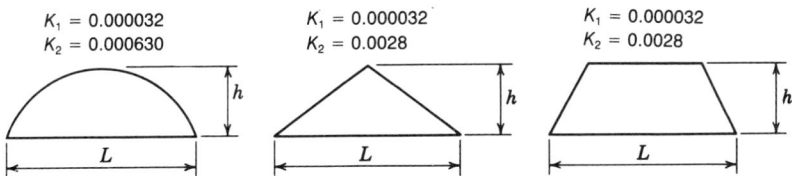

**FIGURE 6.2** Truss camber. Coefficients vary with truss geometry and roof height. See applicable building codes or ASCE 7-88 (3) for specific information on application of wind loads.

$$\text{Camber} = K_1 \frac{L^3}{h} + K_2 \frac{L^2}{h} \tag{6-2}$$

where  $L$ = span (ft),
 $h$ = rise (ft), and
 $K_1, K_2$ = coefficients depending on the truss configuration (see Fig. 6.2) and camber is in inches.

Both lower and upper chords may be cambered, but this is generally done only in flat or parallel chord trusses and in longer triangular trusses. When only the lower chord is cambered, as is usual for bowstring and some triangular or pitched trusses, the effective depth of the truss should be reduced accordingly in stress computations.

The following camber recommendations by the Canadian Standards Association, CSA Standard 086 (1), are provided as a guide and should be modified where unusual conditions exist. The amount of camber built into a truss varies with the use to which the truss is put and the expectancy of full design load. The following camber limitations are suggestions only and are for simple-span trusses only:

(a)  Bowstring trusses: (i) with continuous glued laminated top chord, camber in bottom chord only, $\frac{1}{4}$ in. per 10 ft of span; (ii) segmental overlapping sawn chord trusses, camber in bottom chord only, $\frac{3}{8}$ in. per 10 ft of span.

(b)  Triangular trusses—where possible, camber should be built into both top and bottom chords: (i) with sawn timber or rod and sawn timber trusses, $\frac{1}{2}$ in. per 10 ft of span on bottom chord; at midpoint of half span on top chord, $\frac{3}{8}$ in. per 10 ft of span; (ii) with glued laminated timber or rod and glued laminated timber trusses, $\frac{3}{8}$ in. per 10 ft of span on bottom chord; at midpoint of half span on top chord, $\frac{1}{4}$ in. per 10 ft of span.

(c)  Pratt and Howe trusses—camber both top and bottom chords similarly: (i) for sawn timber or rod and sawn trusses, $\frac{1}{2}$ in. per 10 ft of span; (ii) for glued laminated timber or rod and glued laminated timber trusses, $\frac{3}{8}$ in. per 10 ft of span.

### 6.2.4  General Design Procedure

Given the truss type, span, depth, spacing, and species and grade of lumber to be used, the following general procedure may be used for all truss types. Special design features for each particular truss type must also be considered.

1.   Determine dead loads acting on the truss. Dead load should include weights of all roof and ceiling construction, mechanical and electrical equipment, the estimated truss weight, and other permanently applied loads that will act throughout the service life of the truss.

2.   Determine snow or live loads, wind loads, and any special loads required by applicable building codes and/or good engineering judgment. Apply loading magnification factors such as for snow drifting, water ponding, and impact as may be required. Most building codes and ordinances specify loading combinations to be used. Unbalanced loading conditions frequently control truss designs. If net uplift is possible, special considerations for stress reversals and bottom chord buckling will be required.

### 6.2.5   Preliminary Design

Most timber trusses have chords that are continuous across several panel points and also have distributed loads on the chords. Member forces and architectural considerations generally determine the type of connections to be used and can result in pinned, fixed, or semifixed-type connections. These fixed or semifixed conditions affect the distribution of forces throughout the truss and result in the truss being internally indeterminate. If the overall height of a truss must be held to a given dimension, the geometry of the centerlines of the members will be dependent on the truss member sizes. Consequently, it is usually necessary to conduct a preliminary truss design to determine approximate member sizes and connection types.

For the purposes of a preliminary design, loads are placed at panel points and all joints are assumed pinned. Member forces can then be determined graphically or analytically. Preliminary chord sizes can be selected based on these axial forces plus approximate moments due to any distributed loads or concentrated loads that do not occur at panel points. Preliminary web sizes can also be determined using the axial forces. Refer to design of members for axial forces or bending plus axial forces, which are covered in Section 5.9. Locate splices in chords considering available lengths and minimal axial forces. Design the joint connections considering first the joint or joints carrying the greatest load. Connections should be concentric and hinged wherever possible. Consideration should be given to the effect of possible wood shrinkage on the connection. Member sizes may have to be increased to compensate for loss of section at connections.

### 6.2.6   Final Design

Draw truss to scale using member sizes from the preliminary design, locate centerlines of members, and chord splices, and indicate locations of hinged and fixed connections. Indicate loading exactly as it will be applied to the truss. The designer must now decide the method of analysis to be used. Table 6.2 lists typical truss parameters, along with methods of analysis and results possible.

If no members are continuous but are all pinned at each panel point, a simple mathematical or graphical analysis will give a correct distribution of member forces

**TABLE 6.2**

**Truss Parameters and Methods of Analysis**

| Truss Parameters | Method of Analysis | Design Results |
|---|---|---|
| A. All members discontinuous with pins at every panel point. | Manual analytical, graphical or computer. | Exact[a] |
| B. Chords continuous through panel points; chords pinned at ends and splices and all webs pinned to chords. | Same as above. Stiffness analysis by computer or complex manual method. | Approximate Exact[a] |
| C. Continuous chords with fixed connections at all panel points. | Stiffness analysis by computer or complex manual method. | Exact[a] |

[a]Based on the assumption of average modulus of elasticity in each member and the accuracy of the assumed joint fixity.

(Table 6.2, Type A). If chords are continuous or any connections are not pinned, the distribution of forces will be affected and a more complex method of analysis should be considered. In reality, most truss top and bottom chords have continuity, and may be pinned only at the middle of the truss span and at the ends, although webs are usually pinned to the chords (Table 6.2, Type B).

If the truss span is relatively short, the depth–span ratio is in accordance with Table 6.1 and loads are fairly low, the designer may decide that a manual analytical or graphical analysis will be satisfactory, although not exact. In such cases, the chords should be analyzed as multispan beams in order to determine support reactions at panel points (assume rigid supports and apply loading exactly as it will be transmitted from secondary framing). Apply these reactions to the truss as panel point loads and analyze for axial member forces. Member sizes and connections can then be designed. Connections should be concentric, pinned (in accordance with design assumption), and given consideration as to possible wood shrinkage. The designer must keep in mind that this analysis is not exact. Deflection of the truss under load will cause a redistribution of member forces and possible higher forces and moments in some locations.

As a general rule, if truss members are continuous or any joint is not hinged, a more complex and accurate method of analysis must take into account the relative stiffness of members, deflection of the truss under load, and resulting redistribution of member forces and moments. Likewise, if there are knee braces or other structural elements attached to the truss that can lend support or influence load distribution, they must be included as integral with the truss in the overall analysis.

The design of truss chord members will be governed primarily by axial tension or compression if chords are not continuous and if all secondary framing transmits loads into panel points only. However, if distributed loads are applied to chord members or chord members are continuous over panel points or panel point connections are fixed, then significant bending stresses along with axial stresses could result and a combined stress unity check must be made. Combined loading in beams and lateral stability considerations are discussed in Chapter 5. If chords are curved, moments resulting from eccentricity and axial forces must be included in the design.

Truss web member designs are governed by axial tension or compression loads for all pinned webs. If web connections offer moment-resisting capacity, the webs must be sized to resist the resulting combined loading, the same as for chords. Slenderness ratios of compression webs should be checked as columns using the width as the least dimension and the distance between panel points as the length. Unbalanced, wind, wind uplift, or other special loadings can cause increased stresses or reversal of stresses, and thus control portions of a total design. Net member sections must be used where drilling and dapping for connection hardware is required. The required truss member sizes are determined for appropriate conditions of loading based on the above-stated procedures utilizing the latest available design values for structural glued laminated timber. For members stressed principally in axial tension or axial compression, refer to Table 2, AITC 117—Design (4). If chord members are highly stressed in bending, refer to Table 1, AITC 117—Design. For hardwood glued laminated timber, see AITC 119—94 (5).

Connections are to be designed in accordance with Chapter 7. Truss connections should be concentric; that is, the centerlines of members at each joint should intersect at a common point so that moments are not induced into the members being connected.

Connections should be of the pinned type whenever possible. Truss members tend to rotate in relation to each other as a result of distortion and deflection of the truss under loading. If rotation is restrained by the joints, moments and additional forces in members and connections will result.

Determine the size and type of fastening to be used. When possible, use fastenings of the same size and type throughout the truss. Check spacing, end distance, and edge distance requirements for each fastener. See Chapter 7 on design considerations for fastenings. Table 3, AITC 117—Design (4), contains information on the connector groups to be used. Possible shrinkage of wood members must be considered. For splices in deep members where steel plates are used, it may be necessary to utilize two rows of splice plates. Checking caused by a decrease in moisture content is a possibility in all wood members and should be considered, especially if members are large or if rapid or large changes in moisture content take place. It is generally advisable to stagger connectors.

### 6.2.7  Monochord Trusses

Design of steel straps in compression must be based on actual support conditions for the plates. For plates connecting webs to chords, generally the web does not provide lateral stability for the plate but instead the plate is an extension of the web. Therefore, the end of the plate fastened to the web cannot be considered fixed in determining the plate effective length. When the web plates are pinned at the chord, the spacing between bolts in the web should be large enough to provide stability and prevent splitting of the web. Normal fabrication tolerances must be considered. If steel rods are used as tension elements in a truss, written instruction should be given to the truss assembler regarding tightening of the rods. Otherwise, overtightening could occur, which might overstress truss elements when design loads are applied.

### 6.2.8  Truss Bracing

In structures employing trusses, a system of bracing is required to provide resistance to lateral forces, to hold the trusses true and plumb, and to hold compression elements in line. Both permanent bracing and temporary erection bracing should be designed according to accepted engineering principles to resist all loads that will normally act on the system. Erection bracing is that which is installed during erection to hold the trusses in a safe position until sufficient permanent construction is in place to provide full stability. Permanent bracing is that which forms an integral part of the completed structure. Part or all of the permanent bracing may also act as erection bracing.

Permanent bracing must be provided to resist both transverse and longitudinal forces and may consist of either of the following systems:

1.   A structural diaphragm in the plane of the top chord. The diaphragm, acting as a plate girder, transmits forces to end and side walls. (See 6.7, Structural Diaphragms.) This system is the preferred method of providing truss bracing and is usually most economical.

2.   Horizontal bracing between trusses and in the plane of the bottom and/or top chords. In effect, this is a horizontal truss. It is a positive method of bracing, but it is more costly and should be used only where the strength of system 1 is insufficient. If horizontal bracing is used, it should be designed to resist all the transverse and longitudinal loads that will normally act on the system.

If the roof construction does not provide proper top chord strut action, separate additional members should be provided.

If bottom chords can go into compression under wind uplift or some other load application, additional bracing may be required to satisfy $l/d$ requirements. To provide lateral support to the compression chord of a truss, the bracing system should be designed to withstand a horizontal force equal to at least $2\%$ of the compressive force in the truss chord if the members are in perfect alignment; or, if the members are not aligned, the calculation of the force should be based on the eccentricity of the members due to the misalignment.

In all cases, vertical X-bracing between trusses should be installed in each third or fourth bay at intervals of approximately 35 ft measured parallel to trusses. Bottom chord lateral bracing should be installed perpendicular to the trusses and in line with vertical X-bracing and extend from end wall to end wall.

Bracing is also required to provide overall building stability and transmit forces from the roof to the ground. This bracing typically consists of shear walls, diagonal bracing, buttresses, cantilevered columns, or knee braces.

Use of knee braces between the trusses and columns is a method of bracing particularly adaptable to buildings that have a large length–width ratio. The stresses induced in the trusses and columns can be critical. Lateral loads acting in the long direction of the building are usually resisted by means of diagonal bracing and struts in the end bays.

### 6.2.9    Erection Truss Bracing

Erection truss bracing is that installed to hold trusses true and plumb and in a safe condition until permanent truss bracing and other permanent components, such as joists and sheathing contributing to the rigidity of the complete roof structure, are in place.

Erection truss bracing may consist of struts, ties, cables, guys, shores, or similar items. Joists, purlins, and other permanent elements may be used as part of the erection bracing.

### 6.2.10    Design of Glued Laminated Timber Truss

The design of a glued laminated timber truss is illustrated in the following example:

**Example.**    Design a triangular Howe truss using glued laminated timber members for the following conditions:

Span = 48 ft center to center of supports
Truss spacing = 12 ft center to center
Panel point spacing = 8 ft
Pitch = 6 : 12

Heavy timber decking is attached directly to the top chord. Bottom chord is braced laterally at 16-ft intervals.

Loading:    DL = 15 psf on horizontal projection (no ceiling load)
SL = 30 psf on horizontal projection (2-month duration)
WL = 25 psf applied as shown in the sketch

Use the following tabular design values:

| **Top Chord** | **Bottom Chord and Webs** |
|---|---|
| $F_{bx}$ = 2400 psi | $F_{bx}$ = 1600 psi |
| $E_x = E_y$ = 1,700,000 psi | $F_t$ = 1250 psi |
| $F_c$ = 1600 psi | $F_c$ = 1900 psi |
| $F_{c\perp}$ = 560 psi | $E_x = E_y$ = 1,700,000 psi |
| | $F_{c\perp}$ = 560 psi |

Species: Douglas Fir-Larch. Group B species for connectors and shear plates. Steel plates ASTM A36, bolts ASTM A307.

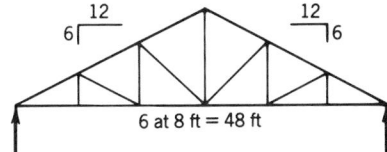

Preliminary size of top chord:

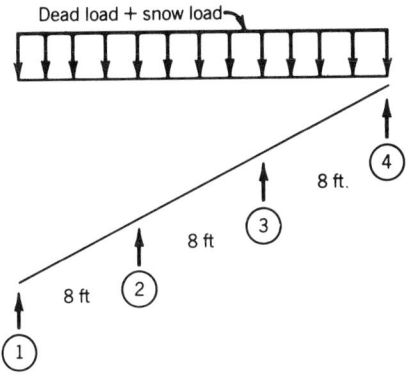

$$w_{DL} = (15)(12) = 180 \text{ plf}$$

$$w_{SL} = (30)(12) = \underline{360} \text{ plf}$$

$$540 \text{ plf}$$

Calculate support reactions and moments using standard engineering equations for a three-span continuous beam.

Dead load plus snow load:

$$R_1 = R_4 = 0.4wl = (0.4)(540)(8) = 1730 \text{ lb}$$

$$R_2 = R_3 = 1.1wl = (1.1)(540)(8) = 4750 \text{ lb}$$

$$M_{1-2} = M_{3-4} = 0.08wL^2 = (0.08)(540)(8)^2(12) = 33,180 \text{ in.-lb}$$

$$M_2 = M_3 = -0.10wL^2 = -(0.10)(540)(8)^2(12) = -41,470 \text{ in.-lb}$$

$$M_{2-3} = 0.025wL^2 = (0.025)(540)(8)^2(12) = 10,370 \text{ in.-lb}$$

where $M_{1-2}$ = maximum moment between points 1 and 2
$M_2$ = moment at point 2
$M_{2-3}$ = maximum moment between points 2 and 3
$M_3$ = moment at point 3
$M_{3-4}$ = maximum moment between points 3 and 4

Dead load plus wind load: (see diagram for application.)

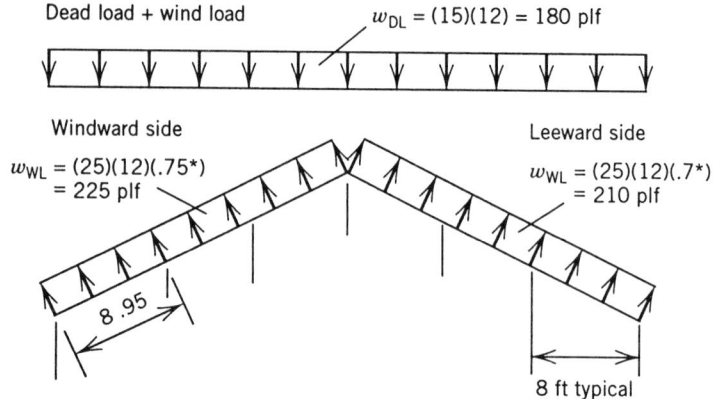

Dead load + wind load    $w_{DL} = (15)(12) = 180$ plf

Windward side
$w_{WL} = (25)(12)(.75^*)$
$= 225$ plf

Leeward side
$w_{WL} = (25)(12)(.7^*)$
$= 210$ plf

8.95

8 ft typical

*Coefficients vary with truss geometry and roof height. See applicable building code or ASCE 7-88 (3) for specific information on application of wind loads.

Maximum moments—windward side:

$$M_{DL} = -0.10w_{DL}L^2 = -(0.10)(180)(8)^2(12) = -13,820 \text{ in.-lb}$$

$$M_{WL} = 0.10w_{WL}L^2 = (0.10)(225)(8.95)^2(12) = \underline{21,630} \text{ in.-lb}$$

$$7810 \text{ in.-lb}$$

Maximum moments—leeward side:

$$M_{DL} = \hspace{4cm} -13,820 \text{ in.-lb}$$

$$M_{WL} = 0.10wL^2 = (0.10)(210)(8.95)^2(12) = \underline{20,190} \text{ in.-lb}$$

$$6370 \text{ in.-lb}$$

Determine preliminary axial member forces graphically as shown in Fig. 6.3 for DL + SL, Fig. 6.4 for DL + unbalanced snow load, and Fig. 6.5 for DL + WL. Results are summarized in Table 6.3.

From the preliminary axial forces, it appears that DL + SL will control. Check compression $P$ and bending $M$ in the top chord.

$$P_{TC} = 25,100 \text{ lb (DL + SL)}$$

$$M_{TC} = 41,470 \text{ in.-lb (DL + SL)}$$

Try $5\frac{1}{8} \times 6$ in., $A = 30.75$ in.$^2$, $S = 30.75$ in.$^3$

$$E'_x = E_x = 1,700,000 \text{ psi}$$

$$F^*_c = F_c C_D = (1600)(1.15) = 1840 \text{ psi}$$

$$F'_b = F_b C_D = (2400)(1.15) = 2760 \text{ psi}$$

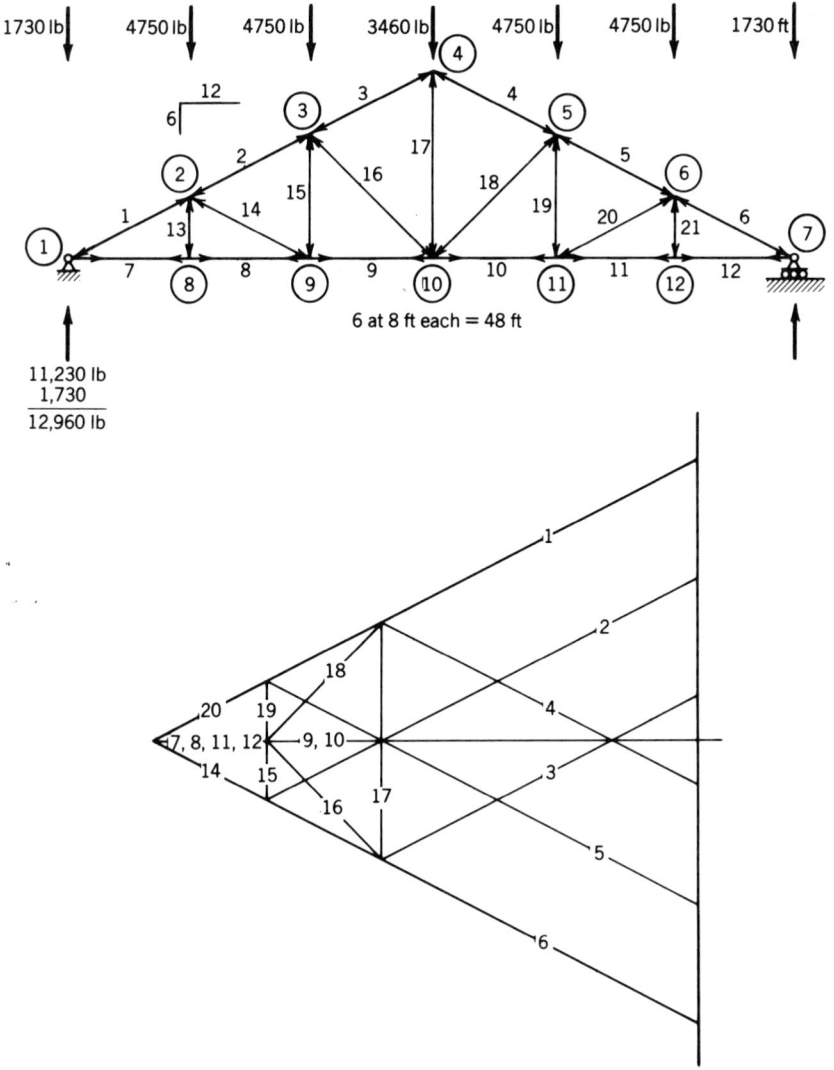

**FIGURE 6.3** Graphical solutions for dead load plus snow load. DL + SL; symmetrical about centerline.

| Member | Force | Member | Force |
|--------|---------|--------|--------|
| 1 | −25,100 | 13 | 0 |
| 2 | −19,900 | 14 | −5300 |
| 3 | −14,700 | 15 | +2300 |
| 7 | +22,500 | 16 | −6700 |
| 8 | +22,500 | 17 | +9400 |
| 9 | +17,800 | | |

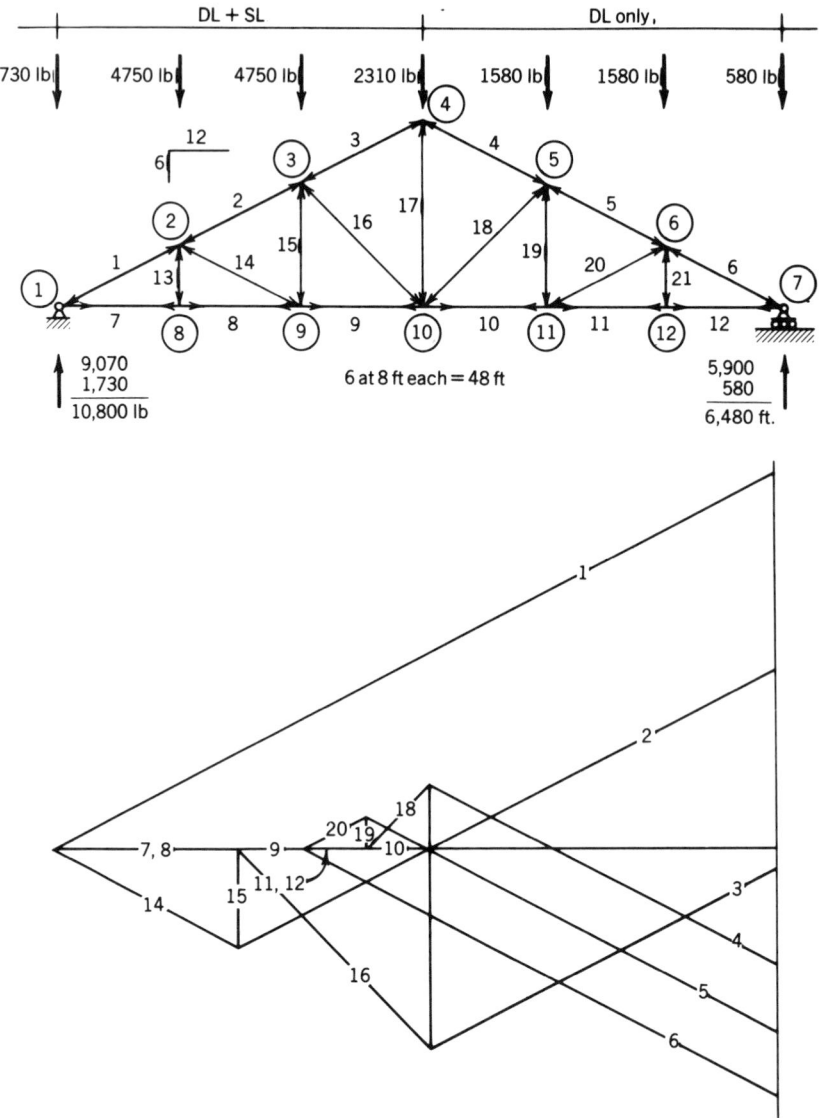

**FIGURE 6.4** Graphical solution for dead load plus unbalanced snow load. DL + USL.

| Member | Force | Member | Force |
|---|---|---|---|
| 1 | −20,300 | 13 | 0 |
| 2 | −15,000 | 14 | −5300 |
| 3 | −9700 | 15 | +2400 |
| 4 | −9700 | 16 | −6700 |
| 5 | −11,400 | 17 | +6300 |
| 6 | −13,200 | 18 | −2200 |
| 7,8 | +18,100 | 19 | +800 |
| 9 | +13,400 | 20 | 0 |
| 10 | +10,200 | 21 | 0 |
| 11,12 | +11,800 | | |

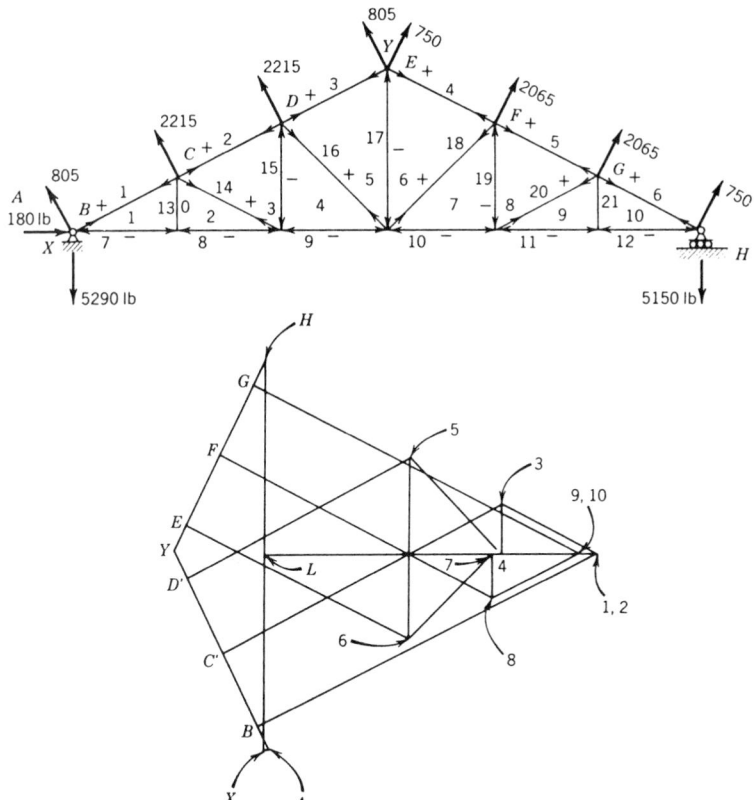

**FIGURE 6.5** Graphical solution for dead load plus wind loads. DL + WL.

Forces (lb)

| Member | WL | DL | WL + DL |
|---|---|---|---|
| 1 | +10,300 | −8500 | +1800 |
| 2 | +8500 | −6700 | +1800 |
| 3 | +6800 | −4900 | +1900 |
| 4 | +6800 | −4900 | +1900 |
| 5 | +8300 | −6700 | +1600 |
| 6 | +9900 | −8500 | +1400 |
| 7 | −9000 | +7600 | −1400 |
| 8 | −9000 | +7600 | −1400 |
| 9 | −6400 | +6000 | −400 |
| 10 | −6100 | +6000 | −100 |
| 11 | −8500 | +7600 | −900 |
| 12 | −8500 | +7600 | −900 |
| 13 | 0 | 0 | 0 |
| 14 | +2900 | −1800 | +1100 |
| 15 | −1300 | +800 | −500 |
| 16 | +3500 | −2200 | +1300 |
| 17 | −4800 | +3100 | −1700 |
| 18 | +3200 | −2200 | +1000 |
| 19 | −1200 | +800 | −400 |
| 20 | +2700 | −1800 | +900 |
| 21 | 0 | 0 | 0 |

## TABLE 6.3

### Summary of Preliminary Member Loads Based on Graphical Analysis of Truss with Pinned Joints[a]

| Loading | DL + SL | | DL + USL[b] | | DL + WL | |
|---|---|---|---|---|---|---|
| Member | $F$ | $M_{max}$ | $F$ | $M_{max}$ | $F$ | $M_{max}$ |
| Top chord | | | | | | |
| 1 | −25.1 | 41.47 | −20.3 | 41.47 | +1.8 | 7.8 |
| 2 | −19.9 | 41.47 | −15.0 | 41.47 | +1.8 | 7.8 |
| 3 | −14.7 | 41.47 | −9.7 | 41.47 | +1.9 | 7.8 |
| 4 | −14.7 | 41.47 | −9.7 | 13.82 | +1.9 | 6.4 |
| 5 | −19.9 | 41.47 | −11.4 | 13.82 | +1.6 | 6.4 |
| 6 | −25.1 | 41.47 | −13.2 | 13.82 | +1.4 | 6.4 |
| Bottom chord | | | | | | |
| 7 | +22.5 | 0 | +18.1 | 0 | −1.4 | 0 |
| 8 | +22.5 | 0 | +18.1 | 0 | −1.4 | 0 |
| 9 | +17.8 | 0 | +13.4 | 0 | −0.4 | 0 |
| 10 | +17.8 | 0 | +10.2 | 0 | −0.1 | 0 |
| 11 | +22.5 | 0 | +11.8 | 0 | −0.9 | 0 |
| 12 | +22.5 | 0 | +11.8 | 0 | −0.9 | 0 |
| Webs | | | | | | |
| 13 | 0 | 0 | 0 | 0 | 0 | 0 |
| 14 | −5.3 | 0 | −5.3 | 0 | +1.1 | 0 |
| 15 | +2.3 | 0 | +2.4 | 0 | −0.5 | 0 |
| 16 | −6.7 | 0 | −6.7 | 0 | +1.3 | 0 |
| 17 | +9.4 | 0 | +6.3 | 0 | −1.7 | 0 |
| 18 | −6.7 | 0 | −2.2 | 0 | +1.0 | 0 |
| 19 | +2.3 | 0 | +0.8 | 0 | −0.4 | 0 |
| 20 | −5.3 | 0 | −1.8 | 0 | +0.9 | 0 |
| 21 | 0 | 0 | 0 | 0 | 0 | 0 |

[a]Units—kips; (+) tension, (−) compression, moment—in.-kips.
[b]USL—unbalanced snow load.

Top chord is laterally braced about $y$-$y$ axis.

$$(l_e/d)_x = \frac{(8.95)(12)}{6} = 17.9 < 50 \qquad \text{O.K.}$$

$$f_c = \frac{P_{TC}}{A} = \frac{25,100}{30.75} = 815 \text{ psi}$$

$$f_b = \frac{M_{TC}}{S} = \frac{41,470}{30.75} = 1350 \text{ psi}$$

Determine column stability factor $C_P$:

From Eq. (5-14), $K_{cE} = 0.418$, $c = 0.9$

$$F_{cE} = \frac{K_{cE} E'}{(l_e/d)^2} = \frac{(0.418)(1{,}700{,}000)}{(17.9)^2} = 2218 \text{ psi}$$

$$C_P = \frac{1 + F_{cE}/F_c^*}{2c} - \sqrt{\left[\frac{1 + F_{cE}/F_c^*}{2c}\right]^2 - \frac{F_{cE}/F_c^*}{c}}$$

$$= \frac{1 + 2218/1840}{(2)(0.9)} - \sqrt{\left[\frac{1 + 2218/1840}{(2)(0.9)}\right]^2 - \frac{2218/1840}{0.9}} = 0.823$$

$$F_c' = F_c^* C_P = (1840)(0.823) = 1510 \text{ psi}$$

$$\left(\frac{f_c}{F_c'}\right)^2 + \frac{f_b}{F_b'[1 - f_c/F_{cE}]} \leq 1$$

$$\left(\frac{815}{1510}\right)^2 + \frac{1350}{2760[1 - 815/2218]} = 1.06 > 1$$

For a preliminary trial size, use $5\frac{1}{8} \times 7\frac{1}{2}$ in. in order to make allowance for a reduction in net section by connectors and to reduce the combined stress to less than one. Note that DL + SL + WL does not control, by observation, for this particular truss design because the wind uplift reduces the DL + SL stresses.

Determine the preliminary size of the bottom chord based on tension:

$$T_{BC} = 22{,}500 \text{ lb (DL + SL)}$$

$$F_t' = (1250)(1.15) = 1440 \text{ psi}$$

$$A = \frac{T_{BC}}{F_t'} = \frac{22{,}500}{1440} = 16 \text{ in.}^2$$

Try a $5\frac{1}{8} \times 7\frac{1}{2}$-in. section (same size as top chord), which is larger than that calculated because of considerations of net section at connectors and moment that will be induced into the bottom chord due to truss deflection.

Determine the preliminary size of web members:

$$P_{max} = 6700 \text{ lb (DL + SL)}$$

$$T_{max} = 9400 \text{ lb (DL + SL)}$$

Try $5\frac{1}{8} \times 6$ in., $A = 30.75$ in.$^2$. Check member 16, Fig. 6.3.

$$(l_e/d)_y = \frac{(11.3)(12)}{5.125} = 26.5$$

$$F_c^* = (1900)(1.15) = 2190 \text{ psi}$$

Determine column stability factor $C_P$:
From Eq. (5-14), $K_{cE} = 0.418$ and $c = 0.9$

$$F_{cE} = \frac{(0.418)(1,700,000)}{(26.5)^2} = 1012 \text{ psi}$$

$$C_P = \frac{1 + 1012/2190}{(2)(0.9)} - \sqrt{\left[\frac{1 + 1012/2190}{(2)(0.9)}\right]^2 - \frac{1012/2190}{0.9}} = 0.430$$

$$F_c' = F_c^* C_P = 941 \text{ psi}$$

$$f_c = \frac{6700}{30.75} = 220 \text{ psi} < 941 \text{ psi} \qquad \text{O.K.}$$

Check member 17 for tension.

$$F_t' = F_t C_D = (1250)(1.15) = 1440 \text{ psi}$$

$$f_t = \frac{9400}{30.75} = 310 \text{ psi} < F_t' \qquad \text{O.K.}$$

Use $5\frac{1}{8} \times 6$ in. as a trial size for all web members considering connections will reduce net sections and a smaller size could present problems in making connections.

Based on these preliminary member sizes, the member forces shown in Table 6.4 were determined using computer analysis.

Design connections on the basis that they will be pinned and will not significantly resist rotations.

Design heel connection. Note that this connection should be as shown in the bottom detail of Fig. 6.6 so as to be considered as a pinned connection and have concentric force lines. The upper detail of Fig. 6.6 does not have concentric force lines and is not a pinned connection. The restraint caused by the four bolts in each chord will tend to cause splitting in the chord.

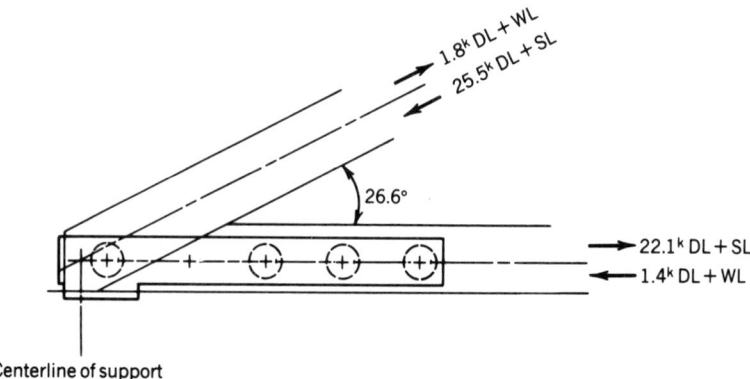

**TABLE 6.4**

**Summary of Forces and Moments**[a]

| Joint Condition | | Pinned | | | | Fixed | | | |
| --- | --- | --- | --- | --- | --- | --- | --- | --- | --- |
| Loading | | DL + SL | | DL + USL[b] | | DL + SL | | DL + USL[b] | |
| | Member | F | M | F | M | F | M | F | M |
| Top chord | 1 | −25.5 | 29.6 | −20.7 | 31.6 | −25.6 | 29.2 | −20.9 | 31.5 |
| | 2 | −20.7 | 39.9 | −15.9 | 40.5 | −20.6 | 33.0 | −15.8 | 35.4 |
| | 3 | −15.6 | 39.9 | −10.8 | 40.5 | −15.7 | 39.8 | −10.9 | 33.4 |
| | 4 | −15.6 | 39.9 | −10.0 | 12.6 | −15.7 | 39.8 | −10.1 | 21.7 |
| | 5 | −20.7 | 39.9 | −11.7 | 12.6 | −20.6 | 33.0 | −11.7 | 8.7 |
| | 6 | −25.5 | 29.6 | −13.3 | 7.8 | −25.6 | 29.2 | −13.3 | 7.5 |
| Bottom chord | 7 | +22.1 | 6.9 | +17.8 | 5.7 | +22.1 | 12.0 | +17.9 | 12.3 |
| | 8 | +22.1 | 6.9 | +17.8 | 5.7 | +22.0 | 5.4 | +17.7 | 4.4 |
| | 9 | +17.7 | 0.5 | +13.4 | 0.4 | +17.6 | 2.8 | +13.3 | 2.4 |
| | 10 | +17.7 | 0.5 | +10.2 | 0.3 | +17.6 | 2.8 | +10.1 | 1.3 |
| | 11 | +22.1 | 6.9 | +11.6 | 3.5 | +22.0 | 5.4 | +11.6 | 2.8 |
| | 12 | +22.1 | 6.9 | +11.6 | 3.5 | +22.1 | 12.0 | +11.6 | 3.8 |
| Webs | 13 | −0.14 | 0 | −0.11 | 0 | −0.04 | 4.6 | −0.06 | 5.1 |
| | 14 | −4.9 | 0 | −5.0 | 0 | −4.9 | 2.7 | −4.9 | 1.8 |
| | 15 | +2.2 | 0 | +2.3 | 0 | +2.2 | 3.8 | +2.2 | 2.2 |
| | 16 | −6.7 | 0 | −6.7 | 0 | −6.1 | 1.2 | −6.3 | 0.4 |
| | 17 | +9.5 | 0 | +6.3 | 0 | +8.7 | 0 | +5.8 | 4.3 |
| | 18 | −6.7 | 0 | −2.2 | 0 | −6.1 | 1.2 | −1.8 | 1.3 |
| | 19 | +2.2 | 0 | +0.73 | 0 | +2.2 | 3.8 | +0.7 | 2.9 |
| | 20 | −4.9 | 0 | −1.6 | 0 | −4.9 | 2.7 | −1.6 | 1.8 |
| | 21 | −0.14 | 0 | −0.07 | 0 | −0.04 | 4.6 | −0.02 | 0.9 |

[a]Units—kips; (+) tension, (−) compression, moments—in.-kips.

[b]USL—unbalanced snow load.

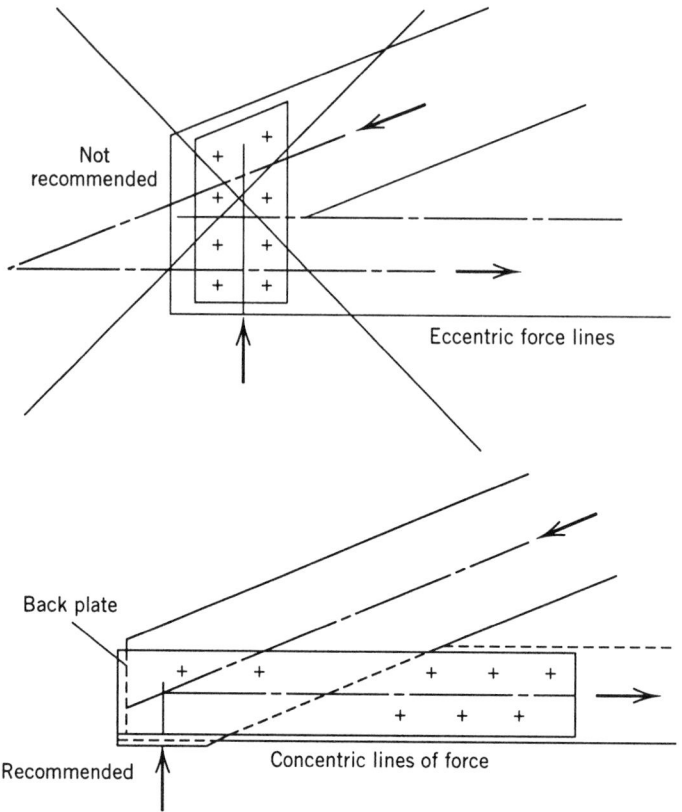

**FIGURE 6.6**    Truss heel connection.

Determine number of connectors in bottom chord. Use 4-in. diameter shear plates. Members are Group B species. From Table 7.32, the design value is 4320 lb for one shear plate. When used with metal side plates, this value is adjusted by the steel side plate factor $C_{st}$ for Group B species. Value can also be obtained directly from Table 7.33.

Number of shear plates required $= T_{BC}/(4320)(1.11)(1.15) = 22,100/(4320)(1.11)(1.15) = 4.1$

Try six shear plates on three $\frac{3}{4}$-in. bolts.

Try $\frac{1}{4} \times 4\frac{1}{2}$-in. side plate. See AISC, *Manual of Steel Construction* (6) for design of steel plates and definition of terms.

$$A_{net} = (0.25)(2)(4.5 - 0.81) = 1.85 \text{ in.}^2$$

$$A_{gross} = (0.25)(2)(4.5) = 2.25 \text{ in.}^2$$

$$\text{Capacity} = 0.50 F_u A_{net} = (29,000)(1.85)$$

$$= 53,650 \text{ lb} > 22,100 \text{ lb} \qquad \text{O.K.}$$

or

$$= 0.60 F_y A_{gross} = (22,000)(2.25)$$

$$= 49,500 \text{ lb} > 22,100 \text{ lb} \qquad \text{O.K.}$$

Check the capacity of the connection considering reduction for fasteners in a row.

$$A_m = (5.125)(7.5) = 38.4 \text{ in.}^2$$

$$A_s = (0.25)(4.5)(2) = 2.25 \text{ in.}^2$$

$$\frac{A_m}{A_s} = \frac{38.4}{2.25} = 17.1$$

From Table 7.6, group action factor, $C_g$, is 0.96 for three shear plates in a row.

Capacity of six 4-in. shear plates $= (6)(4320)(1.11)(1.15)(0.96)$

$$= 31,800 \text{ lb} > 22,100 \text{ lb} \qquad \text{O.K.}$$

Check the end grain bearing of the top chord on the back plate of the heel connection using the Hankinson formula:

$$F'_n = F'_{c-26.6°} = \frac{F'_g F'_{c\perp}}{F'_g \sin^2 26.6° + F'_{c\perp} \cos^2 26.6°}$$

$$F_g = 2300 \text{ psi (from Table A-1, AITC 117—Design)}$$

$$F_{c\perp} = 560 \text{ psi}$$

$$F'_{c-26.6°} = \frac{(2300)(1.15)(560)}{(2300)(1.15)(0.2) + (560)(0.8)} = 1516 \text{ psi}$$

$$f_c = \frac{22,100}{(5.125)(4.5)} = 958 \text{ psi} < 1516 \text{ psi} \qquad \text{O.K.}$$

Check the connection in the top chord for tension loading. Capacity of one $\frac{3}{4}$-in. bolt and one 4-in. shear plate is 4800 lb from Table 7.33.

Determine the reduction for a 5-in. end distance, $C_{\Delta n}$, by interpolation of values from Table 7.35.

| End distance (in.) | Percent of design load |
|---|---|
| 7 | 100 |
| $3\frac{1}{2}$ | 62.5 |

$$C_{\Delta n} = \left[\frac{(1.5)}{(3.5)}\right] 0.375 + 0.625 = 0.79$$

$P$ limited to strength of wood.

$$P = (4800)(1.6)(0.79)(2) = 12,069 \text{ lb} > 1800 \text{ lb (DL + WL)}$$

*P* limited by strength of steel.

$$P = 4800(1.33)(2) = 12,768 \text{ lb (wood controls)}$$

The bottom connection plate should be checked for buckling. If critical, a tie bolt can be added near the end of the bottom plate.

The truss chord to the web connections should be designed as shown in the bottom detail of Fig. 6.7. Web members should be fastened to the side plates with a minimum of two bolts. Determine the number and size of bolts in the ends of web members.

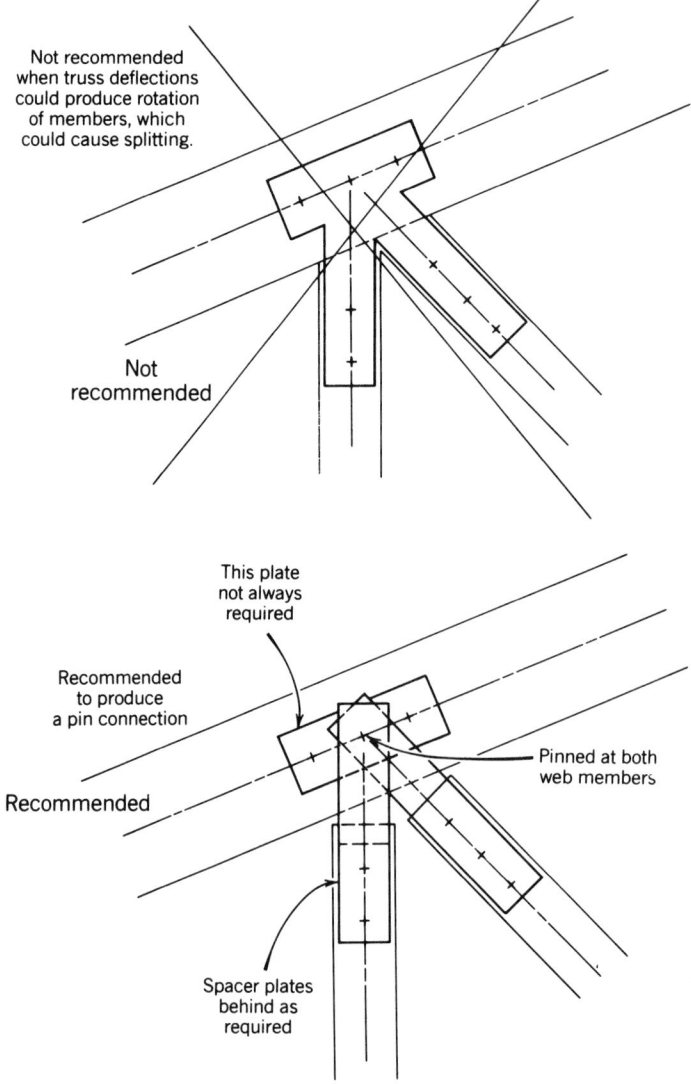

**FIGURE 6.7**    Truss web and chord connection.

Maximum web forces:

$$\text{Tension} = +9500 \text{ lb (DL + SL)}$$

$$\text{Compression} = -6700 \text{ lb (DL + SL)}$$

The capacity of one $\frac{3}{4}$-in. bolt parallel to grain is 3170 lb from Table 7.18. This value can be increased by 1.15 for duration of load for snow.

$$Z' = (3170)(1.15) = 3650 \text{ lb}$$

Two $\frac{3}{4}$-in. bolts are adequate for the maximum forces, except for the ends of member 17, which require three bolts. Use two $\frac{3}{4}$ in. bolts in the end of each web member (three $\frac{3}{4}$-in. bolts for member 17).

Spacing should be 4 times the bolt diameter (3 in.) and minimum end distance should be 7 times the bolt diameter ($5\frac{1}{4}$ in.). Check the connection between webs 15 and 16 and the top chord at point 3.

$$\text{Perpendicular to grain load} = 6360 - 2060 = 4300 \text{ lb}$$

$$\text{Parallel to grain load} = 1030 + 2120 = 3150 \text{ lb}$$

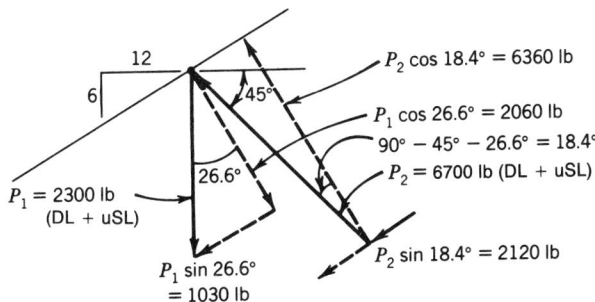

The capacity of one $\frac{3}{4}$-in. bolt perpendicular to grain is 1800 lb. Use three $\frac{3}{4}$-in. bolts in the top chord as shown in the bottom detail of Fig. 6.7. The capacity of one 4-in. shear plate perpendicular to the grain is 3000 lb. Two 4-in. shear plates on a $\frac{3}{4}$-in. bolt = (2)(3000) = 6000 lb. Increase 15% for duration of load,

$$P' = (2)(3000)(1.15) = 6900 > 4300 \text{ lb}$$

An alternate would be to use two 4-in. shear plates on a $\frac{3}{4}$-in. bolt for this connection.

Check the member sizes based on reduced member cross section at connectors. Use member forces based on computer analysis with pinned joints.

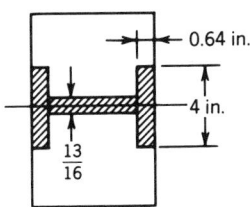

Check member 1:

$$P = 25,500 \text{ lb (DL + SL)}$$

$$M = 29,600 \text{ in.-lb (DL + SL)}$$

$$A_{\text{net}} = 38.44 - 8.2 = 30.2 \text{ in.}^2$$

$$S_{\text{net}} = 1/6[(5.125)(7.5)^2 - (2)(0.64)(4)^2$$

$$- (3.84)(0.81)^2] = 44.2 \text{ in.}^3$$

$$(l_e/d)_x = \frac{(8.95)(12)}{(7.5)} = 14.3$$

Determine column stability factor $C_P$:

$$c = 0.9 \qquad K_{cE} = 0.418 \text{ [from Eq. (5-14)]}$$

$$F_c^* = (1600)(1.15) = 1840 \text{ psi}$$

$$F_{cE} = \frac{K_{cE} E'}{(l_e/d)^2} = \frac{(0.418)(1,700,000)}{(14.3)^2} = 3475 \text{ psi}$$

$$C_P = \frac{1 + 3475/1840}{(2)(0.9)} - \sqrt{\left[\frac{1 + 3475/1840}{(2)(0.9)}\right]^2 - \frac{3475/1840}{0.9}} = 0.914$$

$$F_c' = F_c^* C_P = (1840)(0.914) = 1680 \text{ psi}$$

$$f_c = \frac{25,500}{30.2} = 844 \text{ psi}$$

$$f_{bx} = \frac{29,600}{44.2} = 670 \text{ psi}$$

$$\left(\frac{f_c}{F_c'}\right)^2 + \frac{f_{bx}}{F_{bx}'[1 - (f_c/F_{cE})]} = \left(\frac{844}{1680}\right)^2 + \frac{670}{2760[1 - (844/3475)]} = 0.573$$

$$0.573 < 1 \qquad \text{O.K.}$$

Check member 2:

$$f_c = \frac{20,700}{30.2} = 685 \text{ psi}$$

$$f_{bx} = \frac{39,900}{44.2} = 903 \text{ psi}$$

$$\left(\frac{685}{1680}\right)^2 + \frac{903}{2760[1 - (685/3475)]} = 0.574 < 1 \qquad \text{O.K.}$$

Check member 7:

$$T = 22,100 \text{ lb (DL + SL)}$$

$$M = 6900 \text{ in.-lb (DL + SL)}$$

$$f_t = \frac{22,100}{30.2} = 732 \text{ psi}$$

$$f_b = \frac{6900}{44.2} = 156 \text{ psi}$$

$$\frac{f_t}{F_t'} + \frac{f_b}{F_b^*} \leq 1$$

$$\frac{732}{(1250)(1.15)} + \frac{156}{(1600)(1.15)} = 0.51 + 0.08 = 0.59 < 1.0 \qquad \text{O.K.}$$

$$\frac{f_b - f_t}{F_b^{**}} \leq 1$$

where $F_b^{**} = F_b$ multiplied by all adjustment factors except $C_V$. When $f_b$ is less than $f_t$, $f_b = 156$ psi and $f_t = 732$ psi, this last check is not necessary.

Check compression in bottom chord taking member forces from graphical analysis of DL + WL:

$$P = 1400 \text{ lb (DL + WL)}$$

$$(l_e/d)_y = \frac{(16)(12)}{(5.125)} = 37.5$$

$$F_c^* = (1900)(1.6) = 3040 \text{ lb}$$

Determine column stability factor $C_P$:

$$F_{cE} = \frac{K_{cE}E'}{(l_e/d)^2} = \frac{(0.418)(1,700,000)}{(37.5)^2} = 505 \text{ psi}$$

$$C_P = \frac{1 + 505/3040}{(2)(0.9)} - \sqrt{\left[\frac{1 + 505/3040}{(2)(0.9)}\right]^2 - \frac{505/3040}{0.9}} = 0.163$$

$$F_c' = F_c^* C_P = (3040)(0.163) = 497 \text{ psi}$$

$$f_c = \frac{1400}{30.2} = 46 \text{ psi} < 497 \text{ psi} \qquad \text{O.K.}$$

Check members 16 and 18:

$$P = 6700 \text{ lb (DL + SL)}$$

From preliminary size check, web members of $5\frac{1}{8} \times 6$ in. are O.K.

Check member 17:

$$T = 9500 \text{ lb (DL + SL)}$$

For a $5\frac{1}{8} \times 6$ in. member

$$A_{net} = 30.75 - 8.2 = 22.6 \text{ in.}^2$$

$$F_t' = 1250(1.15) = 1440 \text{ psi}$$

$$f_t = \frac{9500}{22.6} = 420 \text{ psi} < 1440 \text{ psi} \qquad \text{O.K.}$$

All other web members of $5\frac{1}{8} \times 6$ in. are O.K. by comparison with above check.

At this point, adjustment of sizes for minimum design should be considered. If adjustments are made, the stiffness analysis should be recalculated since some redistribution of forces and moments will likely occur. Also, after connection designs are completed, member net sections should be reviewed and compared to those used in member designs. If pinned connections are not used, the forces and moments from the "fixed-joint" analysis must be used in both the design of members and connections.

## 6.3  POLE-TYPE FRAMING

Pole-type frame structures generally consist of preservatively treated round timber poles set in the ground as the main upright supporting members. Pole-type construction should not be confused with "post-frame" construction that uses solid-sawn or laminated timbers and generally uses diaphragm action to carry the bulk of the lateral wind loads. For information on post-frame construction, see Refs. (7)–(10). Preservatively treated sawn timber or glued laminated timber posts may also be used. The hole in which the pole or post is set furnishes both vertical and horizontal support for the structure. Setting the pole in the ground tends to prevent rotation of the bottom of the pole, thereby providing some or all of the required bracing. The roof portion of the structure is usually framed with lumber or glued laminated timber. Preservative treatment of the poles or posts that are set in the ground should be in accordance with the American Wood-Preservers' Association Standards (11). Use AWPA C1 for all timber products, C4 for round poles, C2 for sawn timber posts, and C28 for glued laminated timber posts.

### 6.3.1  General Considerations

General considerations applicable to all pole-type frame structures include the following:

1.  A bracing system can be provided at the top of a pole in order to reduce bending moments at the base of the pole and to distribute loads. The design of buildings supported by poles without bracing requires good knowledge of soil conditions in order to eliminate excessive deflection or sidesway.

2.   Bearing values under butt ends of poles should be checked. It is common practice to backfill holes around poles with well-tamped native soil, sand, or gravel. Backfilling with concrete or soil-cement can develop a more effective pole diameter; consequently, it can be used as a means of reducing required depth of embedment. Concrete backfill also increases the area of the pole for skin friction and this increases the bearing capacity. Skin friction is also effective where uplift due to wind may act on a pole through its connections to the roof framing.

3.   In order to increase bearing capacity under pole butts, concrete footings may be used. If they are used, they should be designed to withstand the punching shear of the pole and load. Concrete footings should be considered for use even in firm soils such as dry hard clay, coarse firm sand, or gravel.

The intended use of the structure largely determines such general features as height, overall length and width, spacing of poles, height at eaves, type of roof framing, and the kind of flooring to be used, as well as any special features such as wide bays, unsymmetrical layouts, or the possible suspending of particular loads from the roof framing. These general design features having been determined, the following procedure may be used.

The design of pole-frame structures as presented in this manual is based on all lateral forces being resisted by embedment of the poles. No diaphragm action of the roof or shear resistance of the end walls is required. However, the use of roof diaphragms and end shear walls will often result in a more economical structure.

The depth of the embedment of poles is based on semiempirical equations. For more precise calculations of pole foundation design, see Post and Pole Foundation Design, ASAE EP 486 (12), or Pole Building Design, American Wood Preservers Institute (13).

The design method included herein is based on structures using a glued laminated beam between poles. When other roof systems are used, it is recommended that Refs. (7) and (8) be considered.

### 6.3.2   Design Values for Poles

The National Design Specification (14) does not contain design values for poles. It is suggested that the design values for piles which are included in the NDS, based on single member use be used for pole design. Design values can also be derived from the tables for MOR, MOE, and COV in ANSI 05.1-1992 (15). For some uses, other design values may be appropriate. [See Refs. (13) and ASTM D 3200 *Standard Specification and Methods for Establishing Recommended Design Stresses for Round Timber Construction Poles* (16).] Pole frame buildings may be used for purposes (such as farm buildings) where the importance of the structure is such that higher design values are justified. The designer should check the applicable building code for the appropriate design values.

Pile design values in Table 6.8 are based on piles being used in a cluster. When piles are used singly, the design value must be modified by the single pile factor $C_{sp}$ shown in Table 6.10 for bending and compression parallel to grain. The single pile factors have already been included in Table 6.5.

## TABLE 6.5

### Design Values for Preservatively Treated Poles[a,b], Wet and Dry Conditions of Use

| Species | Bending $F_b$ (psi) | Compression Parallel to Grain $F_c$ (psi) | Modulus of Elasticity $E$ (psi) | Compression Perpendicular to Grain $F_{c\perp}$ (psi) | Horizontal Shear $F_v$ (psi) | End Grain in Bearing[c] $F_g$ (psi) |
|---|---|---|---|---|---|---|
| Douglas Fir, Coast | 1,850 | 1,000[d] | 1,500,000 | 375 | 115 | 1,200 |
| Jack Pine | 1,500 | 800 | 1,100,000 | 280 | 95 | 870 |
| Lodgepole Pine | 1,350 | 700 | 1,100,000 | 240 | 85 | 870 |
| Northern White Cedar | 1,050 | 525 | 600,000 | 225 | 80 | 670 |
| Ponderosa Pine | 1,300 | 650 | 1,000,000 | 320 | 90 | 820 |
| Red or Norway Pine | 1,450 | 725 | 1,300,000 | 265 | 85 | 790 |

| Species | | | | | |
|---|---|---|---|---|---|
| Southern Pine | 1,700 | 900[d] | 1,500,000 | 320 | 105 | 1,120 |
| Western Red Cedar | 1,350 | 750 | 900,000 | 255 | 95 | 940 |
| Western Hemlock | 1,650 | 900 | 1,300,000 | 245 | 115 | 1,120 |
| Western Larch | 2,050 | 1,075 | 1,500,000 | 375 | 120 | 1,210 |

[a]The design values are based on ASTM D2899-86, *Method for Establishing Design Stresses for Round Timber Piles* (24). The values are for single-member uses of poles and assume that the conditioning prior to treatment was in accordance with ASTM D1760-86a (17). If the poles are conditioned by air or kiln drying only prior to treatment, the design values, with the exception of modulus of elasticity, may be increased by application of the adjustment factor for seasoning conditioning, $C_u$, which is 1.18 for Southern Pine and 1.11 for other species.

[b]The compression perpendicular to grain design values $F_{c\perp}$ are based on design values for sawn lumber under wet conditions of use which have been reduced for conditioning. For dry conditions of use, multiply by 1.5.

[c]End grain bearing design values, $F_g$, are based on the design values for sawn lumber used in a wet location and reduced for conditioning. When used in a dry location, they may be increased by 10% ($C_M = 1.10$).

[d]Design values for compression parallel to grain, $F_c$, in Douglas Fir and Southern Pine may be increased 0.2% for each foot of length from the tip of the pole to the critical section. This increase shall not exceed 10%. The modification factor for location of critical section, $C_{cs} = 1 + 0.002L$, where $L$ = distance (ft) from the tip to the critical section, $C_{cs}$, shall not exceed 1.10.

The design values in Table 6.5 are consistent with other wood design values in regard to safety factors. They include reduction factors for conditioning prior to treatment in accordance with ASTM D 1760 (17), as well as safety factors for single-member use. When air-dried or kiln-dried conditioning only is used prior to treatment, the untreated factor, $C_u$, is applied to all design values except modulus of elasticity. For Southern Pine, the value of $C_u$ is 1.18 and for other species 1.11. The design values for compression perpendicular to grain, $F_{c\perp}$, are calculated by the same method used for sawn lumber based on wet conditions of use and further modified for conditioning prior to treatment. The design values for shear $F_v$ are based on wet conditions of use with the reduction for conditioning prior to treatment applied. The design values for fastenings are the same as for sawn lumber. For general information on poles, see the *Wood Handbook* (18).

### 6.3.3   Design Procedures

1.   Determine loads. The principal load on a pole-type frame is generally the horizontal wind load; therefore, the wind load value required by the governing building code or, in the absence of a governing code, as determined from ASCE 7-88 (3), should be used. In addition to dead load and other loads, roofs will transmit vertical components of wind load to poles. When wind loads control, the duration factor, $C_D$, is applicable to the wood members.

Determine the resultant of the horizontal forces acting on the poles and the height above the ground surface at which it acts. ASCE 7-88 (3) requires determination of windward loading and a leeward loading of the structure. When glued laminated timber or other roofing systems of similar rigidity are used, the tops of the poles deflect approximately the same amount. Because the leeward forces may be less than the windward forces, the leeward pole or poles tend to support the windward pole. When only two poles are used in a bent, the following method can be used to determine the forces acting on the poles. In this method, it is assumed that the wind loads are applied through girts that approximate a uniform load on the poles. If not, the actual loading condition should be used.

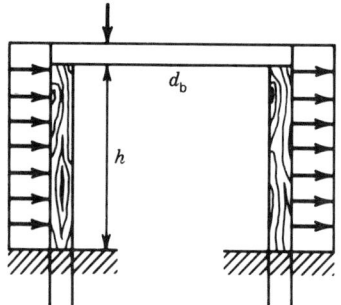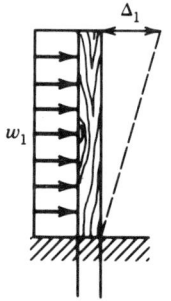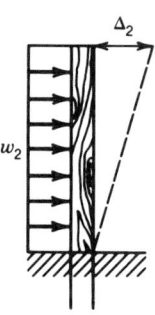

(a)   The top of both poles can be assumed to deflect the same amount. First assume that the poles are not connected by the roof structure and each is allowed to deflect separately. Let the deflection of the windward pole $= \Delta_1$ and the deflection of the leeward pole $= \Delta_2$. Next, calculate the force $P_3$ that will move the windward pole deflection $\Delta_3$ to the left and the leeward pole deflection $\Delta_3$ to the right so that $\Delta_A = \Delta_B$. Next, determine the resultant $P$ on the windward pole of the wind load and its distance $h$ from the ground line. Determine the wind load acting on the structure above the top of the poles, $P_1$ and $P_2$:

$$P_1 = w_1 d_b, \qquad P_2 = w_2 d_b$$

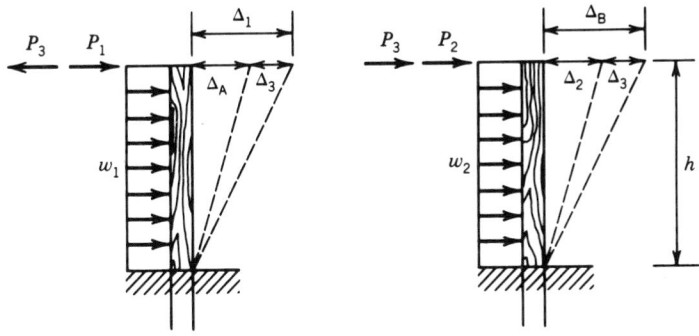

Determine $P_3$:

$$\Delta_1 = \frac{w_1 h^4}{8EI} + \frac{w_1 d_b h^3}{3EI} \qquad (6\text{-}3)$$

$$\Delta_2 = \frac{w_2 h^4}{8EI} + \frac{w_2 d_b h^3}{3EI} \qquad (6\text{-}4)$$

$$\Delta_3 = \frac{P_3 h^3}{3EI} \qquad (6\text{-}5)$$

where   $w_1 =$ wind load on windward side (plf),
$\quad\quad w_2 =$ wind load on leeward side (plf),
$\quad\quad d_b =$ depth of glued laminated beam, plus any additional roof structure above beam (ft),
$\quad\quad EI =$ the stiffness modulus, which is assumed to be the same for each pole, and
$\quad\quad h =$ length of pole below beam (ft).
$$\Delta_A = \Delta_1 - \Delta_3 \qquad \Delta_B = \Delta_2 + \Delta_3 \qquad \Delta_A = \Delta_B$$

therefore,

$$\Delta_1 - \Delta_3 = \Delta_2 + \Delta_3,$$

$$\frac{w_1 h^4}{8EI} + \frac{w_1 d_b h^3}{3EI} - \frac{P_3 h^3}{3EI} = \frac{w_2 h^4}{8EI} + \frac{w_2 d_b h^3}{3EI} + \frac{P_3 h^3}{3EI}$$

Rearranging terms and solving for $P_3$,

$$P_3 = \frac{3h(w_1 - w_2)}{16} + d_b \left( \frac{w_1 - w_2}{2} \right) \tag{6-6}$$

Determine the resultant of the forces on the windward pole, $P$, and the height above ground line, $h'$, at which it acts

$$P = w_1 h + P_1 - P_3$$

$$h' = \frac{(w_1 h^2/2) + (P_1 - P_3)h}{P} \tag{6-7}$$

where notations are those that have been described previously.

(b)   When knee braces are used, the deflection at the points of contraflexture can be assumed to be equal with only a small error. The forces acting on the windward pole at and below the point of contraflexture are calculated in a manner similar to that for poles without knee braces.

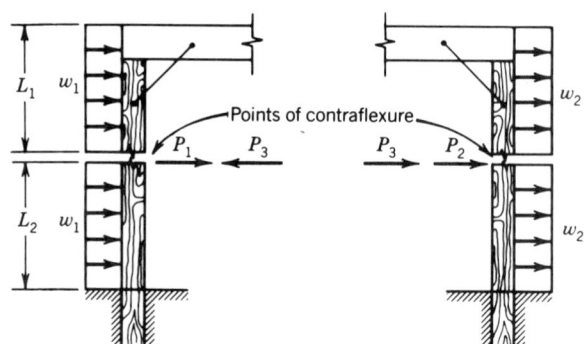

$$P_1 = w_1 L_1$$

$$P_3 = \frac{L_1(w_1 - w_2)}{2} + \frac{3L_2(w_1 - w_2)}{16}$$

$$P = w_1(L_2 + L_1) - P_3$$

$$h' = \frac{(P_1 - P_3)L_2 + (w_1 L_2^2)}{P} \tag{6-8}$$

where notations are those described previously and shown in the sketch.

The following combinations of loads are recommended for roofs unless specified otherwise by the applicable building code: dead load plus snow load, dead load plus wind load, and dead load plus wind load plus one-half snow load.

2.   Determine soil values. If soil tests are not available, visual inspection and a careful estimate of the bearing value of samples of the soil should be made. Build-

ing codes generally contain allowable design values for direct bearing assigned to various soil classifications.

When building codes do not give established allowable lateral passive soil pressure values, the values may be determined by using Rankine's formula:

$$p = \frac{1 + \sin \phi}{1 - \sin \phi} wd \qquad (6\text{-}9)$$

where  $p$ = allowable passive soil pressure (psf per ft of depth),
$\phi$ = angle of internal friction of soil (degrees),
$w$ = weight of soil (pcf), and
$d$ = depth below grade (ft).

Allowable soil pressure values for various classes of soil are also given in Table 6.6.

3.  Estimate the size of pole required. The American National Standards Institute (ANSI) has established certain pole classes. Table 6.7 tabulates the size requirements corresponding to ANSI classes H-6–10 for Douglas Fir and Southern Pine poles. The size requirements for other species may be obtained from ANSI 05.1 (15).

Poles of a given class and length are selected to have approximately the same load-carrying capacity regardless of species. The minimum circumferences specified at 6 ft from the butt in Table 6.7 are based on the maximum fiber stress in bending that will occur at the groundline due to a given horizontal load applied 2 ft from the top of the pole.

The size of a pole is generally influenced more by bending stresses than by compressive stresses. However, these stresses are interactive, and the equation for combined bending and compression stresses should be used in the final design. To estimate the pole size, determine the size based on bending alone, then increase that size based on the effect of the compressive stress.

Circumferences at points other than those tabulated in Table 6.7 may be determined assuming an average taper of 0.25 in. circumference per foot of length of pole for Douglas Fir and Southern Pine.

4.  Determine required embedment of pole. Eq. (6-10) may be used in determining required embedment depth where no constraint (such as a rigid floor or surface pavement) is provided at the ground surface. Note that this procedure requires the assumption of an initial embedment depth for determining lateral soil bearing pressure.

$$d = \frac{A}{2}\left(1 + \sqrt{\frac{1 + 4.36h}{A}}\right) \text{ (see Fig. 6.8)} \qquad (6\text{-}10)$$

where  $d$ = depth of embedment in earth (ft),
$A = 2.34 \, P/S_1 B$,
$P$ = applied horizontal force on pole (lb),

## TABLE 6.6

### Allowable Foundation and Lateral Pressure

| Class of Materials[2] | Allowable Foundation Pressure Lbs./Sq. Ft.[3] | Lateral Bearing Lbs./Sq./Ft./Ft. of Depth Below Natural Grade[4] | Lateral Sliding[1] Coefficients[5] | Lateral Sliding[1] Resistance Lbs./Sq. Ft.[6] |
|---|---|---|---|---|
| 1. Massive Crystalline Bedrock | 4000 | 1200 | .70 | |
| 2. Sedimentary and Foliated Rock | 2000 | 400 | .35 | |
| 3. Sandy Gravel and/or Gravel (GW and GP) | 2000 | 200 | .35 | |
| 4. Sand, Silty Sand, Clayey Sand, Silty Gravel and Clayey Gravel (SW, SP, SM, SC, GM and GC) | 1500 | 150 | .25 | |
| 5. Clay, Sandy Clay, Silty Clay and Clayey Silt (CL, ML, MH and CH) | 1000[7] | 100 | | 130 |

[1]Lateral bearing and lateral sliding resistance may be combined.

[2]For soil classifications OL, OH and PT (i.e., organic clays and peat), a foundation investigation shall be required.

[3]All values of allowable foundation pressure are for footings having a minimum width of 12 inches and a minimum depth of 12 inches into natural grade. Except as in Footnote No. 7 below, increase of 20 percent allowed for each additional foot of width or depth to a maximum value of three times the designated value.

[4]May be increased the amount of the designated value for each additional foot of depth to a maximum of 15 times the designated value. Isolated poles for uses such as flagpoles or signs and poles used to support buildings which are not adversely affected by a $\frac{1}{2}$-inch motion at ground surface due to short-term lateral loads may be designed using lateral bearing values equal to two times the tabulated values.

[5]Coefficient to be multiplied by the dead load.

[6]Lateral sliding resistance value to be multiplied by the contact area. In no case shall the lateral sliding resistance exceed one half the dead load.

[7]No increase for width is allowed.

Source: Reproduced from The Uniform Building Code, 1991 ed. (20), Copyright 1991, with permission of the International Conference of Building Officials.

$S_1$ = allowable lateral soil-bearing pressure as shown in Table 6.6 based on a depth of one-third the depth of embedment but not over 12 ft. For purposes of determining lateral pressure (psf), $S_1 = S_0 d/3$,

$h$ = height above groundline at which force $P$ is applied (ft), and

$B$ = butt diameter of pole, diagonal of square pole, or diameter of concrete casing (ft).

If the pole is restrained at the groundline, such as by a rigid concrete floor, the following equation is used:

$$d = \sqrt{\frac{4.25Ph}{S_3 B}} \qquad (6\text{-}11)$$

where   $S_3$ = allowable lateral soil-bearing pressure as set forth in Table 6.6 based on a depth equal to the depth of embedment (psf) and other terms are as previously defined ($S_3 = S_0 d$).

It is sometimes more convenient to use the equations in the form

$$d = 1.97 \sqrt[3]{\frac{P(h + 0.93d)}{BS_4}} \quad \text{(for unrestrained footings)} \qquad (6\text{-}12)$$

and

$$d = 1.63 \sqrt[3]{\frac{Ph}{BS_4}} \quad \text{(for restrained footings)} \qquad (6\text{-}13)$$

where   $S_4$ = tabular value in Table 6.6 for lateral soil pressure $S_0$ adjusted for allowable movement (psf) and other terms are as previously defined.

Note that when buildings are not adversely affected by a $\frac{1}{2}$-in. movement at the ground surface due to short-term lateral loads, the tabular design value of $S_0$ may be doubled.

For poles having some degree of fixity or restraint at the eave line, such as provided by knee braces, the point at which the applied horizontal force $P$ acts is at the point of contraflexure of the pole. For round tapered poles, the point of contraflexure is assumed to be at two-thirds the distance from the groundline to the point of restraint. For poles with no restraint, $h$ is calculated as shown in Fig. 6.8.

It may be desirable to repeat the procedure for determining embedment with the first determination of embedment as the basis for recalculating the value of $S_1$ or $S_3$.

## TABLE 6.7

### Dimensions of Douglas Fir (Both Types) and Southern Pine Poles$^{a,b}$

| Class | H-6 | H-5 | H-4 | H-3 | H-2 | H-1 | 1 | 2 | 3 | 4 | 5 | 6 | 7 | 9 | 10 |
|---|---|---|---|---|---|---|---|---|---|---|---|---|---|---|---|
| Minimum Circumference at Top (in.) | 39 | 37 | 35 | 33 | 31 | 29 | 27 | 25 | 23 | 21 | 19 | 17 | 15 | 15 | 12 |
| Length of Pole (ft) / Groundline Distance from Butt$^c$ (ft) | | | | | | Minimum Circumference at 6 ft from Butt (in.) | | | | | | | | | |
| 20 / 4.0 | | | | | | | 31.0 | 29.0 | 27.0 | 25.0 | 23.0 | 21.0 | 19.5 | 17.5 | 14.0 |
| 25 / 5.0 | | | | | | | 33.5 | 31.5 | 29.5 | 27.5 | 25.5 | 23.0 | 21.5 | 19.5 | 15.0 |
| 30 / 5.5 | | | | | | | 36.5 | 34.0 | 32.0 | 29.5 | 27.5 | 25.0 | 23.5 | 20.5 | |
| 35 / 6.0 | | | | | 43.5 | 41.5 | 39.0 | 36.5 | 34.0 | 31.5 | 29.0 | 27.0 | 25.0 | | |
| 40 / 6.0 | | | 51.0 | 48.5 | 46.0 | 43.5 | 41.0 | 38.5 | 36.0 | 33.5 | 31.0 | 28.5 | | | |
| 45 / 6.5 | 58.5 | 56.0 | 53.5 | 51.0 | 48.5 | 45.5 | 43.0 | 40.5 | 37.5 | 35.0 | 32.5 | 30.0 | | | |
| 50 / 7.0 | 61.0 | 58.5 | 55.5 | 53.0 | 50.5 | 47.5 | 45.0 | 42.0 | 39.0 | 36.5 | 34.0 | | | | |
| 55 / 7.5 | 63.5 | 60.5 | 58.0 | 55.0 | 52.0 | 49.5 | 46.5 | 43.5 | 40.5 | 38.0 | | | | | |

| | | | | | | | | | | | |
|---|---|---|---|---|---|---|---|---|---|---|---|
| 60 | 8.0 | 65.5 | 62.5 | 59.5 | 57.0 | 54.0 | 51.0 | 48.0 | 45.0 | 42.0 | 39.0 |
| 65 | 8.5 | 67.5 | 64.5 | 61.5 | 58.5 | 55.5 | 52.5 | 49.5 | 46.5 | 43.5 | 40.5 |
| 70 | 9.0 | 69.0 | 66.5 | 63.5 | 60.5 | 57.0 | 54.0 | 51.0 | 48.0 | 45.0 | 41.5 |
| 75 | 9.5 | 71.0 | 68.0 | 65.0 | 62.0 | 59.0 | 55.5 | 52.5 | 49.0 | 46.0 | |
| 80 | 10.0 | 72.5 | 69.5 | 66.5 | 63.5 | 60.0 | 57.0 | 54.0 | 50.5 | 47.0 | |
| 85 | 10.5 | 74.5 | 71.5 | 68.0 | 65.0 | 61.5 | 58.5 | 55.0 | 51.5 | 48.0 | |
| 90 | 11.0 | 76.0 | 73.0 | 69.5 | 66.5 | 63.0 | 59.5 | 56.0 | 53.0 | 49.0 | |
| 95 | 11.0 | 77.5 | 74.5 | 71.0 | 67.5 | 64.5 | 61.0 | 57.0 | 54.0 | | |
| 100 | 11.0 | 79.0 | 76.0 | 72.5 | 69.0 | 65.5 | 62.0 | 58.5 | 55.0 | | |
| 105 | 12.0 | 80.5 | 77.0 | 74.0 | 70.5 | 67.0 | 63.0 | 59.5 | 56.0 | | |
| 110 | 12.0 | 82.0 | 78.5 | 75.0 | 71.5 | 68.0 | 64.5 | 60.5 | 57.0 | | |
| 115 | 12.0 | 83.5 | 80.0 | 76.5 | 72.5 | 69.0 | 65.5 | 61.5 | 58.0 | | |
| 120 | 12.0 | 85.0 | 81.0 | 77.5 | 74.0 | 70.0 | 66.5 | 62.5 | 59.0 | | |
| 125 | 12.0 | 86.0 | 82.5 | 78.5 | 75.0 | 71.0 | 67.5 | 63.5 | 59.5 | | |

[a]This material is reproduced with permission from American National *Specifications and Dimensions for Wood Poles*, ANSI 05.1, copyright 1992 by the American National Standards Institute. Copies of this standard may be purchased from the American National Standards Institute at 1430 Broadway, New York, N.Y. 10018.

[b]Douglas Fir includes Douglas Fir, Interior North, and Douglas Fir, Coastal.

[c]The figures in this column are intended for use only when a definition of groundline is necessary in order to apply requirements relating to scars, straightness, etc.

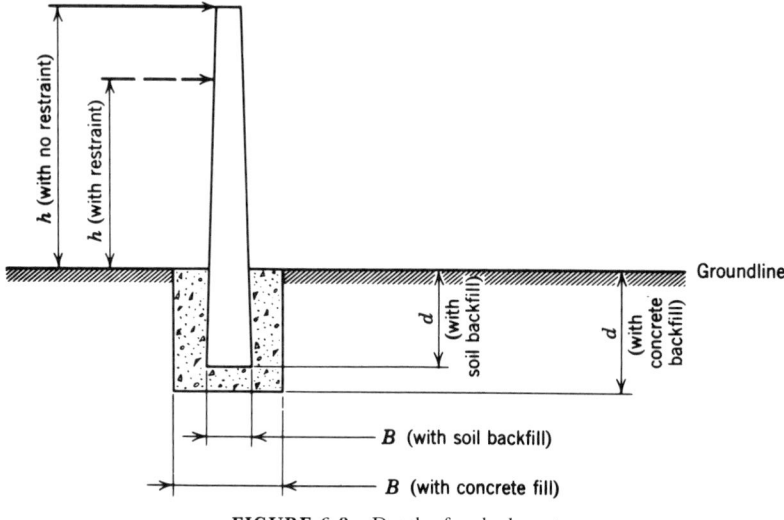

**FIGURE 6.8**    Depth of embedment.

5.    Check pole bending stress. The bending stress in the pole at the point of maximum moment may be determined by the equation

$$f_b = \frac{32\pi^2 M}{C^3} = \frac{32M}{\pi D^3} \tag{6-14}$$

where   $f_b$ = bending stress (psi),
   $C$ = pole circumference at point of maximum moment (in.),
   $D$ = pole diameter at point of maximum moment (in.), and
   $M$ = maximum moment (in.-lb), which, for a pole embedded in earth, is assumed to occur at one-fourth the depth of embedment below groundline; it may be determined if $P$ and $h$, as previously defined, are known. For a pole embedded in concrete, the maximum moment can be assumed to act at the top of the concrete.

The actual bending stress cannot exceed design values in bending in Table 6.5. These tabular design values may be adjusted by 1.6 for wind duration of loading.

When knee braces are used, the bending moment above the pole's point of inflection is maximum at the bottom of the roof bracing. The bending stress at this point should be checked to make sure that it does not exceed the allowable bending stress.

6.    Determine the pole compression stress. The vertical loads on one pole include, in addition to applied live loads and/or snow loads and dead load of the supported structure, the vertical component of wind load and the weight of the pole above the section being analyzed.

The design procedures for long round tapered columns are shown in 5.8.5.

The ratio $l_e/D$ should not exceed 44 for round columns. The unsupported length of long poles may be reduced by providing bracing in both directions at approximately the inflection point of the pole. The actual compression parallel to grain stress at the top, or small end, of the pole should be checked, and it should not exceed design values. Actual compressive stress $f_c$ may be determined by the equation

$$f_c = \frac{P}{A}$$

where   $P$ = total vertical load on the pole (lb) and
$A$ = cross-sectional area at the top of the pole (in.$^2$).

7.   Check combination of bending and axial stresses by use of Eq. (5-23). Assume that only half of the snow load is acting when maximum wind occurs and check combination of stresses at the point of maximum bending stress, which can be assumed to occur at the groundline when concrete backfill is used or one-fourth the depth of embedment below the groundline when soil is used for the backfill. The design should also be checked by the combination of bending and axial loading equation at the critical section of the column using the representative dimensions determined by Eq. (5-16) modified for round columns.

8.   Determine footing requirements. The pole footing requirements can be determined if the kind of soil, the vertical load on the pole, and the size of pole required are known. Soil-bearing pressure (psf) can be determined by the equation

$$\text{Soil bearing pressure} = \frac{P}{A}$$

where   $P$ = total vertical load on pole (lb) and
$A$ = area of butt of pole (ft$^2$).

If this load exceeds the allowable bearing capacity $S_B$ for the type of soil in question, a concrete footing is required.

The required area $A_c$ in square ft of a concrete footing may be determined by the equation

$$A_c = \frac{P}{S_B}$$

where   $S_B$ = allowable bearing capacity (psf)
and other terms are as previously defined.

Concrete footings should be designed in accordance with the recommendations of the American Concrete Institute.

**Example.**   Design the pole for the illustrated typical bent of a pole-type building to meet the stated conditions.

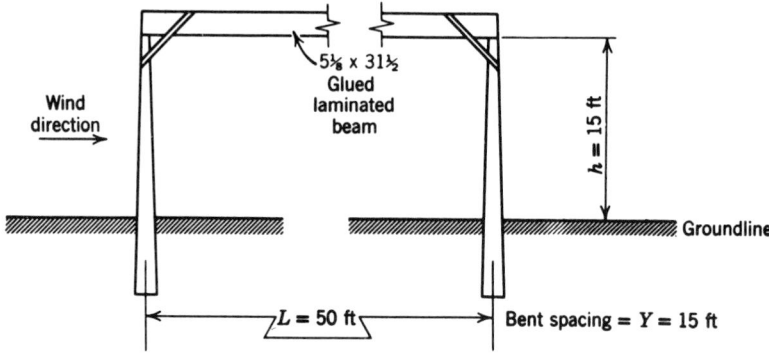

Loads:   DL = 10 psf     Pole design values:   $F_b$ = 1850 psi

SL = 20 psf ($C_D$ = 1.15)                $F_c$ = 1000 psi

WL = 10 psf on windward side plus    $E = E'$ = 1,500,000 psi

5 psf on leeward side

($C_D$ = 1.6)

1.   Determine loads:

$$\text{Vertical load on one pole} = P_{SL} = \frac{YL(DL + SL)}{2}$$

$$P_{SL} = \frac{(15)(50)(10 + 20)}{2} = 11{,}250 \text{ lb (full snow load)}$$

$$P'_{SL} = \frac{(15)(50)(10 + 10)}{2} = 7500 \text{ lb (one-half snow load)}$$

Wind loads:

$$\text{windward side, } w_1 = (10)(15) = 150 \text{ plf}$$

$$\text{leeward side, } w_2 = 75 \text{ plf}$$

Determine the wind load acting on structure above top of the poles. Assume roofing extends as much as $2\frac{1}{2}$ in. above the beam,

$$d_b = 31\frac{1}{2} + 2\frac{1}{2} = 34 \text{ in.}$$

$$P_1 = (34/12)(150) = 425 \text{ lb (windward side)}$$

$$P_2 = (34/12)(75) = 212.5 \text{ lb (leeward side)}$$

Determine the force $P_3$ from Eq. (6-6):

$$P_3 = \frac{3h(w_1 - w_2)}{16} + d_b \frac{(w_1 - w_2)}{2}$$

$$P_3 = \frac{(3)(15)(150 - 75)}{16} + \frac{(34/12)(150 - 75)}{2} = 317 \text{ lb}$$

Determine resultant of the wind forces on the windward pole, $P$, and the height above ground at which it acts.

$$P = w_1 h + P_1 - P_3$$

$$P = (150)(15) + 425 - 317 = 2358 \text{ lb}$$

$$h' = \frac{[w_1 h^2/2] + (P_2 - P_3)h}{P}$$

$$h' = \frac{[(150)(15)^2/2] + (425 - 317)(15)}{2358} = 7.84 \text{ ft}$$

2.   Determine soil values; soil is well-graded sand (Type 3 soil):

$$S_0 = 200 \text{ psf/ft from Table 6.6}$$

Multiply by 2 when $\frac{1}{2}$-in. movement at base permitted.

$$\text{Unrestrained } S_1 = 200 \, d/3$$

$$\text{Restrained } S_3 = 200 \, d$$

3.   Estimate size of pole required.

Determine size of pole required to resist wind load in bending:

$$M = (2358)(7.84)(12) = 221,840 \text{ in.-lb}$$

$$F_b' = (1850)(1.6) = 2960 \text{ psi}$$

$$C = \sqrt[3]{\frac{32\pi^2 M}{F_b'}} = \sqrt[3]{\frac{32\pi^2(221,840)}{2960}} = 28.7 \text{ in.}$$

where   $C$ = circumference (in.) and
$M$ = moment (in.-lb)

Assume that dead load plus $\frac{1}{2}$ snow load is acting at the same time as the maximum wind load.

$$P_{SL}' = 7500 \text{ lb}$$

Assume critical diameter of pole = 10 in.

$$f_c = \frac{P_{SL}'}{A} = \frac{7500}{(\pi)(10)^2/4} = 95 \text{ psi}$$

Assume a slightly larger pole than required to resist bending alone. Assume depth of embedment $d_1 = 9$ ft. A 25-ft class 1 pole has a circumference of 33.5 in. at a distance of 6 ft from the butt (Table 6.7).

4. Determine required embedment. Douglas Fir and Southern Pine poles have a taper of approximately 0.25 in. per ft in circumference.

$$\text{Butt diameter} = B = \frac{33.5 + (6)(0.25)}{\pi} = 11.14 \text{ in.} = 0.928 \text{ ft}$$

Calculated depth of embedment: Using Eq. (6-10), determine depth of embedment assuming a trial depth of 9 ft and movement of pole $= \frac{1}{2}$ in.

$$d = \frac{A}{2}\left(1 + \sqrt{1 + \frac{4.36h'}{A}}\right)$$

$S_0 = 200$ psf (from Table 6.6)

$$S_1 = \frac{(200)(2)(9)}{3} = 1200 \text{ psf}$$

$$A = \frac{2.34P}{S_1 B} = \frac{(2.34)(2358)}{(1200)(0.928)} = 4.95$$

$$d = \frac{4.95}{2}\left(1 + \sqrt{1 + \frac{(4.36)(7.84)}{4.95}}\right) = 9.44 \text{ ft}$$

This is a large depth, and the depth of embedment will be determined for a pole restrained at the floor line by a concrete floor using Eq. (6-11) assuming depth $= 8$ ft.

$$d = \sqrt{\frac{4.25Ph'}{S_3 B}}$$

$S_3 = (200)(8) = 1600$ psf

$$d = \sqrt{\frac{(4.25)(2358)(7.84)}{(1600)(0.928)}} = 7.27 \text{ ft}$$

Recheck depth assuming 7.5 ft.

$$S_3 = (200)(7.5) = 1500 \text{ psf}$$

$$d = \sqrt{\frac{(4.25)(2358)(7.84)}{(1500)(0.928)}} = 7.52 \text{ ft}$$

Use 7.5 ft.

A 25-ft-long pole is required. A 25-ft Class 1 Douglas Fir pole has a minimum circumference at the top of 27 in. and a minimum circumference of 33.5 in. 6 ft from the butt (Table 6.7). (Note: Since the embedment required is 7.5 ft and the pole is 25 ft in length, 2.5 ft of the pole would have to be cut off to maintain a pole height of 15 ft.)

The poles are subjected to a combined axial and bending load and the combination of stresses must be checked by Eq. (5-23) modified for round members. The pole taper and variation of bending stress along its length complicates the design. When the column is embedded in concrete and the concrete floor is capable of resisting the horizontal force of the pole, the point of maximum moment is near enough to the floor surface to assume that the maximum moment occurs at that surface. If the hole for the pole is not filled with concrete and the overturning is resisted by soil, the point of maximum moment can be as much as depth of embedment/3 below the surface.

In this design, the maximum moment is assumed to occur at the floor surface. When the combination of stresses is checked at this point, the design value for compression perpendicular to grain of $F_c^*$ is the tabulated design value multiplied by all applicable adjustment factors except $C_P$.

The combination of bending and axial stresses along the length of the column should also be investigated. Using Eq. (5-16), the representative depth of the column must be determined and the combination of stresses should be checked at the point where this occurs. In some cases, the combination of bending and axial stresses may need to be checked at other points.

   5.   Determine the bending and axial stresses in the pole at the point of maximum bending stress (ground line) and at the critical section for compression load (representative diameter of pole for compression load).

   6.   Determine the effect of dead load plus snow load only and the combined effects of $\frac{1}{2}$ snow load plus dead load and wind load at the ground line.

Determine $f_c$ at ground line (full snow load).

$$D_{\max} \text{ at ground line}$$

$$7.5 - 6 = 1.5 \text{ ft}$$

$$D_{\max} = 33.5/\pi - (1.5)(0.25) = 10.29 \text{ in.}$$

Determine weight of pole above ground line

$$A \text{ 25-ft Class 1 pole weighs 625 lb}$$

$$W = (15/25)(625) = 375 \text{ lb}$$

$$P = P_{\mathrm{SL}} + W = 11{,}250 + 375 = 11{,}625 \text{ lb}$$

$$f_c = \frac{P}{A} = \frac{11{,}625}{\pi/4(10.29)^2} = 140 \text{ psi}$$

$$C_P \text{ at ground line} = 1$$

$$C_{cs} = 1 + 0.002 L,$$

$L$ = the length from the pole tip to the ground line

(see Table 6.5, footnote $d$.)

$F'_c$ at ground line = $F_c C_D C_{cs}$

$$F'_c = (1,000)(1.15)[1 + (0.002)(17.5)] = 1,190 \text{ psi} > 140 \text{ psi} \qquad \text{O.K.}$$

Check combined stresses with 1/2 snow load.

$$F_{cE} = \frac{12 K_{cE} E'}{(l_e/r)^2} = \frac{(12)(0.3)(1,500,000)}{\left[ 378 \middle/ \left( \dfrac{10.29}{4} \right) \right]^2} = 250 \text{ psi}$$

$$F'_c = F_c C_D C_P C_{cs}$$

$$F'_c = (1,000)(1.6)(1.0)[1 + (0.002)(25 - 7.5)] = 1,656 \text{ psi}$$

$f_c$ for 1/2 snow load + dead load

$$P' = 7,000 + 375 = 7,375$$

$$f_c = \frac{P'}{A} = \frac{7,375}{\pi/4(10.29)^2} = 89 \text{ psi}$$

$C_L$ for round section = 1

$$F'_b = (2,960)(1) = 2,960 \text{ psi}$$

Determine $f_b$ at ground line.
The circumference $C$ at the ground line is:

$$C = 33.5 - (7.5 - 6)(0.25) = 33.12 \text{ in.}$$

$$M = 221,840 \text{ in.-lb (Step 3)}$$

$$f_b = \frac{32 \pi^2 M}{C^3} = \frac{32 \pi^2 (221,840)}{33.12^3} = 1928 \text{ psi} < 2960 \text{ psi} \qquad \text{O.K.}$$

Check for combined stresses using Eq. (5-23).

$$\left( \frac{f_c}{F'_c} \right)^2 + \frac{f_b}{F'_b(1 - f_c/F_{cE})}$$

$$\left( \frac{89}{1,656} \right)^2 + \frac{1,928}{2,960(1 - 89/250)} = 1.01 > 1 \qquad \text{Consider O.K.}$$

In the above equation, $F'_c$ was taken as $F^*_c$, the tabular design value multiplied by all applicable adjustment factors except for column stability, because the maximum moment is at the ground surface.

7.   Check stresses at critical point for compression and combination of compression and bending stresses. Determine point at which critical stress occurs in pole by using Eq. (5-16) which was developed for a rectangular cross section. Use the diameters $D_{min}$ and $D_{max}$ to replace $d_{min}$ and $d_{max}$ respectively.

The representative diameter $D$ is obtained as follows:

$$D = D_{min} + (D_{max} - D_{min}) \left[ a - 0.15 \left( \frac{1 - D_{min}}{D_{max}} \right) \right]$$

$$D = 8.79 + (10.29 - 8.79) \left[ 0.70 - 0.15 \left( 1 - \frac{8.79}{10.29} \right) \right] = 9.81 \text{ in.}$$

Critical section occurs

$$\frac{9.81 - 8.79}{10.29 - 8.79} (15) = 10.2 \text{ ft from top}$$

Check compressive stress for full snow load.

Determine weight of pole above critical section:

$$W = (10.2/25)(625) = 255 \text{ lb}$$

$$P = P_{SL} + W = 11,250 + 255 = 11,505 \text{ lb}$$

Determine $F_c^*$. Assume structure unbraced for sidesway. From Table 5.6, $K_e = 2.1$.

$$l_e = lK_e = (15)(2.1) = 31.5 \text{ ft}$$

$$D = 9.81 \text{ in.}$$

$$\frac{l_e}{D} = \frac{(31.5)(12)}{9.81} = 38.5$$

$$F_c^* = F_c C_D C_{cs} = (1000)(1.15)[1 + (0.002)(10.2)] = 1173 \text{ psi}$$

Note that $C_{cs}$ is obtained from Table 6.5.

Determine $C_P$ and $F_c'$.

For round timber poles, $c = 0.85$ $K_{cE} = 0.3$

$$F_{cE} = \frac{12 K_{cE} E'}{\left( \dfrac{l_e}{D/4} \right)^2} = \frac{12(0.3)(1,500,000)}{(378/(9.81/4))^2} = 227 \text{ psi}$$

$$C_P = \frac{1 + F_{cE}/F_c^*}{2c} - \sqrt{\left[ \frac{1 + (F_{cE}/F_c^*)}{2c} \right]^2 - \frac{F_{cE}/F_c^*}{c}}$$

$$C_P = \frac{1 + 227/1173}{(2)(0.85)} - \sqrt{\left[ 1 + \frac{227/1173}{(2)(0.85)} \right]^2 - \frac{227/1173}{0.85}} = 0.193$$

$$F'_c = F^*_c C_P = (1173)(0.193) = 227 \text{ psi}$$

$$C \text{ at critical point} = 9.81\pi = 30.82 \text{ in.}$$

$$f_c = \frac{P}{A} = \frac{11,505}{(30.82)^2/4\pi} = 152 \text{ psi} < 227 \text{ psi} \qquad \text{O.K.}$$

Check combined bending and compressive stresses at point 10.2 ft below top for wind load and $1/2$ snow load.

$$F^*_c = (1000)(1.6)[1 + (0.002)(10.2)] = 1632 \text{ psi}$$

$$\frac{l_e}{D/4} = \frac{(2.1)(15)(12)}{9.81/4} = 154.1$$

$F_{cE} = 227 \text{ psi}$ (from previous calculation).

$$C_P = \frac{1 + F_{cE}/F^*_c}{2c} - \sqrt{\left[\frac{1 + F_{cE}/F^*_c}{2c}\right]^2 - \frac{F_{cE}/F^*_c}{c}}$$

$$C_P = \frac{1 + 227/1632}{(2)(0.85)} - \sqrt{\left[\frac{1 + 227/1632}{(2)(0.85)}\right]^2 - \frac{227/1632}{0.85}} = 0.136$$

$$F'_c = F^*_c C_P = (1632)(0.136) = 222 \text{ psi}$$

$$f_c = \frac{P}{A} = \frac{7500 + 255}{(30.82)^2/4\pi} = 103 \text{ psi}$$

$$M_A = 12\left[\frac{(10.2 + 2.83)^2(150)}{2} - 317(10.2 + 1.31)\right]$$

$$= 107,062 \text{ in.-lb}$$

$$f_b = \frac{32\pi M_A}{C^3} = \frac{32\pi(107,062)}{(30.82)^3} = 367 \text{ psi}$$

$$F^*_b = (1850)(1.6) = 2960 \text{ psi}$$

$$C_L = 1 \text{ (round pole)}$$

$$F'_b = F^*_b = 2960 \text{ psi}$$

$$\left(\frac{f_c}{F'_c}\right)^2 + \frac{f_b}{F'_b(1 - f_c/F_{cE})} \leq 1$$

$$\left(\frac{103}{222}\right)^2 + \frac{367}{2960(1 - 103/227)} = 0.44 < 1 \qquad \text{O.K.}$$

A check 5 ft below top was made and size of pole was satisfactory.

8. Determine footing requirements.

$$\text{Area of pole butt} = A = \frac{[33.5 + (6)(0.25)]^2}{4\pi} = 97.5 \text{ in.}^2 = 0.677 \text{ ft}^2$$

$$\text{Weight of pole} = W = 625(22.5/25) = 563 \text{ lb}$$

$$P = 11{,}250 + 563 = 11{,}813 \text{ lb}$$

From Table 6.6, allowable soil bearing is

$$S_B = (2000)[1 + (0.2)(7.5 - 1)] = 4600 \text{ psf}$$

$$\text{Soil bearing load} = \frac{P}{A} = \frac{11{,}813}{0.677} = 17{,}450 \text{ psf} > 4600 \text{ psf;}$$

therefore, use concrete footing. Required area of concrete footing is calculated as

$$A_c = \frac{P}{S_B} = \frac{11{,}813}{4600} = 2.57 \text{ ft}^2$$

Minimum diameter of hole

$$D_H = \sqrt{\frac{4A_c}{\pi}} = \sqrt{\frac{4(2.57)}{\pi}} = 1.81 \text{ ft}$$

The hole should be backfilled with concrete so that the maximum moment acts near the ground line. For practical reasons, the hole should be large enough to place the concrete backfill. The diameter of the butt of the pole calculates to be slightly less than 1 ft but may actually be larger. Using a 2-ft-diameter hole will allow approximately 6 in. between the side of the hole and the pole, which should be adequate. Another advantage of concrete backfill is that it increases the effective diameter of the pole below the groundline, and by using this in Eq. (6-11), the depth of embedment can be reduced.

Step 4 will be repeated using 2-ft-diameter concrete backfill as the diameter of the pole butt $B$.

Assume depth $d = 6.0$ ft. From Eq. (6-11)

$$d = \sqrt{\frac{(4.25)(2360)(7.84)}{(6)(200)(2.0)}}$$

$$d = 5.72 \text{ ft}$$

$$\text{Length of pole required} = 15 + 5.72 = 20.7 \text{ ft} > 20 \text{ ft}$$

A 25-ft-long pole is still required, but digging the hole will be less costly.

The use of knee braces cuts down the effective length of the column. However, care should be used in designing connections for knee braces because forces in the knee braces may become quite large. The effect of sidesway can also be reduced by use of a roof diaphragm and shear walls.

Additional information on pole type building design is contained in *Wood Engineering* (19).

## 6.4    TIMBER PILES

Recommendations for the use of timber piles in foundations may be found in *Pile Foundations: Know-How* (21). Piles may be specified by the circumference of the nominal butt or nominal tip. Tables specified butt circumference with corresponding tip circumferences and tables of tip circumference with corresponding butt circumference for various size piles are included in ASTM D 25, *Standard Specifications for Round Timber Piles* (22). (These may vary with species.) Piles are generally designed at a critical section governed by the tip size, butt size, or an intermediate section. The ASTM classification allows the designer to specify a pile with adequate dimensions at the critical section. For example, a pile depending on frictional forces along the side of the pile to support the vertical load will generally have a critical section located away from the tip. On the other hand, an end-bearing pile may have the critical section located at the tip.

A number of species are used for piles, with Southern Pine, Douglas Fir, and Oak being the most commonly used. Piles must be relatively straight and possess the strength to resist driving stress and carry the imposed loads. When piles are used below the permanent water table or completely submerged in fresh water, preservative treatment is not necessary. However, for most permanent structures where these conditions do not exist, preservative treatment is necessary.

### 6.4.1    Preservative Treatment of Piles

Foundation piles are most commonly used where some part of the piles are exposed above the permanent water table. Therefore, preservative treatment is required. The preservative treatment should conform to recognized specifications such as Federal Specification TT-W-571 (23) and American Wood-Preservers' Association (AWPA) Standards C1 and C3 (11). Cutoffs at the tops of piles exposing untreated wood should be field treated in accordance with AWPA Standard M4 (11).

Piles used in salt water are subject to attack by marine borers, and special treatment techniques must be used to minimize the problem. The treatment of piles for use in saltwater is covered in Federal Specification TT-W-571 (23) and AWPA Standard C18 (11).

### 6.4.2    Pile Driving

Equipment used for driving timber piles is of special importance. The energy used to drive the piles must be sufficient to drive the pile, but must not impart excessive forces. Pile butts and tips may be damaged severely by sharp blows. For this reason, it is not desirable to drive timber piles with a drop hammer unless a suitable block is employed to dampen the impact. Air, steam, and diesel hammers are commonly used.

Generally, it is desirable to band the butt or driven end of a pile to minimize damage during driving. In some cases where tip damage may occur, a special shoe or fitting is used to protect the tip.

Before selecting a pile foundation, the soil conditions should be investigated. Group action, effective unsupported length, lateral loads, rebound, and other design and construction features should be considered.

### 6.4.3   Design Values for Piles

Design values for piles are published in the *National Design Specification*® (14). They were determined in accordance with ASTM D 2899-86 (24). Table 6.8 contains the design values for commonly used species. These design values apply to both wet and dry use for normal duration of load.

### 6.4.4   Adjustment Factors for Piles

The design values in Table 6.10 must be adjusted for various conditions. See Table 6.9 for applicability of adjustment factors.

Adjustments for load duration, temperature, bearing area, and column stability factor are applied the same as for other wood products.
Untreated Factor. The design values in Table 6.10 are based on steam seasoning or Boltonizing prior to treatment. When piles are seasoned by air drying or are used untreated, all design values in Table 6.10, except modulus of elasticity, may be increased by multiplying by the untreated adjustment

**TABLE 6.8**

**Design Values for Treated Round Timber Piles[a]**

| Species | Compression Parallel to Grain $F_c$ (psi) | Bending $F_b$ (psi) | Shear Parallel to Grain $F_v$ (psi) | Compression Perpendicular to Grain $F_{c\perp}$ (psi) | Modulus of Elasticity $E$ (psi) |
|---|---|---|---|---|---|
| Pacific Coast Douglas Fir[b] | 1250 | 2450 | 115 | 230 | 1,500,000 |
| Red Oak[c] | 1100 | 2450 | 135 | 350 | 1,250,000 |
| Red Pine[d] | 900 | 1900 | 85 | 155 | 1,280,000 |
| Southern Pine[e] | 1200 | 2400 | 110 | 250 | 1,500,000 |

[a]Design values for normal load duration and wet service conditions.

[b]Pacific Coast Douglas Fir values apply to this species as defined in ASTM D 1760-86a, *Standard Specification for Pressure Treatment of Timber Products* (17). For connection design, use Douglas Fir-Larch design values.

[c]Red Oak values apply to Northern and Southern Red Oak.

[d]Red Pine values apply to Red Pine grown in the United States. For fastener design, use Northern Pine design values.

[e]Southern Pine values apply to Loblolly, Longleaf, Shortleaf, and Slash Pines.

Source: ANSI/NF₀PA NDS—1991, *National Design Specification for Wood Construction*, AFPA, 1250 Connecticut Ave., NW, Washington, DC 20036.

## TABLE 6.9

### Applicability of Adjustment Factors for Round Timber Piles

| | | | LOAD DURATION FACTOR | TEMPERA-TURE FACTOR | UNTREATED FACTOR | SIZE FACTOR | SINGLE PILE FACTOR | COLUMN STABILITY FACTOR | CRITICAL SECTION FACTOR | BEARING AREA FACTOR |
|---|---|---|---|---|---|---|---|---|---|---|
| $F_c'$ | = | $(F_c)$ | $(C_D)$ | $(C_t)$ | $(C_u)$ | • | $(C_{sp})$ | $(C_p)$ | $(C_{cs})$ | • |
| $F_b'$ | = | $(F_b)$ | $(C_D)$ | $(C_t)$ | $(C_u)$ | $(C_F)$ | $(C_{sp})$ | • | • | • |
| $F_v'$ | = | $(F_v)$ | $(C_D)$ | $(C_t)$ | $(C_u)$ | • | • | • | • | • |
| $F_{c\perp}'$ | = | $(F_{c\perp})$ | $(C_D)$ | $(C_t)$ | $(C_u)$ | • | • | • | • | $(C_b)$ |
| $E'$ | = | $(E)$ | • | $(C_t)$ | • | • | • | • | • | • |
| $F_g'$ | = | $(F_g)$ | $(C_D)$ | $(C_t)$ | $(C_u)$ | • | • | • | • | • |

Source: ANSI/NF$_o$PA NDS—1991, *National Design Specification for Wood Construction*, AFPA, 1250 Connecticut Ave., NW, Washington, DC 20036.

factor, $C_u$, as follows:

| Species | $C_u$ |
|---|---|
| Pacific Coast Douglas Fir, Red Oak, Red Pine | 1.11 |
| Southern Pine | 1.18 |

Size Factor. The size factor, $C_F$, is applied to the critical section in bending when the pile circumference exceeds 43 in. (diameter = 13.5 in.).

Critical Section Factor. The design values are based on tip of the pile. The compression parallel to grain design value of a pile increases 0.2% for each ft of length from the tip. The critical section factor is determined by the following equation:

$$C_{cs} = 1.0 + (L_c)(0.002)$$

where

$L_c$ = the distance from the tip of the pile to the critical section (ft). $C_{cs}$ should not be greater than 1.10. It is independent of the provisions for tapered columns and both are permitted to be used in design.

Single Pile Factor. The design values in Table 6.8 are based on use of the piles in a group. When piles are used in such a manner that each pile supports a specific load and group action does not exist, the compression parallel to grain and bending design values should be multiplied by the following single pile adjustment factor:

| Design Value | $C_{sp}$ |
|:---:|:---:|
| $F_c$ | 0.80 |
| $F_b$ | 0.77 |

The beam stability factor, $C_L$, does not apply to piles.

When piles or portions of piles are standing unbraced in air, water, or material not capable of lateral support, they should be designed in accordance with the applicable column equation. When piles are subjected to lateral loads, the applicable equations for flexure also apply.

### 6.4.5  Application of Adjustment Factors

The application of adjustment factors for piles is illustrated by the following example.

**Example.**    Determine the design values in bending and compression parallel to grain for a single Southern Pine pile 10 ft from the tip. The pile was conditioned prior to treatment by air drying. The critical section has a 17-in. diameter. The critical stresses are caused by wind. Each pile must support a specific load under wet conditions of use. The unsupported length of the pile is such that $C_P = 0.91$. Compression parallel to grain:

$$F'_c = F_c C_D C_u C_{sp} C_P C_{cs}$$

$$= (1200)(1.6)(1.18)(0.8)(0.91)[1 + (0.002)(10)] = 1682 \text{ psi}$$

Bending:

$$F'_b = F_b C_D C_u C_F C_{sp}$$

Convert round section to equivalent square section.

$$d = \frac{D}{1.13} = \frac{17}{1.13} = 15.04 \text{ in.}$$

$$C_F = \left(\frac{12}{d}\right)^{1/9} = \left(\frac{12}{15.04}\right)^{1/9} = 0.975$$

$$F'_b = (2400)(1.6)(1.18)(0.975)(0.77)$$

$$= 3402 \text{ psi}$$

## 6.5  TIMBER BRIDGES

Timber bridges consist of several basic types, including trestles, girder bridges, truss bridges, and arch bridges. In the design of any of these types, the considerations given under "Moving Loads" in Chapter 4 of this manual should be taken into account. For highway bridges, design loads and their application should be in accordance with the recommendations of the American Association of State Highway and Transportation Officials (AASHTO) (25). For railway bridges, the recommendations of the American Railway Engineering Association (AREA) (26) should be followed.

### 6.5.1   Trestles

The trestle is a simple type of timber bridge. Timber trestles consist of stringers supported by pile or frame bents. The bridge deck is applied to the stringers. Pile and frame bents are capped by timbers 12 × 12 in. or larger, adequately fastened to the tops of the piles or posts.

If pile penetration or height of bent is such that piles longer than those commercially available are required, or if pile bearing values are low and a large number of piles must be driven, posts may be used on top of the pile bents. Frame bents must rest on some type of foundation structure, such as concrete footings or piles. Sway bracing and longitudinal tower bracing, appropriate to the height of the bent, must be provided.

Spacing of bents is determined, in part, by the commercially available lengths of stringers, which are fabricated in even-foot increments. The ends of interior stringers are usually lapped and fastened to the bearing on the caps, whereas exterior stringers are butted to the ends and spliced over the bent caps.

Stringers are designed as simple-span beams under the loadings recommended by AASHTO (25) or AREA (26). Sizes and spacing are determined by the span and loading conditions. Standard sizes of glued laminated or sawn timbers as given in Tables 8.1 and 8.2 should be used. Solid bridging should be provided at the ends of stringers to hold them in line and also to serve as a fire stop. Bridging should also be placed between stringers at midspan or, on long spans, at third-points.

### 6.5.2   Girder Bridges

Girder bridges, consisting of glued laminated or sawn timber girders supporting a bridge deck, may be used for spans that exceed the practical limits of timber trestles, for spans less than those economical for truss bridges, or where a truss bridge is not desirable. Substructures similar to those used for timber trestles can be used for girder bridges, the girder being fastened to the bent caps by means of a fabricated steel girder seat.

Timber girders are designed as beams in accordance with the recommendations in Chapter 5, including those for lateral support (see Table 5.3). Lateral forces acting on the girder will enter into the design of the lateral bracing system. Construction economies will usually result if the standard sizes for glued laminated or sawn timber as given in Tables 8.1 and 8.2 of this manual are used.

### 6.5.3   Truss Bridges

Truss bridges may be of either of two types: deck-truss bridges, in which trusses support the bridge deck and roadway; or through-truss bridges, in which the roadway passes between two parallel trusses forming the bridge structure. The deck-truss type is the more economical since substructures and lateral bracing are narrower. The use of deck trusses may be limited, however, by under-clearance.

Deck trusses may be of the parallel chord type or of the bowstring type with the truss built up to the level of the floor beams. Through-trusses may be of either of these truss types also, but the bowstring is usually more economical. Bowstring trusses particularly are often used as pony truss bridges, which are through-trusses with no overhead bracing above the roadway. Through-trusses have the disadvantage of potential damage from vehicles.

Substructures for truss bridges may be similar to those for timber trestles; however, because the vertical loads are greater and are concentrated at the ends of the trusses, the bents must be capable of carrying a greater load and a system of cribbing is required for bent caps. For longer span and heavier loads, timber, stone, or concrete piers may be required. Lateral forces are greater on truss bridges, and a carefully designed substructure sway bracing system is necessary.

The design of trusses for bridges is similar to that for roof trusses, the length of truss panel being determined by economical spacing of floor beams, minimum number of joints, and commercially available lengths of timber. As in roof truss design, the joint design is an important consideration. Bridge truss joints should be designed to eliminate or minimize pockets, which may tend to collect moisture.

### 6.5.4  Arch Bridges

When site conditions are such that considerable height is required between foundation and roadway or a relatively long clear span is required, an arch bridge may be most economical because of the lesser need for substructure framing. Arch bridges may be of the two-hinged or three-hinged type, two-hinged designs being more frequently used on short spans and three-hinged designs on long spans.

Glued laminated timber arches may be fabricated to the desired shape and the ends built up to the level of the roadway by means of post bents. Post bents may be connected to the arch by means of steel gusset plates, which should be designed for erection loads, possible stress reversals, and lateral forces as well as for the anticipated bridge loads.

### 6.5.5  Bridge Decks

The selection of decks for timber bridges is determined by density of traffic and economics. Plank decks may be used for light traffic or for temporary bridges. Laminated decks can be used for heavier traffic conditions. Asphaltic wearing surfaces may be applied on the decking, although this is not usually done for plank decks.

Composite timber–concrete decks are sometimes used in timber bridge construction. Composite timber–concrete construction combines timber and concrete in such manner that the wood is in tension and the concrete is in compression (except at the supports of continuous spans, where negative bending occurs and these stresses are reversed). Composite timber–concrete construction is of two basic types: T-beams and ''slab'' decks.

T-beams consist of timber stringers, which form the stems, and concrete slabs, which form the flanges of a series of T-shapes. Composite beams of this type are usually simple-span bridges. Slab decks use, as a base for the concrete, a mechanically laminated wooden deck made up of planks set on edge, with alternate planks raised 2 in. to form longitudinal grooves. This grooved surface is usually obtained by using planks of two different widths and alternating them in assembly. This composite type has been used for continuous-span bridges and trestles, but is not very common today.

In both types, a means of shear parallel to grain resistance and a means of preventing separations are needed at the joint between the two materials. In T-beams, resistance to shear parallel to grain is generally provided by a series of notches $\frac{1}{2}$–$\frac{3}{4}$ in. deep cut into the top of the sawn timber stringer and about $1\frac{1}{2}$ in. deep for a glued laminated timber stringer, while nails and spikes partially driven into the top prevent vertical separation of the concrete and timber. Other adequate methods may be used. In slab decks, shear resistance is accomplished either by means of notches $\frac{1}{2}$-in. deep cut into the tops of all laminations, by triangular steel plate "shear developers" driven into precut slots in the channels formed by the raised laminations, or other suitable shear connectors. When the $\frac{1}{2}$-in. notches are used, grooves are milled the full length of both faces of each raised lamination to resist the uplift and separation of the wood and concrete. When the steel shear developers are used, nails or spikes are partially driven into the tops of raised laminations to resist separation.

In T-beam design, secondary shearing stress due to temperature must be considered in designing for shear parallel to grain resistance. These stresses are induced by the thermal expansion or contraction of the concrete, both of which are resisted by the wood, which is assumed to be unaffected by normal temperature changes. Shear connections for temperature change are neglected in slab deck-type composite construction; however, expansion joints should be provided in the concrete slab.

The concrete slab should be reinforced for temperature stress. In continuous spans, steel sufficient to develop negative bending stress is necessary over interior supports.

The dead load of the composite structure is considered to be carried entirely by the timber section. The composite structure carries positive bending moment and, over interior supports in continuous spans, steel reinforcing and the wood act to resist negative bending moment.

In designing a composite structure, if it is assumed that the junction between the two materials is without inelastic deformation and has elastic characteristics in keeping with the materials, the structure can be designed by the transformed-area method, that is, by transforming the composite section into an equivalent homogeneous section. This is accomplished by multiplying the concrete width or depth by the ratio of the moduli of elasticity of the materials.

Glued laminated timber bridges are commonly used for highway bridges, and design examples are included. Because nail-laminated decks and composite timber-concrete decks are not common, design examples for these decks are not included in this manual. For bridges utilizing longitudinal stringers and transverse decks, the following design procedures illustrate the superstructure design with interconnecting steel dowels between the panels and without the dowels. Note that this procedure and examples are for vertical loading from vehicles as described in AASHTO *Standard Specifications for Highway Bridges* (25). The reader should refer to this standard for additional loading requirements on the superstructure components.

### 6.5.6 Longitudinal Stringers and Doweled Deck

The design examples do not show camber for the glued laminated stringers because camber may vary with the particular use. Generally, camber equal to dead-load deflection plus additional amounts for drainage and appearance is sufficient. Glued laminated timber design values are based on AITC 117—Design (4). Wet-use design values are used for the design of deck panels because tests in the field indicate the moisture content is likely to exceed 16% in service. Dry use design values are used for the design of the glued laminated timber stringers, with the exception of compression perpendicular to grain stress, $F_{c\perp}$. The glued laminated timber deck panel, when properly treated and installed, should provide a roof for the stringers, preventing them from exceeding an in-service moisture content of 16%.

Wind-driven moisture can cause a surface wetting on the exposed face of the edge stringers, but this superficial wetting should not affect the design. Localized moisture accumulations can develop, and preservative treatment is recommended for the stringers to protect against possible decay hazards such as at areas of steel connections and at bearing locations.

### Example 1.

Criteria
HS 15-44 live load, two lanes, follow AASHTO Standards
3-in. asphalt wearing surface
Span:   40 ft 0 in.
Width:   34 ft 0 in.
Materials
Decking:   Glued laminated L2 Douglas Fir (Combination 2) or No. 2 MG
            Southern Pine (Combination 47); wet condition of use.
Stringers:   Douglas Fir or Southern Pine; 24F–V4 DF/DF or 24F–V3 SP/SP

### Stringer Design

Assume:   Six girders at 6 ft 7 in. center-to-center
          $5\frac{1}{8}$-in.-thick deck panels, 35 ft long
          Weight of wood, 50 pcf
          Weight of asphalt, 150 pcf

Dead load (one stringer):

$$\text{Deck } \frac{(5.125)(50)(6.58)}{12} = 140 \text{ plf}$$

$$\text{Wearing surface: } \frac{(3)(150)(6.58)}{12} = 247 \text{ plf}$$

$$\text{Stringer (assumed): } = \underline{100 \text{ plf}}$$

$$\text{Total (preliminary): } = 487 \text{ plf}$$

Live load (on one stringer) HS 15-44 (AASHTO, Appendix A)

$$\text{Distribution factor } = \frac{\text{spacing}}{5.0} = \frac{6.58}{5} = 1.32$$

$$\text{Impact factor } = 0 \text{ (none required for timber)}$$

$$\text{Moment } = \frac{(\text{ft-kips/lane})(\text{distribution factor})}{\text{wheel lines/lane}}$$

$$= \frac{(337.4)(1.32)}{2} = 222.7 \text{ ft-kips}$$

Try $6\frac{3}{4} \times 43\frac{1}{2}$-in. girder
Area $= 293.6$ in.$^2$
Section modulus $= 2129$ in.$^3$
Moment of inertia $= 46,300$ in.$^4$
Volume factor for Southern Pine

$$C_V = \left(\frac{5.125}{6.75}\right)^{1/20} \left(\frac{12}{43.5}\right)^{1/20} \left(\frac{21}{40}\right)^{1/20} = 0.895$$

Volume factor for Douglas Fir

$$C_V = \left(\frac{5.125}{6.75}\right)^{1/10} \left(\frac{12}{43.5}\right)^{1/10} \left(\frac{21}{40}\right)^{1/10} = 0.802$$

$$\text{Weight } = \frac{(50)(293.6)}{144} = 102 \text{ plf}$$

Actual dead load $= 140 + 247 + 102 = 489$ plf
Bending:

$$\text{Dead load moment } = \frac{wL^2}{8} = \frac{(0.489)(40)^2}{8} = 97.8 \text{ ft-kips}$$

$$\text{Total load moment } = 97.8 + 222.7 = 320.5 \text{ ft-kips}$$

$$\text{Actual bending stress } f_b = \frac{M}{S} = \frac{(320.5)(12)(1000)}{2129} = 1807 \text{ psi}$$

The deck is fastened to the stringers to provide lateral bracing; therefore, $C_L = 1$.

For Southern Pine, design value in bending:

$$F_b' = F_b C_V C_M = (2400)(0.895)(1.0) = 2148 \text{ psi}$$

For Douglas Fir, design value in bending:

$$F_b' = F_b C_V C_M = (2400)(0.802)(1.0) = 1925 \text{ psi}$$

Shear parallel to grain: points considered
Three times depth from end:

$$\frac{(3)(43.5)}{12} = 10.875 \text{ ft}$$

Span quarter point:

$$\frac{40.0}{4} = 10.0 \text{ ft (controls)}$$

$$V_{DL} = 9.8 - (0.489)(10) = 4.9 \text{ kips}$$

$$V_{LL} = 0.5[(0.6)(13.95) + (1.32)(13.95)] = 13.4 \text{ kips}$$

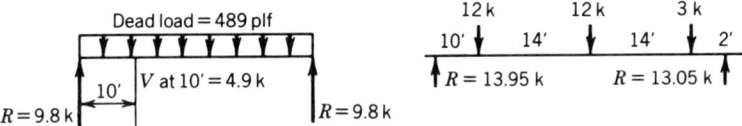

Actual shear parallel to grain stress:

$$f_v = \frac{3V}{2bd} = \frac{(3)(13.4 + 4.9)(1000)}{(2)(6.75)(43.5)} = 93 \text{ psi}$$

A check of the shear at one stringer depth from the end for undistributed live load indicates that the shear stress is

$$f_v = \frac{(3)(18.3 + 8.0)(1000)}{(2)(6.75)(43.5)} = 134 \text{ psi}$$

The design value in shear parallel to grain is 165 psi for Douglas Fir or 200 psi for Southern Pine.

Overload provisions: apply for single lane; distribution factor $= 6.58/6 = 1.10$.

Bending—100% increase in truck or lane load:

$$\text{Moment} = \left(\frac{1.10}{1.32}\right)(222.7)(2) + 97.8 = 468 \text{ ft-kips}$$

$$\text{Actual bending stress} = \frac{M}{S} = \frac{(468)(12)(1000)}{2129} = 2638 \text{ psi}$$

Increase allowable stress by 50%.

For Southern Pine:

Allowable bending stress $= 1.5F_bC_V = (1.5)(2400)(0.895) = 3224$ psi   O.K.

For Douglas Fir:

Allowable bending stress $= 1.5F_bC_V = (1.5)(2400)(0.802) = 2887$ psi   O.K.

Shear parallel to grain with 100% increase in truck or lane load:

Live load shear at $L/4 = (13.95)(2) = 27.9$ kips

Distributed live load shear $= 0.5[(0.6)(27.9) + (1.10)(27.9)] = 23.7$ kips

Actual shear parallel to grain stress:

$$f_v = \frac{3V}{2bd} = \frac{(3)(23.7 + 4.9)(1000)}{(2)(6.75)(43.5)} = 146 \text{ psi}$$

Check shear with overload at one $d$ from end:

$$f_v = \frac{(3)(31.0 + 8.0)(1000)}{(2)(6.75)(43.5)} = 199 \text{ psi}$$

Increase allowable stress by 50%; then allowable shear parallel to grain stress $= (1.5)(165) = 248$ psi for Douglas Fir or $(1.5)(200) = 300$ psi for Southern Pine. Deflection:   The actual calculated deflection of the interior stringer is based on converting the distributed live load moment to an equivalent concentrated load placed at midspan. It should be noted that this calculation of live load deflection is conservative as it does not account for load distribution to adjacent stringers and no allowance is made for composite T-beam action which exists between the decking and stringers.

$$M_{LL} = 222.7 \text{ ft-kips}$$

$$P = \frac{4M_{LL}}{L} = \frac{4(222.7)}{40} = 22.27 \text{ kips}$$

$$\Delta_{LL} = \frac{PL^3}{48EI} = \frac{22.27(40)^3(1000)(1728)}{48(1,800,000)(46,300)} = 0.62 \text{ in.}$$

$$\Delta_{LL} = \frac{L}{300} = \frac{40(12)}{300} = 1.60 \text{ in.} > 0.62 \text{ in.} \qquad \text{O.K.}$$

**Deck Design**

 Basic design equations: (see AASHTO 3.25)

$$M_x = P[(0.51 \log_{10}s) - K]$$

$$R_x = 0.034P$$

$$t = \sqrt{\frac{6M_x}{F_b'}} \text{ or } t = \frac{3R_x}{2F_v'} \text{ (whichever is greater)}$$

where $M_x$ = primary bending moment (in.-lb/in.),
$R_x$ = primary shear (lb/in.),
$P$ = design wheel load = 12,000 lb,
$s$ = effective deck span (in.) = clear span + one half stringer width (75.624 in.) or clear span + assumed deck thickness (77.375 in.) whichever is less,
$t$ = deck thickness (assumed to be $5\frac{1}{8}$ in.),
$K$ = 0.47 for H 15 loading, 0.51 for H 20 loading,
$F_b$ = 1750 psi, Combination 2 Douglas Fir or Combination No. 47 Southern Pine,
$F_v$ = 145 psi (Douglas Fir) or 175 psi (Southern Pine),
$C_{fu}$ = 1.10 for $5\frac{1}{8}$ in. thickness from footnote $f$, Table 2, AITC 117—Design (4),
$F'_b = F_b C_{fu} C_M = (1750)(1.10)(0.80) = 1540$ psi,
$F'_v = F_v C_M = (145)(0.875) = 127$ psi (Douglas Fir),
$F'_v = F_v C_M = (175)(0.875) = 153$ psi (Southern Pine),
$M_x = 12,000[0.51(\log 75.625) - 0.47]$
$\quad = 12,000[(0.51)(1.878665) - 0.47]$
$\quad = 5860$ in.-lb/in.,
$R_x = (0.034)(12,000) = 408$ lb/in., and
$t = \sqrt{6M/F'_b} = \sqrt{(6)(5860)/1540} = 4.78$ in.,

or

$t = 3R/2F'_v = (3)(408)/(2)(127) = 4.82$ in. < 5.125 in.  O.K.

Use $5\frac{1}{8}$-in. glued laminated timber deck.

**Dowel Design** [for basic design equations, see AASHTO 3.25 (25)]

$$n = \frac{1000}{\sigma_{PL}} \left[ \frac{\overline{R}_y}{R_D} + \frac{\overline{M}_y}{M_D} \right]$$

$$\overline{R}_y = \frac{6Ps}{1000} \quad \text{for } s \leq 50 \text{ in.}$$

$$\overline{R}_y = \frac{P}{2s}(s - 20) \quad \text{for } s > 50 \text{ in.}$$

$$\overline{M}_y = \frac{Ps}{1600}(s - 10) \quad \text{for } s \leq 50 \text{ in.}$$

$$\overline{M}_y = \frac{Ps(s - 30)}{20(s - 10)} \quad \text{for } s > 50 \text{ in.}$$

where $R_D, M_D$ = shear and moment capacities in AASHTO 3.25.1.4, and
$\sigma_{PL}$ = 1000 psi for Douglas Fir or Southern Pine.

Try $1\frac{1}{2}$-in. diameter dowels:

$$R_D = 2770 \text{ lb} \quad M_D = 8990 \text{ in.-lb}$$

$$\overline{R}_y = \frac{(12,000)(75.625 - 20)}{(2)(75.625)} = 4410 \text{ lb}$$

$$\overline{M}_y = \frac{(12,000)(75.625)(75.625 - 30)}{(20)(75.625 - 10)} = 31,550 \text{ in.-lb}$$

$$n = \frac{1000}{1000} \left[ \frac{4410}{2770} + \frac{31,550}{8990} \right] = 5.10$$

Use five $1\frac{1}{2}$-in. diameter by $19\frac{1}{2}$-in. long dowels in each span. Check stress in dowels:

$$\sigma = \frac{1}{n} (C_R \overline{R}_y + C_M \overline{M}_y)$$

$$C_R = 3.11 \text{ in.}^2$$

$$C_M = 3.02 \text{ in.}^3$$

$$\sigma = (1/5) [(3.11)(4410) + (3.02)(31,550)] = 21,800 \text{ psi} \qquad \text{O.K.}$$

because this is less than the elastic limit of the dowels.

### 6.5.7  Longitudinal Stringers and Nondoweled Deck

For nondoweled decks carrying HS 20-44 or HS 15-44 loading, the following chart shows the maximum clear span between stringers and the maximum deck overhang beyond the outside stringer.

| Deck Thickness (in.) | Maximum Clear Span | Maximum Clear Overhang |
|---|---|---|
| $3\frac{1}{8}$ | 2 ft 6 in. | 1 ft 3 in. |
| $5\frac{1}{8}$ | 4 ft 6 in. | 2 ft 3 in. |
| $6\frac{3}{4}$ | 6 ft 0 in. | 3 ft 0 in. |

These values are based on experience in constructing non-interconnected (nondoweled) decks that are fastened to the stringers using the bracket system shown in Fig. 6.9 or similar deck connection systems.

**Example 2.**  The following example illustrates the procedures to follow in the design of non-doweled decks on longitudinal stringers.

Criteria
HS 20 loading, two lanes
Asphalt wearing surface, 3 in.
Span = 50 ft 0 in. center to center of bearings
Width = 34 ft 0 in.

**Stringer Design**
Use combination 24F-V4 Douglas Fir-Larch, dry service conditions.
Determine number of stringers, $n$, assuming $10\frac{3}{4}$ in. wide $\times$ 48 in. deep stringers.

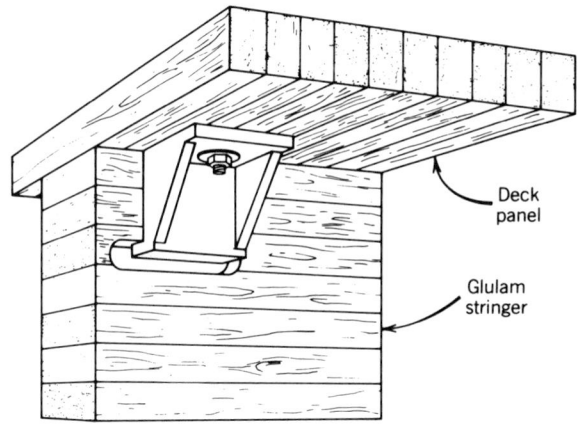

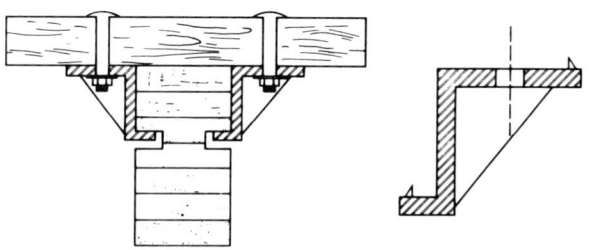

**FIGURE 6.9**   Bracket for fastening deck to glulam stringer.

$$n = \frac{34 - 2(3.0)}{6 + (10.75/12)} = 4.06 \quad \text{(use 5 stringers)}$$

$$\text{Stringer spacing} = \frac{34 - (2)(3.0)}{5} = 5.6 \text{ ft} \quad \text{(use 5 ft 8 in.)}$$

$$F_b = 2400 \text{ psi} \quad F_v = 165 \text{ psi} \quad F_{c\perp} = 650 \text{ psi and} \quad E = 1{,}800{,}000 \text{ psi}$$

$$\text{Dead loads: 3 in. asphalt} \frac{(3)(150)(5.67)}{12} = 213 \text{ plf}$$

$$6\tfrac{3}{4} \text{ in. deck} \frac{(6.75)(50)(5.67)}{12} = 159 \text{ plf}$$

$$\text{Railing, posts, and curb} = 75 \text{ plf}$$

$$\text{Stringer (assume } 10\tfrac{3}{4} \times 48 \text{ in.)} = \underline{179 \text{ plf}}$$

$$\text{Total dead load} = 626 \text{ plf}$$

Live load moment per wheel lane (from AASHTO, Appendix A):

$$\text{Distribution factor} = \frac{\text{spacing}}{5} = \frac{5.67}{5} = 1.134$$

$$M_{LL} = \frac{627,900}{2} (1.134) = 356 \text{ ft-kips}$$

$$M_{DL} = \frac{wL^2}{8} = \frac{(0.626)(50)^2}{8} = 196 \text{ ft-kips}$$

$$M_{LL} + M_{DL} = 356 + 196 = 552 \text{ ft-kips}$$

For a $10\frac{3}{4} \times 48$-in. stringer, $S = 4128$ in.$^3$, $I = 99,070$ in.$^4$

$$f_b = \frac{M}{S} = \frac{(552)(12)(1000)}{4128} = 1605 \text{ psi}$$

Check lateral stability.

The top of the stringer is laterally supported by brackets at 12 in. centers; there-fore, no reduction for lack of lateral stability is required.

$$C_V = \left(\frac{5.125}{10.75}\right)^{1/10} \left(\frac{12}{48}\right)^{1/10} \left(\frac{21}{50}\right)^{1/10} = 0.741$$

$$F_b' = F_b C_V = (2400)(0.741) = 1778 \text{ psi} > 1605 \text{ psi} \qquad \text{O.K.}$$

Check shear:

$$\frac{\text{Span}}{4} = \frac{50}{4} = 12.5 \text{ ft}$$

$$3 \text{ (stringer depth)} = (3)(48/12) = 12.0 \text{ ft (controls)}$$

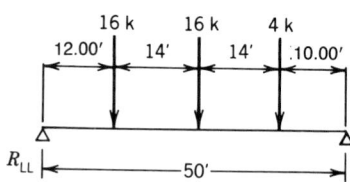

$$R_{LL} = (16) \left(\frac{38}{50}\right) + (16) \left(\frac{24}{50}\right) + (4) \left(\frac{10}{50}\right) = 20.64 \text{ kips}$$

$$V_{LL} = (0.5)[(0.6)(20.64) + (1.2)(20.64)] = 18.6 \text{ kips}$$

$$V_{DL} = \left(\frac{L}{2} - 12\right) w_{DL} = \left[\left(\frac{50}{2}\right) - 12\right] (0.626) = 8.14 \text{ kips}$$

$$V_{LL} + V_{DL} = 18,600 + 8140 = 26,740 \text{ lb}$$

$$f_v = \frac{(3)(26,740)}{(2)(10.75)(48)} = 77 \text{ psi} < F_v = 165 \text{ psi} \quad \text{O.K.}$$

Check shear at 1 times the depth from the end:

$$V_{LL} = 26.4 \text{ kips}$$

For 48 in. deep member:

$$V_{DL} = \frac{0.626(50 - 8)}{2} = 13.1 \text{ kips}$$

$$V_{TL} = 26.4 + 13.1 = 39.5 \text{ kips}$$

$$f_v = \frac{(3)(39,500)}{(2)(10.75)(48)} = 115 \text{ psi} < 165 \text{ psi} \quad \text{O.K.}$$

**Deck Design for Combination 1**

$$F_b = 1450 \text{ psi} \quad F_v = 145 \text{ psi}$$

The wheel load is distributed over a deck length of 15 in. plus the deck thickness, $15 + 6.75 = 21.75$ in.

$$s = \text{Deck span} = 68 - 10.75 + \frac{10.75}{2} = 62.625 \text{ in.}$$

or

$$s = \text{Deck span (maximum)} = 68 - 10.75 + 6.75 = 64 \text{ in.}$$

Use $s = $ smaller of two values calculated (62.625 in.)

$$b_t = \sqrt{(0.01P)(2.5)} = \sqrt{(0.01)(12,000)(2.5)} = 17.3 \text{ in.}$$

$$M_{LL} = P\left[\frac{s}{4} - \frac{b_t}{8}\right] = (12,000)\left[\frac{62.625}{4} - \frac{17.3}{8}\right] = 161,900 \text{ in.-lb}$$

$$w_{DL} = \left[\left(\frac{3}{12}\right)(150) + \left(\frac{6.75}{12}\right)(50)\right]\left(\frac{1}{12}\right)\left(\frac{21.75}{12}\right) = 9.9 \text{ lb/in.}$$

$$M_{DL} = \frac{w_{DL}s^2}{8} = \frac{(9.9)(62.625)^2}{8} = 4850 \text{ in.-lb}$$

Because the deck is continuous over two or more stringers, the moment is 80% of the simple-span moment.

$$M_{TL} = (0.8)(161{,}900 + 4850) = 133{,}400 \text{ in.-lb}$$

$$S = \frac{(21.75)(6.75)^2}{6} = 165.2 \text{ in.}^3$$

$$F_b' = F_b C_{fu} C_M = (1450)(1.07)(0.8) = 1240 \text{ psi}$$

$$f_b = \frac{133{,}400}{165.2} = 808 \text{ psi} < 1240 \text{ psi} \qquad \text{O.K.}$$

Check shear by placing wheel load 15 in. from centerline of stringer.

$$V_{LL} = \left(\frac{47.625}{62.625}\right)(12{,}000) = 9126 \text{ lb}$$

$$V_{DL} = (9.9)\left(\frac{62.625}{2}\right) = 310 \text{ lb}$$

$$V_{TL} = V_{LL} + V_{DL} = 9126 + 310 = 9436 \text{ lb}$$

$$F_v' = F_v C_M = (145)(0.875) = 127 \text{ psi}$$

$$f_v = \frac{3V}{2bd} = \frac{(3)(9436)}{(2)(48)(6.75)} = 44 \text{ psi} < 127 \text{ psi} \qquad \text{O.K.}$$

Design of stringers and decking for Southern Pine is accomplished in a similar manner.

### Rail, Post, and Curb

Load requirements are in accordance with AASHTO. Load $P = 10$ kips applied with the top of the rail 2 ft 3 in. above roadway surface (see Fig. 6.10). The load is applied in an outward direction at post for post design and at midspan for rail design. For post design, a longitudinal load of 5 kips distributed over four posts must be applied simultaneously with the outward load. Each post shall be designed to resist an inward load of 10 kips/4. The attachment of the rail shall be designed to resist a vertical load of 10 kips/4 either upward or downward. The curb loading is 500 lb/ft. Combination 47 Southern Pine will be used in this example. The design of posts, rails, and curbs using Western Species is accomplished in a similar manner. Spacing of posts is 8 ft.

Rail:

Analyze a $6\frac{3}{4}$ in. × 11 in. rail for bending using the AASHTO recommended equation for bending of $Pl/6$ and for shear by locating the load a distance from the face of the post equal to the rail width.

Bending:

$$M = \frac{Pl}{6} = \frac{(10{,}000)(8)(12)}{6} = 160{,}000 \text{ in.-lb}$$

$$S_y = 83.53 \text{ in.}^3 \quad \text{(Table 8.2)}$$

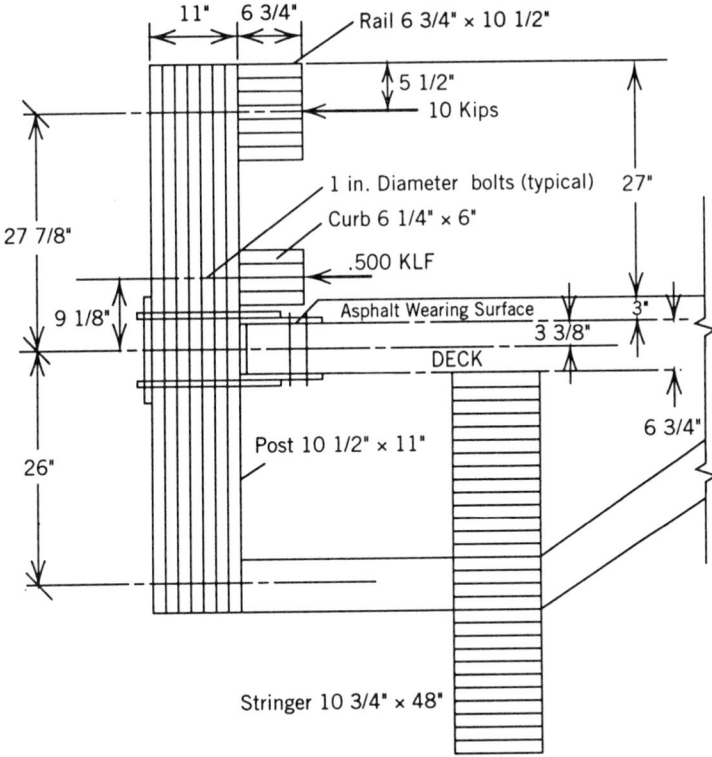

**FIGURE 6.10**   Post and Rail.

$$S_{y(net)} = 67.39 \text{ in.}^3 \quad (2 \text{ bolt holes})$$

$$f_b = \frac{M}{S_{y(net)}} = \frac{160,000}{67.39} = 2374 \text{ psi}$$

For Combination 47, $F_{by} = 1750$ psi; from AASHTO for rail loading, $C_D = 1.65$; from footnote $f$, Table 2, AITC 117—Design, $C_{fu} = 1.07$, $C_M = 0.8$.

$$F_b' = F_b C_{fu} C_D C_M = (1750)(1.07)(1.65)(0.8)$$

$$= 2472 \text{ psi} > 2374 \text{ psi} \qquad \text{O.K.}$$

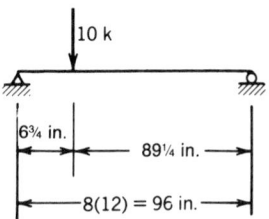

$$V_{max} = \left(\frac{89.25}{96}\right)(10,000) = 9297 \text{ lb}$$

$$f_v = \frac{3V}{2bd} = \frac{(3)(9297)}{(2)(11)(6.75)} = 188 \text{ psi}$$

$$F'_{vy} = F_{vy} C_D C_M = (175)(1.65)(0.875) = 253 \text{ psi} > 188 \text{ psi} \qquad \text{O.K.}$$

Bearing: Check compression perpendicular to grain between rail and post (assume post is 10.5 in. wide).

$$F_{c\perp} = 560 \text{ psi (from Table 2 AITC 117—Design)}$$

$$f_{c\perp} = \frac{P}{A} = \frac{(10,000)}{(11)(10.5)} = 87 \text{ psi}$$

$$F'_{c\perp} = F_{c\perp} C_D C_M = (560)(1.65)(0.53) = 490 \text{ psi} > 87 \text{ psi} \qquad \text{O.K.}$$

Check tension in bolts connecting the rail to the post for a load of one-fourth of the 10-kip load. Two $\frac{3}{4}$-in. diameter A307 bolts provide a tensile capacity of $(8.8)(2) = 17.6$ kips $> 10$ kips O.K.

Check the shear in bolts connecting the rail to the post. Use Eqs. (7-6)–(7-11) to determine the capacity of a $\frac{3}{4}$ in. diameter bolt in single shear.

The nominal design value $Z$ was calculated by use of these equations and Mode IV failure controlled:

$$Z = 1235 \text{ lb}$$

See 7.24 for information on determining nominal bolt design values.

$$\text{The capacity of two bolts} = 2ZC_D C_M = (2)(1235)(1.65)(0.75)$$

$$= 3,050 \text{ lb} > 2500 \text{ lb} \qquad \text{O.K.}$$

Curb:

Check a 6-$\frac{3}{4}$ in. × 5-$\frac{1}{2}$ in. curb for bending and shear.
The curb loading $w = 500$ plf

$$F'_{by} = F_{by} C_{fu} C_D C_M = 1750(1.07)(1.65)(0.8) = 2472 \text{ psi}$$

$$M_y = \frac{wL^2}{8} = \frac{(500)(8)^2(12)}{8} = 48,000 \text{ in.-lb}$$

$$S_y = 41.77 \text{ in.}^3$$

$$f_{by} = \frac{M_y}{S_y} = \frac{48,000}{41.77} = 1149 \text{ psi} < 2472 \text{ psi} \qquad \text{O.K.}$$

$$\text{Shear: } V = \frac{wL}{2} = \frac{(500)(96 - 10.5 - 2(6.75))}{2(12)} = 1500 \text{ lb}$$

$$F_{vy} = 175 \text{ psi}$$

$$F'_{vy} = F_v C_D C_M = (175)(1.65)(0.875) = 253 \text{ psi}$$

$$f_v = \frac{3V}{2bd} = \frac{(3)(1500)}{(2)(6.75)(5.5)} = 61 \text{ psi} < 253 \text{ psi} \qquad \text{O.K.}$$

Post:

Check an 11 in. $\times$ 10-$\frac{1}{2}$ in. post for bending and shear:

$$F_{bx} = 1600 \text{ psi} \quad F_{vx} = 200 \text{ psi} \quad E = 1,400,000 \text{ psi}$$

$$F_{by} = 1750 \text{ psi} \quad F_{vy} = 175 \text{ psi}$$

Bending:

Check biaxial bending in the post with railing and curb loads applied simultaneously.

$$M_x = (10,000)(27.3125) + 500(8)(8.5625) = 307,375 \text{ in.-lb}$$

$$M_y = (5,000/4)(27.3125) = 34,141 \text{ in.-lb}$$

Post is attached to deck with bolts outside of post.

$$S_x = 211.8 \text{ in.}^3$$

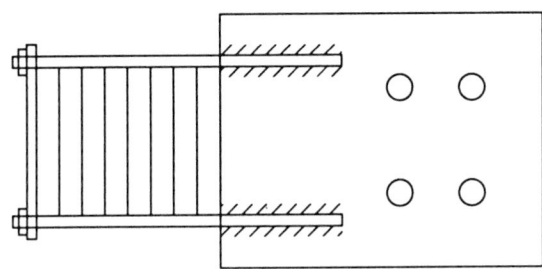

$$C_V = (K_L) \left(\frac{5\frac{1}{8}}{b}\right)^{1/20} \left(\frac{12}{d}\right)^{1/20} \left(\frac{21}{L}\right)^{1/20}$$

$$= (1.09) \left(\frac{5.125}{10.5}\right)^{1/20} \left(\frac{12}{11}\right)^{1/20} \left(\frac{21}{4.4}\right)^{1/20} = 1.14. \quad \text{Use 1.0.}$$

$C_D = 1.65$ for impact load (AASHTO 13.2.5.1)

$C_M = 0.80$

$C_L$ was calculated and determined to be equal to 1.0.

$$F'_{bx} = F_{bx} C_V C_D C_M C_L = (1600)(1.0)(1.65)(0.80)(1.0) = 2112 \text{ psi}$$

$$f_{bx} = \frac{M_x}{S_x} = \frac{307,375}{211.8} = 1451 \text{ psi} < 2112 \text{ psi} \qquad \text{O.K.}$$

$C_{fu} = 1.01$ (Footnote $f$, Table 2 AITC 117—Design)

$$S_y = 202.1 \text{ in.}^3$$

$$F'_{by} = F_{by} C_{fu} C_D C_M = 1750(1.01)(1.65)(.80) = 2333 \text{ psi}$$

$$f_{by} = \frac{M_y}{S_y} = \frac{34,141}{202.1} = 170 \text{ psi}$$

$l_e$ (post in bending) $= 1.11 l_u = (1.11)(53.3) = 59.16$ in.

$$R_B = \sqrt{\frac{l_e d}{b^2}} = \sqrt{\frac{(59.16)(11)}{(10.5)^2}} = 2.43$$

$$K_{bE} = 0.609 \text{ [From Eq. (5-10)]}$$

$$F_{bE} = \frac{K_{bE} E'}{R_B^2} = \frac{(0.609)(1,400,000)(0.833)}{2.43^2} = 120,270 \text{ psi}$$

$$\frac{f_{bx}}{F'_{bx}} + \frac{f_{by}}{F'_{by}(1 - f_{bx}/F_{bE})} = \frac{1451}{2112} + \frac{170}{2333(1 - 1451/120,270)}$$

$$= 0.761 < 1 \qquad \text{O.K.}$$

Shear parallel to grain:

Check shear on net area: Anchor post to deck.

$$A = 115.5 \text{ in.}^2$$

$$f_v = \frac{3V}{2A} = \frac{(3)(14,000)}{(2)(115.5)} = 182 \text{ psi}$$

$$F'_{vy} = F_{vy} C_D C_M = (175)(1.65)(.875) = 252 \text{ psi} > 182 \text{ psi} \qquad \text{O.K.}$$

Check post to deck connection:

$$(4)(35.125) + (10)(53.875) = 26T$$

$$T = 26,125 \text{ lb}$$

Two 1-in. diameter A307 bolts provide $(15.7)(2) = 31.4$ kips capacity for post to steel plate deck bracket connection. For bracket to deck connection, determine number, $n$, of 1-in.-diameter bolts required. From Table 7.18, the bolt value is 5,750 lb, which may be increased by $C_D = 1.65$ for impact, but is reduced for wet service, $C_M = .67$

$$n = \frac{26,125}{(5750)(1.65)(.67)} = 4.1 \text{ (use four bolts)}$$

Design of these members is similar to the preceding analysis and is not detailed here.

The reaction at the bottom of the post must be transferred to the stringer and, in turn, through a diagonal member to the deck.

### 6.5.8   Longitudinal Deck with Transverse Stiffeners

The following example illustrates the procedure for designing a simple span longitudinal deck bridge using glued laminated timber panels.

**Example 3.**   Criteria

Bridge Configuration:
Span:   20 ft 0 in. (simple span)
Roadway width:   32 ft 0 in. (multiples of 4 ft)
Loading Criteria:
HS 15-44 live load, two lanes, in accordance with AASHTO Standards with
   a 3 in. asphalt wearing surface
Materials:
Design values for longitudinal deck (based on bending about the y-y axis) from
   Table 2 of AITC 117—Design (Douglas Fir-Larch, L2)

$$F_{by} = 1800 \text{ psi}$$

$$F_{vy} = 145 \text{ psi}$$

$$F_{c\perp} = 560 \text{ psi}$$

$$E = 1,700,000 \text{ psi}$$

Design values must be reduced by appropriate adjustment, $C_M$, for wet-use conditions when applicable.
Deck Design:
Assume:

$8\frac{3}{4}$ in. thick deck panel $\times$ 48 in. wide $\times$ 20 ft long (interior panel)

$6\text{-}\frac{3}{4}$ in. $\times$ $4\text{-}\frac{1}{2}$ in. $\times$ 34 ft long stiffeners (2 @ 6'-8" cc.)

Weight of glulam $= 50$ pcf   (AASHTO 3.3.6)

Weight of asphalt $= 150$ pcf   (AASHTO 3.3.6)

Dead load per panel:

$$\text{Deck} = \frac{(8.75)(4.0)(50)}{12} = 146 \text{ plf}$$

$$\text{Asphalt} = \frac{(3.0)(4.0)(150)}{12} = \underline{150 \text{ plf}}$$

$$\text{Total dead load} = w_{DL} = 296 \text{ plf}$$

Live load:

$$P = 12,000 \text{ lb} \quad (\text{AASHTO } 3.7.7\text{A})$$

Bending:

The maximum moment for this span and loading condition occurs when a 12,000 lb wheel load is placed at midspan. The wheel load fractions (WLF) for two or more lanes, (Iowa State University report) (Ref. 27) are calculated as:

$$W_p = \text{Width of Panel} = 4 \text{ ft}$$

$$L = \text{Span} = 20 \text{ ft}$$

$$\text{Wheel Load Fraction} = \text{WLF} = \frac{W_p}{3.75 + L/28}$$

$$= \frac{4}{3.75 + 20/28} = 0.896 \text{ (controls)}$$

$$\text{or} = \frac{W_p}{5.00} = \frac{4}{5} = 0.80$$

whichever is greater,

$$M_{DL} = \frac{w_{DL}L^2}{8} = \frac{(0.296)(20)^2}{8} = 14.8 \text{ ft-kips}$$

$$M_{LL} = \frac{PL(WLF)}{4} = \frac{(12)(20)(0.896)}{4} = 53.8 \text{ ft-kips}$$

$$M_{TL} = M_{DL} + M_{LL} = 14.8 + 53.8 = 68.6 \text{ ft-kips}$$

Section Properties for $8\text{-}\frac{3}{4}$ in. $\times$ 48 in. deck, $C_{fu} = 1.04$ (footnote $f$, Table 2 AITC 117—Design)

$$A = \text{Area} = 420 \text{ in.}^2$$

$$S = \text{Section Modulus} = 612.5 \text{ in.}^3$$

$$I = \text{Moment of Inertia} = 2680 \text{ in.}^4$$

Actual bending stress,

$$f_{by} = \frac{M_{TL}}{S} = \frac{(68.6)(12,000)}{612.5} = 1344 \text{ psi}$$

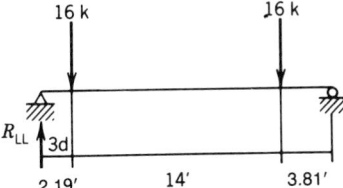

Design value in bending.

$$F'_{by} = F_{by} C_{fu} C_M = 1800(1.04)(0.8) = 1498 \text{ psi} > 1344 \text{ psi} \qquad \text{O.K.}$$

Shear parallel to grain:

Design live load is calculated by placing the wheel load at either three times the depth of the deck from the end or at the quarter span. The value which results in the wheel load being located closest to the support will control.

$$3 \times \text{depth} = 3 \times \frac{8.75}{12} = 2.19 \text{ ft (controls)}$$

$$\text{or at quarter span} = \frac{20}{4} = 5 \text{ ft}$$

$V_{DL}$ at distance "$t$" from the support $= 0.296[(20/2) - 8.75/12] = 2.7$ kips

$$V_{LL} = \left[ \frac{12(20.0 - 2.19)}{20} + \frac{12(20.0 - 16.19)}{20} \right] \text{WLF} = 13.0 \text{ kips}$$

$$\text{where WLF} = W_p/4.0 = 4.0/4.0 = 1$$

$$V_{TL} = V_{DL} + V_{LL} = 2.7 + 13.0 = 15.7 \text{ kips}$$

Actual shear parallel to grain, $f_v = 3V/2A = (3)(15.7)/(2)(420) = 56$ psi
Design value for shear parallel to grain, $F'_{vy} = F_{vy} C_M = (145)(0.875) = 127$ psi $> 56$ psi O.K.

Deflection:

Deflection is determined by simple beam action. For live load deflection, wheel load is positioned as for maximum moment with load distributed by wheel load fraction as determined for moment.

$$\text{Wet-use adjustment factor for } E, \ C_M = 0.833.$$

$$E' = (0.833)(1,700,000) = 1,416,000 \text{ psi}$$

For uniform load:

$$\Delta_{DL} = \frac{5w_{DL}L^4}{384EI} = \frac{(296)(20)^4(1728)}{(384)(1,416,000)(2680)} = 0.28 \text{ in.}$$

For point load at center span:

$$\Delta_{LL} = \frac{PL^3}{48EI}(WLF) = \frac{(12,000)(20)^3(1728)}{(48)(1,416,000)(2680)}(0.896) = 0.82 \text{ in.}$$

For live load, deflection limit is span/300

$$\Delta_{LL} = L/300 = \frac{20(12)}{300} = 0.80 \approx 0.82 \quad \text{O.K.}$$

Stiffeners:

Size of stiffeners is based on a minimum $EI$ value of 80,000 kips-in.$^2$
Use:   $E = 1,700,000$ psi (dry use)
Try:   6-$\frac{3}{4}$ in. × 4-$\frac{1}{2}$ in. glulam

$$I_{xx} = 51.3 \text{ in.}^4$$

$$EI = (1,700,000)(51.3) = 87,210 \text{ kip-in.}^2 > 80,000 \text{ kip-in.}^2 \quad \text{O.K.}$$

Locate one stiffener at third points.

Therefore, stiffeners at 6 ft 8 in. spacing $(20/3 = 6.67 \text{ ft})$

Maximum stiffener spacing is 8 ft $> 6$ ft 8 in.    O.K.

The cross sectional size of the stiffeners may need to be increased depending on the requirements of the railing and post design and their attachment to the deck.

### 6.5.9   Guard Rails for Longitudinal Deck Bridges

Guard rails for longitudinal deck glued laminated timbers bridges have been tested in accordance with performance level 1 (PL-1) specified in AASHTO *Guide Specifications for Bridge Railings* (28). One of the types which passed the crash test for performance level 1 is shown in the following sketch. For further information, see Refs. (29, 30).

## 6.6   SHEATHING, FLOORING, AND DECKING

### 6.6.1   Sheathing and Flooring

#### 6.1.1.1   *Lumber*

Sheathing consisting of lumber up to 1 in. in nominal thickness nailed transversely or diagonally at about 45° to studs or joists is sometimes used in wood frame construction. When subjected to lateral forces such as wind or earthquakes, lumber sheathing and its supporting framework may act as a diaphragm, when properly designed as such, serving to brace the building against the lateral forces and transmitting these forces to the foundations. Diaphragm design procedures

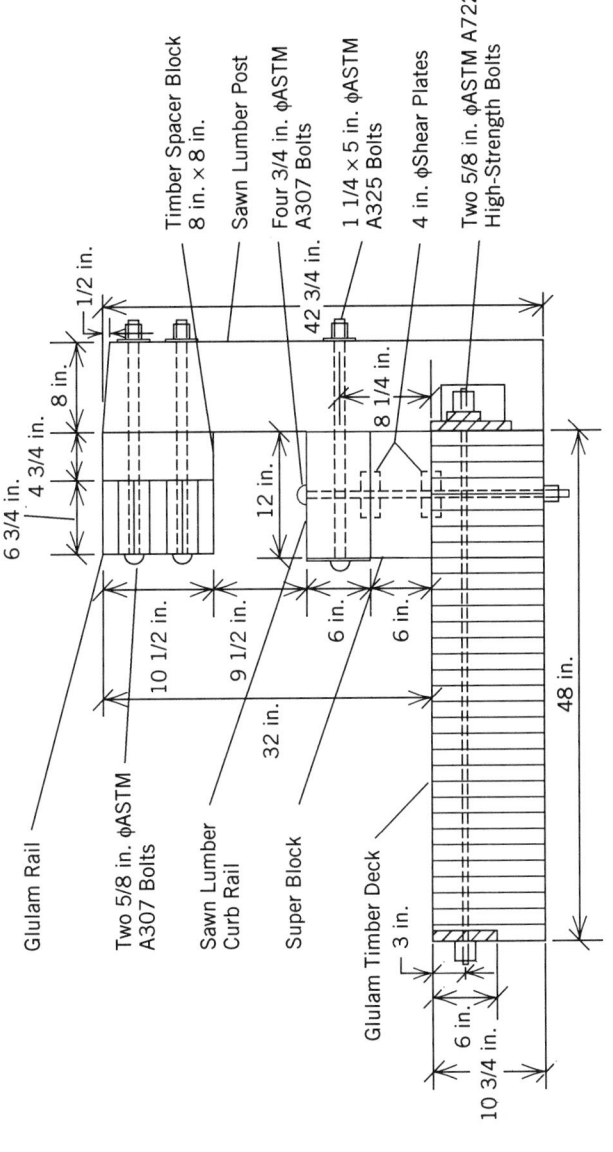

Timber Spacer Block
8 in. × 8 in.

Sawn Lumber Post

Four 3/4 in. φASTM
A307 Bolts

1 1/4 × 5 in. φASTM
A325 Bolts

4 in. φShear Plates

Two 5/8 in. φASTM A722
High-Strength Bolts

Glulam Rail

Two 5/8 in. φASTM
A307 Bolts

Sawn Lumber
Curb Rail

Super Block

Glulam Timber Deck

1/2 in.
8 in.
4 3/4 in.
6 3/4 in.
42 3/4 in.
8 1/4 in.
12 in.
10 1/2 in.
9 1/2 in.
32 in.
6 in.
6 in.
48 in.
3 in.
6 in.
10 3/4 in.

Notes: (1) Post Spacing 6 ft. 3 in.
(2) Many details have been omitted, refer to Ref. (30).

Tested in accordance with the requirements for PL-1 described in *Guide Specifications for Bridge Railings*, AASHTO 1989 (28).

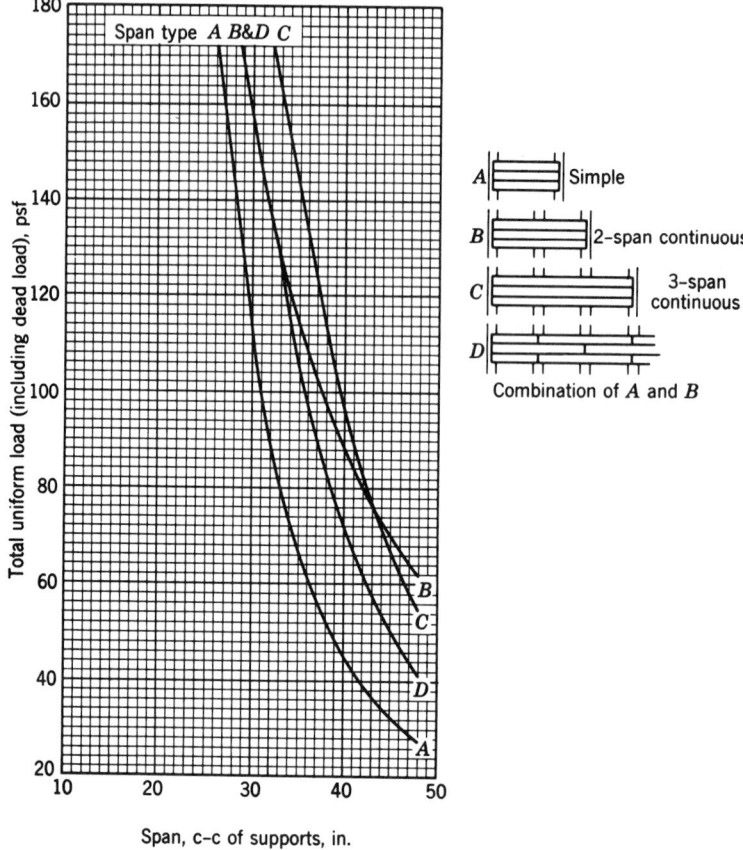

**FIGURE 6.11**    Span-load curves. For 1-in. nominal thickness lumber sheathing. Conditions: Deflection = $l/240$; Modulus of elasticity = 1,800,000 psi; design value in bending = 1200 psi, span types $A$–$D$ illustrated at right of figure.

for lumber sheathing 1 in. in nominal thickness are given in *Western Woods Use Book* (31).

The edges of lumber sheathing may be square, shiplapped, splined, or tongue-and-groove. Sheathing runs should always be spliced over supports unless end matched or scarfed and glued end joints or splice blocks are used. Sheathing used as subflooring in wood frame construction has the effect of a shallow beam, and therefore, its design procedure is similar to that for beams with certain modifications. Fig. 6.11 presents span-load curves for 1 in. nominal thickness lumber sheathing for various types of spans.

### 6.1.1.2    Panels

Panels for construction and industrial applications can be manufactured in a variety of ways: as plywood (cross-laminated wood veneer), as composites (veneer faces bonded to reconstituted wood cores), or as non-veneered panels (including waferboard, oriented strand board, and certain specific classes of particleboard).

Some grades of veneered panels are manufactured under the provisions of Voluntary Product Standard PS1, *Construction and Industrial Plywood* (32). These panels are referred to in this manual as plywood. Other veneered panels, however, as well as other composite and non-veneered panels are manufactured under provisions of Voluntary Product Standard PS2 (33), *Performance Standard for Wood Based Structural Panels*, or American Plywood Association performance standards to establish their performance for specific construction applications. The panels meeting these APA standards are referred to in this manual as performance-rated panels.

Plywood is divided into two basic types by the exposure durability category. Exterior type is manufactured with a waterproof glue line and C grade or better veneers. Interior types may contain D grade veneers and may use a moisture-resistant glue line. However, most Interior-type plywood manufactured today is made with waterproof glue and designated Exposure 1. Performance-rated panels can be manufactured in three exposure durability classifications: Exterior, Exposure 1, and Exposure 2. Panels marked Exterior are for applications subject to continuous exposure to the weather or moisture and are comparable to Exterior-type plywood. Panels with an Exposure 1 designation are intended for protected construction applications where ability to resist the effects of moisture during construction delays or similar exposure conditions is needed.

Exposure 1 panels are comparable to Interior-type plywood with exterior glue. Panels with an Exposure 2 designation are intended for protected construction applications where the ability to resist the effects of high humidity and water leakage is required. Exposure 2 panels are comparable to Interior-type plywood with intermediate glue.

Plywood can be manufactured from over 70 species of wood. These species are divided into five groups by strength and stiffness categories. Group 1, containing Douglas Fir, Western Larch, and Southern Pine, is the strongest group. Group 2 contains the next strongest species, and so on.

Panel products are also categorized by span ratings, which denote the maximum recommended center-to-center spacing in inches of supports over which the panels should be placed in construction applications. A typical span rating for sheathing consists of two numbers separated by a slash, for example, 32/16. The left-hand number (32 in this example) indicates the maximum recommended spacing of supports when used as roof sheathing. The right-hand number indicates the maximum spacing for subflooring. Both of these ratings apply when the long dimension of the panel is across three or more supports.

Table 6.10 contains recommended roof live loads for performance-rated sheathing in conventional applications where the long dimension is perpendicular to the supports. Table 6.11 contains recommended live loads for plywood where the long dimension is parallel to the supports. Table 6.12 contains minimum fastener recommendations for performance-rated panels.

Panel edge support is required to resist concentrated loads in some cases, such

**TABLE 6.10**

**Recommended Uniform Roof Live Load for Performance-Rated Panel Sheathing with Long Dimension Perpendicular to Supports[a,b,c]**

| Panel Span Rating | Panel Thickness (in.) | Maximum Span (in.) | | Allowable Live Loads (psf),[d] Spacing of Supports Center-to-Center (in.) | | | | | | | |
| | | With Edge Support[e] | Without Edge Support | 12 | 16 | 20 | 24 | 32 | 40 | 48 | 60 |
|---|---|---|---|---|---|---|---|---|---|---|---|
| 12/0 | 5/16 | 12 | 12 | 30 | | | | | | | |
| 16/0 | 5/16, 3/8 | 16 | 16 | 70 | 30 | | | | | | |
| 20/0 | 5/16, 3/8 | 20 | 20[f] | 120 | 50 | 30 | | | | | |
| 24/0 | 3/8, 7/16, 1/2 | 24 | 20[f] | 190 | 100 | 60 | 30 | | | | |
| 24/16 | 7/16, 1/2 | 24 | 24 | 190 | 100 | 75 | 40 | | | | |
| 32/16 | 15/32, 1/2, 5/8 | 32 | 28 | 325 | 180 | 120 | 70 | 30 | | | |
| 40/20 | 19/32, 5/8, 3/4, 7/8 | 40 | 32 | | 305 | 205 | 130 | 60 | 30 | | |
| 48/24 | 23/32, 3/4, 7/8 | 48 | 36 | | | 280 | 175 | 95 | 45 | 35 | |
| 48 oc[g] | 1-1/8 | 60 | 48 | | | | 290 | 160 | 100 | 65 | 40 |

[a]Source: American Plywood Association.

[b]Rated sheathing and Structural I rated sheathing.

[c]When roofing is to be guaranteed or warranties, check with roofing manufacturer for acceptable deck.

[d]10-psf dead load assumed.

[e]Tongue-and-groove edges, panel edge clips (one between each support, except two between supports 48 in. on center), lumber blocking, or other.

[f]24 in. for 1/2 in. panels.

[g]Span rating applies to APA-rated Sturd-I-Floor.

as single-layer floors, or where the span–thickness ratio is high in roof sheathing. Such edge supports may be furnished by solid blocking cut in between framing or by tongue-and-groove joints in the plywood. For roof sheathing, specially manufactured H-shaped metal clips may be used. When the panel edges are required to transmit lateral shear, as in some diaphragms, they should be attached to solid blocking or otherwise fastened to resist lateral loading.

### 6.6.2 Decking

Timber decking is commonly used for floor and roof construction in conjunction with timber joists, purlins, beams, arches and trusses. Timber decking may also be used with other structural materials.

### 6.6.3 Two-, Three-, and Four-Inch Heavy Timber Decking

Sawn lumber roof decking. For information on species, sizes, patterns, lengths, moisture content, application, specifications, applicable allowable stress, and allowable loads for 2-, 3-, and 4-in.-nominal-thickness tongue-and-groove heavy timber decking used as roof decking, see *Standard for Tongue-and-Groove Heavy Timber Roof Decking*, AITC 112 (34), in Chapter 8. Lumber grading rules the NDS and building codes also provide specific information for design and construction.

If decking is to be used for floors or other purposes, special care should be given to end joint locations and to deflection limitations.

### 6.6.4 Mechanically Laminated Decking

Mechanically laminated decks consist of square-edged dimension lumber set on edge, wide face-to-wide face, with the pieces connected by nails or other fasteners. If side nails are used, they should be long enough to penetrate approximately two and one-half lamination thicknesses for load transfer. Where deck supports are 4 ft center to center or less, side nails should be spaced not more than 30 in. on centers and staggered one third of the spacing in adjacent laminations. When supports are spaced more than 4 ft center to center, side nails should be spaced approximately 18 in. on centers alternately near top and bottom edge and also staggered one-third of the spacing in adjacent laminations. Two side nails should be used at each end of the butt-joined pieces.

Laminations should be toe-nailed to supports using $20d$ or larger common nails. When the supports are 4 ft center to center or less, alternate laminations should be toe-nailed to alternate supports; when supports are spaced more than 4 ft center to center, alternate laminations should be toe-nailed to every support.

The five arrangements of spans in *Standard for Tongue-and-Groove Heavy Timber Roof Decking*, AITC 112 (34), Chapter 8, are also applicable for mechanically laminated decks. The design equations for mechanically laminated decks are as follows:

Simple span (Type 1):

$$w_b = \frac{8F_b'I}{l^2c} \quad \text{and} \quad w_\Delta = \frac{384\Delta E'I}{5l^4}$$

**TABLE 6.11**

**Recommended Uniform Roof Loads (psf) for APA Rated Sheathing with Long Dimension Parallel to Supports[a,b]**

| Panel Grade | Thickness (in.) | Span Rating | Maximum Span (in.) | Load at Maximum Span | |
|---|---|---|---|---|---|
| | | | | Live Load | Total Load |
| APA Structural I Rated Sheathing | | | | | |
| | 7/16 | 24/16 | 24[d] | 20 | 30 |
| | 15/32 | 32/16 | 24 | 35[a] | 45[a] |
| | 1/2 | 32/16 | 24 | 40[a] | 50[a] |
| | 19/32 and 5/8 | 40/20 | 24 | 70 | 80 |
| | 23/32 and 3/4 | 48/24 | 24 | 90 | 100 |

**APA Rated Sheathing**

| | | | | |
|---|---|---|---|---|
| 7/16[b] | 24/0, 24/16 | 16 | 40 | 50 |
| 15/32 | 32/16 | 24[d] | 20 | 25 |
| 1/2 | 32/16 | 24 | 25 | 30 |
| 19/32 | 40/20 | 24 | 40 | 50 |
| 5/8 | 32/16, 40/20 | 24 | 45 | 55 |
| 23/32 and 3/4 | 40/20, 48/24 | 24 | 60 | 65 |

[a]Source: American Plywood Association.

[b]When roofing is to be guaranteed by a performance bond, check with roofing manufacturer for minimum panel thickness, span, and edge support requirements.

[c]Number of layers equal to number of piles, except 4 ply is 3 layer and 6 ply is 5 layer.

[d]Solid blocking recommended at 24-in. span.

[e]25-psf live and 35-psf total load with solid blocking at panel ends.

6-407

## TABLE 6.12

### Recommended Minimum Fastening Schedule for Performance-Rated Panel Roof Sheathing[a,b]

| Panel Thickness (in.) | Size | Nailing[c] Spacing (in.) Panel Edges | Intermediate |
|---|---|---|---|
| 5/16 | 6d | 6 | 12 |
| 3/8 | 6d | 6 | 12 |
| 7/16, 15/32, 1/2 | 6d | 6 | 12 |
| 19/32, 5/8, 23/32, 3/4, 7/8 | 8d | 6 | 12[f] |
| 1-1/8 | 8d or 10d | 6 | 12[f] |

[a]Source: American Plywood Association.

[b]Closer nail spacing may be required to obtain higher diaphragm shear values.

[c]Use common smooth or deformed shank nails with panels to 1 in. thick. For 1-1/8 in. panels, use 8d deformed shank or 10d common smooth-shank nails.

[f]For stapling asphalt shingles to 5/16 in. and thicker panels, use staples with a 15/16 in. minimum crown width and a 1 in. leg length. Space according to shingle manufacturer's recommendations.

[f]For spans 48 in. or greater, space nails 6 in. at all supports.

Two-span continuous (Type 2):

$$w_b = \frac{8F_b' I}{l^2 c} \quad \text{and} \quad w_\Delta = \frac{185 \Delta E' I}{l^4}$$

Combination simple- and two-span continuous (Type 3):

$$w_b = \frac{8F_b' I}{l^2 c} \quad \text{and} \quad w_\Delta = \frac{109 \Delta E' I}{l^4}$$

Cantilevered pieces intermixed (Type 4):

$$w_b = \frac{20F_b' I}{3l^2 c} \quad \text{and} \quad w_\Delta = \frac{105 \Delta E' I}{l^4}$$

Controlled random layup (Type 5):

$$w_b = \frac{20F_b' I}{3l^2 c} \quad \text{and} \quad w_\Delta \frac{100 \Delta E' I}{l^4}$$

where   $w_b$ = allowable load limited by bending, pli, pounds per lineal inch,
   $w_\Delta$ = allowable load limited by deflection, pli, pounds per lineal inch,

$F_b'$ = design value in bending (psi) modified by applicable adjustment factors,

$E'$ = design value for modulus of elasticity (psi) modified by applicable adjustment factors,

$I$ = moment of inertia (in.$^4$),

$l$ = span (in.),

$c$ = half the depth of the decking (in.), and

$\Delta$ = deflection limitation (in.).

Note: The preceding equations are dimensionally correct with $w_b$ and $w_\Delta$ in lbs per linear in. of a strip 12 in. wide. Both $w_b$ and $w_\Delta$ should be multiplied by 12 to obtain the customary units of pounds per lineal ft (plf) to use in design.

Table 6.13 gives allowable uniformly distributed loads for mechanically laminated decks for various spans as limited by bending and deflection. The table is based on the use of seasoned lumber, normal duration of loading, and loads applied normal to the decking surface. The allowable loads include dead load.

### 6.6.5  Other Decking

Other special decking products are available. They include panelized decking, glued laminated decking, and heavy plywood decking such as "2–4–1". Panelized decking is a decking component made up of splined panels, usually about 2 ft wide and of any specified length up to a maximum, which may vary among manufacturers.

Glued laminated decking is manufactured by laminating two or more pieces of lumber into single decking members. Common nominal lumber thicknesses of the decking are 2, 3, or 4 in., although exact finished sizes may differ between manufacturers.

Plywood panels known as "2–4–1" plywood are $1\frac{1}{8}$ in. thick, for 32- or 48-in. floor spans and may also be used as roof decking in "heavy timber" construction. They are usually supplied with tongue-and-groove joints on the long edge. Manufacturers of these special decking products should be consulted for information concerning their products.

### 6.6.6  Cantilevered Overhangs

Under uniformly distributed loads, cantilevered overhangs act to reduce the deflections in the remaining areas of a deck structure. Thus, span increases are often justified where overhangs are used. The effects of cantilevered overhangs are varied. The span arrangement and the length of overhang effect support reactions, bending stress, and span and overhang deflections. Charts that permit the determination of these factors for simple-span, two-span continuous, and up to five-span continuous arrangements are available from AITC.

The designer should also consider the effects of heating on snow loads for roofs with cantilevered overhangs. Unbalanced loading may be created where melting occurs over heated areas but not over unheated overhangs.

## TABLE 6.13

### Load Capacity of Mechanically Laminated Decks[a]

| Span (ft) | Nominal Thickness (in.) | Load Controlled by Bending $F'_b = 1000$ psi | | Load Controlled by Deflection Limited by $l/240$ $E' = 1,000,000$ psi | | | | |
|---|---|---|---|---|---|---|---|---|
| | | Types 1, 2, and 3[b] | Types 4 and 5[b] | Type 1[c] | Type 2[c] | Type 3[c] | Type 4[c] | Type 5[c] |
| | | Uniform Load $w$, psf | | | | | | |
| 4 | 4 | 1021 | 851 | 1488 | 3587 | 2114 | 2036 | 1939 |
| 5 | 4 | 653 | 544 | 762 | 1836 | 1082 | 1042 | 993 |
| | 6 | 1613 | 1344 | 2958 | 7126 | 4199 | 4045 | 3852 |
| 6 | 4 | 454 | 378 | 441 | 1063 | 626 | 603 | 574 |
| | 6 | 1120 | 934 | 1712 | 4124 | 2430 | 2341 | 2229 |
| | 8 | 1946 | 1622 | 3920 | 9440 | 5566 | 5362 | 5106 |
| 7 | 4 | 333 | 278 | 278 | 669 | 394 | 380 | 362 |
| | 6 | 823 | 686 | 1078 | 2597 | 1530 | 1474 | 1404 |
| | 8 | 1430 | 1192 | 2469 | 5948 | 3505 | 3376 | 3215 |
| 8 | 4 | 255 | 213 | 186 | 448 | 264 | 254 | 242 |
| | 6 | 630 | 525 | 722 | 1740 | 1025 | 988 | 940 |
| | 8 | 1095 | 912 | 1654 | 3985 | 2348 | 2262 | 2154 |
| | 10 | 1782 | 1485 | 3435 | 8276 | 4877 | 4698 | 4474 |
| 9 | 4 | 201 | 168 | 131 | 315 | 186 | 179 | 170 |
| | 6 | 498 | 415 | 507 | 1222 | 720 | 694 | 660 |
| | 8 | 865 | 721 | 1162 | 2799 | 1649 | 1589 | 1513 |
| | 10 | 1408 | 1174 | 2412 | 5813 | 3425 | 3299 | 3142 |
| 10 | 6 | 403 | 336 | 370 | 891 | 525 | 506 | 481 |
| | 8 | 701 | 584 | 847 | 2040 | 1202 | 1158 | 1103 |
| | 10 | 1141 | 951 | 1759 | 4237 | 2497 | 2405 | 2290 |
| 11 | 8 | 579 | 483 | 636 | 1533 | 903 | 870 | 829 |
| | 10 | 943 | 786 | 1321 | 3184 | 1876 | 1807 | 1721 |
| 12 | 8 | 487 | 406 | 490 | 1181 | 696 | 670 | 638 |
| | 10 | 792 | 660 | 1018 | 2452 | 1445 | 1392 | 1326 |
| 13 | 8 | 415 | 346 | 385 | 929 | 547 | 527 | 502 |
| | 10 | 675 | 562 | 800 | 1929 | 1137 | 1095 | 1043 |
| 14 | 8 | 357 | 298 | 309 | 744 | 438 | 422 | 402 |
| | 10 | 582 | 485 | 641 | 1544 | 910 | 877 | 835 |
| 15 | 8 | 311 | 260 | 251 | 605 | 356 | 343 | 327 |
| | 10 | 507 | 422 | 521 | 1256 | 740 | 713 | 679 |

**TABLE 6.13** (*Continued*)

| Span (ft) | Nominal Thickness (in.) | Load Controlled by Bending $F_b' = 1000$ psi | | Load Controlled by Deflection Limited by $l/240$ $E' = 1,000,000$ psi | | | | |
|---|---|---|---|---|---|---|---|---|
| | | Types 1, 2, and 3[b] | Types 4 and 5[b] | Type 1[c] | Type 2[c] | Type 3[c] | Type 4[c] | Type 5[c] |
| | | Uniform Load $w$, psf | | | | | | |
| 16 | 8 | 274 | 228 | 207 | 498 | 294 | 283 | 269 |
| | 10 | 446 | 371 | 429 | 1034 | 610 | 587 | 559 |
| 17 | 8 | 242 | 202 | 172 | 415 | 245 | 236 | 224 |
| | 10 | 395 | 329 | 358 | 862 | 508 | 490 | 466 |
| 18 | 8 | 216 | 180 | 145 | 350 | 206 | 199 | 189 |
| | 10 | 352 | 293 | 302 | 727 | 428 | 412 | 393 |
| 19 | 10 | 316 | 263 | 256 | 618 | 364 | 351 | 334 |
| 20 | 10 | 285 | 238 | 220 | 530 | 312 | 301 | 286 |

[a]For seasoned lumber, normal conditions of loading, and load applied normal to decking surface. Loads for other stress and deflection values can be determined by proportion. Load includes weight of decking, which should be subtracted to determine allowable superimposed load.

[b]Limited by bending $F_b' = 1000$ psi.

[c]Limited by a deflection limit of $l/240$. $E' = 1,000,000$ psi.

## 6.7 STRUCTURAL DIAPHRAGMS

### 6.7.1 General Considerations

Structural diaphragms are relatively thin, usually rectangular, structural elements capable of resisting shear parallel to their edges. A conventional frame roof, wall, or floor will normally function as a structural diaphragm with only slight design modification. They may be used as walls in a vertical position, as roofs or floors in a horizontal position, or as roofs pitched or curved to conform with common truss shapes.

The function of the diaphragm is to brace a structure against lateral forces, such as wind or earthquake loads, and to transmit these forces to the other resisting elements of the structure. Fig. 6.12 illustrates the distribution of lateral forces acting on a simple structure. The lateral loads produced act on the side walls spanning from foundation to roof. The top of the side wall thereby produces horizontal loads against the roof framing. When designed as a diaphragm, the roof framing system acts as a large plate girder, generally with continuous chords resisting

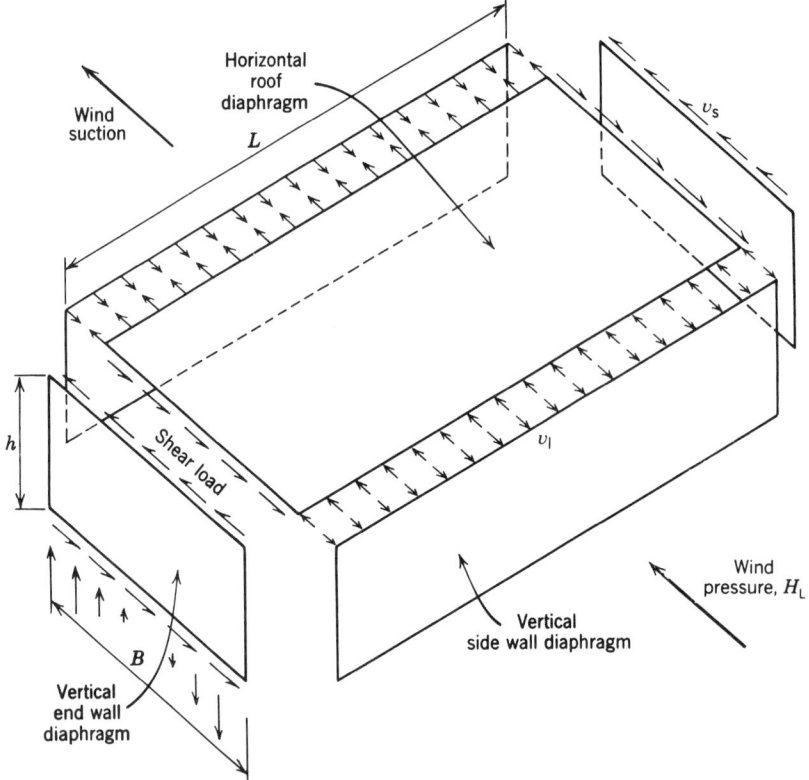

**FIGURE 6.12**    Lateral forces on a simple structure.

bending moment as flanges and with the sheathing itself resisting shear forces as the web. The roof framing system carries the side wall reactions as horizontal loads and spans to the end walls, which acts as a cantilever system extending up from the foundation to provide the necessary horizontal support.

A bibliography on the design and performance of lumber and wood panel diaphragms can be found in Ref. (35).

### 6.7.2  Common Types

Common types of wood diaphragms are the following:

(a)  Plywood—consists of sheets of plywood or other panel products fastened to cross members usually by means of nails, although sometimes by adhesives. For additional information see *Plywood Diaphragm Construction* (36).

(b)  Heavy timber decking—consists of 2-, 3-, or 4-in. heavy timber decking and may be covered with plywood sheathing to improve diaphragm values. Available test data on this system indicate that when sheathed with plywood, it can be designed as equivalent to a blocked plywood diaphragm.

(c) Diagonal sheathing—consists either of 1-in. nominal-thickness boards or 2-in. nominal-thickness lumber, nailed at a 45° angle in a single layer to cross members. Considerably greater strength and stiffness result from this placement than when transverse sheathing is used. Tests verify that although there may be considerable bending in the sheathing, the primary load resistance in an efficient diaphragm is due to the axial stress in the sheathing. The axial stress may be either direct tension or compression. Moment forces are resisted by the continuous chords.

(d) Double diagonal sheathing—consists of two layers of diagonal sheathing, one on top of the other, with the sheathing in one layer at a 90° angle with the sheathing in the outer layer. This type is considerably stiffer and stronger than the types (e) and (c). One layer of sheathing is in axial tension and is counteracted by the other layer, which is in compression; thus, the effects counteract and cancel each other.

(e) Transverse sheathing—consists of either 1-in. nominal boards or 2-in. nominal-thickness lumber, nailed in a single layer at right angles to the direction of cross members, such as joists or studs. This type is suitable when loads are light and when deflection is not important. When used vertically as a wall diaphragm, the load capacity of transverse sheathing is quite low compared to that of other types, and it is recommended that cross bracing be used to increase the strength and stiffness.

### 6.7.3 General Design Procedure

All diaphragms are essentially deep beams with a shear-resistant web. As the web is not considered effective in bending, ordinary beam theory rather than deep beam theory normally is used. The sheathing material acts as the beam web to carry the shear forces. Because the individual pieces of the sheathing are discontinuous, the sheathing connections are one of the most critical items in diaphragm design. The beams, girts, studs, columns, joists, and purlins act as stiffeners and cross ties for the web. Chord members must be provided to carry the flange forces of the beam. Either additional members, a portion of the walls, or some other continuous framing must be provided. For frame buildings with shear resistant walls, some engineers consider that the chord forces are directly transferred to the foundation by the walls in shear, so no separate chord members are required.

Openings in diaphragms must have the shear force on their two sides distributed in proportion to their stiffeners (size). The bending stiffness is calculated from the main diaphragm chords and the secondary chords required at the edges of the opening. The resulting chord forces are the total of the force due to bending on the whole diaphragm with the opening removed plus the secondary bending on each side of the hole loaded with the moment due to the shear on the side times one-half the length of the opening. This secondary bending is of opposite sign at the two ends so it is additive to the main chords. The secondary chords must extend far enough into the diaphragm to anchor the chord forces into the diaphragm.

At both ends of the opening, drag struts are required to transmit the shear force on the cut section to each side. Drag struts or collectors are the members used to collect shear forces from the diaphragm at openings or other discontinuities. Most frequently, they are anchored to shear walls.

For design of lumber diaphragms, see *Western Woods Use Book* (31). For plywood diaphragms, see *Plywood Diaphragm Construction* (36) or *Design of Wood Structures* (37).

## 6.8   REFERENCES

1.  Canadian Standards Association, *Code for Engineering Design in Wood*, CAN3-086, Rexdale, Ontario, 1980.

2.  TECO/Lumberlock, *Design Manual for TECO Timber Connector Construction*, Colliers, WV, 1973.

3.  American Society of Civil Engineers, *Minimum Design Loads for Buildings and Other Structures*, ASCE 7-88, New York, NY, 1988.

4.  American Institute of Timber Construction, *Standard Specifications for Structural Glued Laminated Timber of Softwood Species*, AITC 117—Design, 1993, Englewood, CO, 1993.

5.  American Institute of Timber Construction, *Standard Specification for Hardwood Glued Laminated Timber*, AITC 119, 1994, Englewood, CO, 1994.

6.  American Institute of Steel Construction, *Manual of Steel Construction*, Chicago, IL, 1989.

7.  American Society of Agricultural Engineers (ASAE), *Post-Frame Building Design*. (Eds.) J. N. Walker and F. E. Woeste. ASAE Monograph No. 11, ASAE, St. Joseph, MI, 1993.

8.  Wood Products Center (WPC), *Wood Design Focus*, Volume 3, Number 1, Wood Products Center, 4033 SW Canyon Road, Portland, OR, 1992.

9.  Alumax Building Products, *Technical Data Information for Powerpanel 29-Gauge Roof Diaphragms*. Alumax Building Products, Reidsville, NC, 1991.

10. Alumax Building Products, *Powerpanel Test Data for Post-Frame End Walls with Diaphragm Loading*. Alumax Building Products, Reidsville, NC, 1992.

11. American Wood-Preservers' Association, *Book of Standards*, Stevensville, MD, 1984.

12. American Society of Agricultural Engineers, ASAE, *Post and Pole Foundation Design*, EP486. ASAE, St. Joseph, MI, 1991.

13. American Wood Preservers Institute, *Pole Building Design*. Vienna, VA, 1977.

14. American Forest and Paper Association, *National Design Specification® for Wood Construction*, NDS®, Washington, DC, 1991.

15. American National Standards Institute, *Specifications and Dimensions for Wood Poles*, ANSI 05.1, New York, NY, 1992.

16. American Society for Testing and Materials, *Standard Specifications and Methods for Establishing Recommended Design Stresses for Round Timber Construction Poles*, ASTM D 3200, Philadelphia, PA, 1974 (Reapproved 1986).

17. American Society for Testing and Materials, *Standard Specification for Pressure Treatment of Timber Products*, ASTM D 1760, Philadelphia, PA, 1986.

18. United States Department of Agriculture, Forest Service, Forest Products Laboratory, *Wood Handbook: Wood as an Engineering Material*, Agriculture Handbook No. 72, Madison, WI, 1987.

19. Southern Forest Products Association, Wood Engineering, Kenner, LA, 1973.

20. International Conference of Building Officials, *Uniform Building Code*, 1991.

21. American Wood-Preserver's Institute, *Pile Foundations Know-How*, Vienna, VA, 1969.

22. American Society for Testing and Materials, *Standard Specifications for Round Timber Piles*, ASTM D25, Philadelphia, PA, 1991.

23. United States Government, Federal Specification. *Wood Preservation, Treating Practice*, TT-W-571i (2), 1972.

24. American Society for Testing and Materials, *Standard Method for Establishing Design Stresses for Round Timber Piles*, ASTM D 2899, Philadelphia, PA, 1986.

25. American Association of State Highway and Transportation Officials, *Standard Specifications for Highway Bridges*, Washington, DC, 1983.

26. American Railway Engineering Association, *Manual of Recommended Practice*, Chicago, IL, 1969.

27. *Load Distribution in Glued Laminated Longitudinal Timber Deck Highway Bridges*, Iowa State University, Ames, IA, 1985.

28. American Association of State Highway and Transportation Officials, *Guide Specifications for Bridge Railings*, Washington, DC, 1989.

29. Transportation Research Board, *Performance Level 1 Bridge Railings for Timber Decks*, Washington, DC, 1993.

30. Federal Highway Administration, *Timber Bridge Rail and Transition Rail Testing and Evaluation-Glulam Bridge Rail and Deck on Glulam Stringers*, Report No. FHWA-RD-94, McLean, WV, 1993.

31. Western Wood Products Association, *Western Woods Use Book*, Portland, OR, 1983.

32. United States Department of Commerce, *Construction and Industrial Plywood PS 1*, 1983.

33. United States Department of Commerce, Performance Standard for Wood-Based Structural-Use Panels PS2, Published by American Plywood Association, Tacoma, WA, 1992.

34. American Institute of Timber Construction, *Standard for Tongue-and Groove Heavy Timber Roof Decking*, AITC 112, Englewood, CO, 1993.

35. American Society of Civil Engineers, *Bibliography of Lumber and Wood Diaphragms, Journal of Structural Engineering*, 109 (12) Dec. 1983.

36. American Plywood Association, *Diaphragm Publications* L350, Tacoma, WA, 1992.

37. D. E. Breyer, *Design of Wood Structures*, McGraw-Hill, New York, NY, 1993.

# CHAPTER 7

# FASTENERS

## 7.1 GENERAL CONSIDERATIONS

The following considerations, in general, apply to all the mechanical fasteners for timber joints covered in this section. Considerations applicable to a particular mechanical fastener will be found under the appropriate headings herein. The factors that require consideration in determining design values for mechanically fastened joints are: (1) lumber species (specific gravity), (2) critical section, (3) angle of load to grain, (4) spacing of mechanical fastenings, (5) edge and end distances, (6) conditions of loading, (7) eccentricity, and (8) adjustments to tabular design values.

In addition to being designed for strength to transfer loads, connections should be designed to avoid splitting the members and to permit swelling and shrinkage of the wood. Glued laminated timbers are often larger than sawn lumber, and the loads transferred are also larger; therefore, the effect of increased size should be considered in the design of glued laminated timber connections. For additional information on fasteners, see *National Design Specification for Wood Construction*® NDS® (1).

Examples of good and poor detailing practices for glued laminated timber construction are included in *Typical Construction Details* (2), AITC 104, in Chapter 8.

### 7.1.1 Lumber Species

The design values for mechanical fasteners vary with the specific gravity of species of wood with which they are used. The specific gravities of the commonly used species of woods are shown in the tables for design values for each dowel type fastener, bolts, lag screws, nails, and wood screws. For species not shown, determine the specific gravity ($G$) of the species based on weight and volume when oven dry, and enter the tables for bearing strength with that specific gravity. Design values for connections in a given species applies to all grades of lumber of that species unless indicated otherwise. Species groupings for split rings and shear plate connectors are shown in Table 7.27.

The recommendations herein are based on the assumption that the mechanical fasteners are used in structural glued laminated timber or lumber that meets the requirements for stress-graded lumber, and that proper fabrication practices have been followed in the installation of the fasteners.

### 7.1.2    Nominal Design Values

The nominal lateral resistance, $Z$, for dowel type fasteners is based on the European Yield Theory and can, in most cases, be determined by solving the equations shown for each fastener type. The nominal design value, $Z$, for split rings and shear plate connectors are given in Tables 7.32–7.34. The nominal design value $Z$ is for one fastener. Design values for more than one type of fastener in the same joint have not been determined, and design values for mixed types of fasteners should be determined by tests. However, when a connection contains two or more fasteners of the same type and similar size which exhibit the same yield mode, the total allowable design value of the connection is the sum of the allowable design values for each individual fastener. The nominal design values, $Z$ or $W$, must be multiplied by all applicable adjustment factors listed in Table 7.1 to determine the allowable design values $Z'$ or $W'$.

### 7.1.3    Adjustment Factors for Fasteners

The design values for fasteners in the tables in this section are shown for one fastener in wood when fabricated with full edge and end distance requirements and spacing and then used under the stated moisture content conditions for a normal duration of load.

When other conditions exist, these design values must be multiplied by the appropriate adjustment factors.

These adjustment factors for design values of fasteners are applied in the same manner as the adjustment factors for strength properties of wood.

The adjustment factors for fasteners consist of:

| | |
|---|---|
| $C_D$, Load duration factor | $C_\Delta$, Geometric factor designated as |
| $C_M$, Wet service factor | $C_{\Delta s}$, Spacing factor |
| $C_t$, Temperature factor | $C_{\Delta n}$, End distance factor |
| $C_d$, Penetration depth factor | $C_{\Delta e}$, Edge distance factor |
| $C_g$, Group action factor | $C_{eg}$, End grain factor |
| $C_{st}$, Metal plate factor | $C_{di}$, Diaphragm factor |
| | $C_{tn}$, Toe-nail factor |

Table 7.1 shows the applicability of these adjustment factors to the various types of fasteners.

The nominal tabular lateral design value $Z$ and the nominal withdrawal design value $W$ are multiplied by all applicable adjustment factors as shown in Table 7.1 to obtain the design values $Z'$ and $W'$.

#### 7.1.3.1    *Duration of Load Factor $C_D$*

The same duration of load factors applicable to the strength properties of wood are also applicable to fasteners where the strength of the fastener is controlled by the wood. An exception is for shear plates where the strength is limited by the bearing of the shear plate on the bolt. Where this occurs in Tables 7.32 and 7.33, the exceptions are indicated. (Note: Load duration factor, $C_D$, applies to load perpendicular to grain design values for fasteners, but not to compression perpendicular to grain design values of wood). See Fig. 4.4 or Table 4.7 for load duration adjustment factors. Note the load duration factor should not exceed 1.6.

## TABLE 7.1

### Applicability of Adjustment Factors for Connections

| | | LOAD DURATION FACTOR [1] | WET SERVICE FACTOR [2] | TEMPERA-TURE FACTOR | GROUP ACTION FACTOR | GEOMETRY FACTOR [3] | PENETRATION DEPTH FACTOR [3] | END GRAIN FACTOR [3] | METAL SIDE PLATE FACTOR [3] | DIAPHRAGM FACTOR [3] | TOE-NAIL FACTOR [3] |
|---|---|---|---|---|---|---|---|---|---|---|---|
| BOLTS | $Z' = (Z)$ | $(C_D)$ | $(C_M)$ | $(C_t)$ | $(C_g)$ | $(C_\Delta)$ | • | • | • | • | • |
| LAG SCREWS | $W' = (W)$ | $(C_D)$ | $(C_M)$ | $(C_t)$ | • | • | • | $(C_{eg})$ | • | • | • |
| | $Z' = (Z)$ | $(C_D)$ | $(C_M)$ | $(C_t)$ | $(C_g)$ | $(C_\Delta)$ | $(C_d)$ | $(C_{eg})$ | • | • | • |
| SPLIT RING and SHEAR PLATE CONNECTORS | $P' = (P)$ | $(C_D)$ | $(C_M)$ | $(C_t)$ | $(C_g)$ | $(C_\Delta)$ | $(C_d)$ | • | $(C_{st})$ | • | • |
| | $Q' = (Q)$ | $(C_D)$ | $(C_M)$ | $(C_t)$ | $(C_g)$ | $(C_\Delta)$ | $(C_d)$ | • | • | • | • |
| WOOD SCREWS | $W' = (W)$ | $(C_D)$ | $(C_M)$ | $(C_t)$ | • | • | • | $(C_{eg})$ | • | • | • |
| | $Z' = (Z)$ | $(C_D)$ | $(C_M)$ | $(C_t)$ | • | • | $(C_d)$ | $(C_{eg})$ | • | • | • |
| NAILS and SPIKES | $W' = (W)$ | $(C_D)$ | $(C_M)$ | $(C_t)$ | • | • | • | • | • | $(C_{di})$ | $(C_{tn})$ |
| | $Z' = (Z)$ | $(C_D)$ | $(C_M)$ | $(C_t)$ | • | • | $(C_d)$ | $(C_{eg})$ | • | | $(C_{tn})$ |
| METAL PLATE CONNECTORS | $Z' = (Z)$ | $(C_D)$ | $(C_M)$ | $(C_t)$ | • | • | • | • | • | • | • |
| DRIFT BOLTS and DRIFT PINS | $W' = (W)$ | $(C_D)$ | $(C_M)$ | $(C_t)$ | • | • | • | $(C_{eg})$ | • | • | • |
| | $Z' = (Z)$ | $(C_D)$ | $(C_M)$ | $(C_t)$ | $(C_g)$ | $(C_\Delta)$ | $(C_d)$ | $(C_{eg})$ | • | • | • |
| SPIKE GRIDS | $Z' = (Z)$ | $(C_D)$ | $(C_M)$ | $(C_t)$ | • | $(C_\Delta)$ | • | • | • | • | • |

Source: ANSI/NF$_o$PA NDS—1991, *National Design Specification for Wood Construction*, AFPA, 1250 Connecticut Ave., NW, Washington, DC 20036.

1. The load duration factor, $C_D$, shall not exceed 1.6 for connections (see 7.1.3.1).
2. The wet service factor, $C_M$, shall not apply to toe-nails loaded in withdrawal (see 7.1.3.13).
3. Specific information concerning geometry factors ($C_\Delta$), penetration depth factors ($C_d$), end grain factors ($C_{eg}$), metal side plate factors ($C_{st}$), diaphragm factors ($C_{di}$) and toe-nail factors ($C_{tn}$) is provided in this chapter.

### 7.1.3.2  Wet Service Factor $C_M$

The nominal design values for connections in wood are based on wood that is seasoned to a moisture content of 19% or less and used under continuously dry conditions. When the wood is unseasoned, partially seasoned, or the connection is exposed to wet service conditions, the nominal design value must be multiplied by the wet service factors in Table 7.2. These factors apply to both sawn lumber and glued laminated timber. Sawn lumber may be manufactured in the dry, par-

**TABLE 7.2**

**Wet Service Factors, $C_M$, for Connections**

| Fastener Type | Condition of Wood[1] | | Wet Service Factor $C_M$ |
| --- | --- | --- | --- |
| | At time of fabrication | In service | |
| Split Ring or Shear Plate[2] Connectors | Dry Partially seasoned Wet Dry or wet | Dry Dry Dry Partially seasoned or wet | 1.0 see Footnote 3 0.8 0.67 |
| Bolts or Lag Screws | Dry Partially seasoned or wet Dry or wet Dry or wet | Dry Dry Exposed to weather Wet | 1.0 see Footnote 4 0.75 0.67 |
| Laterally loaded Drift Bolts or Drift Pins | Dry or wet Dry or wet | Dry Partially seasoned or wet, or subject to wetting and drying | 1.0 0.7 |
| Wood Screws | Dry or wet Dry or wet Dry or wet | Dry Exposed to weather Wet | 1.0 0.75 0.67 |
| Common Wire Nails, Box Nails or Common Wire Spikes — Withdrawal Loads | Dry Partially seasoned or wet Partially seasoned or wet Dry | Dry Wet Dry Subject to wetting and drying | 1.0 1.0 0.25 0.25 |
| — Lateral Loads | Dry Partially seasoned or wet Dry | Dry Dry or wet Partially seasoned or wet | 1.0 0.75 0.75 |
| Threaded, Hardened Steel Nails | Dry or wet | Dry or wet | 1.0 |
| Metal Connector Plates | Dry Partially seasoned or wet | Dry Dry or wet | 1.0 0.8 |

Source: ANSI/NF₀PA NDS—1991, *National Design Specification for Wood Construction*, AFPA, 1250 Connecticut Ave., NW, Washington, DC 20036.

1. "Conditions of wood" are defined as follows for determining wet service factors for connections:
   "Dry" wood has a moisture content ≤19%.
   "Wet" wood has a moisture content ≥30% (approximate fiber saturation point).
   "Partially seasoned" wood has 19% < moisture content <30%.
   "Exposed to weather" means that the wood will vary in moisture content from dry to partially seasoned, but it is not expected to reach the fiber saturation point at times when the connection is supporting full design load.
   "Subject to wetting and drying" means that the wood will vary in moisture content from dry to partially seasoned or wet, or vice versa, with consequent effects on the tightness of the connection.

**Footnotes—Continued**

2. For split ring or shear plate connectors, moisture content limitations apply to a depth of 3/4" below the surface of the wood.
3. When split ring or shear plate connectors are installed in wood that is partially seasoned at the time of fabrication, but that will be dry before full design load is applied, proportional intermediate wet service factors shall be permitted to be used.
4. When bolts or lag screws are installed in wood that is wet at the time of fabrication, but that will be dry before full design load is applied, the following wet service factors, $C_M$, shall apply:

| Arrangement of bolts or lag screws | $C_M$ |
|---|---|
| —one fastener only, or<br>—two or more fasteners placed in a single row parallel to grain, or<br>—fasteners placed in two or more rows parallel to grain with separate splice plates<br>    for each row | 1.0 |
| —all other arrangements | 0.4 |

When bolts or lag screws are installed in wood that is partially seasoned at the time of fabrication, but that will be dry before full design load is applied, proportional intermediate wet service factors shall be permitted to be used.

tially seasoned, or wet condition and may be in service in the dry, wet, partially seasoned, or exposed to weather conditions. Glued laminated timber is manufactured dry and may be used under the dry, wet, or exposed to weather conditions.

### 7.1.3.3 *Temperature Factor $C_t$*

The strength of fasteners in wood is generally controlled by the strength of the wood. In those unusual cases where a reduction is required for strength properties of wood due to temperature, the design value for fasteners should also be adjusted by the temperature factor. (See Table 7.3.)

### 7.1.3.4 *Group Action Factor $C_g$*

The design values of bolts, lag screws, connectors, drift bolts, and drift pins are decreased for group action when used in rows. The adjustment factors are shown in Tables 7.4–7.7.

**TABLE 7.3**

**Temperature Factors, $C_t$, for Connections**

| In Service Moisture Conditions[a] | $C_t$ | | |
|---|---|---|---|
| | $T \leq 100°F$ | $100°F < T \leq 125°F$ | $125°F < T \leq 150°F$ |
| Dry | 1.0 | 0.8 | 0.7 |
| Wet | 1.0 | 0.7 | 0.5 |

[a]Wet and dry service conditions for connections are specified in Table 7.2.

Source: ANSI/NF$_o$PA NDS—1991, *National Design Specification for Wood Construction*, AFPA, 1250 Connecticut Ave., NW, Washington, DC 20036.

## TABLE 7.4

### Group Action Factors, $C_g$, for Bolt or Lag Screw Connections with Wood Side Members[2]
#### For $D = 1$ in., $s = 4$ in., $E = 1,400,000$ psi

| $A_s/A_m$[1] | $A_s$[1] in.$^2$ | Number of fasteners in a row | | | | | | | | | | |
|---|---|---|---|---|---|---|---|---|---|---|---|---|
| | | 2 | 3 | 4 | 5 | 6 | 7 | 8 | 9 | 10 | 11 | 12 |
| 0.5 | 5 | 0.98 | 0.92 | 0.84 | 0.75 | 0.68 | 0.61 | 0.55 | 0.50 | 0.45 | 0.41 | 0.38 |
| | 12 | 0.99 | 0.96 | 0.92 | 0.87 | 0.81 | 0.76 | 0.70 | 0.65 | 0.61 | 0.57 | 0.53 |
| | 20 | 0.99 | 0.98 | 0.95 | 0.91 | 0.87 | 0.83 | 0.78 | 0.74 | 0.70 | 0.66 | 0.62 |
| | 28 | 1.00 | 0.98 | 0.96 | 0.93 | 0.90 | 0.87 | 0.83 | 0.79 | 0.76 | 0.72 | 0.69 |
| | 40 | 1.00 | 0.99 | 0.97 | 0.95 | 0.93 | 0.90 | 0.87 | 0.84 | 0.81 | 0.78 | 0.75 |
| | 64 | 1.00 | 0.99 | 0.98 | 0.97 | 0.95 | 0.93 | 0.91 | 0.89 | 0.87 | 0.84 | 0.82 |
| 1 | 5 | 1.00 | 0.97 | 0.91 | 0.85 | 0.78 | 0.71 | 0.64 | 0.59 | 0.54 | 0.49 | 0.45 |
| | 12 | 1.00 | 0.99 | 0.96 | 0.93 | 0.88 | 0.84 | 0.79 | 0.74 | 0.70 | 0.65 | 0.61 |
| | 20 | 1.00 | 0.99 | 0.98 | 0.95 | 0.92 | 0.89 | 0.86 | 0.82 | 0.78 | 0.75 | 0.71 |
| | 28 | 1.00 | 0.99 | 0.98 | 0.97 | 0.94 | 0.92 | 0.89 | 0.86 | 0.83 | 0.80 | 0.77 |
| | 40 | 1.00 | 1.00 | 0.99 | 0.98 | 0.96 | 0.94 | 0.92 | 0.90 | 0.87 | 0.85 | 0.82 |
| | 64 | 1.00 | 1.00 | 0.99 | 0.98 | 0.97 | 0.96 | 0.95 | 0.93 | 0.91 | 0.90 | 0.88 |

Source: ANSI/NF₀PA NDS—1991, *National Design Specification for Wood Construction*, AFPA, 1250 Connecticut Ave., NW, Washington, DC 20036.

1. When $A_s/A_m > 1.0$, use $A_m/A_s$ and use $A_m$ instead of $A_s$.

2. Tabulated group action factors ($C_g$) are conservative for $D < 1''$, $s < 4''$ or $E > 1,400,000$ psi.

# TABLE 7.5

**Group Action Factors, $C_g$, for 4 in. Split Ring or Shear Plate Connectors with Wood Side Members[2]**

**For $s = 9$ in., $E = 1,400,000$ psi**

| $A_s/A_m$[1] | $A_s$[1] in² | Number of fasteners in a row | | | | | | | | | | |
|---|---|---|---|---|---|---|---|---|---|---|---|---|
| | | 2 | 3 | 4 | 5 | 6 | 7 | 8 | 9 | 10 | 11 | 12 |
| 0.5 | 5 | 0.90 | 0.73 | 0.59 | 0.48 | 0.41 | 0.35 | 0.31 | 0.27 | 0.25 | 0.22 | 0.20 |
| | 12 | 0.95 | 0.83 | 0.71 | 0.60 | 0.52 | 0.45 | 0.40 | 0.36 | 0.32 | 0.29 | 0.27 |
| | 20 | 0.97 | 0.88 | 0.78 | 0.69 | 0.60 | 0.53 | 0.47 | 0.43 | 0.39 | 0.35 | 0.32 |
| | 28 | 0.97 | 0.91 | 0.82 | 0.74 | 0.66 | 0.59 | 0.53 | 0.48 | 0.44 | 0.40 | 0.37 |
| | 40 | 0.98 | 0.93 | 0.86 | 0.79 | 0.72 | 0.65 | 0.59 | 0.54 | 0.49 | 0.45 | 0.42 |
| | 64 | 0.99 | 0.95 | 0.91 | 0.85 | 0.79 | 0.73 | 0.67 | 0.62 | 0.58 | 0.54 | 0.50 |
| 1 | 5 | 1.00 | 0.87 | 0.72 | 0.59 | 0.50 | 0.43 | 0.38 | 0.34 | 0.30 | 0.28 | 0.25 |
| | 12 | 1.00 | 0.93 | 0.83 | 0.72 | 0.63 | 0.55 | 0.48 | 0.43 | 0.39 | 0.36 | 0.33 |
| | 20 | 1.00 | 0.95 | 0.88 | 0.79 | 0.71 | 0.63 | 0.57 | 0.51 | 0.46 | 0.42 | 0.39 |
| | 28 | 1.00 | 0.97 | 0.91 | 0.83 | 0.76 | 0.69 | 0.62 | 0.57 | 0.52 | 0.47 | 0.44 |
| | 40 | 1.00 | 0.98 | 0.93 | 0.87 | 0.81 | 0.75 | 0.69 | 0.63 | 0.58 | 0.54 | 0.50 |
| | 64 | 1.00 | 0.98 | 0.95 | 0.91 | 0.87 | 0.82 | 0.77 | 0.72 | 0.67 | 0.62 | 0.58 |

Source: ANSI/NF₀PA NDS—1991, *National Design Specification for Wood Construction*, AFPA, 1250 Connecticut Ave., NW, Washington, DC 20036.

1. When $A_s/A_m > 1.0$, use $A_m/A_s$ and use $A_m$ instead of $A_s$.

2. Tabulated group action factors ($C_g$) are conservative for $2\frac{1}{2}''$ split ring connectors, $2\frac{5}{8}''$ shear plate connectors, $s < 9''$ or $E > 1,400,000$ psi.

## TABLE 7.6

**Group Action Factors, $C_g$, for 4 in. Shear Plate Connectors with Steel Side Plates[1]**
For $s = 9$ in., $E_{wood} = 1,400,000$ psi, $E_{steel} = 30,000,000$ psi

| $A_m/A_s$ | $A_m$ in$^2$ | Number of fasteners in a row | | | | | | | | | | |
|---|---|---|---|---|---|---|---|---|---|---|---|---|
| | | 2 | 3 | 4 | 5 | 6 | 7 | 8 | 9 | 10 | 11 | 12 |
| | 5 | 0.97 | 0.89 | 0.80 | 0.70 | 0.62 | 0.55 | 0.49 | 0.44 | 0.40 | 0.37 | 0.34 |
| | 8 | 0.98 | 0.93 | 0.85 | 0.77 | 0.70 | 0.63 | 0.57 | 0.52 | 0.47 | 0.43 | 0.40 |
| | 16 | 0.99 | 0.96 | 0.92 | 0.86 | 0.80 | 0.75 | 0.69 | 0.64 | 0.60 | 0.55 | 0.52 |
| 12 | 24 | 0.99 | 0.97 | 0.94 | 0.90 | 0.85 | 0.81 | 0.76 | 0.71 | 0.67 | 0.63 | 0.59 |
| | 40 | 1.00 | 0.98 | 0.96 | 0.94 | 0.90 | 0.87 | 0.83 | 0.79 | 0.76 | 0.72 | 0.69 |
| | 64 | 1.00 | 0.99 | 0.98 | 0.96 | 0.94 | 0.91 | 0.88 | 0.86 | 0.83 | 0.80 | 0.77 |
| | 120 | 1.00 | 0.99 | 0.99 | 0.98 | 0.96 | 0.95 | 0.93 | 0.91 | 0.90 | 0.87 | 0.85 |
| | 200 | 1.00 | 1.00 | 0.99 | 0.99 | 0.98 | 0.97 | 0.96 | 0.95 | 0.93 | 0.92 | 0.90 |
| | 5 | 0.99 | 0.93 | 0.85 | 0.76 | 0.68 | 0.61 | 0.54 | 0.49 | 0.44 | 0.41 | 0.37 |
| | 8 | 0.99 | 0.95 | 0.90 | 0.83 | 0.75 | 0.69 | 0.62 | 0.57 | 0.52 | 0.48 | 0.44 |
| | 16 | 1.00 | 0.98 | 0.94 | 0.90 | 0.85 | 0.79 | 0.74 | 0.69 | 0.65 | 0.60 | 0.56 |
| 18 | 24 | 1.00 | 0.98 | 0.96 | 0.93 | 0.89 | 0.85 | 0.80 | 0.76 | 0.72 | 0.68 | 0.64 |
| | 40 | 1.00 | 0.99 | 0.97 | 0.95 | 0.93 | 0.90 | 0.87 | 0.83 | 0.80 | 0.77 | 0.73 |
| | 64 | 1.00 | 0.99 | 0.98 | 0.97 | 0.95 | 0.93 | 0.91 | 0.89 | 0.86 | 0.83 | 0.81 |
| | 120 | 1.00 | 1.00 | 0.99 | 0.98 | 0.97 | 0.96 | 0.95 | 0.93 | 0.92 | 0.90 | 0.88 |
| | 200 | 1.00 | 1.00 | 0.99 | 0.99 | 0.98 | 0.98 | 0.97 | 0.96 | 0.95 | 0.94 | 0.92 |
| | 40 | 1.00 | 0.99 | 0.97 | 0.95 | 0.93 | 0.89 | 0.86 | 0.83 | 0.79 | 0.76 | 0.72 |
| 24 | 64 | 1.00 | 0.99 | 0.98 | 0.97 | 0.95 | 0.93 | 0.91 | 0.88 | 0.85 | 0.83 | 0.80 |
| | 120 | 1.00 | 1.00 | 0.99 | 0.98 | 0.97 | 0.96 | 0.95 | 0.93 | 0.91 | 0.90 | 0.88 |
| | 200 | 1.00 | 1.00 | 0.99 | 0.99 | 0.98 | 0.98 | 0.97 | 0.96 | 0.95 | 0.93 | 0.92 |
| | 40 | 1.00 | 0.98 | 0.96 | 0.93 | 0.89 | 0.85 | 0.81 | 0.77 | 0.73 | 0.69 | 0.65 |
| 30 | 64 | 1.00 | 0.99 | 0.97 | 0.95 | 0.93 | 0.90 | 0.87 | 0.83 | 0.80 | 0.77 | 0.73 |
| | 120 | 1.00 | 0.99 | 0.99 | 0.97 | 0.96 | 0.94 | 0.92 | 0.90 | 0.88 | 0.85 | 0.83 |
| | 200 | 1.00 | 1.00 | 0.99 | 0.98 | 0.97 | 0.96 | 0.95 | 0.94 | 0.92 | 0.90 | 0.89 |
| | 40 | 0.99 | 0.97 | 0.94 | 0.91 | 0.86 | 0.82 | 0.77 | 0.73 | 0.68 | 0.64 | 0.60 |
| 35 | 64 | 1.00 | 0.98 | 0.96 | 0.94 | 0.91 | 0.87 | 0.84 | 0.80 | 0.76 | 0.73 | 0.69 |
| | 120 | 1.00 | 0.99 | 0.98 | 0.97 | 0.95 | 0.92 | 0.90 | 0.88 | 0.85 | 0.82 | 0.79 |
| | 200 | 1.00 | 0.99 | 0.99 | 0.98 | 0.97 | 0.95 | 0.94 | 0.92 | 0.90 | 0.88 | 0.86 |
| | 40 | 0.99 | 0.97 | 0.93 | 0.88 | 0.83 | 0.78 | 0.73 | 0.68 | 0.63 | 0.59 | 0.55 |
| 42 | 64 | 0.99 | 0.98 | 0.95 | 0.92 | 0.88 | 0.84 | 0.80 | 0.76 | 0.72 | 0.68 | 0.64 |
| | 120 | 1.00 | 0.99 | 0.97 | 0.95 | 0.93 | 0.90 | 0.88 | 0.85 | 0.81 | 0.78 | 0.75 |
| | 200 | 1.00 | 0.99 | 0.98 | 0.97 | 0.96 | 0.94 | 0.92 | 0.90 | 0.88 | 0.85 | 0.83 |
| | 40 | 0.99 | 0.96 | 0.91 | 0.85 | 0.79 | 0.74 | 0.68 | 0.63 | 0.58 | 0.54 | 0.51 |
| 50 | 64 | 0.99 | 0.97 | 0.94 | 0.90 | 0.85 | 0.81 | 0.76 | 0.72 | 0.67 | 0.63 | 0.59 |
| | 120 | 1.00 | 0.98 | 0.97 | 0.94 | 0.91 | 0.88 | 0.85 | 0.81 | 0.78 | 0.74 | 0.71 |
| | 200 | 1.00 | 0.99 | 0.98 | 0.96 | 0.95 | 0.92 | 0.90 | 0.87 | 0.85 | 0.82 | 0.79 |

Source: ANSI/NF$_o$PA NDS—1991, *National Design Specification for Wood Construction*, AFPA, 1250 Connecticut Ave., NW, Washington, DC 20036.

1. Tabulated group action factors ($C_g$) are conservative for D < 1" or s < 4".

Tables 7.4 and 7.5 contain group action factors for members with wood side members, and Tables 7.6 and 7.7 contain the factors for wood members with steel side plates. These tables are based on specific sizes and spacing of fasteners and a modulus of elasticity of 1,400,000 psi for the wood. The values obtained from these tables are conservative for smaller fasteners and spacings and larger values of modulus of elasticity. If a more accurate value of $C_g$ is needed, it may be determined by Eq. (7-1).

$$C_g = \left[ \frac{m(1 - m^{2n})}{n[(1 + R_{EA}m^n)(1 + m) - 1 + m^{2n}]} \right] \left[ \frac{1 + R_{EA}}{1 - m} \right] \qquad (7-1)$$

**TABLE 7.7**

**Group Action Factors, $C_g$, for Bolt or Lag Screw Connections with Steel Side Plates[1]**
**For $D = 1$ in., $s = 4$ in., $E_{wood} = 1,400,000$ psi, $E_{steel} = 30,000,000$ psi**

| $A_m/A_s$ | $A_m$ in.² | Number of fasteners in a row | | | | | | | | | | |
|---|---|---|---|---|---|---|---|---|---|---|---|---|
| | | 2 | 3 | 4 | 5 | 6 | 7 | 8 | 9 | 10 | 11 | 12 |
| 12 | 5 | 0.91 | 0.75 | 0.60 | 0.50 | 0.42 | 0.36 | 0.31 | 0.28 | 0.25 | 0.23 | 0.21 |
| | 8 | 0.94 | 0.80 | 0.67 | 0.56 | 0.47 | 0.41 | 0.36 | 0.32 | 0.29 | 0.26 | 0.24 |
| | 16 | 0.96 | 0.87 | 0.76 | 0.66 | 0.58 | 0.51 | 0.45 | 0.40 | 0.37 | 0.33 | 0.31 |
| | 24 | 0.97 | 0.90 | 0.82 | 0.73 | 0.64 | 0.57 | 0.51 | 0.46 | 0.42 | 0.39 | 0.35 |
| | 40 | 0.98 | 0.94 | 0.87 | 0.80 | 0.73 | 0.66 | 0.60 | 0.55 | 0.50 | 0.46 | 0.43 |
| | 64 | 0.99 | 0.96 | 0.91 | 0.86 | 0.80 | 0.74 | 0.69 | 0.63 | 0.59 | 0.55 | 0.51 |
| | 120 | 0.99 | 0.98 | 0.95 | 0.91 | 0.87 | 0.83 | 0.79 | 0.74 | 0.70 | 0.66 | 0.63 |
| | 200 | 1.00 | 0.99 | 0.97 | 0.95 | 0.92 | 0.89 | 0.85 | 0.82 | 0.79 | 0.75 | 0.72 |
| 18 | 5 | 0.97 | 0.83 | 0.68 | 0.56 | 0.47 | 0.41 | 0.36 | 0.32 | 0.28 | 0.26 | 0.24 |
| | 8 | 0.98 | 0.87 | 0.74 | 0.62 | 0.53 | 0.46 | 0.40 | 0.36 | 0.32 | 0.30 | 0.27 |
| | 16 | 0.99 | 0.92 | 0.82 | 0.73 | 0.64 | 0.56 | 0.50 | 0.45 | 0.41 | 0.37 | 0.34 |
| | 24 | 0.99 | 0.94 | 0.87 | 0.78 | 0.70 | 0.63 | 0.57 | 0.51 | 0.47 | 0.43 | 0.39 |
| | 40 | 0.99 | 0.96 | 0.91 | 0.85 | 0.78 | 0.72 | 0.66 | 0.60 | 0.55 | 0.51 | 0.47 |
| | 64 | 1.00 | 0.97 | 0.94 | 0.89 | 0.84 | 0.79 | 0.74 | 0.69 | 0.64 | 0.60 | 0.56 |
| | 120 | 1.00 | 0.99 | 0.97 | 0.94 | 0.90 | 0.87 | 0.83 | 0.79 | 0.75 | 0.71 | 0.67 |
| | 200 | 1.00 | 0.99 | 0.98 | 0.96 | 0.94 | 0.91 | 0.89 | 0.86 | 0.82 | 0.79 | 0.76 |
| 24 | 40 | 1.00 | 0.96 | 0.91 | 0.84 | 0.77 | 0.71 | 0.65 | 0.59 | 0.54 | 0.50 | 0.46 |
| | 64 | 1.00 | 0.98 | 0.94 | 0.89 | 0.84 | 0.78 | 0.73 | 0.68 | 0.63 | 0.58 | 0.54 |
| | 120 | 1.00 | 0.99 | 0.96 | 0.94 | 0.90 | 0.86 | 0.82 | 0.78 | 0.74 | 0.70 | 0.66 |
| | 200 | 1.00 | 0.99 | 0.98 | 0.96 | 0.94 | 0.91 | 0.88 | 0.85 | 0.82 | 0.78 | 0.75 |
| 30 | 40 | 0.99 | 0.93 | 0.86 | 0.78 | 0.70 | 0.63 | 0.57 | 0.52 | 0.47 | 0.43 | 0.40 |
| | 64 | 0.99 | 0.96 | 0.90 | 0.84 | 0.78 | 0.71 | 0.66 | 0.60 | 0.56 | 0.51 | 0.48 |
| | 120 | 0.99 | 0.98 | 0.94 | 0.90 | 0.86 | 0.81 | 0.76 | 0.71 | 0.67 | 0.63 | 0.59 |
| | 200 | 1.00 | 0.98 | 0.96 | 0.94 | 0.91 | 0.87 | 0.83 | 0.79 | 0.76 | 0.72 | 0.68 |
| 35 | 40 | 0.98 | 0.91 | 0.83 | 0.74 | 0.66 | 0.59 | 0.53 | 0.48 | 0.43 | 0.40 | 0.36 |
| | 64 | 0.99 | 0.94 | 0.88 | 0.81 | 0.73 | 0.67 | 0.61 | 0.56 | 0.51 | 0.47 | 0.43 |
| | 120 | 0.99 | 0.97 | 0.93 | 0.88 | 0.82 | 0.77 | 0.72 | 0.67 | 0.62 | 0.58 | 0.54 |
| | 200 | 1.00 | 0.98 | 0.95 | 0.92 | 0.88 | 0.84 | 0.80 | 0.76 | 0.71 | 0.68 | 0.64 |
| 42 | 40 | 0.97 | 0.88 | 0.79 | 0.69 | 0.61 | 0.54 | 0.48 | 0.43 | 0.39 | 0.36 | 0.33 |
| | 64 | 0.98 | 0.92 | 0.84 | 0.76 | 0.69 | 0.62 | 0.56 | 0.51 | 0.46 | 0.42 | 0.39 |
| | 120 | 0.99 | 0.95 | 0.90 | 0.85 | 0.78 | 0.72 | 0.67 | 0.62 | 0.57 | 0.53 | 0.49 |
| | 200 | 0.99 | 0.97 | 0.94 | 0.90 | 0.85 | 0.80 | 0.76 | 0.71 | 0.67 | 0.62 | 0.59 |
| 50 | 40 | 0.95 | 0.86 | 0.75 | 0.65 | 0.56 | 0.49 | 0.44 | 0.39 | 0.35 | 0.32 | 0.30 |
| | 64 | 0.97 | 0.90 | 0.81 | 0.72 | 0.64 | 0.57 | 0.51 | 0.46 | 0.42 | 0.38 | 0.35 |
| | 120 | 0.98 | 0.94 | 0.88 | 0.81 | 0.74 | 0.68 | 0.62 | 0.57 | 0.52 | 0.48 | 0.45 |
| | 200 | 0.99 | 0.96 | 0.92 | 0.87 | 0.82 | 0.77 | 0.71 | 0.66 | 0.62 | 0.58 | 0.54 |

Source: ANSI/NF$_o$PA NDS—1991, *National Design Specification for Wood Construction*, AFPA, 1250 Connecticut Ave., NW, Washington, DC 20036.

1. Tabulated group action factors ($C_g$) are conservative for $2\frac{5}{8}''$ shear plate connectors or $s < 9''$

in which   $n$ = number of fasteners in a row,

$R_{EA}$ = the lesser of $E_s A_s / E_m A_m$ or $E_m A_m / E_s A_s$,

$E_m$ = modulus of elasticity of main member (psi),

$E_s$ = modulus of elasticity of side members (psi),

$A_m$ = gross cross-sectional area of main member (in.²),

$A_s$ = sum of gross cross-sectional areas of side members (in.²),

$m = u - \sqrt{u^2 - 1}$,

$u = 1 + \gamma \dfrac{s}{2} \left[ \dfrac{1}{E_m A_m} + \dfrac{1}{E_s A_s} \right]$,

> $s$ = center-to-center spacing between adjacent fasteners in a row (in.),
> $\gamma$ = load/slip modulus for a connection (lb/in.),
> $\gamma$ = 500,000 lb/in. for 4 in. split ring or shear plate connectors,
> $\gamma$ = 400,000 lb/in. for $2\text{-}\frac{1}{2}$ in. split ring or $2\text{-}\frac{5}{8}$ in. shear plate connectors,
> $\gamma$ = $(180,000)(D^{1.5})$ for bolts or lag screws in wood-to-wood connections,
> $\gamma$ = $(270,000)(D^{1.5})$ for bolts or lag screws in wood-to-metal connections, and
> $D$ = diameter of bolt or lag screw (in.).

The factor is applicable to the group of fasteners acting as a whole, but for convenience of calculation, it may be applied to the single factor value along with other adjustment factors. The total load that can be carried by the joint is the number of fasteners multiplied by the design value of each fastener.

### 7.1.3.5   Geometry Factor $C_\Delta$

The nominal design values for bolts, split rings, shear plates, lag screws, drift bolts, drift pins, and spike grids require specified end and edge distances and spacings for full design value. When these distances are less than required for full design values but no less than a specified minimum, the design value can be obtained by multiplying by the geometry factor $C_\Delta$. The methods for determining the various geometry factors are contained in the subsection for the fastener used.

For convenience of calculation, the geometry factor, $C_\Delta$, can be subdivided into

$C_{\Delta s}$ = Spacing factor
$C_{\Delta e}$ = Edge distance factor
$C_{\Delta n}$ = End distance factor

These factors are not accumulative and the smallest controls in determining $C_\Delta$.

When fasteners are used in a group and any one of the fasteners needs to be adjusted by any of the geometry factors $C_{\Delta e}$, $C_{\Delta n}$, $C_{\Delta s}$, the whole group must be adjusted by the lowest value obtained.

### 7.1.3.6   Geometry Factor for Edge Distance $C_{\Delta e}$

The tabular design values for bolts, lag screws, connectors, drift bolts, and drift pins are based on the full edge distance requirements. When full edge distance is not used in fabrication, the tabular design loads must be modified by the edge distance factor which is listed for each fastener where required.

### 7.1.3.7   Geometry Factor for End Distance $C_{\Delta n}$

The tabular design values for bolts, lag screws, connectors, drift bolts, and drift pins are based on full end distance requirements. When full end distance is not used in fabrication, the tabular design loads must be adjusted by the end distance factor which is listed for each fastener.

### 7.1.3.8   Geometry Factor for Spacing $C_{\Delta s}$

The tabular design values for bolts, lag screws, connectors, drift bolts, and drift pins are based on the full spacing requirements. When full spacing is not used, the tabular design values must be adjusted by the spacing factor which is listed for each fastener.

### 7.1.3.9 Penetration Depth Factor $C_d$

The tabular design values for lateral loads of lag screws, wood screws, nails, drift bolts, and spikes are based on a specific embedment of the fastening in the piece receiving the point. The design value at a smaller embedment is less, and is usually determined by interpolation between the full design value and lower design value at minimum penetration. It is sometimes advantageous to show this decreased value as a decimal fraction of the full load, in which case it is used as the depth of embedment factor, $C_d$.

### 7.1.3.10 End Grain Factor $C_{eg}$

The nominal design values for fasteners are based on the fasteners being used in the side grain of wood. When lag screws, wood screws, nails and spikes, drift pins, and drift bolts are used in end grain, the end grain factor must be applied as shown in Table 7.1. The end grain factor, $C_{eg}$, for lag screws, wood screws, nails, and spikes is 0.67. For spiral dowels and drift bolts, use 0.60. See 7.4.12 for use of connectors in end grain.

### 7.1.3.11 Metal Side Plate Factor $C_{st}$

The design values of 4 in. shear plates loaded parallel to grain may be increased when metal rather than wood side plates are used (see Table 7.1).

Values in Table 7.33 for shear plates have been increased for steel side plates, and no further increase is needed.

### 7.1.3.12 Diaphragm Factor $C_{di}$

When nails or spikes are used for diaphragms, the nominal lateral design values are multiplied by the diaphragm factor $C_{di}$, which is equal to 1.1.

### 7.1.3.13 Toe-Nail Factor $C_{tn}$

When toe-nailed connections are used, the nominal lateral design value is multiplied by the toe-nail factor, $C_{tn}$, which is equal to 0.83, and the nominal withdrawal design value is multiplied by 0.67.

## 7.1.4 Critical Section

The critical section of a wood member in a joint is that section, taken at a right angle to the longitudinal axis of the member, that gives the maximum stress based on the net area. The net area at this section is equal to the full cross-sectional area of the member less the projected area of that portion of the mechanical fastener within the member, including the projected area of associated holes not within the fastener's projected area. See provisions herein for determining net section when a specific type of mechanical fastener is staggered.

## 7.1.5 Angle of Load to Grain

Angle of load to grain is a factor in the determination of the design value on certain types of mechanical fasteners because wood has a greater bearing value parallel to grain than perpendicular to grain. The angle of load to grain is the

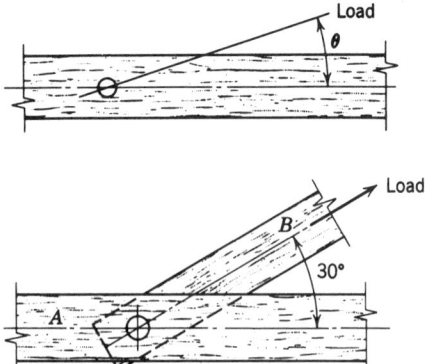

**FIGURE 7.1**    Angle of load to grain.

angle between the resultant of the load exerted by mechanical fastener acting on a member and the longitudinal axis of the member (angle $\theta$ in Fig. 7.1).

The letter $P$ is sometimes used to designate parallel-to-grain tabular design values, $Q$ is used for perpendicular-to-grain tabular design values, and $N$ is used for design values at angles of load to grain between $0°$ and $90°$.

The Hankinson formula is used to determine the design value $N$ of loads at an angle of load to grain between $0°$ and $90°$ as shown in Figure 7.2. $P'$, $Q'$ or $N'$ are used to designate the actual values used in design after multiplying by the applicable adjustment factors.

The angle of load to grain applies only to the particular member under consideration; that is, the angle of load to grain may be different for the various members being connected by the same fastener. For example, in Fig. 7.1, the angle of load to grain with respect to member $A$ is $30°$, whereas the angle of load to grain with respect to member $B$ is zero.

The Hankinson formula is used directly to determine the design value of shear plates and split rings. The adjustment factors for edge distance, end distance, and spacing are based on angle of load to grain, and should not be applied to the design value prior to solution of the Hankinson formula. Therefore, for shear plates and split rings, $N$, the load at an angle of grain, is calculated first and the adjustment factors are applied to $N$. For some other applications, the adjustment factors are applied prior to solution of the formula.

The design values for bolts and lag screws at angles of load to grain between $0°$ and $90°$ cannot be determined directly by using $Z_\parallel$ and $Z_\perp$ in the Hankinson formula. The dowel bearing stress $F_{e\theta}$, however, is determined by the Hankinson formula, and is used to replace the dowel bearing stress for the members loaded at an angle of grain in the equations to determine $Z$ for the various yield modes. A close approximation of the load at an angle to grain other than $0°$ or $90°$ can be obtained, however, by substituting $Z_\parallel$ for $P$ and $Z_\perp$ for $Q$ in the Hankinson formula.

$$N = \frac{PQ}{P \sin^2 \theta + Q \cos^2 \theta} \tag{7-2}$$

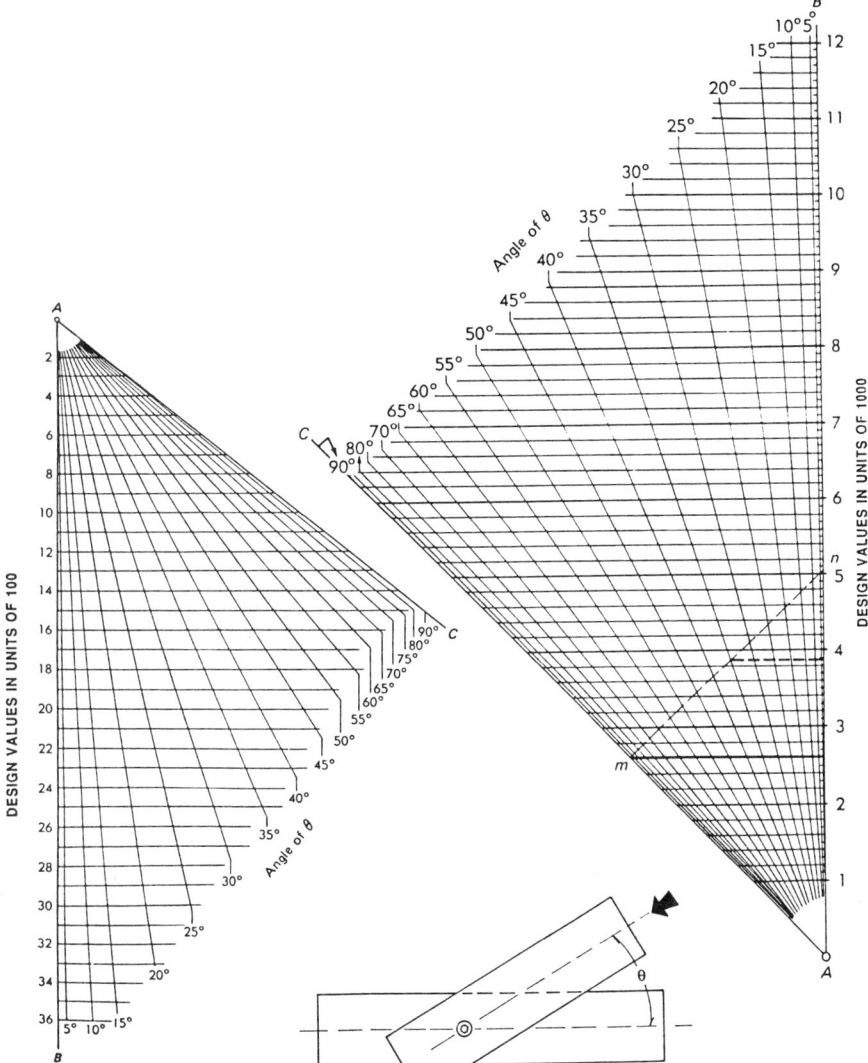

**FIGURE 7.2** Scholten nomograph for solution of Hankinson formula for Split Ring or Shear Plate connection. Example of use: Conditions—$P' = 5030$ lb, $Q' = 2620$ lb, $\theta = 35°$. Locate 5030 lb at $n$ on line $A$-$B$ on right-hand chart. Locate 2620 lb at $m$ on line $A$-$C$, opposite that value on line $A$-$B$. Where line $m$-$n$ intersects the 35° radial line, project to line $A$-$B$ and read allowable load $N' = 3870$ lb. Source: *National Design Specification for Wood Construction* (1).

where $N$ = design value for load acting at an angle $\theta$ with direction of grain (lb),

$P$ = design value for load acting parallel to grain (lb),

$Q$ = design value for load acting perpendicular to grain (lb), and

$\theta$ = angle between the direction of load and the direction of grain (degrees).

The Hankinson formula may be solved graphically through use of the nomographs in Fig. 7.2. The difference between the two nomographs is in their scale. The units on the vertical scales are shown in pounds for design values for fasteners and in pounds per square inch (psi) for design values for strength properties of wood such as bearing at an angle of load to grain between 0° and 90°. When the Hankinson formula is used to determine the design value in compression at an angle of load to grain, the adjustments must be applied prior to use of the formula.

### 7.1.6    Spacing of Mechanical Fasteners

The spacing of mechanical fasteners is the distance between centers of the fasteners measured on a straight line joining their centers. Spacing may also be measured parallel and perpendicular to grain. These measurements are illustrated in Fig. 7.3. Spacing between fasteners in a group should be sufficient to develop the required strength of each fastener.

### 7.1.7    Edge Distance

Edge distance is the distance from the edge of a member to the center of the mechanical fastener closest to that edge measured perpendicular to the edge. The unloaded edge is the edge toward which the load induced by the fastener acts. In Fig. 7.4, $A$ is the loaded edge distance and $B$ is the unloaded edge distance. Edge distance should be sufficient to develop the required strength of fasteners.

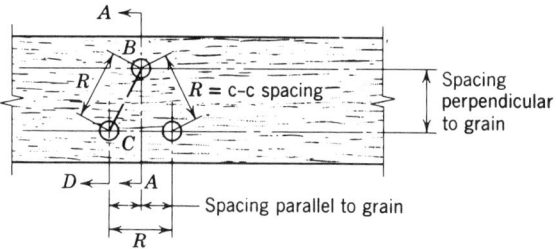

**FIGURE 7.3**    Spacing measurements.

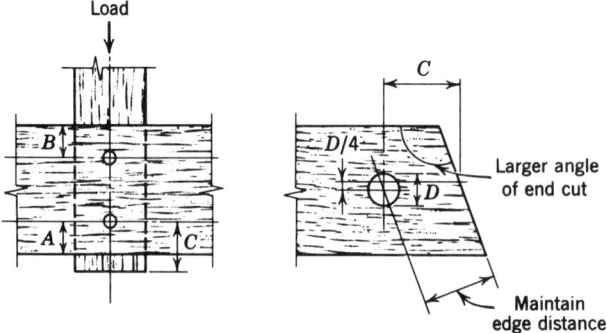

**FIGURE 7.4**    Edge and end distances.

### 7.1.8 End Distance

End distance is the distance, measured parallel to grain, from the center of mechanical fastener to the square-cut end of a member. If the end of the member is cut at an angle, the end distance is measured parallel to the length of the piece on a line that is one-fourth the fastener diameter, $D$, from the center of the connector and on the side of the larger angle of the end cut. The dimensions, $C$, in Fig. 7.4 are end distances. End distances should be sufficient to develop the required strength of fasteners.

### 7.1.9 Effect of Treatment

No reduction in design values is recommended for preservatively treated wood except for fire retardant treatment. (See Fire Retardant Treatment.) Also, check with the manufacturer of the treatment to determine whether or not the fire retardant treatment has a corrosive effect on the fasteners being used.

### 7.1.10 Eccentricity

Eccentric timber connector, bolt, or lag screw joints as shown in Fig. 7.5 should be avoided wherever possible, especially in heavily stressed members. If eccentric joints are used, the effect of the shear, moment, tension perpendicular to grain, and tension stress should be taken into consideration in the design. Concentric joints as shown in Fig. 7.6 should be used.

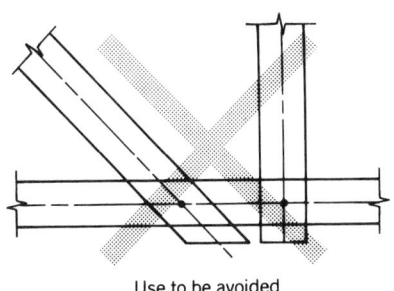

Use to be avoided

**FIGURE 7.5** Truss connection—eccentric joint. When the centerlines of members do not intersect at a common point in a truss, considerable shear, moment, and tension perpendicular to grain may result in the bottom chord. When these are combined with the presumable high-tension stress, the member may be overstressed.

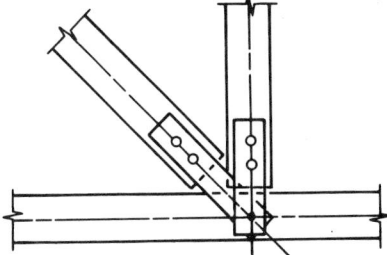

**FIGURE 7.6** Truss connection—concentric joint. For trusses with single-piece upper and lower chords, such as bowstring trusses, the chords and web members should be in the same plane, with straps or gusset plates used for the connection.

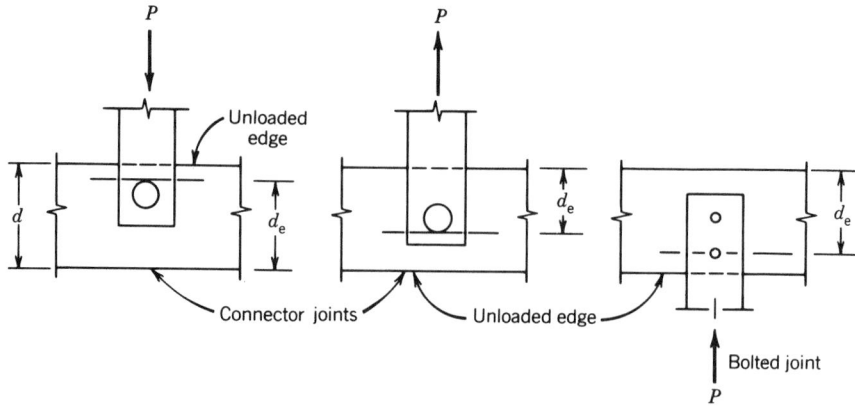

**FIGURE 7.7**    Depth, $d_e$, for members with various fasteners.

### 7.1.11    Shear Stress in Joints at Ends of Members

It is preferable to have the ends of beams supported by resting on another member, column cap, bolster, and so on. However, beams made of sawn lumber and smaller glued laminated beams have been successfully supported by fasteners such as illustrated in Fig. 7.7. When this is done, the same design procedure used for beams notched on the end should be used.

The allowable shear, $V$, is determined by the following equation:

$$V = \frac{2F'_v bd_e^2}{3d} \tag{7-3}$$

where   $V$ = shear at the end of the member (lb) ($V$ = vertical reaction $R_v$ when member is horizontal),

$F'_v$ = design value in shear (psi) adjusted by applicable adjustment factors,

$b$ = width of beam (in.),

$d$ = depth of beam (in.) (see Fig. 7.7), and

$d_e$ = depth from fastener to the loaded edge (in.).

The designer is cautioned against using fasteners to support the ends of large glued laminated beams. See Fig. 7.7 for additional information.

### 7.1.12    Shear Stress in Joints Away from Ends of Members

When the joint is located five times the depth or more from the end of a beam or when a load is applied to a member at least five times the depth of the member from the ends as shown in Fig. 7.8, the design value in shear for joint detail may be increased 50% and the shear is determined as follows:

$$V = (2/3)(1.5F'_v bd_e) = F'_v bd_e \tag{7-4}$$

where terms are as described previously.

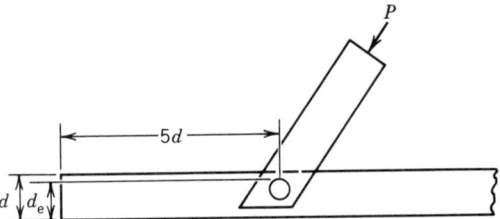

**FIGURE 7.8**  Shear in joint details.

**Example.**  Determine the maximum value of *P* that is limited by shear in joint detail.

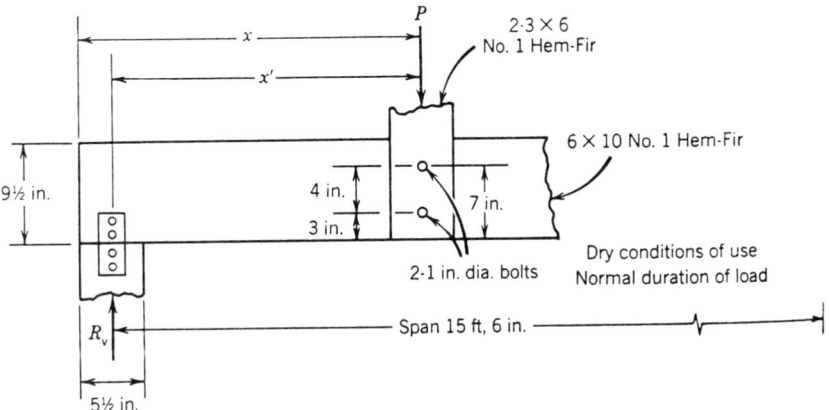

(a)  When $x = 4$ ft, joint is located $(4)(12)/9.5 = 5.05d$ from end; therefore, shear in joint detail may be increased 50%. From Table 8.6, $F_v = 70$ psi, $F'_v = 75$ psi for $x = 4$ ft,

$$x' = (12)(4) - \frac{5.5}{2} = 45.25 \text{ in.}$$

$$R_v = \frac{[(15.5)(12) - (45.25)]\,(P)}{(15.5)(12)} = 0.757P$$

$$V = (2/3)(1.5F'_v bd_e)$$

$$V = R_v = 0.757P$$

$$V = (2/3)(1.5)(70)(5.5)(7) = 2695 \text{ lb}$$

$$P = \frac{2695}{0.757} = 3560 \text{ lb}$$

However, $P$ can be no larger than would be allowed based on use of full cross section with no increase for shear.

$$V = 2/3F'_v bd = (2/3)(70)(5.5)(9.5) = 2438 \text{ lb}$$

$$P = \frac{2438}{0.757} = 3221 \text{ lb} < 3816 \text{ lb}$$

Use $P = 3221$ lb

(b)   When $x = 3$ ft, joint is located $(3)(12)/9.5 = 3.79d$ from end.

Since $3.79d < 5d$, no increase for $F_v$.

$$F'_v = F_v = 70 \text{ psi}$$

$$V = \frac{2F'_v bd^2_e}{3d}$$

$$V = \frac{(2)(70)(5.5)7^2}{(3)(9.5)} = 1324 \text{ lb.}$$

For $x = 3$ ft,

$$R_v = \frac{(15.5)(12) - [(3)(12) - 5.5/2]\,P}{(15.5)(12)} = 0.821P$$

$$R_v = V = 0.821P, \; P = \frac{1324}{0.821} = 1612 \text{ lb} < 3221 \text{ lb}$$

Use $P = 1612$ lb.

## 7.1.13   Group Action of Fasteners

Research has indicated that a load carried by a row of fasteners is not equally divided among the fasteners; that is, end fasteners such as $A$ and $G$ in Fig. 7.9 tend to carry a larger portion of the load than the intermediate fasteners. The distribution of load is determined by the relative stiffness of the main member and the side members.

1.   A group of fasteners consists of one or more rows of fasteners.

2.   A row of fasteners consists of either two or more bolts loaded in single or multiple shear or two or more split rings, shear plates, or lag screws loaded in single shear. The row is aligned with the direction of the load.

When fasteners in adjacent rows are staggered, and the distance, $a$, between

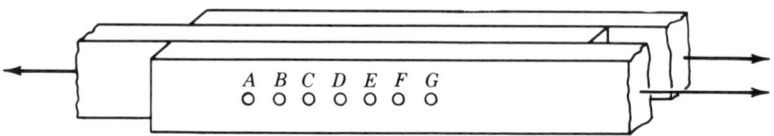

**FIGURE 7.9**   Row of fasteners.

the rows is less than $\frac{1}{4}$ of the spacing, $b$, between the closest fasteners in an adjacent row as shown in Fig. 7.10a, the fasteners in adjacent rows should be considered as one row for the purpose of determining the adjustment factor $C_g$.

3. The load for each row of fasteners is determined by summing the individual loads for each fastener in the row and then multiplying this value by the adjustment factor $C_g$ in Table 7.4, 7.5, 7.6, or 7.7.

For convenience $C_g$ may be applied to individual fastener values prior to summation of values.

4. The design value for the group of fasteners is the sum of the design values of the rows in the group.

5. When a member is loaded perpendicular to grain, its equivalent cross-sectional area is the product of the thickness of the member and the overall width of the fastener group for calculating cross-sectional area ratios. When only one row of fasteners is used, the width is equal to the minimum spacing for full load for the type of fastening used. In general, long rows of fasteners perpendicular to grain should be avoided.

When fasteners in adjacent rows are staggered and the distance between adjacent rows is less than $\frac{1}{4}$ the distance between a fastener in one row and the fastener in an adjacent row, the fasteners in both rows shall be considered to be the same row. (See Fig. 7.10.)

The group action factor, $C_g$, does not apply to nails and wood screws because they can deform and spread the load more evenly than stiffer fasteners.

### 7.1.14 Uplift Loads

Where gravity loads are transferred by bearing perpendicular to grain, adequate fasteners must be provided to carry both horizontal and uplift loads. These loads are normally of a transient nature and are of short duration. The fastener is usually placed toward the lower edge of the member to prevent gravity loads being transferred to the fastener due to shrinkage. The distance $d_e$ shown in Fig. 7.11 must be no less than the distance required for the perpendicular-to-grain edge distance of the type of fastener used. It should also be deep enough so that the calculated shear parallel to grain (horizontal shear), $f_v$, in the following equation (7-5) does not exceed the design value in shear, $F_v'$.

$$f_v = \frac{3Vd}{2bd_e^2} \tag{7-5}$$

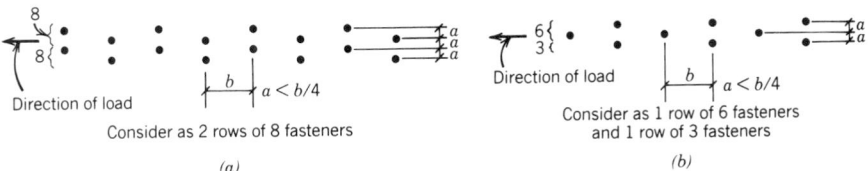

FIGURE 7.10 Group action of fasteners.

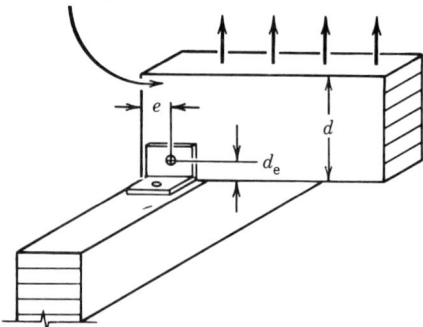

**FIGURE 7.11**   Uplift loading conditions.

where   $f_v$ = calculated shear parallel to grain (psi),
   $V$ = shear (lb), ($V$ = vertical reaction $R_v$ when member is horizontal),
   $b$ = width of member (in.),
   $d_e$ = effective depth of member (in.) (see Fig. 7.11), and
   $d$ = depth of member (in.) (see Fig. 7.11).

The end distance $e$ must also be adequate for the type of fastener used. The uplift loads and horizontal loads are usually small compared to other loads, and frequently only one fastener is required. An adequate number of fasteners, however, should be used to resist the uplift load.

### 7.1.15   Design Values for Dowel Type Fasteners

The design values for lateral strength $Z$ for dowel type fasteners—bolts, lag screws, wood screws, and nails and spikes—are based on the European Yield Model for determining fastener strengths. Fig. 7.12 contains an illustration of the primary yield modes for single shear and double shear connections.

Mode $I_m$ represents the bearing dominated yield of wood bearing on the main member, and Mode $I_s$ is the yield mode for side members.

Mode II yielding occurs only in single shear connections where pivoting takes place, producing limited localized crushing near the faces.

Mode III yielding occurs with bending at one plastic hinge point per shear plane.

Mode IV yielding occurs with 2 plastic hinge points per shear plane.

### 7.1.16   Lateral Strength Dowel Type Fasteners

The nominal design values for lateral strength for each dowel type fastener $Z$ are determined by the appropriate equations for each fastener equation for each applicable yielding mode. The lesser value for $Z$ obtained from these equations controls. Tables of design values of $Z$ have been prepared, and in many cases, the design values can be taken directly from the table.

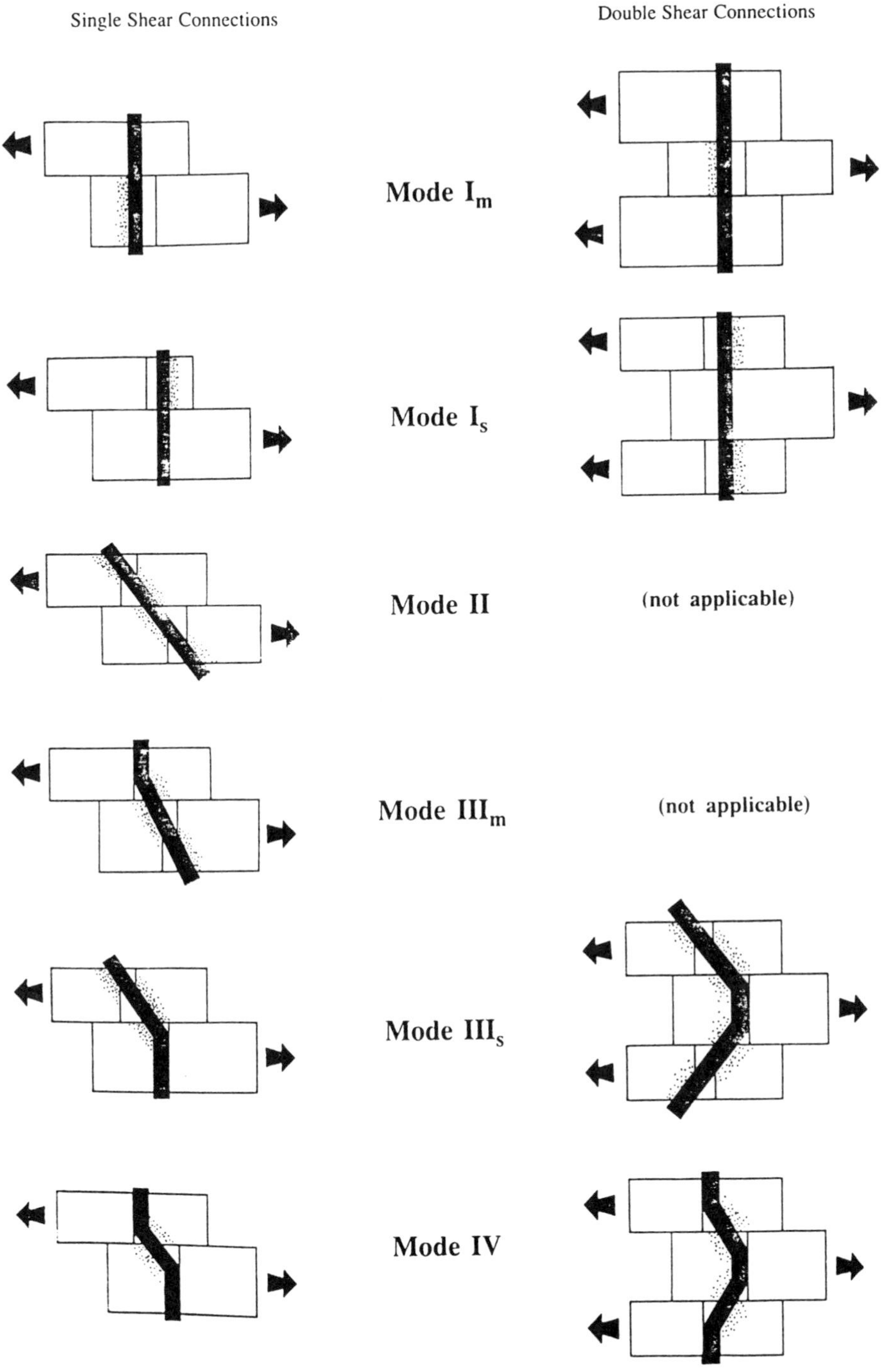

Single Shear Connections

Double Shear Connections

Mode $I_m$

Mode $I_s$

Mode II    (not applicable)

Mode $III_m$    (not applicable)

Mode $III_s$

Mode IV

**FIGURE 7.12**    Connection yield modes.

Source: ANSI/NF₀PA NDS—1991, *National Design Specification for Wood Construction*, AFPA, 1250 Connecticut Ave., NW, Washington, DC 20036.

### 7.1.17   Lateral Strength for Split Ring and Shear Connectors

Split rings and shear plates transfer loads by large circular rings in the faces of the wood pieces being connected, and the yield equations for dowel type connectors do not apply. Design values for loads parallel to grain $P$ and loads perpendicular to grain $Q$ have been determined by testing, and these nominal design values $P$ and $Q$ can be obtained directly from Tables 7.32–7.34.

### 7.1.18   Withdrawal Strength Dowel Type Fasteners

Withdrawal strength for lag screws, wood screws, nails, and spikes have been determined by testing.

Tables for nominal withdrawal per inch of penetration of the threaded portion of lag screws and wood screws and penetration of the point of nails and spikes for commonly used sizes are included.

## 7.2   BOLTS

### 7.2.1   General Considerations

The design values given herein are based on accurate drilling and placement of the bolt holes in both the main member and the side members.

A tight fit requiring forcible driving of bolts is not recommended. Bolt holes should be $\frac{1}{32}$–$\frac{1}{16}$ in. larger than the bolt diameter, depending on the size of the bolt. Careful centering of holes in main members and splice plates is assumed. Washers of proper size or a metal plate or strap should be used between the wood and the bolt head and between the wood and the nut. Nuts should be tightened snugly, but not so tightly as to cause crushing of the wood under the washer or plate. The threaded portion of the nut which will bear on the wood should be kept to a minimum.

Design values are based on the use of bolts $\leq 1$ in. diameter, conforming to ANSI/ASME Standard B18.2.1-1981 (3).

### 7.2.2   Bolt Design Values

The nominal design values of bolts are a function of the specific gravity of the wood, the diameter and length of the bolt and joint configuration. Species with the same specific gravity have the same design values. Table 7.8 contains the specific gravity as well as the dowel bearing strengths, $F_{e\parallel}$ and $F_{e\perp}$, of bolted connections of most species used in construction. If a species is not listed in Table 7.8, the design values for bolts can be determined by first calculating the specific gravity of the species based on oven dried weight and volume. Enter Table 7.8 with this specific gravity to determine dowel bearing stresses $F_{e\parallel}$ and $F_{e\perp}$.

Bolt design values for both single shear and double shear configurations can be determined from the tables or by calculations. When a multiple shear bolted connection with four or more members is used, the design value of the connection is the lowest design value for any shear plane multiplied by the number of shear planes in the connection.

Bolt design values for loads parallel to grain, $Z_\parallel$, and loads perpendicular to grain, $Z_\perp$, can be determined directly using Tables 7.11, 7.12, 7.15, and 7.16 for sawn lumber and Tables 7.13, 7.14, 7.17, and 7.18 for glued laminated timber. These design values are based on using the same species in all members of the connection; however, the species can be mixed if the design value is based on the species with the lowest specific gravity.

Design values may also be determined by using equations shown for the various types of bolt connections. When the load is at an angle to the grain, the tables for bolt design values cannot be used, and the equations for bolt design values at an angle between 0° and 90° must be used.

The design value for bolts at an angle of load to grain between 0° and 90° can be closely approximated, however, by use of the Hankinson's formula; see 7.2.8 for an example of the calculations.

### 7.2.3  Adjustment Factors for Bolts

The applicable adjustment factors shown in Table 7.1 must be applied to the tabular design values that have been adjusted for joint configuration.

Duration of load factor $C_D$ is obtained from Fig. 4.4 or Table 4.7.

Wet service factors, $C_M$, are shown in Table 7.2.

Fire retardant treatment factor should be obtained from the manufacturer of the treatment for the species of lumber being fastened.

Temperature factor, $C_t$, is the same as for wood and is given in Table 7.3.

Group action factor, $C_g$, is explained in 7.1.3.4, Table 7.4, and Table 7.6.

Geometry factor, $C_\Delta$, is subdivided into three parts:

End distance, $C_{\Delta n}$, is explained in 7.2.8 and Table 7.9;

Edge distance, $C_{\Delta e}$, is explained in 7.2.8 and Table 7.9;

Spacing, $C_{\Delta s}$, is explained in 7.2.8, Table 7.9.

### 7.2.4  Single Shear Bolt Connections

The nominal design values for bolts in single shear, $Z_\parallel$ and $Z_\perp$ for the most commonly used species for both wood side plate and steel side plates can be determined directly from Tables 7.11–7.14. (See Fig. 7.13.) When wood side plates are used, the $Z_\perp$ values for perpendicular to grain loading are shown for both the side members and the main member—$Z_{s\perp}$ and $Z_{m\perp}$. When the design is such that the tables cannot be used, the design values must be determined by calculations using the appropriate mode failure equations.

These equations [(7-6)–(7-11)] for determining design values are based on single shear (two members) wood-to-wood connections with end distance, edge distance, and spacing adequate for full design load. The nominal design values, $Z$, for each mode of failure must be determined and the lesser of values controls. For bolts in single shear, the following equations apply:

## TABLE 7.8
### Dowel Bearing Strength for Bolted Connections

| Species Combination[a] | Specific Gravity[b] G | $F_{e\parallel}$ | Dowel bearing strength in pounds per square inch (psi) | | | | |
|---|---|---|---|---|---|---|---|
| | | | $F_{e\perp}$ D=1/2" | $F_{e\perp}$ D=5/8" | $F_{e\perp}$ D=3/4" | $F_{e\perp}$ D=7/8" | $F_{e\perp}$ D=1" |
| Aspen | 0.39 | 4350 | 2200 | 1950 | 1800 | 1650 | 1550 |
| Balsam Fir | 0.36 | 4050 | 1950 | 1750 | 1600 | 1500 | 1400 |
| Beech-Birch-Hickory | 0.71 | 7950 | 5250 | 4700 | 4300 | 3950 | 3700 |
| Coast Sitka Spruce | 0.39 | 4350 | 2200 | 1950 | 1800 | 1650 | 1550 |
| Cottonwood | 0.41 | 4600 | 2350 | 2100 | 1950 | 1800 | 1650 |
| Douglas Fir-Larch | 0.50 | 5600 | 3150 | 2800 | 2600 | 2400 | 2250 |
| Douglas Fir-Larch (North) | 0.49 | 5500 | 3050 | 2750 | 2500 | 2300 | 2150 |
| Douglas Fir-South | 0.46 | 5150 | 2800 | 2500 | 2300 | 2100 | 2000 |
| Eastern Hemlock | 0.41 | 4600 | 2350 | 2100 | 1950 | 1800 | 1650 |
| Eastern Hemlock-Tamarack | 0.41 | 4600 | 2350 | 2100 | 1950 | 1800 | 1650 |
| Eastern Hemlock-Tamarack (North) | 0.47 | 5250 | 2900 | 2600 | 2350 | 2200 | 2050 |
| Eastern Softwoods | 0.36 | 4050 | 1950 | 1750 | 1600 | 1500 | 1400 |
| Eastern Spruce | 0.41 | 4600 | 2350 | 2100 | 1950 | 1800 | 1650 |
| Eastern White Pine | 0.36 | 4050 | 1950 | 1750 | 1600 | 1500 | 1400 |
| Engelmann Spruce-Lodgepole Pine[c] (MSR 1650f and higher grades) | 0.46 | 5150 | 2800 | 2500 | 2300 | 2100 | 2000 |
| Engelmann Spruce-Lodgepole Pine[c] (MSR 1500f and lower grades) | 0.38 | 4250 | 2100 | 1900 | 1750 | 1600 | 1500 |
| Hem-Fir | 0.43 | 4800 | 2550 | 2250 | 2050 | 1900 | 1800 |
| Hem-Fir (North) | 0.46 | 5150 | 2800 | 2500 | 2300 | 2100 | 2000 |
| Mixed Maple | 0.55 | 6150 | 3650 | 3250 | 2950 | 2750 | 2550 |
| Mixed Oak | 0.68 | 7600 | 4950 | 4400 | 4050 | 3750 | 3500 |

| Species | | | | | | | |
|---|---|---|---|---|---|---|---|
| Mixed Southern Pine | 0.51 | 5700 | 3250 | 2900 | 2650 | 2450 | 2300 |
| Mountain Hemlock | 0.47 | 5250 | 2900 | 2600 | 2350 | 2200 | 2050 |
| Northern Pine | 0.42 | 4700 | 2450 | 2200 | 2000 | 1850 | 1750 |
| Northern Red Oak | 0.68 | 7600 | 4950 | 4400 | 4050 | 3750 | 3500 |
| Northern Species | 0.35 | 3900 | 1900 | 1700 | 1550 | 1400 | 1350 |
| Northern White Cedar | 0.31 | 3450 | 1600 | 1400 | 1300 | 1200 | 1100 |
| Ponderosa Pine | 0.43 | 4800 | 2550 | 2250 | 2050 | 1900 | 1800 |
| Red Maple | 0.58 | 6500 | 3900 | 3500 | 3200 | 2950 | 2750 |
| Red Oak | 0.67 | 7500 | 4850 | 4300 | 3950 | 3650 | 3400 |
| Red Pine | 0.44 | 4950 | 2600 | 2350 | 2150 | 2000 | 1850 |
| Redwood, close grain | 0.44 | 4950 | 2600 | 2350 | 2150 | 2000 | 1850 |
| Redwood, open grain | 0.37 | 4150 | 2050 | 1850 | 1650 | 1550 | 1450 |
| Sitka Spruce | 0.43 | 4800 | 2550 | 2250 | 2050 | 1900 | 1800 |
| Southern Pine | 0.55 | 6150 | 3650 | 3250 | 2950 | 2750 | 2550 |
| Spruce-Pine-Fir | 0.42 | 4700 | 2450 | 2200 | 2000 | 1850 | 1750 |
| Spruce-Pine-Fir (South) | 0.36 | 4050 | 1950 | 1750 | 1600 | 1500 | 1400 |
| Western Cedars | 0.36 | 4050 | 1950 | 1750 | 1600 | 1500 | 1400 |
| Western Cedars (North) | 0.35 | 3900 | 1900 | 1700 | 1550 | 1400 | 1350 |
| Western Hemlock | 0.47 | 5250 | 2900 | 2600 | 2350 | 2200 | 2050 |
| Western Hemlock (North) | 0.46 | 5150 | 2800 | 2500 | 2300 | 2100 | 2000 |
| Western White Pine | 0.40 | 4500 | 2300 | 2050 | 1850 | 1750 | 1600 |
| Western Woods | 0.36 | 4050 | 1950 | 1750 | 1600 | 1500 | 1400 |
| White Oak | 0.73 | 8200 | 5450 | 4900 | 4450 | 4150 | 3850 |
| Yellow Poplar | 0.43 | 4800 | 2550 | 2250 | 2050 | 1900 | 1800 |

Source: ANSI/NF,PA NDS—1991, *National Design Specification for Wood Construction*, AFPA, 1250 Connecticut Ave., NW, Washington, DC 20036.

[a]Alaska cedar has a specific gravity of 0.46.

[b]Specific gravity based on weight and volume when oven-dry.

[c]Applies only to Engelmann Spruce-Lodgepole Pine machine stress rated (MSR) structural lumber.

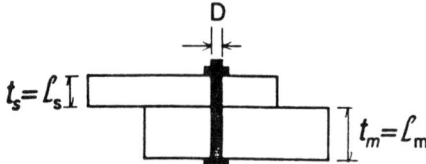

**FIGURE 7.13**    Single shear bolted connection.

equations apply:

Yield Mode

$$Z = \frac{Dt_m F_{em}}{4K_\theta} \qquad \text{Mode I}_m \qquad (7\text{-}6)$$

$$Z = \frac{Dt_s F_{es}}{4K_\theta} \qquad \text{Mode I}_s \qquad (7\text{-}7)$$

$$Z = \frac{k_1 Dt_s F_{es}}{3.6K_\theta} \qquad \text{Mode II} \qquad (7\text{-}8)$$

$$Z = \frac{k_2 Dt_m F_{em}}{3.2(1 + 2R_e)K_\theta} \qquad \text{Mode III}_m \qquad (7\text{-}9)$$

$$Z = \frac{k_3 Dt_s F_{em}}{3.2(2 + R_e)K_\theta} \qquad \text{Mode III}_s \qquad (7\text{-}10)$$

$$Z = \frac{D^2}{3.2K_\theta}\sqrt{\frac{2F_{em}F_{yb}}{3(1 + R_e)}} \qquad \text{Mode IV} \qquad (7\text{-}11)$$

where $\quad k_1 = \dfrac{\sqrt{R_e + 2R_e^2(1 + R_t + R_t^2) + R_t^2 R_e^3} - R_e(1 + R_t)}{1 + R_e}$

$$k_2 = -1 + \sqrt{2(1 + R_e) + \frac{2F_{yb}(1 + 2R_e)D^2}{3F_{em}t_m^2}}$$

$$k_3 = -1 + \sqrt{\frac{2(1 + R_e)}{R_e} + \frac{2F_{yb}(2 + R_e)D^2}{3F_{em}t_s^2}}$$

$R_e = F_{em}/F_{es}$
$R_t = t_m/t_s$
$t_m$ = thickness of main (thicker) member (in.),
$t_s$ = thickness of side (thinner) member (in.),
$F_{em}$ = dowel bearing strength of main (thicker) member (psi),
$F_{es}$ = dowel bearing strength of side (thinner) member (psi),
$F_{yb}$ = bending yield strength of bolt (psi),

$D$ = nominal bolt diameter, (in.),
$K_\theta$ = $1 + (\theta_{max}/360°)$, and
$\theta_{max}$ = maximum angle of load to grain ($0° \leq \theta \leq 90°$) for any member in a connection.

When a member is loaded at an angle to grain, the dowel bearing strength, $F_{e\theta}$ for the member is determined as follows:

$$F_{e\theta} = \frac{F_{e\|} F_{e\perp}}{F_{e\|} \sin^2\theta + F_{e\perp} \cos^2\theta} \tag{7-12}$$

in which   $\theta$ = angle between direction of load and direction of grain (longitudinal axis of member).

The dowel bearing strength for a member loaded at an angle to grain between $0°$ and $90°$, $F_{e\theta}$ is substituted for $F_{es}$ or $F_{em}$ in the preceding equations, depending on which member is loaded at an angle to grain. For example, if the main member is loaded at an angle of load to grain, $F_{e\theta}$ is substituted for $F_{em}$ in the equations. If the side member is loaded at an angle of load to grain, $F_{e\theta}$ is substituted for $F_{es}$.

**Example.**   A horizontal $5\frac{1}{8}$ in. $\times$ 12 in. glued laminated timber member is fastened to an $8\frac{3}{4}$ in. $\times$ 9 in. vertical post as shown. The species is Douglas Fir-Larch. The horizontal member is fastened to the vertical member with two $\frac{3}{4}$ in. bolts. Determine the maximum vertical load that can be carried by the connection assuming a load duration factor of 1.6.

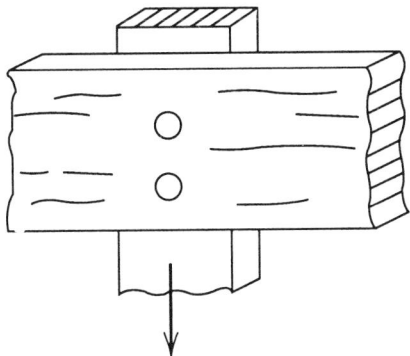

Use Eq. (7-6)–(7-11) to determine the load. The solution of these equations is very time consuming, and the use of a computer program is recommended. For purposes of illustration, the hand calculations are shown:

The input values for the Yield Mode equations are as follows:

$t_m$ = $8\frac{3}{4}$ in.
$t_s$ = $5\frac{1}{8}$ in.
$F_{em}$ = $F_{e\|}$ = 5600 psi (from Table 7.8)
$F_{es}$ = $F_{e\perp}$ = 2600 psi (from Table 7.8)
$F_{yb}$ = 45,000 psi (from Table 7.11)

$D = \frac{3}{4}$ in.

$\theta_{\max} = 90°$

$K_\theta = 1 + \theta_{\max}/360°,\ K_\theta = 1 + 90°/360° = 1.25$

$R_e = F_{em}/F_{es} = 5600/2600 = 2.154$

$R_t = t_m/t_s = 8.75/5.125 = 1.707$

$$k_1 = \frac{\sqrt{R_e + 2R_e^2(1 + R_t + R_t^2) + R_t^2 R_e^3} - R_e(1 + R_t)}{(1 + R_e)}$$

$$k_1 = \frac{\sqrt{2.154 + 2(2.154)^2(1 + 1.707 + 1.707^2) + (1.707)^2(2.154)^3} - (2.154)(1 + 1.707)}{(1 + 2.154)}$$

$k_1 = 1.047$

$$k_2 = -1 + \sqrt{2(1 + R_e) + \frac{2F_{yb}(1 + 2R_e)D^2}{3F_{em}t_m^2}}.$$

$$k_2 = -1 + \sqrt{2(1 + 2.154) + \frac{(2)(45,000)[(1 + (2)(2.154)](0.75)^2}{(3)(5600)(8.75)^2}} = 1.553$$

$$k_3 = -1 + \sqrt{\frac{2(1 + R_e)}{R_e} + \frac{2F_{yb}(2 + R_e)D^2}{3F_{em}t_s^2}}$$

$$k_3 = -1 + \sqrt{\frac{2(1 + 2.154)}{2.154} + \frac{(2)(45,000)(2 + 2.154)(0.75)^2}{(3)(5600)(5.125)^2}} = 0.845$$

Mode I$_m$  $\quad Z = \dfrac{Dt_m F_{em}}{4K_\theta} = \dfrac{(0.75)(8.75)(5600)}{(4)(1.25)} = 7350$ lb

Mode I$_s$  $\quad Z = \dfrac{Dt_s F_{es}}{4K_\theta} = \dfrac{(0.75)(5.125)(2600)}{(4)(1.25)} = 1999$ lb

Mode II  $\quad Z = \dfrac{k_1 Dt_s F_{es}}{3.6K_\theta} = \dfrac{(1.047)(0.75)(5.125)(2600)}{(3.6)(1.25)} = 2326$ lb

Mode III$_m$  $\quad Z = \dfrac{k_2 Dt_m F_{em}}{3.2(1 + 2R_e)K_\theta} = \dfrac{(1.553)(0.75)(8.75)(5600)}{3.2[1 + (2)(2.154)](1.25)} = 2688$ lb

Mode III$_s$  $\quad Z = \dfrac{k_3 Dt_s F_{em}}{3.2(2 + R_e)K_\theta} = \dfrac{(0.845)(0.75)(5.125)(5600)}{3.2(2 + 2.154)(1.25)} = 1095$ lb

Mode IV  $\quad Z = \dfrac{D^2}{3.2K_\theta}\sqrt{\dfrac{2F_{em}F_{yb}}{3(1 + R_e)}}$

$$= \frac{0.75^2}{(3.2)(1.25)}\sqrt{\frac{(2)(5600)(45,000)}{3(1 + 2.154)}} = 1026 \text{ lb}$$

Failure Mode IV controls.

$$P = 2ZC_D = (2)(1026)(1.6) = 3283 \text{ lb}$$

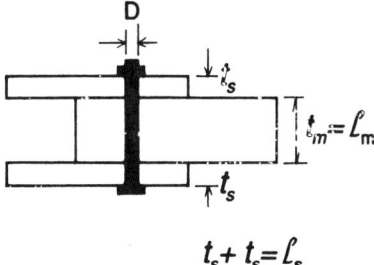

**FIGURE 7.14**   Double shear connection.

### 7.2.4.1   Wood-to-Metal Connections

The design values for wood-to-metal connections for the most commonly used bolt connections are given in Tables 7.12 and 7.14. They can also be determined by using the lesser of the values obtained from Eqs. (7-6), (7-8), (7-9), and (7-10) by setting $F_{es}$ equal to the dowel bearing stress of the metal. The dowel bearing stress of ASTM A36 steel $F_e$ is 58,000 psi.

### 7.2.5   Double Shear Connections

The design values $Z_{\parallel}$ and $Z_{\perp}$ for double shear connections can be determined from Tables 7.15–7.18 for most commonly used species. (See Fig. 7.14.) These equations can be used to determine the design values $Z_{\parallel}$ and $Z_{\perp}$ for any sized members and species provided the specific gravity is known. The dowel bearing strength $F_{e\parallel}$ or $F_{e\perp}$ is determined from Table 7.8 and is used in the equations for mode failure.

The following equations are for three member joints with wood side plates loaded in double shear.

$$\text{Yield Modes}$$

$$Z = \frac{Dt_m F_{em}}{4K_\theta} \qquad \text{Mode I}_m \qquad (7\text{-}13)$$

$$Z = \frac{Dt_s F_{es}}{2K_\theta} \qquad \text{Mode I}_s \qquad (7\text{-}14)$$

$$Z = \frac{k_3 Dt_s F_{em}}{1.6(2 + R_e)K_\theta} \qquad \text{Mode III}_s \qquad (7\text{-}15)$$

$$Z = \frac{D^2}{1.6K_\theta} \sqrt{\frac{2F_{em}F_{yb}}{3(1 + R_e)}} \qquad \text{Mode IV} \qquad (7\text{-}16)$$

$$\text{where} \quad k_3 = -1 + \sqrt{\frac{2(1 + R_e)}{R_e} + \frac{2F_{yb}(2 + R_e)D^2}{3F_{em}t_s^2}},$$

$$R_e = F_{em}/F_{es},$$

$t_m$ = thickness of main (center) member (in.),

$t_s$ = thickness of side member (in.),

$F_{em}$ = dowel bearing strength of main (center) member (psi),

$F_{es}$ = dowel bearing strength of side members (psi),

$F_{yb}$ = bending yield strength of bolt (psi),

$D$ = nominal bolt diameter (in.), and

$K_\theta$ = $1 + (\theta_{max}/360°)$.

$\theta_{max}$ = maximum angle of load to grain ($0° \leq \theta \leq 90°$) for any member in a connection.

The design values obtained are based on a symmetric joint loaded in double shear with wood side members of equal thickness and of identical species. Edge distance, end distance, and spacing should be sufficient to develop full design values. The nominal design values $Z$ are the lesser of the design values obtained using Eq. (7-13)–(7-16).

When members are loaded at an angle of load to grain between $0°$ and $90°$, Eq. (7-12) must be used to determine the dowel bearing strength, $F_{e\theta}$, at the angle of load to grain. This value is substituted for dowel bearing strength of the members. When the main member is loaded at an angle of load to grain, $F_{e\theta}$ is substituted for $F_{em}$. When the side member is loaded at an angle of load to grain, $F_{e\theta}$ is substituted for $F_{es}$. The design value of an angle of load to grain between $0°$ and $90°$ can be closely approximated by use of the Hankinson formula and the values of $Z_{\|}$ and $Z_{\perp}$ taken from the tables. See example in this chapter.

### 7.2.5.1    Wood-to-Metal Connections

Design values $Z_{\|}$ and $Z_{\perp}$ for wood-to-metal bolted connections loaded in double shear for the most commonly used connections are given in Tables 7.14 and 7.16 for sawn lumber and Table 7.18 for glued laminated timber. They can also be calculated by using the lesser of the values determined by Eqs. (7-13), (7-15), and (7-16) by setting the dowel bearing strength of the metal equal to $F_{es}$.

### 7.2.6    Critical or Net Section

The net area at the critical section of a bolted joint is equal to the full cross-sectional area of the timber less the projected area of bolt holes at that section. For parallel-to-grain loading where bolts are staggered, adjacent bolts in a row with parallel-to-grain spacing less than eight times the bolt diameter, the nearest bolt in the next row is considered to occur at the same critical section (see Fig. 7.15).

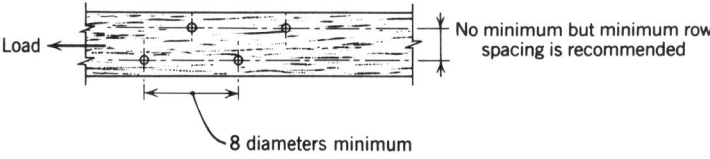

**FIGURE 7.15**    Staggered bolt spacing.

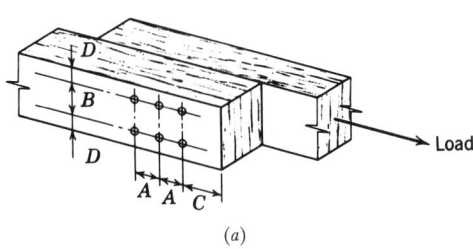
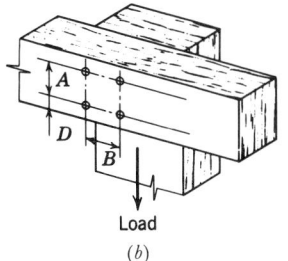

$(a)$                                        $(b)$

**FIGURE 7.16**   Bolt spacing.

The net cross-sectional area, in square inches, required at the critical section is determined by dividing the total load, in pounds, transferred through the critical section of the member by the appropriate unit tensile or compressive design value for the species. In other words, for tension members, divide by $F_t'$, and for compression members, divide by $F_c'$.

### 7.2.7   Spacing of Bolts

A row of bolts is a number of bolts placed in line parallel to the direction of load when the load is parallel or perpendicular to grain. Spacing of bolts in a row is measured in the direction of the load from center to center of bolts [dimension A, Fig. 7.16($a$) and ($b$)]. Row spacing is measured between rows perpendicular to the direction of the load [dimension B, Fig. 7.16($a$) and ($b$)]. Recommended bolt spacing values are given in Table 7.9.

### 7.2.8   End and Edge Distances

Parallel-to-grain loading end distance is the distance from the end of a member to the center of the bolt nearest the end [dimension C, Fig. 7.16($a$)]. Edge distance is the distance from the edge of the member to the center of the nearest bolt [dimension D, Fig. 7.16($a$) and ($b$)]. For perpendicular-to-grain loading, the loaded edge distance is measured from the center of the bolt to the edge of the member in the direction of the load, and the unloaded edge distance is measured to the opposite edge of the member. End and edge distance values are given in Table 7.9. Various types of bolted connections are shown in Table 7.10.

For other than normal duration of loading, the design values from the tables or those calculated by the yield mode equations should be multiplied by the appropriate adjustment factor $C_D$ from Fig. 4.4 or Table 4.7. The full design values may be used for lumber that is installed in seasoned (less than 19% moisture content) wood that will remain dry in service. They apply for bolted joints with wood side members having a single bolt and loaded parallel or perpendicular to grain, or having multiple rows of bolts loaded parallel to grain with separate splice plates for each row. The full design values may also be used for bolted joints with steel

## TABLE 7.9

### Recommended Spacing and Edge and End Distance Values for Bolts[a]

| Dimension[b] | Parallel-to-Grain Loading | Perpendicular-to-Grain Loading |
|---|---|---|
| A, c–c spacing of bolts in a row | Four times bolt diameter for full design value. Minimum 3 times bolt diameter for 75% of full design value. Use straight-line interpolation for design values for intermediate spacing. | Design value limited by spacing requirements of attached member or members (whether of metal or of wood loaded parallel to grain). |
| Staggered bolts | Adjacent bolts are considered to be placed at critical section unless bolts in a row are spaced at a minimum of 8 times the bolt diameter (see Fig. 7.15). | Staggering of bolts is desirable for members loaded perpendicular to grain. |
| B, row spacing | Minimum of $1\frac{1}{2}$ times bolt diameter. | $2\frac{1}{2}$ times bolt diameter for $l/D$[c] ratio $\leq 2$; $(5l + 1DD)/8$ times the bolt diameter when $2 < l/D < 6$; 5 times bolt diameter for $l/D$ ratios $\leq 6$ or more; use straight-line interpolation for $l/D$ ratios between 2 and 6. |
| | Spacing between rows paralleling a member may not exceed 5 in. unless separate splice plates are used for each row. | |
| C, end distance | In tension, 7 times bolt diameter for softwoods and 5 times bolt diameter for hardwoods for full design load. In compression, 4 times bolt diameter for full design load. Minimum end distance is $\frac{1}{2}$ that for full load for which the design value is 50% of that for full end distance. Interpolate for intermediate end distances. | 4 times bolt diameter for full design load, 2 times bolt diameter for 50% of design load. Interpolate for intermediate loads. When members abut at a joint (not illustrated), the strength of the joint shall be evaluated also as a beam supported by fastenings. See 7.1.11 and 7.1.12. |
| D, edge distance | $1\frac{1}{2}$ times the bolt diameter; except that for $l/D$ ratios of more than 6, use $1\frac{1}{2}$ times bolt diameter or one-half the row spacing B, whichever is greater. | Minimum of 4 times bolt diameter at edge toward which load acts; minimum of $1\frac{1}{2}$ times bolt diameter at opposite edge. |

[a]These are minimum distance values for bolt design values from tables or from calculations.
[b]See Fig. 7.16.
[c]Ratio of length of bolt in main member $l$ to diameter of bolt $D$.

**TABLE 7.10**

**Types of Bolted Joints**

| Wood-to-Wood Connections | |
|---|---|
| Type of Joint | Determination of $Z$ |

Three-member joint (Double shear)

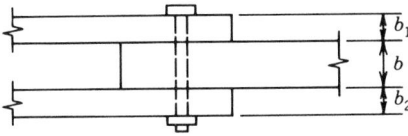

Wood side members

$b_1 = b_2$ — Use tabulated value or calculate.

$b_1 \neq b_2$ (asymmetric) — Calculate each shear plane as a single shear connection using Eqs. (7-6)–(7-11). Multiply lowest value obtained for any shear plane by the number of shear planes.

Different species — Calculate or use tabular value for species of the lowest specific gravity.

Two-member joint (Single shear)

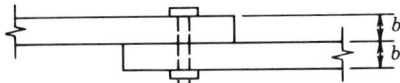

$b_1 = b_2$ — Use tabulated value or calculate.

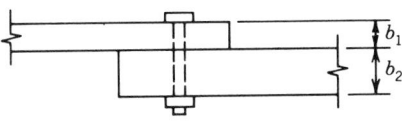

$b_1 \neq b_2$ (asymmeric) — Use tabulated value or calculate.

Different species — Calculate or use tabular value for species with the lower specific gravity.

Multiple-member joints (4 or more members) — Calculate each shear plane as a single shear connection using Eqs. (7-6)–(7-11). Multiply the lowest plane by the number of shear planes.

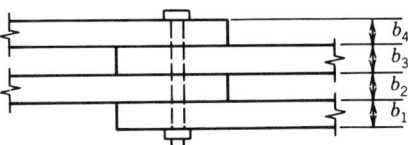

All members of equal or unequal thickness

**TABLE 7.10** *(Continued)*

Wood-to-Metal Connections

| Type of Joint | Determination of Z |
|---|---|

Three-member joint (Double Shear)

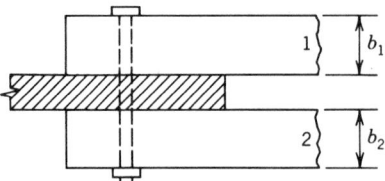

$b_1 = b_2$

$b_1 \neq b_2$ (asymmetric)

Tabular value or calculate using Eqs.
(7-13), (7-15), and (7-16).
Tabular value using thinnest side plate
or calculate using thinnest side plate
with Eqs. (7-13), (7-15), and (7-16).

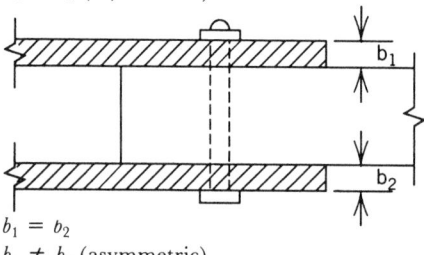

$b_1 = b_2$
$b_1 \neq b_2$ (asymmetric)

Calculate using Eqs. (7-13)–(7-16).
Calculate using procedure for multiple-
member joint.

Two-member joint (single shear)

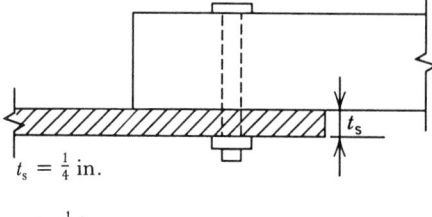

$t_s = \frac{1}{4}$ in.

$t_s > \frac{1}{4}$ in.

Use tabular value or calculate. Use Eqs.
(7-6), (7-8), (7-9), (7-10) and (7-11).
Conservative value obtained using $\frac{1}{4}$ in.
steel side plate or calculate.

Wood-to-Concrete

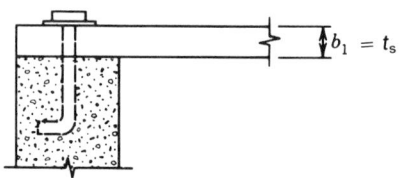

The allowable bolt design value, $Z'$ for
a single shear wood connection
applies with $t_m = 2t_s$ (twice the
thickness of the wood member).
$F_{em} = F_{es}$ dowel bearing strength of
wood member.
Calculate using Eqs. (7-6)–(7-11).

**TABLE 7.10**    (*Continued*)

---

Connection with Members at an Angle

---

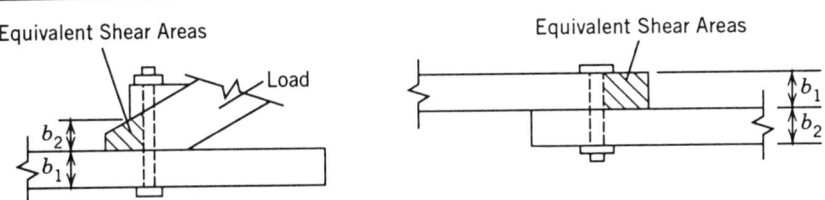

Calculate: The component of load acting at 90° to the bolt axis should not exceed the allowable design value $Z'$ for a connection in which two members at 90° with the bolt axis have thicknesses $t_s = l_s$ and $t_m = l_m$. Sufficient bearing area must be provided under washers or plates to resist the load component acting parallel to the bolt's axis.

---

gusset plates on each side having a single row of bolts parallel to grain in each member and loaded parallel or perpendicular to grain (see Fig. 7.17). For joints in use under other service or seasoning conditions, the design values determined from Tables 7.11–7.18 are multiplied by the appropriate wet service factor from Table 7.2.

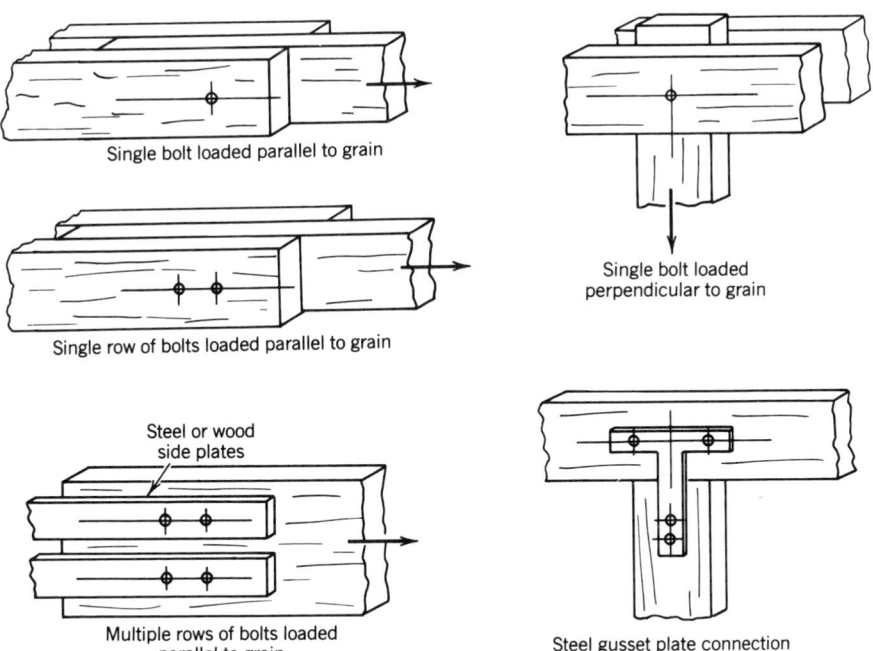

**FIGURE 7.17**    Examples of bolted joints for which full tabular design values may be used (further adjustment may be necessary due to conditions of use).

## TABLE 7.11

### Bolt Design Values (Z) for Single Shear (Two Member) Connections[1,2] for Sawn Lumber with Both Members of Identical Species

| THICKNESS | | | G=0.67 RED OAK | | | G=0.55 MIXED MAPLE SOUTHERN PINE | | | G=0.50 DOUGLAS FIR-LARCH | | | G=0.49 DOUGLAS FIR-LARCH (N) | | | G=0.46 DOUGLAS FIR (S) HEM-FIR (N) | | |
|---|---|---|---|---|---|---|---|---|---|---|---|---|---|---|---|---|---|
| MAIN MEMBER $t_m$ inches | SIDE MEMBER $t_s$ inches | BOLT DIAMETER D inches | $Z_\parallel$ lbs. | $Z_{s\perp}$ lbs. | $Z_{m\perp}$ lbs. | $Z_\parallel$ lbs. | $Z_{s\perp}$ lbs. | $Z_{m\perp}$ lbs. | $Z_\parallel$ lbs. | $Z_{s\perp}$ lbs. | $Z_{m\perp}$ lbs. | $Z_\parallel$ lbs. | $Z_{s\perp}$ lbs. | $Z_{m\perp}$ lbs. | $Z_\parallel$ lbs. | $Z_{s\perp}$ lbs. | $Z_{m\perp}$ lbs. |
| 1-1/2 | 1-1/2 | 1/2 | 650 | 420 | 420 | 530 | 330 | 330 | 480 | 300 | 300 | 470 | 290 | 290 | 440 | 270 | 270 |
| | | 5/8 | 810 | 500 | 500 | 660 | 400 | 400 | 600 | 360 | 360 | 590 | 350 | 350 | 560 | 320 | 320 |
| | | 3/4 | 970 | 580 | 580 | 800 | 460 | 460 | 720 | 420 | 420 | 710 | 400 | 400 | 670 | 380 | 380 |
| | | 7/8 | 1130 | 660 | 660 | 930 | 520 | 520 | 850 | 470 | 470 | 830 | 460 | 460 | 780 | 420 | 420 |
| | | 1 | 1290 | 740 | 740 | 1060 | 580 | 580 | 970 | 530 | 530 | 950 | 510 | 510 | 890 | 480 | 480 |
| 2-1/2 | 1-1/2 | 1/2 | 770 | 480 | 540 | 660 | 400 | 420 | 610 | 370 | 370 | 610 | 360 | 360 | 580 | 340 | 330 |
| | | 5/8 | 1070 | 660 | 630 | 930 | 560 | 490 | 850 | 520 | 430 | 830 | 520 | 420 | 780 | 470 | 390 |
| | | 3/4 | 1360 | 890 | 720 | 1120 | 660 | 560 | 1020 | 590 | 500 | 1000 | 560 | 480 | 940 | 520 | 450 |
| | | 7/8 | 1590 | 960 | 800 | 1300 | 720 | 620 | 1190 | 630 | 550 | 1170 | 600 | 540 | 1090 | 550 | 500 |
| | | 1 | 1820 | 1020 | 870 | 1490 | 770 | 680 | 1360 | 680 | 610 | 1330 | 650 | 590 | 1250 | 600 | 550 |
| 3 | 1-1/2 | 1/2 | 770 | 480 | 470 | 660 | 400 | 470 | 610 | 370 | 420 | 610 | 360 | 410 | 580 | 340 | 380 |
| | | 5/8 | 1070 | 660 | 710 | 940 | 560 | 550 | 880 | 520 | 480 | 870 | 520 | 470 | 830 | 470 | 440 |
| | | 3/4 | 1450 | 890 | 800 | 1270 | 660 | 620 | 1190 | 590 | 550 | 1170 | 560 | 530 | 1090 | 520 | 500 |
| | | 7/8 | 1860 | 960 | 890 | 1520 | 720 | 690 | 1390 | 630 | 610 | 1360 | 600 | 590 | 1280 | 550 | 540 |
| | | 1 | 2120 | 1020 | 970 | 1740 | 770 | 750 | 1590 | 680 | 670 | 1560 | 650 | 650 | 1460 | 600 | 600 |
| 3-1/2 | 1-1/2 | 1/2 | 770 | 480 | 560 | 660 | 400 | 470 | 610 | 370 | 430 | 610 | 360 | 420 | 580 | 340 | 400 |
| | | 5/8 | 1070 | 660 | 760 | 940 | 560 | 620 | 880 | 520 | 540 | 870 | 520 | 530 | 830 | 470 | 490 |
| | | 3/4 | 1450 | 890 | 900 | 1270 | 660 | 690 | 1200 | 590 | 610 | 1190 | 560 | 590 | 1140 | 520 | 550 |
| | | 7/8 | 1890 | 960 | 990 | 1680 | 720 | 770 | 1590 | 630 | 680 | 1570 | 600 | 650 | 1470 | 550 | 600 |
| | | 1 | 2410 | 1020 | 1080 | 2010 | 770 | 830 | 1830 | 680 | 740 | 1790 | 650 | 710 | 1680 | 600 | 660 |
| | 3-1/2 | 1/2 | 830 | 590 | 590 | 750 | 520 | 520 | 720 | 490 | 490 | 710 | 480 | 480 | 690 | 460 | 460 |
| | | 5/8 | 1290 | 880 | 880 | 1170 | 780 | 780 | 1120 | 700 | 700 | 1110 | 690 | 690 | 1070 | 650 | 650 |
| | | 3/4 | 1860 | 1190 | 1190 | 1610 | 960 | 960 | 1610 | 870 | 870 | 1600 | 850 | 850 | 1540 | 800 | 800 |
| | | 7/8 | 2540 | 1410 | 1410 | 2170 | 1160 | 1160 | 1970 | 1060 | 1060 | 1940 | 1040 | 1040 | 1810 | 980 | 980 |
| | | 1 | 3020 | 1670 | 1670 | 2480 | 1360 | 1360 | 2260 | 1230 | 1230 | 2210 | 1190 | 1190 | 2070 | 1110 | 1110 |
| 4-1/2 | 1-1/2 | 5/8 | 1070 | 660 | 760 | 940 | 560 | 640 | 880 | 520 | 590 | 870 | 520 | 590 | 830 | 470 | 560 |
| | | 3/4 | 1450 | 890 | 990 | 1270 | 660 | 840 | 1200 | 590 | 750 | 1190 | 590 | 720 | 1140 | 520 | 670 |
| | | 7/8 | 1890 | 960 | 1210 | 1680 | 720 | 930 | 1590 | 630 | 820 | 1570 | 600 | 790 | 1520 | 550 | 730 |
| | | 1 | 2410 | 1020 | 1310 | 2150 | 770 | 1000 | 2050 | 680 | 890 | 2030 | 650 | 860 | 1930 | 600 | 800 |
| | 3-1/2 | 5/8 | 1290 | 880 | 880 | 1170 | 780 | 780 | 1120 | 700 | 730 | 1110 | 690 | 720 | 1070 | 650 | 690 |
| | | 3/4 | 1860 | 1190 | 1240 | 1690 | 960 | 1090 | 1610 | 870 | 1000 | 1600 | 850 | 980 | 1540 | 800 | 910 |
| | | 7/8 | 2540 | 1410 | 1630 | 2300 | 1160 | 1300 | 2190 | 1060 | 1160 | 2170 | 1040 | 1130 | 2060 | 980 | 1050 |
| | | 1 | 3310 | 1670 | 1830 | 2870 | 1390 | 1440 | 2610 | 1290 | 1290 | 2560 | 1260 | 1250 | 2400 | 1210 | 1170 |
| 5-1/2 | 1-1/2 | 5/8 | 1070 | 660 | 760 | 940 | 560 | 640 | 880 | 520 | 590 | 870 | 520 | 590 | 830 | 470 | 560 |
| | | 3/4 | 1450 | 890 | 990 | 1270 | 660 | 850 | 1200 | 590 | 790 | 1190 | 560 | 780 | 1140 | 520 | 740 |
| | | 7/8 | 1890 | 960 | 1260 | 1680 | 720 | 1090 | 1590 | 630 | 980 | 1570 | 600 | 940 | 1520 | 550 | 860 |
| | | 1 | 2410 | 1020 | 1560 | 2150 | 770 | 1190 | 2050 | 680 | 1060 | 2030 | 650 | 1010 | 1930 | 600 | 940 |
| | 3-1/2 | 5/8 | 1290 | 880 | 880 | 1170 | 780 | 780 | 1120 | 700 | 730 | 1110 | 690 | 720 | 1070 | 650 | 690 |
| | | 3/4 | 1860 | 1190 | 1240 | 1690 | 960 | 1090 | 1610 | 870 | 1030 | 1600 | 850 | 1010 | 1540 | 800 | 970 |
| | | 7/8 | 2540 | 1410 | 1640 | 2300 | 1160 | 1410 | 2190 | 1060 | 1260 | 2170 | 1040 | 1220 | 2060 | 980 | 1130 |
| | | 1 | 3310 | 1670 | 1980 | 2870 | 1390 | 1550 | 2660 | 1290 | 1390 | 2630 | 1260 | 1340 | 2500 | 1210 | 1250 |
| 7-1/2 | 1-1/2 | 5/8 | 1070 | 660 | 760 | 940 | 560 | 640 | 880 | 520 | 590 | 870 | 520 | 590 | 830 | 470 | 560 |
| | | 3/4 | 1450 | 890 | 990 | 1270 | 660 | 850 | 1200 | 590 | 790 | 1190 | 560 | 780 | 1140 | 520 | 740 |
| | | 7/8 | 1890 | 960 | 1260 | 1680 | 720 | 1090 | 1590 | 630 | 1010 | 1570 | 600 | 990 | 1520 | 550 | 950 |
| | | 1 | 2410 | 1020 | 1560 | 2150 | 770 | 1350 | 2050 | 680 | 1270 | 2030 | 650 | 1240 | 1930 | 600 | 1190 |
| | 3-1/2 | 5/8 | 1290 | 880 | 880 | 1170 | 780 | 780 | 1120 | 700 | 730 | 1110 | 690 | 720 | 1070 | 650 | 690 |
| | | 3/4 | 1860 | 1190 | 1240 | 1690 | 960 | 1090 | 1610 | 870 | 1030 | 1600 | 850 | 1010 | 1540 | 800 | 970 |
| | | 7/8 | 2540 | 1410 | 1640 | 2300 | 1160 | 1450 | 2190 | 1060 | 1360 | 2170 | 1040 | 1340 | 2060 | 980 | 1280 |
| | | 1 | 3310 | 1670 | 2090 | 2870 | 1390 | 1830 | 2660 | 1290 | 1630 | 2630 | 1260 | 1570 | 2500 | 1210 | 1470 |

## TABLE 7.11 (Continued)

| $t_m$ in. | $t_s$ in. | D in. | G=0.43 HEM-FIR $Z_\parallel$ | $Z_{s\perp}$ | $Z_{m\perp}$ | G=0.42 SPRUCE-PINE-FIR $Z_\parallel$ | $Z_{s\perp}$ | $Z_{m\perp}$ | G=0.37 REDWOOD (open grain) $Z_\parallel$ | $Z_{s\perp}$ | $Z_{m\perp}$ | G=0.36 EASTERN SOFTWOODS SPRUCE-PINE-FIR(S) WESTERN CEDARS WESTERN WOODS $Z_\parallel$ | $Z_{s\perp}$ | $Z_{m\perp}$ | G=0.35 NORTHERN SPECIES $Z_\parallel$ | $Z_{s\perp}$ | $Z_{m\perp}$ |
|---|---|---|---|---|---|---|---|---|---|---|---|---|---|---|---|---|---|
| 1-1/2 | 1-1/2 | 1/2 | 410 | 250 | 250 | 410 | 240 | 240 | 360 | 210 | 210 | 350 | 200 | 200 | 340 | 200 | 200 |
|  |  | 5/8 | 520 | 300 | 300 | 510 | 290 | 290 | 450 | 250 | 250 | 440 | 240 | 240 | 420 | 240 | 240 |
|  |  | 3/4 | 620 | 350 | 350 | 610 | 340 | 340 | 540 | 290 | 290 | 520 | 280 | 280 | 500 | 270 | 270 |
|  |  | 7/8 | 720 | 390 | 390 | 710 | 380 | 380 | 630 | 330 | 330 | 610 | 320 | 320 | 590 | 310 | 310 |
|  |  | 1 | 830 | 440 | 440 | 810 | 430 | 430 | 720 | 370 | 370 | 700 | 360 | 360 | 670 | 350 | 350 |
| 2-1/2 | 1-1/2 | 1/2 | 550 | 320 | 310 | 540 | 320 | 300 | 500 | 290 | 250 | 490 | 280 | 240 | 470 | 280 | 240 |
|  |  | 5/8 | 730 | 420 | 360 | 710 | 410 | 350 | 630 | 350 | 300 | 610 | 330 | 290 | 590 | 320 | 280 |
|  |  | 3/4 | 870 | 460 | 410 | 850 | 450 | 400 | 750 | 370 | 340 | 740 | 360 | 330 | 710 | 350 | 320 |
|  |  | 7/8 | 1020 | 500 | 450 | 1000 | 490 | 440 | 880 | 410 | 380 | 860 | 390 | 370 | 830 | 370 | 350 |
|  |  | 1 | 1160 | 540 | 500 | 1140 | 530 | 490 | 1010 | 440 | 420 | 980 | 420 | 410 | 940 | 410 | 390 |
| 3 | 1-1/2 | 1/2 | 550 | 320 | 350 | 540 | 320 | 330 | 500 | 290 | 280 | 490 | 280 | 270 | 480 | 280 | 260 |
|  |  | 5/8 | 790 | 420 | 400 | 780 | 410 | 390 | 720 | 350 | 330 | 710 | 330 | 320 | 690 | 320 | 310 |
|  |  | 3/4 | 1020 | 460 | 450 | 1000 | 450 | 440 | 880 | 370 | 370 | 860 | 360 | 360 | 830 | 350 | 350 |
|  |  | 7/8 | 1190 | 500 | 500 | 1160 | 490 | 490 | 1030 | 410 | 420 | 1000 | 390 | 400 | 970 | 370 | 380 |
|  |  | 1 | 1360 | 540 | 550 | 1330 | 530 | 540 | 1170 | 440 | 460 | 1150 | 420 | 440 | 1100 | 410 | 430 |
| 3-1/2 | 1-1/2 | 1/2 | 550 | 320 | 380 | 540 | 320 | 370 | 500 | 290 | 320 | 490 | 280 | 300 | 480 | 280 | 290 |
|  |  | 5/8 | 790 | 420 | 440 | 780 | 410 | 430 | 720 | 350 | 370 | 710 | 330 | 350 | 700 | 320 | 340 |
|  |  | 3/4 | 1100 | 460 | 500 | 1080 | 450 | 480 | 1010 | 370 | 410 | 990 | 360 | 400 | 950 | 350 | 380 |
|  |  | 7/8 | 1370 | 500 | 550 | 1340 | 490 | 540 | 1180 | 410 | 460 | 1160 | 390 | 440 | 1110 | 370 | 420 |
|  |  | 1 | 1570 | 540 | 600 | 1530 | 530 | 590 | 1350 | 440 | 500 | 1320 | 420 | 480 | 1270 | 410 | 470 |
|  | 3-1/2 | 1/2 | 660 | 440 | 440 | 660 | 430 | 430 | 620 | 400 | 400 | 610 | 390 | 390 | 600 | 380 | 380 |
|  |  | 5/8 | 1040 | 600 | 600 | 1020 | 590 | 590 | 960 | 520 | 520 | 950 | 500 | 500 | 930 | 490 | 490 |
|  |  | 3/4 | 1450 | 740 | 740 | 1420 | 730 | 730 | 1250 | 650 | 650 | 1220 | 630 | 630 | 1180 | 620 | 620 |
|  |  | 7/8 | 1690 | 910 | 910 | 1660 | 890 | 890 | 1460 | 770 | 770 | 1430 | 750 | 750 | 1370 | 720 | 720 |
|  |  | 1 | 1930 | 1030 | 1030 | 1890 | 1000 | 1000 | 1670 | 870 | 870 | 1630 | 840 | 840 | 1570 | 810 | 810 |
| 4-1/2 | 1-1/2 | 5/8 | 790 | 420 | 530 | 780 | 410 | 520 | 720 | 350 | 450 | 710 | 350 | 430 | 700 | 320 | 440 |
|  |  | 3/4 | 1100 | 460 | 600 | 1080 | 450 | 590 | 1010 | 370 | 490 | 990 | 360 | 480 | 970 | 350 | 460 |
|  |  | 7/8 | 1460 | 500 | 660 | 1440 | 490 | 640 | 1350 | 410 | 550 | 1520 | 390 | 530 | 1280 | 370 | 500 |
|  |  | 1 | 1800 | 540 | 720 | 1760 | 530 | 710 | 1560 | 440 | 590 | 1520 | 420 | 570 | 1460 | 410 | 550 |
|  | 3-1/2 | 5/8 | 1040 | 600 | 660 | 1020 | 590 | 650 | 960 | 520 | 600 | 950 | 500 | 570 | 930 | 490 | 560 |
|  |  | 3/4 | 1490 | 740 | 840 | 1480 | 730 | 820 | 1390 | 650 | 720 | 1370 | 630 | 700 | 1330 | 620 | 670 |
|  |  | 7/8 | 1950 | 920 | 940 | 1920 | 910 | 940 | 1690 | 820 | 820 | 1650 | 800 | 790 | 1590 | 770 | 750 |
|  |  | 1 | 2240 | 1140 | 1070 | 2190 | 1120 | 1050 | 1940 | 1020 | 900 | 1890 | 980 | 880 | 1820 | 950 | 840 |
| 5-1/2 | 1-1/2 | 5/8 | 790 | 420 | 530 | 780 | 410 | 520 | 720 | 350 | 470 | 710 | 330 | 460 | 700 | 320 | 450 |
|  |  | 3/4 | 1100 | 460 | 700 | 1080 | 450 | 690 | 1010 | 370 | 580 | 990 | 360 | 570 | 970 | 350 | 550 |
|  |  | 7/8 | 1460 | 500 | 780 | 1440 | 490 | 760 | 1350 | 410 | 650 | 1330 | 390 | 630 | 1280 | 370 | 590 |
|  |  | 1 | 1800 | 540 | 860 | 1760 | 530 | 830 | 1560 | 440 | 700 | 1520 | 420 | 680 | 1460 | 410 | 650 |
|  | 3-1/2 | 5/8 | 1040 | 600 | 660 | 1020 | 590 | 650 | 960 | 520 | 610 | 950 | 500 | 590 | 930 | 490 | 580 |
|  |  | 3/4 | 1490 | 740 | 920 | 1480 | 730 | 900 | 1390 | 650 | 770 | 1370 | 630 | 750 | 1330 | 620 | 720 |
|  |  | 7/8 | 1950 | 920 | 1030 | 1920 | 910 | 1010 | 1740 | 820 | 870 | 1710 | 800 | 840 | 1660 | 770 | 800 |
|  |  | 1 | 2370 | 1140 | 1150 | 2330 | 1120 | 1120 | 2120 | 1020 | 960 | 2080 | 980 | 930 | 2030 | 950 | 890 |
| 7-1/2 | 1-1/2 | 5/8 | 790 | 420 | 530 | 780 | 410 | 520 | 720 | 350 | 470 | 710 | 330 | 460 | 700 | 320 | 450 |
|  |  | 3/4 | 1100 | 460 | 700 | 1080 | 450 | 690 | 1010 | 370 | 630 | 990 | 360 | 620 | 970 | 350 | 600 |
|  |  | 7/8 | 1460 | 500 | 900 | 1440 | 490 | 890 | 1350 | 410 | 810 | 1330 | 390 | 800 | 1280 | 370 | 770 |
|  |  | 1 | 1800 | 540 | 1130 | 1760 | 530 | 1110 | 1560 | 440 | 920 | 1520 | 420 | 890 | 1460 | 410 | 860 |
|  | 3-1/2 | 5/8 | 1040 | 600 | 660 | 1020 | 590 | 650 | 960 | 520 | 610 | 950 | 500 | 590 | 930 | 490 | 580 |
|  |  | 3/4 | 1490 | 740 | 920 | 1480 | 730 | 910 | 1390 | 650 | 840 | 1370 | 630 | 820 | 1330 | 620 | 810 |
|  |  | 7/8 | 1950 | 920 | 1210 | 1920 | 910 | 1180 | 1740 | 820 | 1010 | 1710 | 800 | 980 | 1660 | 770 | 920 |
|  |  | 1 | 2370 | 1140 | 1340 | 2330 | 1120 | 1300 | 2120 | 1020 | 1100 | 2080 | 980 | 1070 | 2030 | 950 | 1030 |

1. Tabulated lateral design values (Z) for bolted connections shall be multiplied by all applicable adjustment factors (see Table 7.1).

2. Tabulated lateral design values (Z) are for "full diameter" bolts (see Reference 3) with a bending yield strength ($F_{yb}$) of 45,000 psi.

Source: ANSI/NFₒPA NDS—1991, *National Design Specification for Wood Construction*, AFPA, 1250 Connecticut Ave., NW, Washington, DC 20036.

## TABLE 7.12

### Bolt Design Values ($Z$) for Single Shear (Two Member) Connections[1,2,3] for Sawn Lumber with $\frac{1}{4}$ in. ASTM A36 Steel Side Plate

| THICKNESS MAIN MEMBER $t_m$ inches | STEEL SIDE PLATE $t_s$ inches | BOLT DIAMETER D inches | G=0.67 RED OAK $Z_{\|}$ lbs. | $Z_\perp$ lbs. | G=0.55 MIXED MAPLE SOUTHERN PINE $Z_{\|}$ lbs. | $Z_\perp$ lbs. | G=0.50 DOUGLAS FIR-LARCH $Z_{\|}$ lbs. | $Z_\perp$ lbs. | G=0.49 DOUGLAS FIR-LARCH (N) $Z_{\|}$ lbs. | $Z_\perp$ lbs. | G=0.46 DOUGLAS FIR (S) HEM-FIR (N) $Z_{\|}$ lbs. | $Z_\perp$ lbs. |
|---|---|---|---|---|---|---|---|---|---|---|---|---|
| 1-1/2 | 1/4 | 1/2 | 670 | 380 | 570 | 310 | 530 | 270 | 520 | 270 | 500 | 250 |
|  |  | 5/8 | 840 | 430 | 710 | 350 | 660 | 320 | 650 | 310 | 620 | 290 |
|  |  | 3/4 | 1010 | 490 | 860 | 390 | 800 | 360 | 780 | 350 | 740 | 330 |
|  |  | 7/8 | 1180 | 540 | 1000 | 440 | 930 | 400 | 920 | 390 | 870 | 370 |
|  |  | 1 | 1350 | 580 | 1140 | 480 | 1060 | 440 | 1050 | 430 | 990 | 410 |
| 2-1/2 | 1/4 | 1/2 | 850 | 570 | 780 | 440 | 750 | 390 | 750 | 380 | 730 | 350 |
|  |  | 5/8 | 1270 | 640 | 1100 | 500 | 1010 | 440 | 990 | 430 | 940 | 400 |
|  |  | 3/4 | 1580 | 710 | 1320 | 550 | 1210 | 490 | 1190 | 480 | 1120 | 450 |
|  |  | 7/8 | 1850 | 770 | 1540 | 610 | 1410 | 540 | 1390 | 520 | 1310 | 480 |
|  |  | 1 | 2110 | 830 | 1760 | 650 | 1620 | 590 | 1590 | 560 | 1500 | 530 |
| 3 | 1/4 | 1/2 | 850 | 570 | 780 | 500 | 750 | 450 | 750 | 440 | 730 | 410 |
|  |  | 5/8 | 1270 | 750 | 1170 | 580 | 1130 | 510 | 1120 | 500 | 1090 | 460 |
|  |  | 3/4 | 1800 | 830 | 1560 | 640 | 1430 | 570 | 1410 | 550 | 1330 | 510 |
|  |  | 7/8 | 2200 | 900 | 1830 | 700 | 1670 | 620 | 1650 | 600 | 1550 | 550 |
|  |  | 1 | 2510 | 970 | 2090 | 750 | 1910 | 670 | 1880 | 650 | 1770 | 610 |
| 3-1/2 | 1/4 | 1/2 | 850 | 570 | 780 | 500 | 750 | 470 | 750 | 460 | 730 | 450 |
|  |  | 5/8 | 1270 | 800 | 1170 | 670 | 1130 | 580 | 1120 | 570 | 1090 | 520 |
|  |  | 3/4 | 1800 | 960 | 1650 | 730 | 1580 | 650 | 1570 | 630 | 1530 | 580 |
|  |  | 7/8 | 2420 | 1040 | 2120 | 800 | 1940 | 710 | 1900 | 680 | 1790 | 630 |
|  |  | 1 | 2920 | 1110 | 2420 | 850 | 2210 | 760 | 2180 | 730 | 2050 | 690 |
| 4-1/2 | 1/4 | 5/8 | 1270 | 800 | 1170 | 710 | 1130 | 660 | 1120 | 660 | 1090 | 630 |
|  |  | 3/4 | 1800 | 1090 | 1650 | 920 | 1580 | 820 | 1570 | 790 | 1530 | 730 |
|  |  | 7/8 | 2420 | 1310 | 2220 | 1010 | 2130 | 880 | 2110 | 850 | 2050 | 780 |
|  |  | 1 | 3140 | 1400 | 2880 | 1070 | 2760 | 950 | 2740 | 910 | 2610 | 850 |
| 5-1/2 | 1/4 | 5/8 | 1270 | 800 | 1170 | 710 | 1130 | 660 | 1120 | 660 | 1090 | 630 |
|  |  | 3/4 | 1800 | 1090 | 1650 | 950 | 1580 | 900 | 1570 | 880 | 1530 | 850 |
|  |  | 7/8 | 2420 | 1410 | 2220 | 1220 | 2130 | 1070 | 2110 | 1030 | 2050 | 940 |
|  |  | 1 | 3140 | 1700 | 2880 | 1290 | 2760 | 1150 | 2740 | 1100 | 2660 | 1030 |
| 7-1/2 | 1/4 | 5/8 | 1270 | 800 | 1170 | 710 | 1130 | 660 | 1120 | 660 | 1090 | 630 |
|  |  | 3/4 | 1800 | 1090 | 1650 | 950 | 1580 | 900 | 1570 | 880 | 1530 | 850 |
|  |  | 7/8 | 2420 | 1410 | 2220 | 1240 | 2130 | 1160 | 2110 | 1140 | 2050 | 1090 |
|  |  | 1 | 3140 | 1760 | 2880 | 1540 | 2760 | 1460 | 2740 | 1420 | 2660 | 1380 |
| 9-1/2 | 1/4 | 3/4 | 1800 | 1090 | 1650 | 950 | 1580 | 900 | 1570 | 880 | 1530 | 850 |
|  |  | 7/8 | 2420 | 1410 | 2220 | 1240 | 2130 | 1160 | 2110 | 1140 | 2050 | 1090 |
|  |  | 1 | 3140 | 1760 | 2880 | 1540 | 2760 | 1460 | 2740 | 1420 | 2660 | 1380 |
| 11-1/2 | 1/4 | 7/8 | 2420 | 1410 | 2220 | 1240 | 2130 | 1160 | 2110 | 1140 | 2050 | 1090 |
|  |  | 1 | 3140 | 1760 | 2880 | 1540 | 2760 | 1460 | 2740 | 1420 | 2660 | 1380 |
| 13-1/2 | 1/4 | 1 | 3140 | 1760 | 2880 | 1540 | 2760 | 1460 | 2740 | 1420 | 2660 | 1380 |

## TABLE 7.12  (*Continued*)

| THICKNESS | | BOLT DIAMETER | G=0.43 HEM-FIR | | G=0.42 SPRUCE-PINE-FIR | | G=0.37 REDWOOD (open grain) | | G=0.36 EASTERN SOFTWOODS SPRUCE-PINE-FIR(S) WESTERN CEDARS WESTERN WOODS | | G=0.35 NORTHERN SPECIES | |
|---|---|---|---|---|---|---|---|---|---|---|---|---|
| MAIN MEMBER $t_m$ inches | STEEL SIDE PLATE $t_s$ inches | D inches | $Z_{\parallel}$ lbs. | $Z_{\perp}$ lbs. | $Z_{\parallel}$ lbs. | $Z_{\perp}$ lbs. | $Z_{\parallel}$ lbs. | $Z_{\perp}$ lbs. | $Z_{\parallel}$ lbs. | $Z_{\perp}$ lbs. | $Z_{\parallel}$ lbs. | $Z_{\perp}$ lbs. |
| 1-1/2 | 1/4 | 1/2 | 470 | 240 | 460 | 230 | 420 | 210 | 410 | 200 | 400 | 200 |
|  |  | 5/8 | 590 | 270 | 580 | 270 | 530 | 240 | 520 | 230 | 500 | 230 |
|  |  | 3/4 | 700 | 310 | 690 | 300 | 630 | 270 | 620 | 270 | 600 | 260 |
|  |  | 7/8 | 820 | 340 | 810 | 340 | 740 | 300 | 720 | 300 | 700 | 290 |
|  |  | 1 | 940 | 380 | 920 | 370 | 840 | 340 | 830 | 330 | 800 | 320 |
| 2-1/2 | 1/4 | 1/2 | 700 | 320 | 690 | 310 | 620 | 270 | 600 | 260 | 580 | 260 |
|  |  | 5/8 | 880 | 370 | 860 | 360 | 770 | 310 | 760 | 300 | 730 | 290 |
|  |  | 3/4 | 1050 | 410 | 1030 | 400 | 930 | 340 | 910 | 330 | 880 | 330 |
|  |  | 7/8 | 1230 | 450 | 1210 | 440 | 1080 | 380 | 1060 | 370 | 1020 | 350 |
|  |  | 1 | 1410 | 490 | 1380 | 480 | 1240 | 410 | 1210 | 400 | 1170 | 390 |
| 3 | 1/4 | 1/2 | 710 | 370 | 700 | 360 | 660 | 310 | 660 | 300 | 650 | 290 |
|  |  | 5/8 | 1040 | 420 | 1020 | 410 | 910 | 360 | 890 | 340 | 860 | 330 |
|  |  | 3/4 | 1240 | 470 | 1220 | 460 | 1090 | 390 | 1060 | 380 | 1030 | 370 |
|  |  | 7/8 | 1450 | 510 | 1420 | 500 | 1270 | 430 | 1240 | 420 | 1200 | 400 |
|  |  | 1 | 1660 | 560 | 1630 | 540 | 1450 | 460 | 1420 | 450 | 1370 | 440 |
| 3-1/2 | 1/4 | 1/2 | 710 | 430 | 700 | 410 | 660 | 350 | 660 | 340 | 650 | 330 |
|  |  | 5/8 | 1050 | 480 | 1040 | 470 | 990 | 400 | 980 | 380 | 960 | 370 |
|  |  | 3/4 | 1440 | 530 | 1410 | 520 | 1250 | 440 | 1230 | 420 | 1180 | 410 |
|  |  | 7/8 | 1680 | 570 | 1640 | 560 | 1460 | 480 | 1430 | 470 | 1380 | 440 |
|  |  | 1 | 1910 | 630 | 1880 | 610 | 1670 | 520 | 1630 | 500 | 1580 | 490 |
| 4-1/2 | 1/4 | 5/8 | 1050 | 600 | 1040 | 580 | 990 | 500 | 980 | 470 | 960 | 460 |
|  |  | 3/4 | 1480 | 660 | 1470 | 640 | 1390 | 540 | 1370 | 520 | 1350 | 510 |
|  |  | 7/8 | 1990 | 710 | 1970 | 700 | 1860 | 590 | 1810 | 570 | 1750 | 540 |
|  |  | 1 | 2440 | 770 | 2390 | 750 | 2120 | 640 | 2070 | 620 | 2000 | 600 |
| 5-1/2 | 1/4 | 5/8 | 1050 | 600 | 1040 | 600 | 990 | 550 | 980 | 540 | 960 | 530 |
|  |  | 3/4 | 1480 | 790 | 1470 | 770 | 1390 | 640 | 1370 | 620 | 1350 | 610 |
|  |  | 7/8 | 1990 | 860 | 1970 | 830 | 1860 | 710 | 1840 | 680 | 1810 | 640 |
|  |  | 1 | 2580 | 930 | 2550 | 900 | 2410 | 760 | 2390 | 730 | 2350 | 710 |
| 7-1/2 | 1/4 | 5/8 | 1050 | 600 | 1040 | 600 | 990 | 550 | 980 | 540 | 960 | 530 |
|  |  | 3/4 | 1480 | 810 | 1470 | 800 | 1390 | 730 | 1370 | 720 | 1350 | 710 |
|  |  | 7/8 | 1990 | 1040 | 1970 | 1030 | 1860 | 940 | 1840 | 910 | 1810 | 860 |
|  |  | 1 | 2580 | 1250 | 2550 | 1210 | 2410 | 1010 | 2390 | 980 | 2350 | 940 |
| 9-1/2 | 1/4 | 3/4 | 1480 | 810 | 1470 | 800 | 1390 | 730 | 1370 | 720 | 1350 | 710 |
|  |  | 7/8 | 1990 | 1040 | 1970 | 1030 | 1860 | 950 | 1840 | 930 | 1810 | 900 |
|  |  | 1 | 2580 | 1310 | 2550 | 1290 | 2410 | 1180 | 2390 | 1160 | 2350 | 1140 |
| 11-1/2 | 1/4 | 7/8 | 1990 | 1040 | 1970 | 1030 | 1860 | 950 | 1840 | 930 | 1810 | 900 |
|  |  | 1 | 2580 | 1310 | 2550 | 1290 | 2410 | 1180 | 2390 | 1160 | 2350 | 1140 |
| 13-1/2 | 1/4 | 1 | 2580 | 1310 | 2550 | 1290 | 2410 | 1180 | 2390 | 1160 | 2350 | 1140 |

1. Tabulated lateral design values (Z) for bolted connections shall be multiplied by all applicable adjustment factors (see Table 7.1).

2. Tabulated lateral design values (Z) are for "full diameter" bolts (see Reference 3) with a bending yield strength ($F_{yb}$) of 45,000 psi.

3. Tabulated lateral design values (Z) are based on a dowel bearing strength ($F_e$) of 58,000 psi for ASTM A36 steel.

Source: ANSI/NF$_o$PA NDS—1991, *National Design Specification for Wood Construction*, AFPA, 1250 Connecticut Ave., NW, Washington, DC 20036.

## TABLE 7.13

## Bolt Design Values (Z) for Single Shear (Two Member) Connections[1,2] for Glued Laminated Timber Main Member with Sawn Lumber Side Member of Identical Species

| THICKNESS MAIN MEMBER $t_m$ inches | SIDE MEMBER $t_s$ inches | BOLT DIAMETER D inches | SOUTHERN PINE G=0.55 | | | DOUGLAS FIR-LARCH G=0.50 | | | DOUGLAS FIR (S) G=0.46 | | | HEM-FIR G=0.43 | | | SPRUCE-PINE-FIR G=0.42 | | | SPRUCE-PINE-FIR (S) WESTERN WOODS G=0.36 | | |
|---|---|---|---|---|---|---|---|---|---|---|---|---|---|---|---|---|---|---|---|---|
| | | | $Z_\parallel$ lbs. | $Z_{s\perp}$ lbs. | $Z_{m\perp}$ lbs. | $Z_\parallel$ lbs. | $Z_{s\perp}$ lbs. | $Z_{m\perp}$ lbs. | $Z_\parallel$ lbs. | $Z_{s\perp}$ lbs. | $Z_{m\perp}$ lbs. | $Z_\parallel$ lbs. | $Z_{s\perp}$ lbs. | $Z_{m\perp}$ lbs. | $Z_\parallel$ lbs. | $Z_{s\perp}$ lbs. | $Z_{m\perp}$ lbs. | $Z_\parallel$ lbs. | $Z_{s\perp}$ lbs. | $Z_{m\perp}$ lbs. |
| 2-1/2 | 1-1/2 | 1/2 | - | - | - | 610 | 370 | 370 | 580 | 340 | 330 | 550 | 320 | 310 | 540 | 320 | 300 | 490 | 280 | 240 |
| | | 5/8 | - | - | - | 850 | 520 | 430 | 780 | 470 | 390 | 730 | 420 | 360 | 710 | 410 | 350 | 610 | 330 | 290 |
| | | 3/4 | - | - | - | 1020 | 590 | 500 | 940 | 520 | 450 | 870 | 460 | 410 | 850 | 450 | 400 | 740 | 360 | 330 |
| | | 7/8 | - | - | - | 1190 | 630 | 550 | 1090 | 550 | 500 | 1020 | 500 | 450 | 1000 | 490 | 440 | 860 | 390 | 370 |
| | | 1 | - | - | - | 1360 | 680 | 610 | 1250 | 600 | 550 | 1160 | 540 | 500 | 1140 | 530 | 490 | 980 | 420 | 410 |
| 3 | 1-1/2 | 1/2 | 660 | 400 | 470 | - | - | - | - | - | - | - | - | - | - | - | - | - | - | - |
| | | 5/8 | 940 | 560 | 550 | - | - | - | - | - | - | - | - | - | - | - | - | - | - | - |
| | | 3/4 | 1270 | 660 | 620 | - | - | - | - | - | - | - | - | - | - | - | - | - | - | - |
| | | 7/8 | 1520 | 720 | 690 | - | - | - | - | - | - | - | - | - | - | - | - | - | - | - |
| | | 1 | 1740 | 770 | 750 | - | - | - | - | - | - | - | - | - | - | - | - | - | - | - |
| 3-1/8 | 1-1/2 | 1/2 | - | - | - | 610 | 370 | 430 | 580 | 340 | 390 | 550 | 320 | 360 | 540 | 320 | 340 | 490 | 280 | 280 |
| | | 5/8 | - | - | - | 880 | 520 | 500 | 830 | 470 | 450 | 790 | 420 | 410 | 780 | 410 | 400 | 710 | 330 | 330 |
| | | 3/4 | - | - | - | 1200 | 590 | 570 | 1130 | 520 | 510 | 1060 | 460 | 460 | 1040 | 450 | 450 | 890 | 360 | 370 |
| | | 7/8 | - | - | - | 1440 | 630 | 630 | 1320 | 550 | 560 | 1230 | 500 | 510 | 1210 | 490 | 500 | 1040 | 390 | 410 |
| | | 1 | - | - | - | 1640 | 680 | 690 | 1510 | 600 | 620 | 1410 | 540 | 560 | 1380 | 530 | 550 | 1190 | 420 | 450 |
| 5 | 1-1/2 | 5/8 | 940 | 560 | 640 | - | - | - | - | - | - | - | - | - | - | - | - | - | - | - |
| | | 3/4 | 1270 | 660 | 850 | - | - | - | - | - | - | - | - | - | - | - | - | - | - | - |
| | | 7/8 | 1680 | 720 | 1020 | - | - | - | - | - | - | - | - | - | - | - | - | - | - | - |
| | | 1 | 2150 | 770 | 1100 | - | - | - | - | - | - | - | - | - | - | - | - | - | - | - |
| 5-1/8 | 1-1/2 | 5/8 | - | - | - | 880 | 520 | 590 | 830 | 470 | 560 | 790 | 420 | 530 | 780 | 410 | 520 | 710 | 330 | 460 |
| | | 3/4 | - | - | - | 1200 | 590 | 790 | 1140 | 520 | 740 | 1100 | 460 | 670 | 1080 | 450 | 660 | 990 | 360 | 530 |
| | | 7/8 | - | - | - | 1590 | 630 | 920 | 1520 | 550 | 810 | 1460 | 500 | 740 | 1440 | 490 | 720 | 1330 | 390 | 590 |
| | | 1 | - | - | - | 2050 | 680 | 990 | 1930 | 600 | 890 | 1800 | 540 | 810 | 1760 | 530 | 780 | 1520 | 420 | 640 |
| 6-3/4 | 1-1/2 | 5/8 | 940 | 560 | 640 | 880 | 520 | 590 | 830 | 470 | 560 | 790 | 420 | 530 | 780 | 410 | 520 | 710 | 330 | 460 |
| | | 3/4 | 1270 | 660 | 850 | 1200 | 590 | 790 | 1140 | 520 | 740 | 1100 | 460 | 700 | 1080 | 450 | 690 | 990 | 360 | 620 |
| | | 7/8 | 1680 | 720 | 1090 | 1590 | 630 | 1010 | 1520 | 550 | 950 | 1460 | 500 | 900 | 1440 | 490 | 890 | 1330 | 390 | 750 |
| | | 1 | 2150 | 770 | 1350 | 2050 | 680 | 1270 | 1930 | 600 | 1140 | 1800 | 540 | 1030 | 1760 | 530 | 1000 | 1520 | 420 | 810 |

1. Tabulated lateral design values (Z) for bolted connections shall be multiplied by all applicable adjustment factors (see Table 7.1).

2. Tabulated lateral design values (Z) are for "full diameter" bolts (see Reference 3) with a bending yield strength ($F_{yb}$) of 45,000 psi.

Source: ANSI/NF₀PA NDS—1991, *National Design Specification for Wood Construction*, AFPA, 1250 Connecticut Ave., NW, Washington, DC 20036.

**Example.** Design a tension connection using two $\frac{3}{8}$ in. thick steel side plates to splice the ends of two $6\frac{3}{4}$ in. $\times$ $8\frac{1}{4}$ in. glued laminated tension members made of combination 48 Southern Pine. Load is due to snow, therefore $C_D = 1.15$.

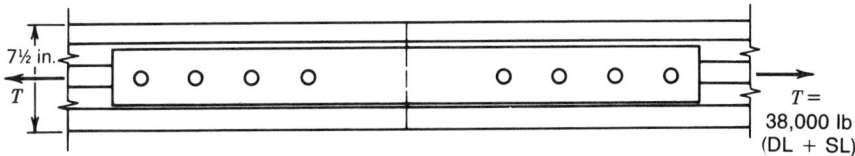

From Table 2, AITC 117—Design (4), combination 48 is all No. 2 Dense Southern Pine.

$$F_t = 1400 \text{ psi}$$

From Table 7.18, $Z_\| = 5750$ lb for a 1-in. diameter bolt with $\frac{1}{4}$ in. steel side plates.

Determine trial $Z^*$.

$$Z^* = Z_\| C_D = (5750)(1.15) = 6610 \text{ lb}$$

$$N = T/Z^* = 38,000/6610 = 5.75$$

Trial number of bolts required $= 5.75$. Try 6 bolts.

The nominal design value 5750 lb obtained from Table 7.18 for $\frac{1}{4}$ in. steel side straps is conservative for other thicknesses and may be used safely with all larger thicknesses of steel side plates. In this example, the design appears close and the nominal bolt design values obtained from Eqs. (7-13), (7-15), and (7-16) will be used.

Main member $6\frac{3}{4} \times 8\frac{1}{4}$ in. Southern Pine
Side plates $\frac{3}{8}$ in. thick steel plate
Bolts 1 in. diameter

The input values for the failure mode equations are:

$F_{em} = 6150$ psi (Table 7.8)
$F_{es} = F_e = 58,000$ psi (Table 7.12)
$F_{yb} = 45,000$ psi (Table 7.4)
$R_e = F_{em}/F_{es} = 6150/58,000 = 0.106$
$t_m = 6.75$ in.
$t_s = 0.375$
$D = 1.0$ in.
$\theta = 0°$
$K_\theta = 1 + \dfrac{0°}{360°} = 1$

$$k_3 = -1 + \sqrt{\frac{2(1 + 0.106)}{0.106} + \frac{2(45,000)(2 + 0.106)(1)^2}{(3)(6150)(0.375)^2}}$$

$$k_3 = 8.691$$

# TABLE 7.14

## Bolt Design Values (Z) for Single Shear (Two Member) Connections[1,2,3] for Glued Laminated Timber with $\frac{1}{4}$ in. ASTM A36 Steel Side Plate

### Steel Side Plate

| MAIN MEMBER $t_m$ inches | STEEL SIDE PLATE $t_s$ inches | BOLT DIAMETER D inches | G=0.55 SOUTHERN PINE $Z_{\parallel}$ lbs. | $Z_{\perp}$ lbs. | G=0.50 DOUGLAS FIR-LARCH $Z_{\parallel}$ lbs. | $Z_{\perp}$ lbs. | G=0.46 DOUGLAS FIR (S) $Z_{\parallel}$ lbs. | $Z_{\perp}$ lbs. | G=0.43 HEM-FIR $Z_{\parallel}$ lbs. | $Z_{\perp}$ lbs. | G=0.42 SPRUCE-PINE-FIR $Z_{\parallel}$ lbs. | $Z_{\perp}$ lbs. | G=0.41 SPRUCE-PINE-FIR (S) WESTERN WOODS $Z_{\parallel}$ lbs. | $Z_{\perp}$ lbs. |
|---|---|---|---|---|---|---|---|---|---|---|---|---|---|---|
| 2-1/2 | 1/4 | 1/2 | - | - | 750 | 390 | 730 | 350 | 700 | 320 | 690 | 310 | 600 | 260 |
|  |  | 5/8 | - | - | 1010 | 440 | 940 | 400 | 880 | 370 | 860 | 360 | 760 | 300 |
|  |  | 3/4 | - | - | 1210 | 490 | 1120 | 450 | 1050 | 410 | 1030 | 400 | 910 | 330 |
|  |  | 7/8 | - | - | 1410 | 540 | 1310 | 480 | 1230 | 450 | 1210 | 440 | 1060 | 370 |
|  |  | 1 | - | - | 1620 | 590 | 1500 | 530 | 1410 | 490 | 1380 | 480 | 1210 | 400 |
| 3 | 1/4 | 1/2 | 780 | 500 | - | - | - | - | - | - | - | - | - | - |
|  |  | 5/8 | 1170 | 580 | - | - | - | - | - | - | - | - | - | - |
|  |  | 3/4 | 1560 | 640 | - | - | - | - | - | - | - | - | - | - |
|  |  | 7/8 | 1830 | 700 | - | - | - | - | - | - | - | - | - | - |
|  |  | 1 | 2090 | 750 | - | - | - | - | - | - | - | - | - | - |
| 3-1/8 | 1/4 | 1/2 | - | - | 750 | 470 | 730 | 420 | 710 | 390 | 700 | 370 | 660 | 310 |
|  |  | 5/8 | - | - | 1130 | 530 | 1090 | 480 | 1050 | 430 | 1010 | 430 | 920 | 350 |
|  |  | 3/4 | - | - | 1490 | 590 | 1380 | 530 | 1290 | 480 | 1270 | 470 | 1100 | 390 |
|  |  | 7/8 | - | - | 1740 | 640 | 1610 | 570 | 1510 | 520 | 1480 | 510 | 1290 | 430 |
|  |  | 1 | - | - | 1990 | 690 | 1840 | 630 | 1720 | 570 | 1690 | 560 | 1470 | 460 |

Tabulated lateral design values, Z (lb), for bolts in single shear — wood member with 1/4 in. ASTM A36 steel side plate.

| Member thickness | 1/4 | D (in.) | | | | | | | | | | | | |
|---|---|---|---|---|---|---|---|---|---|---|---|---|---|---|
| 5 | 1/4 | 5/8 | 1170 | 710 | – | – | – | – | – | – | – | – | – | – |
| | | 3/4 | 1650 | 950 | – | – | – | – | – | – | – | – | – | – |
| | | 7/8 | 2220 | 1110 | – | – | – | – | – | – | – | – | – | – |
| | | 1 | 2880 | 1180 | – | – | – | – | – | – | – | – | – | – |
| 5-1/8 | 1/4 | 5/8 | – | – | 1130 | 660 | 1090 | 630 | 1050 | 600 | 1040 | 600 | 980 | 530 |
| | | 3/4 | – | – | 1580 | 900 | 1530 | 820 | 1480 | 740 | 1470 | 720 | 1370 | 590 |
| | | 7/8 | – | – | 2130 | 1000 | 2050 | 880 | 1990 | 800 | 1970 | 780 | 1840 | 640 |
| | | 1 | – | – | 2760 | 1070 | 2660 | 960 | 2580 | 870 | 2550 | 850 | 2350 | 690 |
| 6-3/4 | 1/4 | 5/8 | 1170 | 710 | 1130 | 660 | 1090 | 630 | 1050 | 600 | 1040 | 600 | 980 | 540 |
| | | 3/4 | 1650 | 950 | 1580 | 900 | 1530 | 850 | 1480 | 810 | 1470 | 800 | 1370 | 720 |
| | | 7/8 | 2220 | 1240 | 2130 | 1160 | 2050 | 1090 | 1990 | 1040 | 1970 | 1010 | 1840 | 830 |
| | | 1 | 2880 | 1540 | 2760 | 1400 | 2660 | 1250 | 2580 | 1130 | 2550 | 1100 | 2390 | 890 |
| 8-1/2 | 1/4 | 3/4 | 1650 | 950 | – | – | – | – | – | – | – | – | – | – |
| | | 7/8 | 2220 | 1240 | – | – | – | – | – | – | – | – | – | – |
| | | 1 | 2880 | 1540 | – | – | – | – | – | – | – | – | – | – |
| 8-3/4 | 1/4 | 3/4 | – | – | 1580 | 900 | 1530 | 850 | 1480 | 810 | 1470 | 800 | 1370 | 720 |
| | | 7/8 | – | – | 2130 | 1160 | 2050 | 1090 | 1990 | 1040 | 1970 | 1030 | 1840 | 930 |
| | | 1 | – | – | 2760 | 1460 | 2660 | 1380 | 2580 | 1310 | 2550 | 1290 | 2390 | 1130 |
| 10-1/2 | 1/4 | 7/8 | 2220 | 1240 | – | – | – | – | – | – | – | – | – | – |
| | | 1 | 2880 | 1540 | – | – | – | – | – | – | – | – | – | – |
| 10-3/4 | 1/4 | 7/8 | – | – | 2130 | 1160 | 2050 | 1090 | 1990 | 1040 | 1970 | 1030 | 1840 | 930 |
| | | 1 | – | – | 2760 | 1460 | 2660 | 1380 | 2580 | 1310 | 2550 | 1290 | 2390 | 1160 |
| 12-1/4 | 1/4 | 7/8 | – | – | 2130 | 1160 | 2050 | 1090 | 1990 | 1040 | 1970 | 1030 | 1840 | 930 |
| | | 1 | – | – | 2760 | 1460 | 2660 | 1380 | 2580 | 1310 | 2550 | 1290 | 2390 | 1160 |
| 14-1/4 | 1/4 | 1 | – | – | 2760 | 1460 | 2660 | 1380 | 2580 | 1310 | 2550 | 1290 | 2390 | 1160 |

1. Tabulated lateral design values (Z) for bolted connections shall be multiplied by all applicable adjustment factors (see Table 7.1).

2. Tabulated lateral design values (Z) are for "full diameter" bolts (see Reference 3) with a bending yield strength ($F_{yb}$) of 45,000 psi.

3. Tabulated lateral design values (Z) are based on a dowel bearing strength ($F_e$) of 58,000 psi for ASTM A36 steel.

Source: ANSI/NF₀PA NDS—1991, *National Design Specification for Wood Construction*, AFPA, 1250 Connecticut Ave., NW, Washington, DC 20036.

## TABLE 7.15

### Bolt Design Values (Z) for Double Shear (Three Member) Connections[1,2] for Sawn Lumber with All Members of Identical Species

| THICKNESS MAIN MEMBER $t_m$ inches | SIDE MEMBER $t_s$ inches | BOLT DIAMETER D inches | G=0.67 RED OAK $Z_\parallel$ lbs. | $Z_{s\perp}$ lbs. | $Z_{m\perp}$ lbs. | G=0.55 MIXED MAPLE SOUTHERN PINE $Z_\parallel$ lbs. | $Z_{s\perp}$ lbs. | $Z_{m\perp}$ lbs. | G=0.50 DOUGLAS FIR-LARCH $Z_\parallel$ lbs. | $Z_{s\perp}$ lbs. | $Z_{m\perp}$ lbs. | G=0.49 DOUGLAS FIR-LARCH (N) $Z_\parallel$ lbs. | $Z_{s\perp}$ lbs. | $Z_{m\perp}$ lbs. | G=0.46 DOUGLAS FIR (S) HEM-FIR (N) $Z_\parallel$ lbs. | $Z_{s\perp}$ lbs. | $Z_{m\perp}$ lbs. |
|---|---|---|---|---|---|---|---|---|---|---|---|---|---|---|---|---|---|
| 1-1/2 | 1-1/2 | 1/2 | 1410 | 960 | 730 | 1150 | 800 | 550 | 1050 | 730 | 470 | 1030 | 720 | 460 | 970 | 680 | 420 |
|  |  | 5/8 | 1760 | 1310 | 810 | 1440 | 1130 | 610 | 1310 | 1040 | 530 | 1290 | 1030 | 520 | 1210 | 940 | 470 |
|  |  | 3/4 | 2110 | 1690 | 890 | 1730 | 1330 | 660 | 1580 | 1170 | 590 | 1550 | 1130 | 560 | 1450 | 1040 | 520 |
|  |  | 7/8 | 2460 | 1920 | 960 | 2020 | 1440 | 720 | 1840 | 1260 | 630 | 1800 | 1210 | 600 | 1690 | 1100 | 550 |
|  |  | 1 | 2810 | 2040 | 1020 | 2310 | 1530 | 770 | 2100 | 1350 | 680 | 2060 | 1290 | 650 | 1930 | 1200 | 600 |
| 2-1/2 | 1-1/2 | 1/2 | 1530 | 960 | 1120 | 1320 | 800 | 910 | 1230 | 730 | 790 | 1210 | 720 | 760 | 1160 | 680 | 700 |
|  |  | 5/8 | 2150 | 1310 | 1340 | 1870 | 1130 | 1020 | 1760 | 1040 | 880 | 1740 | 1030 | 860 | 1660 | 940 | 780 |
|  |  | 3/4 | 2890 | 1770 | 1480 | 2550 | 1330 | 1110 | 2400 | 1170 | 980 | 2380 | 1130 | 940 | 2280 | 1040 | 860 |
|  |  | 7/8 | 3780 | 1920 | 1600 | 3360 | 1440 | 1200 | 3060 | 1260 | 1050 | 3010 | 1210 | 1010 | 2820 | 1100 | 920 |
|  |  | 1 | 4690 | 2040 | 1700 | 3840 | 1530 | 1280 | 3500 | 1350 | 1130 | 3440 | 1290 | 1080 | 3220 | 1200 | 1000 |
| 3 | 1-1/2 | 1/2 | 1530 | 960 | 1120 | 1320 | 800 | 940 | 1230 | 730 | 860 | 1210 | 720 | 850 | 1160 | 680 | 810 |
|  |  | 5/8 | 2150 | 1310 | 1510 | 1870 | 1130 | 1220 | 1760 | 1040 | 1050 | 1740 | 1030 | 1030 | 1660 | 940 | 940 |
|  |  | 3/4 | 2890 | 1770 | 1780 | 2550 | 1330 | 1330 | 2400 | 1170 | 1170 | 2380 | 1130 | 1130 | 2280 | 1040 | 1040 |
|  |  | 7/8 | 3780 | 1920 | 1920 | 3360 | 1440 | 1440 | 3180 | 1260 | 1260 | 3150 | 1210 | 1210 | 3030 | 1100 | 1100 |
|  |  | 1 | 4820 | 2040 | 2040 | 4310 | 1530 | 1530 | 4090 | 1350 | 1350 | 4050 | 1290 | 1290 | 3860 | 1200 | 1200 |
| 3-1/2 | 1-1/2 | 1/2 | 1530 | 960 | 1120 | 1320 | 800 | 940 | 1230 | 730 | 860 | 1210 | 720 | 850 | 1160 | 680 | 810 |
|  |  | 5/8 | 2150 | 1310 | 1510 | 1870 | 1130 | 1290 | 1760 | 1040 | 1190 | 1740 | 1030 | 1170 | 1660 | 940 | 1090 |
|  |  | 3/4 | 2890 | 1770 | 1980 | 2550 | 1330 | 1550 | 2400 | 1170 | 1370 | 2380 | 1130 | 1310 | 2280 | 1040 | 1210 |
|  |  | 7/8 | 3780 | 1920 | 2240 | 3360 | 1440 | 1680 | 3180 | 1260 | 1470 | 3150 | 1210 | 1410 | 3030 | 1100 | 1290 |
|  |  | 1 | 4820 | 2040 | 2380 | 4310 | 1530 | 1790 | 4090 | 1350 | 1580 | 4050 | 1290 | 1510 | 3860 | 1200 | 1400 |
|  | 3-1/2 | 1/2 | 1660 | 1180 | 1180 | 1500 | 1040 | 1040 | 1430 | 970 | 970 | 1420 | 960 | 960 | 1370 | 920 | 920 |
|  |  | 5/8 | 2590 | 1770 | 1770 | 2340 | 1560 | 1420 | 2240 | 1410 | 1230 | 2220 | 1390 | 1200 | 2150 | 1290 | 1090 |
|  |  | 3/4 | 3730 | 2380 | 2070 | 3380 | 1910 | 1550 | 3220 | 1750 | 1370 | 3190 | 1700 | 1310 | 3090 | 1610 | 1210 |
|  |  | 7/8 | 5080 | 2820 | 2240 | 4600 | 2330 | 1680 | 4290 | 2130 | 1470 | 4210 | 2070 | 1410 | 3940 | 1960 | 1290 |
|  |  | 1 | 6560 | 3340 | 2380 | 5380 | 2780 | 1790 | 4900 | 2580 | 1580 | 4810 | 2520 | 1510 | 4510 | 2410 | 1400 |
| 4-1/2 | 1-1/2 | 5/8 | 2150 | 1310 | 1510 | 1870 | 1130 | 1290 | 1760 | 1040 | 1190 | 1740 | 1030 | 1170 | 1660 | 940 | 1110 |
|  |  | 3/4 | 2890 | 1770 | 1980 | 2550 | 1330 | 1690 | 2400 | 1170 | 1580 | 2380 | 1130 | 1550 | 2280 | 1040 | 1480 |
|  |  | 7/8 | 3780 | 1920 | 2520 | 3360 | 1440 | 2170 | 3180 | 1260 | 1890 | 3150 | 1210 | 1810 | 3030 | 1100 | 1650 |
|  |  | 1 | 4820 | 2040 | 3060 | 4310 | 1530 | 2300 | 4090 | 1350 | 2030 | 4050 | 1290 | 1940 | 3860 | 1200 | 1800 |
|  | 3-1/2 | 5/8 | 2590 | 1770 | 1770 | 2340 | 1560 | 1560 | 2240 | 1410 | 1460 | 2220 | 1390 | 1450 | 2150 | 1290 | 1390 |
|  |  | 3/4 | 3730 | 2380 | 2480 | 3380 | 1910 | 1990 | 3220 | 1750 | 1760 | 3190 | 1700 | 1690 | 3090 | 1610 | 1550 |
|  |  | 7/8 | 5080 | 2820 | 2870 | 4600 | 2330 | 2170 | 4390 | 2130 | 1890 | 4350 | 2070 | 1810 | 4130 | 1960 | 1650 |
|  |  | 1 | 6630 | 3340 | 3060 | 5740 | 2780 | 2300 | 5330 | 2580 | 2030 | 5250 | 2520 | 1940 | 4990 | 2410 | 1800 |
| 5-1/2 | 1-1/2 | 5/8 | 2150 | 1310 | 1510 | 1870 | 1130 | 1290 | 1760 | 1040 | 1190 | 1740 | 1030 | 1170 | 1660 | 940 | 1110 |
|  |  | 3/4 | 2890 | 1770 | 1980 | 2550 | 1330 | 1690 | 2400 | 1170 | 1580 | 2380 | 1130 | 1550 | 2280 | 1040 | 1480 |
|  |  | 7/8 | 3780 | 1920 | 2520 | 3360 | 1440 | 2170 | 3180 | 1260 | 2030 | 3150 | 1210 | 1990 | 3030 | 1100 | 1900 |
|  |  | 1 | 4820 | 2040 | 3120 | 4310 | 1530 | 2700 | 4090 | 1350 | 2480 | 4050 | 1290 | 2370 | 3860 | 1200 | 2200 |
|  | 3-1/2 | 5/8 | 2590 | 1770 | 1770 | 2340 | 1560 | 1560 | 2240 | 1410 | 1460 | 2220 | 1390 | 1450 | 2150 | 1290 | 1390 |
|  |  | 3/4 | 3730 | 2380 | 2480 | 3380 | 1910 | 2180 | 3220 | 1750 | 2050 | 3190 | 1700 | 2020 | 3090 | 1610 | 1900 |
|  |  | 7/8 | 5080 | 2820 | 3290 | 4600 | 2330 | 2310 | 4390 | 2130 | 2310 | 4350 | 2070 | 2210 | 4130 | 1960 | 2020 |
|  |  | 1 | 6630 | 3340 | 3740 | 5740 | 2780 | 2810 | 5330 | 2580 | 2480 | 5250 | 2520 | 2370 | 4990 | 2410 | 2200 |
| 7-1/2 | 1-1/2 | 5/8 | 2150 | 1310 | 1510 | 1870 | 1130 | 1290 | 1760 | 1040 | 1190 | 1740 | 1030 | 1170 | 1660 | 940 | 1110 |
|  |  | 3/4 | 2890 | 1770 | 1980 | 2550 | 1330 | 1690 | 2400 | 1170 | 1580 | 2380 | 1130 | 1550 | 2280 | 1040 | 1480 |
|  |  | 7/8 | 3780 | 1920 | 2520 | 3360 | 1440 | 2170 | 3180 | 1260 | 2030 | 3150 | 1210 | 1990 | 3030 | 1100 | 1900 |
|  |  | 1 | 4820 | 2040 | 3120 | 4310 | 1530 | 2700 | 4090 | 1350 | 2530 | 4050 | 1290 | 2480 | 3860 | 1200 | 2390 |
|  | 3-1/2 | 5/8 | 2590 | 1770 | 1770 | 2340 | 1560 | 1560 | 2240 | 1410 | 1460 | 2220 | 1390 | 1450 | 2150 | 1290 | 1390 |
|  |  | 3/4 | 3730 | 2380 | 2480 | 3380 | 1910 | 2180 | 3220 | 1750 | 2050 | 3190 | 1700 | 2020 | 3090 | 1610 | 1940 |
|  |  | 7/8 | 5080 | 2820 | 3290 | 4600 | 2330 | 2890 | 4390 | 2130 | 2720 | 4350 | 2070 | 2670 | 4130 | 1960 | 2560 |
|  |  | 1 | 6630 | 3340 | 4190 | 5740 | 2780 | 3680 | 5330 | 2580 | 3380 | 5250 | 2520 | 3230 | 4990 | 2410 | 3000 |

## TABLE 7.15  (Continued)

| MAIN MEMBER t_m inches | SIDE MEMBER t_s inches | BOLT DIAMETER D inches | G=0.43 HEM-FIR $Z_\parallel$ lbs. | $Z_{s\perp}$ lbs. | $Z_{m\perp}$ lbs. | G=0.42 SPRUCE-PINE-FIR $Z_\parallel$ lbs. | $Z_{s\perp}$ lbs. | $Z_{m\perp}$ lbs. | G=0.37 REDWOOD (open grain) $Z_\parallel$ lbs. | $Z_{s\perp}$ lbs. | $Z_{m\perp}$ lbs. | G=0.36 EASTERN SOFTWOODS SPRUCE-PINE-FIR(S) WESTERN CEDARS WESTERN WOODS $Z_\parallel$ lbs. | $Z_{s\perp}$ lbs. | $Z_{m\perp}$ lbs. | G=0.35 NORTHERN SPECIES $Z_\parallel$ lbs. | $Z_{s\perp}$ lbs. | $Z_{m\perp}$ lbs. |
|---|---|---|---|---|---|---|---|---|---|---|---|---|---|---|---|---|---|
| 1-1/2 | 1-1/2 | 1/2 | 900 | 650 | 380 | 880 | 640 | 370 | 780 | 580 | 310 | 760 | 560 | 290 | 730 | 550 | 290 |
|  |  | 5/8 | 1130 | 840 | 420 | 1100 | 830 | 410 | 970 | 690 | 350 | 950 | 660 | 330 | 910 | 640 | 320 |
|  |  | 3/4 | 1350 | 920 | 460 | 1320 | 900 | 450 | 1170 | 740 | 370 | 1140 | 720 | 360 | 1100 | 700 | 350 |
|  |  | 7/8 | 1580 | 1000 | 500 | 1540 | 970 | 490 | 1360 | 810 | 410 | 1330 | 790 | 390 | 1280 | 740 | 370 |
|  |  | 1 | 1800 | 1080 | 540 | 1760 | 1050 | 530 | 1560 | 870 | 440 | 1520 | 840 | 420 | 1460 | 810 | 410 |
| 2-1/2 | 1-1/2 | 1/2 | 1100 | 650 | 640 | 1080 | 640 | 610 | 990 | 580 | 510 | 980 | 560 | 490 | 950 | 550 | 480 |
|  |  | 5/8 | 1590 | 840 | 700 | 1570 | 830 | 690 | 1450 | 690 | 580 | 1430 | 660 | 550 | 1390 | 640 | 530 |
|  |  | 3/4 | 2190 | 920 | 770 | 2160 | 900 | 750 | 1950 | 740 | 620 | 1900 | 720 | 600 | 1830 | 700 | 580 |
|  |  | 7/8 | 2630 | 1000 | 830 | 2570 | 970 | 810 | 2270 | 810 | 680 | 2210 | 790 | 660 | 2130 | 740 | 610 |
|  |  | 1 | 3000 | 1080 | 900 | 2940 | 1050 | 880 | 2590 | 870 | 730 | 2530 | 840 | 700 | 2440 | 810 | 680 |
| 3 | 1-1/2 | 1/2 | 1100 | 650 | 760 | 1080 | 640 | 740 | 990 | 580 | 620 | 980 | 560 | 590 | 950 | 550 | 570 |
|  |  | 5/8 | 1590 | 840 | 840 | 1570 | 830 | 830 | 1450 | 690 | 690 | 1430 | 660 | 660 | 1390 | 640 | 640 |
|  |  | 3/4 | 2190 | 920 | 920 | 2160 | 900 | 900 | 2010 | 740 | 740 | 1990 | 720 | 720 | 1940 | 700 | 700 |
|  |  | 7/8 | 2920 | 1000 | 1000 | 2880 | 970 | 970 | 2690 | 810 | 810 | 2660 | 790 | 790 | 2560 | 740 | 740 |
|  |  | 1 | 3600 | 1080 | 1080 | 3530 | 1050 | 1050 | 3110 | 870 | 870 | 3040 | 840 | 840 | 2930 | 810 | 810 |
| 3-1/2 | 1-1/2 | 1/2 | 1100 | 650 | 760 | 1080 | 640 | 740 | 990 | 580 | 670 | 980 | 560 | 660 | 950 | 550 | 640 |
|  |  | 5/8 | 1590 | 840 | 980 | 1570 | 830 | 960 | 1450 | 690 | 810 | 1430 | 660 | 770 | 1390 | 640 | 740 |
|  |  | 3/4 | 2190 | 920 | 1080 | 2160 | 900 | 1050 | 2010 | 740 | 870 | 1990 | 720 | 840 | 1940 | 700 | 810 |
|  |  | 7/8 | 2920 | 1000 | 1160 | 2880 | 970 | 1130 | 2690 | 810 | 950 | 2660 | 790 | 920 | 2560 | 740 | 860 |
|  |  | 1 | 3600 | 1080 | 1260 | 3530 | 1050 | 1230 | 3110 | 870 | 1020 | 3040 | 840 | 980 | 2930 | 810 | 950 |
|  | 3-1/2 | 1/2 | 1330 | 880 | 880 | 1310 | 870 | 860 | 1230 | 800 | 720 | 1220 | 780 | 680 | 1200 | 760 | 670 |
|  |  | 5/8 | 2070 | 1190 | 980 | 2050 | 1170 | 960 | 1930 | 1030 | 810 | 1900 | 1000 | 770 | 1870 | 970 | 740 |
|  |  | 3/4 | 2980 | 1490 | 1080 | 2950 | 1460 | 1050 | 2720 | 1290 | 870 | 2660 | 1270 | 840 | 2560 | 1240 | 810 |
|  |  | 7/8 | 3680 | 1840 | 1160 | 3600 | 1810 | 1130 | 3180 | 1640 | 950 | 3100 | 1610 | 920 | 2990 | 1550 | 860 |
|  |  | 1 | 4200 | 2280 | 1260 | 4110 | 2240 | 1230 | 3630 | 2030 | 1020 | 3540 | 1960 | 980 | 3410 | 1890 | 950 |
| 4-1/2 | 1-1/2 | 5/8 | 1590 | 840 | 1050 | 1570 | 830 | 1040 | 1450 | 690 | 940 | 1430 | 660 | 920 | 1390 | 640 | 900 |
|  |  | 3/4 | 2190 | 920 | 1380 | 2160 | 900 | 1350 | 2010 | 740 | 1110 | 1990 | 720 | 1080 | 1940 | 700 | 1050 |
|  |  | 7/8 | 2920 | 1000 | 1500 | 2880 | 970 | 1460 | 2690 | 810 | 1220 | 2660 | 790 | 1180 | 2560 | 740 | 1100 |
|  |  | 1 | 3600 | 1080 | 1620 | 3530 | 1050 | 1580 | 3110 | 870 | 1310 | 3040 | 840 | 1260 | 2930 | 810 | 1220 |
|  | 3-1/2 | 5/8 | 2070 | 1190 | 1270 | 2050 | 1170 | 1240 | 1930 | 1030 | 1040 | 1900 | 1000 | 980 | 1870 | 970 | 960 |
|  |  | 3/4 | 2980 | 1490 | 1380 | 2950 | 1460 | 1350 | 2770 | 1290 | 1110 | 2740 | 1270 | 1080 | 2660 | 1240 | 1050 |
|  |  | 7/8 | 3900 | 1840 | 1500 | 3840 | 1810 | 1460 | 3480 | 1640 | 1220 | 3410 | 1610 | 1180 | 3320 | 1550 | 1100 |
|  |  | 1 | 4730 | 2280 | 1620 | 4660 | 2240 | 1580 | 4240 | 2030 | 1310 | 4170 | 1960 | 1260 | 4050 | 1890 | 1220 |
| 5-1/2 | 1-1/2 | 5/8 | 1590 | 840 | 1050 | 1570 | 830 | 1040 | 1450 | 690 | 940 | 1430 | 660 | 920 | 1390 | 640 | 900 |
|  |  | 3/4 | 2190 | 920 | 1400 | 2160 | 900 | 1380 | 2010 | 740 | 1250 | 1990 | 720 | 1230 | 1940 | 700 | 1210 |
|  |  | 7/8 | 2920 | 1000 | 1800 | 2880 | 970 | 1780 | 2690 | 810 | 1490 | 2660 | 790 | 1440 | 2560 | 740 | 1350 |
|  |  | 1 | 3600 | 1080 | 1980 | 3530 | 1050 | 1930 | 3110 | 870 | 1600 | 3040 | 840 | 1540 | 2930 | 810 | 1490 |
|  | 3-1/2 | 5/8 | 2070 | 1190 | 1320 | 2050 | 1170 | 1310 | 1930 | 1030 | 1210 | 1900 | 1000 | 1180 | 1870 | 970 | 1160 |
|  |  | 3/4 | 2980 | 1490 | 1690 | 2950 | 1460 | 1650 | 2770 | 1290 | 1360 | 2740 | 1270 | 1320 | 2660 | 1240 | 1280 |
|  |  | 7/8 | 3900 | 1840 | 1830 | 3840 | 1810 | 1780 | 3480 | 1640 | 1490 | 3410 | 1610 | 1440 | 3320 | 1550 | 1350 |
|  |  | 1 | 4730 | 2280 | 1980 | 4660 | 2240 | 1930 | 4240 | 2030 | 1600 | 4170 | 1960 | 1540 | 4050 | 1890 | 1490 |
| 7-1/2 | 1-1/2 | 5/8 | 1590 | 840 | 1050 | 1570 | 830 | 1040 | 1450 | 690 | 940 | 1430 | 660 | 920 | 1390 | 640 | 900 |
|  |  | 3/4 | 2190 | 920 | 1400 | 2160 | 900 | 1380 | 2010 | 740 | 1250 | 1990 | 720 | 1230 | 1940 | 700 | 1210 |
|  |  | 7/8 | 2920 | 1000 | 1800 | 2880 | 970 | 1780 | 2690 | 810 | 1630 | 2660 | 790 | 1600 | 2560 | 740 | 1550 |
|  |  | 1 | 3600 | 1080 | 2270 | 3530 | 1050 | 2240 | 3110 | 870 | 2040 | 3040 | 840 | 2010 | 2930 | 810 | 1970 |
|  | 3-1/2 | 5/8 | 2070 | 1190 | 1320 | 2050 | 1170 | 1310 | 1930 | 1030 | 1210 | 1900 | 1000 | 1180 | 1870 | 970 | 1160 |
|  |  | 3/4 | 2980 | 1490 | 1850 | 2950 | 1460 | 1820 | 2770 | 1290 | 1670 | 2740 | 1270 | 1650 | 2660 | 1240 | 1620 |
|  |  | 7/8 | 3900 | 1840 | 2450 | 3840 | 1810 | 2420 | 3480 | 1640 | 2030 | 3410 | 1610 | 1970 | 3320 | 1550 | 1840 |
|  |  | 1 | 4730 | 2280 | 2700 | 4660 | 2240 | 2630 | 4240 | 2030 | 2180 | 4170 | 1960 | 2100 | 4050 | 1890 | 2030 |

1. Tabulated lateral design values (Z) for bolted connections shall be multiplied by all applicable adjustment factors (see Table 7.1).

2. Tabulated lateral design values (Z) are for "full diameter" bolts (see Reference 3) with a bending yield strength ($F_{yb}$) of 45,000 psi.

Source: ANSI/NF$_o$PA NDS—1991, *National Design Specification for Wood Construction*, AFPA, 1250 Connecticut Ave., NW, Washington, DC 20036.

## TABLE 7.16

### Bolt Design Values (Z) for Double Shear (Three Member) Connections[1,2,3] for Sawn Lumber with $\frac{1}{4}$ in. ASTM A36 Steel Side Plates

| MAIN MEMBER $t_m$ inches | STEEL SIDE PLATE $t_s$ inches | BOLT DIAMETER D inches | G=0.67 RED OAK $Z_{\parallel}$ lbs. | $Z_{\perp}$ lbs. | G=0.55 MIXED MAPLE SOUTHERN PINE $Z_{\parallel}$ lbs. | $Z_{\perp}$ lbs. | G=0.50 DOUGLAS FIR-LARCH $Z_{\parallel}$ lbs. | $Z_{\perp}$ lbs. | G=0.49 DOUGLAS FIR-LARCH (N) $Z_{\parallel}$ lbs. | $Z_{\perp}$ lbs. | G=0.46 DOUGLAS FIR (S) HEM-FIR (N) $Z_{\parallel}$ lbs. | $Z_{\perp}$ lbs. |
|---|---|---|---|---|---|---|---|---|---|---|---|---|
| 1-1/2 | 1/4 | 1/2 | 670 | 380 | 570 | 310 | 530 | 270 | 520 | 270 | 500 | 250 |
|  |  | 5/8 | 840 | 430 | 710 | 350 | 660 | 320 | 650 | 310 | 620 | 290 |
|  |  | 3/4 | 1010 | 490 | 860 | 390 | 800 | 360 | 780 | 350 | 740 | 330 |
|  |  | 7/8 | 1180 | 540 | 1000 | 440 | 930 | 400 | 920 | 390 | 870 | 370 |
|  |  | 1 | 1350 | 580 | 1140 | 480 | 1060 | 440 | 1050 | 430 | 990 | 410 |
| 2-1/2 | 1/4 | 1/2 | 850 | 570 | 780 | 440 | 750 | 390 | 750 | 380 | 730 | 350 |
|  |  | 5/8 | 1270 | 640 | 1100 | 500 | 1010 | 440 | 990 | 430 | 940 | 400 |
|  |  | 3/4 | 1580 | 710 | 1320 | 550 | 1210 | 490 | 1190 | 480 | 1120 | 450 |
|  |  | 7/8 | 1850 | 770 | 1540 | 610 | 1410 | 540 | 1390 | 520 | 1310 | 480 |
|  |  | 1 | 2110 | 830 | 1760 | 650 | 1620 | 590 | 1590 | 560 | 1500 | 530 |
| 3 | 1/4 | 1/2 | 850 | 570 | 780 | 500 | 750 | 450 | 750 | 440 | 730 | 410 |
|  |  | 5/8 | 1270 | 750 | 1170 | 580 | 1130 | 510 | 1120 | 500 | 1090 | 460 |
|  |  | 3/4 | 1800 | 830 | 1560 | 640 | 1430 | 570 | 1410 | 550 | 1330 | 510 |
|  |  | 7/8 | 2200 | 900 | 1830 | 700 | 1670 | 620 | 1650 | 600 | 1550 | 550 |
|  |  | 1 | 2510 | 970 | 2090 | 750 | 1910 | 670 | 1880 | 650 | 1770 | 610 |
| 3-1/2 | 1/4 | 1/2 | 850 | 570 | 780 | 500 | 750 | 470 | 750 | 460 | 730 | 450 |
|  |  | 5/8 | 1270 | 800 | 1170 | 670 | 1130 | 580 | 1120 | 570 | 1090 | 520 |
|  |  | 3/4 | 1800 | 960 | 1650 | 730 | 1580 | 650 | 1570 | 630 | 1530 | 580 |
|  |  | 7/8 | 2420 | 1040 | 2120 | 800 | 1940 | 710 | 1900 | 680 | 1790 | 630 |
|  |  | 1 | 2920 | 1110 | 2420 | 850 | 2210 | 760 | 2180 | 730 | 2050 | 690 |
| 4-1/2 | 1/4 | 5/8 | 1270 | 800 | 1170 | 710 | 1130 | 660 | 1120 | 660 | 1090 | 630 |
|  |  | 3/4 | 1800 | 1090 | 1650 | 920 | 1580 | 820 | 1570 | 790 | 1530 | 730 |
|  |  | 7/8 | 2420 | 1310 | 2220 | 1010 | 2130 | 880 | 2110 | 850 | 2050 | 780 |
|  |  | 1 | 3140 | 1400 | 2880 | 1070 | 2760 | 950 | 2740 | 910 | 2610 | 850 |
| 5-1/2 | 1/4 | 5/8 | 1270 | 800 | 1170 | 710 | 1130 | 660 | 1120 | 660 | 1090 | 630 |
|  |  | 3/4 | 1800 | 1090 | 1650 | 950 | 1580 | 900 | 1570 | 880 | 1530 | 850 |
|  |  | 7/8 | 2420 | 1410 | 2220 | 1220 | 2130 | 1070 | 2110 | 1030 | 2050 | 940 |
|  |  | 1 | 3140 | 1700 | 2880 | 1290 | 2760 | 1150 | 2740 | 1100 | 2660 | 1030 |
| 7-1/2 | 1/4 | 5/8 | 1270 | 800 | 1170 | 710 | 1130 | 660 | 1120 | 660 | 1090 | 630 |
|  |  | 3/4 | 1800 | 1090 | 1650 | 950 | 1580 | 900 | 1570 | 880 | 1530 | 850 |
|  |  | 7/8 | 2420 | 1410 | 2220 | 1240 | 2130 | 1160 | 2110 | 1140 | 2050 | 1090 |
|  |  | 1 | 3140 | 1760 | 2880 | 1540 | 2760 | 1460 | 2740 | 1420 | 2660 | 1380 |
| 9-1/2 | 1/4 | 3/4 | 1800 | 1090 | 1650 | 950 | 1580 | 900 | 1570 | 880 | 1530 | 850 |
|  |  | 7/8 | 2420 | 1410 | 2220 | 1240 | 2130 | 1160 | 2110 | 1140 | 2050 | 1090 |
|  |  | 1 | 3140 | 1760 | 2880 | 1540 | 2760 | 1460 | 2740 | 1420 | 2660 | 1380 |
| 11-1/2 | 1/4 | 7/8 | 2420 | 1410 | 2220 | 1240 | 2130 | 1160 | 2110 | 1140 | 2050 | 1090 |
|  |  | 1 | 3140 | 1760 | 2880 | 1540 | 2760 | 1460 | 2740 | 1420 | 2660 | 1380 |
| 13-1/2 | 1/4 | 1 | 3140 | 1760 | 2880 | 1540 | 2760 | 1460 | 2740 | 1420 | 2660 | 1380 |

## TABLE 7.16   (Continued)

| THICKNESS MAIN MEMBER $t_m$ inches | STEEL SIDE PLATE $t_s$ inches | BOLT DIAMETER D inches | G=0.43 HEM-FIR $Z_{||}$ lbs. | $Z_\perp$ lbs. | G=0.42 SPRUCE-PINE-FIR $Z_{||}$ lbs. | $Z_\perp$ lbs. | G=0.37 REDWOOD (open grain) $Z_{||}$ lbs. | $Z_\perp$ lbs. | G=0.36 EASTERN SOFTWOODS SPRUCE-PINE-FIR(S) WESTERN CEDARS WESTERN WOODS $Z_{||}$ lbs. | $Z_\perp$ lbs. | G=0.35 NORTHERN SPECIES $Z_{||}$ lbs. | $Z_\perp$ lbs. |
|---|---|---|---|---|---|---|---|---|---|---|---|---|
| | | 1/2 | 470 | 240 | 460 | 230 | 420 | 210 | 410 | 200 | 400 | 200 |
| | | 5/8 | 590 | 270 | 580 | 270 | 530 | 240 | 520 | 230 | 500 | 230 |
| 1-1/2 | 1/4 | 3/4 | 700 | 310 | 690 | 300 | 630 | 270 | 620 | 270 | 600 | 260 |
| | | 7/8 | 820 | 340 | 810 | 340 | 740 | 300 | 720 | 300 | 700 | 290 |
| | | 1 | 940 | 380 | 920 | 370 | 840 | 340 | 830 | 330 | 800 | 320 |
| | | 1/2 | 700 | 320 | 690 | 310 | 620 | 270 | 600 | 260 | 580 | 260 |
| | | 5/8 | 880 | 370 | 860 | 360 | 770 | 310 | 760 | 300 | 730 | 290 |
| 2-1/2 | 1/4 | 3/4 | 1050 | 410 | 1030 | 400 | 930 | 340 | 910 | 330 | 880 | 330 |
| | | 7/8 | 1230 | 450 | 1210 | 440 | 1080 | 380 | 1060 | 370 | 1020 | 350 |
| | | 1 | 1410 | 490 | 1380 | 480 | 1240 | 410 | 1210 | 400 | 1170 | 390 |
| | | 1/2 | 710 | 370 | 700 | 360 | 660 | 310 | 660 | 300 | 650 | 290 |
| | | 5/8 | 1040 | 420 | 1020 | 410 | 910 | 360 | 890 | 340 | 860 | 330 |
| 3 | 1/4 | 3/4 | 1240 | 470 | 1220 | 460 | 1090 | 390 | 1060 | 380 | 1030 | 370 |
| | | 7/8 | 1450 | 510 | 1420 | 500 | 1270 | 430 | 1240 | 420 | 1200 | 400 |
| | | 1 | 1660 | 560 | 1630 | 540 | 1450 | 460 | 1420 | 450 | 1370 | 440 |
| | | 1/2 | 710 | 430 | 700 | 410 | 660 | 350 | 660 | 340 | 650 | 330 |
| | | 5/8 | 1050 | 480 | 1040 | 470 | 990 | 400 | 980 | 380 | 960 | 370 |
| 3-1/2 | 1/4 | 3/4 | 1440 | 530 | 1410 | 520 | 1250 | 440 | 1230 | 420 | 1180 | 410 |
| | | 7/8 | 1680 | 570 | 1640 | 560 | 1460 | 480 | 1430 | 470 | 1380 | 440 |
| | | 1 | 1910 | 630 | 1880 | 610 | 1670 | 520 | 1630 | 500 | 1580 | 490 |
| | | 5/8 | 1050 | 600 | 1040 | 580 | 990 | 500 | 980 | 470 | 960 | 460 |
| 4-1/2 | 1/4 | 3/4 | 1480 | 660 | 1470 | 640 | 1390 | 540 | 1370 | 520 | 1350 | 510 |
| | | 7/8 | 1990 | 710 | 1970 | 700 | 1860 | 590 | 1810 | 570 | 1750 | 540 |
| | | 1 | 2440 | 770 | 2390 | 750 | 2120 | 640 | 2070 | 620 | 2000 | 600 |
| | | 5/8 | 1050 | 600 | 1040 | 600 | 990 | 550 | 980 | 540 | 960 | 530 |
| 5-1/2 | 1/4 | 3/4 | 1480 | 790 | 1470 | 770 | 1390 | 640 | 1370 | 620 | 1350 | 610 |
| | | 7/8 | 1990 | 860 | 1970 | 830 | 1860 | 710 | 1840 | 680 | 1810 | 640 |
| | | 1 | 2580 | 930 | 2550 | 900 | 2410 | 760 | 2390 | 730 | 2350 | 710 |
| | | 5/8 | 1050 | 600 | 1040 | 600 | 990 | 550 | 980 | 540 | 960 | 530 |
| 7-1/2 | 1/4 | 3/4 | 1480 | 810 | 1470 | 800 | 1390 | 730 | 1370 | 720 | 1350 | 710 |
| | | 7/8 | 1990 | 1040 | 1970 | 1030 | 1860 | 940 | 1840 | 910 | 1810 | 860 |
| | | 1 | 2580 | 1250 | 2550 | 1210 | 2410 | 1010 | 2390 | 980 | 2350 | 940 |
| | | 3/4 | 1480 | 810 | 1470 | 800 | 1390 | 730 | 1370 | 720 | 1350 | 710 |
| 9-1/2 | 1/4 | 7/8 | 1990 | 1040 | 1970 | 1030 | 1860 | 950 | 1840 | 930 | 1810 | 900 |
| | | 1 | 2580 | 1310 | 2550 | 1290 | 2410 | 1180 | 2390 | 1160 | 2350 | 1140 |
| 11-1/2 | 1/4 | 7/8 | 1990 | 1040 | 1970 | 1030 | 1860 | 950 | 1840 | 930 | 1810 | 900 |
| | | 1 | 2580 | 1310 | 2550 | 1290 | 2410 | 1180 | 2390 | 1160 | 2350 | 1140 |
| 13-1/2 | 1/4 | 1 | 2580 | 1310 | 2550 | 1290 | 2410 | 1180 | 2390 | 1160 | 2350 | 1140 |

1. Tabulated lateral design values (Z) for bolted connections shall be multiplied by all applicable adjustment factors (see Table 7.1).

2. Tabulated lateral design values (Z) are for "full diameter" bolts (see Reference 3) with a bending yield strength ($F_{yb}$) of 45,000 psi.

3. Tabulated lateral design values (Z) are based on a dowel bearing strength ($F_e$) of 58,000 psi for ASTM A36 steel.

Source: ANSI/NF$_o$PA NDS—1991, *National Design Specification for Wood Construction*, AFPA, 1250 Connecticut Ave., NW, Washington, DC 20036.

# TABLE 7.17

## Bolt Design Values (Z) for Double Shear (Three Member) Connections[1,2] for Glued Laminated Timber Main Member with Sawn Lumber Side Members of Identical Species

| MAIN MEMBER $t_m$ inches | SIDE MEMBER $t_s$ inches | BOLT DIAMETER D inches | SOUTHERN PINE G=0.55 | | | DOUGLAS FIR-LARCH G=0.50 | | | DOUGLAS FIR (S) G=0.46 | | | HEM-FIR G=0.43 | | | SPRUCE-PINE-FIR G=0.42 | | | SPRUCE-PINE-FIR (S) WESTERN WOODS G=0.36 | | |
|---|---|---|---|---|---|---|---|---|---|---|---|---|---|---|---|---|---|---|---|---|
| | | | $Z_\parallel$ lbs. | $Z_{s\perp}$ lbs. | $Z_{m\perp}$ lbs. | $Z_\parallel$ lbs. | $Z_{s\perp}$ lbs. | $Z_{m\perp}$ lbs. | $Z_\parallel$ lbs. | $Z_{s\perp}$ lbs. | $Z_{m\perp}$ lbs. | $Z_\parallel$ lbs. | $Z_{s\perp}$ lbs. | $Z_{m\perp}$ lbs. | $Z_\parallel$ lbs. | $Z_{s\perp}$ lbs. | $Z_{m\perp}$ lbs. | $Z_\parallel$ lbs. | $Z_{s\perp}$ lbs. | $Z_{m\perp}$ lbs. |
| 2-1/2 | 1-1/2 | 1/2 | – | – | – | 1230 | 730 | 790 | 1160 | 680 | 700 | 1100 | 650 | 640 | 1080 | 640 | 610 | 980 | 560 | 490 |
| | | 5/8 | – | – | – | 1760 | 1040 | 880 | 1660 | 940 | 780 | 1590 | 840 | 700 | 1570 | 830 | 690 | 1430 | 660 | 550 |
| | | 3/4 | – | – | – | 2400 | 1170 | 980 | 2280 | 1040 | 860 | 2190 | 920 | 770 | 2160 | 900 | 750 | 1900 | 720 | 600 |
| | | 7/8 | – | – | – | 3060 | 1260 | 1050 | 2820 | 1100 | 920 | 2630 | 1000 | 830 | 2570 | 970 | 810 | 2210 | 790 | 660 |
| | | 1 | – | – | – | 3500 | 1350 | 1130 | 3220 | 1200 | 1000 | 3000 | 1080 | 900 | 2940 | 1050 | 880 | 2530 | 840 | 700 |
| 3 | 1-1/2 | 1/2 | 1320 | 800 | 940 | – | – | – | – | – | – | – | – | – | – | – | – | – | – | – |
| | | 5/8 | 1870 | 1130 | 1220 | – | – | – | – | – | – | – | – | – | – | – | – | – | – | – |
| | | 3/4 | 2550 | 1330 | 1330 | – | – | – | – | – | – | – | – | – | – | – | – | – | – | – |
| | | 7/8 | 3360 | 1440 | 1440 | – | – | – | – | – | – | – | – | – | – | – | – | – | – | – |
| | | 1 | 4310 | 1530 | 1530 | – | – | – | – | – | – | – | – | – | – | – | – | – | – | – |
| 3-1/8 | 1-1/2 | 1/2 | – | – | – | 1230 | 730 | 860 | 1160 | 680 | 810 | 1100 | 650 | 760 | 1080 | 640 | 740 | 980 | 560 | 610 |
| | | 5/8 | – | – | – | 1760 | 1040 | 1090 | 1660 | 940 | 980 | 1590 | 840 | 880 | 1570 | 830 | 860 | 1430 | 660 | 680 |
| | | 3/4 | – | – | – | 2400 | 1170 | 1220 | 2280 | 1040 | 1080 | 2190 | 920 | 960 | 2160 | 900 | 940 | 1990 | 720 | 750 |
| | | 7/8 | – | – | – | 3180 | 1260 | 1310 | 3030 | 1100 | 1150 | 2920 | 1000 | 1040 | 2880 | 970 | 1010 | 2660 | 790 | 820 |
| | | 1 | – | – | – | 4090 | 1350 | 1410 | 3860 | 1200 | 1250 | 3600 | 1080 | 1130 | 3530 | 1050 | 1090 | 3040 | 840 | 880 |
| 5 | 1-1/2 | 5/8 | 1870 | 1130 | 1290 | – | – | – | – | – | – | – | – | – | – | – | – | – | – | – |
| | | 3/4 | 2550 | 1330 | 1690 | – | – | – | – | – | – | – | – | – | – | – | – | – | – | – |
| | | 7/8 | 3360 | 1440 | 2170 | – | – | – | – | – | – | – | – | – | – | – | – | – | – | – |
| | | 1 | 4310 | 1530 | 2550 | – | – | – | – | – | – | – | – | – | – | – | – | – | – | – |
| 5-1/8 | 1-1/2 | 5/8 | – | – | – | 1760 | 1040 | 1190 | 1660 | 940 | 1110 | 1590 | 840 | 1050 | 1570 | 830 | 1040 | 1430 | 660 | 920 |
| | | 3/4 | – | – | – | 2400 | 1170 | 1580 | 2280 | 1040 | 1480 | 2190 | 920 | 1400 | 2160 | 900 | 1380 | 1990 | 720 | 1230 |
| | | 7/8 | – | – | – | 3180 | 1260 | 2030 | 3030 | 1100 | 1880 | 2920 | 1000 | 1700 | 2880 | 970 | 1660 | 2660 | 790 | 1350 |
| | | 1 | – | – | – | 4090 | 1350 | 2310 | 3860 | 1200 | 2050 | 3600 | 1080 | 1850 | 3530 | 1050 | 1790 | 3040 | 840 | 1440 |
| 6-3/4 | 1-1/2 | 5/8 | 1870 | 1130 | 1290 | 1760 | 1040 | 1190 | 1660 | 940 | 1110 | 1590 | 840 | 1050 | 1570 | 830 | 1040 | 1430 | 660 | 920 |
| | | 3/4 | 2550 | 1330 | 1690 | 2400 | 1170 | 1580 | 2280 | 1040 | 1480 | 2190 | 920 | 1400 | 2160 | 900 | 1380 | 1990 | 720 | 1230 |
| | | 7/8 | 3360 | 1440 | 2170 | 3180 | 1260 | 2030 | 3030 | 1100 | 1900 | 2920 | 1000 | 1800 | 2880 | 970 | 1780 | 2660 | 790 | 1600 |
| | | 1 | 4310 | 1530 | 2700 | 4090 | 1350 | 2530 | 3860 | 1200 | 2390 | 3600 | 1080 | 2270 | 3530 | 1050 | 2240 | 3040 | 840 | 1890 |

1. Tabulated lateral design values (Z) for bolted connections shall be multiplied by all applicable adjustment factors (see Table 7.1).
2. Tabulated lateral design values (Z) are for "full diameter" bolts (see Reference 3) with a bending yield strength ($F_{yb}$) of 45,000 psi.

Source: ANSI/NF.PA NDS—1991, *National Design Specification for Wood Construction*, AFPA, 1250 Connecticut Ave., NW, Washington, DC 20036.

$$\text{Mode } I_m \quad Z = \frac{(1)(6.75)(6150)}{(4)(1)} = 10{,}378 \text{ lb}$$

$$\text{Mode } III_s \quad Z = \frac{(8.691)(1)(0.375)(6150)}{(1.6)(2 + 0.106)(1)} = 5{,}948 \text{ lb}$$

$$\text{Mode } IV \quad Z = \frac{1^2}{(1.6)(1)} \sqrt{\frac{(2)(6150)(45{,}000)}{3(1 + 0.106)}} = 8{,}072 \text{ lb}$$

Mode $III_s$ controls, $Z_\| = 5{,}948$ lb

The calculated value with $\frac{3}{8}$ in. side plates is 5,948 lb which is 198 lb more than the tabular design value based on $\frac{1}{4}$ in. thick steel side plates.

$$Z^* = (Z)(C_D) = (5{,}948)(1.15) = 6{,}840 \text{ lb}$$

Determine group action factor $C_g$.

Assume 4 in. $\times \frac{3}{8}$ in. steel side plates are used.

$A_s = (2)(4)(0.375) = 3 \text{ in.}^2, A_m = (6.75)(8.25) = 55.69 \text{ in.}^2.$
$A_s/A_m = 3/55.69 = 0.0539$
$A_m/A_s = 55.69/3 = 18.56, \text{ use } 18.$
$C_g = 0.94$, by interpolation in Table 7.6. When the design is close, use of Eq.
 (7-1) may yield a higher $C_g$ than that obtained from the tables.
$Z' = (6840)(0.94) = 6430 \text{ lb}$
$N = 38{,}000/6430 = 5.9, \text{ use } 6 \text{ bolts.}$

Check net section.

$$(6.75)(8.25) - (1.0625)(6.75) = 48.52 \text{ in.}^2$$

$$(48.52)(1400)(1.15) = 78{,}100 \text{ lb} > 38{,}000 \text{ lb} \qquad \text{O.K.}$$

Check steel side plate. For two $\frac{3}{8} \times 4$ in. steel plates.

$$(6.75)(8.25) - (1.0625)(6.75) = 48.52 \text{ in.}^2$$

$$(48.52)(1400)(1.15) = 78{,}100 \text{ lb} > 38{,}000 \text{ lb} \qquad \text{O.K.}$$

Check bearing on steel $(2)(0.375)(1) = 0.75 \text{ in.}^2/\text{bolt}, (6)(0.75) = 4.50 \text{ in.}^2$, $38{,}000/4.50 = 8{,}444 \text{ psi} < 35{,}000 \text{ psi. O.K.}$

From Table 7.9 for full design load, the spacing between bolts in a row is $4D = 4$ in., the end distance is $7D = 7$ in., and the edge distance is $1.5D = 1.5$ in. The actual load per bolt is $38{,}000/6 = 6333$ lb.

The spacing between the bolts could be reduced to $(6333/6430)(4) = 3.94$ in., but it is recommended that the spacing of 4 in. for full design load be used unless space is restricted.

The required end distance for full load is $7D$, which also could be reduced to $(6333/6430)(7) = 6.89$ in., but it is recommended that 7 in. end distance for full load be used unless space is restricted.

# TABLE 7.18

## Bolt Design Values (Z) for Double Shear (Three Member) Connections[1,2,3] for Glued Laminated Timber with $\frac{1}{4}$ in. ASTM A36 Steel Side Plates

| THICKNESS MAIN MEMBER $t_m$ inches | STEEL SIDE PLATE $t_s$ inches | BOLT DIAMETER D inches | G=0.55 SOUTHERN PINE | | G=0.50 DOUGLAS FIR-LARCH | | G=0.46 DOUGLAS FIR (S) | | G=0.43 HEM-FIR | | G=0.42 SPRUCE-PINE-FIR | | G=0.36 SPRUCE-PINE-FIR (S) WESTERN WOODS | |
|---|---|---|---|---|---|---|---|---|---|---|---|---|---|---|
| | | | $Z_\parallel$ lbs. | $Z_\perp$ lbs. | $Z_\parallel$ lbs. | $Z_\perp$ lbs. | $Z_\parallel$ lbs. | $Z_\perp$ lbs. | $Z_\parallel$ lbs. | $Z_\perp$ lbs. | $Z_\parallel$ lbs. | $Z_\perp$ lbs. | $Z_\parallel$ lbs. | $Z_\perp$ lbs. |
| 2-1/2 | 1/4 | 1/2 | - | - | 1510 | 790 | 1460 | 700 | 1410 | 640 | 1400 | 610 | 1270 | 490 |
| | | 5/8 | - | - | 2190 | 880 | 2010 | 780 | 1880 | 700 | 1840 | 690 | 1580 | 550 |
| | | 3/4 | - | - | 2630 | 980 | 2410 | 860 | 2250 | 770 | 2200 | 750 | 1900 | 600 |
| | | 7/8 | - | - | 3060 | 1050 | 2820 | 920 | 2630 | 830 | 2570 | 810 | 2210 | 660 |
| | | 1 | - | - | 3500 | 1130 | 3220 | 1000 | 3000 | 900 | 2940 | 880 | 2530 | 700 |
| 3 | 1/4 | 1/2 | 1570 | 1000 | - | - | - | - | - | - | - | - | - | - |
| | | 5/8 | 2350 | 1220 | - | - | - | - | - | - | - | - | - | - |
| | | 3/4 | 3300 | 1330 | - | - | - | - | - | - | - | - | - | - |
| | | 7/8 | 4040 | 1440 | - | - | - | - | - | - | - | - | - | - |
| | | 1 | 4610 | 1530 | - | - | - | - | - | - | - | - | - | - |
| 3-1/8 | 1/4 | 1/2 | - | - | 1510 | 940 | 1460 | 880 | 1410 | 800 | 1400 | 770 | 1310 | 610 |
| | | 5/8 | - | - | 2250 | 1090 | 2170 | 980 | 2110 | 880 | 2090 | 860 | 1960 | 680 |
| | | 3/4 | - | - | 3170 | 1220 | 3020 | 1080 | 2810 | 960 | 2750 | 940 | 2370 | 750 |
| | | 7/8 | - | - | 3830 | 1310 | 3520 | 1150 | 3280 | 1040 | 3210 | 1010 | 2770 | 820 |
| | | 1 | - | - | 4380 | 1410 | 4020 | 1250 | 3750 | 1130 | 3670 | 1090 | 3160 | 880 |

| Size | Bolt dia. | C1 | C2 | C3 | C4 | C5 | C6 | C7 | C8 | C9 | C10 | C11 | C12 |
|------|-----------|------|------|------|------|------|------|------|------|------|------|------|------|
| 5 | 1/4 | 2350 | 1420 | – | – | – | – | – | – | – | – | – | – |
| | | 3300 | 1910 | – | – | – | – | – | – | – | – | – | – |
| | | 4440 | 2410 | – | – | – | – | – | – | – | – | – | – |
| | | 5750 | 2550 | – | – | – | – | – | – | – | – | – | – |
| 5-1/8 | 1/4 | – | – | 2250 | 1330 | 2170 | 1260 | 2110 | 1200 | 2090 | 1190 | 1960 | 1070 |
| | | – | – | 3170 | 1800 | 3060 | 1700 | 2960 | 1580 | 2940 | 1540 | 2750 | 1230 |
| | | – | – | 4260 | 2150 | 4100 | 1880 | 4100 | 1700 | 3940 | 1660 | 3690 | 1350 |
| | | – | – | 5520 | 2310 | 5320 | 2050 | 5150 | 1850 | 5110 | 1790 | 4770 | 1440 |
| 6-3/4 | 1/4 | 2350 | 1420 | 2250 | 1330 | 2170 | 1260 | 2110 | 1200 | 2090 | 1190 | 1960 | 1070 |
| | | 3300 | 1910 | 3170 | 1800 | 3060 | 1700 | 2960 | 1610 | 2940 | 1590 | 2750 | 1440 |
| | | 4440 | 2470 | 4260 | 2320 | 4100 | 2180 | 3980 | 2080 | 3940 | 2060 | 3690 | 1770 |
| | | 5750 | 3090 | 5520 | 2910 | 5320 | 2700 | 5150 | 2430 | 5110 | 2360 | 4770 | 1890 |
| 8-1/2 | 1/4 | 3300 | 1910 | – | – | – | – | – | – | – | – | – | – |
| | | 4440 | 2470 | – | – | – | – | – | – | – | – | – | – |
| | | 5750 | 3090 | – | – | – | – | – | – | – | – | – | – |
| 8-3/4 | 1/4 | – | – | 3170 | 1800 | 3060 | 1700 | 2960 | 1610 | 2940 | 1590 | 2750 | 1440 |
| | | – | – | 4260 | 2320 | 4100 | 2180 | 3980 | 2080 | 3940 | 2060 | 3690 | 1860 |
| | | – | – | 5520 | 2910 | 5320 | 2750 | 5150 | 2620 | 5110 | 2590 | 4770 | 2330 |
| 10-1/2 | 1/4 | 4440 | 2470 | – | – | – | – | – | – | – | – | – | – |
| | | 5750 | 3090 | – | – | – | – | – | – | – | – | – | – |
| 10-3/4 | 1/4 | – | – | 4260 | 2320 | 4100 | 2180 | 3980 | 2080 | 3940 | 2060 | 3690 | 1860 |
| | | – | – | 5520 | 2910 | 5320 | 2750 | 5150 | 2620 | 5110 | 2590 | 4770 | 2330 |
| 12-1/4 | 1/4 | – | – | 4260 | 2320 | 4100 | 2180 | 3980 | 2080 | 3940 | 2060 | 3690 | 1860 |
| | | – | – | 5520 | 2910 | 5320 | 2750 | 5150 | 2620 | 5110 | 2590 | 4770 | 2330 |
| 14-1/4 | 1/4 | – | – | 5520 | 2910 | 5320 | 2750 | 5150 | 2620 | 5110 | 2590 | 4770 | 2330 |

1. Tabulated lateral design values (Z) for bolted connections shall be multiplied by all applicable adjustment factors (see Table 7.1).

2. Tabulated lateral design values (Z) are for "full diameter" bolts (see Reference 3) with a bending yield strength ($F_{yb}$) of 45,000 psi.

3. Tabulated lateral design values (Z) are based on a dowel bearing strength ($F_e$) of 58,000 psi for ASTM A36 Steel.

Source: ANSI/NF₀PA NDS—1991, *National Design Specification for Wood Construction*, AFPA, 1250 Connecticut Ave., NW, Washington, DC 20036.

The required edge distance for full load is $1.5D = (1.5)(1) = 1.5$ in.

Another solution is to use 2 rows of bolts:.
Member size $= 6\frac{3}{4}$ in. $\times$ $8\frac{1}{4}$ in.
Required spacing between rows $= 1.5D = 1.5$ in.
Required edge distance $= 1.5D = 1.5$ in.

With the space available, spacing between rows is set at 3 in. and edge distance is set at $2\frac{5}{8}$ in.
Place the bolts in two rows. $C_g$ will increase.
For 6 bolts in 2 rows, (3 bolts per row) from Table 7.6, $C_g = 0.99$.

$$Z^* = 6840 \text{ lb (previously calculated)}$$

$$Z' = (6840)(0.99) = 6773 \text{ lb}$$

$$\text{Number of bolts required} = \frac{38,000}{6,773} = 5.6 < 6 \qquad \text{O.K.}$$

Check stress at net section of wood member.

$$A_{net} = 55.69 - (1.0625)(6.75)(2) = 41.34 \text{ in.}^2$$

From Table 2, AITC 117—Design, combination 48, $F_t = 1400$ psi:

$$F_t' = F_t C_D = (1400)(1.15) = 1610 \text{ psi}$$

$$f_t = \frac{T}{A_{net}} = \frac{38,000}{41.34} = 919 \text{ psi} < 1610 \text{ psi} \qquad \text{O.K.}$$

Check stresses in steel side plates.

$$A_{gross} = (2)(0.375)(7) = 5.25 \text{ in.}^2$$

$$A_{net} = 5.25 - (1.0625)(0.375)(2) = 4.45 \text{ in.}^2$$

$$f_t \text{ (on gross area)} = \frac{T}{A_{gross}} = \frac{38,000}{5.25} = 7238 \text{ psi} < 22,000 \text{ psi} \qquad \text{O.K.}$$

$$f_t \text{ (on net area)} = \frac{38,000}{4.45} = 8539 \text{ psi} < 29,000 \text{ psi} \qquad \text{O.K.}$$

From Table 7.9, minimum spacing parallel to grain for full load is 4 times the bolt diameter, or 4 in. The minimum end distance for tension loading is 7 in. The minimum edge distance in wood is $1\frac{1}{2}$ times the bolt diameter or $1\frac{1}{2}$ in. The minimum row spacing for the wood is $1\frac{1}{2}$ times the bolt diameter, but this is less than the recommended 3-in. row spacing for the steel used. This results in an edge distance of 2 in. for the steel side plate, which is more than the $1\frac{3}{4}$ in. minimum requirement. The edge distance in the wood is $2\frac{5}{8}$ in., which is more than the minimum of $1\frac{1}{2}$ in.

**Example.**  Determine the capacity $T$ of two 3 in. $\times \frac{1}{4}$ in. steel straps attached to a 5 in. $\times 6\frac{7}{8}$ in. glued laminated timber with one $\frac{3}{4}$ in. bolt loaded in double shear. Combination 48 Southern Pine is used. Steel straps are placed at an angle of 30° with the main member.

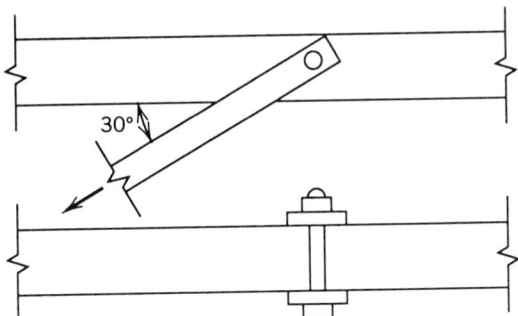

The yield mode equations (7-13), (7-15), and (7-16) will be used, and Eq. (7-12) will be used to determine $F_{e\theta}$ at 30°.

$$F_{em} = F_{e\theta} = \frac{F_{e\parallel}F_{e\perp}}{F_{e\parallel}\sin^2\theta + F_{e\perp}\cos^2\theta}$$

$$F_{e\parallel} = 6150, \quad F_{e\perp} = 2950 \ (\text{Table 7.8})$$

$$F_{e\theta} = \frac{(6150)(2950)}{6150\sin^2 30° + 2950\cos^2 30°} = 4838 \text{ psi}$$

$$F_{es} = F_e = 58,000 \text{ psi (Table 7.12)}$$

$$R_e = \frac{F_{em}}{F_{es}} = \frac{F_{e\theta}}{F_{es}} = \frac{4,838}{58,000} = 0.083$$

$$F_{yb} = 45,000 \text{ psi (Table 7.8)}$$

$$k_3 = -1 + \sqrt{\frac{2(1 + R_e)}{R_e} + \frac{2F_{yb}(2 + R_e)D^2}{3F_{em}t_s^2}}$$

$$k_3 = -1 + \sqrt{\frac{2(1 + 0.083)}{0.083} + \frac{(2)(45,000)(2 + 0.083)(0.75)^2}{(3)(4838)(0.25)^2}} = 10.931$$

$$t_m = 5.0 \text{ in.}$$

$$t_s = 0.25 \text{ in.}$$

$$D = 3/4 \text{ in.}$$

$$\theta = 30°$$

$$K_\theta = 1 + \theta_{max}/360 = 1 + 30/360 = 1.08333$$

$$\text{Mode I,} \quad Z = \frac{(0.75)(5.0)(4838)}{(4)(1.08333)} = 4187 \text{ lb}$$

$$\text{Mode III}_s, \quad Z = \frac{(10.931)(0.75)(0.25)(4,838)}{(1.6)(2 + 0.083)(1.08333)} = 2745 \text{ lb}$$

$$\text{Mode IV,} \quad Z = \frac{0.75^2}{(1.6)(1.08333)} \sqrt{\frac{(2)(4,838)(45,000)}{(3)(1 + 0.083)}} = 3756 \text{ lb}$$

Use $Z_{30°} = 2745$ lb

The value of $Z$ can be closely approximated by using the tabular design values from Table 7.16 in the Hankinson formula:

$$Z_{30°} = \frac{(3300)(1910)}{(3300) \sin^2 30° + 1910 \cos^2 30°} = 2792 \text{ lb}$$

## 7.3   LAG SCREWS

### 7.3.1   General Considerations

Lag screws are sometimes referred to in the industry as "lag bolts" because their size more closely approximates that of bolts than that of wood screws. Sizes range from $\frac{1}{4}$ in. diameter to $1\frac{1}{4}$ in. diameter and with lengths up to 12 in. For special purposes such as reinforcing pitched and tapered curved beams for radial tension, specially made lag screws may be 4 ft or even longer. However, in order for consistency of terminology in publications on wood fasteners, the term *lag screws* will be used in this manual.

The design values determined herein for lag screws apply to lag screws conforming to ANSI/ASME Standard B18.2.1—1981 (3).

Lag screws, except as noted, must be inserted in lead holes. The lead holes in the side member for the shank should be drilled $\frac{1}{32}-\frac{1}{16}$ in. oversize, the same as for bolts. The lead hole for the shank in the main member must be the same diameter as the shank and extend the same depth as the depth of penetration of the unthreaded shank. The lead hole for the threaded portion should have a diameter equal to 65–85% of the shank diameter in wood with $G > 0.60$, 60–75% in wood with $0.50 < G \le 0.60$, and 40–70% for wood with $G \le 0.50$. The larger percentages in each range should apply to the larger diameter lag screws. (See Table 7.19.)

Lead holes are not required for $\frac{3}{8}$ in. diameter and smaller lag screws used in wood with a $G \le 0.50$ provided end distance, edge distance, and spacing are such that unusual splitting does not occur.

Lag screws should be inserted in the lead hole by turning with a wrench, not by driving with a hammer. Soap or other lubricant should be used to facilitate insertion. Washers of proper size or a metal plate or strap should be installed between the wood and the bolt head.

**TABLE 7.19**

**Lead Hole Diameters for Lag Screws**

| Nominal Diameter of Lag Bolt (in.) | Shank (Unthreaded) Portion (in.) | Diameter of Lead Hole (in.) | | |
|---|---|---|---|---|
| | | Threaded Portion | | |
| | | Specific Gravity $G > 0.60$ | Specific Gravity $0.60 \geq G > 0.50$ | Specific Gravity $0.50 \geq G$ |
| $\frac{1}{4}$ | $\frac{1}{4}$ | $\frac{3}{16}$ | $\frac{5}{32}$ | $\frac{3}{32}$ |
| $\frac{5}{16}$ | $\frac{5}{16}$ | $\frac{13}{64}$ | $\frac{3}{16}$ | $\frac{9}{64}$ |
| $\frac{3}{8}$ | $\frac{3}{8}$ | $\frac{1}{4}$ | $\frac{15}{64}$ | $\frac{11}{64}$ |
| $\frac{7}{16}$ | $\frac{7}{16}$ | $\frac{19}{64}$ | $\frac{9}{32}$ | $\frac{13}{64}$ |
| $\frac{1}{2}$ | $\frac{1}{2}$ | $\frac{11}{32}$ | $\frac{5}{16}$ | $\frac{15}{64}$ |
| $\frac{9}{16}$ | $\frac{9}{16}$ | $\frac{13}{32}$ | $\frac{23}{64}$ | $\frac{9}{32}$ |
| $\frac{5}{8}$ | $\frac{5}{8}$ | $\frac{29}{64}$ | $\frac{13}{32}$ | $\frac{5}{16}$ |
| $\frac{3}{4}$ | $\frac{3}{4}$ | $\frac{9}{16}$ | $\frac{1}{2}$ | $\frac{13}{32}$ |
| $\frac{7}{8}$ | $\frac{7}{8}$ | $\frac{43}{64}$ | $\frac{39}{64}$ | $\frac{33}{64}$ |
| $1$ | $1$ | $\frac{51}{64}$ | $\frac{23}{32}$ | $\frac{5}{8}$ |
| $1\frac{1}{4}$ | $1\frac{1}{4}$ | $\frac{59}{64}$ | $\frac{53}{64}$ | $\frac{3}{4}$ |
| $1\frac{1}{4}$ | $1\frac{1}{4}$ | $1\frac{1}{16}$ | $\frac{15}{16}$ | $\frac{7}{8}$ |

### 7.3.2 Species and Specific Gravity

The strength of lag screws in lateral resistance is a function of specific gravity. Table 7.20 contains the specific gravity of the most commonly used species and the dowel bearing strengths for use in determining lateral resistance. For species not included in the table, determine the specific gravity of that species based on oven dry weight and volume and enter tables with value obtained.

### 7.3.3 Critical or Net Section

The net section requirements for lag screw joints are the same as for bolted joints with bolts of a diameter equal to the shank diameter of the lag screw used. Typical dimensions of standard lag screws are given in Table 7.21.

### 7.3.4 Angle of Load to Grain

When the angle of load to grain is between $0°$ and $90°$, the lateral design value at any angle $\theta$ is $Z_\theta$, which is determined by calculations using Eq. (7-12) shown in the section on bolts.

## TABLE 7.20

## Typical Dimensions for Standard Lag Screws

D = unthreaded shank diameter  
Dr = root diameter of threaded portion  
W = width of head across flats  
H = height of head

S = unthreaded shank length  
T = thread length[a]  
E = length of tapered tip  
N = number of threads/inch

| Nominal Length L | | Unthreaded shank diameter, D | | | | | | | | | | |
|---|---|---|---|---|---|---|---|---|---|---|---|---|
| | | 1/4" | 5/16" | 3/8" | 7/16" | 1/2" | 5/8" | 3/4" | 7/8" | 1" | 1-1/8" | 1-1/4" |
| | Dr | 0.173" | 0.227" | 0.265" | 0.328" | 0.371" | 0.471" | 0.579" | 0.683" | 0.780" | 0.887" | 1.012" |
| | E | 5/32" | 3/16" | 7/32" | 9/32" | 5/16" | 13/32" | 1/2" | 19/32" | 11/16" | 25/32" | 7/8" |
| | H | 11/64" | 7/32" | 1/4" | 19/64" | 11/32" | 27/64" | 1/2" | 37/64" | 43/64" | 3/4" | 27/32" |
| | W | 7/16" | 1/2" | 9/16" | 5/8" | 3/4" | 15/16" | 1-1/8" | 1-5/16" | 1-1/2" | 1-11/16" | 1-7/8" |
| | N | 10 | 9 | 7 | 7 | 6 | 5 | 4-1/2 | 4 | 3-1/2 | 3-1/4 | 3-1/4 |
| 1" | S | 1/4" | 1/4" | 1/4" | 1/4" | 1/4" | | | | | | |
| | T | 3/4" | 3/4" | 3/4" | 3/4" | 3/4" | | | | | | |
| | T-E | 19/32" | 9/16" | 17/32" | 15/32" | 7/16" | | | | | | |
| 1-1/2" | S | 1/4" | 1/4" | 1/4" | 1/4" | 1/4" | | | | | | |
| | T | 1-1/4" | 1-1/4" | 1-1/4" | 1-1/4" | 1-1/4" | | | | | | |
| | T-E | 1-3/32" | 1-1/16" | 1-1/32" | 31/32" | 15/16" | | | | | | |
| 2" | S | 1/2" | 1/2" | 1/2" | 1/2" | 1/2" | 1/2" | | | | | |
| | T | 1-1/2" | 1-1/2" | 1-1/2" | 1-1/2" | 1-1/2" | 1-1/2" | | | | | |
| | T-E | 1-11/32" | 1-5/16" | 1-9/32" | 1-7/32" | 1-3/16" | 1-3/32" | | | | | |
| 2-1/2" | S | 3/4" | 3/4" | 3/4" | 3/4" | 3/4" | 3/4" | | | | | |
| | T | 1-3/4" | 1-3/4" | 1-3/4" | 1-3/4" | 1-3/4" | 1-3/4" | | | | | |
| | T-E | 1-19/32" | 1-9/16" | 1-17/32" | 1-15/32" | 1-7/16" | 1-11/32" | | | | | |

| L | | Col 1 | Col 2 | Col 3 | Col 4 | Col 5 | Col 6 | Col 7 | Col 8 | Col 9 |
|---|---|---|---|---|---|---|---|---|---|---|
| 3" | S | 1" | 1" | 1" | 1" | 1" | 1" | 1" | 1" | 1" |
| | T | 2" | 2" | 2" | 2" | 2" | 2" | 2" | 2" | 2" |
| | T-E | 1-27/32" | 1-13/16" | 1-25/32" | 1-23/32" | 1-11/16" | 1-19/32" | 1-13/32" | 1-1/4" | 1-1/8" |
| 4" | S | 1-1/2" | 1-1/2" | 1-1/2" | 1-1/2" | 1-1/2" | 1-1/2" | 1-1/2" | 1-1/2" | 1-1/2" |
| | T | 2-1/2" | 2-1/2" | 2-1/2" | 2-1/2" | 2-1/2" | 2-1/2" | 2-1/2" | 2-1/2" | 2-1/2" |
| | T-E | 2-11/32" | 2-5/16" | 2-9/32" | 2-7/32" | 2-3/16" | 2-3/32" | 1-29/32" | 1-3/4" | 1-5/8" |
| 5" | S | 2" | 2" | 2" | 2" | 2" | 2" | 2" | 2" | 2" |
| | T | 3" | 3" | 3" | 3" | 3" | 3" | 3" | 3" | 3" |
| | T-E | 2-27/32" | 2-13/16" | 2-25/32" | 2-23/32" | 2-11/16" | 2-19/32" | 2-13/32" | 2-1/4" | 2-1/8" |
| 6" | S | 2-1/2" | 2-1/2" | 2-1/2" | 2-1/2" | 2-1/2" | 2-1/2" | 2-1/2" | 2-1/2" | 2-1/2" |
| | T | 3-1/2" | 3-1/2" | 3-1/2" | 3-1/2" | 3-1/2" | 3-1/2" | 3-1/2" | 3-1/2" | 3-1/2" |
| | T-E | 3-11/32" | 3-5/16" | 3-9/32" | 3-7/32" | 3-3/16" | 3-3/32" | 2-29/32" | 2-3/4" | 2-5/8" |
| 7" | S | 3" | 3" | 3" | 3" | 3" | 3" | 3" | 3" | 3" |
| | T | 4" | 4" | 4" | 4" | 4" | 4" | 4" | 4" | 4" |
| | T-E | 3-27/32" | 3-13/16" | 3-25/32" | 3-23/32" | 3-11/16" | 3-19/32" | 3-13/32" | 3-1/4" | 3-1/8" |
| 8" | S | 3-1/2" | 3-1/2" | 3-1/2" | 3-1/2" | 3-1/2" | 3-1/2" | 3-1/2" | 3-1/2" | 3-1/2" |
| | T | 4-1/2" | 4-1/2" | 4-1/2" | 4-1/2" | 4-1/2" | 4-1/2" | 4-1/2" | 4-1/2" | 4-1/2" |
| | T-E | 4-11/32" | 4-5/16" | 4-9/32" | 4-7/32" | 4-3/16" | 4-3/32" | 3-29/32" | 3-3/4" | 3-5/8" |
| 9" | S | 4" | 4" | 4" | 4" | 4" | 4" | 4" | 4" | 4" |
| | T | 5" | 5" | 5" | 5" | 5" | 5" | 5" | 5" | 5" |
| | T-E | 4-27/32" | 4-13/16" | 4-25/32" | 4-23/32" | 4-11/16" | 4-19/32" | 4-13/32" | 4-1/4" | 4-1/8" |
| 10" | S | 4-1/2" | 4-1/2" | 4-1/2" | 4-1/2" | 4-1/2" | 4-1/2" | 4-1/2" | 4-1/2" | 4-1/2" |
| | T | 5-1/2" | 5-1/2" | 5-1/2" | 5-1/2" | 5-1/2" | 5-1/2" | 5-1/2" | 5-1/2" | 5-1/2" |
| | T-E | 5-11/32" | 5-5/16" | 5-9/32" | 5-7/32" | 5-3/16" | 5-3/32" | 4-29/32" | 4-3/4" | 4-5/8" |
| 11" | S | 5" | 5" | 5" | 5" | 5" | 5" | 5" | 5" | 5" |
| | T | 6" | 6" | 6" | 6" | 6" | 6" | 6" | 6" | 6" |
| | T-E | 5-27/32" | 5-13/16" | 5-25/32" | 5-23/32" | 5-11/16" | 5-19/32" | 5-13/32" | 5-1/4" | 5-1/8" |
| 12" | S | 6" | 6" | 6" | 6" | 6" | 6" | 6" | 6" | 6" |
| | T | 6" | 6" | 6" | 6" | 6" | 6" | 6" | 6" | 6" |
| | T-E | 5-27/32" | 5-13/16" | 5-25/32" | 5-23/32" | 5-11/16" | 5-19/32" | 5-13/32" | 5-1/4" | 5-1/8" |

[a]Thread length (T) for intermediate nominal lag screw lengths (L) is 6", or $\frac{1}{2}$ the nominal lag screw length plus 0.5", whichever is less.

Source: ANSI/NF$_{o}$PA NDS—1991, *National Design Specification for Wood Construction*, AFPA, 1250 Connecticut Ave., NW, Washington, DC 20036.

## TABLE 7.21
## Dowel Bearing Strength for Lag Screw Connections

| Species Combination[a] | Specific Gravity[b] G | $F_{c\|\|}$ | Dowel bearing strength in pounds per square inch (psi) | | | | | | | | | | |
|---|---|---|---|---|---|---|---|---|---|---|---|---|---|
| | | | $F_{c\perp}$ D=1/4" | $F_{c\perp}$ D=5/16" | $F_{c\perp}$ D=3/8" | $F_{c\perp}$ D=7/16" | $F_{c\perp}$ D=1/2" | $F_{c\perp}$ D=5/8" | $F_{c\perp}$ D=3/4" | $F_{c\perp}$ D=7/8" | $F_{c\perp}$ D=1" | $F_{c\perp}$ D=1-1/8" | $F_{c\perp}$ D=1-1/4" |
| Aspen | 0.39 | 4350 | 3100 | 2800 | 2550 | 2350 | 2200 | 1950 | 1800 | 1650 | 1550 | 1450 | 1400 |
| Balsam Fir | 0.36 | 4050 | 2750 | 2500 | 2250 | 2100 | 1950 | 1750 | 1600 | 1500 | 1400 | 1300 | 1250 |
| Beech-Birch-Hickory | 0.71 | 7950 | 7400 | 6650 | 6050 | 5600 | 5250 | 4700 | 4300 | 3950 | 3700 | 3500 | 3300 |
| Coast Sitka Spruce | 0.39 | 4350 | 3100 | 2800 | 2550 | 2350 | 2200 | 1950 | 1800 | 1650 | 1550 | 1450 | 1400 |
| Cottonwood | 0.41 | 4600 | 3350 | 3000 | 2750 | 2550 | 2350 | 2100 | 1950 | 1800 | 1650 | 1600 | 1500 |
| Douglas Fir-Larch | 0.50 | 5600 | 4450 | 4000 | 3650 | 3400 | 3150 | 2800 | 2600 | 2400 | 2250 | 2100 | 2000 |
| Douglas Fir-Larch (North) | 0.49 | 5500 | 4350 | 3900 | 3550 | 3300 | 3050 | 2750 | 2500 | 2300 | 2150 | 2050 | 1950 |
| Douglas Fir-South | 0.46 | 5150 | 3950 | 3550 | 3250 | 3000 | 2800 | 2500 | 2300 | 2100 | 2000 | 1850 | 1750 |
| Eastern Hemlock | 0.41 | 4600 | 3350 | 3000 | 2750 | 2550 | 2350 | 2100 | 1950 | 1800 | 1650 | 1600 | 1500 |
| Eastern Hemlock-Tamarack | 0.41 | 4600 | 3350 | 3000 | 2750 | 2550 | 2350 | 2100 | 1950 | 1800 | 1650 | 1600 | 1500 |
| Eastern Hemlock-Tamarack (North) | 0.47 | 5250 | 4100 | 3650 | 3350 | 3100 | 2900 | 2600 | 2350 | 2200 | 2050 | 1900 | 1850 |
| Eastern Softwoods | 0.36 | 4050 | 2750 | 2500 | 2250 | 2100 | 1950 | 1750 | 1600 | 1500 | 1400 | 1300 | 1250 |
| Eastern Spruce | 0.41 | 4600 | 3350 | 3000 | 2750 | 2550 | 2350 | 2100 | 1950 | 1800 | 1650 | 1600 | 1500 |
| Eastern White Pine | 0.36 | 4050 | 2750 | 2500 | 2250 | 2100 | 1950 | 1750 | 1600 | 1500 | 1400 | 1300 | 1250 |
| Engelmann Spruce-Lodgepole Pine[c] (MSR 1650f and higher grades) | 0.46 | 5150 | 3950 | 3550 | 3250 | 3000 | 2800 | 2500 | 2300 | 2100 | 2000 | 1850 | 1750 |
| Engelmann Spruce-Lodgepole Pine[c] (MSR 1500f and lower grades) | 0.38 | 4250 | 3000 | 2700 | 2450 | 2250 | 2100 | 1900 | 1750 | 1600 | 1500 | 1400 | 1350 |
| Hem-Fir | 0.43 | 4800 | 3600 | 3200 | 2950 | 2700 | 2550 | 2250 | 2050 | 1900 | 1800 | 1700 | 1600 |
| Hem-Fir (North) | 0.46 | 5150 | 3950 | 3550 | 3250 | 3000 | 2800 | 2500 | 2300 | 2100 | 2000 | 1850 | 1750 |
| Mixed Maple | 0.55 | 6150 | 5150 | 4600 | 4200 | 3900 | 3650 | 3250 | 2950 | 2750 | 2550 | 2400 | 2300 |

| Species | Specific Gravity[a],[b] | | | | | | | | | | | |
|---|---|---|---|---|---|---|---|---|---|---|---|---|
| Mixed Oak | 0.68 | 7600 | 6950 | 6250 | 5700 | 5250 | 4950 | 4400 | 4050 | 3750 | 3500 | 3300 | 3100 |
| Mixed Southern Pine | 0.51 | 5700 | 4600 | 4100 | 3750 | 3450 | 3250 | 2900 | 2650 | 2450 | 2300 | 2150 | 2050 |
| Mountain Hemlock | 0.47 | 5250 | 4100 | 3650 | 3350 | 3100 | 2900 | 2600 | 2350 | 2200 | 2050 | 1900 | 1850 |
| Northern Pine | 0.42 | 4700 | 3450 | 3100 | 2850 | 2600 | 2450 | 2200 | 2000 | 1850 | 1750 | 1650 | 1550 |
| Northern Red Oak | 0.68 | 7600 | 6950 | 6250 | 5700 | 5250 | 4950 | 4400 | 4050 | 3750 | 3500 | 3300 | 3100 |
| Northern Species | 0.35 | 3900 | 2650 | 2400 | 2150 | 2000 | 1900 | 1700 | 1550 | 1400 | 1350 | 1250 | 1200 |
| Northern White Cedar | 0.31 | 3450 | 2250 | 2000 | 1800 | 1700 | 1600 | 1400 | 1300 | 1200 | 1100 | 1050 | 1000 |
| Ponderosa Pine | 0.43 | 4800 | 3600 | 3200 | 2950 | 2700 | 2550 | 2250 | 2050 | 1900 | 1800 | 1700 | 1600 |
| Red Maple | 0.58 | 6500 | 5550 | 4950 | 4500 | 4200 | 3900 | 3500 | 3200 | 2950 | 2750 | 2600 | 2500 |
| Red Oak | 0.67 | 7500 | 6850 | 6100 | 5550 | 5150 | 4850 | 4300 | 3950 | 3650 | 3400 | 3200 | 3050 |
| Red Pine | 0.44 | 4950 | 3700 | 3300 | 3050 | 2800 | 2600 | 2350 | 2150 | 2000 | 1850 | 1750 | 1650 |
| Redwood, close grain | 0.44 | 4950 | 3700 | 3300 | 3050 | 2800 | 2600 | 2350 | 2150 | 2000 | 1850 | 1750 | 1650 |
| Redwood, open grain | 0.37 | 4150 | 2900 | 2600 | 2350 | 2200 | 2050 | 1850 | 1650 | 1550 | 1450 | 1350 | 1300 |
| Sitka Spruce | 0.43 | 4800 | 3600 | 3200 | 2950 | 2700 | 2550 | 2250 | 2050 | 1900 | 1800 | 1700 | 1600 |
| Southern Pine | 0.55 | 6150 | 5150 | 4600 | 4200 | 3900 | 3650 | 3250 | 2950 | 2750 | 2550 | 2400 | 2300 |
| Spruce-Pine-Fir | 0.42 | 4700 | 3450 | 3100 | 2850 | 2600 | 2450 | 2200 | 2000 | 1850 | 1750 | 1650 | 1550 |
| Spruce-Pine-Fir (South) | 0.36 | 4050 | 2750 | 2500 | 2250 | 2100 | 1950 | 1750 | 1600 | 1500 | 1400 | 1300 | 1250 |
| Western Cedars | 0.36 | 4050 | 2750 | 2500 | 2250 | 2100 | 1950 | 1750 | 1600 | 1500 | 1400 | 1300 | 1250 |
| Western Cedars (North) | 0.35 | 3900 | 2650 | 2400 | 2150 | 2000 | 1900 | 1700 | 1550 | 1400 | 1350 | 1250 | 1200 |
| Western Hemlock | 0.47 | 5250 | 4100 | 3650 | 3350 | 3100 | 2900 | 2600 | 2350 | 2200 | 2050 | 1900 | 1850 |
| Western Hemlock (North) | 0.46 | 5150 | 3950 | 3550 | 3250 | 3000 | 2800 | 2500 | 2300 | 2100 | 2000 | 1850 | 1750 |
| Western White Pine | 0.40 | 4500 | 3250 | 2900 | 2650 | 2450 | 2300 | 2050 | 1850 | 1750 | 1600 | 1500 | 1450 |
| Western Woods | 0.36 | 4050 | 2750 | 2500 | 2250 | 2100 | 1950 | 1750 | 1600 | 1500 | 1400 | 1300 | 1250 |
| White Oak | 0.73 | 8200 | 7750 | 6900 | 6300 | 5850 | 5450 | 4900 | 4450 | 4150 | 3850 | 3650 | 3450 |
| Yellow Poplar | 0.43 | 4800 | 3600 | 3200 | 2950 | 2700 | 2550 | 2250 | 2050 | 1900 | 1800 | 1700 | 1600 |

[a]Alaska cedar (AC) has a specific gravity of 0.46.

[b]Specific gravity based on weight and volume when oven-dry.

[c]Applies only to Engelmann Spruce-Lodgepole Pine Machine Stress Rated (MSR) structural lumber.

Source: ANSI/NF₀PA NDS—1991, *National Design Specification for Wood Construction*, AFPA, 1250 Connecticut Ave., NW, Washington, DC 20036.

### 7.3.5   Spacing, End Distance, and Edge Distance

Spacing, end distance, and edge distance requirements for lag screws are the same as those for bolts of a diameter equal to the shank diameter of the lag screws used.

### 7.3.6   Lag Screw Design Values

Lag screw design values may be determined by using tabular values from Tables 7.22 and 7.23 for lateral loads and Tables 7.24 and 7.25 for withdrawal loads. The design values for lateral loads can also be determined by use of Eqs. (7-17)–(7-19).

**TABLE 7.22**

**Lag Screw Design Values (Z) for Single Shear (Two Member) Connections[1,2] with Both Members of Identical Species**

| SIDE MEMBER THICKNESS $t_s$ inches | LAG SCREW DIAMETER $D$ inches | G=0.67 RED OAK $Z_{\parallel}$ lbs. | $Z_{s\perp}$ lbs. | $Z_{m\perp}$ lbs. | G=0.55 MIXED MAPLE SOUTHERN PINE $Z_{\parallel}$ lbs. | $Z_{s\perp}$ lbs. | $Z_{m\perp}$ lbs. | G=0.50 DOUGLAS FIR-LARCH $Z_{\parallel}$ lbs. | $Z_{s\perp}$ lbs. | $Z_{m\perp}$ lbs. | G=0.49 DOUGLAS FIR-LARCH (N) $Z_{\parallel}$ lbs. | $Z_{s\perp}$ lbs. | $Z_{m\perp}$ lbs. | G=0.46 DOUGLAS FIR (S) HEM-FIR (N) $Z_{\parallel}$ lbs. | $Z_{s\perp}$ lbs. | $Z_{m\perp}$ lbs. |
|---|---|---|---|---|---|---|---|---|---|---|---|---|---|---|---|---|
| 1/2 | 1/4 | 190 | 150 | 150 | 170 | 130 | 130 | 160 | 110 | 120 | 160 | 110 | 120 | 150 | 100 | 120 |
| | 5/16 | 270 | 190 | 210 | 240 | 140 | 180 | 220 | 130 | 170 | 210 | 120 | 170 | 200 | 110 | 160 |
| | 3/8 | 340 | 210 | 250 | 290 | 160 | 220 | 260 | 140 | 200 | 260 | 130 | 200 | 240 | 120 | 190 |
| 5/8 | 1/4 | 210 | 160 | 160 | 180 | 130 | 140 | 170 | 120 | 130 | 170 | 120 | 130 | 160 | 120 | 120 |
| | 5/16 | 290 | 210 | 220 | 250 | 180 | 190 | 240 | 160 | 180 | 240 | 150 | 170 | 230 | 140 | 170 |
| | 3/8 | 350 | 250 | 260 | 310 | 200 | 230 | 290 | 170 | 210 | 290 | 170 | 210 | 280 | 150 | 200 |
| 3/4 | 1/4 | 230 | 170 | 180 | 200 | 140 | 150 | 180 | 130 | 140 | 180 | 130 | 140 | 170 | 120 | 130 |
| | 5/16 | 310 | 220 | 240 | 270 | 190 | 200 | 250 | 170 | 190 | 250 | 170 | 180 | 240 | 160 | 170 |
| | 3/8 | 380 | 260 | 280 | 330 | 220 | 240 | 310 | 210 | 220 | 310 | 200 | 220 | 290 | 180 | 210 |
| 1 | 1/4 | 260 | 200 | 200 | 230 | 160 | 180 | 210 | 150 | 160 | 210 | 150 | 160 | 200 | 140 | 150 |
| | 5/16 | 360 | 250 | 270 | 300 | 210 | 230 | 280 | 190 | 210 | 280 | 190 | 210 | 270 | 170 | 200 |
| | 3/8 | 430 | 290 | 330 | 370 | 240 | 270 | 350 | 220 | 250 | 340 | 220 | 250 | 330 | 210 | 240 |
| 1-1/4 | 1/4 | 260 | 200 | 200 | 230 | 180 | 180 | 220 | 170 | 170 | 220 | 170 | 170 | 210 | 150 | 160 |
| | 5/16 | 370 | 280 | 280 | 340 | 230 | 250 | 320 | 210 | 240 | 320 | 210 | 230 | 300 | 190 | 220 |
| | 3/8 | 470 | 330 | 340 | 420 | 270 | 300 | 390 | 240 | 290 | 390 | 240 | 280 | 370 | 220 | 270 |
| 1-1/2 | 1/4 | 260 | 200 | 200 | 230 | 180 | 180 | 220 | 170 | 170 | 220 | 170 | 170 | 210 | 160 | 160 |
| | 5/16 | 370 | 280 | 280 | 340 | 250 | 250 | 320 | 230 | 240 | 320 | 230 | 230 | 310 | 210 | 220 |
| | 3/8 | 470 | 340 | 340 | 420 | 300 | 300 | 400 | 270 | 290 | 400 | 260 | 280 | 390 | 240 | 270 |
| | 7/16 | 630 | 430 | 460 | 570 | 350 | 400 | 540 | 320 | 380 | 540 | 310 | 380 | 510 | 290 | 360 |
| | 1/2 | 830 | 510 | 590 | 710 | 420 | 500 | 660 | 380 | 460 | 650 | 370 | 450 | 610 | 350 | 430 |
| | 5/8 | 1140 | 670 | 800 | 980 | 570 | 680 | 920 | 530 | 620 | 910 | 520 | 620 | 870 | 470 | 580 |
| | 3/4 | 1510 | 890 | 1040 | 1320 | 660 | 880 | 1240 | 590 | 820 | 1220 | 560 | 800 | 1170 | 520 | 760 |
| | 7/8 | 1940 | 960 | 1300 | 1710 | 720 | 1110 | 1620 | 630 | 1040 | 1600 | 600 | 1020 | 1540 | 550 | 970 |
| | 1 | 2450 | 1020 | 1600 | 2170 | 770 | 1370 | 2060 | 680 | 1290 | 2040 | 650 | 1260 | 1930 | 600 | 1210 |
| | 1-1/8 | 3030 | 1080 | 1930 | 2590 | 810 | 1670 | 2360 | 710 | 1560 | 2320 | 690 | 1540 | 2170 | 620 | 1460 |
| | 1-1/4 | 3520 | 1140 | 2300 | 2880 | 860 | 2000 | 2630 | 750 | 1860 | 2580 | 730 | 1840 | 2410 | 660 | 1750 |
| 2-1/2 | 1/4 | 260 | 200 | 200 | 230 | 180 | 180 | 220 | 170 | 170 | 220 | 170 | 170 | 210 | 160 | 160 |
| | 5/16 | 370 | 280 | 280 | 340 | 250 | 250 | 320 | 240 | 240 | 320 | 230 | 230 | 310 | 220 | 220 |
| | 3/8 | 470 | 340 | 340 | 420 | 300 | 300 | 400 | 290 | 290 | 400 | 280 | 280 | 390 | 270 | 270 |
| | 7/16 | 630 | 460 | 460 | 570 | 400 | 400 | 550 | 380 | 380 | 540 | 380 | 380 | 520 | 360 | 360 |
| | 1/2 | 830 | 590 | 590 | 750 | 520 | 520 | 710 | 480 | 480 | 710 | 480 | 480 | 690 | 450 | 460 |
| | 5/8 | 1290 | 860 | 880 | 1170 | 700 | 780 | 1120 | 630 | 730 | 1110 | 620 | 720 | 1070 | 580 | 690 |
| | 3/4 | 1860 | 1060 | 1240 | 1680 | 870 | 1090 | 1570 | 800 | 1020 | 1550 | 780 | 1010 | 1470 | 740 | 970 |
| | 7/8 | 2460 | 1290 | 1640 | 2100 | 1080 | 1400 | 1950 | 1000 | 1280 | 1930 | 970 | 1260 | 1830 | 920 | 1180 |
| | 1 | 2970 | 1550 | 1980 | 2560 | 1280 | 1660 | 2390 | 1130 | 1530 | 2360 | 1080 | 1500 | 2250 | 1000 | 1430 |
| | 1-1/8 | 3550 | 1800 | 2320 | 3080 | 1350 | 1950 | 2880 | 1180 | 1800 | 2850 | 1150 | 1770 | 2720 | 1040 | 1670 |
| | 1-1/4 | 4190 | 1910 | 2680 | 3650 | 1440 | 2270 | 3430 | 1250 | 2100 | 3390 | 1220 | 2070 | 3250 | 1090 | 1950 |

**TABLE 7.22** *(Continued)*

| SIDE MEMBER THICKNESS $t_s$ inches | LAG SCREW DIAMETER D inches | G=0.43 HEM-FIR | | | G=0.42 SPRUCE-PINE-FIR | | | G=0.37 REDWOOD (open grain) | | | G=0.36 EASTERN SOFTWOODS SPRUCE-PINE-FIR(S) WESTERN CEDARS WESTERN WOODS | | | G=0.35 NORTHERN SPECIES | | |
|---|---|---|---|---|---|---|---|---|---|---|---|---|---|---|---|---|
| | | $Z_{||}$ lbs. | $Z_{s\perp}$ lbs. | $Z_{m\perp}$ lbs. | $Z_{||}$ lbs. | $Z_{s\perp}$ lbs. | $Z_{m\perp}$ lbs. | $Z_{||}$ lbs. | $Z_{s\perp}$ lbs. | $Z_{m\perp}$ lbs. | $Z_{||}$ lbs. | $Z_{s\perp}$ lbs. | $Z_{m\perp}$ lbs. | $Z_{||}$ lbs. | $Z_{s\perp}$ lbs. | $Z_{m\perp}$ lbs. |
| 1/2 | 1/4 | 150 | 90 | 110 | 150 | 90 | 110 | 130 | 70 | 100 | 130 | 70 | 100 | 120 | 70 | 100 |
| | 5/16 | 190 | 100 | 150 | 180 | 100 | 150 | 160 | 80 | 130 | 160 | 80 | 130 | 150 | 80 | 120 |
| | 3/8 | 230 | 110 | 180 | 220 | 110 | 180 | 190 | 90 | 160 | 190 | 80 | 150 | 180 | 80 | 150 |
| 5/8 | 1/4 | 160 | 110 | 120 | 150 | 110 | 110 | 140 | 90 | 100 | 140 | 90 | 100 | 140 | 80 | 100 |
| | 5/16 | 220 | 130 | 160 | 210 | 120 | 160 | 200 | 100 | 140 | 200 | 100 | 140 | 190 | 90 | 140 |
| | 3/8 | 270 | 140 | 190 | 270 | 130 | 190 | 240 | 110 | 170 | 240 | 110 | 170 | 230 | 100 | 160 |
| 3/4 | 1/4 | 160 | 110 | 120 | 160 | 110 | 120 | 150 | 100 | 110 | 150 | 100 | 110 | 140 | 100 | 100 |
| | 5/16 | 230 | 150 | 170 | 220 | 150 | 160 | 210 | 120 | 150 | 200 | 120 | 150 | 200 | 110 | 140 |
| | 3/8 | 280 | 170 | 200 | 280 | 160 | 200 | 260 | 130 | 180 | 250 | 130 | 170 | 250 | 120 | 170 |
| 1 | 1/4 | 190 | 130 | 140 | 190 | 120 | 140 | 170 | 110 | 120 | 170 | 110 | 120 | 160 | 100 | 120 |
| | 5/16 | 250 | 160 | 190 | 250 | 160 | 180 | 230 | 140 | 160 | 220 | 140 | 160 | 220 | 140 | 160 |
| | 3/8 | 310 | 190 | 220 | 300 | 190 | 220 | 280 | 170 | 200 | 270 | 170 | 190 | 270 | 160 | 190 |
| 1-1/4 | 1/4 | 210 | 140 | 150 | 200 | 140 | 150 | 190 | 120 | 140 | 190 | 120 | 140 | 180 | 110 | 130 |
| | 5/16 | 280 | 180 | 210 | 280 | 170 | 210 | 250 | 160 | 180 | 250 | 150 | 180 | 240 | 150 | 170 |
| | 3/8 | 350 | 210 | 250 | 340 | 200 | 250 | 310 | 180 | 220 | 300 | 180 | 210 | 300 | 170 | 210 |
| 1-1/2 | 1/4 | 210 | 150 | 150 | 200 | 150 | 150 | 190 | 140 | 140 | 190 | 130 | 140 | 190 | 130 | 130 |
| | 5/16 | 300 | 200 | 210 | 300 | 190 | 210 | 280 | 170 | 190 | 270 | 160 | 190 | 270 | 160 | 190 |
| | 3/8 | 370 | 230 | 260 | 370 | 220 | 260 | 340 | 190 | 240 | 340 | 190 | 230 | 330 | 180 | 230 |
| | 7/16 | 480 | 270 | 340 | 470 | 270 | 330 | 430 | 240 | 300 | 420 | 230 | 290 | 410 | 230 | 280 |
| | 1/2 | 580 | 330 | 400 | 570 | 320 | 400 | 520 | 290 | 350 | 510 | 290 | 350 | 500 | 280 | 340 |
| | 5/8 | 820 | 420 | 550 | 810 | 410 | 540 | 750 | 350 | 490 | 730 | 330 | 480 | 720 | 320 | 470 |
| | 3/4 | 1120 | 460 | 720 | 1100 | 450 | 710 | 1020 | 370 | 640 | 1010 | 360 | 630 | 980 | 350 | 620 |
| | 7/8 | 1470 | 500 | 920 | 1450 | 490 | 900 | 1350 | 410 | 820 | 1330 | 390 | 810 | 1280 | 370 | 780 |
| | 1 | 1800 | 540 | 1150 | 1760 | 530 | 1130 | 1560 | 440 | 1030 | 1520 | 420 | 1010 | 1460 | 410 | 990 |
| | 1-1/8 | 2030 | 570 | 1400 | 1980 | 560 | 1380 | 1750 | 460 | 1250 | 1710 | 440 | 1230 | 1650 | 420 | 1200 |
| | 1-1/4 | 2250 | 600 | 1670 | 2200 | 580 | 1650 | 1950 | 490 | 1510 | 1900 | 470 | 1480 | 1830 | 450 | 1450 |
| 2-1/2 | 1/4 | 210 | 150 | 150 | 200 | 150 | 150 | 190 | 140 | 140 | 190 | 140 | 140 | 190 | 130 | 130 |
| | 5/16 | 300 | 210 | 210 | 300 | 210 | 210 | 280 | 190 | 190 | 270 | 190 | 190 | 270 | 190 | 190 |
| | 3/8 | 370 | 260 | 260 | 370 | 260 | 260 | 350 | 240 | 240 | 340 | 230 | 230 | 340 | 230 | 230 |
| | 7/16 | 510 | 340 | 340 | 500 | 340 | 340 | 470 | 310 | 310 | 470 | 300 | 310 | 460 | 290 | 300 |
| | 1/2 | 660 | 420 | 440 | 650 | 410 | 430 | 620 | 360 | 400 | 610 | 340 | 390 | 600 | 340 | 390 |
| | 5/8 | 1030 | 540 | 660 | 1020 | 530 | 650 | 960 | 470 | 600 | 940 | 460 | 590 | 910 | 450 | 580 |
| | 3/4 | 1390 | 690 | 910 | 1360 | 680 | 890 | 1240 | 610 | 800 | 1220 | 600 | 780 | 1180 | 580 | 760 |
| | 7/8 | 1740 | 830 | 1110 | 1710 | 810 | 1090 | 1560 | 680 | 980 | 1540 | 660 | 960 | 1500 | 610 | 930 |
| | 1 | 2140 | 900 | 1340 | 2110 | 880 | 1320 | 1940 | 730 | 1180 | 1910 | 700 | 1160 | 1860 | 680 | 1130 |
| | 1-1/8 | 2600 | 960 | 1590 | 2560 | 930 | 1560 | 2360 | 760 | 1400 | 2330 | 730 | 1370 | 2270 | 700 | 1340 |
| | 1-1/4 | 3110 | 1000 | 1850 | 3070 | 970 | 1820 | 2840 | 810 | 1650 | 2800 | 780 | 1620 | 2740 | 750 | 1580 |

1. Tabulated lateral design values (Z) for lag screw connections shall be multiplied by all applicable factors (Table 7.1).

2. Tabulated lateral design values (Z) are for "full diameter" lag screws (see Reference 3) inserted in side grain with lag screw axis perpendicular to wood fibers, and with the following lag screw bending yield strengths ($F_{yb}$):

$F_{yb} = 70,000$ psi for $D = \frac{1}{4}''$

$F_{yb} = 60,000$ psi for $D = \frac{5}{16}''$

$F_{yb} = 45,000$ psi for $D \geq \frac{3}{8}''$

Source: ANSI/NF<sub>o</sub>PA NDS—1991, *National Design Specification for Wood Construction*, AFPA, 1250 Connecticut Ave., NW, Washington, DC 20036.

# TABLE 7.23

## Lag Screw Design Values (Z) for Single Shear (two member) Connections[a,b,c] with $\frac{1}{4}''$ ASTM A36 steel side plate, or ASTM A446, Grade A steel side plate (for $t_s < \frac{1}{4}''$)

| Steel Side Plate $t_s$ inches | Lag Screw Diameter $D$ inches | G = 0.67 Red Oak $Z_\parallel$ lbs. | $Z_\perp$ lbs. | G = 0.55 Mixed Maple Southern Pine $Z_\parallel$ lbs. | $Z_\perp$ lbs. | G = 0.50 Douglas Fir-Larch $Z_\parallel$ lbs. | $Z_\perp$ lbs. | G = 0.49 Douglas Fir-Larch (N) $Z_\parallel$ lbs. | $Z_\perp$ lbs. | G = 0.46 Douglas Fir (S) Hem-Fir (N) $Z_\parallel$ lbs. | $Z_\perp$ lbs. |
|---|---|---|---|---|---|---|---|---|---|---|---|
| $\frac{1}{4}''$ | $\frac{1}{4}$ | 330 | 260 | 310 | 230 | 300 | 220 | 300 | 210 | 290 | 200 |
| | $\frac{5}{16}$ | 450 | 330 | 410 | 290 | 400 | 280 | 400 | 270 | 390 | 260 |
| | $\frac{3}{8}$ | 550 | 390 | 510 | 350 | 490 | 330 | 480 | 320 | 470 | 310 |
| | $\frac{7}{16}$ | 700 | 480 | 650 | 430 | 620 | 400 | 620 | 400 | 600 | 380 |
| | $\frac{1}{2}$ | 870 | 580 | 810 | 520 | 780 | 490 | 770 | 480 | 750 | 460 |
| | $\frac{5}{8}$ | 1290 | 820 | 1190 | 720 | 1140 | 680 | 1140 | 670 | 1100 | 640 |
| | $\frac{3}{4}$ | 1810 | 1100 | 1660 | 960 | 1600 | 910 | 1580 | 890 | 1540 | 860 |
| | $\frac{7}{8}$ | 2420 | 1410 | 2220 | 1240 | 2130 | 1170 | 2120 | 1140 | 2060 | 1100 |
| | 1 | 3130 | 1760 | 2870 | 1540 | 2750 | 1460 | 2730 | 1430 | 2650 | 1380 |
| | $1\frac{1}{8}$ | 3930 | 2150 | 3610 | 1880 | 3460 | 1770 | 3430 | 1750 | 3340 | 1670 |
| | $1\frac{1}{4}$ | 4840 | 2580 | 4440 | 2260 | 4260 | 2120 | 4220 | 2090 | 4100 | 1990 |
| 3 gage $t_s = 0.239''$ | $\frac{1}{4}$ | 300 | 230 | 270 | 210 | 260 | 190 | 260 | 190 | 260 | 180 |
| | $\frac{5}{16}$ | 400 | 300 | 370 | 270 | 360 | 250 | 360 | 250 | 350 | 240 |
| | $\frac{3}{8}$ | 500 | 350 | 460 | 320 | 440 | 300 | 440 | 290 | 430 | 280 |
| 7 gage $t_s = 0.179''$ | $\frac{1}{4}$ | 270 | 210 | 250 | 180 | 240 | 170 | 240 | 170 | 230 | 160 |
| | $\frac{5}{16}$ | 370 | 270 | 340 | 240 | 330 | 230 | 330 | 230 | 320 | 220 |
| | $\frac{3}{8}$ | 460 | 320 | 420 | 290 | 410 | 270 | 400 | 270 | 390 | 260 |
| 10 gage $t_s = 0.134''$ | $\frac{1}{4}$ | 250 | 190 | 230 | 170 | 220 | 160 | 220 | 160 | 210 | 150 |
| | $\frac{5}{16}$ | 350 | 260 | 330 | 230 | 310 | 220 | 310 | 210 | 300 | 210 |
| | $\frac{3}{8}$ | 440 | 310 | 400 | 280 | 390 | 260 | 390 | 260 | 380 | 250 |
| 11 gage $t_s = 0.12''$ | $\frac{1}{4}$ | 250 | 190 | 230 | 170 | 220 | 160 | 220 | 160 | 210 | 150 |
| | $\frac{5}{16}$ | 350 | 260 | 320 | 230 | 310 | 210 | 310 | 210 | 300 | 200 |
| | $\frac{3}{8}$ | 430 | 310 | 400 | 270 | 380 | 260 | 380 | 250 | 370 | 240 |
| 12 gage $t_s = 0.105''$ | $\frac{1}{4}$ | 240 | 190 | 220 | 170 | 210 | 160 | 210 | 150 | 210 | 150 |
| | $\frac{5}{16}$ | 350 | 250 | 320 | 220 | 310 | 210 | 300 | 210 | 300 | 200 |
| | $\frac{3}{8}$ | 430 | 300 | 400 | 270 | 380 | 250 | 380 | 250 | 370 | 240 |
| 14 gage $t_s = 0.075''$ | $\frac{1}{4}$ | 240 | 180 | 220 | 160 | 210 | 150 | 210 | 150 | 200 | 140 |

## TABLE 7.23 (*Continued*)

| Steel Side Plate $t_s$ inches | Lag Screw Diameter $D$ inches | G = 0.43 Hem-Fir $Z_\parallel$ lbs. | $Z_\perp$ lbs. | G = 0.42 Spruce-Pine-Fir $Z_\parallel$ lbs. | $Z_\perp$ lbs. | G = 0.37 Redwood (open grain) $Z_\parallel$ lbs. | $Z_\perp$ lbs. | G = 0.36 Eastern Softwoods Spruce-Pine-Fir(s) Western Cedars Western Woods $Z_\parallel$ lbs. | $Z_\perp$ lbs. | G = 0.35 Northern Species $Z_\parallel$ lbs. | $Z_\perp$ lbs. |
|---|---|---|---|---|---|---|---|---|---|---|---|
| $\frac{1}{4}''$ | $\frac{1}{4}$ | 280 | 200 | 280 | 190 | 260 | 180 | 260 | 170 | 250 | 170 |
| | $\frac{5}{16}$ | 370 | 250 | 370 | 250 | 350 | 230 | 350 | 230 | 340 | 220 |
| | $\frac{3}{8}$ | 460 | 300 | 450 | 290 | 430 | 270 | 430 | 270 | 420 | 260 |
| | $\frac{7}{16}$ | 580 | 370 | 580 | 360 | 550 | 330 | 540 | 330 | 540 | 320 |
| | $\frac{1}{2}$ | 730 | 440 | 720 | 430 | 680 | 400 | 680 | 390 | 670 | 390 |
| | $\frac{5}{8}$ | 1070 | 610 | 1060 | 610 | 1010 | 560 | 1000 | 550 | 980 | 540 |
| | $\frac{3}{4}$ | 1490 | 820 | 1480 | 810 | 1400 | 740 | 1390 | 730 | 1360 | 720 |
| | $\frac{7}{8}$ | 1990 | 1050 | 1980 | 1030 | 1870 | 950 | 1850 | 940 | 1820 | 910 |
| | 1 | 2570 | 1310 | 2550 | 1300 | 2410 | 1190 | 2380 | 1170 | 2340 | 1150 |
| | $1\text{-}\frac{1}{8}$ | 3230 | 1600 | 3200 | 1580 | 3030 | 1430 | 2990 | 1410 | 2940 | 1380 |
| | $1\text{-}\frac{1}{4}$ | 3970 | 1910 | 3930 | 1880 | 3720 | 1730 | 3680 | 1690 | 3610 | 1660 |
| 3 gage $t_s$ = 0.239" | $\frac{1}{4}$ | 250 | 180 | 250 | 170 | 240 | 160 | 230 | 160 | 230 | 160 |
| | $\frac{5}{16}$ | 340 | 230 | 340 | 230 | 320 | 210 | 320 | 210 | 310 | 200 |
| | $\frac{3}{8}$ | 420 | 270 | 410 | 270 | 390 | 250 | 390 | 240 | 380 | 240 |
| 7 gage $t_s$ = 0.179" | $\frac{1}{4}$ | 220 | 160 | 220 | 160 | 210 | 140 | 210 | 140 | 200 | 140 |
| | $\frac{5}{16}$ | 310 | 210 | 310 | 210 | 290 | 190 | 290 | 190 | 280 | 180 |
| | $\frac{3}{8}$ | 380 | 250 | 380 | 250 | 360 | 230 | 360 | 220 | 350 | 220 |
| 10 gage $t_s$ = 0.134" | $\frac{1}{4}$ | 210 | 150 | 210 | 140 | 200 | 130 | 190 | 130 | 190 | 130 |
| | $\frac{5}{16}$ | 290 | 200 | 290 | 190 | 280 | 180 | 270 | 180 | 270 | 170 |
| | $\frac{3}{8}$ | 360 | 240 | 360 | 230 | 340 | 210 | 340 | 210 | 330 | 200 |
| 11 gage $t_s$ = 0.12" | $\frac{1}{4}$ | 200 | 140 | 200 | 140 | 190 | 130 | 190 | 130 | 190 | 130 |
| | $\frac{5}{16}$ | 290 | 190 | 290 | 190 | 270 | 180 | 270 | 170 | 260 | 170 |
| | $\frac{3}{8}$ | 360 | 230 | 360 | 230 | 340 | 210 | 330 | 210 | 330 | 200 |
| 12 gage $t_s$ = 0.105" | $\frac{1}{4}$ | 200 | 140 | 200 | 140 | 190 | 130 | 190 | 130 | 180 | 120 |
| | $\frac{5}{16}$ | 290 | 190 | 280 | 190 | 270 | 170 | 270 | 170 | 260 | 170 |
| | $\frac{3}{8}$ | 360 | 230 | 350 | 230 | 330 | 210 | 330 | 200 | 320 | 200 |
| 14 gage $t_s$ = 0.075" | $\frac{1}{4}$ | 200 | 140 | 190 | 140 | 180 | 120 | 180 | 120 | 180 | 120 |

Source: ANSI/NF₀PA NDS—1991, *National Design Specification for Wood Construction*, AFPA, 1250 Connecticut Ave., NW, Washington, DC 20036.

[a]Tabulated lateral design values (Z) for lag screw connections shall be multiplied by all applicable adjustment factors (Table 7.1).

[b]Tabulated lateral design values (Z) are for "full diameter" lag screws (see Reference 3) inserted in side grain with lag screw axis perpendicular to wood fibers, and with the following lag screw bending yield strengths ($F_{yb}$):

$F_{yb}$ = 70,000 psi for $D = \frac{1}{4}''$
$F_{yb}$ = 60,000 psi for $D = \frac{5}{16}''$
$F_{yb}$ = 45,000 psi for $D \geq \frac{3}{8}''$

[c]Tabulated lateral design values (Z) are based on dowel bearing strengths ($F_e$) of 58,000 psi for ASTM A36 steel, and 45,000 psi for ASTM A446, Grade A steel.

## TABLE 7.24

### Lag Screw Withdrawal Design Values $(W)^1$

Tabulated withdrawal design values (W) are in pounds per inch of thread penetration into side grain of main member. Length of thread penetration in main member shall not include the length of the tapered tip (see Table 7.20).

| Specific Gravity G | Lag Screw Unthreaded Shank Diameter, D | | | | | | | | | | |
|---|---|---|---|---|---|---|---|---|---|---|---|
| | 1/4" | 5/16" | 3/8" | 7/16" | 1/2" | 5/8" | 3/4" | 7/8" | 1" | 1-1/8" | 1-1/4" |
| 0.73 | 397 | 469 | 538 | 604 | 668 | 789 | 905 | 1016 | 1123 | 1226 | 1327 |
| 0.71 | 381 | 450 | 516 | 579 | 640 | 757 | 868 | 974 | 1077 | 1176 | 1273 |
| 0.68 | 357 | 422 | 484 | 543 | 600 | 709 | 813 | 913 | 1009 | 1103 | 1193 |
| 0.67 | 349 | 413 | 473 | 531 | 587 | 694 | 796 | 893 | 987 | 1078 | 1167 |
| 0.58 | 281 | 332 | 381 | 428 | 473 | 559 | 641 | 719 | 795 | 869 | 940 |
| 0.55 | 260 | 307 | 352 | 395 | 437 | 516 | 592 | 664 | 734 | 802 | 868 |
| 0.51 | 232 | 274 | 314 | 353 | 390 | 461 | 528 | 593 | 656 | 716 | 775 |
| 0.50 | 225 | 266 | 305 | 342 | 378 | 447 | 513 | 576 | 636 | 695 | 752 |
| 0.49 | 218 | 258 | 296 | 332 | 367 | 434 | 498 | 559 | 617 | 674 | 730 |
| 0.47 | 205 | 242 | 278 | 312 | 345 | 408 | 467 | 525 | 580 | 634 | 686 |
| 0.46 | 199 | 235 | 269 | 302 | 334 | 395 | 453 | 508 | 562 | 613 | 664 |
| 0.44 | 186 | 220 | 252 | 283 | 312 | 369 | 423 | 475 | 525 | 574 | 621 |
| 0.43 | 179 | 212 | 243 | 273 | 302 | 357 | 409 | 459 | 508 | 554 | 600 |
| 0.42 | 173 | 205 | 235 | 264 | 291 | 344 | 395 | 443 | 490 | 535 | 579 |
| 0.41 | 167 | 198 | 226 | 254 | 281 | 332 | 381 | 428 | 473 | 516 | 559 |
| 0.40 | 161 | 190 | 218 | 245 | 271 | 320 | 367 | 412 | 455 | 497 | 538 |
| 0.39 | 155 | 183 | 210 | 236 | 261 | 308 | 353 | 397 | 438 | 479 | 518 |
| 0.38 | 149 | 176 | 202 | 227 | 251 | 296 | 340 | 381 | 422 | 461 | 498 |
| 0.37 | 143 | 169 | 194 | 218 | 241 | 285 | 326 | 367 | 405 | 443 | 479 |
| 0.36 | 137 | 163 | 186 | 209 | 231 | 273 | 313 | 352 | 389 | 425 | 460 |
| 0.35 | 132 | 156 | 179 | 200 | 222 | 262 | 300 | 337 | 373 | 407 | 441 |
| 0.31 | 110 | 130 | 149 | 167 | 185 | 218 | 250 | 281 | 311 | 339 | 367 |

1. Tabulated withdrawal design values (W) for lag screw connections shall be multiplied by all applicable adjustment factors (see Table 7.1).

Source: ANSI/NF₀PA NDS—1991, *National Design Specification for Wood Construction*, AFPA, 1250 Connecticut Ave., NW, Washington, DC 20036.

Tabular design values in Tables 7.22 and 7.23 apply for one lag screw in a two-member joint under normal duration of loading and dry-use conditions. The total design value for more than one lag screw is the sum of the values for each lag screw, provided that spacings, end distances, and edge distances are sufficient to develop the full strength of each lag screw and that the values are modified for group action in accordance with the recommendations given in 7.1.3.4. For other than normal durations of loading, the design values should be multiplied by the appropriate adjustment factor for duration of load $C_D$ from Fig. 4.4 or Table 4.7. The full design values may be used for lag screws installed in wood that will remain dry in service. For other conditions of service, the appropriate adjustment factor $C_M$ from Table 7.2 should be applied. See Table 7.1 for applicability of adjustment factors for lag screws.

## TABLE 7.25

### Withdrawal Design Values for Lag Screws in Softwood Glued Laminated Timber[a]

| Species[b] | Growth Rate[b] | Lag Bolt Diameter $D$[c] | | | | | | | | | | | |
|---|---|---|---|---|---|---|---|---|---|---|---|---|---|
| | | $\frac{1}{4}$, 0.250 | $\frac{5}{16}$, 0.3125 | $\frac{3}{8}$, 0.375 | $\frac{7}{16}$, 0.4375 | $\frac{1}{2}$, 0.500 | $\frac{9}{16}$, 0.5625 | $\frac{5}{8}$, 0.625 | $\frac{3}{4}$, 0.750 | $\frac{7}{8}$, 0.875 | 1, 1.000 | $1\frac{1}{8}$, 1.125 | $1\frac{1}{4}$, 1.250 |
| Southern Pine | Dense Medium Grain | 260 | 307 | 352 | 395 | 437 | 477 | 516 | 592 | 664 | 734 | 802 | 868 |
| | | 260 | 307 | 352 | 395 | 437 | 477 | 516 | 592 | 664 | 734 | 802 | 868 |
| | Coarse Grain | 167 | 198 | 226 | 254 | 281 | 307 | 332 | 381 | 428 | 473 | 516 | 559 |
| Douglas Fir-Larch | | 225 | 266 | 305 | 342 | 378 | 413 | 447 | 513 | 576 | 636 | 695 | 752 |
| Hem-Fir | | 179 | 212 | 243 | 273 | 302 | 330 | 357 | 409 | 459 | 508 | 554 | 600 |
| Softwood Species | | 137 | 163 | 186 | 209 | 231 | 253 | 273 | 313 | 352 | 389 | 425 | 460 |
| California Redwood | Close Grain | 186 | 220 | 252 | 283 | 312 | 340 | 369 | 423 | 475 | 525 | 574 | 621 |
| Douglas Fir South | | 199 | 235 | 269 | 302 | 334 | 364 | 395 | 453 | 508 | 562 | 613 | 664 |

[a] Normal load duration, dry service conditions. Design values for load in withdrawal in lb/in. of penetration of threaded part into side grain of the laminations receiving the point.

[b] Refer to Table 3, AITC 117—Design (4) for species, rate of growth, and location of lumber used in various combinations. For other species refer to Table 7.24 for withdrawal design values based on specific gravity obtained from Table 7.21.

[c] $D$ = shank diameter.

### 7.3.7  Withdrawal Loads

Table 7.24 may be used to determine the withdrawal design values for lag screws installed in the side grain of a member consisting of either sawn lumber or glued laminated timber. This table is based on specific gravity. Table 7.25 contains withdrawal values for softwood glued laminated timbers based on species used for laminating. If possible, lag screws should not be loaded in withdrawal from end grain. When this condition is unavoidable, the design value as determined from either table is multiplied by 0.75.

The withdrawal design value may not exceed the tensile strength of the lag screw at its net (root) section. A depth of engagement of the lag screw thread in its lead hole of 7 diameters for species with specific gravities of $G > 0.60$, 8 diameters for species with specific gravities of $0.60 \geq G > 0.50$, 10 diameters for species with specific gravities of $0.50 \geq G > 0.40$, and 11 diameters for species with specific gravities of $0.40 \leq G$ where withdrawal load $W$ is in lb per inch based on the depth of the embedment of the threaded portion excluding the tapered gimlet point. The required depth of penetration to develop the allowable tension stress of the lag screw is shown in Table 7.26.

### 7.3.8  Lateral Loads

The design values for the most commonly used connections of lag screws can be obtained from Table 7.22 for wood to wood connections and from Table 7.23 for wood to steel connections.

**TABLE 7.26**

**Depth of Penetration ($p$) of Lag Screw Required to Develop Allowable Tension Stress in Lag Screw, in.[a,b]**

| | Depth of Penetration $p$, in. | | | | | | |
|---|---|---|---|---|---|---|---|
| Diameter in. | Southern Pine | Douglas Fir-Larch | Hem-Fir | Spruce-Pine-Fir | Western Woods | White Oak | Red Oak |
| 1 | 13.0 | 15.0 | 18.8 | 19.5 | 24.6 | 8.5 | 9.7 |
| $\frac{7}{8}$ | 11.0 | 12.7 | 16.0 | 16.5 | 20.8 | 7.2 | 8.2 |
| $\frac{3}{4}$ | 8.9 | 10.3 | 12.9 | 13.3 | 16.8 | 5.8 | 6.6 |
| $\frac{5}{8}$ | 6.8 | 7.8 | 9.8 | 10.1 | 12.8 | 4.4 | 5.0 |
| $\frac{1}{2}$ | 5.0 | 5.7 | 7.2 | 7.4 | 9.4 | 3.2 | 3.7 |
| $\frac{7}{16}$ | 4.3 | 4.9 | 6.2 | 6.4 | 8.1 | 2.8 | 3.2 |
| $\frac{3}{8}$ | 3.1 | 3.6 | 4.5 | 4.7 | 5.9 | 2.1 | 2.3 |
| $\frac{5}{16}$ | 2.6 | 3.0 | 3.8 | 4.0 | 5.0 | 1.7 | 2.0 |
| $\frac{1}{4}$ | 1.8 | 2.1 | 2.6 | 2.7 | 3.4 | 1.2 | 1.4 |

[a]Based on tensile stress in steel of 20,000 psi and diameter of lag screw at the root of the threads $D_r$.

[b]The depth of penetration $p$ does not include the Gimlet Point.

Table 7.22 contains the design values for parallel to grain loading, $Z_\parallel$ for the wood main member with a wood side member. It also contains the design values for loading perpendicular to grain in both the main member $Z_{m\perp}$ and the side member $Z_{s\perp}$. These lateral design values are based on one lag screw installed in side grain of a two member joint comprised of seasoned wood with load applied either parallel or perpendicular to grain. Member thickness and shank and thread penetration in the main member have been taken into consideration in preparing Tables 7.22 and 7.23.

If metal side plates thicker than $\frac{1}{2}$ in. are used, the design values obtained from Table 7.23 should be reduced in proportion to the lesser penetration of the lag screw. The allowable stresses for the metal side members should not be exceeded.

When it is more convenient to use electronic methods of calculation for determining lag screw design values or when the connection is not included in the tables, the lag screw design values can be determined by calculations using Eqs. (7-17)–(7-19).

Design values for loading at an angle of load to grain other than $0°$ or $90°$ cannot be determined by use of the tables and must be determined by calculation using Eq. (7-12).

When the load acts perpendicular to grain and the lag screw is inserted in the end grain of the main member parallel to the wood fibers, the design value must be determined by use of Eqs. (7-17)–(7-19) with $F_{em} = F_{c\perp}$, and the design value thus obtained must be multiplied by the end grain factor $C_{eg} = 0.67$.

$$Z = \frac{Dt_s F_{es}}{4K_\theta} \qquad \text{Mode I}_s \qquad (7\text{-}17)$$

$$Z = \frac{kDt_s F_{em}}{2.8(2 + R_e)K_\theta} \qquad \text{Mode III}_s \qquad (7\text{-}18)$$

$$Z = \frac{D^2}{3K_\theta}\sqrt{\frac{1.75F_{em}F_{yb}}{3(1 + R_e)}} \qquad \text{Mode IV} \qquad (7\text{-}19)$$

where   $k = -1 + \sqrt{\left(\frac{2(1 + R_e)}{R_e} + \frac{F_{yb}(2 + R_e)D^2}{2F_{em}t_s^2}\right)}$,

$R_e = F_{em}/F_{es}$,
$t_s$ = thickness of side member (in.),
$F_{em}$ = dowel bearing strength of main member (member holding point) (psi) (see Table 7.21),
$F_{es}$ = dowel bearing strength of side member (psi) (see Table 7.21),
$F_{yb}$ = bending yield strength of lag screw (psi),
$D$ = unthreaded shank diameter of lag screw (in.)
$K_\theta = 1 + (\theta_{max}/360°)$ (degrees), and
$\theta_{max}$ = maximum angle of load to grain $(0° \le \theta \le 90°)$ for any member in a connection (degrees).

When a member is loaded at an angle to grain, the dowel bearing strength, $F_{e\theta}$, for the member shall be determined from Eq. (7-12) as follows:

$$F_{e\theta} = \frac{F_{e\parallel}F_{e\perp}}{F_{e\parallel}\sin^2\theta + F_{e\perp}\cos^2\theta}$$

where   $\theta$ = angle between direction of load and direction of grain (longitudinal axis of member) (degrees).

The dowel bearing strength for a member loaded at an angle to grain between $0°$ and $90°$, $F_{e\theta}$, is substituted for $F_{es}$ or $F_{em}$ in the preceding equations, depending on which member is loaded at an angle to grain. For example, if the main member (member holding the point) is loaded at an angle to grain, $F_{e\theta}$ is substituted for $F_{em}$.

The nominal lag screw design values for lateral loads are based on a penetration of the point in the main member of $8D$. The minimum lag screw penetration in the main member is $4D$ where $D$ is the shank diameter, $p_{min} = 4D$. When the penetration depth is equal to or greater than $4D$ but less than $8D$, the design value must be multiplied by a depth of penetration factor $C_d$.

$$C_d = \frac{p}{8D} \leq 1 \qquad (7\text{-}20)$$

where   $p$ = the depth of penetration (in.) and
   $D$ = shank diameter (in.).

**Example.**   Lag screws in withdrawal: Determine the capacity of two $\frac{3}{4}$ in. $\times$ 8 in. lag screws in withdrawal from the top of a 5 in. $\times$ 15 in. deep glued laminated beam made of combination 24F–V3 Southern Pine. The member is preservatively treated and is used in a wet location. The load is a wind load.

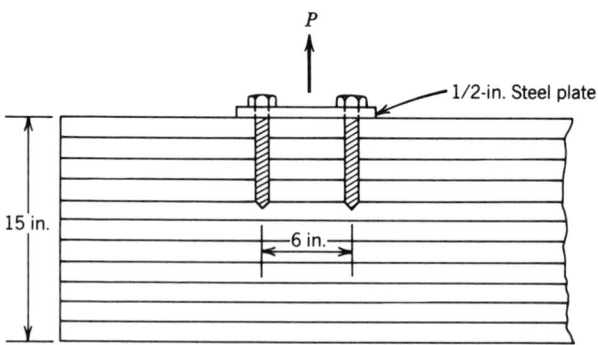

From Table 4.7, $C_D = 1.6$; from Table 7.2, $C_M = 0.67$; from Table 3, AITC 117—Design (4), the core laminations are medium grain Southern Pine for laminations receiving the point. From Table 7.25, the design value of the threaded portion, $W = 592$ lb/in. of penetration. From Table 7.20, the length of the threaded portion of the lag screw $T - E = 4\frac{1}{16}$ in.

$$Wp = W(T - E)C_D C_M = (592)(4.0625)(1.6)(0.67) = 2580 \text{ lb/lag screw}$$

Check tensile strength of lag screw based on area of root diameter. From Table 7.20, $D_r = 0.579$ in.

$$A_n = \frac{\pi D_r^2}{4} = \frac{\pi(0.579)^2}{4} = 0.263 \text{ in.}^2$$

Capacity of lag screw based on steel strength $= A_n F_t = (0.263)(20,000) = 5260$ lb $> 2580$ lb, use 2580 lb.

$$P = 2Wp = (2)(2580) = 5160 \text{ lb}$$

**Example.** Lag screws for lateral wind load: Determine the lateral load that two $\frac{3}{4}$ in. $\times$ 6 in. lag screws can carry when used to attach a $\frac{1}{4}$ in. $\times$ 3 in. strap to a $6 \times 8$ in. Douglas Fir sawn timber as shown: No. 1 Beam and Stringer of Douglas Fir-Larch. The timber is fabricated unseasoned (wet) and will be wet when full load is applied. From Table 7.2, $C_M$ (wet at time of fabrication and wet in service) $= 0.67$. From Table 7.23, $Z_{\|} = 1600$ lb for Douglas Fir-Larch. End distance for full load is

$$7D = (7)(3/4) = 5.25 \text{ in.}$$

Adjustment factor for end distance is

$$C_{\Delta n} = \frac{5.00}{5.25} = 0.952$$

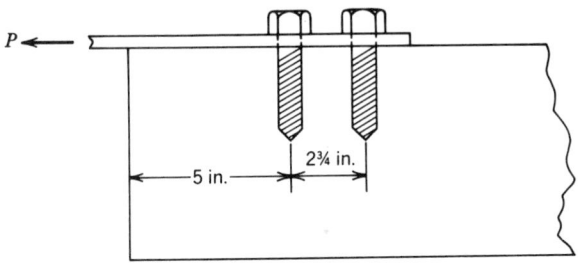

Spacing for full load is

$$4D = (4)(3/4) = 3 \text{ in.}$$

Adjustment factor for spacing is

$$C_{\Delta s} = \frac{2.75}{3} = 0.917 < 0.952$$

Edge distance for full load $= 1.5D = 1\frac{1}{8}$ in.

$$2.75 > 1\frac{1}{8} \text{ in.}, \quad C_{\Delta e} = 1$$

$C_{\Delta s}$ controls ($C_{\Delta n}$ and $C_{\Delta s}$ are not cumulative).
Adjustment factor for depth of penetration is $C_d = 5.1875/6 = 0.864$
Lateral load $= P = 2Z_\| C_D C_M C_{\Delta s} C_d = (2)(1600)(1.6)(0.67)(0.917)(0.864) = 2720$
lb.

### 7.3.9  Combined Lateral and Withdrawal Loads

When a lag screw is subjected to combined lateral and withdrawal loads such as occurs when the lag screw is inserted perpendicular to the wood fiber and the load acts at an angle between $0°$ and $90°$ with the surface, the design value should be determined by use of the following equation:

$$Z'_\alpha = \frac{(W'p)Z'_\|}{Z' \sin^2\alpha + W'p \cos^2\alpha} \tag{7-21}$$

where  $\alpha$ = angle between wood surface and the direction of the load (degrees),
  $p$ = length of thread penetration in the main member (in.),
  $Z'_\|$ = lateral design value for lag screw connections multiplied by appropriate adjustment factors (lb), and
  $W'$ = design value in withdrawal obtained from Table 7.24 and multiplied by appropriate adjustment factors (lb/in.).

The value obtained by this adaptation of the Hankinson formula is conservative.

**Example.**   Determine the load $P_\alpha$ that a $\frac{1}{2}$ in. $\times$ 5 in. lag screw in a $5\frac{1}{8}$ in. $\times$ $6\frac{7}{8}$ in. glued laminated timber can carry. Load is applied through a $\frac{1}{4}$-in. thick metal side plate, which is in the same plane that contains the longitudinal axis and is at an angle of $30°$ with the surface. The glued laminated timber is made with combination 47 (all No. 2 medium grain Southern Pine) which is preservatively treated. The load is primarily a snow load and the fastening is used in a wet location.

From Table 4.7, $C_D = 1.15$. From Table 7.2, $C_M = 0.67$. End distance is greater than $7D$ in both directions, $C_{\Delta n} = 1$.

$$\text{Edge distance} = \frac{5.125}{2} = 2.56 \text{ in.},$$

Perpendicular to grain, edge distance required

$$4D = 4(0.5) = 2.0 \text{ in.} < 2.56 \text{ in.} \quad \text{O.K.}$$

$$C_{\Delta e} = 1$$

From Table 7.21, Southern Pine has a specific gravity of 0.55. From Table 7.23, $Z_\| = 810$ lb; from Table 7.25, $W = 437$ lb/in.; from Table 7.20, $p = T - E = 2\frac{11}{16}$ in.

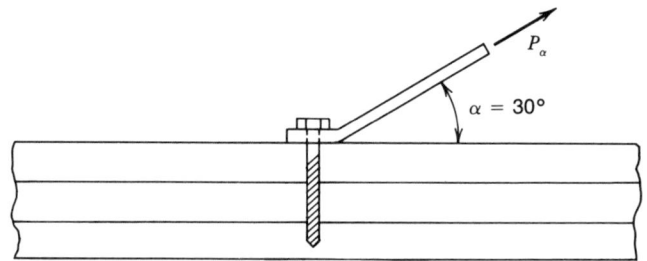

$Wp = 437(2.6875) = 1174$ lb

Depth of penetration in main member $= L - t_s - E = 5 - \frac{1}{4} - \frac{5}{16} = 4\frac{7}{16}$ in.

$8D = (8)(0.5) = 4$ in. $< 4\frac{7}{16}$ in., $C_d = 1$

$Z_\parallel' = Z_\parallel C_D C_M = (810)(1.15)(0.67) = 624$ lb

$W'p = Wp C_D C_M = (1174)(1.15)(0.67) = 905$ lb

Use Eq. (7-20).

$$Z_\alpha' = \frac{(W'p)(Z')}{Z' \sin^2\alpha + W'p \cos^2\alpha} = \frac{(905)(624)}{(624) \sin^2 30° + (905) \cos^2 30°} = 677 \text{ lb}$$

If the direction of the applied load in the above example had not been in the same plane as the longitudinal axis of the piece, the lateral resistance at the angle of load to grain, $Z_\theta$, would need to be determined. For example, if the load $P_\alpha$ is in a plane that is 45° to the longitudinal axis of the piece, the following calculations apply.

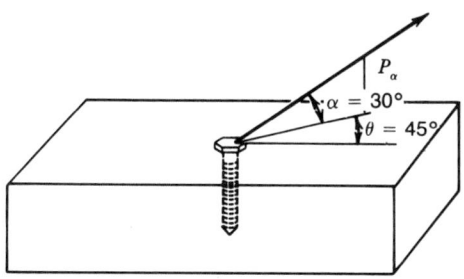

$$\text{Loaded edge distance} = \frac{6.875}{2} = 3.44 \text{ in.} > 4D \qquad \text{O.K.}$$

First, determine the dowel bearing stress at an angle of load to grain of 45°. From Table 7.21 for $\frac{1}{2}$ in. lag screw in Southern Pine,

$$F_{e\parallel} = 6150 \text{ lb}, \quad F_{e\perp} = 3650 \text{ lb}$$

$F_{es}$ for steel side plate ASTM A36 steel $= 58,000$ psi

Substitute $F_{e\theta}$ for $F_{em}$ in Eqs. (7-18) and (7-19).

$$\theta_{max} = 45°$$

$$F_{e\theta} = \frac{F_{e\|}F_{e\perp}}{F_{e\|} \sin^2\theta + F_{e\perp} \cos^2\theta} = \frac{(6150)(3650)}{(6150)(\sin^2 45°) + (3650)(\cos^2 45°)} = 4581 \text{ lb}$$

Use the lowest value of $Z$ obtained from Eqs. (7-17)–(7-19).
Values used in equations:

$$F_{em} = F_{e\theta} = 4581 \text{ psi}$$
$$F_{es} = 58,000 \text{ psi (ASTM A36 steel)}$$
$$R_e = F_{em}/F_{es} = F_{e\theta}/F_{es} = \frac{4581}{58,000} = 0.079$$

$$t_s = \tfrac{1}{4} \text{ in.} = 0.25 \text{ in.}$$
$$F_{yb} = 45,000 \text{ psi}$$
$$D = \tfrac{1}{2} \text{ in.} = 0.5 \text{ in.}$$

$$K_\theta = 1 + \frac{45°}{360°} = 1.125$$

$$k = -1 + \sqrt{\frac{2(1 + R_e)}{R_e} + \frac{F_{yb}(2 + R_e)D^2}{2F_{e\theta}t_s^2}}$$

$$k = -1 + \sqrt{\frac{2(1 + 0.079)}{0.079} + \frac{(45,000)(2 + 0.079)(0.5)^2}{(2)(4581)(0.25)^2}} = 7.256$$

Note that Mode $I_s$ is not applicable to steel side plates.

Yield Mode III$_s$    $Z_{45°} = \dfrac{kDt_sF_{em}}{(2.8)(2 + R_e)K_\theta} = \dfrac{(7.256)(0.5)(0.25)(4581)}{(2.8)(2 + 0.079)(1.125)}$

$$= 634 \text{ lb}$$

Yield Mode IV    $Z_{45°} = \dfrac{D^2}{3K_\theta}\sqrt{\dfrac{1.75F_{em}F_{yb}}{3(1 + R_e)}}$

$$Z_{45°} = \frac{0.5^2}{(3)(1.125)}\sqrt{\frac{(1.75)(4581)(45,000)}{(3)(1 + 0.079)}} = 782 \text{ lb}$$

$$Z'_{45°} = Z_{45°}C_DC_M = (634)(1.15)(0.67) = 488 \text{ lb}$$

$$Z'_\alpha = \frac{Z'_{45°}W'p}{Z'_{45°} \sin^2\alpha + W'p \cos^2\alpha}$$

$$Z'_\alpha = \frac{(488)(905)}{(488)\sin^2 30° + (905)\cos^2 30°} = 551 \text{ lb}$$

## 7.4   TIMBER CONNECTORS—SHEAR PLATES AND SPLIT RINGS

The design values given in Table 7.32 are for two shear plates used back-to-back in the contact faces of a wood-to-wood joint with their bolt in single shear or for one shear plate with its bolt in single shear used in conjunction with a steel strap or shape in a wood-to-metal joint. The design values in Table 7.32 are adjusted by the metal side plate adjustment factors given in Table 7.31 when steel side plates are used. Table 7.33 includes design values for shear plates with the adjustment factor included for metal side plates and no further adjustment is needed. Split rings and shear plates are illustrated in Fig. 7.18. The design values given in Table 7.34 are for one split ring with its bolt in single shear. Projected areas of connectors and bolts for use in determining net sections are given in Table 7.28. Typical dimensions for timber connectors are given in Table 7.29.

When installing timber connectors and bolts, a nut must be placed on each bolt. Washers, not smaller than the size given in Table 7.30, must be placed between the wood member and the bolt head and between the wood member and the nut. When a steel strap or shape is used in conjunction with shear plates, the washer may be omitted, except when desirable to extend the bolt length to prevent the metal from bearing on the threaded portion of the bolt.

### 7.4.1   Lumber Species

The density of the wood affects the allowable loads for timber connectors. The species groupings for connectors in Table 7.27 are based on density and related factors.

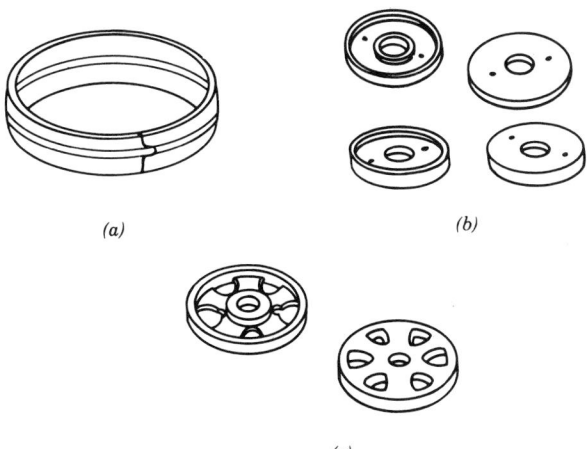

*(a)*          *(b)*

*(c)*

**FIGURE 7.18**   Typical timber connectors. (*a*) Split ring; (*b*) pressed steel shear plates; (*c*) malleable iron shear plates.

Source: TECO/Lumberlok, *Design Manual*, copyright, TECO, TECO/Lumberlok.

**TABLE 7.27**

**Species Groups for Split Ring and Shear Plate Connectors**

| Group A | Group B | Group C | Group D[a] |
|---------|---------|---------|---------|
| Beech-Birch-Hickory | Douglas Fir-Larch | Douglas Fir-South | Aspen |
| Douglas Fir-Larch | Douglas Fir-Larch | Eastern Hemlock- | Balsam Fir |
| (Dense) | (North) | Tamarack (North) | |
| | | Hem-Fir | Coast Sitka Spruce |
| Mixed Oak | Mixed Maple | Hem-Fir (North) | Cottonwood |
| Northern Red Oak | Mixed Southern Pine | Mountain Hemlock | Eastern Hemlock |
| Red Oak | Red Maple | Northern Pine | Eastern Hemlock- |
| Southern Pine | Southern Pine[b] | Ponderosa Pine | Tamarack |
| (Dense) | | Red Pine | Eastern Softwoods |
| White Oak | | Redwood (close | Eastern Spruce |
| | | grain) | Eastern White Pine |
| | | Sitka Spruce | Northern Species |
| | | Spruce-Pine-Fir | Northern White Cedar |
| | | Western Hemlock | Redwood (open grain) |
| | | Western Hemlock | Spruce-Pine-Fir (South) |
| | | (North) | Western Cedars |
| | | Yellow Poplar | Western Cedars (North) |
| | | | Western White Pine |
| | | | Western Woods |

Source: ANSI/NF₀PA NDS—1991, *National Design Specification for Wood Construction*, AFPA, 1250 Connecticut Ave., NW, Washington, DC 20036.

[a]Alaska Cedar is in Group D.

[b]Coarse grain Southern Pine, as used in some glued laminated timber combinations, is in Group C.

### 7.4.2  Connector Design Values

The connector design values in Tables 7.32–7.34 are for one connector unit installed in a seasoned timber joint under normal duration of loading and dry-use conditions. For connectors, lumber is considered to be seasoned if the moisture content is no higher than 19% for a depth of $\frac{3}{4}$ in. These design values must be adjusted if end distances, edge distances, or spacings are between the minimums shown and those required for maximum design value. The reductions due to reduced end distances, $C_{\Delta n}$, edge distances, $C_{\Delta e}$, and spacings, $C_{\Delta s}$, are not cumulative and the lowest value controls. See Table 7.35 for geometry factors. However, any reduction due to moisture content factor $C_M$ is cumulative with other adjustment factors. The adjustment factor for connectors in a row, $C_g$, is also cumulative with other adjustment factors.

The tabular design values in Table 7.33 are based on the capacity of the shear plate in the wood. Values marked with an asterisk (*) may need to be reduced as indicated in footnote *b* when limited by the capacity of the steel. The higher values are included for use in calculating reduced spacing, edge, and end distances for the loads that are limited by the steel.

TABLE 7.28

## TABLE 7.28
### Total Projected Area of Connectors and Bolts (in.$^2$) For Use in Determining Net Sections

| Connector Type | Size (in.) | Bolt Diameter (in.) | Placement of Connectors | Member Thickness (in.) | | | | | | | | | |
|---|---|---|---|---|---|---|---|---|---|---|---|---|---|
| | | | | $1\frac{1}{2}$ | $2\frac{1}{2}$ | $3\frac{1}{8}$ | $3\frac{1}{2}$ | $5\frac{5}{8}$ | $5\frac{1}{2}$ | $6\frac{3}{4}$ | $8\frac{3}{4}$ | $10\frac{3}{4}$ | $12\frac{1}{4}$ |
| **Split Rings** | | | | | | | | | | | | | |
| 1 | $2\frac{1}{2}$ | $\frac{1}{2}$ | 1 face | 1.73 | 2.29 | 2.65 | 2.86 | 3.77 | 3.98 | 4.69 | 5.81 | 6.94 | 7.78 |
| | | $\frac{1}{2}$ | 2 faces | 2.62 | 3.18 | 3.54 | 3.75 | 4.66 | 4.87 | 5.58 | 6.70 | 7.82 | 8.67 |
| 2 | 4 | $\frac{3}{4}$ | 1 face | 3.05 | 3.86 | 4.37 | 4.68 | 6.00 | 6.30 | 7.32 | 8.94 | 10.56 | 11.79 |
| | | $\frac{3}{4}$ | 2 faces | 4.88 | 5.69 | 6.20 | 6.51 | 7.83 | 8.13 | 9.15 | 10.77 | 12.40 | 13.62 |
| **Shear Plates** | | | | | | | | | | | | | |
| 1 | $2\frac{5}{8}$ | $\frac{3}{4}$ | 1 face | 2.03 | 2.85 | 3.35 | 3.66 | 4.98 | 5.28 | 6.30 | 7.92 | 9.55 | 10.77 |
| | | $\frac{3}{4}$ | 2 faces | 2.84 | 3.66 | 4.16 | 4.47 | 5.79 | 6.09 | 7.11 | 8.73 | 10.36 | 11.58 |
| 1-LG[a] | $2\frac{5}{8}$ | $\frac{3}{4}$ | 1 face | 1.91 | 2.72 | 3.23 | 3.53 | 4.85 | 5.16 | 6.18 | 7.80 | 9.43 | 10.64 |
| | | $\frac{3}{4}$ | 2 faces | 2.60 | 3.41 | 3.92 | 4.22 | 5.54 | 5.85 | 6.87 | 8.49 | 10.11 | 11.33 |
| 2 | 4 | $\frac{3}{4}$ | 1 face | 3.26 | 4.07 | 4.58 | 4.89 | 6.21 | 6.51 | 7.53 | 9.15 | 10.78 | 12.00 |
| | | $\frac{3}{4}$ | 2 faces | — | 6.11 | 6.62 | 6.93 | 8.25 | 8.55 | 9.57 | 11.19 | 12.82 | 14.04 |
| 2-A | 4 | $\frac{7}{8}$ | 1 face | 3.37 | 4.30 | 4.89 | 5.24 | 6.77 | 7.12 | 8.29 | 10.16 | 12.04 | 13.45 |
| | | $\frac{7}{8}$ | 2 faces | — | 6.26 | 6.85 | 7.20 | 8.73 | 9.08 | 10.25 | 12.12 | 14.00 | 15.41 |

[a]Light gage.
Source: TECO/Lumberlok, *Design Manual*, copyright, TECO/Lumberlok.

## TABLE 7.29

### Dimensions and Specifications for Connectors[a]

Split Rings[b]

| Size | $2\frac{1}{2}$ in. | 4 in. |
|---|---|---|
| **Split ring** | | |
| Inside diameter at center when closed | 2.500 | 4.000 |
| Thickness of metal at center | 0.163 | 0.193 |
| Depth of metal (width of ring) | 0.750 | 1.000 |
| **Groove** | | |
| Inside diameter | 2.56 | 4.08 |
| Width | 0.18 | 0.21 |
| Depth | 0.375 | 0.50 |
| **Bolt hole** | | |
| Diameter | $\frac{9}{16}$ | $\frac{13}{16}$ |
| **Washers, standard** | | |
| Round, cast or malleable iron, diameter | $2\frac{5}{8}$ | 3 |
| Round, wrought iron (minimum): | | |
|   Diameter | $1\frac{3}{8}$ | 2 |
|   Thickness | $\frac{3}{32}$ | $\frac{5}{32}$ |
| Square plate | | |
|   Length of side | 2 | 3 |
|   Thickness | $\frac{1}{8}$ | $\frac{3}{16}$ |
| **Projected area** | | |
| Portion of one ring within member (in.$^2$) | 1.10 | 2.24 |

Shear Plates[b, c]

| Size | $2\frac{5}{8}$ in. | $2\frac{5}{8}$ in. | 4 in. | 4 in. |
|---|---|---|---|---|
| Shear plate material | Pressed steel | Light gage | Malleable iron | Malleable iron |
| Diameter of plate | 2.62 | 2.62 | 4.03 | 4.03 |
| Diameter of bolt hole | 0.81 | 0.81 | 0.81 | 0.94 |
| Thickness of plate | 0.172 | 0.12 | 0.20 | 0.20 |
| Depth of plate | 0.42 | 0.35 | 0.64 | 0.64 |
| Hole diameter in straps or shapes for bolts | $\frac{13}{16}$ | $\frac{13}{16}$ | $\frac{13}{16}$ | $\frac{15}{16}$ |
| Bolt hole—diameter in timber | $\frac{13}{16}$ | $\frac{13}{16}$ | $\frac{13}{16}$ | $\frac{15}{16}$ |
| Washers, standard | | | | |
|   Round, cast or malleable iron, diameter | 3 | 3 | 3 | $3\frac{1}{2}$ |
|   Round, wrought iron, minimum | | | | |
|     Diameter | 2 | 2 | 2 | $2\frac{1}{4}$ |
|     Thickness | $\frac{5}{32}$ | $\frac{5}{32}$ | $\frac{5}{32}$ | $\frac{11}{64}$ |
|   Square plate | | | | |
|     Length of side | 3 | 3 | 3 | 3 |
|     Thickness | $\frac{1}{4}$ | $\frac{1}{4}$ | $\frac{1}{4}$ | $\frac{1}{4}$ |
| Projected area | | | | |
|   Portion of one shear plate within member (in.$^2$) | 1.18 | 1.00 | 2.58 | 2.58 |

[a] Source: TECO/Lumberlok, *Design Manual*, copyright, TECO/Lumberlok.
[b] Dimensions in inches.
[c] Steel straps or shapes, for use with shear plates, shall be designed in accordance with accepted engineering practices.

It is recommended that, wherever possible, a connector joint be designed with edge distance, end distance, and spacing for maximum design value as shown in Table 7.36 or Figs. 7.20 and 7.21. If space is not available, these dimensions can be reduced provided that the reduced design values of the connectors are capable of carrying the design load. Conversely, if the allowable load is reduced because of a reduced end distance, edge distance, or spacing, the other distances or spacings may be reduced to those resulting in the same allowable load. In determining the total load capacity of a joint, however, the design value for each individual connector is calculated, and the total load that can be carried by the joint is the lowest connector value obtained multiplied by the number of connectors.

The required end distance and spacing requirements may be reduced below those required for 100% of the tabulated load only when the full design load on the joint is reduced in proportion to the reduction in end distance or spacing. Conversely, if the end distance of one connector or the spacing of a pair of connectors of a group of connectors in a joint is less than that required for 100% of full load, the design load of each connector in the group shall not exceed that of the connector with the reduced end distance or spacing.

The design values of shear plates are based on the use of standard size bolts with cut threads. Bolts with rolled threads may have a smaller diameter resulting in a loose fit. See Ref. (3) for information on bolts.

### 7.4.3   Adjustment Factors for Timber Connectors

Adjustment factors for timber connector design values for duration of load are given in Fig. 4.4 or Table 4.7. Adjustment factors for various service and seasoning conditions are given in Table 7.2. A summary of adjustment factors applicable to shear plates and split rings is contained in Table 7.1.

Lag screws may be used instead of bolts in connector units, provided that they have the same shank diameter as the bolt specified for the connector and they meet all other provisions for lag screws as given herein.

When lag screws are used with timber connectors, the design values from Tables 7.32–7.34 are multiplied by the applicable adjustment factor $C_d$ from Table 7.30. Penetration of the lag screw into the member receiving its point should not be less than the minimum shown in Table 7.30. The diameter of the hole for the unthreaded shank must be the same as the diameter of the shank. The diameter of the lead hole for the threaded portion of the lag screw shall be approximately 70% of the shank diameter or as specified in Table 7.19.

The tabular design values given in Table 7.32 for shear plates may be multiplied by the metal plate adjustment factor $C_{st}$ shown in Table 7.31.

For convenience of the designer, Table 7.33 is included which has the metal plate adjustment factor $C_{st}$ included in the tabular values. Do not apply the adjustment factor $C_{st}$ to Table 7.33.

### 7.4.4   Critical or Net Section

The critical section of a timber connector joint will probably be at the centerline of the bolt and connector. The net area of this section is equal to the full cross-

**TABLE 7.30**

**Penetration Depth Factors $C_d$, for Split Ring and Shear Plate Connectors Used with Lag Screws**

| | Side Member | Penetration | Penetration of Lag Screw into Main Member (number of shank diameters) Species Group (see Table 7.27) | | | | Penetration Depth Factor $C_d$ |
|---|---|---|---|---|---|---|---|
| | | | Group A | Group B | Group C | Group D | |
| 2-1/2" Split Ring 4" Split Ring or 4" Shear Plate | Wood or Metal | Minimum for Full Design Value | 7 | 8 | 10 | 11 | 1.0 |
| | | Minimum for Reduced Design Value | 3 | 3-1/2 | 4 | 4-1/2 | 0.75 |
| 2-5/8" Shear Plate | Wood | Minimum for Full Design Value | 4 | 5 | 7 | 8 | 1.0 |
| | | Minimum for Reduced Design Value | 3 | 3-1/2 | 4 | 4-1/2 | 0.75 |
| | Metal | Minimum for Full Design Value | 3 | 3-1/2 | 4 | 4-1/2 | 1.0 |

Source: ANSI/NF₀PA NDS—1991, *National Design Specification for Wood Construction*, AFPA, 1250 Connecticut Ave., NW, Washington, DC 20036.

**TABLE 7.31**

**Metal Side Plate Factors, $C_{st}$, for 4 in. Shear Plate
Connectors Loaded Parallel to Grain**

| Species Grouping | $C_{st}$ |
|:---:|:---:|
| A | 1.18 |
| B | 1.11 |
| C | 1.05 |
| D | 1.00 |

Source: ANSI/NF₀PA NDS—1991, *National Design Specification for Wood Construction*, AFPA, 1250 Connecticut Ave., NW, Washington, DC 20036.

sectional area of the timber less the projected area of the connectors within the member and the projected bolt area (see Table 7.28 for projected areas). When connectors are staggered, adjacent connectors with parallel-to-grain spacing equal to or less than one connector diameter are considered to occur at the same critical section. In Fig. 7.3, section *A–B–C–D* is the critical section when connectors *B* and *C* have a parallel-to-grain spacing equal to or less than one connector diameter. For greater parallel-to-grain spacings, section *A–A* is the critical section.

### 7.4.5    Glued Laminated Timber

Knots occurring at, or near, the critical section are disregarded in determining net section. The cross-sectional area, in square inches, required at the critical section is determined by dividing the total load, in pounds, transferred through the critical section of the member by the design value in tension, $F_t'$, for tension members and the design value in compression, $F_c'$, for compression members.

### 7.4.6    Sawn Lumber

The net cross-sectional area, in square inches, required at the critical section is determined by dividing the total load, in pounds, transferred through the critical section of the member by the design values in tension, $F_t'$, for tension members and the design value in compression, $F_c'$, for compression members. Values for $F_t$ and $F_c$ are contained in Tables 8.3–8.6. These must be adjusted by the appropriate adjustment factors to obtain $F_t'$ and $F_c'$. Conversely, the total load capacity in pounds may be determined by multiplying the net area in square inches by the appropriate design value.

When analyzing an existing structure, the required net area may be determined by dividing total load, in pounds, transferred through the critical section by the appropriate design value in Tables 8.3–8.6, provided no knots occur in the critical section. This recommendation is based on the assumption that the area of the connector and bolt hole will be deducted from the net section. This procedure is not applicable to glued laminated timber.

When the fabrication of connections in sawn lumber containing connectors is subject to supervision of the designer, the design value $F_g$ from Table 8.7 may be used at the net section for both compression and tension members, provided the area of the knots in the cross section as well as the projected area of the connector and bolt is subtracted from the gross area to obtain the net section. When the joint is in compression, the compression stress at the net section shall not exceed the bearing design stress $f_c \leq F_g$. When the joint is in tension, $f_t \leq F_g$.

### 7.4.7 Angle of Load to Grain

The design values of connectors for angles of load to grain other than 0° and 90° are determined by the application of the Hankinson formula Eq. (7-2) to the design values for parallel and perpendicular-to-grain loading given in Tables 7.32–7.34.

### 7.4.8 Angle of Axis of Connectors to Grain

The connector axis is formed by a line joining the centers of any two adjacent connectors located in the same face of a member in a joint. The angle of axis of the connectors is the angle formed by the axis line of the connectors and the longitudinal axis of the member. This angle is a factor in the determination of the required spacing of connectors for a given load as illustrated by angle $\phi$ in Fig. 7.19.

### 7.4.9 Spacing of Connectors

Spacing between connectors must be considered in determining connector loads because it controls the shearing area that develops the connector load. Factors that influence spacing are angle of load to grain and angle of axis of connectors to grain. Table 7.35 contains the spacing required for 100% and 50% of full load for parallel and perpendicular-to-grain loading. The design values for intermediate spacings can be obtained by straight-line interpolation. The spacing in members loaded at an angle to grain with the connector axis at various angles to the grain can be calculated by Eq. (7-22) or obtained graphically from Figs. 7.20 and 7.21. For tension and compression members loaded at an angle of grain, $\theta$, other than 0° and 90° with connector axis and various angles with the grain, $\phi$, other than 0° and 90°, the spacing, $R$, can be determined by the following equation:

$$R = \frac{AB}{\sqrt{A^2 \sin^2\phi + B^2 \cos^2\phi}} \tag{7-22}$$

## TABLE 7.32

Shear Plate Connector Unit Design Values[a,b,c]. Tabulated design values apply to ONE shear plate unit and bolt in single shear.

| Shear plate diameter inches | Bolt diameter inches | Number of faces of member with connectors on same bolt | Net thickness of member inches | Loaded parallel to grain (0°) Design value, P, per connector unit and bolt, lbs. | | | | Loaded perpendicular to grain (90°) Design value, Q, per connector unit and bolt, lbs. | | | |
|---|---|---|---|---|---|---|---|---|---|---|---|
| | | | | Group A species | Group B species | Group C species | Group D species | Group A species | Group B species | Group C species | Group D species |
| 2-5/8 | 3/4 | 1 | 1-1/2" minimum | 3110* | 2670 | 2220 | 2010 | 2170 | 1860 | 1550 | 1330 |
| | | 2 | 1-1/2" minimum | 2420 | 2080 | 1730 | 1500 | 1690 | 1450 | 1210 | 1040 |
| | | | 2" | 3190* | 2730 | 2270 | 1960 | 2220 | 1910 | 1580 | 1370 |
| | | | 2-1/2" or thicker | 3330* | 2860 | 2380 | 2060 | 2320 | 1990 | 1650 | 1440 |

| | | | Thickness of member | | | | | | | | |
|---|---|---|---|---|---|---|---|---|---|---|---|
| 4 | 3/4 or | 1 | 1-1/2" minimum | 4370 | 3750 | 3130 | 2700 | 3040 | 2620 | 2170 | 1860 |
| | | | 1-3/4" or thicker | 5090* | 4360 | 3640 | 3140 | 3540 | 3040 | 2530 | 2200 |
| | 7/8 | 2 | 1-3/4" minimum | 3390 | 2910 | 2420 | 2090 | 2360 | 2020 | 1680 | 1410 |
| | | | 2" | 3790 | 3240 | 2700 | 2330 | 2640 | 2260 | 1880 | 1630 |
| | | | 2-1/2" | 4310 | 3690 | 3080 | 2660 | 3000 | 2550 | 2140 | 1850 |
| | | | 3" | 4830* | 4140 | 3450 | 2980 | 3360 | 2880 | 2400 | 2060 |
| | | | 3-1/2" or thicker | 5030* | 4320 | 3600 | 3110 | 3500 | 3000 | 2510 | 2160 |

Source: ANSI/NF₀PA NDS—1991, *National Design Specification for Wood Construction*, AFPA, 1250 Connecticut Ave., NW, Washington, DC 20036.

*a*Tabulated lateral design values (P, Q) for shear plate connector units shall be multiplied by all applicable adjustment factors (see Table 7.1).

*b*Allowable design values for shear plate connector units shall not exceed the following:

    (a) $2\frac{5}{8}$" shear plate .............................................2900 pounds

    (b) 4" shear plate with $\frac{3}{4}$" bolt.......................4400 pounds

    (c) 4" shear plate with $\frac{7}{8}$" bolt.......................6000 pounds

The design values in Footnote b shall be permitted to be increased in accordance with the American Institute of Steel Construction (AISC) Manual of Steel Construction, 9th edition, Section A5.2 "Wind and Seismic Stresses", except when design loads have already been reduced by load combination factors (see NDS 7.2.3).

*c*Loads followed by an asterisk (*) exceed those permitted by Footnote b, but are needed for determination of design values for other angles of load to grain. Footnote b limitations apply in all cases.

# TABLE 7.33

## Shear Plate Design Values—Steel Side Plates[a,b]

| | | | | | Loaded Parallel to Grain (0°) | | | | Loaded Perpendicular to Grain (90°) | | | | | |
|---|---|---|---|---|---|---|---|---|---|---|---|---|---|---|
| | | | | | Design Value per Connector Unit and Bolt (lb)[c] | | | | Edge Distance (in.) | | Design Value per Connector Unit and Bolt (lb) | | | |
| Shear Plate Diameter (in.) | Bolt Diameter (in.) | Number of Faces of Piece with Connectors on Same Bolt | Net Thickness of Piece (in.) | Minimum Edge Distance (in.) | Group A Woods | Group B Woods | Group C Woods | Group D Woods | Unloaded Edge, min. | Loaded edge[c] | Group A Woods | Group B Woods | Group C Woods | Group D Woods |
| $2\frac{5}{8}$ | $\frac{3}{4}$ | 1 | $1\frac{1}{2}$ min | $1\frac{3}{4}$ | 3110* | 2670 | 2220 | 2010 | $1\frac{3}{4}$ | $1\frac{3}{4}$ min | 1810 | 1550 | 1290 | 1110 |
| | | | | | | | | | | $2\frac{3}{4}$ or more | 2170 | 1860 | 1550 | 1330 |
| | | 2 | $1\frac{1}{2}$ min | $1\frac{3}{4}$ | 2420 | 2080 | 1730 | 1500 | $1\frac{3}{4}$ | $1\frac{3}{4}$ min | 1410 | 1210 | 1010 | 870 |
| | | | | | | | | | | $2\frac{3}{4}$ or more | 1690 | 1450 | 1210 | 1040 |
| | | | 2 | $1\frac{3}{4}$ | 3190* | 2730 | 2270 | 1960 | $1\frac{3}{4}$ | $1\frac{3}{4}$ min | 1850 | 1590 | 1320 | 1140 |
| | | | | | | | | | | $2\frac{3}{4}$ or more | 2220 | 1910 | 1580 | 1370 |
| | | | $2\frac{1}{2}$ or more | $1\frac{3}{4}$ | 3330* | 2860 | 2380 | 2060 | $1\frac{3}{4}$ | $1\frac{3}{4}$ min | 1940 | 1660 | 1380 | 1200 |
| | | | | | | | | | | $2\frac{3}{4}$ or more | 2320 | 1990 | 1650 | 1440 |
| 4 | $\frac{3}{4}$ | 1 | $1\frac{1}{2}$ min | $2\frac{3}{4}$ | 5160 | 4160 | 3290 | 2700 | $2\frac{3}{4}$ | $2\frac{3}{4}$ min | 2540 | 2180 | 1810 | 1550 |
| | | | | | | | | | | $3\frac{3}{4}$ or more | 3040 | 2620 | 2170 | 1860 |
| | or | 2 | $1\frac{3}{4}$ or more | $2\frac{3}{4}$ | 6010* | 4840* | 3820 | 3140 | $2\frac{3}{4}$ | $2\frac{3}{4}$ min | 2950 | 2530 | 2110 | 1810 |
| | | | | | | | | | | $3\frac{3}{4}$ or more | 3540 | 3040 | 2530 | 2200 |

| | | | | | | | | | | | | |
|---|---|---|---|---|---|---|---|---|---|---|---|---|
| $\frac{7}{8}$ | $1\frac{3}{4}$ min | $2\frac{3}{4}$ | 4000 | 3230 | 2540 | 2090 | $2\frac{3}{4}$ | $2\frac{3}{4}$ min | 1970 | 1680 | 1400 | 1250 |
| | | | | | | | | $3\frac{3}{4}$ or more | 2360 | 2020 | 1680 | 1410 |
| | 2 | $2\frac{3}{4}$ | 4470 | 3600 | 2830 | 2330 | $2\frac{3}{4}$ | $2\frac{3}{4}$ min | 2200 | 1880 | 1570 | 1360 |
| | | | | | | | | $3\frac{3}{4}$ or more | 2640 | 2260 | 1880 | 1630 |
| | $2\frac{1}{2}$ | $2\frac{3}{4}$ | 5090* | 4100 | 3230 | 2660 | $2\frac{3}{4}$ | $2\frac{3}{4}$ min | 2500 | 2140 | 1780 | 1540 |
| | | | | | | | | $3\frac{3}{4}$ or more | 3000 | 2550 | 2140 | 1850 |
| | 3 | $2\frac{3}{4}$ | 5700* | 4600* | 3620 | 2980 | $2\frac{3}{4}$ | $2\frac{3}{4}$ min | 2800 | 2400 | 2000 | 1720 |
| | | | | | | | | $3\frac{3}{4}$ or more | 3360 | 2880 | 2400 | 2060 |
| | $3\frac{1}{2}$ or more | $2\frac{3}{4}$ | 5940* | 4800* | 3780 | 3110 | $2\frac{3}{4}$ | $2\frac{3}{4}$ min | 2920 | 2500 | 2090 | 1800 |
| | | | | | | | | $3\frac{3}{4}$ or more | 3500 | 3000 | 2510 | 2160 |

[a]Source: *National Design Specification for Wood Construction*, AFPA, 1250 Connecticut Ave., N.W. Washington, D.C. 20036.

[b]Design values apply to one shear plate unit and bolt in single shear when installed with seasoned wood members that will remain dry in service and be subject to normal loading conditions.

[c]Loads followed by an asterisk (*) exceed those permitted by Note *d*, but are needed for proper determination of loads for other angles of load to grain. Note *d* limitations apply in all cases.

[d]Design loads (lb) on shear plates shall not exceed the following:

| | |
|---|---|
| (1) $2\frac{5}{8}$-in. shear plate | 2900 |
| (2) 4-in. shear plate with $\frac{3}{4}$ in. bolt | 4400 |
| (3) 4-in. shear plate with $\frac{7}{8}$ in. bolt | 6000 |

[e]See Figure 7.4.

## TABLE 7.34

### Split Ring Connector Unit Design Values[a]. Tabulated design values apply to ONE split ring and bolt in single shear.

| Split ring diameter (inches) | Bolt diameter (inches) | Number of faces of member with connectors on same bolt | Net thickness of member (inches) | Loaded parallel to grain (0°) Design value, P, per connector unit and bolt, lbs. Group A species | Group B species | Group C species | Group D species | Loaded perpendicular to grain (90°) Design value, Q, per connector unit and bolt, lbs. Group A species | Group B species | Group C species | Group D species |
|---|---|---|---|---|---|---|---|---|---|---|---|
| 2-1/2 | 1/2 | 1 | 1" minimum | 2630 | 2270 | 1900 | 1640 | 1900 | 1620 | 1350 | 1160 |
| | | | 1-1/2" or thicker | 3160 | 2730 | 2290 | 1960 | 2280 | 1940 | 1620 | 1390 |
| | | 2 | 1-1/2" minimum | 2430 | 2100 | 1760 | 1510 | 1750 | 1500 | 1250 | 1070 |
| | | | 2" or thicker | 3160 | 2730 | 2290 | 1960 | 2280 | 1940 | 1620 | 1390 |
| 4 | 3/4 | 1 | 1" minimum | 4090 | 3510 | 2920 | 2520 | 2840 | 2440 | 2040 | 1760 |
| | | | 1-1/2" | 6020 | 5160 | 4280 | 3710 | 4180 | 3590 | 2990 | 2580 |
| | | | 1-5/8" or thicker | 6140 | 5260 | 4380 | 3790 | 4270 | 3660 | 3050 | 2630 |
| | | 2 | 1-1/2" minimum | 4110 | 3520 | 2940 | 2540 | 2980 | 2450 | 2040 | 1760 |
| | | | 2" | 4950 | 4250 | 3540 | 3050 | 3440 | 2960 | 2460 | 2120 |
| | | | 2-1/2" | 5830 | 5000 | 4160 | 3600 | 4050 | 3480 | 2890 | 2500 |
| | | | 3" or thicker | 6140 | 5260 | 4380 | 3790 | 4270 | 3660 | 3050 | 2630 |

Source: ANSI/NFoPA NDS—1991, *National Design Specification for Wood Construction*, AFPA, 1250 Connecticut Ave., NW, Washington, DC 20036.

[a]Tabulated lateral design values (P, Q) for split ring connector units shall be multiplied to all applicable adjustment factors (see Table 7.1).

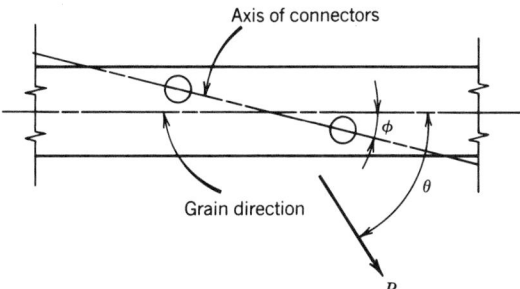

**FIGURE 7.19**   Angle of axis of connectors to grain, $\phi$; angle of load to grain, $\theta$.

where   $R$ = spacing along connector axis as shown in Fig. 7.3 for the required design value (in.),

$A$ = dimension from column 3, Table 7.36, opposite the connector in column 1 and angle of load to grain, $\theta$, in column 2, (in.),

$B$ = dimension from column 4, Table 7.36, opposite the connector in column 1 and angle of load to grain, $\theta$, in column 2 (in.), and

$\phi$ = angle of axis of connectors to grain as shown in Fig. 7.19 (degrees).

Dimension $C$ in column 5, Table 7.36, is the minimum spacing along the connector axis and will permit 50% of the design load for the spacing $R$ determined for full load. For spacings intermediate between $R$ and $C$, the design load is determined by straight-line interpolation.

**Example.**   The following illustrates the use of Fig. 7.20 to determine the required spacing for full load when the angle of axis of connectors is 40° and the angle of load to grain is 45°.

For full design value spacing, enter chart with an angle of axis of 40°. From where this line crosses the angle of load to grain curve at 45°, project downward and read parallel-to-grain spacing of approximately $3\frac{1}{4}$ in. Project horizontally to read perpendicular-to-grain spacing of approximately $2\frac{3}{4}$ in. The center-to-center spacing at 100% of design load is $4\frac{1}{4}$ in. measured from the intersection to the origin. This value can also be determined by calculating the square root of the sum of the squares of the parallel and perpendicular-to-grain spacings. For minimum spacing at 50% of design load, determine where the 40° angle of axis line crosses the 50% of full design value line and project downward to approximately $2\frac{3}{4}$ in. for parallel-to-grain spacing and horizontally to approximately $2\frac{1}{4}$ in. for perpendicular-to-grain spacing. This results in a minimum spacing center to center of approximately $3\frac{1}{2}$ in. For intermediate design values, use straight-line interpolation between spacing at 50 and 100%.

Figs. 7.20 and 7.21 each have five curves representing recommended spacing for full design value at the particular angle of load to grain noted on the curve. For intermediate angles of load to grain, straight-line interpolation may be used. The spacing for full design value is determined by locating the intersection of the

# TABLE 7.35

## Geometry Factors, $C_\Delta$, for Split Ring and Shear Plate Connectors

| | | 2-1/2" Split Ring Connectors & 2-5/8" Shear Plate Connectors | | | | 4" Split Ring Connectors & 4" Shear Plate Connectors | | | |
| | | Parallel to grain loading | | Perpendicular to grain loading | | Parallel to grain loading | | Perpendicular to grain loading | |
| | | Minimum for Reduced Design Value | Minimum for Full Design Value | Minimum for Reduced Design Value | Minimum for Full Design Value | Minimum for Reduced Design Value | Minimum for Full Design Value | Minimum for Reduced Design Value | Minimum for Full Design Value |
|---|---|---|---|---|---|---|---|---|---|
| Edge Distance | Unloaded Edge $C_\Delta$ | 1-3/4" / 1.0 | 1-3/4" / 1.0 | 1-3/4" / 1.0 | 1-3/4" / 1.0 | 2-3/4" / 1.0 | 2-3/4" / 1.0 | 2-3/4" / 1.0 | 2-3/4" / 1.0 |
| Edge Distance | Loaded Edge $C_\Delta$ | 1-3/4" / 1.0 | 1-3/4" / 1.0 | 1-3/4" / 0.83 | 2-3/4" / 1.0 | 2-3/4" / 1.0 | 2-3/4" / 1.0 | 2-3/4" / 0.83 | 3-3/4" / 1.0 |
| End Distance | Tension Member $C_\Delta$ | 2-3/4" / 0.625 | 5-1/2" / 1.0 | 2-3/4" / 0.625 | 5-1/2" / 1.0 | 3-1/2" / 0.625 | 7" / 1.0 | 3-1/2" / 0.625 | 7" / 1.0 |
| End Distance | Compression Member $C_\Delta$ | 2-1/2" / 0.625 | 4" / 1.0 | 2-3/4" / 0.625 | 5-1/2" / 1.0 | 3-1/4" / 0.625 | 5-1/2" / 1.0 | 3-1/2" / 0.625 | 7" / 1.0 |
| Spacing | Spacing parallel to grain $C_\Delta$ | 3-1/2" / 0.5 | 6-3/4" / 1.0 | 3-1/2" / 1.0 | 3-1/2" / 1.0 | 5" / 0.5 | 9" / 1.0 | 5" / 1.0 | 5" / 1.0 |
| Spacing | Spacing perpendicular to grain $C_\Delta$ | 3-1/2" / 1.0 | 3-1/2" / 1.0 | 3-1/2" / 0.5 | 4-1/4" / 1.0 | 5" / 1.0 | 5" / 1.0 | 5" / 0.5 | 6" / 1.0 |

Source: ANSI/NF₀PA NDS—1991, *National Design Specification for Wood Construction*, AFPA, 1250 Connecticut Ave., NW, Washington, DC 20036.

Note: For convenience in calculations, the geometry factor $C_\Delta$ is designated as follows for the various geometric conditions:

Edge Distance $C_{\Delta e}$  End Distance $C_{\Delta n}$  Spacing $C_{\Delta s}$

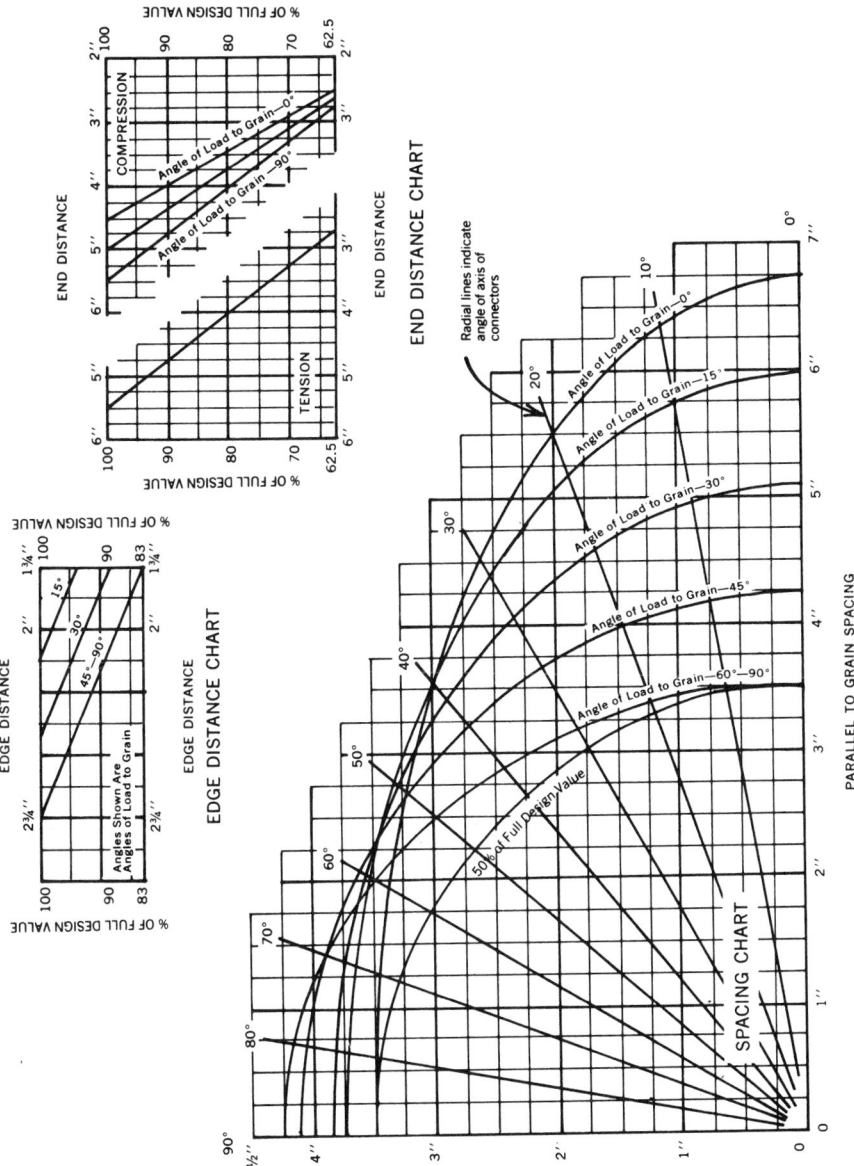

**FIGURE 7.20** Design value data for connectors: $2\frac{5}{8}$-in. shear plates and $2\frac{1}{2}$-in. split rings. Source: TECO/Lumberlok, *Design Manual*, copyright, TECO/Lumberlok.

7-509

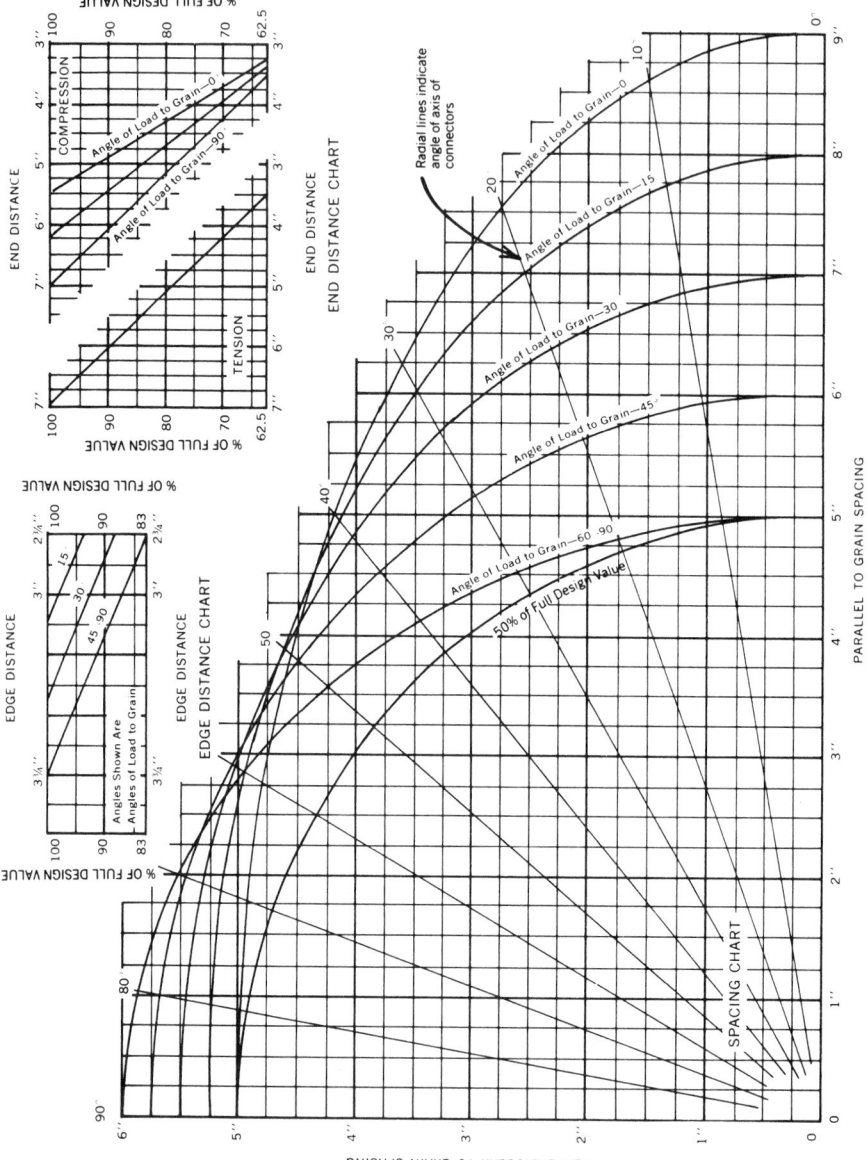

**FIGURE 7.21** Design value data for connector: 4-in. shear plates and 4-in. split rings. Source: TECO/Lumberlok, *Design Manual*, copyright, TECO/Lumberlok.

**TABLE 7.36**

**Values to Use with Eq. $(7\text{-}22)^{a,b}$**

| 1 | 2 | 3 | 4 | 5 |
| --- | --- | --- | --- | --- |
| | Angle of | | | $C$ |
| | Load to | | | (50% |
| Type and Size of | Grain ($\theta$) | $A$ | $B$ | value) |
| Connector | (deg) | (in.) | (in.) | (in.) |
| $2\frac{1}{2}$-in. split ring or | 0 | $6\frac{3}{4}$ | $3\frac{1}{2}$ | $3\frac{1}{2}$ |
| $2\frac{5}{8}$-in. shear plate | 15 | 6 | $3\frac{3}{4}$ | $3\frac{1}{2}$ |
| | 30 | $5\frac{1}{8}$ | $3\frac{7}{8}$ | $3\frac{1}{2}$ |
| | 45 | $4\frac{1}{4}$ | $4\frac{1}{8}$ | $3\frac{1}{2}$ |
| | 60–90 | $3\frac{1}{2}$ | $4\frac{1}{4}$ | $3\frac{1}{2}$ |
| 4-in split ring or | 0 | 9 | 5 | 5 |
| 4–in. shear plate. | 15 | 8 | $5\frac{1}{4}$ | 5 |
| | 30 | 7 | $5\frac{1}{2}$ | 5 |
| | 45 | 6 | $5\frac{3}{4}$ | 5 |
| | 60–90 | 5 | 6 | 5 |

[a]Reproduced from the 1991 edition of the Uniform Building Code, copyright © 1991, with permission of the publisher, the International Conference of Building Officials (8).

[b]Use straight-line interpolation for determining $A$ and $B$ for intermediate angle of loads to grain.

proper angle of load-to-grain curve and the radial line representing angle of axis to grain. The distance from that point of intersection to the origin of the axes (lower left-hand corner of the chart) is the recommended spacing. The parallel-to-grain component of spacing may be read by projecting downward from the point of intersection to the bottom of the chart. The perpendicular-to-grain component of spacing may be read by projecting horizontally from the point of intersection to the left side of the chart.

The quarter circle on each spacing chart represents the spacing for 50% of design value for any angle of load to grain and is the minimum spacing permissible. For design values between 50% and full design value, interpolate radially on the straight line between the 50% curve and full design value curve representing the proper angle of load to grain.

When three or more connectors are used in one face of a member, such as is shown in Fig. 7.3, the spacing between any two connectors should be checked. In the joint shown in Fig. 7.3, the angle of load to grain is the same for all three connectors but the angle of axis of connectors to grain varies with each pair of connectors; therefore, the minimum spacing requirements are determined for each axis.

### 7.4.10   Edge Distance

Edge distances for connectors are given in Table 7.35. They are also shown graphically in Figs. 7.20 and 7.21.

The edge distances shown in Table 7.35 are for the unloaded edge and the loaded edge. The unloaded edge distance is a minimum distance, and is used both for parallel-to-grain loading and for the unloaded edge when the load acts at an angle to the grain. The edge distances for the loaded edge are shown as the minimum distance and the distance required for full load. For intermediate spacings, use straight-line interpolation between the design values at minimum edge distance and full edge distance. For angles of load to grain of 45°–90°, use the edge distance for 90°. For angles of load to grain between 0° and 45°, the required edge distance may be determined by straight-line interpolation. The capacity of a connector at minimum edge distance loaded at an angle of 45°–90° is 83% of full load.

The edge distance charts in Figs. 7.20 and 7.21 may be used to determine a adjustment factor, $C_{\Delta e}$, to apply to the full design load. The minimum edge distance is given in the lower right corner of these charts. The dimension is the minimum edge distance for any direction of loading. Loaded edge distance varies with the angle of load to grain. Percentages of full design loads for various loaded edge distances and angles of load to grain are determined by projecting horizontally to either side from the point of intersection of the proper loaded edge distance line and angle of load-to-grain line. The right side of the edge distance charts is set at the minimum edge distance permitted. This distance allows 100% of full load when the angle of load to grain is 0° and varies to 83% for an angle of load to grain from 45° to 90°.

The edge distances for connectors were determined by test on lumber and design values for $2\frac{1}{2}$ in. and $2\frac{5}{8}$ in. connectors were based on edge distances that allow the use of a connector in nominal 4-in.-wide pieces (net $3\frac{1}{2}$ in.) and the use of 4-in. connectors in nominal 6-in.-wide pieces (net $5\frac{1}{2}$ in.). Glued laminated timbers are slightly smaller in width than sawn lumber of the same nominal width. A review of Technical Bulletin No. 865 *Timber Connector Joints; Their Strength and Design* (5), which contains the test data from which the design values were developed, indicates that a adjustment factor $C_{\Delta e}$ as shown in Table 7.37 should be applied to members where the minimum tabular edge distances shown in Tables 7.33 and 7.35 cannot be obtained. A $2\frac{1}{2}$-in. or $2\frac{5}{8}$-in. connector should not be used in a member less than 3 in. wide and a 4-in. wide connector should not be used in a member less than 5 in. wide. The reduction factors shown in Table 7.37 should be used when connectors are used on the narrow faces of nominal 4-in. and 6-in. wide glued laminated timbers.

### 7.4.11    End Distance

End distances and adjustment factors for less than full end distance, $C_{\Delta n}$, are shown in Table 7.35 and Figs. 7.20 and 7.21. The end distances in Table 7.35 are listed as minimum end distances in both tension and compression for parallel and perpendicular-to-grain loading. When minimum end distance is used, the design value is 62.5% of the full design value. For end distances intermediate between the minimum and that required for full load, the percentage of full design value is determined by straight-line interpolation. End distances for members

**TABLE 7.37**

**Adjustment Factors $C_{\Delta e}$ for Connectors with Reduced Edge Distances, Connectors Centered on Face**

| Connector Size | | Parallel-to-Grain Loading[a] | | | Perpendicular-to-Grain Loading[b] | |
|---|---|---|---|---|---|---|
| $2\frac{1}{2}$-in. split ring or | Width | 3 in. | $3\frac{1}{8}$ in. | | 3 in. | $3\frac{1}{8}$ in. |
| $2\frac{5}{8}$-in. shear plate | | 0.88 | 0.91 | | 0.86 | 0.90 |
| 4-in. split ring or | Width | 5 in. | $5\frac{1}{8}$ in. | | 5 in. | $5\frac{1}{8}$ in. |
| shear plate | | 0.93 | 0.95 | | 0.93 | 0.95 |

[a]The geometry adjustment factor $C_{\Delta e}$ for parallel-to grain loading should be applied to tabular design values in Tables 7.32–7.34.

[b]The geometry adjustment factor $C_{\Delta e}$ for perpendicular-to-grain should be applied to the tabular design values for minimum loaded edge distances in Tables 7.32–7.34.

loaded at angles of grain other than $0°$ or $90°$ may be determined by straight-line interpolation.

End distance may also be determined from Figs. 7.20 and 7.21. These charts are divided into two sections. End distance requirements depend on whether the member is in tension or compression. If the member is in tension, the allowable percentage of full design value is obtained by projecting horizontally from the intersection of the end distance line and the diagonal line. This process can be reversed by starting with the percentage of full design value and determining the required end distance. For members in compression, the angle of load to grain must also be considered. The use of the chart is the same, except that the diagonal line for the proper angle of load to grain should be used. Straight-line interpolation between lines should be used for intermediate angles.

The following example illustrates the design of a tension connection.

**Example.**   For group B species, $F_t = 1200$ psi, $E = 1,700,000$ psi, used in dry location, $4 \times 6$ in. main member, $2 \times 6$ in. side members. Use $2\frac{1}{2}$ in. split ring connectors.

Design load: Dead load $= 5000$ lb, snow load $= 12,000$ lb, $C_D = 1.15$.

$$F'_t = F_t C_D = (1200)(1.15) = 1380 \text{ psi}$$

Check net section of main and side members. From Table 7.28, projected area of one $2\frac{1}{2}$-in. split ring and bolt is 1.73 in.$^2$ for side members.

$$\text{Net area} = (2)(8.25 - 1.73) = 13.04 \text{ in.}^2$$

For two $2\frac{1}{2}$-in. split rings and a bolt in the main member, projected area is 3.75 in.$^2$.

$$\text{Net area} = 19.25 - 3.75 = 15.50 \text{ in.}^2$$

Determine design load controlled by the strength of the wood.

$$\text{Design load} = (13.04)(1380) = 18,000 \text{ lb} > 17,000 \text{ lb} \quad \text{O.K.}$$

Determine connector spacings, end distance, and edge distance for full design value of each connector from Fig. 7.20 or Table 7.35.

End distance $= 5\frac{1}{2}$ in.
Edge distance $= 2\frac{3}{4}$ in.
Spacing $= 6\frac{3}{4}$ in.

Determine number of connectors required. From Table 7.34, full design value for one connector,

$$P = 2730 \text{ lb}$$
$$P' = PC_D = (2730)(1.15) = 3140 \text{ lb}$$
Number of split rings required $= 17,000/3140 = 5.4$

Try 3 split rings in each side member spaced $6\frac{3}{4}$ in. center to center and check for group action of connectors in a row. For this connection, there are two rows of three connectors in each row.

Area of side members $A_s = (2)(8.25) = 16.50$ in.$^2$
Area of main member $A_m = 19.25$ in.$^2$

Use $A_s/A_m = 16.50/19.25 = 0.857$
A conservative value of $C_g$ can be obtained by using Table 7.5. For three connectors in a row and $A_s = 16.5$, interpolate between $A_s/A_m = 0.5$ and $A_s/A_m = 1.0$.

Adjustment factor $C_g = 0.92$

A more precise group action factor can be obtained by using Eq. (7-1) where units are as shown previously.

$$C_g = \left[\frac{m(1 - m^{2n})}{n[(1 + R_{EA}m^n)(1 + m) - 1 + m^{2n}]}\right]\left[\frac{1 + R_{EA}}{1 - m}\right]$$

$\gamma = 400,000$ lb/in. for a $2\frac{1}{2}$ in. split ring.

$$u = 1 + \gamma\frac{s}{2}\left[\frac{1}{E_mA_m} + \frac{1}{E_sA_s}\right]$$

$$= 1 + \left[\left(400,000\ \frac{6.75}{2}\right)\right]\left[\frac{1}{(1,700,000)(19.25)} + \frac{1}{(1,700,000)(16.5)}\right]$$

$$= 1.0984$$

$$m = u - \sqrt{u^2 - 1} = 1.0894 - \sqrt{1.0894^2 - 1} = 0.657$$

$$R_{EA} = \text{lesser of } \frac{E_s A_s}{E_m A_m} \text{ or } \frac{E_m A_m}{E_s A_s}$$

$$\frac{E_s A_s}{E_m A_m} = \frac{1,700,000(16.5)}{1,700,000(19.25)} = 0.857$$

$$\frac{E_m A_m}{E_s A_s} = \frac{1,700,000(19.25)}{1,700,000(16.5)} = 1.167, \quad R_{EA} = 0.857$$

$$C_g = \left[ \frac{0.657(1 - 0.657^{2 \times 3})}{3[(1 + (0.857)(0.657^3)(1 + 0.657) - 1 + 0.657^{2 \times 3}]} \right]$$

$$\cdot \left[ \frac{1 + 0.857}{1 - 0.657} \right] = 0.96$$

Use adjustment factor $C_g = 0.96$

Design value per connector $P' = PC_D C_g = (2730)(1.15)(0.96) = 3014$ lb

Design load capacity $= (6)(3014) = 18,084$ lb $> 17,000$ lb    O.K.

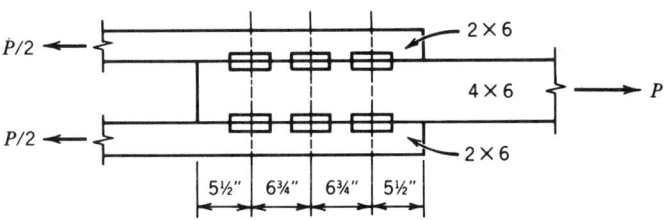

The final joint configuration is as shown in the sketch. Where clearance for the connection is critical, the end distance and spacing could be reduced by taking into account the fact that the connectors are not fully loaded. The ratio of the actual load to design load in this example is $17,000/18,084 = 0.94$. From Fig. 7.21, the required end distance for 94% of full load is $5\frac{1}{4}$ in. and the required spacing is $6\frac{5}{8}$ in. When the end distance and spacing are not critical, it is recommended that the full end distance and spacing be used.

When multiple piece members are used in a joint, it is usually assumed in design that the load in the member is equally divided between the pieces of the member such as in a two piece chord. The difference in modulus of elasticity in the pieces of lumber and twisting may cause uneven loading. When necessary, this should be taken into account by the designer. In the following example the assumption was made that the forces in the multiple piece members were equally divided.

**Example.** The following example with the top chord connection shown in Fig. 7.22 illustrates the transfer of loads, determination of angle of load to grain, angle of axis to grain, and the determination of required end and edge distances using multiple 4 in.-split rings in an existing truss joint.

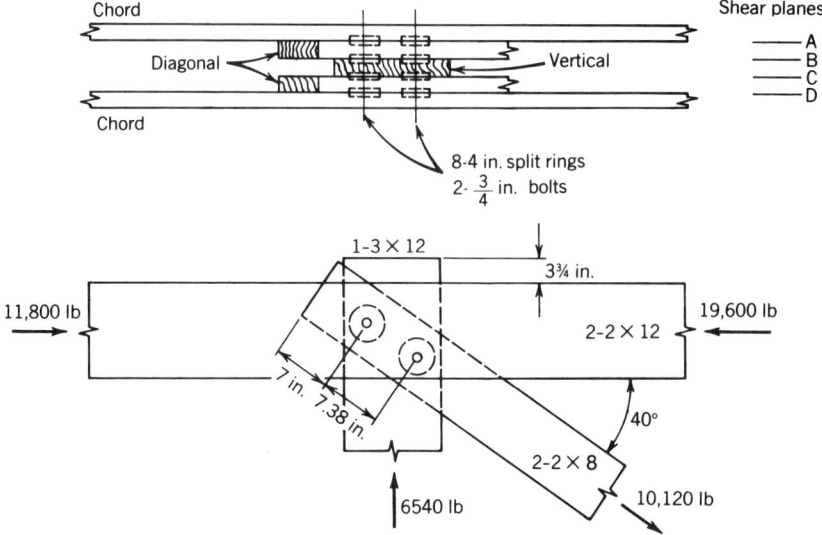

**FIGURE 7.22**    Top chord connection.

Dry condition of use, snow load controls the design ($C_D = 1.15$). Select structural Douglas Fir-Larch (group B species).

Determine the loads transferred in each connector plane and the angle of load and angle of axis to grain of the members on each side of the plane being considered.

There are four different loading conditions that need to be investigated:

1. Shear Planes $A$ and $D$:
   (a) Load from diagonals to chord,
   (b) Load from chords to diagonals.
2. Shear planes $B$ and $C$:
   (a) Load from diagonals to vertical,
   (b) Load from vertical to diagonals.

A 7800 lb (19,600 − 11,800) horizontal force is transferred by the connectors in planes $A$ and $D$ and from the two-member top chord to the two-member diagonal. Assume the load is shared equally by the four connectors. The loads transferred on both planes $A$ and $D$ are as shown in Fig. 7.23.

The angle of load to grain is 0° in the chord and 40° in the diagonal. The angle of axis to grain is 40° in the chord and 0° in the diagonal.

A 6540 lb force is transferred between the diagonal and the vertical member in planes $B$ and $C$. The force is assumed to be divided equally between the four connectors as shown in Fig. 7.24.

The angle of load to grain in the vertical is 0° and angle of axis to grain is 50°.

The angle of load to grain in the diagonal is 50° and angle of axis to grain is 0°.

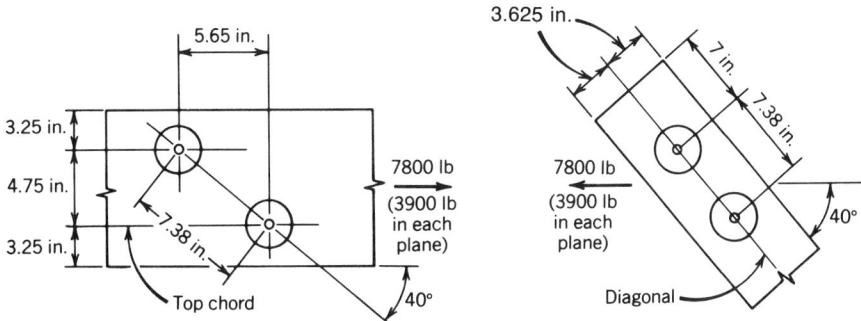

**FIGURE 7.23**   Load transfer in planes $A$ and $D$.

In some instances, the controlling member of the joint can be determined by inspection depending on end, edge distances, or angle of load to grain. In this example, for purposes of illustration, each member of the joint will be analyzed.

Determine spacing from geometry of joint spacing along axis

$$s = \sqrt{4.75^2 + 5.65^2} = 7.38 \text{ in.}$$

Top chord (planes $A$ and $D$): Check load capacity as determined by connectors in chord. From Table 7.34, $P = 5160$ lb (connector in one face of a $1\frac{1}{2}$-in.-thick member). Actual edge distance $= 3.25$ in.

From Table 7.35, required edge distance for full load $= 2\frac{3}{4}$ in., use $C_{\Delta e} = 1.00$. Actual end distance exceeds distance required for full load from Table 7.35 which is 7 in., use $C_{\Delta n} = 1.00$. Actual spacing $= 7.38$ in. at angle of 40° to grain.

From Fig. 7.21, spacing required for full load at angle of axis of connectors of 40° $= 5$ in. parallel to grain and 4.2 in. perpendicular to grain. $s = \sqrt{5^2 + 4.2^2} = 6.53$ in. along connector axis. Because 7.38 in. $> 6.53$ in., $C_{\Delta s} = 1.00$.

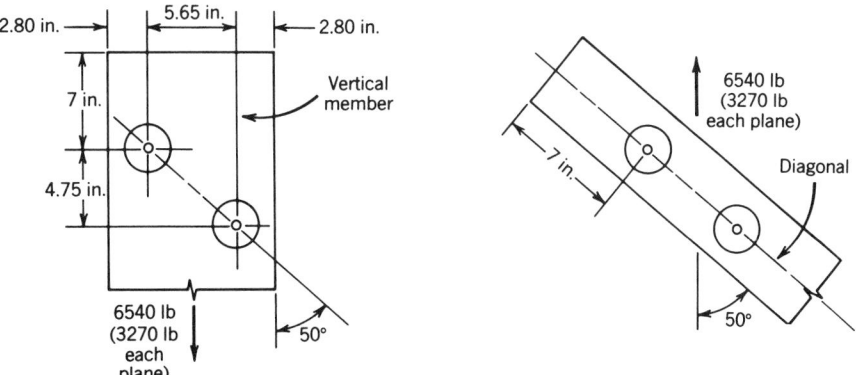

**FIGURE 7.24**   Vertical to diagonal connection.

Since $C_{\Delta e}$, $C_{\Delta n}$, and $C_{\Delta s}$ each $= 1.00$, no adjustment is necessary for edge distance, end distance, or spacing.

$$P' = PC_D = (5160)(1.15) = 5930 \text{ lb}$$

$$\text{Total load} = (4)(5930) = 23,700 \text{ lb} > 7800 \text{ lb} \qquad \text{O.K.}$$

Diagonal (planes $A$ and $D$): Check load capacity of connectors in diagonal as determined by transfer of forces between chord and diagonal. From Table 7.34, $P = 3520$ lb and $Q = 2450$ lb. Actual loaded edge distance $= 3\frac{5}{8}$ in. From Fig. 7.21, minimum edge distance for 100% of full load at 45° angle of load to grain $= 3\frac{1}{4}$ in. Therefore, angle of load to grain is 40°, $C_{\Delta e} = 1.00$.
Actual end distance $= 7$ in.

From Table 7.35, end distance for required full load $= 7$ in.

$$C_{\Delta n} = 1.00$$

Actual spacing $= 7.38$ in. parallel to grain with angle of axis of connectors to grain of 0°.

From Fig. 7.35, spacing for 100% full load at angle of load to grain of 40° is obtained by interpolation of values along the 0° angle of axis of connectors to grain at 30° and 45° angle of load to grain. This results in a required spacing of 6.3 in.

$$7.38 \text{ in.} > 6.3 \text{ in.} \qquad C_{\Delta s} = 1.00$$

Determine split ring capacity $N$ at an angle of load to grain of 40°.

$$N = \frac{PQ}{P' \sin^2 \theta + Q \cos^2 \theta} = \frac{(3520)(2450)}{3520 \sin^2 40° + 2450 \cos^2 40°} = 2982 \text{ lb}$$

$$C_{\Delta e}, C_{\Delta n}, \text{ and } C_{\Delta s} = 1.0$$

$$N' = NC_D = (2982)(1.15) = 3429 \text{ lb}$$

$$\text{Total load} = 4N' = (4)(3429) = 13,717 \text{ lb} > 7800 \text{ lb} \qquad \text{O.K.}$$

Diagonal (planes $B$ and $C$): Check load capacity of the connectors in the diagonals as determined by the transfer of the 6540 lb vertical load between the diagonals and the vertical member.

Check capacity of connectors in diagonal. From Table 7.34 (for connectors in two faces of $1\frac{1}{2}$-in.-thick member),

$$P = 3520 \text{ lb} \qquad Q = 2450 \text{ lb}$$

Actual edge distance $= 3\frac{5}{8}$ in. From Fig. 7.21, the capacity of a 4-in. split ring with minimum edge distance of $3\frac{1}{4}$ in. and loaded at an angle of load to grain of 45°–90° is 100% of full design value. Diagonal is loaded at an angle of load to grain of 50° ($C_{\Delta e} = 1.0$).

Actual end distance $= 7$ in. From Table 7.35, end distance for full loading $= 7$ in. ($C_{\Delta n} = 1.00$).

Actual spacing = 7.38 in. From Fig. 7.21 for angle of load to grain of 50° and angle of axis of connectors to grain of 0°, required spacing is 5.67 in. along axis of connectors (by interpolation). 7.38 in. > 5.67 in. $C_{\Delta s} = 1.00$; therefore, $C_{\Delta s} = 1.0$.

$$N = \frac{(3520)(2450)}{(3520)\ \sin^2 50° + (2450)\ \cos^2 50°} = 2802 \text{ lb}$$

$$N' = NC_D = (2802)(1.15) = 3222 \text{ lb}$$

$$\text{Total load} = 4N' = 4(3222) = 12,888 \text{ lb} > 6540 \text{ lb} \qquad \text{O.K.}$$

Vertical (planes $B$ and $C$): Check capacity of connectors in vertical member. From Table 7.34, $P = 5000$ lb (for connectors in two faces of a $2\frac{1}{2}$-in.-thick member). Actual edge distance = 2.80 in. From Table 7.35, the required edge distance for full load = $2\frac{3}{4}$ in. 2.80 in. > 2.75 in., $C_{\Delta e} = 1.00$.

Actual end distance = 7 in. From Table 7.35, the required end distance for full load = 7 in. $C_{\Delta n} = 1.00$.

Actual spacing = 7.38 in. From Fig. 7.21, spacing for angle of load to grain of 0° and angle of axis of connectors to grain of 50° = 3.9 in. parallel to grain and 4.6 in. perpendicular to grain or 6.0 in. along axis of connectors. Since 7.38 > 6.0, $C_{\Delta s} = 1.00$. No adjustment required for edge distance, end distance, or spacing.

$$P' = PC_D = (5000)(1.15) = 5750 \text{ lb}$$

$$\text{Total load} = 4P' = (4)(5750) = 23,000 \text{ lb} > 6540 \text{ lb} \qquad \text{O.K.}$$

In this example, the vertical and diagonals were shown with full end distances. This is recommended. If space above the top chord had been limited, the end distances could have been cut back. The connectors in the vertical are required to develop $(6540/23,000)(100) = 28\%$ of their capacity; therefore, the minimum end distance of $3\frac{1}{2}$ in., which gives 62.5% of full load, is adequate. The connectors in the diagonal are loaded to $(6540/12,888)(100) = 51\%$ of their capacity.

$$\text{End distance for } 62.5\% \text{ of capacity} = 3.5 \text{ in.}$$

The $3\frac{1}{2}$-in. end distance is measured as shown in Fig. 7.25.
Cut the diagonal horizontally and even with the top of the vertical as shown.
Check net sections: $C_F = 1.0$ for 12 in. wide members

$$F_t = 1000 \text{ psi}, \quad F_t' = (1000)(1.15) = 1150 \text{ psi}$$

$$F_c = 1700 \text{ psi}, \quad F_c' = (1700)(1.15) = 1955 \text{ psi}$$

Top chord $A = 16.88$ in.$^2$.

Because parallel-to-grain spacing is greater than the diameter of one connector, only the area of one split ring and bolt needs to be subtracted from the gross area. From Table 7.28, the projected area of the split rings and bolt within a $1\frac{1}{2}$ in. thick member is 3.05 in.$^2$.

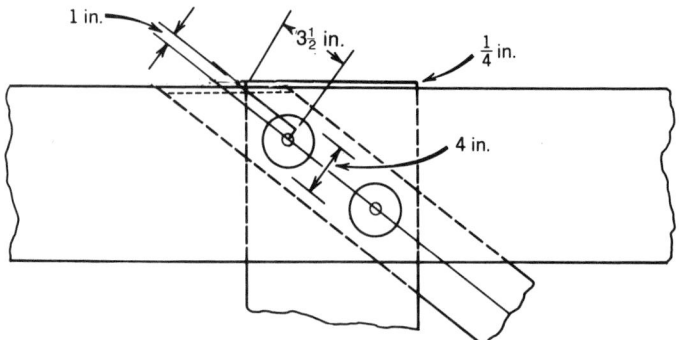

**FIGURE 7.25**    End distance on diagonal.

For a 2 × 12 chord member with connectors in one face,

$$A_{net} = 16.88 - 3.05 = 13.83 \text{ in.}^2$$

For 2 chord members, check compression stress:

$$f_c = \frac{19,600}{(2)(13.83)} = 710 \text{ psi} < F'_c = 1955 \text{ psi} \qquad \text{O.K.}$$

Diagonal

For a 2 × 8 diagonal member with connectors in two faces, $F'_t = F_t C_F C_D = (1000)(1.2)(1.15) = 1380$ psi

$$A = 10.88 \text{ in.}^2 \qquad \text{Area of connectors and bolt} = 4.88 \text{ in.}^2$$

$$A_{net} = 10.88 - 4.88 = 6.0 \text{ in.}^2$$

For 2 diagonal members, check tension stress:

$$f_t = \frac{10,120}{(2)(6.0)} = 843 \text{ psi} < 1380 \text{ psi} \qquad \text{O.K.}$$

Vertical

For the 3 × 12 vertical member with connectors in two faces,

$$A = 28.12 \text{ in.}^2 \qquad \text{Area of connectors and bolts} = 5.69 \text{ in.}^2$$

$$A_{net} = 28.12 - 5.69 = 22.43 \text{ in.}^2$$

$$f_c = \frac{6540}{22.43} = 292 \text{ psi} < F'_c = 1955 \text{ psi} \qquad \text{O.K.}$$

### 7.4.12    Timber Connector Design Values in End Grain

When timber connectors are installed in a surface not parallel to the general direction of the grain of a member, such as the square cut or sloping surface at the end of a member, the design values in Tables 7.32–7.34 must be adjusted. Fig. 7.26 illustrates these conditions of loading.

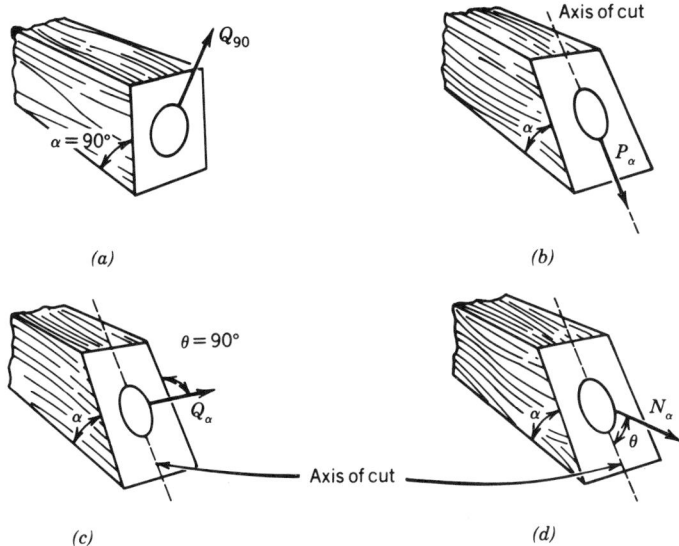

*(a)*                        *(b)*

*(c)*                        *(d)*

**FIGURE 7.26**   Timber connectors in end grain. (*a*) Square cut end, $\alpha = 90°$; (*b*) load parallel to axis of cut, $\theta$; (*c*) load perpendicular to axis of cut, $\theta = 90°$; (*d*) load at angle $\theta$ to axis of cut.

When the end of a member is square cut as shown in Fig. 7.26(*a*), the allowable design value, $Q'_{90}$, is 60% of the perpendicular-to-grain design value regardless of the direction of the load, $Q_{90}$, in the plane of the cut.

$$Q'_{90} = 0.60Q' \tag{7-23}$$

where   $Q'$ = design value for a connector in a side grain surface loaded perpendicular to grain (lb), multiplied by applicable adjustment factors.

When the end of a member is a sloping surface forming an angle $\alpha$ with the general longitudinal grain direction of the member or laminations, the load $P_\alpha$ is applied parallel to the axis of cut and $\theta = 0°$, as shown in Fig. 7.26(*b*); the design value $P_\alpha$ is determined by the Hankinson formula:

$$P'_\alpha = \frac{P'Q'_{90}}{P'\sin^2 \alpha + Q'_{90}\cos^2 \alpha} \tag{7-24}$$

where   $P'_\alpha$ = design value for a connector in a sloping surface, cut at angle $\alpha$ and loaded parallel to the axis of cut (lb), multiplied by applicable adjustment factors,

        $P'$ = design value for a connector in a side grain surface, loaded parallel to grain (lb), multiplied by all applicable adjustment factors,

       $Q'_{90}$ = 60% of the design value for a connector in a side grain surface, $Q'$, loaded perpendicular to grain (lb), multiplied by applicable adjustment factors, and

        $\alpha$ = slope of cut surface (degrees), as shown in Fig. 7.26(*b*).

When the end of a member is a sloping surface forming an angle $\alpha$ with the general longitudinal grain direction of the member or laminations, the load $Q_\alpha$ is applied perpendicular to the axis of the cut and $\theta = 90°$, as shown in Fig. 7.26(c); the design value $Q'_\alpha$ is determined by the Hankinson formula:

$$Q'_\alpha = \frac{Q' Q'_{90}}{Q' \sin^2 \alpha + Q'_{90} \cos^2 \alpha} \tag{7-25}$$

where   $Q'_\alpha$ = design load for a connector in a sloping surface, loaded perpendicular to the axis of cut (lb), multiplied by applicable adjustment factors,

$Q'$ = design value for a connector in a side grain surface, loaded perpendicular to grain (lb), multiplied by applicable adjustment factors,

$Q'_{90}$ = 60% of the design value for a connector in a side grain surface, loaded perpendicular to grain (lb), multiplied by applicable adjustment factors, and

$\alpha$ = slope of cut surface (degrees).

When the end of a member is a sloping surface forming an angle $\alpha$ with the general longitudinal grain direction of the member or laminations, and the load $N_\alpha$ is applied at an angle $\theta$ other than $0°$ or $90°$ to the axis of the cut, as shown in Fig. 7.26(d), the design value $N'_\alpha$ is determined by the Hankinson formula:

$$N'_\alpha = \frac{P'_\alpha Q'_\alpha}{P'_\alpha \sin^2 \theta + Q'_\alpha \cos^2 \theta} \tag{7-26}$$

where   $N'_\alpha$ = design value for connector in a sloping surface, loaded at an angle $\theta$ to the axis of the cut (lb), and

other terms are as defined previously.

The provisions for edge distance, end distance, and spacing shown for connectors in side grain shall be applied to the sloping cuts as follows:

1. For square cut ends, the provisions for perpendicular-to-grain loading apply.

2. For sloping surfaces with angle $\alpha$ from $45°$ to $90°$ loaded in any direction, the provisions for perpendicular-to-grain loading apply.

3. For sloping surfaces with angle $\alpha$ less than $45°$ loaded parallel to the axis of the cut, the provisions for parallel-to-grain loading apply.

4. For sloping surfaces with angle $\alpha$ less than $45°$ loaded perpendicular to the axes of the cut, the provisions for perpendicular-to-grain loading apply.

5. For sloping surfaces with angle $\alpha$ less than $45°$ loaded at angle $\theta$ to axis of cut, the provisions for members loaded at angles of grain other than $0°$ or $90°$ apply.

The design values for connectors installed in end grain should also be checked for reduction in end shear caused by the notched beam effect as given in 7.1.11. Only the component of load acting perpendicular to grain should be used in computing vertical reaction.

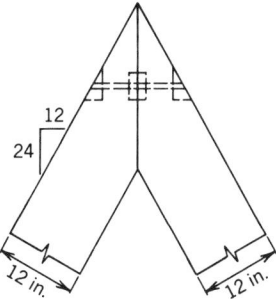

**Example.**   The peak of an A-frame is fastened together with two 4-in. shear plates placed back to back and a $\frac{3}{4}$-in. through bolt with washers counterbored into the arch. Determine the vertical force $P'_\alpha$ which is primarily unbalanced snow load ($C_D = 1.15$) that can be transferred through the joint in either direction. From Table 7.32, the tabular design value for group B species for one 4-in. shear plate with wood side plates is:

parallel-to-grain loading, $P = 4360$ lb
perpendicular-to-grain loading with a loaded edge distance of 6 in., $Q = 3040$ lb.
From Table 7.37, for $5\frac{1}{8}$-in. width (edge distance $= 2.56$ in.), $C_{\Delta e} = 0.95$ for both parallel and perpendicular-to-grain loading.

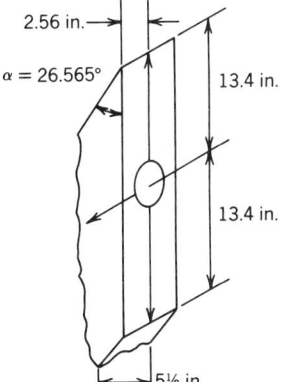

End distance in both directions equals 13.4 in., angle $\alpha > 45°$; therefore, use this distance for parallel-to-grain end distance. From Table 7.35, end distance equals 7 in. for full (100%) load (for both parallel- and perpendicular-to-grain loading).

$$C_{\Delta n} = 1.00 \quad (13.4 \text{ in.} > 7 \text{ in.})$$

When adjustment factors are the same for both parallel- and perpendicular-to-grain loading, the factors can be applied directly to $P$ and $Q$.

$$P' = PC_D C_{\Delta e} = (4360)(1.15)(0.95) = 4760 \text{ lb}$$

$$Q' = QC_D C_{\Delta e} = (3040)(1.15)(0.95) = 3320 \text{ lb}$$

$$Q'_{90} = 0.60 Q' = (0.6)(3320) = 1990 \text{ lb}$$

$$P'_\alpha = \frac{P' Q'_{90}}{P' \sin^2 \alpha + Q'_{90} \cos^2 \alpha}$$

$$= \frac{(4760)(1990)}{(4760) \sin^2 26.565° + (1990) \cos^2 26.565°} = 3720 \text{ lb}$$

## 7.5  WOOD SCREWS

### 7.5.1  General Considerations

The design values given herein are for wood screws conforming to ANSI/ASME Standard B18.6.1—1981 (9). Wood screws require prebored holes to prevent splitting of the wood except as noted. Lead hole diameters for the shank and threaded portions of wood screws are given in Table 7.38. Wood screws should be inserted by turning with a screwdriver and not by driving with a hammer. Soap or other lubricant may be used to aid insertion.

The wood species with the higher specific gravity, $G$, require lead holes to prevent splitting when loaded in withdrawal. Species with $G > 0.60$ should have a lead hole of approximately 90% of the wood screw root diameter and approximately 70% of the wood screw root diameter when $0.50 > G \le 0.60$; when $G \le 0.5$, no lead hole is necessary.

When wood screws are loaded laterally and $G > 0.60$, the part of the lead hole receiving the shank should have approximately the same diameter as the shank, and the part receiving the threaded portion should have approximately the same diameter as the diameter at the root of the thread. When $G \le 0.60$, the part of the lead hole receiving the shank should be approximately $\frac{7}{8}$ the shank diameter, and the portion receiving the threaded portion should be about $\frac{7}{8}$ the diameter of the screw at the root of the thread (1).

See Table 7.41 for shank diameter of wood screws.

Table 7.38 may be used as a guide for the diameter of lead holes.

### 7.5.2  Lumber Species and Specific Gravity

Species and specific gravities for wood screw design values are given in Table 7.39.

**TABLE 7.38**

**Lead Hole Diameters for Wood Screws**

| | | Diameter of Lead Hole (in.) | | | | | |
|---|---|---|---|---|---|---|---|
| | | Withdrawal Loads | | Lateral Loads | | | |
| | | $G > 0.6$ | $0.6 > G \geq 0.5$ | $G > 0.6$ | | $0.6 > G \geq 0.5$ | |
| Gage of Screw | Shank Diameter of Screw (in.) | | | Shank Portion | Threaded Portion | Shank Portion | Threaded Portion |
| 6 | 0.138 | $\frac{5}{64}$ | $\frac{1}{16}$ | $\frac{9}{64}$ | $\frac{3}{32}$ | $\frac{1}{8}$ | $\frac{5}{64}$ |
| 7 | 0.151 | $\frac{3}{32}$ | $\frac{5}{64}$ | $\frac{5}{32}$ | $\frac{7}{64}$ | $\frac{1}{8}$ | $\frac{3}{32}$ |
| 8 | 0.164 | $\frac{7}{64}$ | $\frac{5}{64}$ | $\frac{5}{32}$ | $\frac{7}{64}$ | $\frac{9}{64}$ | $\frac{3}{32}$ |
| 9 | 0.177 | $\frac{7}{64}$ | $\frac{5}{64}$ | $\frac{11}{64}$ | $\frac{1}{8}$ | $\frac{5}{32}$ | $\frac{7}{64}$ |
| 10 | 0.190 | $\frac{7}{64}$ | $\frac{3}{32}$ | $\frac{3}{16}$ | $\frac{1}{8}$ | $\frac{11}{64}$ | $\frac{7}{64}$ |
| 12 | 0.216 | $\frac{9}{64}$ | $\frac{7}{64}$ | $\frac{7}{32}$ | $\frac{9}{64}$ | $\frac{3}{16}$ | $\frac{1}{8}$ |
| 14 | 0.242 | $\frac{5}{32}$ | $\frac{7}{64}$ | $\frac{1}{4}$ | $\frac{5}{32}$ | $\frac{7}{32}$ | $\frac{9}{64}$ |
| 16 | 0.268 | $\frac{11}{64}$ | $\frac{1}{8}$ | $\frac{11}{64}$ | $\frac{3}{16}$ | $\frac{15}{64}$ | $\frac{5}{32}$ |
| 18 | 0.294 | $\frac{3}{16}$ | $\frac{9}{64}$ | $\frac{19}{64}$ | $\frac{13}{64}$ | $\frac{1}{4}$ | $\frac{11}{64}$ |
| 20 | 0.320 | $\frac{13}{64}$ | $\frac{5}{32}$ | $\frac{5}{16}$ | $\frac{7}{32}$ | $\frac{9}{32}$ | $\frac{3}{16}$ |
| 24 | 0.372 | $\frac{15}{64}$ | $\frac{3}{16}$ | $\frac{3}{8}$ | $\frac{1}{4}$ | $\frac{21}{64}$ | $\frac{15}{64}$ |

See Table 7.39 for specific gravity, $G$, of the most commonly used species.

### 7.5.3   Spacing, End Distance, and Edge Distance

Spacing, end distance, and edge distance should be sufficient to prevent unusual splitting of the wood.

### 7.5.4   Wood Screw Design Values

The design values in Tables 7.40 and 7.41 apply for one wood screw in a two-member joint under normal duration of loading and dry-use conditions. The total design value for more than one wood screw is the sum of the values for each wood screw, provided that spacings, end distances, and edge distances are sufficient to develop the full strength of each wood screw. Tabular values apply to both sawn lumber and glued laminated timber. See Table 3, AITC—117 Design (4) for species in glued laminated timber combinations.

For other than normal durations of loading, the design values should be multiplied by the appropriate adjustment factor $C_D$ from Fig. 4.4 or Table 4.7. For wood screw joints under other service or seasoning conditions, the design value determined is multiplied by the appropriate adjustment factor $C_M$ from Table 7.2. See Table 7.1 for applicability of adjustment factors for wood screws.

## TABLE 7.39

### Dowel Bearing Strength for Wood Screw Connections

| Species Combination | Specific[b] Gravity G | Dowel bearing strength in pounds per square inch (psi) $F_e$ |
|---|---|---|
| Aspen | 0.39 | 2950 |
| Balsam Fir | 0.36 | 2550 |
| Beech-Birch-Hickory | 0.71 | 8850 |
| Coast Sitka Spruce | 0.39 | 2950 |
| Cottonwood | 0.41 | 3200 |
| Douglas Fir-Larch | 0.50 | 4650 |
| Douglas Fir-Larch (North) | 0.49 | 4450 |
| Douglas Fir-South | 0.46 | 4000 |
| Eastern Hemlock | 0.41 | 3200 |
| Eastern Hemlock-Tamarack | 0.41 | 3200 |
| Eastern Hemlock-Tamarack (North) | 0.47 | 4150 |
| Eastern Softwoods | 0.36 | 2550 |
| Eastern Spruce | 0.41 | 3200 |
| Eastern White Pine | 0.36 | 2550 |
| Engelmann Spruce-Lodgepole Pine[c] (MSR 1650f and higher grades) | 0.46 | 4000 |
| Engelmann Spruce-Lodgepole Pine[c] (MSR 1500f and lower grades) | 0.38 | 2800 |
| Hem-Fir | 0.43 | 3500 |
| Hem-Fir (North) | 0.46 | 4000 |
| Mixed Maple | 0.55 | 5550 |
| Mixed Oak | 0.68 | 8150 |
| Mixed Southern Pine | 0.51 | 4800 |
| Mountain Hemlock | 0.47 | 4150 |
| Northern Pine | 0.42 | 3350 |
| Northern Red Oak | 0.68 | 8150 |
| Northern Species | 0.35 | 2400 |
| Northern White Cedar | 0.31 | 1900 |
| Ponderosa Pine | 0.43 | 3500 |
| Red Maple | 0.58 | 6100 |
| Red Oak | 0.67 | 7950 |
| Red Pine | 0.44 | 3650 |
| Redwood, close grain | 0.44 | 3650 |
| Redwood, open grain | 0.37 | 2650 |
| Sitka Spruce | 0.43 | 3500 |
| Southern Pine | 0.55 | 5550 |
| Spruce-Pine-Fir | 0.42 | 3350 |
| Spruce-Pine-Fir (South) | 0.36 | 2550 |
| Western Cedars[a] | 0.36 | 2550 |
| Western Cedars (North) | 0.35 | 2400 |
| Western Hemlock | 0.47 | 4150 |
| Western Hemlock (North) | 0.46 | 4000 |
| Western White Pine | 0.40 | 3100 |
| Western Woods | 0.36 | 2550 |
| White Oak | 0.73 | 9300 |
| Yellow Poplar | 0.43 | 3500 |

[a]Alaska Cedar has a specific gravity of 0.46. Use $F_e = 4000$ psi

[b]Specific gravity based on weight and volume when oven-dry.

[c]Applies only to Engelmann Spruce-Lodgepole Pine machine stress rated (MSR) structural lumber.

Source: ANSI/NF$_o$PA NDS—1991, *National Design Specification for Wood Construction*, AFPA, 1250 Connecticut Ave., NW, Washington, DC 20036.

# TABLE 7.40

**Cut Thread or Rolled Thread Wood Screw Withdrawal Design Values (W)[a,b]. Tabulated withdrawal design values (W) are in pounds per inch of thread penetration into side grain of main member. Thread length is approximately 2/3 the total wood screw length (see Reference 9).**

| Specific Gravity G | Wood Screw Gage | | | | | | | | | | |
|---|---|---|---|---|---|---|---|---|---|---|---|
| | 6g | 7g | 8g | 9g | 10g | 12g | 14g | 16g | 18g | 20g | 24g |
| 0.73 | 209 | 229 | 249 | 268 | 288 | 327 | 367 | 406 | 446 | 485 | 564 |
| 0.71 | 198 | 216 | 235 | 254 | 272 | 310 | 347 | 384 | 421 | 459 | 533 |
| 0.68 | 181 | 199 | 216 | 233 | 250 | 284 | 318 | 352 | 387 | 421 | 489 |
| 0.67 | 176 | 193 | 209 | 226 | 243 | 276 | 309 | 342 | 375 | 409 | 475 |
| 0.58 | 132 | 144 | 157 | 169 | 182 | 207 | 232 | 256 | 281 | 306 | 356 |
| 0.55 | 119 | 130 | 141 | 152 | 163 | 186 | 208 | 231 | 253 | 275 | 320 |
| 0.51 | 102 | 112 | 121 | 131 | 141 | 160 | 179 | 198 | 217 | 237 | 275 |
| 0.50 | 98 | 107 | 117 | 126 | 135 | 154 | 172 | 191 | 209 | 228 | 264 |
| 0.49 | 94 | 103 | 112 | 121 | 130 | 147 | 165 | 183 | 201 | 219 | 254 |
| 0.47 | 87 | 95 | 103 | 111 | 119 | 136 | 152 | 168 | 185 | 201 | 234 |
| 0.46 | 83 | 91 | 99 | 107 | 114 | 130 | 146 | 161 | 177 | 193 | 224 |
| 0.44 | 76 | 83 | 90 | 97 | 105 | 119 | 133 | 148 | 162 | 176 | 205 |
| 0.43 | 73 | 79 | 86 | 93 | 100 | 114 | 127 | 141 | 155 | 168 | 196 |
| 0.42 | 69 | 76 | 82 | 89 | 95 | 108 | 121 | 134 | 147 | 161 | 187 |
| 0.41 | 66 | 72 | 78 | 85 | 91 | 103 | 116 | 128 | 141 | 153 | 178 |
| 0.40 | 63 | 69 | 75 | 81 | 86 | 98 | 110 | 122 | 134 | 146 | 169 |
| 0.39 | 60 | 65 | 71 | 77 | 82 | 93 | 105 | 116 | 127 | 138 | 161 |
| 0.38 | 57 | 62 | 67 | 73 | 78 | 89 | 99 | 110 | 121 | 131 | 153 |
| 0.37 | 54 | 59 | 64 | 69 | 74 | 84 | 94 | 104 | 114 | 125 | 145 |
| 0.36 | 51 | 56 | 60 | 65 | 70 | 80 | 89 | 99 | 108 | 118 | 137 |
| 0.35 | 48 | 53 | 57 | 62 | 66 | 75 | 84 | 93 | 102 | 111 | 130 |
| 0.31 | 38 | 41 | 45 | 48 | 52 | 59 | 66 | 73 | 80 | 87 | 102 |

[a]Design value obtained from Equation $W = 2850G^2D$, where G = specific gravity (oven dry weight and volume) and D = diameter of wood screw.

[b]Tabulated withdrawal design values (W) for wood screw connections shall be multiplied by all applicable adjustment factors (see Table 7.1).

Source: ANSI/NF₀PA NDS—1991, *National Design Specification for Wood Construction*, AFPA, 1250 Connecticut Ave:, NW, Washington, DC 20036.

## TABLE 7.41

## Cut Thread Wood Screw Design Values $(Z)^{1,2}$ for Single Shear (Two Member) Connections with Both Members of Identical Species

| SIDE MEMBER THICNESS | WOOD SCREW DIAMETER | WOOD SCREW GAGE | G=0.67 RED OAK | G=0.55 MIXED MAPLE SOUTHERN PINE | G=0.50 DOUGLAS FIR-LARCH | G=0.49 DOUGLAS FIR-LARCH (N) | G=0.46 DOUGLAS FIR (S) HEM-FIR (N) |
|---|---|---|---|---|---|---|---|
| $t_s$ inches | D inches | | Z lbs. | Z lbs. | Z lbs. | Z lbs. | Z lbs. |
| | 0.138 | 6g | 111 | 85 | 75 | 73 | 68 |
| | 0.151 | 7g | 123 | 95 | 84 | 82 | 76 |
| | 0.164 | 8g | 140 | 108 | 96 | 94 | 88 |
| | 0.177 | 9g | 152 | 119 | 107 | 104 | 97 |
| 1/2 | 0.190 | 10g | 156 | 122 | 109 | 106 | 100 |
| | 0.216 | 12g | 173 | 138 | 124 | 121 | 114 |
| | 0.242 | 14g | 183 | 146 | 132 | 128 | 121 |
| | 0.268 | 16g | 212 | 171 | 155 | 151 | 143 |
| | 0.294 | 18g | 235 | 190 | 172 | 168 | 159 |
| | 0.138 | 6g | 127 | 95 | 83 | 80 | 74 |
| | 0.151 | 7g | 140 | 105 | 92 | 89 | 83 |
| | 0.164 | 8g | 157 | 119 | 105 | 101 | 94 |
| | 0.177 | 9g | 170 | 130 | 114 | 111 | 103 |
| 5/8 | 0.190 | 10g | 173 | 133 | 117 | 114 | 106 |
| | 0.216 | 12g | 189 | 147 | 131 | 127 | 119 |
| | 0.242 | 14g | 198 | 155 | 138 | 134 | 125 |
| | 0.268 | 16g | 227 | 179 | 161 | 156 | 147 |
| | 0.294 | 18g | 252 | 199 | 178 | 174 | 163 |
| | 0.138 | 6g | 132 | 106 | 92 | 88 | 81 |
| | 0.151 | 7g | 150 | 117 | 102 | 98 | 90 |
| | 0.164 | 8g | 177 | 132 | 114 | 111 | 102 |
| | 0.177 | 9g | 190 | 142 | 124 | 120 | 111 |
| | 0.190 | 10g | 193 | 145 | 127 | 123 | 114 |
| 3/4 | 0.216 | 12g | 208 | 159 | 140 | 135 | 126 |
| | 0.242 | 14g | 217 | 166 | 146 | 142 | 132 |
| | 0.268 | 16g | 246 | 190 | 169 | 164 | 153 |
| | 0.294 | 18g | 272 | 210 | 187 | 182 | 170 |
| | 0.320 | 20g | 310 | 242 | 216 | 210 | 197 |
| | 0.372 | 24g | 362 | 283 | 252 | 246 | 230 |
| | 0.138 | 6g | 132 | 110 | 101 | 99 | 93 |
| | 0.151 | 7g | 150 | 125 | 115 | 112 | 106 |
| | 0.164 | 8g | 177 | 148 | 135 | 132 | 121 |
| | 0.177 | 9g | 199 | 167 | 147 | 142 | 130 |
| | 0.190 | 10g | 205 | 171 | 150 | 144 | 132 |
| 1 | 0.216 | 12g | 239 | 186 | 161 | 156 | 143 |
| | 0.242 | 14g | 256 | 193 | 168 | 162 | 150 |
| | 0.268 | 16g | 290 | 217 | 190 | 184 | 170 |
| | 0.294 | 18g | 320 | 240 | 210 | 203 | 188 |
| | 0.320 | 20g | 359 | 272 | 239 | 232 | 215 |
| | 0.372 | 24g | 419 | 317 | 279 | 270 | 251 |
| | 0.138 | 6g | 132 | 110 | 101 | 99 | 93 |
| | 0.151 | 7g | 150 | 125 | 115 | 112 | 106 |
| | 0.164 | 8g | 177 | 148 | 135 | 132 | 125 |
| | 0.177 | 9g | 199 | 167 | 152 | 149 | 141 |
| | 0.190 | 10g | 205 | 171 | 157 | 153 | 145 |
| 1-1/4 | 0.216 | 12g | 239 | 200 | 183 | 179 | 164 |
| | 0.242 | 14g | 256 | 213 | 193 | 186 | 170 |
| | 0.268 | 16g | 305 | 249 | 215 | 208 | 191 |
| | 0.294 | 18g | 340 | 275 | 238 | 229 | 211 |
| | 0.320 | 20g | 403 | 308 | 268 | 259 | 238 |
| | 0.372 | 24g | 471 | 359 | 312 | 302 | 278 |
| | 0.138 | 6g | 132 | 110 | 101 | 99 | 93 |
| | 0.151 | 7g | 150 | 125 | 115 | 112 | 106 |
| | 0.164 | 8g | 177 | 148 | 135 | 132 | 125 |
| | 0.177 | 9g | 199 | 167 | 152 | 149 | 141 |
| | 0.190 | 10g | 205 | 171 | 157 | 153 | 145 |
| 1-1/2 | 0.216 | 12g | 239 | 200 | 183 | 179 | 169 |
| | 0.242 | 14g | 256 | 213 | 195 | 191 | 181 |
| | 0.268 | 16g | 305 | 255 | 233 | 228 | 214 |
| | 0.294 | 18g | 340 | 284 | 260 | 254 | 236 |
| | 0.320 | 20g | 403 | 336 | 300 | 289 | 265 |
| | 0.372 | 24g | 471 | 394 | 349 | 337 | 309 |

**TABLE 7.41   (Continued)**

| SIDE MEMBER THICKNESS | WOOD SCREW DIAMETER | WOOD SCREW GAGE | G=0.43 HEM-FIR | G=0.42 SPRUCE-PINE-FIR | G=0.37 REDWOOD (open grain) | G=0.36 EASTERN SOFTWOODS SPRUCE-PINE-FIR(S) WESTERN CEDARS WESTERN WOODS | G=0.35 NORTHERN SPECIES |
|---|---|---|---|---|---|---|---|
| $t_s$ inches | $D$ inches | | $Z$ lbs. | $Z$ lbs. | $Z$ lbs. | $Z$ lbs. | $Z$ lbs. |
| | 0.138 | 6g | 62 | 60 | 52 | 51 | 49 |
| | 0.151 | 7g | 70 | 68 | 59 | 58 | 56 |
| | 0.164 | 8g | 81 | 79 | 68 | 67 | 65 |
| | 0.177 | 9g | 90 | 87 | 77 | 75 | 72 |
| 1/2 | 0.190 | 10g | 92 | 90 | 79 | 77 | 74 |
| | 0.216 | 12g | 105 | 103 | 91 | 89 | 86 |
| | 0.242 | 14g | 112 | 110 | 97 | 95 | 92 |
| | 0.268 | 16g | 133 | 130 | 115 | 113 | 107 |
| | 0.294 | 18g | 148 | 145 | 128 | 125 | 118 |
| | 0.138 | 6g | 67 | 65 | 55 | 54 | 52 |
| | 0.151 | 7g | 75 | 73 | 62 | 61 | 58 |
| | 0.164 | 8g | 86 | 83 | 72 | 70 | 67 |
| | 0.177 | 9g | 95 | 92 | 79 | 78 | 75 |
| 5/8 | 0.190 | 10g | 97 | 94 | 81 | 79 | 77 |
| | 0.216 | 12g | 109 | 106 | 93 | 91 | 88 |
| | 0.242 | 14g | 116 | 113 | 99 | 96 | 93 |
| | 0.268 | 16g | 136 | 132 | 116 | 114 | 110 |
| | 0.294 | 18g | 151 | 147 | 129 | 127 | 123 |
| | 0.138 | 6g | 73 | 71 | 60 | 58 | 55 |
| | 0.151 | 7g | 82 | 79 | 67 | 65 | 62 |
| | 0.164 | 8g | 92 | 90 | 76 | 74 | 71 |
| | 0.177 | 9g | 101 | 98 | 83 | 81 | 78 |
| | 0.190 | 10g | 103 | 100 | 85 | 83 | 80 |
| 3/4 | 0.216 | 12g | 115 | 112 | 96 | 94 | 90 |
| | 0.242 | 14g | 121 | 118 | 102 | 99 | 96 |
| | 0.268 | 16g | 141 | 137 | 119 | 116 | 112 |
| | 0.294 | 18g | 156 | 152 | 132 | 129 | 125 |
| | 0.320 | 20g | 182 | 177 | 155 | 152 | 147 |
| | 0.372 | 24g | 212 | 207 | 181 | 177 | 171 |
| | 0.138 | 6g | 87 | 84 | 69 | 67 | 64 |
| | 0.151 | 7g | 97 | 93 | 77 | 75 | 71 |
| | 0.164 | 8g | 108 | 104 | 87 | 84 | 80 |
| | 0.177 | 9g | 117 | 113 | 94 | 91 | 87 |
| | 0.190 | 10g | 119 | 115 | 96 | 93 | 89 |
| 1 | 0.216 | 12g | 129 | 125 | 106 | 103 | 99 |
| | 0.242 | 14g | 135 | 131 | 111 | 108 | 104 |
| | 0.268 | 16g | 154 | 150 | 128 | 124 | 120 |
| | 0.294 | 18g | 171 | 166 | 141 | 138 | 133 |
| | 0.320 | 20g | 196 | 190 | 163 | 160 | 154 |
| | 0.372 | 24g | 229 | 222 | 191 | 186 | 179 |
| | 0.138 | 6g | 87 | 86 | 76 | 75 | 72 |
| | 0.151 | 7g | 99 | 97 | 86 | 85 | 82 |
| | 0.164 | 8g | 117 | 115 | 99 | 96 | 92 |
| | 0.177 | 9g | 132 | 129 | 107 | 104 | 99 |
| | 0.190 | 10g | 136 | 132 | 109 | 106 | 101 |
| 1-1/4 | 0.216 | 12g | 147 | 142 | 118 | 114 | 109 |
| | 0.242 | 14g | 153 | 147 | 123 | 119 | 114 |
| | 0.268 | 16g | 172 | 166 | 139 | 135 | 130 |
| | 0.294 | 18g | 190 | 184 | 154 | 150 | 144 |
| | 0.320 | 20g | 215 | 208 | 176 | 171 | 164 |
| | 0.372 | 24g | 251 | 243 | 206 | 200 | 192 |
| | 0.138 | 6g | 87 | 86 | 76 | 75 | 72 |
| | 0.151 | 7g | 99 | 97 | 86 | 85 | 82 |
| | 0.164 | 8g | 117 | 115 | 102 | 100 | 97 |
| | 0.177 | 9g | 132 | 129 | 115 | 113 | 110 |
| | 0.190 | 10g | 136 | 133 | 118 | 116 | 113 |
| 1-1/2 | 0.216 | 12g | 159 | 155 | 132 | 127 | 121 |
| | 0.242 | 14g | 170 | 166 | 137 | 132 | 126 |
| | 0.268 | 16g | 192 | 185 | 153 | 149 | 142 |
| | 0.294 | 18g | 212 | 204 | 169 | 164 | 157 |
| | 0.320 | 20g | 238 | 230 | 191 | 186 | 178 |
| | 0.372 | 24g | 277 | 268 | 223 | 217 | 207 |

1. Tabulated lateral design values (Z) for wood screw connections shall be multiplied by all applicable adjustment factors (see Table 7.1).

2. Tabulated lateral design values (Z) are for cut thread wood screws inserted in side grain with wood screw axis perpendicular to wood fibers, and with the following wood screw bending yield strengths ($F_{yb}$):

$F_{yb}$ = 100,000 psi for 6g wood screws          $F_{yb}$ = 70,000 psi for 14g and 16g wood screws

$F_{yb}$ = 90,000 psi for 7g, 8g and 9g wood screws          $F_{yb}$ = 60,000 psi for 18g and 20g wood screws

$F_{yb}$ = 80,000 psi for 10g and 12g wood screws          $F_{yb}$ = 45,000 psi for 24g wood screws

Source: ANSI/NF$_o$PA NDS—1991, *National Design Specification for Wood Construction*, AFPA, 1250 Connecticut Ave., NW, Washington, DC 20036.

### 7.5.5  Withdrawal Loads

If possible, structural designs should be such that wood screws are not loaded in withdrawal. When such loading is unavoidable, the tensile strength of the wood screw at its net (root) section should not be exceeded. Loading in withdrawal from end grain is not permitted.

Table 7.40 may be used to determine withdrawal design values for one wood screw inserted in side grain of the member holding the screw point of a two-member joint. The effective depth of penetration used to determine the design value is the length of the threaded portion of the screw in the member receiving the point. Approximately two-thirds of the length of a wood screw is threaded. The diameter of a wood screw at the root of the thread varies from tip to end of thread and also may vary with the type of screw. The ratio of the diameter at the root of the threads to the nominal diameter also varies with size.

**Example.**  Determine the withdrawal load $Wp$ of a 3-in. No. 12 wood screw fastening a 7 gauge strap to seasoned Douglas Fir used in a dry location. The load is of normal duration ($C_D = 1.0$). The threaded length of the screw is $\frac{2}{3}$ the length $= 2$ in. From Table 7.40, $W = 154$ lb/in. for species with $G = 0.50$.

$$Wp = (154)(2) = 308 \text{ lb.}$$

Check capacity $T_s$ at the net section assuming the diameter at the root of the thread $D_r$ is $\frac{2}{3}$ of the nominal diameter and $F_t = 20,000$ psi.

$D =$ Diameter, No. 12 screw $= 0.216$ in.
Net Section $= \pi D_r^2/4 = \pi[(0.216)(2/3)]^2/4 = 0.01629$ in.$^2$
$T_s = (0.01629)20,000 = 325$ lb $> 308$ lb,
therefore, use withdrawal load of 308 lb.

### 7.5.6  Lateral Loads Wood-to-Wood Connections

The nominal lateral design value $Z$ for wood screws in single shear applies to screws inserted in side grain with the axis of the wood screw perpendicular to the wood fibers. It applies to any angle of load to grain. The depth of penetration $p$ in the main member is seven times the shank diameter ($7D$) for full design load. Minimum penetration is $4D$. When a lesser penetration is obtained but no less than the minimum penetration required, the design value must be adjusted by $C_d$. The adjustment factor $C_d$ to use for intermediate penetration is:

$$C_d = \frac{p}{7D} \le 1.0 \tag{7-27}$$

where  $p =$ length of penetration (in.),
$\quad\quad\;\; D =$ diameter of shank (in.).

The nominal design value $Z$ can be determined from Tables 7.41 and 7.42, or can be calculated using the following equations:

The nominal wood screw design value shall be the lesser of:

Yield Mode

$$Z = \frac{Dt_s F_{es}}{K_D} \qquad \text{Mode } I_s \qquad (7\text{-}28)$$

$$Z = \frac{kDt_s F_{em}}{K_D(2 + R_e)} \qquad \text{Mode } III_s \qquad (7\text{-}29)$$

$$Z = \frac{D^2}{K_D} \sqrt{\frac{1.75 F_{em} F_{yb}}{3(1 + R_e)}} \qquad \text{Mode } IV \qquad (7\text{-}30)$$

where $\quad k = -1 + \sqrt{\dfrac{2(1 + R_e)}{R_e} + \dfrac{F_{yb}(2 + R_e)D^2}{2 F_{em} t_s^2}}$

$R_e = F_{em}/F_{es}$,

$t_s =$ thickness of side member (in.),

$F_{em} =$ dowel bearing strength of main member (member holding point) (psi) (see Table 7.42),

$F_{es} =$ dowel bearing strength of side member (psi) (see Table 7.42),

$F_{yb} =$ bending yield strength of wood screw (psi),

$F_{yb} =$ 100,000 psi for No. 6 wood screws,

$F_{yb} =$ 90,000 psi for Nos. 7, 8, and 9 wood screws,

$F_{yb} =$ 80,000 psi for Nos. 10 and 12 wood screws,

$F_{yb} =$ 70,000 psi for Nos. 14 and 16 wood screws,

$F_{yb} =$ 60,000 psi for Nos. 18 and 20 wood screws,

$F_{yb} =$ 45,000 psi for No. 24 wood screws,

$D =$ unthreaded shank diameter of wood screw (in.),

$K_D =$ 2.2 for $D \le 0.17$ in.,

$K_D =$ $10D + 0.5$, for $0.17$ in. $< D < 0.25$ in., and

$K_D =$ 3.0, for $D \ge 0.25$ in.

### 7.5.7 Wood-to-Metal Connections

Table 7.42 contains design values for lateral loads for wood members with steel side plates. The lateral design loads can also be determined using the lesser of the values obtained from Eqs. (7-29)–(7-30) and using the dowel bearing strength of the metal. The dowel bearing strength of side plates of ASTM A446 Grade A Steel, $F_{es}$ is 45,000 psi.

When a wood screw is inserted in end grain and the load acts perpendicular to the grain, the design values for lateral resistance are two-thirds of those for lateral resistance given in Tables 7.41 and 7.42.

For most designs, the nominal lateral wood screw design values in Tables 7.41 and 7.42 can be used. If different species are used in the main member and side member or if a different thickness side plate is used, Eqs. (7-28)–(7-30) are useful.

**TABLE 7.42**

## Cut Thread Wood Screw Design Values $(Z)^{1,2,3}$ for Single Shear (Two Member) Connections with ASTM A446, Grade A Steel Side Plate

| STEEL SIDE PLATE | WOOD SCREW DIAMETER | WOOD SCREW GAGE | G=0.67 RED OAK | G=0.55 MIXED MAPLE SOUTHERN PINE | G=0.50 DOUGLAS FIR-LARCH | G=0.49 DOUGLAS FIR-LARCH (N) | G=0.46 DOUGLAS FIR (S) HEM-FIR (N) |
|---|---|---|---|---|---|---|---|
| | D inches | | Z lbs. | Z lbs. | Z lbs. | Z lbs. | Z lbs. |
| 3 gage $t_s$=0.239" | 0.242 | 14g | 276 | 241 | 225 | 221 | 211 |
| | 0.268 | 16g | 315 | 275 | 256 | 252 | 241 |
| | 0.294 | 18g | 349 | 304 | 283 | 278 | 267 |
| | 0.320 | 20g | 400 | 348 | 325 | 319 | 305 |
| | 0.372 | 24g | 467 | 407 | 379 | 372 | 356 |
| 7 gage $t_s$=0.179" | 0.190 | 10g | 210 | 184 | 171 | 168 | 161 |
| | 0.216 | 12g | 234 | 204 | 190 | 186 | 178 |
| | 0.242 | 14g | 247 | 215 | 200 | 196 | 188 |
| | 0.268 | 16g | 286 | 248 | 231 | 227 | 217 |
| | 0.294 | 18g | 318 | 276 | 256 | 252 | 241 |
| | 0.320 | 20g | 369 | 320 | 297 | 292 | 279 |
| | 0.372 | 24g | 431 | 374 | 347 | 341 | 326 |
| 10 gage $t_s$=0.134" | 0.138 | 6g | 131 | 114 | 107 | 105 | 100 |
| | 0.151 | 7g | 147 | 128 | 119 | 117 | 112 |
| | 0.164 | 8g | 169 | 147 | 137 | 134 | 129 |
| | 0.177 | 9g | 187 | 162 | 151 | 148 | 142 |
| | 0.190 | 10g | 192 | 166 | 155 | 152 | 145 |
| | 0.216 | 12g | 217 | 188 | 175 | 172 | 164 |
| | 0.242 | 14g | 230 | 199 | 185 | 182 | 174 |
| | 0.268 | 16g | 271 | 234 | 217 | 213 | 204 |
| | 0.294 | 18g | 301 | 260 | 241 | 237 | 226 |
| | 0.320 | 20g | 353 | 304 | 282 | 277 | 265 |
| | 0.372 | 24g | 413 | 356 | 330 | 324 | 309 |
| 11 gage $t_s$=0.12" | 0.138 | 6g | 127 | 110 | 103 | 101 | 96 |
| | 0.151 | 7g | 142 | 124 | 115 | 113 | 108 |
| | 0.164 | 8g | 164 | 143 | 133 | 130 | 125 |
| | 0.177 | 9g | 182 | 158 | 147 | 144 | 138 |
| | 0.190 | 10g | 187 | 162 | 151 | 148 | 141 |
| | 0.216 | 12g | 213 | 184 | 171 | 168 | 160 |
| | 0.242 | 14g | 226 | 196 | 182 | 178 | 170 |
| | 0.268 | 16g | 267 | 230 | 214 | 210 | 200 |
| | 0.294 | 18g | 297 | 256 | 238 | 233 | 223 |
| | 0.320 | 20g | 349 | 301 | 279 | 274 | 261 |
| | 0.372 | 24g | 408 | 352 | 326 | 320 | 306 |
| 12 gage $t_s$=0.105" | 0.138 | 6g | 122 | 106 | 99 | 97 | 93 |
| | 0.151 | 7g | 138 | 119 | 111 | 109 | 104 |
| | 0.164 | 8g | 160 | 138 | 129 | 126 | 121 |
| | 0.177 | 9g | 178 | 154 | 143 | 140 | 134 |
| | 0.190 | 10g | 183 | 158 | 147 | 144 | 138 |
| | 0.216 | 12g | 209 | 181 | 168 | 165 | 157 |
| | 0.242 | 14g | 223 | 192 | 178 | 175 | 167 |
| | 0.268 | 16g | 264 | 227 | 211 | 207 | 197 |
| | 0.294 | 18g | 294 | 253 | 234 | 230 | 219 |
| | 0.320 | 20g | 346 | 298 | 276 | 271 | 258 |
| | 0.372 | 24g | 405 | 348 | 323 | 317 | 302 |
| 14 gage $t_s$=0.075" | 0.138 | 6g | 116 | 100 | 93 | 91 | 87 |
| | 0.151 | 7g | 131 | 113 | 105 | 103 | 98 |
| | 0.164 | 8g | 153 | 132 | 122 | 120 | 115 |
| | 0.177 | 9g | 172 | 148 | 137 | 135 | 128 |
| | 0.190 | 10g | 177 | 152 | 141 | 138 | 132 |
| | 0.216 | 12g | 205 | 176 | 163 | 160 | 152 |
| | 0.242 | 14g | 219 | 188 | 174 | 170 | 163 |
| | 0.268 | 16g | 261 | 223 | 207 | 203 | 193 |
| | 0.294 | 18g | 290 | 249 | 230 | 226 | 215 |
| 16 gage $t_s$=0.06" | 0.138 | 6g | 114 | 98 | 90 | 89 | 85 |
| | 0.151 | 7g | 129 | 111 | 103 | 101 | 96 |
| | 0.164 | 8g | 151 | 130 | 120 | 118 | 113 |
| | 0.177 | 9g | 171 | 146 | 135 | 133 | 127 |
| | 0.190 | 10g | 175 | 150 | 139 | 136 | 130 |
| 18 gage $t_s$=0.048" | 0.138 | 6g | 113 | 97 | 89 | 88 | 84 |
| | 0.151 | 7g | 128 | 110 | 101 | 99 | 95 |
| | 0.164 | 8g | 151 | 129 | 119 | 117 | 112 |

## TABLE 7.42  (*Continued*)

| STEEL SIDE PLATE | WOOD SCREW DIAMETER | WOOD SCREW GAGE | G=0.43 HEM-FIR | G=0.42 SPRUCE-PINE-FIR | G=0.37 REDWOOD (open grain) | G=0.36 EASTERN SOFTWOODS SPRUCE-PINE-FIR(S) WESTERN CEDARS WESTERN WOODS | G=0.35 NORTHERN SPECIES |
|---|---|---|---|---|---|---|---|
| | $D$ inches | | $Z$ lbs. | $Z$ lbs. | $Z$ lbs. | $Z$ lbs. | $Z$ lbs. |
| 3 gage $t_s$=0.239" | 0.242 | 14g | 200 | 197 | 178 | 175 | 171 |
| | 0.268 | 16g | 228 | 224 | 203 | 199 | 194 |
| | 0.294 | 18g | 252 | 248 | 224 | 221 | 215 |
| | 0.320 | 20g | 289 | 283 | 257 | 252 | 246 |
| | 0.372 | 24g | 337 | 331 | 299 | 294 | 287 |
| 7 gage $t_s$=0.179" | 0.190 | 10g | 152 | 150 | 135 | 133 | 130 |
| | 0.216 | 12g | 169 | 166 | 150 | 147 | 144 |
| | 0.242 | 14g | 178 | 174 | 158 | 155 | 151 |
| | 0.268 | 16g | 205 | 201 | 182 | 179 | 174 |
| | 0.294 | 18g | 228 | 223 | 202 | 199 | 193 |
| | 0.320 | 20g | 264 | 259 | 234 | 230 | 224 |
| | 0.372 | 24g | 308 | 302 | 273 | 268 | 261 |
| 10 gage $t_s$=0.134" | 0.138 | 6g | 95 | 93 | 84 | 83 | 81 |
| | 0.151 | 7g | 106 | 104 | 94 | 93 | 90 |
| | 0.164 | 8g | 122 | 119 | 108 | 106 | 103 |
| | 0.177 | 9g | 134 | 132 | 119 | 117 | 114 |
| | 0.190 | 10g | 137 | 135 | 122 | 120 | 117 |
| | 0.216 | 12g | 155 | 152 | 137 | 135 | 131 |
| | 0.242 | 14g | 164 | 161 | 145 | 143 | 139 |
| | 0.268 | 16g | 192 | 188 | 170 | 167 | 163 |
| | 0.294 | 18g | 213 | 209 | 189 | 186 | 181 |
| | 0.320 | 20g | 250 | 245 | 221 | 217 | 211 |
| | 0.372 | 24g | 292 | 286 | 258 | 254 | 247 |
| 11 gage $t_s$=0.12" | 0.138 | 6g | 91 | 89 | 81 | 80 | 77 |
| | 0.151 | 7g | 102 | 100 | 91 | 89 | 87 |
| | 0.164 | 8g | 118 | 115 | 104 | 103 | 100 |
| | 0.177 | 9g | 130 | 128 | 115 | 113 | 110 |
| | 0.190 | 10g | 133 | 131 | 118 | 116 | 113 |
| | 0.216 | 12g | 152 | 149 | 134 | 132 | 128 |
| | 0.242 | 14g | 161 | 158 | 142 | 140 | 136 |
| | 0.268 | 16g | 189 | 185 | 167 | 164 | 160 |
| | 0.294 | 18g | 210 | 206 | 186 | 182 | 177 |
| | 0.320 | 20g | 246 | 242 | 218 | 214 | 208 |
| | 0.372 | 24g | 288 | 283 | 254 | 250 | 243 |
| 12 gage $t_s$=0.105" | 0.138 | 6g | 88 | 86 | 78 | 76 | 74 |
| | 0.151 | 7g | 98 | 97 | 87 | 86 | 84 |
| | 0.164 | 8g | 114 | 112 | 101 | 99 | 97 |
| | 0.177 | 9g | 127 | 124 | 112 | 110 | 107 |
| | 0.190 | 10g | 130 | 127 | 115 | 113 | 110 |
| | 0.216 | 12g | 148 | 145 | 131 | 129 | 125 |
| | 0.242 | 14g | 158 | 155 | 139 | 137 | 133 |
| | 0.268 | 16g | 186 | 182 | 164 | 161 | 157 |
| | 0.294 | 18g | 207 | 203 | 183 | 179 | 175 |
| | 0.320 | 20g | 243 | 239 | 215 | 211 | 205 |
| | 0.372 | 24g | 285 | 279 | 251 | 247 | 240 |
| 14 gage $t_s$=0.075" | 0.138 | 6g | 82 | 80 | 72 | 71 | 69 |
| | 0.151 | 7g | 92 | 91 | 82 | 80 | 78 |
| | 0.164 | 8g | 108 | 106 | 95 | 94 | 91 |
| | 0.177 | 9g | 121 | 119 | 107 | 105 | 102 |
| | 0.190 | 10g | 124 | 122 | 110 | 108 | 105 |
| | 0.216 | 12g | 144 | 141 | 127 | 124 | 121 |
| | 0.242 | 14g | 153 | 150 | 135 | 133 | 129 |
| | 0.268 | 16g | 182 | 178 | 160 | 157 | 153 |
| | 0.294 | 18g | 203 | 199 | 178 | 175 | 170 |
| 16 gage $t_s$=0.06" | 0.138 | 6g | 80 | 78 | 70 | 69 | 67 |
| | 0.151 | 7g | 90 | 89 | 80 | 78 | 76 |
| | 0.164 | 8g | 106 | 104 | 93 | 92 | 89 |
| | 0.177 | 9g | 119 | 117 | 105 | 103 | 100 |
| | 0.190 | 10g | 122 | 120 | 108 | 106 | 103 |
| 18 gage $t_s$=0.048" | 0.138 | 6g | 79 | 77 | 69 | 68 | 66 |
| | 0.151 | 7g | 89 | 88 | 79 | 77 | 75 |
| | 0.164 | 8g | 105 | 103 | 92 | 91 | 88 |

1. Tabulated lateral design values (Z) for wood screw connections shall be multiplied by all applicable adjustment factors (see Table 7.1).

2. Tabulated lateral design values (Z) are for cut thread wood screws inserted in side grain with wood screw axis perpendicular to wood fibers, and with the following wood screw bending yield strengths ($F_{yb}$):

$F_{yb} = 100{,}000$ psi for 6g wood screws

$F_{yb} = 90{,}000$ psi for 7g, 8g and 9g wood screws

$F_{yb} = 80{,}000$ psi for 10g and 12g wood screws

$F_{yb} = 70{,}000$ psi for 14g and 16g wood screws

$F_{yb} = 60{,}000$ psi for 18g and 20g wood screws

$F_{yb} = 45{,}000$ psi for 24g wood screws

3. Tabulated lateral design values (Z) are based on a dowel bearing strength ($F_e$) of 45,000 psi for ASTM A446, Grade A steel.

Source: ANSI/NF$_o$PA NDS—1991, *National Design Specification for Wood Construction*, AFPA, 1250 Connecticut Ave., NW, Washington, DC 20036.

**Example.**   A $\frac{7}{8}$ in. thick piece of Redwood (open grain) is attached to a 4 × 6 Southern Pine member with a No. 24 wood screw. Determine the nominal lateral design value of the connection. Loads act parallel to grain.

Input values for the Yield Mode equations are:

$F_{em}$ = 5550 psi (Southern Pine)
$F_{es}$ = 2650 psi
$R_e = F_{em}/F_{es} = \dfrac{5550}{2650} = 2.094$
$D$ = 0.372 in. (Table 7.42)
$K_D$ = 3.0 [Eq. (7-30)]
$F_{yb}$ = 45,000 psi [Eq. (7-30)]

$$k = -1 + \sqrt{\frac{2(1 + R_e)}{R_e} + \frac{F_{yb}(2 + R_e)D^2}{2F_{em}t_s^2}}$$

$$k = -1 + \sqrt{\frac{2(1 + 2.094)}{2.094} + \frac{45,000(2 + 2.094)(0.372)^2}{(2)(5550)(0.875)^2}} = 1.44$$

Mode $I_s$   $Z = \dfrac{Dt_s F_{es}}{K_D} = \dfrac{(0.372)(0.875)(2650)}{3.0} = 288$ lb

Mode $III_s$   $Z = \dfrac{kDt_s F_{em}}{K_D(2 + R_e)} = \dfrac{(1.44)(0.372)(0.875)(5550)}{3(2 + 2.094)} = 212$ lb

Mode IV   $Z = \dfrac{D^2}{K_D}\sqrt{\dfrac{1.75F_{em}F_{yb}}{3(1 + R_e)}} = \dfrac{0.372^2}{3}\sqrt{\dfrac{(1.75)(5,550)(45,000)}{3(1 + 2.094)}}$

$= 317$ lb

Failure Mode $III_s$ controls.
$Z = 212$ lb.

**Example.**   Determine the design value, $W'p$ in withdrawal of a No. 16 wood screw when the threaded portion penetrates $1\frac{3}{4}$ in. in a Southern Pine glued laminated timber used under dry conditions. The load to be resisted is a snow load.

$$C_M = 1, \quad C_D = 1.15$$

From Table 7.39, Southern Pine has specific gravity of 0.55.
From Table 7.40, the design value, $W$ is 231 lb/in. of penetration.

$$Wp = (231)(1.75) = 404 \text{ lb}$$

$$W'p = WpC_D = (404)(1.15) = 465 \text{ lb}$$

**Example.**   Determine the lateral wind load that four No. 14 wood screws 2-in. long can carry in Hem-Fir with $\frac{1}{4}$-in. steel side plates.
From Table 7.38, diameter of No. 14 gauge screw = 0.242 in.

Depth of penetration required for full load is

$7D = 7(0.242) = 1.69$ in.
$2 - 0.25 = 1.75$ in. $> 1.69$ in.    O.K.
For wind load, $C_D = 1.6$. From Table 7.42, $Z = 200$ lb for No. 3 gauge steel
side plate. Thickness of No. 3 gauge steel plate $= 0.239$ in. $< 0.25$ in. Use
value for No. 3 gauge steel side plate.
$Z' = (200)(1.6) = 320$ lb
$P = 4Z' = (4)(320) = 1280$ lb

**Example.**    Determine the lateral load for one No. 14 wood screw $1\frac{1}{2}$ in. long
attaching a $\frac{1}{4}$ in. steel side plate to seasoned Douglas Fir, normal duration of load.
The fastening is to be used in a wet location. Depth of penetration required for
full load $(7D)$ is $(7)(0.242) = 1.69$ in. Actual penetration $= 1\frac{1}{2} - \frac{1}{4} = 1.25$ in.
Minimum depth of embedment $= 4D$

$4D = (4)(0.242) = 0.97$ in. $< 1.25$ in.    O.K.
Depth of embedment adjustment factor, $C_d = p/7D = 1.25/1.69 = 0.74$.

From Table 7.42, $Z = 225$ lb using No. 3 gauge steel side plate (0.239 in. thick).
From Table 7.2, $C_M = 0.67$.

$$Z' = ZC_d C_M = (225)(0.74)(0.67) = 111 \text{ lb}$$

**Example.**    Determine the lateral resistance $Z'$ of a 10 gauge wood screw 2 in.
long fastening a nominal $1 \times 6$ to a $4 \times 6$. Species is Douglas Fir-Larch.

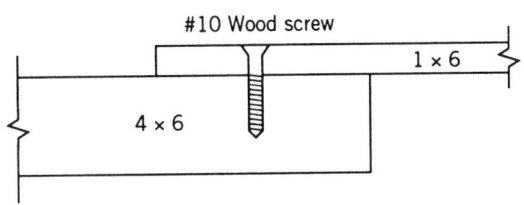

From Table 7.38,

Diameter of 10 gauge wood screw $= 0.190$ in.
Penetration for full load $= 7D = (7)(0.190) = 1.33$ in.
Minimum penetration $= 4D = (4)(0.190) = 0.76$ in.
Actual penetration $= p = 2 - 0.75 = 1.25$ in. $> 4D$
Depth of embedment factor, $C_d = p/7D = 1.25/1.33 = 0.94$

From Table 7.41,
$$Z = 127 \text{ lb}$$

$$Z' = ZC_d = (127)(0.94) = 119 \text{ lb}$$

### 7.5.8    Combined Lateral and Withdrawal Loads

When a load acts to pull a wood screw in withdrawal and exerts a lateral load,
the combined lateral and withdrawal design value for the screw acting at an angle,

$\alpha$, with the surface of the wood is determined as follows:

$$Z'_\alpha = \frac{(W'p)Z'}{Z' \sin^2 \alpha + W'p \cos^2 \alpha}$$

where   $\alpha$ = angle between the wood surface and direction of load (degrees),
       $W'$ = design value in withdrawal (lb/in.), and
       $p$ = length of thread penetration in main member (in.).

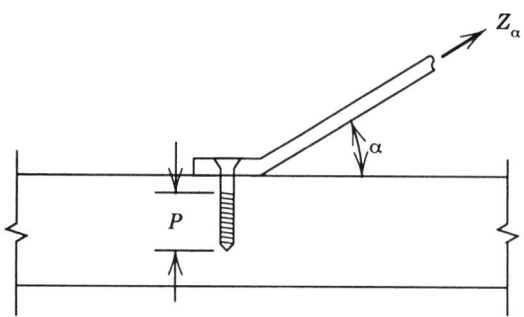

## 7.6  SPIRAL DOWELS

A spiral dowel is a twisted steel rod with spirally grooved ridges along its entire length, the lead point of the spiral thread being sufficient to permit driving by any suitable means. The design values given in Table 7.44 are for any spiral dowel of sufficient strength to cause failure in the wood rather than in the metal. Ranges of diameters and lengths of stock sizes of spiral dowels are listed in Table 7.43. Spiral dowels require prebored lead holes of the diameters given in Table 7.43. The detail illustrated in Fig. 7.27 is recommended for driving long dowels.

### 7.6.1  Lumber Species and Specific Gravity

The design values for spiral dowels are for Douglas Fir-Larch and Southern Pine and other species with a specific gravity of 0.49 or greater.

### 7.6.2  Availability

The designer should check on availability before using spiral dowels in design.

### 7.6.3  Spiral Dowel Design Values

The design values in Table 7.44 are for either sawn lumber or glued laminated timber, and apply for one spiral dowel in a two-member joint under normal load duration and dry-use conditions. The design values given in Table 7.44 are for any spiral dowel of sufficient strength to cause failure in the wood rather than in the metal.

The total design value for more than one spiral dowel is the sum of the values

**TABLE 7.43**

**Dimensions and Lead Hole Diameters for Spiral Dowels in Species with Specific Gravity G ≥ 0.49[a]**

| Outside Diameter of Dowel (in.) | Minimum Length (in.)[b] | Maximum Length (in.)[b] | Maximum Driving Length (in.)[c,d] | Diameter of Lead Hole (in.)[c] |
|---|---|---|---|---|
| 1/4 | 2-1/2 | 6 | | 3/16 |
| 5/16 | 3 | 6-1/2 | | 1/4 |
| 3/8 | 3-1/2 | 10 | | 9/32 |
| 7/16 | 3-1/2 | 12 | 12 | 11/32 |
| 1/2 | 4 | 18 | 12 | 3/8 |
| 5/8 | 4 | 24 | 12 | 15/32 |
| 3/4 | | | 12 | 9/16 |
| 7/8 | Lengths in these three sizes | must be specially ordered | 13 | 21/32 |
| 1 | | | 14 | 3/4 |

[a]For species included, see Table 7.39.
[b]Increments of length of $\frac{1}{2}$ in. up to lengths of 8 in. and 1 in. for lengths over 8 in.
[c]See Fig. 7.27.

for each spiral dowel, provided that spacings, end distances, and edge distances are sufficient to develop the full strength of each spiral dowel. The spacings, end distances, and edge distances required for bolts may be used as a guide.

For other than normal duration of loading, the design values should be multiplied by the appropriate adjustment factor $C_D$ from Fig. 4.4 or Table 4.7. For spiral dowel joints used in unseasoned material or exposed to the weather, the value determined should be multiplied by the factor, $C_M = 0.25$. See Table 7.1 for adjustment factors applicable to spiral dowels.

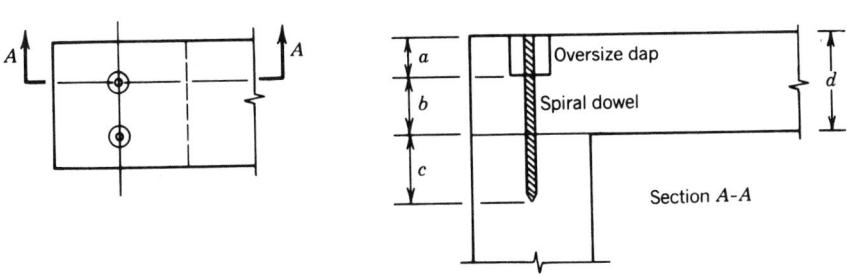

**FIGURE 7.27**  Driving of long spiral dowels. $a$ = overall depth $d$, of member, less $\frac{1}{2}$ maximum driving length: $b = c = \frac{1}{2}$ maximum driving length (see Table 7.43).

**TABLE 7.44**

**Design Values for Spiral Dowels for Southern Pine, Douglas Fir-Larch, and Species with Specific Gravity of 0.49 or Greater[a,b]**

| Outside Diameter of Dowel (in.) | Lateral Load Design Value | Withdrawal Load Design Value[c] (lb/in. of penetration) | |
|---|---|---|---|
| | | Side Grain | End Grain |
| $\frac{1}{4}$ | 111 | 103 | 59 |
| $\frac{5}{16}$ | 174 | 123 | 70 |
| $\frac{3}{8}$ | 249 | 141 | 81 |
| $\frac{7}{16}$ | 340 | 158 | 90 |
| $\frac{1}{2}$ | 444 | 174 | 99 |
| $\frac{5}{8}$ | 692 | 206 | 117 |
| $\frac{3}{4}$ | 998 | 236 | 135 |
| $\frac{7}{8}$ | 1360 | 266 | 151 |
| 1 | 1775 | 293 | 168 |

[a]All applicable provisions under the heading "Spiral Dowel Design Values" must be met to develop these values.

[b]For connections with the thickness of the side member, at least $5D$ and penetration in the main member of at least $7D$.

[c]For spiral dowels installed or used in unseasoned material or exposed to the weather, use 25% of these design values ($C_M = 0.25$).

### 7.6.4    Withdrawal Loads

Withdrawal load design values per inch of penetration in the piece receiving the point for one spiral dowel installed in side or end grain are given in Table 7.44.

### 7.6.5    Lateral Loads

Lateral load design values for one spiral dowel installed in side grain of a two-member joint, when the thickness of the side member is at least 5 times the diameter of the spiral dowel and the dowel penetrates at least 7 diameters into the member receiving the point, are given in Table 7.44.

Lateral strength of spiral dowels in end grain is 60% of the side grain lateral strength, and penetration of 12 diameters into the end grain is required to develop that strength.

## 7.7    DRIFT BOLTS AND PINS

Drift bolts and pins are long unthreaded bolts or pins that are sometimes used to fasten large timbers. Design values for drift bolts and pins have not been studied as much as those for other fasteners and are not usually included in codes. However, an estimate of both the withdrawal design value and the lateral resistance design value can be obtained from information contained in the *Wood Handbook* (6).

The following procedures are based on the *Wood Handbook*.

### 7.7.1   Withdrawal Loads

The ultimate load formula in the *Wood Handbook* (6) can be divided by six and multiplied by 1.10 to obtain an estimated design value for loads of normal duration. This is the same procedure used for nails and spikes. The estimated design value in withdrawal for drift bolts, $P_w$, is

$$P_w = 1200G^2 D \qquad (7-31)$$

where   $P_w$ = design value in withdrawal (lb/in. of penetration),
   $G$ = specific gravity based on oven-dry weight and volume, and
   $D$ = diameter of drift bolt or pin (in.).

The design value is based on use of a predrilled hole $\frac{1}{8}$ in. less in diameter than the pin in seasoned wood.

### 7.7.2   Lateral Resistance

The lateral resistance of drift bolts or pins should be no greater than for bolts of the same diameter. Usually, a lesser value is used because of the lack of nuts and washers or heads on drift pins. The value determined for drift bolts should not exceed 75% of that for the same size bolt. The applicable adjustment factors in Table 7.1 for bolts apply to drift bolts and pins.

Spiral dowels have a greater resistance to withdrawal than drift bolts or pins and their use is recommended where withdrawal is critical.

## 7.8   NAILS AND SPIKES

The design values shown are for common wire nails and spikes, box nails, and threaded, hardened nails and spikes of the sizes conforming to Federal Specification FF-N-105B (10). Commonly used sizes are listed in Table 7.45. Threaded, hardened nails and spikes as covered herein are made of high-carbon steel wire, headed, pointed, and annularly or helically threaded, and heat treated and tempered to provide greater strength than is developed by common wire nails and spikes of corresponding sizes. See Table 7.46 for bending yield strengths for nails and spikes.

When it is necessary to avoid splitting of the wood, a prebored hole of a diameter not exceeding 0.9 of the nail or spike diameter for species with a specific gravity greater than 0.6. The diameter of the prebored hole should not exceed 0.75 of the nail or spike diameter for species with a specific gravity less than 0.6. When these prebored holes are used, the design values for the same size nail or spike taken from Tables 7.50–7.57 may be used.

**TABLE 7.45**

**Nail and Spike Sizes**

| Pennyweight | Length (in.) | Box Nails | Common Wire Nails | Threaded, Hardened Steel Nails | Common Wire Spikes |
|---|---|---|---|---|---|
| | | | **Wire Diameter (in.)** | | |
| 6 $d$ | 2 | 0.099 | 0.113 | 0.120 | |
| 8 $d$ | $2\frac{1}{2}$ | 0.113 | 0.131 | 0.120 | |
| 10 $d$ | 3 | 0.128 | 0.148 | 0.135 | 0.192 |
| 12 $d$ | $3\frac{1}{4}$ | 0.128 | 0.148 | 0.135 | 0.192 |
| 16 $d$ | $3\frac{1}{2}$ | 0.135 | 0.162 | 0.148 | 0.207 |
| 20 $d$ | 4 | 0.148 | 0.192 | 0.177 | 0.225 |
| 30 $d$ | $4\frac{1}{2}$ | 0.148 | 0.207 | 0.177 | 0.244 |
| 40 $d$ | 5 | 0.162 | 0.225 | 0.177 | 0.263 |
| 50 $d$ | $5\frac{1}{2}$ | | 0.244 | 0.177 | 0.283 |
| 60 $d$ | 6 | | 0.263 | 0.177 | 0.283 |
| 70 $d$ | 7 | | | 0.207 | |
| 80 $d$ | 8 | | | 0.207 | |
| 90 $d$ | 9 | | | 0.207 | |
| $\frac{5}{16}$ | 7 | | | | 0.312 |
| $\frac{3}{8}$ | $8\frac{1}{2}$ | | | | 0.375 |

[a]Source: *National Design Specification for Wood Construction* (3).

**TABLE 7.46**

**Bending Yield Strength $F_{yb}$ for Nails and Spikes**

| Type Nail | Diameters, in. | $F_{yb}$ psi |
|---|---|---|
| Box Nail | 0.099, 0.113, 0.128, and 0.135 | 100,000 |
| | 0.148 and 0.162 | 90,000 |
| Common Wire Nail | 0.113 and 0.131 | 100,000 |
| | 0.148 and 0.162 | 90,000 |
| | 0.192, 0.207, and 0.225 | 80,000 |
| | 0.244 and 0.263 | 70,000 |
| Threaded Hardened Steel Nail | 0.12 and 0.135 | 130,000 |
| | 0.148 and 0.177 | 115,000 |
| | 0.207 | 100,000 |
| Common Wire Spike | 0.192, 0.207, and 0.225 | 80,000 |
| | 0.244 and 0.263 | 70,000 |
| | 0.283 and $\frac{5}{6}$ in. dia. | 60,000 |
| | $\frac{3}{8}$ in. dia. | 45,000 |

Source: ANSI/NF$_o$PA NDS—1991, *National Design Specification for Wood Construction*, AFPA, 1250 Connecticut Ave., NW, Washington, DC 20036.

### 7.8.1 Lumber Species and Specific Gravity

The specific gravities of the most commonly used species are listed in Table 7.47. Specific gravities are used in determining both the lateral design value $Z$ and the withdrawal value $W$.

### 7.8.2 Spacing, End Distance, and Edge Distance

Spacing, end distance, and edge distances should be sufficient to avoid unusual splitting of the wood.

### 7.8.3 Nail and Spike Design Values

The design values obtained for both lateral loads and withdrawal loads are for one nail or spike in a two-member joint under normal duration of loading and dry-use conditions. The total design value for more than one nail or spike is the sum of the values for each nail or spike.

For joints in use under other service or seasoning conditions, the values determined from the tables are multiplied by the appropriate adjustment factors from Table 7.2.

### 7.8.4 Withdrawal Loads

If possible, structural designs should be such that nails or spikes are not loaded in withdrawal. When such loading is unavoidable, the withdrawal load design values per inch of penetration for nails or spikes driven in side grain (perpendicular to grain) as determined from Table 7.48 apply. Loading of nails or spikes in withdrawal from end grain is not recommended.

The value $W$ obtained from Table 7.48 is in pounds per inch of penetration $p$ of the piece receiving the point. The withdrawal design load in pounds is determined as follows:

$$\text{Withdrawal design load} = Wp \qquad (7\text{-}32)$$

where  $W$ = withdrawal load (lb/in.) and
  $p$ = penetration in the member receiving the point (in.).

The withdrawal load design values for toe-nailed joints, for all conditions of seasoning, are two-thirds ($C_{tn} = 0.67$) those determined from Table 7.48. The wet service factor $C_M$ does not apply for toe-nails loaded in withdrawal. It is recommended that toe-nails be driven at an angle of approximately 30° to the piece and be started at approximately one-third the nail length from the end of the piece. Toe-nails are usually not permitted to be used in withdrawal when seismic design codes apply.

The withdrawal resistance of clinched, common wire nails is considerably higher than that of unclinched fasteners. The ratio between values for clinched and unclinched nails varies with the moisture content of the wood, the difference in time between when the nail is driven and when the withdrawal load is applied, the species of wood, the size of nail, and the direction of clinch with respect to grain. In dry or green wood, a clinched nail provides from 45 to 170% more with-

## TABLE 7.47

### Dowel Bearing Strength for Nail or Spike Connections

| Species Combination | Specific[c] Gravity G | Dowel bearing strength in pounds per square inch (psi) $F_e$ |
|---|---|---|
| Aspen | 0.39 | 2950 |
| Balsam Fir | 0.36 | 2550 |
| Beech-Birch-Hickory | 0.71 | 8850 |
| Coast Sitka Spruce | 0.39 | 2950 |
| Cottonwood | 0.41 | 3200 |
| Douglas Fir-Larch | 0.50 | 4650 |
| Douglas Fir-Larch (North) | 0.49 | 4450 |
| Douglas Fir-South | 0.46 | 4000 |
| Eastern Hemlock | 0.41 | 3200 |
| Eastern Hemlock-Tamarack | 0.41 | 3200 |
| Eastern Hemlock-Tamarack (North) | 0.47 | 4150 |
| Eastern Softwoods | 0.36 | 2550 |
| Eastern Spruce | 0.41 | 3200 |
| Eastern White Pine | 0.36 | 2550 |
| Engelmann Spruce-Lodgepole Pine[a] (MSR 1650f and higher grades) | 0.46 | 4000 |
| Engelmann Spruce-Lodgepole Pine[a] (MSR 1500f and lower grades) | 0.38 | 2800 |
| Hem-Fir | 0.43 | 3500 |
| Hem-Fir (North) | 0.46 | 4000 |
| Mixed Maple | 0.55 | 5550 |
| Mixed Oak | 0.68 | 8150 |
| Mixed Southern Pine | 0.51 | 4800 |
| Mountain Hemlock | 0.47 | 4150 |
| Northern Pine | 0.42 | 3350 |
| Northern Red Oak | 0.68 | 8150 |
| Northern Species | 0.35 | 2400 |
| Northern White Cedar | 0.31 | 1900 |
| Ponderosa Pine | 0.43 | 3500 |
| Red Maple | 0.58 | 6100 |
| Red Oak | 0.67 | 7950 |
| Red Pine | 0.44 | 3650 |
| Redwood, close grain | 0.44 | 3650 |
| Redwood, open grain | 0.37 | 2650 |
| Sitka Spruce | 0.43 | 3500 |
| Southern Pine | 0.55 | 5550 |
| Spruce-Pine-Fir | 0.42 | 3350 |
| Spruce-Pine-Fir (South) | 0.36 | 2550 |
| Western Cedars[b] | 0.36 | 2550 |
| Western Cedars (North) | 0.35 | 2400 |
| Western Hemlock | 0.47 | 4150 |
| Western Hemlock (North) | 0.46 | 4000 |
| Western White Pine | 0.40 | 3100 |
| Western Woods | 0.36 | 2550 |
| White Oak | 0.73 | 9300 |
| Yellow Poplar | 0.43 | 3500 |

[a]Applies only to Engelmann Spruce-Lodgepole Pine machine stress rated (MSR) structural lumber.

[b]Alaska Cedar has a specific gravity of 0.46. Use $F_e$ = 4000 psi.

[c]Specific gravity based on weight and volume when oven-dry.

Source: ANSI/NF$_o$PA NDS—1991, *National Design Specification for Wood Construction*, AFPA, 1250 Connecticut Ave., NW, Washington, DC 20036.

## TABLE 7.48
### Nail and Spike Withdrawal Design Values (W)[a]

Tabulated withdrawal design values (W) are in pounds per inch of penetration into side grain of main member.

| Specific Gravity G | COMMON WIRE NAILS, BOX NAILS, and COMMON WIRE SPIKES Diameter, D | | | | | | | | | | | | | | | THREADED NAILS Wire Diameter, D | | | | |
|---|---|---|---|---|---|---|---|---|---|---|---|---|---|---|---|---|---|---|---|---|
| | 0.099" | 0.113" | 0.128" | 0.131" | 0.135" | 0.148" | 0.162" | 0.192" | 0.207" | 0.225" | 0.244" | 0.263" | 0.283" | 0.312" | 0.375" | 0.12" | 0.135" | 0.148" | 0.177" | 0.207" |
| 0.73 | 62 | 71 | 80 | 82 | 85 | 93 | 102 | 121 | 130 | 141 | 153 | 165 | 178 | 196 | 236 | 82 | 93 | 102 | 121 | 141 |
| 0.71 | 58 | 66 | 75 | 77 | 79 | 87 | 95 | 113 | 121 | 132 | 143 | 154 | 166 | 183 | 220 | 77 | 87 | 95 | 113 | 132 |
| 0.68 | 52 | 59 | 67 | 69 | 71 | 78 | 85 | 101 | 109 | 118 | 128 | 138 | 149 | 164 | 197 | 69 | 78 | 85 | 101 | 118 |
| 0.67 | 50 | 57 | 65 | 66 | 68 | 75 | 82 | 97 | 105 | 114 | 124 | 133 | 144 | 158 | 190 | 66 | 75 | 82 | 97 | 114 |
| 0.58 | 35 | 40 | 45 | 46 | 48 | 52 | 57 | 68 | 73 | 80 | 86 | 93 | 100 | 110 | 133 | 46 | 52 | 57 | 68 | 80 |
| 0.55 | 31 | 35 | 40 | 41 | 42 | 46 | 50 | 59 | 64 | 70 | 76 | 81 | 88 | 97 | 116 | 41 | 46 | 50 | 59 | 70 |
| 0.51 | 25 | 29 | 33 | 34 | 35 | 38 | 42 | 49 | 53 | 58 | 63 | 67 | 73 | 80 | 96 | 34 | 38 | 42 | 49 | 58 |
| 0.50 | 24 | 28 | 31 | 32 | 33 | 36 | 40 | 47 | 50 | 55 | 60 | 64 | 69 | 76 | 91 | 32 | 36 | 40 | 47 | 55 |
| 0.49 | 23 | 26 | 30 | 30 | 31 | 34 | 38 | 45 | 48 | 52 | 57 | 61 | 66 | 72 | 87 | 30 | 34 | 38 | 45 | 52 |
| 0.47 | 21 | 24 | 27 | 27 | 28 | 31 | 34 | 40 | 43 | 47 | 51 | 55 | 59 | 65 | 78 | 27 | 31 | 34 | 40 | 47 |
| 0.46 | 20 | 22 | 25 | 26 | 27 | 29 | 32 | 38 | 41 | 45 | 48 | 52 | 56 | 62 | 74 | 26 | 29 | 32 | 38 | 45 |
| 0.44 | 18 | 20 | 23 | 23 | 24 | 26 | 29 | 34 | 37 | 40 | 43 | 47 | 50 | 55 | 66 | 23 | 26 | 29 | 34 | 40 |
| 0.43 | 17 | 19 | 21 | 22 | 23 | 25 | 27 | 32 | 35 | 38 | 41 | 44 | 47 | 52 | 63 | 22 | 25 | 27 | 32 | 38 |
| 0.42 | 16 | 18 | 20 | 21 | 21 | 23 | 26 | 30 | 33 | 35 | 38 | 41 | 45 | 49 | 59 | 21 | 23 | 26 | 30 | 35 |
| 0.41 | 15 | 17 | 19 | 19 | 20 | 22 | 24 | 29 | 31 | 33 | 36 | 39 | 42 | 46 | 56 | 19 | 22 | 24 | 29 | 33 |
| 0.40 | 14 | 16 | 18 | 18 | 19 | 21 | 23 | 27 | 29 | 31 | 34 | 37 | 40 | 44 | 52 | 18 | 21 | 23 | 27 | 31 |
| 0.39 | 13 | 15 | 17 | 17 | 18 | 19 | 21 | 25 | 27 | 29 | 32 | 34 | 37 | 41 | 49 | 17 | 19 | 21 | 25 | 29 |
| 0.38 | 12 | 14 | 16 | 16 | 17 | 18 | 20 | 24 | 25 | 28 | 30 | 32 | 35 | 38 | 46 | 16 | 18 | 20 | 24 | 28 |
| 0.37 | 11 | 13 | 15 | 15 | 16 | 17 | 19 | 22 | 24 | 26 | 28 | 30 | 33 | 36 | 43 | 15 | 17 | 19 | 22 | 26 |
| 0.36 | 11 | 12 | 14 | 14 | 14 | 16 | 17 | 21 | 22 | 24 | 26 | 28 | 30 | 33 | 40 | 14 | 16 | 17 | 21 | 24 |
| 0.35 | 10 | 11 | 13 | 13 | 14 | 15 | 16 | 19 | 21 | 23 | 24 | 26 | 28 | 31 | 38 | 13 | 15 | 16 | 19 | 23 |
| 0.31 | 7 | 8 | 9 | 10 | 10 | 11 | 12 | 14 | 15 | 17 | 18 | 19 | 21 | 23 | 28 | 10 | 11 | 12 | 14 | 17 |

[a]Tabulated withdrawal design values (W) for nail or spike connections shall be multiplied by all applicable adjustment factors (see Table 7.1).

Source: ANSI/NF₀PA NDS 1991, *National Design Specification for Wood Construction*, AFPA, 1250 Connecticut Ave., NW, Washington, DC 20036.

drawal resistance than does an unclinched nail withdrawn soon after driving. Nails clinched across the grain have approximately 20% more resistance to withdrawal than do nails clinched long the grain. The proper clinching of nails may be difficult to achieve under field conditions.

### 7.8.5 Lateral Loads

Lateral load $Z$ may be determined by using one of the Tables 7.50–7.57. Select table for the type of fastener and the components—wood or steel side plate. Enter the table selected with the size of fastener, the thickness of the side plate, and the species and read the design value $Z$. When a wood side plate is used, it should be of the same species as the member receiving the point. Otherwise, the design value should be that determined for the species with the lowest specific gravity.

Lateral design values may also be obtained by using Eqs. (7-33)–(7-36). In general, the tabular lateral design values are adequate for most designs. However, use of the yield limit equations allows the designer to determine design values when the main member and side members are of different species and when metals of different bending yield strengths and dowel bearing strengths are used. These equations are based on the nail or spike being driven in the side grain of the main member, with the nail perpendicular to the wood fibers and the depth of penetration of the nail in the main member equal to that required for full load. When a lesser depth of penetration is used but not less than the stated minimum, the design value $Z$ determined by use of these equations must be reduced by the depth of penetration factor $C_d$.

The value of the nominal design value $Z$ is the lesser of the values obtained in Eqs. (7-33)–(7-36).

Yield Mode

$$Z = \frac{Dt_s F_{es}}{K_D} \qquad \text{Mode I}_s \qquad (7\text{-}33)$$

$$Z = \frac{k_1 Dp F_{em}}{K_D(1 + 2R_e)} \qquad \text{Mode III}_m \qquad (7\text{-}34)$$

$$Z = \frac{k_2 Dt_s F_{em}}{K_D(2 + R_e)} \qquad \text{Mode III}_s \qquad (7\text{-}35)$$

$$Z = \frac{D^2}{K_D}\sqrt{\frac{2F_{em}F_{yb}}{3(1 + R_e)}} \qquad \text{Mode IV} \qquad (7\text{-}36)$$

where

$$k_1 = -1 + \sqrt{2(1 + R_e) + \frac{2F_{yb}(1 + 2R_e)D^2}{3F_{em}p^2}}$$

$$k_2 = -1 + \sqrt{\frac{2(1 + R_e)}{R_e} + \frac{2F_{yb}(2 + R_e)D^2}{3F_{em}t_s^2}}$$

where $R_e = F_{em}/F_{es}$,

$p$ = penetration of nail or spike in main member (member holding point) (in.),

$t_s$ = thickness of side member (in.),

$F_{em}$ = dowel bearing strength of main member (member holding point) (psi) (see Table 7.47),

$F_{es}$ = dowel bearing strength of side member (psi) (see Table 7.47),

$F_{yb}$ = bending yield strength of nail or spike (psi),

$D$ = nail or spike diameter (in.) (when annularly threaded nails are used with threads at the shear plane, $D$ = root diameter of threaded portion of nail),

$K_D = 2.2$, for $D \leq 0.17$ in.,

$K_D = 10D + 0.5$, for 0.17 in. $< D < 0.25$ in., and

$K_D = 3.0$ for $D \geq 0.25$ in.

The design values obtained from calculations or tables are for loads in any lateral direction, for one nail or spike driven in the side grain of the main member for depth of penetration of the point in the member receiving the point of $12D$. The minimum permissible penetration is $6D$. The adjustment factor for depth of penetration $C_d$ is determined as follows:

$$C_d = \frac{p}{12D} \leq 1 \tag{7-37}$$

where $p$ = depth of penetration (in.) and

$D$ = diameter of nail or spike (in.).

These values apply when side and main members have approximately the same density. When side and main members have different densities, the lighter-density member controls.

For nails or spikes with a length not exceeding $12D$ which are driven through three members and extending for at least 3 diameters beyond the side members, the lateral load design values may be doubled when the nails are clinched, provided the side members are at least $\frac{3}{8}$-in. thick.

The lateral load design values for toe-nailed joints are five-sixths of those determined for lateral loads from tables or by calculation ($C_{tn} = 0.83$). It is recommended that toe-nails be driven at an angle of approximately 30° to the piece and be started at approximately one-third the nail length from the end of the piece.

Design values for nails and spikes determined by the equations used in this chapter may be increased 10% for use in diaphragm construction ($C_{di} = 1.1$). The diaphragm design values may be further increased by multiplying by the load duration factor $C_D = 1.6$ for wind or earthquake loading. Diaphragm lateral load design values are also subject to the penetration requirements stated previously.

For nails or spikes driven in end grain and loaded laterally, design values are two-thirds of those determined for nails driven into the side grain.

Special nails have been developed for use with strap-type joist and purlin han-

**TABLE 7.49**

**Lateral Load Design Values for Special Nails**[a, b, d]

| Size | Diameter | Length (in.) | Description | Lateral Load Design Value[c] (lb) |
|------|----------|-------------|-------------|-----------------------------------|
| 8d | 11 gage | $1\frac{1}{4}$ | Smooth shank | 85 |
| 10d | 9 gage | $1\frac{1}{2}$ | Smooth shank | 118 |
| 10d | 9 gage | $2\frac{1}{8}$ | Annular thread, stainless steel | 118 |
| 16d | 8 gage | $2\frac{1}{2}$ | Smooth shank | 135 |
| 16d | 0.165 in. | $1\frac{3}{4}$ | Annular thread, stainless steel | 135 |
| 20d | 0.192 in. | $1\frac{3}{4}$ | Annular ring | 174 |
| 20d | 0.192 in. | $2\frac{1}{8}$ | Annular ring | 174 |
| | 0.250 in. | $2\frac{1}{2}$ | Annular ring | 257 |
| | 0.250 in. | 3 | Annular ring | 257 |

[a]See paragraph preceding for definition of special nails.
[b]Used with a steel strap for species with a specific gravity of 0.49 or greater.
[c]Loads have been increased 25% for metal side plates.
[d]For more information on other special nails and staples refer to (11).

gers, tie straps on pipe-type hangers, and other metal ties and straps. Short nails of the same diameter given in Table 7.49 have been developed for use with plywood. Lateral load design values for such special nails used with a metal strap are given in Table 7.49.

### 7.8.6  Wood-to-Metal Connections

Tables 7.54–7.57 contain design lateral design values for wood members with steel side plates. The lateral design values can also be determined by using the lesser value obtained from the Yield Mode Eqs. (7-34)–(7-36) and using the dowel bearing strength $F_e$ of the metal for $F_{es}$. The dowel bearing strength of side plates for ASTM A446 steel, $F_e$, is equal to 45,000 psi.

## 7.9  STAPLES

Because staples and nails are similar in nature, the loads for staples may be determined in a manner similar to that for nails. The design value for one staple of a given diameter equals twice the value for a nail of equal diameter, provided that the staple leg spacing (or crown width) is adequate, and that the penetration of both legs of the staple into the member receiving the points is approximately two-thirds of their length. In general, nail penetration requirements and other provisions regarding seasoning of members, service conditions, and so on apply equally to staples.

## 7.10  LIGHT METAL FRAMING DEVICES

### 7.10.1  Framing Anchors

Framing anchors are metal fittings used to provide a more positive connection between wood members than is obtained by toe-nailing. The several types of manufactured framing anchors are right-angle pieces formed from light gage (typically 18 gage) galvanized sheet steel (see Fig. 7.28).

These may be bent to conform with use conditions. The outstanding legs may be of rectangular, triangular, or other shape, and they are predrilled or prepunched to receive special nails. Nails appropriate to the particular framing anchor should be used in all holes provided in the anchor to develop its full load-carrying capacity. Framing anchors are commonly used in pairs to avoid eccentricity.

Allowable load values for framing anchors are generally determined by test, but an estimate may be based on the lateral resistance of the nails used. Allowable loads for framing anchors are generally given for normal duration of load but may be adjusted for other durations of load. Before making adjustments for duration of load, the manufacturer's literature should be checked to ascertain the duration of load basis on which the recommended values were based and also to make sure that the stress in the anchors or nails are not exceeded by increases for duration of load. The manufacturer's data for the particular type of framing anchor being used should be followed.

### 7.10.2  Joist and Purlin Hangers

Joist and purlin hangers are standard items fabricated by several manufacturers. The allowable loads shown in the manufacturer's literature are usually determined by tests. Several types of hangers are shown in Fig. 7.29. The manufacturer's data for the particular type of hanger being used should be followed.

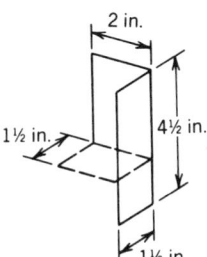

**FIGURE 7.28**  Typical framing anchor dimensions.

## TABLE 7.50

### Box Nail Design Values (Z) for Single Shear (two member) Connections[1,2] with both members of identical species

| SIDE MEMBER THICKNESS | NAIL LENGTH | NAIL DIAMETER | PENNY-WEIGHT | G=0.67 RED OAK | G=0.55 MIXED MAPLE SOUTHERN PINE | G=0.50 DOUGLAS FIR-LARCH | G=0.49 DOUGLAS FIR-LARCH (N) | G=0.46 DOUGLAS FIR (S) HEM-FIR (N) |
|---|---|---|---|---|---|---|---|---|
| $t_s$ inches | L inches | D inches | | Z lbs. | Z lbs. | Z lbs. | Z lbs. | Z lbs. |
| | 2 | 0.099 | 6d | 73 | 55 | 48 | 47 | 43 |
| | 2-1/2 | 0.113 | 8d | 88 | 67 | 59 | 57 | 53 |
| | 3 | 0.128 | 10d | 106 | 82 | 73 | 71 | 66 |
| 1/2 | 3-1/4 | 0.128 | 12d | 106 | 82 | 73 | 71 | 66 |
| | 3-1/2 | 0.135 | 16d | 115 | 89 | 79 | 77 | 72 |
| | 4 | 0.148 | 20d | 129 | 101 | 90 | 87 | 82 |
| | 4-1/2 | 0.148 | 30d | 129 | 101 | 90 | 87 | 82 |
| | 5 | 0.162 | 40d | 149 | 117 | 105 | 102 | 96 |
| | 2 | 0.099 | 6d | 73 | 61 | 55 | 53 | 48 |
| | 2-1/2 | 0.113 | 8d | 94 | 76 | 66 | 64 | 59 |
| | 3 | 0.128 | 10d | 120 | 91 | 79 | 77 | 71 |
| 5/8 | 3-1/4 | 0.128 | 12d | 120 | 91 | 79 | 77 | 71 |
| | 3-1/2 | 0.135 | 16d | 130 | 98 | 86 | 84 | 78 |
| | 4 | 0.148 | 20d | 144 | 110 | 97 | 94 | 87 |
| | 4-1/2 | 0.148 | 30d | 144 | 110 | 97 | 94 | 87 |
| | 5 | 0.162 | 40d | 165 | 126 | 112 | 109 | 101 |
| | 2 | 0.099 | 6d | 73 | 61 | 55 | 54 | 51 |
| | 2-1/2 | 0.113 | 8d | 94 | 79 | 72 | 71 | 65 |
| | 3 | 0.128 | 10d | 121 | 101 | 87 | 84 | 78 |
| 3/4 | 3-1/4 | 0.128 | 12d | 121 | 101 | 87 | 84 | 78 |
| | 3-1/2 | 0.135 | 16d | 135 | 108 | 94 | 91 | 84 |
| | 4 | 0.148 | 20d | 154 | 121 | 105 | 102 | 94 |
| | 4-1/2 | 0.148 | 30d | 154 | 121 | 105 | 102 | 94 |
| | 5 | 0.162 | 40d | 183 | 138 | 121 | 117 | 108 |
| | 2-1/2 | 0.113 | 8d | 94 | 79 | 72 | 71 | 67 |
| | 3 | 0.128 | 10d | 121 | 101 | 93 | 91 | 86 |
| | 3-1/4 | 0.128 | 12d | 121 | 101 | 93 | 91 | 86 |
| 1 | 3-1/2 | 0.135 | 16d | 135 | 113 | 103 | 101 | 96 |
| | 4 | 0.148 | 20d | 154 | 128 | 118 | 115 | 109 |
| | 4-1/2 | 0.148 | 30d | 154 | 128 | 118 | 115 | 109 |
| | 5 | 0.162 | 40d | 184 | 154 | 141 | 137 | 125 |
| | 3 | 0.128 | 10d | 121 | 101 | 93 | 91 | 86 |
| | 3-1/4 | 0.128 | 12d | 121 | 101 | 93 | 91 | 86 |
| 1-1/4 | 3-1/2 | 0.135 | 16d | 135 | 113 | 103 | 101 | 96 |
| | 4 | 0.148 | 20d | 154 | 128 | 118 | 115 | 109 |
| | 4-1/2 | 0.148 | 30d | 154 | 128 | 118 | 115 | 109 |
| | 5 | 0.162 | 40d | 184 | 154 | 141 | 138 | 131 |
| | 3-1/4 | 0.128 | 12d | 121 | 101 | 93 | 91 | 86 |
| | 3-1/2 | 0.135 | 16d | 135 | 113 | 103 | 101 | 96 |
| 1-1/2 | 4 | 0.148 | 20d | 154 | 128 | 118 | 115 | 109 |
| | 4-1/2 | 0.148 | 30d | 154 | 128 | 118 | 115 | 109 |
| | 5 | 0.162 | 40d | 184 | 154 | 141 | 138 | 131 |

**TABLE 7.50**   (*Continued*)

| SIDE MEMBER THICKNESS | NAIL LENGTH | NAIL DIAMETER | PENNY-WEIGHT | G=0.43 HEM-FIR | G=0.42 SPRUCE-PINE-FIR | G=0.37 REDWOOD (open grain) | G=0.36 EASTERN SOFTWOODS SPRUCE-PINE-FIR(S) WESTERN CEDARS WESTERN WOODS | G=0.35 NORTHERN SPECIES |
|---|---|---|---|---|---|---|---|---|
| $t_s$ inches | L inches | D inches | | Z lbs. | Z lbs. | Z lbs. | Z lbs. | Z lbs. |
| | 2 | 0.099 | 6d | 39 | 38 | 33 | 32 | 31 |
| | 2-1/2 | 0.113 | 8d | 49 | 47 | 41 | 40 | 38 |
| | 3 | 0.128 | 10d | 61 | 59 | 51 | 50 | 48 |
| 1/2 | 3-1/4 | 0.128 | 12d | 61 | 59 | 51 | 50 | 48 |
| | 3-1/2 | 0.135 | 16d | 66 | 65 | 56 | 55 | 53 |
| | 4 | 0.148 | 20d | 75 | 73 | 64 | 63 | 61 |
| | 4-1/2 | 0.148 | 30d | 75 | 73 | 64 | 63 | 61 |
| | 5 | 0.162 | 40d | 89 | 87 | 76 | 74 | 72 |
| | 2 | 0.099 | 6d | 44 | 42 | 35 | 34 | 33 |
| | 2-1/2 | 0.113 | 8d | 53 | 52 | 44 | 43 | 41 |
| | 3 | 0.128 | 10d | 65 | 63 | 54 | 53 | 51 |
| 5/8 | 3-1/4 | 0.128 | 12d | 65 | 63 | 54 | 53 | 51 |
| | 3-1/2 | 0.135 | 16d | 71 | 69 | 59 | 58 | 55 |
| | 4 | 0.148 | 20d | 80 | 77 | 67 | 65 | 63 |
| | 4-1/2 | 0.148 | 30d | 80 | 77 | 67 | 65 | 63 |
| | 5 | 0.162 | 40d | 93 | 90 | 78 | 77 | 74 |
| | 2 | 0.099 | 6d | 48 | 47 | 39 | 38 | 36 |
| | 2-1/2 | 0.113 | 8d | 58 | 57 | 47 | 46 | 44 |
| | 3 | 0.128 | 10d | 70 | 68 | 57 | 56 | 54 |
| 3/4 | 3-1/4 | 0.128 | 12d | 70 | 68 | 57 | 56 | 54 |
| | 3-1/2 | 0.135 | 16d | 76 | 74 | 63 | 61 | 58 |
| | 4 | 0.148 | 20d | 85 | 83 | 70 | 69 | 66 |
| | 4-1/2 | 0.148 | 30d | 85 | 83 | 70 | 69 | 66 |
| | 5 | 0.162 | 40d | 99 | 96 | 82 | 80 | 77 |
| | 2-1/2 | 0.113 | 8d | 63 | 61 | 55 | 54 | 51 |
| | 3 | 0.128 | 10d | 80 | 79 | 66 | 64 | 61 |
| | 3-1/4 | 0.128 | 12d | 80 | 79 | 66 | 64 | 61 |
| 1 | 3-1/2 | 0.135 | 16d | 89 | 86 | 71 | 69 | 66 |
| | 4 | 0.148 | 20d | 99 | 96 | 80 | 77 | 74 |
| | 4-1/2 | 0.148 | 30d | 99 | 96 | 80 | 77 | 74 |
| | 5 | 0.162 | 40d | 113 | 109 | 91 | 89 | 85 |
| | 3 | 0.128 | 10d | 80 | 79 | 70 | 69 | 67 |
| | 3-1/4 | 0.128 | 12d | 80 | 79 | 70 | 69 | 67 |
| 1-1/4 | 3-1/2 | 0.135 | 16d | 89 | 88 | 78 | 76 | 74 |
| | 4 | 0.148 | 20d | 102 | 100 | 89 | 87 | 84 |
| | 4-1/2 | 0.148 | 30d | 102 | 100 | 89 | 87 | 84 |
| | 5 | 0.162 | 40d | 122 | 120 | 103 | 100 | 95 |
| | 3-1/4 | 0.128 | 12d | 80 | 79 | 70 | 69 | 67 |
| | 3-1/2 | 0.135 | 16d | 89 | 88 | 78 | 76 | 74 |
| 1-1/2 | 4 | 0.148 | 20d | 102 | 100 | 89 | 87 | 84 |
| | 4-1/2 | 0.148 | 30d | 102 | 100 | 89 | 87 | 84 |
| | 5 | 0.162 | 40d | 122 | 120 | 106 | 104 | 101 |

1. Tabulated lateral design values (Z) for nailed connections shall be multiplied by all applicable adjustment factors (see Table 7.1).
2. Tabulated lateral design values (Z) are for box nails inserted in side grain with nail axis perpendicular to wood fibers, and with the following nail bending yield strengths ($F_{yb}$):

$F_{yb}$ = 100,000 psi for 0.099″, 0.113″, 0.128″ and 0.135″ diameter box nails
$F_{yb}$ = 90,000 psi for 0.148″ and 0.162″ diameter box nails

Source: ANSI/NF$_o$PA NDS—1991, *National Design Specification for Wood Construction*, AFPA, 1250 Connecticut Ave., NW, Washington, DC 20036.

# TABLE 7.51

## Common Wire Nail Design Values (Z) for Single Shear[1,2] (two member) Connections with both members of identical species

| SIDE MEMBER THICKNESS $t_s$ inches | NAIL LENGTH L inches | NAIL DIAMETER D inches | PENNY-WEIGHT | G=0.67 RED OAK Z lbs. | G=0.55 MIXED MAPLE SOUTHERN PINE Z lbs. | G=0.50 DOUGLAS FIR-LARCH Z lbs. | G=0.49 DOUGLAS FIR-LARCH (N) Z lbs. | G=0.46 DOUGLAS FIR (S) HEM-FIR (N) Z lbs. |
|---|---|---|---|---|---|---|---|---|
| 1/2 | 2 | 0.113 | 6d | 88 | 67 | 59 | 57 | 53 |
| | 2-1/2 | 0.131 | 8d | 110 | 85 | 76 | 73 | 69 |
| | 3 | 0.148 | 10d | 129 | 101 | 90 | 87 | 82 |
| | 3-1/4 | 0.148 | 12d | 129 | 101 | 90 | 87 | 82 |
| | 3-1/2 | 0.162 | 16d | 149 | 117 | 105 | 102 | 96 |
| | 4 | 0.192 | 20d | 172 | 137 | 124 | 121 | 114 |
| | 4-1/2 | 0.207 | 30d | 185 | 148 | 134 | 131 | 123 |
| | 5 | 0.225 | 40d | 200 | 162 | 147 | 143 | 135 |
| | 5-1/2 | 0.244 | 50d | 205 | 166 | 151 | 147 | 139 |
| | 6 | 0.263 | 60d | 230 | 188 | 171 | 167 | 158 |
| 5/8 | 2 | 0.113 | 6d | 94 | 76 | 66 | 64 | 59 |
| | 2-1/2 | 0.131 | 8d | 124 | 94 | 82 | 80 | 74 |
| | 3 | 0.148 | 10d | 144 | 110 | 97 | 94 | 87 |
| | 3-1/4 | 0.148 | 12d | 144 | 110 | 97 | 94 | 87 |
| | 3-1/2 | 0.162 | 16d | 165 | 126 | 112 | 109 | 101 |
| | 4 | 0.192 | 20d | 188 | 146 | 130 | 126 | 118 |
| | 4-1/2 | 0.207 | 30d | 199 | 156 | 140 | 136 | 127 |
| | 5 | 0.225 | 40d | 213 | 169 | 151 | 147 | 138 |
| | 5-1/2 | 0.244 | 50d | 218 | 173 | 155 | 151 | 142 |
| | 6 | 0.263 | 60d | 242 | 194 | 175 | 170 | 160 |
| 3/4 | 2-1/2 | 0.131 | 8d | 127 | 104 | 90 | 87 | 80 |
| | 3 | 0.148 | 10d | 154 | 121 | 105 | 102 | 94 |
| | 3-1/4 | 0.148 | 12d | 154 | 121 | 105 | 102 | 94 |
| | 3-1/2 | 0.162 | 16d | 183 | 138 | 121 | 117 | 108 |
| | 4 | 0.192 | 20d | 206 | 157 | 138 | 134 | 125 |
| | 4-1/2 | 0.207 | 30d | 216 | 166 | 147 | 143 | 133 |
| | 5 | 0.225 | 40d | 229 | 178 | 158 | 154 | 144 |
| | 5-1/2 | 0.244 | 50d | 234 | 182 | 162 | 158 | 147 |
| | 6 | 0.263 | 60d | 258 | 203 | 181 | 176 | 165 |
| 1 | 3 | 0.148 | 10d | 154 | 128 | 118 | 115 | 109 |
| | 3-1/4 | 0.148 | 12d | 154 | 128 | 118 | 115 | 109 |
| | 3-1/2 | 0.162 | 16d | 184 | 154 | 141 | 137 | 125 |
| | 4 | 0.192 | 20d | 222 | 183 | 159 | 154 | 142 |
| | 4-1/2 | 0.207 | 30d | 243 | 192 | 167 | 162 | 149 |
| | 5 | 0.225 | 40d | 268 | 202 | 177 | 171 | 159 |
| | 5-1/2 | 0.244 | 50d | 274 | 207 | 181 | 175 | 162 |
| | 6 | 0.263 | 60d | 298 | 227 | 199 | 193 | 179 |
| 1-1/4 | 3-1/4 | 0.148 | 12d | 154 | 128 | 118 | 115 | 109 |
| | 3-1/2 | 0.162 | 16d | 184 | 154 | 141 | 138 | 131 |
| | 4 | 0.192 | 20d | 222 | 185 | 170 | 166 | 157 |
| | 4-1/2 | 0.207 | 30d | 243 | 203 | 186 | 182 | 169 |
| | 5 | 0.225 | 40d | 268 | 224 | 200 | 193 | 177 |
| | 5-1/2 | 0.244 | 50d | 276 | 230 | 204 | 197 | 181 |
| | 6 | 0.263 | 60d | 314 | 256 | 222 | 215 | 198 |
| 1-1/2 | 3-1/2 | 0.162 | 16d | 184 | 154 | 141 | 138 | 131 |
| | 4 | 0.192 | 20d | 222 | 185 | 170 | 166 | 157 |
| | 4-1/2 | 0.207 | 30d | 243 | 203 | 186 | 182 | 172 |
| | 5 | 0.225 | 40d | 268 | 224 | 205 | 201 | 190 |
| | 5-1/2 | 0.244 | 50d | 276 | 230 | 211 | 206 | 196 |
| | 6 | 0.263 | 60D | 314 | 262 | 240 | 235 | 220 |

## TABLE 7.51   (*Continued*)

| SIDE MEMBER THICKNESS | NAIL LENGTH | NAIL DIAMETER | PENNY-WEIGHT | G=0.43 HEM-FIR | G=0.42 SPRUCE-PINE-FIR | G=0.37 REDWOOD (open grain) | G=0.36 EASTERN COTTONWOODS SPRUCE-PINE-FIR(S) WESTERN CEDARS WESTERN WOODS | G=0.35 NORTHERN SPECIES |
|---|---|---|---|---|---|---|---|---|
| $t_s$ inches | L inches | D inches | | Z lbs. | Z lbs. | Z lbs. | Z lbs. | Z lbs. |
| | 2 | 0.113 | 6d | 49 | 47 | 41 | 40 | 38 |
| | 2-1/2 | 0.131 | 8d | 63 | 61 | 53 | 52 | 50 |
| | 3 | 0.148 | 10d | 75 | 73 | 64 | 63 | 61 |
| | 3-1/4 | 0.148 | 12d | 75 | 73 | 64 | 63 | 61 |
| 1/2 | 3-1/2 | 0.162 | 16d | 89 | 87 | 76 | 74 | 72 |
| | 4 | 0.192 | 20d | 105 | 103 | 91 | 89 | 86 |
| | 4-1/2 | 0.207 | 30d | 115 | 112 | 99 | 97 | 94 |
| | 5 | 0.225 | 40d | 126 | 123 | 108 | 104 | 98 |
| | 5-1/2 | 0.244 | 50d | 130 | 127 | 110 | 106 | 100 |
| | 6 | 0.263 | 60d | 147 | 144 | 116 | 112 | 105 |
| | 2 | 0.113 | 6d | 53 | 52 | 44 | 43 | 41 |
| | 2-1/2 | 0.131 | 8d | 67 | 65 | 56 | 55 | 53 |
| | 3 | 0.148 | 10d | 80 | 77 | 67 | 65 | 63 |
| | 3-1/4 | 0.148 | 12d | 80 | 77 | 67 | 65 | 63 |
| 5/8 | 3-1/2 | 0.162 | 16d | 93 | 90 | 78 | 77 | 74 |
| | 4 | 0.192 | 20d | 109 | 106 | 93 | 91 | 88 |
| | 4-1/2 | 0.207 | 30d | 118 | 115 | 100 | 98 | 95 |
| | 5 | 0.225 | 40d | 128 | 125 | 110 | 108 | 105 |
| | 5-1/2 | 0.244 | 50d | 132 | 129 | 113 | 111 | 107 |
| | 6 | 0.263 | 60d | 149 | 145 | 128 | 126 | 122 |
| | 2-1/2 | 0.131 | 8d | 73 | 70 | 60 | 58 | 56 |
| | 3 | 0.148 | 10d | 85 | 83 | 70 | 69 | 66 |
| | 3-1/4 | 0.148 | 12d | 85 | 83 | 70 | 69 | 66 |
| | 3-1/2 | 0.162 | 16d | 99 | 96 | 82 | 80 | 77 |
| 3/4 | 4 | 0.192 | 20d | 114 | 111 | 96 | 93 | 90 |
| | 4-1/2 | 0.207 | 30d | 122 | 119 | 103 | 101 | 97 |
| | 5 | 0.225 | 40d | 132 | 129 | 112 | 110 | 106 |
| | 5-1/2 | 0.244 | 50d | 136 | 132 | 115 | 113 | 109 |
| | 6 | 0.263 | 60d | 152 | 149 | 130 | 127 | 123 |
| | 3 | 0.148 | 10d | 99 | 96 | 80 | 77 | 74 |
| | 3-1/4 | 0.148 | 12d | 99 | 96 | 80 | 77 | 74 |
| | 3-1/2 | 0.162 | 16d | 113 | 109 | 91 | 89 | 85 |
| 1 | 4 | 0.192 | 20d | 128 | 124 | 105 | 102 | 98 |
| | 4-1/2 | 0.207 | 30d | 135 | 131 | 111 | 109 | 104 |
| | 5 | 0.225 | 40d | 144 | 140 | 120 | 117 | 112 |
| | 5-1/2 | 0.244 | 50d | 148 | 143 | 123 | 120 | 115 |
| | 6 | 0.263 | 60d | 164 | 159 | 137 | 134 | 129 |
| | 3-1/4 | 0.148 | 12d | 102 | 100 | 89 | 87 | 84 |
| | 3-1/2 | 0.162 | 16d | 122 | 120 | 103 | 100 | 95 |
| | 4 | 0.192 | 20d | 145 | 140 | 116 | 113 | 108 |
| 1-1/4 | 4-1/2 | 0.207 | 30d | 152 | 147 | 123 | 119 | 114 |
| | 5 | 0.225 | 40d | 160 | 155 | 130 | 127 | 121 |
| | 5-1/2 | 0.244 | 50d | 163 | 158 | 133 | 129 | 124 |
| | 6 | 0.263 | 60d | 179 | 174 | 147 | 143 | 137 |
| | 3-1/2 | 0.162 | 16d | 122 | 120 | 106 | 104 | 101 |
| | 4 | 0.192 | 20d | 147 | 144 | 128 | 126 | 120 |
| 1-1/2 | 4-1/2 | 0.207 | 30d | 161 | 158 | 135 | 131 | 125 |
| | 5 | 0.225 | 40d | 178 | 172 | 143 | 138 | 132 |
| | 5-1/2 | 0.244 | 50d | 181 | 175 | 146 | 141 | 135 |
| | 6 | 0.263 | 60d | 197 | 191 | 159 | 155 | 148 |

1. Tabulated lateral design values (Z) for nailed connections shall be multiplied by all applicable adjustment factors (see Table 7.1).

2. Tabulated lateral design values (Z) are for common wire nails inserted in side grain with nail axis perpendicular to wood fibers, and with the following nail bending yield strength ($F_{yb}$):

   $F_{yb}$ = 100,000 psi for 0.113″ and 0.131″ diameter common wire nails

   $F_{yb}$ =  90,000 psi for 0.148″ and 0.162″ diameter common wire nails

   $F_{yb}$ =  80,000 psi for 0.192″, 0.207″ and 0.225″ diameter common wire nails

   $F_{yb}$ =  70,000 psi for 0.244″ and 0.263″ diameter common wire nails

Source: ANSI/NF₀PA NDS—1991, *National Design Specification for Wood Construction*, AFPA, 1250 Connecticut Ave., NW, Washington, DC 20036.

## TABLE 7.52

### Threaded Hardened-Steel Nail Design Values (Z) for Single Shear[1,2,3] (two member) Connections with both members of identical species

| SIDE MEMBER THICKNESS | NAIL LENGTH | WIRE DIAMETER | PENNY-WEIGHT | G=0.67 RED OAK | G=0.55 MIXED MAPLE SOUTHERN PINE | G=0.50 DOUGLAS FIR-LARCH | G=0.49 DOUGLAS FIR-LARCH (N) | G=0.46 DOUGLAS FIR (S) HEM-FIR (N) |
|---|---|---|---|---|---|---|---|---|
| $t_s$ inches | L inches | D inches | | Z lbs. | Z lbs. | Z lbs. | Z lbs. | Z lbs. |
| 1/2 | 2 | 0.120 | 6d | 103 | 80 | 71 | 69 | 65 |
| | 2-1/2 | 0.120 | 8d | 103 | 80 | 71 | 69 | 65 |
| | 3 | 0.135 | 10d | 124 | 98 | 88 | 85 | 80 |
| | 3-1/4 | 0.135 | 12d | 124 | 98 | 88 | 85 | 80 |
| | 3-1/2 | 0.148 | 16d | 139 | 110 | 98 | 96 | 90 |
| | 4 | 0.177 | 20d | 182 | 147 | 133 | 129 | 122 |
| | 4-1/2 | 0.177 | 30d | 182 | 147 | 133 | 129 | 122 |
| | 5 | 0.177 | 40d | 182 | 147 | 133 | 129 | 122 |
| | 5-1/2 | 0.177 | 50d | 182 | 147 | 133 | 129 | 122 |
| | 6 | 0.177 | 60d | 182 | 147 | 133 | 129 | 122 |
| 5/8 | 2-1/2 | 0.120 | 8d | 116 | 88 | 77 | 75 | 70 |
| | 3 | 0.135 | 10d | 137 | 106 | 93 | 91 | 84 |
| | 3-1/4 | 0.135 | 12d | 137 | 106 | 93 | 91 | 84 |
| | 3-1/2 | 0.148 | 16d | 153 | 118 | 104 | 101 | 95 |
| | 4 | 0.177 | 20d | 196 | 154 | 138 | 134 | 126 |
| | 4-1/2 | 0.177 | 30d | 196 | 154 | 138 | 134 | 126 |
| | 5 | 0.177 | 40d | 196 | 154 | 138 | 134 | 126 |
| | 5-1/2 | 0.177 | 50d | 196 | 154 | 138 | 134 | 126 |
| | 6 | 0.177 | 60d | 196 | 154 | 138 | 134 | 126 |
| 3/4 | 2-1/2 | 0.120 | 8d | 121 | 97 | 84 | 82 | 75 |
| | 3 | 0.135 | 10d | 152 | 115 | 101 | 97 | 90 |
| | 3-1/4 | 0.135 | 12d | 152 | 115 | 101 | 97 | 90 |
| | 3-1/2 | 0.148 | 16d | 169 | 128 | 112 | 109 | 101 |
| | 4 | 0.177 | 20d | 212 | 164 | 145 | 141 | 131 |
| | 4-1/2 | 0.177 | 30d | 212 | 164 | 145 | 141 | 131 |
| | 5 | 0.177 | 40d | 212 | 164 | 145 | 141 | 131 |
| | 5-1/2 | 0.177 | 50d | 212 | 164 | 145 | 141 | 131 |
| | 6 | 0.177 | 60d | 212 | 164 | 145 | 141 | 131 |
| | 7 | 0.207 | 70d | 229 | 178 | 159 | 155 | 145 |
| | 8 | 0.207 | 80d | 229 | 178 | 159 | 155 | 145 |
| | 9 | 0.207 | 90d | 229 | 178 | 159 | 155 | 145 |
| 1 | 2-1/2 | 0.120 | 8d | 121 | 102 | 93 | 91 | 86 |
| | 3 | 0.135 | 10d | 154 | 128 | 118 | 114 | 104 |
| | 3-1/4 | 0.135 | 12d | 154 | 128 | 118 | 114 | 104 |
| | 3-1/2 | 0.148 | 16d | 174 | 145 | 131 | 126 | 116 |
| | 4 | 0.177 | 20d | 241 | 188 | 164 | 159 | 146 |
| | 4-1/2 | 0.177 | 30d | 241 | 188 | 164 | 159 | 146 |
| | 5 | 0.177 | 40d | 241 | 188 | 164 | 159 | 146 |
| | 5-1/2 | 0.177 | 50d | 241 | 188 | 164 | 159 | 146 |
| | 6 | 0.177 | 60d | 241 | 188 | 164 | 159 | 146 |
| | 7 | 0.207 | 70d | 268 | 202 | 177 | 171 | 159 |
| | 8 | 0.207 | 80d | 268 | 202 | 177 | 171 | 159 |
| | 9 | 0.207 | 90d | 268 | 202 | 177 | 171 | 159 |
| 1-1/4 | 3 | 0.135 | 10d | 154 | 128 | 118 | 115 | 109 |
| | 3-1/4 | 0.135 | 12d | 154 | 128 | 118 | 115 | 109 |
| | 3-1/2 | 0.148 | 16d | 174 | 145 | 133 | 130 | 123 |
| | 4 | 0.177 | 20d | 241 | 201 | 184 | 180 | 165 |
| | 4-1/2 | 0.177 | 30d | 241 | 201 | 184 | 180 | 165 |
| | 5 | 0.177 | 40d | 241 | 201 | 184 | 180 | 165 |
| | 5-1/2 | 0.177 | 50d | 241 | 201 | 184 | 180 | 165 |
| | 6 | 0.177 | 60d | 241 | 201 | 184 | 180 | 165 |
| | 7 | 0.207 | 70d | 271 | 227 | 199 | 192 | 177 |
| | 8 | 0.207 | 80d | 271 | 227 | 199 | 192 | 177 |
| | 9 | 0.207 | 90d | 271 | 227 | 199 | 192 | 177 |
| 1-1/2 | 3-1/4 | 0.135 | 12d | 154 | 128 | 118 | 115 | 109 |
| | 3-1/2 | 0.148 | 16d | 174 | 145 | 133 | 130 | 123 |
| | 4 | 0.177 | 20d | 241 | 201 | 184 | 180 | 171 |
| | 4-1/2 | 0.177 | 30d | 241 | 201 | 184 | 180 | 171 |
| | 5 | 0.177 | 40d | 241 | 201 | 184 | 180 | 171 |
| | 5-1/2 | 0.177 | 50d | 241 | 201 | 184 | 180 | 171 |
| | 6 | 0.177 | 60d | 241 | 201 | 184 | 180 | 171 |
| | 7 | 0.207 | 70d | 271 | 227 | 208 | 203 | 193 |
| | 8 | 0.207 | 80d | 271 | 227 | 208 | 203 | 193 |
| | 9 | 0.207 | 90d | 271 | 227 | 208 | 203 | 193 |

## TABLE 7.52 (Continued)

| SIDE MEMBER THICKNESS | NAIL LENGTH | WIRE DIAMETER | PENNY-WEIGHT | G=0.43 HEM-FIR | G=0.42 SPRUCE-PINE-FIR | G=0.37 REDWOOD (open grain) | G=0.36 EASTERN SOFTWOODS SPRUCE-PINE-FIR(S) WESTERN CEDARS WESTERN WOODS | G=0.35 NORTHERN SPECIES |
|---|---|---|---|---|---|---|---|---|
| $t_s$ inches | $L$ inches | $D$ inches | | $Z$ lbs. | $Z$ lbs. | $Z$ lbs. | $Z$ lbs. | $Z$ lbs. |
| | 2 | 0.120 | 6d | 60 | 58 | 51 | 50 | 48 |
| | 2-1/2 | 0.120 | 8d | 60 | 58 | 51 | 50 | 48 |
| | 3 | 0.135 | 10d | 74 | 72 | 63 | 62 | 60 |
| | 3-1/4 | 0.135 | 12d | 74 | 72 | 63 | 62 | 60 |
| 1/2 | 3-1/2 | 0.148 | 16d | 83 | 81 | 72 | 70 | 68 |
| | 4 | 0.177 | 20d | 114 | 111 | 98 | 96 | 94 |
| | 4-1/2 | 0.177 | 30d | 114 | 111 | 98 | 96 | 94 |
| | 5 | 0.177 | 40d | 114 | 111 | 98 | 96 | 94 |
| | 5-1/2 | 0.177 | 50d | 114 | 111 | 98 | 96 | 94 |
| | 6 | 0.177 | 60d | 114 | 111 | 98 | 96 | 94 |
| | 2-1/2 | 0.120 | 8d | 63 | 62 | 53 | 52 | 50 |
| | 3 | 0.135 | 10d | 78 | 75 | 65 | 64 | 62 |
| | 3-1/4 | 0.135 | 12d | 78 | 75 | 65 | 64 | 62 |
| | 3-1/2 | 0.148 | 16d | 87 | 85 | 73 | 72 | 69 |
| 5/8 | 4 | 0.177 | 20d | 116 | 113 | 99 | 97 | 94 |
| | 4-1/2 | 0.177 | 30d | 116 | 113 | 99 | 97 | 94 |
| | 5 | 0.177 | 40d | 116 | 113 | 99 | 97 | 94 |
| | 5-1/2 | 0.177 | 50d | 116 | 113 | 99 | 97 | 94 |
| | 6 | 0.177 | 60d | 116 | 113 | 99 | 97 | 94 |
| | 2-1/2 | 0.120 | 8d | 68 | 66 | 56 | 55 | 52 |
| | 3 | 0.135 | 10d | 82 | 80 | 68 | 67 | 64 |
| | 3-1/4 | 0.135 | 12d | 82 | 80 | 68 | 67 | 64 |
| | 3-1/2 | 0.148 | 16d | 92 | 89 | 77 | 75 | 72 |
| | 4 | 0.177 | 20d | 121 | 117 | 102 | 100 | 96 |
| 3/4 | 4-1/2 | 0.177 | 30d | 121 | 117 | 102 | 100 | 96 |
| | 5 | 0.177 | 40d | 121 | 117 | 102 | 100 | 96 |
| | 5-1/2 | 0.177 | 50d | 121 | 117 | 102 | 100 | 96 |
| | 6 | 0.177 | 60d | 121 | 117 | 102 | 100 | 96 |
| | 7 | 0.207 | 70d | 133 | 130 | 113 | 111 | 107 |
| | 8 | 0.207 | 80d | 133 | 130 | 113 | 111 | 107. |
| | 9 | 0.207 | 90d | 133 | 130 | 113 | 111 | 107 |
| | 2-1/2 | 0.120 | 8d | 80 | 77 | 64 | 62 | 59 |
| | 3 | 0.135 | 10d | 94 | 91 | 76 | 74 | 71 |
| | 3-1/4 | 0.135 | 12d | 94 | 91 | 76 | 74 | 71 |
| | 3-1/2 | 0.148 | 16d | 105 | 101 | 85 | 83 | 79 |
| | 4 | 0.177 | 20d | 133 | 129 | 110 | 107 | 103 |
| 1 | 4-1/2 | 0.177 | 30d | 133 | 129 | 110 | 107 | 103 |
| | 5 | 0.177 | 40d | 133 | 129 | 110 | 107 | 103 |
| | 5-1/2 | 0.177 | 50d | 133 | 129 | 110 | 107 | 103 |
| | 6 | 0.177 | 60d | 133 | 129 | 110 | 107 | 103 |
| | 7 | 0.207 | 70d | 145 | 140 | 120 | 117 | 113 |
| | 8 | 0.207 | 80d | 145 | 140 | 120 | 117 | 113 |
| | 9 | 0.207 | 90d | 145 | 140 | 120 | 117 | 113 |
| | 3 | 0.135 | 10d | 102 | 100 | 86 | 83 | 79 |
| | 3-1/4 | 0.135 | 12d | 102 | 100 | 86 | 83 | 79 |
| | 3-1/2 | 0.148 | 16d | 115 | 113 | 95 | 93 | 88 |
| | 4 | 0.177 | 20d | 149 | 144 | 120 | 117 | 112 |
| | 4-1/2 | 0.177 | 30d | 149 | 144 | 120 | 117 | 112 |
| 1-1/4 | 5 | 0.177 | 40d | 149 | 144 | 120 | 117 | 112 |
| | 5-1/2 | 0.177 | 50d | 149 | 144 | 120 | 117 | 112 |
| | 6 | 0.177 | 60d | 149 | 144 | 120 | 117 | 112 |
| | 7 | 0.207 | 70d | 160 | 155 | 130 | 127 | 121 |
| | 8 | 0.207 | 80d | 160 | 155 | 130 | 127 | 121 |
| | 9 | 0.207 | 90d | 160 | 155 | 130 | 127 | 121 |
| | 3-1/4 | 0.135 | 12d | 102 | 100 | 89 | 87 | 84 |
| | 3-1/2 | 0.148 | 16d | 115 | 113 | 100 | 98 | 95 |
| | 4 | 0.177 | 20d | 160 | 156 | 132 | 129 | 123 |
| | 4-1/2 | 0.177 | 30d | 160 | 156 | 132 | 129 | 123 |
| 1-1/2 | 5 | 0.177 | 40d | 160 | 156 | 132 | 129 | 123 |
| | 5-1/2 | 0.177 | 50d | 160 | 156 | 132 | 129 | 123 |
| | 6 | 0.177 | 60d | 160 | 156 | 132 | 129 | 123 |
| | 7 | 0.207 | 70d | 177 | 171 | 142 | 138 | 132 |
| | 8 | 0.207 | 80d | 177 | 171 | 142 | 138 | 132 |
| | 9 | 0.207 | 90d | 177 | 171 | 142 | 138 | 132 |

1. Tabulated lateral design values (Z) for nail connections shall be multiplied by all applicable adjustment factors (see Table 7.1).

2. Tabulated lateral design values (Z) are for threaded hardened-steel nails inserted in side grain with nail axis perpendicular to wood fibers, and with the following nail bending yield strengths ($F_{yb}$):

   $F_{yb}$ = 130,000 psi for 0.12" and 0.135" diameter threaded hardened-steel nails
   $F_{yb}$ = 115,000 psi for 0.148" and 0.177" diameter threaded hardened-steel nails
   $F_{yb}$ = 100,000 psi for 0.207" diameter threaded hardened-steel nails

3. Tabulated lateral design values (Z) shall not apply for annularly threaded nails when threads occur at the shear plane (see 12.3.1).

Source: ANSI/NF$_o$PA NDS—1991, *National Design Specification for Wood Construction*, AFPA, 1250 Connecticut Ave., NW, Washington, DC 20036.

## TABLE 7.53

### Common Wire Spike Design Values (Z) for Single Shear (two member)[1,2]
### Connections with both members of identical species

| SIDE MEMBER THICKNESS | SPIKE LENGTH | SPIKE DIAMETER | PENNY-WEIGHT | G=0.67 RED OAK | G=0.55 MIXED MAPLE SOUTHERN PINE | G=0.50 DOUGLAS FIR-LARCH | G=0.49 DOUGLAS FIR-LARCH (N) | G=0.46 DOUGLAS FIR (S) HEM-FIR (N) |
|---|---|---|---|---|---|---|---|---|
| $t_s$ inches | L inches | D inches | | Z lbs. | Z lbs. | Z lbs. | Z lbs. | Z lbs. |
| | 3 | 0.192 | 10d | 172 | 137 | 124 | 121 | 114 |
| | 3-1/4 | 0.192 | 12d | 172 | 137 | 124 | 121 | 114 |
| | 3-1/2 | 0.207 | 16d | 185 | 148 | 134 | 131 | 123 |
| 1/2 | 4 | 0.225 | 20d | 200 | 162 | 147 | 143 | 135 |
| | 4-1/2 | 0.244 | 30d | 205 | 166 | 151 | 147 | 139 |
| | 5 | 0.263 | 40d | 230 | 188 | 171 | 167 | 158 |
| | 5-1/2 | 0.283 | 50d | 247 | 201 | 183 | 179 | 169 |
| | 6 | 0.283 | 60d | 247 | 20? | 183 | 179 | 169 |
| | 3 | 0.192 | 10d | 188 | 146 | 130 | 126 | 118 |
| | 3-1/4 | 0.192 | 12d | 188 | 146 | 130 | 126 | 118 |
| | 3-1/2 | 0.207 | 16d | 199 | 156 | 140 | 136 | 127 |
| 5/8 | 4 | 0.225 | 20d | 213 | 169 | 151 | 147 | 138 |
| | 4-1/2 | 0.244 | 30d | 218 | 173 | 155 | 151 | 142 |
| | 5 | 0.263 | 40d | 242 | 194 | 175 | 170 | 160 |
| | 5-1/2 | 0.283 | 50d | 260 | 208 | 187 | 183 | 172 |
| | 6 | 0.283 | 60d | 260 | 208 | 187 | 183 | 172 |
| | 3-1/4 | 0.192 | 12d | 206 | 157 | 138 | 134 | 125 |
| | 3-1/2 | 0.207 | 16d | 216 | 166 | 147 | 143 | 133 |
| | 4 | 0.225 | 20d | 229 | 178 | 158 | 154 | 144 |
| | 4-1/2 | 0.244 | 30d | 234 | 182 | 162 | 158 | 147 |
| 3/4 | 5 | 0.263 | 40d | 258 | 203 | 181 | 176 | 165 |
| | 5-1/2 | 0.283 | 50d | 277 | 217 | 194 | 189 | 177 |
| | 6 | 0.283 | 60d | 277 | 217 | 194 | 189 | 177 |
| | 7 | 0.312 | 5/16 | 325 | 257 | 231 | 225 | 211 |
| | 8-1/2 | 0.375 | 3/8 | 401 | 319 | 287 | 279 | 263 |
| | 3-1/2 | 0.207 | 16d | 243 | 192 | 167 | 162 | 149 |
| | 4 | 0.225 | 20d. | 268 | 202 | 177 | 171 | 159 |
| | 4-1/2 | 0.244 | 30d | 274 | 207 | 181 | 175 | 162 |
| 1 | 5 | 0.263 | 40d | 298 | 227 | 199 | 193 | 179 |
| | 5-1/2 | 0.283 | 50d | 321 | 243 | 214 | 207 | 192 |
| | 6 | 0.283 | 60d | 321 | 243 | 214 | 207 | 192 |
| | 7 | 0.312 | 5/16 | 369 | 283 | 250 | 243 | 226 |
| | 8-1/2 | 0.375 | 3/8 | 452 | 348 | 309 | 300 | 279 |
| | 4 | 0.225 | 20d | 268 | 224 | 200 | 193 | 177 |
| | 4-1/2 | 0.244 | 30d | 276 | 230 | 204 | 197 | 181 |
| | 5 | 0.263 | 40d | 314 | 256 | 222 | 215 | 198 |
| 1-1/4 | 5-1/2 | 0.283 | 50d | 337 | 275 | 239 | 231 | 213 |
| | 6 | 0.283 | 60d | 337 | 275 | 239 | 231 | 213 |
| | 7 | 0.312 | 5/16 | 409 | 316 | 276 | 267 | 247 |
| | 8-1/2 | 0.375 | 3/8 | 512 | 386 | 338 | 327 | 303 |
| | 4-1/2 | 0.244 | 30d | 276 | 230 | 211 | 206 | 196 |
| | 5 | 0.263 | 40d | 314 | 262 | 240 | 235 | 220 |
| 1-1/2 | 5-1/2 | 0.283 | 50d | 337 | 281 | 257 | 252 | 236 |
| | 6 | 0.283 | 60d | 337 | 281 | 257 | 252 | 236 |
| | 7 | 0.312 | 5/16 | 409 | 342 | 305 | 295 | 271 |
| | 8-1/2 | 0.375 | 3/8 | 512 | 428 | 373 | 360 | 331 |
| | 6 | 0.283 | 60d | 337 | 281 | 257 | 252 | 239 |
| 2-1/2 | 7 | 0.312 | 5/16 | 409 | 342 | 313 | 306 | 290 |
| | 8-1/2 | 0.375 | 3/8 | 512 | 428 | 391 | 383 | 363 |
| 3-1/2 | 8-1/2 | 0.375 | 3/8 | 512 | 428 | 391 | 383 | 363 |

## TABLE 7.53  (*Continued*)

| SIDE MEMBER THICKNESS | SPIKE LENGTH | SPIKE DIAMETER | PENNY-WEIGHT | G=0.43 HEM-FIR | G=0.42 SPRUCE-PINE-FIR | G=0.37 REDWOOD (open grain) | G=0.36 EASTERN SOFTWOODS SPRUCE-PINE-FIR(S) WESTERN CEDARS WESTERN WOODS | G=0.35 NORTHERN SPECIES |
|---|---|---|---|---|---|---|---|---|
| $t_s$ inches | L inches | D inches | | Z lbs. | Z lbs. | Z lbs. | Z lbs. | Z lbs. |
| | 3 | 0.192 | 10d | 105 | 103 | 91 | 89 | 86 |
| | 3-1/4 | 0.192 | 12d | 105 | 103 | 91 | 89 | 86 |
| | 3-1/2 | 0.207 | 16d | 115 | 112 | 99 | 97 | 94 |
| 1/2 | 4 | 0.225 | 20d | 126 | 123 | 108 | 104 | 98 |
| | 4-1/2 | 0.244 | 30d | 130 | 127 | 110 | 106 | 100 |
| | 5 | 0.263 | 40d | 147 | 144 | 116 | 112 | 105 |
| | 5-1/2 | 0.283 | 50d | 158 | 155 | 125 | 120 | 113 |
| | 6 | 0.283 | 60d | 158 | 155 | 125 | 120 | 113 |
| | 3 | 0.192 | 10d | 109 | 106 | 93 | 91 | 88 |
| | 3-1/4 | 0.192 | 12d | 109 | 106 | 93 | 91 | 88 |
| | 3-1/2 | 0.207 | 16d | 118 | 115 | 100 | 98 | 95 |
| 5/8 | 4 | 0.225 | 20d | 128 | 125 | 110 | 108 | 105 |
| | 4-1/2 | 0.244 | 30d | 132 | 129 | 113 | 111 | 107 |
| | 5 | 0.263 | 40d | 149 | 145 | 128 | 126 | 122 |
| | 5-1/2 | 0.283 | 50d | 160 | 156 | 138 | 135 | 131 |
| | 6 | 0.283 | 60d | 160 | 156 | 138 | 135 | 131 |
| | 3-1/4 | 0.192 | 12d | 114 | 111 | 96 | 93 | 90 |
| | 3-1/2 | 0.207 | 16d | 122 | 119 | 103 | 101 | 97 |
| | 4 | 0.225 | 20d | 132 | 129 | 112 | 110 | 106 |
| | 4-1/2 | 0.244 | 30d | 136 | 132 | 115 | 113 | 109 |
| 3/4 | 5 | 0.263 | 40d | 152 | 149 | 130 | 127 | 123 |
| | 5-1/2 | 0.283 | 50d | 163 | 159 | 140 | 137 | 132 |
| | 6 | 0.283 | 60d | 163 | 159 | 140 | 137 | 132 |
| | 7 | 0.312 | 5/16 | 196 | 191 | 168 | 165 | 160 |
| | 8-1/2 | 0.375 | 3/8 | 244 | 238 | 210 | 206 | 199 |
| | 3-1/2 | 0.207 | 16d | 135 | 131 | 111 | 109 | 104 |
| | 4 | 0.225 | 20d | 144 | 140 | 120 | 117 | 112 |
| | 4-1/2 | 0.244 | 30d | 148 | 143 | 123 | 120 | 115 |
| 1 | 5 | 0.263 | 40d | 164 | 159 | 137 | 134 | 129 |
| | 5-1/2 | 0.283 | 50d | 176 | 171 | 147 | 143 | 138 |
| | 6 | 0.283 | 60d | 176 | 171 | 147 | 143 | 138 |
| | 7 | 0.312 | 5/16 | 207 | 202 | 175 | 171 | 165 |
| | 8-1/2 | 0.375 | 3/8 | 257 | 250 | 217 | 212 | 205 |
| | 4 | 0.225 | 20d | 160 | 155 | 130 | 127 | 121 |
| | 4-1/2 | 0.244 | 30d | 163 | 158 | 133 | 129 | 124 |
| | 5 | 0.263 | 40d | 179 | 174 | 147 | 143 | 137 |
| 1-1/4 | 5-1/2 | 0.283 | 50d | 193 | 186 | 158 | 154 | 147 |
| | 6 | 0.283 | 60d | 193 | 186 | 158 | 154 | 147 |
| | 7 | 0.312 | 5/16 | 224 | 217 | 185 | 180 | 173 |
| | 8-1/2 | 0.375 | 3/8 | 276 | 268 | 229 | 223 | 214 |
| | 4-1/2 | 0.244 | 30d | 181 | 175 | 146 | 141 | 135 |
| | 5 | 0.263 | 40d | 197 | 191 | 159 | 155 | 148 |
| 1-1/2 | 5-1/2 | 0.283 | 50d | 212 | 205 | 171 | 166 | 159 |
| | 6 | 0.283 | 60d | 212 | 205 | 171 | 166 | 159 |
| | 7 | 0.312 | 5/16 | 244 | 236 | 199 | 193 | 185 |
| | 8-1/2 | 0.375 | 3/8 | 299 | 290 | 244 | 238 | 228 |
| | 6 | 0.283 | 60d | 223 | 219 | 194 | 191 | 185 |
| 2-1/2 | 7 | 0.312 | 5/16 | 271 | 266 | 236 | 232 | 225 |
| | 8-1/2 | 0.375 | 3/8 | 340 | 332 | 296 | 290 | 281 |
| 3-1/2 | 8-1/2 | 0.375 | 3/8 | 340 | 332 | 296 | 290 | 281 |

1. Tabulated lateral design values (Z) for spike connections shall be multiplied by all applicable adjustment factors (see Table 7.1).

2. Tabulated lateral design values (Z) are for common wire spikes inserted in side grain with spike axis perpendicular to wood fibers, and with the following spike bending yield strengths ($F_{yb}$):

   $F_{yb}$ = 80,000 psi for 0.192″, 0.207″ and 0.225″ diameter common wire spikes

   $F_{yb}$ = 70,000 psi for 0.244″ and 0.263″ diameter common wire spikes

   $F_{yb}$ = 60,000 psi for 0.283″ and $\frac{1}{16}$″ diameter common wire spikes

   $F_{yb}$ = 45,000 psi for $\frac{3}{8}$″ diameter common wire spikes

Source: ANSI/NF$_o$PA NDS—1991, *National Design Specification for Wood Construction*, AFPA, 1250 Connecticut Ave., NW, Washington, DC 20036.

## TABLE 7.54

### Box Nail Design Values (Z) for Single Shear (two member) Connections[1,2,3] with ASTM A446, Grade A steel side plate

| STEEL SIDE PLATE | NAIL LENGTH | NAIL DIAMETER | PENNY-WEIGHT | G=0.67 RED OAK | G=0.55 MIXED MAPLE SOUTHERN PINE | G=0.50 DOUGLAS FIR-LARCH | G=0.49 DOUGLAS FIR-LARCH (N) | G=0.46 DOUGLAS FIR (S) HEM-FIR (N) |
|---|---|---|---|---|---|---|---|---|
| | L inches | D inches | | Z lbs. | Z lbs. | Z lbs. | Z lbs. | Z lbs. |
| | 2 | 0.099 | 6d | 84 | 74 | 69 | 67 | 65 |
| | 2-1/2 | 0.113 | 8d | 104 | 90 | 84 | 83 | 79 |
| | 3 | 0.128 | 10d | 127 | 111 | 103 | 101 | 97 |
| 10 gage | 3-1/4 | 0.128 | 12d | 127 | 111 | 103 | 101 | 97 |
| $t_s$=0.134" | 3-1/2 | 0.135 | 16d | 139 | 121 | 113 | 111 | 106 |
| | 4 | 0.148 | 20d | 157 | 137 | 127 | 125 | 119 |
| | 4-1/2 | 0.148 | 30d | 157 | 137 | 127 | 125 | 119 |
| | 5 | 0.162 | 40d | 184 | 160 | 149 | 146 | 140 |
| | 2 | 0.099 | 6d | 80 | 70 | 65 | 64 | 61 |
| | 2-1/2 | 0.113 | 8d | 100 | 87 | 81 | 79 | 76 |
| | 3 | 0.128 | 10d | 123 | 107 | 100 | 98 | 94 |
| 11 gage | 3-1/4 | 0.128 | 12d | 123 | 107 | 100 | 98 | 94 |
| $t_s$=0.12" | 3-1/2 | 0.135 | 16d | 135 | 118 | 109 | 107 | 103 |
| | 4 | 0.148 | 20d | 153 | 133 | 123 | 121 | 116 |
| | 4-1/2 | 0.148 | 30d | 153 | 133 | 123 | 121 | 116 |
| | 5 | 0.162 | 40d | 180 | 156 | 145 | 142 | 136 |
| | 2 | 0.099 | 6d | 77 | 67 | 62 | 61 | 58 |
| | 2-1/2 | 0.113 | 8d | 96 | 83 | 78 | 76 | 73 |
| | 3 | 0.128 | 10d | 120 | 104 | 96 | 95 | 90 |
| 12 gage | 3-1/4 | 0.128 | 12d | 120 | 104 | 96 | 95 | 90 |
| $t_s$=0.105" | 3-1/2 | 0.135 | 16d | 132 | 114 | 106 | 104 | 99 |
| | 4 | 0.148 | 20d | 149 | 129 | 120 | 118 | 112 |
| | 4-1/2 | 0.148 | 30d | 149 | 129 | 120 | 118 | 112 |
| | 5 | 0.162 | 40d | 176 | 152 | 141 | 139 | 133 |
| | 2 | 0.099 | 6d | 70 | 61 | 57 | 56 | 53 |
| | 2-1/2 | 0.113 | 8d | 90 | 78 | 72 | 71 | 68 |
| | 3 | 0.128 | 10d | 114 | 98 | 91 | 89 | 85 |
| 14 gage | 3-1/4 | 0.128 | 12d | 114 | 98 | 91 | 89 | 85 |
| $t_s$=0.075" | 3-1/2 | 0.135 | 16d | 126 | 109 | 101 | 99 | 94 |
| | 4 | 0.148 | 20d | 144 | 124 | 115 | 112 | 107 |
| | 4-1/2 | 0.148 | 30d | 144 | 124 | 115 | 112 | 107 |
| | 5 | 0.162 | 40d | 171 | 147 | 136 | 134 | 128 |
| | 2 | 0.099 | 6d | 68 | 59 | 55 | 54 | 51 |
| | 2-1/2 | 0.113 | 8d | 88 | 76 | 70 | 69 | 66 |
| | 3 | 0.128 | 10d | 112 | 97 | 89 | 88 | 84 |
| 16 gage | 3-1/4 | 0.128 | 12d | 112 | 97 | 89 | 88 | 84 |
| $t_s$=0.06" | 3-1/2 | 0.135 | 16d | 125 | 107 | 99 | 97 | 93 |
| | 4 | 0.148 | 20d | 142 | 122 | 113 | 111 | 106 |
| | 4-1/2 | 0.148 | 30d | 142 | 122 | 113 | 111 | 106 |
| | 5 | 0.162 | 40d | 170 | 146 | 135 | 132 | 126 |
| | 2 | 0.099 | 6d | 67 | 58 | 54 | 53 | 50 |
| | 2-1/2 | 0.113 | 8d | 87 | 75 | 69 | 68 | 65 |
| | 3 | 0.128 | 10d | 112 | 96 | 89 | 87 | 83 |
| 18 gage | 3-1/4 | 0.128 | 12d | 112 | 96 | 89 | 87 | 83 |
| $t_s$=0.048" | 3-1/2 | 0.135 | 16d | 124 | 106 | 98 | 97 | 92 |
| | 4 | 0.148 | 20d | 142 | 121 | 112 | 110 | 105 |
| | 4-1/2 | 0.148 | 30d | 142 | 121 | 112 | 110 | 105 |
| | 5 | 0.162 | 40d | 170 | 145 | 134 | 132 | 125 |
| | 2 | 0.099 | 6d | 67 | 57 | 53 | 52 | 50 |
| 20 gage | 2-1/2 | 0.113 | 8d | 87 | 75 | 69 | 68 | 64 |
| $t_s$=0.036" | 3 | 0.128 | 10d | 112 | 96 | 88 | 87 | 82 |
| | 3-1/4 | 0.128 | 12d | 112 | 96 | 88 | 87 | 82 |
| | 3-1/2 | 0.135 | 16d | 125 | 106 | 98 | 96 | 92 |

**TABLE 7.54**   (*Continued*)

| STEEL SIDE PLATE | NAIL LENGTH | NAIL DIAMETER | PENNY-WEIGHT | G=0.43 HEM-FIR | G=0.42 SPRUCE-PINE-FIR | G=0.37 REDWOOD (open grain) | G=0.36 EASTERN SOFTWOODS SPRUCE-PINE-FIR(S) WESTERN CEDARS WESTERN WOODS | G=0.35 NORTHERN SPECIES |
|---|---|---|---|---|---|---|---|---|
| | L inches | D inches | | Z lbs. | Z lbs. | Z lbs. | Z lbs. | Z lbs. |
| | 2 | 0.099 | 6d | 61 | 60 | 54 | 54 | 52 |
| | 2-1/2 | 0.113 | 8d | 75 | 74 | 67 | 66 | 64 |
| | 3 | 0.128 | 10d | 92 | 90 | 82 | 80 | 78 |
| 10 gage | 3-1/4 | 0.128 | 12d | 92 | 90 | 82 | 80 | 78 |
| $t_s$=0.134" | 3-1/2 | 0.135 | 16d | 100 | 98 | 89 | 88 | 85 |
| | 4 | 0.148 | 20d | 113 | 111 | 100 | 99 | 96 |
| | 4-1/2 | 0.148 | 30d | 113 | 111 | 100 | 99 | 96 |
| | 5 | 0.162 | 40d | 132 | 129 | 117 | 115 | 112 |
| | 2 | 0.099 | 6d | 58 | 57 | 52 | 51 | 50 |
| | 2-1/2 | 0.113 | 8d | 72 | 71 | 64 | 63 | 61 |
| | 3 | 0.128 | 10d | 89 | 87 | 79 | 77 | 75 |
| 11 gage | 3-1/4 | 0.128 | 12d | 89 | 87 | 79 | 77 | 75 |
| $t_s$=0.12" | 3-1/2 | 0.135 | 16d | 97 | 95 | 86 | 85 | 82 |
| | 4 | 0.148 | 20d | 109 | 107 | 97 | 95 | 93 |
| | 4-1/2 | 0.148 | 30d | 109 | 107 | 97 | 95 | 93 |
| | 5 | 0.162 | 40d | 128 | 126 | 114 | 112 | 109 |
| | 2 | 0.099 | 6d | 55 | 54 | 49 | 48 | 47 |
| | 2-1/2 | 0.113 | 8d | 69 | 68 | 61 | 60 | 58 |
| | 3 | 0.128 | 10d | 85 | 84 | 76 | 74 | 72 |
| 12 gage | 3-1/4 | 0.128 | 12d | 85 | 84 | 76 | 74 | 72 |
| $t_s$=0.105" | 3-1/2 | 0.135 | 16d | 94 | 92 | 83 | 82 | 80 |
| | 4 | 0.148 | 20d | 106 | 104 | 94 | 92 | 90· |
| | 4-1/2 | 0.148 | 30d | 106 | 104 | 94 | 92 | 90 |
| | 5 | 0.162 | 40d | 125 | 123 | 111 | 109 | 106 |
| | 2 | 0.099 | 6d | 50 | 49 | 44 | 44 | 43 |
| | 2-1/2 | 0.113 | 8d | 64 | 63 | 56 | 56 | 54 |
| | 3 | 0.128 | 10d | 81 | 79 | 71 | 70 | 68 |
| 14 gage | 3-1/4 | 0.128 | 12d | 81 | 79 | 71 | 70 | 68 |
| $t_s$=0.075" | 3-1/2 | 0.135 | 16d | 89 | 87 | 79 | 77 | 75 |
| | 4 | 0.148 | 20d | 101 | 99 | 89 | 88 | 85 |
| | 4-1/2 | 0.148 | 30d | 101 | 99 | 89 | 88 | 85 |
| | 5 | 0.162 | 40d | 120 | 118 | 106 | 104 | 101 |
| | 2 | 0.099 | 6d | 48 | 47 | 43 | 42 | 41 |
| | 2-1/2 | 0.113 | 8d | 62 | 61 | 55 | 54 | 52 |
| | 3 | 0.128 | 10d | 79 | 77 | 70 | 68 | 66 |
| 16 gage | 3-1/4 | 0.128 | 12d | 79 | 77 | 70 | 68 | 66 |
| $t_s$=0.06" | 3-1/2 | 0.135 | 16d | 87 | 86 | 77 | 76 | 74 |
| | 4 | 0.148 | 20d | 100 | 98 | 88 | 86 | 84 |
| | 4-1/2 | 0.148 | 30d | 100 | 98 | 88 | 86 | 84 |
| | 5 | 0.162 | 40d | 119 | 116 | 105 | 103 | 100 |
| | 2 | 0.099 | 6d | 47 | 46 | 42 | 41 | 40 |
| | 2-1/2 | 0.113 | 8d | 61 | 60 | 54 | 53 | 51 |
| | 3 | 0.128 | 10d | 78 | 76 | 69 | 67 | 66 |
| 18 gage | 3-1/4 | 0.128 | 12d | 78 | 76 | 69 | 67 | 66 |
| $t_s$=0.048" | 3-1/2 | 0.135 | 16d | 87 | 85 | 76 | 75 | 73 |
| | 4 | 0.148 | 20d | 99 | 97 | 87 | 85 | 83 |
| | 4-1/2 | 0.148 | 30d | 99 | 97 | 87 | 85 | 83 |
| | 5 | 0.162 | 40d | 118 | 116 | 104 | 102 | 99 |
| | 2 | 0.099 | 6d | 47 | 46 | 41 | 40 | 39 |
| 20 gage | 2-1/2 | 0.113 | 8d | 61 | 59 | 53 | 52 | 51 |
| $t_s$=0.036" | 3 | 0.128 | 10d | 78 | 76 | 68 | 67 | 65 |
| | 3-1/4 | 0.128 | 12d | 78 | 76 | 68 | 67 | 65 |
| | 3-1/2 | 0.135 | 16d | 86 | 84 | 76 | 74 | 72 |

1. Tabulated lateral design values (Z) for nailed connections shall be multiplied by all applicable adjustment factors (see Table 7.1).
2. Tabulated lateral design values (Z) are for box nails inserted in side grain with nail axis perpendicular to wood fibers, and with the following nail bending yield strengths ($F_{yb}$):
   $F_{yb}$ = 100,000 psi for 0.099", 0.113", 0.128" and 0.135" diameter box nails
   $F_{yb}$ =  90,000 psi for 0.148" and 0.162" diameter box nails
3. Tabulated lateral design values (Z) are based on a dowel bearing strength ($F_e$) of 45,000 psi for ASTM A446, Grade A steel.

Source: ANSI/NF₀PA NDS—1991, *National Design Specification for Wood Construction*, AFPA, 1250 Connecticut Ave., NW, Washington, DC 20036.

## TABLE 7.55

## Common Wire Nail Design Values (Z) for Single Shear[1,2,3]
## (two member) Connections with ASTM A446, Grade A steel side plate

| STEEL SIDE PLATE | NAIL LENGTH | NAIL DIAMETER | PENNY-WEIGHT | G=0.67 RED OAK | G=0.55 MIXED MAPLE SOUTHERN PINE | G=0.50 DOUGLAS FIR-LARCH | G=0.49 DOUGLAS FIR-LARCH (N) | G=0.46 DOUGLAS FIR (S) HEM-FIR (N) |
|---|---|---|---|---|---|---|---|---|
| | L inches | D inches | | Z lbs. | Z lbs. | Z lbs. | Z lbs. | Z lbs. |
| 3 gage t_s=0.239" | 5-1/2 | 0.244 | 50d | 301 | 263 | 245 | 241 | 230 |
| | 6 | 0.263 | 60d | 334 | 291 | 271 | 266 | 255 |
| 7 gage t_s=0.179" | 4 | 0.192 | 20d | 233 | 203 | 189 | 185 | 177 |
| | 4-1/2 | 0.207 | 30d | 249 | 217 | 202 | 198 | 189 |
| | 5 | 0.225 | 40d | 269 | 234 | 217 | 213 | 204 |
| | 5-1/2 | 0.244 | 50d | 276 | 240 | 223 | 219 | 209 |
| | 6 | 0.263 | 60d | 309 | 268 | 249 | 245 | 234 |
| 10 gage t_s=0.134" | 2 | 0.113 | 6d | 104 | 90 | 84 | 83 | 79 |
| | 2-1/2 | 0.131 | 8d | 133 | 115 | 107 | 105 | 101 |
| | 3 | 0.148 | 10d | 157 | 137 | 127 | 125 | 119 |
| | 3-1/4 | 0.148 | 12d | 157 | 137 | 127 | 125 | 119 |
| | 3-1/2 | 0.162 | 16d | 184 | 160 | 149 | 146 | 140 |
| | 4 | 0.192 | 20d | 217 | 188 | 174 | 171 | 164 |
| | 4-1/2 | 0.207 | 30d | 234 | 203 | 188 | 185 | 176 |
| | 5 | 0.225 | 40d | 256 | 221 | 205 | 201 | 192 |
| | 5-1/2 | 0.244 | 50d | 262 | 227 | 210 | 206 | 197 |
| | 6 | 0.263 | 60d | 296 | 256 | 237 | 233 | 222 |
| 11 gage t_s=0.12" | 2 | 0.113 | 6d | 100 | 87 | 81 | 79 | 76 |
| | 2-1/2 | 0.131 | 8d | 128 | 112 | 104 | 102 | 97 |
| | 3 | 0.148 | 10d | 153 | 133 | 123 | 121 | 116 |
| | 3-1/4 | 0.148 | 12d | 153 | 133 | 123 | 121 | 116 |
| | 3-1/2 | 0.162 | 16d | 180 | 156 | 145 | 142 | 136 |
| | 4 | 0.192 | 20d | 213 | 184 | 171 | 168 | 160 |
| | 4-1/2 | 0.207 | 30d | 231 | 199 | 185 | 181 | 173 |
| | 5 | 0.225 | 40d | 252 | 218 | 202 | 198 | 189 |
| | 5-1/2 | 0.244 | 50d | 259 | 224 | 207 | 204 | 194 |
| | 6 | 0.263 | 60d | 294 | 253 | 234 | 230 | 220 |
| 12 gage t_s=0.105" | 2 | 0.113 | 6d | 96 | 83 | 78 | 76 | 73 |
| | 2-1/2 | 0.131 | 8d | 125 | 108 | 100 | 99 | 94 |
| | 3 | 0.148 | 10d | 149 | 129 | 120 | 118 | 112 |
| | 3-1/4 | 0.148 | 12d | 149 | 129 | 120 | 118 | 112 |
| | 3-1/2 | 0.162 | 16d | 176 | 152 | 141 | 139 | 133 |
| | 4 | 0.192 | 20d | 209 | 181 | 168 | 164 | 157 |
| | 4-1/2 | 0.207 | 30d | 228 | 196 | 182 | 178 | 170 |
| | 5 | 0.225 | 40d | 250 | 215 | 199 | 196 | 187 |
| | 5-1/2 | 0.244 | 50d | 257 | 221 | 205 | 201 | 192 |
| | 6 | 0.263 | 60d | 291 | 251 | 232 | 228 | 217 |
| 14 gage t_s=0.075" | 2 | 0.113 | 6d | 90 | 78 | 72 | 71 | 68 |
| | 2-1/2 | 0.131 | 8d | 119 | 103 | 95 | 93 | 89 |
| | 3 | 0.148 | 10d | 144 | 124 | 115 | 112 | 107 |
| | 3-1/4 | 0.148 | 12d | 144 | 124 | 115 | 112 | 107 |
| | 3-1/2 | 0.162 | 16d | 171 | 147 | 136 | 134 | 128 |
| | 4 | 0.192 | 20d | 205 | 176 | 163 | 160 | 153 |
| | 4-1/2 | 0.207 | 30d | 224 | 192 | 178 | 174 | 166 |
| | 5 | 0.225 | 40d | 247 | 212 | 196 | 192 | 183 |
| | 5-1/2 | 0.244 | 50d | 254 | 218 | 202 | 198 | 188 |
| | 6 | 0.263 | 60d | 289 | 248 | 229 | 225 | 214 |
| 16 gage t_s=0.06" | 2 | 0.113 | 6d | 88 | 76 | 70 | 69 | 66 |
| | 2-1/2 | 0.131 | 8d | 118 | 101 | 94 | 92 | 88 |
| | 3 | 0.148 | 10d | 142 | 122 | 113 | 111 | 106 |
| | 3-1/4 | 0.148 | 12d | 142 | 122 | 113 | 111 | 106 |
| | 3-1/2 | 0.162 | 16d | 170 | 146 | 135 | 132 | 126 |
| | 4 | 0.192 | 20d | 204 | 175 | 162 | 159 | 151 |
| | 4-1/2 | 0.207 | 30d | 224 | 191 | 177 | 174 | 165 |
| 18 gage t_s=0.048" | 2 | 0.113 | 6d | 87 | 75 | 69 | 68 | 65 |
| | 2-1/2 | 0.131 | 8d | 117 | 100 | 93 | 91 | 87 |
| | 3 | 0.148 | 10d | 142 | 121 | 112 | 110 | 105 |
| | 3-1/4 | 0.148 | 12d | 142 | 121 | 112 | 110 | 105 |
| | 3-1/2 | 0.162 | 16d | 170 | 145 | 134 | 132 | 125 |
| 20 gage t_s=0.036" | 2 | 0.113 | 6d | 87 | 75 | 69 | 68 | 64 |
| | 2-1/2 | 0.131 | 8d | 117 | 100 | 92 | 91 | 86 |
| | 3 | 0.148 | 10d | 142 | 121 | 112 | 110 | 105 |

**TABLE 7.55**   *(Continued)*

| STEEL SIDE PLATE | NAIL LENGTH | NAIL DIAMETER | PENNY-WEIGHT | G=0.43 HEM-FIR | G=0.42 SPRUCE-PINE-FIR | G=0.37 REDWOOD (open grain) | G=0.36 EASTERN SOFTWOODS SPRUCE-PINE-FIR(S) WESTERN CEDARS WESTERN WOODS | G=0.35 NORTHERN SPECIES |
|---|---|---|---|---|---|---|---|---|
| | L inches | D inches | | Z lbs. | Z lbs. | Z lbs. | Z lbs. | Z lbs. |
| 3 gage $t_s$=0.239" | 5-1/2 | 0.244 | 50d | 218 | 214 | 194 | 191 | 186 |
| | 6 | 0.263 | 60d | 241 | 237 | 214 | 211 | 205 |
| 7 gage $t_s$=0.179" | 4 | 0.192 | 20d | 168 | 165 | 149 | 146 | 143 |
| | 4-1/2 | 0.207 | 30d | 179 | 176 | 159 | 156 | 152 |
| | 5 | 0.225 | 40d | 193 | 189 | 171 | 168 | 164 |
| | 5-1/2 | 0.244 | 50d | 198 | 194 | 175 | 172 | .168 |
| | 6 | 0.263 | 60d | 221 | 217 | 196 | 192 | 187 |
| 10 gage $t_s$=0.134" | 2 | 0.113 | 6d | 75 | 74 | 67 | 66 | 64 |
| | 2-1/2 | 0.131 | 8d | 95 | 94 | 85 | 83 | 81 |
| | 3 | 0.148 | 10d | 113 | 111 | 100 | 99 | 96 |
| | 3-1/4 | 0.148 | 12d | 113 | 111 | 100 | 99 | 96 |
| | 3-1/2 | 0.162 | 16d | 132 | 129 | 117 | 115 | 112 |
| | 4 | 0.192 | 20d | 155 | 152 | 137 | 135 | 131 |
| | 4-1/2 | 0.207 | 30d | 167 | 163 | 147 | 145 | 141 |
| | 5 | 0.225 | 40d | 181 | 178 | 160 | 158 | 153 |
| | 5-1/2 | 0.244 | 50d | 186 | 183 | 165 | 162 | 157 |
| | 6 | 0.263 | 60d | 210 | 206 | 185 | 182 | 177 |
| 11 gage $t_s$=0.12" | 2 | 0.113 | 6d | 72 | 71 | 64 | 63 | 61 |
| | 2-1/2 | 0.131 | 8d | 92 | 90 | 82 | 80 | 78 |
| | 3 | 0.148 | 10d | 109 | 107 | 97 | 95 | 93 |
| | 3-1/4 | 0.148 | 12d | 109 | 107 | 97 | 95 | 93 |
| | 3-1/2 | 0.162 | 16d | 128 | 126 | 114 | 112 | 109 |
| | 4 | 0.192 | 20d | 151 | 148 | 134 | 132 | 128 |
| | 4-1/2 | 0.207 | 30d | 163 | 160 | 145 | 142 | 138 |
| | 5 | 0.225 | 40d | 178 | 175 | 158 | 155 | 151 |
| | 5-1/2 | 0.244 | 50d | 183 | 180 | 162 | 159 | 155 |
| | 6 | 0.263 | 60d | 207 | 203 | 183 | 180 | 175 |
| 12 gage $t_s$=0.105" | 2 | 0.113 | 6d | 69 | 68 | 61 | 60 | 58 |
| | 2-1/2 | 0.131 | 8d | 89 | 87 | 79 | 78 | 75 |
| | 3 | 0.148 | 10d | 106 | 104 | 94 | 92 | 90 |
| | 3-1/4 | 0.148 | 12d | 106 | 104 | 94 | 92 | 90 |
| | 3-1/2 | 0.162 | 16d | 125 | 123 | 111 | 109 | 106 |
| | 4 | 0.192 | 20d | 148 | 145 | 131 | 129 | 125 |
| | 4-1/2 | 0.207 | 30d | 161 | 158 | 142 | 139 | 136 |
| | 5 | 0.225 | 40d | 176 | 173 | 155 | 153 | 148 |
| | 5-1/2 | 0.244 | 50d | 181 | 177 | 160 | 157 | 152 |
| | 6 | 0.263 | 60d | 205 | 201 | 180 | 177 | 172 |
| 14 gage $t_s$=0.075" | 2 | 0.113 | 6d | 64 | 63 | 56 | 56 | 54 |
| | 2-1/2 | 0.131 | 8d | 84 | 83 | 74 | 73 | 71 |
| | 3 | 0.148 | 10d | 101 | 99 | 89 | 88 | 85 |
| | 3-1/4 | 0.148 | 12d | 101 | 99 | 89 | 88 | 85 |
| | 3-1/2 | 0.162 | 16d | 120 | 118 | 106 | 104 | 101 |
| | 4 | 0.192 | 20d | 144 | 141 | 127 | 124 | 121 |
| | 4-1/2 | 0.207 | 30d | 157 | 154 | 138 | 136 | 132 |
| | 5 | 0.225 | 40d | 172 | 169 | 152 | 149 | 145 |
| | 5-1/2 | 0.244 | 50d | 177 | 174 | 156 | 153 | 149 |
| | 6 | 0.263 | 60d | 201 | 198 | 177 | 174 | 169 |
| 16 gage $t_s$=0.06" | 2 | 0.113 | 6d | 62 | 61 | 55 | 54 | 52 |
| | 2-1/2 | 0.131 | 8d | 83 | 81 | 73 | 71 | 70 |
| | 3 | 0.148 | 10d | 100 | 98 | 88 | 86 | 84 |
| | 3-1/4 | 0.148 | 12d | 100 | 98 | 88 | 86 | 84 |
| | 3-1/2 | 0.162 | 16d | 119 | 116 | 105 | 103 | 100 |
| | 4 | 0.192 | 20d | 142 | 140 | 125 | 123 | 120 |
| | 4-1/2 | 0.207 | 30d | 156 | 153 | 137 | 134 | 131 |
| 18 gage $t_s$=0.048" | 2 | 0.113 | 6d | 61 | 60 | 54 | 53 | 51 |
| | 2-1/2 | 0.131 | 8d | 82 | 80 | 72 | 71 | 69 |
| | 3 | 0.148 | 10d | 99 | 97 | 87 | 85 | 83 |
| | 3-1/4 | 0.148 | 12d | 99 | 97 | 87 | 85 | 83 |
| | 3-1/2 | 0.162 | 16d | 118 | 116 | 104 | 102 | 99 |
| 20 gage $t_s$=0.036" | 2 | 0.113 | 6d | 61 | 59 | 53 | 52 | 51 |
| | 2-1/2 | 0.131 | 8d | 81 | 80 | 71 | 70 | 68 |
| | 3 | 0.148 | 10d | 98 | 96 | 86 | 85 | 82 |

1. Tabulated lateral design values (Z) for nailed connections shall be multiplied by all applicable adjustment factors (see Table 7.1).

2. Tabulated lateral design values (Z) are for common wire nails inserted in side grain with nail axis perpendicular to wood fibers, and with the following nail bending yield strengths ($F_{yb}$):

   $F_{yb}$ = 100,000 psi for 0.113" and 0.131" diameter common wire nails

   $F_{yb}$ = 90,000 psi for 0.148" and 0.162" diameter common wire nails

   $F_{yb}$ = 80,000 psi for 0.192", 0.207" and 0.225" diameter common wire nails

   $F_{yb}$ = 70,000 psi for 0.244" and 0.263" diameter common wire nails

3. Tabulated lateral design values (Z) are based on a dowel bearing strength ($F_e$) of 45,000 psi for ASTM A446, Grade A steel.

Source: ANSI/NF$_o$PA NDS—1991, *National Design Specification for Wood Construction*, AFPA, 1250 Connecticut Ave., NW, Washington, DC 20036.

# TABLE 7.56

## Threaded Hardened-Steel Nail Design Values (Z) for Single Shear[1,2,3] (two member) Connections with ASTM A446, Grade A steel side plate

| STEEL SIDE PLATE | NAIL LENGTH | WIRE DIAMETER | PENNY-WEIGHT | G=0.67 RED OAK | G=0.55 MIXED MAPLE SOUTHERN PINE | G=0.50 DOUGLAS FIR-LARCH | G=0.49 DOUGLAS FIR-LARCH (N) | G=0.46 DOUGLAS FIR (S) HEM-FIR (N) |
|---|---|---|---|---|---|---|---|---|
| | L inches | D inches | | Z lbs. | Z lbs. | Z lbs. | Z lbs. | Z lbs. |
| 7 gage $t_s$=0.179" | 7 | 0.207 | 70d | 271 | 235 | 218 | 214 | 205 |
| | 8 | 0.207 | 80d | 271 | 235 | 218 | 214 | 205 |
| | 9 | 0.207 | 90d | 271 | 235 | 218 | 214 | 205 |
| 10 gage $t_s$=0.134" | 2 | 0.120 | 6d | 125 | 109 | 101 | 99 | 95 |
| | 2-1/2 | 0.120 | 8d | 125 | 109 | 101 | 99 | 95 |
| | 3 | 0.135 | 10d | 154 | 133 | 124 | 122 | 116 |
| | 3-1/4 | 0.135 | 12d | 154 | 133 | 124 | 122 | 116 |
| | 3-1/2 | 0.148 | 16d | 172 | 150 | 139 | 137 | 131 |
| | 4 | 0.177 | 20d | 232 | 200 | 186 | 182 | 174 |
| | 4-1/2 | 0.177 | 30d | 232 | 200 | 186 | 182 | 174 |
| | 5 | 0.177 | 40d | 232 | 200 | 186 | 182 | 174 |
| | 5-1/2 | 0.177 | 50d | 232 | 200 | 186 | 182 | 174 |
| | 6 | 0.177 | 60d | 232 | 200 | 186 | 182 | 174 |
| | 7 | 0.207 | 70d | 258 | 223 | 207 | 203 | 194 |
| | 8 | 0.207 | 80d | 258 | 223 | 207 | 203 | 194 |
| | 9 | 0.207 | 90d | 258 | 223 | 207 | 203 | 194 |
| 11 gage $t_s$=0.12" | 2 | 0.120 | 6d | 122 | 106 | 98 | 96 | 92 |
| | 2-1/2 | 0.120 | 8d | 122 | 106 | 98 | 96 | 92 |
| | 3 | 0.135 | 10d | 150 | 130 | 121 | 119 | 113 |
| | 3-1/4 | 0.135 | 12d | 150 | 130 | 121 | 119 | 113 |
| | 3-1/2 | 0.148 | 16d | 169 | 146 | 136 | 133 | 127 |
| | 4 | 0.177 | 20d | 228 | 197 | 183 | 179 | 171 |
| | 4-1/2 | 0.177 | 30d | 228 | 197 | 183 | 179 | 171 |
| | 5 | 0.177 | 40d | 228 | 197 | 183 | 179 | 171 |
| | 5-1/2 | 0.177 | 50d | 228 | 197 | 183 | 179 | 171 |
| | 6 | 0.177 | 60d | 228 | 197 | 183 | 179 | 171 |
| 12 gage $t_s$=0.105" | 2 | 0.120 | 6d | 118 | 102 | 95 | 93 | 89 |
| | 2-1/2 | 0.120 | 8d | 118 | 102 | 95 | 93 | 89 |
| | 3 | 0.135 | 10d | 147 | 127 | 118 | 116 | 111 |
| | 3-1/4 | 0.135 | 12d | 147 | 127 | 118 | 116 | 111 |
| | 3-1/2 | 0.148 | 16d | 166 | 143 | 133 | 130 | 124 |
| | 4 | 0.177 | 20d | 226 | 194 | 180 | 177 | 169 |
| | 4-1/2 | 0.177 | 30d | 226 | 194 | 180 | 177 | 169 |
| | 5 | 0.177 | 40d | 226 | 194 | 180 | 177 | 169 |
| | 5-1/2 | 0.177 | 50d | 226 | 194 | 180 | 177 | 169 |
| | 6 | 0.177 | 60d | 226 | 194 | 180 | 177 | 169 |
| 14 gage $t_s$=0.075" | 2 | 0.120 | 6d | 114 | 98 | 91 | 89 | 85 |
| | 2-1/2 | 0.120 | 8d | 114 | 98 | 91 | 89 | 85 |
| | 3 | 0.135 | 10d | 143 | 123 | 114 | 112 | 106 |
| | 3-1/4 | 0.135 | 12d | 143 | 123 | 114 | 112 | 106 |
| | 3-1/2 | 0.148 | 16d | 161 | 139 | 128 | 126 | 120 |
| | 4 | 0.177 | 20d | 222 | 191 | 176 | 173 | 165 |
| | 4-1/2 | 0.177 | 30d | 222 | 191 | 176 | 173 | 165 |
| | 5 | 0.177 | 40d | 222 | 191 | 176 | 173 | 165 |
| | 5-1/2 | 0.177 | 50d | 222 | 191 | 176 | 173 | 165 |
| | 6 | 0.177 | 60d | 222 | 191 | 176 | 173 | 165 |
| 16 gage $t_s$=0.06" | 2 | 0.120 | 6d | 112 | 97 | 89 | 88 | 84 |
| | 2-1/2 | 0.120 | 8d | 112 | 97 | 89 | 88 | 84 |
| | 3 | 0.135 | 10d | 142 | 122 | 113 | 110 | 105 |
| | 3-1/4 | 0.135 | 12d | 142 | 122 | 113 | 110 | 105 |
| | 3-1/2 | 0.148 | 16d | 160 | 137 | 127 | 125 | 119 |
| | 4 | 0.177 | 20d | 222 | 190 | 176 | 172 | 164 |
| | 4-1/2 | 0.177 | 30d | 222 | 190 | 176 | 172 | 164 |
| | 5 | 0.177 | 40d | 222 | 190 | 176 | 172 | 164 |
| | 5-1/2 | 0.177 | 50d | 222 | 190 | 176 | 172 | 164 |
| | 6 | 0.177 | 60d | 222 | 190 | 176 | 172 | 164 |
| 18 gage $t_s$=0.048" | 2 | 0.120 | 6d | 112 | 96 | 89 | 87 | 83 |
| | 2-1/2 | 0.120 | 8d | 112 | 96 | 89 | 87 | 83 |
| | 3 | 0.135 | 10d | 142 | 121 | 112 | 110 | 105 |
| | 3-1/4 | 0.135 | 12d | 142 | 121 | 112 | 110 | 105 |
| | 3-1/2 | 0.148 | 16d | 160 | 137 | 127 | 124 | 118 |
| 20 gage $t_s$=0.036" | 2 | 0.120 | 6d | 112 | 96 | 88 | 87 | 83 |
| | 2-1/2 | 0.120 | 8d | 112 | 96 | 88 | 87 | 83 |
| | 3 | 0.135 | 10d | 142 | 121 | 112 | 110 | 105 |

## TABLE 7.56 *(Continued)*

| STEEL SIDE PLATE | NAIL LENGTH | WIRE DIAMETER | PENNY-WEIGHT | G=0.43 HEM-FIR | G=0.42 SPRUCE-PINE-FIR | G=0.37 REDWOOD (open grain) | G=0.36 EASTERN SOFTWOODS SPRUCE-PINE-FIR(S) WESTERN CEDARS WESTERN WOODS | G=0.35 NORTHERN SPECIES |
|---|---|---|---|---|---|---|---|---|
| | L inches | D inches | | Z lbs. | Z lbs. | Z lbs. | Z lbs. | Z lbs. |
| 7 gage $t_s$=0.179" | 7 | 0.207 | 70d | 194 | 190 | 172 | 169 | 164 |
| | 8 | 0.207 | 80d | 194 | 190 | 172 | 169 | 164 |
| | 9 | 0.207 | 90d | 194 | 190 | 172 | 169 | 164 |
| | 2 | 0.120 | 6d | 90 | 88 | 80 | 78 | 76 |
| | 2-1/2 | 0.120 | 8d | 90 | 88 | 80 | 78 | 76 |
| | 3 | 0.135 | 10d | 110 | 108 | 98 | 96 | 93 |
| | 3-1/4 | 0.135 | 12d | 110 | 108 | 98 | 96 | 93 |
| | 3-1/2 | 0.148 | 16d | 123 | 121 | 109 | 107 | ·105 |
| 10 gage $t_s$=0.134" | 4 | 0.177 | 20d | 165 | 161 | 146 | 143 | 139 |
| | 4-1/2 | 0.177 | 30d | 165 | 161 | 146 | 143 | 139 |
| | 5 | 0.177 | 40d | 165 | 161 | 146 | 143 | 139 |
| | 5-1/2 | 0.177 | 50d | 165 | 161 | 146 | 143 | 139 |
| | 6 | 0.177 | 60d | 165 | 161 | 146 | 143 | 139 |
| | 7 | 0.207 | 70d | 183 | 179 | 162 | 159 | 155 |
| | 8 | 0.207 | 80d | 183 | 179 | 162 | 159 | 155 |
| | 9 | 0.207 | 90d | 183 | 179 | 162 | 159 | 155 |
| | 2 | 0.120 | 6d | 87 | 85 | 77 | 76 | 74 |
| | 2-1/2 | 0.120 | 8d | 87 | 85 | 77 | 76 | 74 |
| | 3 | 0.135 | 10d | 107 | 105 | 95 | 93 | 91 |
| 11 gage $t_s$=0.12" | 3-1/4 | 0.135 | 12d | 107 | 105 | 95 | 93 | 91 |
| | 3-1/2 | 0.148 | 16d | 120 | 118 | 107 | 105 | 102 |
| | 4 | 0.177 | 20d | 162 | 159 | 143 | 140 | 137 |
| | 4-1/2 | 0.177 | 30d | 162 | 159 | 143 | 140 | 137 |
| | 5 | 0.177 | 40d | 162 | 159 | 143 | 140 | 137 |
| | 5-1/2 | 0.177 | 50d | 162 | 159 | 143 | 140 | 137 |
| | 6 | 0.177 | 60d | 162 | 159 | 143 | 140 | 137 |
| | 2 | 0.120 | 6d | 84 | 83 | 75 | 73 | 71 |
| | 2-1/2 | 0.120 | 8d | 84 | 83 | 75 | 73 | 71 |
| | 3 | 0.135 | 10d | 104 | 102 | 92 | 91 | 88 |
| | 3-1/4 | 0.135 | 12d | 104 | 102 | 92 | 91 | 88 |
| 12 gage $t_s$=0.105" | 3-1/2 | 0.148 | 16d | 117 | 115 | 104 | 102 | 99 |
| | 4 | 0.177 | 20d | 159 | 156 | 140 | 138 | 134 |
| | 4-1/2 | 0.177 | 30d | 159 | 156 | 140 | 138 | 134 |
| | 5 | 0.177 | 40d | 159 | 156 | 140 | 138 | 134 |
| | 5-1/2 | 0.177 | 50d | 159 | 156 | 140 | 138 | 134 |
| | 6 | 0.177 | 60d | 159 | 156 | 140 | 138 | 134 |
| | 2 | 0.120 | 6d | 80 | 79 | 71 | 70 | 68 |
| | 2-1/2 | 0.120 | 8d | 80 | 79 | 71 | 70 | 68 |
| | 3 | 0.135 | 10d | 100 | 98 | 89 | 87 | 85 |
| | 3-1/4 | 0.135 | 12d | 100 | 98 | 89 | 87 | 85 |
| 14 gage $t_s$=0.075" | 3-1/2 | 0.148 | 16d | 113 | 111 | 100 | 98 | 95 |
| | 4 | 0.177 | 20d | 155 | 152 | 137 | 134 | 131 |
| | 4-1/2 | 0.177 | 30d | 155 | 152 | 137 | 134 | 131 |
| | 5 | 0.177 | 40d | 155 | 152 | 137 | 134 | 131 |
| | 5-1/2 | 0.177 | 50d | 155 | 152 | 137 | 134 | 131 |
| | 6 | 0.177 | 60d | 155 | 152 | 137 | 134 | 131 |
| | 2 | 0.120 | 6d | 79 | 77 | 69 | 68 | 66 |
| | 2-1/2 | 0.120 | 8d | 79 | 77 | 69 | 68 | 66 |
| | 3 | 0.135 | 10d | 99 | 97 | 87 | 86 | 83 |
| | 3-1/4 | 0.135 | 12d | 99 | 97 | 87 | 86 | 83 |
| 16 gage $t_s$=0.06" | 3-1/2 | 0.148 | 16d | 112 | 110 | 98 | 97 | 94 |
| | 4 | 0.177 | 20d | 154 | 151 | 136 | 133 | 130 |
| | 4-1/2 | 0.177 | 30d | 154 | 151 | 136 | 133 | 130 |
| | 5 | 0.177 | 40d | 154 | 151 | 136 | 133 | 130 |
| | 5-1/2 | 0.177 | 50d | 154 | 151 | 136 | 133 | 130 |
| | 6 | 0.177 | 60d | 154 | 151 | 136 | 133 | 130 |
| | 2 | 0.120 | 6d | 78 | 76 | 69 | 67 | 66 |
| 18 gage $t_s$=0.048" | 2-1/2 | 0.120 | 8d | 78 | 76 | 69 | 67 | 66 |
| | 3 | 0.135 | 10d | 98 | 97 | 87 | 85 | 83 |
| ₁ | 3-1/4 | 0.135 | 12d | 98 | 97 | 87 | 85 | 83 |
| | 3-1/2 | 0.148 | 16d | 111 | 109 | 98 | 96 | 93 |
| 20 gage $t_s$=0.036" | 2 | 0.120 | 6d | 78 | 76 | 68 | 67 | 65 |
| | 2-1/2 | 0.120 | 8d | 78 | 76 | 68 | 67 | 65 |
| | 3 | 0.135 | 10d | 98 | 96 | 86 | 85 | 82 |

1. Tabulated lateral design values (Z) for nailed connections shall be multiplied by all applicable adjustment factors (see Table 7.1).

2. Tabulated lateral design values (Z) are for threaded hardened-steel nails inserted in side grain with nail axis perpendicular to wood fibers, and with the following nail bending yield strengths ($F_{yb}$):

$F_{yb}$ = 130,000 psi for 0.12" and 0.135" diameter threaded hardened-steel nails

$F_{yb}$ = 115,000 psi for 0.148" and 0.177" diameter threaded hardened-steel nails

$F_{yb}$ = 100,000 psi for 0.207" diameter threaded hardened-steel nails

3. Tabulated lateral design values (Z) are based on a dowel bearing strength ($F_e$) of 45,000 psi for ASTM A446, Grade A steel.

Source: ANSI/NF₀PA NDS—1991, *National Design Specification for Wood Construction*, AFPA, 1250 Connecticut Ave., NW, Washington, DC 20036.

TABLE 7.57

## Common Wire Spike Design Values (Z) for Single Shear[1,2,3] (two member) Connections with ASTM A446, Grade A steel side plate

| STEEL SIDE PLATE | SPIKE LENGTH | SPIKE DIAMETER | PENNY-WEIGHT | G=0.67 RED OAK | G=0.55 MIXED MAPLE SOUTHERN PINE | G=0.50 DOUGLAS FIR-LARCH | G=0.49 DOUGLAS FIR-LARCH (N) | G=0.46 DOUGLAS FIR (S) HEM-FIR (N) |
|---|---|---|---|---|---|---|---|---|
| | L inches | D inches | | Z lbs. | Z lbs. | Z lbs. | Z lbs. | Z lbs. |
| 3 gage $t_s$=0.239" | 4-1/2 | 0.244 | 30d | 301 | 263 | 245 | 241 | 230 |
| | 5 | 0.263 | 40d | 334 | 291 | 271 | 266 | 255 |
| | 5-1/2 | 0.283 | 50d | 359 | 312 | 291 | 286 | 273 |
| | 6 | 0.283 | 60d | 359 | 312 | 291 | 286 | 273 |
| | 7 | 0.312 | 5/16 | 423 | 367 | 342 | 336 | 321 |
| | 8-1/2 | 0.375 | 3/8 | 523 | 454 | 423 | 415 | 397 |
| 7 gage $t_s$=0.179" | 3 | 0.192 | 10d | 233 | 203 | 189 | 185 | 177 |
| | 3-1/4 | 0.192 | 12d | 233 | 203 | 189 | 185 | 177 |
| | 3-1/2 | 0.207 | 16d | 249 | 217 | 202 | 198 | 189 |
| | 4 | 0.225 | 20d | 269 | 234 | 217 | 213 | 204 |
| | 4-1/2 | 0.244 | 30d | 276 | 240 | 223 | 219 | 209 |
| | 5 | 0.263 | 40d | 309 | 268 | 249 | 245 | 234 |
| | 5-1/2 | 0.283 | 50d | 332 | 288 | 267 | 262 | 251 |
| | 6 | 0.283 | 60d | 332 | 288 | 267 | 262 | 251 |
| | 7 | 0.312 | 5/16 | 396 | 343 | 318 | 313 | 299 |
| | 8-1/2 | 0.375 | 3/8 | 492 | 426 | 395 | 388 | 371 |
| 10 gage $t_s$=0.134" | 3 | 0.192 | 10d | 217 | 188 | 174 | 171 | 164 |
| | 3-1/4 | 0.192 | 12d | 217 | 188 | 174 | 171 | 164 |
| | 3-1/2 | 0.207 | 16d | 234 | 203 | 188 | 185 | 176 |
| | 4 | 0.225 | 20d | 256 | 221 | 205 | 201 | 192 |
| | 4-1/2 | 0.244 | 30d | 262 | 227 | 210 | 206 | 197 |
| | 5 | 0.263 | 40d | 296 | 256 | 237 | 233 | 222 |
| | 5-1/2 | 0.283 | 50d | 318 | 274 | 254 | 250 | 238 |
| | 6 | 0.283 | 60d | 318 | 274 | 254 | 250 | 238 |
| | 7 | 0.312 | 5/16 | 383 | 330 | 306 | 300 | 287 |
| | 8-1/2 | 0.375 | 3/8 | 478 | 411 | 381 | 374 | 357 |
| 11 gage $t_s$=0.12" | 3 | 0.192 | 10d | 213 | 184 | 171 | 168 | 160 |
| | 3-1/4 | 0.192 | 12d | 213 | 184 | 171 | 168 | 160 |
| | 3-1/2 | 0.207 | 16d | 231 | 199 | 185 | 181 | 173 |
| | 4 | 0.225 | 20d | 252 | 218 | 202 | 198 | 189 |
| | 4-1/2 | 0.244 | 30d | 259 | 224 | 207 | 204 | 194 |
| | 5 | 0.263 | 40d | 294 | 253 | 234 | 230 | 220 |
| | 5-1/2 | 0.283 | 50d | 315 | 271 | 251 | 247 | 235 |
| | 6 | 0.283 | 60d | 315 | 271 | 251 | 247 | 235 |
| | 7 | 0.312 | 5/16 | 380 | 327 | 303 | 297 | 284 |
| | 8-1/2 | 0.375 | 3/8 | 475 | 408 | 378 | 371 | 354 |
| 12 gage $t_s$=0.105" | 3 | 0.192 | 10d | 209 | 181 | 168 | 164 | 157 |
| | 3-1/4 | 0.192 | 12d | 209 | 181 | 168 | 164 | 157 |
| | 3-1/2 | 0.207 | 16d | 228 | 196 | 182 | 178 | 170 |
| | 4 | 0.225 | 20d | 250 | 215 | 199 | 196 | 187 |
| | 4-1/2 | 0.244 | 30d | 257 | 221 | 205 | 201 | 192 |
| | 5 | 0.263 | 40d | 291 | 251 | 232 | 228 | 217 |
| | 5-1/2 | 0.283 | 50d | 312 | 269 | 249 | 244 | 233 |
| | 6 | 0.283 | 60d | 312 | 269 | 249 | 244 | 233 |
| | 7 | 0.312 | 5/16 | 378 | 325 | 301 | 295 | 281 |
| | 8-1/2 | 0.375 | 3/8 | 473 | 406 | 376 | 368 | 351 |
| 14 gage $t_s$=0.075" | 3 | 0.192 | 10d | 205 | 176 | 163 | 160 | 153 |
| | 3-1/4 | 0.192 | 12d | 205 | 176 | 163 | 160 | 153 |
| | 3-1/2 | 0.207 | 16d | 224 | 192 | 178 | 174 | 166 |
| | 4 | 0.225 | 20d | 247 | 212 | 196 | 192 | 183 |
| | 4-1/2 | 0.244 | 30d | 254 | 218 | 202 | 198 | 188 |
| | 5 | 0.263 | 40d | 289 | 248 | 229 | 225 | 214 |
| | 5-1/2 | 0.283 | 50d | 310 | 266 | 246 | 241 | 230 |
| | 6 | 0.283 | 60d | 310 | 266 | 246 | 241 | 230 |
| | 7 | 0.312 | 5/16 | 377 | 323 | 298 | 292 | 279 |
| | 8-1/2 | 0.375 | 3/8 | 472 | 404 | 373 | 366 | 348 |
| 16 gage $t_s$=0.06" | 3 | 0.192 | 10d | 204 | 175 | 162 | 159 | 151 |
| | 3-1/4 | 0.192 | 12d | 204 | 175 | 162 | 159 | 151 |
| | 3-1/2 | 0.207 | 16d | 224 | 191 | 177 | 174 | 165 |

## TABLE 7.57   (*Continued*)

| STEEL SIDE PLATE | SPIKE LENGTH | SPIKE DIAMETER | PENNY-WEIGHT | G=0.43 HEM-FIR | G=0.42 SPRUCE-PINE-FIR | G=0.37 REDWOOD (open grain) | G=0.36 EASTERN SOFTWOODS SPRUCE-PINE-FIR(S) WESTERN CEDARS WESTERN WOODS | G=0.35 NORTHERN SPECIES |
|---|---|---|---|---|---|---|---|---|
| | L inches | D inches | | Z lbs. | Z lbs. | Z lbs. | Z lbs. | Z lbs. |
| 3 gage ts=0.239" | 4-1/2 | 0.244 | 30d | 218 | 214 | 194 | 191 | 186 |
| | 5 | 0.263 | 40d | 241 | 237 | 214 | 211 | 205 |
| | 5-1/2 | 0.283 | 50d | 259 | 254 | 230 | 226 | 220 |
| | 6 | 0.283 | 60d | 259 | 254 | 230 | 226 | 220 |
| | 7 | 0.312 | 5/16 | 304 | 298 | 270 | 265 | 258 |
| | 8-1/2 | 0.375 | 3/8 | 375 | 368 | 333 | 328 | 319 |
| 7 gage ts=0.179" | 3 | 0.192 | 10d | 168 | 165 | 149 | 146 | 143 |
| | 3-1/4 | 0.192 | 12d | 168 | 165 | 149 | 146 | 143 |
| | 3-1/2 | 0.207 | 16d | 179 | 176 | 159 | 156 | 152 |
| | 4 | 0.225 | 20d | 193 | 189 | 171 | 168 | 164 |
| | 4-1/2 | 0.244 | 30d | 198 | 194 | 175 | 172 | 168 |
| | 5 | 0.263 | 40d | 221 | 217 | 196 | 192 | 187 |
| | 5-1/2 | 0.283 | 50d | 237 | 233 | 210 | 206 | 201 |
| | 6 | 0.283 | 60d | 237 | 233 | 210 | 206 | 201 |
| | 7 | 0.312 | 5/16 | 282 | 277 | 250 | 245 | 239 |
| | 8-1/2 | 0.375 | 3/8 | 350 | 344 | 310 | 305 | 296 |
| 10 gage ts=0.134" | 3 | 0.192 | 10d | 155 | 152 | 137 | 135 | 131 |
| | 3-1/4 | 0.192 | 12d | 155 | 152 | 137 | 135 | 131 |
| | 3-1/2 | 0.207 | 16d | 167 | 163 | 147 | 145 | 141 |
| | 4 | 0.225 | 20d | 181 | 178 | 160 | 158 | 153 |
| | 4-1/2 | 0.244 | 30d | 186 | 183 | 165 | 162 | 157 |
| | 5 | 0.263 | 40d | 210 | 206 | 185 | 182 | 177 |
| | 5-1/2 | 0.283 | 50d | 225 | 221 | 199 | 195 | 190 |
| | 6 | 0.283 | 60d | 225 | 221 | 199 | 195 | 190 |
| | 7 | 0.312 | 5/16 | 270 | 265 | 239 | 234 | 228 |
| | 8-1/2 | 0.375 | 3/8 | 337 | 330 | 297 | 292 | 284 |
| 11 gage ts=0.12" | 3 | 0.192 | 10d | 151 | 148 | 134 | 132 | 128 |
| | 3-1/4 | 0.192 | 12d | 151 | 148 | 134 | 132 | 128 |
| | 3-1/2 | 0.207 | 16d | 163 | 160 | 145 | 142 | 138 |
| | 4 | 0.225 | 20d | 178 | 175 | 158 | 155 | 151 |
| | 4-1/2 | 0.244 | 30d | 183 | 180 | 162 | 159 | 155 |
| | 5 | 0.263 | 40d | 207 | 203 | 183 | 180 | 175 |
| | 5-1/2 | 0.283 | 50d | 222 | 218 | 196 | 193 | 187 |
| | 6 | 0.283 | 60d | 222 | 218 | 196 | 193 | 187 |
| | 7 | 0.312 | 5/16 | 267 | 262 | 236 | 232 | 225 |
| | 8-1/2 | 0.375 | 3/8 | 334 | 327 | 294 | 289 | 281 |
| 12 gage ts=0.105" | 3 | 0.192 | 10d | 148 | 145 | 131 | 129 | 125 |
| | 3-1/4 | 0.192 | 12d | 148 | 145 | 131 | 129 | 125 |
| | 3-1/2 | 0.207 | 16d | 161 | 158 | 142 | 139 | 136 |
| | 4 | 0.225 | 20d | 176 | 173 | 155 | 153 | 148 |
| | 4-1/2 | 0.244 | 30d | 181 | 177 | 160 | 157 | 152 |
| | 5 | 0.263 | 40d | 205 | 201 | 180 | 177 | 172 |
| | 5-1/2 | 0.283 | 50d | 219 | 215 | 193 | 190 | 185 |
| | 6 | 0.283 | 60d | 219 | 215 | 193 | 190 | 185 |
| | 7 | 0.312 | 5/16 | 265 | 260 | 234 | 229 | 223 |
| | 8-1/2 | 0.375 | 3/8 | 331 | 324 | 291 | 286 | 279 |
| 14 gage ts=0.075" | 3 | 0.192 | 10d | 144 | 141 | 127 | 124 | 121 |
| | 3-1/4 | 0.192 | 12d | 144 | 141 | 127 | 124 | 121 |
| | 3-1/2 | 0.207 | 16d | 157 | 154 | 138 | 136 | 132 |
| | 4 | 0.225 | 20d | 172 | 169 | 152 | 149 | 145 |
| | 4-1/2 | 0.244 | 30d | 177 | 174 | 156 | 153 | 149 |
| | 5 | 0.263 | 40d | 201 | 198 | 177 | 174 | 169 |
| | 5-1/2 | 0.283 | 50d | 216 | 212 | 190 | 187 | 182 |
| | 6 | 0.283 | 60d | 216 | 212 | 190 | 187 | 182 |
| | 7 | 0.312 | 5/16 | 262 | 257 | 230 | 226 | 220 |
| | 8-1/2 | 0.375 | 3/8 | 328 | 321 | 288 | 283 | 275 |
| 16 gage ts=0.06" | 3 | 0.192 | 10d | 142 | 140 | 125 | 123 | 120 |
| | 3-1/4 | 0.192 | 12d | 142 | 140 | 125 | 123 | 120 |
| | 3-1/2 | 0.207 | 16d | 156 | 153 | 137 | 134 | 131 |

1. Tabulated lateral design values (Z) for spiked connections shall be multiplied by all applicable adjustment factors (see Table 7.1).

2. Tabulated lateral design values (Z) are for common wire spikes inserted in side grain with spike axis perpendicular to wood fibers, and with the following spike bending yield strengths ($F_{yb}$):

   $F_{yb}$ = 80,000 psi for 0.192", 0.207" and 0.225" diameter common wire spikes

   $F_{yb}$ = 70,000 psi for 0.244" and 0.263" diameter common wire spikes

   $F_{yb}$ = 60,000 psi for 0.283" and $\frac{5}{16}''$ diameter common wire spikes

   $F_{yb}$ = 45,000 psi for $\frac{3}{8}''$ diameter common wire spikes

3. Tabulated lateral design values (Z) are based on a dowel bearing strength ($F_e$) of 45,000 psi for ASTM A446, Grade A steel.

Source: ANSI/NF₀PA NDS—1991, *National Design Specification for Wood Construction*, AFPA, 1250 Connecticut Ave., NW, Washington, DC 20036.

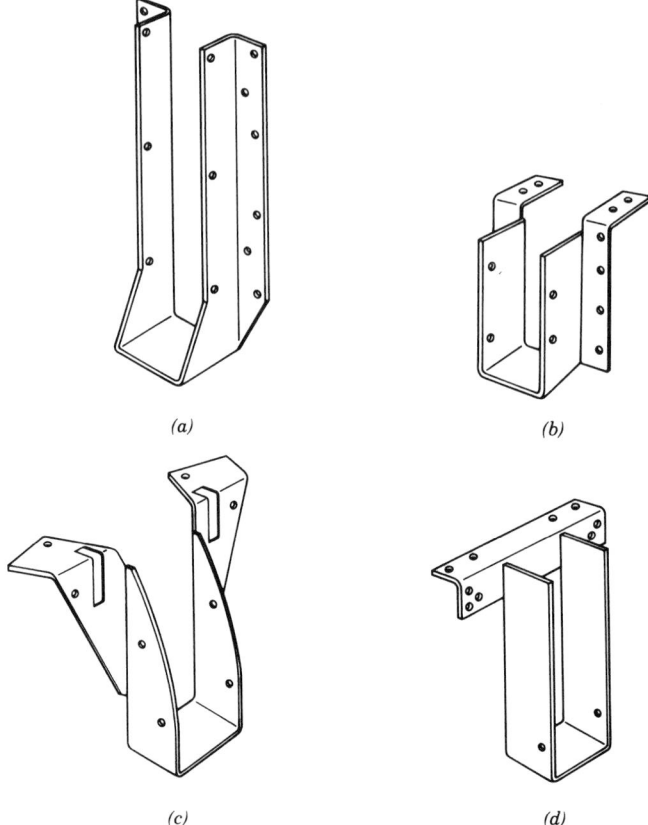

*(a)*    *(b)*

*(c)*    *(d)*

**FIGURE 7.29**    Typical joist and purlin hanger types. (*a*) Joist hanger; (*b*) joist hanger; (*c*) joist and purlin hanger; (*d*) joist and beam hanger.

## 7.11   REFERENCES

1.  American Forest & Paper Association, ANSI/NF₀PA NDS®—1991, *National Design Specification® for Wood Construction*, Washington, DC, 1991.

2.  American Institute of Timber Construction, *Typical Construction Details*, AITC 104, Englewood, CO, 1984.

3.  American Society of Mechanical Engineers, ANSI/ASME Standard B18.2.1—1981, *Square and Hex Bolts and Screws (Inch Series)*, New York, NY.

4.  American Institute of Timber Construction, *Standard Specifications for Structural Glued Laminated Timber of Softwood Species*, AITC 117—Design, Englewood, CO, 1993.

5.  United States Department of Agriculture, Forest Service, Forest Products Laboratory, *Timber-Connector Joints; Their Strength and Design*, Technical Bulletin No. 865, Madison, WI, 1944.

6.  United States Department of Agriculture, Forest Service, Forest Products Laboratory, *Wood Handbook: Wood as an Engineering Material*, Agriculture Handbook No. 72, Madison, WI.

7.  TECO/Lumberlok, *Design Manual for TECO Timber Connector Construction*, Colliers, WV, 1973.

8. International Conference of Building Officials, *Uniform Building Code*, Whittier, CA, 1982.

9. American Society of Mechanical Engineers, ANSI/ASME Standard B18.6.1—1981, *Wood Screws (Inch Series)*, New York, NY, 1981.

10. U.S. Government Services Administration (GSA) Federal Specification FF-N-105B, *Wire, Cut and Wrought Nails, Staples and Spikes*, 1977.

11. National Evaluation Service, NER 272, *Pneumatic of Mechanically Driven Staples, Nails, P-Nails, and Allied Fasteners for Use in All Types of Building Construction*, issued to International Staple, Nail, and Tool Association, Chicago, IL, 1989.

# Part III

# REFERENCE

# CHAPTER 8

# REFERENCE INFORMATION

## 8.1  INTRODUCTION

This chapter contains tables with information useful for design purposes including section properties, design values, span load tables, system diagrams, material weights and English-Metric conversion factors. Also included are selected AITC standards which contain design values for glued laminated timber as well as other design information on wood structures.

## 8.2   SECTION PROPERTIES

### TABLE 8.1

### Section Properties of Standard Dressed (S4S) Sawn Lumber

| Nominal Size b x d | Standard Dressed Size (S4S) b x d inches x inches | Area of Section A in² | X-X AXIS Section Modulus $S_{xx}$ in³ | X-X AXIS Moment of Inertia $I_{xx}$ in⁴ | Y-Y AXIS Section Modulus $S_{yy}$ in³ | Y-Y AXIS Moment of Inertia $I_{yy}$ in⁴ | Approximate weight in pounds per linear foot (lb/ft) of piece when density of wood equals: 25 lb/ft³ | 30 lb/ft³ | 35 lb/ft³ | 40 lb/ft³ | 45 lb/ft³ | 50 lb/ft³ |
|---|---|---|---|---|---|---|---|---|---|---|---|---|
| 1 x 3 | 3/4 x 2-1/2 | 1.875 | 0.781 | 0.977 | 0.234 | 0.088 | 0.326 | 0.391 | 0.456 | 0.521 | 0.586 | 0.651 |
| 1 x 4 | 3/4 x 3-1/2 | 2.625 | 1.531 | 2.680 | 0.328 | 0.123 | 0.456 | 0.547 | 0.638 | 0.729 | 0.820 | 0.911 |
| 1 x 6 | 3/4 x 5-1/2 | 4.125 | 3.781 | 10.40 | 0.516 | 0.193 | 0.716 | 0.859 | 1.003 | 1.146 | 1.289 | 1.432 |
| 1 x 8 | 3/4 x 7-1/4 | 5.438 | 6.570 | 23.82 | 0.680 | 0.255 | 0.944 | 1.133 | 1.322 | 1.510 | 1.699 | 1.888 |
| 1 x 10 | 3/4 x 9-1/4 | 6.938 | 10.70 | 49.47 | 0.867 | 0.325 | 1.204 | 1.445 | 1.686 | 1.927 | 2.168 | 2.409 |
| 1 x 12 | 3/4 x 11-1/4 | 8.438 | 15.82 | 88.99 | 1.055 | 0.396 | 1.465 | 1.758 | 2.051 | 2.344 | 2.637 | 2.930 |
| 2 x 3 | 1-1/2 x 2-1/2 | 3.750 | 1.563 | 1.953 | 0.938 | 0.703 | 0.651 | 0.781 | 0.911 | 1.042 | 1.172 | 1.302 |
| 2 x 4 | 1-1/2 x 3-1/2 | 5.250 | 3.063 | 5.359 | 1.313 | 0.984 | 0.911 | 1.094 | 1.276 | 1.458 | 1.641 | 1.823 |
| 2 x 5 | 1-1/2 x 4-1/2 | 6.750 | 5.063 | 11.39 | 1.688 | 1.266 | 1.172 | 1.406 | 1.641 | 1.875 | 2.109 | 2.344 |
| 2 x 6 | 1-1/2 x 5-1/2 | 8.250 | 7.563 | 20.80 | 2.063 | 1.547 | 1.432 | 1.719 | 2.005 | 2.292 | 2.578 | 2.865 |
| 2 x 8 | 1-1/2 x 7-1/4 | 10.88 | 13.14 | 47.63 | 2.719 | 2.039 | 1.888 | 2.266 | 2.643 | 3.021 | 3.398 | 3.776 |
| 2 x 10 | 1-1/2 x 9-1/4 | 13.88 | 21.39 | 98.93 | 3.469 | 2.602 | 2.409 | 2.891 | 3.372 | 3.854 | 4.336 | 4.818 |
| 2 x 12 | 1-1/2 x 11-1/4 | 16.88 | 31.64 | 178.0 | 4.219 | 3.164 | 2.930 | 3.516 | 4.102 | 4.688 | 5.273 | 5.859 |
| 2 x 14 | 1-1/2 x 13-1/4 | 19.88 | 43.89 | 290.8 | 4.969 | 3.727 | 3.451 | 4.141 | 4.831 | 5.521 | 6.211 | 6.901 |
| 3 x 4 | 2-1/2 x 3-1/2 | 8.750 | 5.104 | 8.932 | 3.646 | 4.557 | 1.519 | 1.823 | 2.127 | 2.431 | 2.734 | 3.038 |
| 3 x 5 | 2-1/2 x 4-1/2 | 11.25 | 8.438 | 18.98 | 4.688 | 5.859 | 1.953 | 2.344 | 2.734 | 3.125 | 3.516 | 3.906 |
| 3 x 6 | 2-1/2 x 5-1/2 | 13.75 | 12.60 | 34.66 | 5.729 | 7.161 | 2.387 | 2.865 | 3.342 | 3.819 | 4.297 | 4.774 |
| 3 x 8 | 2-1/2 x 7-1/4 | 18.13 | 21.90 | 79.39 | 7.552 | 9.440 | 3.147 | 3.776 | 4.405 | 5.035 | 5.664 | 6.293 |
| 3 x 10 | 2-1/2 x 9-1/4 | 23.13 | 35.65 | 164.9 | 9.635 | 12.04 | 4.015 | 4.818 | 5.621 | 6.424 | 7.227 | 8.030 |
| 3 x 12 | 2-1/2 x 11-1/4 | 28.13 | 52.73 | 296.6 | 11.72 | 14.65 | 4.883 | 5.859 | 6.836 | 7.813 | 8.789 | 9.766 |
| 3 x 14 | 2-1/2 x 13-1/4 | 33.13 | 73.15 | 484.6 | 13.80 | 17.25 | 5.751 | 6.901 | 8.051 | 9.201 | 10.35 | 11.50 |
| 3 x 16 | 2-1/2 x 15-1/4 | 38.13 | 96.90 | 738.9 | 15.89 | 19.86 | 6.619 | 7.943 | 9.266 | 10.59 | 11.91 | 13.24 |
| 4 x 4 | 3-1/2 x 3-1/2 | 12.25 | 7.146 | 12.51 | 7.146 | 12.51 | 2.127 | 2.552 | 2.977 | 3.403 | 3.828 | 4.253 |
| 4 x 5 | 3-1/2 x 4-1/2 | 15.75 | 11.81 | 26.58 | 9.188 | 16.08 | 2.734 | 3.281 | 3.828 | 4.375 | 4.922 | 5.469 |
| 4 x 6 | 3-1/2 x 5-1/2 | 19.25 | 17.65 | 48.53 | 11.23 | 19.65 | 3.342 | 4.010 | 4.679 | 5.347 | 6.016 | 6.684 |
| 4 x 8 | 3-1/2 x 7-1/4 | 25.38 | 30.66 | 111.1 | 14.80 | 25.90 | 4.405 | 5.286 | 6.168 | 7.049 | 7.930 | 8.811 |
| 4 x 10 | 3-1/2 x 9-1/4 | 32.38 | 49.91 | 230.8 | 18.89 | 33.05 | 5.621 | 6.745 | 7.869 | 8.993 | 10.12 | 11.24 |
| 4 x 12 | 3-1/2 x 11-1/4 | 39.38 | 73.83 | 415.3 | 22.97 | 40.20 | 6.836 | 8.203 | 9.570 | 10.94 | 12.30 | 13.67 |
| 4 x 14 | 3-1/2 x 13-1/4 | 46.38 | 102.4 | 678.5 | 27.05 | 47.34 | 8.051 | 9.661 | 11.27 | 12.88 | 14.49 | 16.10 |
| 4 x 16 | 3-1/2 x 15-1/4 | 53.38 | 135.7 | 1034 | 31.14 | 54.49 | 9.266 | 11.12 | 12.97 | 14.83 | 16.68 | 18.53 |
| 5 x 5 | 4-1/2 x 4-1/2 | 20.25 | 15.19 | 34.17 | 15.19 | 34.17 | 3.516 | 4.219 | 4.922 | 5.625 | 6.328 | 7.031 |
| 6 x 6 | 5-1/2 x 5-1/2 | 30.25 | 27.73 | 76.26 | 27.73 | 76.26 | 5.252 | 6.302 | 7.352 | 8.403 | 9.453 | 10.50 |
| 6 x 8 | 5-1/2 x 7-1/2 | 41.25 | 51.56 | 193.4 | 37.81 | 104.0 | 7.161 | 8.594 | 10.03 | 11.46 | 12.89 | 14.32 |
| 6 x 10 | 5-1/2 x 9-1/2 | 52.25 | 82.73 | 393.0 | 47.90 | 131.7 | 9.071 | 10.89 | 12.70 | 14.51 | 16.33 | 18.14 |
| 6 x 12 | 5-1/2 x 11-1/2 | 63.25 | 121.2 | 697.1 | 57.98 | 159.4 | 10.98 | 13.18 | 15.37 | 17.57 | 19.77 | 21.96 |
| 6 x 14 | 5-1/2 x 13-1/2 | 74.25 | 167.1 | 1128 | 68.06 | 187.2 | 12.89 | 15.47 | 18.05 | 20.63 | 23.20 | 25.78 |
| 6 x 16 | 5-1/2 x 15-1/2 | 85.25 | 220.2 | 1707 | 78.15 | 214.9 | 14.80 | 17.76 | 20.72 | 23.68 | 26.64 | 29.60 |
| 6 x 18 | 5-1/2 x 17-1/2 | 96.25 | 280.7 | 2456 | 88.23 | 242.6 | 16.71 | 20.05 | 23.39 | 26.74 | 30.08 | 33.42 |
| 6 x 20 | 5-1/2 x 19-1/2 | 107.3 | 348.6 | 3398 | 98.31 | 270.4 | 18.62 | 22.34 | 26.07 | 29.79 | 33.52 | 37.24 |
| 6 x 22 | 5-1/2 x 21-1/2 | 118.3 | 423.7 | 4555 | 108.4 | 298.1 | 20.53 | 24.64 | 28.74 | 32.85 | 36.95 | 41.06 |
| 6 x 24 | 5-1/2 x 23-1/2 | 129.3 | 506.2 | 5948 | 118.5 | 325.8 | 22.44 | 26.93 | 31.41 | 35.90 | 40.39 | 44.88 |
| 8 x 8 | 7-1/2 x 7-1/2 | 56.25 | 70.31 | 263.7 | 70.31 | 263.7 | 9.766 | 11.72 | 13.67 | 15.63 | 17.58 | 19.53 |
| 8 x 10 | 7-1/2 x 9-1/2 | 71.25 | 112.8 | 535.9 | 89.06 | 334.0 | 12.37 | 14.84 | 17.32 | 19.79 | 22.27 | 24.74 |
| 8 x 12 | 7-1/2 x 11-1/2 | 86.25 | 165.3 | 950.5 | 107.8 | 404.3 | 14.97 | 17.97 | 20.96 | 23.96 | 26.95 | 29.95 |
| 8 x 14 | 7-1/2 x 13-1/2 | 101.3 | 227.8 | 1538 | 126.6 | 474.4 | 17.58 | 21.09 | 24.61 | 28.13 | 31.64 | 35.16 |
| 8 x 16 | 7-1/2 x 15-1/2 | 116.3 | 300.3 | 2327 | 145.3 | 544.9 | 20.18 | 24.22 | 28.26 | 32.29 | 36.33 | 40.36 |
| 8 x 18 | 7-1/2 x 17-1/2 | 131.3 | 382.8 | 3350 | 164.1 | 615.2 | 22.79 | 27.34 | 31.90 | 36.46 | 41.02 | 45.57 |
| 8 x 20 | 7-1/2 x 19-1/2 | 146.3 | 475.3 | 4634 | 182.8 | 685.5 | 25.39 | 30.47 | 35.55 | 40.63 | 45.70 | 50.78 |
| 8 x 22 | 7-1/2 x 21-1/2 | 161.3 | 577.8 | 6211 | 201.6 | 755.9 | 27.99 | 33.59 | 39.19 | 44.79 | 50.39 | 55.99 |
| 8 x 24 | 7-1/2 x 23-1/2 | 176.3 | 690.3 | 8111 | 220.3 | 826.2 | 30.60 | 36.72 | 42.84 | 48.96 | 55.08 | 61.20 |
| 10 x 10 | 9-1/2 x 9-1/2 | 90.25 | 142.9 | 678.8 | 142.9 | 678.8 | 15.67 | 18.80 | 21.94 | 25.07 | 28.20 | 31.34 |
| 10 x 12 | 9-1/2 x 11-1/2 | 109.3 | 209.4 | 1204 | 173.0 | 821.7 | 18.97 | 22.76 | 26.55 | 30.35 | 34.14 | 37.93 |
| 10 x 14 | 9-1/2 x 13-1/2 | 128.3 | 288.6 | 1948 | 203.1 | 964.5 | 22.27 | 26.72 | 31.17 | 35.63 | 40.08 | 44.53 |
| 10 x 16 | 9-1/2 x 15-1/2 | 147.3 | 380.4 | 2948 | 233.1 | 1107 | 25.56 | 30.68 | 35.79 | 40.90 | 46.02 | 51.13 |
| 10 x 18 | 9-1/2 x 17-1/2 | 166.3 | 484.9 | 4243 | 263.2 | 1250 | 28.86 | 34.64 | 40.41 | 46.18 | 51.95 | 57.73 |
| 10 x 20 | 9-1/2 x 19-1/2 | 185.3 | 602.1 | 5870 | 293.3 | 1393 | 32.16 | 38.59 | 45.03 | 51.46 | 57.89 | 64.32 |
| 10 x 22 | 9-1/2 x 21-1/2 | 204.3 | 731.9 | 7868 | 323.4 | 1536 | 35.46 | 42.55 | 49.64 | 56.74 | 63.83 | 70.92 |
| 10 x 24 | 9-1/2 x 23-1/2 | 223.3 | 874.4 | 10270 | 353.5 | 1679 | 38.76 | 46.51 | 54.26 | 62.01 | 69.77 | 77.52 |

# TABLE 8.1 (Continued)

| Nominal Size b x d | Standard Dressed Size (S4S) b x d | Area of Section A | X-X AXIS | | Y-Y AXIS | | Approximate weight in pounds per linear foot (lb/ft) of piece when density of wood equals: | | | | | |
|---|---|---|---|---|---|---|---|---|---|---|---|---|
| inches x inches | inches x inches | $in^2$ | Section Modulus $S_{xx}$ $in^3$ | Moment of Inertia $I_{xx}$ $in^4$ | Section Modulus $S_{yy}$ $in^3$ | Moment of Inertia $I_{yy}$ $in^4$ | 25 lb/ft³ | 30 lb/ft³ | 35 lb/ft³ | 40 lb/ft³ | 45 lb/ft³ | 50 lb/ft³ |
| 12 x 12 | 11-1/2 x 11-1/2 | 132.3 | 253.5 | 1458 | 253.5 | 1458 | 22.96 | 27.55 | 32.14 | 36.74 | 41.33 | 45.92 |
| 12 x 14 | 11-1/2 x 13-1/2 | 155.3 | 349.3 | 2358 | 297.6 | 1711 | 26.95 | 32.34 | 37.73 | 43.13 | 48.52 | 53.91 |
| 12 x 16 | 11-1/2 x 15-1/2 | 178.3 | 460.5 | 3569 | 341.6 | 1964 | 30.95 | 37.14 | 43.32 | 49.51 | 55.70 | 61.89 |
| 12 x 18 | 11-1/2 x 17-1/2 | 201.3 | 587.0 | 5136 | 385.7 | 2218 | 34.94 | 41.93 | 48.91 | 55.90 | 62.89 | 69.88 |
| 12 x 20 | 11-1/2 x 19-1/2 | 224.3 | 728.8 | 7106 | 429.8 | 2471 | 38.93 | 46.72 | 54.51 | 62.29 | 70.08 | 77.86 |
| 12 x 22 | 11-1/2 x 21-1/2 | 247.3 | 886.0 | 9524 | 473.9 | 2725 | 42.93 | 51.51 | 60.10 | 68.68 | 77.27 | 85.85 |
| 12 x 24 | 11-1/2 x 23-1/2 | 270.3 | 1058 | 12440 | 518.0 | 2978 | 46.92 | 56.30 | 65.69 | 75.07 | 84.45 | 93.84 |
| 14 x 14 | 13-1/2 x 13-1/2 | 182.3 | 410.1 | 2768 | 410.1 | 2768 | 31.64 | 37.97 | 44.30 | 50.63 | 56.95 | 63.28 |
| 14 x 16 | 13-1/2 x 15-1/2 | 209.3 | 540.6 | 4189 | 470.8 | 3178 | 36.33 | 43.59 | 50.86 | 58.13 | 65.39 | 72.66 |
| 14 x 18 | 13-1/2 x 17-1/2 | 236.3 | 689.1 | 6029 | 531.6 | 3588 | 41.02 | 49.22 | 57.42 | 65.63 | 73.83 | 82.03 |
| 14 x 20 | 13-1/2 x 19-1/2 | 263.3 | 855.6 | 8342 | 592.3 | 3998 | 45.70 | 54.84 | 63.98 | 73.13 | 82.27 | 91.41 |
| 14 x 22 | 13-1/2 x 21-1/2 | 290.3 | 1040 | 11180 | 653.1 | 4408 | 50.39 | 60.47 | 70.55 | 80.63 | 90.70 | 100.8 |
| 14 x 24 | 13-1/2 x 23-1/2 | 317.3 | 1243 | 14600 | 713.8 | 4818 | 55.08 | 66.09 | 77.11 | 88.13 | 99.14 | 110.2 |
| 16 x 16 | 15-1/2 x 15-1/2 | 240.3 | 620.6 | 4810 | 620.6 | 4810 | 41.71 | 50.05 | 58.39 | 66.74 | 75.08 | 83.42 |
| 16 x 18 | 15-1/2 x 17-1/2 | 271.3 | 791.1 | 6923 | 700.7 | 5431 | 47.09 | 56.51 | 65.93 | 75.35 | 84.77 | 94.18 |
| 16 x 20 | 15-1/2 x 19-1/2 | 302.3 | 982.3 | 9578 | 780.8 | 6051 | 52.47 | 62.97 | 73.46 | 83.96 | 94.45 | 104.9 |
| 16 x 22 | 15-1/2 x 21-1/2 | 333.3 | 1194 | 12840 | 860.9 | 6672 | 57.86 | 69.43 | 81.00 | 92.57 | 104.1 | 115.7 |
| 16 x 24 | 15-1/2 x 23-1/2 | 364.3 | 1427 | 16760 | 941.0 | 7293 | 63.24 | 75.89 | 88.53 | 101.2 | 113.8 | 126.5 |
| 18 x 18 | 17-1/2 x 17-1/2 | 306.3 | 893.2 | 7816 | 893.2 | 7816 | 53.17 | 63.80 | 74.44 | 85.07 | 95.70 | 106.3 |
| 18 x 20 | 17-1/2 x 19-1/2 | 341.3 | 1109 | 10810 | 995.3 | 8709 | 59.24 | 71.09 | 82.94 | 94.79 | 106.6 | 118.5 |
| 18 x 22 | 17-1/2 x 21-1/2 | 376.3 | 1348 | 14490 | 1097 | 9602 | 65.32 | 78.39 | 91.45 | 104.5 | 117.6 | 130.6 |
| 18 x 24 | 17-1/2 x 23-1/2 | 411.3 | 1611 | 18930 | 1199 | 10500 | 71.40 | 85.68 | 99.96 | 114.2 | 128.5 | 142.8 |
| 20 x 20 | 19-1/2 x 19-1/2 | 380.3 | 1236 | 12050 | 1236 | 12050 | 66.02 | 79.22 | 92.42 | 105.6 | 118.8 | 132.0 |
| 20 x 22 | 19-1/2 x 21-1/2 | 419.3 | 1502 | 16150 | 1363 | 13280 | 72.79 | 87.34 | 101.9 | 116.5 | 131.0 | 145.6 |
| 20 x 24 | 19-1/2 x 23-1/2 | 458.3 | 1795 | 21090 | 1489 | 14520 | 79.56 | 95.47 | 111.4 | 127.3 | 143.2 | 159.1 |
| 22 x 22 | 21-1/2 x 21-1/2 | 462.3 | 1656 | 17810 | 1656 | 17810 | 80.25 | 96.30 | 112.4 | 128.4 | 144.5 | 160.5 |
| 22 x 24 | 21-1/2 x 23-1/2 | 505.3 | 1979 | 23250 | 1810 | 19460 | 87.72 | 105.3 | 122.8 | 140.3 | 157.9 | 175.4 |
| 24 x 24 | 23-1/2 x 23-1/2 | 552.3 | 2163 | 25420 | 2163 | 25420 | 95.88 | 115.1 | 134.2 | 153.4 | 172.6 | 191.8 |

Source: ANSI/NF₀PA NDS—1991, *National Design Specification (NDS) for Wood Construction*, AFPA (formerly NFPA), Washington, DC, 1991.

# TABLE 8.2
## Section Properties of Structural Glued Laminated Timber Western Species (Based on 1½ in. Thick Laminations)

| Beam Size | | Number of Lams | A Area in.² | x-x Axis | | y-y Axis | | V Volume per ft cf/ft | DF$^a$ 35 pcf Weight per ft plf | HF$^a$ 27 pcf Weight per ft plf |
| --- | --- | --- | --- | --- | --- | --- | --- | --- | --- | --- |
| b Width in. | d Depth in. | | | S$_x$ Section Modulus in.³ | I$_x$ Moment of Inertia in.⁴ | S$_y$ Section Modulus in.³ | I$_y$ Moment of Inertia in.⁴ | | | |
| 2½ × | 3 | 2 | 7.500 | 3.750 | 5.625 | 3.125 | 3.906 | 0.05 | 1.8 | 1.4 |
| 2½ × | 4½ | 3 | 11.25 | 8.438 | 18.98 | 4.688 | 5.859 | 0.08 | 2.7 | 2.1 |
| 2½ × | 6 | 4 | 15.00 | 15.00 | 45.00 | 6.250 | 7.813 | 0.10 | 3.6 | 2.8 |
| 2½ × | 7½ | 5 | 18.75 | 23.44 | 87.89 | 7.813 | 9.766 | 0.13 | 4.6 | 3.5 |
| 2½ × | 9 | 6 | 22.50 | 33.75 | 151.9 | 9.375 | 11.71 | 0.16 | 5.5 | 4.2 |
| 2½ × | 10½ | 7 | 26.25 | 45.94 | 241.2 | 10.93 | 13.67 | 0.18 | 6.4 | 4.9 |
| 2½ × | 12 | 8 | 30.00 | 60.00 | 360.0 | 12.50 | 15.62 | 0.21 | 7.3 | 5.6 |
| 2½ × | 13½ | 9 | 33.75 | 75.94 | 512.6 | 14.06 | 17.58 | 0.23 | 8.2 | 6.3 |
| 2½ × | 15 | 10 | 37.50 | 93.75 | 703.1 | 15.63 | 19.53 | 0.26 | 9.1 | 7.0 |
| 2½ × | 16½ | 11 | 41.25 | 113.4 | 935.9 | 17.19 | 21.48 | 0.29 | 10.0 | 7.7 |
| 2½ × | 18 | 12 | 45.00 | 135.0 | 1215 | 18.75 | 23.44 | 0.31 | 10.9 | 8.4 |
| 3⅛ × | 3 | 2 | 9.375 | 4.688 | 7.031 | 4.883 | 7.629 | 0.07 | 2.3 | 1.8 |
| 3⅛ × | 4½ | 3 | 14.06 | 10.55 | 23.73 | 7.324 | 11.44 | 0.10 | 3.4 | 2.6 |
| 3⅛ × | 6 | 4 | 18.75 | 18.75 | 56.25 | 9.766 | 15.26 | 0.13 | 4.6 | 3.5 |
| 3⅛ × | 7½ | 5 | 23.44 | 29.30 | 109.9 | 12.21 | 19.07 | 0.16 | 5.7 | 4.4 |

| | | | | | | | | | |
|---|---|---|---|---|---|---|---|---|---|
| $3\frac{1}{8}$ × 9 | 6 | 28.13 | 42.19 | 189.8 | 14.65 | 22.89 | 0.20 | 6.8 | 5.3 |
| $3\frac{1}{8}$ × $10\frac{1}{2}$ | 7 | 32.81 | 57.42 | 301.5 | 17.09 | 26.70 | 0.23 | 8.0 | 6.1 |
| $3\frac{1}{8}$ × 12 | 8 | 37.50 | 75.00 | 450.0 | 19.53 | 30.52 | 0.26 | 9.1 | 7.0 |
| $3\frac{1}{8}$ × $13\frac{1}{2}$ | 9 | 42.19 | 94.92 | 640.7 | 21.97 | 34.33 | 0.29 | 10.3 | 7.9 |
| $3\frac{1}{8}$ × 15 | 10 | 46.88 | 117.2 | 878.9 | 24.41 | 38.15 | 0.33 | 11.4 | 8.8 |
| $3\frac{1}{8}$ × $16\frac{1}{2}$ | 11 | 51.56 | 141.8 | 1170 | 26.86 | 41.96 | 0.36 | 12.5 | 9.7 |
| $3\frac{1}{8}$ × 18 | 12 | 56.25 | 168.8 | 1519 | 29.30 | 45.78 | 0.39 | 13.7 | 10.6 |
| $3\frac{1}{8}$ × $19\frac{1}{2}$ | 13 | 60.94 | 198.0 | 1931 | 31.74 | 49.59 | 0.42 | 14.8 | 11.4 |
| $3\frac{1}{8}$ × 21 | 14 | 65.63 | 229.7 | 2412 | 34.18 | 53.41 | 0.46 | 15.9 | 12.3 |
| $3\frac{1}{8}$ × $22\frac{1}{2}$ | 15 | 70.31 | 263.7 | 2966 | 36.62 | 57.22 | 0.49 | 17.1 | 13.2 |
| $3\frac{1}{8}$ × 24 | 16 | 75.00 | 300.0 | 3600 | 39.06 | 61.04 | 0.52 | 18.2 | 14.1 |
| $3\frac{1}{8}$ × $25\frac{1}{2}$ | 17 | 79.69 | 338.7 | 4318 | 41.50 | 64.85 | 0.55 | 19.4 | 14.9 |
| $3\frac{1}{8}$ × 27 | 18 | 84.38 | 379.7 | 5126 | 43.95 | 68.66 | 0.59 | 20.5 | 15.8 |
| $3\frac{1}{8}$ × $28\frac{1}{2}$ | 19 | 89.06 | 423.0 | 6028 | 46.39 | 72.48 | 0.62 | 21.7 | 16.7 |
| $3\frac{1}{8}$ × 30 | 20 | 93.75 | 468.8 | 7031 | 48.83 | 76.29 | 0.65 | 22.8 | 17.6 |
| $5\frac{1}{8}$ × 3 | 2 | 15.38 | 7.688 | 11.53 | 13.13 | 33.65 | 0.11 | 3.7 | 2.9 |
| $5\frac{1}{8}$ × $4\frac{1}{2}$ | 3 | 23.06 | 17.30 | 38.92 | 19.70 | 50.48 | 0.16 | 5.6 | 4.3 |
| $5\frac{1}{8}$ × 6 | 4 | 30.75 | 30.75 | 92.25 | 26.27 | 67.31 | 0.21 | 7.5 | 5.8 |
| $5\frac{1}{8}$ × $7\frac{1}{2}$ | 5 | 38.44 | 48.05 | 180.2 | 32.83 | 84.13 | 0.27 | 9.3 | 7.2 |
| $5\frac{1}{8}$ × 9 | 6 | 46.13 | 69.19 | 311.3 | 39.40 | 101.0 | 0.32 | 11.2 | 8.6 |
| $5\frac{1}{8}$ × $10\frac{1}{2}$ | 7 | 53.81 | 94.17 | 494.4 | 45.96 | 117.8 | 0.37 | 13.1 | 10.1 |
| $5\frac{1}{8}$ × 12 | 8 | 61.50 | 123.0 | 738.0 | 52.53 | 134.6 | 0.43 | 14.9 | 11.5 |
| $5\frac{1}{8}$ × $13\frac{1}{2}$ | 9 | 69.19 | 155.7 | 1051 | 59.10 | 151.4 | 0.48 | 16.8 | 13.0 |
| $5\frac{1}{8}$ × 15 | 10 | 76.88 | 192.2 | 1441 | 65.66 | 168.3 | 0.53 | 18.7 | 14.4 |

## TABLE 8.2 (Continued)
### Western Species (Based on $1\frac{1}{2}$ in. Thick Laminations)

| Beam Size b Width in. | d Depth in. | Number of Lams | A Area in.² | x–x Axis $S_x$ Section Modulus in.³ | $I_x$ Moment of Inertia in.⁴ | y–y Axis $S_y$ Section Modulus in.³ | $I_y$ Moment of Inertia in.⁴ | V Volume per ft cf/ft | DFᵃ 35 pcf Weight per ft plf | HFᵃ 27 pcf Weight per ft plf |
|---|---|---|---|---|---|---|---|---|---|---|
| $5\frac{1}{8}$ × | $16\frac{1}{2}$ | 11 | 84.56 | 232.5 | 1919 | 72.23 | 185.1 | 0.59 | 20.6 | 15.9 |
| $5\frac{1}{8}$ × | 18 | 12 | 92.25 | 276.8 | 2491 | 78.80 | 201.9 | 0.64 | 22.4 | 17.3 |
| $5\frac{1}{8}$ × | $19\frac{1}{2}$ | 13 | 99.94 | 324.8 | 3167 | 85.36 | 218.7 | 0.69 | 24.3 | 18.7 |
| $5\frac{1}{8}$ × | 21 | 14 | 107.6 | 376.7 | 3955 | 91.93 | 235.6 | 0.75 | 26.2 | 20.2 |
| $5\frac{1}{8}$ × | $22\frac{1}{2}$ | 15 | 115.3 | 432.4 | 4865 | 98.50 | 252.4 | 0.80 | 28.0 | 21.6 |
| $5\frac{1}{8}$ × | 24 | 16 | 123.0 | 492.0 | 5904 | 105.1 | 269.2 | 0.85 | 29.9 | 23.1 |
| $5\frac{1}{8}$ × | $25\frac{1}{2}$ | 17 | 130.7 | 555.4 | 7082 | 111.6 | 286.0 | 0.91 | 31.8 | 24.5 |
| $5\frac{1}{8}$ × | 27 | 18 | 138.4 | 622.7 | 8406 | 118.2 | 302.9 | 0.96 | 33.6 | 25.9 |
| $5\frac{1}{8}$ × | $28\frac{1}{2}$ | 19 | 146.1 | 693.8 | 9887 | 124.8 | 319.7 | 1.01 | 35.5 | 27.4 |
| $5\frac{1}{8}$ × | 30 | 20 | 153.8 | 768.8 | 11,530 | 131.3 | 336.5 | 1.07 | 37.4 | 28.8 |
| $5\frac{1}{8}$ × | $31\frac{1}{2}$ | 21 | 161.4 | 847.5 | 13,350 | 137.9 | 353.4 | 1.12 | 39.2 | 30.3 |
| $5\frac{1}{8}$ × | 33 | 22 | 169.1 | 930.2 | 15,350 | 144.5 | 370.2 | 1.17 | 41.1 | 31.7 |
| $5\frac{1}{8}$ × | $34\frac{1}{2}$ | 23 | 176.8 | 1017 | 17,540 | 151.0 | 387.0 | 1.23 | 43.0 | 33.2 |
| $5\frac{1}{8}$ × | 36 | 24 | 184.5 | 1107 | 19,930 | 157.6 | 403.8 | 1.28 | 44.8 | 34.6 |
| $5\frac{1}{8}$ × | $37\frac{1}{2}$ | 25 | 192.2 | 1201 | 22,520 | 164.2 | 420.7 | 1.33 | 46.7 | 36.0 |
| $5\frac{1}{8}$ × | 39 | 26 | 199.9 | 1299 | 25,330 | 170.7 | 437.5 | 1.39 | 48.6 | 37.5 |

| | | | | | | | | | |
|---|---|---|---|---|---|---|---|---|---|
| $5\frac{1}{8} \times 40\frac{1}{2}$ | 27 | 207.6 | 1401 | 28,370 | 177.3 | 454.3 | 1.44 | 50.4 | 38.9 |
| $5\frac{1}{8} \times 42$ | 28 | 215.3 | 1507 | 31,640 | 183.9 | 471.1 | 1.49 | 52.3 | 40.4 |
| $5\frac{1}{8} \times 43\frac{1}{2}$ | 29 | 222.9 | 1616 | 35,150 | 190.4 | 488.0 | 1.55 | 54.2 | 41.8 |
| $5\frac{1}{8} \times 45$ | 30 | 230.6 | 1730 | 38,920 | 197.0 | 504.8 | 1.60 | 56.1 | 43.2 |
| $6\frac{3}{4} \times 3$ | 2 | 20.25 | 10.13 | 15.19 | 22.78 | 76.89 | 0.14 | 4.9 | 3.8 |
| $6\frac{3}{4} \times 4\frac{1}{2}$ | 3 | 30.38 | 22.78 | 51.26 | 34.17 | 115.3 | 0.21 | 7.4 | 5.7 |
| $6\frac{3}{4} \times 6$ | 4 | 40.50 | 40.50 | 121.5 | 45.56 | 153.8 | 0.28 | 9.8 | 7.6 |
| $6\frac{3}{4} \times 7\frac{1}{2}$ | 5 | 50.63 | 63.28 | 237.3 | 56.95 | 192.2 | 0.35 | 12.3 | 9.5 |
| $6\frac{3}{4} \times 9$ | 6 | 60.75 | 91.13 | 410.1 | 68.34 | 230.7 | 0.42 | 14.8 | 11.4 |
| $6\frac{3}{4} \times 10\frac{1}{2}$ | 7 | 70.88 | 124.0 | 651.2 | 79.73 | 269.1 | 0.49 | 17.2 | 13.3 |
| $6\frac{3}{4} \times 12$ | 8 | 81.00 | 162.0 | 972.0 | 91.13 | 307.5 | 0.56 | 19.7 | 15.2 |
| $6\frac{3}{4} \times 13\frac{1}{2}$ | 9 | 91.13 | 205.0 | 1384 | 102.5 | 346.0 | 0.63 | 22.1 | 17.1 |
| $6\frac{3}{4} \times 15$ | 10 | 101.3 | 253.1 | 1898 | 113.9 | 384.4 | 0.70 | 24.6 | 19.0 |
| $6\frac{3}{4} \times 16\frac{1}{2}$ | 11 | 111.4 | 306.3 | 2527 | 125.3 | 422.9 | 0.77 | 27.1 | 20.9 |
| $6\frac{3}{4} \times 18$ | 12 | 121.5 | 364.5 | 3281 | 136.7 | 461.3 | 0.84 | 29.5 | 22.8 |
| $6\frac{3}{4} \times 19\frac{1}{2}$ | 13 | 131.6 | 427.8 | 4171 | 148.1 | 499.8 | 0.91 | 32.0 | 24.7 |
| $6\frac{3}{4} \times 21$ | 14 | 141.8 | 496.1 | 5209 | 159.5 | 538.2 | 0.98 | 34.5 | 26.6 |
| $6\frac{3}{4} \times 22\frac{1}{2}$ | 15 | 151.9 | 569.5 | 6407 | 170.9 | 576.7 | 1.05 | 36.9 | 28.5 |
| $6\frac{3}{4} \times 24$ | 16 | 162.0 | 648.0 | 7776 | 182.3 | 615.1 | 1.13 | 39.4 | 30.4 |
| $6\frac{3}{4} \times 25\frac{1}{2}$ | 17 | 172.1 | 731.5 | 9327 | 193.6 | 653.5 | 1.20 | 41.8 | 32.3 |
| $6\frac{3}{4} \times 27$ | 18 | 182.3 | 820.1 | 11,070 | 205.0 | 692.0 | 1.27 | 44.3 | 34.2 |
| $6\frac{3}{4} \times 28\frac{1}{2}$ | 19 | 192.4 | 913.8 | 13,020 | 216.4 | 730.4 | 1.34 | 46.8 | 36.1 |
| $6\frac{3}{4} \times 30$ | 20 | 202.5 | 1013 | 15,190 | 227.8 | 768.9 | 1.41 | 49.2 | 38.0 |
| $6\frac{3}{4} \times 31\frac{1}{2}$ | 21 | 212.6 | 1116 | 17,580 | 239.2 | 807.3 | 1.48 | 51.7 | 39.9 |

**TABLE 8.2** *(Continued)*

**Western Species (Based on $1\frac{1}{2}$ in. Thick Laminations)**

| Beam Size | | Number of Lams | A Area in.² | x-x Axis | | y-y Axis | | V Volume per ft cf/ft | DFᵃ 35 pcf Weight per ft plf | HFᵃ 27 pcf Weight per ft plf |
| --- | --- | --- | --- | --- | --- | --- | --- | --- | --- | --- |
| b Width in. | d Depth in. | | | $S_x$ Section Modulus in.³ | $I_x$ Moment of Inertia in.⁴ | $S_y$ Section Modulus in.³ | $I_y$ Moment of Inertia in.⁴ | | | |
| $6\frac{3}{4}$ × | 33 | 22 | 222.8 | 1225 | 20,210 | 250.6 | 845.8 | 1.55 | 54.1 | 41.8 |
| $6\frac{3}{4}$ × | $34\frac{1}{2}$ | 23 | 232.9 | 1339 | 23,100 | 262.0 | 884.2 | 1.62 | 56.6 | 43.7 |
| $6\frac{3}{4}$ × | 36 | 24 | 243.0 | 1458 | 26,240 | 273.4 | 922.6 | 1.69 | 59.1 | 45.6 |
| $6\frac{3}{4}$ × | $37\frac{1}{2}$ | 25 | 253.1 | 1582 | 29,660 | 284.8 | 961.1 | 1.76 | 61.5 | 47.5 |
| $6\frac{3}{4}$ × | 39 | 26 | 263.3 | 1711 | 33,370 | 296.2 | 1000 | 1.83 | 64.0 | 49.4 |
| $6\frac{3}{4}$ × | $40\frac{1}{2}$ | 27 | 273.4 | 1845 | 37,370 | 307.5 | 1038 | 1.90 | 66.4 | 51.3 |
| $6\frac{3}{4}$ × | 42 | 28 | 283.5 | 1985 | 41,670 | 318.9 | 1076 | 1.97 | 68.9 | 53.2 |
| $6\frac{3}{4}$ × | $43\frac{1}{2}$ | 29 | 293.6 | 2129 | 46,300 | 330.3 | 1115 | 2.04 | 71.4 | 55.1 |
| $6\frac{3}{4}$ × | 45 | 30 | 303.8 | 2278 | 51,260 | 341.7 | 1153 | 2.11 | 73.8 | 57.0 |
| $6\frac{3}{4}$ × | $46\frac{1}{2}$ | 31 | 313.9 | 2433 | 56,560 | 353.1 | 1192 | 2.18 | 76.3 | 58.9 |
| $6\frac{3}{4}$ × | 48 | 32 | 324.0 | 2592 | 62,210 | 364.5 | 1230 | 2.25 | 78.8 | 60.8 |
| $6\frac{3}{4}$ × | $49\frac{1}{2}$ | 33 | 334.1 | 2757 | 68,220 | 375.9 | 1269 | 2.32 | 81.2 | 62.6 |
| $6\frac{3}{4}$ × | 51 | 34 | 344.3 | 2926 | 74,620 | 387.3 | 1307 | 2.39 | 83.7 | 64.5 |
| $6\frac{3}{4}$ × | $52\frac{1}{2}$ | 35 | 354.4 | 3101 | 81,400 | 398.7 | 1346 | 2.46 | 86.1 | 66.4 |
| $6\frac{3}{4}$ × | 54 | 36 | 364.5 | 3281 | 88,570 | 410.1 | 1384 | 2.53 | 88.6 | 68.3 |
| $6\frac{3}{4}$ × | $55\frac{1}{2}$ | 37 | 374.6 | 3465 | 96,160 | 421.5 | 1422 | 2.60 | 91.1 | 70.2 |
| $6\frac{3}{4}$ × | 57 | 38 | 384.8 | 3655 | 104,200 | 432.8 | 1461 | 2.67 | 93.5 | 72.1 |

| Size | | | | | | | | | |
|---|---|---|---|---|---|---|---|---|---|
| $6\frac{3}{4} \times 58\frac{1}{2}$ | 39 | 394.9 | 3850 | 112,600 | 444.2 | 1499 | 2.74 | 96.0 | 74.0 |
| $6\frac{3}{4} \times 60$ | 40 | 405.0 | 4050 | 121,500 | 455.6 | 1538 | 2.81 | 98.4 | 75.9 |
| $8\frac{3}{4} \times 3$ | 2 | 26.25 | 13.13 | 19.69 | 38.28 | 167.5 | 0.18 | 6.4 | 4.9 |
| $8\frac{3}{4} \times 4\frac{1}{2}$ | 3 | 39.38 | 29.53 | 66.45 | 57.42 | 251.2 | 0.27 | 9.6 | 7.4 |
| $8\frac{3}{4} \times 6$ | 4 | 52.50 | 52.50 | 157.5 | 76.56 | 335.0 | 0.36 | 12.8 | 9.8 |
| $8\frac{3}{4} \times 7\frac{1}{2}$ | 5 | 65.63 | 82.03 | 307.6 | 95.70 | 418.7 | 0.46 | 16.0 | 12.3 |
| $8\frac{3}{4} \times 9$ | 6 | 78.75 | 118.1 | 531.6 | 114.8 | 502.4 | 0.55 | 19.1 | 14.8 |
| $8\frac{3}{4} \times 10\frac{1}{2}$ | 7 | 91.88 | 160.8 | 844.1 | 134.0 | 586.2 | 0.64 | 22.3 | 17.2 |
| $8\frac{3}{4} \times 12$ | 8 | 105.0 | 210.0 | 1260 | 153.1 | 669.9 | 0.73 | 25.5 | 19.7 |
| $8\frac{3}{4} \times 13\frac{1}{2}$ | 9 | 118.1 | 265.8 | 1794 | 172.3 | 753.7 | 0.82 | 28.7 | 22.1 |
| $8\frac{3}{4} \times 15$ | 10 | 131.3 | 328.1 | 2461 | 191.4 | 837.4 | 0.91 | 31.9 | 24.6 |
| $8\frac{3}{4} \times 16\frac{1}{2}$ | 11 | 144.4 | 397.0 | 3276 | 210.5 | 921.1 | 1.00 | 35.1 | 27.1 |
| $8\frac{3}{4} \times 18$ | 12 | 157.5 | 472.5 | 4253 | 229.7 | 1005 | 1.09 | 38.3 | 29.5 |
| $8\frac{3}{4} \times 19\frac{1}{2}$ | 13 | 170.6 | 554.5 | 5407 | 248.8 | 1089 | 1.18 | 41.5 | 32.0 |
| $8\frac{3}{4} \times 21$ | 14 | 183.8 | 643.1 | 6753 | 268.0 | 1172 | 1.28 | 44.7 | 34.5 |
| $8\frac{3}{4} \times 22\frac{1}{2}$ | 15 | 196.9 | 738.3 | 8306 | 287.1 | 1256 | 1.37 | 47.9 | 36.9 |
| $8\frac{3}{4} \times 24$ | 16 | 210.0 | 840.0 | 10,080 | 306.3 | 1340 | 1.46 | 51.0 | 39.4 |
| $8\frac{3}{4} \times 25\frac{1}{2}$ | 17 | 223.1 | 948.3 | 12,090 | 325.4 | 1424 | 1.55 | 54.2 | 41.8 |
| $8\frac{3}{4} \times 27$ | 18 | 236.3 | 1063 | 14,350 | 344.5 | 1507 | 1.64 | 57.4 | 44.3 |
| $8\frac{3}{4} \times 28\frac{1}{2}$ | 19 | 249.4 | 1185 | 16,880 | 363.7 | 1591 | 1.73 | 60.6 | 46.8 |
| $8\frac{3}{4} \times 30$ | 20 | 262.5 | 1313 | 19,690 | 382.8 | 1675 | 1.82 | 63.8 | 49.2 |
| $8\frac{3}{4} \times 31\frac{1}{2}$ | 21 | 275.6 | 1447 | 22,790 | 402.0 | 1759 | 1.91 | 67.0 | 51.7 |
| $8\frac{3}{4} \times 33$ | 22 | 288.8 | 1588 | 26,200 | 421.1 | 1842 | 2.01 | 70.2 | 54.1 |
| $8\frac{3}{4} \times 34\frac{1}{2}$ | 23 | 301.9 | 1736 | 29,940 | 440.2 | 1926 | 2.10 | 73.4 | 56.6 |

**TABLE 8.2** (*Continued*)

**Western Species (Based on $1\frac{1}{2}$ in. Thick Laminations)**

| Beam Size | | Number of Lams | A Area in.² | x–x Axis | | y–y Axis | | V Volume per ft cf/ft | DF^a 35 pcf Weight per ft plf | HF^a 27 pcf Weight per ft plf |
| Width in. | Depth in. | | | $S_x$ Section Modulus in.³ | $I_x$ Moment of Inertia in.⁴ | $S_y$ Section Modulus in.³ | $I_y$ Moment of Inertia in.⁴ | | | |
|---|---|---|---|---|---|---|---|---|---|---|
| $8\frac{3}{4}$ × | 36 | 24 | 315.0 | 1890 | 34,020 | 459.4 | 2010 | 2.19 | 76.6 | 59.1 |
| $8\frac{3}{4}$ × | $37\frac{1}{2}$ | 25 | 328.1 | 2051 | 38,450 | 478.5 | 2094 | 2.28 | 79.8 | 61.5 |
| $8\frac{3}{4}$ × | 39 | 26 | 341.3 | 2218 | 43,250 | 497.7 | 2177 | 2.37 | 82.9 | 64.0 |
| $8\frac{3}{4}$ × | $40\frac{1}{2}$ | 27 | 354.4 | 2392 | 48,440 | 516.8 | 2261 | 2.46 | 86.1 | 66.4 |
| $8\frac{3}{4}$ × | 42 | 28 | 367.5 | 2573 | 54,020 | 535.9 | 2345 | 2.55 | 89.3 | 68.9 |
| $8\frac{3}{4}$ × | $43\frac{1}{2}$ | 29 | 380.6 | 2760 | 60,020 | 555.1 | 2428 | 2.64 | 92.5 | 71.4 |
| $8\frac{3}{4}$ × | 45 | 30 | 393.8 | 2953 | 66,450 | 574.2 | 2512 | 2.73 | 95.7 | 73.8 |
| $8\frac{3}{4}$ × | $46\frac{1}{2}$ | 31 | 406.9 | 3153 | 73,310 | 593.4 | 2596 | 2.83 | 98.9 | 76.3 |
| $8\frac{3}{4}$ × | 48 | 32 | 420.0 | 3360 | 80,640 | 612.5 | 2680 | 2.92 | 102.1 | 78.8 |
| $8\frac{3}{4}$ × | $49\frac{1}{2}$ | 33 | 433.1 | 3573 | 88,440 | 631.6 | 2763 | 3.01 | 105.3 | 81.2 |
| $8\frac{3}{4}$ × | 51 | 34 | 446.3 | 3793 | 96,720 | 650.8 | 2847 | 3.10 | 108.5 | 83.7 |
| $8\frac{3}{4}$ × | $52\frac{1}{2}$ | 35 | 459.4 | 4020 | 105,500 | 669.9 | 2931 | 3.19 | 111.7 | 86.1 |
| $8\frac{3}{4}$ × | 54 | 36 | 472.5 | 4253 | 114,800 | 689.1 | 3015 | 3.28 | 114.8 | 88.6 |
| $8\frac{3}{4}$ × | $55\frac{1}{2}$ | 37 | 485.6 | 4492 | 124,700 | 708.2 | 3098 | 3.37 | 118.0 | 91.1 |
| $8\frac{3}{4}$ × | 57 | 38 | 498.8 | 4738 | 135,000 | 727.3 | 3182 | 3.46 | 121.2 | 93.5 |
| $8\frac{3}{4}$ × | $58\frac{1}{2}$ | 39 | 511.9 | 4991 | 146,000 | 746.5 | 3266 | 3.55 | 124.4 | 96.0 |
| $8\frac{3}{4}$ × | 60 | 40 | 525.0 | 5250 | 157,500 | 765.6 | 3350 | 3.65 | 127.6 | 98.4 |

| | | | | | | | | | |
|---|---|---|---|---|---|---|---|---|---|
| $8\frac{3}{4}$ × $61\frac{1}{2}$ | 41 | 538.1 | 5516 | 169,600 | 784.8 | 3433 | 3.74 | 130.8 | 100.9 |
| $8\frac{3}{4}$ × 63 | 42 | 551.3 | 5788 | 182,300 | 803.9 | 3517 | 3.83 | 134.0 | 103.4 |
| $8\frac{3}{4}$ × $64\frac{1}{2}$ | 43 | 564.4 | 6067 | 195,700 | 823.0 | 3601 | 3.92 | 137.2 | 105.8 |
| $8\frac{3}{4}$ × 66 | 44 | 577.5 | 6353 | 209,600 | 842.2 | 3685 | 4.01 | 140.4 | 108.3 |
| $8\frac{3}{4}$ × $67\frac{1}{2}$ | 45 | 590.6 | 6645 | 224,300 | 861.3 | 3768 | 4.10 | 143.6 | 110.7 |
| $8\frac{3}{4}$ × 69 | 46 | 603.8 | 6943 | 239,500 | 880.5 | 3852 | 4.19 | 146.7 | 113.2 |
| $8\frac{3}{4}$ × $70\frac{1}{2}$ | 47 | 616.9 | 7248 | 255,500 | 899.6 | 3936 | 4.28 | 149.9 | 115.7 |
| $8\frac{3}{4}$ × 72 | 48 | 630.0 | 7560 | 272,200 | 918.8 | 4020 | 4.38 | 153.1 | 118.1 |
| $8\frac{3}{4}$ × $73\frac{1}{2}$ | 49 | 643.1 | 7878 | 289,500 | 937.9 | 4103 | 4.47 | 156.3 | 120.6 |
| $8\frac{3}{4}$ × 75 | 50 | 656.3 | 8203 | 307,600 | 957.0 | 4187 | 4.56 | 159.5 | 123.0 |
| $10\frac{3}{4}$ × 3 | 2 | 32.25 | 16.13 | 24.19 | 57.78 | 310.6 | 0.22 | 7.8 | 6.0 |
| $10\frac{3}{4}$ × $4\frac{1}{2}$ | 3 | 48.38 | 36.28 | 81.63 | 86.67 | 465.9 | 0.34 | 11.8 | 9.1 |
| $10\frac{3}{4}$ × 6 | 4 | 64.50 | 64.50 | 193.5 | 115.6 | 621.1 | 0.45 | 15.7 | 12.1 |
| $10\frac{3}{4}$ × $7\frac{1}{2}$ | 5 | 80.63 | 100.8 | 377.9 | 144.5 | 776.4 | 0.56 | 19.6 | 15.1 |
| $10\frac{3}{4}$ × 9 | 6 | 96.75 | 145.1 | 653.1 | 173.3 | 931.7 | 0.67 | 23.5 | 18.1 |
| $10\frac{3}{4}$ × $10\frac{1}{2}$ | 7 | 112.9 | 197.5 | 1037 | 202.2 | 1087 | 0.78 | 27.4 | 21.2 |
| $10\frac{3}{4}$ × 12 | 8 | 129.0 | 258.0 | 1548 | 231.1 | 1242 | 0.90 | 31.4 | 24.2 |
| $10\frac{3}{4}$ × $13\frac{1}{2}$ | 9 | 145.1 | 326.5 | 2204 | 260.0 | 1398 | 1.01 | 35.3 | 27.2 |
| $10\frac{3}{4}$ × 15 | 10 | 161.3 | 403.1 | 3023 | 288.9 | 1553 | 1.12 | 39.2 | 30.2 |
| $10\frac{3}{4}$ × $16\frac{1}{2}$ | 11 | 177.4 | 487.8 | 4024 | 317.8 | 1708 | 1.23 | 43.1 | 33.3 |
| $10\frac{3}{4}$ × 18 | 12 | 193.5 | 580.5 | 5225 | 346.7 | 1863 | 1.34 | 47.0 | 36.3 |
| $10\frac{3}{4}$ × $19\frac{1}{2}$ | 13 | 209.6 | 681.3 | 6642 | 375.6 | 2019 | 1.46 | 51.0 | 39.3 |
| $10\frac{3}{4}$ × 21 | 14 | 225.8 | 790.1 | 8296 | 404.5 | 2174 | 1.57 | 54.9 | 42.3 |
| $10\frac{3}{4}$ × $22\frac{1}{2}$ | 15 | 241.9 | 907.0 | 10,200 | 433.4 | 2329 | 1.68 | 58.8 | 45.4 |

**TABLE 8.2** (*Continued*)
**Western Species (Based on $1\frac{1}{2}$ in. Thick Laminations)**

| Beam Size | | Number of Lams | A Area in.² | x-x Axis | | y-y Axis | | V Volume per ft cf/ft | DF[a] 35 pcf Weight per ft plf | HF[a] 27 pcf Weight per ft plf |
|---|---|---|---|---|---|---|---|---|---|---|
| b Width in. | d Depth in. | | | $S_x$ Section Modulus in.³ | $I_x$ Moment of Inertia in.⁴ | $S_y$ Section Modulus in.³ | $I_y$ Moment of Inertia in.⁴ | | | |
| $10\frac{3}{4}$ × | 24 | 16 | 258.0 | 1032 | 12,380 | 462.3 | 2485 | 1.79 | 62.7 | 48.4 |
| $10\frac{3}{4}$ × | $25\frac{1}{2}$ | 17 | 274.1 | 1165 | 14,850 | 491.1 | 2640 | 1.90 | 66.6 | 51.4 |
| $10\frac{3}{4}$ × | 27 | 18 | 290.3 | 1306 | 17,630 | 520.0 | 2795 | 2.02 | 70.5 | 54.4 |
| $10\frac{3}{4}$ × | $28\frac{1}{2}$ | 19 | 306.4 | 1455 | 20,740 | 548.9 | 2950 | 2.13 | 74.5 | 57.4 |
| $10\frac{3}{4}$ × | 30 | 20 | 322.5 | 1613 | 24,190 | 577.8 | 3106 | 2.24 | 78.4 | 60.5 |
| $10\frac{3}{4}$ × | $31\frac{1}{2}$ | 21 | 338.6 | 1778 | 28,000 | 606.7 | 3261 | 2.35 | 82.3 | 63.5 |
| $10\frac{3}{4}$ × | 33 | 22 | 354.8 | 1951 | 32,190 | 635.6 | 3416 | 2.46 | 86.2 | 66.5 |
| $10\frac{3}{4}$ × | $34\frac{1}{2}$ | 23 | 370.9 | 2133 | 36,790 | 664.5 | 3572 | 2.58 | 90.1 | 69.5 |
| $10\frac{3}{4}$ × | 36 | 24 | 387.0 | 2322 | 41,800 | 693.4 | 3727 | 2.69 | 94.1 | 72.6 |
| $10\frac{3}{4}$ × | $37\frac{1}{2}$ | 25 | 403.1 | 2520 | 47,240 | 722.3 | 3882 | 2.80 | 98.0 | 75.6 |
| $10\frac{3}{4}$ × | 39 | 26 | 419.3 | 2725 | 53,140 | 751.2 | 4037 | 2.91 | 101.9 | 78.6 |
| $10\frac{3}{4}$ × | $40\frac{1}{2}$ | 27 | 435.4 | 2939 | 59,510 | 780.0 | 4193 | 3.02 | 105.8 | 81.6 |
| $10\frac{3}{4}$ × | 42 | 28 | 451.5 | 3161 | 66,370 | 808.9 | 4348 | 3.14 | 109.7 | 84.7 |
| $10\frac{3}{4}$ × | $43\frac{1}{2}$ | 29 | 467.6 | 3390 | 73,740 | 837.8 | 4503 | 3.25 | 113.7 | 87.7 |
| $10\frac{3}{4}$ × | 45 | 30 | 483.8 | 3628 | 81,630 | 866.7 | 4659 | 3.36 | 117.6 | 90.7 |
| $10\frac{3}{4}$ × | $46\frac{1}{2}$ | 31 | 499.9 | 3874 | 90,070 | 895.6 | 4814 | 3.47 | 121.5 | 93.7 |
| $10\frac{3}{4}$ × | 48 | 32 | 516.0 | 4128 | 99,070 | 924.5 | 4969 | 3.58 | 125.4 | 96.8 |

| | | | | | | | | | |
|---|---|---|---|---|---|---|---|---|---|
| 99.8 | 129.3 | 3.70 | 5124 | 953.4 | 108,650 | 4390 | 532.1 | 33 | $10\frac{3}{4}$ × $49\frac{1}{2}$ |
| 102.8 | 133.3 | 3.81 | 5280 | 982.3 | 118,800 | 4660 | 548.3 | 34 | $10\frac{3}{4}$ × 51 |
| 105.8 | 137.2 | 3.92 | 5435 | 1011 | 129,600 | 4938 | 564.4 | 35 | $10\frac{3}{4}$ × $52\frac{1}{2}$ |
| 108.8 | 141.1 | 4.03 | 5590 | 1040 | 141,100 | 5225 | 580.5 | 36 | $10\frac{3}{4}$ × 54 |
| 111.9 | 145.0 | 4.14 | 5746 | 1069 | 153,100 | 5519 | 596.6 | 37 | $10\frac{3}{4}$ × $55\frac{1}{2}$ |
| 114.9 | 148.9 | 4.26 | 5901 | 1098 | 165,900 | 5821 | 612.8 | 38 | $10\frac{3}{4}$ × 57 |
| 117.9 | 152.9 | 4.37 | 6056 | 1127 | 179,300 | 6132 | 628.9 | 39 | $10\frac{3}{4}$ × $58\frac{1}{2}$ |
| 120.9 | 156.8 | 4.48 | 6211 | 1156 | 193,500 | 6450 | 645.0 | 40 | $10\frac{3}{4}$ × 60 |
| 124.0 | 160.7 | 4.59 | 6367 | 1185 | 208,400 | 6777 | 661.1 | 41 | $10\frac{3}{4}$ × $61\frac{1}{2}$ |
| 127.0 | 164.6 | 4.70 | 6522 | 1213 | 224,000 | 7111 | 677.3 | 42 | $10\frac{3}{4}$ × 63 |
| 130.0 | 168.5 | 4.82 | 6677 | 1242 | 240,400 | 7454 | 693.4 | 43 | $10\frac{3}{4}$ × $64\frac{1}{2}$ |
| 133.0 | 172.4 | 4.93 | 6833 | 1271 | 257,500 | 7805 | 709.5 | 44 | $10\frac{3}{4}$ × 66 |
| 136.1 | 176.4 | 5.04 | 6988 | 1300 | 275,500 | 8163 | 725.6 | 45 | $10\frac{3}{4}$ × $67\frac{1}{2}$ |
| 139.1 | 180.3 | 5.15 | 7143 | 1329 | 294,300 | 8530 | 741.8 | 46 | $10\frac{3}{4}$ × 69 |
| 142.1 | 184.2 | 5.26 | 7298 | 1358 | 313,900 | 8905 | 757.9 | 47 | $10\frac{3}{4}$ × $70\frac{1}{2}$ |
| 145.1 | 188.1 | 5.38 | 7454 | 1387 | 334,400 | 9288 | 774.0 | 48 | $10\frac{3}{4}$ × 72 |
| 148.1 | 192.0 | 5.49 | 7609 | 1416 | 355,700 | 9679 | 790.1 | 49 | $10\frac{3}{4}$ × $73\frac{1}{2}$ |
| 151.2 | 196.0 | 5.60 | 7764 | 1445 | 377,900 | 10,080 | 806.3 | 50 | $10\frac{3}{4}$ × 75 |
| 154.2 | 199.9 | 5.71 | 7920 | 1473 | 401,100 | 10,490 | 822.4 | 51 | $10\frac{3}{4}$ × $76\frac{1}{2}$ |
| 157.2 | 203.8 | 5.82 | 8075 | 1502 | 425,100 | 10,900 | 838.5 | 52 | $10\frac{3}{4}$ × 78 |
| 160.2 | 207.7 | 5.93 | 8230 | 1531 | 450,100 | 11,320 | 854.6 | 53 | $10\frac{3}{4}$ × $79\frac{1}{2}$ |
| 163.3 | 211.6 | 6.05 | 8386 | 1560 | 476,100 | 11,760 | 870.8 | 54 | $10\frac{3}{4}$ × 81 |
| 166.3 | 215.6 | 6.16 | 8541 | 1589 | 503,000 | 12,190 | 886.9 | 55 | $10\frac{3}{4}$ × $82\frac{1}{2}$ |
| 169.3 | 219.5 | 6.27 | 8696 | 1618 | 531,000 | 12,640 | 903.0 | 56 | $10\frac{3}{4}$ × 84 |

**TABLE 8.2** (*Continued*)

**Western Species (Based on 1½ in. Thick Laminations)**

| Beam Size | | Number of Lams | A Area in.² | x-x Axis | | y-y Axis | | V Volume per ft cf/ft | DF[a] 35 pcf Weight per ft plf | HF[a] 27 pcf Weight per ft plf |
|---|---|---|---|---|---|---|---|---|---|---|
| b Width in. | d Depth in. | | | $S_x$ Section Modulus in.³ | $I_x$ Moment of Inertia in.⁴ | $S_y$ Section Modulus in.³ | $I_y$ Moment of Inertia in.⁴ | | | |
| 10¾ × | 85½ | 57 | 919.1 | 13,100 | 559,900 | 1647 | 8851 | 6.38 | 223.4 | 172.3 |
| 10¾ × | 87 | 58 | 935.3 | 13,560 | 589,900 | 1676 | 9007 | 6.49 | 227.3 | 175.4 |
| 10¾ × | 88½ | 59 | 951.4 | 14,030 | 621,000 | 1705 | 9162 | 6.61 | 231.2 | 178.4 |
| 10¾ × | 90 | 60 | 967.5 | 14,510 | 653,100 | 1733 | 9317 | 6.72 | 235.2 | 181.4 |
| 12¼ × | 3 | 2 | 36.75 | 18.38 | 27.56 | 75.03 | 459.6 | 0.26 | 8.9 | 6.9 |
| 12¼ × | 4½ | 3 | 55.13 | 41.34 | 93.02 | 112.5 | 689.3 | 0.38 | 13.4 | 10.3 |
| 12¼ × | 6 | 4 | 73.50 | 73.50 | 220.5 | 150.1 | 919.1 | 0.51 | 17.9 | 13.8 |
| 12¼ × | 7½ | 5 | 91.88 | 114.8 | 430.7 | 187.6 | 1149 | 0.64 | 22.3 | 17.2 |
| 12¼ × | 9 | 6 | 110.3 | 165.4 | 744.2 | 225.1 | 1379 | 0.77 | 26.8 | 20.7 |
| 12¼ × | 10½ | 7 | 128.6 | 225.1 | 1182 | 262.6 | 1608 | 0.89 | 31.3 | 24.1 |
| 12¼ × | 12 | 8 | 147.0 | 294.0 | 1764 | 300.1 | 1838 | 1.02 | 35.7 | 27.6 |
| 12¼ × | 13½ | 9 | 165.4 | 372.1 | 2512 | 337.6 | 2068 | 1.15 | 40.2 | 31.0 |
| 12¼ × | 15 | 10 | 183.8 | 459.4 | 3445 | 375.2 | 2298 | 1.28 | 44.7 | 34.5 |
| 12¼ × | 16½ | 11 | 202.1 | 555.8 | 4586 | 412.7 | 2528 | 1.40 | 49.1 | 37.9 |
| 12¼ × | 18 | 12 | 220.5 | 661.5 | 5954 | 450.2 | 2757 | 1.53 | 53.6 | 41.3 |
| 12¼ × | 19½ | 13 | 238.9 | 776.3 | 7569 | 487.7 | 2987 | 1.66 | 58.1 | 44.8 |

| | | | | | | | | | |
|---|---|---|---|---|---|---|---|---|---|
| 48.2 | 62.5 | 1.79 | 3217 | 525.2 | 9454 | 900.4 | 257.3 | 14 | $12\frac{1}{4} \times 21$ |
| 51.7 | 67.0 | 1.91 | 3447 | 562.7 | 11,630 | 1034 | 275.6 | 15 | $12\frac{1}{4} \times 22\frac{1}{2}$ |
| 55.1 | 71.5 | 2.04 | 3677 | 600.3 | 14,110 | 1176 | 294.0 | 16 | $12\frac{1}{4} \times 24$ |
| 58.6 | 75.9 | 2.17 | 3906 | 637.8 | 16,930 | 1328 | 312.4 | 17 | $12\frac{1}{4} \times 25\frac{1}{2}$ |
| 62.0 | 80.4 | 2.30 | 4136 | 675.3 | 20,090 | 1488 | 330.8 | 18 | $12\frac{1}{4} \times 27$ |
| 65.5 | 84.9 | 2.42 | 4366 | 712.8 | 23,630 | 1658 | 349.1 | 19 | $12\frac{1}{4} \times 28\frac{1}{2}$ |
| 68.9 | 89.3 | 2.55 | 4596 | 750.3 | 27,560 | 1838 | 367.5 | 20 | $12\frac{1}{4} \times 30$ |
| 72.4 | 98.8 | 2.68 | 4825 | 787.8 | 31,910 | 2026 | 385.9 | 21 | $12\frac{1}{4} \times 31\frac{1}{2}$ |
| 75.8 | 98.3 | 2.81 | 5055 | 825.3 | 36,690 | 2223 | 404.3 | 22 | $12\frac{1}{4} \times 33$ |
| 79.2 | 102.7 | 2.93 | 5285 | 862.9 | 41,920 | 2430 | 422.6 | 23 | $12\frac{1}{4} \times 34\frac{1}{2}$ |
| 82.7 | 107.2 | 3.06 | 5515 | 900.4 | 47,630 | 2646 | 441.0 | 24 | $12\frac{1}{4} \times 36$ |
| 86.1 | 111.7 | 3.19 | 5745 | 937.9 | 53,830 | 2871 | 459.4 | 25 | $12\frac{1}{4} \times 37\frac{1}{2}$ |
| 89.6 | 116.1 | 3.32 | 5974 | 975.4 | 60,550 | 3105 | 477.8 | 26 | $12\frac{1}{4} \times 39$ |
| 93.0 | 120.6 | 3.45 | 6204 | 1013 | 67,810 | 3349 | 496.1 | 27 | $12\frac{1}{4} \times 40\frac{1}{2}$ |
| 96.5 | 125.1 | 3.57 | 6434 | 1050 | 75,630 | 3602 | 514.5 | 28 | $12\frac{1}{4} \times 42$ |
| 99.9 | 129.5 | 3.70 | 6664 | 1088 | 84,030 | 3863 | 532.9 | 29 | $12\frac{1}{4} \times 43\frac{1}{2}$ |
| 103.4 | 134.0 | 3.83 | 6893 | 1125 | 93,020 | 4134 | 551.3 | 30 | $12\frac{1}{4} \times 45$ |
| 106.8 | 138.5 | 3.96 | 7123 | 1163 | 102,600 | 4415 | 569.6 | 31 | $12\frac{1}{4} \times 46\frac{1}{2}$ |
| 110.2 | 142.9 | 4.08 | 7353 | 1201 | 112,900 | 4704 | 588.0 | 32 | $12\frac{1}{4} \times 48$ |
| 113.7 | 147.4 | 4.21 | 7583 | 1238 | 123,800 | 5003 | 606.4 | 33 | $12\frac{1}{4} \times 49\frac{1}{2}$ |
| 117.1 | 151.8 | 4.34 | 7813 | 1276 | 135,400 | 5310 | 624.8 | 34 | $12\frac{1}{4} \times 51$ |
| 120.6 | 156.3 | 4.47 | 8042 | 1313 | 147,700 | 5627 | 643.1 | 35 | $12\frac{1}{4} \times 52\frac{1}{2}$ |
| 124.0 | 160.8 | 4.59 | 8272 | 1351 | 160,700 | 5954 | 661.5 | 36 | $12\frac{1}{4} \times 54$ |
| 127.5 | 165.2 | 4.72 | 8502 | 1388 | 174,500 | 6289 | 679.9 | 37 | $12\frac{1}{4} \times 55\frac{1}{2}$ |

**TABLE 8.2** *(Continued)*

Western Species (Based on $1\frac{1}{2}$ in. Thick Laminations)

| Beam Size | | Number of Lams | A Area in.² | x-x Axis | | y-y Axis | | V Volume per ft cf/ft | DF[a] 35 pcf Weight per ft plf | HF[a] 27 pcf Weight per ft plf |
| b Width in. | d Depth in. | | | $S_x$ Section Modulus in.³ | $I_x$ Moment of Inertia in.⁴ | $S_y$ Section Modulus in.³ | $I_y$ Moment of Inertia in.⁴ | | | |
|---|---|---|---|---|---|---|---|---|---|---|
| $12\frac{1}{4}$ × | 57 | 38 | 698.3 | 6633 | 189,100 | 1426 | 8732 | 4.85 | 169.7 | 130.9 |
| $12\frac{1}{4}$ × | $58\frac{1}{2}$ | 39 | 716.6 | 6987 | 204,400 | 1463 | 8962 | 4.98 | 174.2 | 134.4 |
| $12\frac{1}{4}$ × | 60 | 40 | 735.0 | 7350 | 220,500 | 1501 | 9191 | 5.10 | 178.6 | 137.8 |
| $12\frac{1}{4}$ × | $61\frac{1}{2}$ | 41 | 753.4 | 7722 | 237,500 | 1538 | 9421 | 5.23 | 183.1 | 141.3 |
| $12\frac{1}{4}$ × | 63 | 42 | 771.8 | 8103 | 255,300 | 1576 | 9651 | 5.36 | 187.6 | 144.7 |
| $12\frac{1}{4}$ × | $64\frac{1}{2}$ | 43 | 790.1 | 8494 | 273,900 | 1613 | 9881 | 5.49 | 192.0 | 148.1 |
| $12\frac{1}{4}$ × | 66 | 44 | 808.5 | 8894 | 293,500 | 1651 | 10,110 | 5.61 | 196.5 | 151.6 |
| $12\frac{1}{4}$ × | $67\frac{1}{2}$ | 45 | 826.9 | 9302 | 314,000 | 1688 | 10,340 | 5.74 | 201.0 | 155.0 |
| $12\frac{1}{4}$ × | 69 | 46 | 845.3 | 9720 | 335,400 | 1726 | 10,570 | 5.87 | 205.4 | 158.5 |
| $12\frac{1}{4}$ × | $70\frac{1}{2}$ | 47 | 863.6 | 10,150 | 357,700 | 1763 | 10,800 | 6.00 | 209.9 | 161.9 |
| $12\frac{1}{4}$ × | 72 | 48 | 882.0 | 10,580 | 381,000 | 1801 | 11,030 | 6.13 | 214.4 | 165.4 |
| $12\frac{1}{4}$ × | $73\frac{1}{2}$ | 49 | 900.4 | 11,030 | 405,300 | 1838 | 11,260 | 6.25 | 218.8 | 168.8 |
| $12\frac{1}{4}$ × | 75 | 50 | 918.8 | 11,480 | 430,700 | 1876 | 11,490 | 6.38 | 223.3 | 172.3 |
| $12\frac{1}{4}$ × | $76\frac{1}{2}$ | 51 | 937.1 | 11,950 | 457,000 | 1913 | 11,720 | 6.51 | 227.8 | 175.7 |
| $12\frac{1}{4}$ × | 78 | 52 | 955.5 | 12,420 | 484,400 | 1951 | 11,950 | 6.64 | 232.2 | 179.2 |
| $12\frac{1}{4}$ × | $79\frac{1}{2}$ | 53 | 973.9 | 12,900 | 512,900 | 1988 | 12,180 | 6.76 | 236.7 | 182.6 |
| $12\frac{1}{4}$ × | 81 | 54 | 992.3 | 13,400 | 542,500 | 2026 | 12,410 | 6.89 | 241.2 | 186.0 |

| | | | | | | | | |
|---|---|---|---|---|---|---|---|---|
| 12¼ × 82½ | 1011 | 13,900 | 573,200 | 2063 | 12,640 | 7.02 | 245.6 | 189.5 |
| 12¼ × 84 | 1029 | 14,410 | 605,100 | 2101 | 12,870 | 7.15 | 250.1 | 192.9 |
| 12¼ × 85½ | 1047 | 14,930 | 638,000 | 2138 | 13,100 | 7.27 | 254.6 | 196.4 |
| 12¼ × 87 | 1066 | 15,450 | 672,200 | 2176 | 13,330 | 7.40 | 259.0 | 199.8 |
| 12¼ × 88½ | 1084 | 15,990 | 707,600 | 2213 | 13,560 | 7.53 | 263.5 | 203.3 |
| 12¼ × 90 | 1103 | 16,540 | 744,200 | 2251 | 13,790 | 7.66 | 268.0 | 206.7 |
| 12¼ × 91½ | 1121 | 17,090 | 782,000 | 2288 | 14,020 | 7.78 | 272.4 | 210.2 |
| 12¼ × 93 | 1139 | 17,660 | 821,100 | 2326 | 14,250 | 7.91 | 276.9 | 213.6 |
| 12¼ × 94½ | 1158 | 18,230 | 861,500 | 2363 | 14,480 | 8.04 | 281.4 | 217.1 |
| 12¼ × 96 | 1176 | 18,820 | 903,200 | 2401 | 14,710 | 8.17 | 285.8 | 220.5 |
| 12¼ × 97½ | 1194 | 19,410 | 946,200 | 2439 | 14,940 | 8.29 | 290.3 | 223.9 |
| 12¼ × 99 | 1213 | 20,010 | 990,500 | 2476 | 15,170 | 8.42 | 294.8 | 227.4 |
| 12¼ × 100½ | 1231 | 20,620 | 1,036,000 | 2514 | 15,400 | 8.55 | 299.2 | 230.8 |
| 12¼ × 102 | 1250 | 21,240 | 1,083,000 | 2551 | 15,630 | 8.68 | 303.7 | 234.3 |
| 14¼ × 3 | 42.75 | 21.38 | 32.06 | 101.5 | 723.4 | 0.30 | 10.4 | 8.0 |
| 14¼ × 4½ | 64.13 | 48.09 | 108.2 | 152.3 | 1085 | 0.45 | 15.6 | 12.0 |
| 14¼ × 6 | 85.50 | 85.50 | 256.5 | 203.1 | 1447 | 0.59 | 20.8 | 16.0 |
| 14¼ × 7½ | 106.9 | 133.6 | 501.0 | 253.8 | 1809 | 0.74 | 26.0 | 20.0 |
| 14¼ × 9 | 128.3 | 192.4 | 865.7 | 304.6 | 2170 | 0.89 | 31.2 | 24.0 |
| 14¼ × 10½ | 149.6 | 261.8 | 1375 | 355.4 | 2532 | 1.04 | 36.4 | 28.1 |
| 14¼ × 12 | 171.0 | 342.0 | 2052 | 406.1 | 2894 | 1.19 | 41.6 | 32.1 |
| 14¼ × 13½ | 192.4 | 432.8 | 2922 | 456.9 | 3255 | 1.34 | 46.8 | 36.1 |
| 14¼ × 15 | 213.8 | 534.4 | 4008 | 507.7 | 3617 | 1.48 | 52.0 | 40.1 |
| 14¼ × 16½ | 235.1 | 646.6 | 5334 | 558.4 | 3979 | 1.63 | 57.1 | 44.1 |

**TABLE 8.2** (*Continued*)

Western Species (Based on $1\frac{1}{2}$ in. Thick Laminations)

| Beam Size | | Number of Lams | A Area in.² | x-x Axis | | y-y Axis | | V Volume per ft cf/ft | DFᵃ 35 pcf Weight per ft plf | HFᵃ 27 pcf Weight per ft plf |
| Width in. | Depth in. | | | Sₓ Section Modulus in.³ | Iₓ Moment of Inertia in.⁴ | Sᵧ Section Modulus in.³ | Iᵧ Moment of Inertia in.⁴ | | | |
|---|---|---|---|---|---|---|---|---|---|---|
| $14\frac{1}{4}$ × | 18 | 12 | 256.5 | 769.5 | 6926 | 609.2 | 4340 | 1.78 | 62.3 | 48.1 |
| $14\frac{1}{4}$ × | $19\frac{1}{2}$ | 13 | 277.9 | 903.1 | 8805 | 660.0 | 4702 | 1.93 | 67.5 | 52.1 |
| $14\frac{1}{4}$ × | 21 | 14 | 299.3 | 1047 | 11,000 | 710.7 | 5064 | 2.08 | 72.7 | 56.1 |
| $14\frac{1}{4}$ × | $22\frac{1}{2}$ | 15 | 320.6 | 1202 | 13,530 | 761.5 | 5426 | 2.23 | 77.9 | 60.1 |
| $14\frac{1}{4}$ × | 24 | 16 | 342.0 | 1368 | 16,420 | 812.3 | 5787 | 2.38 | 83.1 | 64.1 |
| $14\frac{1}{4}$ × | $25\frac{1}{2}$ | 17 | 363.4 | 1544 | 19,690 | 863.0 | 6149 | 2.52 | 88.3 | 68.1 |
| $14\frac{1}{4}$ × | 27 | 18 | 384.8 | 1731 | 23,370 | 913.8 | 6511 | 2.67 | 93.5 | 72.1 |
| $14\frac{1}{4}$ × | $28\frac{1}{2}$ | 19 | 406.1 | 1929 | 27,490 | 964.5 | 6872 | 2.82 | 98.7 | 76.1 |
| $14\frac{1}{4}$ × | 30 | 20 | 427.5 | 2138 | 32,060 | 1015 | 7234 | 2.97 | 103.9 | 80.2 |
| $14\frac{1}{4}$ × | $31\frac{1}{2}$ | 21 | 448.9 | 2357 | 37,120 | 1066 | 7596 | 3.12 | 109.1 | 84.2 |
| $14\frac{1}{4}$ × | 33 | 22 | 470.3 | 2586 | 42,680 | 1117 | 7958 | 3.27 | 114.3 | 88.2 |
| $14\frac{1}{4}$ × | $34\frac{1}{2}$ | 23 | 491.6 | 2827 | 48,760 | 1168 | 8319 | 3.41 | 119.5 | 92.2 |
| $14\frac{1}{4}$ × | 36 | 24 | 513.0 | 3078 | 55,400 | 1218 | 8681 | 3.56 | 124.7 | 96.2 |
| $14\frac{1}{4}$ × | $37\frac{1}{2}$ | 25 | 534.4 | 3340 | 62,620 | 1269 | 9043 | 3.71 | 129.9 | 100.2 |
| $14\frac{1}{4}$ × | 39 | 26 | 555.8 | 3612 | 70,440 | 1320 | 9404 | 3.86 | 135.1 | 104.2 |
| $14\frac{1}{4}$ × | $40\frac{1}{2}$ | 27 | 577.1 | 3896 | 78,890 | 1371 | 9766 | 4.01 | 140.3 | 108.2 |
| $14\frac{1}{4}$ × | 42 | 28 | 598.5 | 4190 | 87,980 | 1421 | 10,130 | 4.16 | 145.5 | 112.2 |

| Size | | | | | | | | | |
|---|---|---|---|---|---|---|---|---|---|
| 14¼ × 43½ | 29 | 619.9 | 4494 | 97,750 | 1472 | 10,490 | 4.30 | 150.7 | 116.2 |
| 14¼ × 45 | 30 | 641.3 | 4809 | 108,200 | 1523 | 10,850 | 4.45 | 155.9 | 120.2 |
| 14¼ × 46½ | 31 | 662.6 | 5135 | 119,400 | 1574 | 11,210 | 4.60 | 161.1 | 124.2 |
| 14¼ × 48 | 32 | 684.0 | 5472 | 131,300 | 1625 | 11,570 | 4.75 | 166.3 | 128.3 |
| 14¼ × 49½ | 33 | 705.4 | 5819 | 144,000 | 1675 | 11,940 | 4.90 | 171.4 | 132.3 |
| 14¼ × 51 | 34 | 726.8 | 6177 | 157,500 | 1726 | 12,300 | 5.05 | 176.6 | 136.3 |
| 14¼ × 52½ | 35 | 748.1 | 6546 | 171,800 | 1777 | 12,660 | 5.20 | 181.8 | 140.3 |
| 14¼ × 54 | 36 | 769.5 | 6926 | 187,000 | 1828 | 13,020 | 5.34 | 187.0 | 144.3 |
| 14¼ × 55½ | 37 | 790.9 | 7316 | 203,000 | 1878 | 13,380 | 5.49 | 192.2 | 148.3 |
| 14¼ × 57 | 38 | 812.3 | 7716 | 219,900 | 1929 | 13,740 | 5.64 | 197.4 | 152.3 |
| 14¼ × 58½ | 39 | 833.6 | 8128 | 237,700 | 1980 | 14,110 | 5.79 | 202.6 | 156.3 |
| 14¼ × 60 | 40 | 855.0 | 8550 | 256,500 | 2031 | 14,470 | 5.94 | 207.8 | 160.3 |
| 14¼ × 61½ | 41 | 876.4 | 8983 | 276,200 | 2081 | 14,830 | 6.09 | 213.0 | 164.3 |
| 14¼ × 63 | 42 | 897.8 | 9426 | 296,900 | 2132 | 15,190 | 6.23 | 218.2 | 168.3 |
| 14¼ × 64½ | 43 | 919.1 | 9881 | 318,600 | 2183 | 15,550 | 6.38 | 223.4 | 172.3 |
| 14¼ × 66 | 44 | 940.5 | 10,350 | 341,400 | 2234 | 15,920 | 6.53 | 228.6 | 176.3 |
| 14¼ × 67½ | 45 | 961.9 | 10,820 | 365,200 | 2284 | 16,280 | 6.68 | 233.8 | 180.4 |
| 14¼ × 69 | 46 | 983.3 | 11,310 | 390,100 | 2335 | 16,640 | 6.83 | 239.0 | 184.4 |
| 14¼ × 70½ | 47 | 1005 | 11,800 | 416,100 | 2386 | 17,000 | 6.98 | 244.2 | 188.4 |
| 14¼ × 72 | 48 | 1026 | 12,310 | 443,200 | 2437 | 17,360 | 7.13 | 249.4 | 192.4 |
| 14¼ × 73½ | 49 | 1047 | 12,830 | 471,500 | 2488 | 17,720 | 7.27 | 254.6 | 196.4 |
| 14¼ × 75 | 50 | 1069 | 13,360 | 501,000 | 2538 | 18,090 | 7.42 | 259.8 | 200.4 |
| 14¼ × 76½ | 51 | 1090 | 13,900 | 531,600 | 2589 | 18,450 | 7.57 | 265.0 | 204.4 |
| 14¼ × 78 | 52 | 1112 | 14,450 | 563,500 | 2640 | 18,810 | 7.72 | 270.2 | 208.4 |

**TABLE 8.2** (*Continued*)

**Western Species (Based on $1\frac{1}{2}$ in. Thick Laminations)**

| Beam Size b Width in. | Beam Size d Depth in. | Number of Lams | A Area in.² | x–x Axis $S_x$ Section Modulus in.³ | x–x Axis $I_x$ Moment of Inertia in.⁴ | y–y Axis $S_y$ Section Modulus in.³ | y–y Axis $I_y$ Moment of Inertia in.⁴ | V Volume per ft cf/ft | DF[a] 35 pcf Weight per ft plf | HF[a] 27 pcf Weight per ft plf |
|---|---|---|---|---|---|---|---|---|---|---|
| $14\frac{1}{4}$ × | $79\frac{1}{2}$ | 53 | 1133 | 15,010 | 596,700 | 2691 | 19,170 | 7.87 | 275.4 | 212.4 |
| $14\frac{1}{4}$ × | 81 | 54 | 1154 | 15,580 | 631,100 | 2741 | 19,530 | 8.02 | 280.5 | 216.4 |
| $14\frac{1}{4}$ × | $82\frac{1}{2}$ | 55 | 1176 | 16,160 | 666,800 | 2792 | 19,890 | 8.16 | 285.7 | 220.4 |
| $14\frac{1}{4}$ × | 84 | 56 | 1197 | 16,760 | 703,800 | 2843 | 20,260 | 8.31 | 290.9 | 224.4 |
| $14\frac{1}{4}$ × | $85\frac{1}{2}$ | 57 | 1218 | 17,360 | 742,200 | 2894 | 20,620 | 8.46 | 296.1 | 228.4 |
| $14\frac{1}{4}$ × | 87 | 58 | 1240 | 17,980 | 782,000 | 2944 | 20,980 | 8.61 | 301.3 | 232.5 |
| $14\frac{1}{4}$ × | $88\frac{1}{2}$ | 59 | 1261 | 18,600 | 823,100 | 2995 | 21,340 | 8.76 | 306.5 | 236.5 |
| $14\frac{1}{4}$ × | 90 | 60 | 1283 | 19,240 | 865,700 | 3046 | 21,700 | 8.91 | 311.7 | 240.5 |
| $14\frac{1}{4}$ × | $91\frac{1}{2}$ | 61 | 1304 | 19,880 | 909,700 | 3097 | 22,060 | 9.05 | 316.9 | 244.5 |
| $14\frac{1}{4}$ × | 93 | 62 | 1325 | 20,540 | 955,200 | 3147 | 22,430 | 9.20 | 322.1 | 248.5 |
| $14\frac{1}{4}$ × | $94\frac{1}{2}$ | 63 | 1347 | 21,210 | 1,002,000 | 3198 | 22,790 | 9.35 | 327.3 | 252.5 |
| $14\frac{1}{4}$ × | 96 | 64 | 1368 | 21,890 | 1,051,000 | 3249 | 23,150 | 9.50 | 332.5 | 256.5 |
| $14\frac{1}{4}$ × | $97\frac{1}{2}$ | 65 | 1389 | 22,580 | 1,101,000 | 3300 | 23,510 | 9.65 | 337.7 | 260.5 |
| $14\frac{1}{4}$ × | 99 | 66 | 1411 | 23,280 | 1,152,000 | 3351 | 23,870 | 9.80 | 342.9 | 264.5 |
| $14\frac{1}{4}$ × | $100\frac{1}{2}$ | 67 | 1432 | 23,990 | 1,205,000 | 3401 | 24,230 | 9.95 | 348.1 | 268.5 |
| $14\frac{1}{4}$ × | 102 | 68 | 1454 | 24,710 | 1,260,000 | 3452 | 24,600 | 10.09 | 353.3 | 272.5 |

| 14¼ × 103½ | 69 | 1475 | 25,440 | 1,317,000 | 3503 | 24,960 | 10.24 | 358.5 | 276.5 |
| 14¼ × 105 | 70 | 1496 | 26,180 | 1,375,000 | 3554 | 25,320 | 10.39 | 363.7 | 280.5 |
| 14¼ × 106½ | 71 | 1518 | 26,940 | 1,434,000 | 3604 | 25,680 | 10.54 | 368.9 | 284.6 |
| 14¼ × 108 | 72 | 1539 | 27,700 | 1,496,000 | 3655 | 26,040 | 10.69 | 374.1 | 288.6 |
| 14¼ × 109½ | 73 | 1560 | 28,480 | 1,559,000 | 3706 | 26,400 | 10.84 | 379.3 | 292.6 |
| 14¼ × 111 | 74 | 1582 | 29,260 | 1,624,000 | 3757 | 26,770 | 10.98 | 384.5 | 296.6 |
| 14¼ × 112½ | 75 | 1603 | 30,060 | 1,691,000 | 3807 | 27,130 | 11.13 | 389.6 | 300.6 |
| 14¼ × 114 | 76 | 1625 | 30,870 | 1,759,000 | 3858 | 27,490 | 11.28 | 394.8 | 304.6 |
| 14¼ × 115½ | 77 | 1646 | 31,680 | 1,830,000 | 3909 | 27,850 | 11.43 | 400.0 | 308.6 |
| 14¼ × 117 | 78 | 1667 | 32,510 | 1,902,000 | 3960 | 28,210 | 11.58 | 405.2 | 312.6 |
| 14¼ × 118½ | 79 | 1689 | 33,350 | 1,976,000 | 4010 | 28,570 | 11.73 | 410.4 | 316.6 |
| 14¼ × 120 | 80 | 1710 | 34,200 | 2,052,000 | 4061 | 28,940 | 11.88 | 415.6 | 320.6 |

**TABLE 8.2** (*Continued*)

**Southern Pine (Based on 1⅜ in. Thick Laminations)**

| Beam Size b Width in. | d Depth in. | Number of Lams | A Area in.² | x-x Axis Sₓ Section Modulus in.³ | Iₓ Moment of Inertia in.⁴ | y-y Axis Sᵧ Section Modulus in.³ | Iᵧ Moment of Inertia in.⁴ | V Volume per ft cf/ft | SPᵃ 36 pcf Weight per ft plf |
|---|---|---|---|---|---|---|---|---|---|
| 2½ × | 2¾ | 2 | 6.875 | 3.151 | 4.333 | 2.865 | 3.581 | 0.05 | 1.7 |
| 2½ × | 4⅛ | 3 | 10.31 | 7.090 | 14.62 | 4.297 | 5.371 | 0.07 | 2.6 |
| 2½ × | 5½ | 4 | 13.75 | 12.60 | 34.66 | 5.729 | 7.161 | 0.10 | 3.4 |
| 2½ × | 6⅞ | 5 | 17.19 | 19.69 | 67.70 | 7.161 | 8.952 | 0.12 | 4.3 |
| 2½ × | 8¼ | 6 | 20.63 | 28.36 | 117.0 | 8.594 | 10.74 | 0.14 | 5.2 |
| 2½ × | 9⅝ | 7 | 24.06 | 38.60 | 185.8 | 10.03 | 12.53 | 0.17 | 6.0 |
| 2½ × | 11 | 8 | 27.50 | 50.42 | 277.3 | 11.46 | 14.32 | 0.19 | 6.9 |
| 2½ × | 12⅜ | 9 | 30.94 | 63.81 | 394.8 | 12.89 | 16.11 | 0.21 | 7.7 |
| 2½ × | 13¾ | 10 | 34.38 | 78.78 | 541.6 | 14.32 | 17.90 | 0.24 | 8.6 |
| 2½ × | 15⅛ | 11 | 37.81 | 95.32 | 720.9 | 15.76 | 19.69 | 0.26 | 9.5 |
| 2½ × | 16½ | 12 | 41.25 | 113.4 | 935.9 | 17.19 | 21.48 | 0.29 | 10.3 |
| 2½ × | 17⅞ | 13 | 44.69 | 133.1 | 1190 | 18.62 | 23.27 | 0.31 | 11.2 |
| 3 × | 2¾ | 2 | 8.250 | 3.781 | 5.199 | 4.125 | 6.188 | 0.06 | 2.1 |
| 3 × | 4⅛ | 3 | 12.38 | 8.508 | 17.55 | 6.188 | 9.281 | 0.09 | 3.1 |
| 3 × | 5½ | 4 | 16.50 | 15.13 | 41.59 | 8.250 | 12.38 | 0.11 | 4.1 |
| 3 × | 6⅞ | 5 | 20.63 | 23.63 | 81.24 | 10.31 | 15.47 | 0.14 | 5.2 |

| | | | | | | | | |
|---|---|---|---|---|---|---|---|---|
| $3 \times 8\frac{1}{4}$ | 6 | 24.75 | 34.03 | 140.4 | 12.38 | 18.56 | 0.17 | 6.2 |
| $3 \times 9\frac{5}{8}$ | 7 | 28.88 | 46.32 | 222.9 | 14.44 | 21.66 | 0.20 | 7.2 |
| $3 \times 11$ | 8 | 33.00 | 60.50 | 332.8 | 16.50 | 24.75 | 0.23 | 8.3 |
| $3 \times 12\frac{3}{8}$ | 9 | 37.13 | 76.57 | 473.8 | 18.56 | 27.84 | 0.26 | 9.3 |
| $3 \times 13\frac{3}{4}$ | 10 | 41.25 | 94.53 | 649.9 | 20.63 | 30.94 | 0.29 | 10.3 |
| $3 \times 15\frac{1}{8}$ | 11 | 45.38 | 114.4 | 865.0 | 22.69 | 34.03 | 0.32 | 11.4 |
| $3 \times 16\frac{1}{2}$ | 12 | 49.50 | 136.1 | 1123 | 24.75 | 37.13 | 0.34 | 12.4 |
| $3 \times 17\frac{7}{8}$ | 13 | 53.63 | 159.8 | 1428 | 26.81 | 40.22 | 0.37 | 13.4 |
| $3 \times 19\frac{1}{4}$ | 14 | 57.75 | 185.3 | 1783 | 28.88 | 43.31 | 0.40 | 14.5 |
| $3 \times 20\frac{5}{8}$ | 15 | 61.88 | 212.7 | 2193 | 30.94 | 46.41 | 0.43 | 15.5 |
| $3 \times 22$ | 16 | 66.00 | 242.0 | 2662 | 33.00 | 49.50 | 0.46 | 16.5 |
| $3 \times 23\frac{3}{8}$ | 17 | 70.13 | 273.2 | 3193 | 35.06 | 52.59 | 0.49 | 17.5 |
| $3 \times 24\frac{3}{4}$ | 18 | 74.25 | 306.3 | 3790 | 37.13 | 55.69 | 0.52 | 18.6 |
| $3 \times 26\frac{1}{8}$ | 19 | 78.38 | 341.3 | 4458 | 39.19 | 58.78 | 0.54 | 19.6 |
| $3 \times 27\frac{1}{2}$ | 20 | 82.50 | 378.1 | 5199 | 41.25 | 61.88 | 0.57 | 20.6 |
| $3\frac{1}{8} \times 2\frac{3}{4}$ | 2 | 8.594 | 3.939 | 5.416 | 4.476 | 6.994 | 0.06 | 2.1 |
| $3\frac{1}{8} \times 4\frac{1}{8}$ | 3 | 12.89 | 8.862 | 18.28 | 6.714 | 10.49 | 0.09 | 3.2 |
| $3\frac{1}{8} \times 5\frac{1}{2}$ | 4 | 17.19 | 15.76 | 43.33 | 8.952 | 13.99 | 0.12 | 4.3 |
| $3\frac{1}{8} \times 6\frac{7}{8}$ | 5 | 21.48 | 24.62 | 84.62 | 11.19 | 17.48 | 0.15 | 5.4 |
| $3\frac{1}{8} \times 8\frac{1}{4}$ | 6 | 25.78 | 35.45 | 146.2 | 13.43 | 20.98 | 0.18 | 6.4 |
| $3\frac{1}{8} \times 9\frac{5}{8}$ | 7 | 30.08 | 48.25 | 232.2 | 15.67 | 24.48 | 0.21 | 7.5 |
| $3\frac{1}{8} \times 11$ | 8 | 34.38 | 63.02 | 346.6 | 17.90 | 27.97 | 0.24 | 8.6 |
| $3\frac{1}{8} \times 12\frac{3}{8}$ | 9 | 38.67 | 79.76 | 493.5 | 20.14 | 31.47 | 0.27 | 9.7 |
| $3\frac{1}{8} \times 13\frac{3}{4}$ | 10 | 42.97 | 98.47 | 677.0 | 22.38 | 34.97 | 0.30 | 10.7 |

## TABLE 8.2 (Continued)
### Southern Pine (Based on $1\frac{3}{8}$ in. Thick Laminations)

| Beam Size | | Number of Lams | A Area in.² | x–x Axis | | y–y Axis | | V Volume per ft cf/ft | SP[a] 36 pcf Weight per ft plf |
| --- | --- | --- | --- | --- | --- | --- | --- | --- | --- |
| b Width in. | d Depth in. | | | $S_x$ Section Modulus in.³ | $I_x$ Moment of Inertia in.⁴ | $S_y$ Section Modulus in.³ | $I_y$ Moment of Inertia in.⁴ | | |
| $3\frac{1}{8}$ × | $15\frac{1}{8}$ | 11 | 47.27 | 119.1 | 901.1 | 24.62 | 38.46 | 0.33 | 11.8 |
| $3\frac{1}{8}$ × | $16\frac{1}{2}$ | 12 | 51.56 | 141.8 | 1170 | 26.86 | 41.96 | 0.36 | 12.9 |
| $3\frac{1}{8}$ × | $17\frac{7}{8}$ | 13 | 55.86 | 166.4 | 1487 | 29.09 | 45.46 | 0.39 | 14.0 |
| $3\frac{1}{8}$ × | $19\frac{1}{4}$ | 14 | 60.16 | 193.0 | 1858 | 31.33 | 48.96 | 0.42 | 15.0 |
| $3\frac{1}{8}$ × | $20\frac{5}{8}$ | 15 | 64.45 | 221.6 | 2285 | 33.57 | 52.45 | 0.45 | 16.1 |
| $3\frac{1}{8}$ × | 22 | 16 | 68.75 | 252.1 | 2773 | 35.81 | 55.95 | 0.48 | 17.2 |
| $3\frac{1}{8}$ × | $23\frac{3}{8}$ | 17 | 73.05 | 284.6 | 3326 | 38.05 | 59.45 | 0.51 | 18.3 |
| $3\frac{1}{8}$ × | $24\frac{3}{4}$ | 18 | 77.34 | 319.0 | 3948 | 40.28 | 62.94 | 0.54 | 19.3 |
| $3\frac{1}{8}$ × | $26\frac{1}{8}$ | 19 | 81.64 | 355.5 | 4643 | 42.52 | 66.44 | 0.57 | 20.4 |
| $3\frac{1}{8}$ × | $27\frac{1}{2}$ | 20 | 85.94 | 393.9 | 5416 | 44.76 | 69.94 | 0.60 | 21.5 |
| 5 × | $2\frac{3}{4}$ | 2 | 13.75 | 6.302 | 8.665 | 11.46 | 28.65 | 0.10 | 3.4 |
| 5 × | $4\frac{1}{8}$ | 3 | 20.63 | 14.18 | 29.25 | 17.19 | 42.97 | 0.14 | 5.2 |
| 5 × | $5\frac{1}{2}$ | 4 | 27.50 | 25.21 | 69.32 | 22.92 | 57.29 | 0.19 | 6.9 |
| 5 × | $6\frac{7}{8}$ | 5 | 34.38 | 39.39 | 135.4 | 28.65 | 71.61 | 0.24 | 8.6 |
| 5 × | $8\frac{1}{4}$ | 6 | 41.25 | 56.72 | 234.0 | 34.38 | 85.94 | 0.29 | 10.3 |
| 5 × | $9\frac{5}{8}$ | 7 | 48.13 | 77.20 | 371.5 | 40.10 | 100.3 | 0.33 | 12.0 |
| 5 × | 11 | 8 | 55.00 | 100.8 | 554.6 | 45.83 | 114.6 | 0.38 | 13.8 |

| | | | | | | | | | |
|---|---|---|---|---|---|---|---|---|---|
| 5 × 12⅜ | 9 | 61.88 | 127.6 | 789.6 | 51.56 | 128.9 | 0.43 | 15.5 |
| 5 × 13¾ | 10 | 68.75 | 157.6 | 1083 | 57.29 | 143.2 | 0.48 | 17.2 |
| 5 × 15⅛ | 11 | 75.63 | 190.6 | 1442 | 63.02 | 157.6 | 0.53 | 18.9 |
| 5 × 16½ | 12 | 82.50 | 226.9 | 1872 | 68.75 | 171.9 | 0.57 | 20.6 |
| 5 × 17⅞ | 13 | 89.38 | 266.3 | 2380 | 74.48 | 186.2 | 0.62 | 22.3 |
| 5 × 19¼ | 14 | 96.25 | 308.8 | 2972 | 80.21 | 200.5 | 0.67 | 24.1 |
| 5 × 20⅝ | 15 | 103.1 | 354.5 | 3656 | 85.94 | 214.8 | 0.72 | 25.8 |
| 5 × 22 | 16 | 110.0 | 403.3 | 4437 | 91.67 | 229.2 | 0.76 | 27.5 |
| 5 × 23⅜ | 17 | 116.9 | 455.3 | 5322 | 97.40 | 243.5 | 0.81 | 29.2 |
| 5 × 24¾ | 18 | 123.8 | 510.5 | 6317 | 103.1 | 257.8 | 0.86 | 30.9 |
| 5 × 26⅛ | 19 | 130.6 | 568.8 | 7429 | 108.9 | 272.1 | 0.91 | 32.7 |
| 5 × 27½ | 20 | 137.5 | 630.2 | 8,670 | 114.6 | 286.5 | 0.95 | 34.4 |
| 5 × 28⅞ | 21 | 144.4 | 694.8 | 10,030 | 120.3 | 300.8 | 1.00 | 36.1 |
| 5 × 30¼ | 22 | 151.3 | 762.6 | 11,530 | 126.0 | 315.1 | 1.05 | 37.8 |
| 5 × 31⅝ | 23 | 158.1 | 833.5 | 13,180 | 131.8 | 329.4 | 1.10 | 39.5 |
| 5 × 33 | 24 | 165.0 | 907.5 | 14,970 | 137.5 | 343.8 | 1.15 | 41.3 |
| 5 × 34⅜ | 25 | 171.9 | 984.7 | 16,920 | 143.2 | 358.1 | 1.19 | 43.0 |
| 5 × 35¾ | 26 | 178.8 | 1065 | 19,040 | 149.0 | 372.4 | 1.24 | 44.7 |
| 5 × 37⅛ | 27 | 185.6 | 1149 | 21,320 | 154.7 | 386.7 | 1.29 | 46.4 |
| 5 × 38½ | 28 | 192.5 | 1235 | 23,780 | 160.4 | 401.0 | 1.34 | 48.1 |
| 5 × 39⅞ | 29 | 199.4 | 1325 | 26,420 | 166.1 | 415.4 | 1.38 | 49.8 |
| 5 × 41¼ | 30 | 206.3 | 1418 | 29,250 | 171.9 | 429.7 | 1.43 | 51.6 |
| 5 × 42⅝ | 31 | 213.1 | 1514 | 32,270 | 177.6 | 444.0 | 1.48 | 53.3 |
| 5 × 44 | 32 | 220.0 | 1613 | 35,490 | 183.3 | 458.3 | 1.53 | 55.0 |

## TABLE 8.2  (Continued)
### Southern Pine (Based on $1\frac{3}{8}$ in. Thick Laminations)

| Beam Size | | | | x-x Axis | | y-y Axis | | | |
| b Width in. | d Depth in. | Number of Lams | A Area in.² | $S_x$ Section Modulus in.³ | $I_x$ Moment of Inertia in.⁴ | $S_y$ Section Modulus in.³ | $I_y$ Moment of Inertia in.⁴ | V Volume per ft cf/ft | SPᵃ 36 pcf Weight per ft plf |
|---|---|---|---|---|---|---|---|---|---|
| $5\frac{1}{8}$ × | $2\frac{3}{4}$ | 2 | 14.09 | 6.460 | 8.882 | 12.04 | 30.85 | 0.10 | 3.5 |
| $5\frac{1}{8}$ × | $4\frac{1}{8}$ | 3 | 21.14 | 14.53 | 29.98 | 18.06 | 46.27 | 0.15 | 5.3 |
| $5\frac{1}{8}$ × | $5\frac{1}{2}$ | 4 | 28.19 | 25.84 | 71.06 | 24.08 | 61.70 | 0.20 | 7.0 |
| $5\frac{1}{8}$ × | $6\frac{7}{8}$ | 5 | 35.23 | 40.37 | 138.8 | 30.10 | 77.12 | 0.24 | 8.8 |
| $5\frac{1}{8}$ × | $8\frac{1}{4}$ | 6 | 42.28 | 58.14 | 239.8 | 36.12 | 92.55 | 0.29 | 10.6 |
| $5\frac{1}{8}$ × | $9\frac{5}{8}$ | 7 | 49.33 | 79.13 | 380.8 | 42.13 | 108.0 | 0.34 | 12.3 |
| $5\frac{1}{8}$ × | $11$ | 8 | 56.38 | 103.4 | 568.4 | 48.15 | 123.4 | 0.39 | 14.1 |
| $5\frac{1}{8}$ × | $12\frac{3}{8}$ | 9 | 63.42 | 130.8 | 809.4 | 54.17 | 138.8 | 0.44 | 15.9 |
| $5\frac{1}{8}$ × | $13\frac{3}{4}$ | 10 | 70.47 | 161.5 | 1110 | 60.19 | 154.2 | 0.49 | 17.6 |
| $5\frac{1}{8}$ × | $15\frac{1}{8}$ | 11 | 77.52 | 195.4 | 1478 | 66.21 | 169.7 | 0.54 | 19.4 |
| $5\frac{1}{8}$ × | $16\frac{1}{2}$ | 12 | 84.56 | 232.5 | 1919 | 72.23 | 185.1 | 0.59 | 21.1 |
| $5\frac{1}{8}$ × | $17\frac{7}{8}$ | 13 | 91.61 | 272.9 | 2439 | 78.25 | 200.5 | 0.64 | 22.9 |
| $5\frac{1}{8}$ × | $19\frac{1}{4}$ | 14 | 98.66 | 316.5 | 3047 | 84.27 | 215.9 | 0.69 | 24.7 |
| $5\frac{1}{8}$ × | $20\frac{5}{8}$ | 15 | 105.7 | 363.4 | 3747 | 90.29 | 231.4 | 0.73 | 26.4 |
| $5\frac{1}{8}$ × | $22$ | 16 | 112.8 | 413.4 | 4548 | 96.31 | 246.8 | 0.78 | 28.2 |
| $5\frac{1}{8}$ × | $23\frac{3}{8}$ | 17 | 119.8 | 466.7 | 5455 | 102.3 | 262.2 | 0.83 | 29.9 |
| $5\frac{1}{8}$ × | $24\frac{3}{4}$ | 18 | 126.8 | 523.2 | 6475 | 108.3 | 277.6 | 0.88 | 31.7 |

| | | | | | | | | |
|---|---|---|---|---|---|---|---|---|
| $5\frac{1}{8}$ × $26\frac{1}{8}$ | 19 | 133.9 | 583.0 | 7615 | 114.4 | 293.1 | 0.93 | 33.5 |
| $5\frac{1}{8}$ × $27\frac{1}{2}$ | 20 | 140.9 | 646.0 | 8882 | 120.4 | 308.5 | 0.98 | 35.2 |
| $5\frac{1}{8}$ × $28\frac{7}{8}$ | 21 | 148.0 | 712.2 | 10282 | 126.4 | 323.9 | 1.03 | 37.0 |
| $5\frac{1}{8}$ × $30\frac{1}{4}$ | 22 | 155.0 | 781.6 | 11822 | 132.4 | 339.3 | 1.08 | 38.8 |
| $5\frac{1}{8}$ × $31\frac{5}{8}$ | 23 | 162.1 | 854.3 | 13508 | 138.4 | 354.8 | 1.13 | 40.5 |
| $5\frac{1}{8}$ × 33 | 24 | 169.1 | 930.2 | 15348 | 144.5 | 370.2 | 1.17 | 42.3 |
| $5\frac{1}{8}$ × $34\frac{3}{8}$ | 25 | 176.2 | 1009 | 17348 | 150.5 | 385.6 | 1.22 | 44.0 |
| $5\frac{1}{8}$ × $35\frac{3}{4}$ | 26 | 183.2 | 1092 | 19514 | 156.5 | 401.0 | 1.27 | 45.8 |
| $5\frac{1}{8}$ × $37\frac{1}{8}$ | 27 | 190.3 | 1177 | 21853 | 162.5 | 416.5 | 1.32 | 47.6 |
| $5\frac{1}{8}$ × $38\frac{1}{2}$ | 28 | 197.3 | 1266 | 24372 | 168.5 | 431.9 | 1.37 | 49.3 |
| $5\frac{1}{8}$ × $39\frac{7}{8}$ | 29 | 204.4 | 1358 | 27078 | 174.6 | 447.3 | 1.42 | 51.1 |
| $5\frac{1}{8}$ × $41\frac{1}{4}$ | 30 | 211.4 | 1453 | 29977 | 180.6 | 462.7 | 1.47 | 52.9 |
| $5\frac{1}{8}$ × $42\frac{5}{8}$ | 31 | 218.5 | 1552 | 33075 | 186.6 | 478.2 | 1.52 | 54.6 |
| $5\frac{1}{8}$ × 44 | 32 | 225.5 | 1654 | 36381 | 192.6 | 493.6 | 1.57 | 56.4 |
| $6\frac{3}{4}$ × $2\frac{3}{4}$ | 2 | 18.56 | 8.508 | 11.70 | 20.88 | 70.48 | 0.13 | 4.6 |
| $6\frac{3}{4}$ × $4\frac{1}{8}$ | 3 | 27.84 | 19.14 | 39.48 | 31.32 | 105.7 | 0.19 | 7.0 |
| $6\frac{3}{4}$ × $5\frac{1}{2}$ | 4 | 37.13 | 34.03 | 93.59 | 41.77 | 141.0 | 0.26 | 9.3 |
| $6\frac{3}{4}$ × $6\frac{7}{8}$ | 5 | 46.41 | 53.17 | 182.8 | 52.21 | 176.2 | 0.32 | 11.6 |
| $6\frac{3}{4}$ × $8\frac{1}{4}$ | 6 | 55.69 | 76.57 | 315.9 | 62.65 | 211.4 | 0.39 | 13.9 |
| $6\frac{3}{4}$ × $9\frac{5}{8}$ | 7 | 64.97 | 104.2 | 501.6 | 73.09 | 246.7 | 0.45 | 16.2 |
| $6\frac{3}{4}$ × 11 | 8 | 74.25 | 136.1 | 748.7 | 83.53 | 281.9 | 0.52 | 18.6 |
| $6\frac{3}{4}$ × $12\frac{3}{8}$ | 9 | 83.53 | 172.3 | 1066 | 93.97 | 317.2 | 0.58 | 20.9 |
| $6\frac{3}{4}$ × $13\frac{3}{4}$ | 10 | 92.81 | 212.7 | 1462 | 104.4 | 352.4 | 0.64 | 23.2 |
| $6\frac{3}{4}$ × $15\frac{1}{8}$ | 11 | 102.1 | 257.4 | 1946 | 114.9 | 387.6 | 0.71 | 25.5 |

## TABLE 8.2 (*Continued*)
### Southern Pine (Based on $1\frac{3}{8}$ in. Thick Laminations)

| Beam Size | | | | x-x Axis | | y-y Axis | | | $SP^a$ |
| b Width in. | d Depth in. | Number of Lams | A Area in.$^2$ | $S_x$ Section Modulus in.$^3$ | $I_x$ Moment of Inertia in.$^4$ | $S_y$ Section Modulus in.$^3$ | $I_y$ Moment of Inertia in.$^4$ | V Volume per ft cf/ft | 36 pcf Weight per ft plf |
|---|---|---|---|---|---|---|---|---|---|
| $6\frac{3}{4}$ × | $16\frac{1}{2}$ | 12 | 111.4 | 306.3 | 2527 | 125.3 | 422.9 | 0.77 | 27.8 |
| $6\frac{3}{4}$ × | $17\frac{7}{8}$ | 13 | 120.7 | 359.5 | 3213 | 135.7 | 458.1 | 0.84 | 30.2 |
| $6\frac{3}{4}$ × | $19\frac{1}{4}$ | 14 | 129.9 | 416.9 | 4012 | 146.2 | 493.4 | 0.90 | 32.5 |
| $6\frac{3}{4}$ × | $20\frac{5}{8}$ | 15 | 139.2 | 478.6 | 4935 | 156.6 | 528.6 | 0.97 | 34.8 |
| $6\frac{3}{4}$ × | 22 | 16 | 148.5 | 544.5 | 5990 | 167.1 | 563.8 | 1.03 | 37.1 |
| $6\frac{3}{4}$ × | $23\frac{3}{8}$ | 17 | 157.8 | 614.7 | 7184 | 177.5 | 599.1 | 1.10 | 39.5 |
| $6\frac{3}{4}$ × | $24\frac{3}{4}$ | 18 | 167.1 | 689.1 | 8528 | 187.9 | 634.3 | 1.16 | 41.8 |
| $6\frac{3}{4}$ × | $26\frac{1}{8}$ | 19 | 176.3 | 767.8 | 10,030 | 198.4 | 669.6 | 1.22 | 44.1 |
| $6\frac{3}{4}$ × | $27\frac{1}{2}$ | 20 | 185.6 | 850.8 | 11,700 | 208.8 | 704.8 | 1.29 | 46.4 |
| $6\frac{3}{4}$ × | $28\frac{7}{8}$ | 21 | 194.9 | 938.0 | 13,540 | 219.3 | 740.0 | 1.35 | 48.7 |
| $6\frac{3}{4}$ × | $30\frac{1}{4}$ | 22 | 204.2 | 1029 | 15,570 | 229.7 | 775.3 | 1.42 | 51.1 |
| $6\frac{3}{4}$ × | $31\frac{5}{8}$ | 23 | 213.5 | 1125 | 17,790 | 240.2 | 810.5 | 1.48 | 53.4 |
| $6\frac{3}{4}$ × | 33 | 24 | 222.8 | 1225 | 20,210 | 250.6 | 845.8 | 1.55 | 55.7 |
| $6\frac{3}{4}$ × | $34\frac{3}{8}$ | 25 | 232.0 | 1329 | 22,850 | 261.0 | 881.0 | 1.61 | 58.0 |
| $6\frac{3}{4}$ × | $35\frac{3}{4}$ | 26 | 241.3 | 1438 | 25,700 | 271.5 | 916.2 | 1.68 | 60.3 |
| $6\frac{3}{4}$ × | $37\frac{1}{8}$ | 27 | 250.6 | 1551 | 28,780 | 281.9 | 951.5 | 1.74 | 62.7 |

| | | | | | | | | |
|---|---|---|---|---|---|---|---|---|
| $6\frac{3}{4} \times 38\frac{1}{2}$ | 28 | 259.9 | 1668 | 32,100 | 292.4 | 986.7 | 1.80 | 65.0 |
| $6\frac{3}{4} \times 39\frac{7}{8}$ | 29 | 269.2 | 1789 | 35,660 | 302.8 | 1022 | 1.87 | 67.3 |
| $6\frac{3}{4} \times 41\frac{1}{4}$ | 30 | 278.4 | 1914 | 39,480 | 313.2 | 1057 | 1.93 | 69.6 |
| $6\frac{3}{4} \times 42\frac{5}{8}$ | 31 | 287.7 | 2044 | 43,560 | 323.7 | 1092 | 2.00 | 71.9 |
| $6\frac{3}{4} \times 44$ | 32 | 297.0 | 2178 | 47,920 | 334.1 | 1128 | 2.06 | 74.3 |
| $6\frac{3}{4} \times 45\frac{3}{8}$ | 33 | 306.3 | 2316 | 52,550 | 344.6 | 1163 | 2.13 | 76.6 |
| $6\frac{3}{4} \times 46\frac{3}{4}$ | 34 | 315.6 | 2459 | 57,470 | 355.0 | 1198 | 2.19 | 78.9 |
| $6\frac{3}{4} \times 48\frac{1}{8}$ | 35 | 324.8 | 2606 | 62,700 | 365.4 | 1233 | 2.26 | 81.2 |
| $6\frac{3}{4} \times 49\frac{1}{2}$ | 36 | 334.1 | 2757 | 68,220 | 375.9 | 1269 | 2.32 | 83.5 |
| $6\frac{3}{4} \times 50\frac{7}{8}$ | 37 | 343.4 | 2912 | 74,070 | 386.3 | 1304 | 2.38 | 85.9 |
| $6\frac{3}{4} \times 52\frac{1}{4}$ | 38 | 352.7 | 3071 | 80,200 | 396.8 | 1339 | 2.45 | 88.2 |
| $6\frac{3}{4} \times 53\frac{5}{8}$ | 39 | 362.0 | 3235 | 86,700 | 407.2 | 1374 | 2.51 | 90.5 |
| $6\frac{3}{4} \times 55$ | 40 | 371.3 | 3403 | 93,600 | 417.7 | 1410 | 2.58 | 92.8 |
| $8\frac{1}{2} \times 2\frac{3}{4}$ | 2 | 23.38 | 10.71 | 14.73 | 33.11 | 140.7 | 0.16 | 5.8 |
| $8\frac{1}{2} \times 4\frac{1}{8}$ | 3 | 35.06 | 24.11 | 49.72 | 49.67 | 211.1 | 0.24 | 8.8 |
| $8\frac{1}{2} \times 5\frac{1}{2}$ | 4 | 46.75 | 42.85 | 117.9 | 66.23 | 281.5 | 0.32 | 11.7 |
| $8\frac{1}{2} \times 6\frac{7}{8}$ | 5 | 58.44 | 66.96 | 230.2 | 82.79 | 351.8 | 0.41 | 14.6 |
| $8\frac{1}{2} \times 8\frac{1}{4}$ | 6 | 70.13 | 96.42 | 397.7 | 99.34 | 422.2 | 0.49 | 17.5 |
| $8\frac{1}{2} \times 9\frac{5}{8}$ | 7 | 81.81 | 131.2 | 631.6 | 115.9 | 492.6 | 0.57 | 20.5 |
| $8\frac{1}{2} \times 11$ | 8 | 93.50 | 171.4 | 943.0 | 132.5 | 562.9 | 0.65 | 23.4 |
| $8\frac{1}{2} \times 12\frac{3}{8}$ | 9 | 105.2 | 216.9 | 1342 | 149.0 | 633.3 | 0.73 | 26.3 |
| $8\frac{1}{2} \times 13\frac{3}{4}$ | 10 | 116.9 | 267.8 | 1841 | 165.6 | 703.7 | 0.81 | 29.2 |
| $8\frac{1}{2} \times 15\frac{1}{8}$ | 11 | 128.6 | 324.1 | 2451 | 182.1 | 774.1 | 0.89 | 32.1 |
| $8\frac{1}{2} \times 16\frac{1}{2}$ | 12 | 140.3 | 385.7 | 3182 | 198.7 | 844.4 | 0.97 | 35.1 |

## TABLE 8.2 (Continued)
## Southern Pine (Based on $1\frac{3}{8}$ in. Thick Laminations)

| Beam Size | | Number of Lams | A Area in.² | x-x Axis | | y-y Axis | | V Volume per ft cf/ft | SPᵃ 36 pcf Weight per ft plf |
| b Width in. | d Depth in. | | | $S_x$ Section Modulus in.³ | $I_x$ Moment of Inertia in.⁴ | $S_y$ Section Modulus in.³ | $I_y$ Moment of Inertia in.⁴ | | |
|---|---|---|---|---|---|---|---|---|---|
| $8\frac{1}{2}$ × | $17\frac{7}{8}$ | 13 | 151.9 | 452.6 | 4046 | 215.2 | 914.8 | 1.06 | 38.0 |
| $8\frac{1}{2}$ × | $19\frac{1}{4}$ | 14 | 163.6 | 525.0 | 5053 | 231.8 | 985.2 | 1.14 | 40.9 |
| $8\frac{1}{2}$ × | $20\frac{5}{8}$ | 15 | 175.3 | 602.6 | 6215 | 248.4 | 1056 | 1.22 | 43.8 |
| $8\frac{1}{2}$ × | 22 | 16 | 187.0 | 685.7 | 7542 | 264.9 | 1126 | 1.30 | 46.8 |
| $8\frac{1}{2}$ × | $23\frac{3}{8}$ | 17 | 198.7 | 774.1 | 9047 | 281.5 | 1196 | 1.38 | 49.7 |
| $8\frac{1}{2}$ × | $24\frac{3}{4}$ | 18 | 210.4 | 867.8 | 10,740 | 298.0 | 1267 | 1.46 | 52.6 |
| $8\frac{1}{2}$ × | $26\frac{1}{8}$ | 19 | 222.1 | 966.9 | 12,630 | 314.6 | 1337 | 1.54 | 55.5 |
| $8\frac{1}{2}$ × | $27\frac{1}{2}$ | 20 | 233.8 | 1071 | 14,730 | 331.1 | 1407 | 1.62 | 58.4 |
| $8\frac{1}{2}$ × | $28\frac{7}{8}$ | 21 | 245.4 | 1181 | 17,050 | 347.7 | 1478 | 1.70 | 61.4 |
| $8\frac{1}{2}$ × | $30\frac{1}{4}$ | 22 | 257.1 | 1296 | 19,610 | 364.3 | 1548 | 1.79 | 64.3 |
| $8\frac{1}{2}$ × | $31\frac{5}{8}$ | 23 | 268.8 | 1417 | 22,400 | 380.8 | 1618 | 1.87 | 67.2 |
| $8\frac{1}{2}$ × | 33 | 24 | 280.5 | 1543 | 25,460 | 397.4 | 1689 | 1.95 | 70.1 |
| $8\frac{1}{2}$ × | $34\frac{3}{8}$ | 25 | 292.2 | 1674 | 28,770 | 413.9 | 1759 | 2.03 | 73.1 |
| $8\frac{1}{2}$ × | $35\frac{3}{4}$ | 26 | 303.9 | 1811 | 32,360 | 430.5 | 1830 | 2.11 | 76.0 |
| $8\frac{1}{2}$ × | $37\frac{1}{8}$ | 27 | 315.6 | 1953 | 36,240 | 447.0 | 1900 | 2.19 | 78.9 |
| $8\frac{1}{2}$ × | $38\frac{1}{2}$ | 28 | 327.3 | 2100 | 40,420 | 463.6 | 1970 | 2.27 | 81.8 |
| $8\frac{1}{2}$ × | $39\frac{7}{8}$ | 29 | 338.9 | 2253 | 44,910 | 480.2 | 2041 | 2.35 | 84.7 |

| | | | | | | | | |
|---|---|---|---|---|---|---|---|---|
| $8\frac{1}{2}$ × $41\frac{1}{4}$ | 30 | 350.6 | 2411 | 49,720 | 496.7 | 2111 | 2.43 | 87.7 |
| $8\frac{1}{2}$ × $42\frac{5}{8}$ | 31 | 362.3 | 2574 | 54,860 | 513.3 | 2181 | 2.52 | 90.6 |
| $8\frac{1}{2}$ × 44 | 32 | 374.0 | 2743 | 60,340 | 529.8 | 2252 | 2.60 | 93.5 |
| $8\frac{1}{2}$ × $45\frac{3}{8}$ | 33 | 385.7 | 2917 | 66,170 | 546.4 | 2322 | 2.68 | 96.4 |
| $8\frac{1}{2}$ × $46\frac{3}{4}$ | 34 | 397.4 | 3096 | 72,370 | 562.9 | 2393 | 2.76 | 99.3 |
| $8\frac{1}{2}$ × $48\frac{1}{8}$ | 35 | 409.1 | 3281 | 78,900 | 579.5 | 2463 | 2.84 | 102.3 |
| $8\frac{1}{2}$ × $49\frac{1}{2}$ | 36 | 420.8 | 3471 | 85,900 | 596.1 | 2533 | 2.92 | 105.2 |
| $8\frac{1}{2}$ × $50\frac{7}{8}$ | 37 | 432.4 | 3667 | 93,300 | 612.6 | 2604 | 3.00 | 108.1 |
| $8\frac{1}{2}$ × $52\frac{1}{4}$ | 38 | 444.1 | 3868 | 101,000 | 629.2 | 2674 | 3.08 | 111.0 |
| $8\frac{1}{2}$ × $53\frac{5}{8}$ | 39 | 455.8 | 4074 | 109,200 | 645.7 | 2744 | 3.17 | 114.0 |
| $8\frac{1}{2}$ × 55 | 40 | 467.5 | 4285 | 117,800 | 662.3 | 2815 | 3.25 | 116.9 |
| $8\frac{1}{2}$ × $56\frac{3}{8}$ | 41 | 479.2 | 4502 | 126,900 | 678.8 | 2885 | 3.33 | 119.8 |
| $8\frac{1}{2}$ × $57\frac{3}{4}$ | 42 | 490.9 | 4725 | 136,400 | 695.4 | 2955 | 3.41 | 122.7 |
| $8\frac{1}{2}$ × $59\frac{1}{8}$ | 43 | 502.6 | 4952 | 146,400 | 712.0 | 3026 | 3.49 | 125.6 |
| $8\frac{1}{2}$ × $60\frac{1}{2}$ | 44 | 514.3 | 5185 | 156,900 | 728.5 | 3096 | 3.57 | 128.6 |
| $8\frac{1}{2}$ × $61\frac{7}{8}$ | 45 | 525.9 | 5424 | 167,800 | 745.1 | 3167 | 3.65 | 131.5 |
| $8\frac{1}{2}$ × $63\frac{1}{4}$ | 46 | 537.6 | 5667 | 179,200 | 761.6 | 3237 | 3.73 | 134.4 |
| $8\frac{1}{2}$ × $64\frac{5}{8}$ | 47 | 549.3 | 5917 | 191,200 | 778.2 | 3307 | 3.81 | 137.3 |
| $8\frac{1}{2}$ × 66 | 48 | 561.0 | 6171 | 203,600 | 794.8 | 3378 | 3.90 | 140.3 |
| $8\frac{1}{2}$ × $67\frac{3}{8}$ | 49 | 572.7 | 6431 | 216,600 | 811.3 | 3448 | 3.98 | 143.2 |
| $8\frac{1}{2}$ × $68\frac{3}{4}$ | 50 | 584.4 | 6696 | 230,200 | 827.9 | 3518 | 4.06 | 146.1 |
| $10\frac{1}{2}$ × $2\frac{3}{4}$ | 2 | 28.88 | 13.23 | 18.20 | 50.53 | 265.3 | 0.20 | 7.2 |
| $10\frac{1}{2}$ × $4\frac{1}{8}$ | 3 | 43.31 | 29.78 | 61.42 | 75.80 | 397.9 | 0.30 | 10.8 |
| $10\frac{1}{2}$ × $5\frac{1}{2}$ | 4 | 57.75 | 52.94 | 145.6 | 101.1 | 530.6 | 0.40 | 14.4 |

## TABLE 8.2 (Continued)
### Southern Pine (Based on $1\frac{3}{8}$ in. Thick Laminations)

| Beam Size b Width in. | d Depth in. | Number of Lams | A Area in.² | x-x Axis $S_x$ Section Modulus in.³ | $I_x$ Moment of Inertia in.⁴ | y-y Axis $S_y$ Section Modulus in.³ | $I_y$ Moment of Inertia in.⁴ | V Volume per ft cf/ft | SP[a] 36 pcf Weight per ft plf |
|---|---|---|---|---|---|---|---|---|---|
| $10\frac{1}{2}$ × | $6\frac{7}{8}$ | 5 | 72.19 | 82.71 | 284.3 | 126.3 | 663.2 | 0.50 | 18.1 |
| $10\frac{1}{2}$ × | $8\frac{1}{4}$ | 6 | 86.63 | 119.1 | 491.3 | 151.6 | 795.9 | 0.60 | 21.7 |
| $10\frac{1}{2}$ × | $9\frac{5}{8}$ | 7 | 101.1 | 162.1 | 780.2 | 176.9 | 928.5 | 0.70 | 25.3 |
| $10\frac{1}{2}$ × | 11 | 8 | 115.5 | 211.8 | 1165 | 202.1 | 1061 | 0.80 | 28.9 |
| $10\frac{1}{2}$ × | $12\frac{3}{8}$ | 9 | 129.9 | 268.0 | 1658 | 227.4 | 1194 | 0.90 | 32.5 |
| $10\frac{1}{2}$ × | $13\frac{3}{4}$ | 10 | 144.4 | 330.9 | 2275 | 252.7 | 1326 | 1.00 | 36.1 |
| $10\frac{1}{2}$ × | $15\frac{1}{8}$ | 11 | 158.8 | 400.3 | 3028 | 277.9 | 1459 | 1.10 | 39.7 |
| $10\frac{1}{2}$ × | $16\frac{1}{2}$ | 12 | 173.3 | 476.4 | 3931 | 303.2 | 1592 | 1.20 | 43.3 |
| $10\frac{1}{2}$ × | $17\frac{7}{8}$ | 13 | 187.7 | 559.2 | 4997 | 328.5 | 1724 | 1.30 | 46.9 |
| $10\frac{1}{2}$ × | $19\frac{1}{4}$ | 14 | 202.1 | 648.5 | 6242 | 353.7 | 1857 | 1.40 | 50.5 |
| $10\frac{1}{2}$ × | $20\frac{5}{8}$ | 15 | 216.6 | 744.4 | 7677 | 379.0 | 1990 | 1.50 | 54.1 |
| $10\frac{1}{2}$ × | 22 | 16 | 231.0 | 847.0 | 9317 | 404.3 | 2122 | 1.60 | 57.8 |
| $10\frac{1}{2}$ × | $23\frac{3}{8}$ | 17 | 245.4 | 956.2 | 11,180 | 429.5 | 2255 | 1.70 | 61.4 |
| $10\frac{1}{2}$ × | $24\frac{3}{4}$ | 18 | 259.9 | 1072 | 13,270 | 454.8 | 2388 | 1.80 | 65.0 |
| $10\frac{1}{2}$ × | $26\frac{1}{8}$ | 19 | 274.3 | 1194 | 15,600 | 480.0 | 2520 | 1.90 | 68.6 |
| $10\frac{1}{2}$ × | $27\frac{1}{2}$ | 20 | 288.8 | 1323 | 18,200 | 505.3 | 2653 | 2.01 | 72.2 |
| $10\frac{1}{2}$ × | $28\frac{7}{8}$ | 21 | 303.2 | 1459 | 21,070 | 530.6 | 2786 | 2.11 | 75.8 |

| Size | | | | | | | | |
|---|---|---|---|---|---|---|---|---|
| $10\frac{1}{2} \times 30\frac{1}{4}$ | 22 | 317.6 | 1601 | 24,220 | 555.8 | 2918 | 2.21 | 79.4 |
| $10\frac{1}{2} \times 31\frac{5}{8}$ | 23 | 332.1 | 1750 | 27,680 | 581.1 | 3051 | 2.31 | 83.0 |
| $10\frac{1}{2} \times 33$ | 24 | 346.5 | 1906 | 31,440 | 606.4 | 3183 | 2.41 | 86.6 |
| $10\frac{1}{2} \times 34\frac{3}{8}$ | 25 | 360.9 | 2068 | 35,540 | 631.6 | 3316 | 2.51 | 90.2 |
| $10\frac{1}{2} \times 35\frac{3}{4}$ | 26 | 375.4 | 2237 | 39,980 | 656.9 | 3449 | 2.61 | 93.8 |
| $10\frac{1}{2} \times 37\frac{1}{8}$ | 27 | 389.8 | 2412 | 44,770 | 682.2 | 3581 | 2.71 | 97.5 |
| $10\frac{1}{2} \times 38\frac{1}{2}$ | 28 | 404.3 | 2594 | 49,930 | 707.4 | 3714 | 2.81 | 101.1 |
| $10\frac{1}{2} \times 39\frac{7}{8}$ | 29 | 418.7 | 2783 | 55,480 | 732.7 | 3847 | 2.91 | 104.7 |
| $10\frac{1}{2} \times 41\frac{1}{4}$ | 30 | 433.1 | 2978 | 61,420 | 758.0 | 3979 | 3.01 | 108.3 |
| $10\frac{1}{2} \times 42\frac{5}{8}$ | 31 | 447.6 | 3180 | 67,760 | 783.2 | 4112 | 3.11 | 111.9 |
| $10\frac{1}{2} \times 44$ | 32 | 462.0 | 3388 | 74,540 | 808.5 | 4245 | 3.21 | 115.5 |
| $10\frac{1}{2} \times 45\frac{3}{8}$ | 33 | 476.4 | 3603 | 81,740 | 833.8 | 4377 | 3.31 | 119.1 |
| $10\frac{1}{2} \times 46\frac{3}{4}$ | 34 | 490.9 | 3825 | 89,400 | 859.0 | 4510 | 3.41 | 122.7 |
| $10\frac{1}{2} \times 48\frac{1}{8}$ | 35 | 505.3 | 4053 | 97,500 | 884.3 | 4643 | 3.51 | 126.3 |
| $10\frac{1}{2} \times 49\frac{1}{2}$ | 36 | 519.8 | 4288 | 106,100 | 909.6 | 4775 | 3.61 | 129.9 |
| $10\frac{1}{2} \times 50\frac{7}{8}$ | 37 | 534.2 | 4529 | 115,200 | 934.8 | 4908 | 3.71 | 133.6 |
| $10\frac{1}{2} \times 52\frac{1}{4}$ | 38 | 548.6 | 4778 | 124,800 | 960.1 | 5040 | 3.81 | 137.2 |
| $10\frac{1}{2} \times 53\frac{5}{8}$ | 39 | 563.1 | 5032 | 134,900 | 985.4 | 5173 | 3.91 | 140.8 |
| $10\frac{1}{2} \times 55$ | 40 | 577.5 | 5294 | 145,600 | 1011 | 5306 | 4.01 | 144.4 |
| $10\frac{1}{2} \times 56\frac{3}{8}$ | 41 | 591.9 | 5562 | 156,800 | 1036 | 5438 | 4.11 | 148.0 |
| $10\frac{1}{2} \times 57\frac{3}{4}$ | 42 | 606.4 | 5836 | 168,500 | 1061 | 5571 | 4.21 | 151.6 |
| $10\frac{1}{2} \times 59\frac{1}{8}$ | 43 | 620.8 | 6118 | 180,900 | 1086 | 5704 | 4.31 | 155.2 |
| $10\frac{1}{2} \times 60\frac{1}{2}$ | 44 | 635.3 | 6405 | 193,800 | 1112 | 5836 | 4.41 | 158.8 |
| $10\frac{1}{2} \times 61\frac{7}{8}$ | 45 | 649.7 | 6700 | 207,300 | 1137 | 5969 | 4.51 | 162.4 |

**TABLE 8.2** (*Continued*)
Southern Pine (Based on $1\frac{3}{8}$ in. Thick Laminations)

| Beam Size | | Number of Lams | A Area in.² | x–x Axis | | y–y Axis | | V Volume per ft cf/ft | SP[a] 36 pcf Weight per ft plf |
| b Width in. | d Depth in. | | | $S_x$ Section Modulus in.³ | $I_x$ Moment of Inertia in.⁴ | $S_y$ Section Modulus in.³ | $I_y$ Moment of Inertia in.⁴ | | |
|---|---|---|---|---|---|---|---|---|---|
| $10\frac{1}{2}$ × | $63\frac{1}{4}$ | 46 | 664.1 | 7001 | 221,400 | 1162 | 6102 | 4.61 | 166.0 |
| $10\frac{1}{2}$ × | $64\frac{5}{8}$ | 47 | 678.6 | 7309 | 236,200 | 1187 | 6234 | 4.71 | 169.6 |
| $10\frac{1}{2}$ × | $66$ | 48 | 693.0 | 7623 | 251,600 | 1213 | 6367 | 4.81 | 173.3 |
| $10\frac{1}{2}$ × | $67\frac{3}{8}$ | 49 | 707.4 | 7944 | 267,600 | 1238 | 6500 | 4.91 | 176.9 |
| $10\frac{1}{2}$ × | $68\frac{3}{4}$ | 50 | 721.9 | 8271 | 284,300 | 1263 | 6632 | 5.01 | 180.5 |
| $10\frac{1}{2}$ × | $70\frac{1}{8}$ | 51 | 736.3 | 8606 | 301,700 | 1289 | 6765 | 5.11 | 184.1 |
| $10\frac{1}{2}$ × | $71\frac{1}{2}$ | 52 | 750.8 | 8946 | 319,800 | 1314 | 6898 | 5.21 | 187.7 |
| $10\frac{1}{2}$ × | $72\frac{7}{8}$ | 53 | 765.2 | 9294 | 338,600 | 1339 | 7030 | 5.31 | 191.3 |
| $10\frac{1}{2}$ × | $74\frac{1}{4}$ | 54 | 779.6 | 9648 | 358,200 | 1364 | 7163 | 5.41 | 194.9 |
| $10\frac{1}{2}$ × | $75\frac{5}{8}$ | 55 | 794.1 | 10,010 | 378,400 | 1390 | 7295 | 5.51 | 198.5 |
| $10\frac{1}{2}$ × | $77$ | 56 | 808.5 | 10,380 | 399,500 | 1415 | 7428 | 5.61 | 202.1 |
| $10\frac{1}{2}$ × | $78\frac{3}{8}$ | 57 | 822.9 | 10,750 | 421,300 | 1440 | 7561 | 5.71 | 205.7 |
| $10\frac{1}{2}$ × | $79\frac{3}{4}$ | 58 | 837.4 | 11,130 | 443,800 | 1465 | 7693 | 5.82 | 209.3 |
| $10\frac{1}{2}$ × | $81\frac{1}{8}$ | 59 | 851.8 | 11,520 | 467,200 | 1491 | 7826 | 5.92 | 213.0 |
| $10\frac{1}{2}$ × | $82\frac{1}{2}$ | 60 | 866.3 | 11,910 | 491,300 | 1516 | 7959 | 6.02 | 216.6 |
| $10\frac{1}{2}$ × | $83\frac{7}{8}$ | 61 | 880.7 | 12,310 | 516,300 | 1541 | 8091 | 6.12 | 220.2 |
| $10\frac{1}{2}$ × | $85\frac{1}{4}$ | 62 | 895.1 | 12,720 | 542,100 | 1566 | 8224 | 6.22 | 223.8 |

| | | | | | | | | |
|---|---|---|---|---|---|---|---|---|
| $10\frac{1}{2} \times 86\frac{5}{8}$ | 63 | 909.6 | 13,130 | 568,800 | 1592 | 8357 | 6.32 | 227.4 |
| $10\frac{1}{2} \times 88$ | 64 | 924.0 | 13,550 | 596,300 | 1617 | 8489 | 6.42 | 231.0 |
| $10\frac{1}{2} \times 89\frac{3}{8}$ | 65 | 938.4 | 13,980 | 624,700 | 1642 | 8622 | 6.52 | 234.6 |
| $12 \times 2\frac{3}{4}$ | 2 | 33.00 | 15.13 | 20.80 | 66.00 | 396.0 | 0.23 | 8.3 |
| $12 \times 4\frac{1}{8}$ | 3 | 49.50 | 34.03 | 70.19 | 99.00 | 594.0 | 0.34 | 12.4 |
| $12 \times 5\frac{1}{2}$ | 4 | 66.00 | 60.50 | 166.4 | 132.0 | 792.0 | 0.46 | 16.5 |
| $12 \times 6\frac{7}{8}$ | 5 | 82.50 | 94.53 | 325.0 | 165.0 | 990.0 | 0.57 | 20.6 |
| $12 \times 8\frac{1}{4}$ | 6 | 99.00 | 136.1 | 561.5 | 198.0 | 1188 | 0.69 | 24.8 |
| $12 \times 9\frac{5}{8}$ | 7 | 115.5 | 185.3 | 891.7 | 231.0 | 1386 | 0.80 | 28.9 |
| $12 \times 11$ | 8 | 132.0 | 242.0 | 1331 | 264.0 | 1584 | 0.92 | 33.0 |
| $12 \times 12\frac{3}{8}$ | 9 | 148.5 | 306.3 | 1895 | 297.0 | 1782 | 1.03 | 37.1 |
| $12 \times 13\frac{3}{4}$ | 10 | 165.0 | 378.1 | 2600 | 330.0 | 1980 | 1.15 | 41.3 |
| $12 \times 15\frac{1}{8}$ | 11 | 181.5 | 457.5 | 3460 | 363.0 | 2178 | 1.26 | 45.4 |
| $12 \times 16\frac{1}{2}$ | 12 | 198.0 | 544.5 | 4492 | 396.0 | 2376 | 1.38 | 49.5 |
| $12 \times 17\frac{7}{8}$ | 13 | 214.5 | 639.0 | 5711 | 429.0 | 2574 | 1.49 | 53.6 |
| $12 \times 19\frac{1}{4}$ | 14 | 231.0 | 741.1 | 7133 | 462.0 | 2772 | 1.60 | 57.8 |
| $12 \times 20\frac{5}{8}$ | 15 | 247.5 | 850.8 | 8774 | 495.0 | 2970 | 1.72 | 61.9 |
| $12 \times 22$ | 16 | 264.0 | 968.0 | 10,650 | 528.0 | 3168 | 1.83 | 66.0 |
| $12 \times 23\frac{3}{8}$ | 17 | 280.5 | 1093 | 12,770 | 561.0 | 3366 | 1.95 | 70.1 |
| $12 \times 24\frac{3}{4}$ | 18 | 297.0 | 1225 | 15,160 | 594.0 | 3564 | 2.06 | 74.3 |
| $12 \times 26\frac{1}{8}$ | 19 | 313.5 | 1365 | 17,830 | 627.0 | 3762 | 2.18 | 78.4 |
| $12 \times 27\frac{1}{2}$ | 20 | 330.0 | 1513 | 20,800 | 660.0 | 3960 | 2.29 | 82.5 |
| $12 \times 28\frac{7}{8}$ | 21 | 346.5 | 1668 | 24,070 | 693.0 | 4158 | 2.41 | 86.6 |
| $12 \times 30\frac{1}{4}$ | 22 | 363.0 | 1830 | 27,680 | 726.0 | 4356 | 2.52 | 90.8 |

**TABLE 8.2** (*Continued*)

**Southern Pine (Based on $1\frac{3}{8}$ in. Thick Laminations)**

| Beam Size b Width in. | d Depth in. | Number of Lams | A Area in.² | x–x Axis $S_x$ Section Modulus in.³ | $I_x$ Moment of Inertia in.⁴ | y–y Axis $S_y$ Section Modulus in.³ | $I_y$ Moment of Inertia in.⁴ | V Volume per ft cf/ft | SP[a] 36 pcf Weight per ft plf |
|---|---|---|---|---|---|---|---|---|---|
| 12 × | $31\frac{5}{8}$ | 23 | 379.5 | 2000 | 31,630 | 759.0 | 4554 | 2.64 | 94.9 |
| 12 × | 33 | 24 | 396.0 | 2178 | 35,940 | 792.0 | 4752 | 2.75 | 99.0 |
| 12 × | $34\frac{3}{8}$ | 25 | 412.5 | 2363 | 40,620 | 825.0 | 4950 | 2.86 | 103.1 |
| 12 × | $35\frac{3}{4}$ | 26 | 429.0 | 2556 | 45,690 | 858.0 | 5148 | 2.98 | 107.3 |
| 12 × | $37\frac{1}{8}$ | 27 | 445.5 | 2757 | 51,170 | 891.0 | 5346 | 3.09 | 111.4 |
| 12 × | $38\frac{1}{2}$ | 28 | 462.0 | 2965 | 57,070 | 924.0 | 5544 | 3.21 | 115.5 |
| 12 × | $39\frac{7}{8}$ | 29 | 478.5 | 3180 | 63,400 | 957.0 | 5742 | 3.32 | 119.6 |
| 12 × | $41\frac{1}{4}$ | 30 | 495.0 | 3403 | 70,190 | 990.0 | 5940 | 3.44 | 123.8 |
| 12 × | $42\frac{5}{8}$ | 31 | 511.5 | 3634 | 77,400 | 1023 | 6138 | 3.55 | 127.9 |
| 12 × | 44 | 32 | 528.0 | 3872 | 85,200 | 1056 | 6336 | 3.67 | 132.0 |
| 12 × | $45\frac{3}{8}$ | 33 | 544.5 | 4118 | 93,400 | 1089 | 6534 | 3.78 | 136.1 |
| 12 × | $46\frac{3}{4}$ | 34 | 561.0 | 4371 | 102,200 | 1122 | 6732 | 3.90 | 140.3 |
| 12 × | $48\frac{1}{8}$ | 35 | 577.5 | 4632 | 111,500 | 1155 | 6930 | 4.01 | 144.4 |
| 12 × | $49\frac{1}{2}$ | 36 | 594.0 | 4901 | 121,300 | 1188 | 7128 | 4.13 | 148.5 |
| 12 × | $50\frac{7}{8}$ | 37 | 610.5 | 5177 | 131,700 | 1221 | 7326 | 4.24 | 152.6 |
| 12 × | $52\frac{1}{4}$ | 38 | 627.0 | 5460 | 142,600 | 1254 | 7524 | 4.35 | 156.8 |
| 12 × | $53\frac{5}{8}$ | 39 | 643.5 | 5751 | 154,200 | 1287 | 7722 | 4.47 | 160.9 |

| $12 \times$ | | No. | | | | | | | |
|---|---|---|---|---|---|---|---|---|---|
| $55$ | 40 | 660.0 | 6050 | 166,400 | 1320 | 7920 | 4.58 | 165.0 |
| $56\frac{3}{8}$ | 41 | 676.5 | 6356 | 179,200 | 1353 | 8118 | 4.70 | 169.1 |
| $57\frac{3}{4}$ | 42 | 693.0 | 6670 | 192,600 | 1386 | 8316 | 4.81 | 173.3 |
| $59\frac{1}{8}$ | 43 | 709.5 | 6992 | 206,700 | 1419 | 8514 | 4.93 | 177.4 |
| $60\frac{1}{2}$ | 44 | 726.0 | 7321 | 221,400 | 1452 | 8712 | 5.04 | 181.5 |
| $61\frac{7}{8}$ | 45 | 742.5 | 7657 | 236,900 | 1485 | 8910 | 5.16 | 185.6 |
| $63\frac{1}{4}$ | 46 | 759.0 | 8001 | 253,000 | 1518 | 9108 | 5.27 | 189.8 |
| $64\frac{5}{8}$ | 47 | 775.5 | 8353 | 269,900 | 1551 | 9306 | 5.39 | 193.9 |
| $66$ | 48 | 792.0 | 8712 | 287,500 | 1584 | 9504 | 5.50 | 198.0 |
| $67\frac{3}{8}$ | 49 | 808.5 | 9079 | 305,800 | 1617 | 9702 | 5.61 | 202.1 |
| $68\frac{3}{4}$ | 50 | 825.0 | 9453 | 325,000 | 1650 | 9900 | 5.73 | 206.3 |
| $70\frac{1}{8}$ | 51 | 841.5 | 9835 | 344,800 | 1683 | 10,100 | 5.84 | 210.4 |
| $71\frac{1}{2}$ | 52 | 858.0 | 10,220 | 365,500 | 1716 | 10,300 | 5.96 | 214.5 |
| $72\frac{7}{8}$ | 53 | 874.5 | 10,620 | 387,000 | 1749 | 10,490 | 6.07 | 218.6 |
| $74\frac{1}{4}$ | 54 | 891.0 | 11,030 | 409,300 | 1782 | 10,690 | 6.19 | 222.8 |
| $75\frac{5}{8}$ | 55 | 907.5 | 11,440 | 432,500 | 1815 | 10,890 | 6.30 | 226.9 |
| $77$ | 56 | 924.0 | 11,860 | 456,500 | 1848 | 11,090 | 6.42 | 231.0 |
| $78\frac{3}{8}$ | 57 | 940.5 | 12,290 | 481,400 | 1881 | 11,290 | 6.53 | 235.1 |
| $79\frac{3}{4}$ | 58 | 957.0 | 12,720 | 507,200 | 1914 | 11,480 | 6.65 | 239.3 |
| $81\frac{1}{8}$ | 59 | 973.5 | 13,160 | 533,900 | 1947 | 11,680 | 6.76 | 243.4 |
| $82\frac{1}{2}$ | 60 | 990.0 | 13,610 | 561,500 | 1980 | 11,880 | 6.88 | 247.5 |
| $83\frac{7}{8}$ | 61 | 1007 | 14,070 | 590,100 | 2013 | 12,080 | 6.99 | 251.6 |
| $85\frac{1}{4}$ | 62 | 1023 | 14,540 | 619,600 | 2046 | 12,280 | 7.10 | 255.8 |
| $86\frac{5}{8}$ | 63 | 1040 | 15,010 | 650,000 | 2079 | 12,470 | 7.22 | 259.9 |

**TABLE 8.2** (*Continued*)

**Southern Pine (Based on $1\frac{3}{8}$ in. Thick Laminations)**

| Beam Size b Width in. | d Depth in. | Number of Lams | A Area in.² | $S_x$ Section Modulus in.³ | $I_x$ Moment of Inertia in.⁴ | $S_y$ Section Modulus in.³ | $I_y$ Moment of Inertia in.⁴ | V Volume per ft cf/ft | SP[a] 36 pcf Weight per ft plf |
|---|---|---|---|---|---|---|---|---|---|
| 12 × | 88 | 64 | 1056 | 15,490 | 681,500 | 2112 | 12,670 | 7.33 | 264.0 |
| 12 × | $89\frac{3}{8}$ | 65 | 1073 | 15,980 | 713,900 | 2145 | 12,870 | 7.45 | 268.1 |
| 12 × | $90\frac{3}{4}$ | 66 | 1089 | 16,470 | 747,400 | 2178 | 13,070 | 7.56 | 272.3 |
| 12 × | $92\frac{1}{8}$ | 67 | 1106 | 16,970 | 781,900 | 2211 | 13,270 | 7.68 | 276.4 |
| 12 × | $93\frac{1}{2}$ | 68 | 1122 | 17,480 | 817,400 | 2244 | 13,460 | 7.79 | 280.5 |
| 12 × | $94\frac{7}{8}$ | 69 | 1139· | 18,000 | 854,000 | 2277 | 13,660 | 7.91 | 284.6 |
| 12 × | $96\frac{1}{4}$ | 70 | 1155 | 18,530 | 891,700 | 2310 | 13,860 | 8.02 | 288.8 |
| 12 × | $97\frac{5}{8}$ | 71 | 1172 | 19,060 | 930,400 | 2343 | 14,060 | 8.14 | 292.9 |
| 12 × | 99 | 72 | 1188 | 19,600 | 970,300 | 2376 | 14,260 | 8.25 | 297.0 |
| 12 × | $100\frac{3}{8}$ | 73 | 1205 | 20,150 | 1,011,000 | 2409 | 14,450 | 8.36 | 301.1 |
| 12 × | $101\frac{3}{4}$ | 74 | 1221 | 20,710 | 1,053,000 | 2442 | 14,650 | 8.48 | 305.3 |
| 12 × | $103\frac{1}{8}$ | 75 | 1238 | 21,270 | 1,097,000 | 2475 | 14,850 | 8.59 | 309.4 |
| 14 × | $2\frac{3}{4}$ | 2 | 38.50 | 17.65 | 24.26 | 89.83 | 628.8 | 0.27 | 9.6 |
| 14 × | $4\frac{1}{8}$ | 3 | 57.75 | 39.70 | 81.89 | 134.8 | 943.3 | 0.40 | 14.4 |
| 14 × | $5\frac{1}{2}$ | 4 | 77.00 | 70.58 | 194.1 | 179.7 | 1258 | 0.53 | 19.3 |
| 14 × | $6\frac{7}{8}$ | 5 | 96.25 | 110.3 | 379.1 | 224.6 | 1572 | 0.67 | 24.1 |

| | | | | | | | | | | |
|---|---|---|---|---|---|---|---|---|---|---|
| 14 | × | $8\frac{1}{4}$ | 6 | 115.5 | 158.8 | 655.1 | 269.5 | 1887 | 0.80 | 28.9 |
| 14 | × | $9\frac{5}{8}$ | 7 | 134.8 | 216.2 | 1040 | 314.4 | 2201 | 0.94 | 33.7 |
| 14 | × | 11 | 8 | 154.0 | 282.3 | 1553 | 359.3 | 2515 | 1.07 | 38.5 |
| 14 | × | $12\frac{3}{8}$ | 9 | 173.3 | 357.3 | 2211 | 404.3 | 2830 | 1.20 | 43.3 |
| 14 | × | $13\frac{3}{4}$ | 10 | 192.5 | 441.1 | 3033 | 449.2 | 3144 | 1.34 | 48.1 |
| 14 | × | $15\frac{1}{8}$ | 11 | 211.8 | 533.8 | 4037 | 494.1 | 3459 | 1.47 | 52.9 |
| 14 | × | $16\frac{1}{2}$ | 12 | 231.0 | 635.3 | 5241 | 539.0 | 3773 | 1.60 | 57.8 |
| 14 | × | $17\frac{7}{8}$ | 13 | 250.3 | 745.5 | 6663 | 583.9 | 4087 | 1.74 | 62.6 |
| 14 | × | $19\frac{1}{4}$ | 14 | 269.5 | 864.6 | 8322 | 628.8 | 4402 | 1.87 | 67.4 |
| 14 | × | $20\frac{5}{8}$ | 15 | 288.8 | 992.6 | 10,240 | 673.8 | 4716 | 2.01 | 72.2 |
| 14 | × | 22 | 16 | 308.0 | 1129 | 12,420 | 718.7 | 5031 | 2.14 | 77.0 |
| 14 | × | $23\frac{3}{8}$ | 17 | 327.3 | 1275 | 14,900 | 763.6 | 5345 | 2.27 | 81.8 |
| 14 | × | $24\frac{3}{4}$ | 18 | 346.5 | 1429 | 17,690 | 808.5 | 5660 | 2.41 | 86.6 |
| 14 | × | $26\frac{1}{8}$ | 19 | 365.8 | 1593 | 20,800 | 853.4 | 5974 | 2.54 | 91.4 |
| 14 | × | $27\frac{1}{2}$ | 20 | 385.0 | 1765 | 24,260 | 898.3 | 6288 | 2.67 | 96.3 |
| 14 | × | $28\frac{7}{8}$ | 21 | 404.3 | 1945 | 28,090 | 943.3 | 6603 | 2.81 | 101.1 |
| 14 | × | $30\frac{1}{4}$ | 22 | 423.5 | 2135 | 32,290 | 988.2 | 6917 | 2.94 | 105.9 |
| 14 | × | $31\frac{5}{8}$ | 23 | 442.8 | 2334 | 36,900 | 1033 | 7232 | 3.07 | 110.7 |
| 14 | × | 33 | 24 | 462.0 | 2541 | 41,930 | 1078 | 7546 | 3.21 | 115.5 |
| 14 | × | $34\frac{3}{8}$ | 25 | 481.3 | 2757 | 47,390 | 1123 | 7860 | 3.34 | 120.3 |
| 14 | × | $35\frac{3}{4}$ | 26 | 500.5 | 2982 | 53,310 | 1168 | 8175 | 3.48 | 125.1 |
| 14 | × | $37\frac{1}{8}$ | 27 | 519.8 | 3216 | 59,700 | 1213 | 8489 | 3.61 | 129.9 |
| 14 | × | $38\frac{1}{2}$ | 28 | 539.0 | 3459 | 66,580 | 1258 | 8804 | 3.74 | 134.8 |
| 14 | × | $39\frac{7}{8}$ | 29 | 558.3 | 3710 | 73,970 | 1303 | 9118 | 3.88 | 139.6 |

## TABLE 8.2 (*Continued*)
### Southern Pine (Based on $1\frac{3}{8}$ in. Thick Laminations)

| Beam Size | | Number of Lams | A Area in.² | x-x Axis | | y-y Axis | | V Volume per ft cf/ft | SPª 36 pcf Weight per ft plf |
| --- | --- | --- | --- | --- | --- | --- | --- | --- | --- |
| b Width in. | d Depth in. | | | $S_x$ Section Modulus in.³ | $I_x$ Moment of Inertia in.⁴ | $S_y$ Section Modulus in.³ | $I_y$ Moment of Inertia in.⁴ | | |
| 14 × | $41\frac{1}{4}$ | 30 | 577.5 | 3970 | 81,900 | 1348 | 9433 | 4.01 | 144.4 |
| 14 × | $42\frac{5}{8}$ | 31 | 596.8 | 4239 | 90,400 | 1392 | 9747 | 4.14 | 149.2 |
| 14 × | 44 | 32 | 616.0 | 4517 | 99,400 | 1437 | 10,060 | 4.28 | 154.0 |
| 14 × | $45\frac{3}{8}$ | 33 | 635.3 | 4804 | 109,000 | 1482 | 10,380 | 4.41 | 158.8 |
| 14 × | $46\frac{3}{4}$ | 34 | 654.5 | 5100 | 119,200 | 1527 | 10,690 | 4.55 | 163.6 |
| 14 × | $48\frac{1}{8}$ | 35 | 673.8 | 5404 | 130,000 | 1572 | 11,000 | 4.68 | 168.4 |
| 14 × | $49\frac{1}{2}$ | 36 | 693.0 | 5717 | 141,500 | 1617 | 11,320 | 4.81 | 173.3 |
| 14 × | $50\frac{7}{8}$ | 37 | 712.3 | 6039 | 153,600 | 1662 | 11,630 | 4.95 | 178.1 |
| 14 × | $52\frac{1}{4}$ | 38 | 731.5 | 6370 | 166,400 | 1707 | 11,950 | 5.08 | 182.9 |
| 14 × | $53\frac{5}{8}$ | 39 | 750.8 | 6710 | 179,900 | 1752 | 12,260 | 5.21 | 187.7 |
| 14 × | 55 | 40 | 770.0 | 7058 | 194,100 | 1797 | 12,580 | 5.35 | 192.5 |
| 14 × | $56\frac{3}{8}$ | 41 | 789.3 | 7416 | 209,000 | 1842 | 12,890 | 5.48 | 197.3 |
| 14 × | $57\frac{3}{4}$ | 42 | 808.5 | 7782 | 224,700 | 1887 | 13,210 | 5.61 | 202.1 |
| 14 × | $59\frac{1}{8}$ | 43 | 827.8 | 8157 | 241,100 | 1931 | 13,520 | 5.75 | 206.9 |
| 14 × | $60\frac{1}{2}$ | 44 | 847.0 | 8541 | 258,400 | 1976 | 13,830 | 5.88 | 211.8 |
| 14 × | $61\frac{7}{8}$ | 45 | 866.3 | 8933 | 276,400 | 2021 | 14,150 | 6.02 | 216.6 |
| 14 × | $63\frac{1}{4}$ | 46 | 885.5 | 9335 | 295,200 | 2066 | 14,460 | 6.15 | 221.4 |

| | | | | | | | | |
|---|---|---|---|---|---|---|---|---|
| 14 × 64⅝ | 47 | 904.8 | 9745 | 314,900 | 2111 | 14,780 | 6.28 | 226.2 |
| 14 × 66 | 48 | 924.0 | 10,160 | 335,400 | 2156 | 15,090 | 6.42 | 231.0 |
| 14 × 67⅜ | 49 | 943.3 | 10,590 | 356,800 | 2201 | 15,410 | 6.55 | 235.8 |
| 14 × 68¾ | 50 | 962.5 | 11,030 | 379,100 | 2246 | 15,720 | 6.68 | 240.6 |
| 14 × 70⅛ | 51 | 981.8 | 11,470 | 402,300 | 2291 | 16,040 | 6.82 | 245.4 |
| 14 × 71½ | 52 | 1001 | 11,930 | 426,400 | 2336 | 16,350 | 6.95 | 250.3 |
| 14 × 72⅞ | 53 | 1020 | 12,390 | 451,500 | 2381 | 16,660 | 7.09 | 255.1 |
| 14 × 74¼ | 54 | 1040 | 12,860 | 477,600 | 2426 | 16,980 | 7.22 | 259.9 |
| 14 × 75⅝ | 55 | 1059 | 13,340 | 504,600 | 2470 | 17,290 | 7.35 | 264.7 |
| 14 × 77 | 56 | 1078 | 13,830 | 532,600 | 2515 | 17,610 | 7.49 | 269.5 |
| 14 × 78⅜ | 57 | 1097 | 14,330 | 561,700 | 2560 | 17,920 | 7.62 | 274.3 |
| 14 × 79¾ | 58 | 1117 | 14,840 | 591,800 | 2605 | 18,240 | 7.75 | 279.1 |
| 14 × 81⅛ | 59 | 1136 | 15,360 | 622,900 | 2650 | 18,550 | 7.89 | 283.9 |
| 14 × 82½ | 60 | 1155 | 15,880 | 655,100 | 2695 | 18,870 | 8.02 | 288.8 |
| 14 × 83⅞ | 61 | 1174 | 16,420 | 688,400 | 2740 | 19,180 | 8.15 | 293.6 |
| 14 × 85¼ | 62 | 1194 | 16,960 | 722,800 | 2785 | 19,490 | 8.29 | 298.4 |
| 14 × 86⅝ | 63 | 1213 | 17,510 | 758,000 | 2830 | 19,810 | 8.42 | 303.2 |
| 14 × 88 | 64 | 1232 | 18,070 | 795,000 | 2875 | 20,120 | 8.56 | 308.0 |
| 14 × 89⅜ | 65 | 1251 | 18,640 | 833,000 | 2920 | 20,440 | 8.69 | 312.8 |
| 14 × 90¾ | 66 | 1271 | 19,220 | 872,000 | 2965 | 20,750 | 8.82 | 317.6 |
| 14 × 92⅛ | 67 | 1290 | 19,800 | 912,000 | 3009 | 21,070 | 8.96 | 322.4 |
| 14 × 93½ | 68 | 1309 | 20,400 | 954,000 | 3054 | 21,380 | 9.09 | 327.3 |
| 14 × 94⅞ | 69 | 1328 | 21,000 | 996,000 | 3099 | 21,690 | 9.22 | 332.1 |
| 14 × 96¼ | 70 | 1348 | 21,620 | 1,040,000 | 3144 | 22,010 | 9.36 | 336.9 |
| 14 × 97⅝ | 71 | 1367 | 22,240 | 1,086,000 | 3189 | 22,320 | 9.49 | 341.7 |

**TABLE 8.2** (*Continued*)

**Southern Pine (Based on $1\frac{3}{8}$ in. Thick Laminations)**

| Beam Size | | Number of Lams | A Area in.$^2$ | x-x Axis | | y-y Axis | | V Volume per ft cf/ft | SP$^a$ 36 pcf Weight per ft plf |
| --- | --- | --- | --- | --- | --- | --- | --- | --- | --- |
| b Width in. | d Depth in. | | | $S_x$ Section Modulus in.$^3$ | $I_x$ Moment of Inertia in.$^4$ | $S_y$ Section Modulus in.$^3$ | $I_y$ Moment of Inertia in.$^4$ | | |
| 14 × | 99 | 72 | 1386 | 22,870 | 1,132,000 | 3234 | 22,640 | 9.63 | 346.5 |
| 14 × | $100\frac{3}{8}$ | 73 | 1405 | 23,510 | 1,180,000 | 3279 | 22,950 | 9.76 | 351.3 |
| 14 × | $101\frac{3}{4}$ | 74 | 1425 | 24,160 | 1,229,000 | 3324 | 23,270 | 9.89 | 356.1 |
| 14 × | $103\frac{1}{8}$ | 75 | 1444 | 24,810 | 1,279,000 | 3369 | 23,580 | 10.03 | 360.9 |
| 14 × | $104\frac{1}{2}$ | 76 | 1463 | 25,480 | 1,331,000 | 3414 | 23,900 | 10.16 | 365.8 |
| 14 × | $105\frac{7}{8}$ | 77 | 1482 | 26,160 | 1,385,000 | 3459 | 24,210 | 10.29 | 370.6 |
| 14 × | $107\frac{1}{4}$ | 78 | 1502 | 26,840 | 1,439,000 | 3504 | 24,520 | 10.43 | 375.4 |
| 14 × | $108\frac{5}{8}$ | 79 | 1521 | 27,530 | 1,495,000 | 3548 | 24,840 | 10.56 | 380.2 |
| 14 × | 110 | 80 | 1540 | 28,230 | 1,553,000 | 3593 | 25,150 | 10.69 | 385.0 |
| 14 × | $111\frac{3}{8}$ | 81 | 1559 | 28,940 | 1,612,000 | 3638 | 25,470 | 10.83 | 389.8 |
| 14 × | $112\frac{3}{4}$ | 82 | 1579 | 29,660 | 1,672,000 | 3683 | 25,780 | 10.96 | 394.6 |
| 14 × | $114\frac{1}{8}$ | 83 | 1598 | 30,390 | 1,734,000 | 3728 | 26,100 | 11.10 | 399.4 |
| 14 × | $115\frac{1}{2}$ | 84 | 1617 | 31,130 | 1,798,000 | 3773 | 26,410 | 11.23 | 404.3 |
| 14 × | $116\frac{7}{8}$ | 85 | 1636 | 31,870 | 1,863,000 | 3818 | 26,730 | 11.36 | 409.1 |
| 14 × | $118\frac{1}{4}$ | 86 | 1656 | 32,630 | 1,929,000 | 3863 | 27,040 | 11.50 | 413.9 |
| 14 × | $119\frac{5}{8}$ | 87 | 1675 | 33,390 | 1,997,000 | 3908 | 27,350 | 11.63 | 418.7 |

$^a$DF = Douglas Fir-Larch; HF = Hem-Fir; and SP = Southern Pine.

# 8.3 DESIGN VALUES FOR SAWN LUMBER

## TABLE 8.3

### Base Design Values for Visually Graded Dimension Lumber

(All species except Southern Pine—see Table 8.4)

(Tabulated design values are for normal load duration and dry service conditions. See NDS 2.3 for a comprehensive description of design value adjustment factors.)

Use with Table 8.3 Adjustment Factors

| Species and commercial grade | Size classification | Design values in pounds per square inch (psi) | | | | | | Grading Rules Agency |
| | | Bending $F_b$ | Tension parallel to grain $F_t$ | Shear parallel to grain $F_v$ | Compression perpendicular to grain $F_{c\perp}$ | Compression parallel to grain $F_c$ | Modulus of Elasticity E | |
| **ASPEN** | | | | | | | | |
| Select Structural | 2"-4" thick | 875 | 500 | 60 | 265 | 725 | 1,100,000 | |
| No.1 | | 625 | 375 | 60 | 265 | 600 | 1,100,000 | NELMA |
| No.2 | | 600 | 350 | 60 | 265 | 450 | 1,000,000 | NSLB |
| No.3 | 2"& wider | 350 | 200 | 60 | 265 | 275 | 900,000 | WWPA |
| Stud | | 475 | 275 | 60 | 265 | 300 | 900,000 | |
| Construction | 2"-4" thick | 700 | 400 | 60 | 265 | 625 | 900,000 | |
| Standard | | 375 | 225 | 60 | 265 | 475 | 900,000 | |
| Utility | 2"-4" wide | 175 | 100 | 60 | 265 | 300 | 800,000 | |
| **BEECH-BIRCH-HICKORY** | | | | | | | | |
| Select Structural | 2"-4" thick | 1450 | 850 | 100 | 715 | 1200 | 1,700,000 | |
| No.1 | | 1050 | 600 | 100 | 715 | 950 | 1,600,000 | |
| No.2 | | 1000 | 600 | 100 | 715 | 750 | 1,500,000 | NELMA |
| No.3 | 2"& wider | 575 | 350 | 100 | 715 | 425 | 1,300,000 | |
| Stud | | 775 | 450 | 100 | 715 | 475 | 1,300,000 | |
| Construction | 2"-4" thick | 1150 | 675 | 100 | 715 | 1000 | 1,400,000 | |
| Standard | | 650 | 375 | 100 | 715 | 775 | 1,300,000 | |
| Utility | 2"-4" wide | 300 | 175 | 100 | 715 | 500 | 1,200,000 | |

## TABLE 8.3 (Continued)

| Species and commercial grade | Size classification | Design values in pounds per square inch (psi) | | | | | | Grading Rules Agency |
|---|---|---|---|---|---|---|---|---|
| | | Bending $F_b$ | Tension parallel to grain $F_t$ | Shear parallel to grain $F_v$ | Compression perpendicular to grain $F_{c\perp}$ | Compression parallel to grain $F_c$ | Modulus of Elasticity E | |
| **COTTONWOOD** | | | | | | | | |
| Select Structural | 2"-4"thick | 875 | 525 | 65 | 320 | 775 | 1,200,000 | NSLB |
| No.1 | | 625 | 375 | 65 | 320 | 625 | 1,200,000 | |
| No.2 | | 625 | 350 | 65 | 320 | 475 | 1,100,000 | |
| No.3 | 2"& wider | 350 | 200 | 65 | 320 | 275 | 1,000,000 | |
| Stud | | 475 | 275 | 65 | 320 | 300 | 1,000,000 | |
| Construction | 2"-4"thick | 700 | 400 | 65 | 320 | 650 | 1,000,000 | |
| Standard | | 400 | 225 | 65 | 320 | 500 | 900,000 | |
| Utility | 2"-4" wide | 175 | 100 | 65 | 320 | 325 | 900,000 | |
| **DOUGLAS FIR-LARCH** | | | | | | | | |
| Select Structural | 2"-4"thick | 1450 | 1000 | 95 | 625 | 1700 | 1,900,000 | WCLIB WWPA |
| No.1 & Btr | | 1150 | 775 | 95 | 625 | 1500 | 1,800,000 | |
| No.1 | | 1000 | 675 | 95 | 625 | 1450 | 1,700,000 | |
| No.2 | 2"& wider | 875 | 575 | 95 | 625 | 1300 | 1,600,000 | |
| No.3 | | 500 | 325 | 95 | 625 | 750 | 1,400,000 | |
| Stud | | 675 | 450 | 95 | 625 | 825 | 1,400,000 | |
| Construction | 2"-4"thick | 1000 | 650 | 95 | 625 | 1600 | 1,500,000 | |
| Standard | | 550 | 375 | 95 | 625 | 1350 | 1,400,000 | |
| Utility | 2"-4" wide | 275 | 175 | 95 | 625 | 875 | 1,300,000 | |
| **DOUGLAS FIR-LARCH (NORTH)** | | | | | | | | |
| Select Structural | 2"-4"thick | 1300 | 800 | 95 | 625 | 1900 | 1,900,000 | NLGA |
| No.1/No.2 | | 825 | 500 | 95 | 625 | 1350 | 1,600,000 | |
| No.3 | 2"& wider | 475 | 300 | 95 | 625 | 775 | 1,400,000 | |
| Stud | | 650 | 375 | 95 | 625 | 850 | 1,400,000 | |
| Construction | 2"-4"thick | 950 | 575 | 95 | 625 | 1750 | 1,500,000 | |
| Standard | | 525 | 325 | 95 | 625 | 1400 | 1,400,000 | |
| Utility | 2"-4" wide | 250 | 150 | 95 | 625 | 925 | 1,300,000 | |

## DOUGLAS FIR-SOUTH

| Grade | Size | | | | | | | Agency |
|---|---|---|---|---|---|---|---|---|
| Select Structural | | 1300 | 875 | 90 | 520 | 1550 | 1,400,000 | WWPA |
| No.1 | 2"-4" thick | 900 | 600 | 90 | 520 | 1400 | 1,300,000 | |
| No.2 | | 825 | 525 | 90 | 520 | 1300 | 1,200,000 | |
| No.3 | 2" & wider | 475 | 300 | 90 | 520 | 750 | 1,100,000 | |
| Stud | 2"-4" thick | 650 | 425 | 90 | 520 | 825 | 1,100,000 | |
| Construction | 2"-4" thick | 925 | 600 | 90 | 520 | 1550 | 1,200,000 | |
| Standard | | 525 | 350 | 90 | 520 | 1300 | 1,100,000 | |
| Utility | 2"-4" wide | 250 | 150 | 90 | 520 | 875 | 1,000,000 | |

## EASTERN HEMLOCK-TAMARACK

| Grade | Size | | | | | | | Agency |
|---|---|---|---|---|---|---|---|---|
| Select Structural | 2"-4" thick | 1250 | 575 | 85 | 555 | 1200 | 1,200,000 | NELMA NSLB |
| No.1 | | 775 | 350 | 85 | 555 | 1000 | 1,100,000 | |
| No.2 | 2" & wider | 575 | 275 | 85 | 555 | 825 | 1,100,000 | |
| No.3 | | 350 | 150 | 85 | 555 | 475 | 900,000 | |
| Stud | 2"-4" thick | 450 | 200 | 85 | 555 | 525 | 900,000 | |
| Construction | 2"-4" thick | 675 | 300 | 85 | 555 | 1050 | 1,000,000 | |
| Standard | | 375 | 175 | 85 | 555 | 850 | 900,000 | |
| Utility | 2"-4" wide | 175 | 75 | 85 | 555 | 550 | 800,000 | |

## EASTERN SOFTWOODS

| Grade | Size | | | | | | | Agency |
|---|---|---|---|---|---|---|---|---|
| Select Structural | 2"-4" thick | 1250 | 575 | 70 | 335 | 1200 | 1,200,000 | NELMA NSLB |
| No.1 | | 775 | 350 | 70 | 335 | 1000 | 1,100,000 | |
| No.2 | 2" & wider | 575 | 275 | 70 | 335 | 825 | 1,100,000 | |
| No.3 | | 350 | 150 | 70 | 335 | 475 | 900,000 | |
| Stud | 2"-4" thick | 450 | 200 | 70 | 335 | 525 | 900,000 | |
| Construction | 2"-4" thick | 675 | 300 | 70 | 335 | 1050 | 1,000,000 | |
| Standard | | 375 | 175 | 70 | 335 | 850 | 900,000 | |
| Utility | 2"-4" wide | 175 | 75 | 70 | 335 | 550 | 800,000 | |

## EASTERN WHITE PINE

| Grade | Size | | | | | | | Agency |
|---|---|---|---|---|---|---|---|---|
| Select Structural | 2"-4" thick | 1250 | 575 | 70 | 350 | 1200 | 1,200,000 | NELMA NSLB |
| No.1 | | 775 | 350 | 70 | 350 | 1000 | 1,100,000 | |
| No.2 | 2" & wider | 575 | 275 | 70 | 350 | 825 | 1,100,000 | |
| No.3 | | 350 | 150 | 70 | 350 | 475 | 900,000 | |
| Stud | 2"-4" thick | 450 | 200 | 70 | 350 | 525 | 900,000 | |
| Construction | 2"-4" thick | 675 | 300 | 70 | 350 | 1050 | 1,000,000 | |
| Standard | | 375 | 175 | 70 | 350 | 850 | 900,000 | |
| Utility | 2"-4" wide | 175 | 75 | 70 | 350 | 550 | 800,000 | |

**TABLE 8.3** (*Continued*)

| Species and commercial grade | Size classification | Design values in pounds per square inch (psi) | | | | | | Grading Rules Agency |
|---|---|---|---|---|---|---|---|---|
| | | Bending $F_b$ | Tension parallel to grain $F_t$ | Shear parallel to grain $F_v$ | Compression perpendicular to grain $F_{c\perp}$ | Compression parallel to grain $F_c$ | Modulus of Elasticity E | |
| **HEM-FIR** | | | | | | | | |
| Select Structural | 2"-4"thick | 1400 | 900 | 75 | 405 | 1500 | 1,600,000 | WCLIB WWPA |
| No.1 & Btr | | 1050 | 700 | 75 | 405 | 1350 | 1,500,000 | |
| No.1 | | 950 | 600 | 75 | 405 | 1300 | 1,500,000 | |
| No.2 | 2"& wider | 850 | 500 | 75 | 405 | 1250 | 1,300,000 | |
| No.3 | | 500 | 300 | 75 | 405 | 725 | 1,200,000 | |
| Stud | 2"-4"thick | 675 | 400 | 75 | 405 | 800 | 1,200,000 | |
| Construction | | 975 | 575 | 75 | 405 | 1500 | 1,300,000 | |
| Standard | 2"-4" wide | 550 | 325 | 75 | 405 | 1300 | 1,200,000 | |
| Utility | | 250 | 150 | 75 | 405 | 850 | 1,100,000 | |
| **HEM-FIR (NORTH)** | | | | | | | | |
| Select Structural | 2"-4"thick | 1300 | 775 | 75 | 370 | 1650 | 1,700,000 | NLGA |
| No.1/No.2 | | 1000 | 550 | 75 | 370 | 1450 | 1,600,000 | |
| No.3 | 2"& wider | 575 | 325 | 75 | 370 | 850 | 1,400,000 | |
| Stud | 2"-4"thick | 775 | 425 | 75 | 370 | 925 | 1,400,000 | |
| Construction | | 1150 | 625 | 75 | 370 | 1750 | 1,500,000 | |
| Standard | 2"-4" wide | 625 | 350 | 75 | 370 | 1500 | 1,400,000 | |
| Utility | | 300 | 175 | 75 | 370 | 975 | 1,300,000 | |
| **MIXED MAPLE** | | | | | | | | |
| Select Structural | 2"-4"thick | 1000 | 600 | 100 | 620 | 875 | 1,300,000 | NELMA |
| No.1 | | 725 | 425 | 100 | 620 | 700 | 1,200,000 | |
| No.2 | 2"& wider | 700 | 425 | 100 | 620 | 550 | 1,100,000 | |
| No.3 | | 400 | 250 | 100 | 620 | 325 | 1,000,000 | |
| Stud | 2"-4"thick | 550 | 325 | 100 | 620 | 350 | 1,000,000 | |
| Construction | | 800 | 475 | 100 | 620 | 725 | 1,100,000 | |
| Standard | 2"-4" wide | 450 | 275 | 100 | 620 | 575 | 1,000,000 | |
| Utility | | 225 | 125 | 100 | 620 | 375 | 900,000 | |

## MIXED OAK

| Grade | Size | | | | | | Agency |
|---|---|---|---|---|---|---|---|
| Select Structural | 2"-4"thick | 1150 | 675 | 85 | 800 | 1000 | 1,100,000 | NELMA |
| No.1 | | 825 | 500 | 85 | 800 | 825 | 1,000,000 | |
| No.2 | 2"&wider | 800 | 475 | 85 | 800 | 625 | 900,000 | |
| No.3 | | 475 | 275 | 85 | 800 | 375 | 800,000 | |
| Stud | 2"-4"thick | 625 | 375 | 85 | 800 | 400 | 800,000 | |
| Construction | | 925 | 550 | 85 | 800 | 850 | 900,000 | |
| Standard | | 525 | 300 | 85 | 800 | 650 | 800,000 | |
| Utility | 2"-4" wide | 250 | 150 | 85 | 800 | 425 | 800,000 | |

## NORTHERN RED OAK

| Grade | Size | | | | | | Agency |
|---|---|---|---|---|---|---|---|
| Select Structural | 2"-4"thick | 1400 | 800 | 110 | 885 | 1150 | 1,400,000 | NELMA |
| No.1 | | 1000 | 575 | 110 | 885 | 925 | 1,400,000 | |
| No.2 | 2"&wider | 975 | 575 | 110 | 885 | 725 | 1,300,000 | |
| No.3 | | 550 | 325 | 110 | 885 | 425 | 1,200,000 | |
| Stud | 2"-4"thick | 750 | 450 | 110 | 885 | 450 | 1,200,000 | |
| Construction | | 1100 | 650 | 110 | 885 | 975 | 1,200,000 | |
| Standard | | 625 | 350 | 110 | 885 | 750 | 1,100,000 | |
| Utility | 2"-4" wide | 300 | 175 | 110 | 885 | 500 | 1,000,000 | |

## NORTHERN SPECIES

| Grade | Size | | | | | | Agency |
|---|---|---|---|---|---|---|---|
| Select Structural | 2"-4"thick | 950 | 450 | 65 | 350 | 1100 | 1,100,000 | NLGA |
| No.1/No.2 | | 575 | 275 | 65 | 350 | 825 | 1,100,000 | |
| No.3 | 2"&wider | 350 | 150 | 65 | 350 | 475 | 1,000,000 | |
| Stud | 2"-4"thick | 450 | 200 | 65 | 350 | 525 | 1,000,000 | |
| Construction | | 675 | 300 | 65 | 350 | 1050 | 1,000,000 | |
| Standard | | 375 | 175 | 65 | 350 | 850 | 900,000 | |
| Utility | 2"-4" wide | 175 | 75 | 65 | 350 | 550 | 900,000 | |

## NORTHERN WHITE CEDAR

| Grade | Size | | | | | | Agency |
|---|---|---|---|---|---|---|---|
| Select Structural | 2"-4"thick | 775 | 450 | 60 | 370 | 750 | 800,000 | NELMA |
| No.1 | | 575 | 325 | 60 | 370 | 600 | 700,000 | |
| No.2 | 2"&wider | 550 | 325 | 60 | 370 | 475 | 700,000 | |
| No.3 | | 325 | 175 | 60 | 370 | 275 | 600,000 | |
| Stud | 2"-4"thick | 425 | 250 | 60 | 370 | 300 | 600,000 | |
| Construction | | 625 | 375 | 60 | 370 | 625 | 700,000 | |
| Standard | | 350 | 200 | 60 | 370 | 475 | 600,000 | |
| Utility | 2"-4" wide | 175 | 100 | 60 | 370 | 325 | 600,000 | |

# TABLE 8.3 (Continued)

| Species and commercial grade | Size classification | Bending $F_b$ | Tension parallel to grain $F_t$ | Shear parallel to grain $F_v$ | Compression perpendicular to grain $F_{c\perp}$ | Compression parallel to grain $F_c$ | Modulus of Elasticity E | Grading Rules Agency |
|---|---|---|---|---|---|---|---|---|
| **RED MAPLE** | | | | | | | | |
| Select Structural | 2"-4"thick | 1300 | 750 | 105 | 615 | 1100 | 1,700,000 | NELMA |
| No.1 | | 925 | 550 | 105 | 615 | 900 | 1,600,000 | |
| No.2 | 2"& wider | 900 | 525 | 105 | 615 | 700 | 1,500,000 | |
| No.3 | | 525 | 300 | 105 | 615 | 400 | 1,300,000 | |
| Stud | | 700 | 425 | 105 | 615 | 450 | 1,300,000 | |
| Construction | 2"-4"thick | 1050 | 600 | 105 | 615 | 925 | 1,400,000 | |
| Standard | | 575 | 325 | 105 | 615 | 725 | 1,300,000 | |
| Utility | 2"-4" wide | 275 | 150 | 105 | 615 | 475 | 1,200,000 | |
| **RED OAK** | | | | | | | | |
| Select Structural | 2"-4"thick | 1150 | 675 | 85 | 820 | 1000 | 1,400,000 | NELMA |
| No.1 | | 825 | 500 | 85 | 820 | 825 | 1,300,000 | |
| No.2 | 2"& wider | 800 | 475 | 85 | 820 | 625 | 1,200,000 | |
| No.3 | | 475 | 275 | 85 | 820 | 375 | 1,100,000 | |
| Stud | | 625 | 375 | 85 | 820 | 400 | 1,200,000 | |
| Construction | 2"-4"thick | 925 | 550 | 85 | 820 | 850 | 1,200,000 | |
| Standard | | 525 | 300 | 85 | 820 | 650 | 1,100,000 | |
| Utility | 2"-4" wide | 250 | 150 | 85 | 820 | 425 | 1,000,000 | |
| **REDWOOD** | | | | | | | | |
| Clear Structural | | 1750 | 1000 | 145 | 650 | 1850 | 1,400,000 | RIS |
| Select Structural | | 1350 | 800 | 80 | 650 | 1500 | 1,400,000 | |
| Select Structural, open grain | | 1100 | 625 | 80 | 425 | 1100 | 1,100,000 | |
| No.1 | 2"-4"thick | 975 | 575 | 80 | 650 | 1200 | 1,300,000 | |
| No.1, open grain | | 775 | 450 | 80 | 425 | 900 | 1,100,000 | |
| No.2 | 2"& wider | 925 | 525 | 80 | 650 | 950 | 1,200,000 | |
| No.2, open grain | | 725 | 425 | 80 | 425 | 700 | 1,100,000 | |
| No.3 | | 525 | 300 | 80 | 650 | 550 | 1,000,000 | |
| No.3, open grain | | 425 | 250 | 80 | 425 | 400 | 900,000 | |
| Stud | | 575 | 325 | 80 | 425 | 450 | 900,000 | |
| Construction | 2"-4"thick | 825 | 475 | 80 | 425 | 925 | 900,000 | |

Design values in pounds per square inch (psi)

| | | | | | | | | |
|---|---|---|---|---|---|---|---|---|
| Standard | | 450 | 275 | 80 | 425 | 725 | 900,000 | |
| Utility | 2"-4" wide | 225 | 125 | 80 | 425 | 475 | 800,000 | |
| **SPRUCE-PINE-FIR** | | | | | | | | |
| Select Structural | 2"-4"thick | 1250 | 675 | 70 | 425 | 1400 | 1,500,000 | NLGA |
| No.1/No.2 | | 875 | 425 | 70 | 425 | 1100 | 1,400,000 | |
| No.3 | 2"& wider | 500 | 250 | 70 | 425 | 625 | 1,200,000 | |
| Stud | | 675 | 325 | 70 | 425 | 675 | 1,200,000 | |
| Construction | 2"-4"thick | 975 | 475 | 70 | 425 | 1350 | 1,300,000 | |
| Standard | | 550 | 275 | 70 | 425 | 1100 | 1,200,000 | |
| Utility | 2"-4" wide | 250 | 125 | 70 | 425 | 725 | 1,100,000 | |
| **SPRUCE-PINE-FIR (SOUTH)** | | | | | | | | |
| Select Structural | | 1300 | 575 | 70 | 335 | 1200 | 1,300,000 | NELMA NSLB WCLIB WWPA |
| No.1 | 2"-4"thick | 850 | 400 | 70 | 335 | 1050 | 1,200,000 | |
| No.2 | | 750 | 325 | 70 | 335 | 975 | 1,100,000 | |
| No.3 | 2"& wider | 425 | 200 | 70 | 335 | 550 | 1,000,000 | |
| Stud | | 575 | 250 | 70 | 335 | 600 | 1,000,000 | |
| Construction | 2"-4"thick | 850 | 375 | 70 | 335 | 1200 | 1,000,000 | |
| Standard | | 475 | 225 | 70 | 335 | 1000 | 900,000 | |
| Utility | 2"-4" wide | 225 | 100 | 70 | 335 | 650 | 900,000 | |
| **WESTERN CEDARS** | | | | | | | | |
| Select Structural | | 1000 | 600 | 75 | 425 | 1000 | 1,100,000 | WCLIB WWPA |
| No.1 | 2"-4"thick | 725 | 425 | 75 | 425 | 825 | 1,000,000 | |
| No.2 | | 700 | 425 | 75 | 425 | 650 | 1,000,000 | |
| No.3 | 2"& wider | 400 | 250 | 75 | 425 | 375 | 900,000 | |
| Stud | | 550 | 325 | 75 | 425 | 400 | 900,000 | |
| Construction | 2"-4"thick | 800 | 475 | 75 | 425 | 850 | 900,000 | |
| Standard | | 450 | 275 | 75 | 425 | 650 | 800,000 | |
| Utility | 2"-4" wide | 225 | 125 | 75 | 425 | 425 | 800,000 | |
| **WESTERN WOODS** | | | | | | | | |
| Select Structural | | 875 | 400 | 70 | 335 | 1050 | 1,200,000 | WCLIB WWPA |
| No.1 | 2"-4"thick | 650 | 300 | 70 | 335 | 925 | 1,100,000 | |
| No.2 | | 650 | 275 | 70 | 335 | 875 | 1,000,000 | |
| No.3 | 2"& wider | 375 | 175 | 70 | 335 | 500 | 900,000 | |
| Stud | | 500 | 225 | 70 | 335 | 550 | 900,000 | |
| Construction | 2"-4"thick | 725 | 325 | 70 | 335 | 1050 | 1,000,000 | |
| Standard | | 400 | 175 | 70 | 335 | 900 | 900,000 | |
| Utility | 2"-4" wide | 200 | 75 | 70 | 335 | 600 | 800,000 | |

## TABLE 8.3  (Continued)

| Species and commercial grade | Size classification | Design values in pounds per square inch (psi) | | | | | | Grading Rules Agency |
|---|---|---|---|---|---|---|---|---|
| | | Bending F$_b$ | Tension parallel to grain F$_t$ | Shear parallel to grain F$_v$ | Compression perpendicular to grain F$_{c\perp}$ | Compression parallel to grain F$_c$ | Modulus of Elasticity E | |
| **WHITE OAK** | | | | | | | | |
| Select Structural | 2"-4"thick | 1200 | 700 | 110 | 800 | 1100 | 1,100,000 | NELMA |
| No.1 | | 875 | 500 | 110 | 800 | 900 | 1,000,000 | |
| No.2 | | 850 | 500 | 110 | 800 | 700 | 900,000 | |
| No.3 | 2"& wider | 475 | 275 | 110 | 800 | 400 | 800,000 | |
| Stud | | 650 | 375 | 110 | 800 | 450 | 800,000 | |
| Construction | 2"-4"thick | 950 | 550 | 110 | 800 | 925 | 900,000 | |
| Standard | | 525 | 325 | 110 | 800 | 725 | 800,000 | |
| Utility | 2"-4" wide | 250 | 150 | 110 | 800 | 475 | 800,000 | |
| **YELLOW POPLAR** | | | | | | | | |
| Select Structural | 2"-4"thick | 1000 | 575 | 75 | 420 | 900 | 1,500,000 | NSLB |
| No.1 | | 725 | 425 | 75 | 420 | 725 | 1,400,000 | |
| No.2 | 2"& wider | 700 | 400 | 75 | 420 | 575 | 1,300,000 | |
| No.3 | | 400 | 225 | 75 | 420 | 325 | 1,200,000 | |
| Stud | 2"-4"thick | 550 | 325 | 75 | 420 | 350 | 1,200,000 | |
| Construction | | 800 | 475 | 75 | 420 | 750 | 1,300,000 | |
| Standard | 2"-4" wide | 450 | 250 | 75 | 420 | 575 | 1,100,000 | |
| Utility | | 200 | 125 | 75 | 420 | 375 | 1,100,000 | |

Source: ANSI/NF₀PA NDS—1991, *National Design Specification (NDS) for Wood Construction*, AFPA (formerly NFPA), Washington, DC, 1991.

1. **LUMBER DIMENSIONS.** Tabulated design values are applicable to lumber that will be used under dry conditions such as in most covered structures. For 2" to 4" thick lumber the DRY dressed sizes shall be used (see Table 1A of NDS) regardless of the moisture content at the time of manufacture or use. In calculating design values, the natural gain in strength and stiffness that occurs as lumber dries has been taken into consideration as well as the reduction in size that occurs when unseasoned lumber shrinks. The gain in load carrying capacity due to increased strength and stiffness resulting from drying more than offsets the design effect of size reductions due to shrinkage.

2. **STRESS-RATED BOARDS.** Stress-rated boards of nominal 1", 1¼" and 1½" thickness, 2" and wider, of most species, are permitted the design values shown for Select Structural, No. 1 & Btr, No. 1, No. 2, No. 3, Stud, Construction, Standard, Utility, Clear Heart Structural and Clear Structural grades as shown in the 2" to 4" thick categories herein, when graded in accordance with the stress-rated board provisions in the applicable grading rules. Information on stress-rated board grades applicable to the various species is available from the respective grading rules agencies. Information on additional design values may also be available from the respective grading agencies.

**TABLE 8.3—Adjustment Factors**

## SIZE FACTOR, $C_F$

Tabulated bending, tension, and compression parallel to grain design values for dimension lumber 2" to 4" thick shall be multiplied by the following size factors:

### SIZE FACTORS, $C_F$

| Grades | Width | $F_b$ Thickness 2" & 3" | $F_b$ Thickness 4" | $F_t$ | $F_c$ |
|---|---|---|---|---|---|
| Select Structural, No. 1 & Btr. No. 1, No. 2, No. 3 | 2", 3", & 4" | 1.5 | 1.5 | 1.5 | 1.15 |
| | 5" | 1.4 | 1.4 | 1.4 | 1.1 |
| | 6" | 1.3 | 1.3 | 1.3 | 1.1 |
| | 8" | 1.2 | 1.3 | 1.2 | 1.05 |
| | 10" | 1.1 | 1.2 | 1.1 | 1.0 |
| | 12" | 1.0 | 1.1 | 1.0 | 1.0 |
| | 14" & wider | 0.9 | 1.0 | 0.9 | 0.9 |
| Stud | 2", 3", & 4" | 1.1 | 1.1 | 1.1 | 1.05 |
| | 5" & 6" | 1.0 | 1.0 | 1.0 | 1.0 |
| Construction & Standard | 2", 3" & 4" | 1.0 | 1.0 | 1.0 | 1.0 |
| Utility | 4" | 1.0 | 1.0 | 1.0 | 1.0 |
| | 2" & 3" | 0.4 | — | 0.4 | 0.6 |

## TABLE 8.3—Adjustment Factors (*Continued*)

### REPETITIVE MEMBER FACTOR, $C_r$

Bending design values, $F_b$, for dimension lumber 2" to 4" thick shall be multiplied by the repetitive member factor, $C_r = 1.15$, when such members are used as joists, truss chords, rafters, studs, planks, decking or similar members which are in contact or spaced not more than 24" on centers, are not less than 3 in number and are joined by floor, roof or other load distributing elements adequate to support the design load.

### FLAT USE FACTOR, $C_{fu}$

Bending design values adjusted by size factors are based on edgewise use (load applied to narrow face). When dimension lumber is used flatwise (load applied to wide face), the bending design value, $F_b$, shall also be multiplied by the following flat use factors:

#### FLAT USE FACTORS, $C_{fu}$

| Width | Thickness | |
|---|---|---|
| | 2" & 3" | 4" |
| 2" & 3" | 1.0 | --- |
| 4" | 1.1 | 1.0 |
| 5" | 1.1 | 1.05 |
| 6" | 1.15 | 1.05 |
| 8" | 1.15 | 1.05 |
| 10" & wider | 1.2 | 1.1 |

### WET SERVICE FACTOR, $C_M$

When dimension lumber is used where moisture content will exceed 19% for an extended time period, design values shall be multiplied by the appropriate wet service factors from the following table:

#### WET SERVICE FACTORS, $C_M$

| $F_b$ | $F_t$ | $F_v$ | $F_{c\perp}$ | $F_c$ | E |
|---|---|---|---|---|---|
| 0.85* | 1.0 | 0.97 | 0.67 | 0.8** | 0.9 |

\* when $(F_b)(C_F) \leq 1150$ psi, $C_M = 1.0$
\*\* when $(F_c)(C_F) \leq 750$ psi, $C_M = 1.0$

# SHEAR STRESS FACTOR, $C_H$

Tabulated shear design values parallel to grain have been reduced to allow for the occurrence of splits, checks and shakes. Tabulated shear design values parallel to grain, $F_v$, shall be permitted to be multiplied by the shear stress factors specified in the following table when length of split, or size of check or shake is known and no increase in them is anticipated. When shear stress factors are used for Redwood, a tabulated design value of $F_v = 80$ psi shall be assigned for all grades of Redwood dimension lumber. Shear stress factors shall be permitted to be linearly interpolated.

## SHEAR STRESS FACTORS, $C_H$

| Length of split on wide face of 2" (nominal) lumber | $C_H$ | Length of split on wide face of 3" (nominal) and thicker lumber | $C_H$ | Size of shake* in 2" (nominal) and thicker lumber | $C_H$ |
|---|---|---|---|---|---|
| no split ................................ | 2.00 | no split ................................ | 2.00 | no shake ................................ | 2.00 |
| 1/2 x wide face ..................... | 1.67 | 1/2 x narrow face. ................ | 1.67 | 1/6 x narrow face ................. | 1.67 |
| 3/4 x wide face ..................... | 1.50 | 3/4 x narrow face ................. | 1.50 | 1/4 narrow face .................... | 1.50 |
| 1 x wide face ........................ | 1.33 | 1 x narrow face .................... | 1.33 | 1/3 x narrow face ................. | 1.33 |
| 1-1/2 x wide face or more. .... | 1.00 | 1-1/2 x narrow face or more. . | 1.00 | 1/2 x narrow face or more. ..... | 1.00 |
| | | | | *Shake is measured at the end between lines enclosing the shake and perpendicular to the loaded face. | |

## TABLE 8.4

### Design Values for Visually Graded Southern Pine Dimension Lumber

(Tabulated design values are for normal load duration and dry service conditions, unless specified otherwise. See NDS 2.3 for a comprehensive description of design value adjustment factors.)

Use with Adjustment Factors

| Species and commercial grade | Size classification | Design values in pounds per square inch (psi) | | | | | | Grading Rules Agency |
|---|---|---|---|---|---|---|---|---|
| | | Bending $F_b$ | Tension parallel to grain $F_t$ | Shear parallel to grain $F_v$ | Compression perpendicular to grain $F_{c\perp}$ | Compression parallel to grain $F_c$ | Modulus of Elasticity E | |
| **MIXED SOUTHERN PINE** | | | | | | | | |
| Select Structural | 2"-4"thick | 2050 | 1200 | 100 | 565 | 1800 | 1,600,000 | |
| No.1 | | 1450 | 875 | 100 | 565 | 1650 | 1,500,000 | |
| No.2 | | 1300 | 775 | 90 | 565 | 1650 | 1,400,000 | |
| No.3 | 2"-4" wide | 750 | 450 | 90 | 565 | 950 | 1,200,000 | |
| Stud | | 775 | 450 | 90 | 565 | 950 | 1,200,000 | |
| Construction | 2"-4"thick | 1000 | 600 | 100 | 565 | 1700 | 1,300,000 | |
| Standard | | 550 | 325 | 90 | 565 | 1450 | 1,200,000 | |
| Utility | 4" wide | 275 | 150 | 90 | 565 | 950 | 1,100,000 | |
| Select Structural | 2"-4"thick | 1850 | 1100 | 90 | 565 | 1700 | 1,600,000 | |
| No.1 | | 1300 | 750 | 90 | 565 | 1550 | 1,500,000 | SPIB |
| No.2 | | 1150 | 675 | 90 | 565 | 1550 | 1,400,000 | |
| No.3 | 5"-6" wide | 675 | 400 | 90 | 565 | 875 | 1,200,000 | |
| Stud | | 675 | 400 | 90 | 565 | 875 | 1,200,000 | |
| Select Structural | 2"-4"thick | 1750 | 1000 | 90 | 565 | 1600 | 1,600,000 | |
| No.1 | | 1200 | 700 | 90 | 565 | 1450 | 1,500,000 | |
| No.2 | 8" wide | 1050 | 625 | 90 | 565 | 1450 | 1,400,000 | |
| No.3 | | 625 | 375 | 90 | 565 | 850 | 1,200,000 | |
| Select Structural | 2"-4"thick | 1500 | 875 | 90 | 565 | 1600 | 1,600,000 | |
| No.1 | | 1050 | 600 | 90 | 565 | 1450 | 1,500,000 | |
| No.2 | 10" wide | 925 | 550 | 90 | 565 | 1450 | 1,400,000 | |
| No.3 | | 525 | 325 | 90 | 565 | 825 | 1,200,000 | |
| Select Structural | 2"-4"thick | 1400 | 825 | 90 | 565 | 1550 | 1,600,000 | |
| No.1 | | 975 | 575 | 90 | 565 | 1400 | 1,500,000 | |
| No.2 | 12" wide | 875 | 525 | 90 | 565 | 1400 | 1,400,000 | |
| No.3 | | 500 | 300 | 90 | 565 | 800 | 1,200,000 | |

## SOUTHERN PINE

| Grade | Size | | | | | | | |
|---|---|---|---|---|---|---|---|---|
| Dense Select Structural | 2"-4" thick | 2"-4" wide | 3050 | 1650 | 100 | 660 | 2250 | 1,900,000 |
| Select Structural | | | 2850 | 1600 | 100 | 565 | 2100 | 1,800,000 |
| Non-Dense Select Structural | | | 2650 | 1350 | 100 | 480 | 1950 | 1,700,000 |
| No.1 Dense | | | 2000 | 1100 | 100 | 660 | 2000 | 1,800,000 |
| No.1 | | | 1850 | 1050 | 100 | 565 | 1850 | 1,700,000 |
| No.1 Non-Dense | | | 1700 | 900 | 100 | 480 | 1700 | 1,600,000 |
| No.2 Dense | | | 1700 | 875 | 90 | 660 | 1850 | 1,700,000 |
| No.2 | | | 1500 | 825 | 90 | 565 | 1650 | 1,600,000 |
| No.2 Non-Dense | | | 1350 | 775 | 90 | 480 | 1600 | 1,400,000 |
| No.3 | | | 850 | 475 | 90 | 565 | 975 | 1,400,000 |
| Stud | | | 875 | 500 | 90 | 565 | 975 | 1,400,000 |
| Construction | 2"-4" thick | 4" wide | 1100 | 625 | 100 | 565 | 1800 | 1,500,000 |
| Standard | | | 625 | 350 | 90 | 565 | 1500 | 1,300,000 |
| Utility | | | 300 | 175 | 90 | 565 | 975 | 1,300,000 |
| Dense Select Structural | 2"-4" thick | 5"-6" wide | 2700 | 1500 | 90 | 660 | 2150 | 1,900,000 |
| Select Structural | | | 2550 | 1400 | 90 | 565 | 2000 | 1,800,000 |
| Non-Dense Select Structural | | | 2350 | 1200 | 90 | 480 | 1850 | 1,700,000 |
| No.1 Dense | | | 1750 | 950 | 90 | 660 | 1900 | 1,800,000 |
| No.1 | | | 1650 | 900 | 90 | 565 | 1750 | 1,700,000 |
| No.1 Non-Dense | | | 1500 | 800 | 90 | 480 | 1600 | 1,600,000 |
| No.2 Dense | | | 1450 | 775 | 90 | 660 | 1750 | 1,700,000 |
| No.2 | | | 1250 | 725 | 90 | 565 | 1600 | 1,600,000 |
| No.2 Non-Dense | | | 1150 | 675 | 90 | 480 | 1500 | 1,400,000 |
| No.3 | | | 750 | 425 | 90 | 565 | 925 | 1,400,000 |
| Stud | | | 775 | 425 | 90 | 565 | 925 | 1,400,000 |
| Dense Select Structural | 2"-4" thick | 8" wide | 2450 | 1350 | 90 | 660 | 2050 | 1,900,000 |
| Select Structural | | | 2300 | 1300 | 90 | 565 | 1900 | 1,800,000 |
| Non-Dense Select Structural | | | 2100 | 1100 | 90 | 480 | 1750 | 1,700,000 |
| No.1 Dense | | | 1650 | 875 | 90 | 660 | 1800 | 1,800,000 |
| No.1 | | | 1500 | 825 | 90 | 565 | 1650 | 1,700,000 |
| No.1 Non-Dense | | | 1350 | 725 | 90 | 480 | 1550 | 1,600,000 |
| No.2 Dense | | | 1400 | 675 | 90 | 660 | 1700 | 1,700,000 |
| No.2 | | | 1200 | 650 | 90 | 565 | 1550 | 1,600,000 |
| No.2 Non-Dense | | | 1100 | 600 | 90 | 480 | 1450 | 1,400,000 |
| No.3 | | | 700 | 400 | 90 | 565 | 875 | 1,400,000 |
| Dense Select Structural | | | 2150 | 1200 | 90 | 660 | 2000 | 1,900,000 |
| Select Structural | | | 2050 | 1100 | 90 | 565 | 1850 | 1,800,000 |

## TABLE 8.4 (Continued)

| Species and commercial grade | Size classification | Design values in pounds per square inch (psi) | | | | | | Grading Rules Agency |
|---|---|---|---|---|---|---|---|---|
| | | Bending $F_b$ | Tension parallel to grain $F_t$ | Shear parallel to grain $F_v$ | Compression perpendicular to grain $F_{cL}$ | Compression parallel to grain $F_c$ | Modulus of Elasticity E | |
| Non-Dense Select Structural | 2"-4"thick | 1850 | 950 | 90 | 480 | 1750 | 1,700,000 | |
| No.1 Dense | | 1450 | 775 | 90 | 660 | 1750 | 1,800,000 | |
| No.1 | 10" wide | 1300 | 725 | 90 | 565 | 1600 | 1,700,000 | |
| No.1 Non-Dense | | 1200 | 650 | 90 | 480 | 1500 | 1,600,000 | |
| No.2 Dense | | 1200 | 625 | 90 | 660 | 1650 | 1,700,000 | |
| No.2 | | 1050 | 575 | 90 | 565 | 1500 | 1,600,000 | |
| No.2 Non-Dense | | 950 | 550 | 90 | 480 | 1400 | 1,600,000 | |
| No.3 | | 600 | 325 | 90 | 565 | 850 | 1,400,000 | |
| Dense Select Structural | | 2050 | 1100 | 90 | 660 | 1950 | 1,900,000 | |
| Select Structural | | 1900 | 1050 | 90 | 565 | 1800 | 1,800,000 | |
| Non-Dense Select Structural | 2"-4"thick | 1750 | 900 | 90 | 480 | 1700 | 1,700,000 | |
| No.1 Dense | | 1350 | 725 | 90 | 660 | 1700 | 1,800,000 | |
| No.1 | 12" wide | 1250 | 675 | 90 | 565 | 1600 | 1,700,000 | |
| No.1 Non-Dense | | 1150 | 600 | 90 | 480 | 1500 | 1,600,000 | |
| No.2 Dense | | 1150 | 575 | 90 | 660 | 1600 | 1,700,000 | |
| No.2 | | 975 | 550 | 90 | 565 | 1450 | 1,600,000 | |
| No.2 Non-Dense | | 900 | 525 | 90 | 480 | 1350 | 1,400,000 | |
| No.3 | | 575 | 325 | 90 | 565 | 825 | 1,400,000 | |
| **SOUTHERN PINE** | | (Dry service conditions — 19% or less moisture content) | | | | | | |
| Dense Structural 86 | 2"-4"thick | 2600 | 1750 | 155 | 660 | 2000 | 1,800,000 | SPIB |
| Dense Structural 72 | | 2200 | 1450 | 130 | 660 | 1650 | 1,800,000 | |
| Dense Structural 65 | 2"& wider | 2000 | 1300 | 115 | 660 | 1500 | 1,800,000 | |
| **SOUTHERN PINE** | | (Wet service conditions) | | | | | | |
| Dense Structural 86 | 2-1/2"-4"thick | 2100 | 1400 | 145 | 440 | 1300 | 1,600,000 | SPIB |
| Dense Structural 72 | | 1750 | 1200 | 120 | 440 | 1100 | 1,600,000 | |
| Dense Structural 65 | 2-1/2"& wider | 1600 | 1050 | 110 | 440 | 1000 | 1,600,000 | |

Source: ANSI/NF₀PA NDS—1991, *National Design Specification (NDS) for Wood Construction*, AFPA (formerly NFPA), Washington, DC, 1991.

1. **LUMBER DIMENSIONS.** Tabulated design values are applicable to lumber that will be used under dry conditions such as in most covered structures. For 2″ to 4″ thick lumber the DRY dressed sizes shall be used (see Table 1A of NDS) regardless of the moisture content at the time of manufacture or use. In calculating design values, the natural gain in strength and stiffness that occurs as lumber dries has been taken into consideration as well as the reduction in size that occurs when unseasoned lumber shrinks. The gain in load carrying capacity due to increased strength and stiffness resulting from drying more than offsets the design effect of size reductions due to shrinkage.

2. **STRESS-RATED BOARDS.** Information for various grades of Southern Pine stress-rated boards of nominal 1″, $1\frac{1}{4}$ and $1\frac{1}{2}$″ thickness, 2″ and wider, is available from the Southern Pine Inspection Bureau (SPIB) in the "Standard Grading Rules for Southern Pine Lumber."

3. **SPRUCE PINE.** To obtain recommended design values for Spruce Pine graded to SPIB rules, multiply the appropriate design values for Mixed Southern Pine by the corresponding conversion factor shown below and round to the nearest 100,000 psi for E; to the next lower multiple of 5 psi for $F_v$ and $F_{c\perp}$; to the next lower multiple of 50 psi for $F_b$, $F_t$ and $F_c$ if 1000 psi or greater, 25 psi otherwise.

## CONVERSION FACTORS FOR DETERMINING DESIGN VALUES FOR SPRUCE PINE

|  | Bending $F_b$ | Tension parallel to grain $F_t$ | Shear parallel to grain $F_v$ | Compression perpendicular to grain $F_{c\perp}$ | Compression parallel to grain $F_c$ | Modulus of Elasticity E |
|---|---|---|---|---|---|---|
| Conversion Factor | 0.784 | 0.784 | 0.965 | 0.682 | 0.766 | 0.807 |

**TABLE 8.4—Adjustment Factors**

**SIZE FACTOR, $C_F$**

Appropriate size adjustment factors have already been incorporated in the tabulated design values for most thicknesses of Southern Pine and Mixed Southern Pine dimension lumber. For dimension lumber 4" thick, 8" and wider (all grades except Dense Structural 86, Dense Structural 72 and Dense Structural 65), tabulated bending design values, $F_b$, shall be permitted to be multiplied by the size factor, $C_F = 1.1$. For dimension lumber wider than 12" (all grades except Dense Structural 86, Dense Structural 72 and Dense Structural 65), tabulated bending, tension and compression parallel to grain design values for 12" wide lumber shall be multiplied by the size factor, $C_F = 0.9$. When the depth, d, of Dense Structural 86, Dense Structural 72 or Dense Structural 65 dimension lumber exceeds 12", the tabulated bending design value, $F_b$, shall be multiplied by the following size factor:

$$C_F = (12/d)^{1/9}$$

**REPETITIVE MEMBER FACTOR, $C_r$**

Bending design values, $F_b$, for dimension lumber 2" to 4" thick shall be multiplied by the repetitive member factor, $C_r = 1.15$, when such members are used as joists, truss chords, rafters, studs, planks, decking or similar members which are in contact or spaced not more than 24" on centers, are not less than 3 in number and are joined by floor, roof or other load distributing elements adequate to support the design load.

## FLAT USE FACTOR, $C_{fu}$

Bending design values adjusted by size factors are based on edgewise use (load applied to narrow face). When dimension lumber is used flatwise (load applied to wide face), the bending design value, $F_b$, shall also be multiplied by the following flat use factors:

### FLAT USE FACTORS, $C_{fu}$

| Width | Thickness | |
|---|---|---|
| | 2" & 3" | 4" |
| 2" & 3" | 1.0 | -- |
| 4" | 1.1 | 1.0 |
| 5" | 1.1 | 1.05 |
| 6" | 1.15 | 1.05 |
| 8" | 1.15 | 1.05 |
| 10" & wider | 1.2 | 1.1 |

## WET SERVICE FACTOR, $C_M$

When dimension lumber is used where moisture content will exceed 19% for an extended time period, design values shall be multiplied by the appropriate wet service factors from the following table (for Dense Structural 86, Dense Structural 72 and Dense Structural 65 use tabulated design values for wet service conditions without further adjustment):

### WET SERVICE FACTORS, $C_M$

| $F_b$ | $F_t$ | $F_v$ | $F_{c\perp}$ | $F_c$ | E |
|---|---|---|---|---|---|
| 0.85* | 1.0 | 0.97 | 0.67 | 0.8** | 0.9 |

\* when $(F_b)(C_F) \leq 1150$ psi, $C_M = 1.0$
\*\* when $F_c \leq 750$ psi, $C_M = 1.0$

**TABLE 8.4—Adjustment Factors** (*Continued*)

## SHEAR STRESS FACTOR, $C_H$

Tabulated shear design values parallel to grain have been reduced to allow for the occurrence of splits, checks and shakes. Tabulated shear design values parallel to grain, $F_v$, shall be permitted to be multiplied by the shear stress factors specified in the following table when length of split, or size of check or shake is known and no increase in them is anticipated. When shear stress factors are used for Southern Pine and Mixed Southern Pine, a tabulated design value of $F_v = 90$ psi shall be assigned for all grades of Southern Pine and Mixed Southern Pine dimension lumber. Shear stress factors shall be permitted to be linearly interpolated.

### SHEAR STRESS FACTORS, $C_H$

| Length of split on wide face of 2" (nominal) lumber | $C_H$ | Length of split on wide face of 3" (nominal) and thicker lumber | $C_H$ | Size of shake* in 2" (nominal) and thicker lumber | $C_H$ |
|---|---|---|---|---|---|
| no split | 2.00 | no split | 2.00 | no shake | 2.00 |
| 1/2 x wide face | 1.67 | 1/2 x narrow face | 1.67 | 1/6 x narrow face | 1.67 |
| 3/4 x wide face | 1.50 | 3/4 x narrow face | 1.50 | 1/4 narrow face | 1.50 |
| 1 x wide face | 1.33 | 1 x narrow face | 1.33 | 1/3 x narrow face | 1.33 |
| 1-1/2 x wide face or more | 1.00 | 1-1/2 x narrow face or more | 1.00 | 1/2 x narrow face or more | 1.00 |
| | | | | *Shake is measured at the end between lines enclosing the shake and perpendicular to the loaded face. | |

# TABLE 8.5

## Design Values for Mechanically Graded Dimension Lumber

(Tabulated design values are for normal load duration and dry service conditions. See NDS 2.3 for a comprehensive description of design value adjustment factors.)

Use with Adjustment Factors

| Species and commercial grade | Size classification | Design values in pounds per square inch (psi) | | | | Grading Rules Agency |
|---|---|---|---|---|---|---|
| | | Bending $F_b$ | Tension parallel to grain $F_t$ | Compression parallel to grain $F_c$ | Modulus of Elasticity E | |
| **MACHINE STRESS RATED (MSR) LUMBER** | | | | | | |
| 900f-1.0E | | 900 | 350 | 1050 | 1,000,000 | WCLIB, WWPA |
| 1200f-1.2E | | 1200 | 600 | 1400 | 1,200,000 | NLGA, SPIB, WCLIB, WWPA |
| 1350f-1.3E | | 1350 | 750 | 1600 | 1,300,000 | SPIB, WCLIB, WWPA |
| 1450f-1.3E | | 1450 | 800 | 1625 | 1,300,000 | NLGA, WCLIB, WWPA |
| 1500f-1.3E | | 1500 | 900 | 1650 | 1,300,000 | SPIB |
| 1500f-1.4E | | 1500 | 900 | 1650 | 1,400,000 | NLGA, SPIB, WCLIB, WWPA |
| 1650f-1.4E | | 1650 | 1020 | 1700 | 1,400,000 | SPIB |
| 1650f-1.5E | 2" & less in thickness | 1650 | 1020 | 1700 | 1,500,000 | NLGA, SPIB, WCLIB, WWPA |
| 1800f-1.6E | | 1800 | 1175 | 1750 | 1,600,000 | NLGA, SPIB, WCLIB, WWPA |
| 1950f-1.5E | | 1950 | 1375 | 1800 | 1,500,000 | SPIB |
| 1950f-1.7E | | 1950 | 1375 | 1800 | 1,700,000 | NLGA, SPIB, WWPA |
| 2100f-1.8E | 2" & wider | 2100 | 1575 | 1875 | 1,800,000 | NLGA, SPIB, WCLIB, WWPA |
| 2250f-1.6E | | 2250 | 1750 | 1925 | 1,600,000 | SPIB |
| 2250f-1.9E | | 2250 | 1750 | 1925 | 1,900,000 | NLGA, SPIB, WWPA |
| 2400f-1.7E | | 2400 | 1925 | 1975 | 1,700,000 | SPIB |
| 2400f-2.0E | | 2400 | 1925 | 1975 | 2,000,000 | NLGA, SPIB, WCLIB, WWPA |
| 2550f-2.1E | | 2550 | 2050 | 2025 | 2,100,000 | NLGA, SPIB, WWPA |
| 2700f-2.2E | | 2700 | 2150 | 2100 | 2,200,000 | NLGA, SPIB, WCLIB, WWPA |
| 2850f-2.3E | | 2850 | 2300 | 2150 | 2,300,000 | SPIB, WWPA |
| 3000f-2.4E | | 3000 | 2400 | 2200 | 2,400,000 | NLGA, SPIB |
| 3150f-2.5E | | 3150 | 2500 | 2250 | 2,500,000 | SPIB |
| 3300f-2.6E | | 3300 | 2650 | 2325 | 2,600,000 | SPIB |
| 900f-1.2E | | 900 | 350 | 1050 | 1,200,000 | NLGA, WCLIB |

## TABLE 8.5 (Continued)

| Species and commercial grade | Size classification | Design values in pounds per square inch (psi) | | | | Grading Rules Agency |
|---|---|---|---|---|---|---|
| | | Bending $F_b$ | Tension parallel to grain $F_t$ | Compression parallel to grain $F_c$ | Modulus of Elasticity E | |
| **MACHINE STRESS RATED (MSR) LUMBER** | | | | | | |
| 1200f-1.5E | 2" & less in thickness | 1200 | 600 | 1400 | 1,500,000 | NLGA, WCLIB |
| 1350f-1.8E | | 1350 | 750 | 1600 | 1,800,000 | NLGA |
| 1500f-1.8E | 6" & wider | 1500 | 900 | 1650 | 1,800,000 | WCLIB |
| 1800f-2.1E | | 1800 | 1175 | 1750 | 2,100,000 | NLGA, WCLIB |
| **MACHINE EVALUATED LUMBER (MEL)** | | | | | | |
| M-10 | | 1400 | 800 | 1600 | 1,200,000 | |
| M-11 | | 1550 | 850 | 1650 | 1,500,000 | |
| M-12 | | 1600 | 850 | 1700 | 1,600,000 | |
| M-13 | | 1600 | 950 | 1700 | 1,400,000 | |
| M-14 | | 1800 | 1000 | 1750 | 1,700,000 | |
| M-15 | | 1800 | 1100 | 1750 | 1,500,000 | |
| M-16 | 2" & less in thickness | 1800 | 1300 | 1750 | 1,500,000 | |
| M-17 | | 1950 | 1300 | 2050 | 1,700,000 | |
| M-18 | | 2000 | 1200 | 1850 | 1,800,000 | SPIB |
| M-19 | 2" & wider | 2000 | 1300 | 1850 | 1,600,000 | |
| M-20 | | 2000 | 1600 | 2100 | 1,900,000 | |
| M-21 | | 2300 | 1400 | 1950 | 1,900,000 | |
| M-22 | | 2350 | 1500 | 1950 | 1,700,000 | |
| M-23 | | 2400 | 1900 | 2000 | 1,800,000 | |
| M-24 | | 2700 | 1800 | 2100 | 1,900,000 | |
| M-25 | | 2750 | 2000 | 2100 | 2,200,000 | |
| M-26 | | 2800 | 2000 | 2150 | 2,000,000 | |
| M-27 | | 3000 | 2000 | 2400 | 2,100,000 | |

Source: ANSI/NF$_o$PA NDS—1991, *National Design Specification (NDS) for Wood Construction*, AFPA (formerly NFPA), Washington, DC, 1991.

**1. LUMBER DIMENSIONS.** Tabulated design values are applicable to lumber that will be used under dry conditions such as in most covered structures. For 2″ to 4″ thick lumber the DRY dressed sizes shall be used (see Table 1A of NDS) regardless of the moisture content at the time of manufacture or use. In calculating design values, the natural gain in strength and stiffness that occurs as lumber dries has been taken into consideration as well as the reduction in size that occurs when unseasoned lumber shrinks. The gain in load carrying capacity due to increased strength and stiffness resulting from drying more than offsets the design effect of size reductions due to shrinkage.

**2. SHEAR PARALLEL TO GRAIN, $F_v$, AND COMPRESSION PERPENDICULAR TO GRAIN, $F_{c\perp}$.** Design values for shear parallel to grain, $F_v$, and compression perpendicular to grain, $F_{c\perp}$, are identical to the design values given in Tables 8.3 and 8.4 for No. 2 visually graded lumber of the appropriate species.

**3. MODULUS OF ELASTICITY, E.** For any given bending design value, $F_b$, the average modulus of elasticity, E, may vary depending upon species, timber source or other variables. The "E" value included in the "$F_b$-E" grade designations in Table 8.5 are those usually associated with each "$F_b$" level. Grade stamps may show higher or lower "E" values (in increments of 100,000 psi) if machine rating indicates the assignment is appropriate. When the "E" value shown on a grade stamp differs from the "E" value in Table 8.5, the "E" value shown on the grade stamp shall be used for design. The tabulated $F_b$, $F_t$ and $F_c$ values associated with the designated "$F_b$" value shall be used for design.

TABLE 8.5—Adjustment Factors

## REPETITIVE MEMBER FACTOR, $C_r$

Bending design values, $F_b$, for dimension lumber 2" to 4" thick shall be multiplied by the repetitive member factor, $C_r = 1.15$, when such members are used as joists, truss chords, rafters, studs, planks, decking or similar members which are in contact or spaced not more than 24" on centers, are not less than 3 in number and are joined by floor, roof or other load distributing elements adequate to support the design load.

## WET SERVICE FACTOR, $C_M$

When dimension lumber is used where moisture content will exceed 19% for an extended time period, design values shall be multiplied by the appropriate wet service factors from the following table:

### WET SERVICE FACTORS, $C_M$

| $F_b$ | $F_t$ | $F_v$ | $F_{c\perp}$ | $F_c$ | E |
|-------|-------|-------|--------------|-------|-----|
| 0.85* | 1.0 | 0.97 | 0.67 | 0.8 | 0.9 |

* when $F_b \leq 1150$ psi, $C_M = 1.0$

## FLAT USE FACTOR, $C_{fu}$

Bending design values adjusted by size factors are based on edgewise use (load applied to narrow face). When dimension lumber is used flatwise (load applied to wide face), the bending design value, $F_b$, shall also be multiplied by the following flat use factors:

### FLAT USE FACTORS, $C_{fu}$

| Width | Thickness |
|-------|-----------|
| | 2" |
| 2" & 3" | 1.0 |
| 4" | 1.1 |
| 5" | 1.1 |
| 6" | 1.15 |
| 8" | 1.15 |
| 10" & wider | 1.2 |

# SHEAR STRESS FACTOR, $C_H$

Tabulated shear design values parallel to grain have been reduced to allow for the occurrence of splits, checks and shakes. Tabulated shear design values parallel to grain, $F_v$, shall be permitted to be multiplied by the shear stress factors specified in the following table when length of split, or size of check or shake is known and no increase in them is anticipated. Shear stress factors shall be permitted to be linearly interpolated.

## SHEAR STRESS FACTORS, $C_H$

| Length of split on wide face of 2" (nominal) lumber | $C_H$ | Length of split on wide face of 3" (nominal) and thicker lumber | $C_H$ | Size of shake* in 2" (nominal) and thicker lumber | $C_H$ |
|---|---|---|---|---|---|
| no split | 2.00 | no split | 2.00 | no shake | 2.00 |
| 1/2 x wide face | 1.67 | 1/2 x narrow face | 1.67 | 1/6 x narrow face | 1.67 |
| 3/4 x wide face | 1.50 | 3/4 x narrow face | 1.50 | 1/4 narrow face | 1.50 |
| 1 x wide face | 1.33 | 1 x narrow face | 1.33 | 1/3 x narrow face | 1.33 |
| 1-1/2 x wide face or more. | 1.00 | 1-1/2 x narrow face or more. | 1.00 | 1/2 x narrow face or more. | 1.00 |
| | | | | *Shake is measured at the end between lines enclosing the shake and perpendicular to the loaded face. | |

## TABLE 8.6

### Design Values for Visually Graded Timbers (5 in. × 5 in. and Larger)

(Tabulated design values are for normal load duration and dry service conditions, unless specified otherwise. See NDS 2.3 for a comprehensive description of design value adjustment factors.)

Use with Adjustment Factors

| Species and commercial grade | Size classification | Design values in pounds per square inch (psi) | | | | | | Grading Rules Agency |
|---|---|---|---|---|---|---|---|---|
| | | Bending $F_b$ | Tension parallel to grain $F_t$ | Shear parallel to grain $F_v$ | Compression perpendicular to grain $F_{c\perp}$ | Compression parallel to grain $F_c$ | Modulus of Elasticity $E$ | |
| **BALSAM FIR** | | | | | | | | |
| Select Structural | Beams and Stringers | 1350 | 900 | 65 | 305 | 950 | 1,400,000 | |
| No.1 | | 1100 | 750 | 65 | 305 | 800 | 1,400,000 | NELMA |
| No.2 | | 725 | 350 | 65 | 305 | 500 | 1,100,000 | NSLB |
| Select Structural | Posts and Timbers | 1250 | 825 | 65 | 305 | 1000 | 1,400,000 | |
| No.1 | | 1000 | 675 | 65 | 305 | 875 | 1,400,000 | |
| No.2 | | 575 | 375 | 65 | 305 | 400 | 1,100,000 | |
| **BEECH-BIRCH-HICKORY** | | | | | | | | |
| Select Structural | Beams and Stringers | 1650 | 975 | 90 | 715 | 975 | 1,500,000 | |
| No.1 | | 1400 | 700 | 90 | 715 | 825 | 1,500,000 | NELMA |
| No.2 | | 900 | 450 | 90 | 715 | 525 | 1,200,000 | |
| Select Structural | Posts and Timbers | 1550 | 1050 | 90 | 715 | 1050 | 1,500,000 | |
| No.1 | | 1250 | 850 | 90 | 715 | 900 | 1,500,000 | |
| No.2 | | 725 | 475 | 90 | 715 | 425 | 1,200,000 | |
| **COAST SITKA SPRUCE** | | | | | | | | |
| Select Structural | Beams and Stringers | 1150 | 675 | 60 | 455 | 775 | 1,500,000 | |
| No.1 | | 950 | 475 | 60 | 455 | 650 | 1,500,000 | NLGA |
| No.2 | | 625 | 325 | 60 | 455 | 425 | 1,200,000 | |
| Select Structural | Posts and Timbers | 1100 | 725 | 60 | 455 | 825 | 1,500,000 | |
| No.1 | | 875 | 575 | 60 | 455 | 725 | 1,500,000 | |
| No.2 | | 525 | 350 | 60 | 455 | 500 | 1,200,000 | |

## DOUGLAS FIR-LARCH

| Grade | Size Classification | | | | | | | Agency |
|---|---|---|---|---|---|---|---|---|
| Dense Select Structural | Beams and Stringers | 1900 | 1100 | 85 | 730 | 1300 | 1,700,000 | WCLIB |
| Select Structural | | 1600 | 950 | 85 | 625 | 1100 | 1,600,000 | |
| Dense No.1 | | 1550 | 775 | 85 | 730 | 1100 | 1,700,000 | |
| No.1 | | 1350 | 675 | 85 | 625 | 925 | 1,600,000 | |
| No.2 | | 875 | 425 | 85 | 625 | 600 | 1,300,000 | |
| Dense Select Structural | Posts and Timbers | 1750 | 1150 | 85 | 730 | 1350 | 1,700,000 | |
| Select Structural | | 1500 | 1000 | 85 | 625 | 1150 | 1,600,000 | |
| Dense No.1 | | 1400 | 950 | 85 | 730 | 1200 | 1,700,000 | |
| No.1 | | 1200 | 825 | 85 | 625 | 1000 | 1,600,000 | |
| No.2 | | 750 | 475 | 85 | 625 | 700 | 1,300,000 | |
| Dense Select Structural | Beams and Stringers | 1850 | 1100 | 85 | 730 | 1300 | 1,700,000 | WWPA |
| Select Structural | | 1600 | 950 | 85 | 625 | 1100 | 1,600,000 | |
| Dense No.1 | | 1550 | 775 | 85 | 730 | 1100 | 1,700,000 | |
| No.1 | | 1350 | 675 | 85 | 625 | 925 | 1,600,000 | |
| Dense No.2 | | 1000 | 500 | 85 | 730 | 700 | 1,400,000 | |
| No.2 | | 875 | 425 | 85 | 625 | 600 | 1,300,000 | |
| Dense Select Structural | Posts and Timbers | 1750 | 1150 | 85 | 730 | 1350 | 1,700,000 | |
| Select Structural | | 1500 | 1000 | 85 | 625 | 1150 | 1,600,000 | |
| Dense No.1 | | 1400 | 950 | 85 | 730 | 1200 | 1,700,000 | |
| No.1 | | 1200 | 825 | 85 | 625 | 1000 | 1,600,000 | |
| Dense No.2 | | 800 | 550 | 85 | 730 | 550 | 1,400,000 | |
| No.2 | | 700 | 475 | 85 | 625 | 475 | 1,300,000 | |

## DOUGLAS FIR-LARCH (NORTH)

| Grade | Size Classification | | | | | | | Agency |
|---|---|---|---|---|---|---|---|---|
| Select Structural | Beams and Stringers | 1600 | 950 | 85 | 625 | 1100 | 1,600,000 | NLGA |
| No.1 | | 1300 | 675 | 85 | 625 | 925 | 1,600,000 | |
| No.2 | | 875 | 425 | 85 | 625 | 600 | 1,300,000 | |
| Select Structural | Posts and Timbers | 1500 | 1000 | 85 | 625 | 1150 | 1,600,000 | |
| No.1 | | 1200 | 825 | 85 | 625 | 1000 | 1,600,000 | |
| No.2 | | 725 | 475 | 85 | 625 | 700 | 1,300,000 | |

## DOUGLAS FIR-SOUTH

| Grade | Size Classification | | | | | | | Agency |
|---|---|---|---|---|---|---|---|---|
| Select Structural | Beams and Stringers | 1550 | 900 | 85 | 520 | 1000 | 1,200,000 | WWPA |
| No.1 | | 1300 | 625 | 85 | 520 | 850 | 1,200,000 | |
| No.2 | | 825 | 425 | 85 | 520 | 525 | 1,000,000 | |
| Select Structural | Posts and Timbers | 1400 | 950 | 85 | 520 | 1050 | 1,200,000 | |
| No.1 | | 1150 | 775 | 85 | 520 | 925 | 1,200,000 | |
| No.2 | | 650 | 400 | 85 | 520 | 425 | 1,000,000 | |

## TABLE 8.6 (Continued)

| Species and commercial grade | Size classification | Bending $F_b$ | Tension parallel to grain $F_t$ | Shear parallel to grain $F_v$ | Compression perpendicular to grain $F_{c\perp}$ | Compression parallel to grain $F_c$ | Modulus of Elasticity $E$ | Grading Rules Agency |
|---|---|---|---|---|---|---|---|---|
| **EASTERN HEMLOCK** | | | | | | | | |
| Select Structural | Beams and Stringers | 1350 | 925 | 80 | 550 | 950 | 1,200,000 | |
| No.1 | | 1150 | 775 | 80 | 550 | 800 | 1,200,000 | NELMA |
| No.2 | | 750 | 375 | 80 | 550 | 550 | 900,000 | |
| Select Structural | Posts and Timbers | 1250 | 850 | 80 | 550 | 1000 | 1,200,000 | |
| No.1 | | 1050 | 700 | 80 | 500 | 875 | 1,200,000 | NSLB |
| No.2 | | 600 | 400 | 80 | 550 | 400 | 900,000 | |
| **EASTERN HEMLOCK-TAMARACK** | | | | | | | | |
| Select Structural | Beams and Stringers | 1400 | 925 | 80 | 555 | 950 | 1,200,000 | |
| No.1 | | 1150 | 775 | 80 | 555 | 800 | 1,200,000 | NELMA |
| No.2 | | 750 | 375 | 80 | 555 | 500 | 900,000 | |
| Select Structural | Posts and Timbers | 1300 | 875 | 80 | 555 | 1000 | 1,200,000 | |
| No.1 | | 1050 | 700 | 80 | 555 | 875 | 1,200,000 | NSLB |
| No.2 | | 600 | 400 | 80 | 555 | 400 | 900,000 | |
| **EASTERN HEMLOCK-TAMARACK (N)** | | | | | | | | |
| Select Structural | Beams and Stringers | 1450 | 850 | 85 | 555 | 950 | 1,300,000 | |
| No.1 | | 1200 | 600 | 85 | 555 | 800 | 1,300,000 | NLGA |
| No.2 | | 775 | 400 | 85 | 555 | 500 | 1,100,000 | |
| Select Structural | Posts and Timbers | 1350 | 900 | 85 | 555 | 1000 | 1,300,000 | |
| No.1 | | 1100 | 725 | 85 | 555 | 875 | 1,300,000 | |
| No.2 | | 650 | 425 | 85 | 555 | 600 | 1,100,000 | |
| **EASTERN SPRUCE** | | | | | | | | |
| Select Structural | Beams and Stringers | 1050 | 725 | 65 | 390 | 750 | 1,400,000 | |
| No.1 | | 900 | 600 | 65 | 390 | 625 | 1,400,000 | NELMA |
| No.2 | | 575 | 275 | 65 | 390 | 375 | 1,000,000 | |
| Select Structural | Posts and Timbers | 1000 | 675 | 65 | 390 | 775 | 1,400,000 | |
| No.1 | | 800 | 550 | 65 | 390 | 675 | 1,400,000 | NSLB |
| No.2 | | 450 | 300 | 65 | 390 | 300 | 1,000,000 | |

Design values in pounds per square inch (psi)

| Species / Size Classification / Grade | | | | | | | Agency |
|---|---|---|---|---|---|---|---|
| **EASTERN WHITE PINE** | | | | | | | |
| Beams and Stringers | Select Structural | 1050 | 700 | 65 | 350 | 675 | 1,100,000 |
| | No.1 | 875 | 600 | 65 | 350 | 575 | 1,100,000 |
| | No.2 | 575 | 275 | 65 | 350 | 400 | 900,000 |
| | | | | | | | NELMA |
| Posts and Timbers | Select Structural | 975 | 650 | 65 | 350 | 725 | 1,100,000 |
| | No.1 | 800 | 525 | 65 | 350 | 625 | 1,100,000 |
| | No.2 | 450 | 300 | 65 | 350 | 325 | 900,000 |
| | | | | | | | NSLB |
| **HEM-FIR** | | | | | | | |
| Beams and Stringers | Select Structural | 1300 | 750 | 70 | 405 | 925 | 1,300,000 |
| | No.1 | 1050 | 525 | 70 | 405 | 750 | 1,300,000 |
| | No.2 | 675 | 350 | 70 | 405 | 500 | 1,100,000 |
| Posts and Timbers | Select Structural | 1200 | 800 | 70 | 405 | 975 | 1,300,000 |
| | No.1 | 975 | 650 | 70 | 405 | 850 | 1,300,000 |
| | No.2 | 575 | 375 | 70 | 405 | 575 | 1,100,000 |
| | | | | | | | WCLIB |
| Beams and Stringers | Select Structural | 1250 | 725 | 70 | 405 | 925 | 1,300,000 |
| | No.1 | 1050 | 525 | 70 | 405 | 775 | 1,300,000 |
| | No.2 | 675 | 325 | 70 | 405 | 475 | 1,100,000 |
| Posts and Timbers | Select Structural | 1200 | 800 | 70 | 405 | 975 | 1,300,000 |
| | No.1 | 950 | 650 | 70 | 405 | 850 | 1,300,000 |
| | No.2 | 525 | 350 | 70 | 405 | 375 | 1,100,000 |
| | | | | | | | WWPA |
| **HEM-FIR (NORTH)** | | | | | | | |
| Beams and Stringers | Select Structural | 1250 | 725 | 70 | 370 | 900 | 1,300,000 |
| | No.1 | 1000 | 500 | 70 | 370 | 750 | 1,300,000 |
| | No.2 | 675 | 325 | 70 | 370 | 475 | 1,100,000 |
| Posts and Timbers | Select Structural | 1150 | 775 | 70 | 370 | 950 | 1,300,000 |
| | No.1 | 925 | 625 | 70 | 370 | 850 | 1,300,000 |
| | No.2 | 550 | 375 | 70 | 370 | 575 | 1,100,000 |
| | | | | | | | NLGA |
| **MIXED MAPLE** | | | | | | | |
| Beams and Stringers | Select Structural | 1150 | 700 | 90 | 620 | 725 | 1,100,000 |
| | No.1 | 975 | 500 | 90 | 620 | 600 | 1,100,000 |
| | No.2 | 625 | 325 | 90 | 620 | 375 | 900,000 |
| Posts and Timbers | Select Structural | 1100 | 725 | 90 | 620 | 750 | 1,100,000 |
| | No.1 | 875 | 600 | 90 | 620 | 650 | 1,100,000 |
| | No.2 | 500 | 350 | 90 | 620 | 300 | 900,000 |
| | | | | | | | NELMA |

## TABLE 8.6 (Continued)

| Species and commercial grade | Size classification | Bending $F_b$ | Tension parallel to grain $F_t$ | Shear parallel to grain $F_v$ | Compression perpendicular to grain $F_{c\perp}$ | Compression parallel to grain $F_c$ | Modulus of Elasticity $E$ | Grading Rules Agency |
|---|---|---|---|---|---|---|---|---|
| **MIXED OAK** | | | | | | | | |
| Select Structural | Beams and Stringers | 1350 | 800 | 80 | 800 | 825 | 1,000,000 | |
| No.1 | | 1150 | 550 | 80 | 800 | 700 | 1,000,000 | NELMA |
| No.2 | | 725 | 375 | 80 | 800 | 450 | 800,000 | |
| Select Structural | Posts and Timbers | 1250 | 850 | 80 | 800 | 875 | 1,000,000 | |
| No.1 | | 1000 | 675 | 80 | 800 | 775 | 1,000,000 | |
| No.2 | | 575 | 400 | 80 | 800 | 350 | 800,000 | |
| **MIXED SOUTHERN PINE** | | **(Wet Service Conditions)** | | | | | | |
| Select Structural | 5"x 5" & larger | 1500 | 1000 | 110 | 375 | 900 | 1,300,000 | |
| No.1 | | 1350 | 900 | 110 | 375 | 800 | 1,300,000 | SPIB |
| No.2 | | 850 | 550 | 95 | 375 | 525 | 1,000,000 | |
| **MOUNTAIN HEMLOCK** | | | | | | | | |
| Select Structural | Beams and Stringers | 1350 | 775 | 85 | 570 | 875 | 1,100,000 | |
| No.1 | | 1100 | 550 | 85 | 570 | 725 | 1,100,000 | WCLIB |
| No.2 | | 725 | 375 | 85 | 570 | 475 | 900,000 | |
| Select Structural | Posts and Timbers | 1250 | 825 | 85 | 570 | 925 | 1,100,000 | |
| No.1 | | 1000 | 675 | 85 | 570 | 800 | 1,100,000 | |
| No.2 | | 625 | 400 | 85 | 570 | 550 | 900,000 | |
| **NORTHERN PINE** | | | | | | | | |
| Select Structural | Beams and Stringers | 1250 | 850 | 65 | 435 | 850 | 1,300,000 | |
| No.1 | | 1050 | 700 | 65 | 435 | 725 | 1,300,000 | NELMA |
| No.2 | | 675 | 350 | 65 | 435 | 450 | 1,000,000 | |
| Select Structural | Posts and Timbers | 1150 | 800 | 65 | 435 | 900 | 1,300,000 | |
| No.1 | | 950 | 650 | 65 | 435 | 800 | 1,300,000 | NSLB |
| No.2 | | 550 | 375 | 65 | 435 | 375 | 1,000,000 | |

Design values in pounds per square inch (psi)

| Species / Size Classification | Grade | | | | | | | Agency |
|---|---|---|---|---|---|---|---|---|
| **NORTHERN RED OAK** | | | | | | | | |
| Beams and Stringers | Select Structural | 1600 | 950 | 105 | 885 | 950 | 1,300,000 | NELMA |
| | No.1 | 1350 | 675 | 105 | 885 | 800 | 1,300,000 | |
| | No.2 | 875 | 425 | 105 | 885 | 500 | 1,000,000 | |
| Posts and Timbers | Select Structural | 1500 | 1000 | 105 | 885 | 1000 | 1,300,000 | |
| | No.1 | 1200 | 800 | 105 | 885 | 875 | 1,300,000 | |
| | No.2 | 700 | 475 | 105 | 885 | 400 | 1,000,000 | |
| **NORTHERN WHITE CEDAR** | | | | | | | | |
| Beams and Stringers | Select Structural | 900 | 600 | 60 | 370 | 600 | 700,000 | NELMA |
| | No.1 | 750 | 500 | 60 | 370 | 500 | 700,000 | |
| | No.2 | 500 | 250 | 60 | 370 | 325 | 600,000 | |
| Posts and Timbers | Select Structural | 850 | 575 | 60 | 370 | 650 | 700,000 | |
| | No.1 | 675 | 450 | 60 | 370 | 550 | 700,000 | |
| | No.2 | 400 | 250 | 60 | 370 | 250 | 600,000 | |
| **PONDEROSA PINE** | | | | | | | | |
| Beams and Stringers | Select Structural | 1100 | 725 | 65 | 535 | 750 | 1,100,000 | NLGA |
| | No.1 | 925 | 500 | 65 | 535 | 625 | 1,100,000 | |
| | No.2 | 600 | 300 | 65 | 535 | 400 | 900,000 | |
| Posts and Timbers | Select Structural | 1000 | 675 | 65 | 535 | 800 | 1,100,000 | |
| | No.1 | 825 | 550 | 65 | 535 | 700 | 1,100,000 | |
| | No.2 | 475 | 325 | 65 | 535 | 325 | 900,000 | |
| **RED MAPLE** | | | | | | | | |
| Beams and Stringers | Select Structural | 1500 | 875 | 100 | 615 | 900 | 1,500,000 | NELMA |
| | No.1 | 1250 | 625 | 100 | 615 | 750 | 1,500,000 | |
| | No.2 | 800 | 400 | 100 | 615 | 475 | 1,200,000 | |
| Posts and Timbers | Select Structural | 1400 | 925 | 100 | 615 | 950 | 1,500,000 | |
| | No.1 | 1150 | 750 | 100 | 615 | 825 | 1,500,000 | |
| | No.2 | 650 | 425 | 100 | 615 | 375 | 1,200,000 | |
| **RED OAK** | | | | | | | | |
| Beams and Stringers | Select Structural | 1350 | 800 | 80 | 820 | 825 | 1,200,000 | NELMA |
| | No.1 | 1150 | 550 | 80 | 820 | 700 | 1,200,000 | |
| | No.2 | 725 | 375 | 80 | 820 | 450 | 1,000,000 | |
| Posts and Timbers | Select Structural | 1250 | 850 | 80 | 820 | 875 | 1,200,000 | |
| | No.1 | 1000 | 675 | 80 | 820 | 775 | 1,200,000 | |
| | No.2 | 575 | 400 | 80 | 820 | 350 | 1,000,000 | |

## TABLE 8.6  (Continued)

| Species and commercial grade | Size classification | Design values in pounds per square inch (psi) | | | | | | Grading Rules Agency |
|---|---|---|---|---|---|---|---|---|
| | | Bending $F_b$ | Tension parallel to grain $F_t$ | Shear parallel to grain $F_v$ | Compression perpendicular to grain $F_{c\perp}$ | Compression parallel to grain $F_c$ | Modulus of Elasticity E | |
| **RED PINE** | | | | | | | | |
| Select Structural | Beams and Stringers | 1050 | 625 | 65 | 440 | 725 | 1,100,000 | NLGA |
| No.1 | | 875 | 450 | 65 | 440 | 600 | 1,100,000 | |
| No.2 | | 575 | 300 | 65 | 440 | 375 | 900,000 | |
| Select Structural | Posts and Timbers | 1000 | 675 | 65 | 440 | 775 | 1,100,000 | |
| No.1 | | 800 | 550 | 65 | 440 | 675 | 1,100,000 | |
| No.2 | | 475 | 325 | 65 | 440 | 475 | 900,000 | |
| **REDWOOD** | | | | | | | | |
| Clear Heart Structural or Clear Structural | 5"x 5" & larger | 1850 | 1250 | 135 | 650 | 1650 | 1,300,000 | RIS |
| Select Structural | | 1400 | 950 | 95 | 650 | 1200 | 1,300,000 | |
| No.1 | | 1200 | 800 | 95 | 650 | 1050 | 1,300,000 | |
| No.2 | | 975 | 650 | 95 | 650 | 900 | 1,100,000 | |
| **SITKA SPRUCE** | | | | | | | | |
| Select Structural | Beams and Stringers | 1200 | 675 | 70 | 435 | 825 | 1,300,000 | WCLIB |
| No.1 | | 1000 | 500 | 70 | 435 | 675 | 1,300,000 | |
| No.2 | | 650 | 325 | 70 | 435 | 450 | 1,000,000 | |
| Select Structural | Posts and Timbers | 1150 | 750 | 70 | 435 | 875 | 1,300,000 | |
| No.1 | | 925 | 600 | 70 | 435 | 750 | 1,300,000 | |
| No.2 | | 550 | 350 | 70 | 435 | 525 | 1,000,000 | |
| **SOUTHERN PINE** | | (Wet Service Conditions) | | | | | | |
| Dense Select Structural | 5"x 5" | 1750 | 1200 | 110 | 440 | 1100 | 1,600,000 | |
| Select Structural | | 1500 | 1000 | 110 | 375 | 950 | 1,500,000 | |
| No.1 Dense | | 1550 | 1050 | 110 | 440 | 975 | 1,600,000 | |
| No.1 | | 1350 | 900 | 110 | 375 | 825 | 1,500,000 | |

Table of structural lumber design values (headers not shown on this page; columns are the five stress values, modulus of elasticity E, and grading rules agency).

| Group | Classification | Grade | (1) | (2) | (3) | (4) | (5) | E | Agency |
|---|---|---|---|---|---|---|---|---|---|
| | | | 975 | 650 | 100 | 440 | 625 | 1,300,000 | SPIB |
| | | | 850 | 550 | 100 | 375 | 525 | 1,200,000 | |
| | | | 2100 | 1400 | 145 | 440 | 1300 | 1,600,000 | |
| | | | 1750 | 1200 | 120 | 440 | 1100 | 1,600,000 | |
| | | | 1600 | 1050 | 110 | 440 | 1000 | 1,600,000 | |
| | Beams and Stringers | Select Structural | 1100 | 650 | 65 | 425 | 775 | 1,300,000 | NLGA |
| | | No.1 | 900 | 450 | 65 | 425 | 625 | 1,300,000 | |
| | | No.2 | 600 | 300 | 65 | 425 | 425 | 1,000,000 | |
| | Posts and Timbers | Select Structural | 1050 | 700 | 65 | 425 | 800 | 1,300,000 | |
| | | No.1 | 850 | 550 | 65 | 425 | 700 | 1,300,000 | |
| | | No.2 | 500 | 325 | 65 | 425 | 500 | 1,000,000 | |
| **SPRUCE-PINE-FIR (SOUTH)** | Beams and Stringers | Select Structural | 1050 | 625 | 65 | 335 | 675 | 1,200,000 | NELMA / NSLB |
| | | No.1 | 900 | 450 | 65 | 335 | 575 | 1,200,000 | |
| | | No.2 | 575 | 300 | 65 | 335 | 350 | 1,000,000 | |
| | Posts and Timbers | Select Structural | 1000 | 675 | 65 | 335 | 700 | 1,200,000 | WWPA |
| | | No.1 | 800 | 550 | 65 | 335 | 625 | 1,200,000 | |
| | | No.2 | 350 | 225 | 65 | 335 | 225 | 1,000,000 | |
| **WESTERN CEDARS** | Beams and Stringers | Select Structural | 1150 | 675 | 70 | 425 | 875 | 1,000,000 | WCLIB |
| | | No.1 | 975 | 475 | 70 | 425 | 725 | 1,000,000 | |
| | | No.2 | 625 | 325 | 70 | 425 | 475 | 800,000 | |
| | Posts and Timbers | Select Structural | 1100 | 725 | 70 | 425 | 925 | 1,000,000 | |
| | | No.1 | 875 | 600 | 70 | 425 | 800 | 1,000,000 | |
| | | No.2 | 550 | 350 | 70 | 425 | 550 | 800,000 | |
| | Beams and Stringers | Select Structural | 1150 | 700 | 70 | 425 | 875 | 1,000,000 | WWPA |
| | | No.1 | 975 | 475 | 70 | 425 | 725 | 1,000,000 | |
| | | No.2 | 625 | 325 | 70 | 425 | 475 | 800,000 | |
| | Posts and Timbers | Select Structural | 1100 | 725 | 70 | 425 | 925 | 1,000,000 | |
| | | No.1 | 875 | 600 | 70 | 425 | 800 | 1,000,000 | |
| | | No.2 | 500 | 350 | 70 | 425 | 375 | 800,000 | |

**TABLE 8.6** (*Continued*)

| Species and commercial grade | Size classification | Design values in pounds per square inch (psi) | | | | | | Grading Rules Agency |
|---|---|---|---|---|---|---|---|---|
| | | Bending F_b | Tension parallel to grain F_t | Shear parallel to grain F_v | Compression perpendicular to grain F_{c⊥} | Compression parallel to grain F_c | Modulus of Elasticity E | |
| **WESTERN CEDARS (NORTH)** | | | | | | | | |
| Select Structural | Beams and Stringers | 1150 | 675 | 65 | 425 | 850 | 1,000,000 | NLGA |
| No.1 | | 925 | 475 | 65 | 425 | 700 | 1,000,000 | |
| No.2 | | 625 | 300 | 65 | 425 | 450 | 800,000 | |
| Select Structural | Posts and Timbers | 1050 | 700 | 65 | 425 | 900 | 1,000,000 | |
| No.1 | | 875 | 575 | 65 | 425 | 800 | 1,000,000 | |
| No.2 | | 500 | 350 | 65 | 425 | 550 | 800,000 | |
| **WESTERN HEMLOCK** | | | | | | | | |
| Select Structural | Beams and Stringers | 1400 | 825 | 85 | 410 | 1000 | 1,400,000 | WCLIB |
| No.1 | | 1150 | 575 | 85 | 410 | 850 | 1,400,000 | |
| No.2 | | 750 | 375 | 85 | 410 | 550 | 1,100,000 | |
| Select Structural | Posts and Timbers | 1300 | 875 | 85 | 410 | 1100 | 1,400,000 | |
| No.1 | | 1050 | 700 | 85 | 410 | 950 | 1,400,000 | |
| No.2 | | 650 | 425 | 85 | 410 | 650 | 1,100,000 | |
| **WESTERN HEMLOCK (NORTH)** | | | | | | | | |
| Select Structural | Beams and Stringers | 1400 | 825 | 70 | 410 | 1000 | 1,400,000 | NLGA |
| No.1 | | 1150 | 575 | 70 | 410 | 850 | 1,400,000 | |
| No.2 | | 750 | 375 | 70 | 410 | 550 | 1,100,000 | |
| Select Structural | Posts and Timbers | 1300 | 875 | 70 | 410 | 1100 | 1,400,000 | |
| No.1 | | 1050 | 700 | 70 | 410 | 950 | 1,400,000 | |
| No.2 | | 650 | 425 | 70 | 410 | 650 | 1,100,000 | |
| **WESTERN WHITE PINE** | | | | | | | | |
| Select Structural | Beams and Stringers | 1050 | 600 | 60 | 375 | 775 | 1,300,000 | NLGA |
| No. | | 850 | 425 | 60 | 375 | 625 | 1,300,000 | |
| No.2 | | 550 | 275 | 60 | 375 | 400 | 1,000,000 | |
| Select Structural | Posts and Timbers | 975 | 650 | 60 | 375 | 800 | 1,300,000 | |
| No.1 | | 775 | 525 | 60 | 375 | 700 | 1,300,000 | |
| No.2 | | 450 | 300 | 60 | 375 | 500 | 1,000,000 | |

## WESTERN WOODS

| | | | | | | | | |
|---|---|---|---|---|---|---|---|---|
| **Beams and Stringers** | Select Structural | 1050 | 625 | 65 | 335 | 675 | 1,100,000 | |
| | No.1 | 900 | 450 | 65 | 335 | 575 | 1,100,000 | WWPA |
| | No.2 | 575 | 300 | 65 | 335 | 350 | 900,000 | |
| **Posts and Timbers** | Select Structural | 1000 | 675 | 65 | 335 | 700 | 1,100,000 | |
| | No.1 | 800 | 550 | 65 | 335 | 625 | 1,100,000 | |
| | No.2 | 350 | 225 | 65 | 335 | 225 | 900,000 | |

## WHITE OAK

| | | | | | | | | |
|---|---|---|---|---|---|---|---|---|
| **Beams and Stringers** | Select Structural | 1400 | 825 | 105 | 800 | 900 | 1,000,000 | |
| | No.1 | 1200 | 575 | 105 | 800 | 775 | 1,000,000 | NELMA |
| | No.2 | 750 | 375 | 105 | 800 | 475 | 800,000 | |
| **Posts and Timbers** | Select Structural | 1300 | 875 | 105 | 800 | 950 | 1,000,000 | |
| | No.1 | 1050 | 700 | 105 | 800 | 825 | 1,000,000 | |
| | No.2 | 600 | 400 | 105 | 800 | 400 | 800,000 | |

Source: ANSI/NF$_o$PA NDS—1991, *National Design Specification (NDS) for Wood Construction*, AFPA (formerly NFPA), Washington, DC, 1991.

**1. LUMBER DIMENSIONS.** Tabulated design values are applicable to lumber that will be used under dry conditions such as in most covered structures. For 5″ and thicker lumber, the GREEN dressed sizes shall be permitted to be used (see Table 1A of NDS) because design values have been adjusted to compensate for any loss in size by shrinkage which may occur.

**2. SPRUCE PINE.** To obtain recommended design values for Spruce Pine graded to Southern Pine Inspection Bureau (SPIB) rules, multiply the appropriate design values for Mixed Southern Pine by corresponding conversion factor shown below and round to the nearest 100,000 psi for E; to the next lower multiple of 5 psi for $F_v$ and $F_{c\perp}$; to the next lower multiple of 50 psi for $F_b$, $F_t$, and $F_c$ if 1000 psi or greater, 25 psi otherwise.

## CONVERSION FACTORS FOR DETERMINING DESIGN VALUES FOR SPRUCE PINE

| | Bending $F_b$ | Tension parallel to grain $F_t$ | Shear parallel to grain $F_v$ | Compression perpendicular to grain $F_{c\perp}$ | Compression parallel to grain $F_c$ | Modulus of Elasticity E |
|---|---|---|---|---|---|---|
| Conversion Factor | 0.784 | 0.784 | 0.965 | 0.682 | 0.766 | 0.807 |

## TABLE 8.6—Adjustment Factors

### SIZE FACTOR, $C_F$

When the depth, d, of a beam, stringer, post or timber exceeds 12", the tabulated bending design value, $F_b$, shall be multiplied by the following size factor:

$$C_F = (12/d)^{1/9}$$

### WET SERVICE FACTOR, $C_M$

When timbers are used where moisture content will exceed 19% for an extended time period, design values shall be multiplied by the appropriate wet service factors from the following table (for Southern Pine and Mixed Southern Pine use tabulated design values without further adjustment):

#### WET SERVICE FACTORS, $C_M$

| $F_b$ | $F_t$ | $F_v$ | $F_{c\perp}$ | $F_c$ | E |
|-------|-------|-------|--------------|-------|------|
| 1.00  | 1.00  | 1.00  | 0.67         | 0.91  | 1.00 |

# SHEAR STRESS FACTOR, $C_H$

Tabulated shear design values parallel to grain have been reduced to allow for the occurrence of splits, checks and shakes. Tabulated shear design values parallel to grain, $F_v$, shall be permitted to be multiplied by the shear stress factors specified in the following table when length of split, or size of check or shake is known and no increase in them is anticipated. When shear stress factors are used for Redwood, Southern Pine and Mixed Southern Pine, a tabulated design value of $F_v = 80$ psi shall be assigned for all grades of Redwood and a tabulated design value of $F_v = 90$ psi shall be assigned for all grades of Southern Pine and Mixed Southern Pine. Shear stress factors shall be permitted to be linearly interpolated.

## SHEAR STRESS FACTORS, $C_H$

| Length of split on wide face of 5" (nominal) and thicker lumber | $C_H$ | Size of shake* in 5" (nominal) and thicker lumber | $C_H$ |
|---|---|---|---|
| no split | 2.00 | no shake | 2.00 |
| 1/2 x narrow face. | 1.67 | 1/6 x narrow face | 1.67 |
| 3/4 x narrow face | 1.50 | 1/4 narrow face | 1.50 |
| 1 x narrow face | 1.33 | 1/3 x narrow face | 1.33 |
| 1-1/2 x narrow face or more | 1.00 | 1/2 x narrow face or more. | 1.00 |
| | | *Shake is measured at the end between lines enclosing the shake and perpendicular to the loaded face. | |

# TABLE 8.7

## Bearing Design Values Parallel to Grain, $F_g$

(Tabulated design values are for normal load duration. See NDS 2.3 for a comprehensive description of design value adjustment factors.)

| Species Combination | Wet Service Conditions | Dry service conditions | | |
| --- | --- | --- | --- | --- |
| | | Sawn Lumber | | Glued[a,b] Laminated Timber |
| | | 5" x 5" & Larger | 2" - 4" thick | |
| ASPEN | 740 | — | 1110 | — |
| BALSAM FIR | 890 | 980 | 1330 | — |
| BEECH-BIRCH-HICKORY | 1180 | 1300 | 1770 | 2070 |
| COAST SITKA SPRUCE | 950 | 1040 | 1420 | — |
| COAST SPECIES | 950 | — | 1420 | — |
| COTTONWOOD | 760 | — | 1150 | 1340 [a] |
| DOUGLAS FIR-LARCH (Dense) | 1570 | 1730 | 2360 | [a] |
| DOUGLAS FIR-LARCH | 1350 | 1480 | 2020 | — |
| DOUGLAS FIR-LARCH (NORTH) | 1340 | 1480 | 2020 | — |
| DOUGLAS FIR-SOUTH | 1220 | 1340 | 1820 | — |
| EASTERN HEMLOCK | 1140 | 1260 | — | — |
| EASTERN HEMLOCK-TAMARACK | 1150 | 1270 | 1730 | — |
| EASTERN HEMLOCK-TAMARACK (NORTH) | 1160 | 1280 | 1740 | — |
| EASTERN SOFTWOODS | 890 | — | 1340 | — |
| EASTERN SPRUCE | 970 | 1070 | 1460 | — |
| EASTERN WHITE PINE | 900 | 990 | 1360 | — |
| EASTERN WHITE PINE (NORTH) | 960 | — | 1440 | — |
| HEM-FIR | 1110 | 1220 | 1670 | — |
| HEM-FIR (NORTH) | 1160 | 1280 | 1750 | — |
| MIXED MAPLE | 870 | 960 | 1310 | 1520 |
| MIXED OAK | 1010 | 1110 | 1520 | 1770 |
| MIXED SOUTHERN PINE | 1270 | 1390 | 1900 | — |

| Species | | | | |
|---|---|---|---|---|
| MOUNTAIN HEMLOCK | 1070 | 1170 | 1570 | — |
| NORTHERN PINE | 1040 | 1150 | 1730 | — |
| NORTHERN RED OAK | 1150 | 1270 | — | 2010 |
| NORTHERN SPECIES | 880 | — | 1320 | 1540 |
| NORTHERN WHITE CEDAR | 740 | 810 | 1110 | — |
| PONDEROSA PINE | 1050 | 1160 | 1580 | — |
| RED MAPLE | 1100 | 1210 | 1650 | 1930 |
| RED OAK | 1010 | 1110 | 1520 | 1770 |
| RED PINE | 880 | 970 | 1320 | — |
| REDWOOD (Close grain) | 1560 | 1720 | 2340 | — |
| REDWOOD (Open grain) | 1150 | 1270 | 1730 | — |
| SITKA SPRUCE | 990 | 1090 | 1480 | — |
| SOUTHERN PINE (Dense) | 1540 | 1690 | 2310 | — |
| SOUTHERN PINE | 1320 | 1450 | 1970 | — |
| SPRUCE-PINE-FIR | 940 | 1040 | 1410 | 1650 |
| SPRUCE-PINE-FIR (SOUTH) | 810 | 900 | 1220 | 1430 |
| WESTERN CEDARS[c] | 1060 | 1170 | 1590 | — |
| WESTERN CEDARS (NORTH) | 1040 | 1140 | 1560 | — |
| WESTERN HEMLOCK | 1240 | 1360 | 1860 | — |
| WESTERN HEMLOCK (NORTH) | 1240 | 1360 | 1860 | — |
| WESTERN WHITE PINE | 930 | 1030 | 1400 | — |
| WESTERN WOODS | 810 | 900 | 1220 | — |
| WHITE OAK | 1100 | 1210 | 1650 | 1930 |
| YELLOW POPLAR | 890 | — | 1340 | 1560 |

Source: ANSI/NF₀PA NDS—1991, *National Design Specification (NDS) for Wood Construction*, AFPA (formerly NFPA), Washington, DC, 1991.

[a] Wet and dry service conditions are defined in NDS 4.1.4 for sawn lumber and in AITC 117—Design, section 4.3 for glued laminated timber. Wet service factors, $C_M$, have already been applied to sawn lumber tabulated bearing design values to reflect the appropriate moisture service conditions.

[b] Bearing design values for species typically used in structural glued laminated timber are given in AITC 117—Design. Values shown in this column may vary from values used in laminating due to combination of grades, and whether full or partial bearing on end grain.

[c] Values for Alaska Cedar used in glued laminated timber are given in AITC 117—Design.

## 8.4 SPAN LOAD TABLES FOR GLUED LAMINATED TIMBER

### TABLE 8.8
### Structural Glued Laminated Timber[a]
### Simple-Span Beam Sizes

| Span (ft) | Spacing (ft) | Construction Load[b,c] | Roof[c,e] Snow Loads[b,c] | | | | | | Floor[d] Live Load[b,c] |
|---|---|---|---|---|---|---|---|---|---|
| **Duration of Load Factor $C_D$ =** | | 1.25 | 1.15 | | | | | | 1.00 |
| **Total Load =** | | 30 psf | 30 psf | 35 psf | 40 psf | 45 psf | 50 psf | 55 psf | 50 psf |
| 8 | 4 | $*3\frac{1}{8} \times 6$ | $*3\frac{1}{8} \times 6$ | $*3\frac{1}{8} \times 6$ | $*3\frac{1}{8} \times 6$ | $*3\frac{1}{8} \times 6$ | $*3\frac{1}{8} \times 6$ | $*3\frac{1}{8} \times 6$ | $*3\frac{1}{8} \times 6$ |
|  | 6 | $*3\frac{1}{8} \times 6$ | $*3\frac{1}{8} \times 6$ | $*3\frac{1}{8} \times 6$ | $*3\frac{1}{8} \times 6$ | $*3\frac{1}{8} \times 6$ | $*3\frac{1}{8} \times 6$ | $*3\frac{1}{8} \times 6$ | $*3\frac{1}{8} \times 6$ |
|  | 8 | $*3\frac{1}{8} \times 6$ | $*3\frac{1}{8} \times 6$ | $*3\frac{1}{8} \times 6$ | $*3\frac{1}{8} \times 6$ | $*3\frac{1}{8} \times 6$ | $*3\frac{1}{8} \times 6$ | $*3\frac{1}{8} \times 6$ | $*3\frac{1}{8} \times 7\frac{1}{2}$ |
| 10 | 4 | $*3\frac{1}{8} \times 6$ | $*3\frac{1}{8} \times 6$ | $*3\frac{1}{8} \times 6$ | $*3\frac{1}{8} \times 6$ | $*3\frac{1}{8} \times 6$ | $*3\frac{1}{8} \times 6$ | $*3\frac{1}{8} \times 6$ | $*3\frac{1}{8} \times 7\frac{1}{2}$ |
|  | 6 | $*3\frac{1}{8} \times 6$ | $*3\frac{1}{8} \times 6$ | $*3\frac{1}{8} \times 6$ | $*3\frac{1}{8} \times 6$ | $*3\frac{1}{8} \times 6$ | $*3\frac{1}{8} \times 7\frac{1}{2}$ | $*3\frac{1}{8} \times 7\frac{1}{2}$ | $*3\frac{1}{8} \times 7\frac{1}{2}$ |
|  | 8 | $*3\frac{1}{8} \times 6$ | $*3\frac{1}{8} \times 6$ | $*3\frac{1}{8} \times 7\frac{1}{2}$ | $*3\frac{1}{8} \times 7\frac{1}{2}$ | $*3\frac{1}{8} \times 7\frac{1}{2}$ | $*3\frac{1}{8} \times 7\frac{1}{2}$ | $*3\frac{1}{8} \times 7\frac{1}{2}$ | $*3\frac{1}{8} \times 9$ |
|  | 10 | $*3\frac{1}{8} \times 6$ | $*3\frac{1}{8} \times 7\frac{1}{2}$ | $*3\frac{1}{8} \times 7\frac{1}{2}$ | $*3\frac{1}{8} \times 7\frac{1}{2}$ | $*3\frac{1}{8} \times 9$ | $*3\frac{1}{8} \times 9$ | $*3\frac{1}{8} \times 9$ | $*3\frac{1}{8} \times 9$ |

| | | | | | | | | |
|---|---|---|---|---|---|---|---|---|
| 12 | 6 | $*3\frac{1}{8} \times 6$ | $*3\frac{1}{8} \times 7\frac{1}{2}$ | $*3\frac{1}{8} \times 7\frac{1}{2}$ | $*3\frac{1}{8} \times 7\frac{1}{2}$ | $*3\frac{1}{8} \times 9$ | $*3\frac{1}{8} \times 9$ | $*3\frac{1}{8} \times 9$ |
| | 8 | $*3\frac{1}{8} \times 7\frac{1}{2}$ | $*3\frac{1}{8} \times 7\frac{1}{2}$ | $*3\frac{1}{8} \times 7\frac{1}{2}$ | $*3\frac{1}{8} \times 9$ | $*3\frac{1}{8} \times 9$ | $*3\frac{1}{8} \times 9$ | $*3\frac{1}{8} \times 10\frac{1}{2}$ |
| | 10 | $*3\frac{1}{8} \times 9$ | $*3\frac{1}{8} \times 9$ | $*3\frac{1}{8} \times 9$ | $*3\frac{1}{8} \times 9$ | $*3\frac{1}{8} \times 10\frac{1}{2}$ | $*3\frac{1}{8} \times 10\frac{1}{2}$ | $*3\frac{1}{8} \times 12$ |
| | 12 | $*3\frac{1}{8} \times 9$ | $*3\frac{1}{8} \times 9$ | $*3\frac{1}{8} \times 9$ | $*3\frac{1}{8} \times 10\frac{1}{2}$ | $*3\frac{1}{8} \times 10\frac{1}{2}$ | $*3\frac{1}{8} \times 12$ | $*3\frac{1}{8} \times 12$ |
| 14 | 8 | $*3\frac{1}{8} \times 9$ | $*3\frac{1}{8} \times 9$ | $*3\frac{1}{8} \times 9$ | $*3\frac{1}{8} \times 10\frac{1}{2}$ | $*3\frac{1}{8} \times 10\frac{1}{2}$ | $*3\frac{1}{8} \times 10\frac{1}{2}$ | $*3\frac{1}{8} \times 12$ |
| | 10 | $*3\frac{1}{8} \times 9$ | $*3\frac{1}{8} \times 9$ | $*3\frac{1}{8} \times 10\frac{1}{2}$ | $*3\frac{1}{8} \times 10\frac{1}{2}$ | $*3\frac{1}{8} \times 12$ | $*3\frac{1}{8} \times 12$ | $*3\frac{1}{8} \times 13\frac{1}{2}$ |
| | 12 | $*3\frac{1}{8} \times 10\frac{1}{2}$ | $*3\frac{1}{8} \times 10\frac{1}{2}$ | $*3\frac{1}{8} \times 10\frac{1}{2}$ | $*3\frac{1}{8} \times 12$ | $*3\frac{1}{8} \times 13\frac{1}{2}$ | $*3\frac{1}{8} \times 13\frac{1}{2}$ | $*3\frac{1}{8} \times 13\frac{1}{2}$ |
| | 14 | $*3\frac{1}{8} \times 10\frac{1}{2}$ | $*3\frac{1}{8} \times 10\frac{1}{2}$ | $*3\frac{1}{8} \times 12$ | $*3\frac{1}{8} \times 12$ | $*3\frac{1}{8} \times 13\frac{1}{2}$ | $*3\frac{1}{8} \times 15$ | $*3\frac{1}{8} \times 15$ |
| 16 | 8 | $*3\frac{1}{8} \times 10\frac{1}{2}$ | $*3\frac{1}{8} \times 10\frac{1}{2}$ | $*3\frac{1}{8} \times 10\frac{1}{2}$ | $*3\frac{1}{8} \times 10\frac{1}{2}$ | $*3\frac{1}{8} \times 12$ | $*3\frac{1}{8} \times 12$ | $*3\frac{1}{8} \times 13\frac{1}{2}$ |
| | 12 | $*3\frac{1}{8} \times 12$ | $*3\frac{1}{8} \times 12$ | $*3\frac{1}{8} \times 12$ | $*3\frac{1}{8} \times 13\frac{1}{2}$ | $*3\frac{1}{8} \times 15$ | $*3\frac{1}{8} \times 15$ | $*3\frac{1}{8} \times 16\frac{1}{2}$ |
| | 14 | $*3\frac{1}{8} \times 12$ | $*3\frac{1}{8} \times 12$ | $*3\frac{1}{8} \times 13\frac{1}{2}$ | $*3\frac{1}{8} \times 15$ | $*3\frac{1}{8} \times 16\frac{1}{2}$ | $*3\frac{1}{8} \times 16\frac{1}{2}$ | $*3\frac{1}{8} \times 18$ |
| | 16 | $*3\frac{1}{8} \times 13\frac{1}{2}$ | $*3\frac{1}{8} \times 13\frac{1}{2}$ | $*3\frac{1}{8} \times 15$ | $*3\frac{1}{8} \times 15$ | $*3\frac{1}{8} \times 16\frac{1}{2}$ | $*3\frac{1}{8} \times 18$ | $*5\frac{1}{8} \times 15$ |
| 18 | 8 | $*3\frac{1}{8} \times 12$ | $*3\frac{1}{8} \times 12$ | $*3\frac{1}{8} \times 12$ | $*3\frac{1}{8} \times 12$ | $*3\frac{1}{8} \times 13\frac{1}{2}$ | $*3\frac{1}{8} \times 15$ | $*3\frac{1}{8} \times 15$ |
| | 12 | $*3\frac{1}{8} \times 13\frac{1}{2}$ | $*3\frac{1}{8} \times 13\frac{1}{2}$ | $*3\frac{1}{8} \times 13\frac{1}{2}$ | $*3\frac{1}{8} \times 15$ | $*3\frac{1}{8} \times 16\frac{1}{2}$ | $*3\frac{1}{8} \times 18$ | $3\frac{1}{8} \times 16\frac{1}{2}$ |
| | 16 | $*3\frac{1}{8} \times 15$ | $*3\frac{1}{8} \times 15$ | $*3\frac{1}{8} \times 15$ | $*3\frac{1}{8} \times 18$ | $*3\frac{1}{8} \times 19\frac{1}{2}$ | $*3\frac{1}{8} \times 21$ | $*5\frac{1}{8} \times 16\frac{1}{2}$ |
| | 20 | $*3\frac{1}{8} \times 16\frac{1}{2}$ | $*3\frac{1}{8} \times 16\frac{1}{2}$ | $*3\frac{1}{8} \times 16\frac{1}{2}$ | $*3\frac{1}{8} \times 19\frac{1}{2}$ | $*3\frac{1}{8} \times 24$ | $*3\frac{1}{8} \times 25\frac{1}{2}$ | $5\frac{1}{8} \times 16\frac{1}{2}$ |
| 20 | 8 | $*3\frac{1}{8} \times 12$ | $*3\frac{1}{8} \times 12$ | $*3\frac{1}{8} \times 13\frac{1}{2}$ | $*3\frac{1}{8} \times 13\frac{1}{2}$ | $3\frac{1}{8} \times 15$ | $*3\frac{1}{8} \times 16\frac{1}{2}$ | $3\frac{1}{8} \times 16\frac{1}{2}$ |
| | 12 | $*3\frac{1}{8} \times 13\frac{1}{2}$ | $*3\frac{1}{8} \times 15$ | $*3\frac{1}{8} \times 15$ | $*3\frac{1}{8} \times 16\frac{1}{2}$ | $3\frac{1}{8} \times 16\frac{1}{2}$ | $3\frac{1}{8} \times 19\frac{1}{2}$ | $*5\frac{1}{8} \times 16\frac{1}{2}$ |
| | 16 | $*3\frac{1}{8} \times 16\frac{1}{2}$ | $*3\frac{1}{8} \times 16\frac{1}{2}$ | $*3\frac{1}{8} \times 16\frac{1}{2}$ | $*3\frac{1}{8} \times 18$ | $*5\frac{1}{8} \times 16\frac{1}{2}$ | $5\frac{1}{8} \times 16\frac{1}{2}$ | $5\frac{1}{8} \times 18$ |
| | 20 | $*3\frac{1}{8} \times 18$ | $*3\frac{1}{8} \times 16\frac{1}{2}$ | $*5\frac{1}{8} \times 16\frac{1}{2}$ | $*5\frac{1}{8} \times 16\frac{1}{2}$ | $5\frac{1}{8} \times 16\frac{1}{2}$ | $5\frac{1}{8} \times 18$ | $5\frac{1}{8} \times 19\frac{1}{2}$ |

**TABLE 8.8** (*Continued*)

| Span (ft) | Spacing (ft) | Construction Load[b,c] $C_D = 1.25$ 30 psf | Snow Loads[b,c] $C_D = 1.15$ | | | | | | Floor[d] Live Load[b,c] $C_D = 1.00$ 50 psf |
|---|---|---|---|---|---|---|---|---|---|
| | | | 30 psf | 35 psf | 40 psf | 45 psf | 50 psf | 55 psf | |
| 24 | 8 | $*3\frac{1}{8} \times 15$ | $*3\frac{1}{8} \times 15$ | $*3\frac{1}{8} \times 15$ | $*3\frac{1}{8} \times 16\frac{1}{2}$ | $3\frac{1}{8} \times 16\frac{1}{2}$ | $3\frac{1}{8} \times 16\frac{1}{2}$ | $3\frac{1}{8} \times 18$ | $5\frac{5}{8} \times 16\frac{1}{2}$ |
| | 12 | $3\frac{1}{8} \times 16\frac{1}{2}$ | $3\frac{1}{8} \times 16\frac{1}{2}$ | $3\frac{1}{8} \times 16\frac{1}{2}$ | $3\frac{1}{8} \times 18$ | $*5\frac{5}{8} \times 16\frac{1}{2}$ | $5\frac{5}{8} \times 16\frac{1}{2}$ | $5\frac{5}{8} \times 16\frac{1}{2}$ | $5\frac{5}{8} \times 19\frac{1}{2}$ |
| | 16 | $3\frac{1}{8} \times 18$ | $3\frac{1}{8} \times 18$ | $5\frac{5}{8} \times 16\frac{1}{2}$ | $5\frac{5}{8} \times 16\frac{1}{2}$ | $5\frac{5}{8} \times 18$ | $5\frac{5}{8} \times 18$ | $5\frac{5}{8} \times 19\frac{1}{2}$ | $5\frac{5}{8} \times 21$ |
| | 20 | $3\frac{1}{8} \times 19\frac{1}{2}$ | $5\frac{5}{8} \times 16\frac{1}{2}$ | $5\frac{5}{8} \times 18$ | $5\frac{5}{8} \times 18$ | $5\frac{5}{8} \times 19\frac{1}{2}$ | $5\frac{5}{8} \times 21$ | $5\frac{5}{8} \times 22\frac{1}{2}$ | $5\frac{5}{8} \times 22\frac{1}{2}$ |
| | 24 | $3\frac{1}{8} \times 21$ | $5\frac{5}{8} \times 18$ | $5\frac{5}{8} \times 19\frac{1}{2}$ | $5\frac{5}{8} \times 21$ | $5\frac{5}{8} \times 21$ | $5\frac{5}{8} \times 22\frac{1}{2}$ | $5\frac{5}{8} \times 25\frac{1}{2}$ | $5\frac{5}{8} \times 24$ |
| 28 | 8 | $3\frac{1}{8} \times 16\frac{1}{2}$ | $3\frac{1}{8} \times 16\frac{1}{2}$ | $3\frac{1}{8} \times 18$ | $3\frac{1}{8} \times 18$ | $5\frac{5}{8} \times 16\frac{1}{2}$ | $5\frac{5}{8} \times 16\frac{1}{2}$ | $5\frac{5}{8} \times 18$ | $5\frac{5}{8} \times 19\frac{1}{2}$ |
| | 12 | $3\frac{1}{8} \times 19\frac{1}{2}$ | $5\frac{5}{8} \times 16\frac{1}{2}$ | $5\frac{5}{8} \times 16\frac{1}{2}$ | $5\frac{5}{8} \times 18$ | $5\frac{5}{8} \times 18$ | $5\frac{5}{8} \times 19\frac{1}{2}$ | $5\frac{5}{8} \times 19\frac{1}{2}$ | $5\frac{5}{8} \times 21$ |
| | 16 | $3\frac{1}{8} \times 21$ | $5\frac{5}{8} \times 18$ | $5\frac{5}{8} \times 18$ | $5\frac{5}{8} \times 19\frac{1}{2}$ | $5\frac{5}{8} \times 21$ | $5\frac{5}{8} \times 22\frac{1}{2}$ | $5\frac{5}{8} \times 22\frac{1}{2}$ | $5\frac{5}{8} \times 24$ |
| | 20 | $3\frac{1}{8} \times 22\frac{1}{2}$ | $5\frac{5}{8} \times 19\frac{1}{2}$ | $5\frac{5}{8} \times 21$ | $5\frac{5}{8} \times 22\frac{1}{2}$ | $5\frac{5}{8} \times 24$ | $5\frac{5}{8} \times 24$ | $5\frac{5}{8} \times 25\frac{1}{2}$ | $5\frac{5}{8} \times 27$ |
| | 24 | $3\frac{1}{8} \times 25\frac{1}{2}$ | $5\frac{5}{8} \times 19\frac{1}{2}$ | $5\frac{5}{8} \times 22\frac{1}{2}$ | $5\frac{5}{8} \times 24$ | $5\frac{5}{8} \times 25\frac{1}{2}$ | $5\frac{5}{8} \times 27$ | $5\frac{5}{8} \times 30$ | $5\frac{5}{8} \times 28\frac{1}{2}$ |
| | 28 | $5\frac{5}{8} \times 21$ | $5\frac{5}{8} \times 22\frac{1}{2}$ | $5\frac{5}{8} \times 24$ | $5\frac{5}{8} \times 25\frac{1}{2}$ | $5\frac{5}{8} \times 28\frac{1}{2}$ | $5\frac{5}{8} \times 31\frac{1}{2}$ | $5\frac{5}{8} \times 34\frac{1}{2}$ | $5\frac{5}{8} \times 31\frac{1}{2}$ |
| 32 | 8 | $3\frac{1}{8} \times 19\frac{1}{2}$ | $5\frac{5}{8} \times 16\frac{1}{2}$ | $5\frac{5}{8} \times 16\frac{1}{2}$ | $5\frac{5}{8} \times 18$ | $5\frac{5}{8} \times 18$ | $5\frac{5}{8} \times 19\frac{1}{2}$ | $5\frac{5}{8} \times 19\frac{1}{2}$ | $5\frac{5}{8} \times 22\frac{1}{2}$ |
| | 12 | $3\frac{1}{8} \times 21$ | $5\frac{5}{8} \times 18$ | $5\frac{5}{8} \times 19\frac{1}{2}$ | $5\frac{5}{8} \times 19\frac{1}{2}$ | $5\frac{5}{8} \times 21$ | $5\frac{5}{8} \times 22\frac{1}{2}$ | $5\frac{5}{8} \times 22\frac{1}{2}$ | $5\frac{5}{8} \times 25\frac{1}{2}$ |

| | | | | | | | | | |
|---|---|---|---|---|---|---|---|---|---|
| 32 | 16 | $3\frac{1}{8} \times 24$ | $5\frac{1}{8} \times 19\frac{1}{2}$ | $5\frac{1}{8} \times 21$ | $5\frac{1}{8} \times 22\frac{1}{2}$ | $5\frac{1}{8} \times 24$ | $5\frac{1}{8} \times 25\frac{1}{2}$ | $5\frac{1}{8} \times 27$ | $5\frac{1}{8} \times 27$ |
| (cont.) | 20 | $5\frac{1}{8} \times 21$ | $5\frac{1}{8} \times 22\frac{1}{2}$ | $5\frac{1}{8} \times 24$ | $5\frac{1}{8} \times 25\frac{1}{2}$ | $5\frac{1}{8} \times 27$ | $5\frac{1}{8} \times 28\frac{1}{2}$ | $5\frac{1}{8} \times 30$ | $5\frac{1}{8} \times 30$ |
| | 24 | $5\frac{1}{8} \times 22\frac{1}{2}$ | $5\frac{1}{8} \times 24$ | $5\frac{1}{8} \times 25\frac{1}{2}$ | $5\frac{1}{8} \times 28\frac{1}{2}$ | $5\frac{1}{8} \times 30$ | $5\frac{1}{8} \times 31\frac{1}{2}$ | $6\frac{3}{4} \times 28\frac{1}{2}$ | $5\frac{1}{8} \times 33$ |
| | 28 | $5\frac{1}{8} \times 25\frac{1}{2}$ | $5\frac{1}{8} \times 25\frac{1}{2}$ | $5\frac{1}{8} \times 28\frac{1}{2}$ | $5\frac{1}{8} \times 30$ | $6\frac{3}{4} \times 28\frac{1}{2}$ | $6\frac{3}{4} \times 30$ | $6\frac{3}{4} \times 31\frac{1}{2}$ | $6\frac{3}{4} \times 31\frac{1}{2}$ |
| | 32 | $5\frac{1}{8} \times 27$ | $5\frac{1}{8} \times 28\frac{1}{2}$ | $5\frac{1}{8} \times 30$ | $5\frac{1}{8} \times 31\frac{1}{2}$ | $6\frac{3}{4} \times 30$ | $6\frac{3}{4} \times 31\frac{1}{2}$ | $6\frac{3}{4} \times 33$ | $6\frac{3}{4} \times 34\frac{1}{2}$ |
| 36 | 12 | $3\frac{1}{8} \times 24$ | $5\frac{1}{8} \times 21$ | $5\frac{1}{8} \times 21$ | $5\frac{1}{8} \times 22\frac{1}{2}$ | $5\frac{1}{8} \times 24$ | $5\frac{1}{8} \times 25\frac{1}{2}$ | $5\frac{1}{8} \times 25\frac{1}{2}$ | $6\frac{3}{4} \times 25\frac{1}{2}$ |
| | 16 | $5\frac{1}{8} \times 22\frac{1}{2}$ | $5\frac{1}{8} \times 22\frac{1}{2}$ | $5\frac{1}{8} \times 24$ | $5\frac{1}{8} \times 25\frac{1}{2}$ | $5\frac{1}{8} \times 27$ | $5\frac{1}{8} \times 28\frac{1}{2}$ | $5\frac{1}{8} \times 30$ | $6\frac{3}{4} \times 28\frac{1}{2}$ |
| | 20 | $5\frac{1}{8} \times 24$ | $5\frac{1}{8} \times 25\frac{1}{2}$ | $5\frac{1}{8} \times 27$ | $5\frac{1}{8} \times 28\frac{1}{2}$ | $5\frac{1}{8} \times 30$ | $6\frac{3}{4} \times 28\frac{1}{2}$ | $6\frac{3}{4} \times 30$ | $6\frac{3}{4} \times 31\frac{1}{2}$ |
| | 24 | $5\frac{1}{8} \times 25\frac{1}{2}$ | $5\frac{1}{8} \times 27$ | $5\frac{1}{8} \times 30$ | $5\frac{1}{8} \times 31\frac{1}{2}$ | $6\frac{3}{4} \times 30$ | $6\frac{3}{4} \times 31\frac{1}{2}$ | $6\frac{3}{4} \times 33$ | $6\frac{3}{4} \times 33$ |
| | 28 | $5\frac{1}{8} \times 28\frac{1}{2}$ | $5\frac{1}{8} \times 30$ | $5\frac{1}{8} \times 31\frac{1}{2}$ | $6\frac{3}{4} \times 30$ | $6\frac{3}{4} \times 31\frac{1}{2}$ | $6\frac{3}{4} \times 34\frac{1}{2}$ | $6\frac{3}{4} \times 36$ | $6\frac{3}{4} \times 36$ |
| | 32 | $5\frac{1}{8} \times 30$ | $5\frac{1}{8} \times 31\frac{1}{2}$ | $6\frac{3}{4} \times 30$ | $6\frac{3}{4} \times 33$ | $6\frac{3}{4} \times 34\frac{1}{2}$ | $6\frac{3}{4} \times 36$ | $6\frac{3}{4} \times 37\frac{1}{2}$ | $6\frac{3}{4} \times 39$ |
| | 36 | $5\frac{1}{8} \times 31\frac{1}{2}$ | $6\frac{3}{4} \times 30$ | $6\frac{3}{4} \times 31\frac{1}{2}$ | $6\frac{3}{4} \times 34\frac{1}{2}$ | $6\frac{3}{4} \times 36$ | $6\frac{3}{4} \times 39$ | $6\frac{3}{4} \times 42$ | $6\frac{3}{4} \times 42$ |
| 40 | 12 | $5\frac{1}{8} \times 22\frac{1}{2}$ | $5\frac{1}{8} \times 22\frac{1}{2}$ | $5\frac{1}{8} \times 24$ | $5\frac{1}{8} \times 25\frac{1}{2}$ | $5\frac{1}{8} \times 27$ | $5\frac{1}{8} \times 28\frac{1}{2}$ | $6\frac{3}{4} \times 25\frac{1}{2}$ | $6\frac{3}{4} \times 28\frac{1}{2}$ |
| | 16 | $5\frac{1}{8} \times 25\frac{1}{2}$ | $5\frac{1}{8} \times 25\frac{1}{2}$ | $5\frac{1}{8} \times 27$ | $5\frac{1}{8} \times 28\frac{1}{2}$ | $5\frac{1}{8} \times 31\frac{1}{2}$ | $6\frac{3}{4} \times 28\frac{1}{2}$ | $6\frac{3}{4} \times 30$ | $6\frac{3}{4} \times 31\frac{1}{2}$ |
| | 20 | $5\frac{1}{8} \times 27$ | $5\frac{1}{8} \times 28\frac{1}{2}$ | $5\frac{1}{8} \times 30$ | $6\frac{3}{4} \times 28\frac{1}{2}$ | $6\frac{3}{4} \times 30$ | $6\frac{3}{4} \times 31\frac{1}{2}$ | $6\frac{3}{4} \times 33$ | $6\frac{3}{4} \times 34\frac{1}{2}$ |
| | 24 | $5\frac{1}{8} \times 30$ | $5\frac{1}{8} \times 31\frac{1}{2}$ | $6\frac{3}{4} \times 30$ | $6\frac{3}{4} \times 31\frac{1}{2}$ | $6\frac{3}{4} \times 33$ | $6\frac{3}{4} \times 34\frac{1}{2}$ | $6\frac{3}{4} \times 37\frac{1}{2}$ | $6\frac{3}{4} \times 37\frac{1}{2}$ |
| | 28 | $5\frac{1}{8} \times 31\frac{1}{2}$ | $6\frac{3}{4} \times 30$ | $6\frac{3}{4} \times 31\frac{1}{2}$ | $6\frac{3}{4} \times 34\frac{1}{2}$ | $6\frac{3}{4} \times 36$ | $6\frac{3}{4} \times 37\frac{1}{2}$ | $6\frac{3}{4} \times 40\frac{1}{2}$ | $6\frac{3}{4} \times 40\frac{1}{2}$ |
| | 32 | $5\frac{1}{8} \times 34\frac{1}{2}$ | $6\frac{3}{4} \times 31\frac{1}{2}$ | $6\frac{3}{4} \times 34\frac{1}{2}$ | $6\frac{3}{4} \times 36$ | $6\frac{3}{4} \times 39$ | $6\frac{3}{4} \times 40\frac{1}{2}$ | $6\frac{3}{4} \times 42$ | $6\frac{3}{4} \times 43\frac{1}{2}$ |
| | 36 | $5\frac{1}{8} \times 36$ | $6\frac{3}{4} \times 33$ | $6\frac{3}{4} \times 36$ | $6\frac{3}{4} \times 39$ | $6\frac{3}{4} \times 40\frac{1}{2}$ | $6\frac{3}{4} \times 43\frac{1}{2}$ | $8\frac{3}{4} \times 40\frac{1}{2}$ | $6\frac{3}{4} \times 46\frac{1}{2}$ |
| | 40 | $6\frac{3}{4} \times 33$ | $6\frac{3}{4} \times 34\frac{1}{2}$ | $6\frac{3}{4} \times 37\frac{1}{2}$ | $6\frac{3}{4} \times 40\frac{1}{2}$ | $6\frac{3}{4} \times 43\frac{1}{2}$ | $8\frac{3}{4} \times 40\frac{1}{2}$ | $8\frac{3}{4} \times 42$ | $8\frac{3}{4} \times 43\frac{1}{2}$ |
| 44 | 12 | $5\frac{1}{8} \times 25\frac{1}{2}$ | $5\frac{1}{8} \times 25\frac{1}{2}$ | $5\frac{1}{8} \times 25\frac{1}{2}$ | $5\frac{1}{8} \times 28\frac{1}{2}$ | $5\frac{1}{8} \times 30$ | $6\frac{3}{4} \times 27$ | $6\frac{3}{4} \times 28\frac{1}{2}$ | |
| | 16 | $5\frac{1}{8} \times 27\frac{1}{2}$ | $5\frac{1}{8} \times 28\frac{1}{2}$ | $5\frac{1}{8} \times 30$ | $5\frac{1}{8} \times 39\frac{1}{2}$ | $6\frac{3}{4} \times 30$ | $6\frac{3}{4} \times 31\frac{1}{2}$ | $6\frac{3}{4} \times 33$ | |

## TABLE 8.8 (Continued)

| | | Roof$^{d,e}$ | | | | | | Floor$^{d}$ |
| | Construction Load$^{b,c}$ | Snow Loads$^{b,c}$ | | | | | | Live Load$^{b,c}$ |
| Duration of Load Factor $C_D$ = | 1.25 | 1.15 | | | | | | 1.00 |
| Total Load = | 30 psf | 30 psf | 35 psf | 40 psf | 45 psf | 50 psf | 55 psf | 50 psf |
| Span (ft) / Spacing (ft) | | | | | | | | |
| 44 (cont.) 20 | $5\frac{1}{8} \times 30$ | $5\frac{1}{8} \times 31\frac{1}{2}$ | $6\frac{3}{4} \times 30$ | $6\frac{3}{4} \times 31\frac{1}{2}$ | $6\frac{3}{4} \times 34\frac{1}{2}$ | $6\frac{3}{4} \times 36$ | $6\frac{3}{4} \times 37\frac{1}{2}$ | |
| 24 | $5\frac{1}{8} \times 33$ | $6\frac{3}{4} \times 30$ | $6\frac{3}{4} \times 33$ | $6\frac{3}{4} \times 34\frac{1}{2}$ | $6\frac{3}{4} \times 37\frac{1}{2}$ | $6\frac{3}{4} \times 39$ | $6\frac{3}{4} \times 40\frac{1}{2}$ | |
| 28 | $5\frac{1}{8} \times 36$ | $6\frac{3}{4} \times 33$ | $6\frac{3}{4} \times 36$ | $6\frac{3}{4} \times 37\frac{1}{2}$ | $6\frac{3}{4} \times 40\frac{1}{2}$ | $6\frac{3}{4} \times 42$ | $8\frac{3}{4} \times 39$ | |
| 32 | $6\frac{3}{4} \times 33$ | $6\frac{3}{4} \times 34\frac{1}{2}$ | $6\frac{3}{4} \times 37\frac{1}{2}$ | $6\frac{3}{4} \times 40\frac{1}{2}$ | $6\frac{3}{4} \times 43\frac{1}{2}$ | $8\frac{3}{4} \times 40\frac{1}{2}$ | $8\frac{3}{4} \times 42$ | |
| 36 | $6\frac{3}{4} \times 36$ | $6\frac{3}{4} \times 37\frac{1}{2}$ | $6\frac{3}{4} \times 40\frac{1}{2}$ | $6\frac{3}{4} \times 43\frac{1}{2}$ | $8\frac{3}{4} \times 40\frac{1}{2}$ | $8\frac{3}{4} \times 43\frac{1}{2}$ | $8\frac{3}{4} \times 45$ | |
| 40 | $6\frac{3}{4} \times 37\frac{1}{2}$ | $6\frac{3}{4} \times 39$ | $6\frac{3}{4} \times 42$ | $8\frac{3}{4} \times 40\frac{1}{2}$ | $8\frac{3}{4} \times 43\frac{1}{2}$ | $8\frac{3}{4} \times 45$ | $8\frac{3}{4} \times 48$ | |
| 48 12 | $5\frac{1}{8} \times 27$ | $5\frac{1}{8} \times 27$ | $5\frac{1}{8} \times 28\frac{1}{2}$ | $5\frac{1}{8} \times 31\frac{1}{2}$ | $6\frac{3}{4} \times 28\frac{1}{2}$ | $6\frac{3}{4} \times 30$ | $6\frac{3}{4} \times 31\frac{1}{2}$ | |
| 16 | $5\frac{1}{8} \times 30$ | $5\frac{1}{8} \times 31\frac{1}{2}$ | $6\frac{3}{4} \times 30$ | $6\frac{3}{4} \times 31\frac{1}{2}$ | $6\frac{3}{4} \times 33$ | $6\frac{3}{4} \times 34\frac{1}{2}$ | $6\frac{3}{4} \times 37\frac{1}{2}$ | |
| 20 | $5\frac{1}{8} \times 33$ | $6\frac{3}{4} \times 30$ | $6\frac{3}{4} \times 33$ | $6\frac{3}{4} \times 34\frac{1}{2}$ | $6\frac{3}{4} \times 37\frac{1}{2}$ | $6\frac{3}{4} \times 39$ | $6\frac{3}{4} \times 42$ | |
| 24 | $6\frac{3}{4} \times 36$ | $6\frac{3}{4} \times 33$ | $6\frac{3}{4} \times 36$ | $6\frac{3}{4} \times 39$ | $6\frac{3}{4} \times 40\frac{1}{2}$ | $6\frac{3}{4} \times 43\frac{1}{2}$ | $8\frac{3}{4} \times 40\frac{1}{2}$ | |
| 28 | $6\frac{3}{4} \times 34\frac{1}{2}$ | $6\frac{3}{4} \times 36$ | $6\frac{3}{4} \times 39$ | $6\frac{3}{4} \times 42$ | $6\frac{3}{4} \times 43\frac{1}{2}$ | $8\frac{3}{4} \times 42$ | $8\frac{3}{4} \times 43\frac{1}{2}$ | |
| 32 | $6\frac{3}{4} \times 37\frac{1}{2}$ | $6\frac{3}{4} \times 39$ | $6\frac{3}{4} \times 42$ | $6\frac{3}{4} \times 45$ | $6\frac{3}{4} \times 48$ | $8\frac{3}{4} \times 45$ | $8\frac{3}{4} \times 46\frac{1}{2}$ | |
| 36 | $6\frac{3}{4} \times 39$ | $6\frac{3}{4} \times 40\frac{1}{2}$ | $6\frac{3}{4} \times 43\frac{1}{2}$ | $8\frac{3}{4} \times 42$ | $8\frac{3}{4} \times 45$ | $8\frac{3}{4} \times 46\frac{1}{2}$ | $8\frac{3}{4} \times 49\frac{1}{2}$ | |
| 40 | $6\frac{3}{4} \times 40\frac{1}{2}$ | $6\frac{3}{4} \times 43\frac{1}{2}$ | $8\frac{3}{4} \times 42$ | $8\frac{3}{4} \times 45$ | $8\frac{3}{4} \times 46\frac{1}{2}$ | $8\frac{3}{4} \times 49\frac{1}{2}$ | $8\frac{3}{4} \times 52\frac{1}{2}$ | |

| | | D1 | D2 | D3 | D4 | D5 | D6 | D7 |
|---|---|---|---|---|---|---|---|---|
| 52 | 12 | $5\frac{1}{8} \times 30$ | $5\frac{1}{8} \times 30$ | $5\frac{1}{8} \times 31\frac{1}{2}$ | $6\frac{3}{4} \times 30$ | $6\frac{3}{4} \times 31\frac{1}{2}$ | $6\frac{3}{4} \times 33$ | $6\frac{3}{4} \times 34\frac{1}{2}$ |
| | 16 | $5\frac{1}{8} \times 33$ | $6\frac{3}{4} \times 30$ | $6\frac{3}{4} \times 33$ | $6\frac{3}{4} \times 34\frac{1}{2}$ | $6\frac{3}{4} \times 36$ | $6\frac{3}{4} \times 39$ | $6\frac{3}{4} \times 40\frac{1}{2}$ |
| | 20 | $6\frac{3}{4} \times 36$ | $6\frac{3}{4} \times 33$ | $6\frac{3}{4} \times 36$ | $6\frac{3}{4} \times 39$ | $6\frac{3}{4} \times 40\frac{1}{2}$ | $6\frac{3}{4} \times 43\frac{1}{2}$ | $8\frac{3}{4} \times 40\frac{1}{2}$ |
| | 24 | $6\frac{3}{4} \times 34\frac{1}{2}$ | $6\frac{3}{4} \times 36$ | $6\frac{3}{4} \times 39$ | $6\frac{3}{4} \times 42$ | $6\frac{3}{4} \times 45$ | $8\frac{3}{4} \times 42$ | $8\frac{3}{4} \times 43\frac{1}{2}$ |
| | 28 | $6\frac{3}{4} \times 37\frac{1}{2}$ | $6\frac{3}{4} \times 39$ | $6\frac{3}{4} \times 42$ | $8\frac{3}{4} \times 40\frac{1}{2}$ | $8\frac{3}{4} \times 43\frac{1}{2}$ | $8\frac{3}{4} \times 45$ | $8\frac{3}{4} \times 48$ |
| | 32 | $6\frac{3}{4} \times 40\frac{1}{2}$ | $6\frac{3}{4} \times 42$ | $8\frac{3}{4} \times 40\frac{1}{2}$ | $8\frac{3}{4} \times 43\frac{1}{2}$ | $8\frac{3}{4} \times 46\frac{1}{2}$ | $8\frac{3}{4} \times 48$ | $8\frac{3}{4} \times 51$ |
| | 36 | $6\frac{3}{4} \times 43\frac{1}{2}$ | $6\frac{3}{4} \times 45$ | $8\frac{3}{4} \times 43\frac{1}{2}$ | $8\frac{3}{4} \times 46\frac{1}{2}$ | $8\frac{3}{4} \times 49\frac{1}{2}$ | $8\frac{3}{4} \times 51$ | $8\frac{3}{4} \times 54$ |
| | 40 | $6\frac{3}{4} \times 45$ | $8\frac{3}{4} \times 42$ | $8\frac{3}{4} \times 45$ | $8\frac{3}{4} \times 48$ | $8\frac{3}{4} \times 51$ | $8\frac{3}{4} \times 54$ | $8\frac{3}{4} \times 57$ |
| 56 | 12 | $5\frac{1}{8} \times 31\frac{1}{2}$ | $6\frac{3}{4} \times 28\frac{1}{2}$ | $6\frac{3}{4} \times 30$ | $6\frac{3}{4} \times 33$ | $6\frac{3}{4} \times 34\frac{1}{2}$ | $6\frac{3}{4} \times 36$ | $6\frac{3}{4} \times 37\frac{1}{2}$ |
| | 16 | $5\frac{1}{8} \times 34\frac{1}{2}$ | $6\frac{3}{4} \times 33$ | $6\frac{3}{4} \times 34\frac{1}{2}$ | $6\frac{3}{4} \times 37\frac{1}{2}$ | $6\frac{3}{4} \times 39$ | $6\frac{3}{4} \times 42$ | $8\frac{3}{4} \times 39$ |
| | 20 | $6\frac{3}{4} \times 34\frac{1}{2}$ | $6\frac{3}{4} \times 36$ | $6\frac{3}{4} \times 39$ | $6\frac{3}{4} \times 42$ | $8\frac{3}{4} \times 39$ | $8\frac{3}{4} \times 42$ | $8\frac{3}{4} \times 43\frac{1}{2}$ |
| | 24 | $6\frac{3}{4} \times 37\frac{1}{2}$ | $6\frac{3}{4} \times 39$ | $6\frac{3}{4} \times 43\frac{1}{2}$ | $8\frac{3}{4} \times 40\frac{1}{2}$ | $8\frac{3}{4} \times 43\frac{1}{2}$ | $8\frac{3}{4} \times 45$ | $8\frac{3}{4} \times 48$ |
| | 28 | $6\frac{3}{4} \times 40\frac{1}{2}$ | $6\frac{3}{4} \times 43\frac{1}{2}$ | $8\frac{3}{4} \times 42$ | $8\frac{3}{4} \times 43\frac{1}{2}$ | $8\frac{3}{4} \times 46\frac{1}{2}$ | $8\frac{3}{4} \times 49\frac{1}{2}$ | $8\frac{3}{4} \times 52\frac{1}{2}$ |
| | 32 | $6\frac{3}{4} \times 43\frac{1}{2}$ | $8\frac{3}{4} \times 40\frac{1}{2}$ | $8\frac{3}{4} \times 43\frac{1}{2}$ | $8\frac{3}{4} \times 46\frac{1}{2}$ | $8\frac{3}{4} \times 49\frac{1}{2}$ | $8\frac{3}{4} \times 52\frac{1}{2}$ | $8\frac{3}{4} \times 55\frac{1}{2}$ |
| | 36 | $6\frac{3}{4} \times 40\frac{1}{2}$ | $8\frac{3}{4} \times 43\frac{1}{2}$ | $8\frac{3}{4} \times 46\frac{1}{2}$ | $8\frac{3}{4} \times 49\frac{1}{2}$ | $8\frac{3}{4} \times 52\frac{1}{2}$ | $8\frac{3}{4} \times 55\frac{1}{2}$ | $8\frac{3}{4} \times 58\frac{1}{2}$ |
| | 40 | $8\frac{3}{4} \times 43\frac{1}{2}$ | $8\frac{3}{4} \times 45$ | $8\frac{3}{4} \times 49\frac{1}{2}$ | $8\frac{3}{4} \times 52\frac{1}{2}$ | $8\frac{3}{4} \times 55\frac{1}{2}$ | $8\frac{3}{4} \times 58\frac{1}{2}$ | $10\frac{3}{4} \times 57$ |
| 60 | 12 | $5\frac{1}{8} \times 34\frac{1}{2}$ | $6\frac{3}{4} \times 31\frac{1}{2}$ | $6\frac{3}{4} \times 33$ | $6\frac{3}{4} \times 34\frac{1}{2}$ | $6\frac{3}{4} \times 37\frac{1}{2}$ | $6\frac{3}{4} \times 39$ | $6\frac{3}{4} \times 40\frac{1}{2}$ |
| | 16 | $6\frac{3}{4} \times 34\frac{1}{2}$ | $6\frac{3}{4} \times 34\frac{1}{2}$ | $6\frac{3}{4} \times 37\frac{1}{2}$ | $6\frac{3}{4} \times 40\frac{1}{2}$ | $6\frac{3}{4} \times 43\frac{1}{2}$ | $8\frac{3}{4} \times 40\frac{1}{2}$ | $8\frac{3}{4} \times 42$ |
| | 20 | $6\frac{3}{4} \times 37\frac{1}{2}$ | $6\frac{3}{4} \times 39$ | $6\frac{3}{4} \times 42$ | $8\frac{3}{4} \times 40\frac{1}{2}$ | $8\frac{3}{4} \times 43\frac{1}{2}$ | $8\frac{3}{4} \times 45$ | $8\frac{3}{4} \times 48$ |
| | 24 | $6\frac{3}{4} \times 40\frac{1}{2}$ | $6\frac{3}{4} \times 43\frac{1}{2}$ | $8\frac{3}{4} \times 42$ | $8\frac{3}{4} \times 45$ | $8\frac{3}{4} \times 46\frac{1}{2}$ | $8\frac{3}{4} \times 49\frac{1}{2}$ | $8\frac{3}{4} \times 52\frac{1}{2}$ |
| | 28 | $6\frac{3}{4} \times 45$ | $8\frac{3}{4} \times 42$ | $8\frac{3}{4} \times 45$ | $8\frac{3}{4} \times 48$ | $8\frac{3}{4} \times 51$ | $8\frac{3}{4} \times 54$ | $8\frac{3}{4} \times 55\frac{1}{2}$ |
| | 32 | $6\frac{3}{4} \times 48$ | $8\frac{3}{4} \times 45$ | $8\frac{3}{4} \times 48$ | $8\frac{3}{4} \times 51$ | $8\frac{3}{4} \times 54$ | $8\frac{3}{4} \times 57$ | $10\frac{3}{4} \times 54$ |

**TABLE 8.8** *(Continued)*

| Span (ft) | Spacing (ft) | Construction Load[b,c] | Roof[d,e] Snow Loads[b,c] | | | | | | Floor[d] Live Load[b,c] |
|---|---|---|---|---|---|---|---|---|---|
| Duration of Load Factor $C_D =$ | | 1.25 | 1.15 | | | | | | 1.00 |
| Total Load = | | 30 psf | 30 psf | 35 psf | 40 psf | 45 psf | 50 psf | 55 psf | 50 psf |
| 60 (cont.) | 36 | $8\frac{3}{4} \times 45$ | $8\frac{3}{4} \times 46\frac{1}{2}$ | $8\frac{3}{4} \times 51$ | $8\frac{3}{4} \times 54$ | $8\frac{3}{4} \times 57$ | $10\frac{1}{4} \times 55\frac{1}{2}$ | $10\frac{1}{4} \times 58\frac{1}{2}$ | |
| | 40 | $8\frac{3}{4} \times 48$ | $8\frac{3}{4} \times 49\frac{1}{2}$ | $8\frac{3}{4} \times 54$ | $8\frac{3}{4} \times 57$ | $10\frac{1}{4} \times 55\frac{1}{2}$ | $10\frac{1}{4} \times 58\frac{1}{2}$ | $10\frac{1}{4} \times 61\frac{1}{2}$ | |
| 64 | 12 | $5\frac{1}{8} \times 36$ | $6\frac{3}{4} \times 33$ | $6\frac{3}{4} \times 36$ | $6\frac{3}{4} \times 37\frac{1}{2}$ | $6\frac{3}{4} \times 40\frac{1}{2}$ | $6\frac{3}{4} \times 42$ | $6\frac{3}{4} \times 45$ | |
| | 16 | $6\frac{3}{4} \times 37\frac{1}{2}$ | $6\frac{3}{4} \times 37\frac{1}{2}$ | $6\frac{3}{4} \times 40\frac{1}{2}$ | $6\frac{3}{4} \times 43\frac{1}{2}$ | $8\frac{3}{4} \times 42$ | $8\frac{3}{4} \times 43\frac{1}{2}$ | $8\frac{3}{4} \times 45$ | |
| | 20 | $6\frac{3}{4} \times 40\frac{1}{2}$ | $6\frac{3}{4} \times 42$ | $6\frac{3}{4} \times 45$ | $8\frac{3}{4} \times 43\frac{1}{2}$ | $8\frac{3}{4} \times 46\frac{1}{2}$ | $8\frac{3}{4} \times 48$ | $8\frac{3}{4} \times 51$ | |
| | 24 | $6\frac{3}{4} \times 43\frac{1}{2}$ | $8\frac{3}{4} \times 42$ | $8\frac{3}{4} \times 45$ | $8\frac{3}{4} \times 48$ | $8\frac{3}{4} \times 51$ | $8\frac{3}{4} \times 52\frac{1}{2}$ | $8\frac{3}{4} \times 55\frac{1}{2}$ | |
| | 28 | $6\frac{3}{4} \times 48$ | $8\frac{3}{4} \times 45$ | $8\frac{3}{4} \times 48$ | $8\frac{3}{4} \times 51$ | $8\frac{3}{4} \times 54$ | $8\frac{3}{4} \times 57$ | $10\frac{1}{4} \times 55\frac{1}{2}$ | |
| | 32 | $8\frac{3}{4} \times 45$ | $8\frac{3}{4} \times 48$ | $8\frac{3}{4} \times 51$ | $8\frac{3}{4} \times 55\frac{1}{2}$ | $8\frac{3}{4} \times 58\frac{1}{2}$ | $10\frac{1}{4} \times 55\frac{1}{2}$ | $10\frac{1}{4} \times 58\frac{1}{2}$ | |
| | 36 | $8\frac{3}{4} \times 48$ | $8\frac{3}{4} \times 51$ | $8\frac{3}{4} \times 54$ | $8\frac{3}{4} \times 58\frac{1}{2}$ | $10\frac{1}{4} \times 57$ | $10\frac{1}{4} \times 60$ | $10\frac{1}{4} \times 63$ | |
| | 40 | $8\frac{3}{4} \times 51$ | $8\frac{3}{4} \times 52\frac{1}{2}$ | $8\frac{3}{4} \times 57$ | $10\frac{1}{4} \times 55\frac{1}{2}$ | $10\frac{1}{4} \times 60$ | $10\frac{1}{4} \times 63$ | $10\frac{1}{4} \times 63$ | |
| 68 | 12 | $6\frac{3}{4} \times 36$ | $6\frac{3}{4} \times 36$ | $6\frac{3}{4} \times 37\frac{1}{2}$ | $6\frac{3}{4} \times 40\frac{1}{2}$ | $6\frac{3}{4} \times 43\frac{1}{2}$ | $8\frac{3}{4} \times 40\frac{1}{2}$ | $8\frac{3}{4} \times 42$ | |
| | 16 | $6\frac{3}{4} \times 39$ | $6\frac{3}{4} \times 40\frac{1}{2}$ | $6\frac{3}{4} \times 43\frac{1}{2}$ | $8\frac{3}{4} \times 42$ | $8\frac{3}{4} \times 45$ | $8\frac{3}{4} \times 46\frac{1}{2}$ | $8\frac{3}{4} \times 49\frac{1}{2}$ | |
| | 20 | $6\frac{3}{4} \times 43\frac{1}{2}$ | $8\frac{3}{4} \times 40\frac{1}{2}$ | $8\frac{3}{4} \times 43\frac{1}{2}$ | $8\frac{3}{4} \times 46\frac{1}{2}$ | $8\frac{3}{4} \times 49\frac{1}{2}$ | $8\frac{3}{4} \times 52\frac{1}{2}$ | $8\frac{3}{4} \times 54$ | |

A ready-reckoner reference table of dimensions (values expressed in mixed-fraction form). The left-hand index numbers run 24, 28, 32, 36, 40, then a block labelled **72** (12, 16, 20, 24, 28, 32, 36, 40), then a block labelled **76** (12, 16, 20, 24, 28, 32, 36, 40).

**Top block (index 24–40)**

| | | | | | | | |
|---|---|---|---|---|---|---|---|
| 24 | $6\tfrac{3}{4}\times48$ | $8\tfrac{3}{4}\times45$ | $8\tfrac{3}{4}\times48$ | $8\tfrac{3}{4}\times51$ | $8\tfrac{3}{4}\times54$ | $8\tfrac{3}{4}\times57$ | $8\tfrac{3}{4}\times60$ |
| 28 | $8\tfrac{3}{4}\times45$ | $8\tfrac{3}{4}\times48$ | $8\tfrac{3}{4}\times51$ | $8\tfrac{3}{4}\times55\tfrac{1}{2}$ | $8\tfrac{3}{4}\times58\tfrac{1}{2}$ | $10\tfrac{3}{4}\times57$ | $10\tfrac{3}{4}\times58\tfrac{1}{2}$ |
| 32 | $8\tfrac{3}{4}\times49\tfrac{1}{2}$ | $8\tfrac{3}{4}\times51$ | $8\tfrac{3}{4}\times55\tfrac{1}{2}$ | $8\tfrac{3}{4}\times58\tfrac{1}{2}$ | $10\tfrac{3}{4}\times57$ | $10\tfrac{3}{4}\times60$ | $10\tfrac{3}{4}\times63$ |
| 36 | $8\tfrac{3}{4}\times51$ | $8\tfrac{3}{4}\times54$ | $8\tfrac{3}{4}\times58\tfrac{1}{2}$ | $10\tfrac{3}{4}\times57$ | $10\tfrac{3}{4}\times60$ | | |
| 40 | $8\tfrac{3}{4}\times54$ | $8\tfrac{3}{4}\times57$ | $10\tfrac{3}{4}\times57$ | $10\tfrac{3}{4}\times60$ | | | |

**Block 72 (index 12–40)**

| | | | | | | | |
|---|---|---|---|---|---|---|---|
| 12 | $6\tfrac{3}{4}\times37\tfrac{1}{2}$ | $6\tfrac{3}{4}\times37\tfrac{1}{2}$ | $6\tfrac{3}{4}\times40\tfrac{1}{2}$ | $6\tfrac{3}{4}\times40\tfrac{1}{2}$ | $6\tfrac{3}{4}\times43\tfrac{1}{2}$ | $6\tfrac{3}{4}\times43\tfrac{1}{2}$ | $8\tfrac{3}{4}\times45$ |
| 16 | $6\tfrac{3}{4}\times42$ | $6\tfrac{3}{4}\times43\tfrac{1}{2}$ | $8\tfrac{3}{4}\times42$ | $8\tfrac{3}{4}\times42$ | $8\tfrac{3}{4}\times45$ | $8\tfrac{3}{4}\times45$ | $8\tfrac{3}{4}\times52\tfrac{1}{2}$ |
| 20 | $6\tfrac{3}{4}\times46\tfrac{1}{2}$ | $8\tfrac{3}{4}\times48$ | $8\tfrac{3}{4}\times46\tfrac{1}{2}$ | $8\tfrac{3}{4}\times46\tfrac{1}{2}$ | $8\tfrac{3}{4}\times49\tfrac{1}{2}$ | $8\tfrac{3}{4}\times49\tfrac{1}{2}$ | $8\tfrac{3}{4}\times58\tfrac{1}{2}$ |
| 24 | $8\tfrac{3}{4}\times45$ | $8\tfrac{3}{4}\times48$ | $8\tfrac{3}{4}\times51$ | $8\tfrac{3}{4}\times51$ | $8\tfrac{3}{4}\times54$ | $8\tfrac{3}{4}\times54$ | $10\tfrac{3}{4}\times58\tfrac{1}{2}$ |
| 28 | $8\tfrac{3}{4}\times49\tfrac{1}{2}$ | $8\tfrac{3}{4}\times51$ | $8\tfrac{3}{4}\times55\tfrac{1}{2}$ | $8\tfrac{3}{4}\times55\tfrac{1}{2}$ | $8\tfrac{3}{4}\times58\tfrac{1}{2}$ | $8\tfrac{3}{4}\times58\tfrac{1}{2}$ | $10\tfrac{3}{4}\times63$ |
| 32 | $8\tfrac{3}{4}\times52\tfrac{1}{2}$ | $8\tfrac{3}{4}\times54$ | $8\tfrac{3}{4}\times58\tfrac{1}{2}$ | $8\tfrac{3}{4}\times58\tfrac{1}{2}$ | $10\tfrac{3}{4}\times57$ | $10\tfrac{3}{4}\times57$ | |
| 36 | $8\tfrac{3}{4}\times55\tfrac{1}{2}$ | $8\tfrac{3}{4}\times58\tfrac{1}{2}$ | $10\tfrac{3}{4}\times57$ | $10\tfrac{3}{4}\times57$ | $10\tfrac{3}{4}\times61\tfrac{1}{2}$ | $10\tfrac{3}{4}\times61\tfrac{1}{2}$ | |
| 40 | $8\tfrac{3}{4}\times58\tfrac{1}{2}$ | $8\tfrac{3}{4}\times61\tfrac{1}{2}$ | $10\tfrac{3}{4}\times60$ | | | | |

**Block 76 (index 12–40)**

| | | | | | | | |
|---|---|---|---|---|---|---|---|
| 12 | $6\tfrac{3}{4}\times40\tfrac{1}{2}$ | $6\tfrac{3}{4}\times40\tfrac{1}{2}$ | $6\tfrac{3}{4}\times43\tfrac{1}{2}$ | $8\tfrac{3}{4}\times42$ | $8\tfrac{3}{4}\times43\tfrac{1}{2}$ | $8\tfrac{3}{4}\times46\tfrac{1}{2}$ | $8\tfrac{3}{4}\times48$ |
| 16 | $6\tfrac{3}{4}\times43\tfrac{1}{2}$ | $8\tfrac{3}{4}\times42$ | $8\tfrac{3}{4}\times45$ | $8\tfrac{3}{4}\times48$ | $8\tfrac{3}{4}\times51$ | $8\tfrac{3}{4}\times52\tfrac{1}{2}$ | $8\tfrac{3}{4}\times55\tfrac{1}{2}$ |
| 20 | $8\tfrac{3}{4}\times43\tfrac{1}{2}$ | $8\tfrac{3}{4}\times46\tfrac{1}{2}$ | $8\tfrac{3}{4}\times49\tfrac{1}{2}$ | $8\tfrac{3}{4}\times52\tfrac{1}{2}$ | $8\tfrac{3}{4}\times57$ | $8\tfrac{3}{4}\times58\tfrac{1}{2}$ | $10\tfrac{3}{4}\times57$ |
| 24 | $8\tfrac{3}{4}\times48$ | $8\tfrac{3}{4}\times51$ | $8\tfrac{3}{4}\times54$ | $8\tfrac{3}{4}\times58\tfrac{1}{2}$ | $8\tfrac{3}{4}\times61\tfrac{1}{2}$ | $10\tfrac{3}{4}\times60$ | $10\tfrac{3}{4}\times61\tfrac{1}{2}$ |
| 28 | $8\tfrac{3}{4}\times52\tfrac{1}{2}$ | $8\tfrac{3}{4}\times54$ | $8\tfrac{3}{4}\times58\tfrac{1}{2}$ | $10\tfrac{3}{4}\times57$ | $10\tfrac{3}{4}\times61\tfrac{1}{2}$ | | |
| 32 | $8\tfrac{3}{4}\times55\tfrac{1}{2}$ | $8\tfrac{3}{4}\times58\tfrac{1}{2}$ | $10\tfrac{3}{4}\times57$ | $10\tfrac{3}{4}\times61\tfrac{1}{2}$ | | | |
| 36 | $8\tfrac{3}{4}\times58\tfrac{1}{2}$ | $10\tfrac{3}{4}\times57$ | $10\tfrac{3}{4}\times61\tfrac{1}{2}$ | | | | |
| 40 | $8\tfrac{3}{4}\times61\tfrac{1}{2}$ | $10\tfrac{3}{4}\times60$ | | | | | |

## TABLE 8.8  (Continued)

| | | Construction Load[b,c] | Roof[d,e] Snow Loads[b,c] | | | | | | Floor[d] Live Load[b,c] |
|---|---|---|---|---|---|---|---|---|---|
| Duration of Load Factor $C_D$ = | | 1.25 | 1.15 | | | | | | 1.00 |
| Total Load = | | 30 psf | 30 psf | 35 psf | 40 psf | 45 psf | 50 psf | 55 psf | 50 psf |
| Span (ft) | Spacing (ft) | | | | | | | | |
| 80 | 12 | $6\frac{3}{4} \times 42$ | $6\frac{3}{4} \times 43\frac{1}{2}$ | $8\frac{3}{4} \times 42$ | $8\frac{3}{4} \times 43\frac{1}{2}$ | $8\frac{3}{4} \times 46\frac{1}{2}$ | $8\frac{3}{4} \times 49\frac{1}{2}$ | $8\frac{3}{4} \times 51$ | |
| | 16 | $6\frac{3}{4} \times 46\frac{1}{2}$ | $8\frac{3}{4} \times 43\frac{1}{2}$ | $8\frac{3}{4} \times 48$ | $8\frac{3}{4} \times 51$ | $8\frac{3}{4} \times 54$ | $8\frac{3}{4} \times 57$ | $8\frac{3}{4} \times 58\frac{1}{2}$ | |
| | 20 | $8\frac{3}{4} \times 46\frac{1}{2}$ | $8\frac{3}{4} \times 49\frac{1}{2}$ | $8\frac{3}{4} \times 52\frac{1}{2}$ | $8\frac{3}{4} \times 57$ | $8\frac{3}{4} \times 60$ | $10\frac{3}{4} \times 57$ | $10\frac{3}{4} \times 60$ | |
| | 24 | $8\frac{3}{4} \times 51$ | $8\frac{3}{4} \times 54$ | $8\frac{3}{4} \times 58\frac{1}{2}$ | $8\frac{3}{4} \times 61\frac{1}{2}$ | $10\frac{3}{4} \times 60$ | $10\frac{3}{4} \times 63$ | | |
| | 28 | $8\frac{3}{4} \times 55\frac{1}{2}$ | $8\frac{3}{4} \times 58\frac{1}{2}$ | $8\frac{3}{4} \times 63$ | $10\frac{3}{4} \times 61\frac{1}{2}$ | | | | |
| | 32 | $8\frac{3}{4} \times 58\frac{1}{2}$ | $8\frac{3}{4} \times 61\frac{1}{2}$ | $10\frac{3}{4} \times 61\frac{1}{2}$ | | | | | |
| | 36 | $8\frac{3}{4} \times 63$ | $10\frac{3}{4} \times 60$ | | | | | | |
| | 40 | $10\frac{3}{4} \times 60$ | $10\frac{3}{4} \times 63$ | | | | | | |
| 84 | 12 | $6\frac{3}{4} \times 45$ | $8\frac{3}{4} \times 40\frac{1}{2}$ | $8\frac{3}{4} \times 43\frac{1}{2}$ | $8\frac{3}{4} \times 46\frac{1}{2}$ | $8\frac{3}{4} \times 49\frac{1}{2}$ | $8\frac{3}{4} \times 52\frac{1}{2}$ | $8\frac{3}{4} \times 54$ | |
| | 16 | $8\frac{3}{4} \times 45$ | $8\frac{3}{4} \times 46\frac{1}{2}$ | $8\frac{3}{4} \times 51$ | $8\frac{3}{4} \times 54$ | $8\frac{3}{4} \times 57$ | $8\frac{3}{4} \times 60$ | $8\frac{3}{4} \times 63$ | |
| | 20 | $8\frac{3}{4} \times 49\frac{1}{2}$ | $8\frac{3}{4} \times 52\frac{1}{2}$ | $8\frac{3}{4} \times 55\frac{1}{2}$ | $8\frac{3}{4} \times 60$ | $10\frac{3}{4} \times 58\frac{1}{2}$ | $10\frac{3}{4} \times 61\frac{1}{2}$ | | |
| | 24 | $8\frac{3}{4} \times 59\frac{1}{2}$ | $8\frac{3}{4} \times 57$ | $8\frac{3}{4} \times 61\frac{1}{2}$ | $10\frac{3}{4} \times 60$ | $10\frac{3}{4} \times 63$ | | | |

| | Size | | | | | | | |
|---|---|---|---|---|---|---|---|---|
| | 28 | $8\frac{3}{4}\times 58\frac{1}{2}$ | $8\frac{3}{4}\times 61\frac{1}{2}$ | | | | | |
| | 32 | $8\frac{3}{4}\times 63$ | $10\frac{3}{4}\times 60$ | $10\frac{3}{4}\times 60$ | | | | |
| | 36 | $10\frac{3}{4}\times 60$ | $10\frac{3}{4}\times 63$ | | | | | |
| | 40 | $10\frac{3}{4}\times 63$ | | | | | | |
| 88 | 12 | $6\frac{3}{4}\times 46\frac{1}{2}$ | $8\frac{3}{4}\times 43\frac{1}{2}$ | $8\frac{3}{4}\times 46\frac{1}{2}$ | $8\frac{3}{4}\times 49\frac{1}{2}$ | $8\frac{3}{4}\times 52\frac{1}{2}$ | $8\frac{3}{4}\times 55\frac{1}{2}$ | $8\frac{3}{4}\times 57$ |
| | 16 | $8\frac{3}{4}\times 48$ | $8\frac{3}{4}\times 49\frac{1}{2}$ | $8\frac{3}{4}\times 52\frac{1}{2}$ | $8\frac{3}{4}\times 57$ | $8\frac{3}{4}\times 60$ | $10\frac{3}{4}\times 58\frac{1}{2}$ | $10\frac{3}{4}\times 60$ |
| | 20 | $8\frac{3}{4}\times 52\frac{1}{2}$ | $8\frac{3}{4}\times 55\frac{1}{2}$ | $8\frac{3}{4}\times 58\frac{1}{2}$ | $10\frac{3}{4}\times 58\frac{1}{2}$ | $10\frac{3}{4}\times 61\frac{1}{2}$ | | |
| | 24 | $8\frac{3}{4}\times 57$ | $8\frac{3}{4}\times 60$ | $10\frac{3}{4}\times 60$ | $10\frac{3}{4}\times 63$ | | | |
| | 28 | $8\frac{3}{4}\times 61\frac{1}{2}$ | $10\frac{3}{4}\times 60$ | | | | | |
| | 32 | $10\frac{3}{4}\times 60$ | $10\frac{3}{4}\times 63$ | | | | | |
| 92 | 12 | $8\frac{3}{4}\times 46\frac{1}{2}$ | $8\frac{3}{4}\times 45$ | $8\frac{3}{4}\times 49\frac{1}{2}$ | $8\frac{3}{4}\times 52\frac{1}{2}$ | $8\frac{3}{4}\times 55\frac{1}{2}$ | $8\frac{3}{4}\times 58\frac{1}{2}$ | $8\frac{3}{4}\times 60$ |
| | 16 | $8\frac{3}{4}\times 49\frac{1}{2}$ | $8\frac{3}{4}\times 52\frac{1}{2}$ | $8\frac{3}{4}\times 55\frac{1}{2}$ | $8\frac{3}{4}\times 60$ | $10\frac{3}{4}\times 58\frac{1}{2}$ | $10\frac{3}{4}\times 61\frac{1}{2}$ | |
| | 20 | $8\frac{3}{4}\times 55\frac{1}{2}$ | $8\frac{3}{4}\times 58\frac{1}{2}$ | $10\frac{3}{4}\times 57$ | $10\frac{3}{4}\times 61\frac{1}{2}$ | | | |
| | 24 | $8\frac{3}{4}\times 60$ | $10\frac{3}{4}\times 58\frac{1}{2}$ | $10\frac{3}{4}\times 63$ | | | | |
| | 28 | $10\frac{3}{4}\times 60$ | $10\frac{3}{4}\times 63$ | | | | | |
| | 32 | $10\frac{3}{4}\times 63$ | | | | | | |
| 96 | 12 | $8\frac{3}{4}\times 48$ | $8\frac{3}{4}\times 48$ | $8\frac{3}{4}\times 51$ | $8\frac{3}{4}\times 55\frac{1}{2}$ | $8\frac{3}{4}\times 58\frac{1}{2}$ | $8\frac{3}{4}\times 61\frac{1}{2}$ | $10\frac{3}{4}\times 58\frac{1}{2}$ |
| | 16 | $8\frac{3}{4}\times 52\frac{1}{2}$ | $8\frac{3}{4}\times 55\frac{1}{2}$ | $8\frac{3}{4}\times 58\frac{1}{2}$ | $10\frac{3}{4}\times 58\frac{1}{2}$ | $10\frac{3}{4}\times 61\frac{1}{2}$ | | |
| | 20 | $8\frac{3}{4}\times 58\frac{1}{2}$ | $8\frac{3}{4}\times 61\frac{1}{2}$ | $10\frac{3}{4}\times 60$ | | | | |
| | 24 | $8\frac{3}{4}\times 63$ | $10\frac{3}{4}\times 61\frac{1}{2}$ | | | | | |
| | 28 | $10\frac{3}{4}\times 63$ | | | | | | |

## TABLE 8.8 (Continued)

| | | | Roof[d,e] | | | | | | Floor[d] |
| | | Construction Load[b,c] | Snow Loads[b,c] | | | | | | Live Load[b,c] |
| Duration of Load Factor $C_D$ = | | 1.25 | 1.15 | | | | | | 1.00 |
| Total Load = | | 30 psf | 30 psf | 35 psf | 40 psf | 45 psf | 50 psf | 55 psf | 50 psf |
| Span (ft) | Spacing (ft) | | | | | | | | |
|---|---|---|---|---|---|---|---|---|---|
| 100 | 12 | $8\frac{3}{4} \times 49\frac{1}{2}$ | $8\frac{3}{4} \times 51$ | $8\frac{3}{4} \times 54$ | $8\frac{3}{4} \times 57$ | $10\frac{3}{4} \times 55\frac{1}{2}$ | $10\frac{3}{4} \times 58\frac{1}{2}$ | $10\frac{3}{4} \times 61\frac{1}{2}$ | |
| | 16 | $8\frac{3}{4} \times 55\frac{1}{2}$ | $8\frac{3}{4} \times 57$ | $10\frac{3}{4} \times 57$ | $10\frac{3}{4} \times 61\frac{1}{2}$ | | | | |
| | 20 | $8\frac{3}{4} \times 61\frac{1}{2}$ | $10\frac{3}{4} \times 58\frac{1}{2}$ | $10\frac{3}{4} \times 63$ | | | | | |
| | 24 | $10\frac{3}{4} \times 61\frac{1}{2}$ | | | | | | | |

[a]For PRELIMINARY DESIGN PURPOSES ONLY. The beam sizes shown for simple span, straight or cambered glued laminated timber beams. Other beam support systems may be used to meet varying design conditions. All beam sizes are based on continuous lateral bracing of the top (compression side) of the beam. Allowable tabular design values used are as follows:

Combinations: 24F-V4 or 20F-V3 (denoted by asterisk) Western Species

$F_b$ = 2400 psi   Except for beam sizes denoted with an asterisk, which are 2000 psi.

$F_v$ = 165 psi (Douglas Fir-Larch).

$E$ = 1,800,000 psi   Except for beam sizes denoted with an asterisk, which are 1,600,000 psi.

Sizes are based on bending, deflection, and shear only. No wind, earthquake, or other load conditions have been checked.

[b]Loads: Dead load = 10 psf, for all beam types. Beam weight is not included in the dead load, but is used in the analysis.

Floor live loads = 40 psf. Construction load = 20 psf. Snow load varies from 20 psf to 45 psf.

$^c$Deflection limits: for Roof beams, deflection has been analyzed by span/180 for Snow or Construction loads, and span/120 for Total loads. for Floor beams, deflection has been analyzed by span/360 for live loads. Total load deflection does not control design since live load is 80% of total load. See Table 4.4 for other deflection limits for floor beams for commercial and other uses where increased stiffness is desired.

$^d$The tabular design values for bending have been adjusted using the volume factor, $C_v = k[(5.125/b)^{1/x}(12/d)^{1/x}(21/L)^{1/x}]$. The value of $x = 10$ for Western Species has been applied to beam sizes. For Southern Pine, $x = 20$. Sizes for Southern Pine beams may vary compared to sizes shown.

$^e$Roof beam sizes are based on a minimum roof slope. For other roof slopes, additional analysis may be required.

## TABLE 8.9
### Douglas Fir

## Structural Glued Laminated Timber
### *Roof Beams for Construction Load*

Lamination thickness: 1-1/2 in.

Simple Span Beam Table
For Preliminary Design Purposes

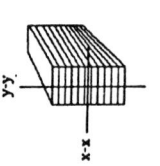

| | | | |
|---|---|---|---|
| $F_b$ 2400 psi | $F_v$ 165 psi | $E$ $1.8 \times 10^6$ psi | $C_D$ 1.25 |

Deflection Limit l/180

| Beam Size | | | Beam Capacity, Uniform Load w, plf | | | | | | | | | | | | | | | | |
|---|---|---|---|---|---|---|---|---|---|---|---|---|---|---|---|---|---|---|---|
| Width b, in. | Depth d, in. | Weight plf | Span, ft 8 | 9 | 10 | 11 | 12 | 13 | 14 | 15 | 16 | 17 | 18 | 19 | 20 | 21 | 22 | 23 | 24 |
| 3-1/8 | 6 | 4.6 | 586 | 412 | 300 | 225 | 174 | 137 | 109 | 89 | 73 | 61 | 51 | 44 | 37 | 32 | 28 | | |
| 3-1/8 | 7-1/2 | 5.7 | 916 | 723 | 586 | 440 | 339 | 267 | 214 | 174 | 143 | 119 | 100 | 85 | 73 | 63 | 55 | 48 | 42 |
| 3-1/8 | 9 | 6.8 | 1190 | 1031 | 844 | 697 | 586 | 461 | 369 | 300 | 247 | 206 | 174 | 148 | 127 | 109 | 95 | 83 | 73 |
| 3-1/8 | 10-1/2 | 8.0 | 1444 | 1245 | 1094 | 949 | 798 | 680 | 586 | 478 | 393 | 327 | 276 | 234 | 201 | 174 | 151 | 132 | 116 |
| 3-1/8 | 12 | 9.1 | 1719 | 1473 | 1289 | 1148 | 1031 | 868 | 765 | 667 | 586 | 488 | 412 | 350 | 300 | 259 | 225 | 197 | 174 |
| 3-1/8 | 13-1/2 | 10.3 | 2018 | 1719 | 1487 | 1326 | 1190 | 1079 | 969 | 844 | 742 | 657 | 586 | 498 | 427 | 369 | 321 | 281 | 247 |
| 3-1/8 | 15 | 11.4 | 2344 | 1983 | 1719 | 1517 | 1357 | 1228 | 1121 | 1031 | 916 | 811 | 723 | 649 | 586 | 506 | 440 | 385 | 339 |
| 3-1/8 | 16-1/2 | 12.5 | 2701 | 2269 | 1956 | 1719 | 1533 | 1383 | 1260 | 1158 | 1070 | 981 | 875 | 766 | 709 | 643 | 586 | 513 | 451 |
| 3-1/8 | 18 | 13.7 | 3094 | 2578 | 2210 | 1934 | 1719 | 1547 | 1406 | 1289 | 1190 | 1105 | 1031 | 935 | 844 | 765 | 697 | 638 | 563 |
| 3-1/8 | 19-1/2 | 14.8 | 3528 | 2914 | 2483 | 2162 | 1915 | 1719 | 1559 | 1426 | 1314 | 1219 | 1136 | 1064 | 990 | 898 | 815 | 743 | 679 |
| 5-1/8 | 6 | 7.5 | 961 | 675 | 492 | 370 | 285 | 224 | 179 | 146 | 120 | 100 | 84 | 72 | 61 | 53 | 46 | 40 | 36 |
| 5-1/8 | 7-1/2 | 9.3 | 1501 | 1188 | 961 | 722 | 556 | 437 | 350 | 285 | 235 | 196 | 165 | 140 | 120 | 104 | 90 | 79 | 70 |
| 5-1/8 | 9 | 11.2 | 1951 | 1691 | 1384 | 1144 | 961 | 756 | 605 | 492 | 405 | 338 | 285 | 242 | 208 | 179 | 156 | 136 | 120 |
| 5-1/8 | 10-1/2 | 13.1 | 2368 | 2041 | 1794 | 1557 | 1308 | 1114 | 961 | 781 | 644 | 537 | 452 | 384 | 330 | 285 | 248 | 217 | 191 |
| 5-1/8 | 12 | 14.9 | 2819 | 2416 | 2114 | 1879 | 1691 | 1456 | 1255 | 1053 | 961 | 801 | 675 | 574 | 492 | 425 | 370 | 323 | 285 |
| 5-1/8 | 13-1/2 | 16.8 | 3309 | 2819 | 2455 | 2174 | 1951 | 1770 | 1588 | 1364 | 1216 | 1077 | 961 | 817 | 701 | 605 | 528 | 481 | 405 |
| 5-1/8 | 15 | 18.7 | 3844 | 3252 | 2819 | 2487 | 2225 | 2013 | 1838 | 1691 | 1501 | 1328 | 1178 | 1052 | 944 | 830 | 722 | 632 | 556 |
| 5-1/8 | 16-1/2 | 20.6 | 4429 | 3721 | 3208 | 2819 | 2514 | 2269 | 2067 | 1898 | 1755 | 1592 | 1412 | 1261 | 1132 | 1022 | 926 | 841 | 740 |

| | | | | | | | | | | | | | | | | | | | |
|---|---|---|---|---|---|---|---|---|---|---|---|---|---|---|---|---|---|---|---|
| 5-1/8 | 18 | 22.4 | 5074 | 4228 | 3624 | 3171 | 2819 | 2537 | 2306 | 2114 | 1951 | 1812 | 1666 | 1487 | 1335 | 1205 | 1093 | 996 | 911 |
| 5-1/8 | 19-1/2 | 24.3 | 5788 | 4780 | 4072 | 3546 | 3141 | 2819 | 2557 | 2339 | 2156 | 1999 | 1863 | 1731 | 1555 | 1403 | 1273 | 1159 | 1060 |
| 5-1/8 | 21 | 26.2 | 6577 | 5381 | 4553 | 3946 | 3482 | 3115 | 2819 | 2574 | 2368 | 2192 | 2041 | 1909 | 1790 | 1615 | 1485 | 1334 | 1220 |
| 5-1/8 | 22-1/2 | 28.0 | 7461 | 6040 | 5074 | 4374 | 3844 | 3428 | 3094 | 2819 | 2589 | 2393 | 2225 | 2079 | 1951 | 1838 | 1670 | 1521 | 1391 |
| 5-1/8 | 24 | 29.9 | 8456 | 6765 | 5638 | 4832 | 4228 | 3758 | 3383 | 3075 | 2819 | 2602 | 2416 | 2255 | 2114 | 1990 | 1879 | 1720 | 1573 |
| 5-1/8 | 25-1/2 | 31.8 | 9584 | 7566 | 6250 | 5324 | 4637 | 4107 | 3686 | 3343 | 3059 | 2819 | 2614 | 2437 | 2282 | 2148 | 2025 | 1917 | 1765 |
| 6-3/4 | 7-1/2 | 12.3 | 1878 | 1563 | 1266 | 951 | 732 | 576 | 461 | 375 | 309 | 258 | 217 | 185 | 158 | 137 | 119 | 104 | 92 |
| 6-3/4 | 9 | 14.8 | 2570 | 2228 | 1823 | 1506 | 1266 | 995 | 797 | 648 | 534 | 445 | 375 | 319 | 273 | 236 | 205 | 180 | 158 |
| 6-3/4 | 10-1/2 | 17.2 | 3119 | 2688 | 2363 | 2050 | 1723 | 1468 | 1266 | 1029 | 848 | 707 | 595 | 506 | 434 | 375 | 326 | 285 | 251 |
| 6-3/4 | 12 | 19.7 | 3713 | 3182 | 2784 | 2475 | 2228 | 1917 | 1653 | 1440 | 1205 | 1055 | 889 | 756 | 648 | 560 | 487 | 426 | 375 |
| 6-3/4 | 13-1/2 | 22.1 | 4358 | 3713 | 3233 | 2864 | 2570 | 2331 | 2092 | 1812 | 1583 | 1393 | 1236 | 1076 | 923 | 797 | 693 | 607 | 534 |
| 6-3/4 | 15 | 24.6 | 5063 | 4284 | 3713 | 3276 | 2931 | 2652 | 2421 | 2214 | 1933 | 1702 | 1510 | 1348 | 1210 | 1092 | 951 | 832 | 732 |
| 6-3/4 | 16-1/2 | 27.1 | 5834 | 4901 | 4225 | 3713 | 3311 | 2988 | 2723 | 2500 | 2312 | 2040 | 1809 | 1615 | 1450 | 1308 | 1187 | 1081 | 975 |
| 6-3/4 | 18 | 29.5 | 6683 | 5569 | 4773 | 4177 | 3713 | 3341 | 3038 | 2784 | 2570 | 2387 | 2135 | 1905 | 1711 | 1544 | 1401 | 1278 | 1167 |
| 6-3/4 | 19-1/2 | 32.0 | 7620 | 6295 | 5363 | 4671 | 4137 | 3713 | 3367 | 3081 | 2839 | 2633 | 2454 | 2218 | 1992 | 1798 | 1631 | 1485 | 1358 |
| 6-3/4 | 21 | 34.5 | 8663 | 7088 | 5997 | 5198 | 4586 | 4103 | 3713 | 3390 | 3119 | 2888 | 2688 | 2515 | 2293 | 2070 | 1877 | 1710 | 1564 |
| 6-3/4 | 22-1/2 | 36.9 | 9827 | 7955 | 6683 | 5761 | 5063 | 4515 | 4075 | 3713 | 3409 | 3152 | 2931 | 2739 | 2570 | 2360 | 2140 | 1949 | 1783 |
| 6-3/4 | 24 | 39.4 | 11138 | 8910 | 7425 | 6364 | 5569 | 4950 | 4455 | 4050 | 3713 | 3427 | 3182 | 2970 | 2784 | 2621 | 2419 | 2204 | 2015 |
| 6-3/4 | 25-1/2 | 41.8 | 12623 | 9965 | 8232 | 7013 | 6108 | 5410 | 4855 | 4403 | 4028 | 3713 | 3443 | 3209 | 3005 | 2826 | 2667 | 2473 | 2261 |
| 6-3/4 | 27 | 44.3 | 14320 | 11138 | 9113 | 7711 | 6683 | 5896 | 5276 | 4773 | 4358 | 4010 | 3713 | 3456 | 3233 | 3038 | 2864 | 2709 | 2521 |
| 6-3/4 | 28-1/2 | 46.8 | 16278 | 12448 | 10077 | 8465 | 7297 | 6413 | 5719 | 5161 | 4703 | 4319 | 3993 | 3713 | 3469 | 3258 | 3067 | 2899 | 2748 |
| 6-3/4 | 30 | 49.2 | 18563 | 13922 | 11138 | 9281 | 7955 | 6961 | 6188 | 5569 | 5063 | 4641 | 4284 | 3978 | 3713 | 3480 | 3276 | 3094 | 2931 |

TABLE SPECIFICATIONS: This table applies to straight, simply supported glued laminated timber beams under dry conditions of use. Beams must be laterally supported at the top along the length of the beam and at the top and bottom at the end bearings. The load carrying capacities tabulated are for total load including the weight of the member. A unit weight of 35 pcf was assumed to determine the tabulated plf beam weights.

DESIGN VALUE MODIFICATIONS: The allowable stress in bending, $F_b$, has been modified by the AITC volume factor, $C_v$. For determination of load carrying capacities governed by shear, loads within a distance "d" (the depth of the beam) from the ends have been neglected.

DEFLECTION LIMITS: For roof beams, a deflection limit of l/180 for total load has been used.

CONTROLLING VALUES: Values above and to the right of the heavy line and/or shaded area indicate load capacity controlled by deflection considerations; whereas, values below and to the left of the heavy line and/or shaded area represent load capacity controlled by shear. Values in the shaded area are those load capacities governed by bending.

SPAN: Span is defined as the length from centerline to centerline of bearing. It is this length that is used in standard engineering equations to calculate deflection, bending and shear.

TABLE 8.9  (Continued)

## Structural Glued Laminated Timber
### Roof Beams for Construction Load

Lamination thickness: 1-1/2 in.

Simple Span Beam Table
For Preliminary Design Purposes

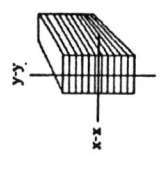

| $F_b$ | $F_v$ | $E$ | $C_D$ | Deflection Limit |
|---|---|---|---|---|
| 2400 psi | 165 psi | $1.8 \times 10^6$ psi | 1.25 | $l/180$ |

| Beam Size | | | Span, ft — Beam Capacity, Uniform Load $w$, plf | | | | | | | | | | | | | | | |
|---|---|---|---|---|---|---|---|---|---|---|---|---|---|---|---|---|---|---|
| Width b, in. | Depth d, in. | Weight plf | 25 | 26 | 27 | 28 | 29 | 30 | 31 | 32 | 33 | 34 | 35 | 36 | 37 | 38 | 39 | 40 |
| 3-1/8 | 6 | 4.6 | | | | | | | | | | | | | | | | |
| 3-1/8 | 7-1/2 | 5.7 | 37 | 33 | 30 | 27 | | | | | | | | | | | | |
| 3-1/8 | 9 | 6.8 | 65 | 58 | 51 | 48 | 42 | 37 | 34 | 31 | 28 | | | | | | | |
| 3-1/8 | 10-1/2 | 8.0 | 103 | 91 | 82 | 73 | 66 | 60 | 54 | 49 | 45 | 41 | 37 | 34 | 32 | 29 | | |
| 3-1/8 | 12 | 9.1 | 154 | 137 | 122 | 109 | 98 | 89 | 81 | 73 | 67 | 61 | 56 | 51 | 47 | 44 | 40 | 37 |
| 3-1/8 | 13-1/2 | 10.3 | 219 | 184 | 174 | 156 | 140 | 127 | 115 | 104 | 95 | 87 | 80 | 73 | 67 | 62 | 58 | 53 |
| 3-1/8 | 15 | 11.4 | 300 | 267 | 238 | 214 | 192 | 174 | 157 | 143 | 130 | 119 | 109 | 100 | 93 | 85 | 79 | 73 |
| 3-1/8 | 16-1/2 | 12.5 | 399 | 355 | 317 | 284 | 256 | 231 | 209 | 190 | 174 | 159 | 146 | 134 | 123 | 114 | 105 | 97 |
| 3-1/8 | 18 | 13.7 | 518 | 461 | 412 | 369 | 332 | 300 | 272 | 247 | 225 | 206 | 189 | 174 | 160 | 148 | 137 | 127 |
| 3-1/8 | 19-1/2 | 14.8 | 623 | 574 | 523 | 469 | 422 | 381 | 346 | 314 | 287 | 262 | 240 | 221 | 203 | 188 | 174 | 161 |
| 5-1/8 | 6 | 7.5 | 31 | 28 | | | | | | | | | | | | | | |
| 5-1/8 | 7-1/2 | 9.3 | 61 | 55 | 49 | 44 | 39 | 36 | 32 | 29 | | | | | | | | |
| 5-1/8 | 9 | 11.2 | 106 | 94 | 84 | 78 | 68 | 61 | 56 | 51 | 46 | 42 | 39 | 38 | 33 | 30 | 28 | |
| 5-1/8 | 10-1/2 | 13.1 | 169 | 150 | 134 | 120 | 108 | 98 | 89 | 80 | 73 | 67 | 61 | 57 | 52 | 48 | 44 | 41 |
| 5-1/8 | 12 | 14.9 | 252 | 224 | 200 | 179 | 161 | 146 | 132 | 120 | 110 | 100 | 92 | 84 | 78 | 72 | 66 | 61 |
| 5-1/8 | 13-1/2 | 16.8 | 359 | 319 | 285 | 255 | 230 | 208 | 188 | 171 | 156 | 143 | 131 | 120 | 111 | 102 | 94 | 88 |
| 5-1/8 | 15 | 18.7 | 492 | 437 | 391 | 350 | 315 | 285 | 256 | 235 | 214 | 198 | 179 | 165 | 152 | 140 | 130 | 120 |
| 5-1/8 | 16-1/2 | 20.6 | 655 | 582 | 520 | 466 | 420 | 379 | 343 | 312 | 285 | 260 | 239 | 219 | 202 | 186 | 172 | 160 |
| 5-1/8 | 18 | 22.4 | 836 | 756 | 675 | 605 | 545 | 492 | 446 | 405 | 370 | 338 | 310 | 285 | 262 | 242 | 224 | 208 |

| b (in) | d (in) | wt (plf) | | | | | | | | | | | | | | | | |
|---|---|---|---|---|---|---|---|---|---|---|---|---|---|---|---|---|---|---|---|
| 5-1/8 | 19-1/2 | 24.3 | 973 | 896 | 828 | 767 | 692 | 626 | 567 | 515 | 470 | 430 | 394 | 362 | 333 | 308 | 285 | 264 |
| 5-1/8 | 21 | 26.2 | 1120 | 1032 | 953 | 883 | 820 | 784 | 708 | 644 | 567 | 537 | 492 | 452 | 418 | 384 | 356 | 330 |
| 5-1/8 | 22-1/2 | 28.0 | 1277 | 1176 | 1086 | 1007 | 935 | 871 | 813 | 760 | 713 | 660 | 605 | 556 | 512 | 473 | 437 | 405 |
| 5-1/8 | 24 | 29.9 | 1444 | 1329 | 1228 | 1138 | 1057 | 984 | 919 | 860 | 808 | 757 | 712 | 671 | 622 | 574 | 531 | 492 |
| 5-1/8 | 25-1/2 | 31.8 | 1620 | 1492 | 1378 | 1277 | 1186 | 1105 | 1031 | 965 | 904 | 849 | 799 | 753 | 711 | 672 | 637 | 590 |
| 6-3/4 | 7-1/2 | 12.3 | 81 | 72 | 64 | 58 | 52 | 47 | 42 | 39 | 35 | 32 | 30 | 27 | | | | |
| 6-3/4 | 9 | 14.8 | 140 | 124 | 111 | 100 | 90 | 81 | 73 | 67 | 61 | 56 | 51 | 47 | 43 | 40 | 37 | 34 |
| 6-3/4 | 10-1/2 | 17.2 | 222 | 198 | 176 | 158 | 142 | 129 | 117 | 106 | 97 | 88 | 81 | 74 | 69 | 63 | 59 | 54 |
| 6-3/4 | 12 | 19.7 | 332 | 295 | 263 | 236 | 213 | 192 | 174 | 158 | 144 | 132 | 121 | 111 | 102 | 94 | 87 | 81 |
| 6-3/4 | 13-1/2 | 22.1 | 472 | 420 | 375 | 336 | 303 | 273 | 248 | 225 | 205 | 188 | 172 | 158 | 146 | 135 | 124 | 115 |
| 6-3/4 | 15 | 24.6 | 648 | 576 | 514 | 461 | 415 | 375 | 340 | 309 | 282 | 258 | 238 | 217 | 200 | 185 | 171 | 158 |
| 6-3/4 | 16-1/2 | 27.1 | 862 | 787 | 685 | 614 | 553 | 499 | 452 | 411 | 375 | 343 | 314 | 289 | 268 | 246 | 227 | 211 |
| 6-3/4 | 18 | 29.5 | 1071 | 998 | 889 | 797 | 717 | 648 | 587 | 534 | 487 | 445 | 408 | 375 | 345 | 319 | 295 | 273 |
| 6-3/4 | 19-1/2 | 32.0 | 1247 | 1148 | 1061 | 983 | 912 | 824 | 747 | 679 | 619 | 566 | 519 | 477 | 439 | 405 | 375 | 348 |
| 6-3/4 | 21 | 34.5 | 1435 | 1322 | 1221 | 1131 | 1051 | 979 | 914 | 848 | 773 | 707 | 648 | 595 | 548 | 506 | 468 | 434 |
| 6-3/4 | 22-1/2 | 36.9 | 1638 | 1507 | 1392 | 1290 | 1198 | 1116 | 1041 | 974 | 913 | 858 | 797 | 732 | 675 | 623 | 576 | 534 |
| 6-3/4 | 24 | 39.4 | 1850 | 1703 | 1574 | 1456 | 1354 | 1261 | 1177 | 1101 | 1032 | 970 | 912 | 880 | 812 | 756 | 699 | 648 |
| 6-3/4 | 25-1/2 | 41.8 | 2075 | 1911 | 1766 | 1636 | 1520 | 1415 | 1321 | 1236 | 1159 | 1088 | 1024 | 965 | 911 | 861 | 816 | 774 |
| 6-3/4 | 27 | 44.3 | 2314 | 2131 | 1968 | 1824 | 1694 | 1578 | 1473 | 1378 | 1291 | 1213 | 1141 | 1078 | 1016 | 960 | 909 | 862 |
| 6-3/4 | 28-1/2 | 46.8 | 2564 | 2361 | 2181 | 2021 | 1877 | 1748 | 1632 | 1527 | 1431 | 1344 | 1265 | 1192 | 1125 | 1064 | 1008 | 958 |
| 6-3/4 | 30 | 49.2 | 2784 | 2603 | 2405 | 2228 | 2069 | 1927 | 1799 | 1683 | 1578 | 1482 | 1394 | 1314 | 1241 | 1173 | 1111 | 1053 |

TABLE SPECIFICATIONS: This table applies to straight, simply supported glued laminated timber beams under dry conditions of use. Beams must be laterally supported at the top along the length of the beam and at the top and bottom at the end bearings. The load carrying capacities tabulated are for total load including the weight of the member. A unit weight of 35 pcf was assumed to determine the tabulated plf beam weights.

DESIGN VALUE MODIFICATIONS: The allowable stress in bending, $F_b$, has been modified by the AITC volume factor, $C_v$. For determination of load carrying capacities governed by shear, loads within a distance "d" (the depth of the beam) from the ends have been neglected.

DEFLECTION LIMITS: For roof beams, a deflection limit of l/180 for total load has been used.

CONTROLLING VALUES: Values above and to the right of the heavy line and/or shaded area indicate load capacity controlled by deflection considerations; whereas, values below and to the left of the heavy line and/or shaded area represent load capacity controlled by shear. Values in the shaded area are those load capacities governed by bending.

SPAN: Span is defined as the length from centerline to centerline of bearing. It is this length that is used in standard engineering equations to calculate deflection, bending and shear.

TABLE 8.9 (*Continued*)

## Structural Glued Laminated Timber
### *Roof Beams for Snow Load*

Lamination thickness: 1-1/2 in.

Simple Span Beam Table
For Preliminary Design Purposes

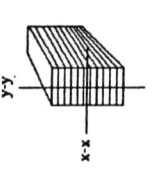

| | | | | | |
|---|---|---|---|---|---|
| $F_b$ | $F_v$ | $E$ | $C_D$ | Deflection Limit | |
| 2400 psi | 165 psi | $1.8 \times 10^6$ psi | 1.15 | $l/180$ | |

| Beam Size | | Weight | Span, ft | | | | | | | | | | | | | | Beam Capacity, Uniform Load $w$, plf | | |
|---|---|---|---|---|---|---|---|---|---|---|---|---|---|---|---|---|---|---|---|---|
| Width b, in. | Depth d, in. | plf | 8 | 9 | 10 | 11 | 12 | 13 | 14 | 15 | 16 | 17 | 18 | 19 | 20 | 21 | 22 | 23 | 24 |
| 3-1/8 | 6 | 4.6 | 539 | 412 | 300 | 225 | 174 | 137 | 109 | 89 | 73 | 61 | 51 | 44 | 37 | 32 | 28 | | |
| 3-1/8 | 7-1/2 | 5.7 | 842 | 666 | 539 | 440 | 339 | 267 | 214 | 174 | 143 | 119 | 100 | 85 | 73 | 63 | 55 | 48 | 42 |
| 3-1/8 | 9 | 6.8 | 1095 | 949 | 776 | 642 | 539 | 459 | 369 | 300 | 247 | 206 | 174 | 148 | 127 | 109 | 95 | 83 | 73 |
| 3-1/8 | 10-1/2 | 8.0 | 1328 | 1145 | 1006 | 873 | 734 | 625 | 539 | 470 | 393 | 327 | 276 | 234 | 201 | 174 | 151 | 132 | 118 |
| 3-1/8 | 12 | 9.1 | 1581 | 1355 | 1186 | 1054 | 949 | 817 | 704 | 613 | 539 | 478 | 412 | 350 | 300 | 259 | 225 | 197 | 174 |
| 3-1/8 | 13-1/2 | 10.3 | 1858 | 1581 | 1377 | 1220 | 1095 | 993 | 891 | 778 | 682 | 604 | 539 | 484 | 427 | 369 | 321 | 281 | 247 |
| 3-1/8 | 15 | 11.4 | 2158 | 1825 | 1581 | 1395 | 1248 | 1129 | 1031 | 949 | 842 | 746 | 666 | 597 | 539 | 489 | 440 | 385 | 339 |
| 3-1/8 | 16-1/2 | 12.5 | 2485 | 2087 | 1799 | 1581 | 1410 | 1273 | 1160 | 1065 | 985 | 903 | 805 | 723 | 652 | 592 | 539 | 493 | 451 |
| 3-1/8 | 18 | 13.7 | 2846 | 2372 | 2033 | 1779 | 1581 | 1423 | 1294 | 1186 | 1095 | 1017 | 949 | 860 | 776 | 704 | 642 | 587 | 537 |
| 3-1/8 | 19-1/2 | 14.8 | 3246 | 2681 | 2284 | 1989 | 1762 | 1581 | 1434 | 1312 | 1209 | 1121 | 1045 | 979 | 911 | 826 | 750 | 683 | 625 |
| 5-1/8 | 6 | 7.5 | 884 | 675 | 492 | 370 | 285 | 224 | 179 | 148 | 120 | 100 | 84 | 72 | 61 | 53 | 46 | 40 | 36 |
| 5-1/8 | 7-1/2 | 9.3 | 1361 | 1091 | 864 | 722 | 556 | 437 | 350 | 285 | 235 | 196 | 165 | 140 | 120 | 104 | 90 | 79 | 70 |
| 5-1/8 | 9 | 11.2 | 1795 | 1556 | 1273 | 1052 | 864 | 753 | 605 | 492 | 405 | 338 | 285 | 242 | 208 | 179 | 156 | 136 | 120 |
| 5-1/8 | 10-1/2 | 13.1 | 2178 | 1878 | 1650 | 1432 | 1203 | 1025 | 884 | 770 | 644 | 537 | 452 | 384 | 330 | 285 | 248 | 217 | 191 |
| 5-1/8 | 12 | 14.9 | 2593 | 2223 | 1945 | 1729 | 1556 | 1339 | 1155 | 1006 | 884 | 783 | 675 | 574 | 492 | 425 | 370 | 323 | 285 |
| 5-1/8 | 13-1/2 | 16.8 | 3044 | 2563 | 2259 | 2001 | 1795 | 1628 | 1461 | 1273 | 1119 | 991 | 884 | 792 | 701 | 605 | 526 | 461 | 405 |
| 5-1/8 | 15 | 18.7 | 3536 | 2992 | 2593 | 2288 | 2047 | 1852 | 1691 | 1556 | 1381 | 1222 | 1084 | 968 | 869 | 784 | 711 | 632 | 556 |
| 5-1/8 | 16-1/2 | 20.6 | 4075 | 3423 | 2951 | 2593 | 2313 | 2087 | 1902 | 1748 | 1615 | 1465 | 1299 | 1160 | 1041 | 940 | 852 | 776 | 710 |
| 5-1/8 | 18 | 22.4 | 4668 | 3890 | 3334 | 2917 | 2593 | 2334 | 2122 | 1945 | 1795 | 1667 | 1533 | 1368 | 1228 | 1109 | 1006 | 916 | 838 |

| Width | Depth | plf | | | | | | | | | | | | | | | | |
|---|---|---|---|---|---|---|---|---|---|---|---|---|---|---|---|---|---|---|
| 5-1/8 | 19-1/2 | 24.3 | 5323 | 4397 | 3746 | 3262 | 2890 | 2593 | 2352 | 2152 | 1983 | 1839 | 1714 | 1593 | 1430 | 1291 | 1171 | 1066 | 975 |
| 5-1/8 | 21 | 26.2 | 6051 | 4951 | 4189 | 3631 | 3203 | 2866 | 2593 | 2368 | 2178 | 2017 | 1878 | 1757 | 1646 | 1486 | 1348 | 1228 | 1123 |
| 5-1/8 | 22-1/2 | 28.0 | 6884 | 5557 | 4668 | 4024 | 3536 | 3154 | 2846 | 2593 | 2382 | 2202 | 2047 | 1913 | 1795 | 1691 | 1537 | 1400 | 1280 |
| 5-1/8 | 24 | 29.9 | 7780 | 6224 | 5187 | 4446 | 3890 | 3458 | 3112 | 2829 | 2593 | 2394 | 2223 | 2075 | 1945 | 1831 | 1729 | 1582 | 1447 |
| 5-1/8 | 25-1/2 | 31.8 | 8817 | 6961 | 5750 | 4898 | 4266 | 3779 | 3391 | 3076 | 2814 | 2593 | 2405 | 2242 | 2099 | 1974 | 1863 | 1763 | 1624 |
| 6-3/4 | 7-1/2 | 12.3 | 1819 | 1436 | 1164 | 951 | 732 | 576 | 461 | 375 | 309 | 258 | 217 | 185 | 158 | 137 | 119 | 104 | 92 |
| 6-3/4 | 9 | 14.8 | 2365 | 2049 | 1677 | 1366 | 1164 | 992 | 797 | 648 | 534 | 445 | 375 | 319 | 273 | 236 | 205 | 180 | 158 |
| 6-3/4 | 10-1/2 | 17.2 | 2869 | 2473 | 2174 | 1866 | 1585 | 1350 | 1164 | 1014 | 848 | 707 | 595 | 508 | 434 | 375 | 326 | 285 | 251 |
| 6-3/4 | 12 | 19.7 | 3415 | 2928 | 2562 | 2277 | 2049 | 1784 | 1521 | 1325 | 1164 | 1025 | 889 | 756 | 648 | 560 | 487 | 426 | 375 |
| 6-3/4 | 13-1/2 | 22.1 | 4009 | 3415 | 2975 | 2635 | 2365 | 2145 | 1925 | 1667 | 1459 | 1282 | 1137 | 1015 | 911 | 797 | 693 | 607 | 534 |
| 6-3/4 | 15 | 24.6 | 4658 | 3941 | 3415 | 3014 | 2696 | 2440 | 2228 | 2037 | 1778 | 1566 | 1388 | 1240 | 1113 | 1005 | 911 | 830 | 732 |
| 6-3/4 | 16-1/2 | 27.1 | 5367 | 4508 | 3887 | 3415 | 3046 | 2749 | 2505 | 2300 | 2127 | 1877 | 1665 | 1486 | 1334 | 1204 | 1092 | 995 | 910 |
| 6-3/4 | 18 | 29.5 | 6148 | 5123 | 4391 | 3842 | 3415 | 3074 | 2795 | 2562 | 2365 | 2196 | 1964 | 1753 | 1574 | 1421 | 1288 | 1174 | 1073 |
| 6-3/4 | 19-1/2 | 32.0 | 7011 | 5782 | 4834 | 4297 | 3806 | 3415 | 3098 | 2834 | 2612 | 2422 | 2258 | 2041 | 1833 | 1654 | 1500 | 1366 | 1250 |
| 6-3/4 | 21 | 34.5 | 7969 | 6520 | 5517 | 4782 | 4219 | 3775 | 3415 | 3118 | 2869 | 2657 | 2473 | 2314 | 2110 | 1904 | 1727 | 1573 | 1439 |
| 6-3/4 | 22-1/2 | 36.9 | 9041 | 7319 | 6148 | 5300 | 4658 | 4154 | 3749 | 3415 | 3137 | 2900 | 2696 | 2520 | 2365 | 2171 | 1969 | 1793 | 1640 |
| 6-3/4 | 24 | 39.4 | 10247 | 8197 | 683* | 5855 | 5123 | 4554 | 4099 | 3726 | 3415 | 3153 | 2928 | 2732 | 2562 | 2411 | 2226 | 2027 | 1854 |
| 6-3/4 | 25-1/2 | 41.8 | 11613 | 9168 | 7573 | 6451 | 5619 | 4977 | 4466 | 4051 | 3726 | 3415 | 3167 | 2952 | 2765 | 2600 | 2453 | 2275 | 2080 |
| 6-3/4 | 27 | 44.3 | 13174 | 10247 | 8384 | 7094 | 6148 | 5425 | 4854 | 4391 | 4009 | 3706 | 3415 | 3180 | 2975 | 2795 | 2635 | 2492 | 2319 |
| 6-3/4 | 28-1/2 | 46.8 | 14976 | 11452 | 9271 | 7787 | 6713 | 5899 | 5262 | 4748 | 4326 | 3973 | 3673 | 3415 | 3192 | 2995 | 2822 | 2667 | 2528 |
| 6-3/4 | 30 | 49.2 | 17078 | 12808 | 10247 | 8539 | 7319 | 6404 | 5683 | 5123 | 4658 | 4269 | 3941 | 3659 | 3415 | 3202 | 3014 | 2846 | 2696 |

TABLE SPECIFICATIONS: This table applies to straight, simply supported glued laminated timber beams under dry conditions of use. Beams must be laterally supported at the top along the length of the beam and at the top and bottom at the end bearings. The load carrying capacities tabulated are for total load including the weight of the member. A unit weight of 35 pcf was assumed to determine the tabulated plf beam weights.

DESIGN VALUE MODIFICATIONS: The allowable stress in bending, $F_b$, has been modified by the AITC volume factor, $C_v$. For determination of load carrying capacities governed by shear, loads within a distance "d" (the depth of the beam) from the ends have been neglected.

DEFLECTION LIMITS: For roof beams, a deflection limit of l/180 for total load has been used.

CONTROLLING VALUES: Values above and to the right of the heavy line and/or shaded area indicate load capacity controlled by deflection considerations; whereas, values below and to the left of the heavy line and/or shaded area represent load capacity controlled by shear. Values in the shaded area are those load capacities governed by bending.

SPAN: Span is defined as the length from centerline to centerline of bearing. It is this length that is used in standard engineering equations to calculate deflection, bending and shear.

TABLE 8.9 (Continued)

## Structural Glued Laminated Timber
### Roof Beams for Snow Load

Lamination thickness: 1-1/2 in.

Simple Span Beam Table
For Preliminary Design Purposes

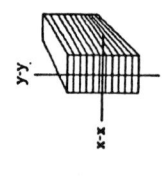

| $F_b$ 2400 psi | $F_v$ 165 psi | $E$ $1.8 \times 10^6$ psi | $C_D$ 1.15 | Deflection Limit l/180 |
|---|---|---|---|---|

**Beam Capacity, Uniform Load w, plf**

| Width b, in. | Depth d, in. | Weight plf | 25 | 26 | 27 | 28 | 29 | 30 | 31 | 32 | 33 | 34 | 35 | 36 | 37 | 38 | 39 | 40 |
|---|---|---|---|---|---|---|---|---|---|---|---|---|---|---|---|---|---|---|
| 3-1/8 | 6 | 4.6 | | | | | | | | | | | | | | | | |
| 3-1/8 | 7-1/2 | 5.7 | 37 | 33 | 30 | 27 | | | | | | | | | | | | |
| 3-1/8 | 9 | 6.8 | 65 | 58 | 51 | 48 | 42 | 37 | 34 | 31 | 28 | | | | | | | |
| 3-1/8 | 10-1/2 | 8.0 | 103 | 91 | 82 | 73 | 66 | 60 | 54 | 49 | 45 | 41 | 37 | 34 | 32 | 29 | | |
| 3-1/8 | 12 | 9.1 | 154 | 137 | 122 | 109 | 98 | 89 | 81 | 73 | 67 | 61 | 56 | 51 | 47 | 44 | 40 | 37 |
| 3-1/8 | 13-1/2 | 10.3 | 219 | 194 | 174 | 156 | 140 | 127 | 115 | 104 | 95 | 87 | 80 | 73 | 67 | 62 | 58 | 53 |
| 3-1/8 | 15 | 11.4 | 300 | 267 | 238 | 214 | 192 | 174 | 157 | 143 | 130 | 119 | 109 | 100 | 93 | 85 | 79 | 73 |
| 3-1/8 | 16-1/2 | 12.5 | 399 | 355 | 317 | 284 | 256 | 231 | 209 | 190 | 174 | 159 | 146 | 134 | 123 | 114 | 105 | 97 |
| 3-1/8 | 18 | 13.7 | 493 | 454 | 412 | 369 | 332 | 300 | 272 | 247 | 225 | 208 | 189 | 174 | 160 | 148 | 137 | 127 |
| 3-1/8 | 19-1/2 | 14.8 | 573 | 528 | 488 | 452 | 420 | 381 | 346 | 314 | 287 | 262 | 240 | 221 | 203 | 188 | 174 | 161 |
| 5-1/8 | 6 | 7.5 | 31 | 28 | | | | | | | | | | | | | | |
| 5-1/8 | 7-1/2 | 9.3 | 61 | 55 | 49 | 44 | 39 | 36 | 32 | 29 | | | | | | | | |
| 5-1/8 | 9 | 11.2 | 106 | 94 | 84 | 76 | 68 | 61 | 56 | 51 | 46 | 42 | 39 | 36 | 33 | 30 | 28 | |
| 5-1/8 | 10-1/2 | 13.1 | 169 | 150 | 134 | 120 | 108 | 98 | 89 | 80 | 73 | 67 | 61 | 57 | 52 | 48 | 44 | 41 |
| 5-1/8 | 12 | 14.9 | 252 | 224 | 200 | 179 | 161 | 146 | 132 | 120 | 110 | 100 | 92 | 84 | 78 | 72 | 66 | 61 |
| 5-1/8 | 13-1/2 | 16.8 | 359 | 319 | 285 | 255 | 230 | 208 | 188 | 171 | 156 | 143 | 131 | 120 | 111 | 102 | 94 | 88 |
| 5-1/8 | 15 | 18.7 | 492 | 437 | 391 | 350 | 315 | 285 | 258 | 235 | 214 | 196 | 179 | 165 | 152 | 140 | 130 | 120 |
| 5-1/8 | 16-1/2 | 20.6 | 652 | 582 | 520 | 466 | 420 | 379 | 343 | 312 | 285 | 260 | 239 | 219 | 202 | 186 | 172 | 160 |
| 5-1/8 | 18 | 22.4 | 769 | 708 | 654 | 605 | 545 | 482 | 446 | 405 | 370 | 338 | 310 | 285 | 262 | 242 | 224 | 208 |

| Width | Depth | Wt (plf) | | | | | | | | | | | | | | | | |
|---|---|---|---|---|---|---|---|---|---|---|---|---|---|---|---|---|---|---|
| 5-1/8 | 19-1/2 | 24.3 | 895 | 824 | 762 | 706 | 655 | 610 | 567 | 515 | 470 | 430 | 394 | 362 | 333 | 308 | 285 | 264 |
| 5-1/8 | 21 | 26.2 | 1030 | 949 | 877 | 812 | 755 | 703 | 656 | 614 | 575 | 537 | 492 | 452 | 416 | 384 | 356 | 330 |
| 5-1/8 | 22-1/2 | 28.0 | 1175 | 1082 | 1000 | 926 | 860 | 801 | 748 | 700 | 656 | 616 | 580 | 546 | 512 | 473 | 437 | 405 |
| 5-1/8 | 24 | 29.9 | 1328 | 1223 | 1150 | 1047 | 972 | 906 | 845 | 791 | 741 | 696 | 655 | 618 | 583 | 551 | 522 | 492 |
| 5-1/8 | 25-1/2 | 31.8 | 1490 | 1372 | 1268 | 1175 | 1091 | 1016 | 949 | 887 | 832 | 781 | 735 | 693 | 654 | 619 | 586 | 555 |
| 6-3/4 | 7-1/2 | 12.3 | 81 | 72 | 64 | 58 | 52 | 47 | 42 | 39 | 35 | 32 | 30 | 27 | | | | |
| 6-3/4 | 9 | 14.8 | 140 | 124 | 111 | 100 | 90 | 81 | 73 | 67 | 61 | 56 | 51 | 47 | 43 | 40 | 37 | 34 |
| 6-3/4 | 10-1/2 | 17.2 | 222 | 198 | 176 | 158 | 142 | 129 | 117 | 106 | 97 | 88 | 81 | 74 | 69 | 63 | 59 | 54 |
| 6-3/4 | 12 | 19.7 | 332 | 295 | 263 | 236 | 213 | 192 | 174 | 158 | 144 | 132 | 121 | 111 | 102 | 94 | 87 | 81 |
| 6-3/4 | 13-1/2 | 22.1 | 472 | 420 | 375 | 336 | 303 | 273 | 248 | 225 | 205 | 188 | 172 | 158 | 146 | 135 | 124 | 115 |
| 6-3/4 | 15 | 24.6 | 648 | 576 | 514 | 461 | 415 | 375 | 340 | 309 | 282 | 258 | 236 | 217 | 200 | 185 | 171 | 158 |
| 6-3/4 | 16-1/2 | 27.1 | 835 | 767 | 685 | 614 | 553 | 499 | 452 | 411 | 375 | 343 | 314 | 289 | 266 | 248 | 227 | 211 |
| 6-3/4 | 18 | 29.5 | 985 | 907 | 838 | 776 | 717 | 648 | 587 | 534 | 487 | 445 | 408 | 375 | 345 | 319 | 295 | 273 |
| 6-3/4 | 19-1/2 | 32.0 | 1147 | 1058 | 976 | 904 | 840 | 782 | 730 | 679 | 619 | 566 | 519 | 477 | 439 | 405 | 375 | 348 |
| 6-3/4 | 21 | 34.5 | 1320 | 1216 | 1123 | 1041 | 967 | 900 | 840 | 786 | 737 | 692 | 648 | 595 | 548 | 508 | 468 | 434 |
| 6-3/4 | 22-1/2 | 36.9 | 1505 | 1386 | 1281 | 1186 | 1102 | 1026 | 958 | 896 | 840 | 789 | 743 | 700 | 661 | 623 | 576 | 534 |
| 6-3/4 | 24 | 39.4 | 1702 | 1567 | 1448 | 1341 | 1248 | 1160 | 1083 | 1013 | 950 | 892 | 839 | 791 | 747 | 706 | 669 | 634 |
| 6-3/4 | 25-1/2 | 41.8 | 1909 | 1758 | 1624 | 1505 | 1398 | 1302 | 1215 | 1137 | 1066 | 1001 | 942 | 888 | 838 | 793 | 750 | 712 |
| 6-3/4 | 27 | 44.3 | 2128 | 1960 | 1811 | 1678 | 1558 | 1451 | 1355 | 1267 | 1188 | 1118 | 1050 | 990 | 934 | 883 | 837 | 793 |
| 6-3/4 | 28-1/2 | 46.8 | 2359 | 2172 | 2007 | 1859 | 1727 | 1608 | 1501 | 1405 | 1317 | 1237 | 1164 | 1097 | 1035 | 979 | 927 | 879 |
| 6-3/4 | 30 | 49.2 | 2562 | 2395 | 2212 | 2050 | 1904 | 1773 | 1655 | 1548 | 1451 | 1363 | 1283 | 1209 | 1141 | 1079 | 1022 | 969 |

TABLE SPECIFICATIONS: This table applies to straight, simply supported glued laminated timber beams under dry conditions of use. Beams must be laterally supported at the top along the length of the beam and at the top and bottom at the end bearings. The load carrying capacities tabulated are for total load including the weight of the member. A unit weight of 35 pcf was assumed to determine the tabulated plf beam weights.

DESIGN VALUE MODIFICATIONS: The allowable stress in bending, $F_b$, has been modified by the AITC volume factor, $C_v$. For determination of load carrying capacities governed by shear, loads within a distance "d" (the depth of the beam) from the ends have been neglected.

DEFLECTION LIMITS: For roof beams, a deflection limit of l/180 for total load has been used.

CONTROLLING VALUES: Values above and to the right of the heavy line and/or shaded area indicate load capacity controlled by deflection considerations; whereas, values below and to the left of the heavy line and/or shaded area represent load capacity controlled by shear. Values in the shaded area are those load capacities governed by bending.

SPAN: Span is defined as the length from centerline to centerline of bearing. It is this length that is used in standard engineering equations to calculate deflection, bending and shear.

TABLE 8.9 (*Continued*)

# Structural Glued Laminated Timber
## *Floor Beams*

**Lamination thickness: 1-1/2 in.**

Simple Span Beam Table
For Preliminary Design Purposes

| $F_b$ | $F_v$ | $E$ | $C_D$ | Deflection Limit |
|---|---|---|---|---|
| 2400 psi | 165 psi | $1.8 \times 10^6$ psi | 1.00 | l/360 |

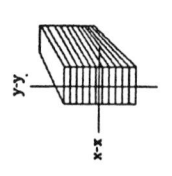

| Beam Size | | | Beam Capacity, Uniform Load w, plf | | | | | | | | | | | | | | | | |
|---|---|---|---|---|---|---|---|---|---|---|---|---|---|---|---|---|---|---|---|
| Width b, in. | Depth d, in. | Weight plf | 8 | 9 | 10 | 11 | 12 | 13 | 14 | 15 | 16 | 17 | 18 | 19 | 20 | 21 | 22 | 23 | 24 |
| 3-1/8 | 8 | 4.6 | 366 | 257 | 187 | 141 | 109 | 85 | 68 | 56 | 46 | | | | | | | | |
| 3-1/8 | 7-1/2 | 5.7 | 715 | 502 | 366 | 275 | 212 | 167 | 133 | 109 | 89 | 75 | 63 | 53 | 46 | | | | |
| 3-1/8 | 9 | 6.8 | 852 | 825 | 633 | 475 | 366 | 288 | 231 | 187 | 154 | 129 | 109 | 92 | 79 | 68 | 59 | 52 | 46 |
| 3-1/8 | 10-1/2 | 8.0 | 1155 | 996 | 875 | 755 | 582 | 457 | 366 | 298 | 245 | 205 | 172 | 147 | 128 | 109 | 94 | 83 | 73 |
| 3-1/8 | 12 | 9.1 | 1375 | 1179 | 1031 | 917 | 825 | 683 | 547 | 444 | 366 | 305 | 257 | 219 | 187 | 162 | 141 | 123 | 109 |
| 3-1/8 | 13-1/2 | 10.3 | 1614 | 1375 | 1198 | 1061 | 952 | 863 | 758 | 633 | 521 | 435 | 366 | 311 | 267 | 231 | 201 | 178 | 154 |
| 3-1/8 | 15 | 11.4 | 1875 | 1587 | 1375 | 1213 | 1086 | 982 | 897 | 825 | 715 | 596 | 502 | 427 | 366 | 316 | 275 | 241 | 212 |
| 3-1/8 | 16-1/2 | 12.5 | 2161 | 1815 | 1565 | 1375 | 1226 | 1107 | 1008 | 928 | 856 | 785 | 689 | 569 | 487 | 421 | 366 | 320 | 282 |
| 3-1/8 | 18 | 13.7 | 2475 | 2063 | 1768 | 1547 | 1375 | 1238 | 1125 | 1031 | 952 | 884 | 825 | 738 | 633 | 547 | 475 | 418 | 366 |
| 3-1/8 | 19-1/2 | 14.8 | 2822 | 2332 | 1988 | 1730 | 1532 | 1375 | 1247 | 1141 | 1051 | 975 | 909 | 851 | 792 | 695 | 604 | 529 | 466 |
| 5-1/8 | 8 | 7.5 | 601 | 422 | 307 | 231 | 178 | 140 | 112 | 91 | 75 | 63 | 53 | 45 | | | | | |
| 5-1/8 | 7-1/2 | 9.3 | 1173 | 824 | 601 | 451 | 348 | 273 | 219 | 178 | 147 | 122 | 103 | 88 | 75 | | | | |
| 5-1/8 | 9 | 11.2 | 1561 | 1353 | 1038 | 780 | 601 | 472 | 378 | 307 | 253 | 211 | 178 | 151 | 130 | 112 | 97 | 85 | 75 |
| 5-1/8 | 10-1/2 | 13.1 | 1894 | 1633 | 1435 | 1238 | 954 | 750 | 601 | 488 | 402 | 335 | 283 | 240 | 206 | 178 | 155 | 135 | 119 |
| 5-1/8 | 12 | 14.9 | 2255 | 1833 | 1691 | 1503 | 1353 | 1120 | 896 | 729 | 601 | 501 | 422 | 359 | 307 | 266 | 231 | 202 | 178 |
| 5-1/8 | 13-1/2 | 16.8 | 2647 | 2255 | 1964 | 1740 | 1561 | 1418 | 1271 | 1038 | 855 | 713 | 601 | 511 | 438 | 378 | 329 | 288 | 253 |
| 5-1/8 | 15 | 18.7 | 3075 | 2602 | 2255 | 1990 | 1780 | 1611 | 1471 | 1353 | 1173 | 978 | 824 | 700 | 601 | 519 | 451 | 395 | 348 |
| 5-1/8 | 16-1/2 | 20.8 | 3544 | 2977 | 2568 | 2255 | 2011 | 1815 | 1654 | 1519 | 1404 | 1274 | 1097 | 932 | 799 | 691 | 601 | 526 | 463 |

| Width | Depth | plf | | | | | | | | | | | | | | | | | |
|---|---|---|---|---|---|---|---|---|---|---|---|---|---|---|---|---|---|---|---|---|
| 5-1/8 | 18 | 22.4 | 4059 | 3383 | 2899 | 2537 | 2255 | 2030 | 1845 | 1691 | 1561 | 1450 | 1333B | 1190B | 1038 | 896 | 780 | 682 | 601 |
| 5-1/8 | 19-1/2 | 24.3 | 4829 | 3824 | 3257 | 2837 | 2513 | 2255 | 2045 | 1871 | 1724 | 1599 | 1491 | 1385B | 1244B | 1123B | 991 | 868 | 764 |
| 5-1/8 | 21 | 26.2 | 5282 | 4305 | 3643 | 3157 | 2786 | 2482 | 2255 | 2059 | 1894 | 1754 | 1633 | 1528 | 1432B | 1292B | 1172B | 1066B | 954 |
| 5-1/8 | 22-1/2 | 28.0 | 5989 | 4832 | 4059 | 3499 | 3075 | 2743 | 2475 | 2255 | 2071 | 1915 | 1780 | 1664 | 1561 | 1471 | 1338B | 1217B | 1113B |
| 5-1/8 | 24 | 29.9 | 6785 | 5412 | 4510 | 3868 | 3383 | 3007 | 2708 | 2460 | 2255 | 2082 | 1933 | 1804 | 1691 | 1592 | 1503 | 1376B | 1258B |
| 5-1/8 | 25-1/2 | 31.8 | 7667 | 6053 | 5000 | 4259 | 3710 | 3286 | 2949 | 2675 | 2447 | 2255 | 2091 | 1949 | 1825 | 1718 | 1620 | 1533 | 1412B |
| 6-3/4 | 7-1/2 | 12.3 | 1545 | 1085 | 791 | 564 | 458 | 360 | 288 | 234 | 193 | 161 | 136 | 115 | 99 | 85 | 74 | 65 | 57 |
| 6-3/4 | 9 | 14.8 | 2056 | 1782 | 1367 | 1027 | 791 | 622 | 498 | 405 | 334 | 278 | 234 | 199 | 171 | 148 | 128 | 112 | 99 |
| 6-3/4 | 10-1/2 | 17.2 | 2495 | 2151 | 1890 | 1631 | 1256 | 988 | 791 | 643 | 530 | 442 | 372 | 316 | 271 | 234 | 204 | 178 | 157 |
| 6-3/4 | 12 | 19.7 | 2970 | 2548 | 2228 | 1980 | 1782 | 1475 | 1181 | 960 | 791 | 659 | 556 | 472 | 405 | 350 | 304 | 266 | 234 |
| 6-3/4 | 13-1/2 | 22.1 | 3487 | 2970 | 2587 | 2291 | 2056 | 1865 | 1674B | 1367 | 1128 | 939 | 791 | 673 | 577 | 498 | 433 | 379 | 334 |
| 6-3/4 | 15 | 24.6 | 4050 | 3427 | 2970 | 2621 | 2345 | 2121 | 1937 | 1771B | 1545 | 1288 | 1085 | 923 | 791 | 683 | 594 | 520 | 458 |
| 6-3/4 | 16-1/2 | 27.1 | 4687 | 3920 | 3380 | 2970 | 2649 | 2390 | 2178 | 2000 | 1849 | 1632B | 1444 | 1228 | 1053 | 909 | 791 | 692 | 609 |
| 6-3/4 | 18 | 29.5 | 5348 | 4455 | 3819 | 3341 | 2970 | 2673 | 2430 | 2228 | 2056 | 1909 | 1708B | 1524B | 1367 | 1181 | 1027 | 899 | 791 |
| 6-3/4 | 19-1/2 | 32.0 | 6098 | 5036 | 4290 | 3736 | 3309 | 2970 | 2694 | 2484 | 2271 | 2106 | 1963 | 1775B | 1594B | 1438B | 1181 | 1143 | 1006 |
| 6-3/4 | 21 | 34.5 | 6930 | 5670 | 4798 | 4158 | 3669 | 3283 | 2970 | 2712 | 2495 | 2310 | 2151 | 2012 | 1834B | 1656B | 1502B | 1368B | 1251B |
| 6-3/4 | 22-1/2 | 36.9 | 7862 | 6364 | 5348 | 4609 | 4050 | 3612 | 3260 | 2970 | 2728 | 2522 | 2345 | 2191 | 2056 | 1888B | 1712B | 1559B | 1426B |
| 6-3/4 | 24 | 39.4 | 8910 | 7128 | 5940 | 5091 | 4455 | 3960 | 3564 | 3240 | 2970 | 2742 | 2546 | 2376 | 2228 | 2096 | 1935B | 1763B | 1612B |
| 6-3/4 | 25-1/2 | 41.8 | 10098 | 7972 | 6588 | 5610 | 4886 | 4328 | 3884 | 3523 | 3223 | 2970 | 2754 | 2567 | 2404 | 2261 | 2133 | 1978B | 1809B |
| 6-3/4 | 27 | 44.3 | 11456 | 8910 | 7290 | 6168 | 5346 | 4717 | 4221 | 3819 | 3487 | 3208 | 2970 | 2765 | 2587 | 2430 | 2291 | 2167 | 2017B |
| 6-3/4 | 28-1/2 | 46.8 | 13022 | 9958 | 8061 | 6772 | 5838 | 5130 | 4575 | 4129 | 3762 | 3455 | 3194 | 2970 | 2775 | 2604 | 2453 | 2319 | 2199 |
| 6-3/4 | 30 | 49.2 | 14850 | 11138 | 8910 | 7425 | 6364 | 5569 | 4950 | 4455 | 4050 | 3713 | 3427 | 3182 | 2970 | 2784 | 2621 | 2475 | 2345 |

TABLE SPECIFICATIONS: This table applies to straight, simply supported glued laminated timber beams under dry conditions of use. Beams must be laterally supported at the top along the length of the beam and at the top and bottom at the end bearings. The load carrying capacities tabulated are for total load including the weight of the member. A unit weight of 35 pcf was assumed to determine the tabulated plf beams weights.

DESIGN VALUE MODIFICATIONS: The allowable stress in bending, $F_b$, has been modified by the AITC volume factor, $C_v$. For determination of load carrying capacities governed by shear, loads within a distance "d" (the depth of the beam) from the ends have been neglected.

DEFLECTION LIMITS: For floor beams, a deflection limit of l/360 for live load has been used. Live load has been assumed to be 80% of the total load as is common for residential floors. For other ratios of live load to total load, the designer should check deflection based on applicable building code provisions.

CONTROLLING VALUES: Values above and to the right of the heavy line and shaded area indicate load capacity controlled by deflection considerations whereas values below and to the left of the heavy line and shaded area represent load capacity controlled by shear. Values in the shaded area are those load capacities governed by bending.

## TABLE 8.9 *(Continued)*

# Structural Glued Laminated Timber
## *Floor Beams*

Simple Span Beam Table
For Preliminary Design Purposes

Lamination thickness: 1-1/2 in.

| $F_b$ | $F_v$ | $E$ | $C_D$ | Deflection Limit |
|---|---|---|---|---|
| 2400 psi | 165 psi | 1.8 × 10⁶ psi | 1.00 | l/360 |

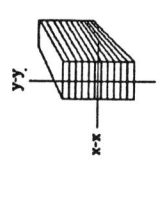

| Beam Size | | | Span, ft — Beam Capacity: Uniform Load w, plf | | | | | | | | | | | | | | | |
|---|---|---|---|---|---|---|---|---|---|---|---|---|---|---|---|---|---|---|---|
| Width b, in. | Depth d, in. | Weight plf | 25 | 26 | 27 | 28 | 29 | 30 | 31 | 32 | 33 | 34 | 35 | 36 | 37 | 38 | 39 | 40 |
| 3-1/8 | 6 | 4.6 | | | | | | | | | | | | | | | | |
| 3-1/8 | 7-1/2 | 5.7 | | | | | | | | | | | | | | | | |
| 3-1/8 | 9 | 6.8 | 48 | | | | | | | | | | | | | | | |
| 3-1/8 | 10-1/2 | 8.0 | 73 | 64 | 57 | 51 | 46 | | | | | | | | | | | |
| 3-1/8 | 12 | 9.1 | 109 | 98 | 85 | 76 | 68 | 62 | 56 | 50 | 46 | | | | | | | |
| 3-1/8 | 13-1/2 | 10.3 | 154 | 137 | 122 | 109 | 97 | 88 | 79 | 72 | 65 | 59 | 54 | 50 | 46 | | | |
| 3-1/8 | 15 | 11.4 | 212 | 187 | 167 | 148 | 133 | 120 | 109 | 98 | 89 | 82 | 75 | 68 | 63 | 58 | 53 | 49 |
| 3-1/8 | 16-1/2 | 12.5 | 282 | 250 | 222 | 198 | 178 | 160 | 144 | 131 | 119 | 109 | 99 | 91 | 84 | 77 | 71 | 68 |
| 3-1/8 | 18 | 13.7 | 366 | 324 | 288 | 257 | 231 | 208 | 187 | 170 | 154 | 141 | 129 | 118 | 109 | 100 | 92 | 85 |
| 3-1/8 | 19-1/2 | 14.8 | 466 | 412 | 366 | 327 | 293 | 264 | 238 | 216 | 196 | 179 | 164 | 150 | 138 | 127 | 117 | 108 |
| 5-1/8 | 6 | 7.5 | | | | | | | | | | | | | | | | |
| 5-1/8 | 7-1/2 | 9.3 | | | | | | | | | | | | | | | | |
| 5-1/8 | 9 | 11.2 | 75 | 66 | 58 | 53 | 47 | | | | | | | | | | | |
| 5-1/8 | 10-1/2 | 13.1 | 119 | 105 | 94 | 84 | 75 | 68 | 61 | 55 | 50 | 46 | | | | | | |
| 5-1/8 | 12 | 14.9 | 178 | 157 | 140 | 125 | 112 | 101 | 91 | 83 | 75 | 68 | 63 | 57 | 53 | 49 | | |
| 5-1/8 | 13-1/2 | 16.8 | 253 | 224 | 199 | 178 | 160 | 144 | 130 | 118 | 107 | 97 | 89 | 82 | 75 | 69 | 64 | 59 |
| 5-1/8 | 15 | 18.7 | 348 | 307 | 273 | 244 | 219 | 197 | 178 | 161 | 147 | 134 | 122 | 112 | 103 | 95 | 88 | 81 |
| 5-1/8 | 16-1/2 | 20.6 | 483 | 409 | 364 | 325 | 291 | 262 | 237 | 215 | 195 | 178 | 163 | 149 | 137 | 126 | 117 | 108 |
| 5-1/8 | 18 | 22.4 | 601 | 531 | 472 | 422 | 378 | 340 | 307 | 279 | 253 | 231 | 211 | 194 | 178 | 164 | 151 | 140 |

| Width | Depth | Weight (plf) | 1 | 2 | 3 | 4 | 5 | 6 | 7 | 8 | 9 | 10 | 11 | 12 | 13 | 14 | 15 | 16 |
|---|---|---|---|---|---|---|---|---|---|---|---|---|---|---|---|---|---|---|
| 5-1/8 | 19-1/2 | 24.3 | 764 | 676 | 601 | 536 | 481 | 433 | 391 | 354 | 322 | 294 | 269 | 248 | 228 | 208 | 192 | 178 |
| 5-1/8 | 21 | 26.2 | 954 | 844 | 750 | 670 | 601 | 541 | 488 | 443 | 402 | 367 | 335 | 307 | 283 | 260 | 240 | 222 |
| 5-1/8 | 22-1/2 | 28.0 | 1113 | 1022 | 923 | 824 | 739 | 665 | 601 | 544 | 495 | 451 | 413 | 378 | 348 | 320 | 296 | 273 |
| 5-1/8 | 24 | 29.9 | 1258 | 1155 | 1064 | 963 | 896 | 807 | 729 | 661 | 601 | 548 | 501 | 459 | 422 | 389 | 359 | 332 |
| 5-1/8 | 25-1/2 | 31.8 | 1412 | 1296 | 1193 | 1102 | 1021 | 949 | 874 | 792 | 720 | 657 | 601 | 551 | 506 | 466 | 430 | 398 |
| 6-3/4 | 7-1/2 | 12.3 | 57 | 51 |  |  |  |  |  |  |  |  |  |  |  |  |  |  |
| 6-3/4 | 9 | 14.8 | 99 | 87 | 78 | 69 | 62 | 56 | 51 | 46 |  |  |  |  |  |  |  |  |
| 6-3/4 | 10-1/2 | 17.2 | 157 | 139 | 123 | 110 | 99 | 89 | 80 | 73 | 66 | 60 | 55 | 51 | 47 |  |  |  |
| 6-3/4 | 12 | 19.7 | 234 | 207 | 184 | 165 | 148 | 133 | 120 | 109 | 99 | 90 | 82 | 76 | 69 | 64 | 59 | 55 |
| 6-3/4 | 13-1/2 | 22.1 | 334 | 295 | 262 | 234 | 210 | 189 | 171 | 155 | 141 | 128 | 117 | 108 | 99 | 91 | 84 | 78 |
| 6-3/4 | 15 | 24.6 | 458 | 405 | 360 | 321 | 288 | 259 | 234 | 212 | 193 | 176 | 161 | 148 | 136 | 125 | 115 | 107 |
| 6-3/4 | 16-1/2 | 27.1 | 609 | 539 | 479 | 428 | 384 | 345 | 312 | 283 | 257 | 234 | 214 | 196 | 181 | 166 | 153 | 142 |
| 6-3/4 | 18 | 29.5 | 791 | 700 | 622 | 556 | 498 | 448 | 405 | 367 | 334 | 304 | 278 | 255 | 234 | 218 | 199 | 184 |
| 6-3/4 | 19-1/2 | 32.0 | 1006 | 890 | 791 | 708 | 633 | 570 | 515 | 467 | 424 | 387 | 354 | 324 | 298 | 274 | 253 | 234 |
| 6-3/4 | 21 | 34.5 | 1251 | 1111 | 988 | 882 | 791 | 712 | 643 | 583 | 530 | 483 | 442 | 405 | 372 | 343 | 316 | 293 |
| 6-3/4 | 22-1/2 | 36.9 | 1428 | 1309 | 1205 | 1085 | 973 | 876 | 791 | 717 | 652 | 594 | 543 | 498 | 458 | 422 | 389 | 360 |
| 6-3/4 | 24 | 39.4 | 1612 | 1490 | 1363 | 1259 | 1166 | 1063 | 960 | 870 | 791 | 721 | 659 | 605 | 556 | 512 | 472 | 437 |
| 6-3/4 | 25-1/2 | 41.8 | 1809 | 1660 | 1529 | 1413 | 1309 | 1216 | 1132 | 1044 | 949 | 865 | 791 | 725 | 662 | 614 | 567 | 524 |
| 6-3/4 | 27 | 44.3 | 2017 | 1851 | 1705 | 1575 | 1459 | 1355 | 1262 | 1178 | 1102 | 1027 | 939 | 861 | 791 | 729 | 673 | 622 |
| 6-3/4 | 28-1/2 | 46.8 | 2199 | 2051 | 1889 | 1745 | 1617 | 1502 | 1399 | 1306 | 1221 | 1145 | 1075 | 1012 | 930 | 857 | 791 | 732 |
| 6-3/4 | 30 | 49.2 | 2345 | 2228 | 2082 | 1924 | 1782 | 1656 | 1542 | 1439 | 1346 | 1262 | 1185 | 1115 | 1051 | 993 | 923 | 853 |

TABLE SPECIFICATIONS: This table applies to straight, simply supported glued laminated timber beams under dry conditions of use. Beams must be laterally supported at the top along the length of the beam and at the top and bottom at the end bearings. The load carrying capacities tabulated are for total load including the weight of the member. A unit weight of 35 pcf was assumed to determine the tabulated plf beams weights.

DESIGN VALUE MODIFICATIONS: The allowable stress in bending, $F_b$, has been modified by the AITC volume factor, $C_v$. For determination of load carrying capacities governed by shear, loads within a distance "d" (the depth of the beam) from the ends have been neglected.

DEFLECTION LIMITS: For floor beams, a deflection limit of l/360 for live load has been used. Live load has been assumed to be 80% of the total load as is common for residential floors. For other ratios of live load to total load, the designer should check deflection based on applicable building code provisions.

CONTROLLING VALUES: Values above and to the right of the heavy line and shaded area indicate load capacity controlled by deflection considerations whereas values below and to the left of the heavy line and shaded area represent load capacity controlled by shear. Values in the shaded area are those load capacities governed by bending.

TABLE 8.9 (Continued)

## Structural Glued Laminated Timber
### Roof Beams for Construction Load

Lamination thickness: 1-3/8 in.

Simple Span Beam Table
For Preliminary Design Purposes

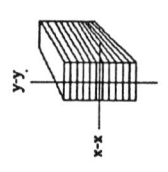

| $F_b$ | $F_v$ | $E$ | $C_D$ | Deflection Limit |
|---|---|---|---|---|
| 2400 psi | 200 psi | $1.8 \times 10^6$ psi | 1.25 | $l/180$ |

| Beam Size | | | Beam Capacity: Uniform Load w, plf | | | | | | | | | | | | | | | | | |
|---|---|---|---|---|---|---|---|---|---|---|---|---|---|---|---|---|---|---|---|---|
| Width b, in. | Depth d, in. | Weight plf | **Span, ft** | | | | | | | | | | | | | | | | |
| | | | 8 | 9 | 10 | 11 | 12 | 13 | 14 | 15 | 16 | 17 | 18 | 19 | 20 | 21 | 22 | 23 | 24 |
| 3 | 5-1/2 | 4.1 | 433 | 304 | 222 | 167 | 128 | 101 | 81 | 66 | 54 | 45 | 38 | 32 | | | | | |
| 3 | 6-7/8 | 5.2 | 739 | 584 | 433 | 326 | 251 | 187 | 158 | 128 | 106 | 88 | 74 | 63 | 54 | 47 | 41 | 38 | 31 |
| 3 | 8-1/4 | 6.2 | 1063 | 840 | 661 | 562 | 433 | 341 | 273 | 222 | 183 | 152 | 128 | 109 | 94 | 81 | 70 | 62 | 54 |
| 3 | 9-5/8 | 7.2 | 1448 | 1144 | 925 | 765 | 643 | 541 | 433 | 352 | 290 | 242 | 204 | 173 | 149 | 128 | 112 | 98 | 86 |
| 3 | 11 | 8.3 | 1784 | 1494 | 1210 | 1000 | 840 | 716 | 617 | 528 | 433 | 361 | 304 | 259 | 222 | 192 | 167 | 148 | 128 |
| 3 | 12-3/8 | 9.3 | 2084 | 1784 | 1531 | 1286 | 1063 | 906 | 781 | 681 | 566 | 514 | 433 | 368 | 316 | 273 | 237 | 208 | 183 |
| 3 | 13-3/4 | 10.3 | 2409 | 2050 | 1784 | 1563 | 1313 | 1119 | 965 | 840 | 739 | 664 | 564 | 505 | 433 | 374 | 326 | 285 | 251 |
| 3 | 15-1/8 | 11.3 | 2760 | 2334 | 2022 | 1784 | 1589 | 1354 | 1167 | 1017 | 884 | 792 | 708 | 634 | 572 | 496 | 433 | 379 | 334 |
| 3 | 16-1/2 | 12.4 | 3143 | 2640 | 2278 | 2000 | 1784 | 1610 | 1389 | 1210 | 1063 | 942 | 840 | 754 | 681 | 617 | 562 | 482 | 433 |
| 3 | 17-7/8 | 13.4 | 3560 | 2969 | 2548 | 2229 | 1982 | 1784 | 1622 | 1420 | 1248 | 1106 | 966 | 885 | 799 | 725 | 660 | 604 | 551 |
| 5 | 5-1/2 | 6.9 | 722 | 507 | 370 | 278 | 214 | 168 | 135 | 110 | 90 | 75 | 63 | 54 | 46 | 40 | 35 | 30 | |
| 5 | 6-7/8 | 8.6 | 1231 | 973 | 722 | 543 | 418 | 329 | 263 | 214 | 176 | 147 | 124 | 105 | 90 | 78 | 68 | 59 | 52 |
| 5 | 8-1/4 | 10.3 | 1772 | 1400 | 1134 | 937 | 722 | 568 | 455 | 370 | 305 | 254 | 214 | 182 | 156 | 135 | 117 | 103 | 90 |
| 5 | 9-5/8 | 12.0 | 2413 | 1906 | 1544 | 1278 | 1072 | 902 | 722 | 587 | 484 | 403 | 340 | 289 | 248 | 214 | 186 | 163 | 143 |
| 5 | 11 | 13.8 | 2973 | 2490 | 2017 | 1667 | 1400 | 1193 | 1029 | 876 | 722 | 602 | 507 | 431 | 370 | 319 | 278 | 243 | 214 |
| 5 | 12-3/8 | 15.5 | 3474 | 2973 | 2552 | 2109 | 1772 | 1510 | 1302 | 1134 | 997 | 857 | 722 | 614 | 528 | 455 | 396 | 346 | 305 |
| 5 | 13-3/4 | 17.2 | 4015 | 3418 | 2973 | 2604 | 2188 | 1865 | 1608 | 1400 | 1231 | 1090 | 973 | 842 | 722 | 624 | 543 | 475 | 418 |
| 5 | 15-1/8 | 18.9 | 4601 | 3891 | 3370 | 2973 | 2648 | 2256 | 1945 | 1695 | 1489 | 1319 | 1174 | 1051 | 948 | 830 | 722 | 632 | 556 |

| Width | Depth | Wt | | | | | | | | | | | | | | | | | |
|---|---|---|---|---|---|---|---|---|---|---|---|---|---|---|---|---|---|---|---|---|
| 5 | 16-1/2 | 20.6 | 5238 | 4400 | 3793 | 3333 | 2973 | 2683 | 2215 | 2017 | 1771 | 1564 | 1391 | 1245 | 1121 | 1014 | 922 | 820 | 722 |
| 5 | 17-7/8 | 22.3 | 5834 | 4948 | 4243 | 3714 | 3303 | 2973 | 2703 | 2382 | 2070 | 1828 | 1626 | 1455 | 1310 | 1185 | 1077 | 984 | 901 |
| 5 | 19-1/4 | 24.1 | 6696 | 5540 | 4724 | 4118 | 3649 | 3277 | 2973 | 2721 | 2391 | 2112 | 1878 | 1681 | 1514 | 1369 | 1245 | 1136 | 1042 |
| 5 | 20-5/8 | 25.8 | 7534 | 6180 | 5238 | 4545 | 4015 | 3595 | 3254 | 2973 | 2736 | 2416 | 2149 | 1923 | 1731 | 1567 | 1424 | 1300 | 1191 |
| 5 | 22 | 27.5 | 8462 | 6875 | 5789 | 5000 | 4400 | 3929 | 3548 | 3235 | 2973 | 2740 | 2437 | 2181 | 1964 | 1777 | 1615 | 1474 | 1351 |
| 5 | 23-3/8 | 29.2 | 9492 | 7633 | 6382 | 5484 | 4807 | 4279 | 3856 | 3508 | 3219 | 2973 | 2743 | 2455 | 2210 | 2000 | 1818 | 1660 | 1521 |
| 6-3/4 | 6-7/8 | 11.6 | 1662 | 1313 | 975 | 732 | 564 | 444 | 355 | 289 | 238 | 198 | 167 | 142 | 122 | 105 | 92 | 80 | 71 |
| 6-3/4 | 8-1/4 | 13.9 | 2393 | 1891 | 1531 | 1296 | 975 | 767 | 614 | 499 | 411 | 343 | 289 | 246 | 211 | 182 | 158 | 138 | 122 |
| 6-3/4 | 9-5/8 | 16.2 | 3257 | 2573 | 2084 | 1723 | 1448 | 1218 | 975 | 793 | 653 | 544 | 459 | 390 | 334 | 289 | 251 | 220 | 194 |
| 6-3/4 | 11 | 18.6 | 4014 | 3391 | 2723 | 2250 | 1891 | 1611 | 1389 | 1183 | 975 | 813 | 685 | 582 | 499 | 431 | 375 | 328 | 289 |
| 6-3/4 | 12-3/8 | 20.9 | 4689 | 4014 | 3446 | 2848 | 2393 | 2039 | 1756 | 1531 | 1344 | 1157 | 975 | 829 | 711 | 614 | 534 | 467 | 411 |
| 6-3/4 | 13-3/4 | 23.2 | 5420 | 4812 | 4014 | 3516 | 2954 | 2517 | 2170 | 1884 | 1650 | 1457 | 1296 | 1137 | 975 | 842 | 732 | 641 | 564 |
| 6-3/4 | 15-1/8 | 25.5 | 6211 | 5252 | 4550 | 4014 | 3574 | 3042 | 2613 | 2268 | 1987 | 1755 | 1561 | 1397 | 1258 | 1121 | 975 | 853 | 751 |
| 6-3/4 | 16-1/2 | 27.8 | 7071 | 5940 | 5121 | 4500 | 4014 | 3604 | 3096 | 2688 | 2355 | 2079 | 1850 | 1655 | 1490 | 1348 | 1226 | 1108 | 975 |
| 6-3/4 | 17-7/8 | 30.2 | 8010 | 6680 | 5728 | 5014 | 4458 | 4014 | 3619 | 3142 | 2752 | 2431 | 2182 | 1835 | 1742 | 1576 | 1433 | 1308 | 1199 |
| 6-3/4 | 19-1/4 | 32.5 | 9039 | 7478 | 6377 | 5559 | 4927 | 4423 | 4014 | 3630 | 3180 | 2806 | 2498 | 2236 | 2013 | 1821 | 1656 | 1511 | 1385 |
| 6-3/4 | 20-5/8 | 34.8 | 10171 | 8343 | 7071 | 6136 | 5420 | 4853 | 4363 | 4014 | 3638 | 3213 | 2858 | 2558 | 2303 | 2063 | 1894 | 1729 | 1585 |
| 6-3/4 | 22 | 37.1 | 11423 | 9281 | 7818 | 6750 | 5940 | 5304 | 4790 | 4368 | 4014 | 3644 | 3241 | 2901 | 2611 | 2353 | 2148 | 1961 | 1797 |
| 6-3/4 | 23-3/8 | 39.4 | 12815 | 10304 | 8618 | 7403 | 6480 | 5777 | 5205 | 4736 | 4345 | 4014 | 3648 | 3265 | 2939 | 2659 | 2418 | 2207 | 2023 |
| 6-3/4 | 24-3/4 | 41.8 | 14371 | 11423 | 9479 | 8100 | 7071 | 6275 | 5639 | 5121 | 4689 | 4325 | 4014 | 3650 | 3286 | 2973 | 2703 | 2467 | 2291 |
| 6-3/4 | 26-1/8 | 44.1 | 16123 | 12652 | 10411 | 8845 | 7688 | 6799 | 6094 | 5522 | 5047 | 4648 | 4308 | 4014 | 3651 | 3304 | 3003 | 2742 | 2512 |
| 6-3/4 | 27-1/2 | 46.4 | 18110 | 14009 | 11423 | 9643 | 8343 | 7351 | 6571 | 5940 | 5420 | 4983 | 4612 | 4292 | 4014 | 3651 | 3319 | 3030 | 2777 |

TABLE SPECIFICATIONS: This table applies to straight, simply supported glued laminated timber beams under dry conditions of use. Beams must be laterally supported at the top along the length of the beam and at the top and bottom at the end bearings. The load carrying capacities tabulated are for total load including the weight of the member. A unit weight of 36 pcf was assumed to determine the tabulated plf beams weights.

DESIGN VALUE MODIFICATIONS: The allowable stress in bending, $F_b$, has been modified by the AITC volume factor, $C_v$. For determination of load carrying capacities governed by shear, loads within a distance "d" (the depth of the beam) from the ends have been neglected.

DEFLECTION LIMITS: For roof beams, a deflection limit of l/180 for total load has been used.

CONTROLLING VALUES: Values above and to the right of the heavy line and shaded area indicate load capacity controlled by deflection considerations whereas values below and to the left of the heavy line and shaded area represent load capacity controlled by shear. Values in the shaded area are those load capacities governed by bending.

# TABLE 8.9 (Continued)

## Structural Glued Laminated Timber
### Roof Beams for Construction Load

Lamination thickness: 1-3/8 in.

Simple Span Beam Table
For Preliminary Design Purposes

| $F_b$ | $F_v$ | $E$ | $C_D$ | Deflection Limit |
|---|---|---|---|---|
| 2400 psi | 155 psi | $1.7 \times 10^6$ psi | 1.25 | l/180 |

y-y.

x-x

| Beam Size | | | Span, ft | | | | | | | | | | | | | | | |
|---|---|---|---|---|---|---|---|---|---|---|---|---|---|---|---|---|---|---|
| Width b, in. | Depth d, in. | Weight plf | 25 | 26 | 27 | 28 | 29 | 30 | 31 | 32 | 33 | 34 | 35 | 36 | 37 | 38 | 39 | 40 |
| | | | Beam Capacity, Uniform Load w, plf | | | | | | | | | | | | | | | |
| 3 | 5-1/2 | 4.1 | | | | | | | | | | | | | | | | |
| 3 | 6-7/8 | 5.2 | 28 | | | | | | | | | | | | | | | |
| 3 | 8-1/4 | 6.2 | 48 | 43 | 38 | 34 | 31 | | | | | | | | | | | |
| 3 | 9-5/8 | 7.2 | 76 | 68 | 60 | 54 | 49 | 44 | 40 | 36 | 33 | 30 | | | | | | |
| 3 | 11 | 8.3 | 114 | 101 | 90 | 81 | 73 | 66 | 60 | 54 | 49 | 45 | 41 | 38 | 35 | 32 | 30 | |
| 3 | 12-3/8 | 9.3 | 162 | 144 | 128 | 115 | 104 | 94 | 85 | 77 | 70 | 64 | 59 | 54 | 50 | 46 | 43 | 39 |
| 3 | 13-3/4 | 10.3 | 222 | 197 | 176 | 158 | 142 | 128 | 118 | 106 | 96 | 88 | 81 | 74 | 68 | 63 | 58 | 54 |
| 3 | 15-1/8 | 11.3 | 295 | 262 | 234 | 210 | 189 | 171 | 155 | 141 | 128 | 117 | 108 | 99 | 91 | 84 | 78 | 72 |
| 3 | 16-1/2 | 12.4 | 383 | 341 | 304 | 273 | 246 | 222 | 201 | 183 | 167 | 152 | 140 | 128 | 118 | 109 | 101 | 84 |
| 3 | 17-7/8 | 13.4 | 487 | 433 | 387 | 347 | 312 | 282 | 256 | 232 | 212 | 194 | 178 | 163 | 150 | 139 | 128 | 119 |
| 5 | 5-1/2 | 6.9 | | | | | | | | | | | | | | | | |
| 5 | 6-7/8 | 8.8 | 46 | | | | | | | | | | | | | | | |
| 5 | 8-1/4 | 10.3 | 80 | 71 | 63 | 57 | 51 | 48 | 42 | 38 | 35 | 32 | | | | | | |
| 5 | 9-5/8 | 12.0 | 127 | 113 | 101 | 90 | 81 | 73 | 67 | 60 | 55 | 50 | 48 | 42 | 39 | 36 | 33 | 31 |
| 5 | 11 | 13.8 | 189 | 168 | 150 | 135 | 121 | 110 | 99 | 90 | 82 | 75 | 69 | 63 | 58 | 54 | 50 | 48 |
| 5 | 12-3/8 | 15.5 | 270 | 240 | 214 | 192 | 173 | 156 | 141 | 129 | 117 | 107 | 98 | 90 | 83 | 77 | 71 | 66 |
| 5 | 13-3/4 | 17.2 | 370 | 329 | 293 | 263 | 237 | 214 | 194 | 178 | 161 | 147 | 135 | 124 | 114 | 105 | 97 | 90 |
| 5 | 15-1/8 | 18.9 | 492 | 437 | 391 | 350 | 315 | 285 | 258 | 235 | 214 | 196 | 179 | 165 | 152 | 140 | 130 | 120 |
| 5 | 16-1/2 | 20.6 | 639 | 568 | 507 | 455 | 409 | 370 | 335 | 305 | 278 | 254 | 233 | 214 | 197 | 182 | 168 | 156 |

| Width | Depth | Wt (plf) | 1 | 2 | 3 | 4 | 5 | 6 | 7 | 8 | 9 | 10 | 11 | 12 | 13 | 14 | 15 | 16 |
|---|---|---|---|---|---|---|---|---|---|---|---|---|---|---|---|---|---|---|
| 5 | 17-7/8 | 22.3 | 198 | 214 | 231 | 251 | 272 | 296 | 323 | 353 | 387 | 428 | 470 | 520 | 578 | 645 | 722 | 812 |
| 5 | 19-1/4 | 24.1 | 248 | 267 | 289 | 313 | 340 | 370 | 403 | 441 | 484 | 532 | 587 | 650 | 722 | 805 | 884 | 958 |
| 5 | 20-5/8 | 25.8 | 305 | 329 | 355 | 385 | 418 | 455 | 496 | 543 | 585 | 654 | 722 | 799 | 869 | 936 | 1011 | 1096 |
| 5 | 22 | 27.5 | 370 | 399 | 431 | 467 | 507 | 552 | 602 | 658 | 722 | 794 | 855 | 917 | 985 | 1061 | 1147 | 1243 |
| 5 | 23-3/8 | 29.2 | 443 | 478 | 517 | 560 | 608 | 662 | 722 | 790 | 843 | 900 | 963 | 1032 | 1109 | 1195 | 1291 | 1399 |
| 6-3/4 | 6-7/8 | 11.6 |  |  |  |  |  |  |  |  | 30 | 33 | 36 | 40 | 44 | 50 | 55 | 62 |
| 6-3/4 | 8-1/4 | 13.9 |  |  | 31 | 33 | 36 | 39 | 43 | 47 | 51 | 57 | 62 | 69 | 77 | 86 | 96 | 108 |
| 6-3/4 | 9-5/8 | 16.2 | 42 | 45 | 49 | 53 | 57 | 62 | 68 | 74 | 82 | 90 | 99 | 110 | 122 | 136 | 152 | 171 |
| 6-3/4 | 11 | 18.6 | 62 | 67 | 73 | 79 | 86 | 93 | 102 | 111 | 122 | 134 | 148 | 164 | 182 | 203 | 227 | 256 |
| 6-3/4 | 12-3/8 | 20.9 | 88 | 96 | 104 | 112 | 122 | 133 | 145 | 158 | 174 | 191 | 211 | 233 | 259 | 289 | 323 | 384 |
| 6-3/4 | 13-3/4 | 23.2 | 122 | 131 | 142 | 154 | 167 | 182 | 198 | 217 | 238 | 262 | 289 | 320 | 355 | 396 | 444 | 499 |
| 6-3/4 | 15-1/8 | 25.5 | 162 | 175 | 189 | 205 | 222 | 242 | 264 | 289 | 317 | 348 | 384 | 428 | 473 | 527 | 591 | 664 |
| 6-3/4 | 16-1/2 | 27.8 | 211 | 227 | 246 | 266 | 289 | 314 | 343 | 375 | 411 | 452 | 499 | 553 | 614 | 685 | 767 | 862 |
| 6-3/4 | 17-7/8 | 30.2 | 268 | 289 | 312 | 338 | 367 | 400 | 436 | 477 | 523 | 575 | 635 | 703 | 781 | 870 | 975 | 1097 |
| 6-3/4 | 19-1/4 | 32.5 | 334 | 361 | 390 | 422 | 459 | 499 | 544 | 595 | 653 | 718 | 793 | 877 | 975 | 1087 | 1175 | 1274 |
| 6-3/4 | 20-5/8 | 34.8 | 411 | 444 | 480 | 520 | 564 | 614 | 670 | 732 | 803 | 884 | 975 | 1075 | 1155 | 1245 | 1345 | 1457 |
| 6-3/4 | 22 | 37.1 | 499 | 539 | 582 | 631 | 685 | 745 | 813 | 889 | 975 | 1063 | 1137 | 1219 | 1310 | 1412 | 1525 | 1653 |
| 6-3/4 | 23-3/8 | 39.4 | 599 | 646 | 698 | 756 | 821 | 894 | 975 | 1053 | 1121 | 1197 | 1290 | 1372 | 1475 | 1589 | 1716 | 1860 |
| 6-3/4 | 24-3/4 | 41.8 | 711 | 767 | 829 | 898 | 975 | 1043 | 1107 | 1177 | 1254 | 1338 | 1431 | 1534 | 1648 | 1776 | 1919 | 2080 |
| 6-3/4 | 26-1/8 | 44.1 | 836 | 902 | 975 | 1034 | 1094 | 1159 | 1230 | 1308 | 1393 | 1487 | 1590 | 1705 | 1832 | 1973 | 2132 | 2311 |
| 6-3/4 | 27-1/2 | 46.4 | 974 | 1026 | 1082 | 1143 | 1209 | 1281 | 1360 | 1445 | 1540 | 1643 | 1757 | 1884 | 2024 | 2181 | 2357 | 2554 |

TABLE SPECIFICATIONS: This table applies to straight, simply supported glued laminated timber beams under dry conditions of use. Beams must be laterally supported at the top and bottom and at the end bearings. The load carrying capacities tabulated are for total load including the weight of the member. A unit weight of 36 pcf was assumed to determine the tabulated plf beams weights.

DESIGN VALUE MODIFICATIONS: The allowable stress in bending, $F_b$, has been modified by the AITC volume factor, $C_v$. For determination of load carrying capacities governed by shear, loads within a distance "d" (the depth of the beam) from the ends have been neglected.

DEFLECTION LIMITS: For roof beams, a deflection limit of $l/180$ for total load has been used.

CONTROLLING VALUES: Values above and to the right of the heavy line and shaded area indicate load capacity controlled by deflection considerations whereas values below and to the left of the heavy line and shaded area represent load capacity controlled by shear. Values in the shaded area are those load capacities governed by bending.

**TABLE 8.9** (*Continued*)

## Structural Glued Laminated Timber
### *Roof Beams for Snow Load*

Lamination thickness: 1-3/8 in.

Simple Span Beam Table
For Preliminary Design Purposes

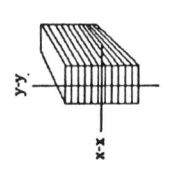

| $F_b$ | $F_v$ | $E$ | $C_D$ | Deflection Limit |
|---|---|---|---|---|
| 2400 psi | 200 psi | $1.8 \times 10^6$ psi | 1.15 | l/180 |

| Beam Size | | | Beam Capacity, Uniform Load w, plf | | | | | | | | | | | | | | | |
|---|---|---|---|---|---|---|---|---|---|---|---|---|---|---|---|---|---|---|
| Width b, in. | Depth d, in. | Weight plf | Span, ft | | | | | | | | | | | | | | | |
| | | | 8 | 9 | 10 | 11 | 12 | 13 | 14 | 15 | 16 | 17 | 18 | 19 | 20 | 21 | 22 | 23 | 24 |
| 3 | 5-1/2 | 4.1 | 433 | 304 | 222 | 167 | 128 | 101 | 81 | 66 | 54 | 45 | 38 | 32 | 28 | | | | |
| 3 | 6-7/8 | 5.2 | 679 | 537 | 433 | 326 | 251 | 197 | 158 | 128 | 106 | 88 | 74 | 63 | 54 | 47 | 41 | 36 | 31 |
| 3 | 8-1/4 | 6.2 | 978 | 773 | 626 | 518 | 433 | 341 | 273 | 222 | 183 | 152 | 128 | 109 | 94 | 81 | 70 | 62 | 54 |
| 3 | 9-5/8 | 7.2 | 1332 | 1052 | 852 | 704 | 592 | 504 | 433 | 352 | 290 | 242 | 204 | 173 | 149 | 128 | 112 | 98 | 86 |
| 3 | 11 | 8.3 | 1641 | 1374 | 1113 | 920 | 773 | 659 | 566 | 495 | 433 | 361 | 304 | 259 | 222 | 192 | 167 | 146 | 128 |
| 3 | 12-3/8 | 9.3 | 1917 | 1641 | 1409 | 1164 | 978 | 834 | 719 | 626 | 550 | 488 | 433 | 368 | 316 | 273 | 237 | 208 | 183 |
| 3 | 13-3/4 | 10.3 | 2216 | 1886 | 1641 | 1438 | 1208 | 1029 | 887 | 773 | 679 | 602 | 537 | 482 | 433 | 374 | 328 | 285 | 251 |
| 3 | 15-1/8 | 11.3 | 2540 | 2148 | 1881 | 1641 | 1462 | 1245 | 1074 | 935 | 822 | 728 | 650 | 583 | 526 | 477 | 433 | 379 | 334 |
| 3 | 16-1/2 | 12.4 | 2891 | 2429 | 2094 | 1840 | 1641 | 1481 | 1278 | 1113 | 978 | 867 | 773 | 694 | 628 | 568 | 518 | 473 | 433 |
| 3 | 17-7/8 | 13.4 | 3275 | 2731 | 2342 | 2050 | 1823 | 1641 | 1492 | 1306 | 1148 | 1017 | 907 | 814 | 735 | 667 | 607 | 556 | 510 |
| 5 | 5-1/2 | 6.9 | 722 | 507 | 370 | 278 | 214 | 168 | 135 | 110 | 90 | 75 | 63 | 54 | 46 | 40 | 35 | 30 | 27 |
| 5 | 6-7/8 | 8.6 | 1132 | 895 | 722 | 543 | 418 | 329 | 263 | 214 | 176 | 147 | 124 | 105 | 90 | 78 | 68 | 59 | 52 |
| 5 | 8-1/4 | 10.3 | 1631 | 1288 | 1044 | 862 | 722 | 568 | 455 | 370 | 305 | 254 | 214 | 182 | 156 | 135 | 117 | 103 | 90 |
| 5 | 9-5/8 | 12.0 | 2220 | 1754 | 1420 | 1174 | 986 | 841 | 722 | 587 | 484 | 403 | 340 | 289 | 248 | 214 | 186 | 163 | 143 |
| 5 | 11 | 13.8 | 2735 | 2291 | 1855 | 1533 | 1288 | 1098 | 947 | 825 | 722 | 602 | 507 | 431 | 370 | 319 | 278 | 243 | 214 |
| 5 | 12-3/8 | 15.5 | 3196 | 2735 | 2348 | 1941 | 1631 | 1389 | 1198 | 1044 | 917 | 813 | 722 | 614 | 528 | 455 | 396 | 346 | 305 |
| 5 | 13-3/4 | 17.2 | 3693 | 3143 | 2735 | 2396 | 2013 | 1715 | 1479 | 1288 | 1132 | 1003 | 895 | 803 | 722 | 624 | 543 | 475 | 418 |
| 5 | 15-1/8 | 18.9 | 4233 | 3579 | 3101 | 2735 | 2436 | 2076 | 1790 | 1559 | 1370 | 1214 | 1080 | 967 | 870 | 787 | 716 | 632 | 556 |
| 5 | 16-1/2 | 20.6 | 4819 | 4048 | 3490 | 3067 | 2735 | 2468 | 2130 | 1855 | 1629 | 1439 | 1279 | 1145 | 1031 | 933 | 848 | 774 | 709 |

| Width | Depth | Wt (plf) | | | | | | | | | | | | | | | | |
|---|---|---|---|---|---|---|---|---|---|---|---|---|---|---|---|---|---|---|---|
| 5 | 17-7/8 | 22.3 | 5459 | 4552 | 3904 | 3417 | 3038 | 2735 | 2487 | 2173 | 1904 | 1682 | 1496 | 1339 | 1205 | 1090 | 991 | 905 | 829 |
| 5 | 19-1/4 | 24.1 | 6160 | 5098 | 4346 | 3788 | 3357 | 3014 | 2735 | 2503 | 2200 | 1943 | 1728 | 1547 | 1392 | 1260 | 1145 | 1046 | 958 |
| 5 | 20-5/8 | 25.8 | 6932 | 5685 | 4819 | 4182 | 3693 | 3307 | 2994 | 2735 | 2517 | 2223 | 1977 | 1770 | 1593 | 1441 | 1310 | 1196 | 1096 |
| 5 | 22 | 27.5 | 7785 | 6325 | 5328 | 4600 | 4048 | 3614 | 3265 | 2976 | 2735 | 2521 | 2242 | 2007 | 1807 | 1635 | 1486 | 1357 | 1243 |
| 5 | 23-3/8 | 29.2 | 8733 | 7022 | 5872 | 5045 | 4423 | 3937 | 3547 | 3228 | 2961 | 2735 | 2523 | 2259 | 2033 | 1840 | 1672 | 1527 | 1399 |
| 6-3/4 | 6-7/8 | 11.6 | 1529 | 1208 | 975 | 732 | 564 | 444 | 355 | 289 | 238 | 198 | 167 | 142 | 122 | 105 | 92 | 80 | 71 |
| 6-3/4 | 8-1/4 | 13.9 | 2201 | 1739 | 1409 | 1164 | 975 | 767 | 614 | 499 | 411 | 343 | 289 | 246 | 211 | 182 | 158 | 138 | 122 |
| 6-3/4 | 9-5/8 | 16.2 | 2995 | 2367 | 1918 | 1585 | 1332 | 1135 | 975 | 793 | 653 | 544 | 459 | 390 | 334 | 289 | 251 | 220 | 194 |
| 6-3/4 | 11 | 18.6 | 3692 | 3092 | 2505 | 2070 | 1739 | 1482 | 1278 | 1113 | 975 | 813 | 685 | 582 | 499 | 431 | 375 | 328 | 289 |
| 6-3/4 | 12-3/8 | 20.8 | 4314 | 3692 | 3170 | 2620 | 2201 | 1876 | 1617 | 1409 | 1236 | 1092 | 971 | 829 | 711 | 614 | 534 | 467 | 411 |
| 6-3/4 | 13-3/4 | 23.2 | 4986 | 4243 | 3692 | 3234 | 2718 | 2316 | 1996 | 1733 | 1518 | 1341 | 1192 | 1067 | 961 | 842 | 732 | 641 | 564 |
| 6-3/4 | 15-1/8 | 25.5 | 5714 | 4832 | 4186 | 3692 | 3289 | 2798 | 2404 | 2087 | 1828 | 1615 | 1436 | 1285 | 1157 | 1047 | 952 | 853 | 751 |
| 6-3/4 | 16-1/2 | 27.8 | 6506 | 5465 | 4711 | 4140 | 3692 | 3316 | 2848 | 2473 | 2166 | 1913 | 1702 | 1523 | 1371 | 1241 | 1128 | 1029 | 943 |
| 6-3/4 | 17-7/8 | 30.2 | 7370 | 6146 | 5270 | 4613 | 4102 | 3692 | 3329 | 2890 | 2532 | 2236 | 1989 | 1780 | 1603 | 1450 | 1318 | 1203 | 1103 |
| 6-3/4 | 19-1/4 | 32.5 | 8316 | 6880 | 5867 | 5114 | 4532 | 4070 | 3692 | 3340 | 2926 | 2584 | 2298 | 2057 | 1852 | 1676 | 1523 | 1390 | 1274 |
| 6-3/4 | 20-5/8 | 34.8 | 9358 | 7675 | 6506 | 5645 | 4986 | 4465 | 4042 | 3692 | 3347 | 2956 | 2629 | 2353 | 2118 | 1917 | 1742 | 1591 | 1458 |
| 6-3/4 | 22 | 37.1 | 10509 | 8539 | 7191 | 6210 | 5465 | 4879 | 4407 | 4018 | 3692 | 3332 | 2962 | 2669 | 2403 | 2174 | 1976 | 1804 | 1653 |
| 6-3/4 | 23-3/8 | 39.4 | 11790 | 9480 | 7927 | 6811 | 5971 | 5315 | 4789 | 4357 | 3997 | 3692 | 3356 | 3004 | 2704 | 2447 | 2224 | 2030 | 1861 |
| 6-3/4 | 24-3/4 | 41.8 | 13221 | 10509 | 8720 | 7452 | 6506 | 5773 | 5188 | 4711 | 4314 | 3979 | 3692 | 3358 | 3023 | 2735 | 2488 | 2270 | 2080 |
| 6-3/4 | 26-1/8 | 44.1 | 14833 | 11640 | 9579 | 8137 | 7073 | 6255 | 5606 | 5080 | 4644 | 4276 | 3963 | 3692 | 3359 | 3039 | 2763 | 2522 | 2311 |
| 6-3/4 | 27-1/2 | 46.4 | 16661 | 12869 | 10509 | 8871 | 7675 | 6763 | 6045 | 5465 | 4986 | 4585 | 4243 | 3949 | 3692 | 3359 | 3053 | 2788 | 2555 |

**TABLE SPECIFICATIONS:** This table applies to straight, simply supported glued laminated timber beams under dry conditions of use. Beams must be laterally supported at the top along the length of the beam and at the top and bottom at the end bearings. The load carrying capacities tabulated are for total load including the weight of the member. A unit weight of 36 pcf was assumed to determine the tabulated plf beam weights.

**DESIGN VALUE MODIFICATIONS:** The allowable stress in bending, $F_b$, has been modified by the AITC volume factor, $C_v$. For determination of load carrying capacities governed by shear, loads within a distance "d" (the depth of the beam) from the ends have been neglected.

**DEFLECTION LIMITS:** For roof beams, a deflection limit of l/180 for total load has been used.

**CONTROLLING VALUES:** Values above and to the right of the heavy line and/or shaded area indicate load capacity controlled by deflection considerations; whereas, values below and to the left of the heavy line and/or shaded area represent load capacity controlled by shear. Values in the shaded area are those load capacities governed by bending.

**SPAN:** Span is defined as the length from centerline to centerline of bearing. It is this length that is used in standard engineering equations to calculate deflection, bending and shear.

TABLE 8.9 (*Continued*)

## Structural Glued Laminated Timber
### *Roof Beams for Snow Load*

Lamination thickness: 1-3/8 in.

Simple Span Beam Table
For Preliminary Design Purposes

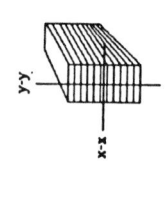

| | | | | Deflection Limit |
|---|---|---|---|---|
| $F_b$ 2400 psi | $F_v$ 200 psi | $E$ $1.8 \times 10^6$ psi | $C_D$ 1.15 | l/180 |

**Beam Capacity, Uniform Load w, plf**

| Width b, in. | Depth d, in. | Weight plf | 25 | 26 | 27 | 28 | 29 | 30 | 31 | 32 | 33 | 34 | 35 | 36 | 37 | 38 | 39 | 40 |
|---|---|---|---|---|---|---|---|---|---|---|---|---|---|---|---|---|---|---|
| 3 | 5-1/2 | 4.1 | 28 | | | | | | | | | | | | | | | |
| 3 | 6-7/8 | 5.2 | 48 | 43 | 38 | 34 | 31 | 28 | | | | | | | | | | |
| 3 | 8-1/4 | 6.2 | 76 | 68 | 60 | 54 | 49 | 44 | 40 | 36 | 33 | 30 | 28 | | | | | |
| 3 | 9-5/8 | 7.2 | 114 | 101 | 90 | 81 | 73 | 66 | 60 | 54 | 49 | 45 | 41 | 38 | 35 | 32 | 30 | 28 |
| 3 | 11 | 8.3 | 162 | 144 | 128 | 115 | 104 | 94 | 85 | 77 | 70 | 64 | 59 | 54 | 50 | 46 | 43 | 39 |
| 3 | 12-3/8 | 9.3 | 222 | 197 | 176 | 158 | 142 | 128 | 116 | 106 | 98 | 88 | 81 | 74 | 68 | 63 | 58 | 54 |
| 3 | 13-3/4 | 10.3 | 295 | 262 | 234 | 210 | 189 | 171 | 155 | 141 | 128 | 117 | 108 | 99 | 91 | 84 | 78 | 72 |
| 3 | 15-1/8 | 11.3 | 383 | 341 | 304 | 273 | 246 | 222 | 201 | 183 | 167 | 152 | 140 | 128 | 118 | 109 | 101 | 94 |
| 3 | 16-1/2 | 12.4 | 489 | 433 | 387 | 347 | 312 | 282 | 256 | 232 | 212 | 194 | 178 | 163 | 150 | 139 | 128 | 119 |
| 3 | 17-7/8 | 13.4 | | | | | | | | | | | | | | | | |
| 5 | 5-1/2 | 6.9 | | | | | | | | | | | | | | | | |
| 5 | 6-7/8 | 8.8 | 46 | 41 | 37 | 33 | 30 | 27 | | | | | | | | | | |
| 5 | 8-1/4 | 10.3 | 80 | 71 | 63 | 57 | 51 | 48 | 42 | 38 | 35 | 32 | 29 | | | | | |
| 5 | 9-5/8 | 12.0 | 127 | 113 | 101 | 90 | 81 | 73 | 67 | 60 | 55 | 50 | 46 | 42 | 39 | 36 | 33 | 31 |
| 5 | 11 | 13.8 | 189 | 168 | 150 | 135 | 121 | 110 | 99 | 90 | 82 | 75 | 69 | 63 | 58 | 54 | 50 | 46 |
| 5 | 12-3/8 | 15.5 | 270 | 240 | 214 | 192 | 173 | 156 | 141 | 129 | 117 | 107 | 98 | 90 | 83 | 77 | 71 | 66 |
| 5 | 13-3/4 | 17.2 | 370 | 329 | 293 | 263 | 237 | 214 | 194 | 176 | 161 | 147 | 135 | 124 | 114 | 105 | 97 | 90 |
| 5 | 15-1/8 | 18.9 | 492 | 437 | 391 | 350 | 315 | 285 | 258 | 235 | 214 | 198 | 179 | 165 | 152 | 140 | 130 | 120 |
| 5 | 16-1/2 | 20.6 | 639 | 568 | 507 | 455 | 409 | 370 | 335 | 305 | 278 | 254 | 233 | 214 | 197 | 182 | 168 | 156 |

| Width | Depth | plf | | | | | | | | | | | | | | | |
|---|---|---|---|---|---|---|---|---|---|---|---|---|---|---|---|---|---|---|
| 5 | 17-7/8 | 22.3 | 763 | 704 | 645 | 578 | 520 | 470 | 426 | 387 | 353 | 323 | 296 | 272 | 251 | 231 | 214 | 198 |
| 5 | 19-1/4 | 24.1 | 881 | 813 | 753 | 699 | 650 | 587 | 532 | 484 | 441 | 403 | 370 | 340 | 313 | 289 | 267 | 248 |
| 5 | 20-5/8 | 25.8 | 1008 | 830 | 881 | 799 | 744 | 694 | 848 | 595 | 543 | 496 | 455 | 418 | 385 | 355 | 329 | 305 |
| 5 | 22 | 27.5 | 1143 | 1055 | 976 | 906 | 843 | 787 | 736 | 689 | 647 | 602 | 552 | 507 | 467 | 431 | 399 | 370 |
| 5 | 23-3/8 | 29.2 | 1287 | 1187 | 1099 | 1020 | 949 | 886 | 828 | 776 | 728 | 685 | 846 | 608 | 560 | 517 | 478 | 443 |
| 6-3/4 | 6-7/8 | 11.6 | 62 | 55 | 50 | 44 | 40 | 36 | 33 | 30 | 27 | | | | | | | |
| 6-3/4 | 8-1/4 | 13.9 | 108 | 96 | 86 | 77 | 69 | 62 | 57 | 51 | 47 | 43 | 39 | 36 | 33 | 31 | 28 | 42 |
| 6-3/4 | 9-5/8 | 16.2 | 171 | 152 | 136 | 122 | 110 | 99 | 90 | 82 | 74 | 68 | 62 | 57 | 53 | 49 | 45 | 62 |
| 6-3/4 | 11 | 18.6 | 256 | 227 | 203 | 182 | 164 | 148 | 134 | 122 | 111 | 102 | 93 | 86 | 79 | 73 | 67 | 89 |
| 6-3/4 | 12-3/8 | 20.9 | 364 | 323 | 289 | 259 | 233 | 211 | 191 | 174 | 158 | 145 | 133 | 122 | 112 | 104 | 96 | 122 |
| 6-3/4 | 13-3/4 | 23.2 | 499 | 444 | 396 | 355 | 320 | 289 | 262 | 238 | 217 | 198 | 182 | 167 | 154 | 142 | 131 | 162 |
| 6-3/4 | 15-1/8 | 25.5 | 684 | 561 | 527 | 473 | 426 | 384 | 348 | 317 | 289 | 264 | 242 | 222 | 205 | 189 | 175 | 211 |
| 6-3/4 | 16-1/2 | 27.8 | 862 | 787 | 685 | 614 | 553 | 499 | 452 | 411 | 375 | 343 | 314 | 289 | 266 | 246 | 227 | 268 |
| 6-3/4 | 17-7/8 | 30.2 | 1014 | 936 | 868 | 781 | 703 | 635 | 575 | 523 | 477 | 436 | 400 | 367 | 338 | 312 | 289 | 334 |
| 6-3/4 | 19-1/4 | 32.5 | 1172 | 1081 | 1001 | 929 | 865 | 783 | 718 | 653 | 595 | 544 | 499 | 459 | 422 | 390 | 381 | 411 |
| 6-3/4 | 20-5/8 | 34.8 | 1341 | 1237 | 1145 | 1063 | 989 | 923 | 863 | 803 | 732 | 670 | 614 | 564 | 520 | 480 | 444 | 499 |
| 6-3/4 | 22 | 37.1 | 1521 | 1403 | 1299 | 1205 | 1122 | 1046 | 978 | 917 | 861 | 810 | 745 | 685 | 631 | 582 | 539 | 599 |
| 6-3/4 | 23-3/8 | 39.4 | 1711 | 1579 | 1462 | 1357 | 1262 | 1178 | 1101 | 1032 | 969 | 911 | 859 | 810 | 756 | 698 | 646 | 711 |
| 8-3/4 | 24-3/4 | 41.8 | 1913 | 1765 | 1634 | 1517 | 1411 | 1317 | 1231 | 1153 | 1083 | 1019 | 960 | 908 | 856 | 811 | 767 | 811 |
| 8-3/4 | 26-1/8 | 44.1 | 2126 | 1962 | 1816 | 1685 | 1568 | 1463 | 1368 | 1282 | 1203 | 1132 | 1067 | 1007 | 952 | 901 | 854 | 811 |
| 8-3/4 | 27-1/2 | 46.4 | 2350 | 2168 | 2007 | 1862 | 1733 | 1617 | 1512 | 1418 | 1330 | 1251 | 1179 | 1113 | 1062 | 996 | 944 | 896 |

TABLE SPECIFICATIONS: This table applies to straight, simply supported glued laminated timber beams under dry conditions of use. Beams must be laterally supported at the top along the length of the beam and at the top and bottom at the end bearings. The load carrying capacities tabulated are for total load including the weight of the member. A unit weight of 36 pcf was assumed to determine the tabulated plf beam weights.

DESIGN VALUE MODIFICATIONS: The allowable stress in bending, $F_b$, has been modified by the AITC volume factor, $C_v$. For determination of load carrying capacities governed by shear, loads within a distance "d" (the depth of the beam) from the ends have been neglected.

DEFLECTION LIMITS: For roof beams, a deflection limit of l/180 for total load has been used.

CONTROLLING VALUES: Values above and to the right of the heavy line and/or shaded area indicate load capacity controlled by deflection considerations; whereas, values below and to the left of the heavy line and/or shaded area represent load capacity controlled by shear. Values in the shaded area are those load capacities governed by bending.

SPAN: Span is defined as the length from centerline to centerline of bearing. It is this length that is used in standard engineering equations to calculate deflection, bending and shear.

## TABLE 8.9 (Continued)

### Structural Glued Laminated Timber
### *Floor Beams*

Simple Span Beam Table
For Preliminary Design Purposes

Lamination thickness: 1-3/8 in.

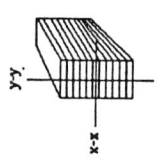

| $F_b$ | $F_v$ | $E$ | $C_D$ | Deflection Limit |
|---|---|---|---|---|
| 2400 psi | 200 psi | $1.8 \times 10^6$ psi | 1.00 | $l/360$ |

**Beam Capacity, Uniform Load w, plf**

| Width b, in. | Depth d, in. | Weight plf | \| | Span, ft 8 | 9 | 10 | 11 | 12 | 13 | 14 | 15 | 16 | 17 | 18 | 19 | 20 | 21 | 22 | 23 | 24 |
|---|---|---|---|---|---|---|---|---|---|---|---|---|---|---|---|---|---|---|---|---|
| 3 | 5-1/2 | 4.1 | | 271 | 190 | 139 | 104 | 80 | 63 | 51 | | | | | | | | | | |
| 3 | 6-7/8 | 5.2 | | 529 | 371 | 271 | 203 | 157 | 123 | 99 | 80 | 66 | 55 | | | | | | | |
| 3 | 8-1/4 | 6.2 | | 851 | 642 | 468 | 352 | 271 | 213 | 171 | 139 | 114 | 95 | 80 | 68 | 58 | 51 | | | |
| 3 | 9-5/8 | 7.2 | | 1158 | 915 | 741 | 558 | 430 | 338 | 271 | 220 | 181 | 151 | 127 | 108 | 93 | 80 | 70 | 61 | 54 |
| 3 | 11 | 8.3 | | 1427 | 1195 | 968 | 800 | 642 | 505 | 404 | 329 | 271 | 226 | 190 | 162 | 139 | 120 | 104 | 91 | 80 |
| 3 | 12-3/8 | 9.3 | | 1667 | 1427 | 1225 | 1013 | 851 | 719 | 578 | 468 | 386 | 321 | 271 | 230 | 197 | 171 | 148 | 130 | 114 |
| 3 | 13-3/4 | 10.3 | | 1927 | 1840 | 1427 | 1250 | 1050 | 895 | 772 | 642 | 529 | 441 | 371 | 316 | 271 | 234 | 203 | 178 | 157 |
| 3 | 15-1/8 | 11.3 | | 2208 | 1868 | 1618 | 1427 | 1271 | 1083 | 934 | 813 | 704 | 587 | 494 | 420 | 360 | 311 | 271 | 237 | 209 |
| 3 | 16-1/2 | 12.4 | | 2514 | 2112 | 1821 | 1600 | 1427 | 1288 | 1111 | 968 | 851 | 754 | 642 | 548 | 468 | 404 | 352 | 308 | 271 |
| 3 | 17-7/8 | 13.4 | | 2848 | 2375 | 2037 | 1783 | 1585 | 1427 | 1296 | 1136 | 998 | 884 | 768 | 684 | 595 | 514 | 447 | 391 | 344 |
| 5 | 5-1/2 | 6.9 | | 451 | 317 | 231 | 174 | 134 | 105 | 84 | 68 | 56 | 47 | | | | | | | |
| 5 | 6-7/8 | 8.6 | | 881 | 619 | 451 | 339 | 281 | 205 | 164 | 134 | 110 | 92 | 77 | 66 | 56 | 49 | | | |
| 5 | 8-1/4 | 10.3 | | 1418 | 1070 | 780 | 586 | 451 | 355 | 284 | 231 | 190 | 159 | 134 | 114 | 97 | 84 | 73 | 64 | 56 |
| 5 | 9-5/8 | 12.0 | | 1930 | 1525 | 1235 | 830 | 717 | 564 | 451 | 367 | 302 | 252 | 212 | 181 | 155 | 134 | 118 | 102 | 90 |
| 5 | 11 | 13.8 | | 2378 | 1992 | 1613 | 1333 | 1070 | 841 | 674 | 548 | 451 | 378 | 317 | 270 | 231 | 200 | 174 | 152 | 134 |
| 5 | 12-3/8 | 15.5 | | 2779 | 2378 | 2042 | 1688 | 1418 | 1198 | 959 | 780 | 643 | 536 | 451 | 384 | 329 | 284 | 247 | 216 | 190 |
| 5 | 13-3/4 | 17.2 | | 3212 | 2733 | 2378 | 2063 | 1751 | 1492 | 1286 | 1070 | 881 | 735 | 619 | 528 | 451 | 380 | 339 | 297 | 261 |
| 5 | 15-7/8 | 18.9 | | 3681 | 3113 | 2696 | 2378 | 2118 | 1805 | 1558 | 1356 | 1173 | 978 | 824 | 701 | 601 | 519 | 451 | 395 | 348 |
| 5 | 16-1/2 | 20.6 | | 4190 | 3520 | 3034 | 2667 | 2378 | 2146 | 1852 | 1613 | 1418 | 1251 | 1070 | 910 | 780 | 674 | 586 | 513 | 451 |

| Width | Depth | Wt | | | | | | | | | | | | | | | | | |
|---|---|---|---|---|---|---|---|---|---|---|---|---|---|---|---|---|---|---|---|---|
| 5 | 17-7/8 | 22.3 | 574 | 652 | 745 | 857 | 992 | 1156 | 1301 | 1462 | 1656 | 1890 | 2163 | 2378 | 2642 | 2971 | 3395 | 3958 | 4747 |
| 5 | 19-1/4 | 24.1 | 717 | 814 | 930 | 1070 | 1211 | 1345 | 1503 | 1690 | 1913 | 2177 | 2378 | 2621 | 2919 | 3294 | 3779 | 4432 | 5357 |
| 5 | 20-5/8 | 25.8 | 881 | 1002 | 1139 | 1253 | 1385 | 1539 | 1719 | 1933 | 2189 | 2378 | 2604 | 2876 | 3212 | 3636 | 4190 | 4944 | 6027 |
| 5 | 22 | 27.5 | 1070 | 1180 | 1292 | 1421 | 1571 | 1745 | 1950 | 2192 | 2378 | 2588 | 2839 | 3143 | 3520 | 4000 | 4632 | 5500 | 6769 |
| 5 | 23-3/8 | 29.2 | 1217 | 1328 | 1454 | 1600 | 1768 | 1964 | 2194 | 2378 | 2575 | 2807 | 3085 | 3423 | 3846 | 4387 | 5106 | 6106 | 7594 |
| 6-3/4 | 6-7/8 | 11.6 |  | 50 | 57 | 66 | 78 | 88 | 104 | 124 | 149 | 181 | 222 | 277 | 353 | 458 | 609 | 836 | 1190 |
| 6-3/4 | 8-1/4 | 13.9 | 76 | 87 | 99 | 114 | 132 | 153 | 181 | 214 | 257 | 312 | 384 | 479 | 609 | 791 | 1053 | 1444 | 1914 |
| 6-3/4 | 9-5/8 | 16.2 | 121 | 137 | 157 | 181 | 209 | 244 | 287 | 340 | 408 | 485 | 609 | 761 | 968 | 1256 | 1668 | 2059 | 2806 |
| 6-3/4 | 11 | 18.6 | 181 | 205 | 234 | 269 | 312 | 364 | 428 | 508 | 609 | 739 | 909 | 1136 | 1444 | 1800 | 2178 | 2689 | 3211 |
| 6-3/4 | 12-3/8 | 20.9 | 257 | 292 | 334 | 384 | 444 | 518 | 609 | 723 | 868 | 1053 | 1295 | 1617 | 1914 | 2278 | 2757 | 3211 | 3752 |
| 6-3/4 | 13-3/4 | 23.2 | 353 | 401 | 458 | 526 | 609 | 711 | 836 | 992 | 1180 | 1444 | 1736 | 2014 | 2363 | 2813 | 3211 | 3689 | 4336 |
| 6-3/4 | 15-1/8 | 25.5 | 469 | 533 | 609 | 701 | 811 | 946 | 1112 | 1320 | 1584 | 1815 | 2090 | 2433 | 2860 | 3211 | 3640 | 4202 | 4969 |
| 6-3/4 | 16-1/2 | 27.8 | 609 | 692 | 791 | 909 | 1053 | 1228 | 1444 | 1664 | 1884 | 2150 | 2477 | 2883 | 3211 | 3600 | 4097 | 4752 | 5657 |
| 6-3/4 | 17-7/8 | 30.2 | 775 | 880 | 1006 | 1156 | 1339 | 1548 | 1730 | 1945 | 2202 | 2513 | 2895 | 3211 | 3567 | 4011 | 4583 | 5344 | 6408 |
| 6-3/4 | 19-1/4 | 32.5 | 968 | 1099 | 1256 | 1444 | 1610 | 1789 | 1998 | 2247 | 2544 | 2904 | 3211 | 3539 | 3941 | 4447 | 5102 | 5983 | 7231 |
| 6-3/4 | 20-5/8 | 34.8 | 1190 | 1352 | 1515 | 1667 | 1842 | 2046 | 2286 | 2570 | 2911 | 3211 | 3515 | 3882 | 4336 | 4909 | 5657 | 6674 | 8137 |
| 6-3/4 | 22 | 37.1 | 1438 | 1568 | 1718 | 1890 | 2069 | 2321 | 2593 | 2915 | 3211 | 3484 | 3832 | 4243 | 4752 | 5400 | 6253 | 7425 | 9138 |
| 6-3/4 | 23-3/8 | 39.4 | 1618 | 1766 | 1934 | 2128 | 2351 | 2612 | 2918 | 3211 | 3476 | 3789 | 4164 | 4622 | 5192 | 5923 | 6893 | 8243 | 10252 |
| 6-3/4 | 24-3/4 | 41.8 | 1809 | 1974 | 2162 | 2378 | 2629 | 2920 | 3211 | 3460 | 3752 | 4097 | 4511 | 5020 | 5657 | 6480 | 7583 | 9138 | 11497 |
| 6-3/4 | 26-1/8 | 44.1 | 2010 | 2183 | 2402 | 2643 | 2921 | 3211 | 3446 | 3719 | 4038 | 4417 | 4875 | 5439 | 6150 | 7076 | 8329 | 10122 | 12898 |
| 6-3/4 | 27-1/2 | 46.4 | 2221 | 2424 | 2655 | 2921 | 3211 | 3434 | 3689 | 3987 | 4336 | 4752 | 5257 | 5881 | 6674 | 7714 | 9138 | 11208 | 14488 |

TABLE SPECIFICATIONS: This table applies to straight, simply supported glued laminated timber beams under dry conditions of use. Beams must be laterally supported at the top along the length of the beam and at the top and bottom at the end bearings. The load carrying capacities tabulated are for total load including the weight of the member. A unit weight of 36 pcf was assumed to determine the tabulated plf beams weights.

DESIGN VALUE MODIFICATIONS: The allowable stress in bending, $F_b$, has been modified by the AITC volume factor, $C_v$. For determination of load carrying capacities governed by shear, loads within a distance "d" (the depth of the beam) from the ends have been neglected.

DEFLECTION LIMITS: For floor beams, a deflection limit of l/360 for live load has been used. Live load has been assumed to be 80% of the total load as is common for residential floors. For other ratios of live load to total load, the designer should check deflection based on applicable building code provisions.

CONTROLLING VALUES: Values above and to the right of the heavy line and shaded area indicate load capacity controlled by deflection considerations whereas values below and to the left of the heavy line and shaded area represent load capacity controlled by shear. Values in the shaded area are those load capacities governed by bending.

TABLE 8.9 (Continued)

## Structural Glued Laminated Timber
### Floor Beams

Lamination thickness: 1-3/8 in.

Simple Span Beam Table
For Preliminary Design Purposes

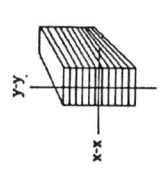

| | $F_b$ | $F_v$ | $E$ | $C_D$ | |
|---|---|---|---|---|---|
| | 2400 psi | 200 psi | $1.8 \times 10^6$ psi | 1.00 | Deflection Limit $l/360$ |

| Beam Size | | | Beam Capacity, Uniform Load w, plf — Span, ft | | | | | | | | | | | | | | |
|---|---|---|---|---|---|---|---|---|---|---|---|---|---|---|---|---|---|
| Width b, in. | Depth d, in. | Weight plf | 25 | 26 | 27 | 28 | 29 | 30 | 31 | 32 | 33 | 34 | 35 | 36 | 37 | 38 | 39 | 40 |
| 3 | 5-1/2 | 4.1 | | | | | | | | | | | | | | | | |
| 3 | 6-7/8 | 5.2 | | | | | | | | | | | | | | | | |
| 3 | 8-1/4 | 6.2 | | | | | | | | | | | | | | | | |
| 3 | 9-5/8 | 7.2 | 48 | | | | | | | | | | | | | | | |
| 3 | 11 | 8.3 | 71 | 63 | 56 | 51 | | | | | | | | | | | | |
| 3 | 12-3/8 | 9.3 | 101 | 90 | 80 | 72 | 65 | 58 | 53 | 48 | | | | | | | | |
| 3 | 13-3/4 | 10.3 | 139 | 123 | 110 | 99 | 89 | 80 | 73 | 66 | 60 | 55 | 51 | | | | | |
| 3 | 15-1/8 | 11.3 | 185 | 164 | 146 | 131 | 118 | 107 | 97 | 88 | 80 | 73 | 67 | 62 | 57 | 53 | 49 | |
| 3 | 16-1/2 | 12.4 | 240 | 213 | 190 | 171 | 153 | 139 | 126 | 114 | 104 | 95 | 87 | 80 | 74 | 68 | 63 | 58 |
| 3 | 17-7/8 | 13.4 | 305 | 271 | 242 | 217 | 195 | 176 | 160 | 145 | 132 | 121 | 111 | 102 | 94 | 87 | 80 | 74 |
| 5 | 5-1/2 | 6.9 | | | | | | | | | | | | | | | | |
| 5 | 6-7/8 | 8.6 | | | | | | | | | | | | | | | | |
| 5 | 8-1/4 | 10.3 | 50 | | | | | | | | | | | | | | | |
| 5 | 9-5/8 | 12.0 | 79 | 70 | 63 | 56 | 51 | 48 | | | | | | | | | | |
| 5 | 11 | 13.8 | 118 | 105 | 94 | 84 | 76 | 68 | 62 | 56 | 51 | 47 | | | | | | |
| 5 | 12-3/8 | 15.5 | 168 | 150 | 134 | 120 | 108 | 97 | 88 | 80 | 73 | 67 | 61 | 56 | 52 | 48 | | |
| 5 | 13-3/4 | 17.2 | 231 | 205 | 183 | 164 | 148 | 134 | 121 | 110 | 100 | 92 | 84 | 77 | 71 | 66 | 61 | 56 |
| 5 | 15-1/8 | 18.9 | 308 | 273 | 244 | 219 | 197 | 178 | 161 | 147 | 134 | 122 | 112 | 103 | 95 | 88 | 81 | 75 |
| 5 | 16-1/2 | 20.8 | 399 | 355 | 317 | 284 | 256 | 231 | 209 | 190 | 174 | 159 | 146 | 134 | 123 | 114 | 105 | 97 |

The following load table gives total load capacities (plf) for glued laminated timber beams. The first three columns give the beam width (in), depth (in), and tabulated weight (plf); the remaining 16 columns give load capacity for successive span conditions (span headers not legible in the source; columns numbered 1–16, shortest to longest span).

| Width | Depth | Wt (plf) | 1 | 2 | 3 | 4 | 5 | 6 | 7 | 8 | 9 | 10 | 11 | 12 | 13 | 14 | 15 | 16 |
|---|---|---|---|---|---|---|---|---|---|---|---|---|---|---|---|---|---|---|
| 5 | 17-7/8 | 22.3 | 508 | 451 | 403 | 381 | 325 | 294 | 266 | 242 | 221 | 202 | 185 | 170 | 157 | 145 | 134 | 124 |
| 5 | 19-1/4 | 24.1 | 634 | 564 | 503 | 451 | 406 | 367 | 333 | 302 | 278 | 252 | 231 | 212 | 196 | 181 | 167 | 155 |
| 5 | 20-5/8 | 25.8 | 780 | 683 | 619 | 555 | 500 | 451 | 409 | 372 | 339 | 310 | 284 | 281 | 241 | 222 | 205 | 190 |
| 5 | 22 | 27.5 | 946 | 841 | 751 | 674 | 606 | 548 | 496 | 451 | 412 | 376 | 345 | 317 | 292 | 270 | 249 | 231 |
| 5 | 23-3/8 | 29.2 | 1119 | 1009 | 901 | 808 | 727 | 657 | 595 | 541 | 494 | 451 | 414 | 380 | 350 | 323 | 299 | 277 |
| 6-3/4 | 6-7/8 | 11.6 | | | | | | | | | | | | | | | | |
| 6-3/4 | 8-1/4 | 13.9 | 67 | 60 | 53 | 48 | | | | | | | | | | | | |
| 6-3/4 | 9-5/8 | 16.2 | 107 | 95 | 85 | 76 | 69 | 62 | 56 | 51 | 47 | | | | | | | |
| 6-3/4 | 11 | 18.6 | 160 | 142 | 127 | 114 | 102 | 92 | 84 | 76 | 69 | 63 | 58 | 53 | 49 | | | |
| 6-3/4 | 12-3/8 | 20.9 | 227 | 202 | 181 | 162 | 146 | 132 | 119 | 108 | 99 | 90 | 83 | 78 | 70 | 65 | 60 | 56 |
| 6-3/4 | 13-3/4 | 23.2 | 312 | 277 | 248 | 222 | 200 | 181 | 164 | 149 | 136 | 124 | 114 | 104 | 96 | 89 | 82 | 78 |
| 6-3/4 | 15-1/8 | 25.5 | 415 | 389 | 330 | 296 | 266 | 240 | 218 | 198 | 181 | 165 | 151 | 139 | 128 | 118 | 109 | 101 |
| 6-3/4 | 16-1/2 | 27.8 | 539 | 479 | 428 | 384 | 345 | 312 | 283 | 257 | 234 | 214 | 196 | 181 | 166 | 153 | 142 | 132 |
| 6-3/4 | 17-7/8 | 30.2 | 685 | 609 | 544 | 488 | 439 | 397 | 359 | 327 | 298 | 272 | 250 | 230 | 211 | 195 | 181 | 167 |
| 6-3/4 | 19-1/4 | 32.5 | 856 | 761 | 680 | 609 | 548 | 495 | 449 | 408 | 372 | 340 | 312 | 287 | 264 | 244 | 225 | 209 |
| 6-3/4 | 20-5/8 | 34.8 | 1053 | 936 | 836 | 749 | 675 | 609 | 552 | 502 | 458 | 419 | 384 | 353 | 325 | 300 | 277 | 257 |
| 6-3/4 | 22 | 37.1 | 1278 | 1136 | 1014 | 909 | 819 | 739 | 670 | 609 | 556 | 508 | 466 | 428 | 394 | 364 | 337 | 312 |
| 6-3/4 | 23-3/8 | 39.4 | 1488 | 1362 | 1217 | 1091 | 982 | 887 | 804 | 731 | 668 | 609 | 559 | 513 | 473 | 436 | 404 | 374 |
| 6-3/4 | 24-3/4 | 41.8 | 1664 | 1535 | 1421 | 1295 | 1166 | 1053 | 954 | 868 | 791 | 723 | 663 | 609 | 561 | 518 | 479 | 444 |
| 6-3/4 | 26-1/8 | 44.1 | 1849 | 1706 | 1579 | 1465 | 1364 | 1238 | 1122 | 1020 | 930 | 851 | 780 | 717 | 660 | 609 | 564 | 522 |
| 6-3/4 | 27-1/2 | 46.4 | 2043 | 1885 | 1745 | 1620 | 1507 | 1406 | 1309 | 1190 | 1085 | 992 | 909 | 836 | 770 | 711 | 657 | 609 |

TABLE SPECIFICATIONS: This table applies to straight, simply supported glued laminated timber beams under dry conditions of use. Beams must be laterally supported at the top along the length of the beam and at the end bearings. The load carrying capacities tabulated are for total load including the weight of the member. A unit weight of 36 pcf was assumed to determine the tabulated plf beams weights.

DESIGN VALUE MODIFICATIONS: The allowable stress in bending, $F_b$, has been modified by the AITC volume factor, $C_v$. For determination of load carrying capacities governed by shear, loads within a distance "d" (the depth of the beam) from the ends have been neglected.

DEFLECTION LIMITS: For floor beams, a deflection limit of $l/360$ for live load has been used. Live load has been assumed to be 80% of the total load as is common for residential floors. For other ratios of live load to total load, the designer should check deflection based on applicable building code provisions.

CONTROLLING VALUES: Values above and to the right of the heavy line and shaded area indicate load capacity controlled by deflection considerations whereas values below and to the left of the heavy line and shaded area represent load capacity controlled by shear. Values in the shaded area are those load capacities governed by bending.

## TABLE 8.9 (Continued)

### Structural Glued Laminated Timber
### *Roof Beams for Construction Load*

Simple Span Beam Table
For Preliminary Design Purposes

Lamination thickness: 1-3/8 in.

| $F_b$ | $F_v$ | $E$ | $C_D$ | Deflection Limit |
|---|---|---|---|---|
| 2400 psi | 200 psi | $1.8 \times 10^6$ psi | 1.25 | l/180 |

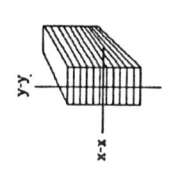

| Beam Size | | | Beam Capacity, Uniform Load w, plf | | | | | | | | | | | | | | | | | |
| Width b, in. | Depth d, in. | Weight plf | Span, ft | | | | | | | | | | | | | | | | | |
| | | | 8 | 9 | 10 | 11 | 12 | 13 | 14 | 15 | 16 | 17 | 18 | 19 | 20 | 21 | 22 | 23 | 24 |
|---|---|---|---|---|---|---|---|---|---|---|---|---|---|---|---|---|---|---|---|
| 3-1/8 | 5-1/2 | 4.3 | 451 | 317 | 231 | 174 | 134 | 105 | 84 | 68 | 56 | 47 | 40 | 34 | 29 | 25 | 22 | 19 | 17 |
| 3-1/8 | 6-7/8 | 5.4 | 769 | 608 | 451 | 339 | 261 | 205 | 164 | 134 | 110 | 92 | 77 | 66 | 56 | 49 | 42 | 37 | 33 |
| 3-1/8 | 8-1/4 | 6.4 | 1108 | 875 | 709 | 586 | 451 | 355 | 284 | 231 | 190 | 159 | 134 | 114 | 97 | 84 | 73 | 64 | 56 |
| 3-1/8 | 9-5/8 | 7.5 | 1506 | 1191 | 965 | 798 | 670 | 564 | 451 | 367 | 302 | 252 | 212 | 181 | 155 | 134 | 116 | 102 | 90 |
| 3-1/8 | 11 | 8.6 | 1858 | 1556 | 1260 | 1042 | 875 | 748 | 643 | 548 | 451 | 376 | 317 | 270 | 231 | 200 | 174 | 152 | 134 |
| 3-1/8 | 12-3/8 | 9.7 | 2171 | 1858 | 1595 | 1318 | 1108 | 944 | 814 | 709 | 623 | 536 | 451 | 384 | 329 | 284 | 247 | 216 | 190 |
| 3-1/8 | 13-3/4 | 10.7 | 2509 | 2135 | 1858 | 1628 | 1368 | 1165 | 1005 | 875 | 769 | 681 | 608 | 526 | 451 | 390 | 339 | 297 | 261 |
| 3-1/8 | 15-1/8 | 11.8 | 2875 | 2432 | 2107 | 1858 | 1655 | 1410 | 1218 | 1059 | 931 | 825 | 735 | 660 | 596 | 519 | 451 | 395 | 348 |
| 3-1/8 | 16-1/2 | 12.9 | 3274 | 2750 | 2371 | 2083 | 1858 | 1677 | 1447 | 1260 | 1108 | 981 | 875 | 788 | 709 | 643 | 586 | 513 | 451 |
| 3-1/8 | 17-7/8 | 14.0 | 3709 | 3093 | 2652 | 2321 | 2064 | 1858 | 1690 | 1479 | 1300 | 1152 | 1027 | 922 | 832 | 755 | 688 | 629 | 574 |
| 5-1/8 | 5-1/2 | 7.0 | 740 | 520 | 379 | 285 | 219 | 172 | 138 | 112 | 93 | 77 | 65 | 55 | 47 | 41 | 36 | 31 | 27 |
| 5-1/8 | 6-7/8 | 8.8 | 1262 | 997 | 740 | 556 | 428 | 337 | 270 | 219 | 181 | 151 | 127 | 108 | 93 | 80 | 70 | 61 | 54 |
| 5-1/8 | 8-1/4 | 10.6 | 1817 | 1435 | 1163 | 981 | 740 | 582 | 466 | 379 | 312 | 260 | 219 | 186 | 160 | 138 | 120 | 105 | 93 |
| 5-1/8 | 8-5/8 | 12.3 | 2473 | 1954 | 1563 | 1308 | 1099 | 924 | 740 | 602 | 496 | 413 | 348 | 296 | 254 | 219 | 191 | 167 | 147 |
| 5-1/8 | 11 | 14.1 | 3047 | 2552 | 2067 | 1708 | 1435 | 1223 | 1055 | 898 | 740 | 617 | 520 | 442 | 379 | 327 | 285 | 249 | 219 |
| 5-1/8 | 12-3/8 | 15.9 | 3561 | 3047 | 2616 | 2162 | 1817 | 1548 | 1335 | 1163 | 1022 | 879 | 740 | 629 | 540 | 466 | 405 | 355 | 312 |
| 5-1/8 | 13-3/4 | 17.6 | 4115 | 3502 | 3047 | 2669 | 2243 | 1911 | 1648 | 1435 | 1262 | 1118 | 997 | 863 | 740 | 639 | 556 | 487 | 428 |
| 5-1/8 | 15-1/8 | 19.4 | 4716 | 3988 | 3455 | 3047 | 2714 | 2312 | 1994 | 1737 | 1527 | 1351 | 1202 | 1075 | 968 | 851 | 740 | 648 | 570 |
| 5-1/8 | 16-1/2 | 21.1 | 5369 | 4510 | 3888 | 3417 | 3047 | 2750 | 2373 | 2067 | 1813 | 1601 | 1424 | 1274 | 1147 | 1038 | 944 | 841 | 740 |

| Width | Depth | Wt (plf) | | | | | | | | | | | | | | | | | |
|---|---|---|---|---|---|---|---|---|---|---|---|---|---|---|---|---|---|---|---|
| 5-1/8 | 17-7/8 | 22.9 | 923 | 1007 | 1103 | 1213 | 1341 | 1490 | 1664 | 1871 | 2119 | 2418 | 2771 | 3047 | 3385 | 3807 | 4349 | 5072 | 6082 |
| 5-1/8 | 19-1/4 | 24.7 | 1066 | 1163 | 1274 | 1402 | 1549 | 1721 | 1923 | 2162 | 2448 | 2789 | 3047 | 3359 | 3741 | 4221 | 4842 | 5678 | 6863 |
| 5-1/8 | 20-5/8 | 26.4 | 1220 | 1331 | 1458 | 1604 | 1773 | 1969 | 2200 | 2473 | 2801 | 3047 | 3336 | 3685 | 4115 | 4659 | 5369 | 6334 | 7723 |
| 5-1/8 | 22 | 28.2 | 1383 | 1509 | 1653 | 1819 | 2010 | 2233 | 2495 | 2805 | 3047 | 3316 | 3637 | 4027 | 4510 | 5125 | 5934 | 7047 | 8673 |
| 5-1/8 | 23-3/8 | 29.9 | 1557 | 1699 | 1861 | 2047 | 2263 | 2513 | 2808 | 3047 | 3299 | 3596 | 3952 | 4386 | 4927 | 5621 | 6542 | 7823 | 9730 |
| 6-3/4 | 6-7/8 | 11.6 | 71 | 80 | 92 | 105 | 122 | 142 | 167 | 198 | 238 | 289 | 355 | 444 | 564 | 732 | 975 | 1313 | 1662 |
| 6-3/4 | 8-1/4 | 13.9 | 122 | 138 | 158 | 182 | 211 | 248 | 289 | 343 | 411 | 499 | 614 | 767 | 975 | 1266 | 1531 | 1891 | 2393 |
| 6-3/4 | 9-5/8 | 16.2 | 194 | 220 | 251 | 289 | 334 | 390 | 459 | 544 | 653 | 793 | 975 | 1218 | 1448 | 1723 | 2084 | 2573 | 3257 |
| 6-3/4 | 11 | 18.6 | 289 | 328 | 375 | 431 | 499 | 582 | 685 | 813 | 975 | 1183 | 1388 | 1611 | 1891 | 2250 | 2723 | 3361 | 4014 |
| 6-3/4 | 12-3/8 | 20.9 | 411 | 467 | 534 | 614 | 711 | 829 | 975 | 1157 | 1344 | 1531 | 1758 | 2039 | 2393 | 2848 | 3446 | 4014 | 4689 |
| 6-3/4 | 13-3/4 | 23.2 | 564 | 641 | 732 | 842 | 975 | 1137 | 1296 | 1457 | 1650 | 1884 | 2170 | 2517 | 2954 | 3516 | 4014 | 4612 | 5420 |
| 6-3/4 | 15-1/8 | 25.5 | 751 | 853 | 975 | 1121 | 1258 | 1397 | 1561 | 1755 | 1987 | 2268 | 2613 | 3042 | 3574 | 4014 | 4550 | 5252 | 6211 |
| 6-3/4 | 16-1/2 | 27.8 | 975 | 1108 | 1226 | 1348 | 1490 | 1655 | 1850 | 2079 | 2355 | 2688 | 3096 | 3604 | 4014 | 4500 | 5121 | 5940 | 7071 |
| 6-3/4 | 17-7/8 | 30.2 | 1199 | 1308 | 1433 | 1576 | 1742 | 1935 | 2162 | 2431 | 2752 | 3142 | 3619 | 4014 | 4458 | 5014 | 5728 | 6680 | 8010 |
| 6-3/4 | 19-1/4 | 32.5 | 1385 | 1511 | 1656 | 1821 | 2013 | 2236 | 2498 | 2809 | 3180 | 3630 | 4014 | 4423 | 4927 | 5559 | 6377 | 7478 | 9039 |
| 6-3/4 | 20-5/8 | 34.8 | 1585 | 1729 | 1894 | 2083 | 2303 | 2556 | 2856 | 3213 | 3638 | 4014 | 4393 | 4853 | 5420 | 6136 | 7071 | 8343 | 10171 |
| 6-3/4 | 22 | 37.1 | 1797 | 1961 | 2148 | 2363 | 2611 | 2901 | 3241 | 3644 | 4014 | 4368 | 4790 | 5304 | 5940 | 6750 | 7816 | 9281 | 11433 |
| 6-3/4 | 23-3/8 | 39.4 | 2023 | 2207 | 2418 | 2659 | 2939 | | 3648 | 4014 | 4345 | 4738 | 5205 | 5777 | 6490 | 7403 | 8616 | 10304 | 12815 |
| 6-3/4 | 24-3/4 | 41.8 | 2261 | 2467 | 2703 | 2973 | 3286 | 3651 | 4014 | 4325 | 4689 | 5121 | 5639 | 6275 | 7071 | 8100 | 9479 | 11423 | 14371 |
| 6-3/4 | 26-1/8 | 44.1 | 2512 | 2742 | 3003 | 3304 | 3651 | 4014 | 4308 | 4648 | 5047 | 5522 | 6094 | 6799 | 7688 | 8845 | 10411 | 12652 | 16123 |
| 6-3/4 | 27-1/2 | 46.4 | 2777 | 3030 | 3319 | 3651 | 4014 | 4292 | 4612 | 4983 | 5420 | 5840 | 6571 | 7351 | 8343 | 9643 | 11423 | 14009 | 18110 |

TABLE SPECIFICATIONS: This table applies to straight, simply supported glued laminated timber beams under dry conditions of use. Beams must be laterally supported at the top along the length of the beam and at the top and bottom at the end bearings. The load carrying capacities tabulated are for total load including the weight of the member. A unit weight of 36 pcf was assumed to determine the tabulated plf beam weights.

DESIGN VALUE MODIFICATIONS: The allowable stress in bending, $F_b$, has been modified by the AITC volume factor, $C_v$. For determination of load carrying capacities governed by shear, loads within a distance "d" (the depth of the beam) from the ends have been neglected.

DEFLECTION LIMITS: For roof beams, a deflection limit of $l/180$ for total load has been used.

CONTROLLING VALUES: Values above and to the right of the heavy line and/or shaded area indicate load capacity controlled by deflection considerations; whereas, values below and to the left of the heavy line and/or shaded area represent load capacity controlled by shear. Values in the shaded area are those load capacities governed by bending.

SPAN: Span is defined as the length from centerline to centerline of bearing. It is this length that is used in standard engineering equations to calculate deflection, bending and shear.

# TABLE 8.9 (Continued)

## Structural Glued Laminated Timber
### *Roof Beams for Construction Load*

Lamination thickness: 1-3/8 in.

Simple Span Beam Table
For Preliminary Design Purposes

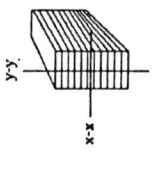

| $F_b$ | $F_v$ | $E$ | $C_D$ | Deflection Limit |
|---|---|---|---|---|
| 2400 psi | 200 psi | $1.8 \times 10^6$ psi | 1.25 | $l/180$ |

| Beam Size | | | Span, ft — Beam Capacity, Uniform Load w, plf | | | | | | | | | | | | | | | |
| Width b, in. | Depth d, in. | Weight plf | 25 | 26 | 27 | 28 | 29 | 30 | 31 | 32 | 33 | 34 | 35 | 36 | 37 | 38 | 39 | 40 |
|---|---|---|---|---|---|---|---|---|---|---|---|---|---|---|---|---|---|---|
| 3-1/8 | 5-1/2  | 4.3  | 29  | 28  |     |     |     |     |     |     |     |     |     |     |     |     |     |     |
| 3-1/8 | 6-7/8  | 5.4  | 50  | 44  | 40  | 36  | 32  | 29  |     |     |     |     |     |     |     |     |     |     |
| 3-1/8 | 8-1/4  | 6.4  | 79  | 70  | 63  | 56  | 51  | 46  | 42  | 38  | 34  | 32  | 29  |     |     |     |     |     |
| 3-1/8 | 9-5/8  | 7.5  | 118 | 105 | 94  | 84  | 78  | 68  | 62  | 56  | 51  | 47  | 43  | 40  | 36  | 34  | 31  | 29  |
| 3-1/8 | 11     | 8.6  | 168 | 150 | 134 | 120 | 108 | 97  | 88  | 80  | 73  | 67  | 61  | 56  | 52  | 48  | 44  | 41  |
| 3-1/8 | 12-3/8 | 9.7  | 231 | 205 | 183 | 164 | 148 | 134 | 121 | 110 | 100 | 92  | 84  | 77  | 71  | 66  | 61  | 56  |
| 3-1/8 | 13-3/4 | 10.7 | 308 | 273 | 244 | 219 | 197 | 178 | 161 | 147 | 134 | 122 | 112 | 103 | 95  | 88  | 81  | 75  |
| 3-1/8 | 15-1/8 | 11.8 | 399 | 355 | 317 | 284 | 256 | 231 | 209 | 190 | 174 | 159 | 146 | 134 | 123 | 114 | 105 | 97  |
| 3-1/8 | 16-1/2 | 12.9 | 508 | 451 | 403 | 361 | 325 | 294 | 266 | 242 | 221 | 202 | 185 | 170 | 157 | 145 | 134 | 124 |
| 3-1/8 | 17-7/8 | 14.0 |     |     |     |     |     |     |     |     |     |     |     |     |     |     |     |     |
| 5-1/8 | 5-1/2  | 7.0  | 47  | 42  | 38  | 34  | 30  | 27  |     |     |     |     |     |     |     |     |     |     |
| 5-1/8 | 6-7/8  | 8.8  | 82  | 73  | 65  | 58  | 52  | 47  | 43  | 39  | 36  | 33  | 30  | 27  |     |     |     |     |
| 5-1/8 | 8-1/4  | 10.6 | 130 | 116 | 103 | 93  | 83  | 75  | 68  | 62  | 57  | 52  | 47  | 44  | 40  | 37  | 34  | 32  |
| 5-1/8 | 9-5/8  | 12.3 | 194 | 172 | 154 | 138 | 124 | 112 | 102 | 93  | 84  | 77  | 71  | 65  | 60  | 55  | 51  | 47  |
| 5-1/8 | 11     | 14.1 | 278 | 248 | 219 | 197 | 177 | 160 | 145 | 132 | 120 | 110 | 101 | 93  | 85  | 79  | 73  | 67  |
| 5-1/8 | 12-3/8 | 15.9 | 379 | 337 | 301 | 270 | 243 | 219 | 199 | 181 | 165 | 151 | 138 | 127 | 117 | 108 | 100 | 93  |
| 5-1/8 | 13-3/4 | 17.8 | 504 | 448 | 400 | 359 | 323 | 292 | 266 | 241 | 219 | 201 | 184 | 169 | 156 | 144 | 133 | 123 |
| 5-1/8 | 15-1/8 | 19.4 | 655 | 582 | 520 | 466 | 420 | 379 | 343 | 312 | 285 | 260 | 239 | 219 | 202 | 186 | 172 | 160 |
| 5-1/8 | 16-1/2 | 21.1 |     |     |     |     |     |     |     |     |     |     |     |     |     |     |     |     |

| Width | Depth | Wt | | | | | | | | | | | | | | | | |
|---|---|---|---|---|---|---|---|---|---|---|---|---|---|---|---|---|---|---|
| 5-1/8 | 17-7/8 | 22.9 | 833 | 740 | 661 | 593 | 533 | 482 | 437 | 397 | 362 | 331 | 303 | 279 | 257 | 237 | 219 | 203 |
| 5-1/8 | 19-1/4 | 24.7 | 981 | 905 | 825 | 740 | 666 | 602 | 545 | 498 | 452 | 413 | 379 | 348 | 321 | 296 | 274 | 254 |
| 5-1/8 | 20-5/8 | 26.4 | 1122 | 1035 | 958 | 889 | 819 | 740 | 671 | 610 | 556 | 508 | 466 | 428 | 395 | 364 | 337 | 312 |
| 5-1/8 | 22 | 28.2 | 1272 | 1174 | 1087 | 1009 | 939 | 876 | 814 | 740 | 675 | 617 | 566 | 520 | 479 | 442 | 409 | 379 |
| 5-1/8 | 23-3/8 | 29.9 | 1432 | 1321 | 1223 | 1135 | 1056 | 985 | 921 | 863 | 810 | 740 | 679 | 624 | 574 | 530 | 490 | 455 |
| 6-3/4 | 6-7/8 | 11.6 | 62 | 55 | 50 | 44 | 40 | 36 | 33 | 30 | 27 | | | | | | | |
| 6-3/4 | 8-1/4 | 13.9 | 108 | 98 | 86 | 77 | 69 | 62 | 57 | 51 | 47 | 43 | 39 | 36 | 33 | 31 | 28 | |
| 6-3/4 | 9-5/8 | 16.2 | 171 | 152 | 136 | 122 | 110 | 99 | 90 | 82 | 74 | 68 | 62 | 57 | 53 | 49 | 45 | 42 |
| 6-3/4 | 11 | 18.8 | 256 | 227 | 203 | 182 | 164 | 148 | 134 | 122 | 111 | 102 | 93 | 86 | 79 | 73 | 67 | 62 |
| 6-3/4 | 12-3/8 | 20.9 | 364 | 323 | 289 | 259 | 233 | 211 | 191 | 174 | 158 | 145 | 133 | 122 | 112 | 104 | 96 | 89 |
| 6-3/4 | 13-3/4 | 23.2 | 499 | 444 | 396 | 355 | 320 | 289 | 262 | 238 | 217 | 198 | 182 | 167 | 154 | 142 | 131 | 122 |
| 6-3/4 | 15-1/8 | 25.5 | 664 | 591 | 527 | 473 | 426 | 384 | 348 | 317 | 289 | 264 | 242 | 222 | 205 | 189 | 175 | 162 |
| 6-3/4 | 16-1/2 | 27.8 | 862 | 767 | 685 | 614 | 553 | 499 | 452 | 411 | 375 | 343 | 314 | 289 | 266 | 246 | 227 | 211 |
| 6-3/4 | 17-7/8 | 30.2 | 1097 | 975 | 870 | 781 | 703 | 635 | 575 | 523 | 477 | 436 | 400 | 367 | 338 | 312 | 289 | 268 |
| 6-3/4 | 19-1/4 | 32.5 | 1274 | 1175 | 1087 | 975 | 877 | 793 | 718 | 653 | 595 | 544 | 499 | 459 | 422 | 390 | 361 | 334 |
| 6-3/4 | 20-5/8 | 34.8 | 1457 | 1345 | 1245 | 1155 | 1075 | 975 | 884 | 803 | 732 | 670 | 614 | 564 | 520 | 480 | 444 | 411 |
| 6-3/4 | 22 | 37.1 | 1653 | 1525 | 1412 | 1310 | 1219 | 1137 | 1063 | 975 | 889 | 813 | 745 | 685 | 631 | 582 | 539 | 499 |
| 6-3/4 | 23-3/8 | 39.4 | 1860 | 1716 | 1589 | 1475 | 1372 | 1280 | 1197 | 1121 | 1053 | 975 | 894 | 821 | 756 | 698 | 646 | 599 |
| 6-3/4 | 24-3/4 | 41.8 | 2080 | 1919 | 1776 | 1648 | 1534 | 1431 | 1338 | 1254 | 1177 | 1107 | 1043 | 975 | 898 | 829 | 767 | 711 |
| 6-3/4 | 26-1/8 | 44.1 | 2311 | 2132 | 1973 | 1832 | 1705 | 1590 | 1487 | 1393 | 1308 | 1230 | 1159 | 1094 | 1034 | 975 | 902 | 836 |
| 6-3/4 | 27-1/2 | 46.4 | 2554 | 2357 | 2181 | 2024 | 1864 | 1757 | 1643 | 1540 | 1445 | 1360 | 1281 | 1209 | 1143 | 1082 | 1026 | 974 |

TABLE SPECIFICATIONS: This table applies to straight, simply supported glued laminated timber beams under dry conditions of use. Beams must be laterally supported at the top along the length of the beam and at the top and bottom at the end bearings. The load carrying capacities tabulated are for total load including the weight of the member. A unit weight of 36 pcf was assumed to determine the tabulated plf beam weights.

DESIGN VALUE MODIFICATIONS: The allowable stress in bending, $F_b$, has been modified by the AITC volume factor, $C_V$. For determination of load carrying capacities governed by shear, loads within a distance "d" (the depth of the beam) from the ends have been neglected.

DEFLECTION LIMITS: For roof beams, a deflection limit of l/180 for total load has been used.

CONTROLLING VALUES: Values above and to the right of the heavy line and/or shaded area indicate load capacity controlled by deflection considerations; whereas, values below and to the left of the heavy line and/or shaded area represent load capacity controlled by shear. Values in the shaded area are those load capacities governed by bending.

SPAN: Span is defined as the length from centerline to centerline of bearing. It is this length that is used in standard engineering equations to calculate deflection, bending and shear.

# TABLE 8.9  (*Continued*)

## Structural Glued Laminated Timber
### *Floor Beams*

Simple Span Beam Table
For Preliminary Design Purposes

Lamination thickness: 1-3/8 in.

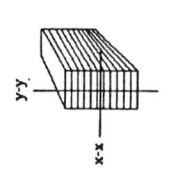

| | | | | | | | |
|---|---|---|---|---|---|---|---|
| $F_b$ | $F_v$ | E | $C_D$ | Deflection Limit | | | |
| 2400 psi | 200 psi | $1.8 \times 10^6$ psi | 1.00 | l/360 | | | |

**Beam Capacity, Uniform Load w, plf**

| Beam Size | | | Span, ft | | | | | | | | | | | | | | | |
|---|---|---|---|---|---|---|---|---|---|---|---|---|---|---|---|---|---|---|
| Width b, in. | Depth d, in. | Weight plf | 8 | 9 | 10 | 11 | 12 | 13 | 14 | 15 | 16 | 17 | 18 | 19 | 20 | 21 | 22 | 23 | 24 |
| 3-1/8 | 5-1/2 | 4.3 | 282 | 198 | 144 | 109 | 84 | 66 | 53 | 43 | | | | | | | | | |
| 3-1/8 | 6-7/8 | 5.4 | 551 | 387 | 282 | 212 | 163 | 128 | 103 | 84 | 69 | 57 | 48 | | | | | | |
| 3-1/8 | 8-1/4 | 6.4 | 866 | 669 | 487 | 366 | 282 | 222 | 178 | 144 | 119 | 99 | 84 | 71 | 61 | 53 | 48 | | |
| 3-1/8 | 9-5/8 | 7.5 | 1209 | 953 | 772 | 562 | 448 | 352 | 282 | 229 | 189 | 158 | 133 | 113 | 97 | 84 | 73 | 64 | 56 |
| 3-1/8 | 11 | 8.6 | 1486 | 1245 | 1008 | 833 | 669 | 528 | 421 | 342 | 282 | 235 | 198 | 168 | 144 | 125 | 109 | 95 | 84 |
| 3-1/8 | 12-3/8 | 9.7 | 1737 | 1486 | 1278 | 1055 | 886 | 749 | 600 | 487 | 402 | 335 | 282 | 240 | 206 | 178 | 154 | 135 | 119 |
| 3-1/8 | 13-3/4 | 10.7 | 2007 | 1708 | 1486 | 1302 | 1094 | 932 | 804 | 669 | 551 | 459 | 387 | 329 | 282 | 244 | 212 | 185 | 163 |
| 3-1/8 | 15-1/8 | 11.8 | 2300 | 1945 | 1685 | 1486 | 1324 | 1128 | 973 | 847 | 733 | 611 | 515 | 438 | 375 | 324 | 282 | 247 | 217 |
| 3-1/8 | 16-1/2 | 12.9 | 2619 | 2200 | 1897 | 1667 | 1486 | 1341 | 1158 | 1008 | 886 | 785 | 669 | 569 | 487 | 421 | 366 | 320 | 282 |
| 3-1/8 | 17-7/8 | 14.0 | 2967 | 2474 | 2122 | 1857 | 1651 | 1486 | 1352 | 1183 | 1040 | 921 | 822 | 723 | 620 | 535 | 466 | 407 | 359 |
| 5-1/8 | 5-1/2 | 7.0 | 463 | 325 | 237 | 178 | 137 | 108 | 86 | 70 | 58 | 48 | | | | | | | |
| 5-1/8 | 6-7/8 | 8.8 | 904 | 635 | 463 | 348 | 268 | 211 | 169 | 137 | 113 | 94 | 79 | 67 | 58 | 50 | 43 | | |
| 5-1/8 | 8-1/4 | 10.6 | 1453 | 1097 | 799 | 601 | 463 | 364 | 291 | 237 | 195 | 163 | 137 | 117 | 100 | 86 | 75 | 66 | 58 |
| 5-1/8 | 9-5/8 | 12.3 | 1978 | 1563 | 1266 | 954 | 735 | 578 | 463 | 376 | 310 | 258 | 218 | 185 | 159 | 137 | 119 | 104 | 92 |
| 5-1/8 | 11 | 14.1 | 2438 | 2042 | 1654 | 1367 | 1097 | 862 | 691 | 561 | 463 | 386 | 325 | 276 | 237 | 205 | 178 | 156 | 137 |
| 5-1/8 | 12-3/8 | 15.9 | 2848 | 2438 | 2093 | 1730 | 1453 | 1228 | 983 | 799 | 659 | 549 | 463 | 393 | 337 | 291 | 253 | 222 | 195 |
| 5-1/8 | 13-3/4 | 17.6 | 3292 | 2801 | 2438 | 2135 | 1794 | 1529 | 1318 | 1097 | 904 | 753 | 635 | 540 | 463 | 400 | 348 | 304 | 268 |
| 5-1/8 | 15-1/8 | 19.4 | 3773 | 3190 | 2764 | 2438 | 2171 | 1850 | 1595 | 1390 | 1203 | 1003 | 845 | 718 | 616 | 532 | 463 | 405 | 356 |
| 5-1/8 | 16-1/2 | 21.1 | 4295 | 3608 | 3110 | 2733 | 2438 | 2200 | 1888 | 1654 | 1450 | 1281 | 1097 | 932 | 799 | 691 | 601 | 526 | 463 |

| Width | Depth | plf | 1 | 2 | 3 | 4 | 5 | 6 | 7 | 8 | 9 | 10 | 11 | 12 | 13 | 14 | 15 | 16 | 17 |
|---|---|---|---|---|---|---|---|---|---|---|---|---|---|---|---|---|---|---|---|
| 5-1/8 | 17-7/8 | 22.9 | 4866 | 4057 | 3480 | 3048 | 2708 | 2438 | 2217 | 1935 | 1695 | 1497 | 1331 | 1185 | 1016 | 878 | 764 | 668 | 588 |
| 5-1/8 | 19-1/4 | 24.7 | 5490 | 4542 | 3874 | 3376 | 2992 | 2687 | 2438 | 2231 | 1959 | 1730 | 1538 | 1377 | 1240 | 1097 | 954 | 835 | 735 |
| 5-1/8 | 20-5/8 | 26.4 | 6178 | 5067 | 4295 | 3727 | 3292 | 2948 | 2669 | 2438 | 2241 | 1979 | 1760 | 1575 | 1418 | 1283 | 1166 | 1027 | 904 |
| 5-1/8 | 22 | 28.2 | 6938 | 5638 | 4747 | 4100 | 3608 | 3221 | 2910 | 2653 | 2438 | 2244 | 1996 | 1787 | 1608 | 1455 | 1323 | 1208 | 1097 |
| 5-1/8 | 23-3/8 | 29.9 | 7784 | 6259 | 5233 | 4497 | 3942 | 3509 | 3162 | 2877 | 2639 | 2438 | 2246 | 2011 | 1810 | 1638 | 1489 | 1359 | 1246 |
| 6-3/4 | 6-7/8 | 11.6 | 1190 | 836 | 609 | 458 | 353 | 277 | 222 | 181 | 149 | 124 | 104 | 89 | 76 | 66 | 57 | 50 | 44 |
| 6-3/4 | 8-1/4 | 13.9 | 1914 | 1444 | 1053 | 791 | 609 | 479 | 384 | 312 | 257 | 214 | 181 | 153 | 132 | 114 | 99 | 87 | 76 |
| 6-3/4 | 9-5/8 | 16.2 | 2606 | 2059 | 1668 | 1256 | 968 | 761 | 609 | 495 | 408 | 340 | 287 | 244 | 209 | 181 | 157 | 137 | 121 |
| 6-3/4 | 11 | 18.6 | 3211 | 2689 | 2178 | 1800 | 1444 | 1136 | 909 | 739 | 609 | 508 | 428 | 364 | 312 | 269 | 234 | 205 | 181 |
| 6-3/4 | 12-3/8 | 20.9 | 3752 | 3211 | 2757 | 2278 | 1914 | 1617 | 1295 | 1053 | 868 | 723 | 609 | 518 | 444 | 384 | 334 | 292 | 257 |
| 6-3/4 | 13-3/4 | 23.2 | 4336 | 3689 | 3211 | 2813 | 2363 | 2014 | 1736 | 1444 | 1190 | 992 | 836 | 711 | 609 | 526 | 458 | 401 | 353 |
| 6-3/4 | 15-1/8 | 25.5 | 4969 | 4202 | 3640 | 3211 | 2860 | 2433 | 2090 | 1815 | 1584 | 1320 | 1112 | 946 | 811 | 701 | 609 | 533 | 469 |
| 6-3/4 | 16-1/2 | 27.8 | 5657 | 4752 | 4097 | 3600 | 3211 | 2883 | 2477 | 2150 | 1884 | 1664 | 1444 | 1228 | 1053 | 909 | 791 | 692 | 609 |
| 6-3/4 | 17-7/8 | 30.2 | 6408 | 5344 | 4583 | 4011 | 3567 | 3211 | 2895 | 2513 | 2202 | 1945 | 1730 | 1548 | 1339 | 1156 | 1006 | 880 | 775 |
| 6-3/4 | 19-1/4 | 32.5 | 7231 | 5983 | 5102 | 4447 | 3941 | 3539 | 3211 | 2904 | 2544 | 2247 | 1998 | 1789 | 1610 | 1444 | 1256 | 1099 | 968 |
| 6-3/4 | 20-5/8 | 34.8 | 8137 | 6674 | 5657 | 4909 | 4336 | 3882 | 3515 | 3211 | 2911 | 2570 | 2288 | 2046 | 1842 | 1667 | 1515 | 1352 | 1190 |
| 6-3/4 | 22 | 37.1 | 9138 | 7425 | 6253 | 5400 | 4752 | 4243 | 3832 | 3494 | 3211 | 2915 | 2593 | 2321 | 2089 | 1890 | 1718 | 1569 | 1438 |
| 6-3/4 | 23-3/8 | 39.4 | 10252 | 8243 | 6893 | 5923 | 5192 | 4622 | 4164 | 3789 | 3476 | 3211 | 2918 | 2612 | 2351 | 2128 | 1934 | 1766 | 1618 |
| 6-3/4 | 24-3/4 | 41.8 | 11497 | 9138 | 7583 | 6480 | 5657 | 5020 | 4511 | 4097 | 3752 | 3460 | 3211 | 2920 | 2629 | 2378 | 2162 | 1974 | 1809 |
| 6-3/4 | 26-1/8 | 44.1 | 12898 | 10122 | 8329 | 7076 | 6150 | 5439 | 4875 | 4417 | 4038 | 3719 | 3446 | 3211 | 2921 | 2643 | 2402 | 2193 | 2010 |
| 6-3/4 | 27-1/2 | 46.4 | 14488 | 11208 | 9138 | 7714 | 6674 | 5881 | 5257 | 4752 | 4336 | 3987 | 3689 | 3434 | 3211 | 2921 | 2655 | 2424 | 2221 |

TABLE SPECIFICATIONS: This table applies to straight, simply supported glued laminated timber beams under dry conditions of use. Beams must be laterally supported at the top along the length of the beam and at the top and bottom at the end bearings. The load carrying capacities tabulated are for total load including the weight of the member. A unit weight of 36 pcf was assumed to determine the tabulated plf beam weights.

DESIGN VALUE MODIFICATIONS: The allowable stress in bending, $F_b$, has been modified by the AITC volume factor, $C_v$. For determination of load carrying capacities governed by shear, loads within a distance "d" (the depth of the beam) from the ends have been neglected.

DEFLECTION LIMITS: For floor beams, a deflection limit of l/360 for live load has been used. Live load has been assumed to be 80% of the total load as is common for residential floors. For other ratios of live load to total load, the designer should check deflection based on applicable building code provisions.

CONTROLLING VALUES: Values above and to the right of the heavy line and/or shaded area indicate load capacity controlled by deflection considerations; whereas, values below and to the left of the heavy line and/or shaded area are those load capacities governed by bending.

SPAN: Span is defined as the length from centerline to centerline of bearing. It is this length that is used in standard engineering equations to calculate deflection, bending and shear.

# TABLE 8.9  (Continued)

## Structural Glued Laminated Timber
### Floor Beams

Lamination thickness: 1-3/8 in.

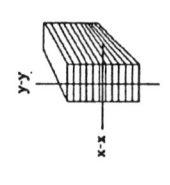

Simple Span Beam Table
For Preliminary Design Purposes

| $F_b$ 2400 psi | $F_v$ 200 psi | $E$ 1.8 x 10⁶ psi | $C_D$ 1.00 | Deflection Limit l/360 |
|---|---|---|---|---|

Beam Capacity: Uniform Load w, plf

| Width b, in. | Depth d, in. | Weight plf | Span, ft 25 | 26 | 27 | 28 | 29 | 30 | 31 | 32 | 33 | 34 | 35 | 36 | 37 | 38 | 39 | 40 |
|---|---|---|---|---|---|---|---|---|---|---|---|---|---|---|---|---|---|---|
| 3-1/8 | 5-1/2 | 4.3 | | | | | | | | | | | | | | | | |
| 3-1/8 | 6-7/8 | 5.4 | | | | | | | | | | | | | | | | |
| 3-1/8 | 8-1/4 | 6.4 | | | | | | | | | | | | | | | | |
| 3-1/8 | 9-5/8 | 7.5 | 50 | 44 | | | | | | | | | | | | | | |
| 3-1/8 | 11 | 8.8 | 74 | 66 | 59 | 53 | 47 | | | | | | | | | | | |
| 3-1/8 | 12-3/8 | 9.7 | 105 | 94 | 84 | 75 | 67 | 61 | 55 | 50 | 46 | | | | | | | |
| 3-1/8 | 13-3/4 | 10.7 | 144 | 128 | 115 | 103 | 93 | 84 | 76 | 69 | 63 | 57 | 53 | 48 | | | | |
| 3-1/8 | 15-1/8 | 11.8 | 192 | 171 | 153 | 137 | 123 | 111 | 101 | 92 | 84 | 76 | 70 | 64 | 59 | 55 | 51 | 47 |
| 3-1/8 | 16-1/2 | 12.9 | 250 | 222 | 198 | 178 | 160 | 144 | 131 | 119 | 109 | 99 | 91 | 84 | 77 | 71 | 66 | 61 |
| 3-1/8 | 17-7/8 | 14.0 | 317 | 282 | 252 | 226 | 203 | 184 | 166 | 151 | 138 | 126 | 116 | 106 | 98 | 90 | 84 | 77 |
| 5-1/8 | 5-1/2 | 7.0 | | | | | | | | | | | | | | | | |
| 5-1/8 | 6-7/8 | 8.8 | | | | | | | | | | | | | | | | |
| 5-1/8 | 8-1/4 | 10.6 | 51 | 45 | | | | | | | | | | | | | | |
| 5-1/8 | 9-5/8 | 12.3 | 81 | 72 | 64 | 58 | 52 | 47 | | | | | | | | | | |
| 5-1/8 | 11 | 14.1 | 121 | 108 | 98 | 86 | 78 | 70 | 64 | 58 | 53 | 48 | | | | | | |
| 5-1/8 | 12-3/8 | 15.9 | 173 | 153 | 137 | 123 | 111 | 100 | 91 | 82 | 75 | 69 | 63 | 58 | 53 | 49 | | |
| 5-1/8 | 13-3/4 | 17.6 | 237 | 211 | 188 | 169 | 152 | 137 | 124 | 113 | 103 | 94 | 86 | 79 | 73 | 67 | 62 | 58 |
| 5-1/8 | 15-1/8 | 19.4 | 315 | 280 | 250 | 224 | 202 | 182 | 165 | 150 | 137 | 125 | 115 | 106 | 97 | 90 | 83 | 77 |
| 5-1/8 | 16-1/2 | 21.1 | 409 | 364 | 325 | 291 | 262 | 237 | 215 | 195 | 178 | 163 | 149 | 137 | 126 | 117 | 108 | 100 |
| 5-1/8 | 17-7/8 | 22.9 | 520 | 463 | 413 | 370 | 333 | 301 | 273 | 248 | 226 | 207 | 190 | 174 | 161 | 148 | 137 | 127 |

| Width | Depth | Weight (plf) | | | | | | | | | | | | | | | | |
|---|---|---|---|---|---|---|---|---|---|---|---|---|---|---|---|---|---|---|
| 5-1/8 | 19-1/4 | 24.7 | 650 | 578 | 516 | 463 | 416 | 376 | 341 | 310 | 283 | 258 | 237 | 218 | 200 | 185 | 171 | 159 |
| 5-1/8 | 20-5/8 | 26.4 | 799 | 711 | 635 | 569 | 512 | 463 | 419 | 381 | 348 | 318 | 291 | 268 | 247 | 228 | 211 | 195 |
| 5-1/8 | 22 | 28.2 | 970 | 862 | 770 | 691 | 622 | 561 | 509 | 463 | 422 | 386 | 354 | 325 | 299 | 278 | 256 | 237 |
| 5-1/8 | 23-3/8 | 29.9 | 1148 | 1034 | 924 | 828 | 746 | 673 | 610 | 555 | 506 | 463 | 424 | 390 | 359 | 331 | 307 | 284 |
| 6-3/4 | 6-7/8 | 11.6 | 67 | 60 | 53 | 48 | | | | | | | | | | | | |
| 6-3/4 | 8-1/4 | 13.9 | 107 | 95 | 85 | 76 | 69 | 62 | 56 | 51 | 47 | | | | | | | |
| 6-3/4 | 9-5/8 | 16.2 | 160 | 142 | 127 | 114 | 102 | 92 | 84 | 76 | 69 | 63 | 58 | 53 | 49 | | | |
| 6-3/4 | 11 | 18.8 | 227 | 202 | 181 | 162 | 146 | 132 | 119 | 108 | 99 | 90 | 83 | 76 | 70 | 65 | 60 | 56 |
| 6-3/4 | 12-3/8 | 20.9 | 312 | 277 | 248 | 222 | 200 | 181 | 164 | 149 | 136 | 124 | 114 | 104 | 96 | 89 | 82 | 76 |
| 6-3/4 | 13-3/4 | 23.2 | 415 | 369 | 330 | 296 | 266 | 240 | 218 | 198 | 181 | 165 | 151 | 139 | 128 | 118 | 109 | 101 |
| 6-3/4 | 15-1/8 | 25.5 | 539 | 479 | 428 | 384 | 345 | 312 | 283 | 257 | 234 | 214 | 196 | 181 | 166 | 153 | 142 | 132 |
| 6-3/4 | 16-1/2 | 27.8 | 685 | 609 | 544 | 488 | 439 | 397 | 359 | 327 | 298 | 272 | 250 | 230 | 211 | 195 | 181 | 167 |
| 6-3/4 | 17-7/8 | 30.2 | 858 | 761 | 680 | 609 | 548 | 495 | 449 | 408 | 372 | 340 | 312 | 287 | 264 | 244 | 225 | 209 |
| 6-3/4 | 19-1/4 | 32.5 | 1053 | 936 | 836 | 749 | 675 | 609 | 552 | 502 | 458 | 419 | 384 | 353 | 325 | 300 | 277 | 257 |
| 6-3/4 | 20-5/8 | 34.8 | 1278 | 1136 | 1014 | 909 | 819 | 739 | 670 | 609 | 556 | 508 | 466 | 428 | 394 | 364 | 337 | 312 |
| 6-3/4 | 22 | 37.1 | 1488 | 1362 | 1217 | 1091 | 982 | 887 | 804 | 731 | 666 | 609 | 559 | 513 | 473 | 436 | 404 | 374 |
| 6-3/4 | 23-3/8 | 39.4 | 1664 | 1535 | 1421 | 1295 | 1166 | 1053 | 954 | 868 | 791 | 723 | 663 | 609 | 561 | 518 | 479 | 444 |
| 6-3/4 | 24-3/4 | 41.8 | 1849 | 1706 | 1579 | 1465 | 1364 | 1238 | 1122 | 1020 | 930 | 851 | 780 | 717 | 660 | 609 | 564 | 522 |
| 6-3/4 | 26-1/8 | 44.1 | 2043 | 1885 | 1745 | 1620 | 1507 | 1406 | 1309 | 1190 | 1085 | 992 | 909 | 836 | 770 | 711 | 657 | 609 |
| 6-3/4 | 27-1/2 | 46.4 | | | | | | | | | | | | | | | | |

TABLE SPECIFICATIONS: This table applies to straight, simply supported glued laminated timber beams under dry conditions of use. Beams must be laterally supported at the top along the length of the beam and at the top and bottom at the end bearings. The load carrying capacities tabulated are for total load including the weight of the member. A unit weight of 36 pcf was assumed to determine the tabulated plf beam weights.

DESIGN VALUE MODIFICATIONS: The allowable stress in bending, $F_b$, has been modified by the AITC volume factor, $C_v$. For determination of load carrying capacities governed by shear, loads within a distance "d" (the depth of the beam) from the ends have been neglected.

DEFLECTION LIMITS: For floor beams, a deflection limit of l/360 for live load has been used. Live load has been assumed to be 80% of the total load as is common for residential floors. For other ratios of live load to total load, the designer should check deflection based on applicable building code provisions.

CONTROLLING VALUES: Values above and to the right of the heavy line and/or shaded area indicate load capacity controlled by deflection considerations; whereas, values below and to the left of the heavy line and/or shaded area represent load capacity controlled by shear. Values in the shaded area are those load capacities governed by bending.

SPAN: Span is defined as the length from centerline to centerline of bearing. It is this length that is used in standard engineering equations to calculate deflection, bending and shear.

## TABLE 8.10

### Structural Glued Laminated Timber Cantilever Beam Span-Load Table[a]

| Span[b], ft. | Dead Load[c], psf | Live Load, psf | TWO-SPAN SYSTEM[d] Suspended Beam | TWO-SPAN SYSTEM[d] Cantilevered Beam | THREE-SPAN SYSTEM[e] Suspended Beam | THREE-SPAN SYSTEM[e] Cantilevered Beam | THREE-SPAN SYSTEM[f] Suspended Beam | THREE-SPAN SYSTEM[f] Cantilevered Beam | THREE-SPAN SYSTEM[f] Double Cantilevered Beam |
|---|---|---|---|---|---|---|---|---|---|
| 32 | 10 | 12 | $*5\frac{1}{8} \times 16\frac{1}{2}$ | $5\frac{1}{8} \times 16\frac{1}{2}$ | $*5\frac{1}{8} \times 10\frac{1}{2}$ | $5\frac{1}{8} \times 16\frac{1}{2}$ | $*5\frac{1}{8} \times 16\frac{1}{2}$ | $5\frac{1}{8} \times 16\frac{1}{2}$ | $5\frac{1}{8} \times 16\frac{1}{2}$ |
|  |  | 20 | $5\frac{1}{8} \times 18$ | $5\frac{1}{8} \times 21$ | $*5\frac{1}{8} \times 12$ | $5\frac{1}{8} \times 21$ | $5\frac{1}{8} \times 18$ | $5\frac{1}{8} \times 21$ | $5\frac{1}{8} \times 19\frac{1}{2}$ |
|  |  | 30 | $5\frac{1}{8} \times 19\frac{1}{2}$ | $5\frac{1}{8} \times 25\frac{1}{2}$ | $*5\frac{1}{8} \times 13\frac{1}{2}$ | $5\frac{1}{8} \times 24$ | $5\frac{1}{8} \times 21$ | $5\frac{1}{8} \times 24$ | $5\frac{1}{8} \times 24$ |
|  | 12 | 12 | $*5\frac{1}{8} \times 16\frac{1}{2}$ | $5\frac{1}{8} \times 16\frac{1}{2}$ | $*5\frac{1}{8} \times 10\frac{1}{2}$ | $5\frac{1}{8} \times 16\frac{1}{2}$ | $*5\frac{1}{8} \times 16\frac{1}{2}$ | $5\frac{1}{8} \times 16\frac{1}{2}$ | $5\frac{1}{8} \times 16\frac{1}{2}$ |
|  |  | 20 | $5\frac{1}{8} \times 18$ | $5\frac{1}{8} \times 21$ | $*5\frac{1}{8} \times 12$ | $5\frac{1}{8} \times 21$ | $5\frac{1}{8} \times 18$ | $5\frac{1}{8} \times 21$ | $5\frac{1}{8} \times 19\frac{1}{2}$ |
|  |  | 30 | $5\frac{1}{8} \times 21$ | $5\frac{1}{8} \times 27$ | $*5\frac{1}{8} \times 13\frac{1}{2}$ | $5\frac{1}{8} \times 24$ | $5\frac{1}{8} \times 21$ | $5\frac{1}{8} \times 24$ | $5\frac{1}{8} \times 25\frac{1}{2}$ |
|  | 15 | 20 | $5\frac{1}{8} \times 19\frac{1}{2}$ | $5\frac{1}{8} \times 22\frac{1}{2}$ | $*5\frac{1}{8} \times 12$ | $5\frac{1}{8} \times 21$ | $5\frac{1}{8} \times 19\frac{1}{2}$ | $5\frac{1}{8} \times 21$ | $5\frac{1}{8} \times 21$ |
|  |  | 30 | $5\frac{1}{8} \times 21$ | $5\frac{1}{8} \times 28\frac{1}{2}$ | $*5\frac{1}{8} \times 13\frac{1}{2}$ | $5\frac{1}{8} \times 25\frac{1}{2}$ | $5\frac{1}{8} \times 22\frac{1}{2}$ | $5\frac{1}{8} \times 25\frac{1}{2}$ | $5\frac{1}{8} \times 27$ |
| 36 | 10 | 12 | $5\frac{1}{8} \times 16\frac{1}{2}$ | $5\frac{1}{8} \times 18$ | $*5\frac{1}{8} \times 12$ | $5\frac{1}{8} \times 18$ | $5\frac{1}{8} \times 16\frac{1}{2}$ | $5\frac{1}{8} \times 18$ | $5\frac{1}{8} \times 16\frac{1}{2}$ |
|  |  | 20 | $5\frac{1}{8} \times 19\frac{1}{2}$ | $5\frac{1}{8} \times 22\frac{1}{2}$ | $*5\frac{1}{8} \times 13\frac{1}{2}$ | $5\frac{1}{8} \times 22\frac{1}{2}$ | $5\frac{1}{8} \times 21$ | $5\frac{1}{8} \times 22\frac{1}{2}$ | $5\frac{1}{8} \times 21$ |
|  |  | 30 | $5\frac{1}{8} \times 22\frac{1}{2}$ | $5\frac{1}{8} \times 28\frac{1}{2}$ | $5\frac{1}{8} \times 15$ | $5\frac{1}{8} \times 27$ | $5\frac{1}{8} \times 24$ | $5\frac{1}{8} \times 27$ | $5\frac{1}{8} \times 27$ |
|  | 12 | 12 | $5\frac{1}{8} \times 16\frac{1}{2}$ | $5\frac{1}{8} \times 19\frac{1}{2}$ | $*5\frac{1}{8} \times 12$ | $5\frac{1}{8} \times 19\frac{1}{2}$ | $5\frac{1}{8} \times 18$ | $5\frac{1}{8} \times 19\frac{1}{2}$ | $5\frac{1}{8} \times 18$ |
|  |  | 20 | $5\frac{1}{8} \times 19\frac{1}{2}$ | $5\frac{1}{8} \times 24$ | $*5\frac{1}{8} \times 13\frac{1}{2}$ | $5\frac{1}{8} \times 24$ | $5\frac{1}{8} \times 21$ | $5\frac{1}{8} \times 24$ | $5\frac{1}{8} \times 21$ |
|  |  | 30 | $5\frac{1}{8} \times 22\frac{1}{2}$ | $5\frac{1}{8} \times 30$ | $*5\frac{1}{8} \times 15$ | $5\frac{1}{8} \times 28\frac{1}{2}$ | $5\frac{1}{8} \times 24$ | $5\frac{1}{8} \times 28\frac{1}{2}$ | $5\frac{1}{8} \times 28\frac{1}{2}$ |

| | | | | | | | |
|---|---|---|---|---|---|---|---|
| 15 | | 20 | $5\frac{1}{8} \times 21$ | $5\frac{1}{8} \times 25\frac{1}{2}$ | $*5\frac{1}{8} \times 13\frac{1}{2}$ | $5\frac{1}{8} \times 24$ | $5\frac{1}{8} \times 22\frac{1}{2}$ | $5\frac{1}{8} \times 24$ |
| | | 30 | $5\frac{1}{8} \times 24$ | $5\frac{1}{8} \times 33$ | $*5\frac{1}{8} \times 16\frac{1}{2}$ | $5\frac{1}{8} \times 28\frac{1}{2}$ | $5\frac{1}{8} \times 25\frac{1}{2}$ | $5\frac{1}{8} \times 31\frac{1}{2}$ |
| 10 | | 12 | $5\frac{1}{8} \times 18$ | $5\frac{1}{8} \times 21$ | $*5\frac{1}{8} \times 12$ | $5\frac{1}{8} \times 21$ | $5\frac{1}{8} \times 19\frac{1}{2}$ | $5\frac{1}{8} \times 19\frac{1}{2}$ |
| | | 20 | $5\frac{1}{8} \times 22\frac{1}{2}$ | $5\frac{1}{8} \times 25\frac{1}{2}$ | $*5\frac{1}{8} \times 15$ | $5\frac{1}{8} \times 25\frac{1}{2}$ | $5\frac{1}{8} \times 22\frac{1}{2}$ | $5\frac{1}{8} \times 24$ |
| | | 30 | $5\frac{1}{8} \times 25\frac{1}{2}$ | $5\frac{1}{8} \times 31\frac{1}{2}$ | $*5\frac{1}{8} \times 16\frac{1}{2}$ | $5\frac{1}{8} \times 30$ | $5\frac{1}{8} \times 27$ | $5\frac{1}{8} \times 30$ |
| 12 | 40 | 12 | $5\frac{1}{8} \times 19\frac{1}{2}$ | $5\frac{1}{8} \times 21$ | $*5\frac{1}{8} \times 12$ | $5\frac{1}{8} \times 21$ | $5\frac{1}{8} \times 19\frac{1}{2}$ | $5\frac{1}{8} \times 19\frac{1}{2}$ |
| | | 20 | $5\frac{1}{8} \times 22\frac{1}{2}$ | $5\frac{1}{8} \times 27$ | $*5\frac{1}{8} \times 15$ | $5\frac{1}{8} \times 27$ | $5\frac{1}{8} \times 24$ | $5\frac{1}{8} \times 24$ |
| | | 30 | $5\frac{1}{8} \times 25\frac{1}{2}$ | $5\frac{1}{8} \times 33$ | $*5\frac{1}{8} \times 18$ | $5\frac{1}{8} \times 31\frac{1}{2}$ | $5\frac{1}{8} \times 27$ | $5\frac{1}{8} \times 31\frac{1}{2}$ |
| 15 | | 20 | $5\frac{1}{8} \times 24$ | $5\frac{1}{8} \times 28\frac{1}{2}$ | $*5\frac{1}{8} \times 16\frac{1}{2}$ | $5\frac{1}{8} \times 27$ | $5\frac{1}{8} \times 24$ | $5\frac{1}{8} \times 27$ |
| | | 30 | $5\frac{1}{8} \times 27$ | $5\frac{1}{8} \times 36$ | $5\frac{1}{8} \times 18$ | $5\frac{1}{8} \times 31\frac{1}{2}$ | $5\frac{1}{8} \times 28\frac{1}{2}$ | $5\frac{1}{8} \times 34\frac{1}{2}$ |
| 10 | | 12 | $5\frac{1}{8} \times 19\frac{1}{2}$ | $5\frac{1}{8} \times 22\frac{1}{2}$ | $*5\frac{1}{8} \times 13\frac{1}{2}$ | $5\frac{1}{8} \times 22\frac{1}{2}$ | $5\frac{1}{8} \times 21$ | $5\frac{1}{8} \times 21$ |
| | | 20 | $5\frac{1}{8} \times 24$ | $5\frac{1}{8} \times 28\frac{1}{2}$ | $*5\frac{1}{8} \times 16\frac{1}{2}$ | $5\frac{1}{8} \times 28\frac{1}{2}$ | $5\frac{1}{8} \times 25\frac{1}{2}$ | $5\frac{1}{8} \times 27$ |
| | | 30 | $5\frac{1}{8} \times 27$ | $5\frac{1}{8} \times 34\frac{1}{2}$ | $5\frac{1}{8} \times 16\frac{1}{2}$ | $5\frac{1}{8} \times 34\frac{1}{2}$ | $5\frac{1}{8} \times 30$ | $5\frac{1}{8} \times 33$ |
| 12 | 44 | 12 | $5\frac{1}{8} \times 21$ | $5\frac{1}{8} \times 24$ | $*5\frac{1}{8} \times 13\frac{1}{2}$ | $5\frac{1}{8} \times 24$ | $5\frac{1}{8} \times 22\frac{1}{2}$ | $5\frac{1}{8} \times 22\frac{1}{2}$ |
| | | 20 | $5\frac{1}{8} \times 24$ | $5\frac{1}{8} \times 30$ | $*5\frac{1}{8} \times 16\frac{1}{2}$ | $5\frac{1}{8} \times 28\frac{1}{2}$ | $5\frac{1}{8} \times 25\frac{1}{2}$ | $5\frac{1}{8} \times 27$ |
| | | 30 | $5\frac{1}{8} \times 28\frac{1}{2}$ | $5\frac{1}{8} \times 36$ | $5\frac{1}{8} \times 18$ | $5\frac{1}{8} \times 34\frac{1}{2}$ | $5\frac{1}{8} \times 30$ | $5\frac{1}{8} \times 34\frac{1}{2}$ |
| 15 | | 20 | $5\frac{1}{8} \times 25\frac{1}{2}$ | $5\frac{1}{8} \times 31\frac{1}{2}$ | $5\frac{1}{8} \times 16\frac{1}{2}$ | $5\frac{1}{8} \times 30$ | $5\frac{1}{8} \times 27$ | $5\frac{1}{8} \times 30$ |
| | | 30 | $5\frac{1}{8} \times 30$ | $6\frac{3}{4} \times 31\frac{1}{2}$ | $5\frac{1}{8} \times 18$ | $5\frac{1}{8} \times 36$ | $5\frac{1}{8} \times 31\frac{1}{2}$ | $6\frac{3}{4} \times 30$ |
| 10 | | 12 | $5\frac{1}{8} \times 22\frac{1}{2}$ | $5\frac{1}{8} \times 25\frac{1}{2}$ | $*5\frac{1}{8} \times 15$ | $5\frac{1}{8} \times 25\frac{1}{2}$ | $5\frac{1}{8} \times 22\frac{1}{2}$ | $5\frac{1}{8} \times 24$ |
| | | 20 | $5\frac{1}{8} \times 27$ | $5\frac{1}{8} \times 31\frac{1}{2}$ | $5\frac{1}{8} \times 16\frac{1}{2}$ | $5\frac{1}{8} \times 31\frac{1}{2}$ | $5\frac{1}{8} \times 28\frac{1}{2}$ | $5\frac{1}{8} \times 30$ |
| | | 30 | $5\frac{1}{8} \times 31\frac{1}{2}$ | $6\frac{3}{4} \times 33$ | $5\frac{1}{8} \times 18$ | $6\frac{3}{4} \times 33$ | $5\frac{1}{8} \times 33$ | $5\frac{1}{8} \times 36$ |

**TABLE 8.10** (*Continued*)

| Span[b], ft. | Dead Load[c], psf | Live Load, psf | TWO-SPAN SYSTEM[a] | | THREE-SPAN SYSTEM[e] | | THREE-SPAN SYSTEM[f] | |
|---|---|---|---|---|---|---|---|---|
| | | | Suspended Beam | Cantilevered Beam | Suspended Beam | Cantilevered Beam | Suspended Beam | Double Cantilevered Beam |
| 48 | 12 | 12 | $5\frac{1}{8} \times 22\frac{1}{2}$ | $5\frac{1}{8} \times 25\frac{1}{2}$ | *$5\frac{1}{8} \times 15$ | $5\frac{1}{8} \times 27$ | $5\frac{1}{8} \times 24$ | $5\frac{1}{8} \times 24$ |
| | | 20 | $5\frac{1}{8} \times 27$ | $5\frac{1}{8} \times 33$ | $5\frac{1}{8} \times 16\frac{1}{2}$ | $5\frac{1}{8} \times 33$ | $5\frac{1}{8} \times 28\frac{1}{2}$ | $5\frac{1}{8} \times 30$ |
| | | 30 | $5\frac{1}{8} \times 31\frac{1}{2}$ | $6\frac{3}{4} \times 33$ | $5\frac{1}{8} \times 19\frac{1}{2}$ | $6\frac{3}{4} \times 33$ | $5\frac{1}{8} \times 33$ | $6\frac{3}{4} \times 31\frac{1}{2}$ |
| | 15 | 20 | $5\frac{1}{8} \times 28\frac{1}{2}$ | $5\frac{1}{8} \times 34\frac{1}{2}$ | $5\frac{1}{8} \times 18$ | $5\frac{1}{8} \times 33$ | $5\frac{1}{8} \times 30$ | $5\frac{1}{8} \times 33$ |
| | | 30 | $5\frac{1}{8} \times 33$ | $6\frac{3}{4} \times 34\frac{1}{2}$ | $5\frac{1}{8} \times 19\frac{1}{2}$ | $6\frac{3}{4} \times 34\frac{1}{2}$ | $5\frac{1}{8} \times 34\frac{1}{2}$ | $6\frac{3}{4} \times 31\frac{1}{2}$ |
| | 10 | 12 | $5\frac{1}{8} \times 24$ | $5\frac{1}{8} \times 27$ | *$5\frac{1}{8} \times 16\frac{1}{2}$ | $5\frac{1}{8} \times 28\frac{1}{2}$ | $5\frac{1}{8} \times 25\frac{1}{2}$ | $5\frac{1}{8} \times 25\frac{1}{2}$ |
| | | 20 | $5\frac{1}{8} \times 30$ | $6\frac{3}{4} \times 30$ | $5\frac{1}{8} \times 18$ | $6\frac{3}{4} \times 31\frac{1}{2}$ | $5\frac{1}{8} \times 30$ | $5\frac{1}{8} \times 31\frac{1}{2}$ |
| | | 30 | $5\frac{1}{8} \times 34\frac{1}{2}$ | $6\frac{3}{4} \times 36$ | $5\frac{1}{8} \times 19\frac{1}{2}$ | $6\frac{3}{4} \times 36$ | $5\frac{1}{8} \times 36$ | $6\frac{3}{4} \times 34\frac{1}{2}$ |
| 52 | 12 | 12 | $5\frac{1}{8} \times 25\frac{1}{2}$ | $5\frac{1}{8} \times 28\frac{1}{2}$ | *$5\frac{1}{8} \times 16\frac{1}{2}$ | $5\frac{1}{8} \times 28\frac{1}{2}$ | $5\frac{1}{8} \times 25\frac{1}{2}$ | $5\frac{1}{8} \times 27$ |
| | | 20 | $5\frac{1}{8} \times 30$ | $6\frac{3}{4} \times 31\frac{1}{2}$ | $5\frac{1}{8} \times 18$ | $6\frac{3}{4} \times 31\frac{1}{2}$ | $5\frac{1}{8} \times 31\frac{1}{2}$ | $5\frac{1}{8} \times 33$ |
| | | 30 | $5\frac{1}{8} \times 34\frac{1}{2}$ | $6\frac{3}{4} \times 36$ | $5\frac{1}{8} \times 21$ | $6\frac{3}{4} \times 37\frac{1}{2}$ | $5\frac{1}{8} \times 36$ | $6\frac{3}{4} \times 34\frac{1}{2}$ |
| | 15 | 20 | $5\frac{1}{8} \times 31\frac{1}{2}$ | $6\frac{3}{4} \times 33$ | $5\frac{1}{8} \times 19\frac{1}{2}$ | $6\frac{3}{4} \times 33$ | $5\frac{1}{8} \times 33$ | $5\frac{1}{8} \times 36$ |
| | | 30 | $5\frac{1}{8} \times 36$ | $6\frac{3}{4} \times 37\frac{1}{2}$ | $5\frac{1}{8} \times 21$ | $6\frac{3}{4} \times 37\frac{1}{2}$ | $6\frac{3}{4} \times 33$ | $6\frac{3}{4} \times 34\frac{1}{2}$ |
| | 10 | 12 | $5\frac{1}{8} \times 25\frac{1}{2}$ | $5\frac{1}{8} \times 30$ | *$5\frac{1}{8} \times 16\frac{1}{2}$ | $5\frac{1}{8} \times 30$ | $5\frac{1}{8} \times 27$ | $5\frac{1}{8} \times 27$ |
| | | 20 | $5\frac{1}{8} \times 31\frac{1}{2}$ | $6\frac{3}{4} \times 33$ | $5\frac{1}{8} \times 19\frac{1}{2}$ | $6\frac{3}{4} \times 33$ | $5\frac{1}{8} \times 33$ | $5\frac{1}{8} \times 34\frac{1}{2}$ |
| | | 30 | $6\frac{3}{4} \times 33$ | $6\frac{3}{4} \times 39$ | $5\frac{1}{8} \times 22\frac{1}{2}$ | $6\frac{3}{4} \times 39$ | $6\frac{3}{4} \times 34\frac{1}{2}$ | $6\frac{3}{4} \times 37\frac{1}{2}$ |

| Span | Spacing | Load | C1 | C2 | C3 | C4 | C5 | C6 |
|---|---|---|---|---|---|---|---|---|
| 56 | 12 | 12 | $5\frac{1}{8} \times 27$ | $5\frac{1}{8} \times 31\frac{1}{2}$ | $5\frac{1}{8} \times 16\frac{1}{2}$ | $5\frac{1}{8} \times 31\frac{1}{2}$ | $5\frac{1}{8} \times 28\frac{1}{2}$ | $5\frac{1}{8} \times 28\frac{1}{2}$ |
|  |  | 20 | $5\frac{1}{8} \times 33$ | $6\frac{3}{4} \times 34\frac{1}{2}$ | $5\frac{1}{8} \times 19\frac{1}{2}$ | $6\frac{3}{4} \times 34\frac{1}{2}$ | $5\frac{1}{8} \times 34\frac{1}{2}$ | $5\frac{1}{8} \times 34\frac{1}{2}$ |
|  |  | 30 | $6\frac{3}{4} \times 33$ | $6\frac{3}{4} \times 39$ | $5\frac{1}{8} \times 22\frac{1}{2}$ | $6\frac{3}{4} \times 40\frac{1}{2}$ | $6\frac{3}{4} \times 34\frac{1}{2}$ | $6\frac{3}{4} \times 37\frac{1}{2}$ |
|  | 15 | 20 | $5\frac{1}{8} \times 28\frac{1}{2}$ | $5\frac{1}{8} \times 33$ | $5\frac{1}{8} \times 21$ | $6\frac{3}{4} \times 36$ | $5\frac{1}{8} \times 36$ | $6\frac{3}{4} \times 31\frac{1}{2}$ |
|  |  | 30 | $5\frac{1}{8} \times 34\frac{1}{2}$ | $6\frac{3}{4} \times 36$ | $5\frac{1}{8} \times 24$ | $6\frac{3}{4} \times 40\frac{1}{2}$ | $6\frac{3}{4} \times 36$ | $6\frac{3}{4} \times 37\frac{1}{2}$ |
| 60 | 10 | 12 | $5\frac{1}{8} \times 30$ | $5\frac{1}{8} \times 33$ | $5\frac{1}{8} \times 16\frac{1}{2}$ | $5\frac{1}{8} \times 33$ | $5\frac{1}{8} \times 30$ | $5\frac{1}{8} \times 30$ |
|  |  | 20 | $6\frac{3}{4} \times 33$ | $6\frac{3}{4} \times 37\frac{1}{2}$ | $5\frac{1}{8} \times 21$ | $6\frac{3}{4} \times 36$ | $5\frac{1}{8} \times 36$ | $6\frac{3}{4} \times 33$ |
|  |  | 30 | $6\frac{3}{4} \times 36$ | $8\frac{3}{4} \times 39$ | $5\frac{1}{8} \times 24$ | $6\frac{3}{4} \times 42$ | $6\frac{3}{4} \times 36$ | $6\frac{3}{4} \times 40\frac{1}{2}$ |
|  | 12 | 12 | $5\frac{1}{8} \times 30$ | $5\frac{1}{8} \times 36$ | $5\frac{1}{8} \times 18$ | $5\frac{1}{8} \times 34\frac{1}{2}$ | $5\frac{1}{8} \times 30$ | $5\frac{1}{8} \times 31\frac{1}{2}$ |
|  |  | 20 | $6\frac{3}{4} \times 33$ | $6\frac{3}{4} \times 39$ | $5\frac{1}{8} \times 21$ | $6\frac{3}{4} \times 37\frac{1}{2}$ | $6\frac{3}{4} \times 33$ | $6\frac{3}{4} \times 33$ |
|  |  | 30 | $6\frac{3}{4} \times 37\frac{1}{2}$ | $8\frac{3}{4} \times 40\frac{1}{2}$ | $5\frac{1}{8} \times 24$ | $8\frac{3}{4} \times 39$ | $6\frac{3}{4} \times 37\frac{1}{2}$ | $6\frac{3}{4} \times 40\frac{1}{2}$ |
|  | 15 | 20 | $6\frac{3}{4} \times 33$ | $6\frac{3}{4} \times 37\frac{1}{2}$ | $5\frac{1}{8} \times 22\frac{1}{2}$ | $6\frac{3}{4} \times 39$ | $6\frac{3}{4} \times 34\frac{1}{2}$ | $6\frac{3}{4} \times 34\frac{1}{2}$ |
|  |  | 30 | $6\frac{3}{4} \times 37\frac{1}{2}$ | $8\frac{3}{4} \times 39$ | $5\frac{1}{8} \times 25\frac{1}{2}$ | $8\frac{3}{4} \times 39$ | $6\frac{3}{4} \times 39$ | $8\frac{3}{4} \times 36$ |
| 64 | 10 | 12 | $5\frac{1}{8} \times 30$ | $5\frac{1}{8} \times 34\frac{1}{2}$ | $5\frac{1}{8} \times 18$ | $5\frac{1}{8} \times 34\frac{1}{2}$ | $5\frac{1}{8} \times 31\frac{1}{2}$ | $5\frac{1}{8} \times 31\frac{1}{2}$ |
|  |  | 20 | $6\frac{3}{4} \times 33$ | $6\frac{3}{4} \times 39$ | $5\frac{1}{8} \times 22\frac{1}{2}$ | $6\frac{3}{4} \times 39$ | $6\frac{3}{4} \times 34\frac{1}{2}$ | $6\frac{3}{4} \times 36$ |
|  |  | 30 | $6\frac{3}{4} \times 37\frac{1}{2}$ | $8\frac{3}{4} \times 39$ | $5\frac{1}{8} \times 25\frac{1}{2}$ | $8\frac{3}{4} \times 40\frac{1}{2}$ | $6\frac{3}{4} \times 39$ | $8\frac{3}{4} \times 37\frac{1}{2}$ |
|  | 12 | 12 | $5\frac{1}{8} \times 31\frac{1}{2}$ | $5\frac{1}{8} \times 36$ | $5\frac{1}{8} \times 19\frac{1}{2}$ | $5\frac{1}{8} \times 36$ | $5\frac{1}{8} \times 33$ | $5\frac{1}{8} \times 33$ |
|  |  | 20 | $6\frac{3}{4} \times 34\frac{1}{2}$ | $6\frac{3}{4} \times 40\frac{1}{2}$ | $5\frac{1}{8} \times 22\frac{1}{2}$ | $6\frac{3}{4} \times 40\frac{1}{2}$ | $6\frac{3}{4} \times 36$ | $6\frac{3}{4} \times 36$ |
|  |  | 30 | $6\frac{3}{4} \times 39$ | $8\frac{3}{4} \times 42$ | $5\frac{1}{8} \times 25\frac{1}{2}$ | $8\frac{3}{4} \times 42$ | $6\frac{3}{4} \times 40\frac{1}{2}$ | $8\frac{3}{4} \times 39$ |
|  | 15 | 20 | $6\frac{3}{4} \times 36$ | $6\frac{3}{4} \times 42$ | $5\frac{1}{8} \times 24$ | $6\frac{3}{4} \times 42$ | $6\frac{3}{4} \times 37\frac{1}{2}$ | $6\frac{3}{4} \times 37\frac{1}{2}$ |
|  |  | 30 | $6\frac{3}{4} \times 40\frac{1}{2}$ | $8\frac{3}{4} \times 42$ | $5\frac{1}{8} \times 27$ | $8\frac{3}{4} \times 42$ | $6\frac{3}{4} \times 42$ | $8\frac{3}{4} \times 39$ |

*Footnotes to Table 8.10*

[a]This preliminary beam design table applies for straight, cantilevered and laminated timber roof beams manufactured from Douglas Fir-Larch. Member sizes are governed by either bending or shear. **Where building code requirements apply, the member size must be checked.** A minimum roof slope of $\frac{1}{4}$ʺ per foot is required to provide proper drainage. Beams are sized for vertical loads only and no wind or earthquake loading have been included in the analysis. Lateral bracing in areas of negative moment is also required.

Specifications and design values:

Beam spacing: 20'-0" (20' bays)

Bending, $F_b$ = 2,400 psi except those marked * in which cases $F_b$ = 2,000 psi. Tabular values are based on modifying allowable bending stresses by volume factor, $C_V$.

Shear parallel to the grain, $F_v$ = 165 psi.

Compression perpendicular to grain, $F_{c\perp}$ = 560 psi.

Duration of load factor: $C_D$ = 1.25 for 12 psf construction loads, and $C_D$ = 1.15 for 20 and 30 psf snow loads.

Member sizes are checked for full and unbalanced live loading.

[b]Main supports are columns or bearing walls. Beam sizes are based on spans as shown in the Table.

[c]Glulam beam weights have been included in the analysis; however, the dead load value does not include the beam weight.

[d]Two-span cantilever system: Cantilevered beam extends over the center support with the length of cantilever, L' equal to approximately 0.20 multiplied by the span.

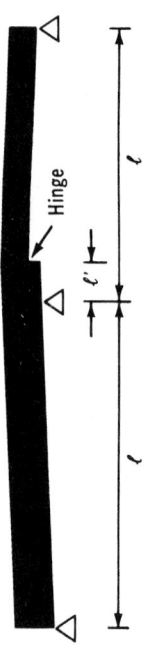

<sup>e</sup>Three-span cantilever system: End members are cantilevered over the intermediate column supports and carry the suspended beam. Length of the cantilever, L', is equal to approximately 0.25 multiplied by the span.

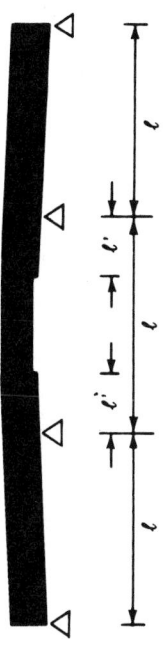

<sup>f</sup>Three-span cantilever system: The center member cantilevers over the intermediate column supports and carrys the suspended wall beams. The length of cantilevers, L', is equal to approximately 0.17 multiplied by the span.

TABLE 8.11

## Structural Glued Laminated Timber Three-Hinged Tudor Arch Span-Load Table[a]

| Loading | Roof Pitch | Wall Ht (ft) | 30' Span Width | Base | Lower Tang. | Upper Tang. | Crown | 35' Span Width | Base | Lower Tang. | Upper Tang. | Crown |
|---|---|---|---|---|---|---|---|---|---|---|---|---|
| Vertical dead + live load = 400 lb/ft | 3/12 | 10 | $3\frac{3}{8}$ | $8\frac{1}{4}$ | 12 | $10\frac{3}{4}$ | $7\frac{1}{2}$ | $3\frac{3}{8}$ | $10\frac{1}{2}$ | $13\frac{1}{4}$ | 12 | $7\frac{1}{2}$ |
| | | 12 | $5\frac{5}{8}$ | $7\frac{1}{2}$ | 11 | $10\frac{3}{4}$ | $7\frac{1}{2}$ | $5\frac{5}{8}$ | $7\frac{1}{2}$ | 12 | 12 | $7\frac{1}{2}$ |
| | | 14 | $5\frac{5}{8}$ | $7\frac{1}{2}$ | 12 | 12 | $7\frac{1}{2}$ | $5\frac{5}{8}$ | $7\frac{1}{2}$ | $13\frac{1}{2}$ | $13\frac{1}{4}$ | $7\frac{1}{2}$ |
| | | 16 | $5\frac{5}{8}$ | $7\frac{1}{2}$ | $13\frac{1}{4}$ | 13 | $7\frac{1}{2}$ | $5\frac{5}{8}$ | $7\frac{1}{2}$ | $14\frac{3}{4}$ | $14\frac{1}{2}$ | $7\frac{1}{2}$ |
| | | 18 | $5\frac{5}{8}$ | $7\frac{1}{2}$ | $14\frac{1}{4}$ | $14\frac{1}{4}$ | $7\frac{1}{2}$ | $5\frac{5}{8}$ | $7\frac{1}{2}$ | 16 | $15\frac{3}{4}$ | $7\frac{1}{2}$ |
| | 4/12 | 10 | $3\frac{3}{8}$ | $7\frac{1}{2}$ | $11\frac{3}{4}$ | $12\frac{3}{4}$ | $7\frac{1}{2}$ | $3\frac{3}{8}$ | $9\frac{3}{4}$ | $13\frac{1}{4}$ | $12\frac{3}{4}$ | $7\frac{1}{2}$ |
| | | 12 | $5\frac{1}{8}$ | $7\frac{1}{2}$ | $10\frac{3}{4}$ | $10\frac{3}{4}$ | $7\frac{1}{2}$ | $5\frac{1}{8}$ | $7\frac{1}{2}$ | $11\frac{3}{4}$ | $11\frac{3}{4}$ | $7\frac{1}{2}$ |
| | | 14 | $5\frac{1}{8}$ | $7\frac{1}{2}$ | 12 | 12 | $7\frac{1}{2}$ | $5\frac{1}{8}$ | $7\frac{1}{2}$ | $13\frac{1}{4}$ | 13 | $7\frac{1}{2}$ |
| | | 16 | $5\frac{1}{8}$ | $7\frac{1}{2}$ | $13\frac{1}{4}$ | 13 | $7\frac{1}{2}$ | $5\frac{1}{8}$ | $7\frac{1}{2}$ | $14\frac{1}{4}$ | $14\frac{1}{2}$ | $7\frac{1}{2}$ |
| | | 18 | $5\frac{1}{8}$ | $7\frac{1}{2}$ | $14\frac{1}{4}$ | 14 | $7\frac{1}{2}$ | $5\frac{1}{8}$ | $7\frac{1}{2}$ | $15\frac{3}{4}$ | $15\frac{1}{2}$ | $7\frac{1}{2}$ |
| | 6/12 | 12 | $5\frac{5}{8}$ | $7\frac{1}{2}$ | $10\frac{1}{2}$ | $10\frac{1}{2}$ | $7\frac{1}{2}$ | $5\frac{5}{8}$ | $7\frac{1}{2}$ | $11\frac{1}{2}$ | $11\frac{1}{4}$ | $7\frac{1}{2}$ |
| | | 14 | $5\frac{5}{8}$ | $7\frac{1}{2}$ | $11\frac{3}{4}$ | $11\frac{3}{4}$ | $7\frac{1}{2}$ | $5\frac{5}{8}$ | $7\frac{1}{2}$ | $12\frac{3}{4}$ | $12\frac{3}{4}$ | $7\frac{1}{2}$ |
| | | 16 | $5\frac{5}{8}$ | $7\frac{1}{2}$ | 13 | $12\frac{3}{4}$ | $7\frac{1}{2}$ | $5\frac{5}{8}$ | $7\frac{1}{2}$ | 14 | 14 | $7\frac{1}{2}$ |
| | | 18 | $5\frac{5}{8}$ | $7\frac{1}{2}$ | 14 | 14 | $7\frac{1}{2}$ | $5\frac{5}{8}$ | $7\frac{1}{2}$ | $15\frac{1}{4}$ | $15\frac{1}{4}$ | $7\frac{1}{2}$ |
| | 8/12 | 12 | $5\frac{1}{8}$ | $7\frac{1}{2}$ | $10\frac{1}{4}$ | 10 | $7\frac{1}{2}$ | $5\frac{1}{8}$ | $7\frac{1}{2}$ | 11 | $10\frac{1}{2}$ | $7\frac{1}{2}$ |
| | | 14 | $5\frac{1}{8}$ | $7\frac{1}{2}$ | $11\frac{1}{2}$ | $11\frac{1}{4}$ | $7\frac{1}{2}$ | $5\frac{1}{8}$ | $7\frac{1}{2}$ | $12\frac{1}{4}$ | 12 | $7\frac{1}{2}$ |
| | | 16 | $5\frac{1}{8}$ | $7\frac{1}{2}$ | $12\frac{3}{4}$ | $12\frac{1}{2}$ | $7\frac{1}{2}$ | $5\frac{1}{8}$ | $7\frac{1}{2}$ | $13\frac{1}{2}$ | $13\frac{1}{2}$ | $7\frac{1}{2}$ |
| | | 18 | $5\frac{1}{8}$ | $7\frac{1}{2}$ | $13\frac{3}{4}$ | $13\frac{1}{2}$ | $7\frac{1}{2}$ | $5\frac{1}{8}$ | $7\frac{1}{2}$ | $14\frac{3}{4}$ | $14\frac{3}{4}$ | $7\frac{1}{2}$ |
| Vertical dead + live load = 600 lb/ft | 3/12 | 10 | $3\frac{3}{8}$ | 12 | $14\frac{1}{2}$ | $12\frac{3}{4}$ | $7\frac{1}{2}$ | $5\frac{1}{8}$ | $9\frac{3}{4}$ | 12 | $12\frac{3}{4}$ | $7\frac{1}{2}$ |
| | | 12 | $5\frac{1}{8}$ | $7\frac{1}{2}$ | 12 | 12 | $7\frac{1}{2}$ | $5\frac{1}{8}$ | $8\frac{1}{2}$ | $13\frac{1}{4}$ | 13 | $7\frac{1}{2}$ |
| | | 14 | $5\frac{1}{8}$ | $7\frac{1}{2}$ | $13\frac{1}{4}$ | $13\frac{1}{4}$ | $7\frac{1}{2}$ | $5\frac{1}{8}$ | $7\frac{3}{4}$ | $15\frac{1}{2}$ | $14\frac{1}{4}$ | $7\frac{1}{2}$ |
| | | 16 | $5\frac{1}{8}$ | $7\frac{1}{2}$ | $14\frac{3}{4}$ | $14\frac{3}{4}$ | $7\frac{1}{2}$ | $5\frac{1}{8}$ | $7\frac{1}{2}$ | $17\frac{1}{4}$ | $15\frac{1}{4}$ | $7\frac{1}{2}$ |
| | | 18 | $5\frac{1}{8}$ | $7\frac{1}{2}$ | 16 | 16 | $7\frac{1}{2}$ | $5\frac{1}{8}$ | $7\frac{1}{2}$ | $18\frac{3}{4}$ | $16\frac{1}{4}$ | $7\frac{1}{2}$ |
| | 4/12 | 10 | $3\frac{3}{8}$ | 11 | 15 | $15\frac{1}{2}$ | $7\frac{1}{2}$ | $5\frac{1}{8}$ | 9 | $10\frac{1}{4}$ | $15\frac{1}{2}$ | $7\frac{1}{2}$ |
| | | 12 | $5\frac{1}{8}$ | $7\frac{1}{2}$ | 12 | 12 | $7\frac{1}{2}$ | $5\frac{1}{8}$ | 8 | $13\frac{3}{4}$ | $12\frac{1}{4}$ | $7\frac{1}{2}$ |
| | | 14 | $5\frac{1}{8}$ | $7\frac{1}{2}$ | $13\frac{1}{2}$ | $13\frac{1}{4}$ | $7\frac{1}{2}$ | $5\frac{1}{8}$ | $7\frac{1}{2}$ | $15\frac{1}{2}$ | $13\frac{3}{4}$ | $7\frac{1}{2}$ |
| | | 16 | $5\frac{1}{8}$ | $7\frac{1}{2}$ | $14\frac{3}{4}$ | $14\frac{3}{4}$ | $7\frac{1}{2}$ | $5\frac{1}{8}$ | $7\frac{1}{2}$ | 17 | 15 | $7\frac{1}{2}$ |
| | | 18 | $5\frac{1}{8}$ | $7\frac{1}{2}$ | 16 | 16 | $7\frac{1}{2}$ | $5\frac{1}{8}$ | $7\frac{1}{2}$ | $18\frac{1}{4}$ | $16\frac{1}{4}$ | $7\frac{1}{2}$ |
| | 6/12 | 12 | $5\frac{1}{8}$ | $7\frac{1}{2}$ | 12 | $11\frac{1}{4}$ | $7\frac{1}{2}$ | $5\frac{5}{8}$ | $7\frac{1}{2}$ | $13\frac{3}{4}$ | $11\frac{1}{4}$ | $7\frac{1}{2}$ |
| | | 14 | $5\frac{1}{8}$ | $7\frac{1}{2}$ | $13\frac{1}{2}$ | $12\frac{3}{4}$ | $7\frac{1}{2}$ | $5\frac{1}{8}$ | $7\frac{1}{2}$ | $15\frac{1}{2}$ | $12\frac{1}{2}$ | $7\frac{1}{2}$ |
| | | 16 | $5\frac{1}{8}$ | $7\frac{1}{2}$ | $14\frac{3}{4}$ | 14 | $7\frac{1}{2}$ | $5\frac{1}{8}$ | $7\frac{1}{2}$ | 17 | 14 | $7\frac{1}{2}$ |
| | | 18 | $5\frac{1}{8}$ | $7\frac{1}{2}$ | 16 | $15\frac{1}{4}$ | $7\frac{1}{2}$ | $5\frac{1}{8}$ | $7\frac{1}{2}$ | $18\frac{1}{4}$ | $15\frac{1}{2}$ | $7\frac{1}{2}$ |
| | 8/12 | 12 | $5\frac{1}{8}$ | $7\frac{1}{2}$ | 12 | $10\frac{1}{2}$ | $7\frac{1}{2}$ | $5\frac{1}{8}$ | $7\frac{1}{2}$ | $13\frac{1}{2}$ | $10\frac{1}{4}$ | $7\frac{1}{2}$ |
| | | 14 | $5\frac{1}{8}$ | $7\frac{1}{2}$ | $13\frac{1}{4}$ | $12\frac{1}{4}$ | $7\frac{1}{2}$ | $5\frac{1}{8}$ | $7\frac{1}{2}$ | $15\frac{1}{4}$ | $11\frac{1}{2}$ | $7\frac{1}{2}$ |
| | | 16 | $5\frac{1}{8}$ | $7\frac{1}{2}$ | $14\frac{1}{2}$ | $13\frac{1}{2}$ | $7\frac{1}{2}$ | $5\frac{1}{8}$ | $7\frac{1}{2}$ | $16\frac{1}{4}$ | $13\frac{1}{4}$ | $7\frac{1}{2}$ |
| | | 18 | $5\frac{1}{8}$ | $7\frac{1}{2}$ | $15\frac{3}{4}$ | $14\frac{3}{4}$ | $7\frac{1}{2}$ | $5\frac{1}{8}$ | $7\frac{1}{2}$ | 18 | $14\frac{1}{4}$ | $7\frac{1}{2}$ |

| 40′ Span | | | | | 50′ Span | | | | |
| --- | --- | --- | --- | --- | --- | --- | --- | --- | --- |
| Width | Base | Lower Tang. | Upper Tang. | Crown | Width | Base | Lower Tang. | Upper Tang. | Crown |
| $3\frac{1}{8}$ | $13\frac{1}{4}$ | $14\frac{1}{2}$ | $13\frac{1}{4}$ | $7\frac{1}{2}$ | $5\frac{1}{8}$ | $11\frac{3}{4}$ | $14$ | $13\frac{3}{4}$ | $7\frac{1}{2}$ |
| $5\frac{1}{8}$ | $7\frac{1}{2}$ | $13\frac{1}{4}$ | $13$ | $7\frac{1}{2}$ | $5\frac{1}{8}$ | $10\frac{1}{2}$ | $14\frac{3}{4}$ | $14\frac{1}{2}$ | $7\frac{1}{2}$ |
| $5\frac{1}{8}$ | $7\frac{1}{2}$ | $14\frac{3}{4}$ | $14\frac{1}{2}$ | $7\frac{1}{2}$ | $5\frac{1}{8}$ | $9\frac{1}{2}$ | $16\frac{3}{4}$ | $16\frac{1}{4}$ | $7\frac{1}{2}$ |
| $5\frac{1}{8}$ | $7\frac{1}{2}$ | $16$ | $16$ | $7\frac{1}{2}$ | $5\frac{1}{8}$ | $8\frac{3}{4}$ | $18\frac{3}{4}$ | $17\frac{1}{2}$ | $7\frac{1}{2}$ |
| $5\frac{1}{8}$ | $7\frac{1}{2}$ | $17\frac{1}{2}$ | $17\frac{1}{4}$ | $7\frac{1}{2}$ | $5\frac{1}{8}$ | $8$ | $20\frac{1}{2}$ | $19$ | $7\frac{1}{2}$ |
| $3\frac{3}{8}$ | $12$ | $15\frac{1}{4}$ | $12\frac{3}{4}$ | $7\frac{1}{2}$ | $5\frac{1}{8}$ | $10\frac{1}{2}$ | $13$ | $12\frac{3}{4}$ | $7\frac{1}{2}$ |
| $5\frac{1}{8}$ | $7\frac{1}{2}$ | $12\frac{3}{4}$ | $12\frac{1}{2}$ | $7\frac{1}{2}$ | $5\frac{1}{8}$ | $9\frac{1}{2}$ | $14\frac{1}{2}$ | $13\frac{1}{2}$ | $7\frac{1}{2}$ |
| $5\frac{1}{8}$ | $7\frac{1}{2}$ | $14\frac{1}{4}$ | $14\frac{1}{4}$ | $7\frac{1}{2}$ | $5\frac{1}{8}$ | $8\frac{3}{4}$ | $16\frac{3}{4}$ | $15$ | $7\frac{1}{2}$ |
| $5\frac{1}{8}$ | $7\frac{1}{2}$ | $15\frac{3}{4}$ | $15\frac{1}{2}$ | $7\frac{1}{2}$ | $5\frac{1}{8}$ | $8$ | $18\frac{1}{2}$ | $16\frac{1}{2}$ | $7\frac{1}{2}$ |
| $5\frac{1}{8}$ | $7\frac{1}{2}$ | $17$ | $17$ | $7\frac{1}{2}$ | $5\frac{1}{8}$ | $7\frac{1}{2}$ | $20\frac{1}{4}$ | $18$ | $7\frac{1}{2}$ |
| $5\frac{1}{8}$ | $7\frac{1}{4}$ | $12\frac{1}{4}$ | $11\frac{1}{2}$ | $7\frac{1}{2}$ | $5\frac{1}{8}$ | $8$ | $14\frac{3}{4}$ | $11\frac{3}{4}$ | $7\frac{1}{2}$ |
| $5\frac{1}{8}$ | $7\frac{1}{2}$ | $13\frac{3}{4}$ | $13\frac{1}{4}$ | $7\frac{1}{2}$ | $5\frac{1}{8}$ | $7\frac{1}{2}$ | $16\frac{1}{2}$ | $12\frac{3}{4}$ | $7\frac{1}{2}$ |
| $5\frac{1}{8}$ | $7\frac{1}{2}$ | $15$ | $15$ | $7\frac{1}{2}$ | $5\frac{1}{8}$ | $7\frac{1}{2}$ | $18\frac{1}{4}$ | $14\frac{1}{4}$ | $7\frac{1}{2}$ |
| $5\frac{1}{8}$ | $7\frac{1}{2}$ | $16\frac{1}{2}$ | $16\frac{1}{4}$ | $7\frac{1}{2}$ | $5\frac{1}{8}$ | $7\frac{1}{2}$ | $19\frac{3}{4}$ | $16$ | $7\frac{1}{2}$ |
| $5\frac{1}{8}$ | $7\frac{1}{2}$ | $12$ | $10\frac{1}{4}$ | $7\frac{1}{2}$ | $5\frac{1}{8}$ | $7\frac{1}{2}$ | $14\frac{1}{4}$ | $11\frac{1}{2}$ | $7\frac{1}{2}$ |
| $5\frac{1}{8}$ | $7\frac{1}{2}$ | $13\frac{1}{2}$ | $11\frac{3}{4}$ | $7\frac{1}{2}$ | $5\frac{1}{8}$ | $7\frac{1}{2}$ | $16$ | $12\frac{1}{4}$ | $7\frac{1}{2}$ |
| $5\frac{1}{8}$ | $7\frac{1}{2}$ | $14\frac{3}{4}$ | $13\frac{1}{2}$ | $7\frac{1}{2}$ | $5\frac{1}{8}$ | $7\frac{1}{2}$ | $17\frac{3}{4}$ | $13$ | $7\frac{1}{2}$ |
| $5\frac{1}{8}$ | $7\frac{1}{2}$ | $16$ | $15$ | $7\frac{1}{2}$ | $5\frac{1}{8}$ | $7\frac{1}{2}$ | $19\frac{1}{4}$ | $14$ | $7\frac{1}{2}$ |
| $5\frac{1}{8}$ | $12\frac{1}{4}$ | $14$ | $13\frac{3}{4}$ | $12\frac{1}{4}$ | $5\frac{1}{8}$ | $17\frac{1}{2}$ | $17\frac{1}{2}$ | $15\frac{1}{2}$ | $12\frac{1}{4}$ |
| $5\frac{1}{8}$ | $10\frac{3}{4}$ | $15$ | $13\frac{1}{2}$ | $7\frac{1}{2}$ | $5\frac{1}{8}$ | $15\frac{1}{2}$ | $17\frac{1}{2}$ | $16$ | $7\frac{1}{2}$ |
| $5\frac{1}{8}$ | $9\frac{3}{4}$ | $17\frac{1}{2}$ | $14\frac{1}{2}$ | $7\frac{1}{2}$ | $5\frac{1}{8}$ | $14$ | $21$ | $16\frac{1}{2}$ | $7\frac{1}{2}$ |
| $5\frac{1}{8}$ | $8\frac{3}{4}$ | $19\frac{1}{4}$ | $15\frac{3}{4}$ | $7\frac{1}{2}$ | $5\frac{1}{8}$ | $12\frac{3}{4}$ | $23\frac{1}{2}$ | $17$ | $7\frac{1}{2}$ |
| $5\frac{1}{8}$ | $8$ | $21\frac{1}{4}$ | $16\frac{3}{4}$ | $7\frac{1}{2}$ | $6\frac{3}{4}$ | $9$ | $20\frac{1}{2}$ | $20\frac{1}{4}$ | $7\frac{1}{2}$ |
| $5\frac{1}{8}$ | $11$ | $11\frac{1}{4}$ | $15\frac{1}{2}$ | $12$ | $5\frac{1}{8}$ | $15\frac{1}{2}$ | $15\frac{1}{2}$ | $15\frac{1}{2}$ | $12$ |
| $5\frac{1}{8}$ | $9\frac{3}{4}$ | $15\frac{1}{4}$ | $12\frac{3}{4}$ | $7\frac{1}{2}$ | $5\frac{1}{8}$ | $14$ | $18$ | $15\frac{1}{4}$ | $7\frac{1}{2}$ |
| $5\frac{1}{8}$ | $9$ | $17\frac{1}{2}$ | $13\frac{3}{4}$ | $7\frac{1}{2}$ | $5\frac{1}{8}$ | $12\frac{3}{4}$ | $21$ | $16$ | $7\frac{1}{2}$ |
| $5\frac{1}{8}$ | $8\frac{1}{4}$ | $19\frac{1}{4}$ | $15\frac{1}{4}$ | $7\frac{1}{2}$ | $5\frac{1}{8}$ | $11\frac{3}{4}$ | $23\frac{1}{2}$ | $16\frac{1}{2}$ | $7\frac{1}{2}$ |
| $5\frac{1}{8}$ | $7\frac{1}{2}$ | $21$ | $16\frac{1}{2}$ | $7\frac{1}{2}$ | $5\frac{1}{8}$ | $10\frac{3}{4}$ | $25\frac{1}{2}$ | $16\frac{3}{4}$ | $7\frac{1}{2}$ |
| $5\frac{1}{8}$ | $8\frac{1}{2}$ | $15\frac{1}{4}$ | $11$ | $7\frac{1}{2}$ | $5\frac{1}{8}$ | $11\frac{1}{4}$ | $18\frac{1}{4}$ | $14\frac{1}{2}$ | $7\frac{3}{4}$ |
| $5\frac{1}{8}$ | $7\frac{3}{4}$ | $17\frac{1}{4}$ | $12\frac{1}{2}$ | $7\frac{1}{2}$ | $5\frac{1}{8}$ | $10\frac{3}{4}$ | $20\frac{3}{4}$ | $15\frac{1}{4}$ | $7\frac{1}{2}$ |
| $5\frac{1}{8}$ | $7\frac{1}{2}$ | $19$ | $14$ | $7\frac{1}{2}$ | $5\frac{1}{8}$ | $10$ | $23$ | $15\frac{3}{4}$ | $7\frac{1}{2}$ |
| $5\frac{1}{8}$ | $7\frac{1}{2}$ | $20\frac{1}{2}$ | $15\frac{1}{4}$ | $7\frac{1}{2}$ | $5\frac{1}{8}$ | $9\frac{1}{2}$ | $25$ | $16\frac{1}{2}$ | $7\frac{1}{2}$ |
| $5\frac{1}{8}$ | $7\frac{1}{2}$ | $15\frac{1}{4}$ | $11$ | $7\frac{1}{2}$ | $5\frac{1}{8}$ | $10$ | $18$ | $14$ | $9\frac{1}{2}$ |
| $5\frac{1}{8}$ | $7\frac{1}{2}$ | $17$ | $11\frac{3}{4}$ | $7\frac{1}{2}$ | $5\frac{1}{8}$ | $9\frac{1}{4}$ | $20\frac{1}{4}$ | $15$ | $8$ |
| $5\frac{1}{8}$ | $7\frac{1}{2}$ | $18\frac{1}{2}$ | $12\frac{3}{4}$ | $7\frac{1}{2}$ | $5\frac{1}{8}$ | $8\frac{3}{4}$ | $22\frac{1}{4}$ | $15\frac{3}{4}$ | $7\frac{1}{2}$ |
| $5\frac{1}{8}$ | $7\frac{1}{2}$ | $20\frac{1}{4}$ | $13\frac{3}{4}$ | $7\frac{1}{2}$ | $5\frac{1}{8}$ | $8\frac{1}{4}$ | $24\frac{1}{4}$ | $16\frac{1}{4}$ | $7\frac{1}{2}$ |

(*continued*)

TABLE 8.11 *(Continued)*

| Loading | Roof Pitch | Wall Ht (ft) | 60' Span | | | | | 70' Span | | | | |
|---|---|---|---|---|---|---|---|---|---|---|---|---|
| | | | Width | Base | Lower Tang. | Upper Tang. | Crown | Width | Base | Lower Tang. | Upper Tang. | Crown |
| Vertical dead + live load = 400 lb/ft | 3/12 | 12 | $5\frac{1}{8}$ | $27\frac{1}{4}$ | $27\frac{1}{4}$ | 23 | $12\frac{1}{4}$ | $6\frac{3}{4}$ | $26\frac{3}{4}$ | $26\frac{3}{4}$ | $24\frac{1}{4}$ | $12\frac{1}{4}$ |
| | | 14 | $6\frac{3}{4}$ | $19\frac{1}{4}$ | $22\frac{1}{2}$ | $21\frac{1}{4}$ | $12\frac{1}{4}$ | $6\frac{3}{4}$ | $24\frac{1}{2}$ | $24\frac{1}{2}$ | $25\frac{3}{4}$ | $12\frac{1}{4}$ |
| | | 16 | $6\frac{3}{4}$ | $17\frac{1}{2}$ | $25\frac{3}{4}$ | $21\frac{3}{4}$ | $12\frac{1}{4}$ | $6\frac{3}{4}$ | $22\frac{1}{2}$ | $28\frac{3}{4}$ | $26\frac{1}{4}$ | $12\frac{1}{4}$ |
| | | 18 | $6\frac{3}{4}$ | $16\frac{1}{4}$ | 28 | $22\frac{1}{4}$ | $12\frac{1}{4}$ | $6\frac{3}{4}$ | 21 | $31\frac{1}{4}$ | $26\frac{3}{4}$ | $12\frac{1}{4}$ |
| | | 20 | $6\frac{3}{4}$ | 15 | $30\frac{1}{4}$ | $22\frac{1}{2}$ | $12\frac{1}{4}$ | $8\frac{3}{4}$ | $15\frac{1}{4}$ | $28\frac{1}{4}$ | $24\frac{3}{4}$ | $12\frac{1}{4}$ |
| | 4/12 | 12 | $5\frac{1}{8}$ | $24\frac{1}{4}$ | $24\frac{1}{4}$ | 22 | 12 | $6\frac{3}{4}$ | $23\frac{1}{2}$ | $23\frac{1}{2}$ | 23 | 12 |
| | | 14 | $6\frac{3}{4}$ | $17\frac{1}{4}$ | $22\frac{3}{4}$ | $20\frac{1}{4}$ | 12 | $6\frac{3}{4}$ | $21\frac{3}{4}$ | $25\frac{1}{4}$ | $24\frac{1}{4}$ | 12 |
| | | 16 | $6\frac{3}{4}$ | 16 | $25\frac{1}{2}$ | 21 | 12 | $6\frac{3}{4}$ | $20\frac{1}{4}$ | $28\frac{3}{4}$ | 25 | 12 |
| | | 18 | $6\frac{3}{4}$ | $14\frac{3}{4}$ | $27\frac{3}{4}$ | $21\frac{1}{2}$ | 12 | $6\frac{3}{4}$ | 19 | $31\frac{1}{2}$ | $25\frac{3}{4}$ | 12 |
| | | 20 | $6\frac{3}{4}$ | $13\frac{3}{4}$ | $29\frac{3}{4}$ | 22 | 12 | $6\frac{3}{4}$ | $17\frac{3}{4}$ | $33\frac{3}{4}$ | $26\frac{1}{4}$ | 12 |
| | 6/12 | 12 | $5\frac{1}{8}$ | 20 | $24\frac{1}{4}$ | $20\frac{1}{4}$ | 12 | $6\frac{3}{4}$ | 19 | 22 | 21 | 13 |
| | | 14 | $6\frac{3}{4}$ | $14\frac{1}{4}$ | $22\frac{1}{2}$ | $18\frac{3}{4}$ | $11\frac{1}{4}$ | $6\frac{3}{4}$ | $17\frac{3}{4}$ | $25\frac{1}{4}$ | 22 | $11\frac{1}{4}$ |
| | | 16 | $6\frac{3}{4}$ | $13\frac{1}{2}$ | $24\frac{3}{4}$ | $19\frac{3}{4}$ | $11\frac{1}{4}$ | $6\frac{3}{4}$ | $16\frac{3}{4}$ | 28 | 23 | $11\frac{1}{4}$ |
| | | 18 | $6\frac{3}{4}$ | $12\frac{3}{4}$ | $26\frac{3}{4}$ | $20\frac{1}{2}$ | $11\frac{1}{4}$ | $6\frac{3}{4}$ | 16 | $30\frac{1}{2}$ | 24 | $11\frac{1}{4}$ |
| | 8/12 | 12 | $5\frac{1}{8}$ | 17 | $24\frac{1}{2}$ | 19 | $14\frac{3}{4}$ | $6\frac{3}{4}$ | 16 | 22 | $19\frac{1}{4}$ | 16 |
| | | 14 | $6\frac{3}{4}$ | $12\frac{1}{4}$ | 22 | 18 | $10\frac{3}{4}$ | $6\frac{3}{4}$ | $15\frac{1}{4}$ | $24\frac{1}{2}$ | $20\frac{3}{4}$ | 14 |
| | | 16 | $6\frac{3}{4}$ | $11\frac{1}{2}$ | 24 | 19 | $10\frac{1}{2}$ | $6\frac{3}{4}$ | $14\frac{1}{2}$ | 27 | 22 | $12\frac{1}{4}$ |
| | | 18 | $6\frac{3}{4}$ | 11 | 26 | 20 | $10\frac{1}{2}$ | $6\frac{3}{4}$ | $13\frac{3}{4}$ | $29\frac{1}{4}$ | 23 | 11 |
| Vertical dead + live load = 600 lb/ft | 3/12 | 12 | $6\frac{3}{4}$ | 26 | 26 | $22\frac{1}{2}$ | $12\frac{1}{4}$ | $6\frac{3}{4}$ | $32\frac{3}{4}$ | $32\frac{3}{4}$ | $26\frac{1}{2}$ | $12\frac{1}{4}$ |
| | | 14 | $6\frac{3}{4}$ | $23\frac{1}{2}$ | 25 | $23\frac{3}{4}$ | $12\frac{1}{4}$ | $6\frac{3}{4}$ | 30 | 30 | $28\frac{1}{2}$ | $12\frac{1}{4}$ |
| | | 16 | $6\frac{3}{4}$ | $21\frac{1}{4}$ | $28\frac{3}{4}$ | $24\frac{1}{4}$ | $12\frac{1}{4}$ | $6\frac{3}{4}$ | $27\frac{3}{4}$ | $32\frac{1}{4}$ | $29\frac{1}{4}$ | $12\frac{1}{4}$ |
| | | 18 | $6\frac{3}{4}$ | 20 | $31\frac{1}{2}$ | $24\frac{3}{4}$ | $12\frac{1}{4}$ | $8\frac{3}{4}$ | $20\frac{1}{4}$ | $29\frac{3}{4}$ | $26\frac{1}{2}$ | $12\frac{1}{4}$ |
| | | 20 | $8\frac{3}{4}$ | $14\frac{1}{2}$ | 28 | $23\frac{1}{2}$ | $12\frac{1}{4}$ | $8\frac{3}{4}$ | $18\frac{3}{4}$ | $31\frac{3}{4}$ | 27 | $12\frac{1}{4}$ |
| | 4/12 | 12 | $6\frac{3}{4}$ | 23 | 23 | $21\frac{1}{2}$ | 12 | $6\frac{3}{4}$ | 29 | 29 | $25\frac{1}{4}$ | 12 |
| | | 14 | $6\frac{3}{4}$ | $21\frac{1}{4}$ | $25\frac{1}{2}$ | $22\frac{1}{2}$ | 12 | $6\frac{3}{4}$ | $26\frac{3}{4}$ | 28 | 27 | 12 |
| | | 16 | $6\frac{3}{4}$ | $19\frac{3}{4}$ | $28\frac{3}{4}$ | $23\frac{1}{4}$ | 12 | $6\frac{3}{4}$ | 25 | $32\frac{1}{4}$ | $27\frac{3}{4}$ | 12 |
| | | 18 | $6\frac{3}{4}$ | $18\frac{1}{4}$ | $31\frac{1}{4}$ | 24 | 12 | $8\frac{3}{4}$ | $18\frac{1}{4}$ | $29\frac{1}{4}$ | $25\frac{1}{2}$ | 12 |
| | | 20 | $6\frac{3}{4}$ | $17\frac{1}{4}$ | $33\frac{3}{4}$ | $24\frac{1}{4}$ | 12 | $8\frac{3}{4}$ | $17\frac{1}{4}$ | $31\frac{1}{4}$ | 26 | 12 |
| | 6/12 | 12 | $6\frac{3}{4}$ | 19 | 22 | $19\frac{3}{4}$ | $11\frac{1}{2}$ | $6\frac{3}{4}$ | $23\frac{1}{2}$ | $24\frac{1}{4}$ | $23\frac{1}{4}$ | $14\frac{3}{4}$ |
| | | 14 | $6\frac{3}{4}$ | $17\frac{3}{4}$ | $25\frac{1}{4}$ | 21 | $11\frac{1}{4}$ | $6\frac{3}{4}$ | 22 | $28\frac{1}{4}$ | $24\frac{1}{2}$ | 13 |
| | | 16 | $6\frac{3}{4}$ | $16\frac{1}{2}$ | 28 | 22 | $11\frac{1}{4}$ | $6\frac{3}{4}$ | $20\frac{3}{4}$ | $31\frac{1}{2}$ | $25\frac{3}{4}$ | $11\frac{1}{4}$ |
| | | 18 | $6\frac{3}{4}$ | $15\frac{3}{4}$ | $30\frac{1}{2}$ | $22\frac{3}{4}$ | $11\frac{1}{4}$ | $8\frac{3}{4}$ | $15\frac{1}{4}$ | $28\frac{1}{4}$ | $23\frac{3}{4}$ | $11\frac{1}{4}$ |
| | 8/12 | 12 | $6\frac{3}{4}$ | $16\frac{1}{4}$ | 22 | $18\frac{3}{4}$ | 14 | $6\frac{3}{4}$ | 20 | $24\frac{1}{2}$ | $21\frac{1}{2}$ | 18 |
| | | 14 | $6\frac{3}{4}$ | $15\frac{1}{4}$ | $24\frac{3}{4}$ | $20\frac{1}{4}$ | $12\frac{1}{4}$ | $6\frac{3}{4}$ | $18\frac{3}{4}$ | $27\frac{3}{4}$ | $23\frac{1}{4}$ | $15\frac{3}{4}$ |
| | | 16 | $6\frac{3}{4}$ | $14\frac{1}{4}$ | $27\frac{1}{4}$ | $21\frac{1}{4}$ | $10\frac{3}{4}$ | $6\frac{3}{4}$ | 18 | $30\frac{1}{2}$ | $24\frac{1}{2}$ | 14 |
| | | 18 | $6\frac{3}{4}$ | $13\frac{3}{4}$ | $29\frac{1}{2}$ | $22\frac{1}{4}$ | $10\frac{1}{2}$ | $6\frac{3}{4}$ | 17 | $33\frac{1}{4}$ | $25\frac{3}{4}$ | $12\frac{1}{2}$ |

| | 80′ Span | | | | | 90′ Span | | | |
|---|---|---|---|---|---|---|---|---|---|
| Width | Base | Lower Tang. | Upper Tang. | Crown | Width | Base | Lower Tang. | Upper Tang. | Crown |
| $6\frac{3}{4}$ | $32\frac{1}{2}$ | $32\frac{1}{2}$ | $27\frac{3}{4}$ | $12\frac{1}{4}$ | $6\frac{3}{4}$ | $38\frac{3}{4}$ | $38\frac{3}{4}$ | $31$ | $12\frac{1}{2}$ |
| $6\frac{3}{4}$ | $30$ | $30$ | $29\frac{1}{2}$ | $12\frac{1}{4}$ | $8\frac{3}{4}$ | $28\frac{1}{4}$ | $28\frac{1}{4}$ | $29\frac{3}{4}$ | $12\frac{1}{4}$ |
| $6\frac{3}{4}$ | $27\frac{3}{4}$ | $31\frac{3}{4}$ | $30\frac{3}{4}$ | $12\frac{1}{4}$ | $8\frac{3}{4}$ | $26\frac{1}{4}$ | $29$ | $30\frac{3}{4}$ | $12\frac{1}{4}$ |
| $8\frac{3}{4}$ | $20\frac{1}{4}$ | $29\frac{1}{2}$ | $27\frac{3}{4}$ | $12\frac{1}{4}$ | $8\frac{3}{4}$ | $24\frac{1}{2}$ | $32\frac{1}{4}$ | $31\frac{3}{4}$ | $12\frac{1}{4}$ |
| $8\frac{3}{4}$ | $19$ | $31\frac{1}{4}$ | $28\frac{1}{2}$ | $12\frac{1}{4}$ | $8\frac{3}{4}$ | $23$ | $34\frac{3}{4}$ | $32\frac{1}{4}$ | $12\frac{1}{4}$ |
| $6\frac{3}{4}$ | $28\frac{1}{2}$ | $28\frac{1}{2}$ | $26$ | $12\frac{1}{2}$ | $6\frac{3}{4}$ | $33\frac{3}{4}$ | $33\frac{3}{4}$ | $29\frac{1}{4}$ | $15\frac{1}{4}$ |
| $6\frac{3}{4}$ | $26\frac{1}{2}$ | $27\frac{1}{2}$ | $28$ | $12$ | $6\frac{3}{4}$ | $31\frac{1}{2}$ | $31\frac{1}{2}$ | $31\frac{1}{4}$ | $12\frac{1}{4}$ |
| $6\frac{3}{4}$ | $24\frac{3}{4}$ | $31\frac{3}{4}$ | $28\frac{3}{4}$ | $12$ | $8\frac{3}{4}$ | $23$ | $29$ | $29$ | $12$ |
| $8\frac{3}{4}$ | $18\frac{1}{4}$ | $29$ | $26\frac{1}{2}$ | $12$ | $8\frac{3}{4}$ | $21\frac{3}{4}$ | $31\frac{3}{4}$ | $29\frac{3}{4}$ | $12$ |
| $8\frac{3}{4}$ | $17$ | $31$ | $27\frac{1}{4}$ | $12$ | $8\frac{3}{4}$ | $20\frac{1}{2}$ | $34$ | $30\frac{3}{4}$ | $12$ |
| $6\frac{3}{4}$ | $23$ | $23\frac{3}{4}$ | $23\frac{3}{4}$ | $16\frac{1}{4}$ | $6\frac{3}{4}$ | $27$ | $27$ | $26\frac{1}{4}$ | $19\frac{3}{4}$ |
| $6\frac{3}{4}$ | $21\frac{1}{2}$ | $27\frac{3}{4}$ | $25$ | $14\frac{1}{4}$ | $6\frac{3}{4}$ | $25\frac{1}{2}$ | $30$ | $28$ | $17\frac{1}{2}$ |
| $6\frac{3}{4}$ | $20\frac{1}{4}$ | $31$ | $26\frac{1}{4}$ | $12\frac{1}{2}$ | $6\frac{3}{4}$ | $24$ | $33\frac{3}{4}$ | $29\frac{1}{2}$ | $15\frac{1}{4}$ |
| $6\frac{3}{4}$ | $19\frac{1}{4}$ | $33\frac{3}{4}$ | $27\frac{1}{2}$ | $11\frac{1}{4}$ | $8\frac{3}{4}$ | $17\frac{3}{4}$ | $30\frac{1}{4}$ | $27\frac{1}{4}$ | $11\frac{3}{4}$ |
| $6\frac{3}{4}$ | $19\frac{1}{4}$ | $24$ | $21\frac{1}{2}$ | $19\frac{3}{4}$ | $6\frac{3}{4}$ | $22\frac{1}{2}$ | $25\frac{3}{4}$ | $23\frac{3}{4}$ | $23\frac{3}{4}$ |
| $6\frac{3}{4}$ | $18\frac{1}{4}$ | $27\frac{1}{4}$ | $23\frac{1}{4}$ | $17\frac{1}{2}$ | $6\frac{3}{4}$ | $21\frac{1}{4}$ | $29\frac{1}{2}$ | $25\frac{3}{4}$ | $21$ |
| $6\frac{3}{4}$ | $17\frac{1}{4}$ | $30$ | $24\frac{3}{4}$ | $15\frac{1}{2}$ | $6\frac{3}{4}$ | $20\frac{1}{4}$ | $32\frac{3}{4}$ | $27\frac{1}{4}$ | $19$ |
| $6\frac{3}{4}$ | $16\frac{1}{2}$ | $32\frac{1}{2}$ | $26$ | $13\frac{3}{4}$ | $8\frac{3}{4}$ | $15\frac{1}{4}$ | $29$ | $25\frac{1}{2}$ | $14\frac{1}{2}$ |
| $6\frac{3}{4}$ | $40\frac{1}{2}$ | $40\frac{1}{2}$ | $30\frac{3}{4}$ | $17$ | $8\frac{3}{4}$ | $37\frac{3}{4}$ | $37\frac{3}{4}$ | $30\frac{3}{4}$ | $17$ |
| $6\frac{3}{4}$ | $37\frac{1}{4}$ | $37\frac{1}{4}$ | $32\frac{3}{4}$ | $17$ | $8\frac{3}{4}$ | $35$ | $35$ | $32\frac{3}{4}$ | $17$ |
| $8\frac{3}{4}$ | $27$ | $29\frac{3}{4}$ | $30\frac{1}{2}$ | $17$ | $8\frac{3}{4}$ | $32\frac{1}{2}$ | $32\frac{1}{2}$ | $34\frac{1}{2}$ | $17$ |
| $8\frac{3}{4}$ | $25\frac{1}{4}$ | $33$ | $31$ | $17$ | $8\frac{3}{4}$ | $30\frac{1}{4}$ | $36$ | $35\frac{1}{2}$ | $17$ |
| $8\frac{3}{4}$ | $23\frac{1}{2}$ | $35\frac{1}{2}$ | $31\frac{3}{4}$ | $17$ | $8\frac{3}{4}$ | $28\frac{1}{2}$ | $39$ | $36$ | $17$ |
| $6\frac{3}{4}$ | $35\frac{1}{4}$ | $35\frac{1}{4}$ | $29\frac{1}{4}$ | $16\frac{3}{4}$ | $8\frac{3}{4}$ | $32\frac{3}{4}$ | $32\frac{3}{4}$ | $28\frac{3}{4}$ | $16\frac{3}{4}$ |
| $6\frac{3}{4}$ | $33$ | $33$ | $31$ | $16\frac{3}{4}$ | $8\frac{3}{4}$ | $30\frac{1}{2}$ | $30\frac{1}{2}$ | $31$ | $16\frac{3}{4}$ |
| $8\frac{3}{4}$ | $24$ | $29\frac{3}{4}$ | $28\frac{1}{2}$ | $16\frac{3}{4}$ | $8\frac{3}{4}$ | $28\frac{3}{4}$ | $32$ | $32\frac{1}{4}$ | $16\frac{3}{4}$ |
| $8\frac{3}{4}$ | $22\frac{1}{2}$ | $32\frac{1}{2}$ | $29\frac{1}{2}$ | $16\frac{3}{4}$ | $8\frac{3}{4}$ | $27$ | $35\frac{1}{2}$ | $33\frac{1}{4}$ | $16\frac{3}{4}$ |
| $8\frac{3}{4}$ | $21\frac{1}{4}$ | $35$ | $30\frac{1}{4}$ | $16\frac{3}{4}$ | $8\frac{3}{4}$ | $25\frac{1}{2}$ | $38\frac{1}{4}$ | $34\frac{1}{4}$ | $16\frac{3}{4}$ |
| $6\frac{3}{4}$ | $28\frac{1}{2}$ | $28\frac{1}{2}$ | $26\frac{1}{4}$ | $18\frac{1}{2}$ | $6\frac{3}{4}$ | $33\frac{1}{4}$ | $33\frac{1}{4}$ | $29$ | $22\frac{3}{4}$ |
| $6\frac{3}{4}$ | $26\frac{3}{4}$ | $31$ | $28$ | $16\frac{1}{4}$ | $8\frac{3}{4}$ | $24\frac{3}{4}$ | $28\frac{1}{2}$ | $27\frac{3}{4}$ | $16\frac{3}{4}$ |
| $8\frac{3}{4}$ | $19\frac{3}{4}$ | $29$ | $26$ | $15\frac{3}{4}$ | $8\frac{3}{4}$ | $23\frac{1}{4}$ | $31\frac{1}{2}$ | $29\frac{1}{4}$ | $15\frac{3}{4}$ |
| $8\frac{3}{4}$ | $18\frac{3}{4}$ | $31\frac{1}{4}$ | $27$ | $15\frac{3}{4}$ | $8\frac{3}{4}$ | $22\frac{1}{4}$ | $34\frac{1}{4}$ | $30\frac{1}{4}$ | $15\frac{3}{4}$ |
| $6\frac{3}{4}$ | $23\frac{3}{4}$ | $26\frac{3}{4}$ | $24\frac{1}{4}$ | $22\frac{1}{4}$ | $6\frac{3}{4}$ | $27\frac{3}{4}$ | $28\frac{3}{4}$ | $26\frac{3}{4}$ | $26\frac{3}{4}$ |
| $6\frac{3}{4}$ | $22\frac{1}{2}$ | $30\frac{3}{4}$ | $26$ | $19\frac{3}{4}$ | $6\frac{3}{4}$ | $26\frac{1}{2}$ | $33\frac{1}{4}$ | $28\frac{3}{4}$ | $24$ |
| $8\frac{3}{4}$ | $16\frac{3}{4}$ | $28$ | $24\frac{1}{4}$ | $15$ | $8\frac{3}{4}$ | $19\frac{3}{4}$ | $30\frac{1}{2}$ | $27$ | $18\frac{1}{2}$ |
| $8\frac{3}{4}$ | $16$ | $30$ | $25\frac{3}{4}$ | $14\frac{3}{4}$ | $8\frac{3}{4}$ | $18\frac{3}{4}$ | $32\frac{3}{4}$ | $28\frac{1}{2}$ | $16\frac{1}{2}$ |

(*continued*)

TABLE 8.11 (*Continued*)

| Loading | Roof Pitch | Wall Ht (ft) | 30' Span | | | | | 35' Span | | | | |
|---|---|---|---|---|---|---|---|---|---|---|---|---|
| | | | Width | Base | Lower Tang. | Upper Tang. | Crown | Width | Base | Lower Tang. | Upper Tang. | Crown |
| Vertical dead + live load = 800 lb/ft | 3/12 | 10 | $5\frac{1}{8}$ | 10 | $12\frac{1}{4}$ | $12\frac{1}{4}$ | $12\frac{1}{4}$ | $5\frac{1}{8}$ | 13 | $13\frac{3}{4}$ | $13\frac{3}{4}$ | $12\frac{1}{4}$ |
| | | 12 | $5\frac{1}{8}$ | $8\frac{3}{4}$ | $13\frac{3}{4}$ | 12 | $7\frac{1}{2}$ | $5\frac{1}{8}$ | $11\frac{1}{4}$ | $15\frac{3}{4}$ | $12\frac{1}{2}$ | $7\frac{1}{2}$ |
| | | 14 | $5\frac{1}{8}$ | $7\frac{3}{4}$ | $15\frac{3}{4}$ | $12\frac{3}{4}$ | $7\frac{1}{2}$ | $5\frac{1}{8}$ | 10 | $18\frac{1}{4}$ | $13\frac{1}{4}$ | $7\frac{1}{2}$ |
| | | 16 | $5\frac{1}{8}$ | $7\frac{1}{2}$ | $17\frac{1}{2}$ | $13\frac{3}{4}$ | $7\frac{1}{2}$ | $5\frac{1}{8}$ | $9\frac{1}{4}$ | $20\frac{1}{4}$ | 14 | $7\frac{1}{2}$ |
| | | 18 | $5\frac{1}{8}$ | $7\frac{1}{2}$ | $19\frac{1}{4}$ | $14\frac{1}{4}$ | $7\frac{1}{2}$ | $5\frac{1}{8}$ | $8\frac{1}{4}$ | $22\frac{1}{4}$ | $14\frac{3}{4}$ | $7\frac{1}{2}$ |
| | 4/12 | 10 | $5\frac{1}{8}$ | $9\frac{1}{4}$ | $9\frac{3}{4}$ | $18\frac{1}{2}$ | 12 | $5\frac{1}{8}$ | $11\frac{3}{4}$ | $11\frac{3}{4}$ | $18\frac{1}{2}$ | 12 |
| | | 12 | $5\frac{1}{8}$ | 8 | 14 | $11\frac{1}{2}$ | $7\frac{1}{2}$ | $5\frac{1}{8}$ | $10\frac{1}{2}$ | 16 | $11\frac{3}{4}$ | $7\frac{1}{2}$ |
| | | 14 | $5\frac{1}{8}$ | $7\frac{1}{2}$ | 16 | $12\frac{1}{2}$ | $7\frac{1}{2}$ | $5\frac{1}{8}$ | $9\frac{1}{2}$ | $18\frac{1}{4}$ | $12\frac{3}{4}$ | $7\frac{1}{2}$ |
| | | 16 | $5\frac{1}{8}$ | $7\frac{1}{2}$ | $17\frac{1}{2}$ | $13\frac{1}{2}$ | $7\frac{1}{2}$ | $5\frac{1}{8}$ | $8\frac{1}{2}$ | $20\frac{1}{4}$ | $13\frac{3}{4}$ | $7\frac{1}{2}$ |
| | | 18 | $5\frac{1}{8}$ | $7\frac{1}{2}$ | 19 | $14\frac{1}{2}$ | $7\frac{1}{2}$ | $5\frac{1}{8}$ | $7\frac{3}{4}$ | 22 | $14\frac{3}{4}$ | $7\frac{1}{2}$ |
| | 6/12 | 12 | $5\frac{1}{8}$ | $7\frac{1}{2}$ | $14\frac{1}{4}$ | $10\frac{3}{4}$ | $7\frac{1}{2}$ | $5\frac{1}{8}$ | 9 | $16\frac{1}{4}$ | $10\frac{1}{2}$ | $7\frac{1}{2}$ |
| | | 14 | $5\frac{1}{8}$ | $7\frac{1}{2}$ | 16 | $11\frac{3}{4}$ | $7\frac{1}{2}$ | $5\frac{1}{8}$ | $8\frac{1}{4}$ | $18\frac{1}{4}$ | $11\frac{3}{4}$ | $7\frac{1}{2}$ |
| | | 16 | $5\frac{1}{8}$ | $7\frac{1}{2}$ | $17\frac{1}{2}$ | 13 | $7\frac{1}{2}$ | $5\frac{1}{8}$ | $7\frac{1}{2}$ | 20 | 13 | $7\frac{1}{2}$ |
| | | 18 | $5\frac{1}{8}$ | $7\frac{1}{2}$ | 19 | 14 | $7\frac{1}{2}$ | $5\frac{1}{8}$ | $7\frac{1}{2}$ | $21\frac{3}{4}$ | 14 | $7\frac{1}{2}$ |
| | 8/12 | 12 | $5\frac{1}{8}$ | $7\frac{1}{2}$ | $14\frac{1}{4}$ | $9\frac{3}{4}$ | $7\frac{1}{2}$ | $5\frac{1}{8}$ | 8 | 16 | $10\frac{3}{4}$ | $7\frac{1}{2}$ |
| | | 14 | $5\frac{1}{8}$ | $7\frac{1}{2}$ | $15\frac{3}{4}$ | $11\frac{1}{4}$ | $7\frac{1}{2}$ | $5\frac{1}{8}$ | $7\frac{1}{2}$ | 18 | $11\frac{1}{2}$ | $7\frac{1}{2}$ |
| | | 16 | $5\frac{1}{8}$ | $7\frac{1}{2}$ | $17\frac{1}{4}$ | $12\frac{1}{2}$ | $7\frac{1}{2}$ | $5\frac{1}{8}$ | $7\frac{1}{2}$ | $19\frac{3}{4}$ | 12 | $7\frac{1}{2}$ |
| | | 18 | $5\frac{1}{8}$ | $7\frac{1}{2}$ | $18\frac{3}{4}$ | $13\frac{1}{2}$ | $7\frac{1}{2}$ | $5\frac{1}{8}$ | $7\frac{1}{2}$ | $21\frac{1}{2}$ | 13 | $7\frac{1}{2}$ |
| Vertical dead + live load = 1000 lb/ft | 3/12 | 10 | $5\frac{1}{8}$ | $12\frac{1}{2}$ | $12\frac{1}{2}$ | $18\frac{1}{2}$ | $12\frac{1}{4}$ | $5\frac{1}{8}$ | 16 | 16 | $18\frac{1}{2}$ | $12\frac{1}{4}$ |
| | | 12 | $5\frac{1}{8}$ | $10\frac{3}{4}$ | $15\frac{1}{2}$ | $11\frac{1}{4}$ | $7\frac{1}{2}$ | $5\frac{1}{8}$ | $13\frac{3}{4}$ | $17\frac{1}{2}$ | 13 | $7\frac{1}{2}$ |
| | | 14 | $5\frac{1}{8}$ | $9\frac{1}{2}$ | 18 | $11\frac{3}{4}$ | $7\frac{1}{2}$ | $5\frac{1}{8}$ | $12\frac{1}{2}$ | $20\frac{1}{2}$ | $12\frac{3}{4}$ | $7\frac{1}{2}$ |
| | | 16 | $5\frac{1}{8}$ | $8\frac{3}{4}$ | 20 | $12\frac{1}{4}$ | $7\frac{1}{2}$ | $5\frac{1}{8}$ | $11\frac{1}{4}$ | 23 | $12\frac{3}{4}$ | $7\frac{1}{2}$ |
| | | 18 | $5\frac{1}{8}$ | 8 | 22 | $12\frac{3}{4}$ | $7\frac{1}{2}$ | $5\frac{1}{8}$ | $10\frac{1}{4}$ | $25\frac{1}{4}$ | $13\frac{1}{4}$ | $7\frac{1}{2}$ |
| | 4/12 | 10 | $5\frac{1}{8}$ | $11\frac{1}{2}$ | $11\frac{1}{2}$ | 23 | 12 | $5\frac{1}{8}$ | $14\frac{1}{2}$ | $14\frac{1}{2}$ | 23 | 12 |
| | | 12 | $5\frac{1}{8}$ | 10 | $15\frac{3}{4}$ | 11 | $7\frac{1}{2}$ | $5\frac{1}{8}$ | $12\frac{3}{4}$ | 18 | 13 | $7\frac{1}{2}$ |
| | | 14 | $5\frac{1}{8}$ | 9 | 18 | $11\frac{3}{4}$ | $7\frac{1}{2}$ | $5\frac{1}{8}$ | $11\frac{1}{2}$ | $20\frac{3}{4}$ | $12\frac{3}{4}$ | $7\frac{1}{2}$ |
| | | 16 | $5\frac{1}{8}$ | $8\frac{1}{4}$ | 20 | $12\frac{1}{4}$ | $7\frac{1}{2}$ | $5\frac{1}{8}$ | $10\frac{1}{2}$ | 23 | $12\frac{1}{2}$ | $7\frac{1}{2}$ |
| | | 18 | $5\frac{1}{8}$ | $7\frac{1}{2}$ | $21\frac{1}{4}$ | $13\frac{1}{4}$ | $7\frac{1}{2}$ | $5\frac{1}{8}$ | $9\frac{3}{4}$ | $25\frac{1}{4}$ | 13 | $7\frac{1}{2}$ |
| | 6/12 | 12 | $5\frac{1}{8}$ | $8\frac{3}{4}$ | $16\frac{1}{4}$ | 10 | $7\frac{1}{2}$ | $5\frac{1}{8}$ | 11 | $18\frac{1}{4}$ | $11\frac{3}{4}$ | $7\frac{1}{2}$ |
| | | 14 | $5\frac{1}{8}$ | 8 | 18 | $11\frac{1}{4}$ | $7\frac{1}{2}$ | $5\frac{1}{8}$ | $10\frac{1}{4}$ | $20\frac{3}{4}$ | $12\frac{1}{4}$ | $7\frac{1}{2}$ |
| | | 16 | $5\frac{1}{8}$ | $7\frac{1}{2}$ | 20 | 12 | $7\frac{1}{2}$ | $5\frac{1}{8}$ | $9\frac{1}{2}$ | $22\frac{3}{4}$ | $12\frac{1}{2}$ | $7\frac{1}{2}$ |
| | | 18 | $5\frac{1}{8}$ | $7\frac{1}{2}$ | $21\frac{1}{4}$ | $12\frac{3}{4}$ | $7\frac{1}{2}$ | $5\frac{1}{8}$ | $8\frac{3}{4}$ | 25 | 13 | $7\frac{1}{2}$ |
| | 8/12 | 12 | $5\frac{1}{8}$ | $7\frac{3}{4}$ | 16 | $9\frac{3}{4}$ | $7\frac{1}{2}$ | $5\frac{1}{8}$ | $9\frac{3}{4}$ | $18\frac{1}{4}$ | 12 | $7\frac{1}{2}$ |
| | | 14 | $5\frac{1}{8}$ | $7\frac{1}{2}$ | 18 | $10\frac{1}{2}$ | $7\frac{1}{2}$ | $5\frac{1}{8}$ | 9 | $20\frac{1}{4}$ | $12\frac{3}{4}$ | $7\frac{1}{2}$ |
| | | 16 | $5\frac{1}{8}$ | $7\frac{1}{2}$ | $19\frac{3}{4}$ | $11\frac{1}{2}$ | $7\frac{1}{2}$ | $5\frac{1}{8}$ | $8\frac{1}{2}$ | $22\frac{1}{2}$ | $13\frac{1}{4}$ | $7\frac{1}{2}$ |
| | | 18 | $5\frac{1}{8}$ | $7\frac{1}{2}$ | $21\frac{1}{2}$ | $12\frac{1}{4}$ | $7\frac{1}{2}$ | $5\frac{1}{8}$ | 8 | $24\frac{1}{2}$ | $13\frac{1}{4}$ | $7\frac{1}{2}$ |

| | 40′ Span | | | | | 50′ Span | | | |
|---|---|---|---|---|---|---|---|---|---|
| Width | Base | Lower Tang. | Upper Tang. | Crown | Width | Base | Lower Tang. | Upper Tang. | Crown |
| $5\frac{1}{8}$ | 16 | 16 | $14\frac{1}{2}$ | $12\frac{1}{4}$ | $5\frac{1}{8}$ | $22\frac{3}{4}$ | $22\frac{3}{4}$ | $18\frac{1}{2}$ | $12\frac{1}{4}$ |
| $5\frac{1}{8}$ | 14 | $17\frac{1}{4}$ | $13\frac{1}{4}$ | $7\frac{1}{2}$ | $5\frac{1}{8}$ | $20\frac{1}{2}$ | $20\frac{1}{2}$ | $18\frac{1}{2}$ | $12\frac{1}{4}$ |
| $5\frac{1}{8}$ | $12\frac{3}{4}$ | $20\frac{1}{2}$ | $13\frac{3}{4}$ | $7\frac{1}{2}$ | $5\frac{1}{8}$ | $18\frac{1}{2}$ | $24\frac{1}{2}$ | 19 | $12\frac{1}{4}$ |
| $5\frac{1}{8}$ | $11\frac{1}{2}$ | $22\frac{3}{4}$ | $14\frac{1}{2}$ | $7\frac{1}{2}$ | $6\frac{3}{4}$ | 13 | 22 | $18\frac{3}{4}$ | $12\frac{1}{4}$ |
| $5\frac{1}{8}$ | $10\frac{1}{2}$ | 25 | $15\frac{1}{4}$ | $7\frac{1}{2}$ | $6\frac{3}{4}$ | 12 | 24 | 20 | $12\frac{1}{4}$ |
| $5\frac{1}{8}$ | $14\frac{1}{2}$ | $14\frac{1}{2}$ | $18\frac{1}{2}$ | 12 | $5\frac{1}{8}$ | $20\frac{1}{2}$ | $20\frac{1}{2}$ | $18\frac{1}{2}$ | 12 |
| $5\frac{1}{8}$ | $12\frac{3}{4}$ | $17\frac{3}{4}$ | 13 | $7\frac{1}{2}$ | $5\frac{1}{8}$ | $18\frac{1}{2}$ | $20\frac{3}{4}$ | $17\frac{1}{2}$ | 12 |
| $5\frac{1}{8}$ | $11\frac{3}{4}$ | $20\frac{1}{2}$ | $13\frac{1}{2}$ | $7\frac{1}{2}$ | $5\frac{1}{8}$ | $16\frac{3}{4}$ | $24\frac{3}{4}$ | $18\frac{1}{4}$ | 12 |
| $5\frac{1}{8}$ | $10\frac{3}{4}$ | $22\frac{3}{4}$ | $13\frac{3}{4}$ | $7\frac{1}{2}$ | $6\frac{3}{4}$ | 12 | 22 | $17\frac{3}{4}$ | 12 |
| $5\frac{1}{8}$ | $9\frac{3}{4}$ | 25 | $14\frac{1}{2}$ | $7\frac{1}{2}$ | $6\frac{3}{4}$ | 11 | $23\frac{3}{4}$ | $19\frac{1}{4}$ | 12 |
| $5\frac{1}{8}$ | 11 | 18 | $12\frac{1}{2}$ | $7\frac{1}{2}$ | $5\frac{1}{8}$ | $15\frac{1}{4}$ | $21\frac{1}{2}$ | $16\frac{1}{2}$ | $11\frac{1}{4}$ |
| $5\frac{1}{8}$ | 10 | $20\frac{1}{4}$ | $13\frac{1}{4}$ | $7\frac{1}{2}$ | $5\frac{1}{8}$ | $14\frac{1}{4}$ | $24\frac{1}{2}$ | $17\frac{1}{4}$ | $11\frac{1}{4}$ |
| $5\frac{1}{8}$ | $9\frac{1}{4}$ | $22\frac{1}{2}$ | $13\frac{3}{4}$ | $7\frac{1}{2}$ | $6\frac{3}{4}$ | $10\frac{1}{4}$ | $21\frac{1}{2}$ | 16 | $11\frac{1}{4}$ |
| $5\frac{1}{8}$ | $8\frac{3}{4}$ | $24\frac{1}{2}$ | $14\frac{1}{4}$ | $7\frac{1}{2}$ | $6\frac{3}{4}$ | $9\frac{1}{2}$ | 23 | $17\frac{1}{2}$ | $11\frac{1}{4}$ |
| $5\frac{1}{8}$ | $9\frac{1}{2}$ | $17\frac{3}{4}$ | $12\frac{1}{2}$ | 8 | $5\frac{1}{8}$ | $13\frac{1}{4}$ | $21\frac{1}{4}$ | 16 | 11 |
| $5\frac{1}{8}$ | 9 | 20 | $13\frac{1}{2}$ | $7\frac{1}{2}$ | $5\frac{1}{8}$ | $12\frac{1}{2}$ | 24 | 17 | $10\frac{1}{2}$ |
| $5\frac{1}{8}$ | $8\frac{1}{4}$ | 22 | 14 | $7\frac{1}{2}$ | $6\frac{3}{4}$ | 9 | $20\frac{3}{4}$ | 16 | $10\frac{1}{2}$ |
| $5\frac{1}{8}$ | $7\frac{3}{4}$ | 24 | $14\frac{1}{2}$ | $7\frac{1}{2}$ | $6\frac{3}{4}$ | $8\frac{1}{2}$ | $22\frac{1}{2}$ | $16\frac{1}{2}$ | $10\frac{1}{2}$ |
| $5\frac{1}{8}$ | $19\frac{3}{4}$ | $19\frac{3}{4}$ | $18\frac{1}{2}$ | $12\frac{1}{4}$ | $5\frac{1}{8}$ | $28\frac{1}{4}$ | $28\frac{1}{4}$ | 23 | 17 |
| $5\frac{1}{8}$ | $17\frac{1}{2}$ | 19 | 16 | $12\frac{1}{4}$ | $5\frac{1}{8}$ | 25 | 25 | $22\frac{1}{4}$ | $12\frac{1}{4}$ |
| $5\frac{1}{8}$ | $15\frac{3}{4}$ | 23 | $15\frac{3}{4}$ | $12\frac{1}{4}$ | $6\frac{3}{4}$ | $17\frac{1}{2}$ | 22 | $18\frac{3}{4}$ | $12\frac{1}{4}$ |
| $6\frac{3}{4}$ | 11 | $20\frac{1}{2}$ | $17\frac{1}{4}$ | $12\frac{1}{4}$ | $6\frac{3}{4}$ | 16 | $24\frac{3}{4}$ | 19 | $12\frac{1}{4}$ |
| $6\frac{3}{4}$ | 10 | $22\frac{1}{4}$ | $18\frac{1}{2}$ | $12\frac{1}{4}$ | $6\frac{3}{4}$ | $14\frac{3}{4}$ | 27 | $19\frac{1}{2}$ | $12\frac{1}{4}$ |
| $5\frac{1}{8}$ | $17\frac{3}{4}$ | $17\frac{3}{4}$ | 23 | 12 | $5\frac{1}{8}$ | $25\frac{1}{4}$ | $25\frac{1}{4}$ | 23 | $16\frac{3}{4}$ |
| $5\frac{1}{8}$ | 16 | $19\frac{3}{4}$ | $15\frac{3}{4}$ | 12 | $5\frac{1}{8}$ | $22\frac{1}{2}$ | $22\frac{1}{2}$ | $21\frac{3}{4}$ | 12 |
| $5\frac{1}{8}$ | $14\frac{1}{2}$ | $23\frac{1}{4}$ | $15\frac{1}{2}$ | 12 | $6\frac{3}{4}$ | 16 | $22\frac{1}{4}$ | 18 | 12 |
| $6\frac{3}{4}$ | $10\frac{1}{4}$ | $20\frac{1}{2}$ | $16\frac{3}{4}$ | 12 | $6\frac{3}{4}$ | $14\frac{3}{4}$ | $24\frac{3}{4}$ | $18\frac{1}{2}$ | 12 |
| $6\frac{3}{4}$ | $9\frac{1}{2}$ | 22 | $18\frac{1}{4}$ | 12 | $6\frac{3}{4}$ | $13\frac{3}{4}$ | $26\frac{3}{4}$ | 19 | 12 |
| $5\frac{1}{8}$ | $13\frac{3}{4}$ | $20\frac{1}{2}$ | $14\frac{1}{4}$ | $11\frac{1}{4}$ | $5\frac{1}{8}$ | 19 | 24 | 19 | $11\frac{1}{4}$ |
| $5\frac{1}{8}$ | $12\frac{1}{2}$ | $23\frac{1}{4}$ | $14\frac{1}{2}$ | $11\frac{1}{4}$ | $6\frac{3}{4}$ | $13\frac{1}{2}$ | 22 | $17\frac{1}{4}$ | $11\frac{1}{4}$ |
| $5\frac{1}{8}$ | $11\frac{3}{4}$ | $25\frac{3}{4}$ | 15 | $11\frac{1}{4}$ | $6\frac{3}{4}$ | $12\frac{3}{4}$ | $24\frac{1}{4}$ | $17\frac{3}{4}$ | $11\frac{1}{4}$ |
| $6\frac{3}{4}$ | $8\frac{1}{4}$ | $21\frac{3}{4}$ | $17\frac{1}{4}$ | $11\frac{1}{4}$ | $6\frac{3}{4}$ | $11\frac{3}{4}$ | $26\frac{1}{4}$ | $18\frac{1}{2}$ | $11\frac{1}{4}$ |
| $5\frac{1}{8}$ | 12 | $20\frac{1}{4}$ | 14 | $10\frac{1}{2}$ | $5\frac{1}{8}$ | $16\frac{1}{4}$ | 24 | $17\frac{3}{4}$ | $12\frac{1}{4}$ |
| $5\frac{1}{8}$ | 11 | $22\frac{3}{4}$ | $14\frac{3}{4}$ | $10\frac{1}{2}$ | $6\frac{3}{4}$ | $11\frac{3}{4}$ | $21\frac{1}{2}$ | $16\frac{3}{4}$ | $10\frac{1}{2}$ |
| $5\frac{1}{8}$ | $10\frac{1}{4}$ | $25\frac{1}{4}$ | $15\frac{1}{2}$ | $10\frac{1}{2}$ | $6\frac{3}{4}$ | 11 | $23\frac{1}{2}$ | $17\frac{3}{4}$ | $10\frac{1}{2}$ |
| $6\frac{3}{4}$ | $7\frac{1}{2}$ | 21 | $16\frac{1}{2}$ | $10\frac{1}{2}$ | $6\frac{3}{4}$ | $10\frac{1}{2}$ | $25\frac{1}{4}$ | $18\frac{1}{2}$ | $10\frac{1}{2}$ |

(*continued*)

| Loading | Roof Pitch | Wall Ht (ft) | 60′ Span | | | | | 70′ Span | | | | |
|---|---|---|---|---|---|---|---|---|---|---|---|---|
| | | | Width | Base | Lower Tang. | Upper Tang. | Crown | Width | Base | Lower Tang. | Upper Tang. | Crown |
| Vertical dead + live load = 400 lb/ft 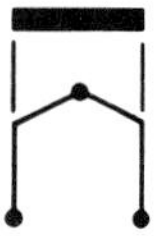 | 3/12 | 12 | $5\frac{1}{8}$ | 14 | $16\frac{1}{4}$ | $16\frac{3}{4}$ | $7\frac{1}{2}$ | $5\frac{1}{8}$ | $17\frac{3}{4}$ | $17\frac{3}{4}$ | 20 | $7\frac{1}{2}$ |
| | | 14 | $5\frac{1}{8}$ | $12\frac{3}{4}$ | $19\frac{1}{2}$ | $17\frac{1}{2}$ | $7\frac{1}{2}$ | $5\frac{1}{8}$ | $16\frac{1}{4}$ | 22 | $20\frac{3}{4}$ | $7\frac{1}{2}$ |
| | | 16 | $5\frac{1}{8}$ | $11\frac{3}{4}$ | 22 | $17\frac{3}{4}$ | $7\frac{1}{2}$ | $5\frac{1}{8}$ | 15 | 25 | $21\frac{1}{4}$ | $7\frac{1}{2}$ |
| | | 18 | $5\frac{1}{8}$ | $10\frac{3}{4}$ | 24 | $19\frac{1}{4}$ | $7\frac{1}{2}$ | $6\frac{3}{4}$ | $10\frac{3}{4}$ | 22 | $21\frac{1}{4}$ | $7\frac{1}{2}$ |
| | | 20 | $6\frac{3}{4}$ | $7\frac{3}{4}$ | 22 | 22 | $7\frac{1}{2}$ | $6\frac{3}{4}$ | 10 | $23\frac{1}{2}$ | $23\frac{1}{4}$ | $7\frac{1}{2}$ |
| | 4/12 | 12 | $5\frac{1}{8}$ | $12\frac{1}{2}$ | $16\frac{3}{4}$ | $15\frac{3}{4}$ | $7\frac{1}{2}$ | $5\frac{1}{8}$ | $15\frac{3}{4}$ | $18\frac{1}{2}$ | $18\frac{3}{4}$ | $7\frac{3}{4}$ |
| | | 14 | $5\frac{1}{8}$ | $11\frac{1}{2}$ | $19\frac{1}{2}$ | $16\frac{1}{2}$ | $7\frac{1}{2}$ | $5\frac{1}{8}$ | $14\frac{1}{2}$ | 22 | $19\frac{1}{2}$ | $7\frac{1}{2}$ |
| | | 16 | $5\frac{1}{8}$ | $10\frac{1}{2}$ | $21\frac{3}{4}$ | 17 | $7\frac{1}{2}$ | $5\frac{1}{8}$ | $13\frac{1}{2}$ | $24\frac{1}{2}$ | $20\frac{1}{4}$ | $7\frac{1}{2}$ |
| | | 18 | $5\frac{1}{8}$ | 10 | $23\frac{3}{4}$ | $17\frac{3}{4}$ | $7\frac{1}{2}$ | $6\frac{3}{4}$ | $9\frac{3}{4}$ | $21\frac{1}{2}$ | $19\frac{1}{2}$ | $7\frac{1}{2}$ |
| | | 20 | $5\frac{1}{8}$ | $9\frac{1}{4}$ | $25\frac{1}{2}$ | $19\frac{1}{4}$ | $7\frac{1}{2}$ | $6\frac{3}{4}$ | 9 | 23 | $21\frac{1}{2}$ | $7\frac{1}{2}$ |
| | 6/12 | 12 | $5\frac{1}{8}$ | $10\frac{1}{4}$ | $16\frac{3}{4}$ | $14\frac{1}{4}$ | $8\frac{1}{4}$ | $5\frac{1}{8}$ | $12\frac{3}{4}$ | $18\frac{3}{4}$ | $16\frac{3}{4}$ | $10\frac{1}{2}$ |
| | | 14 | $5\frac{1}{8}$ | $9\frac{1}{2}$ | 19 | $15\frac{1}{4}$ | $7\frac{1}{2}$ | $5\frac{1}{8}$ | 12 | $21\frac{1}{2}$ | $17\frac{3}{4}$ | 9 |
| | | 16 | $5\frac{1}{8}$ | 9 | 21 | 16 | $7\frac{1}{2}$ | $5\frac{1}{8}$ | $11\frac{1}{4}$ | $23\frac{3}{4}$ | $18\frac{3}{4}$ | $7\frac{3}{4}$ |
| | | 18 | $5\frac{1}{8}$ | $8\frac{1}{2}$ | 23 | $16\frac{1}{2}$ | $7\frac{1}{2}$ | $6\frac{3}{4}$ | 8 | $20\frac{3}{4}$ | $17\frac{1}{4}$ | $7\frac{1}{2}$ |
| | 8/12 | 12 | $5\frac{1}{8}$ | $8\frac{3}{4}$ | $16\frac{1}{2}$ | $13\frac{1}{2}$ | 10 | $5\frac{1}{8}$ | $10\frac{3}{4}$ | $18\frac{1}{2}$ | $15\frac{1}{2}$ | $12\frac{3}{4}$ |
| | | 14 | $5\frac{1}{8}$ | $8\frac{1}{4}$ | $18\frac{1}{2}$ | $14\frac{1}{2}$ | $8\frac{3}{4}$ | $5\frac{1}{8}$ | $10\frac{1}{4}$ | $20\frac{3}{4}$ | $16\frac{3}{4}$ | $11\frac{1}{4}$ |
| | | 16 | $5\frac{1}{8}$ | $7\frac{3}{4}$ | $20\frac{1}{2}$ | $15\frac{1}{2}$ | $7\frac{1}{2}$ | $5\frac{1}{8}$ | $9\frac{1}{2}$ | 23 | $17\frac{3}{4}$ | $9\frac{3}{4}$ |
| | | 18 | $5\frac{1}{8}$ | $7\frac{1}{2}$ | $22\frac{1}{4}$ | $16\frac{1}{4}$ | $7\frac{1}{2}$ | $5\frac{1}{8}$ | $9\frac{1}{4}$ | $25\frac{1}{4}$ | $18\frac{3}{4}$ | $8\frac{3}{4}$ |
| Vertical dead + live load = 600 lb/ft  | 3/12 | 12 | $5\frac{1}{8}$ | $20\frac{1}{2}$ | $20\frac{1}{2}$ | $20\frac{1}{4}$ | $7\frac{1}{2}$ | $5\frac{1}{8}$ | $26\frac{1}{4}$ | $26\frac{1}{4}$ | 24 | $12\frac{1}{4}$ |
| | | 14 | $5\frac{1}{8}$ | $18\frac{3}{4}$ | 24 | 21 | $7\frac{1}{2}$ | $6\frac{3}{4}$ | $18\frac{1}{2}$ | 22 | $22\frac{1}{4}$ | $12\frac{1}{4}$ |
| | | 16 | $6\frac{3}{4}$ | $13\frac{1}{4}$ | 22 | 19 | $7\frac{1}{2}$ | $6\frac{3}{4}$ | $17\frac{1}{4}$ | 25 | 23 | $12\frac{1}{4}$ |
| | | 18 | $6\frac{3}{4}$ | $12\frac{1}{4}$ | 24 | $20\frac{1}{4}$ | $7\frac{1}{2}$ | $6\frac{3}{4}$ | 16 | $27\frac{1}{4}$ | $23\frac{1}{2}$ | $12\frac{1}{4}$ |
| | | 20 | $6\frac{3}{4}$ | $11\frac{1}{2}$ | $25\frac{3}{4}$ | $21\frac{3}{4}$ | $7\frac{1}{2}$ | $6\frac{3}{4}$ | $14\frac{3}{4}$ | $29\frac{1}{2}$ | 24 | $12\frac{1}{4}$ |
| | 4/12 | 12 | $5\frac{1}{8}$ | $18\frac{1}{4}$ | $20\frac{1}{4}$ | $19\frac{1}{4}$ | $7\frac{1}{2}$ | $5\frac{1}{8}$ | $23\frac{1}{4}$ | $23\frac{1}{4}$ | $22\frac{3}{4}$ | 12 |
| | | 14 | $5\frac{1}{8}$ | 17 | $24\frac{1}{4}$ | 20 | $7\frac{1}{2}$ | $6\frac{3}{4}$ | $16\frac{1}{2}$ | 22 | $21\frac{1}{4}$ | 12 |
| | | 16 | $6\frac{3}{4}$ | 12 | $21\frac{3}{4}$ | $18\frac{1}{4}$ | $7\frac{1}{2}$ | $6\frac{3}{4}$ | $15\frac{1}{2}$ | $24\frac{3}{4}$ | $21\frac{3}{4}$ | 12 |
| | | 18 | $6\frac{3}{4}$ | $11\frac{1}{4}$ | $23\frac{1}{2}$ | $18\frac{3}{4}$ | $7\frac{1}{2}$ | $6\frac{3}{4}$ | $14\frac{1}{2}$ | 27 | $22\frac{1}{2}$ | 12 |
| | | 20 | $6\frac{3}{4}$ | $10\frac{1}{2}$ | $25\frac{1}{4}$ | $20\frac{1}{2}$ | $7\frac{1}{2}$ | $6\frac{3}{4}$ | $13\frac{1}{2}$ | $28\frac{3}{4}$ | 23 | 12 |
| | 6/12 | 12 | $5\frac{1}{8}$ | $15\frac{1}{4}$ | 21 | $17\frac{1}{2}$ | $10\frac{1}{4}$ | $5\frac{1}{8}$ | $18\frac{3}{4}$ | $23\frac{1}{4}$ | $20\frac{1}{2}$ | $13\frac{1}{4}$ |
| | | 14 | $5\frac{1}{8}$ | 14 | 24 | $18\frac{1}{2}$ | $8\frac{3}{4}$ | $6\frac{3}{4}$ | $13\frac{1}{2}$ | $21\frac{3}{4}$ | $19\frac{1}{4}$ | $11\frac{1}{4}$ |
| | | 16 | $6\frac{3}{4}$ | $10\frac{1}{4}$ | $21\frac{1}{4}$ | $17\frac{1}{4}$ | $7\frac{1}{2}$ | $6\frac{3}{4}$ | $12\frac{3}{4}$ | 24 | 20 | $11\frac{1}{4}$ |
| | | 18 | $6\frac{3}{4}$ | $9\frac{1}{2}$ | $22\frac{3}{4}$ | $17\frac{3}{4}$ | $7\frac{1}{2}$ | $6\frac{3}{4}$ | 12 | 26 | 21 | $11\frac{1}{4}$ |
| | 8/12 | 12 | $5\frac{1}{8}$ | 13 | $20\frac{3}{4}$ | $16\frac{1}{2}$ | $12\frac{1}{2}$ | $5\frac{1}{8}$ | 16 | $23\frac{1}{4}$ | 19 | 16 |
| | | 14 | $5\frac{1}{8}$ | $12\frac{1}{4}$ | $23\frac{1}{4}$ | $17\frac{3}{4}$ | $10\frac{3}{4}$ | $6\frac{3}{4}$ | $11\frac{1}{2}$ | 21 | 18 | 12 |
| | | 16 | $6\frac{3}{4}$ | $8\frac{3}{4}$ | $20\frac{1}{2}$ | $16\frac{1}{2}$ | 8 | $6\frac{3}{4}$ | 11 | 23 | $19\frac{1}{4}$ | $10\frac{1}{2}$ |
| | | 18 | $6\frac{3}{4}$ | $8\frac{1}{4}$ | 22 | $17\frac{1}{2}$ | $7\frac{1}{2}$ | $6\frac{3}{4}$ | $10\frac{1}{2}$ | 25 | 20 | $10\frac{1}{2}$ |

| 80' Span | | | | | 90' Span | | | | |
|---|---|---|---|---|---|---|---|---|---|
| Width | Base | Lower Tang. | Upper Tang. | Crown | Width | Base | Lower Tang. | Upper Tang. | Crown |
| 5⅛ | 21¾ | 21¾ | 23 | 7½ | 6¾ | 20 | 20 | 22¾ | 8 |
| 5⅛ | 20 | 24 | 24 | 7½ | 6¾ | 18½ | 21½ | 24 | 7½ |
| 6¾ | 14¼ | 22½ | 22 | 7½ | 6¾ | 17¼ | 24¾ | 24¾ | 7½ |
| 6¾ | 13¼ | 24½ | 22¾ | 7½ | 6¾ | 16 | 27 | 25¾ | 7½ |
| 6¾ | 12½ | 26½ | 23 | 7½ | 6¾ | 15 | 29¼ | 26 | 7½ |
| 5⅛ | 19¼ | 20 | 21½ | 9½ | 5⅛ | 22¾ | 22¾ | 24¼ | 12 |
| 5⅛ | 17¾ | 24¼ | 22¾ | 8¼ | 6¾ | 16¼ | 21½ | 22¾ | 8¼ |
| 6¾ | 12¼ | 22 | 20¾ | 7½ | 6¾ | 15¼ | 24¼ | 23¼ | 7½ |
| 6¾ | 12 | 24 | 21¼ | 7½ | 6¾ | 14¼ | 26¼ | 24¼ | 7½ |
| 6¾ | 11¼ | 25¾ | 22 | 7½ | 6¾ | 13½ | 28½ | 24¾ | 7½ |
| 5⅛ | 15¼ | 20½ | 19 | 13¼ | 5⅛ | 18 | 22¼ | 21¼ | 16 |
| 5⅛ | 14½ | 23¾ | 20¼ | 11½ | 6¾ | 13 | 21 | 19¾ | 12 |
| 6¾ | 10½ | 21¼ | 18¾ | 8½ | 6¾ | 12¼ | 23¾ | 21 | 10½ |
| 6¾ | 9¾ | 23 | 19½ | 7½ | 6¾ | 11¾ | 25¼ | 22 | 9¼ |
| 5⅛ | 12¾ | 20¼ | 17¼ | 16 | 5⅛ | 15 | 22 | 19¼ | 19¼ |
| 5⅛ | 12¼ | 23 | 18¾ | 14 | 5⅛ | 14¼ | 25¼ | 20¾ | 17 |
| 5⅛ | 11½ | 25¾ | 20 | 12½ | 6¾ | 10¼ | 22¼ | 19¼ | 12¾ |
| 6¾ | 8½ | 22 | 18½ | 9¼ | 6¾ | 10 | 24 | 20½ | 11½ |
| 6¾ | 25 | 25 | 24½ | 12¼ | 6¾ | 29¾ | 29¾ | 27½ | 12¼ |
| 6¾ | 23 | 23½ | 26 | 12¼ | 6¾ | 27½ | 27½ | 29¼ | 12¼ |
| 6¾ | 21¼ | 27½ | 26¾ | 12¼ | 6¾ | 25½ | 30 | 30¼ | 12¼ |
| 6¾ | 19¾ | 30½ | 27¼ | 12¼ | 6¾ | 23¾ | 33½ | 31 | 12¼ |
| 6¾ | 18½ | 33 | 28 | 12¼ | 8¾ | 17½ | 30 | 28¼ | 12¼ |
| 6¾ | 21¾ | 21¾ | 23 | 12 | 6¾ | 25¾ | 25¾ | 25½ | 12¾ |
| 6¾ | 20¼ | 24¼ | 24¼ | 12 | 6¾ | 24 | 26 | 27¼ | 12 |
| 6¾ | 18¾ | 27¼ | 25 | 12 | 6¾ | 22½ | 29¾ | 28¼ | 12 |
| 6¾ | 17¾ | 30 | 26 | 12 | 6¾ | 21¼ | 32¾ | 29¼ | 12 |
| 6¾ | 16¾ | 32¼ | 26½ | 12 | 8¾ | 15½ | 19¼ | 26¾ | 12 |
| 5⅛ | 22¾ | 25½ | 23½ | 16¼ | 6¾ | 20½ | 22¼ | 22¾ | 17 |
| 6¾ | 16¼ | 23¾ | 22 | 12 | 6¾ | 19¼ | 26 | 24¼ | 14¾ |
| 6¾ | 15½ | 26½ | 23 | 11¼ | 6¾ | 18¼ | 29 | 25½ | 13 |
| 6¾ | 14¾ | 28¾ | 23¾ | 11¼ | 6¾ | 17¼ | 31½ | 26¾ | 11¾ |
| 5⅛ | 19 | 25½ | 21¼ | 19¾ | 6¾ | 17 | 22¼ | 20½ | 20¼ |
| 6¾ | 13¾ | 23¾ | 20¼ | 14¾ | 6¾ | 16 | 25¼ | 22¼ | 18 |
| 6¾ | 13 | 25½ | 21½ | 13¼ | 6¾ | 15½ | 27¾ | 23¾ | 16 |
| 6¾ | 12½ | 27½ | 22½ | 11¾ | 6¾ | 14¾ | 30¼ | 25 | 14½ |

(*continued*)

TABLE 8.11 (*Continued*)

| Loading | Roof Pitch | Wall Ht (ft) | 30' Span | | | | | 35' Span | | | | |
|---|---|---|---|---|---|---|---|---|---|---|---|---|
| | | | Width | Base | Lower Tang. | Upper Tang. | Crown | Width | Base | Lower Tang. | Upper Tang. | Crown |
| Vertical dead = 240 lb/ft horizontal wind 320 lb/ft | 10/12 | 8 | $5\frac{1}{8}$ | $7\frac{1}{2}$ | $7\frac{1}{2}$ | $11$ | $7\frac{1}{2}$ | $5\frac{1}{8}$ | $7\frac{1}{2}$ | $7\frac{1}{2}$ | $12$ | $10\frac{1}{4}$ |
| | | 10 | $5\frac{1}{8}$ | $7\frac{1}{2}$ | $9\frac{1}{4}$ | $12\frac{1}{2}$ | $12\frac{1}{2}$ | $5\frac{1}{8}$ | $7\frac{1}{2}$ | $10$ | $13\frac{1}{4}$ | $8$ |
| | | 12 | $5\frac{1}{8}$ | $7\frac{1}{2}$ | $11\frac{1}{4}$ | $13\frac{3}{4}$ | $13\frac{3}{4}$ | $5\frac{1}{8}$ | $7\frac{1}{2}$ | $12$ | $14\frac{3}{4}$ | $7\frac{1}{2}$ |
| | 12/12 | 8 | $5\frac{1}{8}$ | $7\frac{1}{2}$ | $7\frac{1}{2}$ | $12\frac{1}{2}$ | $8\frac{3}{4}$ | $5\frac{1}{8}$ | $7\frac{1}{2}$ | $8\frac{1}{4}$ | $13\frac{1}{2}$ | $12$ |
| | | 10 | $5\frac{1}{8}$ | $7\frac{1}{2}$ | $9\frac{3}{4}$ | $13\frac{3}{4}$ | $9$ | $5\frac{1}{8}$ | $7\frac{1}{2}$ | $10\frac{1}{2}$ | $15$ | $10\frac{1}{4}$ |
| | | 12 | $5\frac{1}{8}$ | $7\frac{1}{2}$ | $11\frac{1}{4}$ | $15\frac{1}{4}$ | $15\frac{1}{4}$ | $5\frac{1}{8}$ | $7\frac{1}{2}$ | $12\frac{1}{2}$ | $16\frac{1}{2}$ | $9$ |
| | 14/12 | 8 | $5\frac{1}{8}$ | $7\frac{1}{2}$ | $8\frac{3}{4}$ | $13\frac{3}{4}$ | $10\frac{1}{2}$ | $5\frac{1}{8}$ | $7\frac{1}{2}$ | $9$ | $15$ | $13\frac{3}{4}$ |
| | | 10 | $5\frac{1}{8}$ | $7\frac{1}{2}$ | $11\frac{1}{4}$ | $15\frac{1}{4}$ | $8\frac{1}{4}$ | $5\frac{1}{8}$ | $7\frac{1}{2}$ | $11\frac{1}{2}$ | $16\frac{1}{2}$ | $12\frac{1}{4}$ |
| | | 12 | $5\frac{1}{8}$ | $7\frac{1}{2}$ | $13\frac{1}{4}$ | $17$ | $17$ | $5\frac{1}{8}$ | $7\frac{1}{2}$ | $14$ | $18$ | $8\frac{1}{4}$ |
| | 16/12 | 8 | $5\frac{1}{8}$ | $7\frac{1}{2}$ | $10$ | $15$ | $12\frac{1}{2}$ | $5\frac{1}{8}$ | $7\frac{1}{2}$ | $10\frac{1}{2}$ | $16\frac{1}{4}$ | $15\frac{3}{4}$ |
| | | 10 | $5\frac{1}{8}$ | $7\frac{1}{2}$ | $12\frac{1}{4}$ | $16\frac{1}{2}$ | $9\frac{1}{2}$ | $5\frac{1}{8}$ | $7\frac{1}{2}$ | $13$ | $18$ | $14\frac{1}{4}$ |
| | | 12 | $5\frac{1}{8}$ | $7\frac{1}{2}$ | $14\frac{3}{4}$ | $18\frac{1}{4}$ | $7\frac{1}{2}$ | $5\frac{1}{8}$ | $7\frac{1}{2}$ | $15\frac{1}{2}$ | $19\frac{3}{4}$ | $11\frac{1}{2}$ |
| Vertical dead = 320 lb/ft horizontal wind 320 lb/ft | 10/12 | 8 | $5\frac{1}{8}$ | $7\frac{1}{2}$ | $7\frac{1}{2}$ | $10\frac{3}{4}$ | $9$ | $5\frac{1}{8}$ | $7\frac{1}{2}$ | $8\frac{1}{4}$ | $11\frac{1}{4}$ | $11\frac{1}{4}$ |
| | | 10 | $5\frac{1}{8}$ | $7\frac{1}{2}$ | $10$ | $12$ | $7\frac{1}{2}$ | $5\frac{1}{8}$ | $7\frac{1}{2}$ | $10\frac{3}{4}$ | $13$ | $10\frac{1}{2}$ |
| | | 12 | $5\frac{1}{8}$ | $7\frac{1}{2}$ | $12$ | $13\frac{1}{4}$ | $13\frac{1}{4}$ | $5\frac{1}{8}$ | $7\frac{1}{2}$ | $12\frac{3}{4}$ | $14\frac{1}{4}$ | $8\frac{1}{2}$ |
| | 12/12 | 8 | $5\frac{1}{8}$ | $7\frac{1}{2}$ | $8\frac{1}{4}$ | $12\frac{1}{4}$ | $10\frac{1}{4}$ | $5\frac{1}{8}$ | $7\frac{1}{2}$ | $9$ | $13\frac{1}{4}$ | $13\frac{1}{4}$ |
| | | 10 | $5\frac{1}{8}$ | $7\frac{1}{2}$ | $10\frac{1}{2}$ | $13\frac{1}{2}$ | $9$ | $5\frac{1}{8}$ | $7\frac{1}{2}$ | $11\frac{1}{4}$ | $14\frac{3}{4}$ | $12\frac{1}{4}$ |
| | | 12 | $5\frac{1}{8}$ | $7\frac{1}{2}$ | $12\frac{1}{2}$ | $15$ | $9$ | $5\frac{1}{8}$ | $7\frac{1}{2}$ | $13\frac{1}{4}$ | $16$ | $10\frac{1}{4}$ |
| | 14/12 | 8 | $5\frac{1}{8}$ | $7\frac{1}{2}$ | $8\frac{3}{4}$ | $13\frac{1}{2}$ | $12$ | $5\frac{1}{8}$ | $7\frac{1}{2}$ | $9\frac{1}{4}$ | $14\frac{3}{4}$ | $14\frac{3}{4}$ |
| | | 10 | $5\frac{1}{8}$ | $7\frac{1}{2}$ | $10\frac{3}{4}$ | $15$ | $10\frac{1}{4}$ | $5\frac{1}{8}$ | $7\frac{1}{2}$ | $11\frac{1}{4}$ | $16\frac{1}{4}$ | $14\frac{1}{4}$ |
| | | 12 | $5\frac{1}{8}$ | $7\frac{1}{2}$ | $12\frac{3}{4}$ | $16\frac{1}{2}$ | $8\frac{1}{4}$ | $5\frac{1}{8}$ | $7\frac{1}{2}$ | $13\frac{3}{4}$ | $17\frac{3}{4}$ | $12\frac{1}{2}$ |
| | 16/12 | 8 | $5\frac{1}{8}$ | $7\frac{1}{2}$ | $9\frac{1}{2}$ | $15$ | $13\frac{3}{4}$ | $5\frac{1}{8}$ | $7\frac{1}{2}$ | $9\frac{3}{4}$ | $16\frac{1}{4}$ | $16\frac{1}{4}$ |
| | | 10 | $5\frac{1}{8}$ | $7\frac{1}{2}$ | $11\frac{3}{4}$ | $16\frac{1}{2}$ | $12\frac{1}{4}$ | $5\frac{1}{8}$ | $7\frac{1}{2}$ | $12\frac{3}{4}$ | $18$ | $16\frac{1}{4}$ |
| | | 12 | $5\frac{1}{8}$ | $7\frac{1}{2}$ | $14\frac{1}{4}$ | $18$ | $7\frac{3}{4}$ | $5\frac{1}{8}$ | $7\frac{1}{2}$ | $14\frac{3}{4}$ | $19\frac{1}{2}$ | $14\frac{1}{2}$ |
| Vertical dead = 480 lb/ft horizontal wind = 320 lb/ft | 10/12 | 8 | $5\frac{1}{8}$ | $7\frac{1}{2}$ | $9$ | $10\frac{1}{2}$ | $10\frac{1}{2}$ | $5\frac{1}{8}$ | $7\frac{1}{2}$ | $9\frac{3}{4}$ | $12$ | $12$ |
| | | 10 | $5\frac{1}{8}$ | $7\frac{1}{2}$ | $11\frac{1}{4}$ | $11\frac{1}{2}$ | $10\frac{1}{4}$ | $5\frac{1}{8}$ | $7\frac{1}{2}$ | $12\frac{1}{4}$ | $12\frac{1}{2}$ | $12\frac{1}{2}$ |
| | | 12 | $5\frac{1}{8}$ | $7\frac{1}{2}$ | $13\frac{1}{2}$ | $12\frac{3}{4}$ | $8\frac{3}{4}$ | $5\frac{1}{8}$ | $7\frac{1}{2}$ | $14\frac{3}{4}$ | $13\frac{1}{2}$ | $12\frac{1}{2}$ |
| | 12/12 | 8 | $5\frac{1}{8}$ | $7\frac{1}{2}$ | $9\frac{1}{4}$ | $12$ | $12$ | $5\frac{1}{8}$ | $7\frac{1}{2}$ | $10\frac{1}{4}$ | $13\frac{1}{2}$ | $13\frac{1}{2}$ |
| | | 10 | $5\frac{1}{8}$ | $7\frac{1}{2}$ | $11\frac{3}{4}$ | $13\frac{1}{4}$ | $11\frac{1}{2}$ | $5\frac{1}{8}$ | $7\frac{1}{2}$ | $12\frac{3}{4}$ | $14\frac{1}{4}$ | $14\frac{1}{4}$ |
| | | 12 | $5\frac{1}{8}$ | $7\frac{1}{2}$ | $13\frac{3}{4}$ | $14\frac{1}{2}$ | $10$ | $5\frac{1}{8}$ | $7\frac{1}{2}$ | $15$ | $15\frac{1}{2}$ | $14$ |
| | 14/12 | 8 | $5\frac{1}{8}$ | $7\frac{1}{2}$ | $9\frac{3}{4}$ | $13\frac{1}{4}$ | $13\frac{1}{4}$ | $5\frac{1}{8}$ | $7\frac{1}{2}$ | $10\frac{1}{2}$ | $15$ | $15$ |
| | | 10 | $5\frac{1}{8}$ | $7\frac{1}{2}$ | $12$ | $14\frac{3}{4}$ | $13$ | $5\frac{1}{8}$ | $7\frac{1}{2}$ | $13$ | $16$ | $16$ |
| | | 12 | $5\frac{1}{8}$ | $7\frac{1}{2}$ | $14\frac{1}{4}$ | $16$ | $11\frac{1}{2}$ | $5\frac{1}{8}$ | $7\frac{1}{2}$ | $15\frac{1}{2}$ | $17\frac{1}{2}$ | $15\frac{1}{2}$ |
| | 16/12 | 8 | $5\frac{1}{8}$ | $7\frac{1}{2}$ | $10$ | $14\frac{3}{4}$ | $14\frac{3}{4}$ | $5\frac{1}{8}$ | $7\frac{1}{2}$ | $10\frac{3}{4}$ | $16\frac{3}{4}$ | $16\frac{3}{4}$ |
| | | 10 | $5\frac{1}{8}$ | $7\frac{1}{2}$ | $12\frac{1}{4}$ | $16\frac{1}{4}$ | $14\frac{1}{4}$ | $5\frac{1}{8}$ | $7\frac{1}{2}$ | $13\frac{1}{4}$ | $17\frac{3}{4}$ | $17\frac{3}{4}$ |
| | | 12 | $5\frac{1}{8}$ | $7\frac{1}{2}$ | $14\frac{1}{2}$ | $17\frac{3}{4}$ | $13\frac{3}{4}$ | $5\frac{1}{8}$ | $7\frac{1}{2}$ | $15\frac{3}{4}$ | $19\frac{1}{4}$ | $17\frac{1}{2}$ |

| | 40' Span | | | | | 50' Span | | | |
|---|---|---|---|---|---|---|---|---|---|
| Width | Base | Lower Tang. | Upper Tang. | Crown | Width | Base | Lower Tang. | Upper Tang. | Crown |
| $5\frac{1}{8}$ | $7\frac{1}{2}$ | $8\frac{1}{4}$ | $13$ | $12\frac{3}{4}$ | $5\frac{1}{8}$ | $7\frac{1}{2}$ | $9\frac{1}{4}$ | $15\frac{1}{2}$ | $15\frac{1}{2}$ |
| $5\frac{1}{8}$ | $7\frac{1}{2}$ | $10\frac{3}{4}$ | $14\frac{1}{4}$ | $11\frac{3}{4}$ | $5\frac{1}{8}$ | $7\frac{1}{2}$ | $12$ | $16\frac{1}{4}$ | $16\frac{1}{4}$ |
| $5\frac{1}{8}$ | $7\frac{1}{2}$ | $12\frac{3}{4}$ | $15\frac{3}{4}$ | $9\frac{3}{4}$ | $5\frac{1}{8}$ | $7\frac{1}{2}$ | $14\frac{1}{4}$ | $17\frac{1}{2}$ | $16$ |
| $5\frac{1}{8}$ | $7\frac{1}{2}$ | $8\frac{3}{4}$ | $14\frac{3}{4}$ | $14\frac{1}{2}$ | $5\frac{1}{8}$ | $7\frac{1}{2}$ | $9\frac{3}{4}$ | $17\frac{1}{2}$ | $17\frac{1}{2}$ |
| $5\frac{1}{8}$ | $7\frac{1}{2}$ | $11$ | $16\frac{1}{4}$ | $13\frac{3}{4}$ | $5\frac{1}{8}$ | $7\frac{1}{2}$ | $12\frac{1}{2}$ | $18\frac{1}{2}$ | $18\frac{1}{2}$ |
| $5\frac{1}{8}$ | $7\frac{1}{2}$ | $13\frac{1}{4}$ | $17\frac{1}{2}$ | $11\frac{1}{4}$ | $5\frac{1}{8}$ | $7\frac{1}{2}$ | $14\frac{3}{4}$ | $20$ | $18\frac{1}{4}$ |
| $5\frac{1}{8}$ | $7\frac{1}{2}$ | $9\frac{1}{2}$ | $16\frac{1}{4}$ | $16\frac{1}{4}$ | $5\frac{1}{8}$ | $7\frac{1}{2}$ | $10\frac{1}{4}$ | $19\frac{1}{2}$ | $19\frac{1}{2}$ |
| $5\frac{1}{8}$ | $7\frac{1}{2}$ | $12\frac{1}{2}$ | $18$ | $15\frac{3}{4}$ | $5\frac{1}{8}$ | $7\frac{3}{4}$ | $13\frac{1}{4}$ | $20\frac{1}{2}$ | $20\frac{1}{2}$ |
| $5\frac{1}{8}$ | $7\frac{1}{2}$ | $14\frac{1}{2}$ | $19\frac{1}{2}$ | $14\frac{1}{4}$ | $5\frac{1}{8}$ | $8\frac{1}{4}$ | $16$ | $22$ | $20\frac{3}{4}$ |
| $5\frac{1}{8}$ | $7\frac{1}{2}$ | $11$ | $18$ | $18$ | $5\frac{1}{8}$ | $9$ | $11$ | $21\frac{1}{2}$ | $21\frac{1}{2}$ |
| $5\frac{1}{8}$ | $7\frac{3}{4}$ | $14$ | $19\frac{1}{2}$ | $18$ | $5\frac{1}{8}$ | $9\frac{1}{4}$ | $14\frac{3}{4}$ | $22\frac{3}{4}$ | $22\frac{3}{4}$ |
| $5\frac{1}{8}$ | $8\frac{1}{4}$ | $16\frac{1}{4}$ | $21\frac{1}{4}$ | $16\frac{3}{4}$ | $5\frac{1}{8}$ | $9\frac{3}{4}$ | $17\frac{1}{2}$ | $24$ | $23\frac{1}{2}$ |
| $5\frac{1}{8}$ | $7\frac{1}{2}$ | $9$ | $13$ | $13$ | $5\frac{1}{8}$ | $8$ | $10$ | $15\frac{3}{4}$ | $15\frac{3}{4}$ |
| $5\frac{1}{8}$ | $7\frac{1}{2}$ | $11\frac{1}{2}$ | $14$ | $13\frac{1}{2}$ | $5\frac{1}{8}$ | $7\frac{3}{4}$ | $13$ | $16\frac{1}{2}$ | $16\frac{1}{2}$ |
| $5\frac{1}{8}$ | $7\frac{1}{2}$ | $13\frac{3}{4}$ | $15\frac{1}{4}$ | $12\frac{1}{2}$ | $5\frac{1}{8}$ | $7\frac{3}{4}$ | $15\frac{3}{4}$ | $17\frac{1}{4}$ | $17\frac{1}{4}$ |
| $5\frac{1}{8}$ | $7\frac{1}{2}$ | $9\frac{1}{2}$ | $14\frac{3}{4}$ | $14\frac{3}{4}$ | $5\frac{1}{8}$ | $7\frac{3}{4}$ | $10\frac{1}{2}$ | $17\frac{3}{4}$ | $17\frac{3}{4}$ |
| $5\frac{1}{8}$ | $7\frac{1}{2}$ | $12$ | $15\frac{3}{4}$ | $15\frac{1}{4}$ | $5\frac{1}{8}$ | $7\frac{3}{4}$ | $13\frac{1}{2}$ | $18\frac{3}{4}$ | $18\frac{3}{4}$ |
| $5\frac{1}{8}$ | $7\frac{1}{2}$ | $14\frac{1}{4}$ | $17\frac{1}{4}$ | $14\frac{1}{4}$ | $5\frac{1}{8}$ | $7\frac{3}{4}$ | $16$ | $19\frac{1}{2}$ | $19\frac{1}{2}$ |
| $5\frac{1}{8}$ | $7\frac{1}{2}$ | $10$ | $16\frac{1}{2}$ | $16\frac{1}{2}$ | $5\frac{1}{8}$ | $7\frac{3}{4}$ | $10\frac{1}{2}$ | $19\frac{3}{4}$ | $19\frac{3}{4}$ |
| $5\frac{1}{8}$ | $7\frac{1}{2}$ | $12\frac{1}{2}$ | $17\frac{1}{2}$ | $17\frac{1}{4}$ | $5\frac{1}{8}$ | $7\frac{3}{4}$ | $13\frac{3}{4}$ | $20\frac{3}{4}$ | $20\frac{3}{4}$ |
| $5\frac{1}{8}$ | $7\frac{1}{2}$ | $14\frac{3}{4}$ | $19\frac{1}{4}$ | $16\frac{1}{2}$ | $5\frac{1}{8}$ | $8$ | $16\frac{1}{2}$ | $21\frac{3}{4}$ | $21\frac{3}{4}$ |
| $5\frac{1}{8}$ | $7\frac{1}{2}$ | $10\frac{1}{4}$ | $18\frac{1}{4}$ | $18\frac{1}{4}$ | $5\frac{1}{8}$ | $8\frac{1}{4}$ | $10\frac{3}{4}$ | $22$ | $22$ |
| $5\frac{1}{8}$ | $7\frac{1}{2}$ | $13$ | $19\frac{1}{4}$ | $19\frac{1}{4}$ | $5\frac{1}{8}$ | $8\frac{1}{2}$ | $14$ | $23$ | $23$ |
| $5\frac{1}{8}$ | $7\frac{1}{2}$ | $15\frac{3}{4}$ | $21$ | $18\frac{3}{4}$ | $5\frac{1}{8}$ | $9$ | $16\frac{3}{4}$ | $24\frac{1}{4}$ | $24\frac{1}{4}$ |
| $5\frac{1}{8}$ | $8$ | $10\frac{1}{2}$ | $13\frac{1}{2}$ | $13\frac{1}{2}$ | $5\frac{1}{8}$ | $10\frac{1}{4}$ | $11\frac{1}{4}$ | $16\frac{1}{2}$ | $16\frac{1}{2}$ |
| $5\frac{1}{8}$ | $7\frac{3}{4}$ | $13\frac{1}{4}$ | $14$ | $14$ | $5\frac{1}{8}$ | $10$ | $15\frac{1}{4}$ | $17$ | $17$ |
| $5\frac{1}{8}$ | $7\frac{3}{4}$ | $16$ | $14\frac{3}{4}$ | $14\frac{3}{4}$ | $5\frac{1}{8}$ | $9\frac{3}{4}$ | $18$ | $17\frac{1}{2}$ | $17\frac{1}{2}$ |
| $5\frac{1}{8}$ | $7\frac{3}{4}$ | $11$ | $15\frac{1}{4}$ | $15\frac{1}{4}$ | $5\frac{1}{8}$ | $9\frac{3}{4}$ | $11\frac{3}{4}$ | $18\frac{1}{2}$ | $18\frac{1}{2}$ |
| $5\frac{1}{8}$ | $7\frac{3}{4}$ | $13\frac{3}{4}$ | $15\frac{3}{4}$ | $15\frac{3}{4}$ | $5\frac{1}{8}$ | $9\frac{3}{4}$ | $15\frac{1}{2}$ | $19$ | $19$ |
| $5\frac{1}{8}$ | $7\frac{3}{4}$ | $16\frac{1}{4}$ | $16\frac{3}{4}$ | $16\frac{3}{4}$ | $5\frac{1}{8}$ | $9\frac{1}{2}$ | $18\frac{1}{4}$ | $19\frac{3}{4}$ | $19\frac{3}{4}$ |
| $5\frac{1}{8}$ | $7\frac{3}{4}$ | $11\frac{1}{4}$ | $17$ | $17$ | $5\frac{1}{8}$ | $9\frac{1}{2}$ | $12$ | $20\frac{1}{2}$ | $20\frac{1}{2}$ |
| $5\frac{1}{8}$ | $7\frac{3}{4}$ | $14$ | $17\frac{3}{4}$ | $17\frac{3}{4}$ | $5\frac{1}{8}$ | $9\frac{1}{2}$ | $15\frac{1}{2}$ | $21\frac{1}{4}$ | $21\frac{1}{4}$ |
| $5\frac{1}{8}$ | $7\frac{3}{4}$ | $16\frac{1}{2}$ | $18\frac{3}{4}$ | $18\frac{3}{4}$ | $5\frac{1}{8}$ | $9\frac{1}{2}$ | $18\frac{1}{2}$ | $22\frac{1}{4}$ | $22\frac{1}{4}$ |
| $5\frac{1}{8}$ | $7\frac{3}{4}$ | $11\frac{1}{2}$ | $18\frac{3}{4}$ | $18\frac{3}{4}$ | $5\frac{1}{8}$ | $9\frac{1}{2}$ | $12$ | $22\frac{3}{4}$ | $22\frac{3}{4}$ |
| $5\frac{1}{8}$ | $7\frac{3}{4}$ | $14\frac{1}{4}$ | $19\frac{1}{2}$ | $19\frac{1}{2}$ | $5\frac{1}{8}$ | $9\frac{1}{2}$ | $15\frac{1}{2}$ | $23\frac{1}{2}$ | $23\frac{1}{2}$ |
| $5\frac{1}{8}$ | $7\frac{3}{4}$ | $17$ | $20\frac{3}{4}$ | $20\frac{3}{4}$ | $5\frac{1}{8}$ | $9\frac{1}{2}$ | $18\frac{3}{4}$ | $24\frac{1}{2}$ | $24\frac{1}{2}$ |

**Footnote to Table 8.11**

[a] Values for preliminary design purposes only. Sizes are based on Douglas Fir laminated timber, developing an allowable shear stress of 165 psi and with a bending radius of 9 ft 4 in. For Southern Pine laminated timber, an allowable shear stress of 200 psi and a bending radius of 7 ft 0 in. may be used. For roof pitches less than $\frac{10}{12}$, the critical loading is generally the combined dead and live load on the horizontal projection of the full span. For roof pitches of $\frac{10}{12}$ or greater, the critical loading is generally a combination of dead load and horizontal wind load. Sizes shown are determined from a uniformly distributed wind load applied on the vertical projection of the roof arm with a concentrated wind load equal to one-half the total wind load on the wall height acting at the haunch. In the combined stress analysis, it was assumed that the bending portion of the loading exceeded the axial compression portion. The section sizes shown in this table are in inches and are based on the following design criteria: (1) Uniform loading. (2) Radius of curvature at the haunch = 9 ft 4 in. (3) Allowable stresses: Bending stress, $F_b$ = 2400 psi (reduced by curvature factor when applicable); shear stress, $F_v$ = 165 psi; compression parallel to grain stress, $F_c$ = 1500 psi (adjusted for $l/d$ ratio); modulus of elasticity, $E$ = 1,600,000 psi. These stresses were increased for snow loading and wind loading when applicable. (4) Vertical arch legs are laterally supported. (5) Dead load equal to one-third of the total vertical load. Note: When arch deflection is a concern, arch sizes should be checked for deflection at the time of size selection.

## TABLE 8.12

### Controlled Random Layup for Glued Laminated Decking

ALLOWABLE UNIFORMLY DISTRIBUTED TOTAL ROOF LOAD LIMITED BY DEFLECTION

| Species | Actual Size (in.) | Span (ft) 8 | | 9 | | 10 | | 11 | | 12 | | 13 | | 14 | | 15 | | 16 | | 17 | | 18 | | 19 | | 20 | |
|---|---|---|---|---|---|---|---|---|---|---|---|---|---|---|---|---|---|---|---|---|---|---|---|---|---|---|---|
| | | *l*/180 | *l*/240 | *l*/180 | *l*/240 | *l*/180 | *l*/240 | *l*/180 | *l*/240 | *l*/180 | *l*/240 | *l*/180 | *l*/240 | *l*/180 | *l*/240 | *l*/180 | *l*/240 | *l*/180 | *l*/240 | *l*/180 | *l*/240 | *l*/180 | *l*/240 | *l*/180 | *l*/240 |
| GLUED LAMINATED TIMBER DECKING—CONTROLLED RANDOM LAYUP[a-f] | | | | | | | | | | | | | | | | | | | | | | | | | |
| Douglas Fir/ Larch and Southern Pine[c] | $3\frac{21}{32} \times 5\frac{3}{8}$ | | | | | | | | | | | | | 160 | 120 | 130 | 97 | 107 | 80 | 89 | 67 | 75 | 56 | 64 | 48 | 55 | 41 |
| | $2\frac{7}{8} \times 5\frac{3}{8}$ | | | | | | | | | 124 | 93 | 98 | 73 | 78 | 59 | 63 | 48 | 53 | 40 | 44 | 33 | | | | | | |
| | $2\frac{3}{16} \times 5\frac{3}{8}$ | 181 | 136 | 127 | 96 | 93 | 70 | 70 | 52 | 54 | 40 | 42 | 32 | | | | | | | | | | | | | | |
| Idaho White Pine and Inland White Fir | $3\frac{21}{32} \times 5\frac{3}{8}$ | | | | | | | | | | | | | | | 107 | 80 | 88 | 66 | 74 | 55 | 62 | 46 | 53 | 40 | 45 | 34 |
| | $2\frac{7}{8} \times 5\frac{3}{8}$ | | | | | | | | | 102 | 77 | 80 | 60 | 64 | 48 | 52 | 39 | 43 | 32 | 36 | 27 | | | | | | |
| | $2\frac{3}{16} \times 5\frac{3}{8}$ | 151 | 113 | 106 | 80 | 77 | 58 | 58 | 44 | 45 | 34 | 35 | 26 | | | | | | | | | | | | | | |

## TABLE 8.12 (Continued)

Span (ft) — Deflection Limit

| Species | Actual Size (in.) | 8 l/180 | 8 l/240 | 9 l/180 | 9 l/240 | 10 l/180 | 10 l/240 | 11 l/180 | 11 l/240 | 12 l/180 | 12 l/240 | 13 l/180 | 13 l/240 | 14 l/180 | 14 l/240 | 15 l/180 | 15 l/240 | 16 l/180 | 16 l/240 | 17 l/180 | 17 l/240 | 18 l/180 | 18 l/240 | 19 l/180 | 19 l/240 | 20 l/180 | 20 l/240 |
|---|---|---|---|---|---|---|---|---|---|---|---|---|---|---|---|---|---|---|---|---|---|---|---|---|---|---|---|
| Ponderosa Pine | 3¾ × 7⅛ | | | | | | | | | 192 | 144 | 151 | 113 | 120 | 90 | 98 | 74 | 81 | 61 | 68 | 51 | 57 | 43 | 48 | 36 | 41 | 31 |
| | 3 × 7⅛ | | | | | | | | | 95 | 71 | 75 | 56 | 60 | 45 | 49 | 37 | 40 | 30 | 33 | | 28 | | | | | |
| | 2¼ × 7⅛ | | 101 | 95 | 71 | 69 | 52 | 52 | 39 | 40 | 30 | 31 | 23 | 25 | | 20 | | | | | | | | | | | |
| Inland Red Cedar | 3 21/32 × 5⅜ | | | | | | | | | | | | | | | 86 | 64 | 71 | 53 | 59 | 44 | 50 | 37 | 42 | 32 | 36 | 27 |
| | 2⅞ × 5⅜ | | | | | | | | | 82 | 61 | 64 | 48 | 51 | 39 | 42 | 31 | 34 | 26 | 29 | 22 | | | | | | |
| | 2 3/16 × 5⅜ | 121 | 91 | 85 | 64 | 62 | 46 | 47 | 35 | 36 | 27 | 28 | 21 | 22 | 17 | 18 | 14 | 15 | 11 | | | | | | | | |

**DECKING SPECIES DATA**[g,h]

|  | Laminated Timber Decking[i] | | | |
| Species | Modulus of Elasticity, $E$ (psi) | Actual Size (in.) | Weight (psf) | Coverage Factor[a,b] (bd ft/sq ft) |
| --- | --- | --- | --- | --- |
| Douglas Fir/ Larch and Southern Pine[j] | 1,800,000 | $3\frac{21}{32} \times 5\frac{5}{8}$ $2\frac{7}{8} \times 5\frac{5}{8}$ $2\frac{3}{16} \times 5\frac{5}{8}$ | 10.5 8.1 6.5 | 5.58 3.35 3.35 |
| Idaho White Pine and Inland White Fir | 1,500,000 | $3\frac{21}{32} \times 5\frac{5}{8}$ $2\frac{7}{8} \times 5\frac{5}{8}$ $2\frac{3}{16} \times 5\frac{5}{8}$ | 9.5 7.3 5.0 | 5.58 3.35 3.35 |
| Ponderosa Pine | 1,300,000 | $3\frac{1}{4} \times 7\frac{1}{8}$ $3 \times 7\frac{1}{8}$ $2\frac{1}{4} \times 7\frac{1}{8}$ | 8.8 7.0 5.2 | 5.61 4.49 3.37 |
| Inland Red Cedar | 1,200,000 | $3\frac{21}{32} \times 5\frac{5}{8}$ $2\frac{7}{8} \times 5\frac{5}{8}$ $2\frac{3}{16} \times 5\frac{5}{8}$ | 7.5 5.8 4.5 | 5.58 3.35 3.35 |

[a] Allowable uniformly distributed total roof load in pounds per square foot of roof surface for flat roofs, consisting of live load and dead load, including weight of deck (see Decking Species Data at end of Table 7.9). These roof loads do not exceed bending loads allowable under recognized bending formulas. Loads are for dry condition of use.

[b] All load values assume installation conforming to manufacturer's recommendation.

[c] Load values are based on manufacturer's recommendations. In all cases, data is subject to special requirements of local building codes.

[d] Actual size for Southern pine is $2\frac{1}{4}$ in. $\times 5\frac{5}{8}$ in.: To calculate total loads for Southern pine for this size, multiply the listed valued by 1.075.

[e] Values may vary among manufacturers. See manufacturer's literature for more detailed information.

[f] To estimate board feet of decking required, multiply square feet of area to be covered by the coverage factor. Add for job site trimming and waste for irregular areas.

[g] For 8-in. nominal widths ($7\frac{1}{8}$ in. net), coverage factors are: $3\frac{1}{4}$ and $3\frac{15}{16}$ in. — 5.61; 3 in. — 4.49; $2\frac{7}{8}$ and $2\frac{1}{4}$ in. —3.37.

[h] Laminated decking sizes may vary between manufacturers. The designer should check with the supplier to determine actual sizes and load-carrying capacities.

[j] Actual size for Southern Pine is $2\frac{1}{4} \times 5\frac{5}{8}$ in.

# BEAM SYSTEM DIAGRAMS

## TABLE 8.13

### Beam Diagrams and Formulas[a]

---

**1.   SIMPLE BEAM—UNIFORMLY DISTRIBUTED LOAD**

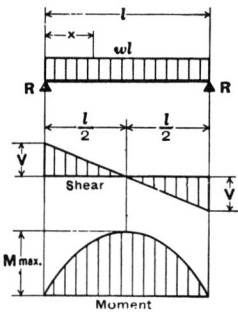

$R = V$ . . . . . . . . . $= \dfrac{wl}{2}$

$V_x$ . . . . . . . . . $= w\left(\dfrac{l}{2} - x\right)$

M max. $\left(\text{at center}\right)$ . . . . $= \dfrac{wl^2}{8}$

$M_x$ . . . . . . . . . $= \dfrac{wx}{2}(l - x)$

$\Delta$max. $\left(\text{at center}\right)$ . . . . $= \dfrac{5\,wl^4}{384\,EI}$

$\Delta_x$ . . . . . . . . . $= \dfrac{wx}{24EI}(l^3 - 2lx^2 + x^3)$

---

**2.   SIMPLE BEAM—LOAD INCREASING UNIFORMLY TO ONE END**

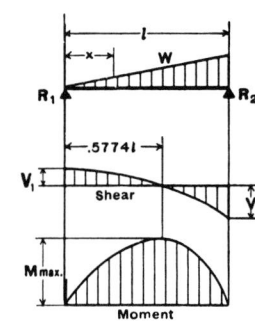

$R_1 = V_1$ . . . . . . . . $= \dfrac{W}{3}$

$R_2 = V_2$ max. . . . . . . . $= \dfrac{2W}{3}$

$V_x$ . . . . . . . . . $= \dfrac{W}{3} - \dfrac{Wx^2}{l^2}$

M max. $\left(\text{at } x = \dfrac{l}{\sqrt{3}} = .5774l\right)$ . . $= \dfrac{2Wl}{9\sqrt{3}} = .1283\,Wl$

$M_x$ . . . . . . . . . $= \dfrac{Wx}{3l^2}(l^2 - x^2)$

$\Delta$max. $\left(\text{at } x = l\sqrt{1 - \sqrt{\dfrac{8}{15}}} = .5193l\right) = .01304\dfrac{Wl^3}{EI}$

$\Delta_x$ . . . . . . . . . $= \dfrac{Wx}{180EI\,l^2}(3x^4 - 10l^2x^2 + 7l^4)$

---

**3.   SIMPLE BEAM—LOAD INCREASING UNIFORMLY TO CENTER**

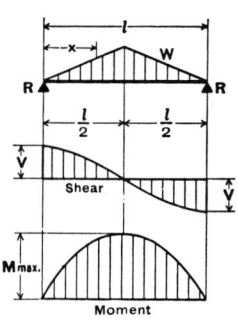

$R = V$ . . . . . . . . . $= \dfrac{W}{2}$

$V_x$ $\left(\text{when } x < \dfrac{l}{2}\right)$ . . . . $= \dfrac{W}{2l^2}(l^2 - 4x^2)$

M max. $\left(\text{at center}\right)$ . . . . $= \dfrac{Wl}{6}$

$M_x$ $\left(\text{when } x < \dfrac{l}{2}\right)$ . . . . $= Wx\left(\dfrac{1}{2} - \dfrac{2x^2}{3l^2}\right)$

$\Delta$max. $\left(\text{at center}\right)$ . . . . $= \dfrac{Wl^3}{60EI}$

$\Delta_x$ . . . . . . . . . $= \dfrac{Wx}{480\,EI\,l^2}(5l^2 - 4x^2)^2$

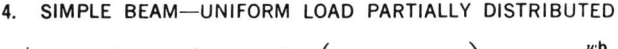

**TABLE 8.13** (*Continued*)

## 4. SIMPLE BEAM—UNIFORM LOAD PARTIALLY DISTRIBUTED

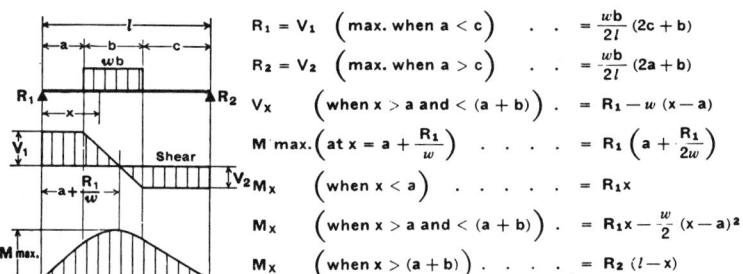

$R_1 = V_1 \left( \text{max. when } a < c \right) \quad \cdots \quad = \dfrac{wb}{2l}(2c + b)$

$R_2 = V_2 \left( \text{max. when } a > c \right) \quad \cdots \quad = \dfrac{wb}{2l}(2a + b)$

$V_x \left( \text{when } x > a \text{ and } < (a+b) \right) . \quad = R_1 - w(x - a)$

$M \text{ max.} \left( \text{at } x = a + \dfrac{R_1}{w} \right) \quad \cdots \quad = R_1\left(a + \dfrac{R_1}{2w}\right)$

$M_x \left( \text{when } x < a \right) \quad \cdots \quad = R_1 x$

$M_x \left( \text{when } x > a \text{ and } < (a+b) \right) . \quad = R_1 x - \dfrac{w}{2}(x - a)^2$

$M_x \left( \text{when } x > (a+b) \right) \quad \cdots \quad = R_2(l - x)$

## 5. SIMPLE BEAM—UNIFORM LOAD PARTIALLY DISTRIBUTED AT ONE END

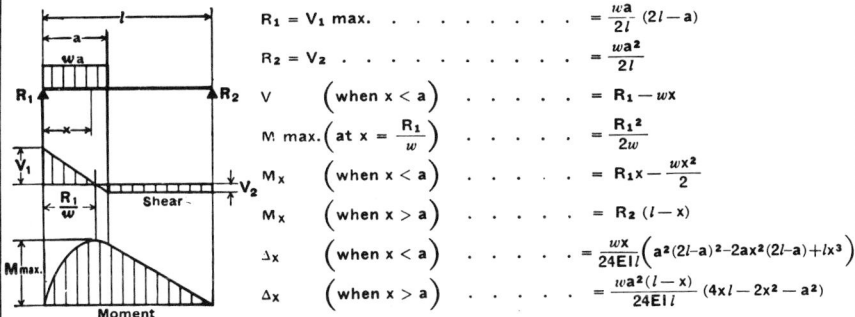

$R_1 = V_1 \text{ max.} \quad \cdots \quad = \dfrac{wa}{2l}(2l - a)$

$R_2 = V_2 \quad \cdots \quad = \dfrac{wa^2}{2l}$

$V \left( \text{when } x < a \right) \quad \cdots \quad = R_1 - wx$

$M \text{ max.} \left( \text{at } x = \dfrac{R_1}{w} \right) \quad \cdots \quad = \dfrac{R_1{}^2}{2w}$

$M_x \left( \text{when } x < a \right) \quad \cdots \quad = R_1 x - \dfrac{wx^2}{2}$

$M_x \left( \text{when } x > a \right) \quad \cdots \quad = R_2(l - x)$

$\Delta_x \left( \text{when } x < a \right) \quad \cdots \quad = \dfrac{wx}{24EIl}\left(a^2(2l-a)^2 - 2ax^2(2l-a) + lx^3\right)$

$\Delta_x \left( \text{when } x > a \right) \quad \cdots \quad = \dfrac{wa^2(l-x)}{24EIl}(4xl - 2x^2 - a^2)$

## 6. SIMPLE BEAM—UNIFORM LOAD PARTIALLY DISTRIBUTED AT EACH END

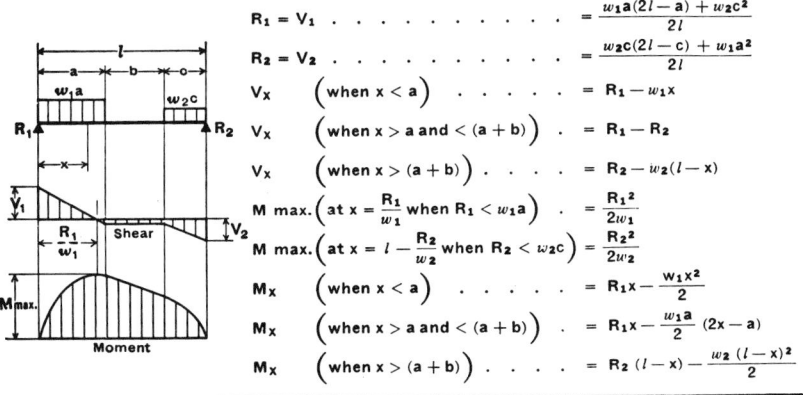

$R_1 = V_1 \quad \cdots \quad = \dfrac{w_1 a(2l - a) + w_2 c^2}{2l}$

$R_2 = V_2 \quad \cdots \quad = \dfrac{w_2 c(2l - c) + w_1 a^2}{2l}$

$V_x \left( \text{when } x < a \right) \quad \cdots \quad = R_1 - w_1 x$

$V_x \left( \text{when } x > a \text{ and } < (a+b) \right) . \quad = R_1 - R_2$

$V_x \left( \text{when } x > (a+b) \right) \quad \cdots \quad = R_2 - w_2(l - x)$

$M \text{ max.} \left( \text{at } x = \dfrac{R_1}{w_1} \text{ when } R_1 < w_1 a \right) \quad = \dfrac{R_1{}^2}{2w_1}$

$M \text{ max.} \left( \text{at } x = l - \dfrac{R_2}{w_2} \text{ when } R_2 < w_2 c \right) = \dfrac{R_2{}^2}{2w_2}$

$M_x \left( \text{when } x < a \right) \quad \cdots \quad = R_1 x - \dfrac{w_1 x^2}{2}$

$M_x \left( \text{when } x > a \text{ and } < (a+b) \right) . \quad = R_1 x - \dfrac{w_1 a}{2}(2x - a)$

$M_x \left( \text{when } x > (a+b) \right) \quad \cdots \quad = R_2(l - x) - \dfrac{w_2(l-x)^2}{2}$

## TABLE 8.13    (*Continued*)

### 7.    SIMPLE BEAM—CONCENTRATED LOAD AT CENTER

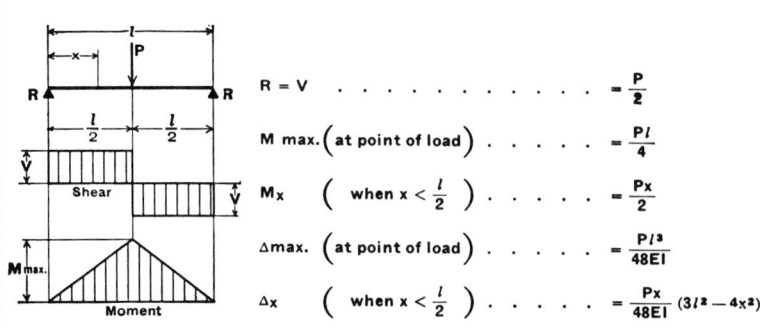

$$R = V \quad \ldots \ldots \ldots \ldots = \frac{P}{2}$$

$$M \text{ max.} \left( \text{at point of load} \right) \ldots \ldots = \frac{Pl}{4}$$

$$M_x \quad \left( \text{ when } x < \frac{l}{2} \right) \ldots \ldots = \frac{Px}{2}$$

$$\Delta \text{max.} \left( \text{at point of load} \right) \ldots \ldots = \frac{Pl^3}{48EI}$$

$$\Delta_x \quad \left( \text{ when } x < \frac{l}{2} \right) \ldots \ldots = \frac{Px}{48EI}(3l^2 - 4x^2)$$

### 8.    SIMPLE BEAM—CONCENTRATED LOAD AT ANY POINT

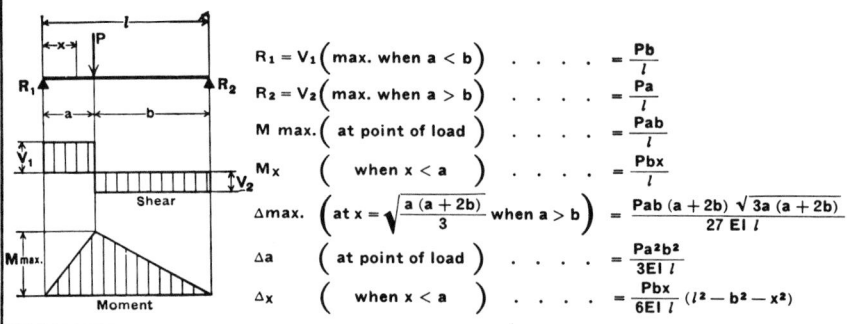

$$R_1 = V_1 \left( \text{max. when } a < b \right) \ldots \ldots = \frac{Pb}{l}$$

$$R_2 = V_2 \left( \text{max. when } a > b \right) \ldots \ldots = \frac{Pa}{l}$$

$$M \text{ max.} \left( \text{at point of load} \right) \ldots \ldots = \frac{Pab}{l}$$

$$M_x \quad \left( \text{ when } x < a \right) \ldots \ldots = \frac{Pbx}{l}$$

$$\Delta \text{max.} \left( \text{at } x = \sqrt{\frac{a(a+2b)}{3}} \text{ when } a > b \right) = \frac{Pab(a+2b)\sqrt{3a(a+2b)}}{27\,EI\,l}$$

$$\Delta_a \quad \left( \text{at point of load} \right) \ldots \ldots = \frac{Pa^2b^2}{3EI\,l}$$

$$\Delta_x \quad \left( \text{ when } x < a \right) \ldots \ldots = \frac{Pbx}{6EI\,l}(l^2 - b^2 - x^2)$$

### 9.    SIMPLE BEAM—TWO EQUAL CONCENTRATED LOADS
### SYMMETRICALLY PLACED

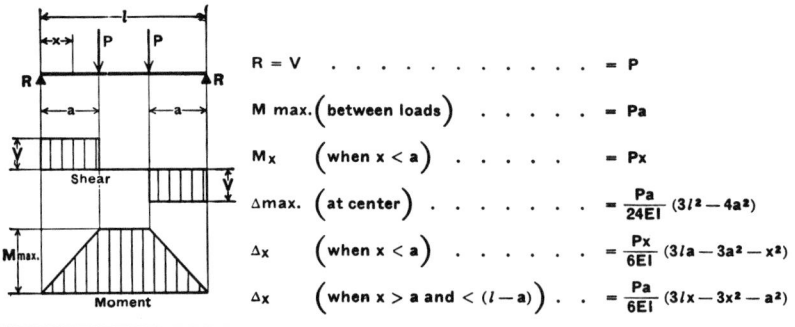

$$R = V \quad \ldots \ldots \ldots \ldots = P$$

$$M \text{ max.} \left( \text{between loads} \right) \ldots \ldots = Pa$$

$$M_x \quad \left( \text{when } x < a \right) \ldots \ldots = Px$$

$$\Delta \text{max.} \left( \text{at center} \right) \ldots \ldots = \frac{Pa}{24EI}(3l^2 - 4a^2)$$

$$\Delta_x \quad \left( \text{when } x < a \right) \ldots \ldots = \frac{Px}{6EI}(3la - 3a^2 - x^2)$$

$$\Delta_x \quad \left( \text{when } x > a \text{ and } < (l-a) \right) \ldots = \frac{Pa}{6EI}(3lx - 3x^2 - a^2)$$

**TABLE 8.13** (*Continued*)

### 10. SIMPLE BEAM—TWO EQUAL CONCENTRATED LOADS UNSYMMETRICALLY PLACED

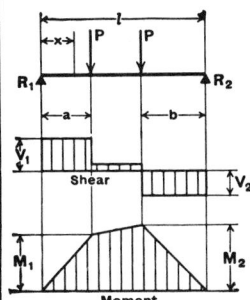

$$R_1 = V_1 \left( \text{max. when } a < b \right) \quad \cdots \cdots = \frac{P}{l}(l - a + b)$$

$$R_2 = V_2 \left( \text{max. when } a > b \right) \quad \cdots \cdots = \frac{P}{l}(l - b + a)$$

$$V_x \quad \left( \text{when } x > a \text{ and } < (l-b) \right) \cdots = \frac{P}{l}(b - a)$$

$$M_1 \quad \left( \text{max. when } a > b \right) \quad \cdots \cdots = R_1 a$$

$$M_2 \quad \left( \text{max. when } a < b \right) \quad \cdots \cdots = R_2 b$$

$$M_x \quad \left( \text{when } x < a \right) \quad \cdots \cdots = R_1 x$$

$$M_x \quad \left( \text{when } x > a \text{ and } < (l-b) \right) \cdots = R_1 x - P(x-a)$$

### 11. SIMPLE BEAM—TWO UNEQUAL CONCENTRATED LOADS UNSYMMETRICALLY PLACED

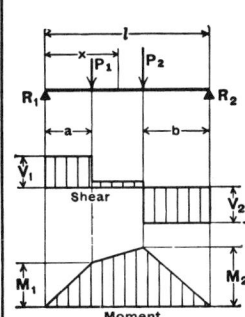

$$R_1 = V_1 \cdots \cdots \cdots \cdots = \frac{P_1(l-a) + P_2 b}{l}$$

$$R_2 = V_2 \cdots \cdots \cdots \cdots = \frac{P_1 a + P_2(l-b)}{l}$$

$$V_x \quad \left( \text{when } x > a \text{ and } < (l-b) \right) \cdots = R_1 - P_1$$

$$M_1 \quad \left( \text{max. when } R_1 < P_1 \right) \quad \cdots \cdots = R_1 a$$

$$M_2 \quad \left( \text{max. when } R_2 < P_2 \right) \quad \cdots \cdots = R_2 b$$

$$M_x \quad \left( \text{when } x < a \right) \quad \cdots \cdots = R_1 x$$

$$M_x \quad \left( \text{when } x > a \text{ and } < (l-b) \right) \cdots = R_1 x - P_1(x-a)$$

### 12. BEAM FIXED AT ONE END, SUPPORTED AT OTHER— UNIFORMLY DISTRIBUTED LOAD

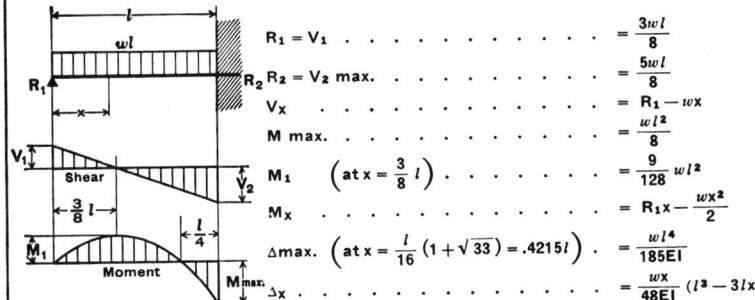

$$R_1 = V_1 \cdots \cdots \cdots \cdots \cdots = \frac{3wl}{8}$$

$$R_2 = V_2 \text{ max.} \cdots \cdots \cdots \cdots = \frac{5wl}{8}$$

$$V_x \cdots \cdots \cdots \cdots \cdots = R_1 - wx$$

$$M \text{ max.} \cdots \cdots \cdots \cdots \cdots = \frac{wl^2}{8}$$

$$M_1 \quad \left( \text{at } x = \frac{3}{8}l \right) \cdots \cdots \cdots = \frac{9}{128}wl^2$$

$$M_x \cdots \cdots \cdots \cdots \cdots = R_1 x - \frac{wx^2}{2}$$

$$\Delta \text{max.} \left( \text{at } x = \frac{l}{16}(1 + \sqrt{33}) = .4215l \right) \cdots = \frac{wl^4}{185EI}$$

$$\Delta_x \cdots \cdots \cdots \cdots \cdots = \frac{wx}{48EI}(l^3 - 3lx^2 + 2x^3)$$

## TABLE 8.13    *(Continued)*

### 13.    BEAM FIXED AT ONE END, SUPPORTED AT OTHER—CONCENTRATED LOAD AT CENTER

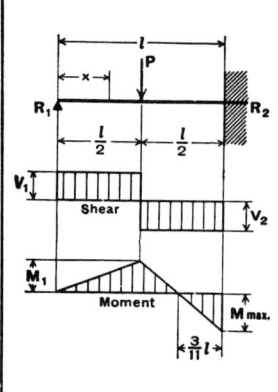

$R_1 = V_1$ . . . . . . . . . . $= \dfrac{5P}{16}$

$R_2 = V_2$ max. . . . . . . . . $= \dfrac{11P}{16}$

M max. $\left(\text{at fixed end}\right)$ . . . . . $= \dfrac{3Pl}{16}$

$M_1$    $\left(\text{at point of load}\right)$ . . . . $= \dfrac{5Pl}{32}$

$M_x$    $\left(\text{when } x < \dfrac{l}{2}\right)$ . . . . . $= \dfrac{5Px}{16}$

$M_x$    $\left(\text{when } x > \dfrac{l}{2}\right)$ . . . . . $= P\left(\dfrac{l}{2} - \dfrac{11x}{16}\right)$

$\Delta$max. $\left(\text{at } x = l\sqrt{\dfrac{1}{5}} = .4472l\right)$ . . $= \dfrac{Pl^3}{48EI\sqrt{5}} = .009317 \dfrac{Pl^3}{EI}$

$\Delta_x$    $\left(\text{at point of load}\right)$ . . . . $= \dfrac{7Pl^3}{768EI}$

$\Delta_x$    $\left(\text{when } x < \dfrac{l}{2}\right)$ . . . . . $= \dfrac{Px}{96EI} (3l^2 - 5x^2)$

$\Delta_x$    $\left(\text{when } x > \dfrac{l}{2}\right)$ . . . . . $= \dfrac{P}{96EI} (x-l)^2 (11x - 2l)$

### 14.    BEAM FIXED AT ONE END, SUPPORTED AT OTHER—CONCENTRATED LOAD AT ANY POINT

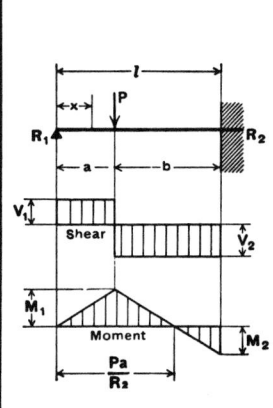

$R_1 = V_1$ . . . . . . . . . . $= \dfrac{Pb^2}{2l^3} (a + 2l)$

$R_2 = V_2$ . . . . . . . . . . $= \dfrac{Pa}{2l^3} (3l^2 - a^2)$

$M_1$    $\left(\text{at point of load}\right)$ . . . . $= R_1 a$

$M_2$    $\left(\text{at fixed end}\right)$ . . . . . $= \dfrac{Pab}{2l^2} (a + l)$

$M_x$    $\left(\text{when } x < a\right)$ . . . . . $= R_1 x$

$M_x$    $\left(\text{when } x > a\right)$ . . . . . $= R_1 x - P(x - a)$

$\Delta$max. $\left(\text{when } a < .414l \text{ at } x = l\dfrac{l^2 + a^2}{3l^2 - a^2}\right) = \dfrac{Pa}{3EI} \dfrac{(l^2 - a^2)^3}{(3l^2 - a^2)^2}$

$\Delta$max. $\left(\text{when } a > .414l \text{ at } x = l\sqrt{\dfrac{a}{2l+a}}\right) = \dfrac{Pab^2}{6EI} \sqrt{\dfrac{a}{2l + a}}$

$\Delta_a$    $\left(\text{at point of load}\right)$ . . . . $= \dfrac{Pa^2 b^3}{12EIl^3} (3l + a)$

$\Delta_x$    $\left(\text{when } x < a\right)$ . . . . . $= \dfrac{Pb^2 x}{12EIl^3} (3al^2 - 2lx^2 - ax^2)$

$\Delta_x$    $\left(\text{when } x > a\right)$ . . . . . $= \dfrac{Pa}{12EIl^3} (l-x)^2 (3l^2 x - a^2 x - 2a^2 l)$

**TABLE 8.13**   *(Continued)*

---

### 15.   BEAM FIXED AT BOTH ENDS—UNIFORMLY DISTRIBUTED LOADS

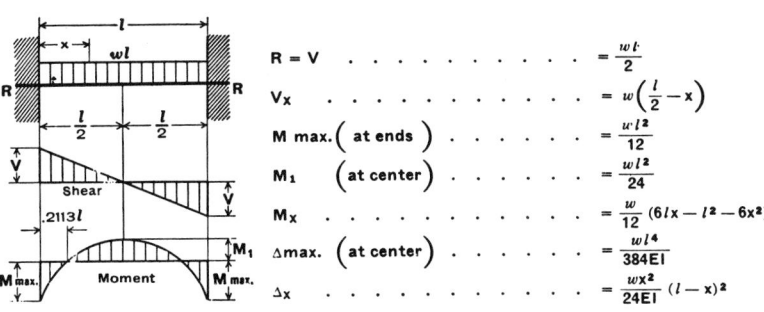

$R = V$ . . . . . . . . . . . $= \dfrac{wl}{2}$

$V_x$ . . . . . . . . . . . $= w\left(\dfrac{l}{2} - x\right)$

$M$ max. $\left(\text{at ends}\right)$ . . . . . $= \dfrac{wl^2}{12}$

$M_1$ $\left(\text{at center}\right)$ . . . . . . $= \dfrac{wl^2}{24}$

$M_x$ . . . . . . . . . . . $= \dfrac{w}{12}(6lx - l^2 - 6x^2)$

$\Delta$max. $\left(\text{at center}\right)$ . . . . . $= \dfrac{wl^4}{384EI}$

$\Delta_x$ . . . . . . . . . . . $= \dfrac{wx^2}{24EI}(l-x)^2$

---

### 16   BEAM FIXED AT BOTH ENDS—CONCENTRATED LOAD AT CENTER

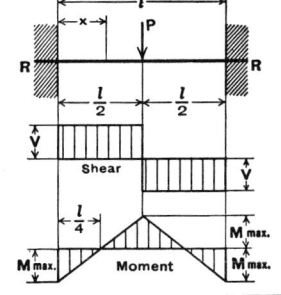

$R = V$ . . . . . . . . . . $= \dfrac{P}{2}$

$M$ max. $\left(\text{at center and ends}\right)$ . . . $= \dfrac{Pl}{8}$

$M_x$ $\left(\text{when } x < \dfrac{l}{2}\right)$ . . . . . $= \dfrac{P}{8}(4x - l)$

$\Delta$max. $\left(\text{at center}\right)$ . . . . . . $= \dfrac{Pl^3}{192EI}$

$\Delta_x$ . . . . . . . . . . . $= \dfrac{Px^2}{48EI}(3l - 4x)$

---

### 17.   BEAM FIXED AT BOTH ENDS—CONCENTRATED LOAD AT ANY POINT

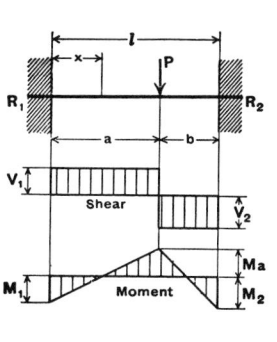

$R_1 = V_1\left(\text{max. when } a < b\right)$ . . . $= \dfrac{Pb^2}{l^3}(3a + b)$

$R_2 = V_2\left(\text{max. when } a > b\right)$ . . . $= \dfrac{Pa^2}{l^3}(a + 3b)$

$M_1$ $\left(\text{max. when } a < b\right)$ . . . $= \dfrac{Pab^2}{l^2}$

$M_2$ $\left(\text{max. when } a > b\right)$ . . . $= \dfrac{Pa^2b}{l^2}$

$M_a$ $\left(\text{at point of load}\right)$ . . . $= \dfrac{2Pa^2b^2}{l^3}$

$M_x$ $\left(\text{when } x < a\right)$ . . . . . $= R_1x - \dfrac{Pab^2}{l^2}$

$\Delta$max. $\left(\text{when } a > b \text{ at } x = \dfrac{2al}{3a + b}\right)$ . $= \dfrac{2Pa^3b^2}{3EI(3a + b)^2}$

$\Delta_a$ $\left(\text{at point of load}\right)$ . . . $= \dfrac{Pa^3b^3}{3EIl^3}$

$\Delta_x$ $\left(\text{when } x < a\right)$ . . . . $= \dfrac{Pb^2x^2}{6EIl^3}(3al - 3ax - bx)$

## TABLE 8.13    (*Continued*)

### 18.    CANTILEVER BEAM—LOAD INCREASING UNIFORMLY TO FIXED END

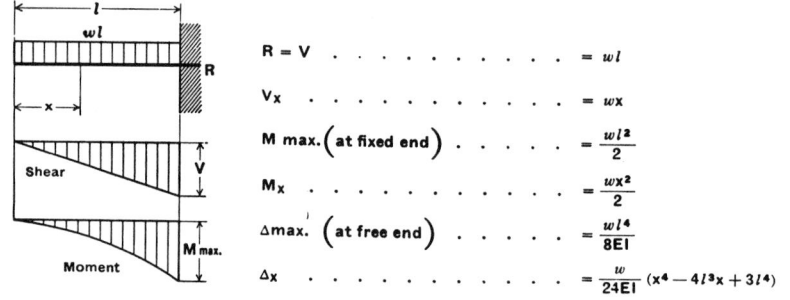

$$R = V \quad \ldots \ldots \ldots \ldots = W$$

$$V_x \quad \ldots \ldots \ldots \ldots = W\frac{x^2}{l^2}$$

$$M \text{ max.} \left( \text{at fixed end} \right) \ldots \ldots = \frac{Wl}{3}$$

$$M_x \quad \ldots \ldots \ldots \ldots = \frac{Wx^3}{3l^2}$$

$$\Delta \text{max.} \left( \text{at free end} \right) \ldots \ldots = \frac{Wl^3}{15EI}$$

$$\Delta_x \quad \ldots \ldots \ldots \ldots = \frac{W}{60EIl^2}(x^5 - 5l^4x + 4l^5)$$

### 19.    CANTILEVER BEAM—UNIFORMLY DISTRIBUTED LOAD

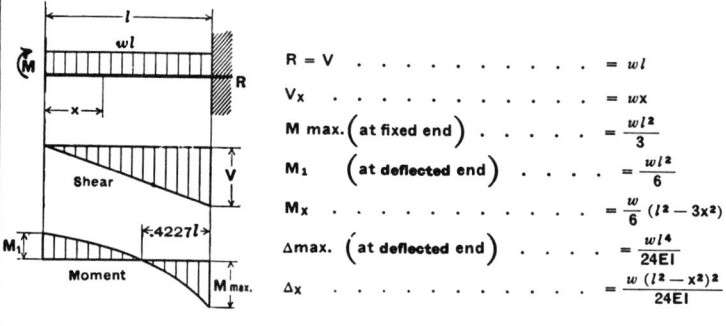

$$R = V \quad \ldots \ldots \ldots \ldots = wl$$

$$V_x \quad \ldots \ldots \ldots \ldots = wx$$

$$M \text{ max.} \left( \text{at fixed end} \right) \ldots \ldots = \frac{wl^2}{2}$$

$$M_x \quad \ldots \ldots \ldots \ldots = \frac{wx^2}{2}$$

$$\Delta \text{max.} \left( \text{at free end} \right) \ldots \ldots = \frac{wl^4}{8EI}$$

$$\Delta_x \quad \ldots \ldots \ldots \ldots = \frac{w}{24EI}(x^4 - 4l^3x + 3l^4)$$

### 20.    BEAM FIXED AT ONE END, FREE TO DEFLECT VERTICALLY BUT NOT ROTATE AT OTHER—UNIFORMLY DISTRIBUTED LOAD

$$R = V \quad \ldots \ldots \ldots \ldots = wl$$

$$V_x \quad \ldots \ldots \ldots \ldots = wx$$

$$M \text{ max.} \left( \text{at fixed end} \right) \ldots \ldots = \frac{wl^2}{3}$$

$$M_1 \left( \text{at deflected end} \right) \ldots \ldots = \frac{wl^2}{6}$$

$$M_x \quad \ldots \ldots \ldots \ldots = \frac{w}{6}(l^2 - 3x^2)$$

$$\Delta \text{max.} \left( \text{at deflected end} \right) \ldots \ldots = \frac{wl^4}{24EI}$$

$$\Delta_x \quad \ldots \ldots \ldots \ldots = \frac{w(l^2 - x^2)^2}{24EI}$$

**TABLE 8.13**    *(Continued)*

## 21.  CANTILEVER BEAM—CONCENTRATED LOAD AT ANY POINT

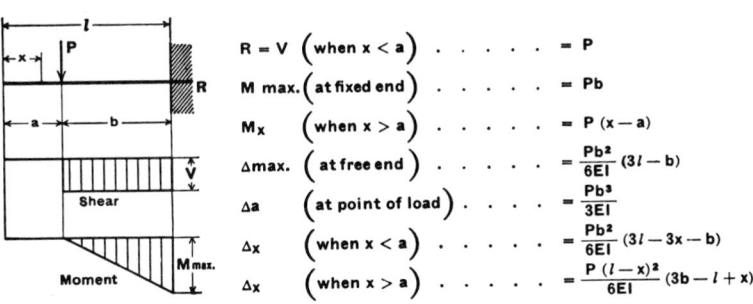

$R = V$ $\left(\text{when } x < a\right)$ . . . . . $= P$

$M$ max. $\left(\text{at fixed end}\right)$ . . . . . $= Pb$

$M_x$ $\left(\text{when } x > a\right)$ . . . . . $= P(x-a)$

$\Delta$max. $\left(\text{at free end}\right)$ . . . . . $= \dfrac{Pb^2}{6EI}(3l-b)$

$\Delta a$ $\left(\text{at point of load}\right)$ . . . . $= \dfrac{Pb^3}{3EI}$

$\Delta x$ $\left(\text{when } x < a\right)$ . . . . . $= \dfrac{Pb^2}{6EI}(3l-3x-b)$

$\Delta x$ $\left(\text{when } x > a\right)$ . . . . . $= \dfrac{P(l-x)^2}{6EI}(3b-l+x)$

## 22.  CANTILEVER BEAM—CONCENTRATED LOAD AT FREE END

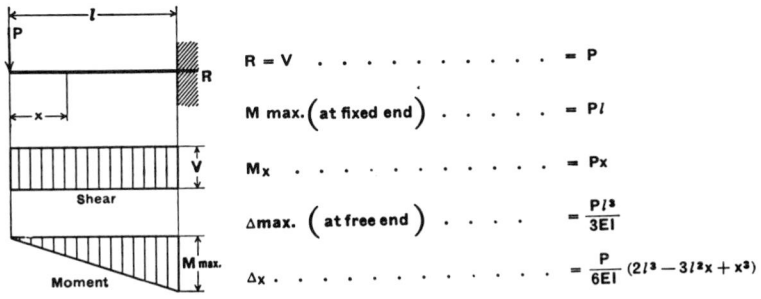

$R = V$ . . . . . . . . . . . $= P$

$M$ max. $\left(\text{at fixed end}\right)$ . . . . . $= Pl$

$M_x$ . . . . . . . . . . . $= Px$

$\Delta$max. $\left(\text{at free end}\right)$ . . . . $= \dfrac{Pl^3}{3EI}$

$\Delta x$ . . . . . . . . . . . . . . $= \dfrac{P}{6EI}(2l^3 - 3l^2x + x^3)$

## 23.    BEAM FIXED AT ONE END, FREE TO DEFLECT VERTICALLY BUT NOT ROTATE AT OTHER—CONCENTRATED LOAD AT DEFLECTED END

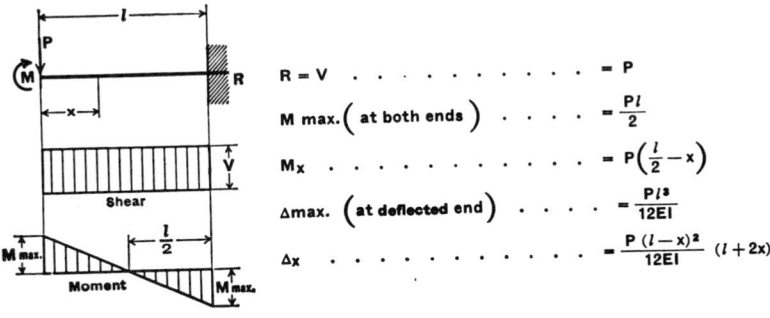

$R = V$ . . . . . . . . . . . $= P$

$M$ max. $\left(\text{at both ends}\right)$ . . . . $= \dfrac{Pl}{2}$

$M_x$ . . . . . . . . . . . . $= P\left(\dfrac{l}{2}-x\right)$

$\Delta$max. $\left(\text{at deflected end}\right)$ . . . . $= \dfrac{Pl^3}{12EI}$

$\Delta x$ . . . . . . . . . . . . . $= \dfrac{P(l-x)^2}{12EI}(l+2x)$

**TABLE 8.13**    *(Continued)*

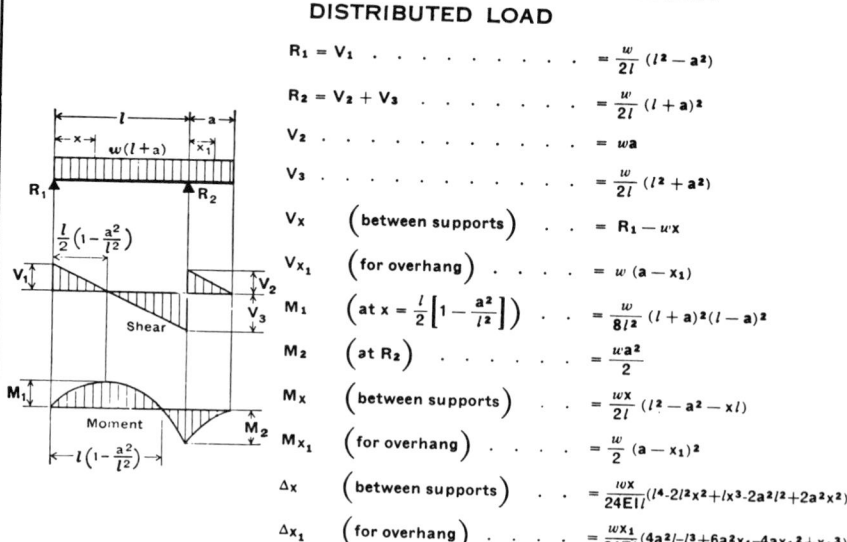

### 24.  BEAM OVERHANGING ONE SUPPORT—UNIFORMLY DISTRIBUTED LOAD

$$R_1 = V_1 \quad\ldots\ldots\ldots\ldots = \frac{w}{2l}(l^2 - a^2)$$

$$R_2 = V_2 + V_3 \quad\ldots\ldots\ldots = \frac{w}{2l}(l + a)^2$$

$$V_2 \quad\ldots\ldots\ldots\ldots\ldots = wa$$

$$V_3 \quad\ldots\ldots\ldots\ldots\ldots = \frac{w}{2l}(l^2 + a^2)$$

$$V_x \quad \left(\text{between supports}\right)\ldots = R_1 - wx$$

$$V_{x_1} \quad \left(\text{for overhang}\right)\ldots\ldots = w(a - x_1)$$

$$M_1 \quad \left(\text{at } x = \frac{l}{2}\left[1 - \frac{a^2}{l^2}\right]\right)\ldots = \frac{w}{8l^2}(l + a)^2(l - a)^2$$

$$M_2 \quad \left(\text{at } R_2\right)\ldots\ldots\ldots = \frac{wa^2}{2}$$

$$M_x \quad \left(\text{between supports}\right)\ldots = \frac{wx}{2l}(l^2 - a^2 - xl)$$

$$M_{x_1} \quad \left(\text{for overhang}\right)\ldots\ldots = \frac{w}{2}(a - x_1)^2$$

$$\Delta_x \quad \left(\text{between supports}\right)\ldots = \frac{wx}{24EIl}(l^4 - 2l^2x^2 + lx^3 - 2a^2l^2 + 2a^2x^2)$$

$$\Delta_{x_1} \quad \left(\text{for overhang}\right)\ldots\ldots = \frac{wx_1}{24EI}(4a^2l - l^3 + 6a^2x_1 - 4ax_1^2 + x_1^3)$$

### 25.  BEAM OVERHANGING ONE SUPPORT—UNIFORMLY DISTRIBUTED LOAD ON OVERHANG

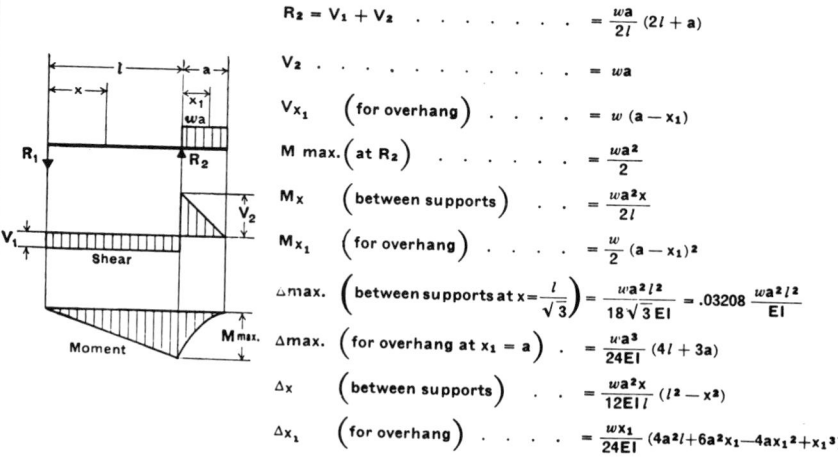

$$R_1 = V_1 \quad\ldots\ldots\ldots\ldots = \frac{wa^2}{2l}$$

$$R_2 = V_1 + V_2 \quad\ldots\ldots\ldots = \frac{wa}{2l}(2l + a)$$

$$V_2 \quad\ldots\ldots\ldots\ldots\ldots = wa$$

$$V_{x_1} \quad \left(\text{for overhang}\right)\ldots\ldots = w(a - x_1)$$

$$M \text{ max.} \left(\text{at } R_2\right)\ldots\ldots\ldots = \frac{wa^2}{2}$$

$$M_x \quad \left(\text{between supports}\right)\ldots = \frac{wa^2x}{2l}$$

$$M_{x_1} \quad \left(\text{for overhang}\right)\ldots\ldots = \frac{w}{2}(a - x_1)^2$$

$$\Delta \text{max.} \left(\text{between supports at } x = \frac{l}{\sqrt{3}}\right) = \frac{wa^2l^2}{18\sqrt{3}\,EI} = .03208\frac{wa^2l^2}{EI}$$

$$\Delta \text{max.} \left(\text{for overhang at } x_1 = a\right)\ldots = \frac{wa^3}{24EI}(4l + 3a)$$

$$\Delta_x \quad \left(\text{between supports}\right)\ldots = \frac{wa^2x}{12EIl}(l^2 - x^2)$$

$$\Delta_{x_1} \quad \left(\text{for overhang}\right)\ldots\ldots = \frac{wx_1}{24EI}(4a^2l + 6a^2x_1 - 4ax_1^2 + x_1^3)$$

**TABLE 8.13** *(Continued)*

## 26. BEAM OVERHANGING ONE SUPPORT—CONCENTRATED LOAD AT END OF OVERHANG

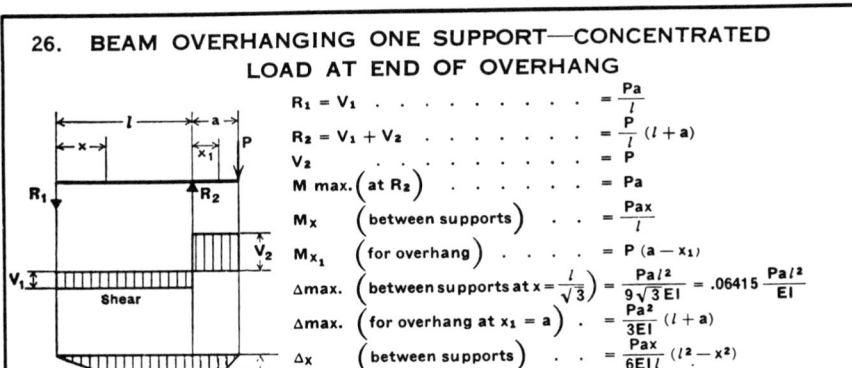

$R_1 = V_1$ . . . . . . . . . $= \dfrac{Pa}{l}$

$R_2 = V_1 + V_2$ . . . . . . . $= \dfrac{P}{l}(l+a)$

$V_2$ . . . . . . . . . $= P$

$M \text{ max.} \left( \text{at } R_2 \right)$ . . . . . . $= Pa$

$M_x \left( \text{between supports} \right)$ . . $= \dfrac{Pax}{l}$

$M_{x_1} \left( \text{for overhang} \right)$ . . . . $= P(a - x_1)$

$\Delta \text{max.} \left( \text{between supports at } x = \dfrac{l}{\sqrt{3}} \right) = \dfrac{Pal^2}{9\sqrt{3}\,EI} = .06415 \dfrac{Pal^2}{EI}$

$\Delta \text{max.} \left( \text{for overhang at } x_1 = a \right)$ . $= \dfrac{Pa^2}{3EI}(l+a)$

$\Delta_x \left( \text{between supports} \right)$ . . $= \dfrac{Pax}{6EIl}(l^2 - x^2)$

$\Delta_{x_1} \left( \text{for overhang} \right)$ . . . . $= \dfrac{Px_1}{6EI}(2al + 3ax_1 - x_1^2)$

## 27. BEAM OVERHANGING ONE SUPPORT—UNIFORMLY DISTRIBUTED LOAD BETWEEN SUPPORTS

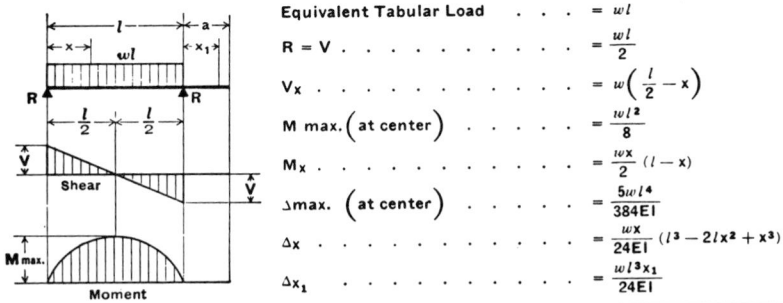

Equivalent Tabular Load . . . $= wl$

$R = V$ . . . . . . . . . $= \dfrac{wl}{2}$

$V_x$ . . . . . . . . . $= w\left( \dfrac{l}{2} - x \right)$

$M \text{ max.} \left( \text{at center} \right)$ . . . . . $= \dfrac{wl^2}{8}$

$M_x$ . . . . . . . . . $= \dfrac{wx}{2}(l - x)$

$\Delta \text{max.} \left( \text{at center} \right)$ . . . . . $= \dfrac{5wl^4}{384EI}$

$\Delta_x$ . . . . . . . . . $= \dfrac{wx}{24EI}(l^3 - 2lx^2 + x^3)$

$\Delta_{x_1}$ . . . . . . . . . $= \dfrac{wl^3 x_1}{24EI}$

## 28. BEAM OVERHANGING ONE SUPPORT—CONCENTRATED LOAD AT ANY POINT BETWEEN SUPPORTS

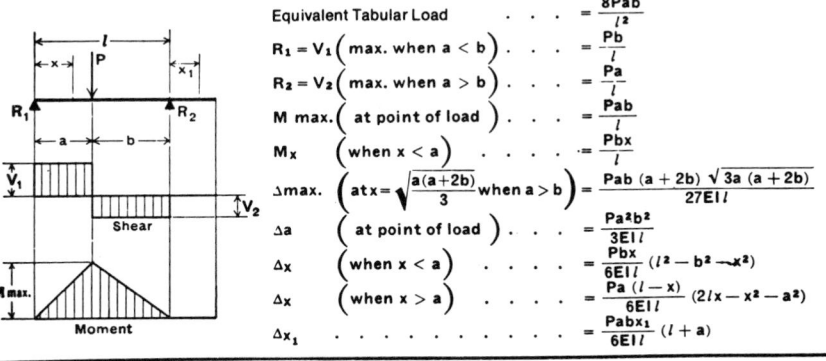

Equivalent Tabular Load . . . $= \dfrac{8Pab}{l^2}$

$R_1 = V_1 \left( \text{max. when } a < b \right)$ . . . $= \dfrac{Pb}{l}$

$R_2 = V_2 \left( \text{max. when } a > b \right)$ . . . $= \dfrac{Pa}{l}$

$M \text{ max.} \left( \text{at point of load} \right)$ . . . $= \dfrac{Pab}{l}$

$M_x \left( \text{when } x < a \right)$ . . . . . $= \dfrac{Pbx}{l}$

$\Delta \text{max.} \left( \text{at } x = \sqrt{\dfrac{a(a+2b)}{3}} \text{ when } a > b \right) = \dfrac{Pab(a+2b)\sqrt{3a(a+2b)}}{27EIl}$

$\Delta a \left( \text{at point of load} \right)$ . . . $= \dfrac{Pa^2 b^2}{3EIl}$

$\Delta_x \left( \text{when } x < a \right)$ . . . . . $= \dfrac{Pbx}{6EIl}(l^2 - b^2 - x^2)$

$\Delta_x \left( \text{when } x > a \right)$ . . . . . $= \dfrac{Pa(l-x)}{6EIl}(2lx - x^2 - a^2)$

$\Delta_{x_1}$ . . . . . . . . . $= \dfrac{Pabx_1}{6EIl}(l+a)$

## TABLE 8.13   (*Continued*)

### 29. CONTINUOUS BEAM – TWO EQUAL SPANS – UNIFORM LOAD ON BOTH SPANS

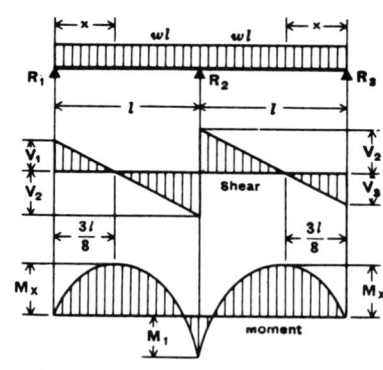

$$R_1 = V_1 = R_3 = V_3 \ldots = \frac{3}{8} wl$$

$$R_2 = 2V_2 \ldots \ldots = \frac{10}{8} wl$$

$$V_2 \ldots \ldots \ldots = \frac{5}{8} wl$$

$$M_x \ldots \ldots \ldots = R_1 x - \frac{wx^2}{2}$$

$$M_x \left( \text{at } x = \frac{3l}{8} \right) \ldots = \frac{9}{128} wl^2$$

$$M_1 \ (\text{at support } R_2) \ldots = -\frac{wl^2}{8}$$

$$\Delta \text{ Max. } (0.4215l \text{ from } R_1 \text{ or } R_3) = wl^4/185EI$$

$$\Delta_x = \frac{wx}{48EI} (l^3 - 3lx^2 + 2x^3)$$

### 30. CONTINUOUS BEAM – TWO EQUAL SPANS – UNIFORM LOAD ON ONE SPAN

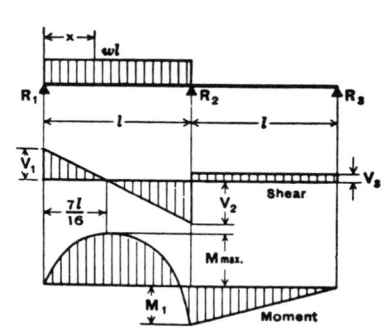

$$R_1 = V_1 \ldots \ldots = \frac{7}{16} wl$$

$$R_2 = V_2 + V_3 \ldots = \frac{5}{8} wl$$

$$R_3 = V_3 \ldots \ldots = -\frac{1}{16} wl$$

$$V_2 \ldots \ldots \ldots = \frac{9}{16} wl$$

$$M \text{ Max.} \left( \text{at } x = \frac{7}{16} l \right) = \frac{49}{512} wl^2$$

$$M_1 \ (\text{at support } R_2) \ldots = \frac{1}{16} wl^2$$

$$M_x \ (\text{when } x < l) \ldots = \frac{wx}{16} (7l - 8x)$$

$$\Delta \text{ Max. } (0.472l \text{ from } R_1) = wl^4/109EI$$

**TABLE 8.13** (*Continued*)

## 31. CONTINUOUS BEAM—TWO EQUAL SPANS—CONCENTRATED LOAD AT CENTER OF ONE SPAN

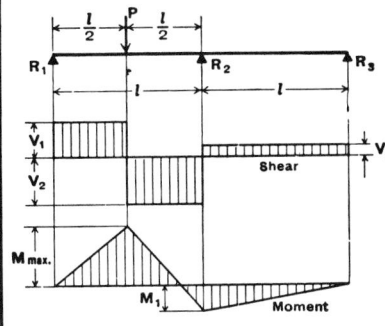

$R_1 = V_1 \quad \cdots \cdots \cdots \quad = \dfrac{13}{32} P$

$R_2 = V_2 + V_3 \quad \cdots \cdots \quad = \dfrac{11}{16} P$

$R_3 = V_3 \quad \cdots \cdots \cdots \quad = -\dfrac{3}{32} P$

$V_2 \quad \cdots \cdots \cdots \cdots \quad = \dfrac{19}{32} P$

$M \text{ Max.} \left( \text{at point of load} \right) \quad = \dfrac{13}{64} Pl$

$M_1 \quad \left( \text{at support } R_2 \right) \quad = \dfrac{3}{32} Pl$

$\Delta \text{ Max. } (0.480\ l \text{ from } R_1) \quad = 0.015\ Pl^3/EI$

## 32. CONTINUOUS BEAM—TWO EQUAL SPANS—CONCENTRATED LOAD AT ANY POINT

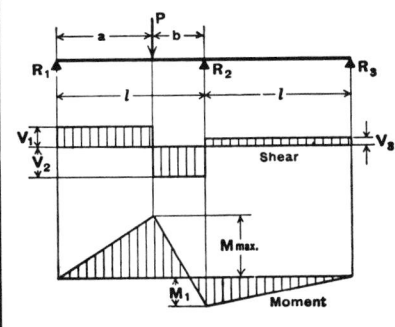

$R_1 = V_1 \quad \cdots \cdots \cdots \quad = \dfrac{Pb}{4l^3} \left( 4l^2 - a(l+a) \right)$

$R_2 = V_2 + V_3 \quad \cdots \cdots \quad = \dfrac{Pa}{2l^3} \left( 2l^2 + b(l+a) \right)$

$R_3 = V_3 \quad \cdots \cdots \cdots \quad = -\dfrac{Pab}{4l^3} (l+a)$

$V_2 \quad \cdots \cdots \cdots \cdots \quad = \dfrac{Pa}{4l^3} \left( 4l^2 + b(l+a) \right)$

$M \text{ max.} \left( \text{at point of load} \right) . \quad = \dfrac{Pab}{4l^3} \left( 4l^2 - a(l+a) \right)$

$M_1 \quad \left( \text{at support } R_2 \right) \quad = \dfrac{Pab}{4l^2} (l+a)$

## TABLE 8.13    *(Continued)*

**33.   BEAM—UNIFORMLY DISTRIBUTED LOAD AND VARIABLE END MOMENTS**

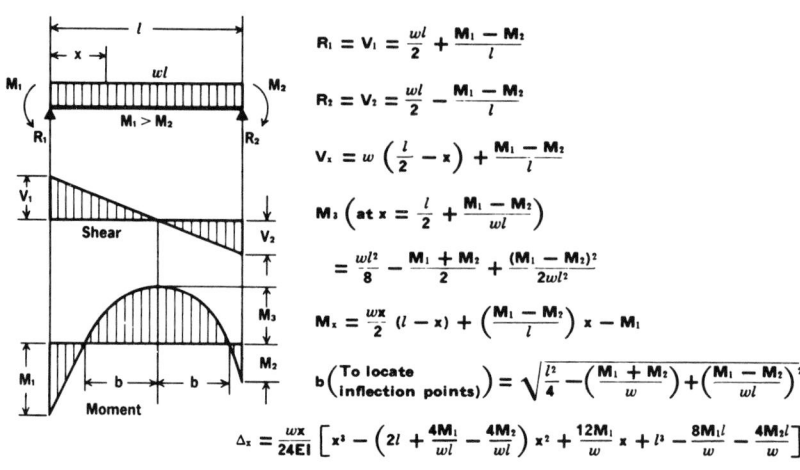

$$R_1 = V_1 = \frac{wl}{2} + \frac{M_1 - M_2}{l}$$

$$R_2 = V_2 = \frac{wl}{2} - \frac{M_1 - M_2}{l}$$

$$V_x = w\left(\frac{l}{2} - x\right) + \frac{M_1 - M_2}{l}$$

$$M_3\left(\text{at } x = \frac{l}{2} + \frac{M_1 - M_2}{wl}\right)$$

$$= \frac{wl^2}{8} - \frac{M_1 + M_2}{2} + \frac{(M_1 - M_2)^2}{2wl^2}$$

$$M_x = \frac{wx}{2}(l - x) + \left(\frac{M_1 - M_2}{l}\right)x - M_1$$

$$b\left(\begin{matrix}\text{To locate}\\ \text{inflection points}\end{matrix}\right) = \sqrt{\frac{l^2}{4} - \left(\frac{M_1 + M_2}{w}\right) + \left(\frac{M_1 - M_2}{wl}\right)^2}$$

$$\Delta_x = \frac{wx}{24EI}\left[x^3 - \left(2l + \frac{4M_1}{wl} - \frac{4M_2}{wl}\right)x^2 + \frac{12M_1}{w}x + l^3 - \frac{8M_1l}{w} - \frac{4M_2l}{w}\right]$$

**34.   BEAM—CONCENTRATED LOAD AT CENTER AND VARIABLE END MOMENTS**

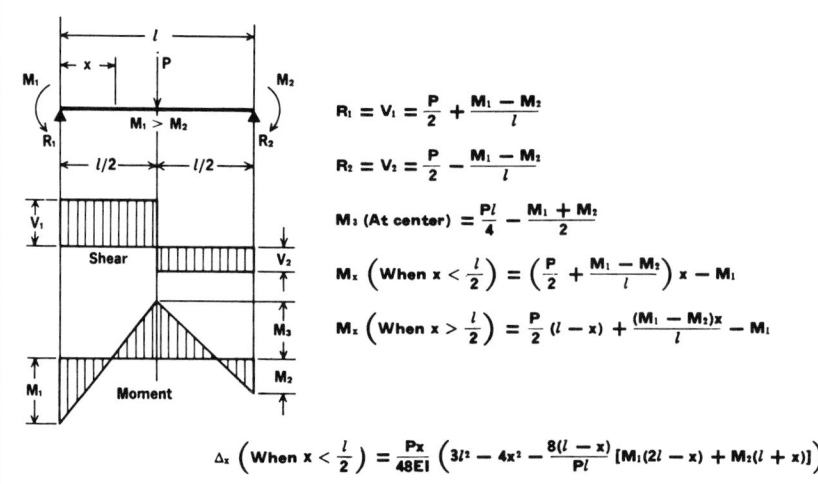

$$R_1 = V_1 = \frac{P}{2} + \frac{M_1 - M_2}{l}$$

$$R_2 = V_2 = \frac{P}{2} - \frac{M_1 - M_2}{l}$$

$$M_3 \text{ (At center)} = \frac{Pl}{4} - \frac{M_1 + M_2}{2}$$

$$M_x\left(\text{When } x < \frac{l}{2}\right) = \left(\frac{P}{2} + \frac{M_1 - M_2}{l}\right)x - M_1$$

$$M_x\left(\text{When } x > \frac{l}{2}\right) = \frac{P}{2}(l - x) + \frac{(M_1 - M_2)x}{l} - M_1$$

$$\Delta_x\left(\text{When } x < \frac{l}{2}\right) = \frac{Px}{48EI}\left(3l^2 - 4x^2 - \frac{8(l - x)}{Pl}[M_1(2l - x) + M_2(l + x)]\right)$$

**TABLE 8.13** (*Continued*)

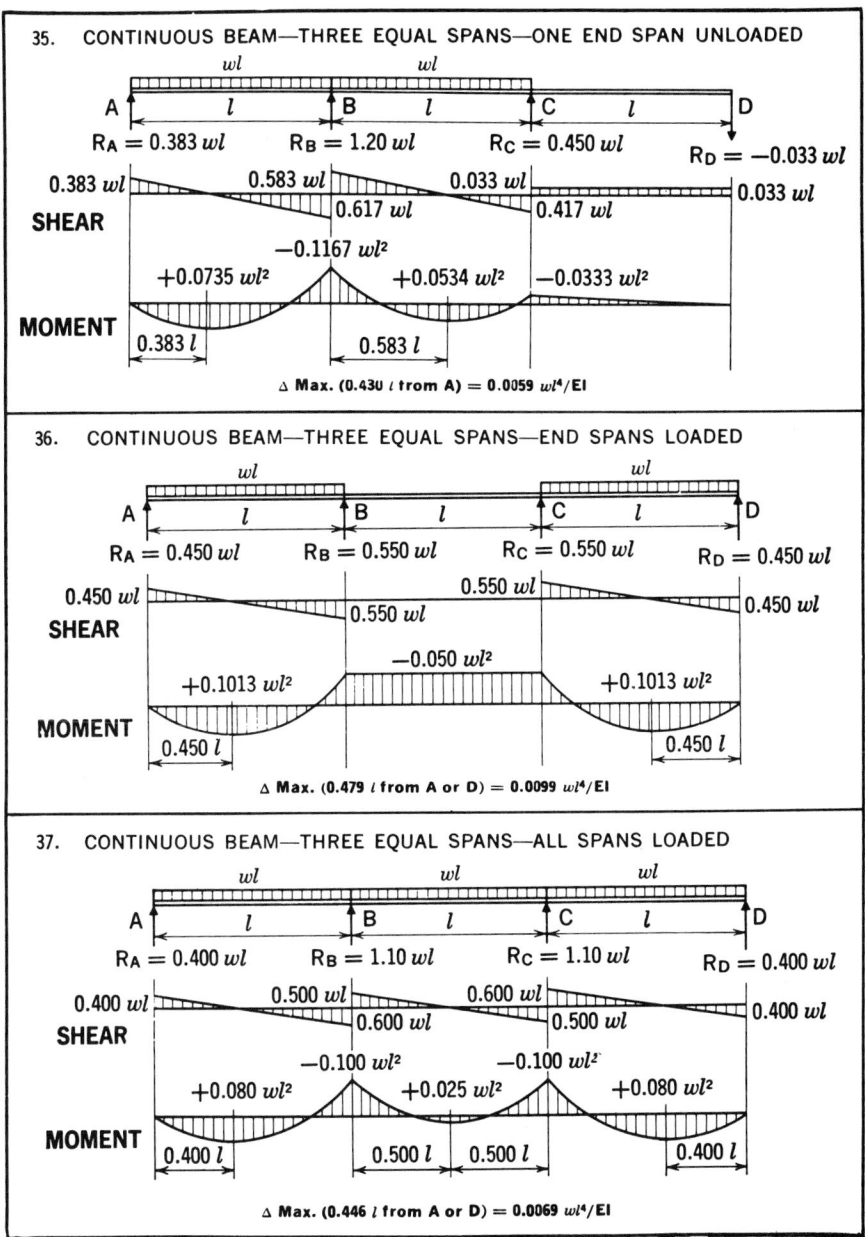

35. CONTINUOUS BEAM—THREE EQUAL SPANS—ONE END SPAN UNLOADED

$R_A = 0.383\,wl$  $R_B = 1.20\,wl$  $R_C = 0.450\,wl$  $R_D = -0.033\,wl$

SHEAR: $0.383\,wl$, $0.583\,wl$, $0.617\,wl$, $0.033\,wl$, $0.417\,wl$, $0.033\,wl$

MOMENT: $-0.1167\,wl^2$, $+0.0735\,wl^2$, $+0.0534\,wl^2$, $-0.0333\,wl^2$, $0.383\,l$, $0.583\,l$

Δ Max. (0.430 *l* from A) = 0.0059 $wl^4/EI$

36. CONTINUOUS BEAM—THREE EQUAL SPANS—END SPANS LOADED

$R_A = 0.450\,wl$  $R_B = 0.550\,wl$  $R_C = 0.550\,wl$  $R_D = 0.450\,wl$

SHEAR: $0.450\,wl$, $0.550\,wl$, $0.550\,wl$, $0.450\,wl$

MOMENT: $+0.1013\,wl^2$, $-0.050\,wl^2$, $+0.1013\,wl^2$, $0.450\,l$, $0.450\,l$

Δ Max. (0.479 *l* from A or D) = 0.0099 $wl^4/EI$

37. CONTINUOUS BEAM—THREE EQUAL SPANS—ALL SPANS LOADED

$R_A = 0.400\,wl$  $R_B = 1.10\,wl$  $R_C = 1.10\,wl$  $R_D = 0.400\,wl$

SHEAR: $0.400\,wl$, $0.500\,wl$, $0.600\,wl$, $0.600\,wl$, $0.500\,wl$, $0.400\,wl$

MOMENT: $-0.100\,wl^2$, $-0.100\,wl^2$, $+0.080\,wl^2$, $+0.025\,wl^2$, $+0.080\,wl^2$, $0.400\,l$, $0.500\,l$, $0.500\,l$, $0.400\,l$

Δ Max. (0.446 *l* from A or D) = 0.0069 $wl^4/EI$

## TABLE 8.13 *(Continued)*

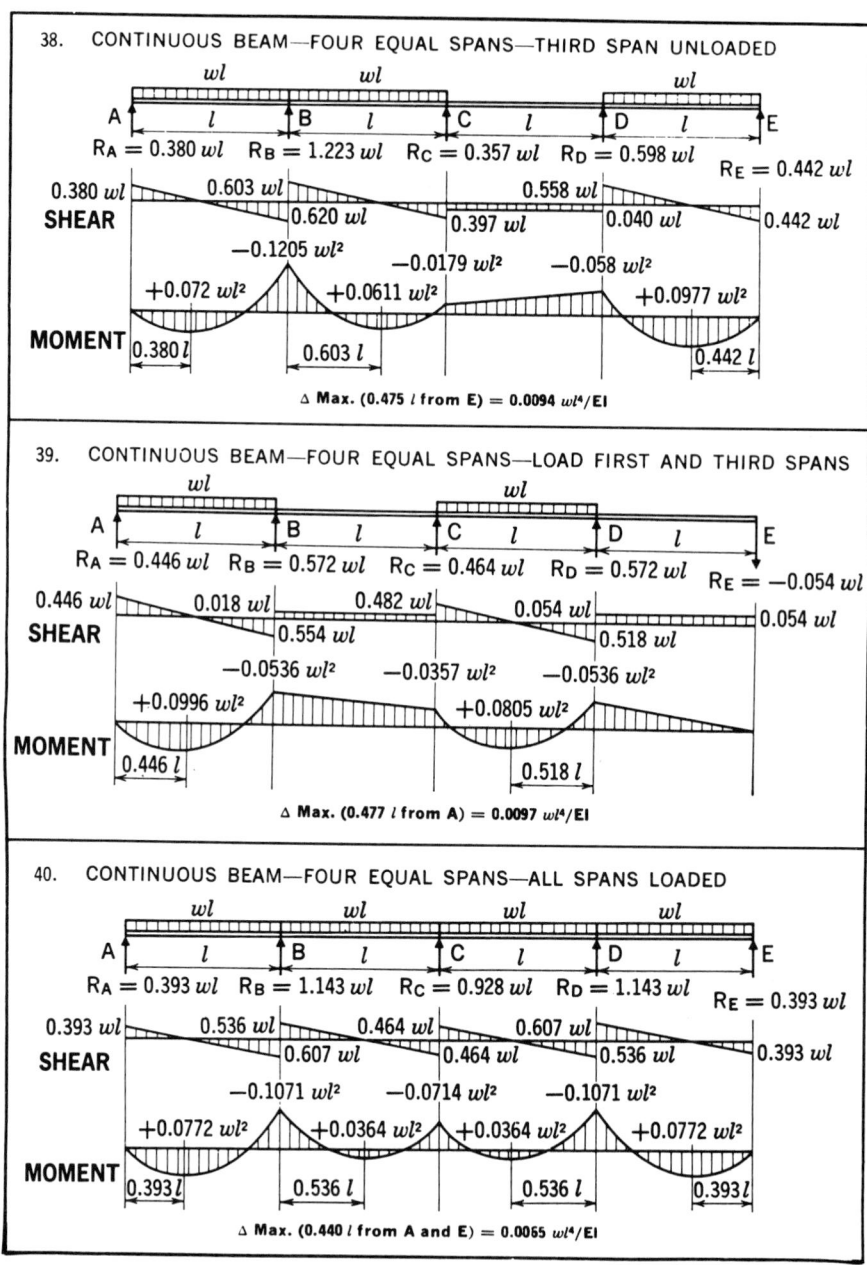

**38.  CONTINUOUS BEAM—FOUR EQUAL SPANS—THIRD SPAN UNLOADED**

$R_A = 0.380\,wl$   $R_B = 1.223\,wl$   $R_C = 0.357\,wl$   $R_D = 0.598\,wl$   $R_E = 0.442\,wl$

SHEAR: $0.380\,wl$   $0.603\,wl$   $0.620\,wl$   $0.558\,wl$   $0.397\,wl$   $0.040\,wl$   $0.442\,wl$

MOMENT: $-0.1205\,wl^2$   $-0.0179\,wl^2$   $-0.058\,wl^2$   $+0.072\,wl^2$   $+0.0611\,wl^2$   $+0.0977\,wl^2$   $0.380\,l$   $0.603\,l$   $0.442\,l$

Δ Max. (0.475 *l* from E) = 0.0094 *wl⁴*/EI

**39.  CONTINUOUS BEAM—FOUR EQUAL SPANS—LOAD FIRST AND THIRD SPANS**

$R_A = 0.446\,wl$   $R_B = 0.572\,wl$   $R_C = 0.464\,wl$   $R_D = 0.572\,wl$   $R_E = -0.054\,wl$

SHEAR: $0.446\,wl$   $0.018\,wl$   $0.482\,wl$   $0.054\,wl$   $0.054\,wl$   $0.554\,wl$   $0.518\,wl$

MOMENT: $-0.0536\,wl^2$   $-0.0357\,wl^2$   $-0.0536\,wl^2$   $+0.0996\,wl^2$   $+0.0805\,wl^2$   $0.446\,l$   $0.518\,l$

Δ Max. (0.477 *l* from A) = 0.0097 *wl⁴*/EI

**40.  CONTINUOUS BEAM—FOUR EQUAL SPANS—ALL SPANS LOADED**

$R_A = 0.393\,wl$   $R_B = 1.143\,wl$   $R_C = 0.928\,wl$   $R_D = 1.143\,wl$   $R_E = 0.393\,wl$

SHEAR: $0.393\,wl$   $0.536\,wl$   $0.464\,wl$   $0.607\,wl$   $0.607\,wl$   $0.464\,wl$   $0.536\,wl$   $0.393\,wl$

MOMENT: $-0.1071\,wl^2$   $-0.0714\,wl^2$   $-0.1071\,wl^2$   $+0.0772\,wl^2$   $+0.0364\,wl^2$   $+0.0364\,wl^2$   $+0.0772\,wl^2$   $0.393\,l$   $0.536\,l$   $0.536\,l$   $0.393\,l$

Δ Max. (0.440 *l* from A and E) = 0.0065 *wl⁴*/EI

*For meaning of symbols, see General Nomenclature. (Source: American Institute of Steel Construction, Inc.)

## 8.5  CANTILEVER BEAM COEFFICIENTS

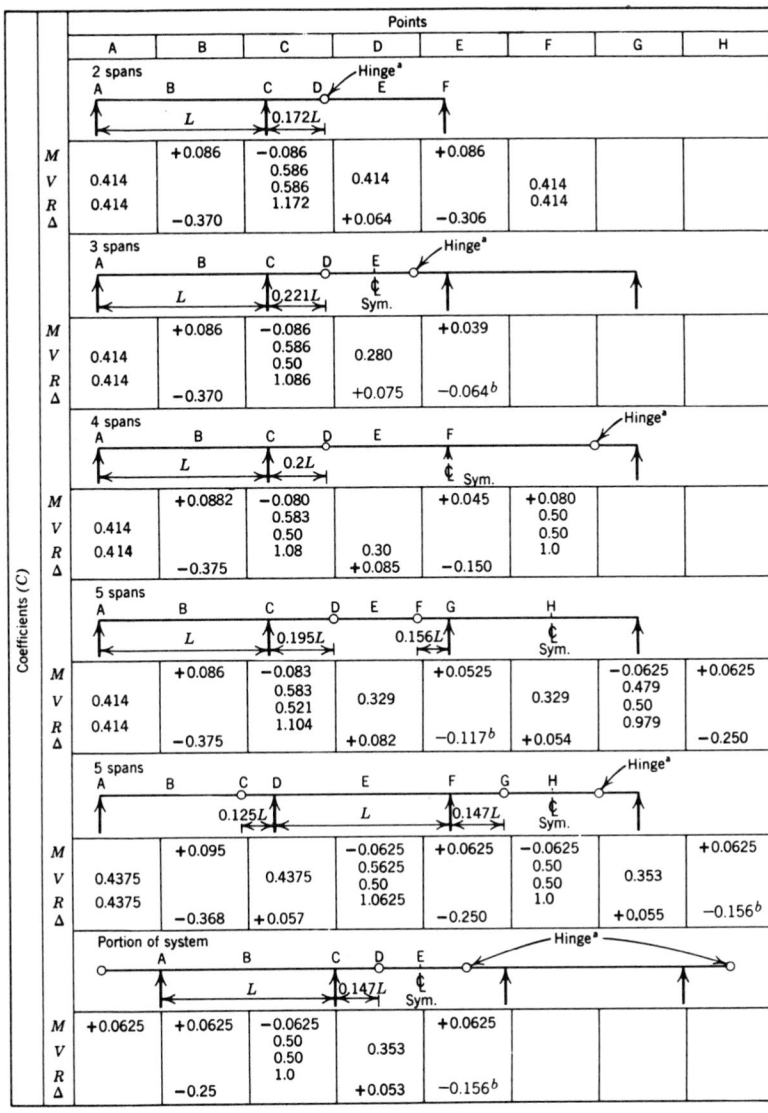

**FIGURE 8.1.**  Cantilever beam coefficients for balanced loading conditions. All spans equal, uniformly distributed load.

$$\text{Moment} = M = CwL^2 \qquad \text{Reaction} = R = CwL$$

$$\text{Shear} = V = CwL$$

$$\text{Deflection} = \Delta = \frac{CwL^4}{48EI}$$

$$w = \text{load (plf)}$$

$$L = \text{span (ft)}$$

[a] For typical hinge details, see *Typical Construction Details*, AITC 104.

[b] Deflecting of suspended beam span in from the ends of the simple-span beam between hinges, but the span $L$ is used in making the calculation.

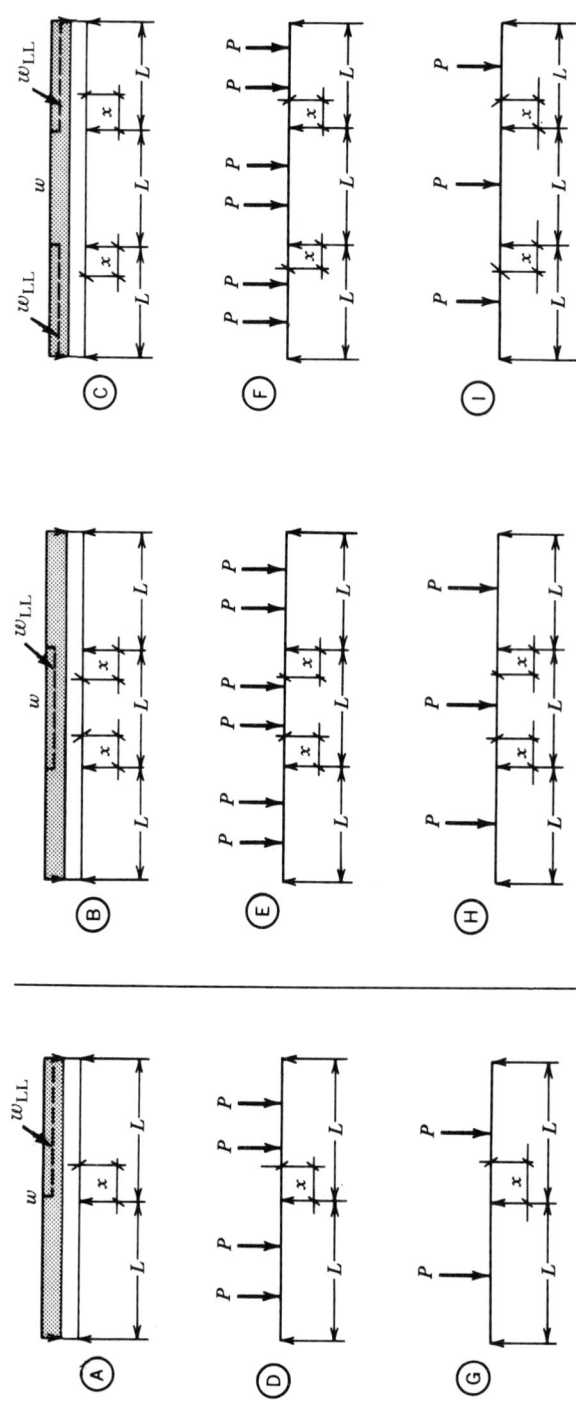

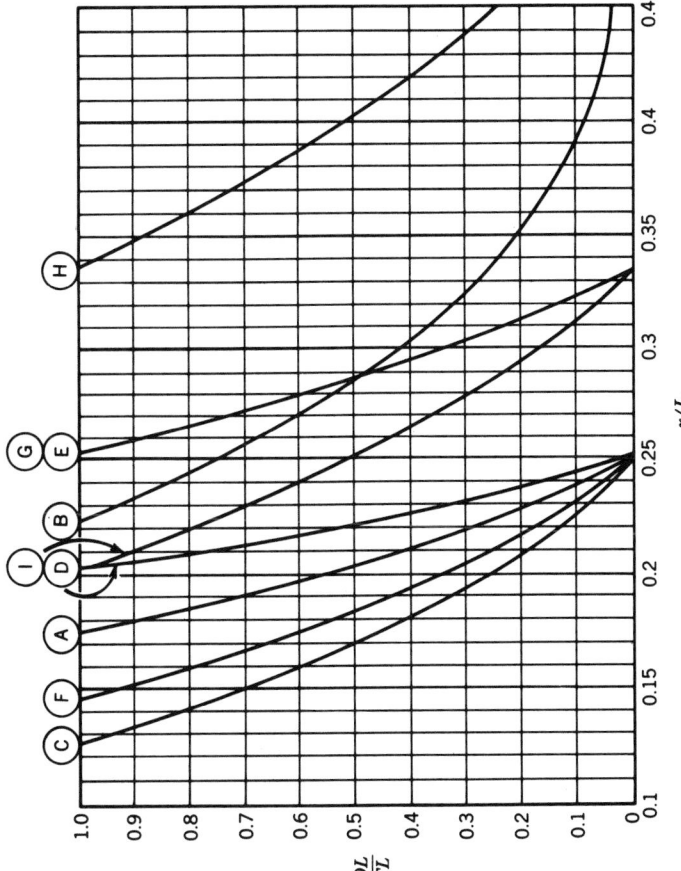

**FIGURE 8.2.** Cantilever beam coefficients (for unbalanced loading conditions). All spans equal, equal maximum positive, and negative moments.

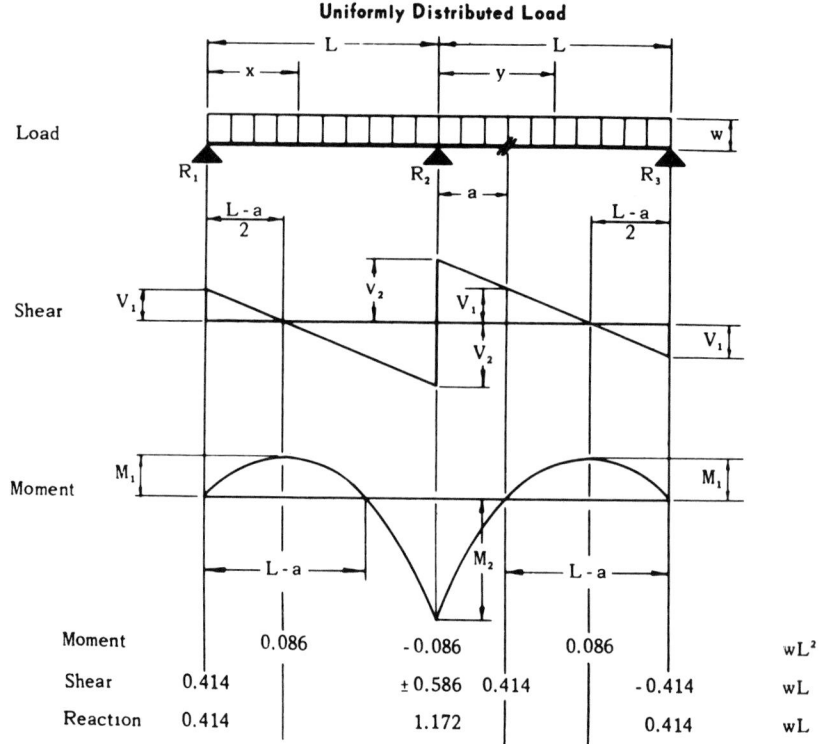

**FIGURE 8.3.** Cantilevered beam diagram. Two equal spans.

For maximum positive moment equal to maximum negative moment, $a = 0.172L$, and the above coefficients may be applied to $wL^2$ and $wL$ to find the respective critical values of moment, shear, and reaction. Maximum deflection in either span will be

$$\Delta = 13.31 \frac{wL^4}{EI} \text{ in.}$$

General formulas are:

$$R_1 = R_3 = \frac{w}{2}(L - a) \qquad V_y = \frac{w}{2}(L + a - 2y)$$

$$R_2 = w(L + a)$$

$$M_1 = \frac{w}{8}(L - a)^2$$

$$V_1 = \pm \frac{w}{2}(L - a)$$

$$M_2 = -\frac{wLa}{2}$$

$$V_2 = \pm \frac{w}{2}(L + a)$$

$$M_x = \frac{wx}{2}(L - a - x)$$

$$V_x = \frac{w}{2}(L - a - 2x)$$

$$M_y = \frac{w}{2}(y - a)(L - y)$$

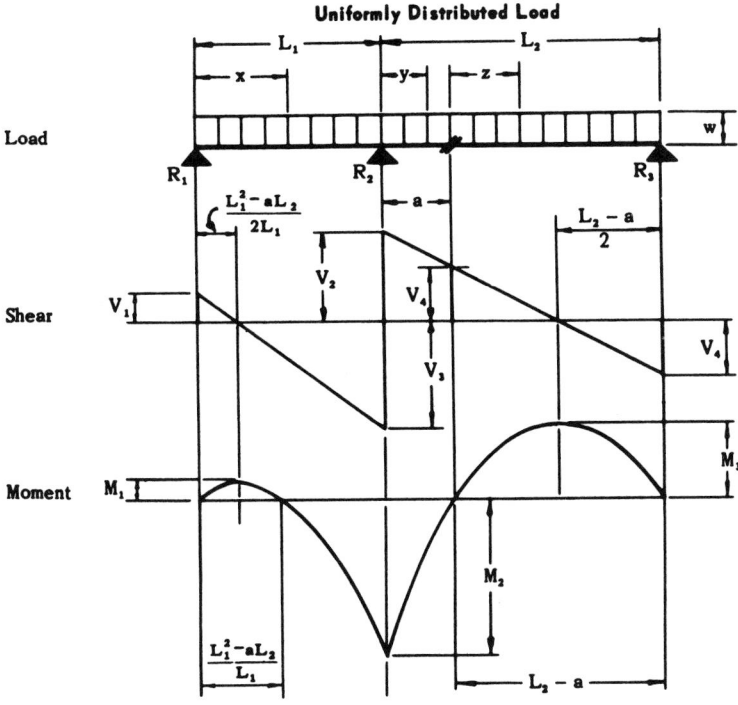

**FIGURE 8.4.**   Cantilevered beam diagram. Two unequal spans.

For maximum positive and negative moments in the cantilevered portion to be equal, $a = 0.172L_1^2/L_2$. Under these conditions, $M_1 = M_2 = 0.086wL_1^2$, $R_1 = V_1 = 0.414wL_1$, and $V_3 = -0.586wL_1$. Other coefficients can be determined for the above or other values of $a$ from the general formulas following:

$$R_1 = \frac{w}{2L_1}(L_1^2 - aL_2) \qquad V_3 = -\frac{w}{2L_1}(L_1^2 - aL_2) \qquad M_2 = -\frac{wL_2a}{2}$$

$$R_2 = \frac{w}{2L_1}(L_1 + a)(L_1 + L_2) \qquad V_4 = \pm\frac{w}{2}(L_2 - a) \qquad M_3 = \frac{w}{8}(L_2 - a)^2$$

$$R_3 = \frac{w}{2}(L_2 - a) \qquad V_x = \frac{w}{2L_1}(L_1^2 - aL_2) - wx \qquad M_x = \frac{wx}{2L_1}(L_1^2 - xL_1 - aL_2)$$

$$V_1 = \frac{w}{2L_1}(L_1^2 - aL_2) \qquad V_y = \frac{w}{2}(L_2 - a) - wy \qquad M_y = \frac{w}{2}(L_2 - y)(y - a)$$

$$V_2 = \frac{w}{2}(L_2 + a) \qquad V_z = \frac{w}{2}(L_2 - a) - wz \qquad M_z = \frac{w}{2}(L_2 - a - z)$$

$$M_1 = \frac{w}{8L_1^2}(L_1^2 - aL_2)^2$$

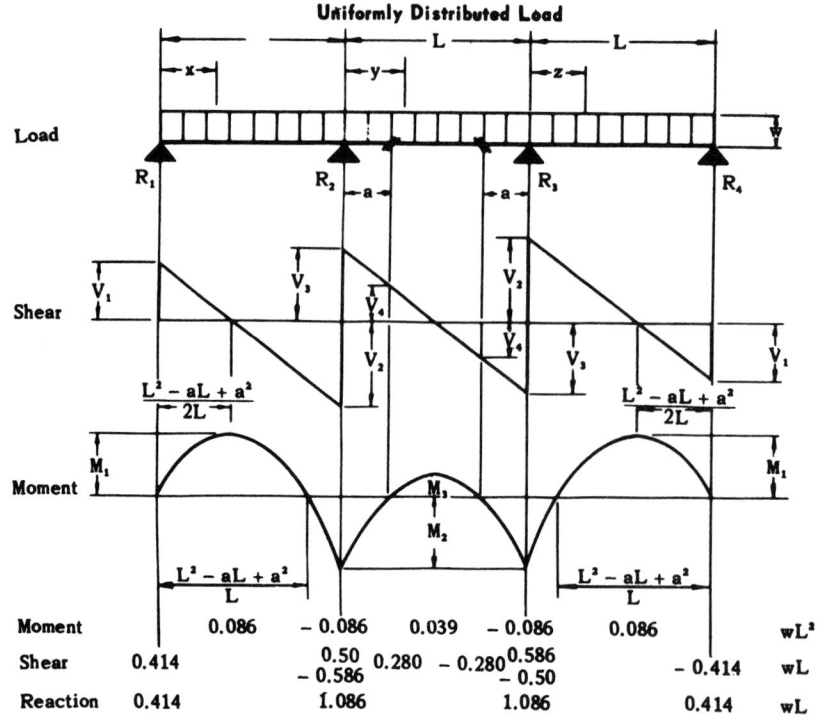

Uniformly Distributed Load

| Moment | | 0.086 | −0.086 | 0.039 | −0.086 | 0.086 | | $wL^2$ |
| Shear | 0.414 | | $\begin{array}{c}0.50\\−0.586\end{array}$ | 0.280 | $−0.280\begin{array}{c}0.586\\−0.50\end{array}$ | | −0.414 | $wL$ |
| Reaction | 0.414 | | 1.086 | | 1.086 | | 0.414 | $wL$ |

**FIGURE 8.5.** Cantilevered beam diagram. Three equal spans—single cantilever each end.

For the maximum positive and negative moments of the cantilevered portions of the beam to be equal, $a = 0.220L$, and the above coefficients may be applied to $wL^2$ and $wL$ to find the respective critical values of moment, shear, and reaction. Maximum deflection in end spans will be

$$\Delta = 13.31 \frac{wL_4}{EI} \text{ in.}$$

General formulas are:

$$R_1 = R_4 = \frac{w}{2L}(L^2 - aL + a^2) \qquad M_1 = -\frac{w}{8L^2}(L^2 - aL + a^2)^2$$

$$R_2 = R_1 = \frac{w}{2L}(2L^2 + aL - a^2) \qquad M_2 = -\frac{w}{2}(aL - a^2)$$

$$V_1 = \pm\frac{w}{2L}(L^2 - aL + a^2) \qquad M_3 = \frac{w}{8}(L - 2a)^2$$

$$V_2 = \pm\frac{w}{2L}(L^2 + aL - a^2) \qquad V_x = \frac{w}{2L}(L^2 - aL + a^2) - wx$$

$$V_3 = \pm\frac{wL}{2} \qquad\qquad V_y = \frac{w}{2}(L - 2y)$$

$$V_4 = \pm\frac{w}{2}(L - 2a) \qquad V_z = \frac{w}{2L}(L^2 + aL - a^2) - wz$$

$$M_x = \frac{wx}{2L}(L^2 - aL + a^2) - \frac{wx^2}{2} \qquad M_y = \frac{w}{2}(y - a)(L - y - a)$$

$$M_z = \frac{w}{2}(L - z)\left(\frac{a^2}{L} + z - a\right)$$

**Uniformly Distributed Load**

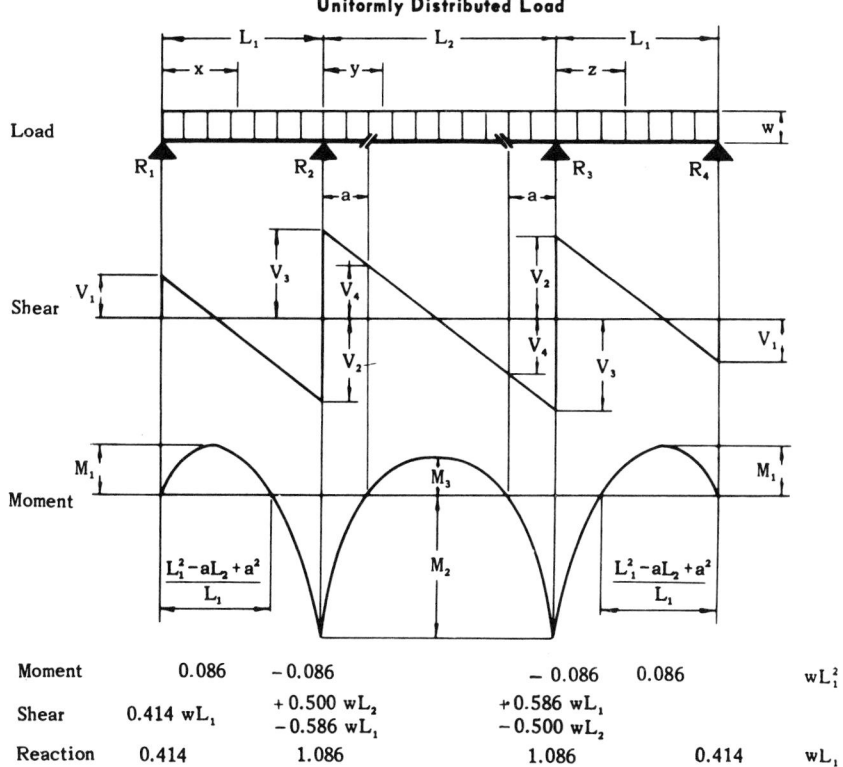

**FIGURE 8.6.**   Cantilevered beam diagrams. Three spans—end spans equal—single cantilever each end.

For the maximum positive and negative moments of the cantilevered portions of the beam to be equal, $a = \frac{1}{2}(L_2 - \sqrt{L_2^2 - 0.688L_1^2})$ and the above coefficients may be applied to find the respective critical values of moment, shear, and reaction. Coefficients are omitted when calculation using the general formula is simpler. General formulas are:

$$R_1 = R_4 = \frac{w}{2L_1}(L_1^2 - aL_2 + a^2) \qquad V_4 = \pm\frac{w}{2}(L_2 - 2a) \qquad M_2 = -\frac{w}{2}(aL_2 - a^2)$$

$$R_2 = R_3 = \frac{w}{2L_1}(L_1 + a)(L_1 + L_2 - a) \qquad M_1 = \frac{w}{8L_1^2}(L_1^2 - aL_2 + a^2)^2 \qquad M_3 = \frac{w}{8}(L_2 - 2a)^2$$

$$V_1 = \pm\frac{w}{2L_1}(L_1^2 - aL_2 - a^2) \qquad V_x = \frac{w}{2L_1}(L_1^2 - aL_2 + a^2) - wx \qquad M_x = \frac{wx}{2L_1}(L_1^2 - aL_2 + a^2) - \frac{wx^2}{2}$$

$$V_2 = \pm\frac{w}{2L_1}(L_1^2 - aL_2 + a^2) \qquad V_y = \frac{w}{2}(L_2 - 2y) \qquad M_y = \frac{w}{2}(y - a)(L_2 - y - a)$$

$$V_3 = \pm\frac{wL_2}{2} \qquad V_z = \frac{w}{2L_1}(L_1^2 - aL_2 + a^2) - wz \qquad M_z = \frac{w}{2L_1}(L_1z - aL_2)(L_1 - z)$$

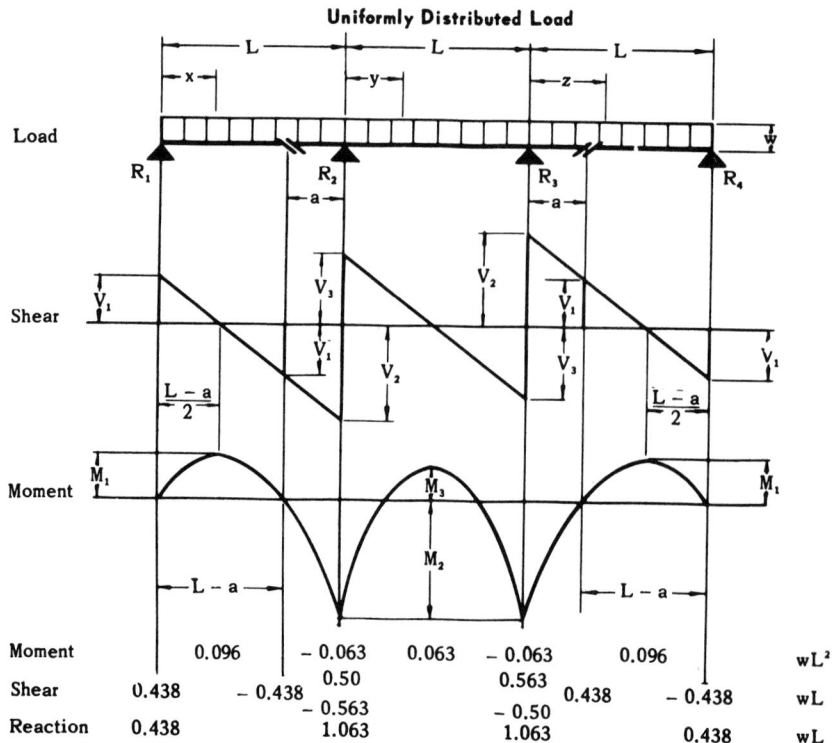

| | | | | |
|---|---|---|---|---|
| Moment | 0.096 | − 0.063    0.063 | − 0.063    0.096 | | $wL^2$ |
| Shear | 0.438 | − 0.438<br>0.50<br>− 0.563 | 0.563<br>0.438    − 0.438<br>− 0.50 | | $wL$ |
| Reaction | 0.438 | 1.063 | 1.063    0.438 | | $wL$ |

**FIGURE 8.7.**   Cantilevered beam diagram. Three equal spans—double cantilever.

For the maximum positive and negative moments of the cantilevered portion of the beam to be equal, $a = L/8$, and the above coefficients may be applied to $wL^2$ and $wL$ to find the respective critical values of moment, shear, and reaction. Maximum deflection in center span will be

$$\Delta = 8.99 \frac{wL^4}{EI}$$

General formulas are

$$R_1 = R_4 = \frac{w}{2}(L - a) \qquad V_x = \frac{w}{2}(L - a - 2x) \qquad M_3 = \frac{w}{8}(L^2 - 4aL)$$

$$R_2 = R_3 = \frac{w}{2}(2L + a) \qquad V_y = \frac{w}{2}(L - 2y) \qquad M_x = \frac{wx}{2}(L - a - x)$$

$$V_1 = \pm\frac{w}{2}(L - a) \qquad V_z = \frac{w}{2}(L + a - 2z) \qquad M_y = \frac{w}{2}(L_2 y - y^2 - La)$$

$$V_2 = \pm\frac{w}{2}(L + a) \qquad M_1 = \frac{w}{8}(L - a)^2 \qquad M_z = \frac{w}{2}(L - z)(z - a)$$

$$V_3 = \pm\frac{wL}{2} \qquad M_2 = -\frac{wLa}{2}$$

Points of zero moment in center span occur at

$$y = \frac{L_2 \pm \sqrt{L_2^2 - 4aL_1}}{2}$$

**Uniformly Distributed Load**

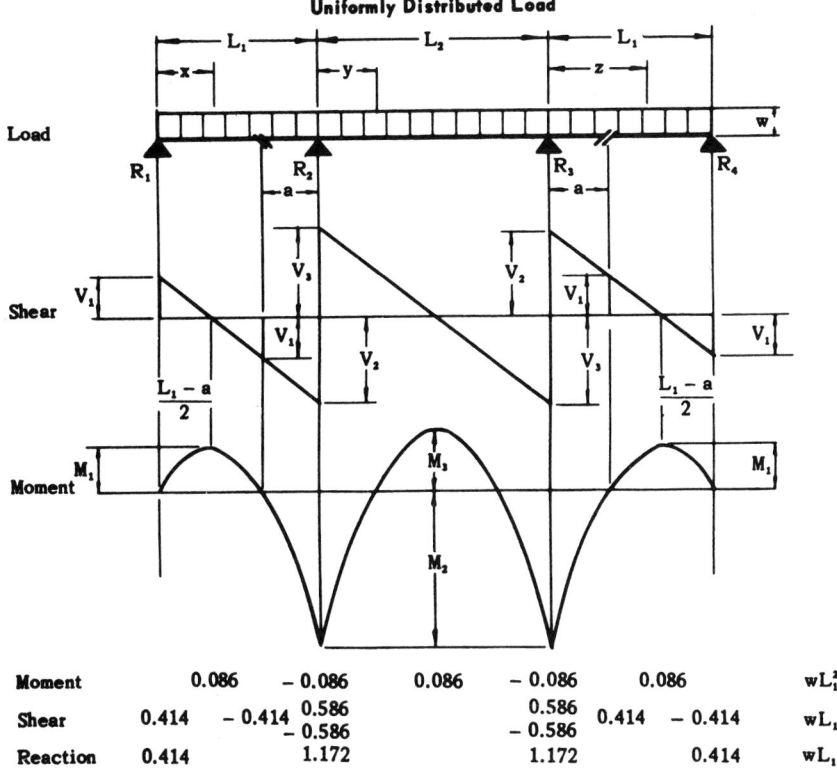

| Moment | 0.086 | − 0.086 | 0.086 | − 0.086 | 0.086 | $wL_1^2$ |
|---|---|---|---|---|---|---|
| Shear | 0.414 | − 0.414 | 0.586<br>− 0.586 | 0.586<br>− 0.586 | 0.414   − 0.414 | $wL_1$ |
| Reaction | 0.414 | | 1.172 | 1.172 | 0.414 | $wL_1$ |

**FIGURE 8.8.**   Cantilevered beam diagram. Three spans—end spans equal—double cantilever.
  For the maximum positive and negative moments of the cantilevered portion of the beam to be equal, $a = 0.125L_2^2/L_1$. For the special case where all maximum positive and negative moments are equal, that is, $M_1 = M_2 = M_3$, $a = 0.172L_1$ and $L_2 = 1.172L_1$, and the above coefficients may be applied to $wL_1^2$ and $wL_1$ to find the respective critical values of moment, shear, and rejection. General formulas are

$$R_1 = R_4 = \frac{w}{2}(L_1 - a) \qquad V_x = \frac{w}{2}(L_1 - a - 2x) \qquad M_3 = \frac{w}{8}(L_2^2 - 4aL_1)$$

$$R_2 = R_3 = \frac{w}{2}(L_1 + L_2 + a) \qquad V_y = \frac{w}{2}(L_2 - 2y) \qquad M_x = \frac{wx}{2}(L_1 - a - x)$$

$$V_1 = \pm(L_1 - a)$$

$$V_z = \frac{w}{2}(L_1 + a - 2z) \qquad M_y = \frac{w}{2}(L_2 y - y^2 - aL_1)$$

$$V_2 = \pm\frac{w}{2}(L_1 + a)$$

$$M_1 = \frac{w}{8}(L_1 - a)^2 \qquad M_z = \frac{w}{2}(z - L_1)(a - z)$$

$$V_3 = \pm\frac{wL_2}{2}$$

$$M_2 = -\frac{wL_1 a}{2}$$

Points of zero moment in center span occur at

$$y = \frac{L_2 \pm \sqrt{L_2^2 - 4aL_1}}{2}$$

## 8.6 PANELIZED ROOF GRID SYSTEMS—EFFICIENCY

The selection of an efficient column grid system is facilitated by the determination of an efficiency factor. The efficiency factor is the ratio of the board footage (fbm) of glued laminated timber used in the primary framing system to the square feet ($ft^2$) of space enclosed under roof cover. In comparing glulam framing systems, the lower the ratio, the more efficient the system.

Fig. 8.9 illustrates the calculation of the efficiency factors for a selected building size and framing system.

Glulam and purlin sizes and corresponding efficiency factors are given in each layout for two different live loads. (*Note:* Regular type refers to one LL factor, with italic type for the other. See legend for complete explanation.) Member sizes shown are governed by either bending or shear. Sawn purlin sizes are based on the use of a No. 1 Douglas Fir grade. In all cases, the data are subject to specific design requirements of local codes or special conditions, including applicable de-

Note: Glued laminated timber sizes are for illustration only.

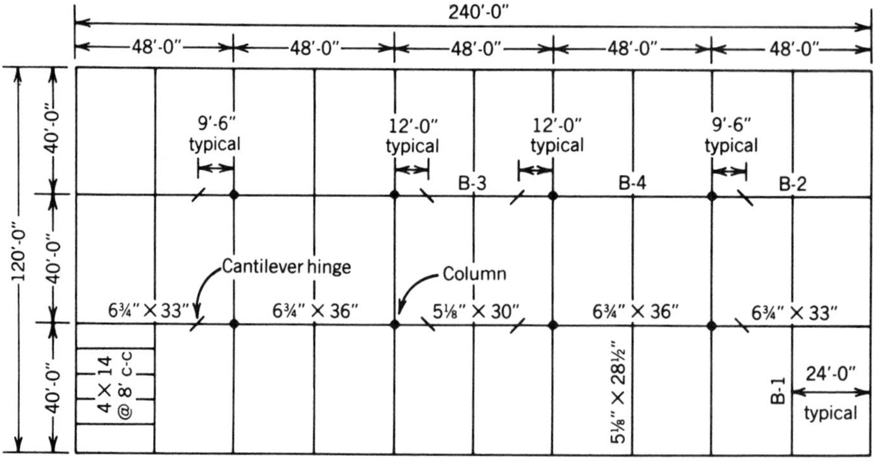

**FIGURE 8.9.**   Calculation of efficiency factors:

B-1 =    760 fbm/beam × 27 beams = 20,520 fbm
B-2 = 1129 fbm/beam ×  4 beams =    4517 fbm
B-3 =    480 fbm/beam ×  2 beams =      960 fbm
B-4 = 2224 fbm/beam ×  4 beams =    8896 fbm
                                    Total = 34,893 fbm

Area enclosed = 28,800 $ft^2$

Efficiency factor = 34,893/28,800 = 1.21 fbm glued laminated/$ft^2$

Solid sawn purlins; = 112 fbm/purlin × 120 purlins = 13,440 fbm

Efficiency factor = 13,440/28,800 = 0.47 fbm sawn/$ft^2$

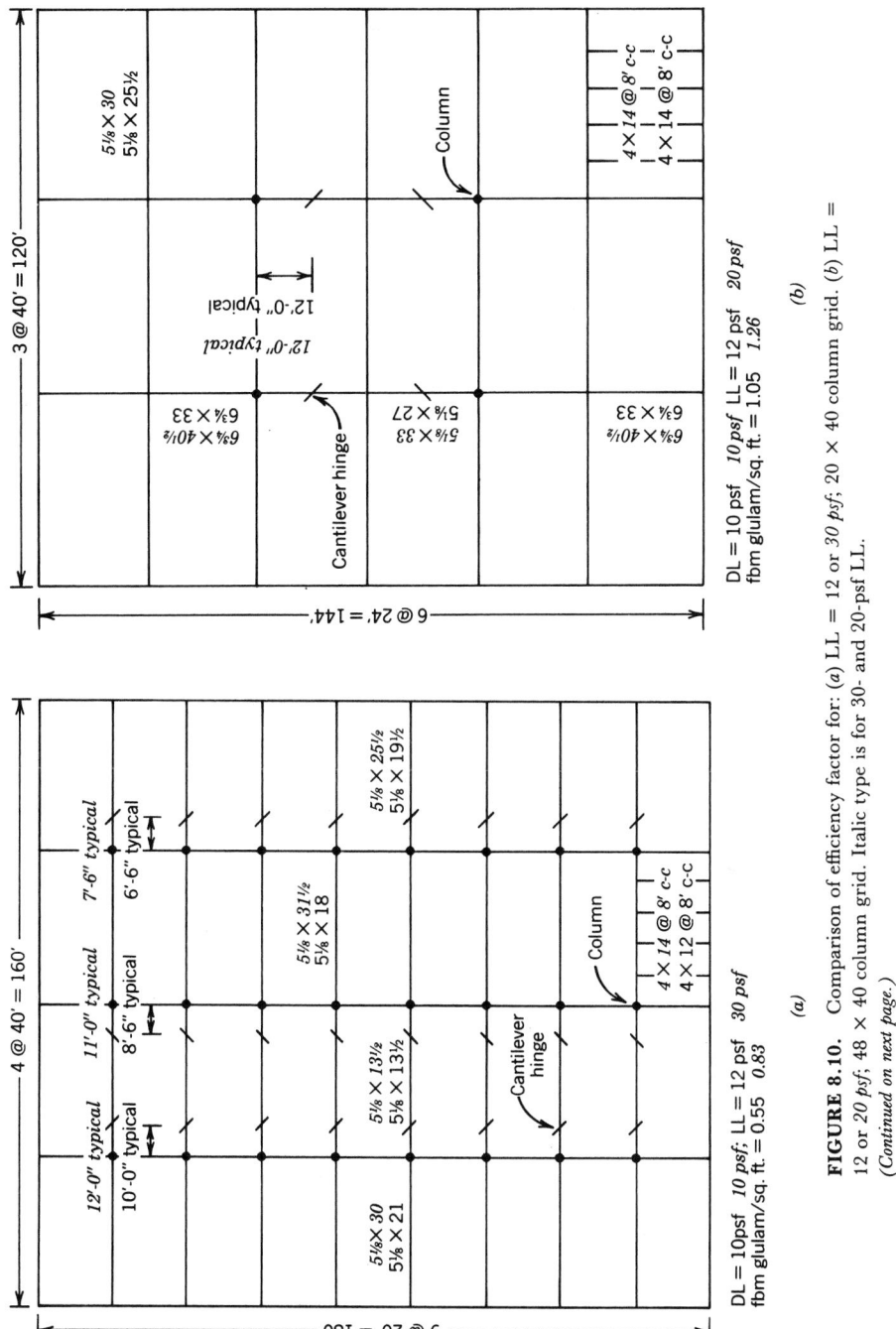

DL = 10psf *10 psf*; LL = 12 psf *30 psf*
fbm glulam/sq. ft. = 0.55  *0.83*

(a)

DL = 10 psf *10 psf*; LL = 12 psf *20 psf*
fbm glulam/sq. ft. = 1.05  *1.26*

(b)

**FIGURE 8.10.** Comparison of efficiency factor for: (*a*) LL = 12 or 30 *psf*; 20 × 40 column grid. (*b*) LL = 12 or 20 *psf*; 48 × 40 column grid. Italic type is for 30- and 20-psf LL. (*Continued on next page.*)

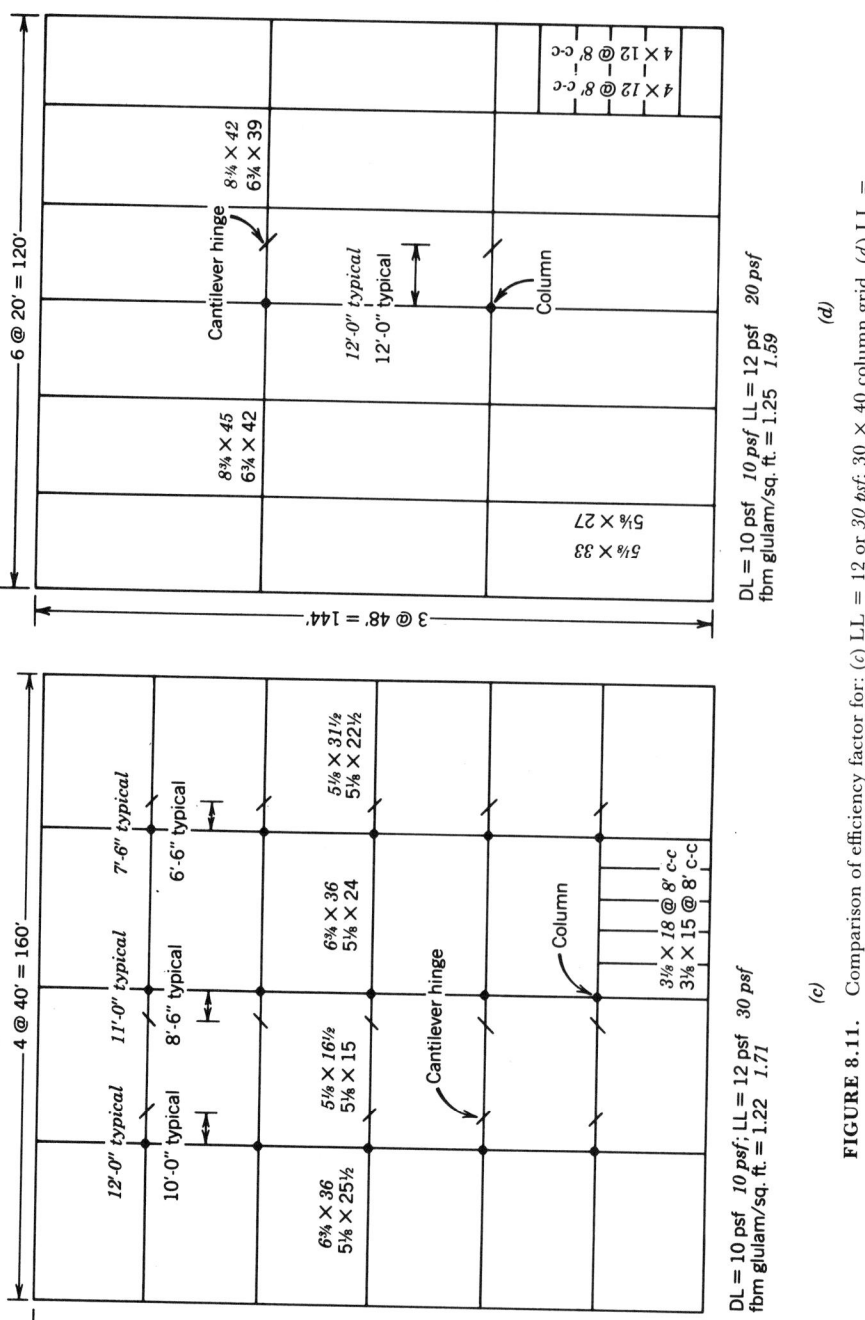

**FIGURE 8.11.** Comparison of efficiency factor for: (*c*) LL = 12 or *30 psf*; 30 × 40 column grid. (*d*) LL = 12 or *20 psf*; 48 × 60 column grid. Italic type is for 30- and 20-psf LL.

DL = 10 psf  *10 psf*  LL = 12 psf  *20 psf*
fbm glulam/sq. ft. = 1.25  *1.59*

*(d)*

DL = 10 psf  *10 psf*; LL = 12 psf  *30 psf*
fbm glulam/sq. ft. = 1.22  *1.71*

*(c)*

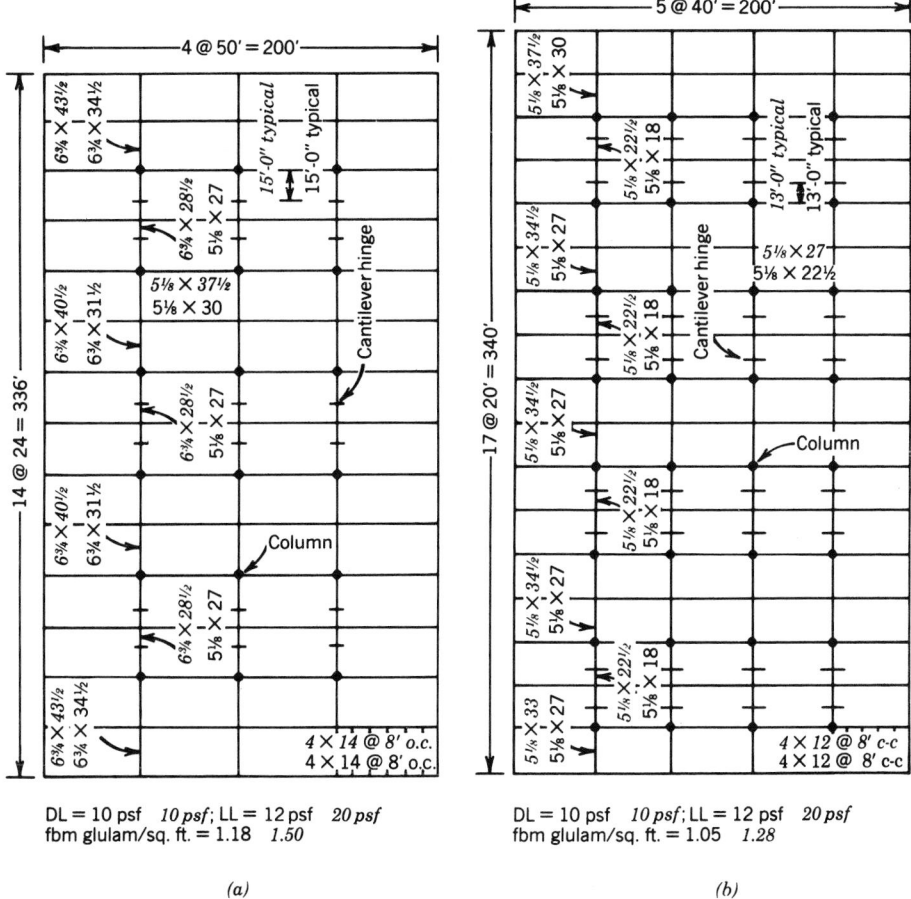

DL = 10 psf  *10 psf*; LL = 12 psf  *20 psf*
fbm glulam/sq. ft. = 1.18  *1.50*

DL = 10 psf  *10 psf*; LL = 12 psf  *20 psf*
fbm glulam/sq. ft. = 1.05  *1.28*

(a)

(b)

**FIGURE 8.12.** Comparison of efficiency of column grids for LL = 12 or *20 psf*; (*a*) 48 × 50 column grid; (*b*) 40 × 40 column grid. Italic type is for 20-psf LL.

Design criteria and allowable design values used in the calculation of the bending member sizes shown for the examples: Fig. 8.9–8.12:

Member sizes are governed by either bending or shear.
Full balanced or unbalanced live load, whichever controls
Dead load does not include weight of glulam
Bending design value, $F_b$ = 2400 psi
Shear design value, $F_v$ = 165 psi
Compression perpendicular to grain stress, $F_{c\perp}$ = 560 psi
Modulus of elasticity, $E$ = 1,800,000 psi
Duration-of-load factor: $C_D$ = 1.25 for 12-psf live loads; $C_D$ = 1.15 for 20- and 30-psf snow loads

flection requirements. A minimum roof slope of $\frac{1}{4}$ in./ft should be provided in addition to camber to minimize water ponding. The examples shown are illustrative only, and a complete design check should be made by competent engineering personnel.

## 8.7 MATERIAL WEIGHTS

### TABLE 8.14

### Minimum Design Dead Loads[a] (psf)

| | | | |
|---|---|---|---|
| **Ceilings** | | **Coverings, roof, and wall (cont.)** | |
| Acoustical fiber tile | 1 | Rigid insulation, $\frac{1}{2}$ in. | 0.75 |
| Gypsum board (per $\frac{1}{8}$ in. thick) | 0.55 | Skylight, metal frame, $\frac{3}{8}$ in. | |
| Mechanical duct allowance | 4 | wire glass | 8 |
| Plaster on tile or concrete | 5 | Slate, $\frac{3}{16}$ in. | 7 |
| Plaster on wood lath | 8 | Slate, $\frac{1}{4}$ in. | 10 |
| Suspended steel channel system | 2 | Waterproofing membranes: | |
| Suspended metal lath and | | Bituminous, gravel covered | 5.5 |
| cement plaster | 15 | Bituminous, smooth surface | 1.5 |
| Suspended metal lath and | | Liquid applied | 1.0 |
| gypsum plaster | 10 | Single-ply, sheer | 0.7 |
| Wood furring suspension | | Wood sheathing (per in. | |
| system | 2.5 | thickness) | 3 |
| **Coverings: roof and wall** | | Wood shingles | 3 |
| Asbestos–cement shingles | 4 | **Floor fill** | |
| Asphalt shingles | 2 | Cinder concrete, per in. | 9 |
| Clay tile | 16 | Lightweight concrete, per in. | 8 |
| Clay tile (for mortar add 10 lb): | | Sand, per in. | 8 |
| Book tile, 2 in. | 12 | Stone concrete, per in. | 12 |
| Book tile, 3 in. | 20 | **Floors and floor finishes** | |
| Ludowici | 10 | Asphalt block (2 in.), $\frac{1}{2}$ in. | |
| Roman | 12 | mortar | 30 |
| Spanish | 19 | Cement finish (1 in.) on stone- | |
| Composition: | | concrete fill | 32 |
| Three-ply ready roofing | 1 | Ceramic or quarry tile ($\frac{3}{4}$ in.) | |
| Four-ply felt and gravel | 5.5 | on $\frac{1}{2}$ in. mortar bed | 16 |
| Five-ply felt and gravel | 6 | Ceramic or quarry tile ($\frac{3}{4}$ in.) | |
| Copper or tin | 1 | on 1 in. mortar bed | 23 |
| Corrugated asbestos–cement | | Concrete fill finish (per in. | |
| roofing | 4 | thick) | 12 |
| Deck, metal, 20 gage | 2.5 | Hardwood flooring, $\frac{7}{8}$ in. | 4 |
| Deck, metal, 18 gage | 3 | Linoleum or asphalt tile, $\frac{1}{4}$ in. | 1 |
| Decking, 2 in. wood (Douglas | | Marble and mortar on stone- | |
| fir) | 5 | concrete fill | 33 |
| Decking, 3 in. wood (Douglas | | Slate (per in. thickness) | 15 |
| fir) | 8 | Solid flat tile on 1 in. | |
| Fiberboard, $\frac{1}{2}$ in. | 0.75 | mortar base | 23 |
| Gypsum sheathing, $\frac{1}{2}$ in. | 2 | Subflooring, $\frac{3}{4}$ in. | 3 |
| Insulation, roof boards (per in. | | Terrazzo ($1\frac{1}{2}$ in.) directly on slab | 19 |
| thickness): | | Terrazzo (1 in.) on stone- | |
| Cellular glass | 0.7 | concrete fill | 32 |
| Fibrous glass | 1.1 | Terrazzo (1 in.) on 2 in. | |
| Fiberboard | 1.5 | stone concrete | 32 |
| Perlite | 0.8 | Wood block (3 in.) on mastic | |
| Polystyrene foam | 0.2 | no fill | 10 |
| Urethane foam with skin | 0.5 | Wood block (3 in.) on $\frac{1}{2}$ in. | |
| Plywood (per $\frac{1}{8}$ in. thickness) | 0.4 | mortar base | 16 |

**TABLE 8.14** (*Continued*)

Floors, wood-joist (no plaster)
  Double wood floor

| Joist sizes, in. | 12 in. spacing (psf) | 16 in. spacing (psf) | 24 in. spacing (psf) |
|---|---|---|---|
| 2 × 6 | 6 | 5 | 5 |
| 2 × 8 | 6 | 6 | 5 |
| 2 × 10 | 7 | 6 | 6 |
| 2 × 12 | 8 | 7 | 6 |

Frame partitions
  Movable steel partitions — 4
  Wood or steel studs, $\frac{1}{2}$ in.
    gypsum board each side — 8
  Wood studs, 2 × 4, unplastered — 4
  Wood studs, 2 × 4, plastered 1
    side — 12
  Wood studs, 2 × 4, plastered 2
    sides — 20
Frame walls
  Exterior stud walls:
    2 × 4 at 16 in., $\frac{5}{8}$ in.
    gypsum, insulated, $\frac{3}{8}$ in.
    siding — 11
    2 × 6 at 16 in., $\frac{5}{8}$ in.
    siding — 12
  Exterior stud walls with brick
    veneer — 48
  Windows, glass frame, and
    sash — 8
Masonry partitions
  Concrete block, light aggregate:
    4 in. — 20
    6 in. — 28

Masonry partitions (cont.)
  Concrete block, light aggregate:
    8 in. — 38
    12 in. — 55
Masonry walls
  Clay brick wythes
    4 in. — 39
    8 in. — 79
    $12\frac{1}{2}$ in. — 115
    17 in. — 155
Hollow concrete masonry unit wythes:

| Wythe thickness (in in.) | 4 | 6 | 8 | 10 | 12 |
|---|---|---|---|---|---|
| Unit Percent Solid | 70 | 55 | 52 | 50 | 48 |
| **Light Weight Units (105 pcf):** | | | | | |
| No grout | 22 | 27 | 35 | 42 | 49 |
| 48 o.c. | | | 31 | 40 | 49 | 58 |
| 40 o.c. | | | 33 | 43 | 53 | 63 |
| 32 o.c. (Grout | | | 34 | 45 | 56 | 66 |
| 24 o.c. (Spacing | | | 37 | 49 | 61 | 72 |
| 16 o.c. | | | 42 | 56 | 70 | 84 |
| Full Grout | | | 57 | 77 | 98 | 119 |
| **Normal Weight Units (135 pcf):** | | | | | |
| No Grout | 29 | | 35 | 45 | 54 | 63 |
| 48 o.c. | | | 33 | 50 | 61 | 72 |
| 40 o.c. (Grout | | | 36 | 53 | 65 | 77 |
| 32 o.c. (Spacing | | | 38 | 55 | 68 | 80 |
| 24 o.c. | | | 41 | 59 | 73 | 86 |
| 16 o.c. | | | 47 | 66 | 82 | 98 |
| Full Grout | | | 64 | 87 | 110 | 133 |

Solid Concrete Masonry Unit Wythes
(incl. Concrete brick):

| Wythe thickness (in. in.) | 4 | 6 | 8 | 10 | 12 |
|---|---|---|---|---|---|
| Light Weight Units (105 pcf): | 32 | 49 | 67 | 84 | 102 |
| Normal Weight Units (135 pcf): | 41 | 63 | 86 | 108 | 131 |

[a]Source: This material is reproduced with permission from American Society of Civil Engineers, ANSI/ASCE 7-88 (formerly A58.1), *Minimum Design Loads for Buildings and Other Structures*, copyright 1990 by the American National Standards Institute. Copies of this standard may be purchased from the American National Standards Institute at 11 West 42nd Street, New York, NY 10036.

[b]Weights of masonry include mortar but not plaster. For plaster, add 5 psf for each face plastered. Values given represent averages. In some cases, there is a considerable range of weight for the same construction.

## TABLE 8.15

### Minimum Design Loads[a] (pcf)

| | | | |
|---|---|---|---|
| Bituminous products | | Silt, moist, loose | 78 |
| Asphaltum | 81 | Silt, moist, packed | 96 |
| Graphite | 135 | Silt, flowing | 108 |
| Paraffin | 56 | Sand and gravel, dry, loose | 100 |
| Petroleum, crude | 55 | | |
| Petroleum, refined | 50 | Sand and gravel, dry, packed | 110 |
| Petroleum, benzine | 46 | | |
| Petroleum, gasoline | 42 | Sand and gravel, wet | 120 |
| Pitch | 69 | Earth (submerged): | |
| Tar | 75 | Clay | 80 |
| Brass | 526 | Soil | 70 |
| Bronze | 552 | River mud | 90 |
| Cast-stone masonry (cement, stone, sand) | 144 | Sand and gravel | 60 |
| | | Sand or gravel, and clay | 65 |
| Cement, Portland, loose | 90 | Gravel, dry | 104 |
| Ceramic tile | 150 | Gypsum, loose | 70 |
| Charcoal | 12 | Gypsum wallboard | 50 |
| Cinders, dry, in bulk | 45 | Ice | 57 |
| Cinder fill | 57 | Iron, cast | 450 |
| Coal, anthracite, piled | 52 | Iron, wrought | 480 |
| Coal, bituminous, piled | 47 | Lead | 710 |
| Coal, lignite, piled | 47 | Lime, hydrated, loose | 32 |
| Coal, peat, dry, piled | 23 | Lime, hydrated, compacted | 45 |
| Concrete, plain: | | Masonry, ashlar: | |
| Cinder | 108 | Granite | 165 |
| Expanded-slag aggregate | 100 | Limestone, crystalline | 165 |
| Haydite (burned-clay aggregate) | 90 | Limestone, oolitec | 135 |
| | | Marble | 173 |
| Slag | 132 | Sandstone | 144 |
| Stone (including gravel) | 144 | Masonry, brick: | |
| Vermiculite and perlite aggregate, nonload-bearing | 25–50 | Hard (low absorption) | 130 |
| | | Medium (medium absorption) | 115 |
| Other light aggregate load-bearing | 70–105 | Soft (high absorption) | 100 |
| | | Masonry, rubble stone: | |
| Concrete, reinforced: | | Granite | 153 |
| Cinder | 111 | Limestone, crystalline | 147 |
| Slag | 138 | Limestone, oolitec | 138 |
| Stone (including gravel) | 150 | Marble | 156 |
| Copper | 556 | Sandstone | 137 |
| Cork, compressed | 14.4 | Mortar | |
| Earth (not submerged) | | Cement | 130 |
| Clay, dry | 63 | Lime | 130 |
| Clay, damp | 110 | Particleboard | 45 |
| Clay and gravel, dry | 100 | | |

**TABLE 8.15  (Continued)**

| | | | |
|---|---|---|---|
| Plywood | 36 | Terra cotta, architectural: | |
| Riprap (not submerged) | | Voids filled | 120 |
| Limestone | 83 | Voids unfilled | 72 |
| Sandstone | 90 | Tin | 459 |
| Sand, clean and dry | 90 | Water, fresh | 62 |
| Sand, river, dry | 106 | Water, sea | 64 |
| Slag, bank | 70 | Wood, seasoned[b] | |
| Slag, bank screenings | 108 | Ash, White | 41 |
| Slag, machine | 96 | Cypress | 34 |
| Slag, sand | 52 | Fir, Douglas | 34 |
| Slate | 172 | Hem-Fir | 28 |
| Steel, cold-drawn | 492 | Oak, red and white | 47 |
| Stone, quarried, piled: | | Pine, Southern | 37 |
| Basalt, granite, gneiss | 96 | Redwood | 28 |
| Limestone, marble, quartz | 95 | Spruce, red, white and | |
| Sandstone | 82 | Sitka | 29 |
| Shale | 92 | Western Hemlock | 32 |
| Greenstone, hornblende | 107 | Zinc, rolled, sheet | 449 |

[a] Source: This material is reproduced with permission from American Society of Civil Engineers, ANSI/ASCE 7-88 (formerly A58.1), *Minimum Design Loads for Buildings and Other Structures*, copyright 1990 by the American National Standards Institute. Copies of this standard may be purchased from the American National Standards Institute at 11 West 42nd Street, New York, NY 10036.

[b] For additional information on weights of commercial lumber species, see Table 2.2

## 8.8  CONVERSION FACTORS

**TABLE 8.16**

**English/Metric Conversion Factors**

| Multiply | English to Metric By | To Obtain |
|---|---|---|
| Length: | | |
| Inches | 25.4 | Millimeters |
| Feet | 0.3048 | Meters |
| Yards | 0.9144 | Meters |
| Miles, statute | 1.609344 | Kilometers |
| Miles, nautical | 1.852 | Kilometers |
| | | |
| Area: | | |
| Square inches | 645.16 | Square millimeters |
| Square feet | 0.09290304 | Square meters |
| Square yards | 0.83612736 | Square meters |

## TABLE 8.16    (*Continued*)

| Multiply | English to Metric<br>By | To Obtain |
|---|---|---|
| Length: | | |
| Square miles | 2.58999776 | Square kilometers |
| Acres | 0.0040469 | Square kilometers |
| | | |
| Volume: | | |
| Cubic inches | 16,387 | Cubic millimeters |
| Cubic feet | 0.02831684 | Cubic meters |
| Cubic yards | 0.76455485 | Cubic meters |
| | | |
| Capacity (U.S. Liquid Measure): | | |
| Ounces | 0.0295735 | Liters |
| Quarts | 0.9463529 | Liters |
| Gallons | 3.785412 | Liters |
| | | |
| Weight (Mass): | | |
| Ounces (avdp) | 28.34952 | Grams |
| Pounds | 453.59243 | Grams |
| Tons, short | 907.1847 | Kilograms |
| | | |
| Force (Weight at Sea Level, Latitude 45°): | | |
| Pounds | 4.44822 | Newtons |
| Tons, short | 8.89644 | Kilonewtons |
| | | |
| Pressure (force): | | |
| Pounds (force) per square inch | 6894.757 | Pascals |
| Pounds (force) per square foot | 47.88026 | Pascals |
| Pounds (force) per square inch | 6.894757 | Megapascals |
| Pounds (force) per square foot | 0.04788026 | Megapascals |
| | | |
| Velocity: | | |
| Feet per second | 1.09728 | Kilometers/hr |
| Miles per hour | 1.609347 | Kilometers/hr |
| | | |
| Temperature: | | |
| Degrees, Fahrenheit (Less 32°) | 0.5556 | Degrees, Celsius |

| Multiply | Metric to English<br>By | To Obtain |
|---|---|---|
| Length: | | |
| Millimeters | 0.039370 | Inches |
| Meters | 3.280840 | Feet |
| Meters | 1.093613 | Yards |

**TABLE 8.16** (*Continued*)

| Multiply | Metric to English<br>By | To Obtain |
|---|---|---|
| Length: | | |
| Kilometers | 0.62137 | Miles, statute |
| Kilometers | 0.5399568 | Miles, nautical |
| | | |
| Area: | | |
| Square millimeters | 0.00155 | Square inches |
| Square meters | 10.76391 | Square feet |
| Square meters | 1.195990 | Square yards |
| Square kilometers | 0.386101 | Square miles |
| Square kilometers | 247.1045 | Acres |
| | | |
| Volume: | | |
| Cubic millimeters | 0.0000610237 | Cubic inches |
| Cubic meters | 35.31467 | Cubic feet |
| Cubic meters | 1.307951 | Cubic yards |
| | | |
| Capacity: | | |
| Liters | 33.81406 | Ounces |
| Liters | 1.056688 | Quarts |
| Liters | 0.2641720 | Gallons |
| | | |
| Weight (Mass): | | |
| Grams | 0.035274 | Ounces (avdp) |
| Grams | 0.0022046 | Pounds |
| Kilograms | 0.0011023 | Tons, short |
| | | |
| Force (Weight at Sea Level, Latitude 45°): | | |
| Newtons | 0.2248090 | Pounds |
| Kilonewtons | 0.11240 | Tons, short |
| | | |
| Pressure (force): | | |
| Pascals | 0.00014504 | Pounds (force) per square inch |
| Pascals | 0.0208854 | Pounds (force) per square foot |
| Megapascals | 0.14504 | Pounds (force) per square inch |
| Megapascals | 20.89 | Pounds (force) per square foot |
| | | |
| Velocity: | | |
| Kilometers/hr | 0.9113444 | Feet per second |
| Kilometers/hr | 0.621370 | Miles per hour |
| | | |
| Temperature: | | |
| Degrees, Celsius | (C × 1.8) + 32 | Degrees, Fahrenheit |

## 8.9 SELECTED AITC STANDARDS

On the following pages are reprints of selected AITC standards pertinent to this timber construction manual. The editions of the standards current at the time of this edition of the manual are included; however, the standards are frequently updated. Please check to determine the latest edition of the standards by writing or calling:

> American Institute of Timber Construction
> 7012 S. Revere Parkway, Suite 140
> Englewood, Colorado 80112
> Phone: (303) 792-9559

Additional AITC standards, not reprinted in this manual, are also available from the same address. They include:

> AITC 117—Manufacturing, Standard Specifications for
> Structural Glued Laminated Timber of Softwood Species
> AITC 114, Structural Glued Laminated Timber for Electric
> Utility Framing and Crossarms
> AITC 119, Standard Specifications for Hardwood Glued
> Laminated Timber

AMERICAN INSTITUTE
7012 South Revere Parkway
Suite 140

TIMBER CONSTRUCTION
Englewood, Colorado 80112
Telephone 303-792-9559

# AITC 104-84
# TYPICAL CONSTRUCTION DETAILS

*Adopted as Recommendations July 18, 1984*
*Copyright 1984 by American Institute of Timber Construction*
*Second Printing—June 1990*

## CONTENTS

## 1. INTRODUCTION

**1.1**  These typical construction details are intended as guides for architects and engineers. They have been developed and used by the engineered timber construction industry and, being based on judgment and experience, will help to assure a high quality of construction.

**1.2  Warning.**  Because the details are to be used only as guides, dimensions have not been included and the drawings should not be scaled. Quantities and sizes of bolts, connectors and other fastening hardware are illustrative only. The actual quantities and sizes required will depend on the loads to be carried and the member sizes. End and edge distances, as well as spacing between fasteners, should be in accordance with the *National Design Specification for Wood Construction* by the National Forest Products Association. Sufficient clearance must be provided between sides of steel connection hardware and wood members to permit installation. This clearance should not exceed the member width plus $\frac{1}{4}$ in.

**1.3  Designing for Strength.**  Connection details must effectively transfer loads, utilize durable materials and be as free from maintenance as possible. The strength of wood is different in the parallel and perpendicular to grain directions. Wood also has much less strength in tension perpendicular to grain than in compression perpendicular to grain. These facts influence design details.

Vertical loads should be transferred so as to take advantage of the high compression perpendicular to grain strength of wood. For example, a beam should bear on the top of a column or wall or be seated in a shoe or hanger. Such a detail is preferred to the support of a beam by bolts at its end, particularly where there are large numbers of bolts.

Beams should be anchored at the ends in order to carry induced horizontal and vertical loads. Vertical loads may be either gravity loads or net uplift loads. The connections typically shown in this standard are primarily for vertical gravity loads. Provisions should be made to resist uplift or lateral loads as required. The bolts or fasteners at the beam ends must be located near the bottom bearing of the beam to minimize the effect of shrinkage of the wood between the bottom of the beam and the fasteners.

In many cases, individual details do not include all structural elements such as lateral bracing ties to connect all the components of the building together.

Loads suspended from glued laminated timber beams or girders should preferably be suspended from the top of the member or at least above the neutral axis.

**1.4  Consideration of Shrinkage and Swelling.**  In addition to designing connections to transfer loads, effort should be made to avoid splitting the member due to expansion and contraction of the wood. Consideration must be given to wood swelling and shrinking due to moisture content changes in service similar to the consideration given to details in metal construction which must accommodate the expanding and contracting metal due to changes in temperature.

Because wood swells and shrinks (primarily in the perpendicular to grain direction) due to moisture content changes, connections should not restrain this movement. Figure 1.1 illustrates typical shrinkage in a sawn member when drying from green to 8% moisture content and a glued laminated timber drying from 12% to 8% moisture content. Even in covered structures, large laminated timbers may shrink after installation due to moisture loss from low relative humidity conditions. Long rows of bolts perpendicular to grain fastened to a single cover plate should be avoided. Although relatively dry at time of manufacture, glued laminated timber can still shrink to reach equilibrium moisture content in service.

When possible, designers should avoid joint details that could loosen in service due to wood shrinking or that could cause problems when wood expands due to increased moisture content. Machine bolts should be used rather than lag bolts whenever possible. When lag bolts are used, correct lead hole sizes are important. Connections should be detailed to avoid loading lag bolts in withdrawal whenever possible.

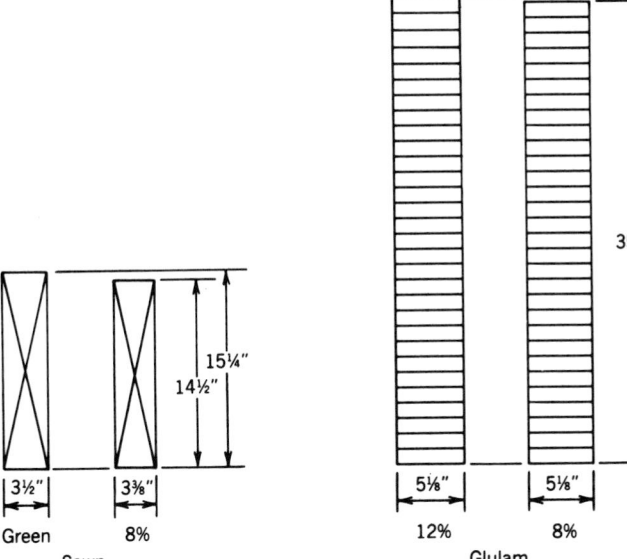

**FIGURE 1.1**  Shrinkage due to moisture loss.

**1.5  Designing to Avoid Tension Perpendicular to Grain Stresses.**  Whenever possible, joints should be designed to avoid causing tension perpendicular to grain stresses in wood members. Examples of connections that induce tension perpendicular to grain stresses are simple beams which have been notched at the ends on the tension side (see Detail A5).

Avoid long lines of fasteners spaced close together along the grain, particularly if the bolts are in tightly drilled holes. These types of connections may induce tension perpendicular to grain stresses due to prying actions from secondary moments.

**1.6  Consideration of Decay.**  When proper construction details are used and other good design and construction practices are followed, wood is a permanent construction material. Moisture barriers, flashings and other protective features should be used to avoid moisture or free water being trapped (see Section 11, "Details to Protect Against Decay"). Preservative treatments are recommended when wood is fully exposed to the weather without roof cover. Provide adequate site drainage, protection during construction and protect metals from corrosion by use of corrosive-resistant metals or resistant coatings or platings. Do not embed wood columns or arch bases below finished concrete floor levels.

**1.7**  Other considerations are also necessary in the design of structures employing engineered timber construction. These are outlined in *Spec-Data Sheet for Structural Glued Laminated Timber*, and other AITC standards and recommendations.

**1.8 End Rotation of Beams.** Consideration should be given to end rotation of beams resulting from vertical load deflection. Location of fasteners that tend to create end fixity should be avoided. Splitting at fasteners can result unless such a connection is designed to develop a fixed end moment sufficient to resist end rotation due to deflection.

## 2. BEAM TO MASONRY ANCHORAGES

Figure 2.1 illustrates a common type of beam seat to resist both uplift and horizontal forces as well as vertical loads. In the case of uplift forces, the notched beam effect must be checked. The seat may be anchored in the concrete or masonry with one or more anchors. The beam may be fastened to the tabs on the seat with one or more bolts or, where forces are greater, with bolts and shear plates.

An important point illustrated in Figure 2.1, as well as other figures in this section, is that the wood member is separated from the masonry or concrete at the bearing surface by the bearing plate and along the vertical surfaces by a minimum of a $\frac{1}{2}$ in. clearance.

For beams sloped where a $\frac{1}{8}$ in. or greater gap might occur, the beam bearing should be detailed to obtain full contact between bearing surfaces (see Figures 2.6 and 2.7). Seat cuts at the top end of the slope should be checked for notched beam effect.

Figures 2.2 through 2.5 illustrate various beam-to-masonry anchorages.

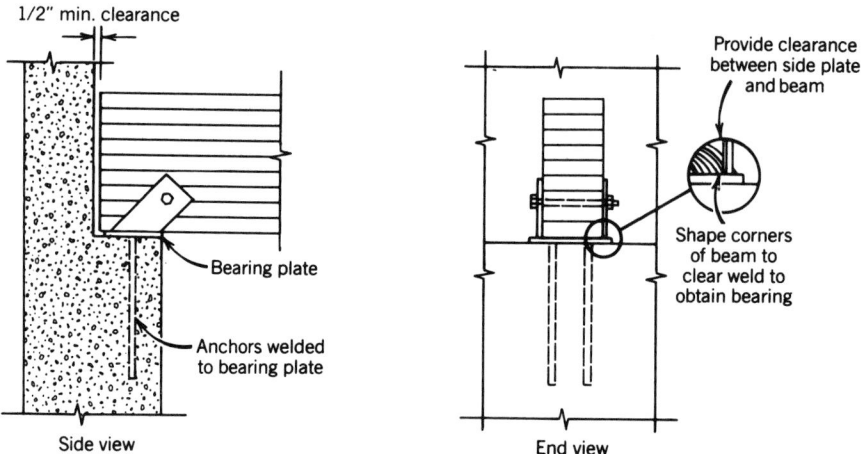

**FIGURE 2.1**

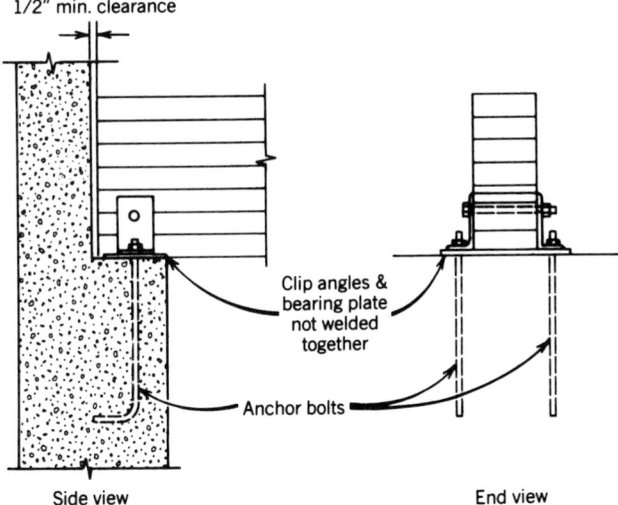

**FIGURE 2.2** Shows steel angles with a separate bearing plate with only the anchor bolts being cast in place.

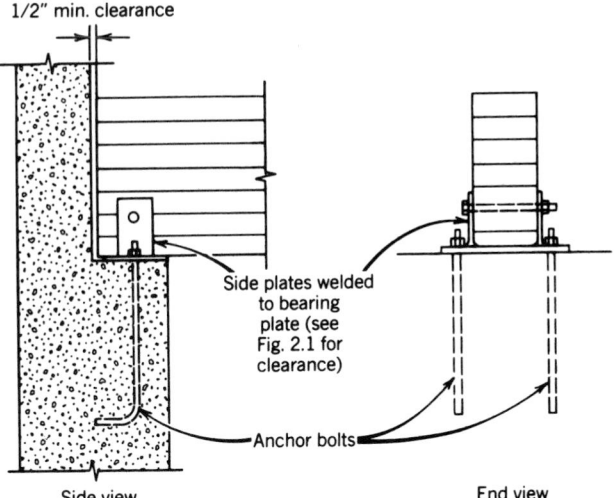

**FIGURE 2.3** Similar to the detail in Figure 2.1 except the side plates are vertical which provides less end distance for the bolts and may lessen their resistance to horizontal forces.

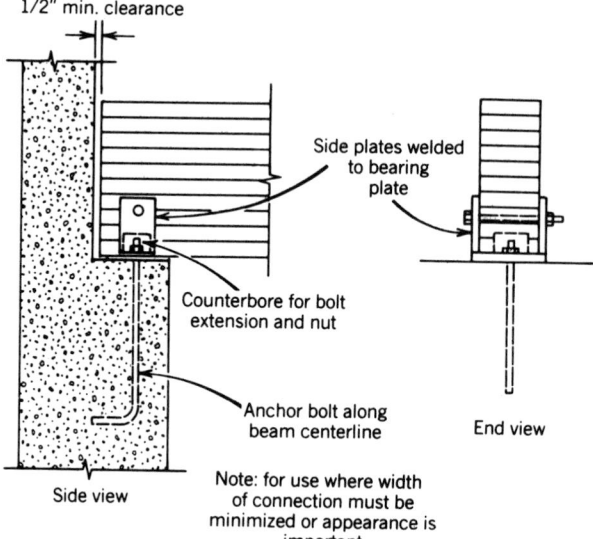

**FIGURE 2.4**    This detail may be used when the pilaster is not wide enough for outside anchor bolts. The anchor(s) may be welded to the underside of the bearing plate or the bolts and nuts may be located in holes counterbored into the bottom of the beam.

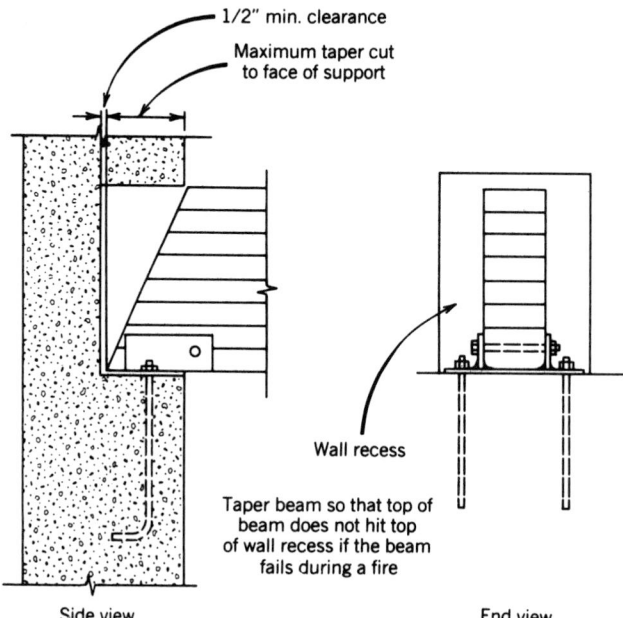

**FIGURE 2.5**    This detail illustrates a typical taper end cut sometimes referred to as a fire cut.

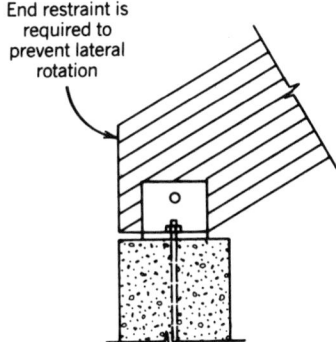

**FIGURE 2.6** Sloped Beam—Lower End. The taper cut beam should be in bearing contact with the bearing plate. See Detail A6.

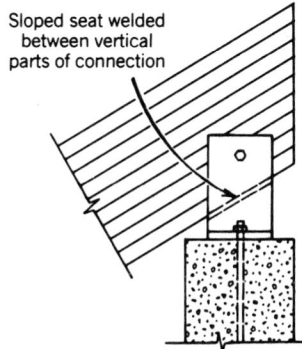

**FIGURE 2.7** Sloped Beam—Upper End. The support at the top end of a sloped member should be designed with a sloping seat rather than a notched end. The bolt must be designed to resist the parallel-to-grain component of the vertical beam reaction. See Detail A7.

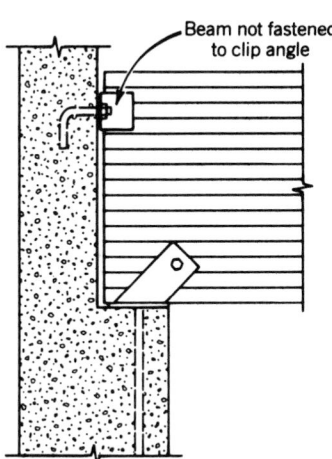

**FIGURE 2.8** Lateral support of the ends of beams can be provided with clip angles anchored to the wall but without a connection to the beam. This will not restrain vertical movement due to shrinkage or horizontal movement due to end rotation. See Detail A8.

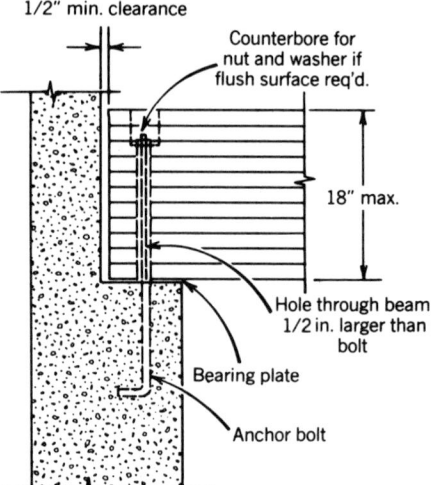

**FIGURE 2.9**   Simple Beam Anchorage. Resists uplift and small horizontal forces. Bearing plate or moisture barrier is recommended. Provide $\frac{1}{2}$ in. minimum clearance from all wall contact surfaces, ends, sides and tops (if masonry exists above beam end).

**FIGURE 2.10**   Curved or Pitched Beam Anchorage—Typical Slip Joint. Slotted or oversize holes at one or both ends of beam permit horizontal movement under lateral deflection or deformation. Length of slotted or oversized hole is based on calculated maximum horizontal deflection. Position bolt in slot to allow for anticipated movement. Bolt should be hand tightened only to permit movement. See Detail A6.

In certain seismic zones where beam provides lateral support for wall, this joint detail is not recommended.

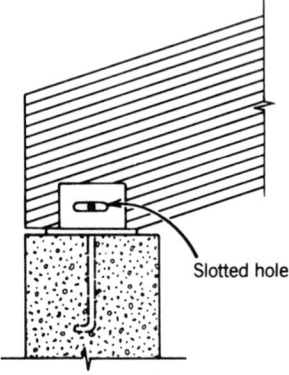

## 3. CANTILEVER BEAM CONNECTIONS

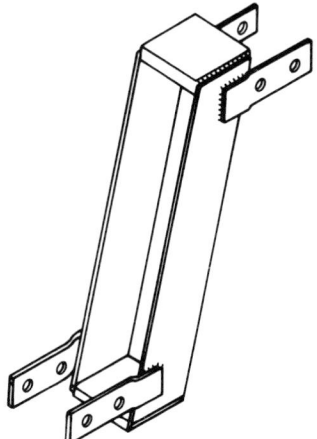

**FIGURE 3.1** Cantilever Hinge Connection. See Detail A9. This is a common type cantilever beam connector. The details that follow are examples using this connector.

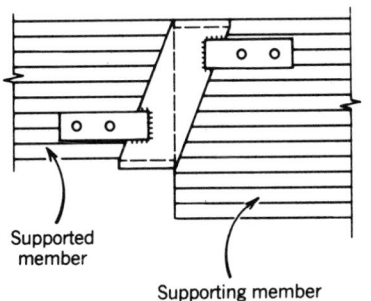

The vertical reaction of the supported member is carried by the side plates and transferred in bearing perpendicular to grain to the supporting member. The rotation due to the eccentric loading is resisted by the bolts through the tabs at the top and bottom. The connector may be installed with the top (and bottom) bearing plates dapped into the members to obtain a flush surface or may be installed without daps. Notching on the tension side should be minimized.

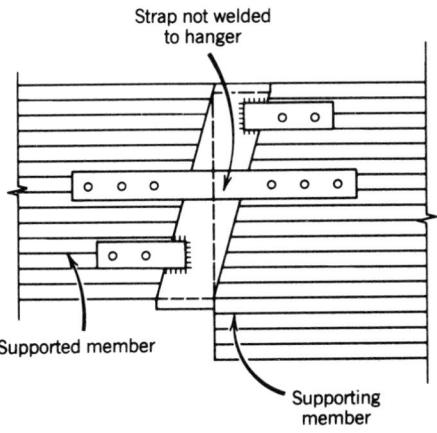

Where horizontal forces must be resisted by a hinge connection, loose tension ties may be installed on both sides of the beam. The tie shown is not fastened to the cantilever hanger.

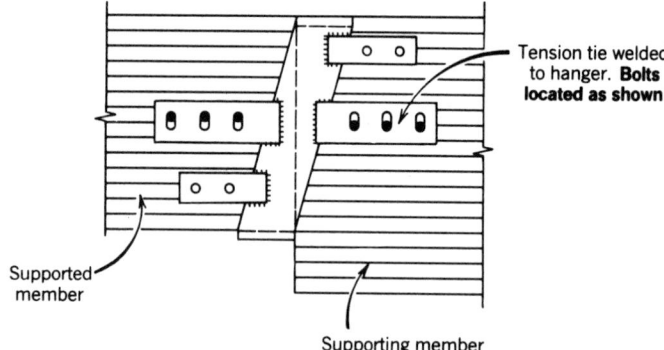

If tension ties are fastened to the cantilever hanger, vertically slotted holes are required in the tie and careful location of the bolts in the end of the slot farthest from the bearing seat is required to prevent splitting due to shrinkage and seating deformations. Bolts in slotted holes should be hand tightened only to permit movement.

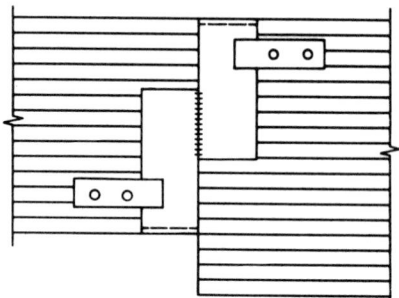

**FIGURE 3.2**   Bent Plate Type Cantilever Hinge Connection. This hinge connection is similar to the one shown in Figure 3.1 but is usually used with smaller members.

## 4.   BEAM AND PURLIN HANGERS FOR ROOF SYSTEMS

In Figure 4.1 and similar details, locate fasteners as close as practical to the bearing surface to minimize splitting due to shrinkage (see Section 1.4). For floor systems, additional restrictions may be necessary to minimize the effects of differential shrinkage of connected members. See Detail A1.

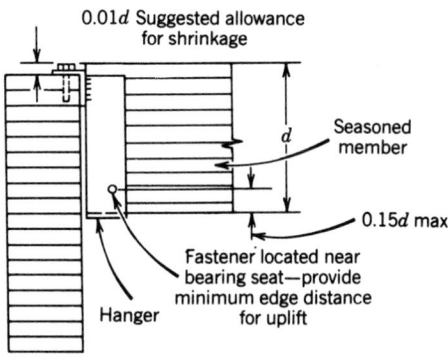

**FIGURE 4.1**   Seasoned Members. When supported members are of seasoned material, the top of the supported member may be set approximately flush with the top of the supporting member.

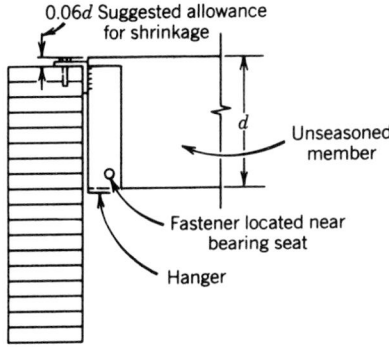

**FIGURE 4.2**  Unseasoned Members. When supported members are of unseasoned material, the hangers should be so dimensioned that the top edge of the supported member is raised above the top of the supporting member or the top of the hanger strap to allow for shrinkage as the members season in place. For supported members with moisture content at or above fiber saturation point when installed, the distance raised should be about 6% of the member's depth above its bearing point.

NOTE: For main members loaded on one side as in Figures 4.1, 4.2 and 4.3, provide a tie between the beam and purlin to restrain potential rotation of the beam due to eccentricity of the hanger load.

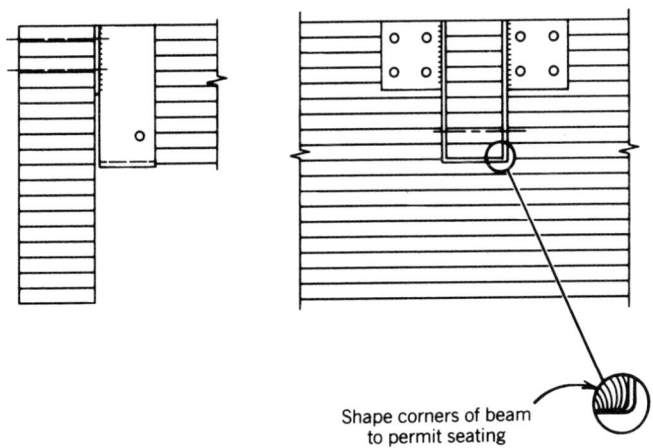

**FIGURE 4.3**  Welded Face Hanger. See Detail A2.

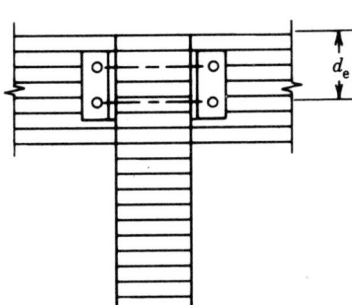

**FIGURE 4.4**  A clip angle connection without a bearing seat may be used for small beams and light loads. The connection should be designed for gravity loads as a notched beam using $d_c$ as shown in the notched beam formula. The distance between bolts should be checked for possible effects of shrinkage. A bearing connection as shown in Figure 4.1 is preferred to the support of the beam by bolts. See Detail A3.

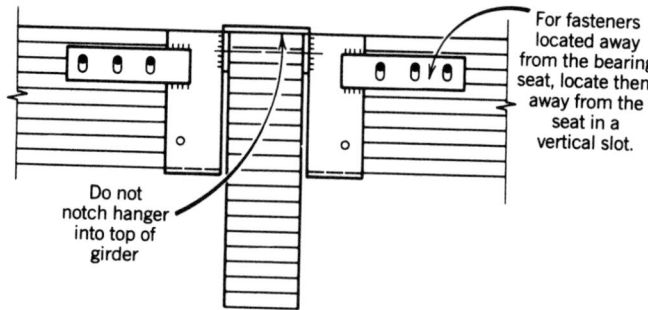

**FIGURE 4.5**   Welded and Bent Strap Hanger. A separate tension tie may be used across the top in lieu of the tabs to resist lateral forces.

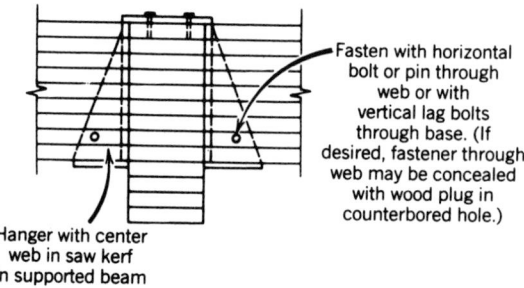

**FIGURE 4.6**   Partially Concealed Type. For moderate loads. Base may be let in flush with bottoms of purlins. See Detail A4.

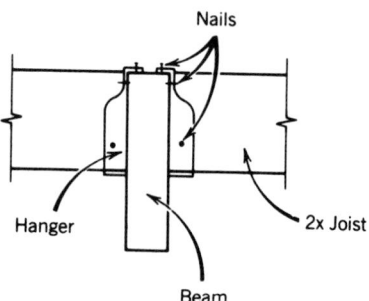

**FIGURE 4.7**   Stamped Joist Hanger. For light loads. Stamped from light-gage metal.

# 5. BEAM TO COLUMN CONNECTIONS

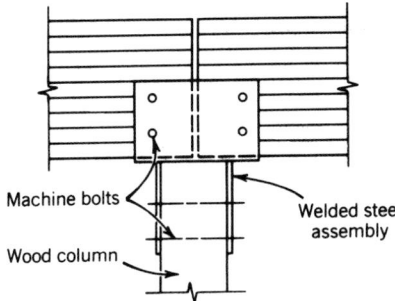

**FIGURE 5.1** Beams to Wood Column—U-Plate. Welded steel assembly passes under abutting wood beams and is welded to steel side plate bolted to wood column.

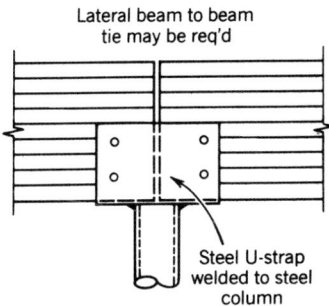

**FIGURE 5.2** Beams to Steel Column. Similar to Figure 5.1.

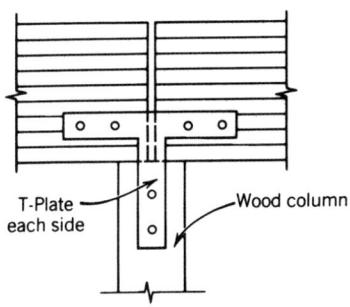

**FIGURE 5.3** Beams to Wood Column—T-Plates. Steel T-plate is bolted to abutting wood beams and to wood column. Loose bearing plate may be used where column cross-sectional area is insufficient to provide bearing for beams in compression perpendicular to grain.

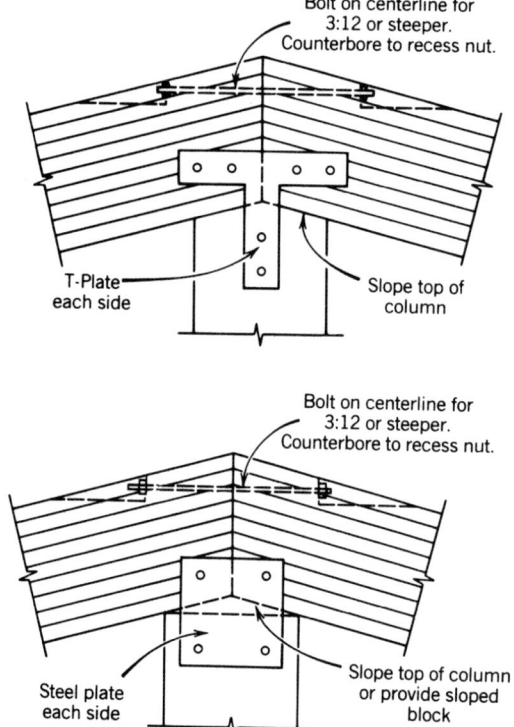

**FIGURE 5.4**   Shed Roof Type End Detail. For beam slopes of 1 : 12 maximum, beams may be notched at bottom to rest on column. Beams may be pitched away from both sides, or from only one side of column. See also Beam-to-Masonry Anchorages, Section 2.

**FIGURE 5.5**   Beam to Wood Column. Connection provides for uplift. Metal bearing plate may be used where column cross-sectional area is insufficient to provide bearing for beam in compression perpendicular to grain.

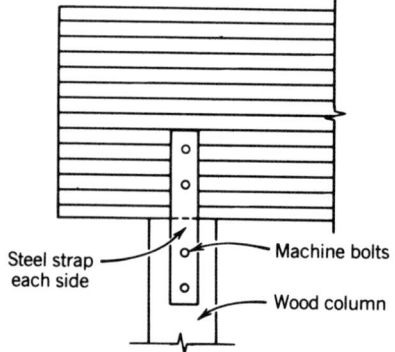

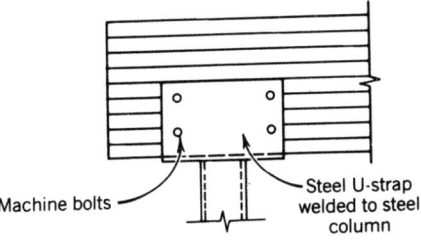

Machine bolts — Steel U-strap welded to steel column

**FIGURE 5.6** Beam to Steel Column. Steel U-strap passes under timber beam and is welded to top of steel column.

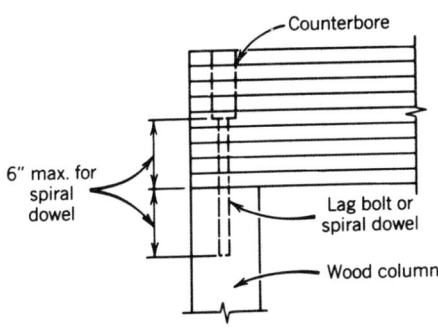

Counterbore

6" max. for spiral dowel

Lag bolt or spiral dowel

Wood column

**FIGURE 5.7** Concealed Type—Beam to Wood Column.

## 6. COLUMN ANCHORAGES

Column bearing elevation should be raised above finished floors which may be subjected to high moisture conditions. In locations where column base anchorages are subject to damage by moving vehicles, protection of the columns from such damage should be considered. Columns exposed to the weather should be treated in accordance with *Standard for Preservative Treatment of Structural Glued Laminated Timber*, AITC 109.

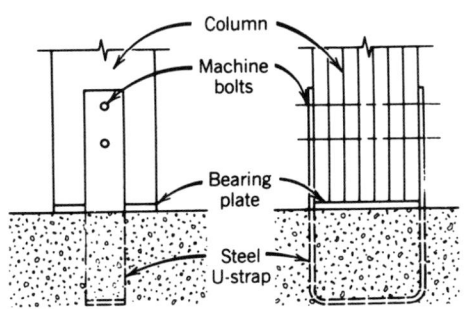

Column

Machine bolts

Bearing plate

Steel U-strap

**FIGURE 6.1** U-Strap Anchorage. Resists both horizontal forces and uplift. Bearing plate or moisture barrier is required. May be used with shear plates.

Do not place column below finished concrete floor level.

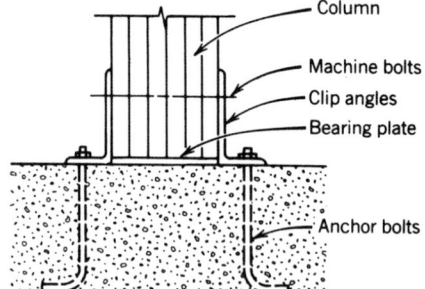

**FIGURE 6.2**    Clip Angle Anchorage to Concrete Base. Resists both horizontal forces and uplift. Bearing plate or moisture barrier is required.

Do not place column below finished concrete floor level.

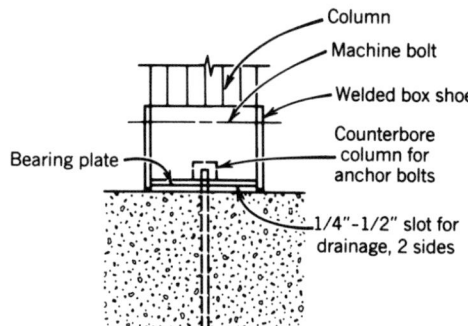

**FIGURE 6.3**    Box Shoe. For use when bottom of box shoe is flush with top of concrete floor.

Do not place column below finished concrete floor level.

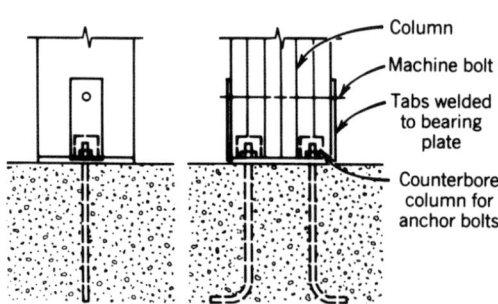

**FIGURE 6.4**    Semi-Concealed Column Anchorage. For use where concrete support area is limited in size. Resists both horizontal forces and uplift.

Do not place column below finished concrete floor level.

## 7.  ARCH ANCHORAGES

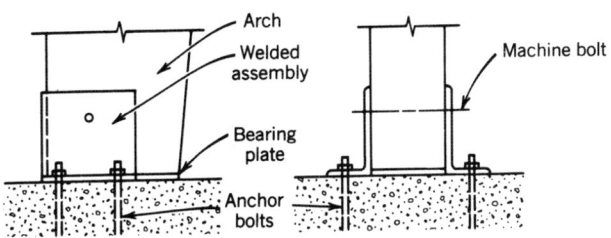

**FIGURE 7.1**    Arch Shoe With Exposed Anchor Bolts.

Do not place arch below finished concrete floor level.

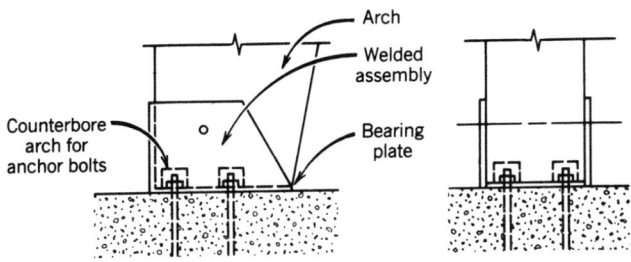

**FIGURE 7.2**    Arch Shoe With Concealed Anchor Bolts. Counterbores are provided in arch base for anchor bolt projections.

Do not place arch below finished concrete floor level.

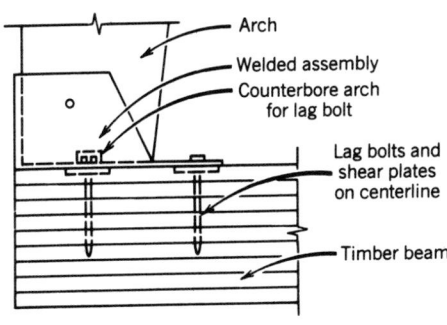

**FIGURE 7.3**    Arch Anchorage to Timber Beam. Vertical load is taken directly by bearing into timber beam. Vertical uplift and thrust are taken by the lag bolts and shear plates into the beam tie.

Do not place arch below finished concrete floor level.

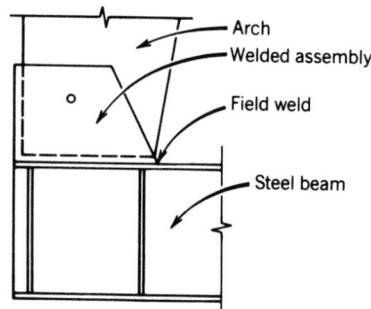

**FIGURE 7.4**   Arch Anchorage to Steel Girder.

Do not place arch below finished concrete floor level.

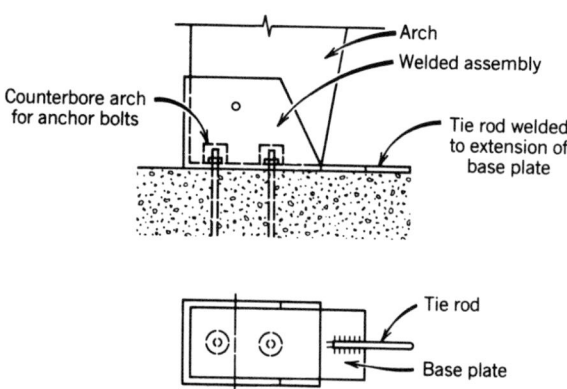

**FIGURE 7.5**   Tie Rod to Arch Shoe. Thrust due to vertical load is taken directly by the tie rod welded to the arch shoe. This detail is intended for use with a raised joist floor where the tie rod can be concealed.

Do not place arch below finished concrete floor level.

**FIGURE 7.6**   Tie Rod Arch. Thrust due to vertical load is taken directly by the tie rod. For use where raised joist floor will conceal the tie rod.

Do not place arch below finished concrete floor level.

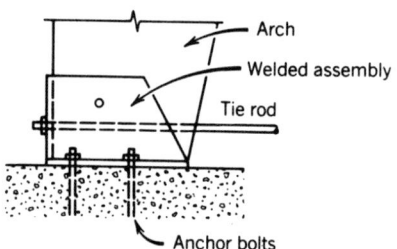

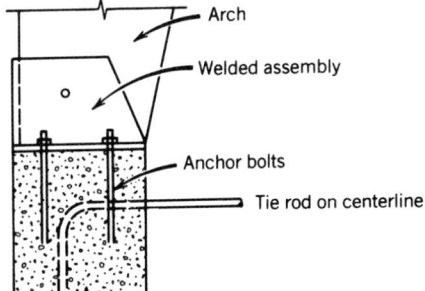

**FIGURE 7.7** Tie Rod in Concrete. Thrust is taken by anchor bolts in shear into the concrete foundation and tie rod.

Do not place arch below finished concrete floor level.

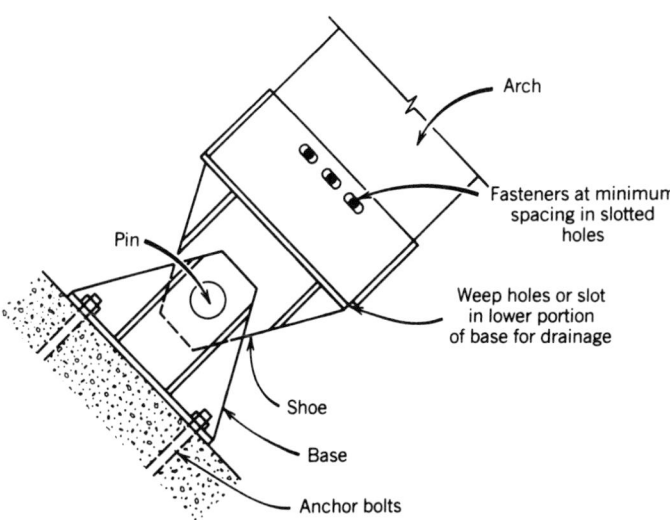

**FIGURE 7.8** True Hinge Anchorage for Arches. Recommended for arches where true hinge action is desired (see Figure 11.6 for protection considerations). Do not place arch below finished concrete floor level.

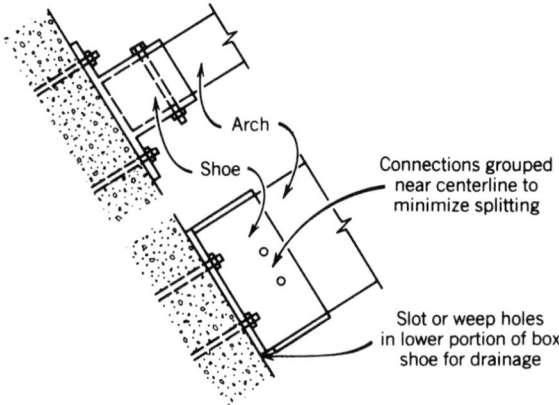

**FIGURE 7.9**    Arch Anchorage Where True Hinge Is Not Required. Recommended for arches where a true hinge is not required. Base shoe is anchored directly to buttress (see Figure 11.6 for protection considerations). Do not embed arch in concrete floor.

## 8.  ARCH CONNECTIONS

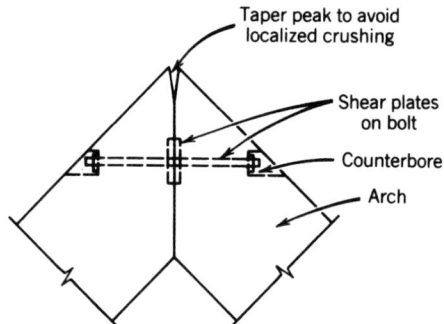

**FIGURE 8.1**    Arch Peak. For arches with slopes of 3 : 12 and greater. This connection will transfer both vertical forces (shear) and horizontal forces (tension and compression).

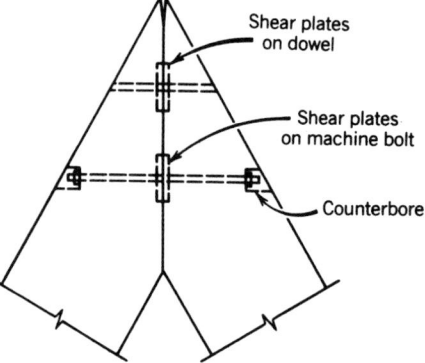

**FIGURE 8.2**    Arch Peak. When the vertical shear is too great for one pair of shear plates, or when deep sections would require extra shear plates for alignment, additional pairs of shear plates centered on dowels or machine bolts may be used.

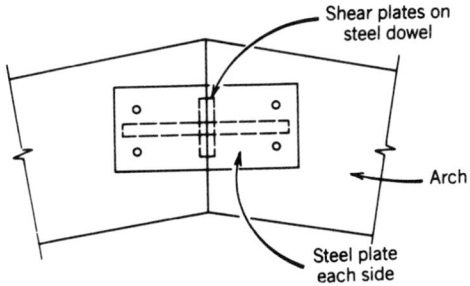

**FIGURE 8.3** Low Pitched Arches. For arches with slopes that would require excessively long through bolts; shear plates back-to-back centered on a dowel are used in conjunction with a tie plate and through bolts.

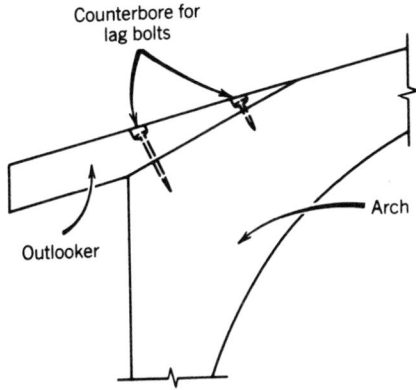

**FIGURE 8.4** Outlooker Connection to Haunched Arch. Lag bolts used in this connection should be long enough so that the withdrawal resistance of the threads is in the main section of the arch. Connection must be designed to resist any cantilever action of the outlooker. Lag bolts may be counterbored when decking is applied.

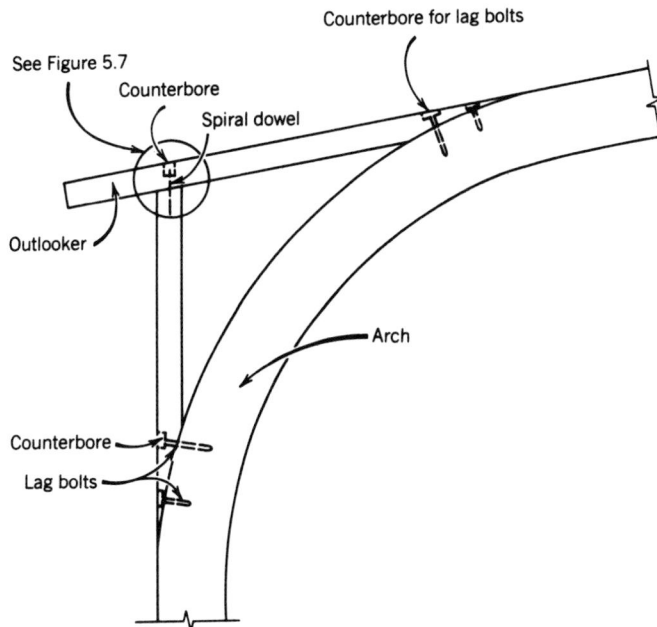

**FIGURE 8.5** Outlooker Connection to Open Haunched Arch.

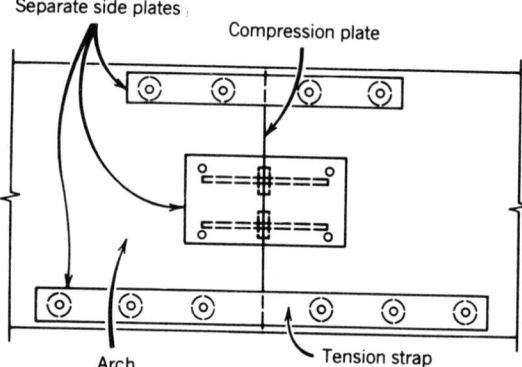

**FIGURE 8.6**    Arch Moment Splice. Drawing shows a typical moment splice. Compression stress is taken in bearing on the wood through a steel compression plate. Tension is taken across the splice by means of steel straps and shear plates. Side plates and straps are used to hold sides and tops of members in position. Shear is taken by shear plates in end grain.

## 9.   TRUSS CONNECTIONS

When unseasoned lumber is used in truss construction, periodic inspection is recommended along with retightening of hardware and connections as necessary.

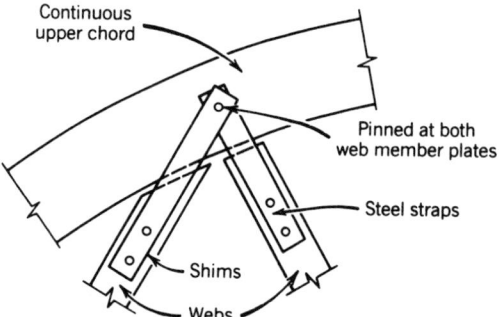

**FIGURE 9.1**    Monochord-Steel Straps. For trusses with continuous upper chord. Provide clearance between web ends and chord. Provide shims at web to prevent bending of plates.

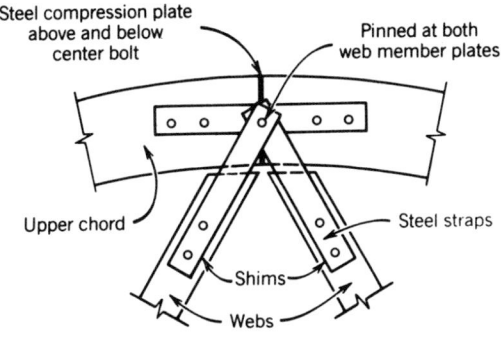

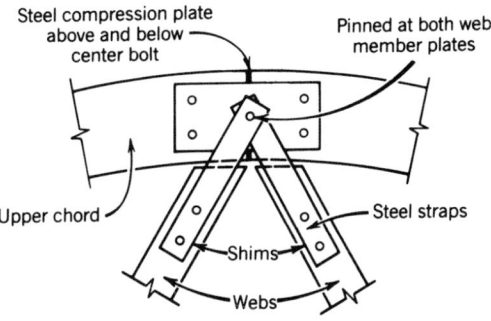

**FIGURE 9.2** Monochord-Steel Strap Assembly. Similar to Figure 9.1. For use at ridge for upper chord splice. Provide shims at webs to prevent bending of plates.

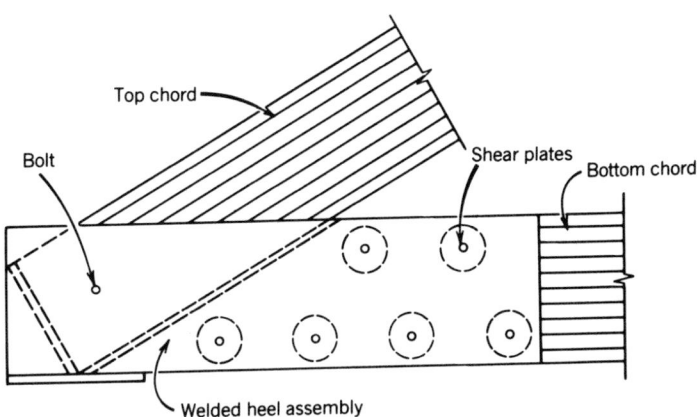

**FIGURE 9.3** Truss Heel Connection. If substantial cross grain shrinkage is anticipated, double steel straps may be used in place of single plate along bottom chord.

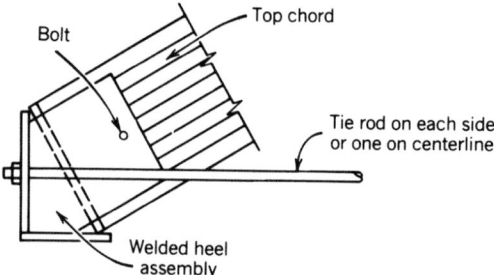

**FIGURE 9.4**    Rod-Tied Arch Heel Connection.

## 10.  SUSPENDED LOADS

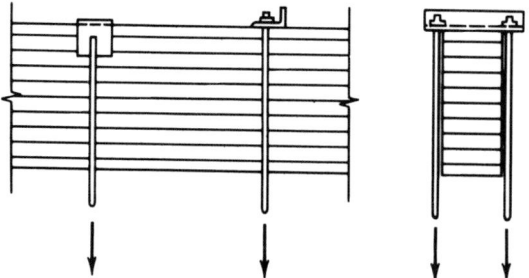

**FIGURE 10.1**    Loads suspended from glued laminated timber beams should be resisted from the top of the member or at least above the neutral axis.

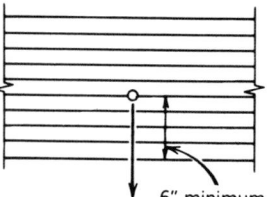

**FIGURE 10.2**    Light loads such as small conduit may be suspended with small fasteners near the bottom of glued laminated timber beams as shown.

# 11.  DETAILS TO PROTECT AGAINST DECAY

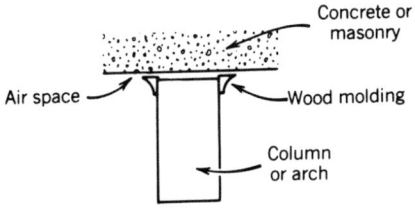

**FIGURE 11.1**  Wood Member Against Continuous Masonry Wall. Minimum of $\frac{1}{2}$ in. air space between member and wall or adequate moisture barrier must be provided. For arches, additional space may be required to permit outward deflection of the arch leg.

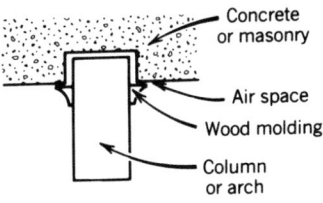

**FIGURE 11.2**  Wood Member Set in Masonry Wall Pocket. Minimum of $\frac{1}{2}$ in. air space between member and wall pocket or adequate moisture barrier must be provided. For arches, additional space may be required to permit outward deflection of the arch leg.

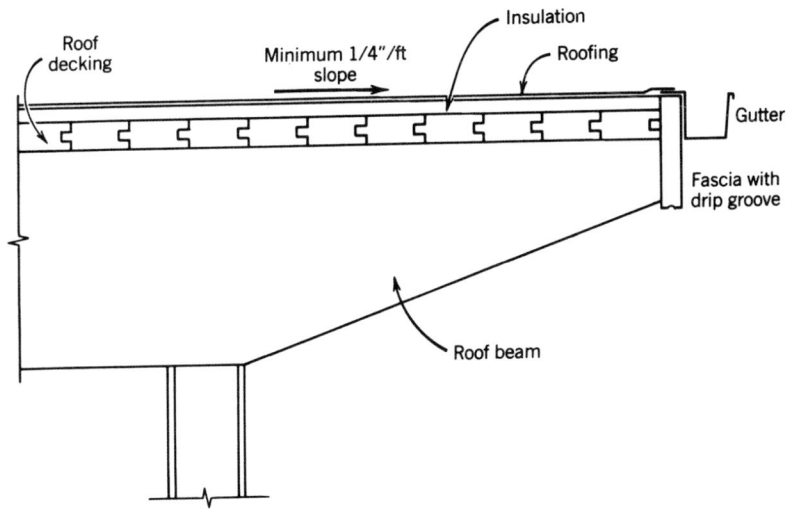

**FIGURE 11.3**  Protection Considerations for Building with Covered Overhang. Beam is protected from direct exposure to weather by fascia. Roof should be sloped for drainage or designed to prevent ponding of water. Fascia should be preservatively treated or made from decay-resistant species. Taper cut should be sealed.

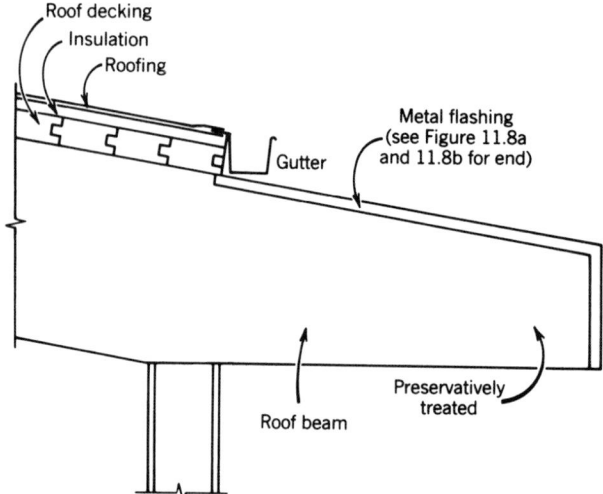

**FIGURE 11.4**    Protection Considerations for Building with Uncovered Overhang. Portion of beam extending outside of building should be protected by metal cap and preservative treatment. Periodic refinishing of the surfaces exposed to the weather may be required to maintain appearance.

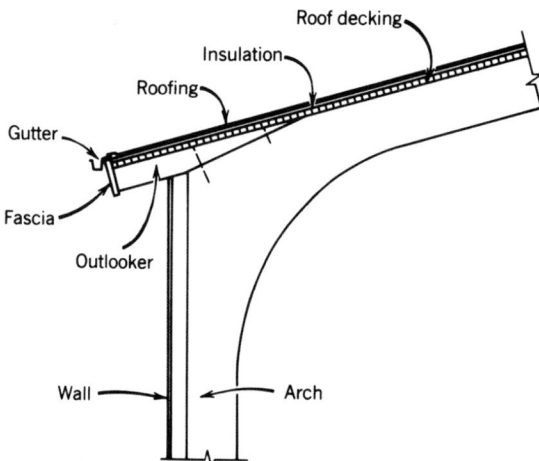

**FIGURE 11.5**    Protection Considerations for Arch Outlooker Overhang. Outlooker is protected from direct exposure to weather. Arch is protected by the wall from direct exposure to the weather.

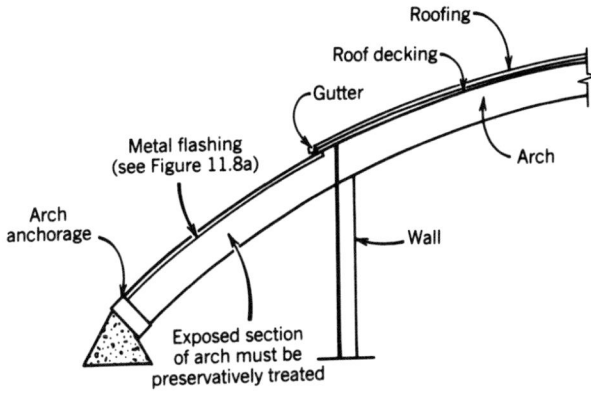

**FIGURE 11.6**  Protection Considerations for Partially Exposed Arches. Portion of arch leg extending outside of building should be protected by metal flashing and preservative treatment. At least 12 in. clearance must be provided between arch base and grade. Preservative treatment in accordance with *Standard for Preservative Treatment of Structural Glued Laminated Timber*, AITC 109, must be used for exposed portion of arch.

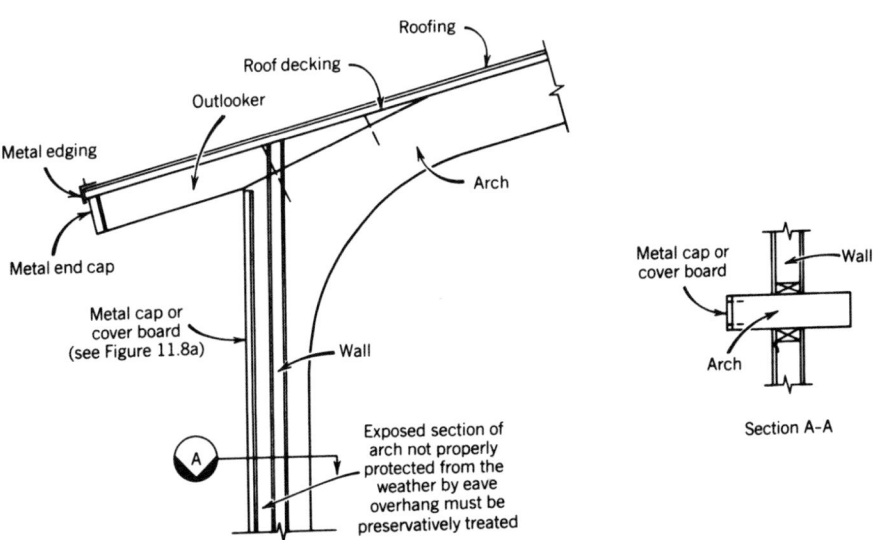

**FIGURE 11.7**  Arch Leg Protection. Metal end cap or treated cover board on edge of exterior portion of arch used in conjunction with preservative treatment of the arch leg. Metal cap is as illustrated in Figure 11.8. Cover board should be vertical grain material set in building sealant and attached with weatherproof nails or screws. All wood with exterior exposure must be adequately protected and maintained.

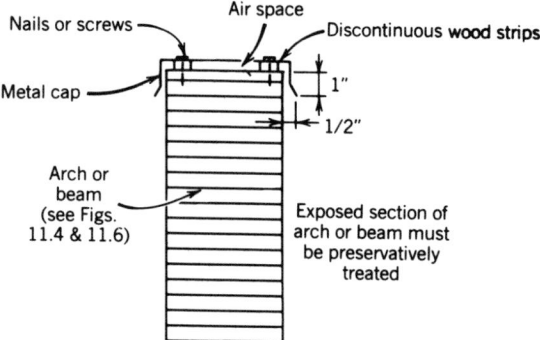

(*a*)   Top cap for horizontal or sloped members.

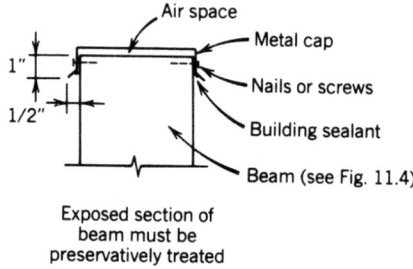

(*b*)   End cap for exposed beams or vertical members.

**FIGURE 11.8**   Protective Metal Cap or Flashing Details. Caps or flashings are made of 20-gage minimum thickness weatherproof metal. Nails or screws are weatherproofed, and heads are sealed with building sealant or neoprene washers. A minimum of $\frac{1}{2}$ in. air space must be provided between cap and the face of the wood section. For vertical use conditions, a continuous bead of building sealant is required.

# APPENDIX TO AITC 104
# CONNECTION DETAILS TO BE AVOIDED

The following are some examples of poor detailing practice and suggestions for improvement of the poor details.

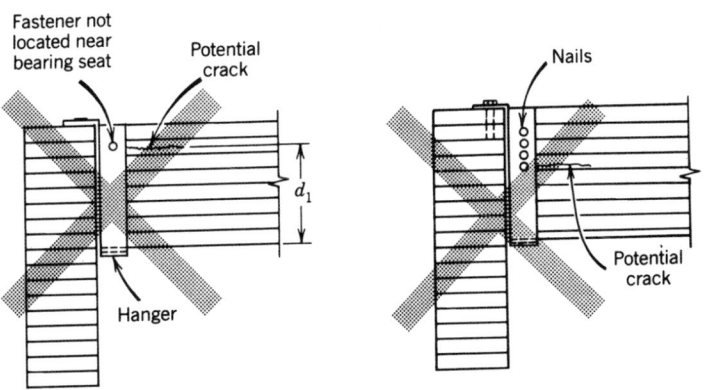

**DETAIL A1**   Glued laminated timbers, although relatively dry at the time of manufacture, may shrink as they reach equilibrium moisture content in service. When fasteners are not located near the bearing seat but in the upper portion of the beam, shrinkage in the beam over the depth, d, can cause the beam reaction to be carried by the fasteners rather than in bearing on the hanger. This induces notch shear and tension perpendicular to grain stresses that can cause splitting along the beam as shown.

**SUGGESTED REVISION**   Detail the connection as shown in Figure 4.1 with the fasteners located near the bearing seat, or slot the hole in the steel hanger and place the fastener in the top of the slot.

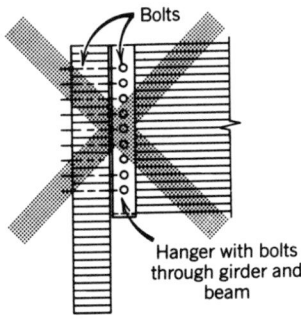

**DETAIL A2**   This detail is similar to Detail A1 in that shrinkage in the beam can result in the bearing being carried by the fasteners rather than in bearing on the hanger seat which, in turn, results in notch shear and tension perpendicular to grain stresses. Also, see Detail A3 concerning long rows of fasteners perpendicular to grain as it relates to the hanger-to-girder connection. Section 1.8 discusses the effect of end rotation on the fasteners in this type of connection.

**SUGGESTED REVISION**   Detail the connection as illustrated in Figure 4.3 with the fasteners in the beam located near the bearing seat and the fasteners in the girder grouped near the top of the girder.

**DETAIL A3** End connections which include long rows of fasteners perpendicular to grain through steel side members should be avoided. Shrinkage of the wood will be restrained by the steel and can result in notch shear and tension perpendicular to grain stresses at the end of the beam which may result in splitting of the member. See also Section 1.4.

**SUGGESTED REVISION** Change to a bearing connection as shown in Figure 4.1 with the beam being supported in bearing and the fasteners located only near the bearing seat.

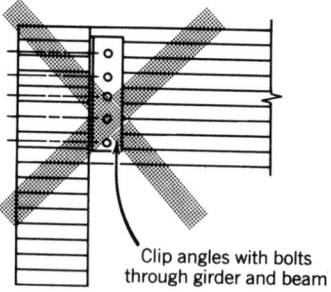

Clip angles with bolts through girder and beam

**DETAIL A4** This detail is similar to Detail A3 except that the beam is supported by bolts through a plate located in a saw kerf in the center of the beam. See also Section 1.4.

**SUGGESTED REVISION** Detail the connection as shown in Figure 4.6 where a bearing seat has been added and the bolts away from the bearing seat have been omitted.

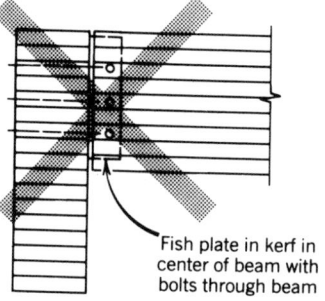

Fish plate in kerf in center of beam with bolts through beam

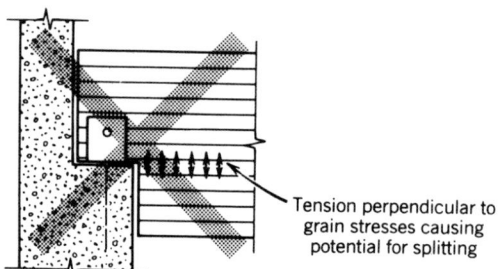

Tension perpendicular to grain stresses causing potential for splitting

**DETAIL A5** An abrupt notch in the end of a wood member creates two problems. One is that the effective shear strength of the member is reduced because of the end notch. The other is that the exposure of end grain in the notch will permit a more rapid migration of moisture in the upper portion of the member causing the indicated split.

**SUGGESTED REVISION** Detail the connection as shown in Figure 2.1 without the end notch. Notches are not recommended on the tension side of glued laminated timber, but if used, they should not exceed 10% of the depth and should be checked by the notched beam formula.

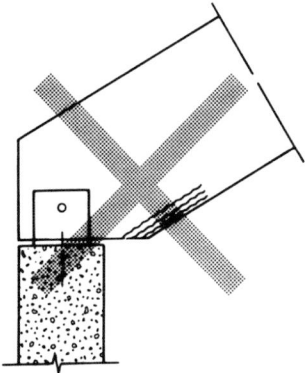

**DETAIL A6**    This condition is similar to that shown in Detail A5. The shear strength of the end of the member is reduced, tension perpendicular to grain stresses are induced and the exposed end grain may result in splitting because of rapid drying.

**SUGGESTED REVISION**    Revise the taper cut to provide bearing as shown in Figure 2.6.

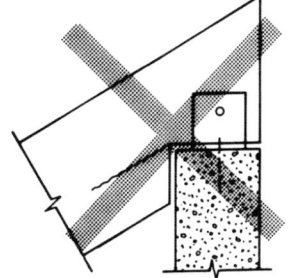

**DETAIL A7**    This detail at the upper end of a sloped beam is similar to the notched beam detail shown in Detail A5.

**SUGGESTED REVISION**    Provide a sloping seat as shown in Figure 2.7.

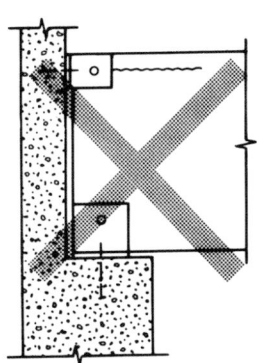

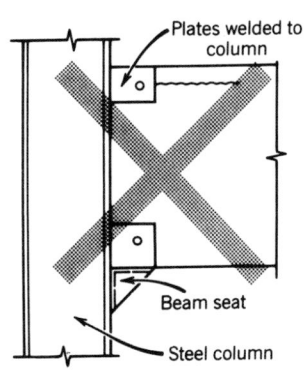

**DETAIL A8**    In this situation, the beam is bearing on the beam seat and the top is laterally supported by clip angles or similar hardware. In a deep beam, the shrinkage due to drying reduces the depth of the beam and will create a split at the upper connection.

**SUGGESTED REVISION**    Provide restraint against lateral rotation without restraining the member against shrinkage as shown in Figure 2.8.

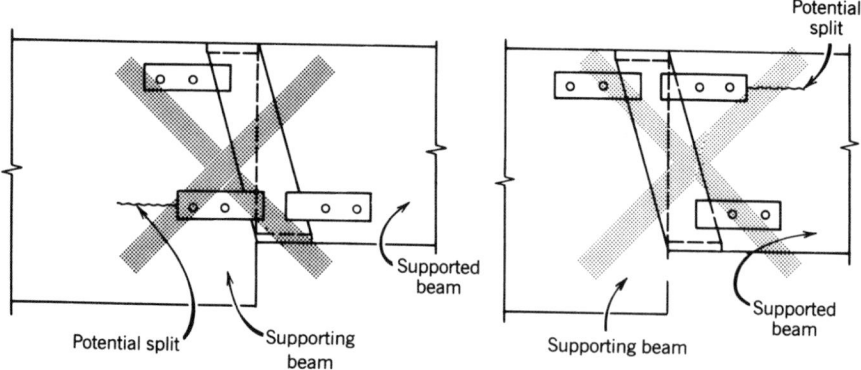

**DETAIL A9**    When a tension connection across a cantilever beam hanger is designed using integral tabs either at the top or bottom of the hanger, splitting may occur due to shrinkage between the bearing point of the hanger and the bolts as shown.

**SUGGESTED REVISION**    Detail as shown in Figure 3.1 for suggested cantilever hanger with a loose tension tie. If tabs are provided as shown, detail vertically slotted holes in the tabs and locate the bolts in the end of the slot farthest from the bearing seat. Bolts in slotted holes should be hand tightened only to permit movement.

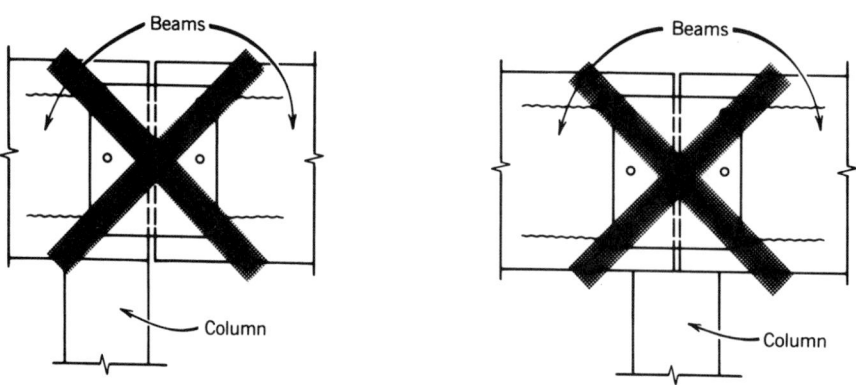

**DETAIL A10**    This situation is similar to Detail A3 where deep splice plates are applied to both faces of the beam. This may be a splice over a column or a situation where one beam is supporting the next one. As the wood shrinks, the steel side plates resist the shrinkage effect causing splits in the beams. This condition is particularly hazardous if one beam is supporting the next one as shown on the right or as a cantilever connection because the splits at the bolt holes will reduce effective strength of the beam.

**SUGGESTED REVISION**    See details in Section 5 for recommended column connections and Figure 8.6 for recommended moment splice if continuity over column is desired. Moment splices are not recommended for beams.

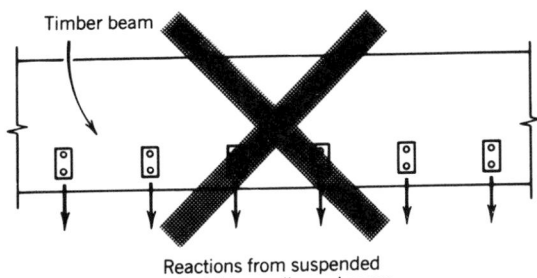

Reactions from suspended
loads, joists, purlins or beams

**DETAIL A11** Loads suspended from beams as shown induce tension perpendicular to grain stresses.
**SUGGESTED REVISION** See Section 10 for recommendations for supporting loads from beams.

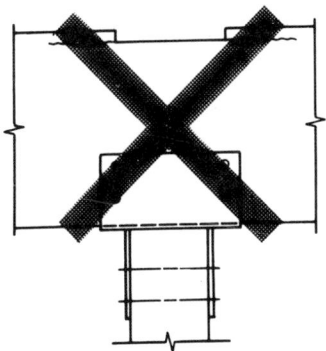

**DETAIL A12** This shows a general condition where, particularly in continuous framing, the top tension fibers have been cut to provide for a recessed hardware connection or for the passage of conduit or other elements over the top of the beam. This is particularly serious in glulam construction since the tension laminations are critical to the proper performance of the structure.
**SUGGESTED REVISION** Detail the connection as shown in Figure 4.5 without the notch in the tension side of the cantilever member. If recessed hardware is required, provide mechanically fastened blocking.

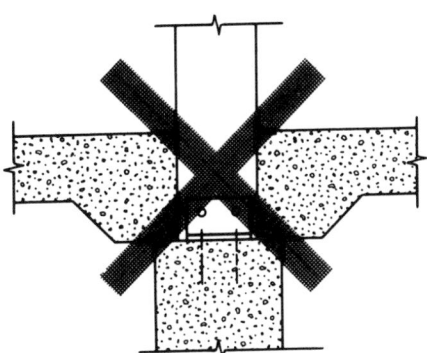

**DETAIL A13** Some designers try to conceal the base of a column or an arch by placing concrete around the connection. Moisture may migrate into the lower portion of the wood and cause decay.
**SUGGESTED REVISION** Detail column bases as shown in Section 6 or arch anchorages as shown in Section 7.

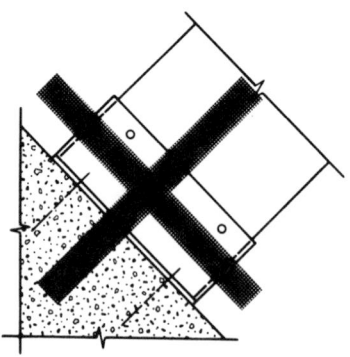

**DETAIL A14**   Similar to Detail A13 in that the base of an arch or a column is placed in a closed steel box where moisture may accumulate and cause decay.
**SUGGESTED REVISION**   Detail arch base as shown in Figure 7.9 with connections grouped near the center of the arch and drainage provided to prevent collection of water in the shoe.

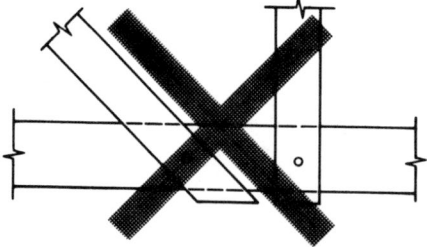

**DETAIL A15**   When the centerlines of members do not line up at a common point in a truss, considerable shear and moment stresses may result in the bottom chord. When these are combined with the presumably high tension stress in the member, failure may occur.
**SUGGESTED REVISION**   Detail web to chord connections to be concentric as shown in Figure 9.1.

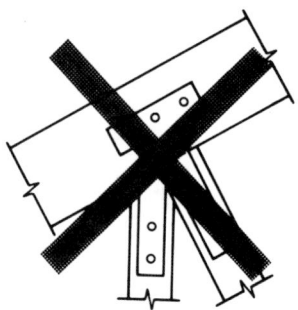

**DETAIL A16**   Truss chord to web connection made from plates welded rigidly together. Not recommended when truss deflections could produce rotation of members, which could cause splitting.
**SUGGESTED REVISION**   Detail as shown in Figure 9.1 with pinned connection at the web and chord, and clearance provided between webs and chord.

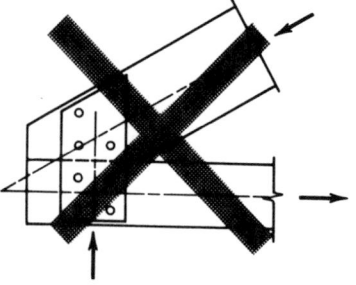

**DETAIL A17**   Truss heel connection with eccentric force lines that cause prying action which may result in splitting of the members.
**SUGGESTED REVISION**   Detail connection similar to Figure 9.3.

AMERICAN INSTITUTE
7012 South Revere Parkway
Suite 140

TIMBER CONSTRUCTION
Englewood, Colorado 80112
Telephone 303-792-9559

# AITC 108-93
# STANDARD FOR HEAVY TIMBER
# CONSTRUCTION

*Adopted as Recommendations April 21, 1993*
*Copyright 1993 by American Institute of Timber Construction*

## CONTENTS

## 1. HEAVY TIMBER CONSTRUCTION

**1.1** ''Heavy timber'' construction is that type in which fire resistance is attained by placing limitations on the minimum size, thickness, or composition of all load-carrying wood members as given in this section; by avoiding concealed spaces under floors or roofs; by using approved fastenings, construction details, and adhesive; and by providing the required degree of fire resistance in exterior and interior walls.

## 2. HEAVY TIMBER FRAMING

### 2.1 Columns

**2.1.1** Wood columns may be sawn or glued laminated and shall be not less than 8 in., nominal, in any dimension when supporting floor loads and not less than 6 in., nominal, in width and not less than 8 in., nominal, in depth when supporting roof and ceiling loads only.

**2.1.2** Columns shall be continuous or superimposed by means of reinforced concrete or metal caps with brackets, or shall be connected by properly designed steel or iron caps, with pintles and base plates, or by timber splice plates affixed to the columns by means of metal connectors housed within the contact faces, or by other approved methods.

## 2.2   Floor Framing

*2.2.1*   Beams and girders of wood may be sawn or glued laminated and shall be not less than 6 in., nominal, in width and not less than 10 in., nominal in depth.

*2.2.2*   Framed or glued laminated arches which spring from grade or the floor line and support floor loads shall not be less than 8 in., nominal, in any dimension.

*2.2.3*   Framed timber trusses supporting floor loads shall have members of not less than 8 in., nominal, in any dimension.

## 2.3   Roof Framing

*2.3.1*   Framed or glued laminated arches for roof construction which spring from grade or the floor line and do not support floor loads shall have members not less than 6 in., nominal, in width and not less than 8 in., nominal, in depth for the lower half of the height and not less than 6 in., nominal, in depth for the upper half.

*2.3.2*   Spaced members may be composed of two or more pieces not less than 3 in., nominal, in thickness when blocked solidly throughout their intervening spaces or when such spaces are tightly closed by a continuous wood cover plate of not less than 2 in., nominal, in thickness, secured to the underside of the members. Splice plates shall not be less than 3 in., nominal, in thickness. When protected by approved automatic sprinklers under the roof deck, framing members shall be not less than 3 in., nominal, in width. Structural non-sawn or glued laminated timber, framed or glued laminated arches which spring from the top of the walls or wall abutments, framed timber trusses, and other wood framing members which do not support floor loads, shall have members not less than 4 in., nominal, in width and not less than 6 in., nominal in depth.

## 3.   HEAVY TIMBER FLOORS

**3.1**   Floors shall be of sawn or glued laminated: (1) planks splined or tongue-and-groove, not less than 3 in., nominal, in thickness covered with 1 in., nominal, dimension tongue-and-groove flooring laid crosswise or diagonally to the plank or with other approved wearing surfaces, or (2) planks, not less than 4 in., nominal, in width set on edge close together and well spiked, and covered as for 3 in. thick plank. The planks shall be laid so that there is no continuous line of end joints except at points of support. Floors shall not extend closer than $\frac{1}{2}$ in. to walls to provide an expansion joint, but the joint shall be covered at top or bottom to avoid flue action.

## 4.   HEAVY TIMBER ROOF DECKS

**4.1**   Roof decks shall be of sawn or glued laminated, (1) planks, splined or tongue-and-groove, not less than 2 in., nominal, in thickness, covered with $1\frac{1}{8}$ in. thick tongue-and-groove interior plywood (exterior glue), or (2) planks, not less than 3 in., nominal, in width, set on edge close together and laid as required for floors. Other wood and/or wood-fiber based decking or other types of decking may be used if noncombustible.

## 5. WALLS

**5.1 Load Bearing Walls.** Load bearing portions of exterior and interior walls shall be of approved noncombustible material and shall have a fire resistance rating of not less than two hours except that, where a horizontal separation of 3 ft or less is provided, load bearing portions of exterior walls shall have a fire resistance rating of not less than three hours.

**5.2 Nonload Bearing Walls.** Nonload bearing portions of exterior walls shall be of approved noncombustible materials except as otherwise noted, and:

*5.2.1* Where a horizontal separation of 3 ft or less is provided, nonbearing exterior walls shall have a fire resistance rating of not less than three hours.

*5.2.2* Where a horizontal separation of more than 3 ft but less than 20 ft is provided, nonbearing exterior walls shall have a fire resistance rating of not less than two hours.

*5.2.3* Where a horizontal separation of 20 to 30 ft is provided, nonbearing exterior walls shall have a fire resistance rating of not less than one hour.

*5.2.4* Where a horizontal separation of 30 ft or more is provided, no fire resistance rating is required.

*5.2.5* Where a horizontal separation of 20 ft or more is provided, wood columns, arches, beams and roof decks conforming with heavy timber sizes may be used externally.

## 6. CONSTRUCTION DETAILS

**6.1** Wall plate boxes of self-releasing type or approved hangers shall be provided where beams and girders enter masonry. An air space of 1 in. shall be provided at the top, end, and sides of the member unless approved durable or treated wood is used.

**6.2** Girders and beams shall be closely fitted around columns, and adjoining ends shall be cross tied to each other, or intertied by caps or ties, to transfer horizontal loads across the joint. Wood bolsters may be placed on top of columns which support roof loads only.

**6.3** Where intermediate beams are used to support a floor, they shall be supported; (1) on the top of the girders, or (2) by ledgers or blocks securely fastened to the sides of the girders, or (3) by approved metal hangers into which the ends of the beam shall be closely fitted.

**6.4** Wood beams and girders supported by walls required to have a fire resistance rating of two hours or more shall have not less than 4 in. of solid masonry between their ends and the outside face of the wall and between adjacent beams.

**6.5**  Columns, beams, girders, arches, trusses, and floors of materials other than wood shall have a fire resistance rating of not less than one hour.

**6.6**  Floors and roof decks shall be without concealed spaces, except that building service equipment may be enclosed provided the spaces between the equipment and enclosures are fire stopped or protected by other acceptable means.

**6.7**  Adequate roof anchorage shall be provided.

## 7.  STANDARD DIMENSIONS FOR HEAVY TIMBER

**7.1**  Excellent fire resistance is achieved with "heavy timber" construction. Minimum sawn lumber sizes have been long established. They are expressed in nominal dimensions and assume surfacing to *American Lumber Standard* net sizes.

**7.2**  For "heavy timber" construction, the net width of glued laminated structural members shall be the standard glued laminated net width for the nominal sawn width specified, and the net depth of glued laminated structural members shall be equal to or greater than the net finished depth specified by the following table.

| | | Minimum Glued Laminated Net Size | | | |
| --- | --- | --- | --- | --- | --- |
| Minimum Nominal Size | | $1\frac{1}{2}$ in. thick laminations | | $1\frac{3}{8}$ in. thick laminations | |
| Width, in. | Depth, in. | Width, in. | Depth, in. | Width, in. | Depth, in. |
| 8 $\times$ 8 | | $6\frac{3}{4}$ $\times$ 9 | | $6\frac{3}{4}$ $\times$ | $8\frac{1}{4}$ |
| 6 $\times$ 10 | | $5\frac{1}{8}$ $\times$ $10\frac{1}{2}$ | | 5 or $5\frac{1}{8}$ $\times$ | 11 |
| 6 $\times$ 8 | | $5\frac{1}{8}$ $\times$ 9 | | 5 or $5\frac{1}{8}$ $\times$ | $8\frac{1}{4}$ |
| 6 $\times$ 6 | | $5\frac{1}{8}$ $\times$ 9 | | 5 or $5\frac{1}{8}$ $\times$ | $6\frac{7}{8}$ |
| 4 $\times$ 6 | | $3\frac{1}{8}$ $\times$ $7\frac{1}{2}$ | | 3 or $3\frac{1}{8}$ $\times$ | $6\frac{7}{8}$ |

AMERICAN INSTITUTE
7012 South Revere Parkway
Suite 140

TIMBER CONSTRUCTION
Englewood, Colorado 80112
Telephone 303-792-9559

# AITC 109-90
# STANDARD FOR PRESERVATIVE TREATMENT OF STRUCTURAL GLUED LAMINATED TIMBER

*Adopted as Recommendations November 2, 1990*
*Copyright 1990 by American Institute of Timber Construction*

## CONTENTS

## 1. PREFACE

**1.1** While the information contained in this standard is based on what is believed to be the most accurate and reliable technical data currently available, it should not be used or relied upon for any general or specific application without competent professional design consideration regarding need for and suitability of the use of preservatively treated wood. By publication of this document, AITC intends no representation or warranty, expressed or implied, that the information contained herein is suitable for any general or specific use or is free from infringement of any patent or copyright. Any user of this information assumes all risk and liability arising from such use.

**1.2**  This standard covers preservative treatment of structural glued laminated timber made from species for which allowable stresses are published in AITC 117-87, Design, *Standard Specifications for Structural Glued Laminated Timber of Softwood Species*, and for which values for treatment penetrations and retentions are published in AWPA C28, *Standard for Preservative Treatment of Structural Glued Laminated Members and Laminations Before Gluing of Southern Pine, Pacific Coast Douglas Fir, Hem-Fir, and Western Hemlock by Pressure Process*. These are Douglas Fir, Hem-Fir, and Southern Pine. It does not cover fire-retardant treatments.

**1.3**  Wood structures properly designed and constructed have performed satisfactorily for centuries. When the use of recognized design and construction principles do not eliminate the hazards of decay, insect, or marine borer attack, then the wood must be pressure preservatively treated to insure durability. The effectiveness of treatment is dependent upon the use of an approved treating chemical and adequate penetration and retention of the chemical used. In some design situations glued laminated timber members, manufactured from the heartwood of naturally decay resistant species, may be used in above ground applications where there is full or partial exposure to the weather without roof cover, or in high humidity situations within a building.

**1.4**  The Environmental Protection Agency (EPA) has established restrictions on the uses of creosote, pentachlorophenol, and inorganic arsenicals as given in Position Document (PD) No. 4—see Reference 13.8. This document contains use and handling precautions on these three types of treatments for all types of wood products.

*1.4.1  Nonhabitable Structure Use.*  Any of the recommended treatments included in this document which meet both the required retentions for the intended end use and the appropriate EPA regulations can be used for structures that are not inhabited and do not carry or store drinking water. Typical applications are pilings, bridges, piers, conveyor structures, culverts, etc.

*1.4.2  Habitable Structure Use.*  Restrictions on the use of preservatively treated glued laminated timber inside buildings and habitable structures are as follows:

*1.4.2.1  Creosote.*  Creosote should only be used for industrial building components which are in ground contact and subject to decay or insect attack and where two coats of an appropriate sealer, as recommended in the EPA Consumer Information Sheet, have been applied to the treated wood. Creosote is basically recommended for exterior use only and should not be used in residential interiors nor in farm building interiors where there may be direct contact with humans or animals. Do not use creosote solutions for structures intended for storing food or where the treated wood may come into contact with drinking water.

*1.4.2.2 Pentachlorophenol.* Pentachlorophenol should not be used in residential, industrial, or commercial structure interiors except for laminated members or building components which are in ground contact and subject to decay or insect attack and where two coats of an appropriate sealer, as recommended in the EPA Consumer Information Sheet, have been applied to the treated wood. Do not use for building interiors where there may be direct contact with humans or animals. Do not use for structures intended for storing food or where the treated wood may come into contact with drinking water.

*1.4.2.3 Inorganic Arsenicals.* Inorganic arsenicals may be used in residential, commercial, or industrial interiors provided that all dust is removed from the wood surface after installation. Do not use for structures used for storing food or where the treated wood may come into contact with drinking water.

*1.4.2.4* Further information on the use and handling of preservatively treated wood products may be found in Consumer Information Sheets (CIS) accompanying shipments or in EPA PD No. 4 (Ref. 13.8).

**1.5** While there is the possibility of exudation of creosote and pentachlorophenol solutions from treated members when the specified retentions are used, the specification of retentions in excess of those shown in Tables 2 and 3 are more likely to result in exudation. Such exudation may continue for a long period of time.

## 2. GENERAL

**2.1 Decay.** Decay is caused by primitive forms of plant life (fungi) that do not contain chlorophyll, but must derive their food from organic matter—in this case wood. Four things are required for decay fungi to grow—food (wood), water, oxygen, and favorable temperature. Decay fungi spores are in the air and only need to find a hospitable environment in order to grow. Prevention of fungal invasion is extremely important since strength losses occur at very early stages of attack. The infected wood may appear sound and not yet be discolored, while already having experienced structural degradation. Prevention of decay damage can be achieved by permanently keeping the wood in-service moisture content to less than 20%. Other means of control include making the food source toxic to decay fungi by preservative treatment, elimination of oxygen by continuous submersion in water, and keeping the temperature of the wood outside the generally accepted decay range—i.e., below about 35°F (2°C) or above about 105°F (40°C).

Active decay fungi attack can be stopped most easily by removing the moisture source. In some cases it may be arrested by field preservative treatment. Any damage done to the wood while decay fungi are active will remain unchanged once the activity has stopped, and whatever wood strength still exists will remain. When the moisture content is reduced below 20% the decay fungi become dormant; however, the decay fungi attack will resume if conditions again become favorable unless the fungi have been killed by an acceptable field treatment. Frequently insect attack will follow attack by decay fungi.

***2.1.1   Installations Where Treating Is Not Required.***   Examples of installations where wood structural members can either be kept below 20% moisture content through proper design, construction details, and maintenance, or wet enough to exclude oxygen are as follows:

*2.1.1.1*   Enclosed buildings with good water tight roof membranes, proper roof maintenance, good joint details, adequate flashing and drainage, good ventilation, properly located insulation and vapor or moisture retarders, and a well-drained building site should assure an in-service wood moisture content below 20%.

*2.1.1.2*   Any usage where wood is permanently and totally submerged in water. However, attack by marine organisms makes the use of submerged, untreated wood in salt water impractical.

***2.1.2   Installations Where Treating Is Required.***   Examples of installations where wood structural members require preservative treatment to be permanent are as follows:

*2.1.2.1*   Building constructions in which portions of wood structural members extend beyond the walls and the protection of the roof, and where adequate weather protection cannot be provided to insure an in-service moisture content of less than 20%.

*2.1.2.2*   Enclosed buildings housing ''wet processes'' or other high humidity environments where, despite ventilation, there is sufficient moisture remaining in the atmosphere to maintain the equilibrium moisture content in any of the wood elements at a level above 20%.

*2.1.2.3*   Outdoor exposures where there is no protective cover for wood structural members from the weather, such as in bridges, piers and wharfs, towers, and electrical utility structures.

*2.1.2.4*   Wood elements in direct contact with the ground or with water, such as in retaining walls, bulkheads, and poles or piles where the wood extends out of the water rather than being permanently and totally submerged.

***2.1.3   Installations Where Treatment May Be Required.***   Examples of installations where wood structural members can only be protected from decay with difficulty and where particular attention to details and connections is required are given in the following subparagraphs. The assurance of a reliable long term maintenance program is needed in these applications to maintain an in-service moisture content of less than 20%. In the absence of these construction details, the members must be treated.

*2.1.3.1*   Certain buildings, such as indoor swimming pools and ice skating rinks in which wood structural members are subject to sustained high humidity and/or moisture condensation. Protection may be provided by means of proper use of vapor and moisture retarders, insulation, and adequate ventilation used collectively. If the designer is sure that an adequate ventilation system will be installed and the system will remain operable during the life of the structure, untreated timbers and dry-use design values may be used, provided the moisture content of the wood is maintained at less than 16%.

*2.1.3.2* Constructions in which wood structural members are adjacent to masonry or concrete, protection may be provided by the use of ventilated air spaces, vapor retarders, or other moisture barrier separations which prevent direct contact of the wood with masonry or concrete.

**2.2 Insect Attack.** Wood structural members that have been preservatively treated to definite retentions and penetrations are resistant to insect attack. The degree of resistance offered will depend upon the type of preservative used as well as the retentions and penetrations obtained.

**2.3 Marine Organisms.** Protection of glued laminated timber members against attack by marine organisms can be provided by pressure treating processes using recommended preservatives with retentions and penetrations as set forth in the current versions of AWPA Standard C2, *Lumber, Timbers, Bridge Ties and Mine Ties—Preservative Treatment by Pressure Processes* and C18, *Standard for Pressure Treated Material in Marine Construction.*

## 3. DESIGN CONSIDERATIONS

**3.1 Materials Properties.** The allowable design values for preservatively treated wood, as included in the *National Design Specification for Wood Construction* (NDS), apply. The NDS requires no reduction in design values for glued laminated timber or other wood products pressure-impregnated by processes and preservatives approved by the American Wood Preservers' Association (AWPA). When the moisture content of the wood in service is less than 16%, untreated timbers and dry-use design values may be used. When the moisture content of the wood in service is 16 to 20%, wet-use design values should be used. When the moisture content of the wood in service is 20% or more, glued laminated timbers should be preservatively treated and wet-use design values used. See Section 7 on Incising if the wood requires incising prior to treatment.

**3.2 Treatment Prior to or After Gluing.** Glued laminated timber may be manufactured with lumber treated prior to gluing, or it may be treated after gluing. While many species may be successfully treated prior to gluing, laminators may choose not to glue pre-treated lumber due to toxic waste disposal problems.

**3.3 Treating Facility Limitations.** The size of available pressure-treating cylinders must be considered when specifying pressure preservative treatments. Due to these cylinder limitations, it may not be possible to treat certain large or curved members after gluing. If treatment is required for these members, treatment prior to gluing should be specified. It is important to contact the supplier prior to initiating the design of large members which require treatment, since special handling and treating problems may exist.

## 4. SPECIES

**4.1** Lumber for glued laminated timber that is to be treated before or after gluing includes the following species: Douglas Fir, Hem-Fir, and Southern Pine for which allowable stresses are published in AITC 117-93 Design and for which treatment penetration and retentions are published in AWPA C28.

**4.2**   Other species may be included if covered by the AWPA standards and the glued laminated timber lay-up combinations are approved by AITC and/or appropriate regulatory agencies.

## 5.   TYPES OF PRESERVATIVE TREATMENTS

**5.1**   Various characteristics of preservative treatments affect their use with glued laminated timber. Some treatments are suitable for use in treating after gluing, and some are suitable for use in treating before gluing. All preservative treatments to be used before gluing should meet the specifications of the current version of AWPA Standard C28. Preservative treatments to be used may be as agreed upon between the buyer and seller, but the treatment process shall be in accordance with applicable AWPA Standards. Preservative treatments are described in the following subsections. A summary of uses of these treatments is given in Table 1. The specifier must select the preservative treatment best suited for a particular job. Availability of treatment types may vary with geographic area and suppliers.

*5.1.1   Creosote and Creosote/Coal Tar Solutions.*   Creosote is a coal tar product containing a multitude of chemical compounds toxic to decay fungi, insects and most marine organisms. Creosote treated material is suitable for the most severe exposure conditions. It has a dark, oily surface appearance which generally cannot be stained or painted. It possesses an odor. Creosote solutions should not be used in contact with materials subject to staining such as plaster, wallboard, etc., or in direct contact with roofing felt. As a practical consideration, the treatment is used only after gluing.

*5.1.2   Oil-Borne Treatments.*   Oil-borne preservative treatments are toxic to decay fungi and insects and utilize various hydrocarbon solvents as carriers. Pentachlorophenol is the most common oil-borne preservative; however, other preservatives such as copper naphthenate are also used. The solvents have various effects on finished products and are classified into the following types in AWPA Standard P9, *Standards for Solvents and Formulations for Organic Preservative Systems.*

*5.1.2.1   Type A—Petroleum Distillates, or a Blend of Petroleum Distillates and Co-Solvents.*   This treatment may become blotchy when exposed to the weather, although this condition diminishes with time. Type A treatment should not be used with wood in contact with material subject to staining, such as wallboard or plaster and does not result in a readily paintable surface. It should not be used in direct contact with roofing felt, but should be separated from the roofing membrane by a barrier of rosin-sized sheathing paper or similar material which is resistant to the treating solution. As a practical consideration, treatment with this type of solvent is used only after gluing.

*5.1.2.2   Type C—Light Hydrocarbon Solvent with Auxiliary Solvent.*   This treatment can leave a natural wood-appearing surface if the theater is advised of appearance requirements and is capable of meeting them. The surface of treated wood can generally be stained or painted after the light solvent volatiles have evaporated.

*5.1.2.3   Type D—Chlorinated Hydrocarbon Solvent-Inhibited Grade of Methylene Chloride.*   The finished surface is slightly darker than untreated wood. The surface can be stained or painted. The solvent recovery process may result in a blotchy appearance, raised grain, and checking of the wood.

**5.1.3   Water-Borne Treatments.**   Water-borne preservative treatments conforming to AWPA Standard P5, *Standards for Waterborne Preservatives*, utilize water soluble preservative chemicals which become fixed in the wood during the treating process. They are toxic to decay fungi, insects and most marine organisms.

*5.1.3.1*   These treatments are typically used to treat laminations prior to gluing. They tend to leave the wood a light green, grey–green, or brown color, depending on the chemical formulation used. When the treated wood surface is dry, the member can be stained or painted in accordance with the coating manufacturer's recommendations. When used in high moisture applications where the in-service moisture content of the wood will be greater than 16%, any metal hardware in contact with the wood should be corrosion resistant in accordance with the chemical treatment manufacturer's recommendations.

*5.1.3.2*   Treating of glued laminated timber members with water-borne preservatives after gluing is not generally recommended. If glued laminated timbers are treated after gluing, dimensional changes caused by saturation of the wood with the water-borne preservatives and their carrier followed by subsequent re-drying may result in excessive warping, checking, or splitting. The use of water-borne treated glued laminated timber members without adequate re-drying of the timbers prior to installation can also result in excessive deflections as well as checking, splitting and warping as previously mentioned as the members ''season'' in-service.

*5.1.3.3*   Treatability of species may vary with different water-borne treatments. It is important to verify with the treater that the preservative specified is compatible with the species used.

*5.1.3.4*   When glued laminated timber members are fabricated with water-borne treated wood and used in chemical or industrial environments, care must be taken to insure compatibility of the treated material with the environment to prevent chemical degradation of the members and their connections. This also applies to other water-borne treated wood members.

## 6.   RETENTION, PENETRATION, CERTIFICATION AND MARKING REQUIREMENTS

**6.1**   Retention and penetration recommendations for the various preservatives and species covered by this standard are given in Tables 2 and 3. The retention and penetration requirements shown are those included in the latest edition of AWPA Standard C28 and should be specified. Members ''Treated to Refusal'' should not be considered as meeting the requirements of this standard unless specified retentions and penetrations are met. Western Red Cedar and the Western Woods should be treated to the equivalent of Hem-Fir treatments.

**6.2**   If laminations are treated prior to gluing, a certificate shall be furnished by the treater or a qualified third party inspection agency to the laminator identifying the type of treatment used and that the treatment has been performed in accordance with, and conforms to, the requirements of AWPA Standard C28. For members treated after gluing, a similar certificate shall be provided to the laminator and/or the end user. The treater should maintain records of assays for retention and penetration which are made to verify conformance to AWPA Standard C28 and these should be available to the laminator and/or the end user. Each bundle or load should be identified by the treater and this identification should be by agreement between the laminator and the treater.

**6.3**   When glued laminated timber is retreated for any reason, consideration must be made for the possible effects of treatment on the strength and/or appearance of the members. A re-treatment may result in some bleeding through of the preservative chemical resulting in a possibly unacceptable appearance of the members. Such re-treatment should be documented in the treating records and shown on the applicable treating certificate supplied by the treater.

## 7.   INCISING

**7.1**   Incising is not required for Southern Pine. However, incising is recommended for the other species listed in Section 4.1 in order to achieve optimum treating results. If incising requirements are waived for appearance or other reasons, penetration and retention requirements still apply. Failure to incise may require re-treatment and can lead to bleeding problems.

*7.1.1*   Nonincised members of species for which incising is recommended should be individually inspected. Each section of a lamination between end joints may be considered as a separate piece of lumber for sampling purposes. Each member should be considered a separate lot and sampled for evaluation of penetration and retention according to the appropriate sections of AWPA Standards C2 & M2, *Standard for Inspection of Treated Timber Products.*

*7.1.2*   Consideration should be given by the designer to potential bending and tension stress reductions caused by incising. The treater doing the incising should provide the necessary information to the designer and laminator on the strength reducing effects of the incising process so that the overall effect on the strength of the member can be assessed.

*7.1.3*   Laminations to be treated prior to gluing shall not be incised on the mating faces to be glued.

## 8.   INDIVIDUAL LAMINATIONS TO BE TREATED PRIOR TO GLUING

**8.1**   The treating of individual laminations with solutions of creosote, creosote-coal tar, creosote–petroleum or pentachlorophenol in heavy petroleum oil prior to gluing is not generally recommended.

**8.2**  Thickness of lumber to be treated should be not more than $\frac{1}{4}$ in. over the thickness of the finished laminations when blanked lumber is used, and $\frac{3}{8}$ in. over when rough lumber is used. The width of either blanked or rough lumber should be not more than $\frac{1}{2}$ in. over the net finished width of the glued laminated member. Lumber treated prior to gluing should be dried after treatment to a moisture content not to exceed the maximum moisture content permitted for production of glued laminated timber in accordance with American National Standard ANSI/ AITC A190.1-1983, *Structural Glued Laminated Timber*.

**8.3**  Lumber treated prior to gluing shall have mating faces resurfaced after treatment and just prior to gluing. The resurfacing of laminations treated prior to gluing should remove as little wood as practical while making the surface clean, planed, and uniform in thickness suitable for gluing. The maximum time between resurfacing and gluing of pre-treated lumber shall be established at the time of plant qualification.

## 9.  ADHESIVES AND PROCESSES

**9.1**  Glued laminated timbers to be treated by pressure processes or pre-treated laminations to be made into laminated timber must be glued with wet-use adhesives conforming to Section 4.5.1.2 of ANSI/AITC A190.1-1983.

**9.2**  In general, longer curing times or higher temperatures are required for gluing pre-treated laminations than for gluing untreated laminations to obtain glue bonds of comparable quality. Treatment may affect adhesive spread requirements and assembly times.

**9.3**  In any process, different combinations of lumber species, pre-treatment preservative, and adhesive type may not produce the same quality of glue bond as obtained with untreated lumber, even though fabricated using the same procedures and conditions. It is, therefore, important that the use of any particular combination of species, pre-treatment preservative, and adhesive be supported by adequate gluing data for the individual laminator's procedures. Each plant should qualify the maximum retention of the treatments it intends to use for each species to be glued.

## 10.  FABRICATION AND MACHINING

**10.1**  Where possible, all fabrication and machining which cuts through the treating envelope should be performed prior to treatment. When fabrication and machining are done after treatment, additional preservative treatment should be applied in accordance with AWPA Standard M4, *Standard for the Care of Preservative-Treated Wood Products*.

## 11.  CARE AFTER TREATMENT AND FIELD TREATMENT

**11.1**  Protect treated wood from mechanical injury when handling. Cutting of treated material should be avoided whenever possible. When handling damage occurs, or cuts, daps, or other machining is done in the field, the cut or damaged surfaces should be field treated as specified in AWPA Standard M4. The application of oil or water-borne preservatives to treated or untreated members in the field is not as effective as and can not replace pressure preservative treatment; therefore, it is important to do as much machining prior to treatment as possible.

**11.2**   Fumigants are specialized preservative chemicals in liquid or solid form that are placed in pre-bored holes to arrest internal decay in untreated or inadequately treated woods. The designer is referred to references 13.5 and 13.10 for additional information on the use of fumigants.

## 12.   EXUDATION OF NATURAL WOOD RESIN

**12.1**   Most species of softwood commonly used in glued laminated timbers contain resins in some portions of the wood. Resins may be present in the form of pitch, pitch streaks, or pitch pockets. Generally, resin is set during drying and remains fixed during the service life of the product. Occasionally, some resin may exude from localized areas or from pitch streaks which were not set during the drying process.

**12.2**   Wood which has been treated with preservatives dissolved in a solvent that also dissolves the resin may tend to have more exudation of resins during the service life than untreated wood. This will vary with species, the type of treatment, and the amount of preservative solvent remaining in the wood. Exudation of resin usually decreases with time, and the resin tends to harden as the solvents evaporate.

## 13.   REFERENCES

13.1   Timber Construction Manual, Third Edition, AITC, Englewood, Colorado 80112.

13.2   AWPA Book of Standards, American Wood Preservers' Association, Stevensville, Maryland 21666.

13.3   The NRCA Roofing and Waterproofing Manual, The National Roofing Contractors Association, Rosemont, Illinois 60018.

13.4   Marine Wood Maintenance Manual: A Guide For Proper Use of Douglas-Fir in Marine Exposures, Research Bulletin 48, Forest Research Laboratory, Oregon State Univ., Corvallis, Oregon 97331.

13.5   Controlling Wood Deterioration With Fumigants: A Review, J. Morrell and M. Corden, Forest Products Journal 36(10):26–34.

13.6   National Design Specification for Wood Construction, National Forest Products Assoc., Washington, DC 20036.

13.7   American National Standard for Wood Products—Structural Glued Laminated Timber, ANSI/AITC A190.1-1983, American National Standards Institute, New York, NY 10018.

13.8   Wood Preservative Pesticides: Creosote, Pentachlorophenol, and Inorganic Arsenicals, EPA Position Document No. 4, PB 84-241538, National Technical Information Service, U.S. Dept. of Commerce, Springfield, Virginia 22161.

13.9   Standard Specifications for Structural Glued Laminated Timber of Softwood Species, 117-93, AITC, Englewood, Colorado 80112.

13.10  In-Service Inspection, Maintenance and Repair of Glued Laminated Timber Subject to Decay Conditions, Technical Note No. 13, AITC, Englewood, Colorado 80112.

**TABLE 1—Uses and Characteristics of Preservative Treatments for Glued Laminated Timbers[1,2]**

| Treatment Type | Visual/Physical Appearance | When Architectural Appearance is Important | Paintability | Treatment Before Gluing | Treatment After Gluing |
|---|---|---|---|---|---|
| Creosote and Creosote/Coal Tar Solutions | Dark, oily | NR | Cannot be painted or stained | NR | A |
| Oil Borne Penta Treatments | | | | | |
| Type A[5] | Oily, may become blotchy when exposed to elements | NR | Not readily paintable | NR | A |
| Type C[5] | Can be natural appearing if treater is advised of finished appearance requirements | A | Can be painted or stained after surface preparation as recommended by treater | A | A |
| Type D[5] | Slightly darker than untreated wood | L | Can be painted or stained | A[2] | L |
| Water-Borne Treatments | Light green, gray-green or brown in color depending upon chemicals used | L | Can be painted or stained when surface is dry and prepared in accordance with coating manufacturer's recommendations | A[2] | L |

A   — Generally acceptable.

L   — Acceptable only with limitations. See comments under "Limitations" in Section 6.

NR — Not recommended

1. Some preservative treatments may cause irritation when in direct human contact. See Reference 13.8 for handling precautions.

2. Availability of treatments for use with specific species and in specific geographical regions should be verified before specifying.

3. See Table 3 for limitations for certain treatments.

4. For potable water, other restrictions may apply. See Reference 13.8 for use site precautions.

5. See Section 5.1.2 for a description of these treatment types.

| Ground Contact or Embedded in Concrete | Exposed to Wetting or in Contact with Masonry | Contact with Fresh Water[4] | Contact with Salt Water | Limitations |
|---|---|---|---|---|
| A[3] | A | A | A | Odor may be objectionable. Should not be used in direct contact with roofing felt. Should not be used in contact with materials subject to staining such as plaster, wallboard, etc. |
| A | A | A | NR | Should not be used in direct contact with roofing felt. Should not be used in contact with materials subject to staining, such as plaster, wallboard, etc. |
| A | A | A | NR | Softwood species used in laminating may exude resin after treatment, although condition diminishes with time. |
| A | A | A | NR | Solvent recovery process may result in raised grain and checking of wood surface. Softwood species used in laminating may exude resin after treatment, although condition diminishes with time. |
| A[3] | A | A | A | Wetting and redrying process associated with treatment may result in dimensional changes, warping, checking, or splitting of the members when treated after gluing. When used in high moisture conditions, all metallic connections should be of corrosive-resistant metal per the treater's recommendations. |

## TABLE 2

### Retention and Penetration for Structural Glued Laminated Timber Treated *After* Gluing

| | From AWPA Standard C28 | | | |
| --- | --- | --- | --- | --- |
| | Southern Pine | | Pacifc Coast Douglas Fir, Hemfir or Western Hemlock | |
| | Above Ground | Soil Contact | Above Ground | Soil[2] Contact |
| Retention-By Assay[1] | | | | |
| Sampling zone for assay-inches from surface | 0–0.60 | 0–0.60 | 0–0.60 | 0–0.60 |
| Number of borings per charge | | | | |
| Creosote & Creosote Solutions | 48 | 48 | 48 | 48 |
| Oil-Borne Preservative | 20 | 20 | 20 | 20 |
| Pcf-Min.: | | | | |
| Creosote & Creosote Solutions | | | | |
| Creosote | 8.0 | 10.0 | 8.0 | 10.0 |
| Creosote–coal tar | 8.0 | 10.0 | 8.0 | 10.0 |
| Creosote–petroleum | 8.0 | 10.0 | 8.0 | 10.0 |
| Oil-Borne Preservatives Pentachlorophenol | 0.30 | 0.60 | 0.30 | 0.60 |
| Penetration in inches of wood and/or percent of sapwood-Min. (AWPA C1, 3.21) | 2.50 or 85 | | Under 5" thick 0.40 and 90. 5" and thicker 0.50 and 90. | |
| Penetration | A borer core shall be taken from 20 pieces in each charge. If 90% of the borings meet the penetration requirements and the remaining pieces have at least 50% of their required depth of penetration, the charge will be accepted. | | A borer core shall be taken from the incised faces of 20 pieces in each charge. If 80% of the borings meet the penetration requirements the charge shall be accepted. | |
| Incising Recommendation | No | No | Yes | Yes |

*Footnotes to Table 2*

[1]More than one boring may be taken from the same piece, but not more then one from the same lamination unless there is an end joint separation or there are an insufficient number of separate laminations. Using an increment borer core, 0.20 inches in diameter, twenty 0.60 inch long borings are required for assay of pentachlorophenol. A minimum of forty-eight 0.60 inch long borings is required for creosote and creosote solutions. A 0.40 inch plug center may be used to increase the size of the sample taken for analysis or to reduce the number of borings required below forty-eight.

The treated wood surface should be lightly scraped prior to taking a sample in order to remove surface deposits of preservative. Samples are to be taken from sapwood only of Southern Pine for determination of both retention and penetration. Samples are to be taken randomly for determination of penetration and from heartwood only for determination of retention of other species.

The retention requirement for pentachlorophenol is based on an assay using the lime ignition method of analysis. When the copper pyridine method of analysis is used, multiply the result by 1.1 to convert to the lime ignition basis.

[2]Soil Contact Use. For members more than 75 square inches in cross section at the groundline, every member shall be bored for penetration. For members 75 square inches or less in cross section at the groundline, 20 members per charge shall be bored for penetration. Should the charge contain less than 20 members, each member shall be bored. When inspecting Southern pine laminated timbers for penetration, borings shall be taken from two different laminations for each member. When boring other species listed in the Table for penetration, a minimum of one boring shall be taken from each of two face laminations and one boring from each of two different interior laminations in each member.

If any boring taken from any member fails to meet the penetration requirement that member shall be rejected. If 90 percent or more of the members bored meet the above requirements for either size category, the charge shall be accepted. If less than 80 percent of the above requirements for either size category, the charge shall be rejected.

Above Ground Use. One boring from each of 20 members in a charge shall be taken for penetration. Should the charge contain less than 20 members, each member shall be bored. If any boring fails to meet the penetration requirement, that member shall be rejected. If 80 percent or more of the members bored meet the above requirement, the charge shall be accepted. If less than 80 percent of the members bored meet the above requirements, the charge shall be rejected.

## TABLE 3

### Retention and Penetration for Structural Glued Laminated Timber Treated *Before* Gluing

| | From AWPA Standard C28 | | | |
| | Southern Pine | | Pacifc Coast Douglas Fir, Hemlock and Western Hemlock | |
| | Above Ground | Soil Contact | Above Ground | Soil[2] Contact |
|---|---|---|---|---|
| Sampling zone for assay[1,2] | | | | |
| Inches from edge of | | | | |
| Interior Laminations | 0.5–1.0 | 0.5–1.0 | 0.5–1.0 | 0.5–1.0 |
| Minimum Number of | | | | |
| Borings per lot | 20 | 20 | 20 | 20 |

## TABLE 3 (*Continued*)

| | From AWPA Standard C28 | | | |
| | Southern Pine | | Pacifc Coast Douglas Fir, Hemlock and Western Hemlock | |
| | Above Ground | Soil Contact | Above Ground | Soil[2] Contact |
|---|---|---|---|---|
| Retention by Assay-pcf (minimum) | | | | |
| Creosote | 8.0 | 10.0 | 8.0 | 10.0 |
| Creosote–Coal Tar Solution | 8.0 | 10.0 | NR | NR |
| Creosote–Petroleum | NR | NR | 8.0 | 10.0 |
| Pentachlorophenol | 0.30 | 0.60 | 0.30 | 0.60 |
| Waterborne Preservatives | | | | |
| ACC | 0.25 | 0.50 | 0.25 | 0.50 |
| ACA | 0.25 | 0.40 | 0.25 | 0.40 |
| ACZA | 0.25 | 0.40 | 0.25 | 0.40 |
| CCA | 0.25 | 0.40 | 0.25 | 0.40 |
| CZC | 0.45 | NR | 0.45 | NR |
| Determination of Penetration[3] Inches from Edge 20 of 20 | 3.00 or 90% | 3.00 or 90% | 1.00 | 1.25 |
| Incising Recommendation | No | No | Yes | Yes |

[1]Using an increment borer core 0.20 inches in diameter, twenty 0.50 inch long borings will provide an adequate sample for assay of pentachlorophenol and the waterborne preservatives. The method of analysis used for assaying retention of the waterborne preservatives should be based on the analytical methods of Standard A2. Forty-eight 0.50 inch long borings are required for creosote and creosote solutions. A 0.40 inch plug cutter may be used to increase the size of the sample taken for analysis or to reduce the number of borings required below forty-eight. Because the assay zone is the second $\frac{1}{2}$ inch, the cores must be 1 inch long.

The retention requirement for pentachlorophenol is based on an assay using the lime ignition method of analysis. When the copper pyridine method of analysis is used, multiply the result by 1.1 to convert to the lime ignition basis.

[2]Laminated beams manufactured from material treated to meet the above requirements can be assayed by changing the sample zone from 0.50 to 1.00 inches, as indicated above, to 0 to 0.50 inches. Results of assay must meet 90% of the retention specified above.

[3]Laminated beams manufactured from material treated to meet the above requirements can be tested for penetration by taking samples from the edges of the individual laminations. The penetration required is 0.50 inch less than that required above on 18 out of 20 samples.

AMERICAN INSTITUTE  TIMBER CONSTRUCTION

7012 South Revere Parkway         Englewood, Colorado 80112

Suite 140                  Telephone 303-792-9559

# AITC 110-84
# STANDARD APPEARANCE GRADES FOR
# STRUCTURAL GLUED LAMINATED TIMBER

*Adopted as Recommendations July 18, 1984*
*Copyright 1984 by American Institute of Timber Construction*

## CONTENTS

## 1. INTRODUCTION

**1.1** Various grades of appearance are desired for different uses. These appearance grades apply to the surfaces of glued laminated members and include such items as growth characteristics, void filling and surfacing operations but not laminating procedures, stains, varnishes, or other finishes, nor wrappings, cratings, or other protective coverings. The appearance grades do not modify the design stresses, fabrication controls, grades of lumber used and other provisions of the standards for structural glued laminated timber.

*1.1.1* These appearance grades are for the guidance of the designer so that a product consistent with the use of the structure may be specified and provide a suitable appearance at appropriate cost. The designer should specify the desired appearance grade to give a clear understanding between buyer and seller. Requirements given in appearance descriptions are intended to achieve a general and distinctive uniformity of appearance, and reasonable tolerance is permitted.

Appearance grading should reflect good judgment. It should be kept in mind that often the natural growth characteristics of the wood enhance the beauty of the member and avoid an artificial appearance.

***1.1.2***   Three appearance grades—Industrial, Architectural and Premium— are applicable to all species or mixtures of species used in laminating. Some combinations in *Standard Specifications for Structural Glued Laminated Timber of Softwood Species*, AITC 117, permit the mixing of species within a given member. With this mixing of species, the potential for differences in color or grain of adjacent laminations must be recognized. For those appearance applications where such possible differences in color or grain might be important, the designer may specify a single species or species group with similar characteristics. In some cases, this may restrict availability.

***1.1.3***   When members containing wood filler are stained, the filler may not accept the stain in the same manner as the adjacent wood. In addition, filler exposed to the elements may weather differently than the adjacent wood. The buyer should check with the supplier concerning past experience in these matters before specifying.

***1.1.4***   Preservative treatment may affect the appearance of the finished laminated timber. See *Standard for Preservative Treatment of Structural Glued Laminated Timber*, AITC 109, for detailed information.

***1.1.5***   When cover boards are not used, a sloping end cut exposed to view shall have the same appearance requirements for loose knots, knot holes, etc., that are required for the sides of members.

***1.1.6***   Unless otherwise specified appearance grade requirements shall apply to three exposed surfaces for beams and arches and to all four exposed surfaces for columns.

## 2.   TEXTURED SURFACES

**2.1**   When specified by the designer, a textured surface may be used. Textured surfaces are produced by a variety of methods and the buyer should check with the supplier before specifying.

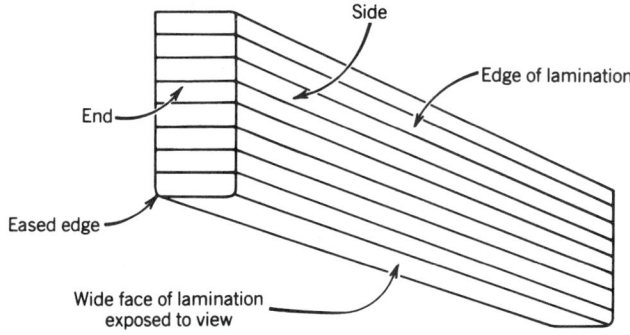

**FIGURE 1.**   Illustration of terms related to appearance grades.

**2.2**   Textured surfaces may change the net finished sizes and tolerances given in *Standard for Dimensions of Glued Laminated Structural Members*, AITC 113. Depending upon the degree of texturing, it may be necessary for the designer to compensate for the resulting loss of cross section.

**2.3**   Texturing will change the appearance grade specified. Additional voids and splintering may result. Textured surfaces will readily absorb stain, but adjacent areas of wood filler may not accept stain in the same manner. Eased edges will not normally be furnished.

### 3.   INDUSTRIAL APPEARANCE GRADE

**3.1   Application.**   Industrial appearance grade is ordinarily suitable for construction in industrial plants, warehouses, garages, and for other uses where appearance is not of primary concern.

**3.2   Description**

*3.2.1*   Laminations may possess the natural growth characteristics of the lumber grade.

*3.2.2*   Voids appearing on the edge of laminations need not be filled.

*3.2.3*   Loose knots and open knot holes in the wide face of laminations exposed to view shall be filled. This restriction does not apply for glued laminated timber truss members. Edge joints appearing on the wide face of laminations exposed to view need not be filled.

*3.2.4*   Members are required to be surfaced on two sides only; appearance requirements apply to these sides. Occasional misses, low laminations or wane (limited to a maximum of 1.4 in. measured across the width) are permitted on a cumulative basis. The cumulative depth of the misses, low laminations or wane shall not exceed 10% of the width of the member at any glue line. The frequency of occurrence shall not exceed one in 10 pieces of lumber used. The maximum area of low laminations shall not exceed 5% of the surface area of a side; and no more than two low laminations shall be adjacent to one another.

*3.2.5*   In accordance with provisions in 3.2.4, wane (limited to $\frac{1}{4}$ in. measured across the width) is permitted in all combinations and is not limited in length. Wane permitted in specific laminating combinations up to $\frac{1}{6}$ the lumber width on each side is not limited in length. Occasional wane approximately one foot in length and not exceeding the permissible depth of a low lamination shall be permitted in all combinations without regard to the cumulative effects indicated in 3.2.4.

### 4.   ARCHITECTURAL APPEARANCE GRADE

**4.1   Application.**   Architectural appearance grade is ordinarily suitable for construction where appearance is an important requirement. Any voids under $\frac{3}{4}$ in. in diameter shall be filled by others that the fabricator if the final decorative finish so requires.

## 4.2  Description

*4.2.1*  Laminations may possess the natural growth characteristics of the lumber grade.

*4.2.2*  In exposed surfaces, knot holes and other voids measuring over $\frac{3}{4}$ in. shall be filled by the fabricator with a wood-tone colored filler which reasonably blends with the final product or with clear wood inserts. When inserts are used, 9they shall be selected with reasonable care for similarity of the grain and color of the wood insert to the adjacent wood.

*4.2.2.1*  For appearance grading purposes, measurement of knot holes and other voids shall be in the direction of the length of the lamination and shall not exceed $\frac{3}{4}$ in. except that a void may be longer than $\frac{3}{4}$ in. if its area does not exceed $\frac{1}{2}$ sq. in. Void measurement limitations apply only to the surfaces of the member exposed in the final structure. All characteristics must be considered, however, with respect to their effects on general appearance.

*4.2.3*  The wide face of laminations exposed to view shall be free of loose knots. Open knot holes shall be filled. Voids greater than $\frac{1}{16}$ in. wide in edge joints appearing on the wide face of laminations exposed to view shall be filled.

*4.2.4*  Exposed faces shall be surfaced smooth. Misses are not permitted.

*4.2.5*  The corners of the wide face of laminations exposed to view in the final structure shall be eased. Current industry practice for eased edges is for a radius between $\frac{1}{8}$ and $\frac{1}{2}$ in. Other radii for eased edges may be agreed upon between buyer and seller.

## 5.  PREMIUM APPEARANCE GRADE

**5.1  Application.**  Premium appearance grade is the highest standard appearance grade.

## 5.2  Description

*5.2.2*  In exposed surfaces, knot holes and other voids shall be filled by the fabricator with a wood-tone colored filler which reasonably blends with the final product or with clear wood inserts. When inserts are used, they shall be selected with reasonable care for similarity of the grain and color of the wood insert to the adjacent wood.

*5.2.3*  The wide face of laminations exposed to view shall be selected for appearance and shall be free of loose knots. Voids shall be filled. Knot size shall be limited to 20% of the net face width of the lamination. Not over two maximum size knots or their equivalent shall occur in a 6 ft length. Voids greater than $\frac{1}{16}$ in. wide in edge joints appearing on the wide face of laminations exposed to view shall be filled.

*5.2.4*  Exposed faces shall be surfaced smooth. Misses are not permitted.

*5.2.5*  The corners of the wide face of laminations exposed to view in the final structure shall be eased. Current industry practice for eased edges is for a radius between $\frac{1}{8}$ and $\frac{1}{2}$ in. Other radii for eased edges may be agreed upon between buyer and seller.

## 6. SPECIAL APPEARANCE REQUIREMENTS

**6.1**  For special applications, the buyer may specify other requirements. Such requirements may limit availability.

*6.1.1*  Industrial Special appearance grade, indicated on the non-customer Quality Inspected Stamp as IND-S, is applicable to nominal 4 in. wide glued laminated timbers surfaced "hit or miss" to $3\frac{1}{2}$ in. All requirements in Section 3.2 apply except the limitations on misses or low laminations and the unsurfaced lamination may exhibit glue smears and discolorations associated with the production process.

## 7. SUMMARY OF APPEARANCE GRADE SPECIFICATIONS

| Description item | Industrial Appearance Grade | Architectural Appearance Grade | Premium Appearance Grade |
|---|---|---|---|
| Natural growth characteristics of lumber grade | Allowed | Allowed | Allowed |
| Paragraph reference | 3.2.1 | 4.2.1 | 5.2.1 |
| Filling of voids on edge of laminations | Not required | Required for voids over $\frac{3}{4}''$ | Required for all voids |
| Paragraph reference | 3.2.2 | 4.2.2 & 4.2.2.1 | 5.2.2 |
| Wide face of laminations exposed to view | Void filling required except for trusses | Free of loose knots, void filling required | Selected for appearance, free of loose knots, void filling required, knot sizes limited |
| Paragraph reference | 3.2.3 | 4.2.3 | 5.2.3 |
| Edge joints appearing on wide faces of laminations exposed to view | Filling not required | Filling required for voids over $\frac{1}{16}''$ wide | Filling required for voids over $\frac{1}{16}''$ wide |
| Paragraph reference | 3.2.3 | 4.2.3 | 5.2.3 |
| Surfacing of sides | Required. Limited amounts of misses, low laminations and wane permitted. | Required. Misses not permitted | Required. Misses not permitted |
| Paragraph reference | 3.2.4 & 3.2.5 | 4.2.4 | 5.2.4 |
| Surfacing of wide face of laminations exposed to view | Not required | Required. Misses not permitted | Required. Misses not permitted |
| Paragraph reference | 3.2.4 | 4.2.4 | 5.2.4 |
| Eased edges | Not required | Required | Required |
| Paragraph reference | — | 4.2.5 | 5.2.5 |

AMERICAN INSTITUTE
7012 South Revere Parkway
Suite 140

TIMBER CONSTRUCTION
Englewood, Colorado 80112
Telephone 303-792-9559

# AITC 111-79
# RECOMMENDED PRACTICE FOR PROTECTION OF STRUCTURAL GLUED LAMINATED TIMBER DURING TRANSIT, STORAGE AND ERECTION

*Adopted as Recommendations March 4, 1979*
*Copyright 1979 by American Institute of Timber Construction*

## CONTENTS

## 1. INTRODUCTION

**1.1**  Protection of glued laminated timber structural members includes end sealers, surface sealers, primer coats and wrapping applied for the protection of members. End sealers, surface sealers, primer coats and wrappings offer a degree of protection, but they do not necessarily preclude damage resulting from negligence and other factors beyond the control of the laminator during shipment, handling, storage and placement.

**1.2**  The protection specified should be commensurate with the end use and final finish of the member. It may also vary with the method of shipment and with exposure to climatic and other conditions before construction is completed.

**1.3**  These recommended specifications are for the guidance of the designer so that the product may have protection consistent with the intended use of the member at appropriate cost. The designer should specify the desired protection in a way that establishes a clear understanding between the buyer and the seller. *This standard contains alternatives from which the specifier must make certain selections to suit the particular job.*

**1.4**  Experience has shown that the following protection methods have a sufficient range to fulfill normal requirements. The designer should select the methods of protection best suited to the particular job and include them in the job specification.

## 2.  END SEALERS

**2.1**  End sealers retard moisture transmission and minimize end checking when use conditions are such that end checking is a major consideration.

**2.2  Recommended Specifications**

*2.2.1*  A coat of sealer should be applied to the fresh-ends of all members after end trimming.

*2.2.2*  A colorless sealer shall be used on ends exposed to view in the completed structure.

## 3.  SURFACE SEALERS

**3.1**  Surface sealers increase resistance to soiling, control grain raising, minimize checking and serve as a moisture retardant. Surface sealers fall into the two following classifications.

*3.1.1  Translucent Penetrating Sealers.*  Translucent penetrating sealers have low solid content. They provide limited protection and are suitable for use when final finish requires staining.

*3.1.1.1  Recommended Specifications.*  A penetrating sealer shall be applied to all surfaces before shipment.

*3.1.2  Primer and Nonpenetrating Sealer Coats.*  Primer and nonpenetrating sealer coats have higher solid content than penetrating sealers and provide maximum protection by sealing the surface of the wood. Primer and non-penetrating sealer coats should not be specified when final finish requires a natural or stained finish.

*3.1.2.1  Recommended Specifications*
(a) A sealer (or primer) coat shall be applied to all surfaces before shipment.
(b) A nonpenetrating sealer (or primer) shall have a minimum solid content of 25%.

## 4.  WRAPPING

**4.1**  Wrapping the member with water-resistant covering or its equivalent for shipment provides additional protection from moisture, soiling and damage in handling. Wrapping is usually recommended when appearance is of prime importance and the additional protection is desired. Bundle or load wrapping may be specified in lieu of individual wrapping when further utilization of wrap after delivery is not desired. Time of removal of factory wrap is optional, but, it must be emphasized, factory-applied wrapping provides additional protection from damage in handling and in transit only. If further utilization of the wrap is desired for protection after shipment, the members should be inspected and provided with additional protection as necessary.

**4.2**  Water-resistant covering used for in-transit protection of individually wrapped members may be left in place until the members are enclosed within the building. If wrapping has to be removed at certain connection points during erection, it should be replaced after connection is made in order to prevent sun bleaching or water staining of parts of the member. If it is impractical to replace the wrapping, all of it should be removed. Individual wrapping should be slit or punctured on the lower side if there is evidence of moisture inside the wrapping.

**4.3  Recommended Specifications**

**4.3.1  *Individual Wrapping***

*4.3.1.1*  Members shall be individually wrapped, covering all surfaces, with water-resistant paper, opaque polyethylene or their equivalent.

*4.3.1.2*  Wrapping shall be secured to the member by staples, tape or other suitable fastenings that do not damage exposed surfaces.

*4.3.1.3*  Seams of wrapping shall inhibit the passage of moisture.

**4.3.2  *Bundle Wrapping***

*4.3.2.1*  Members shall be bundle wrapped, totally inclosing the bundle, with water-resistant paper, opaque polyethylene of their equivalent.

*4.3.2.2*  Convenience in handling shall determine the size of the bundle.

*4.3.2.3*  Wrapping shall be secured to the bundle by staples, tape or other suitable fastenings that do not damage exposed surface. Wherever possible, staples should be located in the bottom of the bundle.

*4.3.2.4*  Seams of wrapping shall inhibit the passage of moisture.

**4.3.3  *Load Wrapping***

*4.3.3.1*  Members shall be load-wrapped, enclosing the top, sides and ends, with water-resistant paper, opaque polyethylene or their equivalent.

*4.3.3.2*  Wrapping shall extend to the bottom of the members included in the load.

*4.3.3.3*  Wrapping shall be secured to the load by staples, tape or other suitable fastenings that do not damage the exposed surfaces.


**5.  PROTECTION FOR PRESERVATIVELY TREATED MEMBERS**

**5.1**  Preservative treatment of glued laminated timber can be divided into three general categories: a) heavy treatments of members after gluing, such as creosote or pentachlorophenol (penta) in heavy oil; b) light treatments of members after gluing such as penta in light solvent (penta in LPG and water-borne treatments are not recommended for treating glued laminated timber after gluing); and c) treatments of laminations prior to gluing as penta in LPG, penta light solvent or waterborne salts.

*5.1.1*   Protection of glued laminated members to receive heavy treatments is generally limited to end sealers. Surface sealers may be recommended if the member are to be stored in an arid climate prior to treatment. If protection during transit and storage is desired, load wrapping should be satisfactory for members to receive heavy treatments. Where the time between fabrication and treatment is short, no special protection may be required.

*5.1.2*   Protection of glued laminated members to receive light treatments is usually more critical since they are generally intended for use where appearance is a factor. End sealers should be applied at the time the end cuts are made *and* following treatment to minimize end checking. Surface sealers may be recommended if the members are to be stored in an arid climate prior to treatment. Surface sealers may also be applied to the members after treatment for the same reasons they are applied to untreated members.

*5.1.3*   Protection of glued laminated members made from treated laminations is generally the same as for untreated members.

## 6.   UNLOADING AND HANDLING
### 6.1   Recommended Specifications

*6.1.1*   Laminated members shall not be dragged or dropped. Care shall be taken in handling to prevent damage to finished surface. Cable slings or chokers should not be used to handle laminated materials unless adequate blocking is provided between the cable and the wood member. Web belting-type slings are recommended. Protection cleats or blocking shall be applied at pickup points to protect corners.

*6.1.2*   Spreader bars of suitable length should be used in lifting long members to reduce the probability of damage. The method of erection and handling should not overstress the member.

*6.1.3*   Whenever possible, members should be lifted on edge.

*6.1.4*   Extreme care should be taken to minimize impact forces during lifting.

## 7.   JOB SITE STORAGE
### 7.1   Recommended Specifications

*7.1.1*   Laminated material stored at the job site shall be treated with care. A level area is required to avoid warpage. Members shall be supported with blocking so spaced as to provide uniform and adequate support. If covered storage is not available, the material shall be blocked well off the ground at a well drained location. Stored members shall be separated with stripping arranged vertically over the supports so that air circulated around all four sides of each member; the top and all sides shall be covered with moisture-resistant covering. If a paved surface is unavailable, the ground under the material shall be covered with polyethylene film. Clear polyethylene film shall not be used. Individual wrappings shall be slit full length or punctured on the lower side to permit drainage of water. Slats inserted between the member and wrapping minimize the marking of the wood. Long members shall be stored on edge.

## 8. ERECTION

### 8.1 Recommended Specifications

*8.1.1* Padded or nonmarring slings shall be used, and corners shall be protected with wood blocking. (See Section 6 for additional information.)

**NOTE:** Heat should not be fully turned on as soon as the structure is enclosed; otherwise, excessive checking may occur due to rapid lowering of the relative humidity in the building. A gradual seasoning period at moderate temperature should be provided.

AMERICAN INSTITUTE
7012 South Revere Parkway
Suite 140

TIMBER CONSTRUCTION
Englewood, Colorado 80112
Telephone 303-792-9559

## AITC 112-93
## STANDARD FOR TONGUE-AND-GROOVE HEAVY TIMBER ROOF DECKING

*Adopted as Recommendations April 21, 1993*
*Copyright 1993 American Institute of Timber Construction*

## CONTENTS

## 1. INTRODUCTION

**1.1**  This Standard applies to sawn tongue-and-groove decking only and does not apply to laminated, panelized or other special decking systems. This standard covers species, sizes, patterns, lengths, moisture content, application, specifications, weights, applicable unit stresses, allowable loads and slope conversion values for heavy timber roof decking in nominal 2, 3, and 4 inch thickness, using single or double tongues and grooves.

**1.2**  Heavy timber roof decking is a specialty lumber product, constituting an important part of modern timber construction, that can be used for many applications to provide an all wood appearance. Nominal three and four inch thick roof decking is especially well adapted for use with glued laminated arches and girders and is easily and quickly erected. To be suitable for purposes intended, heavy timber roof decking must be well manufactured to a low moisture content as described herein.

**1.3**  The lumber used in heavy timber roof decking shall be graded in accordance with the grading rules under which the species is customarily graded. The standard grading and dressing rules referenced in this Standard are:

(a)  *Standard Grading Rules for Northeastern Lumber*, 1991, Northeastern Lumber Manufacturers Association, 272 Tuttle Rd., PO Box 87A, Cumberland Center, ME 04021 (NELMA)

(b)  *Standard Specifications for Grades of California Redwood Lumber*, April 1992, Edition, Redwood Inspection Service, 405 Enfrente Dr., Suite 200, Novato, CA 94949 (RIS)

(c)  *Standard Grading Rules for Southern Pine Lumber*, 1991, Southern Pine Inspection Bureau, 4709 Scenic Highway, Pensacola, FL 32504 (SPIB)

(d)  *Standard Grading Rules for West Coast Lumber, No. 17*, Effective September 1, 1991, West Coast Lumber Inspection Bureau, P.O. Box 23145, Portland, OR 97223 (WCLIB)

(e)  *Standard Grading Rules for Western Lumber*, Effective September 1, 1991, Western Wood Products Association, 522 SW Fifth, Yeon Building, Portland, OR 97204 (WWPA)

(f)  *NLGA Standard Grading Rules for Canadian Lumber*, Effective September 1, 1991, National Lumber Grades Authority, 260-1055 W. Hastings St., Vancouver, B.C. V6E 2E9, Canada (a Canadian Agency) (NLGA)

Copies of these grading rules may be obtained from the respective grading rule agencies.

**1.4**  Moisture content requirements of the regional lumber grading rules may differ from this Standard. Unless their Standard is followed in all requirements, the product will not conform with this Standard.

## 2.  SPECIES

**2.1**  The species usually available and currently used in this product, as well as the regional inspection agencies under which decking lumber is ordinarily graded, are given in Table 1.

## 3.  SIZES AND PATTERNS

**3.1  Two Inch Decking.**  The standard size is 2 × 6 inch and 2 × 8 inch, nominal, dressed at the moisture content specified herein to the actual size and V-grooved pattern shown in Figure 1. Other thicknesses and widths are also available. See regional grading rules listed in paragraph 1.3 for dimensions for individual species.

**3.2  Three and Four Inch Decking.**  Standard sizes are 3 × 6 inch and 4 × 6 inch, nominal, at the moisture content specified herein. Figures 2 and 3 provide typical dimensions for 3 × 6 inch and 4 × 6 inch nominal decking, respectively, illustrating a V-joint pattern. Other thicknesses and widths may be available.

# TABLE 1

## Heavy Timber Deck Species

| Species | Grading Rules Under Which Graded | Paragraph Number of Grading Rules Under Which Graded[a] | |
|---|---|---|---|
| | | Select Quality[b] | Commercial Quality[c] |
| Cedar, Northern White | NELMA | 15.1 | 15.2 |
| Cedars, Western | WWPA, WCLIB | 55.11, 127-b | 55.12, 127-c |
| Cedars, Western (North) | NLGA (Canadian) | 127-b | 127-c |
| Coast Species | NLGA (Canadian) | 127-b | 127-c |
| Douglas Fir-Larch | WWPA, WCLIB | 55.11, 127-b | 55.12, 127-c |
| Douglas Fir-Larch (North) | NLGA (Canadian) | 127-b | 127-c |
| Douglas Fir (South) | WWPA | 55.11 | 55.12 |
| Fir, Balsam | NELMA | 15.1 | 15.2 |
| Hem-Fir | WWPA, WCLIB | 55.11, 127-b | 55.12, 127-c |
| Hem-Fir (North) | NLGA (Canadian) | 127-b | 127-c |
| Hemlock, Eastern-Tamarack | NELMA | 15.1 | 15.2 |
| Hemlock, Eastern-Tamarack (North) | NLGA (Canadian) | 127-b | 127-c |
| Hemlock, Western | WCLIB | 55.11, 127-b | 55.12, 127-c |
| Hemlock, Western (North) | NLGA (Canadian) | 127-b | 127-c |
| Northern Species | NLGA (Canadian) | 127-b | 127-c |
| Pine, Eastern White | NELMA | 15.1 | 15.2 |
| Pine, Eastern White (North) | NLGA (Canadian) | 127-b | 127-c |
| Pine, Northern | NELMA | 15.1 | 15.2 |
| Pine, Ponderosa | NLGA (Canadian) | 127-b | 127-c |
| Pine, Red | NLGA (Canadian) | 127-b | 127-c |
| Pine, Southern[d,e] | SPIB | 412 | 413 |
| Pine, Western White | NLGA (Canadian) | 127-b | 127-c |
| Redwood, California | RIS | 315 | 316 |
| SPF[s] | NELMA, WWPA | 15.1, 55.11 | 15.2, 55.12 |
| Spruce, Coast Sitka | NLGA (Canadian) | 127-b | 127-c |
| Spruce, Eastern | NELMA | 15.1 | 15.2 |
| Spruce, Pine-Fir | NLGA (Canadian) | 127-b | 127-c |
| Spruce, Sitka | WCLIB | 127-b | 127-c |
| Western Woods | WWPA | 55.11 | 55.12 |

[a] When species may be graded under WCLIB and WWPA rules, the first paragraph number is for WWPA and the second for WCLIB rules.

[b] Select quality grades are as follows for the grading rules indicated:

| | | | |
|---|---|---|---|
| WCLIB; | Select Dex | SPIB; | Select Decking |
| WWPA; | Selected Decking | RIS; | Select Decking |
| NELMA; | Selected Decking | NLGA; | Select Decking |

[c] Commercial quality grades are as follows for the grading rules indicated:

| | | | |
|---|---|---|---|
| WCLIB; | Commercial Dex | SPIB; | Commercial Decking |
| WWPA; | Commercial Decking | RIS; | Commercial Decking |
| NELMA; | Commercial Decking | NLGA; | Commercial Decking |

[d] Southern Pine decking is also available in the following grades: Dense Standard Decking, para. 411; Dense Select Decking, para. 412.1; and Dense Commercial Decking, para. 413.1

[e] Southern Pine is limited to the botanical species of longleaf, slash, shortleaf and loblolly. Lumber cut from trees of this species is classified as "Southern Pine" in the SPIB Grading Rules.

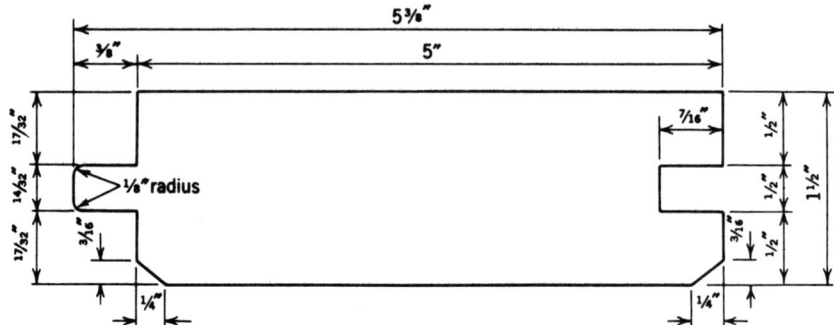

**FIGURE 1.**   2× nominal V-joint pattern. (See regional grading rules listed in paragraph 1.3 for dimensions for individual species.)

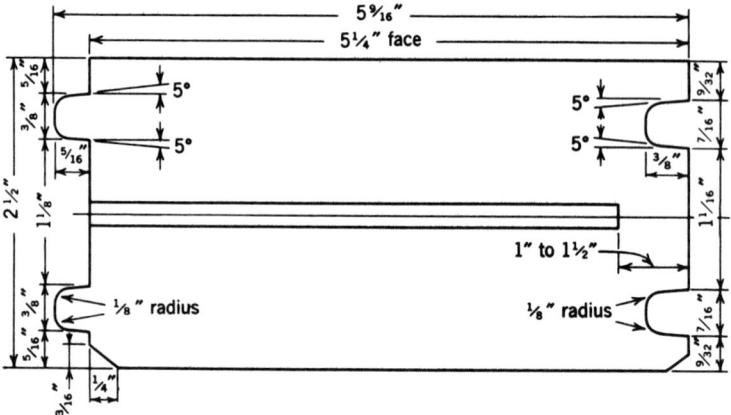

**FIGURE 2.**   3 × 6 inch, nominal V-joint pattern. Note: Profile dimensions apply to all patterns. (See regional grading rules in paragraph 1.3 for dimensions for individual species.)

**3.3**   Other patterns are available, including grooved, striated and eased joint, and the regional grading rules agencies indicated in **paragraph 1.3** should be contacted for further details concerning specific patterns and sizes.

## 4.   LENGTHS

**4.1**   Decking pieces may be of specified length or may be random length. All layup arrangements except controlled random layup require that the specifier indicate the required lengths.

**4.2**   If pieces are for controlled random layup, odd or even lengths are permitted, and the minimum lengths based on fbm percentages shall be as follows:

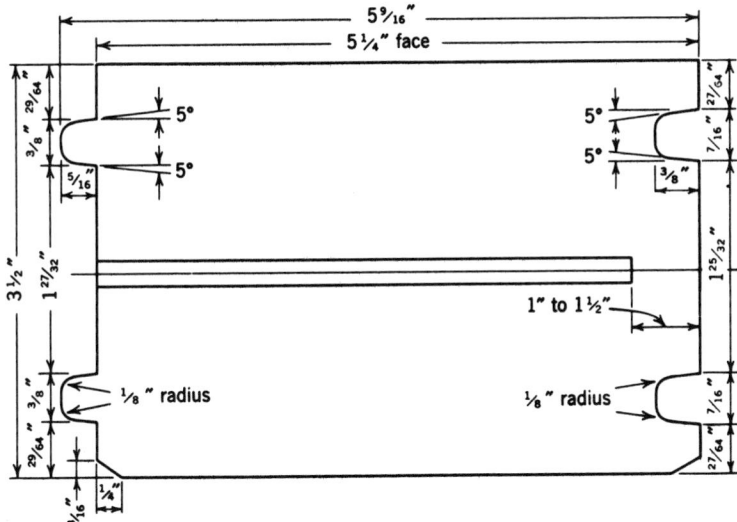

**FIGURE 3.** 4 × 6 inch nominal V-joint pattern. Note: Profile dimensions apply to all patterns. (See regional grading rules in paragraph 1.3 for dimensions for individual species.)

### 4.2.1 Two Inch Decking
* Not less than 40% to be 14 ft and longer
* Not over 10% to be less than 10 ft
* Not over 1% to be 4 to 5 ft
* Minimum length is limited to 75% of the span length (i.e., for 8 ft support spacing, 6 ft)

### 4.2.2 Three Inch Decking
* Not less than 40% to be 14 ft and longer with at least 20% equal to or greater in length than the maximum span.
* Not over 10% to be less than 10 ft
* Not over 1% to be 4 to 5 ft

### 4.2.3 Four Inch Decking
* Not less than 25% to be 16 ft and longer with at least 20% equal to or greater in length than the maximum span
* Not less than 50% to be 15 ft and longer
* Not over 10% to be 5 to 10 ft
* Not over 1% to be 4 to 5 ft

## 5. MOISTURE CONTENT

**5.1 Two Inch Decking.** The maximum moisture content shall be 15%.

**5.2 Three and Four Inch Decking.** The maximum moisture content shall be 19%.

**5.3**  Moisture content shall be determined by such methods as will assure these limitations.

## 6.  APPLICATION

**6.1**  Tongue-and-groove wood decking is to be installed with tongues up on sloped or pitched roofs, and outward in direction of laying on flat roofs. It is to be laid with pattern faces down and exposed on the underside.

**6.2**  Each piece shall be square end trimmed. When random lengths are furnished, each piece must be square end trimmed across the face so that at least 90% of the pieces will be within $\frac{3}{64}$ inches of square. The vertical end cut may vary from square to the bevel cut shown in Figure 4.

**6.3  Nailing Schedules**

   *6.3.1  Two Inch Decking.*    Each piece shall be toenailed through the tongue and face nailed with one nail per support, using 16d common nails.

   *6.3.2  Three and Four Inch Decking.*    Each piece should be toenailed at each support with one 40d nail and face nailed with one 60d nail. Courses shall be spiked to each other with 8 inch spikes at intervals not to exceed 30 inches through predrilled edge holes and with one spike at a distance not exceeding 10 inches from each piece. See Figure 5 for boring details.

**6.4**  Heavy timber decking may be laid in any of the following arrangements:

   *6.4.1  Simple Span.*    All pieces on two supports.

   *6.4.2  Controlled Random Layup.*    This arrangement is applicable to 4 or more supports (3 or more spans). (With less than 4 supports, a special pattern requiring specified lengths must be used.) Joints in the same general line (within 6 inches of being in line each way) shall be separated by at least two intervening courses. In the end bays each piece must rest on at least one support and must continue over the first inner support for at least 2 ft. For 3 and 4 inch decking in the interior bays, occasional pieces not resting over a support may occur provided the ends of the adjacent pieces in the same course are continued for at least

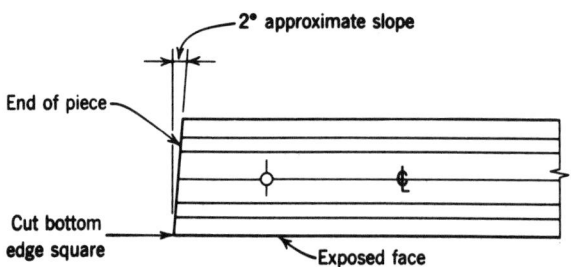

**FIGURE 4.**    Beveled end cut. (Beveled end cut is optional.)

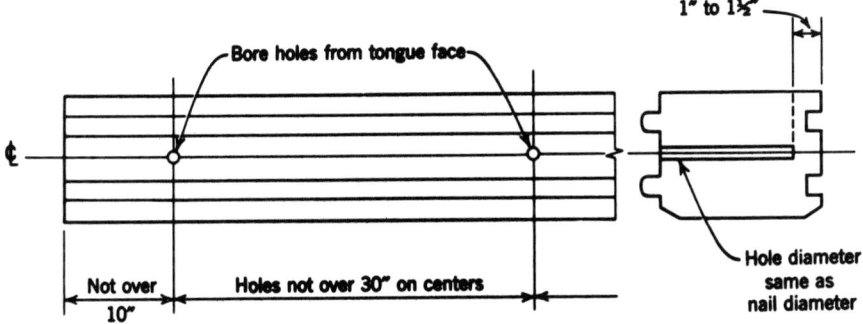

**FIGURE 5.** Boring detail. Locate end holes not over 10 inches from end of piece.

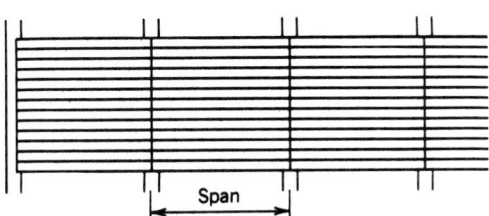

**FIGURE 6.** Simple span layup.

2 ft over the next support. This condition shall not occur more than once in every 6 courses in each interior bay.

*6.4.2.1 Two Inch Decking.* There shall be a minimum distance of 2 ft between end joints in adjacent courses. To provide lateral restraint for the supporting member, the pieces in at least the first and second courses must bear on at least two supports with end joints in these two courses occurring in alternate supports. A maximum of seven intervening courses is allowed before this pattern is repeated. If some other provision, such as plywood overlayment, is made to provide continuity, this pattern is not necessary.

*6.4.2.2 Three and Four Inch Decking.* There shall be a minimum distance of 4 ft between end joints in adjacent courses.

*6.4.3 Cantilever Spans with Controlled Random Layup.* When the overhang does not exceed $1\frac{1}{2}$ ft, 2 ft and 3 ft for nominal 2 inch, 3 inch, and 4 inch thick decking, no special considerations for layup are necessary. The maximum cantilever length for controlled random layup is limited to 0.3 times the length of the first adjacent interior span. For cantilever overhangs exceeding the normal overhang, but not exceeding the maximum, a structural fascia should be fastened

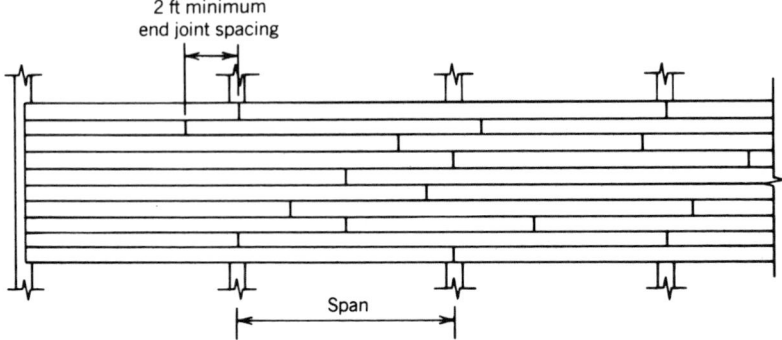

**FIGURE 7.**   Controlled random layup. (Two inch decking.)

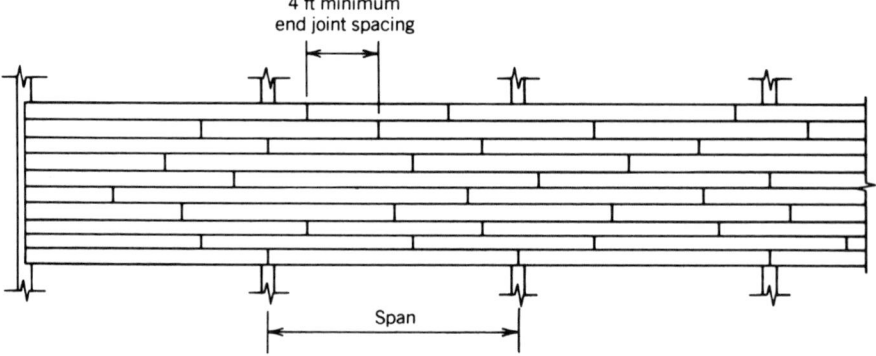

**FIGURE 8.**   Controlled random layup. (Three and four inch decking.)

to each decking piece to maintain a continuously straight roof line. Also, there shall be no end joints in the cantilevered portion or within $\frac{1}{2}$ the span $(L/2)$ of the outer support.

**6.4.4   *Cantilevered Pieces Intermixed.*** This arrangement is applicable to 4 or more supports (3 or more spans). Pieces in the starter course and every third course are simple span. Pieces in other courses are cantilevered over the supports with end joints at alternate quarter or third points of the spans, and each piece rests on at least one support. A tie between supports is provided by the simple span courses of the arrangement.

**6.4.5   *Combination Simple and Two-Span Continuous.*** Alternate pieces in end spans are simple span; adjacent pieces are two-span continuous. End joints are staggered in adjacent courses and occur over support.

**6.4.6   *Two-Span Continuous.*** All pieces bear on three supports. All end joints occur in line on every other support.

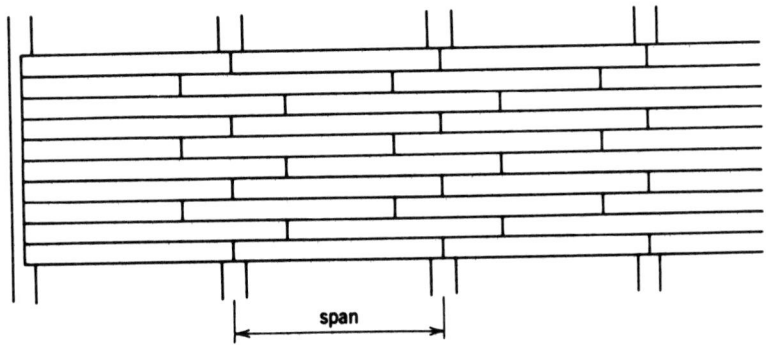

**FIGURE 9.**    Cantilevered pieces intermixed layup.

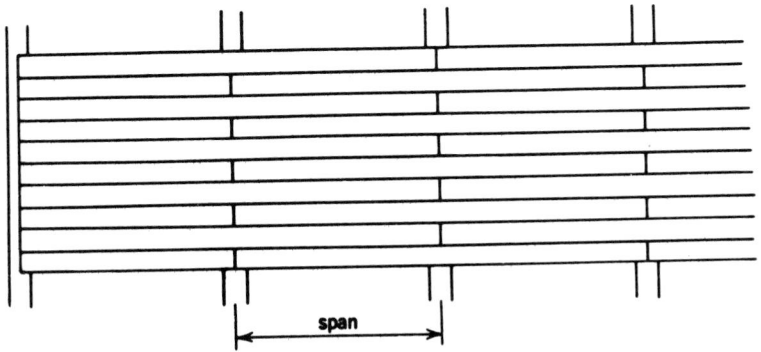

**FIGURE 10.**    Combination simple and two-span continuous layup.

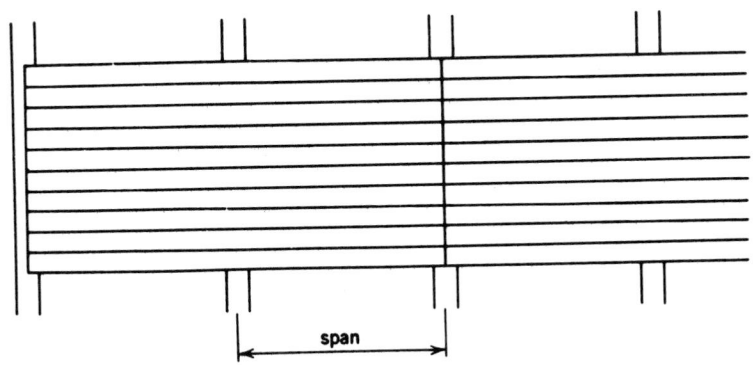

**FIGURE 11.**    Two-span continuous layup.

## 7.  SPECIFICATIONS

**7.1**  The specifications for tongue-and-groove decking for the various species as well as inspection and shipping provisions shall be as specified in the standard grading rules under which the species is graded and shall be subject to such other provisions of the standard grading rules as may be applicable. (See paragraph 1.3.)

**7.2    Select Quality.**    Decking of this quality is recommended for construction for which good strength and fine appearance are desired. Knots and other natural characteristics of specified limitations are permitted.

**7.3    Commercial Quality.**    Decking of this quality is recommended and customarily used for the same purposes served by the higher quality when appearance requirements are less critical.


## 8.  WEIGHTS OF INSTALLED DECKING    (See Table 2, page 8-827.)


## 9.  ALLOWABLE LOADS

**9.1**    Allowable loads for heavy timber decking may be determined by entering Tables 4 through 8 with the appropriate bending stress and modulus of elasticity values, and using the lower of the tabulated load values from the tables for the nominal thickness and span under consideration. Bending stress and modulus of elasticity values for wood decking species, as recommended by the regional lumber rules-writing agency by which the species is graded, are given in Table 3.

**9.2**    Allowable loads given in Tables 4 through 8 are for the simple span and controlled random layup arrangements illustrated under Figs. 6–8.

**9.3    Controlled Random Layup Load Values**

   *9.3.1    Two Inch Decking.*    The allowable loads for controlled random layup, limited by bending, for 2 inch nominal thickness decking as given in Table 4, are based on the standard engineering formula for a three-equal-span continuous, uniformly-loaded member; however, only $\frac{2}{3}$ of the moment of inertia for the cross section was used in calculating the loads. Loads limited by deflection as given in Table 5, are for the maximum deflections in the end spans.

   *9.3.2    Three and Four Inch Decking.*    The allowable loads for controlled random layup of 3 and 4 inch nominal thickness decking as given in Tables 6 through 8, are based on the standard engineering formula for a three-equal-span, continuous, uniformly-loaded member; however, only 80% of the moment of inertia for the cross section was used in calculating the loads. Loads limited by deflection, as given in Tables 7 and 8, are for the maximum deflections in the end spans.

   *9.3.3*    The percentage adjustments in moment of inertia discussed in 9.3.1 and 9.3.2 take into account the differences between continuous decking without joints and the controlled random layup of decking as specified herein. The factors

## TABLE 2

### Weights of Installed Heavy Timber Decking in Pounds per Square Foot of Roof Surface[a]

| Species | 1-½ in. net[b] (2 in. nom.) | 2-½ in. net[c] (3 in. nom.) | 3-½ in. net (4 in. nom.) | Agency[d] |
|---|---|---|---|---|
| Cedar, Northern White | 2.7 | 4.5 | 6.3 | 1 |
| Cedars, Western[d] | 3.0 | 4.9 | 6.9 | 3, 4 |
| Cedars, Western (North)[e] | 2.9 | 4.8 | 6.7 | 2 |
| Coast Species[e] | 3.9 | 6.4 | 9.0 | 2 |
| Douglas Fir-Larch[d] | 4.3 | 7.2 | 10.1 | 3, 4 |
| Douglas Fir-Larch (North)[e] | 4.4 | 7.3 | 10.3 | 2 |
| Douglas Fir (South) | 4.1 | 6.9 | 9.6 | 3 |
| Fir, Balsam | 3.2 | 5.4 | 7.5 | 1 |
| Hem-Fir[e] | 3.7 | 6.1 | 8.6 | 3, 4 |
| Hem-Fir (North)[e] | 3.8 | 6.3 | 8.8 | 2 |
| Hemlock, Eastern-Tamarack[e] | 3.8 | 6.3 | 8.8 | 1 |
| Hemlock, Eastern-Tamarack (North)[d,e] | 4.0 | 6.7 | 9.4 | 2 |
| Hemlock, Western | 4.0 | 6.7 | 9.4 | 4 |
| Northern Species[f] | 2.9–5.3 | 4.8–8.8 | 6.7–12.4 | 2 |
| Pine, Eastern White | 3.3 | 5.5 | 7.7 | 1 |
| Pine, Eastern White (North) | 3.4 | 5.7 | 8.0 | 2 |
| Pine, Northern[f] | 3.8–4.5 | 6.3–7.5 | 8.8–10.5 | 1 |
| Pine, Ponderosa | 4.1 | 6.9 | 9.6 | 2 |
| Pine, Red | 3.7 | 6.1 | 8.6 | 2 |
| Pine, Southern[e] | 4.6 | 7.6 | 10.7 | 5 |
| Pine, Western White | 3.4 | 5.7 | 8.0 | 2 |
| Redwood, California | 3.7 | 6.1 | 8.6 | 6 |
| SPF[s] | 3.7 | 6.1 | 8.6 | 1, 3 |
| Spruce, Coast Sitka | 3.3 | 5.5 | 7.7 | 2 |
| Spruce, Eastern | 3.6 | 6.0 | 8.4 | 1 |
| Spruce-Pine-Fir[f] | 3.0–4.0 | 5.1–6.7 | 7.1–9.4 | 2 |
| Spruce, Sitka | 3.6 | 6.0 | 8.4 | 4 |
| Western Woods[f] | 2.9–4.6 | 4.8–7.6 | 6.7–10.7 | 33 |

[a] All weights shown in Table 2 are based on volume at 14% moisture content; rounded to the nearest 0.1 lb. These weights may be reduced by 2% where 15 maximum moisture content is specified (which is an average of 12% M.C.).

[b] For a net thickness of $1\frac{7}{16}$ in., multiply tabulated weights by a factor of 0.958.

[c] For a net thickness of $2\frac{5}{8}$ in., multiply tabulated weights by a factor of 1.05.

[d] Species listed are as graded by the following grading rules agencies: NELMA (1), NLGA (Canadian) (2), WWPA (3), WCLIB (4), SPIB (5), and RIS (6).

[e] Weights shown for this species grouping are based on the weighted average of the standing timber volume. Lumber from some areas or species within the group may vary slightly from the average.

[f] Weights shown for this species grouping are the range of weights for species that could be included.

**TABLE 3**

Bending Stress and Modulus of Elasticity Values for Heavy Timber Decking Species[a]

| Species | Select Quality Bending Stress[b] psi | Select Quality Modulus of Elasticity[c] psi | Commercial Quality Bending Stress[b] psi | Commercial Quality Modulus of Elasticity[c] psi | Agency[d] |
|---|---|---|---|---|---|
| Cedar, Northern White | 1100 | 800,000 | 950 | 700,000 | 1 |
| Cedars, Western | 1450 | 1,100,000 | 1200 | 1,000,000 | 3, 4 |
| Cedars, Western (North) | 1400 | 1,100,000 | 1200 | 1,000,000 | 2 |
| Coast Species | 1450 | 1,500,000 | 1200 | 1,400,000 | 2 |
| Douglas Fir-Larch | 2000 | 1,800,000 | 1650 | 1,700,000 | 3, 4 |
| Douglas Fir-Larch (North) | 2000 | 1,800,000 | 1650 | 1,700,000 | 2 |
| Douglas Fir (South) | 1900 | 1,400,000 | 1600 | 1,300,000 | 3 |
| Fir, Balsam | 1650 | 1,500,000 | 1400 | 1,300,000 | 1 |
| Hem-Fir | 1600 | 1,500,000 | 1350 | 1,400,000 | 3, 4 |
| Hem-Fir (North) | 1500 | 1,500,000 | 1300 | 1,400,000 | 2 |
| Hemlock, Eastern-Tamarack | 1700 | 1,300,000 | 1450 | 1,100,000 | 1 |
| Hemlock, Eastern-Tamarack (North) | 1700 | 1,300,000 | 1450 | 1,100,000 | 2 |
| Hemlock, Western | 1750 | 1,600,000 | 1450 | 1,400,000 | 4 |
| Hemlock, Western (North) | 1750 | 1,600,000 | 1450 | 1,400,000 | 2 |
| Northern Species | 1050 | 1,100,000 | 875 | 1,000,000 | 2 |
| Pine, Eastern White | 1300 | 1,200,000 | 1100 | 1,100,000 | 1 |
| Pine, Eastern White (North) | 1050 | 1,200,000 | 875 | 1,100,000 | 2 |
| Pine, Northern | 1550 | 1,400,000 | 1300 | 1,300,000 | 1 |
| Pine, Ponderosa | 1450 | 1,300,000 | 1250 | 1,100,000 | 2 |
| Pine, Red | 1350 | 1,300,000 | 1100 | 1,200,000 | 2 |
| Pine, Southern | 1650 | 1,600,000 | 1650 | 1,600,000 | 5 |
| Pine, Western White | 1300 | 1,400,000 | 1050 | 1,300,000 | 2 |
| Redwood, California | 1700 | 1,100,000 | 1350 | 1,000,000 | 6 |
| SPF[s] | 1350 | 1,400,000 | 1100 | 1,200,000 | 1, 3 |
| Spruce, Coast Sitka | 1450 | 1,700,000 | 1200 | 1,500,000 | 2 |
| Spruce, Eastern | 1300 | 1,500,000 | 1100 | 1,400,000 | 1 |
| Spruce-Pine-Fir | 1400 | 1,500,000 | 1150 | 1,300,000 | 2 |
| Spruce, Sitka | 1500 | 1,500,000 | 1250 | 1,300,000 | 4 |
| Western Woods | 1300 | 1,200,000 | 1100 | 1,100,000 | 3 |

[a]The design values in bending ($F_b$), except for Redwood, are based on decking 4 in. thick. For other thicknesses, multiply by the size factor, $C_f$, as follows:

| Thickness | $C_F$ |
|---|---|
| 2 in. | 1.10 |
| 3 in. | 1.04 |

 Design values for visually graded decking are those recommended by the regional lumber rules writing agencies.

 These values are based on decking that is used where the moisture content in-service will not exceed 19%. When the moisture content in-service exceeds 19% for an extended period of time, the tabular design values shall be multiplied by the wet service factor, $C_m$ as follows:

| | $C_M$ | | |
|---|---|---|---|
| $F_b$ | $F_{c\perp}$ | $E$ | *When $(F_b) \times (C_F) < 1150$ psi, $C_M = 1.0$ for bending. |
| 0.85* | 0.67 | 0.9 | |

[b] Repetitive member use values.

[c] The tabulated values for modulus of elasticity are the average for the species grouping. For information concerning coefficient of variation of modulus of elasticity, see the appropriate grading rules for the species.

[d] Stresses listed are as assigned by the following grading rules agencies: NELMA (1), NLGA (Canadian) (2), WWPA (3), WCLIB (4), SPIB (5) and RIS (6).

[e] If specified as "close grain," California Redwood select decking is assigned a bending stress value of 1850 psi and a modulus of elasticity value of 1,400,000 psi when used at 19% M.C.

## TABLE 4

### Two Inch Nominal Thickness[a] Allowable Roof Load Limited by Bending

| Bending Stress, psi | Allowable Uniformly Distributed Total Roof Load[b], psi | | | | | | | | | | | | | |
|---|---|---|---|---|---|---|---|---|---|---|---|---|---|---|
| | Simple Span, ft | | | | | | | Controlled Random Layup Span, ft | | | | | | |
| | 6 | 7 | 8 | 9 | 10 | 11 | 12 | 6 | 7 | 8 | 9 | 10 | 11 | 12 |
| 875 | 73 | 54 | 41 | 32 | 26 | 22 | 18 | 61 | 45 | 34 | 27 | 22 | 18 | 15 |
| 950 | 79 | 58 | 44 | 35 | 28 | 24 | 20 | 66 | 48 | 37 | 30 | 24 | 20 | 16 |
| 1000 | 83 | 61 | 47 | 37 | 30 | 25 | 21 | 69 | 51 | 39 | 31 | 25 | 21 | 17 |
| 1050 | 88 | 64 | 49 | 39 | 32 | 26 | 22 | 73 | 54 | 41 | 33 | 26 | 22 | 18 |
| 1100 | 92 | 67 | 52 | 41 | 33 | 27 | 23 | 76 | 56 | 43 | 34 | 28 | 23 | 19 |
| 1150 | 96 | 70 | 54 | 42 | 34 | 28 | 24 | 80 | 59 | 45 | 36 | 29 | 24 | 20 |
| 1200 | 100 | 73 | 56 | 44 | 36 | 30 | 25 | 83 | 61 | 47 | 38 | 30 | 25 | 21 |
| 1250 | 104 | 76 | 58 | 46 | 38 | 31 | 26 | 87 | 64 | 49 | 39 | 31 | 26 | 22 |
| 1300 | 108 | 80 | 61 | 48 | 39 | 32 | 27 | 90 | 66 | 51 | 41 | 32 | 27 | 22 |
| 1350 | 112 | 83 | 63 | 50 | 40 | 33 | 28 | 94 | 69 | 53 | 42 | 34 | 28 | 23 |
| 1400 | 117 | 86 | 66 | 52 | 42 | 35 | 29 | 97 | 71 | 55 | 44 | 35 | 29 | 24 |
| 1450 | 121 | 89 | 68 | 54 | 44 | 36 | 30 | 101 | 74 | 57 | 45 | 36 | 30 | 25 |
| 1500 | 125 | 92 | 70 | 56 | 45 | 37 | 31 | 104 | 76 | 58 | 47 | 38 | 31 | 26 |
| 1550 | 129 | 95 | 73 | 57 | 46 | 38 | 32 | 108 | 79 | 60 | 48 | 39 | 32 | 27 |
| 1600 | 133 | 98 | 75 | 59 | 48 | 40 | 33 | 111 | 82 | 62 | 50 | 40 | 33 | 28 |
| 1650 | 138 | 101 | 77 | 61 | 50 | 41 | 34 | 114 | 84 | 64 | 52 | 41 | 34 | 29 |
| 1700 | 142 | 104 | 80 | 63 | 51 | 42 | 35 | 118 | 87 | 66 | 53 | 42 | 35 | 30 |
| 1750 | 146 | 107 | 82 | 65 | 52 | 43 | 36 | 122 | 89 | 68 | 55 | 44 | 36 | 30 |
| 1900 | 158 | 116 | 89 | 70 | 57 | 47 | 40 | 132 | 97 | 74 | 60 | 48 | 39 | 33 |
| 2000 | 167 | 122 | 94 | 74 | 60 | 50 | 42 | 139 | 102 | 78 | 63 | 50 | 41 | 35 |

[a] Based on $1\frac{1}{2}$ in. net thickness. To determine allowable loads for $1\frac{7}{16}$ in. net thickness, multiply tabulated values by 0.918.

[b] To determine allowable uniformly distributed total roof loads for other span conditions, use simple span load values for combination simple and two-span continuous, and two-span continuous layups; and use controlled random layup load values for cantilevered pieces intermixed layup.

[c] Duration of load, $C_D = 1.00$ used in this table. For other duration of load, adjust by the appropriate factor.

[d] No increase for size effect has been applied ($C_F = 1.00$). $F_b$ values have been previously adjusted.

[e] Dry condition of use.

## TABLE 5

### Two Inch Nominal Thickness[a] Allowable Roof Load Limited by Deflection

| Modulus of Elasticity psi | Deflection Limit | Allowable Uniformly Distributed Total Roof Load[c], psf | | | | | | | | | | |
|---|---|---|---|---|---|---|---|---|---|---|---|---|
| | | Simple Span, ft | | | | | Controlled Random Layup Span, ft | | | | | |
| | | 6 | 7 | 8 | 9 | 10 | 6 | 7 | 8 | 9 | 10 | 11 | 12 |
| 700,000 | $l/180$ | 32 | 20 | 14 | 10 | 7 | 42 | 27 | 18 | 12 | 9 | 7 | 5 |
| | $l/240$ | 24 | 15 | 10 | 7 | 5 | 32 | 20 | 13 | 9 | 7 | 5 | 4 |
| 800,000 | $l/180$ | 37 | 23 | 16 | 11 | 8 | 48 | 30 | 20 | 14 | 10 | 8 | 6 |
| | $l/240$ | 28 | 17 | 12 | 8 | 8 | 36 | 23 | 15 | 11 | 8 | 6 | 4 |
| 900,000 | $l/180$ | 42 | 26 | 18 | 12 | 9 | 54 | 34 | 23 | 16 | 12 | 9 | 7 |
| | $l/240$ | 31 | 20 | 13 | 9 | 7 | 41 | 26 | 17 | 12 | 9 | 7 | 5 |
| 1,000,000 | $l/180$ | 46 | 29 | 20 | 14 | 10 | 60 | 38 | 25 | 18 | 13 | 10 | 7 |
| | $l/240$ | 35 | 22 | 15 | 12 | 8 | 45 | 19 | 19 | 13 | 10 | 7 | 6 |
| 1,100,000 | $l/180$ | 51 | 32 | 21 | 15 | 11 | 66 | 42 | 28 | 20 | 14 | 11 | 8 |
| | $l/240$ | 38 | 24 | 16 | 11 | 8 | 50 | 31 | 21 | 15 | 11 | 8 | 6 |
| 1,200,000 | $l/180$ | 56 | 35 | 23 | 16 | 12 | 72 | 46 | 30 | 21 | 16 | 12 | 9 |
| | $l/240$ | 42 | 26 | 18 | 12 | 9 | 54 | 34 | 23 | 16 | 12 | 9 | 7 |
| 1,300,000 | $l/180$ | 60 | 38 | 25 | 18 | 13 | 78 | 49 | 33 | 23 | 17 | 13 | 10 |
| | $l/240$ | 45 | 28 | 19 | 13 | 10 | 59 | 37 | 25 | 17 | 13 | 10 | 7 |
| 1,400,000 | $l/180$ | 65 | 41 | 27 | 19 | 14 | 84 | 53 | 36 | 25 | 18 | 14 | 10 |
| | $l/240$ | 49 | 31 | 20 | 14 | 10 | 63 | 40 | 27 | 19 | 13 | 10 | 8 |
| 1,500,000 | $l/180$ | 69 | 44 | 29 | 20 | 15 | 90 | 57 | 38 | 27 | 20 | 15 | 11 |
| | $l/240$ | 52 | 33 | 22 | 15 | 11 | 68 | 43 | 29 | 20 | 15 | 11 | 8 |
| 1,600,000 | $l/180$ | 74 | 47 | 31 | 22 | 16 | 96 | 61 | 41 | 28 | 21 | 16 | 12 |
| | $l/240$ | 56 | 35 | 23 | 16 | 12 | 72 | 46 | 30 | 21 | 16 | 12 | 9 |
| 1,700,000 | $l/180$ | 79 | 50 | 33 | 23 | 17 | 102 | 64 | 43 | 30 | 22 | 17 | 13 |
| | $l/240$ | 59 | 37 | 25 | 17 | 13 | 77 | 48 | 32 | 23 | 17 | 12 | 10 |
| 1,800,000 | $l/180$ | 83 | 52 | 35 | 25 | 18 | 108 | 68 | 46 | 32 | 23 | 18 | 14 |
| | $l/240$ | 62 | 39 | 26 | 18 | 14 | 81 | 51 | 34 | 24 | 18 | 13 | 10 |

[a] Based on a $1\frac{1}{2}$ in. net thickness. To determine allowable loads for $1\frac{7}{16}$ in. net thickness, multiply tabulated value by 0.880.

[b] For a deflection limit $l/360$, use $\frac{1}{2}$ the tabulated value for a deflection limit of $l/180$.

[c] To determine allowable uniformly distributed total roof loads for other span conditions, multiply controlled random layup load values by the following factors:

| | |
|---|---|
| Cantilevered pieces intermixed; | 1.05 |
| Combination simple span and two-span continuous; | 1.31 |
| Two-span continuous; | 1.85 |

[d] Duration of load, $C_D = 1.00$ used in this table. For other duration of load, adjust by the appropriate factor.

[e] No increase for size effect has been applied ($C_F = 1.00$). $F_b$ values have been previously adjusted.

[f] Dry condition of use.

# TABLE 6

## Three and Four Inch Nominal Thickness Allowable Roof Load Limited by Bending Simple Span and Controlled Random Layups (3 or more spans)[a]

Uniformly Distributed Total Roof Load[d], psf

| Bending Stress psi | 3 inch Nominal Thickness[b], Span, ft[c] | | | | | | | | | | | | | 4 inch Nominal Thickness[d], Span, ft[c] | | | | | | | | | | | | |
|---|---|---|---|---|---|---|---|---|---|---|---|---|---|---|---|---|---|---|---|---|---|---|---|---|---|---|
| | 8 | 9 | 10 | 11 | 12 | 13 | 14 | 15 | 16 | 17 | 18 | 19 | 20 | 8 | 9 | 10 | 11 | 12 | 13 | 14 | 15 | 16 | 17 | 18 | 19 | 20 |
| 875 | 114 | 90 | 73 | 60 | 51 | 43 | 37 | 32 | 28 | 25 | 22 | 20 | 18 | 223 | 176 | 143 | 118 | 99 | 84 | 73 | 64 | 56 | 49 | 44 | 40 | 36 |
| 950 | 124 | 98 | 79 | 65 | 55 | 47 | 42 | 35 | 31 | 27 | 24 | 22 | 20 | 242 | 192 | 155 | 128 | 108 | 92 | 79 | 69 | 61 | 54 | 48 | 43 | 39 |
| 1000 | 130 | 103 | 83 | 69 | 58 | 49 | 42 | 37 | 32 | 29 | 26 | 23 | 21 | 255 | 202 | 163 | 135 | 113 | 97 | 83 | 72 | 64 | 56 | 50 | 45 | 41 |
| 1050 | 137 | 108 | 88 | 72 | 61 | 52 | 45 | 39 | 34 | 30 | 27 | 24 | 22 | 262 | 212 | 172 | 142 | 119 | 101 | 88 | 76 | 67 | 59 | 53 | 48 | 43 |
| 1100 | 143 | 113 | 92 | 76 | 64 | 54 | 47 | 41 | 36 | 32 | 28 | 25 | 23 | 281 | 222 | | 148 | 125 | 106 | 92 | 80 | 70 | 62 | 55 | 50 | 45 |
| 1150 | 150 | 118 | 96 | 79 | 66 | 57 | 49 | 42 | 37 | 33 | 30 | 26 | 24 | 293 | 232 | 188 | 155 | 130 | 111 | 96 | 83 | 73 | 65 | 58 | 52 | 47 |
| 1200 | 156 | 123 | 100 | 83 | 69 | 59 | 51 | 44 | 39 | 35 | 31 | 28 | 25 | 306 | 242 | 196 | 162 | 136 | 116 | 100 | 87 | 76 | 68 | 60 | 54 | 49 |
| 1250 | 163 | 129 | 104 | 86 | 72 | 62 | 53 | 46 | 41 | 36 | 32 | 29 | 26 | 319 | 252 | 204 | 169 | 142 | 121 | 104 | 91 | 80 | 71 | 63 | 56 | 51 |
| 1300 | 169 | 134 | 108 | 90 | 75 | 64 | 55 | 48 | 42 | 37 | 33 | 30 | 27 | 332 | 262 | 212 | 175 | 147 | 126 | 108 | 94 | 83 | 73 | 66 | 59 | 53 |
| 1350 | 176 | 139 | 112 | 93 | 78 | 66 | 57 | 50 | 44 | 39 | 35 | 31 | 28 | 344 | 272 | 220 | 182 | 153 | 130 | 112 | 98 | 86 | 76 | 68 | 61 | 55 |
| 1400 | 182 | 144 | 117 | 96 | 81 | 69 | 60 | 52 | 46 | 40 | 36 | 32 | 29 | 357 | 282 | 229 | 189 | 159 | 135 | 117 | 102 | 89 | 79 | 70 | 63 | 57 |
| 1450 | 189 | 149 | 121 | 100 | 84 | 71 | 62 | 54 | 47 | 42 | 37 | 33 | 30 | 370 | 292 | 237 | 196 | 164 | 140 | 121 | 105 | 92 | 82 | 73 | 66 | 59 |
| 1500 | 195 | 154 | 125 | 103 | 87 | 74 | 64 | 56 | 49 | 43 | 38 | 35 | 31 | 383 | 302 | 245 | 202 | 170 | 145 | 125 | 109 | 96 | 85 | 76 | 68 | 61 |
| 1550 | 202 | 159 | 129 | 107 | 90 | 76 | 66 | 57 | 50 | 45 | 41 | 36 | 32 | 396 | 312 | 253 | 209 | 176 | 150 | 129 | 112 | 99 | 88 | 78 | 70 | 63 |
| 1600 | 208 | 165 | 133 | 110 | 92 | 79 | 68 | 59 | 52 | 46 | 41 | 37 | 33 | 408 | 323 | 261 | 216 | 181 | 155 | 133 | 116 | 102 | 90 | 81 | 72 | 65 |
| 1650 | 215 | 170 | 138 | 114 | 95 | 81 | 70 | 61 | 54 | 48 | 42 | 38 | 34 | 421 | 333 | 270 | 223 | 187 | 159 | 138 | 120 | 105 | 93 | 83 | 75 | 67 |
| 1700 | 221 | 175 | 142 | 117 | 98 | 84 | 72 | 63 | 55 | 49 | 44 | 39 | 35 | 434 | 343 | 378 | 229 | 193 | 164 | 142 | 123 | 108 | 96 | 86 | 77 | 69 |
| 1750 | 228 | 180 | 146 | 120 | 101 | 86 | 74 | 65 | 57 | 50 | 45 | 40 | 36 | 447 | 353 | 286 | 236 | 198 | 169 | 146 | 127 | 112 | 99 | 88 | 79 | 71 |
| 1900 | 247 | 195 | 158 | 131 | 110 | 94 | 81 | 70 | 62 | 55 | 49 | 44 | 40 | 485 | 383 | 310 | 256 | 216 | 184 | 158 | 138 | 121 | 107 | 96 | 86 | 78 |
| 2000 | 260 | 206 | 167 | 138 | 116 | 99 | 85 | 74 | 65 | 58 | 51 | 46 | 42 | 510 | 403 | 327 | 270 | 227 | 193 | 167 | 145 | 128 | 113 | 101 | 90 | 82 |

[a] These load values may also be used for cantilevered pieces intermixed, combination simple span and two-span continuous, and two-span continuous layups.

[b] $2\frac{1}{2}$ in. net thickness. To determine allowable loads for $2\frac{5}{8}$ in. net thickness, multiply tabulated loads by 1.10.

[c] All spans to the right of the double line require special ordering of additional long lengths to assure that at least 20% of the decking is equal to the span length or longer.

[d] $3\frac{1}{2}$ in. net thickness.

[e] Duration of load, $C_D$ = 1.00 used in this table. For other duration of load, adjust by the appropriate factor.

[f] No increase for size effect has been applied ($C_F$ = 1.00). $F_b$ values have been previously adjusted.

[g] Dry condition of use.

## TABLE 7

### Three and Four Inch Nominal Thickness Allowable Roof Load Limited by Deflection Simple Span Layup

Allowable Uniformly Distributed Total Roof Load, psf

| Modulus of Elasticity psi | Deflection Limit[a] | 3 inch Thickness[b], Span, ft | | | | | | | | | 4 inch Thickness[c], Span, ft | | | | | | | | | | | | |
|---|---|---|---|---|---|---|---|---|---|---|---|---|---|---|---|---|---|---|---|---|---|---|---|
| | | 8 | 9 | 10 | 11 | 12 | 13 | 14 | 15 | 16 | 8 | 9 | 10 | 11 | 12 | 13 | 14 | 15 | 16 | 17 | 18 | 19 | 20 |
| 700,000 | $l/180$ | 63 | 44 | 32 | 24 | 19 | 15 | 12 | 10 | 8 | 174 | 122 | 89 | 67 | 51 | 40 | 32 | 26 | 22 | 18 | 15 | 13 | 11 |
| | $l/240$ | 47 | 33 | 24 | 18 | 14 | 11 | 9 | 7 | 6 | 130 | 91 | 67 | 50 | 38 | 30 | 24 | 20 | 16 | 14 | 11 | 10 | 8 |
| 800,000 | $l/180$ | 72 | 51 | 737 | 28 | 21 | 17 | 13 | 11 | 9 | 198 | 139 | 102 | 76 | 59 | 46 | 37 | 30 | 25 | 21 | 17 | 15 | 13 |
| | $l/240$ | 54 | 38 | 28 | 21 | 16 | 13 | 10 | 8 | 7 | 149 | 104 | 76 | 57 | 44 | 35 | 28 | 22 | 19 | 16 | 13 | 11 | 10 |
| 900,000 | $l/180$ | 81 | 57 | 42 | 31 | 24 | 19 | 15 | 12 | 10 | 223 | 157 | 114 | 86 | 66 | 52 | 42 | 34 | 28 | 23 | 20 | 17 | 14 |
| | $l/240$ | 61 | 43 | 31 | 23 | 18 | 14 | 11 | 9 | 8 | 167 | 118 | 86 | 64 | 50 | 39 | 31 | 25 | 21 | 17 | 15 | 13 | 11 |
| 1,000,000 | $l/180$ | 90 | 64 | 46 | 35 | 27 | 21 | 17 | 14 | 11 | 248 | 174 | 127 | 95 | 74 | 58 | 46 | 38 | 31 | 26 | 22 | 19 | 16 |
| | $l/240$ | 68 | 48 | 35 | 26 | 20 | 16 | 13 | 10 | 8 | 186 | 131 | 95 | 72 | 55 | 43 | 35 | 28 | 23 | 19 | 16 | 14 | 12 |
| 1,100,000 | $l/180$ | 99 | 70 | 51 | 38 | 29 | 23 | 19 | 15 | 12 | 273 | 192 | 140 | 105 | 81 | 64 | 51 | 41 | 34 | 28 | 24 | 20 | 17 |
| | $l/240$ | 75 | 52 | 38 | 29 | 22 | 17 | 14 | 11 | 9 | 205 | 144 | 105 | 79 | 61 | 48 | 38 | 31 | 26 | 21 | 18 | 15 | 13 |

| $E$ | Defl. | | | | | | | | | | | | | | | | | | | | | | |
|---|---|---|---|---|---|---|---|---|---|---|---|---|---|---|---|---|---|---|---|---|---|---|---|
| 1,200,000 | $l/180$ | 108 | 76 | 56 | 42 | 32 | 25 | 20 | 16 | 14 | 298 | 209 | 153 | 114 | 88 | 69 | 56 | 45 | 37 | 31 | 26 | 22 | 19 |
| | $l/240$ | 81 | 57 | 42 | 31 | 24 | 19 | 15 | 12 | 10 | 223 | 157 | 114 | 86 | 66 | 52 | 42 | 34 | 28 | 23 | 20 | 17 | 14 |
| 1,300,000 | $l/180$ | 117 | 83 | 60 | 45 | 35 | 27 | 22 | 18 | 15 | 322 | 227 | 165 | 124 | 96 | 75 | 60 | 49 | 40 | 34 | 28 | 24 | 21 |
| | $l/240$ | 88 | 62 | 45 | 34 | 26 | 21 | 16 | 13 | 11 | 242 | 170 | 124 | 93 | 72 | 56 | 45 | 37 | 30 | 25 | 21 | 18 | 15 |
| 1,400,000 | $l/180$ | 127 | 89 | 65 | 49 | 38 | 30 | 24 | 19 | 16 | 347 | 244 | 178 | 134 | 103 | 81 | 65 | 53 | 43 | 36 | 30 | 26 | 22 |
| | $l/240$ | 95 | 67 | 49 | 37 | 28 | 22 | 18 | 14 | 12 | 261 | 183 | 133 | 100 | 77 | 61 | 49 | 40 | 33 | 27 | 23 | 19 | 17 |
| 1,500,000 | $l/180$ | 136 | 95 | 69 | 52 | 40 | 32 | 25 | 21 | 17 | 372 | 261 | 191 | 143 | 110 | 87 | 69 | 56 | 47 | 39 | 33 | 28 | 24 |
| | $l/240$ | 102 | 71 | 52 | 39 | 30 | 24 | 19 | 15 | 13 | 279 | 196 | 143 | 107 | 83 | 65 | 52 | 42 | 35 | 29 | 25 | 21 | 18 |
| 1,600,000 | $l/180$ | 145 | 102 | 74 | 56 | 43 | 34 | 27 | 22 | 18 | 397 | 279 | 203 | 153 | 118 | 93 | 74 | 60 | 50 | 41 | 35 | 30 | 25 |
| | $l/240$ | 109 | 76 | 56 | 42 | 32 | 25 | 20 | 16 | 14 | 298 | 209 | 152 | 115 | 88 | 69 | 56 | 45 | 37 | 31 | 26 | 22 | 19 |
| 1,700,000 | $l/180$ | 154 | 108 | 79 | 59 | 46 | 36 | 29 | 23 | 19 | 422 | 296 | 216 | 162 | 125 | 98 | 79 | 64 | 53 | 44 | 37 | 31 | 27 |
| | $l/240$ | 115 | 81 | 59 | 44 | 34 | 27 | 22 | 17 | 14 | 316 | 222 | 162 | 122 | 94 | 74 | 59 | 48 | 40 | 33 | 28 | 24 | 20 |
| 1,800,000 | $l/180$ | 163 | 114 | 83 | 63 | 48 | 38 | 30 | 25 | 20 | 446 | 314 | 229 | 172 | 132 | 104 | 83 | 68 | 56 | 47 | 39 | 33 | 29 |
| | $l/240$ | 122 | 86 | 62 | 47 | 36 | 28 | 23 | 19 | 15 | 335 | 235 | 172 | 129 | 99 | 78 | 62 | 51 | 42 | 35 | 29 | 25 | 21 |

[a] For a deflection limit of $l/360$, use $\frac{1}{2}$ the tabulated value for a deflection limit of $l/180$.

[b] $2\frac{1}{2}$ in. net thickness. To determine allowable loads for $2\frac{5}{8}$ in. net thickness, multiply tabulated loads by 1.16.

[c] $3\frac{1}{2}$ in. net thickness.

[d] Duration of load, $C_D$ = 1.00 used in this table. For other duration of load, adjust by the appropriate factor.

[e] No increase for size effect has been applied ($C_F$ = 1.00). $F_b$ values have been previously adjusted.

[f] Dry condition of use.

## TABLE 8

### Three and Four Inch Nominal Thickness Allowable Roof Load Limited by Deflection Controlled Random Layup[a] (3 or more spans)

| Modulus of Elasticity psi | Deflection Limit[b] | Allowable Uniformly Distributed Total Roof Load, psf | | | | | | | | | | | | | | | | | | | | | | | | | |
| --- | --- | --- | --- | --- | --- | --- | --- | --- | --- | --- | --- | --- | --- | --- | --- | --- | --- | --- | --- | --- | --- | --- | --- | --- | --- | --- | --- |
| | | 3 inch Nominal Thickness[c], Span, ft[d] | | | | | | | | | | | | | 4 inch Nominal Thickness[c], Span, ft[d] | | | | | | | | | | | | |
| | | 8 | 9 | 10 | 11 | 12 | 13 | 14 | 15 | 16 | 17 | 18 | 19 | 20 | 8 | 9 | 10 | 11 | 12 | 13 | 14 | 15 | 16 | 17 | 18 | 19 | 20 |
| 700,000 | $l/180$ | 96 | 67 | 49 | 37 | 28 | 22 | 18 | 14 | 12 | 10 | 8 | 7 | 6 | 262 | 184 | 134 | 101 | 78 | 61 | 49 | 40 | 33 | 27 | 23 | 20 | 17 |
| | $l/240$ | 72 | 50 | 37 | 28 | 21 | 17 | 13 | 11 | 9 | 7 | 6 | 5 | 4 | 197 | 138 | 100 | 76 | 58 | 46 | 37 | 30 | 24 | 20 | 17 | 15 | 12 |
| 800,000 | $l/180$ | 109 | 77 | 56 | 42 | 32 | 25 | 20 | 16 | 14 | 11 | 10 | 8 | 7 | 300 | 210 | 154 | 115 | 89 | 70 | 56 | 45 | 37 | 31 | 26 | 22 | 19 |
| | $l/240$ | 82 | 58 | 42 | 32 | 24 | 19 | 15 | 12 | 10 | 8 | 7 | 6 | 5 | 225 | 162 | 115 | 86 | 67 | 52 | 42 | 34 | 28 | 23 | 20 | 17 | 14 |
| 900,000 | $l/180$ | 123 | 86 | 63 | 47 | 36 | 29 | 23 | 19 | 15 | 13 | 11 | 9 | 8 | 337 | 237 | 173 | 130 | 100 | 79 | 63 | 51 | 42 | 35 | 30 | 25 | 22 |
| | $l/240$ | 92 | 65 | 47 | 35 | 27 | 21 | 17 | 14 | 12 | 10 | 8 | 7 | 6 | 253 | 178 | 129 | 97 | 75 | 59 | 47 | 38 | 32 | 26 | 22 | 19 | 16 |
| 1,000,000 | $l/180$ | 136 | 96 | 70 | 52 | 40 | 32 | 25 | 21 | 17 | 14 | 12 | 10 | 9 | 374 | 263 | 192 | 144 | 111 | 87 | 70 | 57 | 47 | 39 | 33 | 28 | 24 |
| | $l/240$ | 102 | 72 | 52 | 39 | 30 | 24 | 19 | 16 | 13 | 11 | 9 | 8 | 7 | 281 | 197 | 144 | 108 | 83 | 65 | 52 | 43 | 35 | 29 | 25 | 21 | 18 |
| 1,100,000 | $l/180$ | 150 | 105 | 77 | 58 | 44 | 35 | 28 | 23 | 19 | 16 | 13 | 11 | 10 | 412 | 289 | 211 | 158 | 122 | 96 | 77 | 63 | 52 | 43 | 36 | 31 | 26 |
| | $l/240$ | 113 | 79 | 58 | 43 | 33 | 26 | 21 | 17 | 14 | 13 | 10 | 8 | 7 | 309 | 217 | 158 | 119 | 92 | 72 | 58 | 47 | 39 | 32 | 27 | 23 | 20 |
| 1,200,000 | $l/180$ | 164 | 115 | 84 | 63 | 49 | 38 | 31 | 25 | 20 | 17 | 14 | 12 | 10 | 449 | 316 | 230 | 173 | 133 | 105 | 84 | 68 | 56 | 47 | 39 | 34 | 29 |
| | $l/240$ | 123 | 86 | 63 | 47 | 36 | 29 | 23 | 19 | 15 | 13 | 11 | 9 | 8 | 337 | 237 | 173 | 130 | 100 | 79 | 63 | 51 | 42 | 35 | 30 | 25 | 22 |

| | | | | | | | | | | | | | | | | | | | | | | | | | | |
|---|---|---|---|---|---|---|---|---|---|---|---|---|---|---|---|---|---|---|---|---|---|---|---|---|---|---|
| 1,300,000 | $l/180$ | 177 | 125 | 91 | 68 | 53 | 41 | 33 | 27 | 22 | 18 | 16 | 13 | 11 | 487 | 342 | 249 | 187 | 144 | 114 | 91 | 74 | 61 | 51 | 43 | 36 | 31 |
| | $l/240$ | 133 | 93 | 68 | 51 | 39 | 31 | 25 | 20 | 17 | 14 | 12 | 10 | 9 | 365 | 256 | 187 | 140 | 108 | 85 | 68 | 55 | 46 | 38 | 32 | 27 | 23 |
| 1,400,000 | $l/180$ | 191 | 134 | 98 | 73 | 57 | 45 | 36 | 29 | 24 | 20 | 17 | 14 | 12 | 524 | 368 | 269 | 202 | 155 | 122 | 98 | 80 | 66 | 55 | 46 | 39 | 34 |
| | $l/240$ | 143 | 101 | 73 | 55 | 42 | 33 | 27 | 22 | 18 | 15 | 13 | 11 | 9 | 393 | 276 | 201 | 151 | 117 | 92 | 73 | 60 | 49 | 41 | 35 | 29 | 25 |
| 1,500,000 | $l/180$ | 205 | 144 | 105 | 79 | 61 | 48 | 38 | 31 | 26 | 21 | 18 | 15 | 13 | 562 | 395 | 288 | 216 | 166 | 131 | 105 | 85 | 70 | 59 | 49 | 42 | 36 |
| | $l/240$ | 154 | 108 | 79 | 59 | 46 | 36 | 29 | 23 | 19 | 16 | 13 | 11 | 10 | 421 | 296 | 216 | 162 | 125 | 98 | 79 | 64 | 53 | 44 | 37 | 31 | 27 |
| 1,600,000 | $l/180$ | 218 | 153 | 112 | 84 | 65 | 51 | 41 | 33 | 27 | 23 | 19 | 16 | 14 | 599 | 421 | 307 | 230 | 178 | 140 | 112 | 91 | 75 | 62 | 53 | 45 | 38 |
| | $l/240$ | 164 | 115 | 84 | 63 | 49 | 38 | 31 | 25 | 20 | 17 | 14 | 12 | 10 | 449 | 316 | 230 | 173 | 133 | 105 | 84 | 68 | 56 | 47 | 39 | 34 | 29 |
| 1,700,000 | $l/180$ | 232 | 163 | 119 | 89 | 69 | 54 | 43 | 35 | 29 | 24 | 20 | 17 | 15 | 636 | 447 | 326 | 245 | 189 | 148 | 119 | 97 | 80 | 66 | 56 | 48 | 41 |
| | $l/240$ | 174 | 122 | 89 | 67 | 52 | 41 | 32 | 26 | 22 | 18 | 15 | 13 | 11 | 478 | 335 | 245 | 184 | 142 | 111 | 89 | 72 | 60 | 50 | 42 | 36 | 31 |
| 1,800,000 | $l/180$ | 246 | 173 | 126 | 94 | 73 | 57 | 46 | 37 | 31 | 26 | 22 | 18 | 16 | 674 | 474 | 345 | 259 | 200 | 157 | 126 | 102 | 84 | 70 | 59 | 50 | 43 |
| | $l/240$ | 184 | 129 | 94 | 71 | 55 | 43 | 34 | 28 | 23 | 19 | 16 | 14 | 12 | 506 | 355 | 259 | 195 | 150 | 118 | 94 | 77 | 63 | 53 | 44 | 38 | 32 |

[a]To determine allowable uniformly distributed total roof for other span conditions, multiply controlled random layup load values by the following factors: Cantilevered pieces intermixed—0.90; Combination simple span and two-span continuous—1.13; Two-span continuous—1.59.

[b]For a deflection limit of $l/360$, use $\frac{1}{2}$ the tabulated value for a deflection limit of $l/180$.

[c]$2\frac{1}{2}$ in. net thickness. To determine allowable loads for $2\frac{5}{8}$ in. net thickness, multiply tabulated loads by 1.16.

[d]All spans to the right of the double line require special ordering of additional long lengths to assure that at least 20% of the decking is equal to the span length or longer.

[e]$3\frac{1}{2}$ in. net thickness.

of $\frac{2}{3}$ for 2 inch and 80% for 3 and 4 inch decking were selected after careful evaluation of tests and previous experience.

*9.3.4*    When controlled random layup as specified herein is used for unequal spans, non-uniform loading, cantilever action, or conditions other than covered herein by the tabulated values, the same adjustment factors should be applied to the moment of inertia used in standard engineering formulas representing the actual conditions of load and span.

**9.4**    The allowable loads given in Tables 4 and 5 are based on a maximum moisture content of 15% for 2 inch decking. The allowable loads given in Tables 6 through 8 are based on a maximum moisture content of 19% for 3 and 4 inch decking. If the maximum moisture content is limited to 15% for 3 and 4 inches decking, the allowable bending stress valued given in Table 7 may be multiplied by 1.08 and the modulus elasticity values in Tables 7 and 8 may be multiplied by 1.05.

**9.5**    Allowable load values given in Tables 4 and 6 are based on normal duration of loading. If decking is used for purposes where other durations of load control, increase the tabulated values by multiplying by 2 appropriate duration of load factor $C_D$ as follows:

|   |   |
|---|---|
| 0.9 | for permanent load; |
| 1.15 | for 2 months duration, as for snow; |
| 1.25 | for 7 days duration; |
| 1.6 | for wind or earthquake; or |
| 2.0 | for impact |

These increases are not cumulative.

**9.6**    The allowable load tables are for total uniformly distributed vertical loads, including dead and live, in pounds per square foot on a horizontal roof surface. When roofs have only a moderate slope (3 in 12 or less), dead and live load may be added together without adjustment for slope of roof.

**9.7**    For steeper sloping roofs, it is customary to adjust the load so as to express them in terms of square feet of roof surface. (See Figures 12 and 13.) For example, 10 lb dead load (6.7 lb for deck and 3.3 lb for roofing) is the vertical load of one square foot of sloping roof surface. Snow load is usually expressed in pounds per square foot of the horizontal projection of the sloping roof surface. Therefore, the vertical snow load must be converted to the vertical psf load of sloping roof surface. For example, a 60 psf snow load on the horizontal projection is equivalent to a vertical load of 46 psf on a 10 in 12 sloping roof surface. This combined with 10 psf dead load results in a total vertical load of 56 psf on the 10 in 12 sloping roof surface. The 56 psf total vertical load may then be converted to two components, one perpendicular or normal to the roof surface, the vertical load of 56 psf is equivalent to a component perpendicular to the roof of 43 psf and a component parallel to the roof of 37 psf.

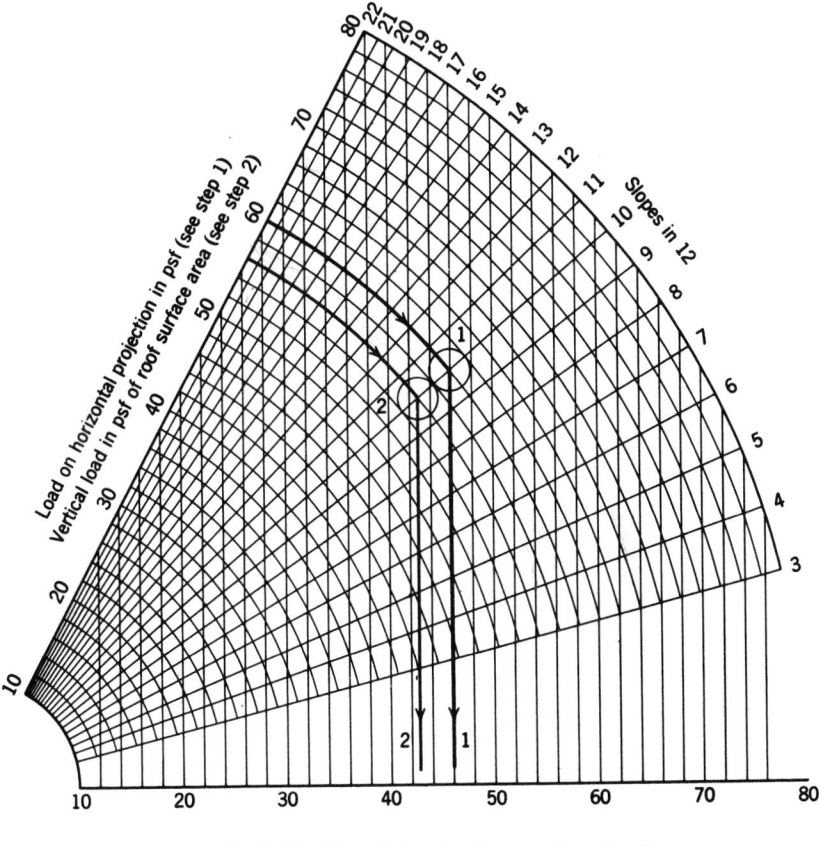

**FIGURE 12.** Load conversion. Example: 60 psf live load and 10 psf dead load on 10 in 12 slope. Step 1: 60 psf live load on horizontal projection equals 46 psf of roof surface area vertical load on 10 in 12 roof slope. Step 2: 10 psf of roof surface area dead load plus 46 psf of roof surface area live load equals 56 psf of roof surface area combined load acting vertically; 56 psf of roof surface area vertical total load equal 43 psf normal to roof causing bending and deflection.

**9.8** Where decking is laid with the longitudinal axis parallel to the slope, the component perpendicular to the roof surface will produce bending and deflection; the parallel component will produce compression. The design value for compression parallel to grain may be taken as that of No. 2 structural joists and planks grade for the species. The decking must be designed for bending and axial stresses as well as deflection.

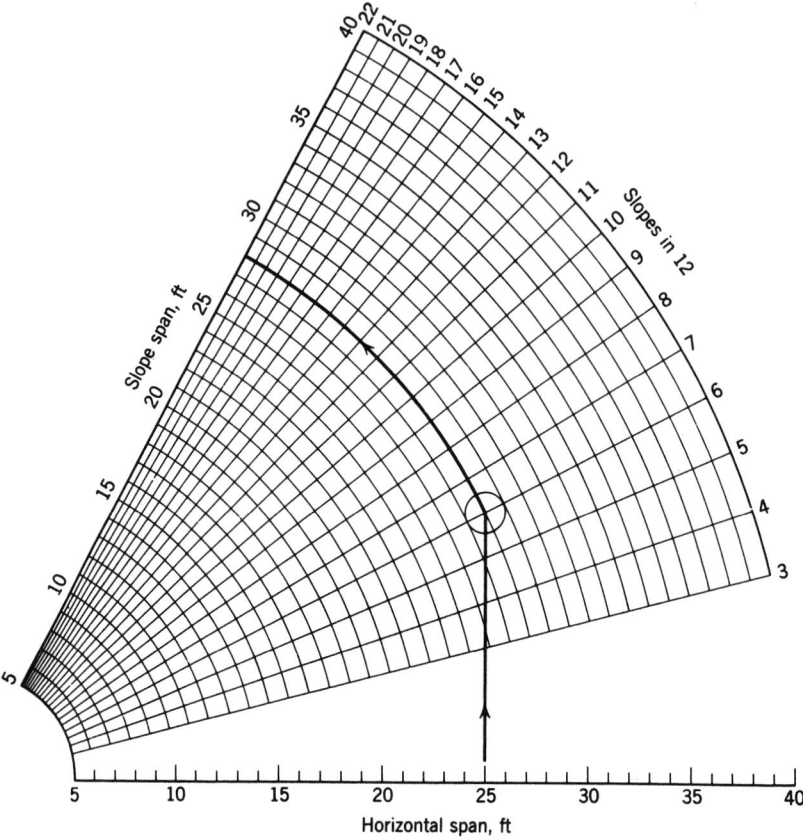

**FIGURE 13.**   Span conversion. Example: 25 ft horizontal spans equals 28 ft slope open when slope is 6 in 12. Use 28 ft in determining footage.

**9.9**   Where decking is laid with the longitudinal axis perpendicular to the slope, the component perpendicular to the roof surface produces bending and deflection; the parallel component, as may be induced by wind forces, is taken by diaphragm action.

AMERICAN INSTITUTE
7012 South Revere Parkway
Suite 140

TIMBER CONSTRUCTION
Englewood, Colorado 80112
Telephone 303-792-9559

# AITC 113-93
# STANDARD FOR DIMENSIONS OF
# STRUCTURAL GLUED LAMINATED TIMBER

*Adopted as Recommendations April 21, 1993*
*Copyright 1993 by American Institute of Timber Construction*

## CONTENTS

## 1. PREFACE

**1.1** The most efficient and economical production of glued laminated structural members results when standard lumber sizes are used for the laminations. Industry recommended practice uses nominal 2 in. thick lumber of standard nominal width to produce straight members and curved members where the radius of curvature is within the bending radius limits for that thickness of the species. Nominal 1 in. thick boards are normally used when the bending radius is too sharp to permit use of nominal 2 in. thick laminations. These are standard practices subject to deviation to conform with specific job requirements and plant procedures. The use of nominal 1 in. and 2 in. thick laminations will generally be the most economical, and therefore, conformance with this standard is recommended for all normal uses. Exceptions should be made only when the shape of the structure requires nonstandard laminations. Textured surfaces for glued laminated timber are permitted in *Standard Appearance Grades for Structural Glued Laminated Timber*, AITC 110, in lieu of the surfaces specified in the AITC appearance grades. When textured surfaces are used, the net finished sizes and tolerances given herein and in AITC 117—Design and ANSI/AITC 190.1-1992, *Structural Glued Laminated Timber*, may not be applicable. Depending upon the degree of texturing, it may be necessary for the designer to compensate for the resulting loss of cross section.

## 2. STANDARD DEPTHS OF MEMBERS

**2.1**   Proper gluing procedures require surfaces planed uniformly smooth to exact thickness. Normal standard practice is to surface nominal 2 in. laminations to a net $1\frac{3}{8}$ in. or $1\frac{1}{2}$ in. thickness, and nominal 1 in. laminations to a net $\frac{3}{4}$ in. thickness. Finished depths of members are thus increments of these net thicknesses.

| | Net Depth of Member, in. | | |
|---|---|---|---|
| | | Nominal 2 in. Laminations | |
| No. of Laminations | Nominal 1 in. Laminations | $1\frac{1}{2}$ in.[1] | $1\frac{3}{8}$ in.[1] |
| 4 | 3 | 6 | $5\frac{1}{2}$ |
| 5 | $3\frac{3}{4}$ | $7\frac{1}{2}$ | $6\frac{7}{8}$ |
| 6 | $4\frac{1}{2}$ | 9 | $8\frac{1}{4}$ |
| 7 | $5\frac{1}{4}$ | $10\frac{1}{2}$ | $9\frac{5}{8}$ |
| 8 | 6 | 12 | 11 |
| Etc. | Etc. | Etc. | Etc. |

[1]$1\frac{1}{2}$ in. thick laminations are normal for Western softwoods; $1\frac{3}{8}$ in. thick laminations are normal for Southern Pine.

**2.2**   The use of laminations of special thicknesses because of bending radius or the mixing of thicknesses for special purposes results in net finished depths other than those shown in the table.

## 3. STANDARD WIDTHS OF MEMBERS

**3.1**   It is necessary to surface the wide faces of members to remove the glue squeezeout and provide a uniformly smooth surface. Therefore, the net finished width of the glued laminated member is less than the net finished width of industry standard boards and dimension.

**3.2**   The normal standard net finished widths for glued laminated structural members are as follows. Other finished widths may be used to meet the size requirements of a design or to meet other special requirements.

| Nominal Width, in. | 3 | 4 | 6 | 8 | 10 | 12 | 14 | 16 |
|---|---|---|---|---|---|---|---|---|
| Net Finished width, in. $1\frac{1}{2}$ in. thick laminations | $2\frac{1}{8}$ | $3\frac{1}{8}$ | $5\frac{1}{8}$ | $6\frac{3}{4}$ | $8\frac{3}{4}$ | $10\frac{3}{4}$ | $12\frac{1}{4}$ | $14\frac{1}{4}$ |
| $1\frac{3}{8}$ in. thick laminations | $2\frac{1}{8}$ | 3 or $3\frac{1}{8}$ | 5 or $5\frac{1}{8}$ | $6\frac{3}{4}$ | $8\frac{1}{2}$ | $10\frac{1}{2}$ | 12 | 14 |

Note: $3\frac{1}{8}$ in. and $5\frac{1}{8}$ in. widths are normal for Western softwoods. 3 in. and 5 in. widths are normal for Southern Pine.

## 4. STANDARD DIMENSIONS FOR HEAVY TIMBER

**4.1** Excellent fire resistance is achieved with "heavy timber" construction (see *Standard for Heavy Timber Construction*, AITC 108). Minimum sawn lumber sizes have been long established and are expressed in nominal dimensions and assume surfacing to *American Lumber Standard* net sizes.

**4.2** For "heavy timber" construction, the net width of glued laminated structural members shall be the standard glued laminated net width for the nominal sawn width specified, and the net depth of glued laminated structural members shall be equal to or greater than the net finished depth specified in the following table.

| | | Minimum Glued Laminated Net Size | | | |
|---|---|---|---|---|---|
| Minimum Nominal Size | | $1\frac{1}{2}$ in. thick laminations | | $1\frac{3}{8}$ in. thick laminations | |
| Width, in. | Depth, in. | Width, in. | Depth, in. | Width, in. | Depth, in. |
| 8 × 8 | | $6\frac{3}{4}$ × 9 | | $6\frac{3}{4}$ × $8\frac{1}{4}$ | |
| 6 × 10 | | $5\frac{1}{8}$ × $10\frac{1}{2}$ | | 5 or $5\frac{1}{8}$ × 11 | |
| 6 × 8 | | $5\frac{1}{8}$ × 9 | | 5 or $5\frac{1}{8}$ × $8\frac{1}{4}$ | |
| 6 × 6 | | $5\frac{1}{8}$ × 6 | | 5 or $5\frac{1}{8}$ × $6\frac{7}{8}$ | |
| 4 × 6 | | $3\frac{1}{8}$ × $7\frac{1}{2}$ | | 3 or $3\frac{1}{8}$ × $6\frac{7}{8}$ | |

## 5. TOLERANCES

**5.1 Dimensions** The tolerances permitted at the time of manufacture shall be as follows:

Width—Plus or minus $\frac{1}{16}$ in.

Depth—Plus $\frac{1}{8}$ in. per ft of depth. Minus $\frac{1}{8}$ in., or $\frac{1}{16}$ in. per ft of depth, whichever is larger.

Length—Up to 20 ft, plus or minus $\frac{1}{16}$ in. Over 20 ft, plus or minus $\frac{1}{16}$ in. per 20 ft of length.

**5.2 Camber or Straightness** The tolerances are applicable to the time of manufacture without allowance for dead load deflection. Up to 20 ft, the tolerance is plus or minus $\frac{1}{4}$ in. Over 20 ft, increase tolerance $\frac{1}{8}$ in. per each additional 20 ft or fraction thereof, but not to exceed $\frac{3}{4}$ in.

The tolerances are intended for use with straight or slightly cambered members and are not applicable to curved members such as arches.

**5.3 Squareness of Cross Section** The tolerance shall be within plus or minus $\frac{1}{8}$ in. per ft of specified depth unless a specially shaped section is specified. Squareness is measured by placing one side of a square along a top or bottom face and determining the offset from the other side of the square to the side of the member.

AMERICAN INSTITUTE
7012 South Revere Parkway
Suite 140

TIMBER CONSTRUCTION
Englewood, Colorado 80112
Telephone 303-792-9559

# AITC 117-93—DESIGN
# STANDARD SPECIFICATIONS FOR
# STRUCTURAL GLUED LAMINATED TIMBER
# OF SOFTWOOD SPECIES

*Adopted as Recommendations, April 21, 1993*
*Copyright 1987, 1993 by American Institute of Timber Construction*

## Contents

## PREFACE

These specifications consolidate, expand and update previously issued laminating specifications and supplements related to specific species or mixtures of species. They represent the latest research available from the U.S. Forest Products Laboratory, various colleges and universities and the American Institute of Timber Construction. With these specifications a designer can specify the required stress levels for a glued laminated timber member. It is the responsibility of a glued laminated timber manufacturer to produce a member with design values that meet or exceed those requirements. When the design stress level allows a choice, manufacturers will select laminating combinations to fit their varying new material supplies, thus better utilizing available forest resources.

A separate publication for the manufacture of glued laminated timber, AITC 117-92—*MANUFACTURING* (Reference 1), has been developed based on ASTM D 3737, *Standard Method for Establishing Stresses for Structural Glued Laminated Timber (Glulam)* (Reference 2), as modified by subsequent research and by American National Standard (ANSI/AITC A 190.1-1992, *Structural Glued Laminated Timber* (Reference 3).

These specifications contain data relating to design values and modification of stresses for the design of glued laminated timber members. They are, however, neither a design manual nor an engineering textbook. For additional design information see the AITC *Timber Construction Manual* (Reference 4).

## 1. GENERAL

### 1.1 Structural Glued Laminated Timber

***1.1.1*** The term *structural glued laminated timber* as employed herein refers to an engineered, stress-rated product of a timber laminating plant, comprising assemblies of suitably selected and prepared wood laminations bonded together with adhesives. The grain of all laminations is approximately parallel longitudinally.

***1.1.2*** The individual laminations shall not exceed 2 in. in net thickness. They may be comprised of pieces end joined to form any length, of pieces placed or glued edge to edge to make any width, or of pieces bent to curved form during gluing.

***1.1.3*** These specifications are applicable to glued laminated timbers with the numbers of laminations indicated in Tables 1 and 2.

***1.1.4*** The production of structural glued laminated timber under these specifications shall be in accordance with AITC 117—*MANUFACTURING*, and American National Standard ANSI/AITC A 190.1-1992, *Structural Glued Laminated Timber.*

***1.1.5*** End joints in laminated timber combinations listed herein may be plain scarf joints, finger joints or other types which qualify for the design values in accordance with the procedures in American National Standard ANSI/AITC A 190.1-1992, *Structural Glued Laminated Timber*, and AITC 117—*MANUFACTURING.*

*1.1.6* The design of glued laminated members and their fastenings should be in accordance with the provisions of these specifications and the *Timber Construction Manual.*

## 1.2 Design Values

*1.2.1* Some of the design values contained herein have been developed by AITC using procedures developed with analytical studies confirmed by full-scale load tests.

## 1.3 Species

*1.3.1* The softwood species most commonly used for laminating are included in these specifications. These are the Western species: Douglas Fir-Larch (DF), Douglas Fir South (DFS), Hem-Fir (HF), Eastern Spruce (ES), Alaska Cedar (AC) and Softwood Species (WW), and the Southern Pine species (SP). Softwood Species and Eastern Spruce are included in the general category of Western species although Eastern Spruce and some Softwood Species are produced in other areas. Canadian Spruce-Pine (CSP) is also used. For species names included in the softwood species (WW), see AITC 117—Mfg., Section 2.

## 1.4 Specification of Design Values

*1.4.1 Principal Stress—Bending.* Table 1 is applicable to members consisting of 4 or more laminations stressed primarily in bending with the load applied perpendicular to the wide faces of the laminations. Table 1 contains 2 groups of species—Western species and Southern Pine species. The table includes combinations manufactured from visually graded lumber and combinations manufactured from $E$-rated lumber. There are four groupings of bending stress ($F$) levels with a number of options within groupings to give the same bending stress, but with some variations in the other design values shown for each option. Many designs can utilize more than one of the options listed with an $F_b$ grouping. Where these other design values ($F_t$, $F_c$, $F_{c\perp}$, $F_v$, and $E$) become critical in design, the designer should specify the stresses as required by design. Obviously, the specifying of values that are much higher than actually required will eliminate certain combinations and may result in a member that is not as readily available as would otherwise be the case. The arbitrary selection of the highest possible design values in all stress categories may result in a member impossible to manufacture under these specifications. It is also possible for the designer to specify a given combination that meets the design requirements, but this may limit availability. Note that increased values for horizontal shear and compression perpendicular to the grain for some combinations can be obtained by so specifying. See the footnotes following Table 1.

*1.4.1.1* The design values in Table 1 are for loads applied perpendicular to the wide faces of the laminations which is the most common direction of loading for glued laminated timbers. For convenience, the design values for loads applied perpendicular to the wide faces of the laminations causing bending about the $X$-$X$ axis are designated in the table by the subscript $X$. Two columns of design values are shown in Table 1 for bending with the load applied perpendicular to the wide faces of the laminations ($F_{bx}$). The first (column 3) is for the most common use of bending members where the tension portion of the bending stress occurs on the face of the member containing the tension zone laminations. The second (column 4) is for use where the face of the member containing the compression zone laminations is stressed in tension, such as short overhang on a simple beam. For continuous beams or beams cantilevered over a support where high tensile stresses can exist on both the top and bottom of a member, see 1.4.1.3.

*1.4.1.2*  The design values for members stressed in bending about the *Y-Y* axis (loads applied parallel to the wide faces of the laminations) and members axially loaded are shown in Table 1. The design values for loads applied parallel to the wide faces of the laminations causing bending about the *Y-Y* axis are designated by the subscript *Y*. Neither the *X* nor *Y* subscripts are commonly used in wood references or textbooks.

*1.4.1.3*  The design values in bending with the load applied perpendicular to the wide faces of the laminations ($F_{bx}$) listed in column 3, Table 1, are for the most common installation of the member as a simple beam. This implies compressive stress occurring at the top of the member and tensile stress occurring at the bottom or soffit of the member (positive moment). For conditions where the beam support configuration and/or loading pattern produce negative moment which becomes significant and the resulting tensile stress on the top of the member exceeds the minimum design values listed in column 4 for the compression zone in tension (1500 psi for 30*F*; 1400 psi for 28*F*; 1300 psi for 26*F*; 1200 psi for 24*F* combinations, 1100 psi for 22*F* combinations), *Tension Zone* grade requirements, including ending joint spacing, must be applied to the top zones of the member so that the basic design values for bending listed in column 3, Table 1, may be utilized. A bending tensile stress in the negative moment area 200 psi higher than that tabulated in column 4 can be obtained by applying only tension zone end joining spacing restrictions to both top and bottom of the member. When specified with *Tension Zone* requirements both top and bottom, the design values in bending ($F_{bx}$) listed in column 3, Table 1, apply to either positive or negative moment locations that the values listed in column 4, Table 1 (1200 psi for 24*F*; 1100 psi for 22*F*; 1000 psi for 20*F*; and 800 psi for 16*F* combinations) should be identified by the designer. The manufacturer will then provide *Tension Zone* laminations in this area as required by the designer.

*1.4.1.4*  Balanced combinations for bending members which have equal or nearly equal positive and negative bending moments are included in Table 1.

*1.4.1.5*  The combinations in Table 2 are usually best suited for members with bending stresses caused by loads applied parallel to the wide faces of the laminations. Design values are also shown for members loaded perpendicular to the wide faces of the laminations. In addition, Table 2 also contains combinations for members with 2 or 3 laminations. These combinations are applicable to members loaded either perpendicular or parallel to the wide faces of the laminations.

*1.4.1.6*  The design values in bending about the *X-X* axis ($F_{bx}$) in Column 3, Table 1 and Column 17, Table 2 are based on the use of special tension laminations for most combinations in Table 1 and all combinations in Table 2 when these combinations are used for bending members. Special tension laminations are not required for arches. Special tension laminations for Table 1 combinations may be omitted from bending members provided the tabular design values for bending ($F_{bxx}$) are multiplied by 0.75 for members greater than 15 inches in depth, or by 0.85 for members up to 15 inches in depth. For Table 2 combinations, bending members 15 inches and less in depth, use design values for bending in Column 16, Table 2, or for members greater than 15 inches in depth, multiply the design value in bending by 0.75.

*1.4.2 Principle Stress—Axial.* Table 2 contains combinations for members stressed primarily in axial tension or compression.

*1.4.3* Members subjected to combined axial and bending stresses—When a combination of axial and bending stresses exists in a member, they should be checked by the interaction formula as shown in the *Timber Construction Manual.* The designer should specify the required tabular design values in bending, $F_b$, and compression parallel to grain, $F_c$, or tension parallel to grain, $F_t$; however, the stresses specified should be available in a single combination. When the predominant stress is bending, the combinations in Table 1 may be more appropriate. The required tabular design values for axial and bending stresses should be specified regardless of whether the combination has been specified.

## 2. LUMBER

**2.1** Lumber shall be of any species or grade shown herein.

## 3. ADHESIVES

**3.1** Adhesives used shall comply with the specifications contained in American National Standard ANSI/AITC A190.1-1992, *Structural Glued Laminated Timber.*

**3.2** Wet-use adhesives may be specified for all moisture conditions but are required when the moisture content exceeds 16% for repeated or prolonged periods of service or when the wood is preservatively treated either before or after gluing.

## 4. DESIGN VALUES

### 4.1 General

*4.1.1* For the design values given herein, or modifications thereof, lumber of the grades required shall be assembled in accordance with the zone requirements indicated in AITC 117-92—*MANUFACTURING.*

*4.1.2* The design values given herein and the modifications required for other conditions of use and loading are also applicable to structural glued laminated timbers that have been pressure impregnated by an approved preservative process in accordance with *Standard for Preservative Treatment of Structural Glued Laminated Timber,* AITC 109 (Reference 5).

*4.1.3* The design values for fire retardant treated glued laminated timber, treated before or after gluing, are dependent upon the species and treatment combinations involved. The effect on strength must be determined for each treatment; however, indications are that a 10 to 25% reduction in bending stress is applicable. The manufacturer of the treatment should be contacted for more specific information on stress adjustments for all design values.

*4.1.4* The design values given herein are for normal durations of loading. Modifications for other durations of loading are given in 4.4.1.

*4.1.5* The design values in bending, $F_b$, given herein apply to a 12 in. deep member, uniformly loaded with a span to depth ratio of 21. Modifications for other durations of loading are given in 4.4.2 and in footnote f to Table 2.

*4.1.6* The modulus of elasticity, $E$, values herein are the average values for the combination shown and reflect the effect of grade. The modulus of elasticity of wood of a given species is somewhat variable. The coefficient of variation (C.O.V.) of visually graded lumber of the same species is approximately 0.25 for species used in laminating. Tests and experience have shown that this variability is considerably reduced by the laminating effect. For glued laminated timber made from 4 laminations of visually graded lumber, the C.O.V. is approximately 0.15, for 10 laminations 0.10 and for 16 or more laminations 0.08. The variation in modulus of elasticity is especially important in designs where stiffness is of prime importance such as in the design of long columns, lateral stability calculations or in calculations for ponding.

A standard deviation is the average value multiplied by the coefficient of variation. In a normal frequency distribution, about $\frac{2}{3}$ of the individual values will be within one standard deviation (above and below) the average value. Also about 95% of the individual values will be within two standard deviations of the average value. Thus, if a combination of glued laminated timber has been an average $E$ of 1,700,000 psi and the coefficient of variation is 0.10, $\frac{2}{3}$ of the members could be expected to have values between 1,360,000 psi and 2,040,000 psi.

In a case where only the lower portion of the variation in $E$ is of engineering importance, similar useful interpretations are possible. In a normal frequency distribution, $\frac{5}{6}$ of the individual values lie above a value located at one standard deviation below the mean (1,530,000 psi in the above example). In the same distribution, 95% of the individual values lie above a value located at 1.645 standard deviations below the mean (1,420,000 psi in the above example).

*4.1.6.1* The tabulated $E$ values shown for bending about the $X-X$ axis of members in Table 1 are higher than those tabulated for bending about the $Y-Y$ axis because the laminations in the outer zones have higher $E$ values than those in the inner zones.

*4.1.6.2* The modulus of elasticity values for bending members listed in Tables 1 and 2 are based on a span to depth ratio of approximately 21 and include an adjustment for shear deflection. These $E$ values can be used for determining deflection for most designs without the necessity of calculating the shear deflection.

*4.1.7* The tabulated compression perpendicular to grain design values in Tables 1 and 2 are based on a deformation limit of 0.04 in. obtained when testing in accordance with the standard method ASTM D 143 for compression perpendicular to grain. In special applications where deformation may be critical, use of a reduced compression perpendicular to grain design value may be appropriate. The following equation may be used for a deformation of 0.02 in. which is 50% of that associated with the values tabulated in Tables 1 and 2.

$$F_{c\perp(0.02)} = 0.73 F_{c\perp}$$

where

$F_{c\perp(0.02)}$ = compression perpendicular to grain at 50% of deformation limit associated with tabulated $F_{c\perp}$ values (0.02 in.)

and

$F_{c\perp}$ = compression perpendicular to grain at 0.04 in. deformation limit.

### 4.2 Radial Tension or Compression

**4.2.1** When a curved member is loaded in bending, radial stresses are induced.

**4.2.2** When the bending moment ($M$) is in the direction tending to increase curvature (decrease the radius), the radial stress is compression across the grain ($F_{rc}$). The design value in radial compression ($F_{rc}$) is equal to the design value in compression perpendicular to grain ($F_{c\perp}$) of the grade and species being used. (For Douglas Fir-Larch, Douglas Fir South, and Southern Pine, use $F_{rc} = 560$ psi; for Hem-Fir use $F_{rc} = 375$ psi; for Eastern Spruce use $F_{rc} = 300$ psi; for Alaska Cedar use $F_{rc} = 470$ psi; for Canadian Spruce-Pine use $F_{rc} = 560$ psi; and for Softwood Species (WW) use $F_{rc} = 255$ psi.) These radial compression values are not subject to the duration of load modifications in 4.4.1.

**4.2.3** When $M$ is in the direction tending to decrease curvature (increase the radius), the radial stress is tension across the grain. The design value in radial tension perpendicular to grain ($F_{rt}$) shall be limited to $\frac{1}{3}$ the design value in horizontal shear for Southern Pine for all load conditions and for Douglas Fir-Larch, Douglas Fir South, Hem-Fir, Eastern Spruce, Alaska Cedar, Canadian Spruce-Pine, and Softwood Species (WW) for wind or earthquake loads. The limit shall be 15 psi for Douglas Fir-Larch, Hem-Fir, Eastern Spruce, Alaska Cedar, Canadian Spruce-Pine, and Softwood Species (WW) for other types of loading. These values are subject to modifications for duration of load and wet conditions of use. For wet conditions of use, the wet-use factor for radial tension is 0.875. If these values are exceeded, mechanical reinforcing shall be used and shall be sufficient to resist all radial tension stresses. For Douglas Fir-Larch, Douglas Fir South, Hem-Fir, Eastern Spruce, Alaska Cedar, Canadian Spruce-Pine, and Softwood Species (WW) where mechanical reinforcement is provided to resist all radial tension stresses, the calculated radial tension stress shall not exceed $\frac{1}{3}$ the design value in horizontal shear. When mechanical reinforcing is used, the maximum moisture content of the laminations at the time of manufacture shall not exceed 12% for dry conditions of use.

### 4.3 Conditions of Use

**4.3.1** Dry conditions of use design values shall be applicable when the moisture content in service is less than 16%, as in most covered structures.

**4.3.2** Wet conditions of use design values shall be applicable when the moisture content in service is 16% or more, as may occur in members directly exposed to precipitation or in covered locations of high relative humidity.

### 4.4 Adjustment of Design Values

#### 4.4.1 Duration of Load

**4.4.1.1** Normal load duration contemplates fully stressing a member to the design value by the application of the full design load for a duration of approximately 10 years (applied either continuously or cumulatively). Tabular design values are based on normal load duration ($C_D = 1.0$).

**4.4.1.2** When the duration of load is other than that for normal load duration the tabular design values, except for modulus of elasticity $E$ and compression perpendicular to grain $F_{c\perp}$, are adjusted by the duration of load factor $C_D$ as shown in 4.4.1.3.

**4.4.1.3** The duration of load factors $C_D$ are shown below:

| Load Duration | $C_D$ | Typical Design Loads |
|---|---|---|
| Permanent | 0.9 | Dead Load |
| Ten Years | 1.0 | Occupany Live Load |
| Two Moneths | 1.15 | Snow Load |
| Seven Days | 1.25 | Construction Load |
| Ten Minutes | 1.6 | Wind/Earthquake Load |
| Impact* | 2.0 | Impact Load |

*The impact load duration factor shall not apply to glued laminated timber members preservatively treated with waterborne preservatives to the heavy retention required for marine exposure, nor to members pressure treated with fire retardant chemicals.

#### 4.4.2 Volume Factor, $C_v$

**4.4.2.1** The volume factor, $C_v$, as shown in AITC Tech. Note 21 (Reference 6) shall be used. The tabular design values in bending about the $x$–$x$ axis are based on a simple span member 12 in. deep, $5\text{-}\frac{1}{8}$ in. wide, 21 ft in length and loaded with a uniform load. When a different size member is used or a different loading condition exists, the tabular design value $F_{bx}$ is multiplied by the volume factor $C_v$ calculated as follows:

$$C_v = K_L[(5.125/b)^{1/x}(12/d)^{1/x}(21/L)^{1/x}] \leq 1.0$$

in which

$K_L$ = loading condition coefficient (see table shown below),
$b$ = width (breadth) of bending member, inches. For multiple piece width layups, $b$ = width of widest piece used in the layup. Thus, $b \leq 10.75$ in.,

$d$ = depth of bending member, in.,
$L$ = length of bending member between points of zero moment, ft,
$x$ = 20 for Southern Pine,
$x$ = 10 for Western Species

| Single Span Beam | $K_L$ |
|---|---|
| Concentrated load at midspan | 1.09 |
| Uniformly distributed load | 1.00 |
| Two equal concentrated loads at $\frac{1}{3}$ points of span | 0.96 |
| Continuous Beam or Cantilever | |
| All loading conditions | 1.00 |

For more information regarding the volume factor and its application to the design of bending members, see AITC Technical Note No. 21.

### 4.4.3  Lateral Stability

*4.4.3.1*   The design values for bending contained in these specifications are applicable to members which are adequately braced. When deep, slender members not adequately braced are used, a reduction to the tabulated design values in bending must be applied based on a computation of the slenderness factor of the member. In the check of lateral stability, the slenderness factor shall be applied in design as shown in the *Timber Construction Manual.*

*4.4.3.2*   A reduction in the design value in bending determined by applying the slenderness factor is not cumulative with a reduction in design value due to the application of the volume factor. In no case shall the design value in bending exceed the stress as determined by applying the volume factor or slenderness factor, whichever governs.

### 4.4.4  Curvature Factor

*4.4.4.1*   For the curved portion of members, the design value in bending ($F_b$) shall be modified by multiplying it by the following curvature factor:

$$C_c = 1 - 2000 \left( \frac{t}{R} \right)^2$$

where  $t$ = thickness of lamination (in.)
       $R$ = radius of curvature of lamination (in.)

No curvature factor need be applied to the design value in the straight portion of an assembly, regardless of curvature elsewhere.

## 5. CONNECTIONS AND FASTENERS

**5.1**   The design values for connections and fasteners for glued laminated timber are contained in the *Timber Construction Manual* and the *National Design Specification* (Reference 7).

**5.2**   The design values for fasteners used in glued laminated timber vary depending upon species and growth rate. The face of the member in which the fastener will be placed (i.e., tension, compression or side face) will determine which species and growth rate. Table 3 contains this information for all combinations.

**5.3**   For additional information on connections, see AITC 104, *Typical Construction Details* (Reference 8).

## 6. DIMENSIONS

**6.1   Standard Sizes.**   American National Standard ANSI/AITC A190.1-1992 permits the use of any width or depth of glued laminated timber. The use of standard finished sizes, however, constitutes recommended practice to the extent that other considerations will permit. The laminator may use any thickness of lumber to develop the specified depth provided the volume of the higher grades of lumber equals or exceeds that specified in laminating combinations which are based on laminations of equal thickness. The depth and width of the glued laminated timber should be as agreed upon by buyer and seller.

### 6.2   Depth and Width

*6.2.1*   Straight and curved members shall be furnished in accordance with the width and depth dimensions required by the design.

*6.2.2*   The standard net finished widths are as follows:

| Nominal Width, in. | Net Finished Width, in. | |
|:---:|:---:|:---:|
| | Western Species | Southern Pine |
| 3 | $2\frac{1}{8}$ | — |
| 4 | $3\frac{1}{8}$ | 3 or $3\frac{1}{8}$ |
| 6 | $5\frac{1}{8}$ | 5 or $5\frac{1}{8}$ |
| 8 | $6\frac{3}{4}$ | $6\frac{3}{4}$ |
| 10 | $8\frac{3}{4}$ | $8\frac{1}{2}$ |
| 12 | $10\frac{3}{4}$ | $10\frac{1}{2}$ |
| 14 | $12\frac{1}{4}$ | — |
| 16 | $14\frac{1}{4}$ | — |

Other finished widths may be used to meet the size requirements of a design or to meet other special requirements.

## 6.3 Radius of Curvature

*6.3.1* The ability to bend laminations is dependent upon many factors relating to both wood properties and manufacturing techniques, and it may be advisable to consult with the laminator prior to specifying. Two prime considerations are thickness of laminations, $t$, and bending radii, $R$. The $t/R$ ratio should not exceed $\frac{1}{100}$ for Southern Pine nor $\frac{1}{125}$ for Douglas Fir-Larch and other softwoods.

*6.3.2* Tudor arches utilize the majority of the most sharply curved laminations and usually include $\frac{5}{8}$ in. to 1 in. actual thickness laminations. The standard radius of curvature used in the industry for these sharply curved members is 9 ft 4 in., but this may be reduced based on the lamination thickness or species. For less sharply curved members fabricated from $1\frac{1}{4}$ in. to $1\frac{5}{8}$ in. thick laminations, the normal radius of curvature is 27 ft 6 in.

## 7. APPEARANCE GRADES

**7.1** Appearance grades shall be in accordance with the current *Standard Appearance Grades for Structural Glued Laminated Timber*, AITC 110 (Reference 9), unless otherwise specified on drawings or specifications.

**7.2** For those combinations permitting the mixing of species, the potential for differences in color or grain of adjacent laminations must be recognized. For those architectural appearance applications where such possible differences in color or grain might be in important, the designer may specify a combination symbol which will restrict the laminations to a single species or group of species with similar characteristics. In some cases, this may restrict availability.

## 8. INSPECTION AND QUALITY CONTROL

**8.1** The assurance that quality material and workmanship are used in structural glued laminated timber members shall be vested in the laminator's day-to-day quality control operations. Visual inspections and physical tests of samples of production are also required to assure conformance with AITC 117—*MANUFACTURING*, American National Standard ANSI/AITC A190.1-1992, *Structural Glued Laminated Timber*, and these specifications.

## 9. MARKING

**9.1** The laminating combinations in Table 1 were developed primarily to resist bending loads. The grades of lumber in laminations on the compression side may not be the same as those on the tension side. Therefore, straight or slightly cambered glued laminated timber bending members shall be stamped **TOP** with letters approximately 2 in. high on the top at both ends of the member. Axially loaded members or bending members which are fabricated in such a manner that they cannot be installed upside down need not be marked with the **TOP** stamp.

# TABLE 1
## Design Values for Structural Glued Laminated Timber
*For normal duration of load and dry conditions of use*[a,b,c]

| Combi-nation Symbol[d] | Species-Outer Lamina-tions/Core Lamina-tions[a] | Bending About X-X Axis (Loaded Perpendicular to Wide Faces of Laminations) | | | | | | Bending About Y-Y Axis (Loaded Parallel to Wide Faces of Laminations) | | | | | Axially Loaded | | |
|---|---|---|---|---|---|---|---|---|---|---|---|---|---|---|---|
| | | Extreme Fiber in Bending, $F_{bx}$ | | Compression Perpendicular to Grain, $F_{c\perp x}^{u}$ | | Shear Parallel to Grain (Horizontal) $F_{vx}$ | Modulus of Elasticity, $E_x$ | Extreme Fiber in Bending,[t,r] $F_{by}$ | Com-pression Perpen-dicular to Grain, $F_{c\perp y}^{v}$ | Shear Parallel to Grain (Horizontal) $F_{vy}$ | Shear Parallel to Grain (Horizontal) (For members with multiple piece laminations which are not edge glued),[t] $F_{vy}$ | Modulus of Elasticity, $E_y$ | Tension Parallel to Grain, $F_t$ | Com-pression Parallel to Grain, $F_c$ | Modulus of Elasticity, $E$ |
| | | Tension Zone Stressed In Tension[t,v] psi | Com-pression Zone Stressed in Tension[g] psi | Tension Face psi | Com-pression Face psi | psi | Million psi | psi | psi | psi | psi | Million psi | psi | psi | Million psi |
| 1 | 2 | 3 | 4 | 5 | 6 | 7 | 8 | 9 | 10 | 11 | 12 | 13 | 14 | 15 | 16 |
| **Visually Graded Western Species** | | | | | | | | | | | | | | | |
| The following three combinations are NOT BALANCED and are for either dry or wet use. | | | | | | | | | | | | | | | |
| 16F-V1 | DF/WW | 1600 | 800 | 560[N] | 560[N] | 140[s,w] | 1.3[x] | 950 | 255 | 130[s,w] | 65[s,w] | 1.1[x] | 675 | 975 | 1.1[x] |
| 16F-V2 | HF/HF | 1600 | 800 | 500 | 375[j] | 155 | 1.4 | 1250 | 375 | 135 | 70 | 1.3 | 875 | 1300 | 1.3 |
| 16F-V3 | DF/DF | 1600 | 800 | 560[N] | 560 | 165 | 1.5 | 1450 | 560 | 145 | 75 | 1.5 | 950 | 1550 | 1.5 |
| The following combination is NOT BALANCED and is intended for straight or slightly cambered members for dry use and industrial appearance[k]. | | | | | | | | | | | | | | | |
| 16F-V4 | DF/WW | 1600 | 800 | 650 | 560[o] | 90[s,w] | 1.5[x] | 900 | 255 | 130[s,w] | 65[s,w] | 1.3[x] | 650 | 600 | 1.3[x] |

| | | | | | | | | | | | | | |
|---|---|---|---|---|---|---|---|---|---|---|---|---|---|
| The following two combinations are BALANCED and are intended for members continuous or cantilevered over supports and provide equal capacity in both positive and negative bending. | | | | | | | | | | | | | |
| 16F-V6 | DF/DF | 1600 | 1600 | $560^{h,j}$ | $560^{h}$ | 165 | 1.5 | 1450 | 560 | 145 | 75 | 1.4 | 950 | 1550 | 1.5 |
| 16F-V7 | HF/HF | | | $375^{j}$ | $375^{j}$ | 155 | 1.4 | 1200 | 375 | 135 | 70 | 1.3 | 850 | 1350 | 1.3 |
| The following five combinations are NOT BALANCED and are for either dry or wet use. | | | | | | | | | | | | | |
| 20F-V1 | DF/WW | | | 650 | $560^{h}$ | $140^{s,w}$ | $1.4^{x}$ | 1000 | 255 | $130^{s,w}$ | $65^{s,w}$ | $1.2^{x}$ | 750 | 1000 | $1.2^{x}$ |
| 20F-V2 | HF/HF | | | $500^{j}$ | $375^{j}$ | 155 | 1.5 | 1200 | 375 | 135 | 70 | 1.4 | 950 | 1350 | 1.4 |
| 20F-V3 | DF/DF | 2000 | 1000 | 650 | $560^{h}$ | 165 | 1.6 | 1450 | 560 | 145 | 75 | 1.5 | 1000 | 1550 | 1.5 |
| 20F-V10 | DF/HF | | | 650 | 560 | 155 | 1.5 | 1300 | 375 | 135 | 70 | 1.4 | 950 | 1500 | 1.4 |
| 20F-V12 | AC/AC | | | 560 | 560 | 190 | 1.5 | 1200 | 470 | 165 | 80 | 1.4 | 900 | 1500 | 1.4 |
| The following three combinations are BALANCED and are intended for members continuous or cantilevered over supports and provide equal capacity in both positive and negative bending. | | | | | | | | | | | | | |
| 20F-V7 | DF/DF | | | 650 | 650 | 165 | 1.6 | 1450 | 560 | 145 | 75 | 1.6 | 1000 | 1600 | 1.6 |
| 20F-V8 | DF/DF | 2000 | 2000 | $590^{h,i}$ | $590^{h,i}$ | 165 | 1.7 | 1450 | 560 | 145 | 75 | 1.6 | 1000 | 1600 | 1.6 |
| 20F-V9 | HF/HF | | | $500^{j}$ | $500^{j}$ | 155 | 1.5 | 1400 | 375 | 135 | 70 | 1.4 | 975 | 1400 | 1.4 |
| The following three combinations are NOT BALANCED and are for either dry or wet use. | | | | | | | | | | | | | |
| 22F-V1 | DF/WW | * | | 650 | $560^{h}$ | $140^{s,w}$ | $1.6^{x}$ | 1050 | 255 | $130^{s,w}$ | $65^{s,w}$ | $1.3^{x}$ | 850 | 1100 | $1.3^{x}$ |
| 22F-V3 | DF/DF | 2200 | 1100 | 650 | $560^{j}$ | 165 | 1.7 | 1450 | 560 | 145 | 75 | 1.6 | 1050 | 1500 | 1.6 |
| 22F-V10 | DF/DFS | | | 650 | $560^{h}$ | 165 | 1.6 | 1600 | 500 | 145 | 75 | 1.3 | 1000 | 1400 | 1.3 |
| The following combination is BALANCED and is intended for members continuous or cantilevered over supports and provides equal capacity in both positive and negative bending. | | | | | | | | | | | | | |
| 22F-V8 | DF/DF | 2200 | 2200 | $590^{h,i}$ | $590^{h,i}$ | 165 | 1.7 | 1450 | 560 | 145 | 75 | 1.6 | 1050 | 1650 | 1.6 |
| The following five combinations are NOT BALANCED and are for either dry or wet use. | | | | | | | | | | | | | |
| 24F-V1 | DF/WW | | | 650 | 650 | $140^{s,w}$ | $1.7^{x}$ | 1250 | 255 | $135^{s,w}$ | $70^{s,w}$ | $1.4^{x}$ | 950 | 1300 | $1.4^{x}$ |
| 24F-V2 | HF/HF | | | $500^{j}$ | $500^{j}$ | 155 | 1.5 | 1250 | 375 | 135 | 70 | 1.4 | 950 | 1300 | 1.4 |
| 24F-V4 | DF/DF | 2400 | 1200 | 650 | 650 | 165 | 1.8 | 1500 | 560 | 145 | 75 | 1.6 | 1100 | 1600 | 1.6 |
| 24F-V5 | DF/HF | | | 650 | 650 | 155 | 1.7 | 1350 | 375 | 140 | 70 | 1.5 | 1100 | 1450 | 1.5 |
| 24F-V11 | DF/DFS | | | 650 | $560^{h}$ | 165 | 1.7 | 1600 | 500 | 145 | 75 | 1.4 | 1150 | 1700 | 1.4 |
| The following two combinations are BALANCED and are intended for members continuous or cantilevered over supports and provide equal capacity in both positive and negative bending. | | | | | | | | | | | | | |
| 24F-V8 | DF/DF | 2400 | 2400 | 650 | 650 | 165 | 1.8 | 1450 | 560 | 145 | 75 | 1.6 | 1100 | 1650 | 1.6 |
| 24F-V10 | DF/HF | | | 650 | 650 | 155 | 1.8 | 1400 | 375 | 140 | 70 | 1.6 | 1150 | 1600 | 1.6 |
| Wet-use factors[b] | | 0.8 | 0.8 | 0.53 | 0.53 | 0.875 | 0.833 | 0.8 | 0.53 | 0.875 | 0.875 | 0.833 | 0.8 | 0.73 | 0.833 |

Note: Some combinations previously shown in this table have been removed because they are not typically used. However, if those combinations are specified, they may be used. The specifier shall consult with the manufacturer for availability.

## TABLE 1  (Continued)

| Combination Symbol[d] | Species-Outer Laminations/Core Laminations[a] | Bending About X-X Axis — Loaded Perpendicular to Wide Faces of Laminations — Extreme Fiber in Bending, $F_{bx}$ — Tension Zone Stressed in Tension[t,v] (psi) | Compression Zone Stressed in Tension[g] (psi) | Compression Perpendicular to Grain, $F_{c\perp x}$ — Tension Face (psi) | Compression Face (psi) | Shear Parallel to Grain (Horizontal) $F_{vx}$ (psi) | Modulus of Elasticity, $E_x$ (Million psi) | Bending About Y-Y Axis — Loaded Parallel to Wide Faces of Laminations — Extreme Fiber in Bending, $F_{by}$[t,r] (psi) | Compression Perpendicular to Grain, $F_{c\perp y}$[v] (psi) | Shear Parallel to Grain (Horizontal), $F_{vy}$ (psi) | Shear Parallel to Grain (Horizontal) (For members with multiple piece laminations, which are not edge glued)[t] $F_{vy}$ (psi) | Modulus of Elasticity, $E_y$ (Million psi) | Axially Loaded — Tension Parallel to Grain, $F_t$ (psi) | Compression Parallel to Grain, $F_c$ (psi) | Modulus of Elasticity, $E$ (Million psi) |
|---|---|---|---|---|---|---|---|---|---|---|---|---|---|---|---|
| 1 | 2 | 3 | 4 | 5 | 6 | 7 | 8 | 9 | 10 | 11 | 12 | 13 | 14 | 15 | 16 |

### E-Rated Western Species

The following three combinations are NOT BALANCED and are for either dry or wet use.

| Combination Symbol[d] | Species-Outer Laminations/Core Laminations[a] | Tension Zone Stressed in Tension[t,v] (psi) | Compression Zone Stressed in Tension[g] (psi) | Tension Face (psi) | Compression Face (psi) | $F_{vx}$ (psi) | $E_x$ (Million psi) | $F_{by}$[t,r] (psi) | $F_{c\perp y}$[v] (psi) | $F_{vy}$ (psi) | $F_{vy}$ (psi) | $E_y$ (Million psi) | $F_t$ (psi) | $F_c$ (psi) | $E$ (Million psi) |
|---|---|---|---|---|---|---|---|---|---|---|---|---|---|---|---|
| 16F-E1 | WW/WW | 1600 | 800 | 255[h] | 255[h] | 140[h,w] | 1.3[x] | 1050 | 255 | 125[h,w] | 65[h,w] | 1.2[x] | 725 | 925 | 1.2[x] |
| 16F-E2[p] | HF/HF | 1600 | 800 | 500[p] | 500[p] | 155 | 1.4 | 1250 | 375 | 135 | 70 | 1.3 | 825 | 1200 | 1.3 |
| 16F-E3 | DF/DF | 1600 | 800 | 650 | 650 | 165 | 1.8 | 1450 | 560 | 145 | 75 | 1.5 | 975 | 1600 | 1.5 |

The following two combinations are BALANCED and are intended for members continuous or cantilevered over supports and[c] provide equal capacity in both positive and negative bending.

| | | | | | | | | | | | | | | | |
|---|---|---|---|---|---|---|---|---|---|---|---|---|---|---|---|
| 16F-E6 | DF/DF | 1600 | 1600 | 650 | 650 | 165 | 1.6 | 1500 | 560 | 145 | 75 | 1.5 | 1000 | 1600 | 1.5 |
| 16F-E7 | HF/HF | 1600 | 1600 | 500[p] | 500[p] | 155 | 1.4 | 1250 | 375 | 135 | 70 | 1.3 | 850 | 1150 | 1.3 |

The following three combinations are NOT BALANCED and are for either dry or wet use.

| Combination | Species | 1 | 2 | 3 | 4 | 5 | 6 | 7 | 8 | 9 | 10 | 11 | 12 | 13 | 14 |
|---|---|---|---|---|---|---|---|---|---|---|---|---|---|---|---|
| 20F-E1 | WW/WW | 2000 | 1000 | $255^n$ | $255^n$ | $140^{λw}$ | $1.6^x$ | 1100 | 255 | $125^{λw}$ | $65^{λw}$ | $1.3^x$ | 800 | 1050 | $1.3^x$ |
| 20F-E2° | HF/HF | 2000 | 1000 | 500° | 500° | 155 | 1.6 | 1400 | 375 | 135 | 70 | 1.4 | 925 | 1550 | 1.4 |
| 20F-E3 | DF/DF | 2000 | 1000 | 650 | 650 | 165 | 1.7 | 1550 | 560 | 145 | 75 | 1.6 | 1050 | 1650 | 1.6 |

The following two combinations are BALANCED and provide equal capacity in both positive and negative bending.

| Combination | Species | 1 | 2 | 3 | 4 | 5 | 6 | 7 | 8 | 9 | 10 | 11 | 12 | 13 | 14 |
|---|---|---|---|---|---|---|---|---|---|---|---|---|---|---|---|
| 20F-E6° | DF/DF | 2000 | 2000 | 650 | 650 | 165 | 1.7 | 1600 | 560 | 145 | 75 | 1.6 | 1150 | 1650 | 1.6 |
| 20F-E7° | HF/HF | 2000 | 2000 | 500° | 500° | 155 | 1.6 | 1500 | 375 | 135 | 70 | 1.4 | 1050 | 1550 | 1.4 |

The following eight combinations are NOT BALANCED and are for either dry or wet use.

| Combination | Species | 1 | 2 | 3 | 4 | 5 | 6 | 7 | 8 | 9 | 10 | 11 | 12 | 13 | 14 |
|---|---|---|---|---|---|---|---|---|---|---|---|---|---|---|---|
| 24F-E1 | DF/DF | 2400 | 1200 | 650 | 650 | 165 | 1.8 | 1550 | 560 | 145 | 75 | 1.6 | 1100 | 1600 | 1.6 |
| 24F-E2° | HF/HF | 2400 | 1200 | 500° | 500° | 155 | 1.7 | 1300 | 375 | 135 | 70 | 1.5 | 850 | 1400 | 1.5 |
| 24F-E3 | DF/HF | 2400 | 1200 | 650° | 500° | 155 | 1.8 | 1500 | 375 | 135 | 70 | 1.5 | 1050 | 1550 | 1.5 |
| 24F-E4 | DF/DF | 2400 | 1200 | 650 | 650 | 165 | 1.8 | 1650 | 560 | 145 | 75 | 1.7 | 1100 | 1700 | 1.7 |
| 24F-E5 | DF/DF | 2400 | 1200 | 650 | 650 | 165 | 1.8 | 1650 | 560 | 145 | 75 | 1.6 | 1100 | 1550 | 1.6 |
| 24F-E6° | HF/WW | 2400 | 1200 | 500° | 500° | $140^{λw}$ | $1.8^x$ | 1100 | 255 | $130^{λw}$ | $65^{λw}$ | $1.4^x$ | 750 | 1250 | $1.4^x$ |
| 24F-E14 | DF/DF | 2400 | 1200 | 650 | 650 | 165 | 1.8 | 1450 | 560 | 145 | 75 | 1.6 | 950 | 1600 | 1.6 |
| 24F-E15 | HF/HF | 2400 | 1200 | 500 | 500 | 155 | 1.8 | 1300 | 375 | 135 | 70 | 1.5 | 950 | 1200 | 1.5 |

The following five combinations are BALANCED and provide equal capacity in both positive and negative bending.

| Combination | Species | 1 | 2 | 3 | 4 | 5 | 6 | 7 | 8 | 9 | 10 | 11 | 12 | 13 | 14 |
|---|---|---|---|---|---|---|---|---|---|---|---|---|---|---|---|
| 24F-E10 | DF/DF | 2400 | 2400 | 650 | 650 | 165 | 1.9 | 1850 | 560 | 145 | 75 | 1.7 | 1300 | 1750 | 1.7 |
| 24F-E11° | HF/HF | 2400 | 2400 | 500° | 500° | 155 | 1.8 | 1600 | 375 | 135 | 70 | 1.5 | 1150 | 1550 | 1.5 |
| 24F-E13 | DF/DF | 2400 | 2400 | 650 | 650 | 165 | 1.8 | 1950 | 560 | 145 | 75 | 1.7 | 1250 | 1700 | 1.7 |
| 24F-E17 | HF/WW | 2400 | 2400 | 500° | 500° | $140^{λw}$ | $1.8^x$ | 1100 | 255 | $130^{λw}$ | $65^{λw}$ | $1.4^x$ | 750 | 1250 | $1.4^x$ |
| 24F-E18 | DF/DF | 2400 | 2400 | 650 | 650 | 165 | 1.8 | 1450 | 560 | 145 | 75 | 1.6 | 950 | 1600 | 1.6 |

The following combination is BALANCED and is for either dry or wet use.

| Combination | Species | 1 | 2 | 3 | 4 | 5 | 6 | 7 | 8 | 9 | 10 | 11 | 12 | 13 | 14 |
|---|---|---|---|---|---|---|---|---|---|---|---|---|---|---|---|
| 24F-E20 | CSP/CSP | 2400 | 2400 | 560 | 560 | 160 | 1.6 | 1150 | 350 | 135 | 60 | 1.5 | 850 | 1800 | 1.5 |
| Wet-use factors[D] | | 0.8 | 0.8 | 0.53 | 0.53 | 0.875 | 0.833 | 0.8 | 0.53 | 0.875 | 0.875 | 0.833 | 0.8 | 0.73 | 0.833 |

Note: Some combinations previously shown in this table have been removed because they are not typically used. However, if those combinations are specified, they may be used. The specifier shall consult with the manufacturer for availability.

## TABLE 1  (Continued)

| Combination Symbol[d] | Species- Outer Laminations/ Core Laminations[e] | Bending About X-X Axis — Loaded Perpendicular to Wide Faces of Laminations — Extreme Fiber in Bending, $F_{bx}$ — Tension Zone Stressed in Tension[t,v], psi | Compression Zone Stressed in Tension[g], psi | Compression Perpendicular to Grain, $F_{c\perp x}$[u] — Tension Face, psi | Compression Face, psi | Shear Parallel to Grain (Horizontal) $F_{vx}$, psi | Modulus of Elasticity, $E_x$, Million psi | Bending About Y-Y Axis — Loaded Parallel to Wide Faces of Laminations — Extreme Fiber in Bending, $F_{by}$[f,r], psi | Compression Perpendicular to Grain, $F_{c\perp y}$[v], psi | Shear Parallel to Grain (Horizontal), $F_{vy}$, psi | Shear Parallel to Grain (Horizontal) (For members with multiple piece laminations which are not edge glued)[t] $F_{vy}$, psi | Modulus of Elasticity, $E_y$, Million psi | Axially Loaded — Tension Parallel to Grain, $F_t$, psi | Compression Parallel to Grain, $F_c$, psi | Modulus of Elasticity, $E$, Million psi |
|---|---|---|---|---|---|---|---|---|---|---|---|---|---|---|---|
| 1 | 2 | 3 | 4 | 5 | 6 | 7 | 8 | 9 | 10 | 11 | 12 | 13 | 14 | 15 | 16 |
| | | | | | | | | **Visually Graded Southern Pine** | | | | | | | |
| The following two combinations are NOT BALANCED and are for either dry or wet use. | | | | | | | | | | | | | | | |
| 16F-V2 | SP/SP | 1600 | 800 | 560[N] | 560[h] | 200 | 1.4 | 1600 | 560 | 175 | 90 | 1.4 | 1000 | 1550 | 1.4 |
| 16F-V3 | SP/SP | | | 650 | 650 | 200 | 1.4 | 1450 | 560 | 175 | 90 | 1.3 | 975 | 1450 | 1.3 |
| The following combination is balanced and is intended for members continuous or cantilevered over supports and provides equal capacity in both positive and negative bending. | | | | | | | | | | | | | | | |
| 16F-V5 | SP/SP | 1600 | 1600 | 560[N] | 560[N] | 200 | 1.4 | 1600 | 560 | 175 | 90 | 1.4 | 1000 | 1550 | 1.4 |
| The following two combinations are NOT BALANCED and are for either dry or wet use. | | | | | | | | | | | | | | | |
| 20F-V2 | SP/SP | 2000 | 1000 | 650 | 650 | 200 | 1.6 | 1450 | 560 | 175 | 90 | 1.4 | 1050 | 1550 | 1.4 |
| 20F-V3 | SP/SP | | | 560[N] | 560[h] | 200 | 1.4 | 1600 | 560 | 175 | 90 | 1.4 | 1000 | 1500 | 1.4 |

| Combination symbol | Species | | | | | | | | | | | | | | |
|---|---|---|---|---|---|---|---|---|---|---|---|---|---|---|---|
| *The following combination is NOT BALANCED and is intended for straight or slightly cambered members for dry use and industrial appearance.[k]* | | | | | | | | | | | | | | | |
| 20F-V4 | SP/SP | 2000 | 1000 | 650 | 560[h] | 90[q] | 1.5 | 1100 | 470 | 150 | 75 | 1.3 | 725 | 950 | 1.3 |
| *The following combination is BALANCED and is intended for members continuous or cantilevered over supports and provides equal capacity in both positive and negative bending.* | | | | | | | | | | | | | | | |
| 20F-V5 | SP/SP | 2000 | 2000 | 650 | 650 | 200 | 1.6 | 1450 | 560 | 175 | 90 | 1.4 | 1050 | 1550 | 1.4 |
| *The following three combinations are NOT BALANCED and are for either dry or wet use.* | | | | | | | | | | | | | | | |
| 22F-V1 | SP/SP | 2200 | 1100 | 650 | 650 | 200 | 1.6 | 1600 | 560 | 175 | 90 | 1.5 | 1050 | 1650 | 1.5 |
| 22F-V2 | SP/SP | 2200 | 1100 | 650 | 560[h,i] | 200 | 1.4 | 1600 | 560 | 175 | 90 | 1.4 | 1000 | 1500 | 1.4 |
| 22F-V3 | SP/SP | 2200 | 1100 | 650 | 560[h] | 200 | 1.6 | 1500 | 560 | 175 | 90 | 1.4 | 1050 | 1500 | 1.4 |
| *The following combination is NOT BALANCED and is intended for straight or slightly cambered members for dry use and industrial appearance.[k]* | | | | | | | | | | | | | | | |
| 22F-V4 | SP/SP | 2200 | 1100 | 650 | 560[h] | 90[q] | 1.6 | 1250 | 470 | 155 | 80 | 1.4 | 825 | 1000 | 1.4 |
| *The following combination is BALANCED and is intended for members continuous or cantilevered over supports and provides equal capacity in both positive and negative bending.* | | | | | | | | | | | | | | | |
| 22F-V5 | SP/SP | 2200 | 2200 | 650 | 650 | 200 | 1.6 | 1600 | 560 | 175 | 90 | 1.5 | 1050 | 1600 | 1.5 |
| *The following two combinations are NOT BALANCED and are for either dry or wet use.* | | | | | | | | | | | | | | | |
| 24F-V1 | SP/SP | 2400 | 1200 | 650 | 560[h] | 200 | 1.7 | 1500 | 560 | 175 | 90 | 1.5 | 1100 | 1350 | 1.5 |
| 24F-V3 | SP/SP | 2400 | 1200 | 650 | 650 | 200 | 1.8 | 1600 | 560 | 175 | 90 | 1.6 | 1150 | 1700 | 1.6 |
| *The following combination is NOT BALANCED and is intended for straight or slightly cambered members for dry use and industrial appearance.[k]* | | | | | | | | | | | | | | | |
| 24F-V4 | SP/SP | 2400 | 1200 | 650 | 560[h] | 90[q] | 1.7 | 1250 | 470 | 155 | 80 | 1.4 | 850 | 1050 | 1.4 |
| *The following combination is BALANCED and is intended for members continuous or cantilevered over supports and provides equal capacity in both positive and negative bending.* | | | | | | | | | | | | | | | |
| 24F-V5 | SP/SP | 2400 | 2400 | 650 | 650 | 200 | 1.7 | 1600 | 560 | 175 | 90 | 1.5 | 1150 | 1700 | 1.5 |
| *The following three combinations are NOT BALANCED and are for either dry or wet use.* | | | | | | | | | | | | | | | |
| 26F-V1 | SP/SP | 2600 | 1300 | 650 | 650 | 200 | 1.8 | 1900 | 560 | 175 | 90 | 1.6 | 1150 | 1600 | 1.6 |
| 26F-V2 | SP/SP | 2600 | 1300 | 650 | 650 | 200 | 1.9 | 2200 | 650 | 175 | 90 | 1.8 | 1200 | 1650 | 1.8 |
| 26F-V3 | SP/SP | 2600 | 1300 | 650 | 650 | 200 | 1.9 | 2100 | 560 | 175 | 90 | 1.8 | 1150 | 1600 | 1.8 |
| *The following combination is BALANCED and is intended for members continuous or cantilevered over supports and provides equal capacity in both positive and negative bending.* | | | | | | | | | | | | | | | |
| 26F-V4 | SP/SP | 2600 | 2600 | 650 | 650 | 200 | 1.9 | 2100 | 560 | 175 | 90 | 1.8 | 1150 | 1600 | 1.8 |
| Wet-use factors[b] | | 0.8 | 0.8 | 0.53 | 0.53 | 0.875 | 0.833 | 0.8 | 0.53 | 0.875 | 0.875 | 0.833 | 0.8 | 0.73 | 0.833 |

Note: Some combinations previously shown in this table have been removed because they are not typically used. However, if those combinations are specified, they may be used. The specifier shall consult with the manufacturer for availability.

## TABLE 1 (Continued)

| Combination Symbol[d] | Species-Outer Laminations/Core Laminations[a] | Bending About X-X Axis — Loaded Perpendicular to Wide Faces of Laminations | | | | | | Bending About Y-Y Axis — Loaded Parallel to Wide Faces of Laminations | | | | | Axially Loaded | | |
|---|---|---|---|---|---|---|---|---|---|---|---|---|---|---|---|
| | | Extreme Fiber in Bending, $F_{bx}$ | | Compression Perpendicular to Grain, $F_{c \perp x}$[u] | | Shear Parallel to Grain (Horizontal) $F_{vx}$ | Modulus of Elasticity, $E_x$ | Extreme Fiber in Bending, $F_{by}$[f,r] | Compression Perpendicular to Grain, $F_{c \perp y}$[v] | Shear Parallel to Grain (Horizontal) $F_{vy}$ | Shear Parallel to Grain (Horizontal) (For members with multiple piece laminations which are not edge glued)[t] $F_{vy}$ | Modulus of Elasticity, $E_y$ | Tension Parallel to Grain, $F_t$ | Compression Parallel to Grain, $F_c$ | Modulus of Elasticity, E |
| | | Tension Zone Stressed in Tension[t,v] | Compression Zone Stressed in Tension[g] | Tension Face | Compression Face | | | | | | | | | | |
| | | psi | psi | psi | psi | psi | Million psi | psi | psi | psi | psi | Million psi | psi | psi | Million psi |
| 1 | 2 | 3 | 4 | 5 | 6 | 7 | 8 | 9 | 10 | 11 | 12 | 13 | 14 | 15 | 16 |
| **E-Rated Southern Pine** | | | | | | | | | | | | | | | |
| The following combination is NOT BALANCED and is for either dry or wet use. | | | | | | | | | | | | | | | |
| 16F-E1 | SP/SP | 1600 | 800 | 650 | 650 | 200 | 1.6 | 1550 | 560 | 175 | 90 | 1.5 | 1050 | 1600 | 1.5 |
| The following combination is BALANCED and is intended for members continuous or cantilevered over supports and provides equal capacity in both positive and negative bending. | | | | | | | | | | | | | | | |
| 16F-E3 | SP/SP | 1600 | 1600 | 650 | 650 | 200 | 1.6 | 1700 | 560 | 175 | 90 | 1.5 | 1100 | 1650 | 1.5 |
| The following combination is NOT BALANCED and is for either dry or wet use. | | | | | | | | | | | | | | | |
| 20F-E1 | SP/SP | 2000 | 1000 | 650 | 650 | 200 | 1.7 | 1600 | 560 | 175 | 90 | 1.5 | 1050 | 1600 | 1.5 |
| The following combination is BALANCED and is intended for members continuous or cantilevered over supports and provides equal capacity in both positive and negative bending. | | | | | | | | | | | | | | | |
| 20F-E3 | SP/SP | 2000 | 2000 | 650 | 650 | 200 | 1.7 | 1800 | 560 | 175 | 90 | 1.5 | 1150 | 1700 | 1.5 |

The following combination is NOT BALANCED and is for either dry or wet use.

| Combination | Species | | | | | | | | | | | | | |
|---|---|---|---|---|---|---|---|---|---|---|---|---|---|---|
| 22F-E1 | SP/SP | 2200 | 1100 | 650 | 650 | 200 | 1.7 | 1600 | 560 | 175 | 90 | 1.5 | 1050 | 1650 | 1.5 |

The following combination is BALANCED and is intended for members continuous or cantilevered over supports and provides equal capacity in both positive and negative bending.

| Combination | Species | | | | | | | | | | | | | |
|---|---|---|---|---|---|---|---|---|---|---|---|---|---|---|
| 22F-E3 | SP/SP | 2200 | 2200 | 650 | 650 | 200 | 1.7 | 1750 | 560 | 175 | 90 | 1.5 | 1150 | 1650 | 1.5 |

The following two combinations are NOT BALANCED and are for either dry or wet use.

| Combination | Species | | | | | | | | | | | | | |
|---|---|---|---|---|---|---|---|---|---|---|---|---|---|---|
| 24F-E1 | SP/SP | | | 650 | 650 | 200 | 1.8 | 1600 | 560 | 175 | 90 | 1.6 | 1100 | 1750 | 1.6 |
| 24F-E2 | SP/SP | 2400 | 1200 | 650 | 650 | 200 | 1.9 | 1700 | 560 | 175 | 90 | 1.6 | 1150 | 1700 | 1.6 |

The following combination is BALANCED and is intended for members continuous or cantilevered over supports and provides equal capacity in both positive and negative bending.

| Combination | Species | | | | | | | | | | | | | |
|---|---|---|---|---|---|---|---|---|---|---|---|---|---|---|
| 24F-E4 | SP/SP | 2400 | 2400 | 650 | 650 | 200 | 1.8 | 2000 | 560 | 175 | 90 | 1.8 | 1250 | 1750 | 1.6 |

The following combination is NOT BALANCED and is for either dry or wet use.

| Combination | Species | | | | | | | | | | | | | |
|---|---|---|---|---|---|---|---|---|---|---|---|---|---|---|
| 28F-E1 | SP/SP | 2800 | 1400 | 650 | 650 | 200 | 2.0 | 1600 | 560 | 175 | 90 | 1.7 | 1300 | 1850 | 1.7 |

The following combination is BALANCED and is for either dry or wet use.

| Combination | Species | | | | | | | | | | | | | |
|---|---|---|---|---|---|---|---|---|---|---|---|---|---|---|
| 28F-E2 | SP/SP | 2800 | 2800 | 650 | 650 | 200 | 2.0 | 1600 | 560 | 175 | 90 | 1.7 | 1300 | 1850 | 1.7 |

The following combination is NOT BALANCED, is only for nominal widths, 6 in. or less, and is for either dry or wet use.

| Combination | Species | | | | | | | | | | | | | |
|---|---|---|---|---|---|---|---|---|---|---|---|---|---|---|
| 30F-E1 | SP/SP | 3000 | 1500 | 650 | 650 | 200 | 2.0 | 1750 | 560 | 175 | 90 | 1.7 | 1250 | 1750 | 1.7 |

The following combination is BALANCED, is only for nominal widths, 6 in. or less, and is for either dry or wet use.

| Combination | Species | | | | | | | | | | | | | |
|---|---|---|---|---|---|---|---|---|---|---|---|---|---|---|
| 30F-E2 | SP/SP | 3000 | 3000 | 650 | 650 | 200 | 2.0 | 1750 | 560 | 175 | 90 | 1.7 | 1250 | 1750 | 1.7 |
| Wet-use factors[b] | | 0.8 | 0.8 | 0.53 | 0.53 | 0.875 | 0.833 | 0.8 | 0.53 | 0.875 | 0.875 | 0.833 | 0.8 | 0.73 | 0.833 |

Note: Some combinations previously shown in this table have been removed because they are not typically used. However, if those combinations are specified, they may be used. The specifier shall consult with the manufacturer for availability.

**Footnotes—Table 1—Design**—Not all combinations are shown in these tables. Other combinations may be used.

[a] The combinations in this table are applicable to members consisting of 4 or more laminations are intended primarily for members stressed in bending due to loads applied perpendicular to the wide faces of the laminations. Design values are tabulated, however, for loading both perpendicular and parallel to the wide faces of the laminations. For combinations and design values applicable to members loaded primarily axially or parallel to the wide faces of the laminations, see Table 2. For members of 2 or 3 laminations, see Table 2.

[b] The tabulated design values are for dry conditions of use. To obtain wet-use design values, multiply the tabulated values by the factors shown at the end of the table.

[c] The tabulated design values are for normal duration of loading. For other durations of loading, see 4.4.1.3.

[d] The combinations symbols relate to a specific combination of grades and species in AITC 117—*MANUFACTURING* that will provide the design values shown for the combination. The first two numbers in the combination symbol correspond to the design value in bending shown in Column 3. The letter in the combination symbol (either a "V" or an "E") indicates whether the combination is made from visually graded (V) or E-rated (E) lumber in the outer zones.

[e] The symbols used for species are DF = Douglas Fir-Larch, DFS = Douglas Fir South, HF = Hem-Fir, WW = Softwood Species, ES = Eastern Spruce, AC = Alaska Cedar, CSP = Canadian Spruce-Pine, and SP = Southern Pine. (N3 refers to No. 3 structural joists and planks or structural light framing grade.) Softwood Species (WW) and Eastern Spruce are included in the general category of Western species although Eastern Spruce and some Softwood Species are produced in other areas.

[f] The tabulated design values for bending about the x–x axis in this table are applicable to a member 12 in. deep, $5\frac{1}{8}$ in. wide, 21 ft. long, uniformly loaded and used for a simple span. When other conditions exist, the requirements of 4.4.2 apply.

[g] Design values in this column are for extreme fiber stress in bending when the member is loaded such that the compression zone laminations are subjected to tensile stresses. For more information, see 1.4.1.3. The values in this column may be increased 200 psi where end joint spacing restrictions are applied to the compression zone when stressed in tension.

[h] Where specified, this value may be increased to 650 psi by providing in the bearing area at least one dense 2 in. nominal thickness lamination of Douglas Fir-Larch for Western species combinations, or Southern Pine for Southern Pine combinations. These dense laminations must be backed by a medium grain lamination of the same species.

[i] For bending members greater than 15 in. in depth, the design value for compression stress perpendicular to grain is 650 psi on the tension face.

[j] Where specified, this value may be increased by providing at least two 2 in. nominal thickness Douglas Fir-Larch laminations in the bearing area. The compression perpendicular to grain design values for Douglas Fir-Larch are 560 psi for medium grain and 650 psi for dense.

[k] These combinations are for dry conditions of use only because they may contain wane. They are recommended for industrial appearance grade and for straight or slightly cambered members only. If wane is omitted these restrictions do not apply.

[l] Where specified, this value may be increased to 140 psi for Softwood Species (WW) and to 155 psi for Hem-Fir by prohibiting wane on both sides of the member; or to 115 psi for Softwood Species (WW) and to 130 psi for Hem-Fir by prohibiting wane on one side of the member.

[m] Where specified, this value may be increased to 110 psi by prohibiting coarse grain material; to 140 psi either by prohibiting wane on both sides of the member or by prohibiting both coarse grain material and wane on one side of the member; or to 165 psi by prohibiting both coarse grain material and wane on both sides of the member.

[n] The compression perpendicular to grain design value of 255 psi is based on the lowest strength species of the Softwood Species (WW) group. If at least one 2 in. nominal thickness lamination of E-rated Hem-Fir with the same E value, or E-rated Douglas Fir-Larch 200,000 psi higher in modulus of elasticity ($E$) than that specified is used in the bearing area on the face of the member subjected to the compression perpendicular to grain stress, $F_{c\perp}$ may be increased to 376 psi. If at least two 2 in. nominal thickness laminations of E-rated Hem-Fir with the same E value, or E-rated Douglas Fir-Larch 200,000 psi higher in modulus of elasticity than that specified are used in the bearing area on the face of the member subjected to the compression perpendicular to grain stress, $F_{c\perp}$ may be increased to 500 psi.

[o] Where specified, this value may be increased to 650 psi by providing in the bearing area at least one 2 in. nominal thickness lamination of Douglas Fir-Larch for Western species combinations, or one 2 in. nominal thickness lamination of Southern Pine for Southern Pine combinations having a modulus of elasticity ($E$) value 200,000 psi higher than the $E$ value specified.

[p] E-rated Douglas Fir-Larch 200,000 psi higher in modulus of elasticity may be substituted for the specified E-rated Hem-Fir.

[q] Where specified, this value may be increased to 140 psi either by prohibiting coarse grain material or by prohibiting wane on both sides of the member; to 165 psi by prohibiting both coarse grain material and wane on one side of the member or to 200 psi by prohibiting both coarse grain material and wane on both sides of the member.

[r] Footnote f to Table 2 also applies.

[s] Where Douglas Fir South is used in place of all of the Softwood Species (WW) laminations required in western species combinations 16F–V1, 16F–V4, 20F–V1, 22F–V1, 24F–V1, 16F–E1, 20F–E1, and 24F–E6 the design value for horizontal shear is the same as for combinations using all Douglas Fir-Larch. ($F_{vx}$ = 165 psi and $F_{vy}$ = 145 psi for L3 and $F_{vx}$ = 90 psi.)

[t] These values for horizontal shear, $F_{vy}$, apply to members manufactured using multiple piece laminations with unbonded edge joints. For members manufactured using single piece laminations or using multiple piece laminations with bonded edge joints the horizontal shear values in column 11 apply. The values in column 12 do not apply to members with 5, 7, or 9 laminations when unbonded edge joints occur in alternate laminations at mid-depth of the member with no edge joints in adjacent laminations and the outside lamination contains unbonded edge joints. The values in column 12 shall be reduced by 16% for members containing 5 laminations when unbonded edge joints occur in each lamination forming a staggered pattern in the member with no edge joint closer than one in. to the mid-depth of the member.

[u] The compression perpendicular to grain design values in this table are not subject to the duration of load modifications in 4.4.1.

[v] The design values in bending about the X–X axis ($F_{bx}$) in this column for bending members shall be multiplied by 0.75 when the member is manufactured without the required special tension lamination(s) (see 1.4.1.6) and 0.85 for beams <15 inches in depth.

[w] The following species may be used for Softwood Species (WW) provided the design values in horizontal shear in Column 7 ($F_{vx}$) and in Column 11 ($F_{vy}$) are reduced by 5 psi: Coast Sitka Spruce, Coast Species, Eastern White Pine (North) and Western White Pine.

[x] The following species may be used for Softwood Species (WW) provided the design values in modulus of elasticity ($E_x$ and $E_y$) in Columns 8, 13, and 16 are reduced by 100,000 psi: Western Cedars, Western Cedars (North), White Woods (Western Woods) and California Redwood—open grain.

## TABLE 2

### Design Values for Structural Glued Laminated Timber

*For normal duration of load and dry conditions of use[a,b,c]*

| | | | All Loading | | Axially Loaded | | | Loaded Parallel to Wide Faces of Laminations | |
|---|---|---|---|---|---|---|---|---|---|
| | | | | | Tension Parallel to Grain, $F_t$ | Compression Parallel to Grain, $F_c$ | | | |
| Combination Symbol | Species[d] | Grade[e] | Modulus of Elasticity, E Million | Compression Perpendicular to Grain[n], $F_{c\perp}$ | | | | Extreme Fiber in Bending[f], $F_{by}$ | |
| | | | | | 2 or More Lams | 4 or More Lams | 2 or 3 Lams | 4 or More Lams | |
| | | | psi | psi | psi | psi | psi | psi | |
| 1 | 2 | 3 | 4 | 5 | 6 | 7 | 8 | 9 | |
| Visually Graded Western Species | | | | | | | | | |
| 1 | | L3 | 1.5 | 560[k] | 900 | 1550 | 1200 | 1450 | |
| 2 | | L2 | 1.7 | 560[k] | 1250 | 1900 | 1600 | 1800 | |
| 3 | DF | L2D | 1.8 | 650 | 1450 | 2300 | 1850 | 2100 | |
| 4 | | L1CL | 1.9 | 590[k] | 1400 | 2100 | 1900 | 2200 | |
| 5 | | L1 | 2.0 | 650 | 1600 | 2400 | 2100 | 2400 | |
| 14 | | L3 | 1.3 | 375[k] | 800 | 1100 | 975 | 1200 | |
| 15 | HF | L2 | 1.4 | 375[k] | 1050 | 1350 | 1300 | 1500 | |
| 16 | | L1 | 1.6 | 375[k] | 1200 | 1500 | 1450 | 1750 | |
| 17 | | L1D | 1.7 | 500 | 1400 | 1750 | 1700 | 2000 | |
| 22 | WW | L3 | 1.0 | 255[o] | 525 | 850 | 675 | 800 | |
| 69 | | L3 | 1.3 | 470 | 700 | 1150 | 1150 | 1000 | |
| 70 | AC | L2 | 1.4 | 470 | 1000 | 1450 | 11550 | 1250 | |
| 71 | | L1D | 1.7 | 560 | 1250 | 1900 | 2050 | 1650 | |
| 72 | | L1S | 1.7 | 560 | 1250 | 1900 | 2050 | 1650 | |
| E-Rated Western Species | | | | | | | | | |
| 27 | | 1.9E-1/2 | 1.8 | 650 | 900 | 1750 | 1200 | 1450 | |
| 28 | | 2.1E-1/2 | 2.0 | 650 | 1100 | 2000 | 1400 | 1450 | |
| 29 | | 2.3E-1/2 | 2.2 | 650 | 1250 | 2300 | 1550 | 1650 | |
| 30 | DF | 1.9E-1/6 | 1.8 | 650 | 1550 | 2100 | 1700 | 2400 | |
| 31 | | 2.1E-1/6 | 2.0 | 650 | 1800 | 2400 | 1900 | 2400 | |
| 32 | | 2.3E-1/6 | 2.2 | 650 | 1800 | 2400 | 2100 | 2400 | |
| 62 | | 2.2E-1/2 | 2.1 | 650 | 1150 | 2200 | 1500 | 1550 | |
| 63 | | 2.2E-1/6 | 2.1 | 650 | 1800 | 2400 | 2000 | 2400 | |
| Wet-use factors[b] | | | 0.833 | 0.53 | 0.8 | 0.73 | 0.73 | 0.8 | |

| | Bending about Y-Y Axis | | | | | Bending About X-X Axis | | |
|---|---|---|---|---|---|---|---|---|
| | | | | | | Loaded Perpendicular to Wide Faces of Laminations | | |
| | | Shear Parallel to Grain (Horizontal), $F_{vy}$ | | | | Extreme Fiber in Bending[h], $F_{bx}$ | | Shear Parallel to Grain[g], (Horizontal) $F_{vx}$ |
| 3 Lams | 2 Lams | 4 or More Lams (For members with multiple piece lams)[m] | 4 or More Lams | 3 Lams | 2 Lams | 2 Lams to 15 in. Deep[l] | 4 or More Lams[l,q] | 2 or more Lams |
| psi | psi | psi | psi | psi | psi | psi | psi | psi |
| 10 | 11 | 12 | 13 | 14 | 15 | 16 | 17 | 18 |
| 1250 | 1000 | 75 | 145 | 135 | 125 | 1250 | 1500 | 165 |
| 1600 | 1300 | 75 | 145 | 135 | 125 | 1700 | 2000 | 165 |
| 1850 | 1550 | 75 | 145 | 135 | 125 | 2000 | 2300 | 165 |
| 2000 | 1650 | 75 | 145 | 135 | 125 | 1900 | 2200 | 165 |
| 2100 | 1800 | 75 | 145 | 135 | 125 | 2200 | 2400 | 165 |
| 1050 | 850 | 70 | 135 | 130 | 115 | 1100 | 1300 | 155 |
| 1350 | 1100 | 70 | 135 | 130 | 115 | 1450 | 1700 | 155 |
| 1550 | 1300 | 70 | 135 | 130 | 115 | 1600 | 1900 | 155 |
| 1850 | 1550 | 70 | 135 | 130 | 115 | 1900 | 2200 | 155 |
| 700 | 550 | 60[p] | 120 | 115 | 105 | 725 | 850 | 140 |
| 875 | 700 | 80 | 165 | 160 | 140 | 1000 | 1150 | 190 |
| 1100 | 925 | 80 | 165 | 160 | 140 | 1350 | 1550 | 190 |
| 1500 | 1250 | 80 | 165 | 160 | 140 | 1700 | 2000 | 190 |
| 1500 | 1250 | 80 | 165 | 160 | 140 | 1700 | 2000 | 190 |
| 1250 | 1000 | 75 | 145 | 135 | 125 | 1250 | 1500 | 165 |
| 1250 | 1000 | 75 | 145 | 135 | 125 | 1500 | 1750 | 165 |
| 140 | 1150 | 75 | 145 | 135 | 125 | 170 | 2000 | 165 |
| 2400 | 2100 | 75 | 145 | 135 | 125 | 1800 | 2100 | 165 |
| 2400 | 2400 | 75 | 145 | 135 | 125 | 2100 | 2400 | 165 |
| 2400 | 2400 | 75 | 145 | 135 | 125 | 2300 | 2400 | 165 |
| 1350 | 1100 | 75 | 145 | 135 | 125 | 1600 | 1900 | 165 |
| 2400 | 2400 | 75 | 145 | 135 | 125 | 2200 | 2400 | 165 |
| 0.8 | 0.8 | 0.875 | 0.875 | 0.875 | 0.875 | 0.8 | 0.8 | 0.875 |

## TABLE 2 *(Continued)*

| Combination Symbol | Species[d] | Grade[e] | All Loading | | Axially Loaded | | | | Loaded Parallel to Wide Faces of Laminations |
|---|---|---|---|---|---|---|---|---|---|
| | | | | | Tension Parallel to Grain, $F_t$ | Compression Parallel to Grain, $F_c$ | | | |
| | | | Modulus of Elasticity, E Million psi | Compression Perpendicular to Grain[n], $F_{c\perp}$ psi | 2 or More Lams psi | 4 or More Lams psi | 2 or 3 Lams psi | 4 or More Lams psi | Extreme Fiber in Bending[f], $F_{by}$ psi |
| 1 | 2 | 3 | 4 | 5 | 6 | 7 | 8 | 9 | |
| 33 | HF | 1.6E-1/2 | 1.5 | 500 | 800 | 1050 | 950 | 1200 | |
| 34 | | 1.9E-1/2 | 1.8 | 500 | 900 | 1300 | 1200 | 1450 | |
| 35 | | 2.1E-1/2 | 2.0 | 500 | 1100 | 1550 | 1400 | 1450 | |
| 36 | | 1.6E-1/4 | 1.5 | 500 | 1200 | 1450 | 1300 | 2100 | |
| 37 | | 1.9E-1/6 | 1.8 | 500 | 1550 | 1950 | 1700 | 2400 | |
| 38 | | 2.1E-1/6 | 2.0 | 500 | 1800 | 2400 | 1900 | 2400 | |
| 39 | WW | 1.6E-1/2 | 1.5 | 255 | 800 | 1200 | 950 | 1200 | |
| 40 | | 1.9E-1/2 | 1.8 | 255 | 900 | 1500 | 1200 | 1450 | |
| 41 | | 2.1E-1/2 | 2.0 | 255 | 1100 | 1750 | 1400 | 1450 | |
| 42 | | 1.6E-1/4 | 1.5 | 255 | 1200 | 1550 | 1300 | 2100 | |
| 43 | | 1.9E-1/6 | 1.8 | 255 | 1550 | 1950 | 1700 | 2400 | |
| 44 | | 2.1E-1/6 | 2.0 | 255 | 1800 | 2200 | 1900 | 2400 | |

**Visually Graded Southern Pine**

| | | | | | | | | | |
|---|---|---|---|---|---|---|---|---|---|
| 47 | SP | N2M[l] | 1.4 | 560[k] | 1200 | 1900 | 1150 | 1750 | |
| 48 | | N2D[l] | 1.7 | 650 | 1400 | 2200 | 1350 | 2000 | |
| 49 | | N1M[l] | 1.7 | 560[k] | 1350 | 2100 | 1450 | 1950 | |
| 50 | | N1D[l] | 1.9 | 650 | 1550 | 2300 | 1700 | 2300 | |

**E-Rated Southern Pine**

| | | | | | | | | | |
|---|---|---|---|---|---|---|---|---|---|
| 53 | SP | 1.9E-1/2 | 1.8 | 650 | 900 | 1900 | 1200 | 1450 | |
| 54 | | 2.1E-1/2 | 2.0 | 650 | 1100 | 2300 | 1400 | 1450 | |
| 55 | | 2.3E-1/2 | 2.2 | 650 | 1250 | 2400 | 1550 | 1650 | |
| 56 | | 1.9E-1/6 | 1.8 | 650 | 1550 | 1850 | 1700 | 2400 | |
| 57 | | 2.1E-1/6 | 2.0 | 650 | 1800 | 2400 | 1900 | 2400 | |
| 58 | | 2.1E-1/6 | 2.2 | 650 | 1800 | 2400 | 2100 | 2400 | |
| Wet-use factors[b] | | | 0.833 | 0.53 | 0.8 | 0.73 | 0.73 | 0.8 | |

| | Bending about Y-Y Axis | | | | | Bending About X-X Axis | | |
| --- | --- | --- | --- | --- | --- | --- | --- | --- |
| | | | | | | Loaded Perpendicular to Wide Faces of Laminations | | |
| | | Shear Parallel to Grain (Horizontal), $F_{vy}$ | | | | Extreme Fiber in Bending[h], $F_{bx}$ | | Shear Parallel to Grain[g], (Horizontal) $F_{vx}$ |
| 3 Lams | 2 Lams | 4 or More Lams (For members with multiple piece lams)[m] | 4 or More Lams | 3 Lams | 2 Lams | 2 Lams to 15 in. Deep[l] | 4 or More Lams[l,q] | 2 or more Lams |
| psi | psi | psi | psi | psi | psi | psi | psi | psi |
| 10 | 11 | 12 | 13 | 14 | 15 | 16 | 17 | 18 |
| 1050 | 850 | 70 | 135 | 130 | 115 | 1100 | 1300 | 155 |
| 1250 | 1000 | 70 | 135 | 130 | 115 | 1250 | 1500 | 155 |
| 1250 | 1000 | 70 | 135 | 130 | 115 | 1500 | 1750 | 155 |
| 1900 | 1700 | 70 | 135 | 130 | 115 | 1400 | 1650 | 155 |
| 2400 | 2100 | 70 | 135 | 130 | 115 | 1800 | 2100 | 155 |
| 2400 | 2400 | 70 | 135 | 130 | 115 | 2100 | 2400 | 155 |
| 1050 | 850 | 60[p] | 120 | 115 | 105 | 1100 | 1300 | 140 |
| 1250 | 1000 | 60[p] | 120 | 115 | 105 | 1250 | 1500 | 140 |
| 1250 | 1000 | 60[p] | 120 | 115 | 105 | 1500 | 1750 | 140 |
| 1900 | 1700 | 60[p] | 120 | 115 | 105 | 1400 | 1650 | 140 |
| 2400 | 2100 | 60[p] | 120 | 115 | 105 | 1800 | 2100 | 140 |
| 2400 | 2400 | 60[p] | 120 | 115 | 105 | 2100 | 2400 | 140 |
| 1550 | 1300 | 90 | 175 | 165 | 150 | 1400 | 1600 | 200 |
| 1800 | 1500 | 90 | 175 | 165 | 150 | 1600 | 1900 | 200 |
| 1750 | 1500 | 90 | 175 | 165 | 150 | 1800 | 2100 | 200 |
| 2100 | 1750 | 90 | 175 | 165 | 150 | 2100 | 2400 | 200 |
| 1250 | 1000 | 90 | 175 | 165 | 150 | 1250 | 1500 | 200 |
| 1250 | 1000 | 90 | 175 | 165 | 150 | 1500 | 1750 | 200 |
| 1400 | 1150 | 90 | 175 | 165 | 150 | 1700 | 2000 | 200 |
| 2400 | 2100 | 90 | 175 | 165 | 150 | 1800 | 2100 | 200 |
| 2400 | 2400 | 90 | 175 | 165 | 150 | 2100 | 2400 | 200 |
| 2400 | 2400 | 90 | 175 | 165 | 150 | 2300 | 2400 | 200 |
| 0.8 | 0.8 | 0.875 | 0.875 | 0.875 | 0.875 | 0.8 | 0.8 | 0.875 |

Note: Some combinations previously shown in this table have been removed because they are not typically used. However, if those combinations are specified, they may be used. The specifier shall consult with the manufacturer.

## Footnotes—Table 2—Design

[a] The combinations in this table are intended primarily for members loaded either axially or in bending with the loads acting parallel to the wide faces of the laminations. Design values for bending due to loading applied perpendicular to the wide faces of the laminations are also included, however, the combinations in Table 1 are usually better suited for this condition of loading. The design values for bending about the $X$-$X$ axis ($F_{bx}$) shown in Column 16 are for members from 2 laminations to 15 in. deep without tension laminations. Design values approximately 15% higher for members with 4 or more laminations are shown in Column 17. These higher design values, however, require special tension laminations which may not be readily available.

[b] The tabulated design values are for dry conditions of use. To obtain wet-use design values, multiply the tabulated values by the factors shown at the end of the table.

[c] The tabulated design values are for normal duration of loading. For other durations of loading, see 4.4.1.3.

[d] The symbols used for species are DF = Douglas Fir-Larch, DFS = Douglas Fir South, HF = Hem-Fir, WW = Softwood Species, ES = Eastern Spruce, AC = Alaska Cedar, and SP = Southern Pine.

[e] Grade designations are as follows:

### Visually Graded Western Species

L1 is L1 laminating grade (dense for Douglas Fir-Larch and Douglas Fir South).

L1D is L1 dense laminating grade for Hem-Fir and Alaska Cedar.

L1CL is L1 close grain laminating grade.

L1S is a special grade of Alaska Cedar, see AITC 117—*MANUFACTURING*.

L2D is L2 laminating grade (dense).

L2 is L2 laminating grade (medium grade).

L3 is L3 laminating grade (medium grain for Douglas Fir-Larch, Douglas Fir South and Hem-Fir).

### Visually Graded Southern Pine

SSD is dense select structural, structural joists and planks, or structural light framing grade (dense).

SSM is select structural, structural joists and planks, or structural light framing grade (medium grain).

N1D is No. 1 dense structural joists and planks, or structural light framing grade or No. 1 boards graded as dense.

N1M is No. 1 structural joists and planks, or structural light framing grade or No. 2 boards graded as dense.

N2D is No. 2 dense structural joists and planks, or structural light framing grade or No. 2 boards graded as dense.

N2M is No. 2 strutural joists and planks, or structural light framing grade or No. 2 boards all with a medium grain rate of growth.

N3M is No. 3 structural joists and planks, or structural light framing grade or No. 3 boards all with a medium grain rate of growth.

N3C is No. 3 structural joists and planks, or structural light framing grade or No. 3 boards all with a coarse grain rate of growth.

### E-Rated Grades—All Species (Except Eastern Spruce)

$2.3E$-$\frac{1}{6}$ has $\frac{1}{6}$ edge characteristic with $2.3E$.

$2.2E$-$\frac{1}{6}$ has $\frac{1}{6}$ edge characteristic with $2.3E$.

$2.1E$-$\frac{1}{6}$ has $\frac{1}{6}$ edge characteristic with $2.1E$.

$1.9E$-$\frac{1}{6}$ has $\frac{1}{6}$ edge characteristic with $1.9E$.

$1.6E$-$\frac{1}{6}$ has $\frac{1}{4}$ edge characteristic with $1.6E$.

$2.3E$-$\frac{1}{2}$, $2.2E$-$\frac{1}{2}$, $2.1E$-$\frac{1}{2}$, $1.9E$-$\frac{1}{2}$, $1.6E$-$\frac{1}{2}$ rate $E$-rated grades with edge characteristics occupying up to $\frac{1}{2}$ of cross section.

Softwood species and Eastern Spruce are included in the general category of Western Species although Eastern Spruce and some Softwood Species are produced in other areas.

[f] The values of $F_{by}$ were calculated based on members 12 in. in depth (bending about $Y$-$Y$ axis). When the depth is less than 12 in., the values of $F_{by}$ can be increased by multiplying by the flat use factor $C_{fu}$ for glued laminated timber.

| | Depth, in. | Flat Use $C_{fu}$ |
|---|---|---|
| Western Species | Southern Pine | Factor |
| 10.75 | 10.50 | 1.01 |
| 8.75 | 8.50 | 1.04 |
| 6.75 | 6.75 | 1.07 |
| 5.125 | 5.00 or 5.125 | 1.10 |
| 3.125 | 3.00 or 3.125 | 1.16 |

$^g$ The design values in horizontal shear contained in this table are based on members without wane.

$^h$ The tabulated design values for bending about the x-x axis in this table are applicable to a member 12 in. deep, $5\frac{1}{8}$ in. wide, 21 ft. long, uniformly loaded and used for a simple span. When other conditions exist, the requirements of 4.4.2 apply.

$^i$ The design values are for members of from 2 laminations to 15 in. in depth without tension laminations.

$^j$ The design values are for members of 4 or more laminations in depth and require special tension laminations. When these values are used in design and the member is specified by combination symbol, the designer should also specify the required design value in bending.

$^k$ When tension laminations are used to obtain the design value for $F_{bx}$ shown in column 16, the compression perpendicular to grain value, $F_{c\perp}$, for the tension face may be increased to 650 psi for Douglas Fir-Larch, Douglas Fir South and Southern Pine, and to 500 psi for Hem-Fir because the tension laminations are required to be dense.

$^l$ Combinations 47, 48, 49, and 50 have more restrictive slope of grain requirements than the basic slope of grain of the grades of lumber used in order to obtain higher tension parallel to grain values and design values in bending when loaded perpendicular to the wide faces of the laminations. The slopes of grain used to calculate the design values in Table 2 were: Combination 47, 1:14, Combination 48, 1:14; Combination 49, 1:16; and Combination 50, 1:14. When design stresses are lower than the design values shown, or when a less restrictive slope of grain provides the same design value, a less restrictive slope of grain may be used. The following table gives the design values of these combinations for various slopes of grain: Values of $F_{bx}$ in column 5 are for members of 2 laminations to 15 in. depth without tension laminations and values in column 6 are for members of 4 or more laminations with tension laminations.

| Slope of Grain. | Comb. No. | Tension Parallel to Grain, $F_t$ 2 or More Lams. psi | Comp. Parallel to Grain, $F_c$ 2 or 3 Lams. psi | Comp. Parallel to Grain, $F_c$ 4 or More Lams. psi | Bending About the X-X Axis, $F_{bx}$ | | Bending About the Y-Y Axis, ($F_{by}$) | | |
|---|---|---|---|---|---|---|---|---|---|
| | | | | | 2 Lams to 15 in. psi | 4 or More Lams. psi | 2 Lams. psi | 3 Lams. psi | 4 or More Lams. psi |
| 1 | | 2 | 3 | 4 | 5 | 6 | 7 | 8 | 9 |
| 1:12 | 47 | 1200 | 1150 | 1900 | 1400 | 1600 | 1300 | 1550 | 1750 |
| | 48 | 1400 | 1350 | 2200 | 1600 | 1900 | 1500 | 1800 | 2000 |
| | 49 | 1300 | 1450 | 1900 | 1750 | 2100 | 1500 | 1750 | 1950 |
| | 50 | 1550 | 1700 | 2200 | 2100 | 2400 | 1750 | 2100 | 2300 |
| 1:10 | 47 | 1150 | 1150 | 1700 | 1400 | 1600 | 1300 | 1550 | 1750 |
| | 48 | 1350 | 1350 | 2000 | 1600 | 1900 | 1500 | 1800 | 2000 |
| | 49 | 1150 | 1450 | 1700 | 1550 | 1850 | 1500 | 1750 | 1850 |
| | 50 | 1350 | 1700 | 2000 | 1800 | 2100 | 1750 | 2100 | 2100 |
| 1:8 | 47 | 1000 | 1150 | 1500 | 1350 | 1600 | 1300 | 1550 | 1600 |
| | 48 | 1150 | 1350 | 1750 | 1600 | 1850 | 1500 | 1800 | 1850 |
| | 49 | — | — | — | — | — | — | — | — |
| | 50 | — | — | — | — | — | — | — | — |

$^m$ These values for shear parallel to grain, $F_{vy}$, apply to members manufactured using multiple piece laminations with unbonded edge joints. For members using single piece laminations or using multiple piece laminations with bonded edge joints the shear parallel to grain values tabulated in columns 13, 14, and 15 apply. The values in column 12 do not apply to members with 5, 7 or 9 laminations when unbonded edge joints occur in alternate laminations at mid-depth of the member with no edge joints in adjacent laminations and the outside lamination contains unbonded edge joints. The values in column 12 shall be reduced by 16% for members containing 5 laminations when unbonded edge joints occur in each lamination forming a staggered pattern in the member with no edge joint closer than 1 in. to the mid-depth of the member.

$^n$ The compression perpendicular to grain design values in this table are not subject to the duration of load modifications in 4.4.1.

$^o$ The following may be used for Softwood Species (WW) provided the modulus of elasticity ($E$) is reduced by 100,000 psi: Western Cedars, Western Cedars (North), White Woods (Western Woods) and California Redwood—open grain.

$^p$ The following may be used for Softwood Species (WW) provided the design values in shear parallel to grain in Column 12 ($F_{vy}$) are reduced by 5 psi and the design values in shear parallel to grain in Columns 13, 14, and 15 ($F_{vy}$) and in Column 18 ($F_{vx}$) are reduced by 10 psi: Coast Sitka Spruce, Coast Species, Western White Pine, and Eastern White Pine.

$^q$ When special tension laminations are not used, the design values in bending about the X-X axis ($F_{bx}$) shall be multiplied by 0.75 for bending members over 15 in. deep. For bending members 15 in. and less in depth, use the design values in Column 16 (see 1.4.1.6).

## TABLE 3
### Design–Species Grouping for Fasteners Used on the Faces of Glued Laminated Timbers

**Table 1   Combinations[a,b,c]**

| Combination Symbol | Tension Face[d] | | | Side Face[e] | | | Compression Face | | |
|---|---|---|---|---|---|---|---|---|---|
| | Species[b] | Growth Rate[c] | Timber Conn. Group | Species[b] | Growth Rate[c] | Timber Conn. Group | Species[b] | Growth Rate[c] | Timber Conn. Group |
| **Visually Graded Western Species** | | | | | | | | | |
| 16F-V1 | DF | M | B | WW | — | D | DF | M | B |
| 16F-V2 | HF | — | C | HF | — | C | HF | — | C |
| 16F-V3 | DF | M | B | DF | M | B | DF | M | B |
| 16F-V4 | DF | D | A | WW | — | D | DF | M | B |
| 16F-V6 | DF | M | B | DF | M | B | DF | M | B |
| 16F-V7 | HF | — | C | HF | — | C | HF | — | C |
| 20F-V1 | DF | D | A | WW | — | D | DF | M | B |
| 20F-V2 | HF | — | C | HF | — | C | HF | — | C |
| 20F-V3 | DF | D | A | DF | M | B | DF | M | B |
| 20F-V7 | DF | D | A | DDF | M | B | DF | D | A |
| 20F-V8 | DF | CL | B | DF | M | B | DF | CL | B |
| 20F-V9 | HF | — | C | HF | — | C | HF | — | C |
| 20F-V10 | DF | D | A | HF | — | C | DF | M | B |
| 20F-V12 | AC | — | D | AC | — | D | AC | — | D |
| 22F-V1 | DF | D | A | WW | — | D | DF | M | B |
| 22F-V3 | DF | D | A | DF | M | B | DF | M | B |
| 22F-V7 | DF | D | A | DF | M | B | DF | D | A |
| 22F-V8 | DF | CL | B | DF | M | B | DF | CL | B |
| 22F-V10 | DF | D | A | DFS | — | C | DF | D | A |

| Combination Symbol | Species | | Grade | E-rated Western Species | | | Species | | Grade |
|---|---|---|---|---|---|---|---|---|---|
| | | | | Species | | Grade | | | |
| 24F-V1 | DF | D | A | WW | — | D | DF | D | A |
| 24F-V2 | HF | — | C | HF | M | C | HF | — | C |
| 24F-V4 | DF | D | A | DF | — | B | DF | D | A |
| 24F-V5 | DF | D | A | HF | M | C | DF | D | A |
| 24F-V8 | DF | D | A | DF | — | B | DF | D | A |
| 24F-V10 | DF | D | A | HF | — | C | DF | D | A |
| 24F-V11 | DF | D | A | DFS | — | C | DF | D | A |
| 16F-E1 | WW | — | D | WW | — | D | WW | — | D |
| 16F-E2 | HF | — | C | HF | — | C | HF | — | C |
| 16F-E3 | DF | — | B | DF | M | B | DF | — | B |
| 16F-E6 | DF | — | B | DF | M | B | DF | — | B |
| 16F-E7 | HF | — | C | HF | — | C | HF | — | C |
| 20F-E1 | WW | — | D | WW | — | D | WW | — | D |
| 20F-E2 | HF | — | C | HF | — | C | HF | — | C |
| 20F-E3 | DF | — | B | DF | M | B | DF | — | B |
| 20F-E6 | DF | — | B | DF | M | B | DF | — | B |
| 20F-E7 | HF | — | C | HF | — | C | HF | — | C |
| 24F-E1 | DF | — | A | DF | M | B | DF | — | A |
| 24F-E2 | HF | — | C | HF | — | C | HF | — | C |
| 24F-E3 | HF | — | C | HF | M | C | HF | — | C |
| 24F-E4 | DF | — | A | DF | M | B | DF | — | A |
| 24F-E5 | DF | — | B | DF | — | D | DF | — | B |
| 24F-E6 | HF | — | C | WW | M | B | HF | — | C |
| 24F-E10 | DF | — | A | HF | — | C | DF | — | A |
| 24F-E11 | DF | — | C | DF | M | B | DF | — | C |
| 24F-E13 | HF | — | A | DF | M | C | HF | — | A |
| 24F-E14 | HF | — | B | HF | — | C | HF | — | B |
| 24F-E15 | DF | — | C | WW | M | D | DF | — | C |
| 24F-E17 | HF | — | C | DF | — | B | HF | — | C |
| 24F-E18 | DF | — | A | DF | M | C | DF | — | A |
| 24F-E20 | CSP | — | C | CSP | — | C | CSP | — | C |

**TABLE 3** *(Continued)*

| Combination Symbol | Tension Face[d] Species[b] | Growth Rate[c] | Timber Conn. Group | Side Face[e] Species[b] | Growth Rate[c] | Timber Conn. Group | Compression Face Species[b] | Growth Rate[c] | Timber Conn. Group |
|---|---|---|---|---|---|---|---|---|---|
| | | | | **Visually Graded Southern Pine** | | | | | |
| 16F-V2 | SP | M | B | SP | M | B | SP | M | B |
| 16F-V3 | SP | D | A | SP | M | B | SP | D | A |
| 16F-V5 | SP | M | B | SP | M | B | SP | M | B |
| 20F-V2 | SP | D | A | SP | M | B | SP | M | B |
| 20F-V3 | SP | M | B | SP | M | B | SP | M | B |
| 20F-V4 | SP | D | A | SP | C | C | SP | M | B |
| 20F-V5 | SP | D | A | SP | M | B | SP | D | A |
| 22F-V1 | SP | D | A | SP | M | B | SP | D | A |
| 22F-V2 | SP | D | A | SP | M | B | SP | M | B |
| 22F-V3 | SP | D | A | SP | M | B | SP | M | B |
| 22F-V4 | SP | D | A | SP | C | C | SP | M | B |
| 22F-V5 | SP | D | A | SP | M | B | SP | D | A |
| 24F-V1 | SP | D | A | SP | M | B | SP | M | B |
| 24F-V3 | SP | D | A | SP | M | B | SP | D | A |
| 24F-V4 | SP | D | A | SP | C | C | SP | M | B |
| 24F-V5 | SP | D | A | SP | M | B | SP | D | A |
| 26F-V1 | SP | D | A | SP | M | B | SP | D | A |
| 26F-V2 | SP | D | A | SP | D | A | SP | D | A |
| 26F-V3 | SP | D | A | SP | M | B | SP | D | A |
| 26F-V4 | SP | D | A | SP | M | B | SP | D | A |

**E-Rated Southern Pine**

| | | | | | | | | | |
|---|---|---|---|---|---|---|---|---|---|
| 16F-E1 | SP | — | B | SP | M | B | SP | — | B |
| 16F-E3 | SP | — | B | SP | M | B | SP | — | B |
| 20F-E1 | SP | — | B | SP | M | B | SP | — | B |
| 20F-E3 | SP | — | B | SP | M | B | SP | — | B |
| 22F-E1 | SP | — | A | SP | M | B | SP | — | B |
| 22F-E3 | SP | — | A | SP | M | B | SP | — | A |
| 24F-E1 | SP | — | A | SP | M | B | SP | — | B |
| 24F-E2 | SP | — | A | SP | M | B | SP | — | A |
| 24F-E4 | SP | — | A | SP | M | B | SP | — | A |
| 28F-E1 | SP | D | A | SP | M | B | SP | D | A |
| 28F-E2 | SP | D | A | SP | M | B | SP | D | A |
| 30F-E1 | SP | D | A | SP | M | B | SP | D | A |
| 30F-E2 | SP | D | A | SP | M | B | SP | D | A |

## TABLE 3

### Design–Species Grouping for Fasteners Used on the Faces of Glued Laminated Timbers

Table 2 Combinations[a,c]

| Combination Symbol[f] | Any Face | |
|---|---|---|
| | Growth Rate[c] | Timber Connector Group |
| **Visually Graded Douglas Fir** | | |
| 1 | M | B |
| 2 | M | B |
| 3 | D | A |
| 4 | CL | B |
| 5 | D | A |
| **Visually Graded Hem-Fir** | | |
| 14 | — | C |
| 15 | — | C |
| 16 | — | C |
| 17 | — | C |
| **Visually Graded Softwood Species** | | |
| 22 | - | D |

| Combination Symbol[f] | Any Face | |
|---|---|---|
| | Growth Rate[c] | Timber Connector Group |
| **E-Rated Hem-Fir** | | |
| 33 | - | C |
| 34 | - | C |
| 35 | - | C |
| 36 | - | C |
| 37 | - | C |
| 38 | - | C |
| **E-Rated Softwood Species** | | |
| 39 | - | D |
| 40 | - | D |
| 41 | - | D |
| 42 | - | D |
| 43 | - | D |
| 44 | - | D |

| Visually Graded Alaska Cedar | | |
|---|---|---|
| 69 | - | D |
| 70 | - | D |
| 71 | - | D |
| 72 | - | D |

| E-Rated Douglas Fir | | |
|---|---|---|
| 27 | - | B |
| 28 | - | A |
| 29 | - | A |
| 30 | - | B |
| 31 | - | A |
| 32 | - | A |
| 62 | - | A |
| 63 | - | A |

| Visually Graded Southern Pine | | |
|---|---|---|
| 47 | M | B |
| 48 | D | A |
| 49 | M | B |
| 50 | D | A |

| E-Rated Southern Pine | | |
|---|---|---|
| 53 | - | B |
| 54 | - | A |
| 55 | - | A |
| 56 | - | B |
| 57 | - | A |
| 58 | - | A |

[a] For bolts, use the species indicated in this table with tables in the *Timber Construction Manual* or the *National Design Specification*. For Softwood Species (WW) use bolt design values for *Northern Species* as given in the *NDS* or *TCM*. Where coarse grain (C) rate of growth is indicated use bolt design values for Hem-Fir as given in the *NDS* or *TCM*.

[b] The symbols used for species are DF = Douglas Fir-Larch, DFS = Douglas Fir South, HF = Hem-Fir, WW = Softwood Species, ES = Eastern Spruce, AC = Alaska Cedar, Canadian Spruce Pine (CSP), and SP = Southern Pine. Softwood Species and Eastern Spruce are included in the general category of Western species although Eastern Spruce and some Softwood Species are produced in other areas.

[c] The symbols used for growth rate are $D$ = dense, $CL$ = close grain, $M$ = medium grain, and $C$ = coarse grain.

[d] The growth rate for the tension face is dense for visually graded Douglas Fir-Larch, Douglas Fir South, Hem-Fir, and Southern Pine for members greater than 15 in. deep. Where the growth rate of dense for the tension face of shallower members is known to the designer, the design values for dense may be used.

[e] The species and growth rate for the side faces are for the core of the member. Where the species and growth rate in the outer zones are known to the designer and the fastenings are used only within those zones, the appropriate fastener design values may be used.

[f] The numerical combinations symbols shown apply only to the combinations listed in Table 2. When dense tension laminations are used for combinations in Table 2, the appropriate fastener design values for dense wood may be used on the tension face.

## 10. PROTECTION DURING SHIPPING AND FIELD HANDLING

**10.1**   End sealers, surface sealers, primer coats and wrappings may be applied for the protection of the members. However, they do not necessarily preclude damage resulting from negligence and other factors beyond the control of the laminator during shipment, handling, storing and placing of the members. The protection specified should be commensurate with the end use and final finish of the member. It may also vary with the method of shipment and with exposure to climatic and other details. See *Recommended Practice for Protection of Structural Glued Laminated Timber During Transit, Storage and Erection*, AITC 111 (Reference 10).

## ANNEX A   DESIGN VALUES FOR END GRAIN IN BEARING PARALLEL TO GRAIN, $F_g$

**A-1**   The following table lists the design values for end grain in bearing parallel to grain, $F_g$.

**A-1.1**   These design values apply to the net area in bearing and are subject to adjustments for duration of load.

**A-1.2**   When the stress in end grain bearing exceeds 75% of the adjusted design value, bearing should be on a metal plate, strap or other durable, rigid, homogenous material of adequate strength.

**A-1.3**   These values are based on dry conditions of use. Multiply the tabulated values by 0.57 to obtain wet conditions of use design values.

**A-1.4**   The design values for end grain in bearing apply to end-to-end bearing of compression members provided there is an adequate lateral support and the end cuts are accurately squared and parallel. When a rigid insert is required by A-1.2, it shall be of not less than 20 gauge metal, or equivalent, inserted with a snug fit between abutting ends.

**A-1.5**   When the load in bearing is at an angle to grain, the maximum bearing value shall be determined by the Hankinson formula using the design value for end grain in bearing parallel to grain from this Annex, and the design value in compression perpendicular to grain as provided in Tables 1 or 2.

**A-1.6**   The design values in the column *Bearing on Full Cross Section* are the average design values of the laminations in the combination. The design values in the column *Bearing on Partial Cross Section* are the design values for the core laminations. These design values should be used for tapered members or members with end bearing on only part of the cross section.

## TABLE A-1

**Design Values for End Grain in Bearing Parallel to Grain, $F_g$**

*Design values are for normal load duration, dry conditions of use, and Table 1 Combinations[a,b]*

| Combination Symbol | Bearing On Full Cross Section, psi | Bearing On Partial Cross Section, psi |
|---|---|---|
| **Visually Graded Western Species** | | |
| 16F-V1 | 1690 | 1430 |
| 16F-V2 | 1960 | 1940 |
| 16F-V3 | 2350 | 2360 |
| 16F-V4 | 1830 | 1430 |
| 16F-V6 | 2350 | 2360 |
| 16F-V7 | 1940 | 1940 |
| 20F-V1 | 1720 | 1430 |
| 20F-V2 | 2010 | 1940 |
| 20F-V3 | 2370 | 2360 |
| 20F-V7 | 2390 | 2360 |
| 20F-V8 | 2370 | 2360 |
| 20F-V9 | 2010 | 1940 |
| 20F-V10 | 2040 | 1940 |
| 20F-V12 | 2170 | 2100 |
| 22F-V1 | 1830 | 1430 |
| 22F-V3 | 2390 | 2360 |
| 22F-V8 | 2380 | 2360 |
| 22F-V10 | 2250 | 2130 |

| Combination Symbol | Bearing On Full Cross Section, psi | Bearing On Partial Cross Section, psi |
|---|---|---|
| **Visually Graded Western Species (cont.)** | | |
| 24F-V1 | 2080 | 1430 |
| 24F-V2 | 2020 | 1940 |
| 24F-V4 | 2430 | 2340 |
| 24F-V5 | 2140 | 1940 |
| 24F-V8 | 2430 | 2360 |
| 24F-V10 | 2180 | 1940 |
| 24F-V11 | 2370 | 2130 |
| **E-Rated Western Species** | | |
| 16F-E1 | 1620 | 1430 |
| 16F-E2 | 1920 | 1800 |
| 16F-E3 | 2360 | 2360 |
| 16F-E6 | 2360 | 2360 |
| 16F-E7 | 1920 | 1800 |
| 20F-E1 | 1770 | 1430 |
| 20F-E2 | 2030 | 1800 |
| 20F-E3 | 2370 | 2360 |
| 20F-E6 | 2370 | 2360 |
| 20F-E7 | 2020 | 1800 |

TABLE A-1 (Continued)

| Combination Symbol | Bearing On Full Cross Section, psi | Bearing On Partial Cross Section, psi |
|---|---|---|
| Visually Graded Southern Pine (cont.) | | |
| 22F-V1 | 2380 | 2300 |
| 22F-V2 | 2300 | 2300 |
| 22F-V3 | 2340 | 2300 |
| 22F-V4 | 2170 | 1840 |
| 22F-V5 | 2340 | 2300 |
| 24F-V1 | 2360 | 2300 |
| 24F-V3 | 2480 | 2300 |
| 24F-V4 | 2170 | 1840 |
| 24F-V5 | 2420 | 2300 |

| Combination Symbol | Bearing On Full Cross Section, psi | Bearing On Partial Cross Section, psi |
|---|---|---|
| E-Rated Western Speices (cont.) | | |
| 24F-E1 | 2460 | 2360 |
| 24F-E2 | 2460 | 1940 |
| 24F-E3 | 2230 | 1940 |
| 24F-E4 | 2450 | 2360 |
| 24F-E5 | 2440 | 2360 |
| 24F-E6 | 1990 | 1430 |
| 24F-E10 | 2510 | 2360 |
| 24F-E11 | 2210 | 1940 |
| 24F-E13 | 2450 | 2360 |

| Combination Symbol | | |
|---|---|---|
| 26F-V1 | 2480 | 2300 |
| 26F-V2 | 2690 | 2690 |
| 26F-V3 | 2500 | 2300 |
| 26F-V4 | 2500 | 2300 |
| **E-Rated Southern Pine** | | |
| 16F-E1 | 2310 | 2300 |
| 16F-E3 | 2320 | 2300 |
| 20F-E1 | 2330 | 2300 |
| 20F-E3 | 2330 | 2300 |
| 22F-E1 | 2360 | 2300 |
| 22F-E3 | 2360 | 2300 |
| 24F-E1 | 2380 | 2300 |
| 24F-E2 | 2480 | 2300 |
| 24F-E4 | 2420 | 2300 |
| 28F-E1 | 2690 | 2600 |
| 28F-E2 | 2690 | 2600 |
| 30F-E1 | 2690 | 2600 |
| 30F-E2 | 2690 | 2600 |

| Combination Symbol | | |
|---|---|---|
| 24F-E14 | 2450 | 2360 |
| 24F-E15 | 2070 | 1940 |
| 24F-E17 | 1850 | 1430 |
| 24F-E18 | 2460 | 2360 |
| 24F-E20 | 1900 | 1820 |
| **Visually Graded Southern Pine** | | |
| 16F-V2 | 2300 | 2300 |
| 16F-V3 | 2340 | 2300 |
| 16F-V5 | 2300 | 2300 |
| 20F-V2 | 2380 | 2300 |
| 20F-V3 | 2300 | 2300 |
| 20F-V4 | 2110 | 1840 |
| 20F-V5 | 2420 | 2300 |

[a]For loads of other duration, see 4.4.1.3.

[b]For wet conditions of use, multiply tabulated values by 0.57.

## TABLE A-2

### Design Values for End Grain in Bearing Parallel to Grain, $F_g$

*Design values are for normal load duration, dry conditions of use and Table 2 Combinations[a,b,c]*

| Combination Symbol | Bearing On Full Cross Section, psi | Combination Symbol | Bearing On Full Cross Section, psi |
|---|---|---|---|
| **Visually Graded Douglas Fir** | | **E-Rated Hem-Fir** | |
| 1 | 2360 | 33 | 1940 |
| 2 | 2360 | 34 | 2270 |
| 3 | 2750 | 35 | 2270 |
| 4 | 2510 | 36 | 1940 |
| 5 | 2750 | 37 | 2270 |
| **Visually Graded Hem-Fir** | | 38 | 2270 |
| 14 | 1940 | **E-Rated Softwood Species** | |
| 15 | 1940 | 39 | 1430 |
| 16 | 1940 | 40 | 1430 |
| 17 | 2270 | 41 | 1430 |
| **Visually Graded Softwood Species** | | 42 | 1430 |
| | | 43 | 1430 |
| 22 | 1430 | 44 | 1430 |
| **Visually Graded Alaska Cedar** | | **Visually Graded Southern Pine** | |
| 69 | 2100 | 47 | 2300 |
| 70 | 2100 | 48 | 2690 |
| 71 | 2460 | 49 | 2300 |
| 72 | 2460 | 50 | 2690 |
| **E-Rated Douglas Fir** | | **E-Rated Southern Pine** | |
| 27 | 2360 | 53 | 2300 |
| 28 | 2750 | 54 | 2690 |
| 29 | 2750 | 55 | 2690 |
| 30 | 2360 | 56 | 2300 |
| 31 | 2750 | 57 | 2690 |
| 32 | 2750 | 58 | 2690 |
| 62 | 2750 | | |
| 63 | 2750 | | |

[a]For loads of other duration, see 4.4.1.3.

[b]For wet conditions of use, multiply tabulated values by 0.57.

[c]Apply to partial bearing also.

# ANNEX B   CALIFORNIA REDWOOD

## B-1   General

*B-1.1*   Glued laminated timbers manufactured from California Redwood are used for special purposes where appearance or natural resistance to decay is a prime consideration. They may not be generally available and the designer should check on availability prior to specifying.

*B-1.2*   All sections of AITC 117—*DESIGN* are applicable to California Redwood except as modified or revised herein.

## B-2   Lumber

*B-2.1*   Lumber shall be of the grades specified herein.

## B-3   Adhesives

*B-3.1*   Wet-use adhesives conforming to American National Standard ANSI/AITC A190.1-1992, *Structural Glued Laminated Timber* are required for all moisture conditions of service in laminating California Redwood.

## B-4   Design Stresses

*B-4.1*   The information on design stresses in Section 4 of the main body of these specifications is also applicable to California Redwood except as modified in B-4.2.

### B-4.2   *Radial Tension or Compression*

*B-4.2.1*   The design value for radial compression $F_{rc}$, is 315 psi.

## TABLE B-1

## Design Values for Visually California Redwood Glued Laminated Timber

*For normal duration of load, dry conditions of use[a,b,c]*

| 1 | 2 | Bending About XX Axis | | | | | | Bending About YY Axis | | | | Axially Loaded | | |
|---|---|---|---|---|---|---|---|---|---|---|---|---|---|---|
| | | Loaded Parallel to the Wide Faces of the Laminations | | | | | | Loaded Perpendicular to Wide Faces of Lamination | | | | | | |
| | | Extreme Fiber in Bending | | Compression Perp. to Grain | | Shear Parallel to Grain (Horizontal) $F_{vx}$ | Modulus of Elasticity $E_x$ | Extreme Fiber in Bending $F_{by}$ | Compression Perpendicular to Grain Side Faces[g] $F_{c\perp y}$ | Shear Parallel to Grain (Horizontal) $F_{vy}$ | Modulus of Elasticity $E_y$ | Tension Parallel To Grain $F_t$ | Compression Parallel to Grain $F_t$ | Modulus of Elasticity $E$ |
| | | Tension Zone Stressed in Tension[e] $F_{bx}$ | Compression Zone Stressed in Tension[f] $F_{bx}$ | Tension Face[g] $F_{c\perp}$ | Compression Face[g] $F_{c\perp}$ | | | | | | | | | |
| Combination Symbol | Species Outer Laminations/Core Laminations[d] | psi | psi | psi | psi | psi | Million psi | psi | psi | psi | Million psi | psi | psi | Million psi |
| 1 | 2 | 3 | 4 | 5 | 6 | 7 | 8 | 9 | 10 | 11 | 12 | 13 | 14 | 15 |
| B-16F-V1 | CR/CR | 1600 | 800 | 315 | 315 | 125 | 1.1 | 1400 | 315 | 110 | 1.1 | 900 | 1400 | 1.1 |
| Wet-Use factors[b] | | 0.8 | 0.8 | 0.53 | 0.53 | 0.875 | 0.833 | 0.8 | 0.53 | 0.875 | 0.833 | 0.8 | 0.73 | 0.833 |

[a] The combinations in this table are intended primarily for members stressed in bending due to loads applied perpendicular to the wide faces of the laminations for members with 4 or more laminations. For loading, however, for loading both perpendicular and parallel to the wide faces of the laminations. For combinations and stresses applicable to members loaded primarily axially or parallel to the wide faces of the laminations, see Table B-2. For members of 2 or 3 laminations, see Table B-2.

[b] The tabulated design values are for dry conditions of use. To obtain wet-use design values, multiply the tabulated values by the factors shown at the end of the table.

[c] The tabulated design values are for normal duration of loading. For other durations of loading, see 4.4.1 in the main body of these specifications.

[d] CR = California Redwood

[e] The tabulated design values in bending are applicable to members 12 in. deep, 5⅛ in. wide, 21 ft long loaded uniformly as a simple span. When other conditions exist, the requirements of 4.4.2 in the main body of these specifications apply.

[f] Design values in this column are for extreme fiber stress in bending when the member is loaded such that the compression zone laminations are subjected to tensile stresses. For more information, see 1.4.1.2 in the main body of these specifications. The values in this column may be increased to 1200 psi when end joint spacing restrictions are applied to the compression zone when stressed in tension.

[g] The compression perpendicular to grain design values in this table are not subject to the duration of load modifications in 4.4.1.

## TABLE B-2
### Design Values for Visually Graded California Redwood Glued Laminated Timber for Normal Loading Duration, Dry Conditions of Use, psi$^{a,b,c}$

| Combination Symbol | Species$^d$ | Grade$^e$ | Modulus of Elasticity, E Million psi | Comp. Perp. to Grain$^f$ $F_{c\perp}$ psi | Axially Loaded — Tension Parallel to Grain 2 or More Lams $F_t$ psi | Axially Loaded — Compression Parallel to Grain 4 or More Lams $F_c$ psi | Axially Loaded — Compression Parallel to Grain 2 or 3 Lams $F_c$ psi | Bending About Y-Y Axis — Extreme Fiber in Bending 4 or More Lams $F_{by}$ psi | Extreme Fiber in Bending 3 Lams $F_{by}$ psi | Extreme Fiber in Bending 2 Lams $F_{by}$ psi | Shear Parallel to Grain (Horizontal) 4 or More Lams $F_{vy}$ psi | Shear Parallel to Grain 3 Lams $F_{vy}$ psi | Shear Parallel to Grain 2 Lams $F_{vy}$ psi | Bending About X-X Axis — Extreme Fiber in Bending 2 Lams to 15" deep $F_{bx}$ psi | Shear Parallel to Grain (Horizontal) 2 or More Lams $F_{vx}$ psi |
|---|---|---|---|---|---|---|---|---|---|---|---|---|---|---|---|
| 1 | 2 | 3 | 4 | 5 | 6 | 7 | 8 | 9 | 10 | 11 | 12 | 13 | 14 | 15 | 16 |
| B-1 | CR | L5 | 1.0 | 315 | 875 | 1350 | 1350 | 1450 | 1300 | 1100 | 110 | 105 | 95 | 1200 | 125 |
| B-2 | CR | L4 | 1.0 | 315 | 875 | 1350 | 1350 | 1450 | 1300 | 1100 | 110 | 105 | 95 | 1200 | 125 |
| B-3 | CR | L3 | 1.2 | 315 | 1000 | 1550 | 1550 | 1450 | 1300 | 1100 | 110 | 105 | 95 | 1350 | 125 |
| B-4 | CR | L2 | 1.2 | 315 | 1000 | 1600 | 1600 | 1500 | 1350 | 1150 | 110 | 105 | 95 | 1350 | 125 |
| B-5 | CR | L1 | 1.2 | 315 | 1000 | 1600 | 1600 | 1600 | 1500 | 1250 | 110 | 105 | 95 | 1350 | 125 |
| Wet-Use factors$^b$ | | | 0.833 | 0.53 | 0.8 | 0.73 | 0.73 | 0.8 | 0.8 | 0.8 | 0.875 | 0.875 | 0.875 | 0.8 | 0.875 |

$^a$The tabulated combinations in this table are intended primarily for members loaded either axially or in bending with the loads acting parallel to the wide faces of the laminations. Design values for bending due to loading applied perpendicular to the wide faces of the laminations are also included; however, the combination in Table B-1 is usually better suited for this condition of loading for members with 4 or more laminations.

$^b$The tabulated design values are for dry conditions of use. To obtain wet-use design values, multiply the tabulated values by the factors shown at the end of the table.

$^c$The tabulated values are for normal duration of loading. For other durations of loading, see 4.4.1.3 of the main body of these specifications.

$^d$CR = California Redwood.

Grade designations are as follows:

Visually Graded—California Redwood.

L1 is L1 laminating grade (close grain).  L4 is L4 laminating grade (close grain).
L2 is L2 laminating grade (close grain).  L5 is L5 laminating grade (close grain).
L3 is L3 laminating grade (close grain).

[f] The values of $F_{by}$ were calculated based on members 12 in. in depth (bending about $Y-Y$ axis). When the depth is less than 12 in., the values of $F_{by}$ can be increased by multiplying by the following flat use factor, $C_{fu}$:

| Depth, in. | Multiplying Factor, $C_{fu}$ |
|---|---|
| 10.75 | 1.01 |
| 8.75 | 1.04 |
| 6.75 | 1.07 |
| 5.125 | 1.10 |
| 3.125 | 1.16 |

[g] The tabulated design values for bending are applicable to members 12 in. or less depth. For members greater than 12 in. in depth, the requirements of 4.4.2 in the main body of these specifications apply.

[h] The combinations in this table are not intended for deep bending members when loaded perpendicular to the wide faces of the laminations. However, if members over 15 in. in depth are necessary, AITC 302-24 tension laminations are required and the designer must specify that the member is for use in bending about the $X-X$ axis; in which case, the design value $F_{bx}$ is 1400 psi for combinations B-1 and B-2 and 1600 psi for B-3, B-4 and B-5.

[i] The compression perpendicular to grain design values in this Table are not subject to the duration of load adjustments in 4.4.1.

## TABLE B-4

## Design Values for End Grain in Bearing Parallel to Grain, $C_g^1$

*Design values are for normal duration of load, dry conditions of use.*

| Combination Symbol | Bearing on Full Cross Section, psi | Bearing on Partial Cross Section, psi |
|---|---|---|
| B-16F-V1 | 2020 | 2020 |
| B1 | 2020 | 2020 |
| B2 | 2020 | 2020 |
| B3 | 2020 | 2020 |
| B4 | 2020 | 2020 |
| B5 | 2020 | 2020 |

[1] For wet conditions of use, multiply tabulated values by 0.57.

## TABLE B-3

## Design Species Groupings for Fasteners Used on the Faces of Visually Graded California Redwood Glued Laminated Timber[a,b,c]

| Combination Symbol | Any Face | | |
|---|---|---|---|
| | Species | Growth Rate | Timber Connector Group |
| All Combinations | CR | CL | C |

[a] CL = close grain growth rate.
[b] CR = California Redwood.

# APPENDIX I   GUIDE FOR SPECIFYING GLUED LAMINATED TIMBER

## Contents

### I-1   General

*I-1.1*   AITC 117—*DESIGN* and its references include all of the information necessary to specify structural glued laminated timber. Design values for glued laminated timber are given in Tables 1 and 2.

Table 3 contains the information necessary to determine species groupings for use in the design of fastenings.

Annex A contains design values for end grain in bearing.

### *I-1.2   Table 1*

*I-1.2.1*   Table 1 contains combinations of species and grades which are best suited for members resisting bending loads applied perpendicular to the wide faces of the lamination. The minimum depth of members made with all combinations in Table 1 is 4 laminations.

The combination symbol indicates the design value in bending about the $X$-$X$ axis. For instance, the 24F combinations have a design value of 2400 psi in bending ($F_{bx}$). Table 1 contains 16 columns. The first column (1) contains the combination designations, and the second column (2) indicates the species. In column 2, for most combinations, the first species designation indicates the species used in the outer zones, and the second indicates the species used in the core or inner zones. For example, DF/WW indicates Douglas Fir-Larch in the outer laminations and Softwood Species (WW) in the core.

The next group of columns (3–8) contains the design values for members loaded in bending about the $X$-$X$ axis caused by loads applied perpendicular to the wide faces of the laminations—the most common loading condition. These are bending about the $X$-$X$ axis ($F_{bx}$), compression perpendicular to grain ($F_{c\perp}$) on both the tension and compression faces, shear parallel to grain ($F_{vx}$) [horizontal], and modulus of elasticity ($E_x$).

When members made with Table 1 combinations are stressed in bending about the $Y$-$Y$ axis (load applied parallel to the wide faces of the laminations), the corresponding design values are generally lower than those for the same members stressed in bending about the $X$-$X$ axis.

The next group of columns (9–13) contains the design values when the members are loaded in bending about the $Y$-$Y$ axis caused by loads applied parallel to the wide faces of the laminations. These values are listed for use primarily when a member is loaded in bending about both axes. When the principal load in bending is about the *Y–Y* axis, the combinations in Table 2 may be more suitable.

The last group of columns (14–16) lists axial stresses. These strength properties are listed primarily for members which are loaded in both bending and axial compression or tension. When the load is primarily compression or tension, the combinations in Table 2 may be more suitable. However, the combinations in Table 1 are generally used for arches.

### I-1.3   Table 2

*I-1.3.1*   Table 2 contains combinations that are best suited for axially loaded members (either in tension or compression) and members stressed in bending about the *Y–Y* axis (loads applied parallel to the wide faces of the laminations). It is divided into two parts according to species—Western species and Southern Pine species. It is further subdivided into visually graded combinations and *E*-rated combinations.

These grades are listed to provide greater flexibility. Each combination contains only one grade of lumber. The design values listed are for dry conditions of use. The line at the end of the table contains multiplying factors to apply to the tabulated values to obtain wet conditions of use values.

Table 2 contains 18 columns divided into 5 groups. The first group (1–3) indicates the combination symbol (1), the species (2), and the grade (3).

The second group of columns (4–5) contains the modulus of elasticity (4) and the compression perpendicular to grain design values (5) which are applicable to all directions of loading.

The third group of columns (6–8) contains strength properties for axially loaded members.

The fourth of columns (9–15) contains design values for bending about the *Y–Y* axis when loads are applied parallel to the wide faces of the laminations. Note that strength properties may vary with the number of laminations.

The fifth group of columns (16–18) contains design values for members loaded in bending about the *X–X* axis. These design values are for members loaded in bending about both axes or for axially loaded members also loaded in bending about the *X–X* axis. For members loaded primarily in bending about the *X–X* axis, the combinations in Table 1 are more suitable than those shown in this table.

The design values shown for bending about the *X–X* axis ($F_{bx}$) in column 16, Table 2—DESIGN are for members from 2 laminations to 15 in. in depth and should suffice for most purposes. Column 17 contains design values for members of 4 or more laminations loaded in bending about the *X–X* axis when special tension laminations are used. These values are approximately 15% higher than the Column 16 values.

Note that Table 2 combinations are generally not as well suited for bending about the *X–X* axis as Table 1 combinations and bending members requiring special tension laminations may not be readily available. Column 18 contains the design values in shear ($F_{vx}$) when the members are loaded in bending about the *X–X* axis.

## I-2   Guide for Specifying Structurally Glued Laminated Timber

### *I-2.1   Principal Stress in Bending with Loads Applied Perpendicular to Wide Faces of Laminations.*

*I-2.1.1*   Glued laminated timber may be specified in several ways. It is recommended that the specifier list the lowest required stresses for each strength property. This will allow the manufacturer to select the most readily available grades and species to meet the design requirements. It is important that the designer check the design values very carefully and not arbitrarily specify the highest possible values in all categories since this will tend to reduce the manufacturing options and in some cases, result in a member impossible to manufacture.

*I-2.1.2*   A suggested form of specification when the principal stress in bending is as follows:

*Glued laminated timber shall be manufactured from species and grades of lumber which will produce design values equal to or exceeding the following when loaded perpendicular to the wide faces of the laminations:*

*Bending ($F_b$) = _____ psi*
*Shear Parallel to grain ($F_v$) = _____ psi*
*Modulus of elasticity ($E$) = _____ psi*
*Compression perpendicular to grain (tension face) ($F_{c\perp}$) = _____ psi*
*Compression perpendicular to grain (compression face) ($F_{c\perp}$) = _____ psi*
*Compression parallel to grain ($F_c$) = _____ psi*
*Tension parallel to grain ($F_t$) = _____ psi*

All of these values may not be necessary for design, and when they are not, they need not be specified. For example, tension parallel to grain ($F_t$) and compression parallel to grain ($F_c$) may not be needed in the design of a beam. For other designs, additional design values may need to be specified such as for a member loaded in bending about the $Y$–$Y$ axis as well as the $X$–$X$ axis, in which case the design value in bending ($F_{by}$) must be specified.

Glued laminated timbers may also be specified by species or by species combinations. This method is not recommended since it tends to limit the availability.

Where California Redwood is desired for appearance or special use conditions, it should be specified by species. See Annex B for details.

*I-2.1.3*   A second alternative is for the designer to specify a species group. The species groupings are arranged in combinations using visually graded lumber or *E*-rated lumber. It is recommended that the manufacturer have the option of supplying combinations made either of visually graded lumber or *E*-rated lumber. Specifying by species groups can be accomplished by adding a notation such as *Western species* or *Southern Pine species* to the suggested specification in I.2.1.2.

*I-2.1.4*   Glued laminated timbers may also be obtained by specifying the combinations in Table 1, such as 24*F*–*V*1 DF/WW or 24*F*–*V*1 SP/SP. This procedure can limit availability and should be used only where there is a specific reason for so doing. Both visually graded and *E*-rated combinations are shown in Table 1. Visually graded combinations are generally available; however, the designer should check with suppliers on availability of *E*-rated combinations prior to specifying.

*I-2.1.5* California Redwood can be specified by either method since only one combination is listed in Table B-1 (B-16*F-V*1 (CR/CR)). If *all heart* redwood is desired, the designer should add *all heart* to the specification. For special appearance purposes, higher grade combinations listed in Table B-2 may also be used.

*I-2.1.6* Design values for small members of 2 or 3 laminations with loads applied perpendicular to the wide faces of the laminations are listed in Table 2. It is recommended that the designer specify the design values needed and allow the manufacturer to pick the grade and species to meet the design requirements. If it is necessary to specify species, the manufacturer should be allowed the option of selecting the grade in order to increase availability.

### I-2.2  Principal Stress in Bending with Loads Applied Parallel to the Wide Faces of the Laminations

*I-2.2.1* When members are loaded parallel to the wide faces of the laminations, the design values in bending are designed in Tables 1 and 2 as $F_{by}$. When the primary bending stresses of a member are in this direction, the combinations in Table 2 are more suitable. It is recommended, however, that the designer specify the design values required and let the manufacturer select the combination of species and grades of lumber. If specification of a species is necessary, the manufacturer should be allowed the option of selecting the grade of lumber to increase availability.

### I-2.3  Axially Loaded Members

*I-2.3.1* The combinations in Table 2 are generally best suited for members where the primary load is axial—either tension or compression. It is recommended that the designer specify the required axial design value as well as the other required values, including the bending design value in both the *X–X* and *Y–Y* directions if needed. The manufacturer will then select species and grades from available stock to manufacture a member which will meet or exceed the design requirements.

### I-2.4  Preservatively Treated Glued Laminated Timber

*I-2.4.1* The designer should refer to AITC 109 for information on treatments and treatability of species. The species specified should be those recommended for preservative treatments, and the type of preservative and retention should also be specified.

## I-3  Examples of Design and Specification of Glued Laminated Timber

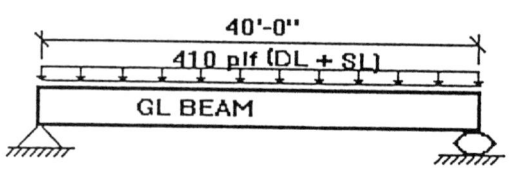

**Example:**   Design a roof beam to support the specified loads. The beam is laterally supported continuously by decking applied directly to the top of the beam and the ends are restrained from rotation. Determine the proper size and specify the proper stresses. (Roof slope equals $\frac{1}{4}$ in. per foot.)

*Select:*
$F_{bx} = 2000$ psi
$C_D = 1.15$ (for snow load)
$F'_{bx} = (2000)(1.15) = 2300$ psi
Deflection limit $= L/180 = (40)(12)/180 = 2.67$ in.

**Load:**
Uniform $w = 410$ plf
(Loads induce stresses about the $X$-$X$ axis)

*Bending:*
$M = wL^2/8 = (410)(40)^2(12)/8 = 984,000$ in.-lb.
$S = M/F_{bx'} = 984,000/2,300 = 428$ in.$^3$

**For Western Species:**
Select a beam based on Section Modulus, $S_x$, $5\frac{1}{8}$ in. $\times$ 24 in.
$S_x = 492$ in.$^3 \geq 428$ in.$^3$
$A_x = 123$ in.$^2$, $I_x = 5,904$ in.$^4$

**Determine Volume Factor, $C_v$**
For Western Species, use exponent of $1/10$
$C_v = (5.125/b)^{1/10}(12/d)^{1/10}(21/L)^{1/10} = (5.125/5.125)^{1/10}(12/29)^{1/10}$
$\quad (21/40)^{1/10} = 0.874$
$f_{bx} = M/S\,C_v = 984,000/(492)(0.874) = 2288$ psi
$F_{bx}(\text{REQ'D}) = 2288/1.15 = 1990$ psi

*Shear:*
Neglect all loads within a distance equal to depth of beam from the end.
$V_x = w(L - 2d/2)/2$
$V = (410)(40 - 4)/2 = 7380$ lb
$f_{vx} = 3V/2bd = (3)(7380)/(2)(5.125)(24) = 90$ psi
$F_{vx}(\text{REQ'D}) = 90/1.15 = 78$ psi

*Bearing:*
Compression perpendicular to grain on tension face. Assume 6 in. bearing at each end.
$R_v = (410)(40)/2 = 8200$ lb.
$f_{c\perp x} = 8200/(5.125)(6) = 267$ psi
$F_{c\perp x}(\text{REQ'D}) = 267$ psi

**Note:**   No special specification is needed for the compression face because the roof decking is applied directly to the top of the beam and the lowest value listed for $F_{c\perp}$ is 255 psi. When purlin hangers or other connectors are used on the top of the beam, the compression perpendicular to grain on the top of the member should be checked and the designer should specify the required design value of $F_{c_\perp}$ for the top of the member.

*Deflection:*

$I_x = 5904$ in.$^4$

$$E(\text{REQ'D}) = \frac{5wL4}{384(L/180)(I_x)}$$

$$= \frac{(5)(410)(40)^4(12)^3}{(384)(2.67)(5904)}$$

$$= 1{,}498{,}000 \text{ psi}$$

Note that bending controls the design of this member.

The recommended method of specifying this member is:

*Glued laminated timber shall be manufactured from species and grades of lumber which will produce design values equal to or exceeding the following when loaded perpendicular to the wide faces of the laminations:*

Bending ($F_{bx}$) = 2000 psi

Shear Parallel to Grain ($F_{vx}$) = 90 psi

Compression perpendicular to grain on the tension face ($F_{c_\perp x}$) = 375 psi

Modulus of elasticity ($E_x$) = 1,500,000 psi.

Ordinarily the bending about the $Y$–$Y$ axis, axial compression, or axial tension are not significant stresses in the design of a roof beam. When the building code requires other loading conditions, the stresses should be checked and the design values specified.

**Note:**   A complete specification should also cover such items as camber, adhesives, appearance grades, protection, quality marks and certificates, and preservative treatment (if required). Refer to the AITC *Timber Construction Manual* for more information.

## I-4   Southern Pine Design Example

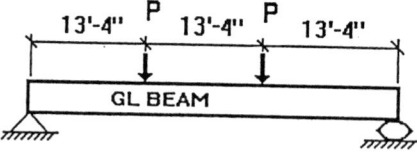

**Loads:**

Dead Load = DL = 10 psf

Live Load = LL = <u>12</u> psf (construction load)

Total Load = TL = 22 psf

Purlins transfer loads a third points. (For example purposes, other loads such as wind and earthquake are not analyzed.) Beam is laterally supported continuously by roof sheathing applied directly to the top of the beam, and the ends are restrained from rotation. Roof slope equals a minimum of $\frac{1}{4}$ in. per foot. Dry conditions of use.

**Select Glulam Beam Properties:**
Bending Design Value $= F_{bx} = 2000$ psi
$F_{bx}^* = F_{bx} C_D = 2000(1.25) = 2500$ psi

**Design Limitation:**
Deflection limited to span$/180 = 40(12)/180 = 2.67$ in.

**Bending Analysis:**
Simple span beam, two equal concentrated loads, symmetrically placed, bending about the $X$–$X$ axis.

$$M(\text{max.}) = Pa$$

$$P = 14(22)(13.33) = 4107 \text{ lb.}$$

$$M = (4107)(13.33) = 657,000 \text{ in.-lb.}$$

$$S(\text{REQ'D}) = M/F_{bx}^* = 657,000/2,500 = 262.8 \text{ in.}^3$$

**Trial Beam Size:**
From section properties table, select beam size according to Section Modulus, $S$ greater than $S(\text{REQ'D})$.
Find $S_x = 403.3$ for a 5 in. $\times$ 22 in. beam.

**Determine Volume Factor, $C_V$, Section 4.4.2**
For Southern Pine use exponent of $\frac{1}{20}$.
Then

$$C_V = K_L \left[ \left( \frac{5.125}{b} \right)^{1/20} \left( \frac{12}{d} \right)^{1/20} \left( \frac{21}{L} \right)^{1/20} \right]$$

$K_L = 0.96$ (see page 8-867)

$$C_V = 0.96 \left[ \left( \frac{5.125}{5.0} \right)^{1/20} \left( \frac{12}{22} \right)^{1/20} \left( \frac{21}{40} \right)^{1/20} \right] = 0.903\cdot$$

$$f_{bx} = M/S_x C_V C_D = 656,796/403.3(0.903)(1.25) = 1.443 \text{ psi}$$

$$F_{bx}(\text{REQ'D}) \geq f_{bx} \geq 1443$$

$$2000 \geq 1443 \qquad \text{O.K.}$$

**Shear:**
$V = P = 4107$ lb.
$f_{vx} = 3V/2bdC_D = 3(4107)/2(5)(22)(1.25) = 45$ psi
$F_{vx}(\text{REQ'D}) \geq f_{vx} = 200$ psi for S.P.
Specify $F_{vx} = 200$ psi

**Bearing:**

Bearing support on bottom of beam is 3 in. × 5 in.

$A = (3)(5) = 15$ in.$^2$

$R_v = P = 4107$ lb.

$f_{c\perp x} = R_v/A = 4107/15 = 274$ psi

$F_{c\perp x} \geq f_{c\perp x} \geq 274$ psi

specify $F_{c\perp x}$(REQ'D) = 275 psi

(Southern Pine $F_{c\perp x}$min. = 560 psi)

**Deflection:**

Determine the $E$(REQ'D) Modulus of Elasticity

$$\Delta = \frac{Pa}{24EI}(3L^2 - 4a^2)$$

$\Delta = 2.67$ inches per deflection limitation, $I = 4437$ in.$^4$

$$E_x(\text{REQ'D}) = \frac{Pa}{24(L/180)I}(3L^2 - 4a^2)$$

$$E_x(\text{REQ'D}) = \frac{4107(13.33)}{24(2.67)(4437)}[3(40)^2 - 4(13.33)^2]$$

$$= 1{,}362{,}000 \text{ psi}$$

$$\text{Specify } E_x = 1{,}400{,}000 \text{ psi}$$

Summary of Design Values to Specify (bending about the $X$–$X$ axis)

Bending = $F_{bx}$ = 2000 psi

Shear Parallel to Grain = $F_{vx}$ = 200 psi

Compression Perpendicular to Grain (tension face) = $F_{c\perp x}$ = 560 psi

Modulus of Elasticity = $E_x$ = 1,400,000 psi

*Note:*   IN THIS EXAMPLE, DEFLECTION CONTROLS.

The beam size can be reduced by using a combination with a greater MOE ($E_x$).

## REFERENCES

1.  AITC 117-93—*MANUFACTURING, Standard Specifications for Glued Laminated Timber of Softwood Species*, American Institute of Timber Construction.

2.  ASTM D 3737-92, *Standard Method for Establishing Stresses for Structural Glued Laminated Timber (Glulam)*, American Society for Testing and Materials.

3.  American National Standard ANSI/AITC A190.1-92, *Structural Glued Laminated Timber*, American National Standards Institute.

4.  *Timber Construction Manual*, American Institute of Timber Construction, published by John Wiley & Sons, Inc.

5.  AITC 109-90, *Standard for Preservative Treatment of Structural Glued Laminated Timber*, American Institute of Timber Construction.

6.  AITC Technical Note No. 21, October 1, 1992, *Use of a Volume Effect Factor in the Design of Glued Laminated Timber Beams*.

7.  *National Design Specification for Wood Construction*, 1991, American Forest & Paper Association.

8.  AITC 104-84, *Typical Construction Details*, American Institute of Timber Construction.

9.  AITC 110-84, *Standard Appearance Grades for Structural Glued Laminated Timber*, American Institute of Timber Construction.

10. AITC 111-79, *Recommended Practice for Protection of Structural Glued Laminated Timber During Transit, Storage and Erection*, American Institute of Timber Construction.

11. AITC 500-91, *Determination of Design Values for Structural Glued Laminated Timber in Accordance with ASTM D3737-89a*, American Institute of Timber Construction.

Cost efficiency in the design of structural glued laminated timber requires that the stress values as determined by the design be specified rather than a particular species or stress combination. The designer should specify the primary stress used in the design, such as extreme fiber in bending, together with the actual computed design values such as those for modulus of elasticity, compression perpendicular to grain and shear. This method of specifying provides timely availability and economy of material and allows the manufacturer the most latitude in selecting raw material—thus promoting better utilization of available forest resources. Tables of design values are conveniently arranged to aid the designer in specifying required stress values. Not all design values as shown in the tables are readily available in all areas. Please check for availability before specifying.

With this edition, previously shown glued laminated timber combinations have been removed from Tables 1–4, in an effort to simplify the tables. The majority of these combinations may be specified and used if confirmed with the manufacturer. AITC has also developed procedures which will allow new combinations to be used and approved based on AITC 500-91, *Determination of Design Values for Structural Glued Laminated Timber in Accordance with ASTM D3737-89a* (Reference 11). This Standard and AITC's computer program determines design values for combinations not previously manufactured, and allows Laminators to develop more efficient layups that better utilize the wood resource. This procedure has been submitted to and approved by the ICBO Evaluation Service as NER 466. Although AITC 117—Manufacturing does not include design values, it is influenced by the AITC 500-91 in that variations of grades and amounts used in glued laminated timber can now be evaluated very quickly, thus allowing more flexibility in using available laminating lumber.

# INDEX